Advances in
CONCURRENT ENGINEERING
CE2000

Information Modeling
Management and Organization
Collaborative Decision Making
Enterprise Engineering
CE in Virtual Environment
Multi-Agents Architectures
Collaboration Technologies
Design Technologies
CE in Construction Industry
Knowledge-Based Concurrent Engineering
Industry Applications in CE
Design for Manufacturing Applications
Training and Education in CE
Standards in CE

PRESENTED AT
SEVENTH ISPE INTERNATIONAL CONFERENCE ON CONCURRENT ENGINEERING: RESEARCH AND APPLICATIONS
Lyon Claude Bernard University, France
July 17–20, 2000

SPONSORED BY
THE INTERNATIONAL SOCIETY FOR PRODUCTIVITY ENHANCEMENT (ISPE)
CERA JOURNAL
THE IEE MANUFACTURING DIVISION
LYON CLAUDE BERNARD UNIVERSITY
INSA OF LYON

TECHNOMIC
PUBLISHING CO., INC.
LANCASTER • BASEL

Advances in Concurrent Engineering—CE2000

aTECHNOMIC®publication

Technomic Publishing Company, Inc.
851 New Holland Avenue, Box 3535
Lancaster, Pennsylvania 17604 U.S.A.

Printed in the United States of America
10 9 8 7 6 5 4 3 2 1

Main entry under title:
   Advances in Concurrent Engineering—CE2000

A Technomic Publishing Company book
Bibliography: p.
Includes index p. 861

Library of Congress Catalog Card No. 00-105557
ISBN No. 1-58716-033-1

HOW TO ORDER THIS BOOK
BY PHONE: 800-233-9936 or 717-291-5609, 8AM–5PM Eastern Time

BY FAX: 717-295-4538

BY MAIL: Order Department
Technomic Publishing Company, Inc.
851 New Holland Avenue, Box 3535
Lancaster, PA 17604, U.S.A.

BY CREDIT CARD: American Express, VISA, MasterCard

BY WWW SITE: http://www.techpub.com

Contents

Preface

It is our great pleasure to introduce this collection of papers presented at the 7th ISPE International Conference on Concurrent Engineering: Research and Applications, being held on July 17 through July 20, 2000 at Lyon Claude Bernard University, France.

Concurrent Engineering is considered one of the key concepts that enables companies to improve the quality of their products, reduce the development time, reduce the cost, create the new concepts, and attain the globalization. Using this approach, everyone contributing to the development of product, from definition of needs, conceptual design to manufacturing, maintenance and recycling. This conference brought together contributions in theory, tools and applications of this area.

We would like to gratefully acknowledge the institutional support and encouragement that we have received from our sponsors: the ISPE, the IEE Manufacturing Division, Lyon Town Hall, Région Rhône-Alpes, Pôle Productique de la Région Rhône-Alpes, the Lyon Claude Bernard University and INSA of Lyon.

We would also like to thank all those who submitted papers for consideration. All papers were refereed by two reviewers. This year, CE2000 represents 100 contributions.

We are grateful to the colleagues who helped in all aspects of organizing the conference. In particular, we would like to thank the members of the CE2000 Committee.

We would also like to thank Technomic Publishing, the Managing Editor of CERA Journal for sponsoring this Conference and the Series Editor of this volume, Dr. Biren Prasad, for their support in turning the individual contributions into an excellent product.

PARISA GHODOUS
LIGIM Lyon, France

D. VANDORPE
LIGIM Lyon, France

Advances in Concurrent Engineering—CE2000

SERIES EDITOR, BIREN PRASAD, Ph.D.

Global competitiveness is of paramount concern to the business community in the developing countries. Establishing Manufacturing Competitiveness is considered equivalent to being a world leader in "world-class manufacturing." In other words, manufacturing competitiveness means sustained growth and earnings through building customer loyalty—by creating high value products—in very dynamic global markets. For decades American and European business and industries focused their efforts primarily on their end products quality rather than on the processes used that created these quality products and services. During the last five years this has changed. A number of modernization programs have been launched by many industries, whose focus has been on business processes and on information usage in the manufacturing environment. Programs, such as enterprise re-engineering, multi-dimensional integration of engineering, manufacturing, and logistics practices, have resulted in far reaching changes and profits. The changes we are witnessing now are a movement from the "age of control" to the "age of flexibility." These changes describe a set of dominant trends in the business environment that have influenced the competitiveness of the companies in the 1990s. These changes could not have been possible without process tools like Concurrent Engineering (CE), CAD/CAM, PDM (Product Data Management) and Knowledge-based Engineering in action. Attributes such as empowerment, flexibility, total quality management, agility, fast-to-market, accountability, teamwork, and integration are inherent in the notion of CE.

Concurrent Engineering is considered one of the key concepts that enable companies to attain world-class stature. Today, it is an organizational keyword. The term refers to the same organizational process described as "simultaneous engineering," "life-cycle engineering," "parallel engineering," and sometimes as the "multi-disciplinary team approach" or "integrated product and process development" (IPPD). Here, everyone contributing to the final product, from conceptual design to marketing teams, is required to participate in the project from its very inception. Some industrial analysts claim that this systematic approach is revolutionizing the manufacturing industries and could be just the boost that will enable firms to survive in the competitive global marketplace.

This is the fifth volume of this series published by Technomic Publishing. The first volume was published in 1996 entitled *Advances in Concurrent Engineering—CE96*. The second volume (CE97) included papers from a conference held in Oakland University, Rochester Hills, MI, August 20–22, 1997. The third (CE98) and fourth (CE99) volumes were published in 1998 and 1999, respectively. This is the fifth volume (CE2000) of this series.

The papers included in this series volume is encompassed by CE2000—the 7th ISPE International Conference on Concurrent Engineering: Research and Applications, held on July 17 through July 20, 2000 at Lyon Claude Bernard University, France. The last five volumes and the forthcoming volumes of this series yet to be issued will broaden the understanding and applicability of the CE concepts to new industrial challenges that we face every day. As we gain a better understanding of CE theory, concepts and its usage to our domain, we will see an increasing impact on our state of competitiveness and our ability to be a world leader in "world-class manufacturing."

This is the fifth volume of the series of books published under the *Advances in Concurrent Engineering* title. The series takes you beyond the current state of developments, focusing in on new ideas, new theories and new practical applications that you can apply to your job for achieving both long-term and short-term productivity and efficiency gains. The advice comes from top names and leaders in the field around the world. Each forthcoming book in the series will be written and reviewed by the leaders themselves, and each book will have the stamp of approval of the international editorial board and the series editor, Biren Prasad, the managing editor of CERA Journal and authors of the most successful two-volume textbooks on Concurrent Engineering Fundamentals, published by Prentice Hall in 1997. Refer to http://rassp.scra.org/newsletter/archive/ for information on *Concurrent Engineering Wheels*.

All of the books in the CE series have a uniform *cover design*. The *cover design* of this CE2000 volume depicts seven elements of cooperation philosophy (called 7Cs in Prasad's *CE Fundamentals Textbook*, Volume I, See

Reference>>http://vig.prenhall.com/acadbook/0,2581,0131474634,00.html) Page 222, Figure 5.2. The cooperation has been and is often the key lynchpin of achieving teamwork and enabling concurrent engineering. Concurrent engineering uses the collective experience/knowledge of project teams to develop design and manufacturing concepts so that, if circumstances change, the prior decisions can be altered quickly. The *cooperation scenario* of the cover design thus shows a virtual setup where, collectively, teams come up with a reasonable set of specifications and objectives that are feasible and fully understood by all parties before they are finally deployed. The success of rapid product realization will depend upon the concurrent team's ability to handle change. Our mission for this series is to provide CE tools, methods, concepts, and case histories to help you manage change. Concurrent teams must manage change carefully, whether it occurs upstream (for instance, during strategic planning), or downstream (for instance, during process execution) levels. *Advances in Concurrent Engineering—henceforth called the CE series* are expected to provide a forum for early publication of Concurrent Engineering developments—made available to the research community for the first time. Many of these advancements in CE are new and being explored by the industries and academia for the *purpose of improving quality, competitiveness, responsiveness, improving customer satisfaction and reducing overall cost.*

Biren Prasad, Ph.D.
Unigraphics Solutions
Knowledge-based Engineering—Product Business Unit
http://www.ugsolutions.com/
Tel: (714) 952-5562   Fax: (714) 952-5758
Email: <prasadb@ugsolutions.com>

ISPE/CE2000 Conference
Organizing Committee

Conference Chairs

Dr. Parisa Ghodous *and* Professor Denis Vandorpe
  Laboratory of CAD/CAM and Modeling (LIGIM), University of Lyon I, Bat 710-43, Bd du 11 Novembre 1918— 69622 Villeurbanne Cedex, France

  • Tel: +33-4-72448000 ext: 4006   •   Fax: +33-4-72431312   •   Email: ghodous@ligim.univ-lyon1.fr • Url: http://www710.univ-lyon1.fr/~ghodous/

Conference International Chairs

Professor Shuichi Fukuda, *Tokyo Metropolitan Institute of Technology, Japan*

Professor S. N. Dwivedi, *University of Louisiana, USA*

Professor Andrew Kusiak, *University of Iowa, USA*

Dr. Pravir Chawdhry, *University of Bath, UK*

Program Chair

Dr. Parisa Ghodous
  LIGIM, University of Lyon I

Proceeding Series Editor

Dr. Biren Prasad
  Unigraphics Solutions, KBE, CA, USA

Program Area Chairs

H. Adeli, *USA*

A. Al-Ashaab, *Mexico*

A. Balbontin-Posadas, *UK*

L. Blessing, *USA*

B.-C. Bjork, *Sweden*

A. F. Cutting-Decelle, *France*

S. N. Dwivedi, *USA*

I. S. Fan, *UK*

T. Fernando, *UK*

R. Goncalves, *Portugal*

T. Horgen, *USA*

P. Ikamonov, *Japan*

H. Ismail, *UK*

B. Prasad, *USA*

M. Pratt, *USA*

R. Roy, *UK*

A. Taleb-Bendiab, *UK*

A. Trappey, *Taiwan*

K. Wallace, *USA*

P. M. Wognum, *Netherlands*

A. Zarli, *France*

Local Organization Committee

President: M. Martinez, *France*

M. Belaziz, F. Biennier, A. Bouras, R. Chaine, A. Dussauchoy, J. Favrel, Y. Gardan, P. Ghodous, S. Tichkiewitch, E. Tosan, D. Vandorpe, *France*

Secretary: S. Croze, *University of Lyon I, France*

ISPE Conference Representatives

Dr. Biren Prasad, *Unigraphics Solutions, Cypress, CA, USA*

Professor Shuichi Fukuda, *Tokyo Metropolitan Institute of Technology, Japan*

Professor S. N. Dwivedi, *University of Louisiana, USA*

Dr. Pravir Chawdhry, *University of Bath, UK*

CERA Journal & Conference Web Sites

Dr. Biren Prasad
  Director, CERA Institute, P.O. Box 3882, Tustin, CA 92782, USA
   • Tel: (714) 952-5562   •   Fax: (71) 505-0663   •   Email: prasadb@ugsolutions.com

ISPE Web Page: http://www.secs.oakland.edu/SECS_prof_orgs/ISPE/CE97.htm

CE'9X Web Page: http://www.bath.ac.uk/Departments/Eng/CE98/
                http://www.bath.ac.uk/Departments/Eng/CE99/

CERA J. Web Page: http://cs.wpi.edu/~dcb/CERA/Jnl-descr.html

Technomic Publishing Co. Web Page: http://www.techpub.com/

CE99 Tutorial: http://www.bath.ac.uk/Departments/Eng/CE99/tutorials.html

CE2000 Technical Program: http://bat710.univ-lyon1.fr/ligim/CE2000/

Best Paper Awards: http://www.bath.ac.uk/Departments/Eng/CE99/awards.html

CE Books: http://vig.prenhall.com/acadbook/0,2581,0131474634,00.html

CE Wheel: http://rassp.scra.org/newsletter/archive/

Towards Informationally-Complete Product Models of Complex Arrangements for Concurrent Engineering: Modeling Design-Constraints Using Virtual Solids

Nickolas S. Sapidis*   Gabriel Theodosiou

Ship Design Laboratory
Department of Naval Architecture and Marine Engineering
National Technical University of Athens
Zografou 157-73, Athens, Greece
{*sapidis,gabriel*}@deslab.ntua.gr

Abstract

Complex mechanical arrangements involve a large number of constraints referring to either properties of individual components (of the arrangement) or relationships between components. Current practices approximate these constraints with simplified "dimensional constraints" aiming at formulating the design problem as a system of (in)equalities to be solved automatically by an appropriate technique. This paper identifies limitations of this approach, leading to inaccurate descriptions of problems that disallow development of a collaborative engineering environment. The work emphasizes solid-modeling aspects of design constraints, and proposes information models leading to complete descriptions of design problems.

1 Introduction

Research/development and industrial application of CAD/CAE systems, during the last twenty years, has established the need for informationally-complete descriptions of products/structures as a prerequisite for any type of work and especially for concurrent or simultaneous or internet-based engineering [7] [11]. When the subject is highly-complex arrangements with thousands of components, like a plant, a ship's engine-room or a complex product, then one is faced also with the vital issue of recording, managing, and satisfying the related "design constraints". Traditionally, these constraints are expressed as "dimensional constraints", i.e., constraints on design parameters,

*Also with *Marine Technology Development Co. (P. Faliro)* and *Elefsis Shipyards S.A. (Elefsis), Greece.*

specifying geometric features or the position of components. This approach has been adopted by the majority of CAD researchers and system vendors as it has the potential of expressing a design problem as a system of mathematical relations which allows employment of powerful tools from numerical analysis and optimization theory. Indeed, promising results have appeared on solving simple 2D design-problems by constraint analysis. This has motivated researchers to attempt applying similar techniques to 3D problems, ignoring the fact that the underlying philosophy is *"a design is fully described by a prescribed number of parameters"*, which might be appropriate for simple 2D problems but not so for complex ones and surely not for 3D arrangements.

This paper considers "object layout" in complex mechanical arrangements and focuses on modeling/processing the various constraints involved in the problem. Two case studies are analyzed establishing the assertion that most of the related constraints are best described in terms of volumes, for which current solid modeling systems offer an ideal environment for construction and processing. §3 presents the first case-study, "required free spaces in a ship's engine-room", and establishes that free-space definitions may be complex and extending beyond their geometric characteristics. Appropriate solid-modeling techniques to define free spaces are presented followed by description of an *extended product model* recording also nongeometric properties of these objects. §4 considers the second case-study, "component layout in product design", identifies limitations in existing approaches, and discusses advantages from employing a solid-modeling based approach.

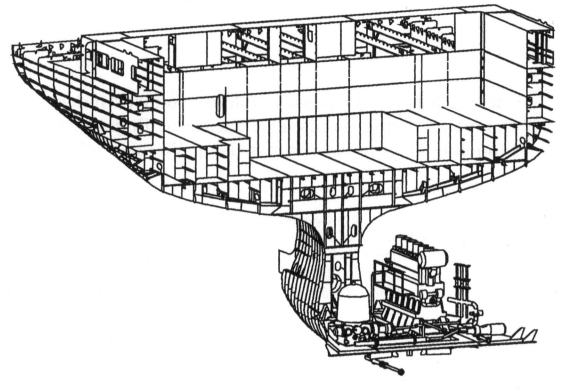

Figure 1: CAD model of a ship's part with approximately 3500 objects (courtesy of Applicon, GmbH).

2 Modeling Complex Mechanical Arrangements

It has long been established that an informationally complete 3D product model offers significant advantages, compared to simplified 2D or 3D models, even for industries dealing with highly complex products and systems, where development of a complete model may be costly [7]. Indeed, in shipbuilding, aircraft manufacturing, and the power and process industries, where it is not unlikely for a CAD-model to include up to half-a-million parts (Fig. 1), CAD vendors promote development of informationally-complete 3D models as these lead to major time-saving benefits in design verification, visual evaluation, and in manufacturability analysis, as well as to improved communication between collaborating teams [11]. In the field of ship design, and in particular, engine-room layout, such an effort has been documented in the recent series of articles published by the Engine Room Arrangement Modeling (ERAM) Team [4]. Surely, engine-room layout is one of the most complex arrangement-problems as it incorporates additional difficulties compared to similar tasks: the available space is very limited, positioning of major components must take under con-

sideration "stability" of the whole vessel, etc. The ERAM publications describe an evolving design effort which adopts the realistic approach of considering an existing design and aiming at specific improvements. Initially the core design-team limits CAD work to 2D drawings and discourages development of a 3D model. Subsequently, that team starts using the 3D model (developed only as an efficient tool for drawing production) for visualization and collision detection. At the concluding stages of the project, the core team fully adopts use of the 3D product model and even attempts to interactively "optimize" parts of the arrangements.

2.1 Research in Complex System Design: State of the Art

Extending the discussion presented in [14], we identify the following subtasks in complex system design: (i) object modeling, (ii) object layout, (iii) system visualization, (iv) system analysis & validation.

The abundance of published geometric modeling techniques [6][16], employing splines, constructive solid geometry, boundary representations, etc, satisfactorily facilitate the first subtask offering a vari-

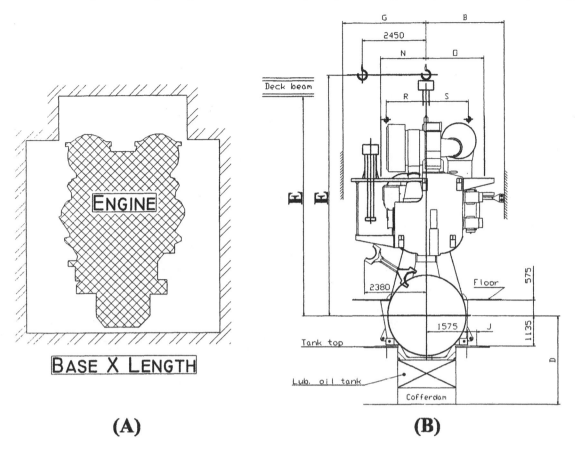

Figure 2: Required free space of case 1 (A) and of case 2 (B).

ety of approaches to shape creation. Regarding the second subtask, "object layout", the literature offers many techniques based on the *assumption* that all constraints and relations, involved in the layout problem, may be expressed as a set of dimensional constraints which is solved using search methods and/or mathematical optimization and/or artificial intelligence techniques. This approach has many disadvantages (see also §4): (a) The underlying assumption is, in general, erroneous and should be applied only to 2D layouts with simple geometric objects. (b) The produced techniques often lead to "black box" software-tools, i.e., they allow no user involvement. (c) These methods are unable to take advantage of evolving solid modeling and CAD technologies. Very recently, a few researchers have started considering this problem in a "solid-modeling framework" producing concepts and tools applicable also to 3D problems. The papers [1] and [14] correctly emphasize how difficult is to manipulate an object in 3D space using standard equipment, since moving (the image of) an object with a mouse on the screen does not specify uniquely the object's

new location. Advanced control devices with many degrees-of-freedom do give more choices to the designer, yet positioning of objects is still based on sight information and thus continues to be cumbersome and time-consuming. The work [1] describes an approach to disambiguate 2D cursor motion using a combination of realistic-looking pseudo-physical behavior and goal-oriented properties, called "object associations". The technique in [14] consists of certain "layout operations" which are performed using physical simulations and a collision-detection algorithm based on software z-buffering. Finally, regarding the subtasks (iii) and (iv), the literature and the existing CAD/CAE systems offer satisfactory solutions based on creating approximate faceted models and applying acceleration techniques from computer graphics [14].

2.2 Modeling Arrangement Constraints Using Volumes

Our research focuses on "object layout" and in particular on informationally-complete descriptions of re-

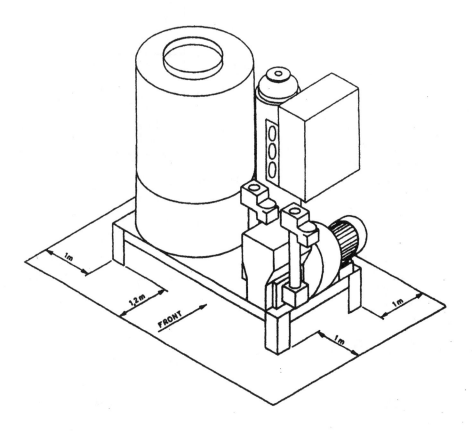

Figure 3: Required free space of case 3: working area around a steam generator [3].

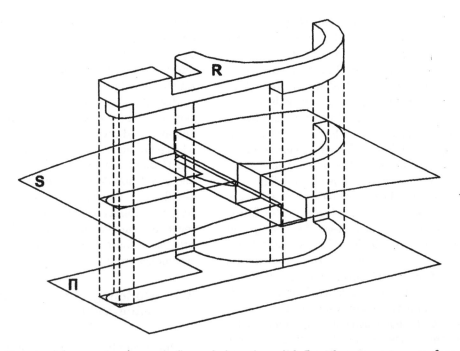

Figure 4: Required free space of case 4: Space below the solid R with respect to a surface S.

lated constraints, which is a prerequisite for a modern concurrent-engineering design approach. The case studies presented below document the assertion that most of the arrangement constraints are best described in terms of volumes, for which a solid modeling system offers the ideal environment for construction and processing. This is exactly the crux of our proposal: development of special-purpose "virtual solid" models which are combined with current object technology to produce an *informationally complete description of arrangement problems*.

3   Case Study I: Required Free Spaces in a Ship's Engine Room

The equipment arranged in a ship's engine-room (engines, pumps, coolers, etc) require free spaces (FS) for maintenance, parts-replacement, inspection, etc, as described in the manufacturers' documentation. Also, there must be sufficient free floor-area for temporary storage of parts, timing shafts, and for the maintenance personnel. A preliminary examination of the engine room's (ER's) major components has produced the following classification regarding FSs and their definition:

Case 1. Complete geometric definition of a FS as a generalized cylinder of given *base* (a polygon, or more generally, an arbitrary planar domain) and *height/length*. *Example:* Fig. 2(A) presents the FS for maintenance of an engine; both the base and the length of the FS are fully defined by the manufacturer.

Case 2. Partial definition of a FS by a "height" or "distance" from a solid's face or "principal plane". Here, the manufacturer's description is, in general, incomplete and must be complemented with additional information by the designer [13]. *Example:* We copy Fig. 2(B) from [9] showing that the space required for removal of piston/liner of the engine is specified by the height E from the crankshaft centerline to the lower edge of the deck beam (above the engine).

Case 3. Incomplete geometric definition of a FS, surrounding a part of the ER, by a given collection of distances from specific faces or "principal planes" (e.g., faces of the bounding box) of that part. *Example:* See steam generator [3] in Fig. 3.

Case 4. Incomplete definition of a FS, associated to a part P positioned above a surface S, as the *"space below P"*. *Example:* The manufacturer of the steam generator, shown in Fig. 3, requires that the space below the generator's base is free of piping; see [3].

Case 5. Complete geometric definition of a FS by specifying movement of an object along a given trajectory. *Example:* See Fig. 2(B) and Fig. 5.

Case 6. Indirect definition of FSs by specifying ER parts that must be removable and/or service/maintenance operations that must be feasible. Here, the system manufacturer provides no explicit information about the required FSs. *Example:* In relation to the steam generator of Fig. 3, the guidelines in [3], regarding stack arrangements and visual inspection, indicate flue pipes that must be removable without mentioning any required FSs.

3.1   Properties of Free Spaces

The above classification makes obvious that FSs are finite manifold 3D objects with complex geometric features, for which a description based solely on "dimensional parameters" is not sufficient. FSs should be modeled using exactly the same solid modeling techniques and data structures used for defining, in a CAD system, all parts of the ER. In fact, FSs have additional geometric and non-geometric properties, described also in the manufacturers' documentation, that make them more complicated than regular solids representing ER parts. These properties are:

1. *Precedence Relationships:* Often a particular FS presumes certain component configuration, or availability of some other FSs, or completion of a specific procedure. *Example:* The FS of Fig. 5 (removal of piston of an engine) presumes availability of the FSs required for removing first the cylinder head and related parts of the engine.

2. *Shape Variations:* There are cases where a particular task may be served by two or more alternative procedures. Then, the corresponding FS is defined differently depending on the selected procedure; e.g., see the height E in Fig. 2(B).

3. *Associations to ER Parts:* Although for most FSs it is true that each one is associated to a single ER component, there are cases where a FS is associated to two or more components (e.g., the required FS between two electrical generators), and also there exist FSs associated to components as well as structural elements of the ER (see Fig. 2(B)).

3.2   "Virtual Solid" Models for Free Spaces

Current solid modeling systems offer a variety of techniques to model a 3D object [6], which are purely geometrical and thus directly applicable also to FSs as solids with zero mass-density. We will call such

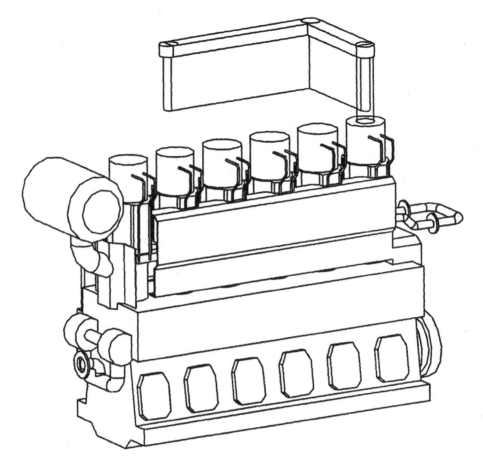

Figure 5: Required free space of case 5 for piston removal.

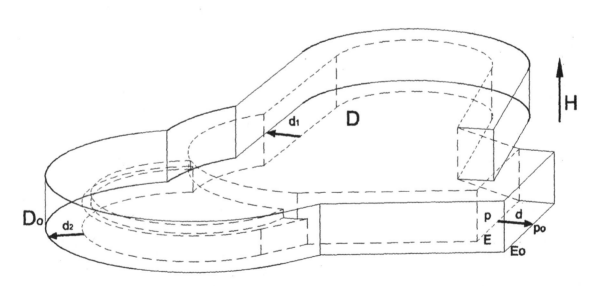

Figure 6: Construction of a case-3 FS model.

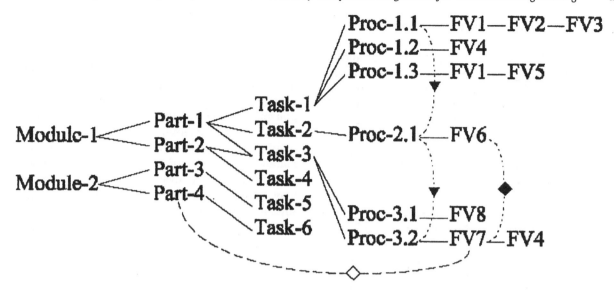

Figure 7: EXTENDED PRODUCT MODEL: the data structure.

a 3D object a "virtual solid". Regarding the modeling techniques appropriate for FS modeling, it is evident that a FS of case 1 or 2 may be defined as an *extruded solid* [6][12]. Regarding a FS of case 3, all examples identified by the authors (including that of Fig. 3) allow transfer of the given distance requirements, associated to faces of the ER component, to appropriate edges of a planar domain D which is the "base" of the component; see Fig. 6. Thus, the space described by the given distances may be constructed by a process based on (a) offsetting edges of D, (b) extracting a new offset-base D_O from these offset edges, and (c) extruding D_O to an appropriate distance H; see Fig. 6.

Constructing a case-4 FS, i.e., (see Fig. 4) the solid below the 3D object R with respect to the surface S, requires solution of complex geometric-modeling problems, which will be discussed elsewhere. Finally, regarding a case-5 FS, this object requires use of the "sweep" operator, which defines a 3D object by moving another object along a trajectory [16].

3.3 The EXTENDED PRODUCT MODEL of an Arrangement

Section 3.1 makes clear that an informationally-complete arrangement model must include all geometric and non-geometric properties of FSs. The *Extended*

Product Model (EPM), described below, extends the standard CAD-model of an arrangement into a comprehensive "functional model" satisfying this requirement.

The main component of the EPM is an *array of tasks* (see Fig. 7), whose elements are well-defined by current ship technology, e.g., "main engine: piston removal", "auxiliary engine: pulling the charge air cooler element", etc. Each task carries pointer(s) to the ER *part(s)* it refers to. Although the strict majority of tasks can be performed in exactly one way, for few of them some manufacturers offer two or three alternatives. Typical example is the task "main engine: piston removal"; see Fig. 2(B). As the designer must record in EPM all alternatives, we associate each task to a list of alternative *procedures*; e.g., in Fig. 7, Task-1 is associated to three procedures and Task-3 to two. Each procedure corresponds to a list of FSs all of which are required for implementing this procedure. Since some procedures are possible only after one or more other procedures are performed, we must add to the "procedure" data structure also two kinds of pointers: incoming pointers from other procedures that are required for the present procedure to be applicable, and outgoing pointers to procedures presuming the present one; e.g., in Fig. 7, Procedure 2.1 carries two such pointers.

Regarding the data structure describing each FS, this must include pointers to parts that are allowed to interfere with this FS. The principal reason for allowing a few interferences is simplification of the geometric description of FSs. Finally, the EPM must record exceptions to the rule *"each FS may be intersected by any other FS"* which generally holds in ER layout. These exceptions identify procedures that must be executed simultaneously and are recorded as pointers added to each FS pointing to FSs that must not intersect it; see Fig. 7.

3.4 Prototype Implementation of EXTENDED PRODUCT MODEL in AutoCAD

The techniques of §3.2 and the model of §3.3 are currently being implemented in the AutoCAD system and in particular in ObjectARX [10], the object-oriented programming environment where also Auto-CAD is developed. Fig. 5 depicts an example of a free space created by the present implementation. This is the FS corresponding to piston removal for the engine of Fig. 2(B). Here, we do not define this FS according to case 2, but rather we treat it as a case-5 FS defined by the exact movement of the piston, which is fully described in [9].

4 Case Study II: Component Layout in Product Design

A mechanical product includes a number of components whose position must satisfy various design constraints. [15] considers the related 2D design problem, and classifies these constraints into three groups: *dimensional constraints, regional constraints*, and *interference constraints*. The first group includes distance and angular dimensions, like those appearing in engineering drawings, that constrain the size, geometric form, location and orientation and of components. Regional constraints restrict the region (area) where a component may lie, and finally, interference constraints specify components and regions that must not overlap.

[15] presents an approach to solving these constraints that systematizes past works. Indeed, the fundamental strategy of most published approaches is clarified as consisting of the following two steps:

- All constraints are translated into dimensional constraints.

- Dimensional constraints are viewed as either a system of algebraic (in)equalities or as a graph,

and a solution is produced by numeric or symbolic or graph-theoretic method.

This approach has obvious disadvantages of which the most important are: (a) it can handle only 2D problems (and some trivial 3D problems where, e.g., components are oriented and move along orthogonal axes) involving only elementary geometric objects, and (b) it requires that linear edges of polygons are approximated with ellipses of a small "width" h, which not only makes the analysis "h-dependent" but also it may lead to erroneous point/object classification results and thus incorrect solutions.

"Design with dimensional constraints" is one of the very early approaches investigated by CAD pioneers, more than twenty years ago, which is based on the assumption, implied by standard engineering drawings, that "a design is fully described by a prescribed number of parameters". This assumption is true only for arrangements including very simple geometric objects with trivial topology. Thus, it is not a surprise to anyone that despite the research efforts of twenty years — which considerably advanced the available tools for treating constraints — the scope of the method is still limited to 2D and trivial 3D problems.

Our proposal, "solidified constraints", implies exactly the inverse approach: (i) translate all design constraints into properties of real and/or virtual solids and relationships between these solids, (ii) treat constraints by procedures operating on these solids. Obviously, regional and interference constraints are "solidified constraints" according to (i), thus we focus on dimensional constraints. Those describing features of a solid may be directly incorporates into the explicit or parametric description of the solid. In the case a parametric solid modeler is not available, then one faces the problem of incorporating inequality constraints like "the height H of the object X must be less than h_0". In this case, one must use virtual solids and translate the dimensional constraint into a regional one. For the above example, the geometric description of X and the value h_0 uniquely define an object X_0 (defined as virtual solid) allowing statement of the above constraint as "X is a subset of X_0". Finally, dimensional constraints describing relations between components are incorporated into the solid model using appropriate local coordinate systems.

It must be clarified that our approach does not aim at eliminating dimensional constraints from the design process but rather placing them at a correct perspective in view of today's object-oriented solid modelers. We propose viewing these as constraints on properties of (real or virtual) solids and processing them

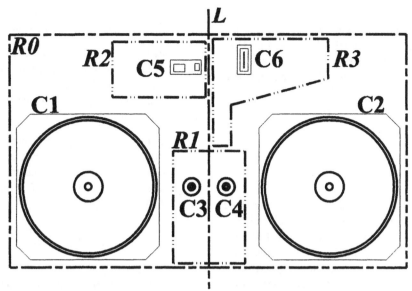

Figure 8: Component Layout Problem [15]: (1) All components inside *R0*, (2) *C1*, *C2* at symmetric positions, (3) *C3*, *C4* at a given distance and inside *R1*, (4) *C5* inside *R2*, etc.

in a solid-modeling environment. This allows solvers to take advantage of the rapidly evolving CAD and solid-modeling technology as opposed to traditional "design with constraints" techniques that view layout problems as systems of equations and thus rely primarily on analytic and numerical methods. Indeed, a huge variety of approaches to "design with constraints" is available, where the differentiating factors are: "amount of interactiveness", "level of compartmentization", "level of optimization", etc [5]. The two extreme cases, and their relevance to "solidified constraints", are discussed briefly below.

4.1   Interactive Component Layout

Very seldom an industrial designer adopts a global approach to constraint solving, simply because the corresponding tools operate in a "black box" mode, i.e., the only input is the constraints, and in a single step, a (usually "unique" and/or "optimal") solution is produced, to which the designer sees often obvious improvements. For this reason, designers prefer the interactive approach where, the design problem is subdivided into smaller sub-problems, which are treated individually in an interactive manner. This means that, to each design modification, the software must respond with a report on constraints satisfied/violated. In this case, solidified constraints are clearly superior to standard constraint modeling. Let us consider Example 1 of [15]: here the user must place non-convex polygons inside a rectangular region. The technique of [15], which suffers from the drawbacks highlighted in

the second paragraph of this section, can only inform the user about violations of constraints. Employing "solidified constraints" would allow use of any current solid modeling system that may offer appropriate measures for "closeness to constraint violation" and even allows biasing the solution using "constraint volumes" [2]. Furthermore, treating constraints in a solid-modeling environment guarantees that the system can handle any number of constraints for arbitrarily complex geometric objects. Indeed, our research is implemented in the AutoCAD system, as this includes a robust implementation of 2D solids, called "regions", that offers numerous interrogation tools and handles analytic and spline geometry also allowing nonmanifold topology.

4.2   Global Optimization of a Layout

In this approach, all design constraints are assembled in a system of (in)qualities treated by an optimization procedure that operates on the whole set of design parameters and produces a solution also minimizing an appropriate "cost function". This is the classical version of the "design with parameters" approach, discussed above, which in principle is incompatible with the new proposal of "solidified constraints". However, if the constraint solver can collaborate with external procedures, then it is possible that materializing certain constraints as solid-modeling queries may improve the scope and the robustness of the method. For example, revisiting the method [15], we observe that regional constraints are implemented in the form of a procedure employing "point/polygon classification", which is the correct approach as established by standard solid-modeling theory. However, use of this classification operation is not sufficient to fully solve the problem. Instead, the solver should collaborate with a solid-modeling library that offers robust solutions to regional and other constraints.

5   Summary

Complex mechanical arrangements are comprised of a large number of components each associated to a varying number of constraints. Current techniques, for producing valid arrangements, view these constraints only as algebraic relations among variables and rely on numerical and/or graph-theoretic tools that ignore the geometric/topological nature of the problem. This paper has proposed the idea of "solidifying constraints", i.e., expressing them in terms of volumes using standard solid-modeling theory and software. The approach has been applied to two cases of "design with constraints": (1) modeling a complex mechanical arrangement with required free-spaces, and (2) component layout in product design.

Acknowledgements

This work was funded by the Hellenic Ministry of Development (grant PENED'95/880 awarded to P. Kaklis), and the Ship Design Laboratory of NTUA. G. Theodosiou was also supported by NTUA's Institute of Communication and Computer Systems through a Graduate Scholarship. Technical information and valuable advice were provided by H. Nowacki (TU Berlin), J. Ioannidis, A. Papanikolaou, N. Kyrtatos, P. Kaklis (NTUA), C. Ventouris (consultant), S. Buchwald, M. Zaika (Applicon, GmbH), J. Tzagarakis, Th. Papakonstantinou (Hellenic Shipyards, Co), and MAN B&W.

References

[1] Bukowski, R.W., Sequin, C.H. (1995), "Object Associations: A Simple and Practical Approach to Virtual 3D Manipulation", Proc. Symp. Interactive 3D Graphics, pp. 131-138.

[2] Chang, H., Li, T.Y., (1995), "Assembly Maintainability Using Motion Planning", Proc. IEEE Int'nal Conf. Robotics & Automation.

[3] Clayton of Belgium (1998), *Steam Generator E-models: Installation Manual* (191.076[rev.B]).

[4] Engine Room Arrangement Modeling (ERAM) Team (1997-98), "ERAM: A Revolutionary, International Approach To Engine Room Design", *Maritime Reporter*, series of four articles: Sept. '97, Jan. '98, July '98, Oct. '98.

[5] Fujita, K., Akagi, S., Nakatogawa, T. (1994), "Hybrid Approach to Plant Layout Design Using Constraint-Directed Search and an Optimization Technique", *Journal of Mechanical Design*, Vol. 116, pp. 1026-1033.

[6] Hoffmann, C. (1989), *Geometric and Solid Modeling*, Morgan Kaufmann.

[7] Johansson, K. (1996), "The Product Model as a Central Information Source in a Shipbuilding Environment", *Journal of Ship Production*, Vol. 12, pp. 99-106.

[8] MAN B&W Diesel A/S, Holeby (1997), *Project Guide for Marine GenSets L16/24* (Ed. 03).

[9] MAN B&W Diesel A/S (1996, 1998[update]), *S60MC-C Project Guide*, www.manbw.dk.

[10] Autodesk (1999), *ObjectARX Developer's Guide*, www.autodesk.com/develop/index.htm.

[11] Potter, C.D. (1999), "A New Role for 3D Plant Models", *Computer Graphics World*, Vol. 22, No. 4, pp. 37-42.

[12] Sapidis, N., Theodosiou, G. (1999), "Planar Domain Simplification for Modeling Virtual-Solids in Plant and Machinery Layout", *CAD*, Vol. 31, pp. 597-610.

[13] Sapidis, N., Theodosiou, G. (1999), "Operational Constraints in Computer-Aided Ship Design: Modeling the Required 'Virtual Solids'", Proc. 10th Intern. Conf. Computer Applications in Shipbuilding, Vol. I, MIT, USA, pp. 83-96.

[14] Shinya, M., Forgue, M.C. (1995), "Laying Out Objects with Geometric and Physical Constraints", *Visual Computer*, Vol. 11, pp. 188-201.

[15] Suzuki, H., Ito, T., Ando, H., Kikkawa, K., Kimura, F. (1997), "Solving Regional Constraints in Components Layout Design based on Geometric Gadgets", *AIEDAM*, Vol. 11, pp. 343-353.

[16] Zeid, I. (1991), *CAD/CAM Theory and Practice*, McGraw-Hill.

Digital Approach for Product and Process

J-F CUGY / C. ROUCHON
Dassault Aviation, CC-CFAO, 78 Quai M. Dassault
Cedex 300 - 92552 SAINT-CLOUD - FRANCE
Tel : +33(1) 47115221 Fax : +33(1)47115244

1 A Concurrent Engineering Approach

The gains necessary for our survival cannot be obtained merely by optimizing the existing industrial structures. A new way of working must take place, it is often referred to as CONCURRENT ENGINEERING.

A typical definition of Concurrent Engineering is :

" A systematic approach to creating a product design that considers in parallel all elements of the product lifecycle from conception of the design to disposal of the product, and in so doing, defines the product, its manufacturing processes, and all other required lifecycle processes such as logistic support.

These activities must be started before all prerequisites are frozen and hence must be adjusted afterwards. In this way, it is possible to do much work in parallel with the main goal to shorten the elapsed time. By powerful computer and communication network support Concurrent Engineering also opens the possibility to test a number of alternative solutions.

Achieving this, the ultimate effort of Concurrent Engineering is to integrate product and process design."

2 Information Technology Tools

With the CE approach and the use of "up to date" hardware and software, it is now possible to deliver the right piece of information, at the right time, to the right person, giving to everybody a coherent vision of one project.

Since 1979, CATIA has been used at Dassault Aviation for design and manufacturing activities. At first the problem was to define a single part (mechanical, sheet metal...). We have defined optimized product lines. A product line is characterized as a data flow between activities associated to an aeronautical part category, from design to manufacturing, including quality control inspection and customer services.

Benefits of this global optimization are definitely higher than pure local adjustment or automatization of isolated tasks. Results have been achieved on one hand with a clear settlement of our CAD/CAM use in our business processes (the dissemination of standard rules and procedures among all partners -internally or outside the company) and on the other hand with the development of many dedicated software (integration of our know-how) on the same CATIA platform. We are used to call this approach a vertical integration

Vertical integration is today under control in our company. We are enhancing it with an industrial exploitation of "design by features" which allows us to encapsulate information in entities (features) of the digital definition [3]. Afterwards, during the manufacturing phase, automatic routines are based on features recognition. Today, "design by feature" is a reality for sheet metal parts at Dassault Aviation.

For the Falcon 2000, RAFALE and the future airplanes, our company has taken the very decisive choice to replace the "physical mock-up" (PMU) by a "digital mock-up" (DMU). Today, DMU specific applications for design, manufacturing and support activities are running on Dassault Aviation sites.

By mean of a large scale digital assembly application, every designer can (as frequently as necessary) search in the database for parts located in a given area. By this way, design development is an iterative cycle starting with the creation of 3D models then checking, revising and sharing the assembly until this one is achieved. This application is based on the ENOVIA and CATIA capabilities in a Relational Data Base environment.

At the design office level, all parts of an aircraft (more than 20.000) are created and DMU is used by every designer who is checking that his layouts fit with those which are concurrently defined by other teams (structure, hydraulics, wire bundles). The user is in position to check for collisions or to analyze accessibility and assembly-disassembly methods without having to rely on physical mock-up. With VR techniques, we are improving our capabilities in visualization and space navigation. It is particularly important when we are working in group (more than ten persons) to validate the definition of some areas. For this occasion, they are persons representing quality control department and some other from the different disciplines involved in the particular area.

3 Digital Mock-up and Manufacturing Process

Assembly process planning and manufacturing instructions

At the manufacturing engineering level, we have deployed a specific CATIA application (called SOFIA) which allows the user to build the "as-planned-for_manufacturing" view of the product from the "as-designed" view released in the DMU.

Today, we are exploiting these information for installation datasets, process plans and working instructions for assembly, certification and customer documentation.

Advanced Virtual Manufacturing

Current developments of Dassault Systemes integrate the Product Model; the Industrial Resources Model and a Process Model.

Data models are based on STEP. Process model includes product flow, control flow and state diagrams representations.

This integration is a decisive issue for Manufacturing Virtual Execution.

Assembly process and Tooling (jig)

We are developing CAD/CAM applications which connect aircraft product design and preliminary design of assembly jigs.

A jig knowledge base is developed (degrees of freedom of mating features, accessibility criteria...). Kinematics simulation (relative extraction directions of the product after assembly) and interference checking are implemented. This allows an early validation of assembly feasibility.

Review / Global release

This case is quite different. New information interface technologies are unavoidable because "mock-up reviews" which were formerly achieved on a physical mock-up at workfloor must be now carried out in dedicated rooms at design office.

The name has been changed from "mock-up review" to "digital assembly review" but the principle to "navigate" in the environment still remains. The project manager is asking for the expertise of the different specialists during common navigation in order to get consensus and validation.

This activity requires a few rooms fitted out with big screens, so investments can be noteworthy. Pure visual feedback will have to be completed with force feedback in order to validate, for instance, assembly or disassembly of heavy equipment.

However, decisions at this step are not immediately executed; they usually presuppose long interventions of CAD/CAM specialists. So, the visualization database may be different from the CAD/CAM database provided reliable updating.

Our action in this domain is still an optimization of CAD/CAM solutions (especially visualization time delay) and furthermore an integration of new simulation tools (datagloves, masterarm, headmounted display...).

4 Virtual Product Management

This leads to continuously manage through DMU the configuration of each physical airplane "as designed", "as-planned for manufacturing" and "as really built". This management is organized for long term (30 years).

Concerning any extension of our data model, we are focusing on ISO 10303-203 recommendations (STEP AP 203) for product structure and configuration.

To be more efficient we are combining the PDM and the CAD/CAM functionalities in our information system.

We are used to calling this approach an horizontal and vertical integration, because this is dealing both with the different product components and the different stages in the product life cycle.

Our vision for the future of design and manufacturing engineering activities is what we call "an integrated space for definition". The output will be numerical definition of a product and the related manufacturing processes. This data will be the result of cooperation between multiple partners (some of them in Dassault Aviation premises some others outside, in France or abroad). To analyze efficiently this huge amount of data, different techniques will be used, but for the geometrical aspect, the "navigation" with the associated VR techniques will be decisive.

5 Conclusion

CE techniques use hardware and software as means of communication to help men and women to ensure their mission to sell, design, manufacture and support aircraft. VR techniques enrich this tools.

It is wrong to think that these techniques will reduce the exchanges between persons. On the contrary, they will break some organizational frontiers and decrease the distance effects allowing people to work together.

Many progress are still necessary in DMU and associated Tools and Methods but today it is a reality at Dassault Aviation and will be a competitive advantage for the future.

References

[1]   G. Burdea, P. Coiffet : La Réalité Virtuelle - Editions Hermès

[2]   P. Chedmail, T. Damay, C. Rouchon : Integrated Design and Artificial Reality - Accessibility and path planning

[3]   BREU 033 Th. Hardy - Final Report February 96

[4]   ESPRIT III Project N° 6562 - Final Report - 95

Concurrent Engineering in the Construction Industry

Hojjat Adeli

Department of Civil and Environmental Engineering and Geodetic Science
The Ohio State University
470 Hitchcock Hall, 2070 Neil Avenue, Columbus, Ohio, 43210 U.S.A.

The goal of concurrent or simultaneous engineering is to move a product concept based on a market need to a manufactured and marketable product in the shortest possible time and with minimum cost. Several recent technological advances should help expedite this process. They include (Adeli, 1999)

- increasingly powerful personal computers and workstations at decreasing prices,

- development of fundamentally different approaches to problem solving and new computing paradigms such as neurocomputing, evolutionary computing, and fuzzy logic (Adeli and Hung, 1995; Adeli and Park, 1998)

- advances in sensor technology,

- more refined visualization tools such as solid modeling and virtual reality, and

- significant and continuous improvement in network infrastructure in most enterprises (Adeli and Kumar, 1999).

Integration of design, engineering, and manufacturing (CAD/CAE/CAM) is the critical issue in concurrent engineering. Design of a product requires significant cost-sensitive choices in how the materials are made and how parts are assembled. Automation of engineering design has alluded engineers for decades as this problem has been brushed aside as open-ended and *ill-defined*. But, design automation is prerequisite for its integration with manufacturing. And design automation has to be considered in the context of optimization of resources. With recent development of robust design automation and optimization algorithms and other aforementioned advances all the elements needed to perform effective concurrent engineering are all within reach. Replacing physical models with virtual prototypes in a concurrent processing environment can reduce the cost and lead time of product development substantially (Adeli, 1999).

Concurrent engineering technology has been developed substantially in automotive and other manufacturing applications. It is now finding applications in the construction industry in both U.S. (El-Bibany and Paulson, 1999; Pena-Mora and Hussein, 1999) and Japan (Kaneta et al., 1999). The current prevailing practice is to complete the design before the construction is started. But, changes in the design might be necessary to improve the product or project even after the construction has already begun. For concurrent engineering in the construction industry it is necessary to integrate the construction management and scheduling with the design process.

Successful application of concurrent engineering in the construction industry should be based on effective integration of design and construction. Two essential prerequisites for such an integration are a tool to automate the complex process of engineering design and a tool for construction scheduling, cost optimization and change order management.

Automation of design of large one-of-a-kind civil engineering systems is a challenging problem due partly to the open-ended nature of the problem and partly to the highly nonlinear constraints that can baffle optimization algorithms (Adeli, 1994). Optimization of large and complex engineering systems is particularly challenging in terms of convergence, stability, and efficiency. Recently, Adeli and Park (1998) developed a neural dynamics model for automating the complex process of engineering design through adroit integration of a novel neurocomputing model (Adeli and Hung, 1995), mathematical optimization (Adeli, 1994), and massively parallel computer architecture (Adeli, 1992a&b, Adeli and Soegiarso, 1999).

The computational models have been applied to fully automated minimum weight design of high-rise and superhighrise building structures of arbitrary size and configuration, including a very large 144-story superhighrise building structure with 20,096 members. The structure is subjected to dead, live, and multiple wind loading conditions applied in three different directions according to the Uniform Building Code (UBC, 1997). Optimization of such a large structure subjected to the highly nonlinear and implicit constraints of actual design codes such as the AISC LRFD code (AISC, 1998) where nonlinear second order effects have to be taken into account has never been reported before. The patented neural dynamics model of Adeli and Park finds the minimum weight design for this very large structure subjected to multiple dead, live, and wind loadings in different directions automatically.

Adeli and Karim (1997) present a general mathematical formulation for scheduling of construction projects. An optimization formulation is presented for the construction project scheduling problem with the goal of minimizing the direct construction cost. The nonlinear optimization problems is then solved by the patented neural dynamics model of Adeli and Park (1998). For any given construction duration, the model yields the optimum construction schedule for minimum construction cost automatically. By varying the construction duration, one can solve the cost-duration trade-off problem and obtain the global optimum schedule and the corresponding minimum construction cost. The new construction scheduling model is particularly suitable for studying the effects of change order on the construction cost.

Recently, Karim and Adeli (1999a) presented an object-oriented (OO) information model for construction scheduling, cost optimization, and change order management based on the new construction scheduling model, with the objective of laying the foundation for a new generation of flexible, powerful, maintainable, and reusable software systems for the construction scheduling problem. The model is presented as a domain-specific development *framework* using the Microsoft Foundation Class (MFC) library and utilizing the software reuse feature of the *framework*. The OO information model for construction scheduling and cost management can be integrated into a concurrent engineering model for the construction industry. The information and computational models have been implemented into a new generation prototype software system called CONSCOM (for CONstruction Scheduling, Cost Optimization, and Management) (Karim and Adeli, 1999b).

CONSCOM includes a superset of all currently available models such as CPM plus new features such as (Karim and Adeli, 1999c):

- Integrated construction scheduling and minimum cost model based on the patented robust and powerful neural dynamics optimization model of Adeli and Park that provides reliable cost minimization of the construction plan, time-cost trade-off analyses, and change order management.
- Support for a hierarchical work breakdown structure with tasks, crews, and segments of work.
- Capability to handle multiple-crew strategies.
- Support for location (distance) modeling of work breakdown structures (very useful for modeling linear projects such as highway construction).
- A mechanism to handle varying job conditions.
- Nonlinear and piecewise linear cost modeling capability for work crews.
- Capability to handle time and distance buffer constraints in addition to all the standard precedence relationships.
- Ability to provide construction plan milestone tracking.

Concurrent engineering processes will be refined further and advanced in several directions in the coming years. The current concurrent engineering technology relies primarily on management of large data bases (Adeli, 1999). One refinement will be integration of artificial intelligence techniques with virtual visual models. Knowledge management and knowledge engineering techniques have the potential to enhance concurrent engineering processes significantly. This can take several forms from natural language processing and speech recognition, to machine vision technology, to machine learning techniques (Adeli, 1990a,b, 1997) to strategic use of smart sensors and robots (Adeli and Saleh, 1999).

Acknowledgment

This Keynote Lecture is based upon work sponsored by the U.S. *National Science Foundation* under Grant No. MSS-9222114, *American Iron and Steel Institute, American Institute of Steel Construction, Ohio Department of Transportation,* and *Federal Highway Administration.* Supercomputing time was provided by the *Ohio Supercomputer Center* and *National Center for Supercomputing Applications* at the University of Illinois at Urbana-Champaign. Part of the work resulted in a United States Patent entitled *Method and apparatus for efficient design automation and optimization, and structure produced thereby.* The patent was issued by the *U.S. Patent and Trademark Office* on September 29, 1998 (Patent 5,815,394). The inventors are Hojjat Adeli and H.S. Park.

References

Adeli, H., Ed. (1990a), *Knowledge Engineering – Volume One – Fundamentals*, McGraw-Hill Book Company, New York.

Adeli, H., Ed. (1990b), *Knowledge Engineering – Volume Two – Applications*, McGraw-Hill Book Company, New York.

Adeli, H., Ed. (1992a), *Supercomputing in Engineering Analysis*, Marcel Dekker, New York.

Adeli, H., Ed. (1992b), *Parallel Processing in Computational Mechanics*, Marcel Dekker, New York.

Adeli, H., Ed. (1994), *Advances in Design Optimization*, Chapman and Hall, London.

Adeli, H., Ed (1997), *Intelligent Information Systems*, IEEE Computer Society, Los Alamitos, California.

Adeli, H. (1999), "Competitive Edge and Environmentally-conscious Design Through Concurrent Engineering", Assembly Automation, Vol. 19, No. 2, pp. 92-94.

Adeli, H. and Hung, S.-L. (1995), *Machine Learning – Neural Networks, Genetic Algorithms, and Fuzzy Systems*, John Wiley & Sons, New York.

Adeli, H. and Karim, A. (1997), "Scheduling/Cost Optimization and Neural Dynamics Model for Construction", Journal of Construction Engineering and Management, Vol. 123, Np. 4, pp. 450-458

Adeli, H. and Kumar, S. (1999), *Distributed Computer-Aided Engineering for Analysis, Design, and Visualization*, CRC Press, Boca Raton, Florida.

Adeli, H. and Park, H.S. (1998), *Neurocomputing for Design Automation*, CRC Press, Boca Raton, Florida.

Adeli, H. and Saleh, A (1999), *Control, Optimization, and Smart Structures - High-Performance Bridges and Buildings of the Future*, John Wiley & Sons, New York.

Adeli, H. and Soegiarso, R. (1999), *High-Performance Computing in Structural Engineering*, CRC Press, Boca Raton, Florida.

AISC (1998), *Manual of Steel Construction – Load and Resistance Factor Design – Volume 1 Structural Members, Specifications, & Codes*, American Institute of Steel Construction, Chicago, IL

El-Bibany, H. and Paulson, B.C. (1999) "A Parametric Architecture for Design, Management, and Coordination in a Collaborative AEC Environment", *Computer-Aided Civil and Infrastructure Engineering*, Vol. 14, No. 1, pp. 1-13.

Kaneta, T., Furusaka, S., Nagaoka, H., Kimoto, K. and Okamoto, H. (1999), "Process Model of Design and Construction Activities of a Building", *Computer-Aided Civil and Infrastructure Engineering*, Vol. 14, No. 1, pp. 45-54.

Karim, A. and Adeli, H. (1999a), "OO Information Model for Construction Project Management", *Journal of Construction Engineering and Management*, ASCE, Vol. 125, No. 5, pp. 361-367.

Karim, A. and Adeli, H. (1999b), "CONSCOM: An OO Construction Scheduling and Change Management System", *Journal of Construction Engineering and Management*, ASCE, Vol. 125, No. 5, pp. 368-376.

Karim, A. and Adeli, H. (1999c), "A New Generation Software for Construction Scheduling and Management", *Engineering, Construction, and Architectural Management*, Vol. 6, No. 4, pp. 380-390.

Pena-Mora, F. and Hussein, K. (1999) "Interaction Dynamics in Collaborative Design Discourse", *Computer-Aided Civil and Infrastructure Engineering*, Vol. 14, No. 3, pp. 171-185.

UBC (1997), *Uniform Building Code - Volume 2 - Structural Engineering Design Provisions*, International Conference of Building Officials, Whittier, California.

CHAPTER 1

Enterprise Engineering

An Engineering Workflow System for the Management of Engineering Processes Across Company Borders

Kamel Rouibah & Kevin Caskey
Department of Information Technology
Faculty of Technology Management
Technische Universiteit Eindhoven
P. O. Box 513, NL-5600 MB, Eindhoven, The Netherlands

Abstract

The most four common views used in business are the DATA, FUNCTIONAL, PROCESS, AND ORGANISATIONAL view (see, e.g. AMICE [1.]). Most research deals with one of these views or a combination. This paper presents another view based upon a parameter approach.

Most available workflow systems, stand alone or incorporated as a module in PDM systems, focus upon administrative workflow and offer little support for simultaneous engineering work. This paper deals with the specification of an engineering workflow (ewf) methodology, suitable for simultaneous engineering, based on the parameter approach and its implementation within the PDM system CADIM/EDB. The parameter approach is based on the concepts of hardness grades, transition statuses and their relationships. Engineering workflow is a combination of classical workflow and a parameter based-approach. The engineering workflow methodology consists of: the control procedure for upgrading parameters, the control procedure roles, the work list associated to activities performed, and the data associated to ewf.

1. Introduction

The changing business and competitive environment requires firms to introduce new products more frequently and in shorter time. This also means that organisations have to be agile as well as responsive to the changing needs of customers. This has lead to new reorganisation approaches such as Business Process Reengineering (BPR). BPR involves a fundamental renewal of corporate activities, often aimed at creating customer order driven business processes. BPR also implies the restructuring of internal processes and the flow of information between different actors. Therefore BPR and workflow occur together [2.]. They are both used when automating the complex processes found in office or industry. Available workflow software more often supports office activities. There is a need for a new generation of workflow systems that could offer support for complex business processes such as product development.

1.1. Characteristics of engineering activities during complex processes of product development

In engineering activities, such as product development, most processes are not fixed but highly networked, and extremely dynamic. The development of a new railcar bogy, for example, consists of the simultaneous development of several components, including "Running Gear", "Drive Unit" and "Brake System". This procedure requires a clear definition of interfaces and constant communication and consideration of the respective requirements. Development activities are not completed in a single pass and are often subject to events that lead to many iterations. The mentioned example serves to show that an advanced degree of simultaneity and networking already exists. The more detailed the analysis of the product structure, the more complex is the interplay of the various development processes.

Analyses of engineering activities [3.][4.] show that a very high share of engineering hours are spent on organisational matters rather than on "productive"' tasks. Major factors identified as decreasing engineering efficiency include:

- poor engineering process management, both in definition and monitoring,
- missing direct access to product data among the engineering partners,
- heterogeneous communication and systems infrastructure,
- lack of engineer involvement in early stages of the project.

For efficient customer driven engineering process development, information content and deployment is relevant, but the timing of the gathering and deployment process is more important. This has resulted in deployment models such as simultaneous engineering [5.]. Although many people think of simultaneous engineering as a concept where development activities are performed in parallel, increased efficiency can only be reached when the following requirements are fulfilled:

- Designers Co-operate, not only within a company but also across company.
- Designers maintain a high level of communication during engineering activities.
- Potential problems and risks are identified and resolved as early as possible in the design process.
- Actual capabilities of product (functionality, quality, etc.) are validated as soon as possible in the design and development process.
- Relevant actors within the organisation or network check and validate the evolved information.
- Customer specified changes should be integrated continuously and as early as possible.

All these actions affect the downstream issues of manufacturing, quality, product final cost, product introduction time, and the marketplace success of the product.

The SIMNET[1] project develops an approach to satisfy these requirements.

1.2. Focus of SIMNET

Several projects, within European Commission Research Programmes [6.][7.][8.][9.] are developing methodologies and tools to locate, organise, transfer, and use the information and expertise distributed throughout organisations.

The focus of SIMNET is on the engineering workflow for complex engineering processes that are order or project related, the customer order driven engineering process. These processes generate the product data that are required for the control and realisation of the production process. Customer order driven engineering processes are characterised [10.] by:

- *High product and process uncertainty*: the more innovative the project the higher the uncertainty in the technology used and the processes required.
- *Frequent changes and disturbances*: during the project the requirements can change for many reasons. Customer's requesting a change of a specific function is one of the most common.
- *Iterative nature*: Due to the nature of the design and engineering process the outcome of an activity will not be stable after the first pass.

SIMNET is to specify an engineering workflow methodology suitable for complex simultaneous engineering across company borders. This methodology aims to link product structure to workflow management. This methodology will be applied to enhance the workflow module of the PDM system CADIM/EDP in order to conform with the STEP and WfMC standards and to offer flexibility to engineering activities.

The project will capture the industrial needs, exploit existing technology and develop new solutions, and to demonstrate the value in an actual industrial environment. This paper is focused on the SIMNET approach to "engineering workflow" which is an emerging concept [11.]. This approach is focused not only on the product but also on the learning capability of a company based upon information. Essential considerations of this approach are:

- the level of information detail needed to perform an activity,
- the relevant people in the network receive information,
- the time required to obtain and deploy this information.

In the following section we will characterise the concept of "engineering workflow".

2. Classical workflow and engineering workflow

According to the WfMC[2] [12.] workflow is defined as: *the automation of a business process, in whole or part, during which documents, information or tasks are passed from one participant to another for action, according to a set of procedural rules*. Thus workflow is characterised by a predefined, fixed, and stable process description and most of the procedural rules are defined in advance. Workflows are based on events and tasks which are combined via different rules (AND, OR, XOR). Each activity has starting (respectively completing) events that trigger (respectively end) that activity. The starting event could be triggered by (1) a clock alarm, (2) a document query, (3) and external event (such as the arrival of an email) etc. A specific work list (list of work items) is assigned to each user who fulfils a specific organisational role. At any time, the workflow participant, i.e. a person who fulfils a role, knows what he must do and which tasks to perform and complete within a defined duration.

[1] See acknowledgement

[2] WfMC– the Workflow Management Coalition is a non profit organisation with the objectives of advancing the opportunities for the exploitation of workflow technology through the development of common terminology and standards.

2.1. Types of workflows

Workflow can be classified into [12.]:
- administrative (or "classical") workflow,
- ad-hoc workflow.

Administrative workflows are represented by structured, predefined, and time invariant process definitions. The triggering events, the activities to be performed, their sequence, as well as the workflow participants (usually defined in terms of roles) are fixed. Then, it is possible to support the process execution by informing the workflow participant about tasks to be done and by providing the information needed for the execution of these tasks. Administrative workflows are instantiated without modification each time they are performed and are therefore suitable for standardised and/or highly repetitive processes (e.g. processing tax forms or insurance claims).

Ad-hoc workflows are represented by unstructured and highly variant processes. The triggering events, the activities to be performed, and the workflow participants are not fixed in advance. The order of activities forming ad-hoc processes is defined only at runtime. Therefore, ad-hoc workflows are defined at runtime.

2.2. Definition of the engineering workflow (SIMNET workflow)

In the engineering domain, processes are complex and networked and both types of workflow occur. Document approval, release, and version control procedures as well as Engineering Change Management (processing of Engineering Change Requests and Orders) are typical examples of well structured, highly repetitive, and time invariant processes whose participants can be pre-defined in terms of roles prior to the workflow execution. In engineering, such administrative workflows are primarily meant to support paperless quality assurance procedures in distributed environments.

In [14.], we have shown that the characteristics of technical (i.e. non-administrative) engineering processes depend very much on "the specific circumstances" in a given engineering situation. Normally, neither the occurring workflow events, the activities to be performed, their sequences, the workflow structure nor the workflow participants can be pre-defined. The workflow is therefore purely ad-hoc and would have to be specifically designed for each particular engineering situation. This is of course not feasible.

The SIMNET approach to engineering workflow allows the formalisation of engineering situations in such a way that at least the occurring events, the activities to be performed, and (to some extent) the process structure can be represented by a single, standardised process definition.

The final structure of the workflow and specification of the workflow participants are generated during the workflow runtime. These are based on the type of product data which is to be communicated as part of the workflow. These product data depict the mentioned "specific circumstances in a given engineering situation" which trigger and control the technical engineering workflow.

We define engineering workflow (*ewf*) as [14.]: *a combination of classical and product data controlled workflow*. Ewf aims to reduce the time spent in developing activities, to monitor processes developed, to ensure the quality of work in progress, and to allow a greater transparency of engineering work. Ewf can be found in highly networked, iterative, and extremely dynamic processes such as product development, prototype engineering, distributed development, and pilot production.

2.3. Restrictions of current WfMS in the support of *ewf*

Current workflow management solutions (including those integrated into PDM systems designed to support engineering processes), cover the administrative workflow only and are not sufficient for dealing with *ewf* [15.][16.]. These workflow systems mainly define and control the process of reviewing and approving changes to product data, and help workflow participants identify tasks to be done, and provide information needed to perform those tasks. These systems are well suited to supporting recurring administrative operations with static structures However, they allow little flexibility and do not support highly networked and dynamic processes subject to ad-hoc influences, which characterise the work of designers and engineers. Indeed engineering work is not primarily described by process and document reference [13.]. An engineer's thinking and creative process works back and forth between imagination and analysis. It is necessary to create systems that support *ewf*.

Some other systems (given that they provide convenient process modelling functionality) may allow rapid process modelling which may afterwards be executed as ad-hoc workflows. However, due to the large number of processes required, this approach to the management of technical workflows is hardly feasible.

In order to fully support ewf as defined above, current systems require significant enhancement. The following section describes a new approach for providing this enhancement.

3. Modelling the engineering process from a parameter perspective

The SIMNET approach considers complex product development as a form of parameter processing. Therefore, the engineering process is approached as a network of activities that uses and produces parameters. An activity can start when all required input parameters are available and ends when the target parameters are stable. The SIMNET parameter based approach consists of three concepts [13.]. First, the *parameter checklist* provides engineering activity transparency (the parameters are available in each design phase). Second, the parameter checklist is linked to a simultaneous engineering workflow. Third, the parameter network is logged to support Engineering Change Management. These three steps constitute the backbone of the engineering workflow methodology to be described below.

3.1. Definition of parameters

SIMNET defines parameters as the specific circumstances in a given engineering situation including the involvement of suppliers and engineering partners. Parameters, and their interactions, represent the variables within engineering design functions. A parameter refers to forces, accelerations, movements, etc. For example, parameters related to design of a railcar bogie include:
- maximum axle diameter,
- maximum static axle load,
- bearing distance,
- track gauge,
- gear transmission ratio,
- worn wheel diameter,
- clearance to upper surface of the rail,
- distance between wheel axle and motor axle,
- number of gear steps.

These parameters share complex relationships. For instance, the parameter "maximum axle diameter" is directly dependent upon the parameters "maximum static axle load", "bearing distance", and "track gauge". Whereas the parameters "distance between "wheel axle" and "motor axle" are only indirectly dependent upon the above parameters but directly dependent upon the parameter "maximum axle diameter".

3.2. Parameter life cycle

- SIMNET defines 6 parameter statuses:
- "Predefined" is parameter whose value is specified by customers according to internal functional requirements.

- "Un-worked" is parameter whose value is not yet specified but will be specified, checked, approved and released.
- "In work" is parameter whose value is under construction (preliminary estimation) or is still in specification and check. However, it is neither approved nor released.
- "In approval" is a parameter whose value has been estimated and needs to be fixed.
- "In release" is parameter whose value has been estimated and fixed.
- "Released" is parameter whose value has been accepted by all parties.

3.3. Hardness grades

The concept of hardness grade (HG) has been initiated by McKinsey and adopted by SGP's engineers in order to express how secure and stable the value of a specific parameter can be considered. We define HG both as a quality measure and reliability measure of a parameter specification during the design process. This allows greater transparency and control of the engineering activities. The HG is linked to design life cycle. In SIMNET, we have defined 5 HGs, which indicate the maturity of parameter's setting, in the view of all engineers involved. For example, HG 2 means that the parameter value is:
- specified, based on more precise calculations in co-operation with other engineering groups (others than the one creating the first value, e.g. mechanical engineering),
- checked and preliminarily approved by suppliers (involved in the design process),
- checked and preliminarily approved by internal organisational units different from engineering (e.g. operations planners or purchasers),
- checked and released by the responsible project or product manager.

3.4. Roles

Five roles are defined to handle parameters:

- "Creator" depicts an engineer who is clearly appointed to be in charge of the parameter (assigning values). If there is a group of engineers, the one appointed by the group to be in charge would be the creator; the other team members would be the editors.
- "Editor" depicts a group of engineers who are directly involved in the parameter value creation or modification, i.e. engineers who may need to work with or change an estimated or calculated parameter value. If the team decides that a new value should be

assigned to the parameter, the creator is in charge of actually assigning it.

- "Reviewer". In addition to the above-mentioned involved engineers, reviewers are persons from other areas such as operations planning, purchase, etc., who may need to approve a parameter value.
- "Supervisor" depicts the engineer who only defines the release time and makes sure that whatever required documentation is available before releasing the parameter. He may then change the status of parameter from "approved" to "released" and move the parameter to a higher HG.
- "Subscriber" depicts engineers who want to be informed about parameter values, HGs, and statuses. While they are not assigned to the parameters, they sign up to the parameter themselves.

3.5. Steps toward the application of *ewf*

Before using the parameter based approach toward ewf, several preliminary steps need to be performed in order to assure engineering transparency (see figure 1).

- define the life cycle development process (define the design phases) and milestones,
- assign parameters to each phase of design (sales & marketing, and the other phases),
- initiate a new project (designing a specific bogie).

After initiating a new project with its parameters, engineers then need to identify parameters, which are "predefined" and "un-worked". Then, the engineering process can begin.

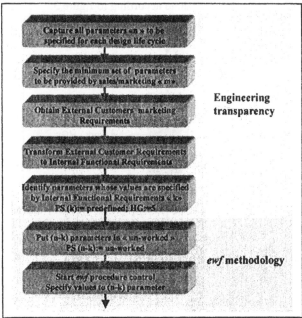

Figure 1 Steps toward the application of ewf

4. Specification of the ewf methodology

The specification of an *ewf* methodology necessitates four basic elements (Figure 2). These will be discussed in the following subsections.

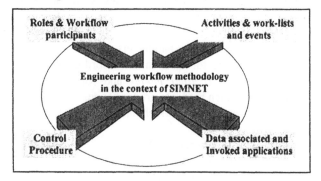

Figure 2. Basic elements for an *ewf* methodology

4.1. Duties and privileges related to roles

The following paragraph gives further clarification of roles related to the ewf methodology.

Users are assigned to roles, categories, and parameters. Users can be assigned to more than one role. In addition, each person could be in charge of more than one parameter. Roles and their duties include:

Creator
- is technically responsible,
- creates the parameter,
- changes status (and hardness grade in a simple approval procedure),
- must
  - obtain approval,
  - inform people about changes,
  - monitor progress,
  - push for consensus (a consensus is required),
  - set and keep deadlines.

Editors
- are directly involved in parameter elaboration,
- represent different backgrounds,
- have its own values to maintain,
- have no write permission to other values,
- have access to discussion,
- must explicitly approve a hardness grade increase (by increasing the status from "in approval" to "in release").
- Reviewer
- must explicitly approve a hardness grade (by increasing the status from "in approval" to "in release"),

- can set request for information according to each status change *(default mode)* and/ or each value change,
- may grant approval up to a certain hardness grade (simplified approval) and for a value within a certain range.

Subscriber

- can freely subscribe to parameters,
- can set request for information according to - each status change *(default mode)* and/ or each value change and each hardness grade change,
- may raise an objection which has to be dealt with.

Supervisor

- is organisationally responsible,
- releases parameters at a certain time,
- co-ordinates the overall project,
- must deal with rejected objections on demand,
- must explicitly release a parameter to increase the hardness grade,
- may reject release due to technical or organisational reasons.

4.2. Control procedure for upgrading parameters

The control procedure for parameters has three steps:

1. Control procedure (triggered by):
   - the creator
   - a request modification stated by editor, reviewer and the subscriber,
   - an expired deadline,
2. collective simple or complex release procedure,
3. parameter update.

During the collective release, there are two phases in upgrading parameter values, statuses, and hardness grades:

- by using "informal messages"
- by setting a "formal approval procedure" workflow.

First phase, using an informal message. Prior to approval, the value must be stable. All those affected are informed of a change, and they communicate until each person accepts the value. In this case the parameter goes from "in work" to "in approval" status. The messages may be supported by meetings, phone calls etc. One way to end this procedure is to assign a deadline to each parameter value and to allow each person to define his proper interval for each parameter value. Feedback from a person (engineer) involved in the "informal message" could be: a silent response (no feedback) or, an explicit approval (agreement) without justification or, a veto (disagreement) with justification.

Second phase, using the "formal approval procedure" workflow. This is used when a parameter value is agreed by the participants in the simultaneous engineering process (phase 1), and needs to be stable. The parameter value fixed in the phase 1 will also be subject to the approval of reviewers. This represents an additional formal procedure for approving and releasing the value. In this case, upgrading parameters is done according to an automatic execution of a single classical workflow by means of a workflow engine. Workflow participants are users who fulfil the role of reviewers, who must approve the parameter value. Feedback of engineers involved in the formal approval procedure should be: an explicit approval (agreement) without justification or, a veto (disagreement) with justification.

According to the SIMNET approach, it is possible to use two release procedures. A "complex release procedure" is a combination of phase 1 and 2. All parameter statuses are used within each HG. A "simple release procedure" is used in order to reduce the e-mails related to the use of "formal approval procedure". In this case, upgrading the hardness grade takes place only according to "informal messages" and parameter statuses involved are only "in work" and "released".

Identification of persons (engineers) who need to be informed, in the case of parameter upgrading, is done by checking persons who were assigned, for example by the supervisor, or freely subscribed to parameters. This is a part of the intelligence behind the SIMNET solution. Upgrading parameters from a current hardness grade i to hardness grade $i+1$ is done according to figure 3.

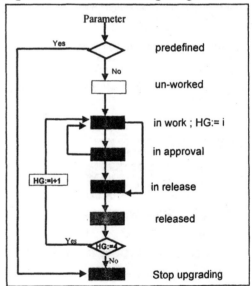

Figure 3. Upgrading parameter from HG i to HG $i+1$ through the *ewf*

A parameter marked *pre-defined* by a customer requirement is not subject to status transition and does not need to be released. Other parameters can be in different statuses. In the same hardness grade, these parameters have different statuses. All the parameters, except predefined, are subject to transition from "un-worked" to "release" statuses.

4.3.　Activities related to *ewf* methodology

In SIMNET, we have identified many activities within the ewf methodology. These can be manual (performed by workflow participants), automatic (performed either by the WfMS or by invoked/external applications), or a combination of both manual and automatic. Each activity has a name, a role associated, a resource (or user) that executes the activity, an event (either external or internal) that triggers the activity, a short description, and a list of instructions.

For example the activity "set initial value" consists of fixing an initial value to an "un-worked" parameter. This activity is triggered by the external event "*predefined parameters have been determined and the project of a bogie development has started*". This activity is performed by a specific user designated the creator. Executing this activity consists of the following steps:

- The creator accesses the system (in our case the PDM system).
- the system verifies identification.
- the creator accesses the parameter with status "un-worked".
- the system verifies permission.
- the creator assigns a minimum, maximum or both values.
- the system checks the writing permission of the creator and puts the parameter in "in work" status and HG:= i.
- the system generates an email, identifies recipients (different users that would be informed with appropriate information), adds a folder and sends the email.
- Users who fulfil the editor role are free to set their own ranges for further discussions.

4.4.　Data associated with *ewf* methodology and invoked applications

Each parameter stored in the database of the PDM system has three types of data:

1. Parameter attributes (e.g.)
   - parameter name,
   - maximum and minimum value,
   - creator name,

2. Data history of parameters that records all changes made through parameter upgrading (e.g.)
   - created by,
   - created date,
   - remark (any kind of memo or explanatory text),
   - remark stated by,

3. Data related to emails or notification services (e.g.)
   - current and old value, status, hardness grade,
   - response to "Do you agree with the current value?"

If information is exchanged between engineers by email only, it is easy to keep the parameter history. However, if the exchanged information includes other communication channels, such as phone calls, meetings, faxes, the discussion requires a mechanism to store history information in the database.

As stated above, some external applications called "invoked applications" are necessary to perform some or part of specific activities. These applications include:

- An external e-mail system (e.g. MS Outlook) used as a communication (notification) tool in order to exchange e-mails between engineers during parameter upgrading.
- A process modelling system (e.g. ProView) required to model and execute the process definitions that are used by the "formal approval procedure" and interpreted as single classical workflow.
- CAD systems (e.g. Pro/Engineer) used allow an engineer access to additional information during the design process.

5.　SIMNET application

To illustrate the *ewf* within the SIMNET project, we are using as an example the design of bogies (the wheel-motor-brake units) of passenger railcars. A bogie is a complex product made of different components (running gear, drive unit, brake system, etc. In the case of SIMNET these components are designed within SGP and Knorr-Bremse.

The design process life cycle moves through 5 phases: sales/marketing, conceptual design, system embodiment design, component design, and details design. SIMNET focuses on two phases of bogie development: "System Embodiment Design" and "Component Design". In the "System Embodiment Design" phase, the number of parameters to be handled is about 300 parameters, with about 3000 parameters in the "Component Design" phase. However, the SIMNET solution will only concentrate on the approximately 300 parameters already taken into account in the "System Embodiment Design" phase.

About 100 engineers both from SGP and Knorr will deal with these parameters.

The SGP engineers have designated 5 HGs and linked parameter hardness grades to engineering milestones.

Hardness grade transition 1-2 (3-4) depicts a preliminary specification of parameter value within the "System Embodiment Design" ("Component Design") phase. The specification uses "informal messages" exchanged between engineers involved in the parameter specification. Hardness grade transition 2-3 (4-5) consists of making further specification of parameter values within the "System Embodiment Design" ("Component Design") phase. Further specification uses both "informal messages" and a "formal approval procedure".

When parameters reach the hardness grade 5, they are quite stable. In order to change the parameter values, it is necessary to launch (request) an Engineering Change Order (ECO).

Besides the direct involvement of Siemens SGP and Knorr Bremse, workshops have been held with the automotive and aerospace industries. These workshops have shown several advantages of the SIMNET workflow:

- Companies have expressed great interest in the SIMNET engineering workflow approach.

- SIMNET is fully applicable within the automotive industry and aeronautics sector. It can contribute to the management of interfaces between components for which different suppliers, engineering partners, or the customers are responsible.

- The parameter based approach in connection with HG definition is considered a very useful approach to better quality assurance in engineering

- The SIMNET approach should be considered a means for quality assurance rather than a communication support tool.

- SGP and Knorr are in the moving toward adopting the SIMNET approach,

- The SIMNET approach is considered the new generation of engineering workflows.

6.   Architecture of *ewf* within the SIMNET

Considering the user requirements specified by SGP's and Knorr's engineers, functional requirements have been established and addressed The architecture [14.] which is under implementation within the workflow module of the PDM system CADIM/EDB is based upon these functional requirements,.

6.1.   Functional requirements of the *ewf*

The following paragraph focuses on the future interface that CADIM/EDB should contain in order to support the engineering workflow methodology as defined above. CADIM/EDB is an independent PDM system, which is available as an independent product. This system may be integrated with a large number of other applications (such as CAD/CAM).

The established functional requirements of the *ewf* methodology are compared with the existing functionality of the CADIM/EDB workflow module. This identifies which functional requirements are already covered by the current functionality of CADIM/EDB, enhance its existing functionality and, develop new functionality for CADIM/EDB. Thus, we have identified three kinds of actions related to CADIM/EDB.

1. Customising: a user of CADIM/EDB is able to fulfil the functional requirement. This action does not need the intervention or support of Eigner & Partner but can use the development tools of CADIM/EDB.
2. Tailoring: functionality is already available based on CADIM/EDB functionality and/or tools. Little or no programming is required.
3. Actions requiring programming: are those related to notification services (email send and received, analyse their content, etc.).

6.2.   Architecture of the system

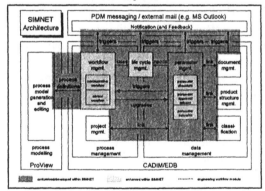

Figure 4. The SIMNET engineering workflow

Ewf serves as a backbone for the product development process to which other tools (such as EDM/PDM /PPC/CAx) are linked. A clear distinction between these systems and ewf-system cannot be provided within this paper. Neither can *ewf*-systems be regarded as unrelated to other systems, as *ewf* is based on data provided by the above-mentioned systems.

The main functions of the resulting *ewf* system are:
- Control the product development process by means of parameters, causal relationships between parameters, responsibilities, deadlines, and, depending on the process status, by blocking process steps or actions. For example, after an "engineer" in charge of a current parameter has entered a certain value for a parameter, he/she will not be able to change it except with prior consultation with other engineers.
- Monitor deadlines and notify when deadlines have been kept or missed. If deadlines are missed, then the effects on following processes can easily be judged from the continuously modelled total process.
- Improve the transparency of the engineering work and ensuring the quality assurance of the engineering work which enables the control of the project progress and project risk. Therefore, it is possible at any time to obtain information about the status of activities (tasks, operations) and the progress of development.
- Manage the co-operation and communication of concurrent local activities between engineers from various companies in a distributed environment (several dispersed production sites). Therefore, it is possible to handle duplication of centralised data and ensure its security.

7. Conclusion

The paradigm of *ewf* is a new area of research. This paper has presented the concept of engineering workflow (*ewf*) and the specification of an *ewf* methodology suitable for simultaneous engineering across company borders. The *ewf* methodology is developed on the basis of engineering parameters. We have defined engineering workflow (*ewf*) as a combination of classical workflow and the parameter based approach in order to support the work of an engineer. This methodology aims to decrease the time spent in developing activities, to monitor processes developed, to ensure the quality of the work in progress and to allow a greater transparency of the progress of engineering work. This *ewf* methodology consists of four basic elements:
- the control procedure of upgrading parameters,
- the roles associated to such procedure,
- the activities performed by roles,
- the data associated and invoked applications.

The first item is the most important because it constitutes the main difference between "classical workflow" and engineering workflow. The *ewf* specification has been transformed into a list of functional requirements. This has been compared to the definition of functional modules of the CADIM/EDB workflow module. Based on this comparison we have identified three categories:
- functionality already provided by the standard modules of CADIM/EDB,
- functionality achieved through enhancement of existing modules of CADIM/EDB and,
- functionality to be developed.

The SIMNET approach to *ewf* has several benefits for companies:
- reducing time to market,
- shortening engineering throughput time,
- reducing product cost by improving communication between engineers,
- improving product quality through the integration of the engineering units involved in product development.

8. Acknowledgements

This research has been partially funded by the ESPRIT programme of the European Commission(EP26780: from 1/11/1998 to 30/4/2001).
Partners:
- Siemens SGP Verkehrstechnik GmbH, Austria (Co-ordinator)
- EIGNER + PARTNER AG, Germany
- Knorr-Bremse Systeme für Schienenfahrzeuge GmbH, Germany
- Mission Critical SA, Belgium
- BETA at Eindhoven University of Technology, The Netherlands
- Technische Universität Clausthal, Germany
- IPS Ingeniería de Productos, Procesos y Sistemas Integrados S.L., Spain

9. References

[1.] AMICE, 1993 CIMOSA: Open Systems Architecture for CIM, 2nd revised and extended version. Springer-Verlag, Berlin.

[2.] Zunkunft O., and Rump F., 1996, From business process modeling to workflow management: an integration approach. In Business Process Modelling by Bernd cholz-Reiter Scholz (Eds.) Springer-publishing, ISBN : 3-540-61707, pp.1-20.

[3.] Harmon R. L., 1996; Das management der neuen fabrik.

[4.] Holmes C. and Yazdani B.,1999, Internal drivers for concurrent engineering industrial case study. 5th International Conference on Concurrent Enterprising (ICE'99), pp. 455-464.

[5.] Parsaei R.H., and Sullivan G.W., 1993; Concurrent engineering : contemporary issues and modern design tools, Chapman & Hall, London.

[6.] Esprit 29364- ENHANCE (Enhanced Aeronautical Concurrent Engineering).

[7.] Esprit 26892- RapidPDM (Development and validation of innovative methods and tools that makes implementation of Product Data Management systems in SMEs faster, easier and more effective).

[8.] INCO-COP 960234- CONFLOW (Concurrent engineering workflow).

[9.] iViP
http://www.egd.igd.fhg.de/zgdv/zr2/projekte/apm_ivi p.htm.l

[10.]   Munstlag, D.R., 1993, Managing customer order driven engineering, an interdisciplinary and design oriented   approach,   PhD-Thesis,   CIP-DATA Koniklijke Bibliotheek, Den Haag.

[11.]   Sihn W., and Kallmeyer O., 1999, Engineering workflow – The emergence of a new concept, 7 p. CAD-CAM-Conference in Neuchatel/ Switzerland on 22-24/02/1999.

[12.]   WfMC.   Workflow   Management   Coalition Terminology   and   Glossary   (WfMC-TC-1011). Technical report, Workflow Management Coalition, Brussels, 1996.

[13.]   Schmitt R., 1999, Target concept with respect to the   industrial   requirements   and   demands   on simultaneous engineering workflow, Confidential report submitted to EC Brussels (120 p).

[14.]   Rouibah K., M. Goltz, M. Zagel et R. Schmitt, 1999; The Implemented Engineering Workflow Management Mechanism and the Methodology for Linking Engineering Workflows to Product Data. Confidential report submitted to EC Brussels (87 p.).

[15.]   Metis http://www.metis.no/.

[16.]   Anaxagora http://www.anaxagoras.com/.

Re-engineering of the bid preparation process in an aircraft modification business

I-S Fan, J. M. García-Fornieles, A. Perez, K. Sehdev, C. Wainwright
Department of Enterprise Integration, SIMS, Cranfield University, Bedford MK43 0AL, UK

Abstract

Aircraft modification business covers many different types of aircraft. Modification work required spans from small maintenance activities to large scale re-fitting of the aircraft for missions not planned in the original design. The preparation of a bid for tender is a critical part of the business as it starts the formal commitment of the company to its customer on the cost of work. Throughout the years, the case study company has developed different processes to prepare bids to match the requirements of specific jobs. The multiple bid processes become an obstacle for corporate learning in bid preparation and is expensive to operate. The availability of network computing provides an opportunity to integrate these processes into a single workflow. A re-engineering exercise was carried out to define the process and streamline the decision-making procedure. The new process takes a holistic approach and manages th]e different types and sizes of bids in the company.

This paper reports on the bid re-engineering process and the business benefits it brings to the company.

1 Introduction

A bid proposal is a plan to carry out a work in order to fulfil a customer need. It can also be considered as the initial part of a project where the planning is done.

An engineer must be concerned not only with the best technical solution that he/she can provide to solve a problem or satisfy a need, he/she also has to analyse the economical impact of that solution.

Cost estimating consists of assessing the economic impact of an engineering design in the most accurate way or in another words [1] "the evaluation of a design expressed as cost". Another definition for estimating can be, according to the Society of Cost Estimating and Analysis, "the art of approximating the probable worth or

cost of an activity based on information available at the time" [2].

Cost estimating is one of the most important issues in any type of business. To obtain a job in a competitive environment, it is necessary to win a bid. It is also necessary to know the cost as accurately as possible to gain business advantage. To achieve maximum profitability, two issues have to be accomplished: to have a cost estimation as accurate as possible; and to achieve the cost objectives within the margins established.

Cost estimating is a vital part in decision-making for projects.

The precision of an estimate is mainly related to four basic factors [3]:

- Available data
- Time spent preparing the estimate
- Estimating method
- Estimating skills

Most of today's contracts for aircraft modifications are fixed price, therefore it is very important to have a well-defined and efficient estimating process.

2 Characteristics of aircraft modification

The aircraft modification business can perform modification and service of any type and size of aircraft. Aircraft may have very long service life. After some time in service the original product may be changed and the level of alteration may be significant.

The aircraft modification business is growing rapidly during the last few years. The modification or reconfiguration of old aircraft can be done at a fraction of the cost of new aircraft and give a new lease of life to old airframe. Military expenditures are being reduced, and the option of upgrading the operational life or capabilities of aircraft has become a valuable option.

Modifications are carried out to cover many different types of needs. Some of the most important are: comply

13

with regulations, incorporate operational improvements, increase reliability and extend life [4]. They can also be applied to new aircraft to change the original functionality of a determined type of aircraft such as converting passenger aircraft into early warning systems.

Aircraft modifications usually involve one or more of the following tasks:

- Design
- Manufacturing
- Installation

What makes aircraft modification different from new aircraft manufacture is that there are uncertainties that cannot be ascertained until the aircraft has been disassembled. Unforeseen modifications may be required due to a *"worst than expected"* condition of the aircraft. Availability of original drawings is not always assured and the original tooling may not be available. Aircraft structures may also deviate from drawings for each specific aircraft.

Another very important point in the aircraft modification business is the diversity of the types of aircraft, from large airliners to small aircraft. Modification can also vary greatly from large modifications like airliner conversion to freighter to simple changes of cabin decor. The type of work is also very variable and can affect different areas such as structures, systems, avionics or others. The combination of all the variety makes the generalisation of aircraft modification business processes very difficult.

3   Types of estimates

Cost estimates can be classified in different ways. A simple distinction is between preliminary and detailed estimates. The factors to make this differentiation are: the type of information available to carry out the estimation; time available; the project stage when it is carried out and the accuracy required from the estimate.

- **Preliminary estimates**
  They are required at the initial stages of a project, when there is not much information available about the design. At this stage there are no engineering drawings and the cost engineer has to calculate the cost from specifications or other types of measurable units. Historical data may be used to compare the proposed work with other projects carried out previously. These estimates are often used as feasibility studies, as they are inexpensive and quick. It is used in selecting

alternative design concepts and as a checking tool for more complex estimates. A preliminary estimate may also be used to establish the commitment of the customer and the contractor to carry out a job. If the initial price satisfies the price range expected by the customer then both sides may commit to work closely and carry out a detailed estimation.

This type of estimates are also known as rough, conceptual, screening, guesstimate, scope or order-of-magnitude estimate.

- **Detailed estimates**
  Usually it is the preliminary estimate what leads to the detailed estimation. In this case the estimator requires a fair amount of design information. The resources and time required in carrying out the estimation increases, therefore increasing the cost of the estimate preparation. For this type of estimates, engineering data such as specifications, basic drawings, detailed sketches or equipment quotations is necessary. The cost of making estimates rises rapidly with the accuracy required from the estimate or the level of detail. The cost and time necessary to carry out this type of estimates are high, and this type of estimates often requires a dedicated team. Detailed estimates are used for the presentation of complex bids or the authorisation of project funds.

This type of estimates is also called definitive, bid, proposal or appropriation estimate.

Preliminary estimates are quick and cheap and require less numerical analysis than detailed ones.

Estimations are built based on available data. If there is no data available there will be no estimate. Also if actual cost is known, then it is not an estimate, it is just a calculation. The accuracy of an estimate varies with the quality of the design information and the time available to carry out the estimation. The range of time and data available differentiate between preliminary and detail estimates. Data can be obtained from different sources such as:

- In-house historical data
- Estimator's knowledge
- Published data
- Prototype modelling

The AACE International has a classification of the types of estimates depending on their accuracy [5]:

Type of estimate	Accuracy range (%)		
Order of magnitude	-30	to	+50
Budget	-15	to	+30
Definitive	-5	to	+15

In this case the Budget estimate is an intermediate case between the preliminary and the definitive estimate that is usually prepared after a part of the design is done. The budget estimate may be used for approval of funds to further develop a project.

Depending on the type of work activity being carried out the estimate can be developed in different manners. Estimates could then be classified depending on the type of work output as:

- **Process or operating estimates**
  A manufacturing design includes the part drawing(s) and a process plan. Process is the combination of tools and persons to produce a change in the value of raw material. Process estimations are used to calculate the cost of labour, materials and tools, fixtures or test equipment used to produce an object. Manufacturing operation estimates are required for product estimating.

- **Product estimates**
  A product estimate is the essential element used to determine the price of a product along with other factors from the sales, marketing and operation estimates. Cash flow, rate-of-return and profit/loss analysis are factors that needed to be considered together with the cost estimate to determine the product price. The bill of materials is the key document required for product estimating.

- **Project estimates**
  A project is a one-of-a-kind end item. Projects require initial capital investment. Capital investments can affect business fitness, therefore project estimating is an important issue for companies. In project estimating, different skills are required as they are usually multidisciplinary activities. Operation and product estimates are part of project estimation.
  A project is divided into work packages, these are then subdivided into smaller elements. The Work Breakdown Structure (WBS) is the key element in project estimation.

- **Service or system estimates**
  Company reputation and customer satisfaction are very important in service estimations. In this case the objective is the effectiveness of the system. A system estimation could be considered as a combination of operation, product and project estimations. System design configurations are commonly based on performance, general requirements, equipment specifications and their compatibility.

4   Estimating methods

As seen in the previous section the quality of an estimate depends very clearly on the quality of the data and time available. Another important factor for achieving good estimates is to choose an appropriate estimating method. An estimating method can be defined as a systematic and consistent approach to predict or estimate the cost and schedule impact of overall job execution. It is recommended that a company set up a methods group that develops and maintains in-house estimating methods. They can then tailor the estimating methods to the company's needs. Good estimating methods that allow the estimator to take advantage of a good knowledge base of estimates and hence prepare a good estimate. Some of the reasons to justify the effort to create methods are [6]:

- Consistency: Since the methods developed will be used by all the company estimators, the estimates will have a consistent format and approach.
- Accuracy: The actual costs can be analysed easily and with the proper feedback the accuracy of the estimates can be improved.
- Time and money saving: If the methods are predetermined then the estimator has to spend less time carrying out the estimate.
- Availability: The factor of finishing the estimate earlier gives advantage for planning and scheduling.
- Confidence and morale: If the method is good and management understands it, the important and the trivial can be analysed, such as finding if an overrun is because of a poor estimate or other problems.
- Training: The methods used give a quick introduction to the company's estimating approach for new estimators.

Estimating methods can vary from judgement to mathematical methods. Some of the most common estimating methods are given below:

- **Opinion**
  This is usually the case when there is no data or the time available to do the estimate is very short. It is the engineer's experience, common sense and knowledge what is used in this case.

- **Team decision**
  With this method people from different departments are put together. Every engineer has specialised knowledge in his/her area and each determines a cost figure independently of the other departments. These figures are put together and the estimator usually adds overheads and profit.

- **Comparison, analogy or similarity**
  This is similar to the two previous methods as there is some judgement involved but there is some logic included. It consists of finding similar tasks or descriptions and then interpolates the similarities between the design that is required and the ones that have a known cost. Caution has to be applied, as it requires that comparisons are done with up-to-date costs, the processes applied have to be similar and factors such as production quantities have to be considered.

- **Unit, average, base unit or module estimating**
  This method estimates the unit cost of a defined unit and aggregates the total to the required quantity. It gives an average value based on similar historical data. Economy of scale saving cannot be represented in this method.

- **Standard time data**
  Standard time data are a catalogue of standard tasks used for estimating a variety of works. The method is very cheap and it offers consistency.

- **Factor, ratio or percentage methods**
  It is similar to the unit method but different factors can be applied to different cost items therefore increasing the accuracy. The factor method is used mainly to carry out estimations of new designs applying the factors to previous known estimates and designs. Factors can be found in catalogues, handbooks, internet databases and they need to be updated regularly.

- **Cost Estimating Relationships (CER's)**
  This estimating method consists of building mathematical models to estimate cost. The models are statistical regression analysis to estimate cost based on parameters that are independent variables. These parameters are usually performance or physical characteristics of the design. The parameters required are known at an early stage of the design and therefore it is a good method for preliminary cost estimates. The CER estimates are based on historical data and thus they establish a relationship between the characteristics required from the design and previous designs costs. The independent parameters used to calculate the cost are also known as cost drivers.

- **Learning curve**
  This method is based on the appreciation that a person requires less time and resources to repeat similar activities. The graphical representation of the mathematical models developed to represent this phenomenon are the learning curves.

- **Probability approaches or Bayesian analysis**
  A given point estimate represents an average value and it uses deterministic information. With the probabilistic analysis, the probability of expected values is obtained instead of an exact deterministic value. Estimation involves elements of uncertainty by nature. If there is uncertainty, probability analysis could be used and the cost estimation could be expressed in terms of range.

- **Engineering or detail estimating**
  The engineering method divides the work output into different tasks. These tasks are further broken down into lower levels. This structure is known as the Work breakdown Structure. Costs such as materials and labour are then calculated for the lowest elements of the WBS. To achieve the total cost, the lowest elements of the WBS are then added up to find the immediate level elements cost. This is then repeated until the highest element is reached.

The engineering method is usually considered a **"bottom-up"** approach as it is based on a detailed assessment of the work required for a task and builds from the lower elements until it arrives to the top. The other methods have usually been applied from a **"top-down"** point of view, which is based on global properties, such as the parametric method. The 'top-down' approach requires factors and does not require neither the breakdown nor the labour or material estimates but knowing the cost drivers. The combination of them, like using parametric analysis in the lower parts of the WBS, should provide good results as advantages from different methods can be exploited such as the estimate visibility of the engineering method and the quick and cheap use of parametric analysis.

5   The bid preparation

A bid proposal can be defined as a description of work to be performed that provides enough information for a customer to make a purchase decision [7]. In the completion of a bid there are some elements that are necessary. The basic elements required in any bid preparation are:

- Skills: It can be an estimator, usually in small bids, or in most cases a multidisciplinary team. To achieve good results, the bidding team must have the necessary mix of skills to suit the nature of the project. Skills such as marketing, business, engineering, manufacturing, production planning or technical publication are usually required.
- Methodical approach: The application of a standard methodology will cut ambiguities and provide a robust bid. The utilisation of computerised tools is very important as it can reduce greatly the time required to execute the method and present the bids.
- Information: The quality and accessibility to data related to the different processes involved in the project are key issues. The re-use of the data is another very important aspect that has to be considered.

The time spent to create a detailed estimation can vary from 8 to 18% [2] of total project time while the cost can range from 0.05 to 2% of the total project cost [3]. If the estimation process is well-defined, the time and cost involved in developing the estimate could be minimised.

Different authors have adopted different presentations of the steps followed in bid preparation [7, 8, 9, 10]. For the completion of a bid, it is possible to generalise the process into the following steps:

- Enquiry assessment
- Bid project planning
- Product design
- Cost estimation
- Product scheduling
- Price determination and payment conditions
- Bid document compilation and presentation

Depending on the way bid preparation is carried out it is possible to classify bid preparation as [11]:

- **Centralised bid preparation**
  In this case a project engineer executes most of the bid-related activities.

- **Sequential bid preparation**
  The development of most complex products involves different skills at different stages of the bidding. In this method, the steps mentioned above are carried out in a sequential manner, passing from department to department until the bid is finished.

- **Simultaneous bid preparation**
  This concept is based on the formation of multidisciplinary, temporary bidding teams set up specifically for a project. Bidding team meetings are an essential part of this method. The use of computer tools facilitating concurrent engineering is very important as it can enhance team co-ordination and information exchange.

6   The business process re-engineering methodology

The academic partner carried out the analysis of the bidding proposal preparation in the collaborating company. The following methodology was used to carry out the Business Process Re-engineering (BPR) exercise.

1. Research: This was carried out to define the characteristics of the bidding process and to classify knowledge about the aircraft modification business and its point of application.
2. Plan: Activities that needed to be performed were acknowledged and scheduled. Company culture was studied and key personnel involved in the bidding process were identified. Finally, case studies were selected.
3. Capture the process knowledge: The company bidding processes were captured through the use of semi-structured interviews with the staff involved in the bid preparation and through the analysis of company documents and computer files.
4. Map the "as-is process": After the relevant data was compiled and analysed the different processes were represented in a model.
5. Analyse the model: The models were studied to find possible weakness in systems infrastructure and integration, non-value-added or redundant

tasks, illogical sequencing of tasks, missing tasks and inputs/outputs.

6. Compare with state of the art models: Some state of the art reference models were analysed in order to establish best practices in other sectors.

7. Develop the "to-be" model: Based on the best practice analysed in the previous step and the "as-is" model, a "to-be" model was develop for aircraft modification. The new reference model was designed as the workflow of the software that will be developed to carry out the estimation.

8. Validate "to-be" model: Once the new model was finished it was presented to the staff involved in the bid preparation. The model was then refined and presented to top-management for approval for implementation.

9. Implement "to-be" model: The "to-be" model was used as the core for the development of an estimating software system.

10. Continuous model improvement: The model should be under regular revision to accommodate possible changes in the future.

7   Process modelling

Different modelling techniques such as IDEF0 [12], Object-Orientated approach [13] and GRAI [14] were investigated as tools to map the bid process in the aircraft modification business.

The modelling technique should be able to:

*   Identify functional parts and interfaces
*   Use appropriate descriptive techniques to describe both the information and the processing systems
*   Identify time based elements of decisional activity

IDEF0 was considered the most appropriate methodology, as it was easy to understand by the staff interviewed in the company. Therefore IDEF0 was used for the conceptual model representing the estimation process in the aircraft modification business. Advantages and disadvantages of the technique are discussed below.

IDEF0

IDEF0 is used to produce a function model that is a structured representation of a system and the flow paths of information and objects that inter-relate those functions. It is a top-down approach that models the system as a whole at the highest level and then decompose this model level by level to describe each of the sub-systems within the

system hierarchy. It can be expanded to any level of detail.

IDEF0 is a static representation of the model. The flow paths linking function blocks do not indicate a specific sequence over time.

Some major advantages are:

*   It permits an effective, standardised systems communication method whereby system analyst can properly communicate their concepts.
*   It permits the description of a system as detailed as desired.
*   It provides a mechanism for decomposing a function into smaller sub-functions, allowing individuals to work on different aspects of the total system and be consistent in terms of final system integration.
*   It has the potential to be used as an industry standard for manufacturing system design.

Some major disadvantages are the learning time involved, ambiguity of function specification and probably most importantly the static nature.

8   The original bid process

As an "external" party not related to the estimation process within the industrial collaborator, the academic partner was able to bring insights that are beyond the normal practice of the company staff. Typical to many organisations, the company has defined procedures and document control systems that require authorisation and signature at the appropriate control points.

Detail working procedures evolves as staff develops different methods and shortcuts. In time, these become adopted procedure and become entrenched in some cases and remembered by staff as "it has always being done like that". As different parts of the organisation develop their own methods, gaps in communications developed and many members involved in the bid process did not have an understanding of what the other parts were doing. The bid process becomes clouded and this becomes the main obstacle towards the "internal" improvement of the company procedure.

Initial analysis of the company bid process found that there was not a unified and explicitly defined way of carrying out estimates. Non-explicit methods or shortcuts existed depending on factors such as:

*   Size of the project: Small projects had shortcuts to avoid lengthy processes, while large projects seemed to follow a more systematic approach. Large projects tend to have dedicated bid teams.

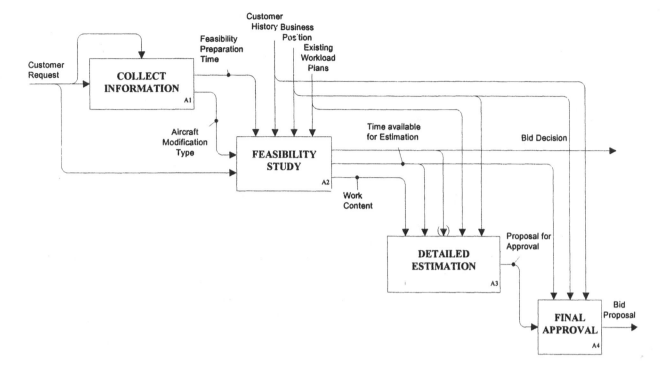

Figure 1 "As-is" bid process

- Customer: Knowledge and understanding of the customer was a factor that influenced the bid process taken. These shortcuts were often facilitated by personal knowledge of key persons in the customer organisation. Constraints imposed by the customer in the submission of the estimates also influenced the choice of process.

- Knowledge of the product: Familiarity of aircraft type or ownership of the design authority for particular aircraft type allow the estimate to follow simplified routes, for example the Product Support Centre (PSC) procedures.

- The type of work to be carried out: Jobs with a dominant engineering, or manufacturing, or production content would use a different process according to the nature of work.

As the process was executed manually without an integrated information system, most intentionally developed shortcuts have the positive effect of minimising resources and saving lead-time for the bid preparation. The disadvantage in taking these shortcuts in the paper-based system was probably that knowledge about different bids was not incorporated into the company's system. Without this feedback the problems mentioned in

Section 3 caused by the lack of methodology could be developed.

From the different processes observed, a dominant one that represented most of the cases in the company was mapped. This process was analysed in detail to look for possible improvement areas.

Figure 1 shows the top level of the "As-is" model for the bid preparation process. The four steps in the process are:

- **Information collection**: This is the gathering of data about the job to be carried out.

- **Feasibility study**: In this step preliminary estimates are carried out to assess the feasibility of the job or to check that customer cost expectations match with the job cost. Usually opinion, conference, comparison or CER estimating methods are used at this stage. The bid/no-bid decision is taken at this stage.

- **Detailed estimation**: Once the decision to bid has been taken, the company then proceeds to develop the bid and engineering estimation is carried out.

- **Final approval & bid submission.** This is the final approval for the bid to go ahead and its submission to the customer.

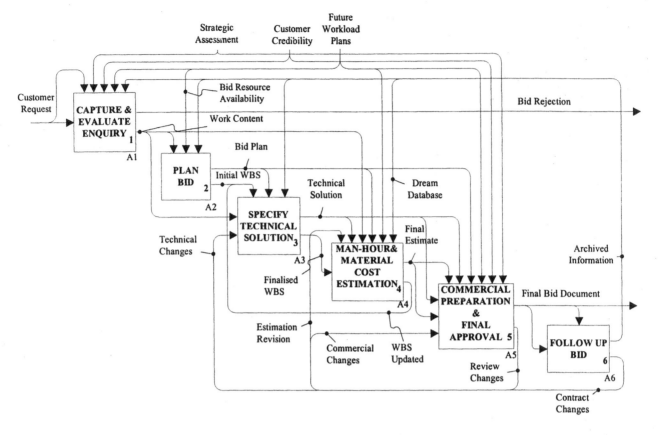

Figure 2 "To-be" bid process

The following points were observed with regards to the process:

- It was found that the bid followed a sequential flow between different departments during the estimation of design man-hours. This was actually identified as the sequential bid preparation type indicated in Section 5.
- Senior design engineers carry out the engineering man-hour estimates as part of their engineering task. Estimates have to compete for the engineers' time.
- Estimating planning was systematic with multiple variations. Some estimates have budgets assigned and some not.
- No explicit risk assessment procedure was found. De-risking was achieved by safety factors built into each stage of the estimation.

The use of knowledge within the process has the following characteristics:

- Rules of thumb used were different and not explicit, dependant on the individual carrying the estimate.

- In the design office, engineers who made the estimate may be different to the ones who carry out the design if the bid is successful.
- Working data on previous estimates were in personal files - not shared.
- Medium and top management factored estimates and the reasons were not feedback to engineers as learning.

Access and re-use of old project expenditure data was limited and historical data was captured without detail classification of low level detail.

9   A unified bid process

Availability of network computing provides an opportunity to maximise the re-use of knowledge and data in the estimating process. A knowledge model that captures the product and project tasks information was developed to enable the communication between members of the bid team and provide access to historical information to enhance the quality of the estimate. This is implemented within a networked software solution. A unified workflow for the estimation process needs to be designed as the working model for the software.

To define a new process that accommodates all the characteristics of the modification business is a major challenge. The process has to be generic enough so that all the different variations in the estimating methodology can become part of the model; and at the same time specific enough to ensure a unique workflow. The use of computer software ensures that the workflow is always maintained coherently.

The use of IT systems also facilitate concurrent decision-making. Sequential processes could now be carried out in near parallel time by simultaneous access to the working files and information by the bid team.

The top level of the new IDEF0 model for the bidding process is given in Figure 2. The main activities at the top level can be decomposed into:

- **Capture & evaluate enquiry**: This activity corresponds to the contact information and feasibility study steps in the "as-is" model.

- **Bid plan**: In this activity the team responsible for the bid is assigned and the estimation budget is set.

- **Specify technical solution**: The technical work for the project is defined. This activity takes a long time and uses a lot of resources.

- **Man-hour & materials estimation**: With a technical solution defined, the cost is assessed and the schedule is arranged.

- **Commercial preparation & final approval**: This is the compilation of all the relevant documentation. Commercial analysis such as payments and prices are added to the technical cost and the bid is presented to the company directors for approval. The bid documentation is prepared to the customer's required format and submitted.

- **Follow up bid**: The decision process of the customer is tracked. From the customer's decision, the reasons for rejecting or accepting a bid by the customer are analysed. All the relevant information is logged for future reuse in the company.

A second level index for the "to-be" model is given in table 1.

[A0] CREATE BID PROPOSAL
    [A1] CAPTURE & EVALUATE ENQUIRY
        [A11] COLLECT INFORMATION
        [A12] SCREANING OF OPPORTUNITY
    [A2] PLAN BID
        [A21] ASSIGN BID RESPONSIBILITY
        [A22] SCOPE JOB
        [A23] DETERMINE ESTIMATION BUDGET
        [A24] DETERMINE BID RESOURCES
    [A3] SPECIFY TECHNICAL SOLUTION
        [A31] IDENTIFY SIMILAR SOLUTIONS
        [A32] CREATE DESIGN CONCEPT
        [A33] GENERATE WBS
        [A34] DOCUMENT TECHNICAL SOLUTION
    [A4] MAN-HOUR & MATERIAL COST ESTIMATION
        [A41] CARRY OUT ESTIMATION
        [A44] SCHEDULE JOB
        [A45] CHECK ESTIMATION
    [A5] COMMERCIAL PREPARATION & FINAL APPROVAL
        [A51] PRICING & COMMERCIAL PREPARATION
        [A52] DOCUMENT PREPARATION
        [A53] BID REVIEW
        [A54] SUBMIT BID
    [A6] FOLLOW UP BID
        [A61] GET CUSTOMER FEED-BACK
        [A62] CONTRACT NEGOTIATION
        [A63] RECORD BID

Table 1 "To-be" model index

10 Benefits of the unified process

The unified process brings many benefits. The new process is unique and as such is enforcing a single workflow for the bid process. By establishing a common methodology, the difficulties in managing multiple processes could be eliminated. There will be a consistent way of carrying estimates. The accuracy will improve, as learning is possible across different bids. The unified process and the use of a common WBS forms the basis for estimates to be compared with historical estimation data as well as actual production data. The bid preparation time will be shortened as people will always follow the same known steps.

The new estimation software automatically captures information for future reuse. By capturing the knowledge and managing it, the company can build up the data in a structure that can be used to improve the accuracy in the estimates.

Concurrent engineering by the use of the new process together with the information technology implementation will shorten the time taken up by waiting for the sequential flow of the paper documents.

Traceability and visibility of the estimates will be immediate.

The new system provides the required infrastructure for managing bid changes during contract negotiation or design updates for developing new solutions.

Comparing the model presented for bid preparation in the aircraft modification business with reference models [7, 8, 9, 10], there are similarity in the top level of the bid preparation across industrial sectors. The major difference is that the model developed in this study introduces the use of WBS at the top level in the process. WBS provides the link between different activities and provides a common understanding of the work to be carried out, from defining the technical solution to carrying the cost estimation or scheduling the job.

11 Conclusions

Aircraft modification projects can vary greatly in size, type of aircraft and tasks involved in the project. This business environment encourages the development of multiple processes for bid preparation depending on the specific project being carried out.

A generic (unified) model has been developed for the aircraft modification bid preparation. This model can be applied to any type of projects in this business. By achieving a common methodology for developing bid proposals, ambiguities can be reduced, improving the efficiency of the process as well as reducing the bid preparation time. Accuracy can also be improved by the systematic use of the methodology as it facilitates knowledge capture and reuse, thus providing feedback for future use of information.

The new process has been developed to be implemented by networked IT solutions. The use of the new estimation process together with the use of IT systems supports the concurrent engineering approach to the bid preparation and shortens the elapsed time for bid preparation.

Acknowledgements

This research has been carried out within the EPSRC IMI Aerospace DREAM (Defining knowledge models for reconfiguration and modification processes in aerospace) project (GR/M43470). The authors would like to thank Marshall of Cambridge Aerospace and the EPSRC for supporting this research.

References

[1] Ostwald, P. F. "Engineering cost estimating", 3rd ed. Prentice-Hall Inc., 1992.

[2] Stewart, R. D. "Cost estimating". John Wiley & Sons Inc., 1991.

[3] Ahuja, H. N. and Campbell, W. J. "Estimating: from concept to completion". Prentice-Hall Inc., 1988.

[4] Garcia-Fornieles, J. M., Fan, I.-S., *et al.* "Practical considerations in aircraft modification and its impact in cost estimation". In *15th National conference on manufacturing research (NCMR)*, Bramley, A.N., *et al.* (Eds.), 1999, Bath. Professional Engineering Publishing Limited, pp. 337-341.

[5] Humphreys, K. K. and Wellman, P. "Basic cost engineering", 3rd ed. Marcel Dekker Inc., 1996.

[6] Clark, F. D. and Lorenzoni, A. B. "Applied cost engineering", 2nd ed. Marcel Dekker Inc., 1985.

[7] Stewart, R. D. and Stewart, A. L. "Proposal preparation", 2nd ed. John Wiley & Sons Inc., 1992.

[8] Kingsman, B. and Souza, A. A. d. "A knowledge-based decision support system for cost estimation and pricing decisions in versatile manufacturing companies". *International Journal of Production Economics*, Vol. 53 (2), 1997, pp. 119-139.

[9] Krömker, M., Thoben, K.-D., *et al.* "An integrated system for simultaneous bid preparation". In *IEEE International Conference on Systems, Man and Cybernetics*, 12-15 October, 1997, Orlando, Florida.

[10] Krömker, M., Weber, F., *et al.* "A reference model and software support for bid preparation in the supply chains in the construction industry". In *Proceedings of 9th International Workshop on Database and Expert Systems Applications*, Tjoa, A.M. (Ed.), 26-28 August, 1998, Vienna, Austria.

[11] Krömker, M., Thoben, K. D. and Wickner, A. "An infrastructure to support concurrent engineering in bid preparation". *Computers in Industry*, Vol. 33, 1997, pp. 201-208.

[12] U.S. Air Force "Integrated Computer-Aided Manufacturing (ICAM) Architecture Part II - Volume IV. Function Modelling Manual (IDEF0)". Air Force Materials Laboratory, Wright-Patterson Air Force Base, Ohio 45433, June, 1981.

[13] Booch, G. "Object-orientated analysis and design", 2nd ed. Addison-Wesley Pub. Co., 1994.

[14] Doumeingts, G., Vallespir, B., *et al.* "Design methodology for advanced manufacturing systems". *Computers in industry*, Vol. 9 (4), 1987, pp. 271-296.

Risk analysis of parametric cost estimates within a concurrent engineering environment

Rajkumar Roy, Sara Forsberg, Sara Kelvesjo, Christopher Rush

Department of Enterprise Integration, SIMS, Cranfield University, Cranfield, Bedford, MK43 0AL, United Kingdom.
Tel: 44 (0) 1234 754072. Email: r.roy@cranfield.ac.uk or c.rush@cranfield.ac.uk

Abstract

In industries where project time scales are long and the investment capital required is high, it is essential to minimise the risk involved with the development of a new product. Aerospace manufacturers are an excellent example of this kind of industry.

Parametric estimating is a method used for predicting cost based on historical relationships between cost and one or more predictor variables. This method is commonly used during the conceptual stages of design where potential risk is at its highest. There is an increasing trend within parametric cost estimating to combine parametric analysis with statistical risk analysis methods.

All cost estimates have a degree of uncertainty associated with them. The objective of a cost risk analysis is to predict the amount of uncertainties involved during the estimate of future projects. This paper presents the results of a risk analysis study conducted on a recently developed set of cost estimating relationships (CERs).

The first case study considers the risks involved within a parametric cost estimate. This particular cost risk methodology can be used when only conceptual information is available. The main risk involved is related to the independent variable and its probability of change during the design process. The second case study provides the estimator with a possible range of costs for designing a part. This method can be used when there is little information available concerning a part.

1. Introduction

Within the aerospace industry, where project time scales are long and the capital required is very high, it is essential to minimise the risks involved, especially during the early stages of a project's lifecycle. The objective of a cost risk analysis is to predict the amount of uncertainties involved in the cost estimate of future projects.

There will always be uncertainties, i.e. risks, involved in a project. If these uncertainties can be identified and quantified, effort can be made to successfully deal with the impact of them occurring. Risk analysis is a very broad term, meaning the study of any situation, which is controlled in one way or another by uncertainty. The outcome of a risk analysis study may help to make extreme "yes or no" decisions, but it can also show all solutions between these extremes.

By looking at the uncertain variables within a situation, a risk analysis can show which ones have the most effect on the solution, and pinpoint where most effort should be targeted. The risk analysis makes sure that uncertainty within the variable can be accounted for before committing the project. Therefore, the outcome of the analysis can be used as a decision tool for the designer, that is, if the designer understands the risks involved with certain cost drivers, he can choose a different approach to lower the risk. Thus, when using risk assessment and risk analysis it ensures that the consequences of risks to a programme cost and schedule are understood and taken account of for the commercial bid on programme price and duration [1].

This paper begins by highlighting the need for cost risk reduction initiatives within a concurrent engineering environment. It then examines different aspects of parametric costing and risk analysis with respect to their use for cost risk analysis.

Section four describes the first of two risk assessment methodologies that can be used with a parametric cost estimate. Section five demonstrates how to use this methodology by applying it to a recently produced cost estimating relationship (CER). Section six describes the second risk analysis methodology and section seven demonstrates its practical application. The results of both methodologies are based on a CER that was previously developed within a European aerospace company [2, 3, 4]. In summary, the risk analysis methods examine whether the independent variables, or cost drivers, included in the CER change as the product matures through the design process. The results demonstrate how changes affect the accuracy of the cost estimate. The paper concludes the main findings of this research in section eight.

2. Aerospace concurrent engineering environment

The company sponsoring this research embraces a concurrent engineering philosophy. Figure 1 illustrates a typical structure of an integrated project team by a family tree type representation. The tree illustrates the team members operating within the Technical and Engineering activities on the project under scrutiny [4].

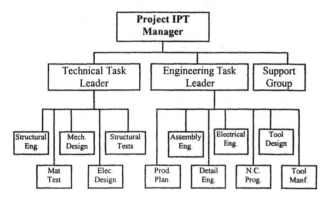

Figure 1. Typical representation of IPT members

Figure 2 represents a diagrammatic form of the IPT environment. The company recognised that new cost estimating relationships were required to reflect the activities of their concurrent engineering environment. The CER used for the risk assessment in this paper was developed around the activities within phase D. The reasons behind this were that data was available for the study within this area [13].

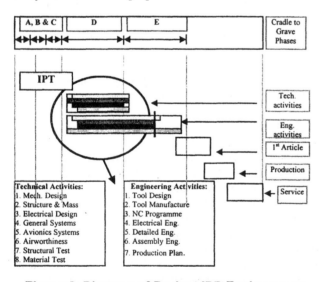

Figure 2. Diagram of Project IPD Environment

2.1. Risk analysis within a concurrent engineering environment

The developed CER was designed for use during the early stages of design to predict the future effort of a new product development. The design of a new product goes through many phases. From the concept stage a SET (simultaneous engineering team) produces a project plan. From this plan initial schemes are produced until a mature concept is ready. The designs then pass through a freeze gate to prevent any further changes. Components are then separated from the Scheme and modelled on CATIA. These models then pass through a freeze gate before going on to the 2D stage (see Figure 3).

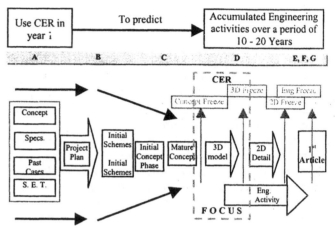

Figure 3: Design activities

An optimised concurrent engineering environment provides an opportunity to substantially reduce the total risk of a project. Because, integrated product teams (IPTs) containing members of various skilled disciplines, enable a simultaneous contribution to an early product development and definition. Therefore, within a fully integrated product development (IPD) cycle, multidisciplinary teams working together reduce the likelihood of product failure by avoiding costly alterations later in the design process.

However, up to 70-80% of a product cost is committed during the concept phase of product development [5, 6, 7, 8]. Making a poor decision at this stage can be extremely costly. This is because product modifications and process alterations are more expensive the later they occur in the development cycle. Since the developed CER is intended to predict the cost of a product 20 years henceforth the sponsoring company wanted to understand the level of risk involved with the CER result. The remainder of this paper describes the process developed for this purpose.

3. Related research

The following discussion captures different aspects of parametric costing and risk analysis within the conceptual stages of project development. This section provides fundamental knowledge concerning the tools and techniques currently used within the area of costing and risk analysis.

3.1. Parametric costing

Parametric cost estimating (PCE) is a technique commonly used to estimate the cost of future systems. It provides a technique for predicting cost based on historical relationships between cost and one or more predictor variables i.e. cost estimating relationships (CERs). The method uses a statistical approach, and is commonly used during the conceptual design stages [9].

Mileham *et al.* [8] state that there is a growing need within the concurrent engineering environment to provide the designer with a simple, accurate method of estimating product costs during the conceptual stage of design. Their developed methodology is based on the basic information available to the designer during the conceptual design stage, and a set of data converters, which are used to calculate the values of cost-driving parameters.

Parametric estimating can be used throughout the product life cycle. Both industry and Government accept the techniques. Many authors commend its usefulness [5, 6, 8, 10]. However, PCE does have its limitations, for example, CERs are sometimes too simplistic to forecast costs. If ill considered a CER could provide a completely misleading result. A broad outline of the CER development process is described below.

3.1.1.   CER development process

CERs can range from simple rules of thumb to complex relationships involving multiple variables. The principal function of CERs is to provide equations or graphs that summarise historical cost or resource data in a manner that will allow the equations or graphs to be used to estimate a future cost [5].

A general methodology for developing CERs includes activities such as data collection, testing a CERs logic, statistical analysis, CER significance tests and validation. Figure 4 illustrates this sequential process.

The collection of data is often a very critical and time-consuming activity. Typically more effort is devoted to assembling a quality database than to any other task in the CER development process. It is often said that the estimate is only as good as the data input to the cost model; therefore it is essential to collect accurate data.

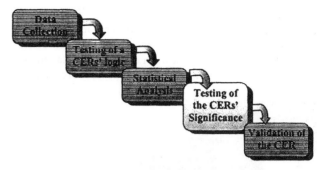

Figure 4. A general process of developing CERs

After a database is developed the first step is to hypothesise and then test the mathematical form of the CER. The analyst must determine and test a proposed CER, in order to determine its logic. The work involves discussions with engineers to identify potential cost driving variables, scrutiny of the technical and cost proposals, and identification of cost relationships. Only with an understanding of estimating requirements can an analyst attempt to hypothesise a forecasting model necessary to develop a CER.

In order to test and validate a CER statistical analysis is used. Multiple regression is the most common method used to test hypotheses [11].

Although widely accepted PCE is primarily based on statistical assumptions concerning the cost driver relationships to cost, and therefore, estimators should not completely rely upon statistical analysis techniques. Hypotheses, common sense and engineering knowledge should come first, and then the relationship should be tested with statistical analysis.

Because of the identified limitations within PCE and CERs there is an increasing trend with parametric cost estimators to combine the statistical techniques of parametric analysis with statistical risk analysis methods. Parametric estimating offers the cost analyst the advantage of being able to quantify the risk. This helps to restore confidence within the results and improves decision-making.

3.2. Risk Analysis

Edmonds [1] discusses the advantages of using risk management within the aerospace industry. Stating that the use of risk assessment and risk analysis ensures that the consequences of risks to programme cost and schedule are understood provided they are taken account of during the commercial bid of a programmes price, and duration. A cost risk analysis generates a range of costs for a project, and assigns a probability level to each cost value within the range.

The introduction of risk assessment and risk analysis ensures that the goals of the producer and consumer materialise and that they both benefit. It provides confidence concerning the final product and identifies actions needed to keep cost and schedule on target. There are five key steps to follow in the risk management process [12]. Figure 5 illustrates this process more clearly.

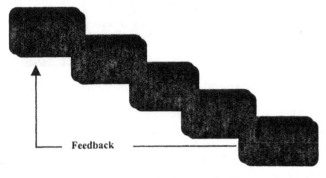

Figure 5: Risk Management Process

One of the most important benefits of using risk assessment is to generate a distribution/range of costs i.e. to move away from single point estimating, since a range of costs are much easier to estimate than a single cost [13]. Furthermore, once a risk analysis has been conducted the analyst can consider ways to reduce the risk e.g. by avoidance, deflection or contingency and then plan accordingly to control the reduction process.

Statistical theories are well established and there are many software packages available to perform such calculations. However, for risk assessment the situation is very different and software is not so readily available. One computer model that can, and is used for the risk analysis of cost estimating is named SAM [14] (Stochastic Aggregation Model). SAM is a Monte Carlo simulation program designed to help the cost analyst quantify the uncertainty associated with a parametric cost estimate. The areas of cost risk that SAM evaluates are:

- cost estimating relationships (CER);
- independent variable uncertainty;
- complexity factor uncertainty, and;
- CER statistical uncertainty.

Crossland *et al.* [15] recognised that in the early stages of a design project, it was necessary to evaluate the design to determine its feasibility in order to focus attention with particular areas of the design. They developed a tool-set called RiTo (RIsk TOol) to fulfil this purpose. RiTo, is an objective-oriented risk modelling tool that was developed to support the development of design from the earliest stage by building models for risk assessment.

By combining risk analysis with the normal pricing process, the estimator gets a direct measure of the risk at the same time the estimate is formed, and therefore can allow for a contingency value to be quantified. This contingency provides a better understanding of the correlation between items, which can have large combined effect on the overall distribution [16].

The remainder of this paper discusses two different approaches to the cost risk analysis process. Each approach is tested on case studies in order to validate their applicability.

4. Cost risk methodology one

4.1. Identification

The first step in handling risks is identification. This is readily available from the results of the cost estimating relationship. The risk consists of the independent variables; included in the CERs (see below), and their probability of changing throughout the process. The independent variables, also called cost drivers, are selected through statistical analysis and form the basis of the CER.

$Y = C_0 + C_1 (Mass) + C_2 (Surface\ related)$
Where:
Y = The dependent variable, time;
C_0 = The constant;
C_1 = The constant multiplied with the value of mass, and;
C_2 = The constant multiplied with the surface related.

Because the independent variables are prone to change as the product matures through the design process the accuracy of the cost estimate is affected. The higher the probability of the change occurring, and its impact, the less accurate the cost estimate will be.

4.2. Assessment

Having identified possible sources of risk, included in the cost estimate, the analyst then needs to calculate their impact on the cost. This is the risk assessment. The risk can be defined as:

$Risk = p * c$,
Where:
p is the probability of the event occurring, and;
c is the impact of the risk on the estimate.

This means that both the probability of the risk occurring, and the impacts are assessed.

The quantification of the cost risks can be made through a probability distribution. One way of representing this is through a triangular distribution. To achieve this the minimum, most likely, and maximum cost are required, see Figure 6 below.

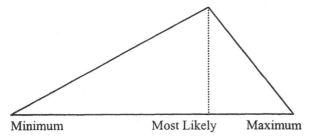

Figure 6: Triangular distribution

The values of the impact are then taken from the coefficient of the independent variable of the CER and multiplied with the values gathered from the risk assessment. As discussed earlier one of the most important benefits of using risk assessment is to generate a distribution of costs, and move away from single point estimating.

4.3. Analysis

After the risk assessment is performed, the analysis can be carried through. This is completed using simulation, such as Monte Carlo or Latin Hypercube. The outcome of the risk analysis provides the estimator with a range of costs, instead of a single one, from the CERs. In this way, it can be assumed that the cost will not exceed a specific value, with a certain probability; normally an 85 % probability is used. This is presented using a cumulative probability curve, the S curve, see Figure 7.

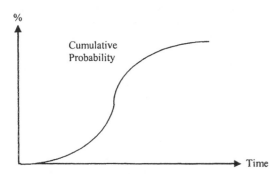

Figure 7: A cumulative probability curve

4.4. Mitigation

If the risk involved is shown to be very high, a mitigation plan should be made. The mitigation plan should include discussion with experts concerning the area under consideration and estimators who have experience in risk management.

5. Case Study One

A case study was conducted where the risks involved from a parametric cost estimate were considered The estimate, a Cost Estimating Relationship (CER), was developed to capture the design time of producing a specific part [2, 3, 4]. This cost risk analysis methodology can be used during the conceptual design stage when there is limited information and uncertainty concerning the independent variable. That is, there is a probability of the independent variable values changing during the design process.

5.1. Identification

In this study, the risk incorporates the independent variables from the CER i.e. "mass" and "surface related". The value of these independent variables can change during the design process, therefore, they are considered as risks.

5.2. Assessment

With the assistance of expert opinions an assessment of the risks, included from the CER, were conducted. The identification and evaluation of the two independent variables are illustrated in Table 1.

	Mass risk	Surface related risk
Description	Risk – An increase in the mass (within the time span that the estimate covers).	Risk – A change of the product whether it is surface related or not.
Probability	50%	0%
MIN	5% of the mass	-
Most Likely	20% of the mass	-
MAX	50% of the mass	-

Table 1: The identification and assessment of the different risks involved in the cost estimate

According to the experts, the surface related variable did not have any probability of changing during the design process. This meant that the variable was no longer considered as a risk, and it was therefore excluded from further analysis. The percentages of the likely changes of the mass figure were gathered from the experts. The independent variable, mass, and the impact, c, were then distributed using the triangular distribution. Since there was a large range within the mass of the parts that the CER was based upon, the impacts were grouped into five different weight categories. The averages of these weight categories were then multiplied with the risk percentages and the coefficient of the CER, see example below. This sum was then used as the impact of the risk described in a triangular distribution.

(min % of change in mass, most likely % of change in mass, max % of change in mass)
*(average weight of the specific category)
*(coefficient of the variable from the CER)
= Triangular distribution (impact) of the cost estimate.

Worked example:

(5%, 20%, 50%)*(800 gram)*0.05 = (2 hours, 8 hours, 20 hours)

The probability of the risk occurring can be shown in a discrete distribution. This type of distribution is used when there is a known amount of outcomes, i.e. the risk will occur or not.

Discrete ({0,1}, 50%, 50%})

The probability of the risk occurring and the impact of the risk are then multiplied to compute the actual risk. This is achieved by multiplying the discrete distribution and the triangular distribution together.

5.3. Analysis

During the identification and evaluation of the risks a tool named "PREDICT!" was used to perform the analysis. This tool is generic and not specifically designed for cost risk analysis. The analysis was made using a Monte Carlo simulation.

5.4. Results

The output produced from the analysis is illustrated using a cumulative probability curve. This plot shows that there is a probability of 80 % that the extra design time, if mass increase during the design process, will not exceed 10 hours. See Figure 8 below:

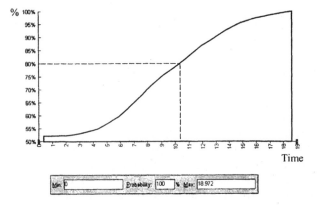

Figure 8: Cumulative probability curve

6. Risk analysis methodology two

To predict the range of a CER, when limited information is available about the product i.e. a new concept/idea, another cost risk analysis is required. This provides the analyst with a technique that can predict the maximum outcome of the cost estimate, with a specific probability.

6.1. Identification

As previously mentioned, the risks involved that can cause changes within the CER range, are the independent variables. These independent variables are readily available from the CER since they form the basis of the estimate.

6.2. Assessment

The probability distributions of the independent variables are also readily available from the data that was used for the development of the CER. These data can be distributed as a triangular distribution with a minimum, most likely and maximum value and used for the assessment of the probability, see Figure 6 above.

6.3. Analysis

To perform the risk analysis, a software package can be used. The input data is taken from the identification and evaluation of the risks, or the independent variables included within the estimate. Because probability distribution data are readily available from CERs, parametric costing is sometimes combined with cost risk analysis techniques involving Monte Carlo simulation. The simulation performs several thousand iterations to form a frequency distribution of cost [5].

6.4. Results

The outcome of this type of cost risk analysis is a range of costs for the total CER. This type of analysis is used when there is only a limited amount of information available. The analyst can say to a specific probability, that the cost of a particular type of part, used for the cost estimate, will have a certain cost range. This can be extremely useful for early estimates, when little or no information about a part is available.

7. Case Study Two

The same CER, used within case study one, was used for this case study. This cost risk analysis provides a range of total costs, to a certain probability, for designing a specific part.

7.1. Identification

The CER used two independent variables, mass and surface related that could cause changes within the CERs range. From the data set used for the CER development, the variation of the independent variables, mass and surface related, were identified.

7.2. Assessment

The range of mass, which was readily available from the data set, was assessed for the analysis. A minimum, most likely and maximum weight was captured. The other independent variable, surface related had only two possible outcomes, surface related or not.

7.3. Analysis

An analysis was conducted on the independent variable mass. A triangular distribution, which included the impact mass had on the CER, was used in the simulation tool PREDICT.

7.4. Result

The impact that mass had on the CER is illustrated in the graph, Figure 9. The graph illustrates with an 80% certainty the impact mass has on the CER will not exceed 67 hours.

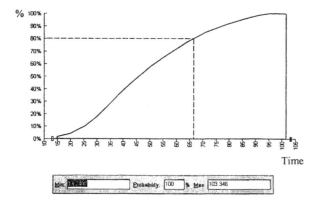

Figure 9: Cumulative probability plot

This result is then summarised with the outcome from the CER. The maximum occurring time is added to the independent variable mass:

Total time = C0 + [C1 (mass)+Maximum added time if the weight change] + C2 (surface related)

Thus, the total time will not exceed 213 hours, with the probability 80%. This is the absolute maximum value that the estimate can be. The second independent variable is chosen to find the maximum impact of the total cost.

8. Discussion and Conclusions

This paper has presented two different methodologies for risk analysis of CERs within a concurrent engineering environment. The risk identification and assessment is very much dependent on the environment. When conceptual information is available and a cost estimate is completed using a CER, a risk analysis like the one used in case study one, will provide the analyst with the extra amount of time the design process will take if the independent variable changes, within a certain probability.

In some cases when the conceptual information is not totally defined a risk analysis can be very useful for the estimator. The analysis will indicate the upper and lower boundary of cost that a part will obtain as it is being designed.

These case studies are based on a CER that was developed for a specific type of part; a small sample size was used for the CER development. Therefore, the results from these case studies can only be used for the same type of part. Nevertheless, the results indicate the possibility of using the two approaches as predictive tools for quantifying the risks involved when using CERs within a concurrent engineering environment.

Acknowledgements

The authors would like to thank Ian Taylor of BAE SYSTEMS who helped to develop and supervise this research with Cranfield University.

References

[1] EDMONDS, R. J. *A case study illustrating the risk assessment and risk analysis process at the bid phase of a project.* British Aerospace Defence Limited, Dynamics division. Ch. 2.

[2] ROY R., BENDALL D., TAYLOR J. P., JONES P., MADARIAGA A. P., CROSSLAND J., HAMEL J., TAYLOR I. *Development of airframe engineering*

CERs for military aerostructures. MSc group project, Cranfield University, UK, 1999.

[3] ROY R., BENDALL D., TAYLOR J.P., JONES P., MADARIAGA A. P., CROSSLAND J., HAMEL J., TAYLOR I. M. *Development of airframe engineering CERs for military aerostructures.* Second World Manufacturing Congress (WMC'99), Durham (UK), 27-30th Sep., 1999.

[4] ROY R., BENDALL D., TAYLOR J.P., JONES P., MADARIAGA A. P., CROSSLAND J., HAMEL J., TAYLOR I. M. *Identifying and capturing the qualitative cost drivers within a concurrent engineering environment.* Advances in Concurrent Engineering, Chawdhry, P.K., Ghodous, P., Vandorpe, D. (Eds), Technomic Publishing Co. Inc., Pennsylvania (USA), pp. 39-50, 1999.

[5] STEWART, R., WYSKIDSA, R., JOHANNES, J., Cost Estimator's Reference Manual, 2nd ed., Wiley Interscience, 1995.

[6] DEPARTMENT OF DEFENCE. *Parametric Estimating Handbook*, 2nd Ed., DoD, http://www.ispa-cost.org/PEIWeb/cover.htm, 1999.

[7] TAYLOR, I. M. *Cost engineering-a feature based approach.* In: 85th Meeting of the AGARD Structures and Material Panel, Aalborg, Denmark, October 13-14, 14:1-9, 1997.

[8] MILEHAM, R. A., CURRIE, C. G., MILES, A. W., BRADFORD, D. T. A parametric approach to cost estimating at the conceptual stage of design. *Journal of Engineering Design,* 4 (2): pp. 117-125, 1993.

[9] RUSH, C. & ROY, R. *Analysis of cost estimating processes used within a concurrent engineering environment throughout a product life cycle.* Proceedings, 7th International Conference on Concurrent Engineering, University Lyon 1, France, July 17-20th, 2000.

[10] PUGH, P. Working top-down: cost estimating before development begins. *Journal of Aerospace Engineering*, Part G, Vol. 206, pp. 143-151, 1992.

[11] NORUSIS, M.J. *SPSS 8.0 A guide to data analysis.* Prentece-Hall, Inc, New Jersey, 1998

[12] TURNER, R: J. *The handbook of project-based management.* McGraw-Hill International (UK) Limited, 1993.

[13] FORSBERG, S., KELVESJÖ, P. S. *CER development for airframe engineering.* Masters Thesis, Cranfield University, UK, 1999.

[14] HAMAKER, J. W. *SAM user Manual.* Huntsville, AL, Cited in: Stewart, R. D., Wyskida, R. M., Johannes, J. D. (ed). *Cost estimator's reference manual*, ch 8, 1980.

[15] CROSSLAND, R., SIMS WILLIAMS, J. H., MCMAHON, C. A. *An object - oriented design model incorporating uncertainty for early risk assessment.* In: International Computers in Engineering Conference, Boston, MA, 1995.

[16] HULL, K. AEPS/ETG, Ministry of Defence, Procurement Executive. *Risk analysis techniques in defence procurement.* (unpublished), 1991.

Developing an Integrated Approach to Design and Manufacturing Cost Modelling

Rajkumar Roy and Phil Jones

Department of Enterprise Integration, SIMS, Cranfield University, Cranfield, Bedford, MK43 OAL, United Kingdom.
Tel: +44 (0) 1234 754072, Fax: +44 (0)1234 750852, Email: r.roy@cranfield.ac.uk

Abstract

The world manufacturing industry is becoming increasingly cost conscious. As a result, a lack of integration between cost estimating activities, at the various project lifecycle stages have been noticed. Integration is seen as the next step towards accurate prediction of the manufacturing cost at the design stage. Furthermore, as Supply Chain Management and Customer Relationship Management are pushed to the fore, transparent costing practices are becoming a prerequisite. This means cost estimating is required and must be presented in a formal manner.

This paper presents the development of an integrated approach to cost estimating. The research utilises two market leading cost estimating software: KAPES and CostAdvantage. The aim is to develop a framework for the software packages to be integrated. This would then create a system that links manufacturing cost knowledge with the design function. A prototype to validate the model is then created. The prototype shows some of the limitations of the proposed system.

The paper also looks at using parametric methods to achieve the same aim. It suggests how some of the processing in updating cost estimating relationships can be automated. The study compares the two systems and defines which areas each method would be more effective in. It then analyses the business benefits of the integrated costing approach and how they can be used in a corporate costing strategy.

1. Introduction

Cost of manufacture is now one of the most important criteria during decision making at the design phase. This project is concerned with developing an integrated cost estimating approach that will feedback data to allow more informed decisions in the design process. The aim of this project is to develop a tool that can accurately predict the cost of a manufactured product from the design stage. Clear visibility of manufacturing costs would enable them to make important tactical and strategic changes at all stages in the supply chain. The research is based around two software packages CostAdvantage and KAPES.

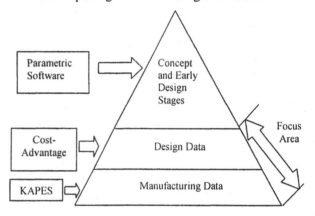

Figure 1: Focus of the research

KAPES (Knowledge Aided Planning & Estimating System) is a manufacturing cost planning system which can be used to control purchasing and outsourcing costs, the generation of bid cost estimates, and the improvement of shop floor efficiency. Access to the KAPES central knowledge base by estimators, manufacturing engineers and purchasing staff, provides a complete solution to many manufacturing engineering requirements. The development engineers essentially determine manufacturing costs of new products. The definitions of section geometry, the material choice of the sections as well as the module structure establish the costs of a total system.

In order to be able to control the cost structures of a new product as early as possible CostAdvantage was developed. Methods for managing the manufacturing costs of a new product are common and effective at the manufacturing stage with engineers knowing the capabilities of the production facilities. Often though the concept and design decision phase are left as side issues and simply ignored. This results in large inaccuracies when a prototype is put into production. If one considers that product manufacturing costs are usually between 80-90% of total product lifecycle costs then forethought to these costs from earlier stages becomes important. By using CostAdvantage to compare the geometry or features of a part with a knowledge-based cost structure system the costs of a component or a module can be derived directly.

CostAdvantage is an object-oriented expert system, in which specific material and process costs can be stored. Based on these structures geometry or feature models are analysed and the costs are calculated. Common and recurring geometry items or features can be defined by keyboard entry and analysed immediately.

With complex geometry it is inevitable to illustrate the sections in one coherent Modeller. With the CAD/CAM interface a part is called directly from Pro/ENGINEER to CostAdvantage. Additionally, the design features in CostAdvantage can be directly changed through Pro/ENGINEER. Using CostAdvantage could be described as " Concurrent Cost engineering " - at each redraw or development of a part a cost model can be used to track cost implications. As well as differing geometry's and dimensions, various materials and different processes can be analysed. Should you mill a flange directly from the billet? Or is the section to be first poured and processed? - Which version is cheaper? A more expensive material, which permits smaller wall thickness'? Or a cheaper material, with which I the walls must be thicker? These and similar questions are answered with CostAdvantage quickly and reliably to give the technical designer the ability to achieve his cost targets.

The question of cost becomes more important as international competition becomes an ever more pressing issue. Development cycles become more compact, therefore manufacturing costs become more and more important. The enterprise that puts an efficient and economical product on the market will emerge as at the very least a cost leader, if not market leader. CostAdvantage supports developers in achieving this critical requirement. Development needs to be carefully considered and planned. Recent research demonstrates that companies unable to provide detailed, meaningful cost estimates, at the early development phases, have a significant higher percentage of programs behind schedule with higher development costs, than those that can provide completed cost estimates [1]. Therefore, it is essential that the cost of a new project development be understood before it actually begins. It could mean the difference between success and failure.

This paper begins with an overview of related research to provide the foundation for the research. The problems associated with linking manufacturing cost data to designers for decision making are also discussed. Sections 3 and 4 detail the main objectives of the research. and describe in detail the functionality and characteristics of both KAPES and CostAdvantage tools. This analysis clearly describes how these tools are used and emphasises their advantages and limitations, which also illustrates the attractiveness of integrating the two systems.

Section 5 introduces the case study on which the integration of the two systems is based. Within Section 6, two integration methods are proposed and tested. The first is concerned with providing a direct cost estimating link between available raw data and the design inputs. CostAdvantage is used to read in data from a CAD/CAM system and then linked with KAPES via a transfer file. This system provides the designer with a cost estimate during the design process. The second proposal describes a method for creating a software tool that is based on parametric estimating principles. The main aim of this proposal is to provide a CER link between CAD/CAM design features and historical data that is specific to the company.

Section 7 provides a detailed discussion concerning the integration of KAPES and CostAdvantage and how the combination of their strengths could be used to improve industry cost prediction capability from the early stages of product design. Before concluding the results of the research within Section 8, the technical limitations of the two proposals are discussed, which will assist further development into providing a cost estimating link from detail manufacturing data to the early design stages.

2. Related Research

Cost estimating can be broken down into a variety of areas. This paper looks at parametric costing, costing at the design stage, costing from standards, and feature based costing.

Parametric costing is a method of estimation based upon mathematical equations that relate cost to the physical or performance measures associated with the product or project being estimated [1][2]. Parametric estimating is sometimes called statistical estimating. This is because the method relies on statistical equations to relate cost to input variables. These are generally called CERs (Cost Estimating Relationships) with cost as a dependant variable and required performance or physical dimensions as independent variables [3]. It is through statistical analysis of the variables and the identification of important variables (cost drivers) that CERs may be developed to allow accurate early phase cost estimation. The statistical analysis is based upon past cases and data from previous projects or products of a similar nature. In this respect parametric estimating requires lots of relevant data for the estimation. Roy et. al. [4] describes two types of CERs: quantitative and qualitative. Quantitative CERs are based on measurable and thus quantitative cost drivers. On the other hand, a qualitative CER represents the subjective part of cost estimating and is based on 'difficult to measure' subjective cost drivers.

It is widely acknowledged that collection of data in industry situations is a great deal more challenging than the statistical analysis [5]. Data must be checked for consistency and integrity. Any analysis based on poor data sets can only hope to produce poor estimating relationships. It also worth considering that data and models need continual maintenance to remain useful. Cost estimating relationships based on outdated processes or with old data will produce inaccurate cost forecasts. They either need adapting to suit the new process or a new model or relationship must be formed.

An easy way to represent manufacturing to design engineers is by showing the cost implications of their actions in real time. Value analysis can add the cost issue as a design driver. It provides a disciplined and logical approach to reducing unnecessary production costs without sacrificing functionality or other quality [6]. A comprehensive discussion about different approaches to cost estimating within a concurrent engineering environment is presented in Rush and Roy [7].

Manufacturing cost is partially estimated based on the operations involved, and standard time required for each operation [8]. The standard labour times are based on motion study standards, such as MTM (Methods Time Management) and MOST (Maynard's Operational Sequencing Times). From these systems standard-costing systems evolved. These would incorporate the standard operation times along with material costs, and sometimes including overheads, into a cost estimate. These cost estimates could be used in bidding for work or measuring performance. The advantage of using standard costing systems is that little historic data is needed from the manufacturing system or product in question. The main drawback to standard costing is that it can only be applied to processes that are typical and conform to an industry standard.

Feature based costing is used at a lower level than parametric costing. There are different types of features for instance: form features, tolerance features, functional features, material features and assembly features.

Currently there is much work being undertaken to standardise what a feature is and create some commonality between functions, specifically design and manufacturing [9]. One must assign a cost to each feature or group of features, and the size of these features depends on the product, the cost estimator's point of view, etc. This assignment is not always possible and can cause inaccuracy. Manufacturing cost is determined by shape, complexity, product precision and tooling process. If these can be obtained at the early design stage an estimate can be formed. The aim of feature based costing is to provide a framework for a designer to cost estimate with little knowledge of the manufacturing process. This should reduce costs downstream. An example of a feature based costing system is put forward in Developing an Integrated Framework for Feature-Based Early Manufacturing Cost Estimation [10]. The proposed system has three modules: CAD module, reference library and analysis module. The CAD module supports the development of designs using feature-based modelling. The common features are drawn from a library. The analysis module then looks at the design checks for manufacturability using CAM techniques and estimates the costs using rules associated with each of the features created from the library. Feature based models offer considerable advantages over other methods. They have the potential to close the gap between design on the one side and process planning and cost information on the other side. A feature is a partial form or a product characteristic that is considered as a unit and that has a meaning in design, process planning, manufacture, cost estimation and other engineering disciplines. The key advantage of using features is that they match the levels of abstraction on which engineers think; they are entities in engineering reasoning processes. They can also serve as units for the storage of product data.

One of the main problems with the use of features is that CAD and CAM systems use a different protocol. Because designers work with lines and circles, the language of CAD has been one of geometric entities. CAM, with its focus on production, needs a different language, one that addresses shop floor functions [11]. Functions such as computer aided process planning (CAPP) understand part features, not part geometry. Typical part features are not recognisable within the CAD database. Furthermore, the database contains no computer-intelligible information about features. Therefore, it is observed that though several attempts have been made to provide manufacturing cost information to designers for their decision making, this is still a difficult area to address. This paper aims to bridge the gap by studying possible integration between KAPES (mostly used for manufacturing cost estimation) and

CostAdvantage (an expert system used to calculate the total cost of a product).

3. The objective

The objective of this research is to create a link between designers and the knowledge of manufacturing engineers for cost estimating purposes. This will allow designers to make better-informed decisions on the manufacturing implications of their designs. More specifically the project is concerned with developing a framework for the integration of CostAdvantage and KAPES and detailing how a prototype could be created.

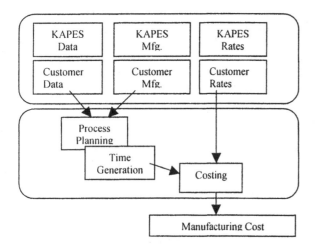

Figure 2: Diagram Illustrating the operation of KAPES

4. KAPES and CostAdvantage

4.1 KAPES Functionality's and architecture

Knowledge aided planning and estimating system (KAPES) consists of many functions covering areas such as line balancing and route selection to product and operation costing. It is the latter on which this project concentrates. The costing functions contained within KAPES cover three main areas: the interface with the supplier while determining a purchase cost, the interface with the customer while determining product cost and finally the in house manufacturing cost. KAPES holds all its data within a central reference library. This contains detailed data at the lowest level such as manufacturing times and material characteristics. It also collects rules and details that embody knowledge of how parts are manufactured. For example milling a slot may consist of several operations in a particular order. This would be the knowledge. The times on each operation and the rates would be the data. By using a centralised data repository a high degree of consistency can be achieved across functions such as route planning and costing. If this data can be used throughout the company then there would be higher compatibility well amongst the various areas of the company.

This project seeks to further this aim so designers can access some of the data contained within KAPES via CostAdvantage. In this way they can make more informed design decisions with respect to cost, leading to greater profitability. Figure 2 illustrates the pattern of operation that KAPES goes through when producing a cost estimate. The core data structure of KAPES is shown in Figure 3 below. All KAPES's tools use this core data structure which is based around parts records, operation records and bills of materials to store information. Parts records are the key element in the data structure of KAPES. It contains a huge array of parts records that can be used in cost estimates. To begin with his library of parts records must be populated. Once this is done KAPES can be used to process plan, generate times and cost effectively.

KAPES uses MTM to generate times for assembly and manual labour. This involves breaking work down into a series of small discrete elements. The user can add synthetic times for any unusual tasks carried out on the shopfloor or alter values according to their work practice. Machining times are calculated using standard algorithms. The user is requested details on speed, feed rate and number of cuts. The inputs are combined with information contained within the data library and all the associated machine operations have their time calculated. There is a standard set of times contained within KAPES along with the possibility of the user adding their own specific data and times for operations.

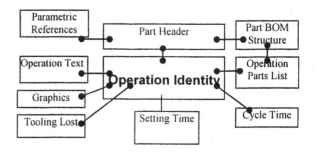

Figure 3: Core Data Structure of KAPES

The machining times and the manual tasks are kept separate to account for concurrent working within the manufacturing system. The KAPES data structure allows cost estimates to be created at any given level of the bill of material. It will drill down into any given assembly and calculate the cost of every part contained within it. The cost estimate is broken down into five estimates:

- Raw Materials
- Purchased Parts Costs
- Labour Costs
- Facility Costs
- Non Recurring Costs

Raw materials refers to the cost of the basic material used to produce a component. Purchased parts cost accounts for any costs incurred outside the company or profit centre. Labour costs are given per hour and are kept in sections for skilled, semiskilled and unskilled work. There is also a facility to include non production staff such as quality engineers and have them factored into cost estimates. The facility costs includes depreciation, power usage and rent/cost of the shop floor area. Non-recurring costs is used as a catchall for any one off costs this includes tooling costs.

Knowledge is contained within the KAPES system in the form of what is referred to as parametric references. These capture manufacturing engineer's knowledge without the need for complex knowledge rules or expert system programming. It does this by storing part characteristics as parameter's that are attached to the route structure. These references are variables that the system requests from the user. Once a part is created it can be copied and the parametric references changed to define the new part.

4.2 The CostAdvantage System

" CostAdvantage is a knowledge based software system that provides expert level design guidance and can analyse manufacturing alternatives, producibility, and predictive cost analysis." – Cost Advantage User's Guide, Cognition.

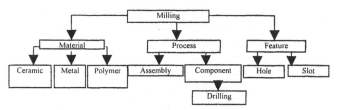

Figure 4: Diagram showing cost model structure

There are two modes of operation. The first involves the user entering data about the part into the model manually. The second uses CostAdvantage's feature based costing facility. This can read in data directly from any feature based CAD system. Each of these features is then broken down into a set of inputs for the model. Once the model has the required inputs it can produce a cost estimate. The model works by assigning each a cost to each feature. It then assigns a material cost and a tool handling cost for the part as a whole. The model consists of a set of design equations and design rules used to calculate the cost of a part. It also has functions to check the producibility of the part. These include warnings if the invalid dimensions are chosen. It can even take into account the machine availability and scheduling difficulties. Cost models have three main cost drivers (material, process and feature) as shown in Figure 4. CostAdvantage provides a user interface that is easy to configure and makes data entry simple. It can be tuned to the user's specifications making it flexible enough to cross industrial boundaries. The aim of CostAdvantage is to provide designer's with the knowledge of an expert manufacturing engineer. The software attempts to capture knowledge about manufacturing processes and costs. These can then be called on at the design stage to make improved decisions.

5. Case Study: A snubbered pin fix compressor blade

This model is concerned with the estimation of the manufacturing process for a compressor blade. The compressor blade was decided upon because of the relatively high availability of data points (there are numerous compressor blades in any one engine!). The material choice is between titanium, aluminium, steel or nickel with titanium being the most common at present.

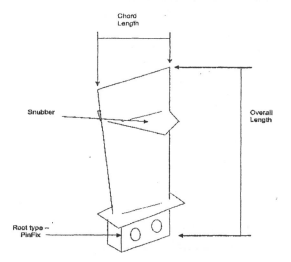

Figure 5: A typical snubbered pin fix type compressor blade.

Each compressor blade consists of a root that can have one of three forms: dovetail, firtree or pin fixing. There is also the possibility that the blade may be snubbered. This is an additional piece of metal that ensures a constant gap is kept between the blades when assembled. Figure 5.0 shows a typical compressor blade configuration. The processes in the production of the compressor blades comprise of drilling, milling, broaching, forging and trimming with some ancillary tasks. The root feature is broached unless it is a pin fix in which case milling and drilling are used. The data collected included past costs, operation times on various machines and materials uses.

Building a model in CostAdvantage is a process that requires recording manufacturing knowledge in terms of values, equations and design rules in a manner that allows calculation of cost estimates. A cost model must be created that incorporates the best estimating equations. The model must represent the part in such a way that a user can easily create a cost estimate. Figure 6 shows some of the equations used in the CostAdvantage model.

6. The Proposed Integration

This section investigates the integration of KAPES and CostAdvantage. It looks at an alternative strategy for providing a cost estimating link between raw data and designer inputs. The proposal is based around parametric methods and uses the CostAdvantage (CA) software.

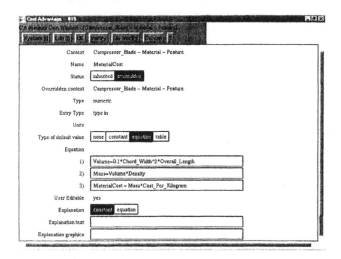

Figure 6: Screenshot showing equations entered for material cost calculation.

6.1 Direct Integration: A data centric view

The aim of the integration is to produce a software system that combines the strong points of the two software packages. This would make use of CostAdvantage's ability to read in data directly from CAD/CAM systems and be controlled using an intuitive graphical user interface (GUI) whilst calculations use KAPES's broad data repository. The basic idea is illustrated in Figure 7.0.

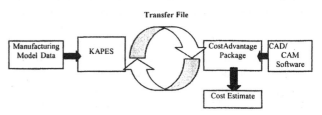

Figure 7: Illustrating the operation of the CA KAPES integration proposal

The entire system should be controlled through the CostAdvantage GUI and read in data from CAD/CAM (or be manually input). These features can then be sent to KAPES as a text ASCII file. This file will contain the name of the feature and details of dimensions surface finishes etc. Each feature is then identified in the text file, its dimensions are copied into the KAPES operation of the same name. KAPES then performs detailed calculations on the production cost of the part. This production cost is entered into the ASCII text file. Once all the features have been costed in this manner the text file is returned for CostAdvantage to read in. This will give CA a read out of the features in the part and their associated costs according to the data in KAPES.

6.1.1 An Example

CAD data from a system such as pro engineer is fed directly into CostAdvantage for the model presented in section 5. This will be in the form of a set of features used to create the part and their associated dimensions. A feature list with all associated dimensions and details are passed to KAPES, and processed. The features are then broken down to the lowest level, and also the manufacturing processes and set-up steps are identified. These lead to a cost breakdown for each feature, which is then transferred to CostAdvantage using a text file. The entry in the Text transfer file for the pin fix feature would be of the form shown in Figure 8 with the processed text file looking similar to Figure 9.

Feature Pin Fix Root
Quantity 1 length .08 width .035 depth .02 hole-radius
0.005
Material cost 0.00
Process cost 0.00
Tooling cost 0.00

Figure 8: Shows the transfer file as it exits from CostAdvanatge

Feature Pin Fix Root
Quantity 1 length .08 width .035 depth .02 hole-radius
0.005
Material cost 0.00
Process cost 0. 1 8
Tooling cost 0.07

Figure 9: Shows the processed transfer file for entry into CostAdvantage from KAPES

6.2 Parametric Integration Proposal

This section contains details on a proposed method for creating a software tool that is based on parametric estimating principles. The proposal operates on using a computer program to provide the link between data collection from the shop floor and the CostAdvantage Package. This would give a complete link between CAD/CAM design features and historical data that is specific to the company. The historical data about manufacturing can be spooled from the KAPES database. This system is illustrated in Figure 10.

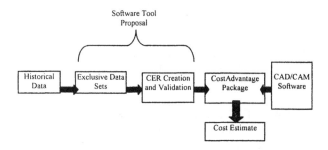

Figure 10: A flow diagram for the parametric integration approach.

To begin with the software would contain a database containing all the historical data that exists within the company. This would be in the form of a part's features and production costs (this may include inflation etc to make it relevant with today). Once this data is collected a sensitivity analysis would be performed to identify what the effects of various features have been on the production cost. This would only give an estimate of the average cost of a feature. It is observed that depending on the repetitions, cost of manufacture for a feature can be lowered. Simply averaging these will give an unrealistic cost estimate and lead to designers and engineers making poor decisions based on this information. It is for this reason that the software must be intelligent enough to identify features both in terms of their description and in terms of their cost history pattern. This means a system must be created that involves identifying subsets of parts, isolating features, and forming a cost estimating relationship (CER) for each feature when created on a particular part. Figure 11 shows a CER for the snubbered type of compressor blade. Jones [2] also presents several other CERs developed for the blade.

For historical information an effective means of data entry would be to use an identical interface as that used within CostAdvantage. This would allow the program to quickly and easily identify if a group of parts already existed to which the part could belong. It would do this by comparing the part type, number and sort of features and material type characteristics. A comparison can then be made between the parts actual cost and that which the CER forecasts. If the part's actual cost is within pre-set limits of the forecast then the parts data can be added into that data-set and the CER refined accordingly. If the forecast does not match then the system should flag up both the part and the relationship for further investigation by a cost engineer.

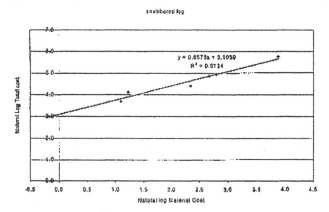

Figure 11: Cost estimating relationship for snubbered blades.

7. Discussion

The main area for incurring costs is typically found in the manufacturing sector. If a business has the ability to control manufacturing costs both by superior operations management and by intelligent design then it has a competitive advantage in the marketplace. KAPES is a package that deals with detailed costing and is more suited to low level costing on the manufacturing floor. It is designed to produce actual cost estimates that represent reality and give an accurate indication of the cost of production. It also provides other tools such as process planing and routing. CostAdvantage on the other hand is better suited to use at the design stage where less data is known and parametric methods can be used. The function of this package leans more towards comparing design options and their cost implications rather than detailed cost estimates. KAPES is at a level of detail where considerations such as set-up costs and indirect labour are of importance whereas these data are not readily available at the design stage. For example it is unlikely that the batch size will be known unless orders have been taken before design. When considering cost issues such as batch size, design functionality and market issues then interdependency is created at the design stage. This is particularly significant in consumer goods. If a decision is made between a luxurious or up market design and a more mundane one then batch size will cause problems. If a luxury design is preferred then the market niche aimed for is likely to be smaller than that of a more standard design. This will have impact on the batch size that will then cause change in the set-up costs. This in turn will influence the overall cost of production. Altering the cost in this way could influence the decision to choose a luxury or standard design. This is why it is difficult to include any consideration for set-up costs in a system dealing with a consumer product.

Therefore, creating a system that allows manufacturing knowledge and expertise to be used further up the product development process is of great value. The KAPES CostAdvantage integration proposal allows data to be accessed in KAPES through CostAdvantage. This is a potential solution to the problem faced by cost estimators. However, there are some technical limitations with the proposed integration. Alerting the software packages to the existence of a new transfer file is a difficult task. As a transfer file is used all features pertaining to a part must be sent to KAPES. This means that calculations are made for every feature every time the file is sent. This is fine for the first instance but when the designers is making slight alterations and wishes to view the cost implications of changes to the design the entire file is sent to KAPES and all features are recalculated whether or not they have changed. This means a considerable wait for calculation time occurs even when the design is changed only slightly, These problems mean the integration of CostAdvantage and KAPES in this manner is not commercially viable. The fact that designer's could be forced to wait for calculations to take place even when

administering only slight changes is a limitation of the approach. The problem can be addressed by automating the file transfer process. On the other hand the parametric integration approach requires a lot of past data about a product or product features. This is often difficult to obtain within an industrial environment.

This project has highlighted the recent change in the costing environment. It has looked at the current situation in terms of software solutions and costing strategies. It has then gone on to look at the ways in which an integrated approach could be created to cover all the costing functions. This included looking at both detailed and parametric costing. The end result is the specification of a method for creating a complete cost estimate.

8. Conclusions

The research has highlighted some of the benefits and issues related to an integrated approach to costing. It has suggested two approaches for solving the problem. It has also looked at the integration issues of two market leading software packages for standard costing. Finally it has taken into account the business benefits of each method and identified their pros and cons. The data centric integration is feasible with further automation, whereas, if past data is available the parametric integration approach can be used for the total cost estimation.

References

[1]   DEPARTMENT OF DEFENCE, *Parametric Estimating Handbook*, 2nd Ed., DoD, http://www.ispa-cost.org/PEIWeb/cover.htm, (1999).

[2]   JONES, P., *Developing an Integrated Approach to Cost Modelling,.* MSc Thesis, Cranfield University, 1999.

[3]   STEWART, R., WYSKIDSA, R., JOHANNES, J., Cost Estimator's Reference Manual, 2nd ed., Wiley Interscience, 1995.

[4]   ROY R., BENDALL D., TAYLOR J.P., JONES P., MADARIAGA A. P., CROSSLAND J., HAMEL J., TAYLOR I. M. *Development of Airframe Engineering CERs for Military Aerostructures.* Second World Manufacturing Congress (WMC'99), Durham (UK), 27-30th Sep., pp. 838-844, 1999.

[5] DEWHURST, P.and BOOTHROYD, G., *Early cost Estimation in product design,* Journal of Manufacturing Systems, 7(3), pp. 183 -191, 1988.

[6] MILES, L. D., Techniques of value Analysis and Engineering, McGraw Hill Book Company, 1972.

[7] RUSH, C, and ROY, R., *Analysis of cost estimating processes used within a concurrent engineering environment throughout a product life cycle,* Accepted for CE2000 conference, Lyon (France), 17-20 July, 2000.

[8] OTSWALD, P.F., Cost Estimating, 2nd ed., Prentice Hall, Englewood Cliffs, NJ, 1984.

[9] WIERDA, L. S., *Linking Design, Process Planning and Cost Information by Feature-based Modelling,* Journal of Engineering Design, Vol. 2, No. 19, 1991

[10] OU-YANG, C. and Lin, T. S., *Developing an Integrated Framework for Feature-Based Early Manufacturing Cost Estimation,* The International Journal of Advanced Manufacturing Technology, No. 13, pp. 618-629, 1997.

[11] MOILEY, E., YANG, H. and BROWNE, J., *Feature Based Modelling in design for Assembly,* International Journal of Computer integrated Manufacturing, No. 6 1721, pp. 119 -125,1993.

A Model-based Methodology for Extended Enterprise Engineering

Sobah A. Petersen

Dept. of Computer & Information Science, Norwegian University of Science & Technology,
N-7491 Trondheim, Norway, sap@idi.ntnu.no

Orsolya Szegheo

Dept. of Production & Quality Engineering, Norwegian University of Science & Technology,
N-7491 Trondheim, Norway, orsolya@ipk.ntnu.no

Abstract

This paper describes an enterprise model-based methodology for Extended Enterprise Engineering. The aim of the model-based methodology is to provide industrial practitioners guidelines and an operational environment for Extended Enterprise Engineering. The enterprise modelling capabilities enable the extension of the methodology from that of a guideline to a knowledge repository. The basic approach is process-oriented, where the activities that take place during the lifecycle are modelled. The methodology not only identifies the inputs, constraints, resources and the outputs for each activity, but it also provides access to them. Thus, the methodology model plays the role of an extended enterprise knowledge management platform as well as helps industrial practitioners to set up the technical and the organisational infrastructure that is central to an Extended Enterprise. The uniqueness of the methodology is achieved through graphical representation and by providing access to information and to other resources in a distributed working environment.

1 Introduction

Recent changes in the market can best be described by terms such as globalisation and customisation. New business practices have emerged to cope with the new challenges and new terms have been created to depict them. Preiss uses the term interprises for enterprises, which are international and interactive in the culture of the Internet, [1]. Davidow and Malone define virtual corporation as a temporary network of independent companies, suppliers, customers, even erstwhile rivals linked by IT to share skills, costs and access to one another's markets, [2]. Wiendahl and Helms identify different types of networks, such as strategic network, virtual enterprise, regional network and operational network, [3].

One of the major factors affecting the way business is conducted is, no doubt, the recent developments in IT; in particular, distributed information systems and the Internet. Enabled by new business practices and technology, enterprises have gone beyond the geographical and sociocultural boundaries and have become entities that compete in a global market by forming international alliances. Due to the possibility of distributed working, new forms of organisations have emerged. The concepts of Extended and Virtual Enterprises have emerged as central to the ideas of forming alliances and networks of enterprises.

1.1 Extended and Virtual Enterprises

There are several definitions of an Extended Enterprise, some of which are listed below:
- A conceptual business unit or system that consists of a purchasing company and suppliers who collaborate closely in such a way as to maximise the returns to each partner, [4].
- The formation of closer coordination in the design, development, costing and the coordination of the respective manufacturing schedules of cooperating individual manufacturing enterprises and related suppliers, [5].
- An enterprise mostly made of functions provided by other enterprises and that it relies heavily on the use of standards, computer communications and electronic data interchange, [6].

Some authors describe only the Extended Enterprise, others only the Virtual Enterprise, and there are some that describe both and define the relationship between them. In our work, we have distinguished between the two concepts. We define the Extended Enterprise as:

A partnership among enterprises, where the goal is to achieve competitive advantages by forming formal or informal links and maintaining distributed co-operation throughout the partnership.

For example, in the manufacturing industry, it can be described as a partnership among manufacturing enterprises. It also includes very close collaboration between the manufacturer and the customer and the supplier. The collaboration among the partners is supported by ICT. In the Extended Enterprise, the collaborating enterprises are encouraged to focus on activities in which they have special competence.

The Virtual Enterprise can be viewed as *a temporary alliance of enterprises participating in the Extended Enterprise*. They join to take advantage of a market opportunity. The Virtual Enterprise, compared to the Extended Enterprise, lasts a shorter period of time, and has less formal partnerships. The Virtual Enterprise is reconfigured for each occurring window of opportunity, [7].

1.2 Extended Enterprise Engineering

In our work, we have addressed the concepts of Extended as well as Virtual Enterprises and investigated their main features. We have tried to define the activities that take place in such enterprises, in particular the activities that are involved in setting up the enterprises and operating them efficiently. We have given special emphasis to the collaboration activities and interactions that take place among the partners, such as establishing the operating rules.

We propose an *Extended Enterprise Engineering Framework and Methodology*. By definition, the activity of systematic design and specification of enterprise business processes is called *Enterprise Engineering*, [6]. So, consequently, the design and specification of the business processes of the Extended Enterprise is called *Extended Enterprise Engineering*.

Our aim is to create a computerised, dynamic, enterprise model-based methodology that describes the Extended Enterprise and, at the same time, provide an Extended Enterprise engineering environment that enables the integration of tools and supports knowledge management.

1.3 The need for a Methodology Model

This research is necessitated by the fact that all the existing enterprise modelling and engineering architectures, frameworks and methodologies have initially been created for individual enterprises and have not taken into consideration the special characteristics of the Extended and Virtual Enterprises, (e.g. GERAM, PERA, CIMOSA, (see [6]). Though some initiatives have been made to apply them for the Extended Enterprise (e.g. CIMOSA, [8]), they are not adjusted yet in every detail to the peculiarities of the Extended Enterprise.

The success of an Extended Enterprise depends on the level of collaboration among the partners. Hence, it is important to address explicitly the issue of collaboration. There is a need to identify the collaborations among the partners, the context and type of the collaboration, the resources that are needed to support the collaboration and the information and materials that are exchanged among the partners. Unless we address these issues explicitly, we fail to support Extended Enterprises in their most important characteristic.

Another important issue in Extended Enterprise Engineering is that of Knowledge Management and the reusability of knowledge and experience. The flexibility that is desired in our business environment is achieved by being able to set up a Virtual Enterprise from enterprises participating in the Extended Enterprise, very quickly, so that it is ready for operation as soon as possible. Similarly, the management of competencies and skills are of equal importance because the Extended and the Virtual Enterprises are primarily focussed on getting together complementing competency profiles. We believe that enterprise models can play a significant role in resolving these two issues.

In our work, we aim to provide an Extended Enterprise Engineering Methodology, in the form of an enterprise model that can be used by industrial practitioners to create and operate Extended and Virtual Enterprises. The enterprise modelling capabilities enables the extension of the methodology from that of a guideline to a knowledge repository. The methodology has a sound theoretical foundation as well as value for industrial use because it can be customised for further use. The methodology is, thus, formulated in such a way that it is easy for industrial practitioners to apply it.

The main audience of our methodology is Extended Enterprise engineers and enterprise architects. The approach is based on Active Knowledge Modelling (AKM) and uses graphical visualisation, [7]. The computerised, model-based representation provides several advantages. It is easy to navigate in the model and simple to extend, modify and customise it to individual requirements. And most importantly, we provide a customised, graphical user interface and a navigation guide to the model.

The rest of the paper is organised as follows. Section 2 provides a background to our work by summarising some of the main ideas of the Globeman 21 project; Section 3 describes how Extended Enterprises support Concurrent Engineering; Section 4 discusses the importance of enterprise modelling for Extended Enterprises and our modelling approach; Section 5 describes the methodology model; Section 6 describes how the model could be used by industrial practitioners and Section 7 summarises this work.

2 Background

While there has been a lot discussion about the concepts of Extended and Virtual Enterprises, (e.g. in [4], [5], [6]), there has been very little work done in terms of clarifying these concepts and in providing guidelines for the industry to operate in this manner. A significant contribution has been made by the Globeman 21 project by identifying some of the activities that take place within an Extended and a Virtual Enterprise. Our work is strongly influenced by the ideas developed in the Globeman 21 project.

2.1 Extended and Virtual Enterprises in Globeman 21

The Globeman 21 project defined an Extended Enterprise Framework, which includes the three concepts; the Network, the Virtual Enterprise and the Product. The Network concept implies a network of enterprises that collaborate. Their relationship may be formalised, (e.g. by a set of operating rules) or informal, (e.g. they might know one another through joint projects in the past). From this network or cluster of enterprises, a Virtual Enterprise is formed to meet a customer request, where each enterprise contributes one or more competencies. The deliverable of the Virtual Enterprise is a product, [9].

In the Globeman 21 project, the relationships among the three entities, the Network, the Virtual Enterprise and the product, are shown using the concept of a lifecycle, as shown in Figure 1. The Extended Enterprise Framework is based on GERAM (Generalised Enterprise Reference Architecture and Methodology), [10], [11].

The Extended Enterprise Framework shows the complete life cycle of the Network, the Virtual Enterprise and the Product. It says that when the Network gets operationalised, a Virtual Enterprise is created. Similarly, in the operational phase of the Virtual Enterprise, the Product is created, [12].

In addition to the Extended Enterprise Framework, the Globeman 21 Concept also identified different management positions and "project groups", or work groups that are formed within an Extended and a Virtual Enterprise to perform the work. An example of a management position is the network manager while an example of a project group is the IT infrastructure group. They also identified some of the deliverables that these managers and project groups must produce.

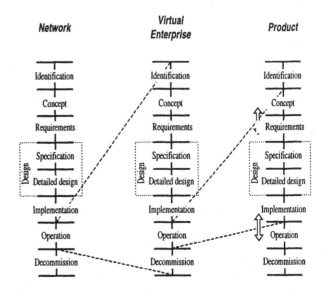

Figure 1: The Globeman21 Extended Enterprise Framework

2.2 Extended Enterprise Engineering Framework

A model-based framework for Extended Enterprise Engineering was proposed in [7]. The framework consists of an Extended Enterprise methodology, a test case and a supporting environment. This framework is based on the Globeman21 Extended Enterprise Framework and the methodology reflects the three lifecycles; the Extended and Virtual Enterprises and the Product. (Note that, in the Globeman 21 framework, these three enterprise entities were referred to as the Network, the Virtual Enterprise and the Product).

A supporting environment is modelled to provide additional descriptive detail of the methodology part. The test case provides example models of Extended and Virtual Enterprises, and can be considered as a library of models. In this paper, we focus on the methodology part of the framework.

3 Extended Enterprises and Concurrent Engineering

The close collaboration among the partners in the Extended Enterprise and the core competence specialisation facilitate Concurrent Engineering. Some examples of how Concurrent Engineering can be achieved in the Extended Enterprise are described below:

- *Manufacturing*: In the manufacturing industry, the customer, the manufacturer and the suppliers collaborate to form the Extended Enterprise. The extensive communication among the three partners means that while the manufacturer is agreeing with the customer, he/she can also ensure that the suppliers are able to deliver the materials that are necessary for manufacturing.
- *Engineering Design*: The different modules or parts of the product may be designed by different designers or teams. Since the different designers collaborate with one another, they can agree on the interfaces of each module and design them concurrently. Thus, mutual designing is achieved.
- *Software development*: The different components of the software, such as the GUI or the database, may be developed by the different partners of the Extended Enterprise. Since the partners collaborate, they can test parts of the modules as they become available, thus tracking bugs and interface inconsistencies earlier on in the development process.
- *Building and Construction*: There are a lot of similarities to both manufacturing and engineering design. Since the Extended Enterprise is focussed on specialising in core competencies, each group (e.g. the electricians or the plumbers) can work in parallel and the communication among them ensure that each partner's work is a part of a bigger job.

4 Enterprise Modelling and Modelling Environments

The subject of enterprise modelling has been addressed by several authors, [13], [6]. The role of enterprise modelling is central to understanding enterprises and is concerned with designing and representing enterprise activities, structure and behaviour. "An enterprise model is a computational representation of the structure, activities, processes, information, resources, people, behaviour, goals and constraints of a business, government or other enterprises", [13]. The distributed and flexible nature of the Extended Enterprise introduces additional challenges in understanding it's full potential, and hence, enterprise modelling can play a significant role in providing an overview and creating a common understanding among the partners in an Extended Enterprise.

In the past, most methodologies and architectures have been described as documents with figures, tables, and other forms of illustrations, (e.g. GERAM, [10]), and based on these descriptions, a list of requirements for a modelling tool is drawn up. Within the context of an Extended Enterprise, we have realised the importance of an Extended Enterprise *modelling environment*, rather than a modelling tool, where the geographically distributed enterprises can realise modelling.

Another reason for the need for a modelling environment is that modelling is no longer an isolated activity within an enterprise. An enterprise model can be a part of the operation of the enterprise, where the model is used to assist people in their daily work. For example, people can be assigned work through the model, they can access their homepages through the model and see their task lists for the day. An example of using a model actively is to access other information sources and software tools. Most importantly, in an Extended Enterprise, there is a need to transfer model information from one model to another and from one enterprise to another. Our modelling endeavour convinced us that there is a need for a modelling environment, which is adjusted, to the requirements of the Extended Enterprise.

4.1 Requirements for an Extended Enterprise Modelling Environment

The main categories of requirements that we have identified for a modelling environment are as follows:
- *Technical requirements*: These requirements address the capabilities of the underlying technology that support the modelling. For example, the modelling environment must support distribution of models over a network and provide capabilities for supporting collaborative modelling in a multi-user environment.
- *Model building capabilities*: These requirements address the model building capabilities that are provided by the modelling environment. In an Extended Enterprise, the responsibility for the complete enterprise model may be shared among several partners. Hence, capabilities such as sub-modelling, where a single model is composed of several sub-models, becomes extremely important.

In order to support knowledge management and experience reuse, it must be possible to support a repository of models.

- *Model usage and operation capabilities*: These requirements address the capabilities that can be applied to use the model after it has been created. Since the model may be used by several people from different enterprises, it must be possible to set access rights to the different parts of the model.
- *Other requirements*: These requirements address general capabilities that are desired in an extended enterprise modelling environment such as the support for the lifecycle of the extended enterprise.

While there are no Extended Enterprise modelling environments that meet all the requirements, we believe that AKM and METIS go a long way in meeting the modelling requirements listed above.

4.2 Active Knowledge Models and METIS

We have chosen AKM as the modelling concept and METIS 2.1 as our modelling environment. An AKM is defined as a "visual model of externalised knowledge represented as flows and structures that can be visualised, traversed, studied, analysed, simulated and executed", [14]. This approach advocates that an enterprise model must contain 4 main aspects of an enterprise; processes and activities, products and services, organisation and people, systems and tools.

We have developed our model in METIS, which is an enterprise modelling software that is based on AKM. More details on AKM and the METIS enterprise modelling environment can be found in [7], [15].

5 The Methodology Model

In the model based Extended Enterprise Engineering approach, the basic idea is that when a group of enterprises want to engineer their collaboration, they model their processes or activities. The methodology covers the three entities, the Extended and the Virtual Enterprises and the Product. Therefore, we have modelled the activities that take place in the different phases of the lifecycles of these three entities.

The lifecycle phases are represented as process models. The top level of the process model addresses the lifecycle of the Extended Enterprise; the lifecycle of the Virtual Enterprises and their Products are modelled in the lower levels. Each process can be decomposed into a set of sub-processes. METIS provides a process modelling capability, where the inputs, outputs, controls

or constraints and mechanisms or resources (collectively referred to as ICOMs), can be defined for each process and sub-process, as shown in

Figure 2. Similarly, it is possible to show the dependencies among the processes in terms of their ICOMs; e.g. the output of one process can be the input, the control and/or the mechanism for another process.

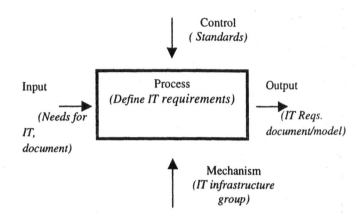

Figure 2: A Process and ICOMs

The modelling was conducted as follows:

1 Model the top-level processes, based on the lifecycle phases.
2 Identify and model the ICOMs for the processes.
3 Decompose and detail the process hierarchy.
4 Describe the ICOMs by modelling them in the supporting environment and providing access to them wherever possible.

5.1 The Lifecycle Phases

The main processes of the Extended Enterprise engineering methodology are shown as the top level of the process model. Each process can be decomposed into its sub-processes, which are shown graphically as small boxes within the main process. These boxes, in fact, represent process objects in the model. It is possible to decompose all the processes in a similar manner. The main processes of the lifecycle were described in [7]. Figure 3 shows the main processes of the three entities and the relationships among the three lifecycles. (Note that we are using figures from the Globeman 21 Methodology model. Hence, the term EEE Network has been used instead of Extended Enterprise.) The processes are, of course, related to each other. For the sake of simplicity, the relationships among the processes are not included in the figure.

During the operational phase of the Extended Enterprise, a Virtual Enterprise is created to meet a customer request. The Extended Enterprise practices and procedures are customised to the needs of the Virtual Enterprise. The customer request is analysed. The Virtual Enterprise is further decomposed into the Product life cycle. In Figure 3, the arrows between the lifecycles indicate this hierarchical decomposition.

5.2 Modelling the Processes

In our methodology, we have detailed the activities of the project groups and the different managerial roles that were identified in the Globeman 21 project. Using these ideas and our experience from other enterprise models, we defined a set of processes and their ICOMs. For example, the IT infrastructure group must produce the IT requirements and architectural specifications. Translating this information into our modelling language, we have a process "define IT requirements" that is performed by the resource "IT infrastructure group" with an output, a document, "IT requirements". This example is shown in

Figure 2.

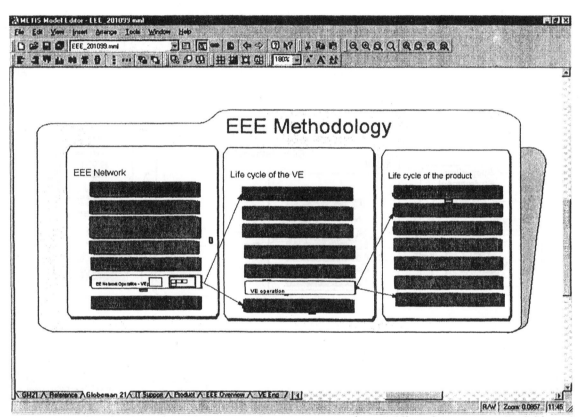

Figure 3: Extended Enterprise Engineering Processes

While modelling the processes within each phase in the lifecycle, we tried to distinguish between the processes that were pertaining to the management of the entity and the processes that were a part of the activities that were conducted by the partners in actually fulfilling a customer request. For example, in the Extended Enterprise, the management activities include the initial selection of partners and setting operational rules of the enterprise, while the partner activities include identifying business opportunities and sending in a bid to compete for such an opportunity. Similarly, in the Virtual Enterprise, the management activities include the setting up of a communications infrastructure and drawing up workplans, while the partner activities include specifying the customer requirements.

Knowledge management plays a major role in the management activities. We have tried to address this issue by modelling a process in each phase of the lifecycle, where some form of documentation of the work that has been done must be reported. This process is included in each phase in order to avoid this being done only at the dissemination phase; but rather, to capture the experience as it happens. Also, wherever possible, we have tried to include other, more automatic means of capturing the experience. Incorporating a model-based approach encourages the modelling of information wherever possible. For example, building models of the competencies of the partners and building product and information models is one way of capturing experience in a structured manner. Also, these models can be reused as well as be kept in a model library for reference purposes.

5.3 Methodology and the Supporting Environment

The identification of the ICOMs in the model necessitated the creation of a dedicated place where all the items defined in the ICOMs could be stored and could be accessed. ICOMs might represent information sources, but they might be supporting tools as well. Thus, the supporting environment in the Extended Enterprise engineering framework is where all the information identified in the ICOMs are modelled, and where all the tools that are needed to execute the activities are accessible. Figure 4 shows a simplified picture of how the supporting environment and the processes are related.

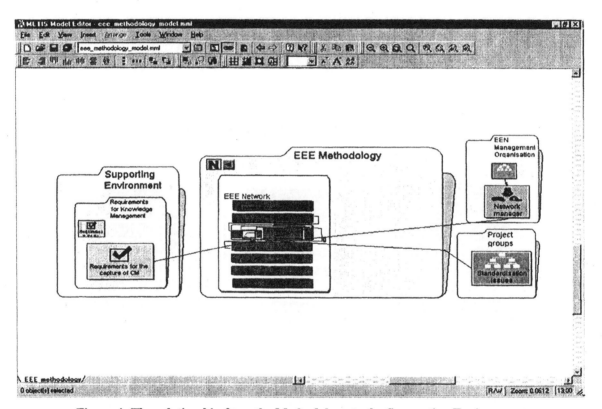

Figure 4: The relationship from the Methodology to the Supporting Environment

For example, if the methodology part contains a process that has the output (or the input) "Requirements for the Capture of Corporate Memory", this output could be modelled in the supporting environment as an on-line document. In the METIS modelling environment, the object type "on-line document" has the capability to spawn another process using the application name and the filename that is given as properties or attributes of that object. Hence, it is possible to open the document directly from the model. Similarly, software applications, other enterprise models and web browsers can be accessed directly from our model. Also, the person (or the group of people), e.g. the network manager, that is responsible for a process can be shown in the supporting environment and we can have links to the person's homepage and their CVs.

5.4 Visualisation and Navigation

When all the information pertaining to an Extended and a Virtual Enterprise is modelled, the model becomes very big and complex. At a first glance, it may appear to be very complicated. In order to make it easier to view and use the contents of the model, we have used the following visualisation capabilities in the model. Some of these are described briefly:

- Selective hiding and displaying of relationships in the model: if the user wants to focus on a particular object in the model and just wants to see the relationships from that object only, it is possible to hide all other relationships in the model and display just ones that are of interest.

- If the user wants to select one or a few objects and all the objects that are related to these objects, it is possible to use a capability called criteria and generate a special view with these objects and relationships. An example of such a view is shown in Figure 4.

- In order to simplify the model, we have created several views of the model. For example, we have special views just showing the lifecycle of the Virtual Enterprise, an overview of IT resources that are required and an overview of the management of the Extended Enterprise. These are just graphical views extracted from the main model and can be edited in either view without making the model inconsistent.

METIS also provides several navigation capabilities. We have used the action button (hot buttons for navigation) capability extensively to provide easy access to the relevant part of the model. These buttons can be placed anywhere in the model and they have a target object and dedicated operations. By double-clicking on this button, it is possible to move to the target area of the model. We have used this capability to provide the following guides to the user:

- A navigation guide showing an overview of the model contents structured according to the GERA model content views [10]. In addition to the four view described in GERA, we have added a view to show partner collaboration.

- A model guide and scenario, which is a simple guide to use the model, with dedicated capabilities to perform simple operations in the model. **Figure 5** shows this view of the model.

5.5 The Graphical User Interface

METIS has a graphical user interface with drag-and-drop modelling capabilities and a symbol editor. The objects and relationships in the model are represented as graphical symbols, which can be customised by the user. Graphical images make models more vivid and lifelike, and less threatening for the user. They help in associating the model to reality. Figure 5 shows an example of utilising the modelling capabilities and the graphical user interface to provide in-line documentation on using the model. We have used the "pushpin" object, which is analogous to sticking post-it notes for comments, to provide instructions to the user.

6 Model usage

This section of the paper describes how the methodology model can be used in the industry.

- *As a guideline for industrial practitioners*
The model provides an overview of the activities that take place in the lifecycle of an Extended and a Virtual Enterprise and a Product. This is based on existing enterprise modelling reference models and architectures, industrial experience and best practice methods. The model also describes operational procedures, the inputs that are needed to perform these activities and the resources that are required. Also, the competencies of the people that are required to perform these activities are highlighted. Similarly, the model also provide the IT requirements for performing the activities. Since the information is structured in the model, the user can easily obtain the specific information that he needs. For example, if the user wants to look at the IT resources that are required in a specific phase of the lifecycle, the model will show this information. Thus, the model acts as a guideline or a "cookbook" for Extended Enterprise engineers.

- *As an operational environment*
Since the model is created in a modelling environment, it is possible to copy the parts of the model that are of interest to the user and instantiate it for the current purpose, thus, creating an enterprise model of the current Extended Enterprise.

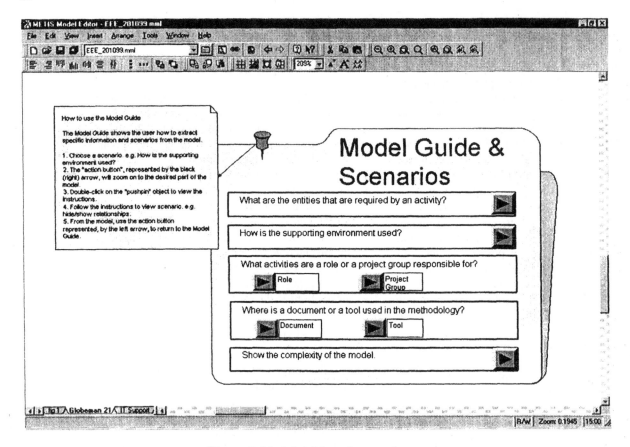

Figure 5: Model guide and scenario

In addition to creating the model, the user is able to use the analytical capabilities that are supported by the METIS modelling environment. The user can traverse through the relationships to see the dependencies among the processes and the other objects in the model, analyse critical paths and automatically generate reports (in several formats, such as ascii text and HTML) from the contents of the model.

Most importantly, it is possible to access external sources such as documents and software applications directly from the model. For example, a specific process in the model involves the editing of a document. The user can open the document directly from the model, edit and save it, while continuing to work in the modelling environment. Similarly, if the user wants to run a simulation using some data in the model, the simulation software can be executed directly from the model. Hence, the model acts as a "worktop" for the user by enabling the user to execute his tasks via the model. This facilitates work optimisation and coordination in an Extended Enterprise. Also, by providing the possibility to represent an overview of all the external resources and access them via the model, the model becomes an integrator or a high level infrastructure for the Extended Enterprise.

- *As a knowledge repository*

The model of the Extended Enterprise grows and is enriched throughout the lifecycle. It becomes a model of the actual work that was conducted in the Extended Enterprise and is a reflection of the life history of the enterprise. The model itself consists of several models, e.g. the IT requirements model, the competency model, etc. These models can be reused in other Extended or Virtual Enterprises as well as be used as reference models. Creating such a knowledge repository where the different groups can access distributed information also supports concurrent engineering.

In the model, knowledge is stored in a structured way, which makes navigation and knowledge acquisition simple. Learning and reflection on experiences are essential concepts of the methodology and feedback loops are built in for knowledge management support. We consider it important to make industrial practitioners conscious about the importance of describing, analysing and evaluating experiences and integrating the gained knowledge into the knowledge base of the Extended Enterprise.

7 Summary

This paper describes an enterprise model-based methodology for Extended Enterprise Engineering. The aim of the methodology model is to provide industrial practitioners guidelines as well as an operational environment for Extended Enterprise Engineering. The enterprise modelling capabilities enables the extension of the methodology from that of a guideline to a knowledge repository. The uniqueness of the methodology model is achieved through graphical representation and by providing access to information and to other resources in a distributed working environment.

By describing the methodology as an enterprise model, we were able to provide an environment that can be made operational as well as one that captures the experience in a reusable, structured format. By building an enterprise model, we also realised the importance of addressing the interactions among the partners and the new requirements this imposes on the modelling environment.

8 Acknowledgements

This work has been conducted in collaboration with NCR METIS and we express our gratitude to Frank Lillehagen for fruitful discussions on Extended Enterprise Engineering. We also wish to thank the Globeman21 Concept team, in particular Johan Vesterager, Jens Dahl Pedersen and Martin Tølle, of the Technical University of Denmark, for interesting discussions.

9 References

[1] Preiss, K., 1997, "The Emergence of the Interprise". In *Organizing the Extended Enterprise* Eds. Paul Schonsleben and Alfred Buchel, Chapman & Hall 1998, IFIP Tc5/WG5.7 International working Conference on Organizing the Extended Enterprise, Ascona, Switzerland.

[2] Davidow, W.H., and M.S. Malone, 1992. *The virtual corporation*, HarperCollins, New York.

[3] Wiendahl, H. P., K. Helms, 1997, "Variable Production networks- Successful Operating" in *An Alliance of the Best*. In *Organizing the Extended Enterprise* Eds. Paul Schonsleben and Alfred Buchel, Chapman & Hall 1998, IFIP Tc5/WG5.7 International working Conference on Organizing the Extended Enterprise, Ascona, Switzerland.

[4] Childe, S. J., 1998, "The extended enterprise- a concept of co-operation", Production Planning & Control, Vol.9, No.3, p. 320-327.

[5] Jagdev, H. S., and Browne, J., 1998, "The extended enterprise- a context for manufacturing", Production Planning & Control, Vol.9, No.3, p. 216-229.

[6] Vernadet, F. B., 1996, *Enterprise Modeling and Integration Principles and Applications*, Chapman and Hall.

[7] Szegheo, O. and Petersen, S. A., 1999, "Extended Enterprise Engineering – A Model-based Approach", *Advances in Concurrent Engineering*, CE99, Eds. P.K. Chawdry et. al., Technomic Publishing Co., Inc., p. 3-10.

[8] Zelm, M., 1997, "CIMOSA and the Enterprise Organisation" ICEIMT'97. Eds. Kosanke et. al., Springer Verlag., p. 539-547.

[9] Globeman 21: Project Homepage available from http://ims.toyo-eng.co.jp/

[10] IFIP-IFAC Task Force, 1998, "GERAM: Generalised Enterprise Reference Architecture and Methodology", Version 1.6.2, available from http://www.cit.gu.au/~bernus/

[11] Bernus, P. and G. Schmidt, 1998. "Architectures of information systems" in P. Bernus, K. Mertins, G Schmidt (Eds.), *Handbook on Architectures of Information Systems*, Springer Verlag.

[12] Vesterager, J., Larsen, L.B. and Gobbi, C, 1999, "Architecture and methodology for creating virtual enterprises – results from Globeman 21", presented at the IMS Globeman 21 Open day, March, Tokyo, Japan.

[13] Fox, M.S. and Gruninger, M., 1998. "Enterprise Modelling". In: AI Magazine, AAAI Press, Fall, p. 109-121.

[14] Lillehagen, F. and Karlsen, D., 1999. "Visual Extended Enterprise Engineering embedding Knowledge Management, Systems Engineering and Work Execution". In: Proceedings of IFIP International Enterprise Modelling Conference IEMC '99, Verdal, Norway, June.

[15] METIS : METIS product information available: http://www.metis.no

Introduction To The Concept Of Enterprise Virtualization

Stephen Chi-fai Chan

Department of Computing, The Hong Kong Polytechnic University, Hong Kong
Email: csschan@comp.polyu.edu.hk

Li Zhang

Institute of Manufacture Systems, Beijing University of Aeronautics & Astronautics, Beijing, China
Email: lizhang@public.fhnet.cn.net

Abstract

*Enterprise Integration and Virtual Enterprise can be considered two different stages in enterprise development. Enterprise Integration focuses on information exchange and sharing within an enterprise. A Virtual Enterprise is a temporary alliance of enterprises, focuses on resource sharing and interoperation on a global scale, and can be considered a later stage of enterprise development. We developed the concept of **Enterprise Virtualization** as an intermediate stage between Enterprise Integration and Virtual Enterprise. This paper presents an overview of the concept: its definition, architecture and basic activities. It aims to map an Integrated Enterprise into the Virtual Space of the Internet and associated standard communication protocols. A prototype Virtualized Enterprise has been successfully developed, in which an Internet-based design and manufacturing system was implemented through resource sharing.*

1 Introduction

The focus of Enterprise Integration is information exchange and sharing across internal boundaries in traditional enterprises. One of the popular techniques of which is to use the STEP standard [9] to define the Global Product Data Model over all product life-cycles so that product data can be exchanged and shared among different operation platforms and different product development phases. Along with the development of computer networks and information technology, and rapid changes of the market and user requirements, manufacturing is becoming a global undertaking. The production mode, organizational structure and range of activities of traditional enterprises face severe challenges, and a variety of techniques have been tried in responding to these challenges [4, 12, 13, 114].

Virtual Enterprise (VE) is a comparatively newer research area which is attracting more and more attention [7, 3]. CIMOSA [2] defined an extensive architectural framework for enterprises consisting of four views: functional, information, resource, and organizational; a three-step instantiation process from generic to partial to particular; and a three-level approach consisting of models from requirements definition to design specification to implementation. Since then, some projects have concentrated on the methodologies for building the Virtual Enterprises [6, 8, 1]. The National Industrial Information Infrastructure Protocols (NIIIP) consortium defines a *Virtual Enterprise* as a temporary consortium or alliance of companies formed to share costs and skills and to exploit fast-changing opportunities, and developed a Reference Architecture [10] for it. The concept of Virtual Enterprise emphasizes that an enterprise must break through both internal and external boundaries, and it should be created and reconfigured on the world-wide scope in flexible and optimized mode in order to meet the requirement of the rapidly changing market.

In this paper we present the narrower concept of Enterprise Virtualization. In our model, the process of enterprise development can be divided into three stages. The First stage is the Integrated Enterprise which focuses on breaking through internal boundaries (such as operation system boundary, language boundary and information boundary), and one can say that an Integrated Enterprise is built in reality space. The Second, or intermediate stage, is the **Virtualized Enterprise** (VLE), where the main objective is to map a traditional Integrated Enterprise from reality space to the virtual space of the Internet. External boundaries are removed when an enterprise is virtualized. The Virtual Enterprise is a higher stage of enterprise development, which focuses on the mechanisms and protocols for building a temporary consortium or alliance of Virtualized Enterprises quickly. So, one can say that a

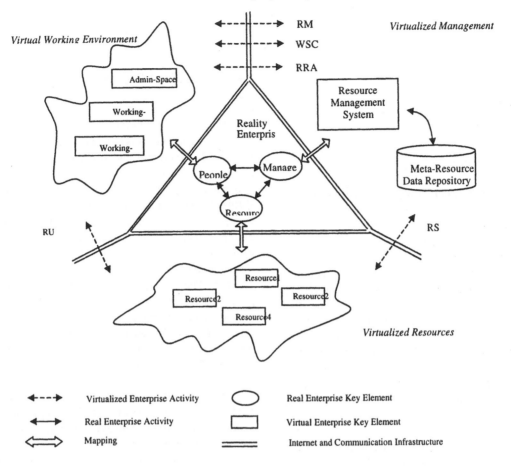

Figure 1. The Architecture of a Virtualized Enterprise.

Virtual Enterprise is built on the base of Virtualized Enterprises. The paper provides an overview of the Virtualized Enterprise, including definition, architecture and basic activities. A prototype Virtualized Enterprise will be described.

2 Basic Concept

Following are some key definitions:

1. *Virtual Space* (**VS**): Unbounded information process field built on networks (Intranet/Internet/extranet) and standard communication protocols (TCP/IP, HTTP, HTMP, CORBA etc.).
2. *Virtualization*: Mapping process through which inner structure and implementation of an entity in reality space are hidden and only outer functional features are mapped to the VS so that the entity becomes visible, operable, manageable and portable in the VS.
3. *Enterprise Virtualization*: A set of mapping processes which map key elements of an enterprise

from the limited Reality Space to the unbounded Virtual Space. The result is a **Virtualized Enterprise (VLE)** which is visible, operable, manageable, and reconfigurable through the Virtual Space.

2.1 Virtualized Enterprise Architecture

The architecture of VLE is shown in Figure 1. VLE is built in VS and it divides VS into three sub-spaces:

1. *Virtualized Resource* sub-space which is the set of all the Virtualized Resources registered in the Meta Resource Data Repository (MRDR).
2. *Virtual Working Environment* sub-space which is the set of all the Administrator Spaces and Working Spaces,
3. *Virtualized Management* sub-space that includes the **Resource Management System (RMS)** and the MRDR.

These three sub-spaces are key elements of the VLE, correspond to the three key elements of a traditional integrated enterprise.

2.2   Basic Activities

The activities that are likely to take place in a VLE can be classified into these five types:

Resource Management (RM): Interactions between administrators and the RMS, such as registering new resources, querying meta-resource data, maintaining resources and supervising resources.

Working-Space Configuration (WSC): Interactions between users and the Resource Management System (RMS). For example, a user signs in the RMS; the RMS sets up the use authority and offers an initial WorkSpace (WS); the user enters the initial WS and subscribes to a Virtualized Resource from the Meta-Resource Data Repository. When the user enters the WS next time, he/she will get a special working platform which contains the Resource Graphical User Interfaces (RGUIs) of the ordered resources.

Resource Request and Assignment (RRA): Interactions between users and the RMS. When a user wants to use a resource through the RGUI, he/she needs to apply from the RMS. The RMS will make a decision based on the current state of the resource. The decision can be: approval, rejection, or wait. If an application is approved, the RMS will notify the applicant and send him/her the operation-handle of the resource server.

Resource Use (RU): Interactions between users and the used resources. When a user gets use authority of a resource, he/she can transparently operate the resource through the RGUI in the WS in client/server mode. The user firstly launches resource server by the RGUI. Then, the server checks the user's authority of access and operation. Next, server begins to provide services. Finally, the server returns service results to the user.

Resource Supervision (RS): Collaboration between Virtualized Resources and the Resource Management System. For example, resources may notify RMS of changes of state and RMS may modify access and operation rules of resources.

3   Functional Model of VLE

3.1.   Virtualized Resources

From the viewpoint of wrapping technology, the resources in an enterprise can be divided into two classes: **wrappable resources** and **unwrappable resources**. For example, software systems and databases can be wrapped - to be accessible as a "black box" object through the Virtual Space, but machines, tools and people cannot be wrapped - often because human interaction is required. Figure 2 depicts the functional model of Virtualized Resources. The Event Channel receives service requests from the Internet or sends information packages to the Internet. The resource server interface consists of two parts: the resource operation interface and the resource management interface. Accordingly, the information flows are also divided into two branches, one involving the resource operation services, and the second involving the other five common management services.

The resource operation interface contains four modules: *access process module, operation process module, translation process module, and interaction module.* Access process searches for the requested object server from the server repository and checks the user's access authority. Operation process checks the user's operating authority. If the resource is "wrappable", a *translation process* translates the standard object invoking operations into specific operation commands of the application resource function interface. If the resource is "unwrappable", the *interaction module* can provide an interactive window through which the user's operation commands are displayed and the resource states are sent to the management system.

The resource management interface includes two modules: the *object service control module* and the *rule management module*. *The object service control module* is responsible for checking the user's access and operation authority and sending messages to the Resource Management System when resource states are changed. The *rule management module* allows the RMS to query and modify current resource rules in the rule repository.

3.2.   Virtualized Resource Management

The functional model of Virtualized Resource Management is shown in Figure 2. It consists of two parts:

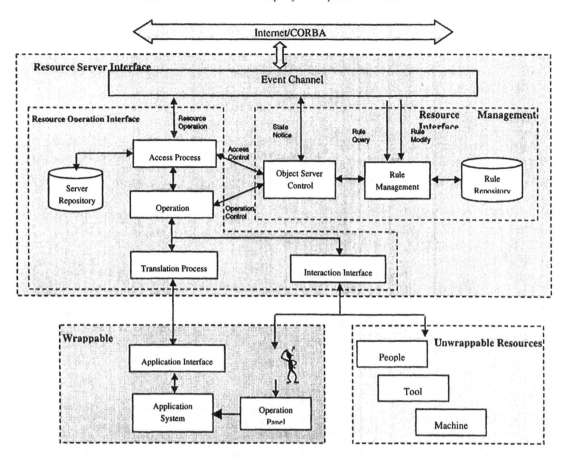

Figure 2 Virtualized Resource Function Model

Resource Management System (RMS) and Meta-Resource Data Repository (MRDR). The MRDR contains a Meta-Resource Database and an Operation Interface. The RMS is provided with four function modules: *Meta-Resource Management, Working-Space Management, Resource Request / Assignment, and Resource Supervision.*

3.2.1 Meta-Resource Management

The Mete-Resource Management module is designed to supporting resource management activities in the VLE. It helps administrators to extend and maintain the Meta-Resource Data Repository (MRDR). The basic functions of Mete-Resource Management module are:

- Meta-Resource Registry: Each Virtualized Resource must be registered in the MRDR.
- MRDR Maintenance: Modifying or deleting a registered resource data may affect other resources associated with it. For example, if a resource subscribed by a customized WS is deleted, the configuration of that WS should be modified accordingly.
- General Functions: These include the general

database operation functions, such as meta resource data querying, displaying, printing and so on.

3.2.2 Working-Space (WS) Management

The Working-Space Management module supports the *Working-Space Configuration* activity in the VLE. It helps user to register, configure and operate special WSs.

- WS registration: In the VLE, the WS is also a special resource. Each WS belongs to only one user. When a user wants to register a customized WS, a default configuration will be provided.
- WS configuration: In the default WS, the user can view and subscribe to all the resources that have been registered in the MRDR. The RMS will store user subscription information so that when the user enters the same WS again, the RGUI of the subscribed resource can be downloaded to the WS.
- WS download: Each WS must be downloaded to the user's computer platform. A user-specific WS consists of two parts: One contains default common software tools that can be directly downloaded from a special directory. The other contains the RGUI of

the subscribed resource. All the RGUIs are platform-independent and their execution codes are copied to the special directory when the resource is registered. The RMS can locate these execution codes according to the user subscription information.

3.2.3 Resource Request and Assignment

Sometimes, more than one applicant may apply for the same resource at the same time. In this case, some rules would be specified for deciding which application will be granted. The process of resource assignment is as follows:

- Request identification: Extract the information related to the resource applicant, applicant's authority and the requested resource from the resource request message.
- Rule query: Get the rules related to resource assignment from the MRDR according to the ID of the requested resource.
- State query: Get current states of the requested resource (working-state and using-state) from the MRDR.
- Making decision: Possible outcomes of the decision are to accept, reject, or to ask the applicant to wait.
- Request reply: If the user's application is accepted, an operation handle to the requested resource server will be sent to the applicant. If the result of decision is "wait", the information related to the applicant should be added to a waiting list.

3.2.4 Resource Supervision

The Resource Supervision Module controls the collaborative work process between the Virtualized Resources and the RMS. It has the following function requirements:

- Control for access: When a resource is assigned to a resource applicant, the RMS should send appropriate messages to notify the assigned resource.
- Control for operation: Sometimes a user is permitted to operate only a limited subset of the Virtualized Resource, and relevant operation rules should be sent to the resource server.
- Response to state change: The RMS may need to respond to state changes of the resource. For example, if the working state of a resource is changed from "occupied" to "unoccupied", the resource may be assigned to the first applicant in the waiting list.

4   Prototype Implementation

A prototype of a Virtualized Enterprise (VLE/1.0) has been successfully developed. Through this prototype, Internet-based design and manufacturing has been implemented by resource sharing and reconfiguration. The VLE/1.0 includes four parts: Virtual Space, Resource Management System, Virtual Working Environment and Virtualized Resource.

4.1   Building The Virtual Space

Virtual Space is built on the information communication network based on the Internet and TCP/IP Protocol. Web browsers such as Netscape or Microsoft Internet Explorer can be used as working platform for users to enter VLE/1.0. Four types of Internet Servers are used. We use a Microsoft Internet Information Server (IIS) as an Internet WWW Server to dynamically create a specialized WS by taking advantage of the Active Service Page (ASP) mechanism. CORBA-based OrbixWeb daemons, Orbix Java daemons and Orbix C++ daemons are used as Internet Monitors to support distributed object operations based on CORBA/IIOP. As shown in Figure 3, four types of communication channels are built into VLE/1.0.

- Channel 1 is based on Internet/HTTP protocol and HTML/ASP mechanism. It is used in the communication between Internet Explorer (IE) Browser and IIS in order to implement some resource management functions, such as resource registry, resource query, WS subscription and WS download.
- Channel 2 is also built between the VWE and the RMS, and is used for communication between IE Browser and OrbixWeb daemon for the other resource management activities related to CORBA-based object operation, such as resource request and assignment.
- Channel 3 acts as a connection between the RMS and the Resource Server for carrying out resource control and supervision.
- The final resource operations between the VWE and Resource Servers are established through Channel 4. Channels 2, 3, 4 all are based on CORBA/IIOP.

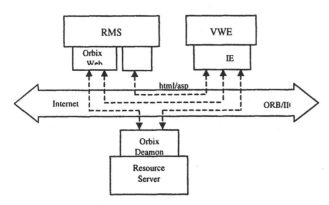

Figure 3   The four types of communication channels in VLE/1.0

4.2   CORBA-Based Resource Wrapping

We use a practical example to explain CORBA-based resource wrapping and distributed object operation. The resource to be wrapped is a controller of a Rapid Prototyping Machine (FDM1600). One of the functions of this controller is to send a machining program (smlfile) to the FDM1600 through RS232 port (A) [11]. This function is operated by a DOS command: "Ssend smlfile portname". In order to wrap this function by the CORBA-based method, at least the following five steps need to be performed.

(1) Define standard interface in CORBA/IDL according to the function to be wrapped.
(2) Compile this IDL interface file using IDL compiler in order to generate four kinds of program codes (Java or C++): Stub Head files, Stub files, Skeleton Head files and Skeleton files.
(3) Write a server program that creates CORBA server ("SendServer") and implements the server function of the resource in it.
(4) Write a client application program to bind this CORBA Server ("SendServer") and operate implementation object through IDL interface.
(5) Register this CORBA server ("SendServer") in the server repository so that the Orbix daemon can launch this server automatically when it is invoked.

Before invoking practical operations on a CORBA server object, an Orbix daemon should be run first. The operation process will have two phases. Firstly, the client program invokes the server object through the ORB stub, which will send a command "bind (ServerName, HostName)" to Internet. This message will be received by the Orbix daemon, which will search for the specific

server (SendServer) in the server repository and launch it automatically if it exists. Then, the Orbix daemon will reply to the client with the message: "CORBA_Orbix_imp_is_ready". Finally, actual inter-operation will begin through the ORB stub and the ORB skeleton.

4.3   Resource Management System and Virtual Working Environment

The kernel of VLE/1.0 is the RMS, which was built on the Windows NT platform. Two Internet monitor systems are run simultaneously. One is the Microsoft IIS server that is in charge of listening for and processing information exchange between the RMS and the VWE. Another Internet server is the OrbixWeb daemon responsible for supervising CORBA-based object service requests which come from the VWE. The RMS contains three types of repositories: Meta-Resource Data Repository, Object Server Repository and ASP File Repository. The principle and operation processes between the RMS and the VWE are as follows:

Firstly, a person enters the homepage of VLE/1.0 via an URL address (http://158.132.10.251/work/index.asp). This login message is received by the IIS server. The IIS server will locate the file index.asp from the ASP File Repository and produce a corresponding homepage.html file, and then sends it back to the user. The browser at the user end will display this homepage.

Secondly, the person can enter VWE by submitting username and password. The IIS server automatically distinguishes between administrators and users and sends back the corresponding Admini-Space.html or Working-Space.html file to the client.

Thirdly, if the person is an administrator, the Admini-Space page will be displayed on the client platform. An administrator can execute management operations, such as resource registration, resource query and resource maintenance. Since resource registration and query operations only involve general database operations, they can be processed by the IIS server through directly accessing and operating the Meta-Resource Repository. However, other resource operations, such as updating and management, involve rule-based decision management, they must be processed by the OrbixWeb daemon. When the OrbixWeb daemon receives an object request event, it will search in the Object Server Repository to find the matched object server. Once found, the object server will be launched by the daemon. This object server also needs to access the Meta Resource Data

Repository through an ODBC driver. Based on the rules and states of the resource, the object server will make a decision and then perform the related resource maintenance operations.

If the client is a general user, a special Working-Space page will be downloaded and displayed on the user's platform. This Working-Space page contains some common software tools and user-ordered Resource Graphical User Interfaces (RGUIs). Requests from common software tools will be processed by the IIS server. Requests from RGUIs, however, will be processed by the OrbixWeb daemon. If the request is accepted, real interactive operation between the user and the virtualized resource will begin.

One customized WorkingSpace has subscribed to five Virtualized Resources: Lonicera-NCP, Lonicera-MDA, FDM1600-RPM, QuickSlice-RPP and CSM Session Manager.

- Lonicera-NCP is a numerical control (NC) programming system that generates NC milling machine tool paths.
- Lonicera-MDA is a mechanical design assistant that includes a number of basic computer-aided design (CAD) modules, such as two-dimensional (2D) engineering drawing, three-dimensional (3D) parametric model, advanced surface model and geometric model data conversion.
- Quickslice is a powerful slicing program that creates Stratasys Modeling Language (SML) codes to drive the Fused Deposition Modeling (FDM) system.
- FDM1600 is a rapid prototyping machine made by Stratasys Inc.
- CSM is an experimental collaborative solid modeling system that allows multiple designers to collaborate in real-time on the Web [4, 5].

In Figure 4, QuickSlice-RPP's RGUI is opened, through which the user can do remote programming for a rapid prototyping machine, such as uploading a STL file from other resource server to QuickSlice-RPP server, setting control parameters, checking operation processes, viewing and modifying the outputs, and so on. Another RGUI, FDM1600-PRM, is illustrated in Figure 5. Some remote machining controls can be implemented through this RGUI. For example, the user can upload a machining program from any other resource server to the FDM1600-PRM server, send it to the RP machine and start the machining process.

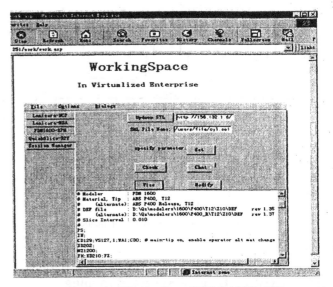

Figure 4.   A customized Working-Space: RGUI-QuickSlice-RPP

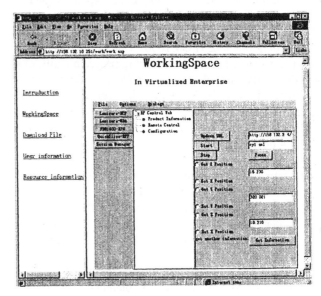

Figure 5. A customized Working-Space: RGUI-FDM1600-RPM

4.4   Resource Wrapping and Resource Registry

Five resources have been wrapped and registered in the RMS. The whole processes of resource wrapping is roughly divided into four steps:

1. Analyze the basic characteristics of resources, such as resource location, platform, language, role and main functions.
2. Analyze the operation interfaces provided by the resource; select the wrapping method; define the unified function interface.

3. Implement resource wrapping: define and implement the RGUI of the wrapped resource and implement the management functions of the wrapped resource which are common to all the Virtualized Resource.

4. Register the wrapped resource into the RMS.

5 Summary and Conclusions

The key issue in the implementation of a Virtualized Enterprise is "Virtualization". It can be divided into three sub-processes: Resource Virtualization, Environment Virtualization, and Management Virtualization. The main components of Virtualization technology include Object-oriented-Wrapping, platform-independent Resource Graphical User Interface, Meta Resource Data Model, Meta Resource Data Repository, Rule-based Resource Management, and Reconfigurable Working-Spaces. A prototype of a Virtualized Enterprise has been developed. The relevant theory and methodology presented in this paper has been partly tested and verified.

This paper has concentrated on virtualization of an enterprise. A Virtual Enterprise is built in Virtual Space. More research on how to build the Virtual Space to support Virtualized Enterprises are required. This will involve the methods and technologies of the communication infrastructure, standard protocols, security, real-time operations, open systems, and so on. On the other hand, how to ally multiple Virtualized Enterprises to form the alliance of a Virtual Enterprise involves a different set of issues.

Acknowledgement

The work described in this paper was partially supported by Hong Kong UGC/RGC/CERG Research Project PolyU5102/97E and China's State 863/CIMS Research Project 863-511-9503-001.

References

[1] Barry, J., Aparicio, M., Durniak, T., Gilman, C., Ramnath, R., 1998. NIIIP-SMART: An Investigation of Distributed Object Approaches to support MES Development in a Virtual Enterprise. *The Second International Enterprise Distributed Computing Workshop (EDOC98)*. http://smart.npo.org/public-forum/.

[2] ESPRIT Consortium AMICE. *CIM-OSA: Open Systems Architecture for CIM*. Berlin: Springer Verlag, 1989.

[3] Dewey et al., 1996. The Impact of NIIIP Virtual Enterprise Technology on Next Generation Manufacturing. *Proceeding of Conference on Agile and Intelligent Manufacturing Systems*, Troy, NY.

[4] Chan, S., Wong, M., and Ng., V., 1999. "Collaborative solid Modeling on the WWW", *1999 ACM symposium on Applied Computing - Special Track on World Wide Web Applications*, February 28 ■ March 2, 1999, San Antonio, Texas, USA, pp. 598-602.

[5] Chan, S., Ng, V., Yu, K. M. and Au, A., 1999. "An Internet-Integrated Manufacturing System Prototype", *Proceedings, 1999 International Computer Science Congress*, 13-15 Dec., 1999, Hong Kong, pp. 409 - 414.

[6] Chen, Y. M., Liao, C. C. and Prasad, B., 1998. "A System Approach of Virtual Enterprising Through Knowledge Management Techniques", *Concurrent Engineering: Research and Applications*. Vol. 6, No. 3, pp. 225 - 244.

[7] Erkes, J. W., Kenny, K. B., Lewis, J. W., Sarachan, B. D., Sobolewski, M. W. and Sum, R. N., 1996. "Implementing shared Manufacturing Services on the World Wide Web", *Communications of the ACM*. Vol.39, No.2, pp34-45.

[8] Gilman, C., Aparicio, M., Barry, J., Durniak, T., Lam, H., and Ramnath, R., 1997. "Integration of Design and Manufacturing in a virtual enterprise using enter rules, intelligent agents, STEP, and workflow. Architecture, Network and Intelligent System for Integrated Manufacturing", *Proceeding, Society of Photo-Optical Instrumentation Engineers (SPIE)*, http://smart.npo.org/public-forum/.

[9] ISO/IS 10303-1, 1994. Industrial Automation Systems - Product Data Representation and Exchange: Part I, Overview and Fundamental Principles.

[10] NIIIP Inc., 1998. *Guide to the NIIIP Reference Archjitecture Model. NIIIP Reference Architecture, Book 1*.

[11] Stratasys Inc., 1998. *QuickSlice Manual Release 6.0*.

[12] Rajagopalan, S., Pinilla, J. M., Losleben, P., 1998. "Integrated design and rapid manufacturing over the Internet", *Proceedings DETC98 1998 ASME Design Engineering Technical Conference*, Sept. 13-16, Atlanta, GA.

[13] Zhang, S., Chen, B., 1997. "A Concept of Virtual Global Manufacturing based on Autonomous Manufacturing Islands", *CIRP International Symposium - Advanced Design and Manufacturing in Global Manufacturing Era*, August 21 - 22, Hong Kong.

[14] Yung, C. K., Grier, C., and Lin, I., 1998. "Development of collaborative CAD/CAM system", *Robotics and Computer-Integrated Manufacturing*, Vol. 14, pp. 55 - 68.

Analysis of cost estimating processes used within a concurrent engineering environment throughout a product life cycle

Christopher Rush and Rajkumar Roy,

Department of Enterprise Integration, SIMS, Cranfield University, Cranfield, Bedford, MK43 OAL, United Kingdom.
Tel: +44 (0) 1254 765261. Email: c.rush@cranfield.ac.uk
Tel: +44 (0) 1234 754072. Email: r.roy@cranfield.ac.uk

Abstract

Concurrent engineering environments affect the cost estimating and engineering capability of an organisation. Cost estimating tools become outdated and need changing in order to reflect the new environment. This is essential, since cost estimating is the start of the cost management process and influences the 'go', 'no go' decisions concerning a new product development. This paper examines both traditional and more recent developments in order to highlight their advantages and limitations. The analysis includes parametric estimating, feature based costing, artificial intelligence, and cost management techniques. This study was deemed necessary because recent investigations carried out by Cranfield University highlighted that many concurrent engineering companies are not making efficient, wide spread use of existing estimating and cost management tools. In order to promote more efficient use of the discussed estimating processes within the twenty first century, this paper highlights the work of a leading European aerospace manufacturer and their efforts to develop a more seamless estimating environment. Furthermore, a matrix is developed that illustrates particular concurrent engineering environments to which each technique is aptly suited.

1. Introduction

Cost is perhaps the most influential factor in the outcome of a product or service within many of today's industries. More often than not, reducing cost is essential for survival. To compete and qualify, companies are increasingly required to improve their quality, flexibility, product variety, and novelty while consistently maintaining or reducing their costs. In short, customers expect higher quality at an ever-decreasing cost. Not surprisingly, cost reduction initiatives are essential within today's highly competitive market place. Concurrent engineering is one such initiative. Since cost has become such an important factor of success, project development needs to be carefully considered and planned. Recent research demonstrates that companies unable to provide detailed, meaningful cost estimates, at the early development phases, have a significant higher percentage of programs behind schedule with higher development costs, than those that can provide completed cost estimates [1]. Therefore, it is essential that the cost of a new project development be understood before it actually begins. It could mean the difference between success and failure.

This article is divided into three broad sections. The first highlights the increasing need for effective cost estimating and cost management techniques within a concurrent engineering environment. Cost estimating being defined as the process of predicting the cost/outcome of an as yet undefined project, and cost management being defined as a technique for managing the development processes in order to achieve the estimate.

The second section of the paper discusses several available estimating and cost management techniques, in order to provide a broad overview and to better understand where and when to use them within a project life cycle. Furthermore, it promotes awareness concerning the traditional and more state of the art techniques that have emerged over the last decade.

The final section presents a snapshot view of several leading concurrent engineering companies, which demonstrates how estimating and cost management techniques are being utilised within industry. This study highlights a general lack of structure and order concerning the use of current estimating techniques within concurrent engineering environments. In an attempt to counter this problem a matrix is developed to assist the choice of applying an estimating technique at different stages of a product lifecycle.

2. The need for cost estimating/engineering

Cost estimating helps companies with decision-making, cost management, and budgeting with respect to product development. It is a methodology used for predicting/forecasting the cost of a work activity or output

[2]. It is the start of the cost management process. Cost estimates during the early stages of product development are crucial. They influence the go, no go decision concerning a new development. If an estimate is too high it could mean the loss of business to a competitor. If the estimate is too low it could mean the company is unable to produce the product and make a reasonable profit.

Many authors agree that 70-80% of a product cost is committed during the concept phase [2, 3, 4, 5]. Making a wrong decision at this stage is extremely costly further down the development process (see Figure 1). Product modifications and process alterations are more expensive the later they occur in the development cycle. Thus, cost estimators need to approximate the *true* cost of producing a product, based on empirical data, with the purpose of satisfying both the customer and company.

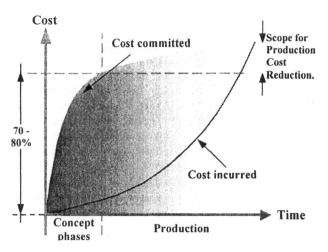

Figure 1: Cost commitment curve

The difficulties of estimating at the conceptual design phase are well recognised [6, 7, 8, 9]. The major obstacles estimators need to address are:
- Working with a limited amount of available data concerning the new development;
- Accounting for step changes within technology over the life span of a product development; a more pronounced problem within the aerospace industry;
- The requirements to show how cost estimates were derived including the assumptions and risks, and;
- The estimates need to be accurate.

Therefore estimators/engineers need company-wide co-operation and support, to assist them with their decision making. Concurrent engineering is an excellent initiative to assist this process; however, it does present a new set of challenges as outlined below.

3. Cost estimating within a concurrent engineering environment

An optimised concurrent engineering environment provides an opportunity to substantially reduce the total cost of a project. This is because, integrated product teams (IPTs) containing members of various skilled disciplines, enable a simultaneous contribution to an early product development and definition. Therefore, within a fully integrated product development (IPD) cycle, multidisciplinary teams working together increase the likelihood of a reduced lifecycle cost by avoiding costly alterations later in the design process. With this view in mind, concurrent engineering is a great step forward when compared to an 'over the wall mentality', where each department works in 'isolation'.

However, a concurrent engineering environment presents many new challenges to cost estimators whom, it could be argued, are more used to predicting the cost of an 'over the wall' environment. The impacts from adopting a concurrent engineering philosophy are substantial and often require significant changes to long-standing working practices. The whole culture begins to change. Existing costing methods and systems soon become outdated and require updating to reflect the new environment. Thus, estimators find it extremely difficult to predict cost within this new environment with their existing tools. This is not all bad because it offers an opportunity to introduce new approaches to old and possibly outdated working practices. This could cause difficulty for some, since advances in technology and techniques have grown rapidly over the last decade. This period of change could be a daunting prospect unless practitioners have had the opportunity to follow recent trends and developments. The remainder of this paper attempts to highlight and dispel some of the mystique behind several state-of-the-art-estimating techniques, in order to make aware the choices that are currently available.

4. Cost estimating methods

4.1. Traditional cost estimating

In traditional costing there are two main estimates: a "first sight" estimate, which is done early in the cost stage, and a detailed estimate, done to calculate costs precisely. The former of these cost-estimating methods is largely based around the experience of the estimator. For example it is not uncommon for a "first sight" project estimate to be based upon a past similar project or purely on experience in costing. However, to attain this level of experience takes years of apprenticeship and considerable oversight from senior estimators. Although useful for a rough order of magnitude estimate, this type of estimating is too subjective in today's cost conscious culture and more quantified and justified estimates are required [10, 11].

For detailed estimates, cost is based upon the number of operations, time per operation, labour cost, material cost and overhead costs. Much of the information in a detailed estimate is based upon the internal synthetics (times or costs based upon expected rates of work for any particular task) of the company. To generate these estimates, it is necessary to have an understanding of the product, the methods of manufacture/process and relationships between processes. Detailed estimating goes through several iterations, since feedback from the relevant departments enables the estimates to be reviewed and improved. Thus, detailed estimating can be achieved only when a product is well defined and understood.

Activity based costing (ABC) is a process for measuring the cost of the activities of an organisation [12, 13]. It is a quantitative technique used to measure the cost and performance of activities e.g. inspection, production processes and administration. Each activity within an organisation is first identified and then an average cost is associated. Once this is achieved it is then possible to estimate the amount of activity a product is likely to need and then associate the relative costs. This makes ABC appealing, since it combines estimates with hard data. This method follows similar processes to detailed estimating, and also requires a detailed understanding of the product definition. Thus, both detailed and ABC techniques are not useful during the conceptual phase of project development. In order to estimate a project during this stage other approaches are required which are discussed below.

4.2.   Parametric estimating

A widely used method for estimating product cost at the early stages of development is known as parametric estimating (PE). To illustrate this concept more clearly the following example will suffice. Typically, for aircraft development, mass relates to the cost of production. That is, as the weight of the aircraft increases, so does the cost of producing it. What's more, this particular relationship is often described as linear, as illustrated in Figure 2 below.

In this hypothetical example the points of the graph represent the relationship of cost to mass for different aircraft. The line traversing the points represents a linear relationship i.e. as the mass increases so does the cost. Using relatively simple algebra it is possible to derive a formula to determine a mathematical relationship for cost to mass. For the above graph the equation, $y = ax + b$ is used to describe the line of best fit between the points. With the relationship described it is then possible to use the formula to predict the cost of a future aircraft based on its weight alone. Within the field of cost estimating this relationship is known as a cost estimating relationship (CER).

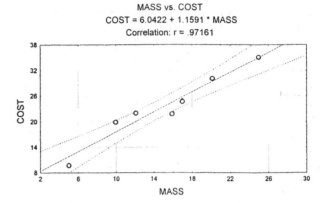

Figure 2: Simple linear equation

This is a rather simplistic illustration describing the main principals of parametric estimating. Nonetheless, variations of this approach are a widely used method within industry to predict the cost of a product under development and throughout the life cycle. As CERs become more complex involving several variables, more complex mathematical equations are used to describe the relationships. When CERs become too complex for mathematical equations to solve, cost algorithms are developed [3].

4.2.1.   Using parametric estimating

Parametric estimating can be used throughout the product life cycle. However, it is mainly used during the early stages of development and for trade studies e.g. within design to cost (DTC) analyses (see Section 5.2). Both industry and Government accept the techniques. Many authors commend its usefulness [2, 3, 5, 6].

However, PE does have its downsides, for example, CERs are sometimes too simplistic to forecast costs. Furthermore, PE is primarily based on statistical assumptions concerning the cost driver relationships to cost, and estimators should not completely rely upon statistical analysis techniques. Hypotheses, common sense and engineering knowledge should come first, and then the relationship should be tested with statistical analysis. Most CER literature describes the process for estimating quantitative issues but not qualitative/judgmental issues. Cranfield University is currently researching this area and early work demonstrates the validity of this innovative approach [10, 11].

In summary parametric estimating is an excellent predictor of cost when procedures are followed, data is meaningful and accurate, and assumptions are clearly identified and carefully documented. A relatively new form of PE is that of feature based costing. This has become popular due to the rise and sophistication of CAD tools. The implications of FBC are discussed below.

4.3. Feature Based Costing

The growth of CADCAM technology and that of 3D modelling tools have largely influenced the development of feature based costing (FBC). Researchers are investigating the integration of design, process planning and manufacturing for cost engineering purposes using a feature based modelling approach [14, 15, 16, 17].

FBC has not yet been fully established or developed with respect to cost engineering. Nonetheless, there are several good reasons for examining the use of features as a basis for costing during the design phase. Products can essentially be described as a number of associated features i.e. holes, flat faces, edges, folds etc (see Figure 3). It follows that each product feature has cost implications during production, since the more features a product has the more manufacturing and planning it will require [18]. Therefore, choices regarding the inclusion or omission of a feature impact the downstream costs of a part, and eventually the life cycle costs of the product [19].

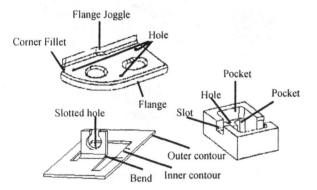

Figure 3: Examples of different views on features

Other reasons for using FBC are that the same features appear in many different parts and products; therefore, the basic cost information prepared for a class of features can be used comparatively often. Furthermore, manufacturers will have numerous past geometric data that can be related to features. Another reason developers explore whether costs should be assigned to individual design features is that it would provide the designer with a tool to visualise the relation between costs, and aspects of the design that s/he can influence in real time as the product is developed. Furthermore, it is possible engineering intent can be encapsulated within features such as, product functionality, performance, manufacturing processes, and behaviour characteristics.

4.3.1. FBC Issues

Although feature based costing is gaining popularity, there are limitations for using them for the costing process. There is no widely accepted consensus on what a feature is across the disciplines of an organisation. This problem is magnified when viewed across companies and industries.

With respect to this problem, companies are faced with producing their own feature definitions. Table 1 shows an example of how one cost engineering group, categorised features for the purpose of costing [4].

Feature type	Examples
Geometric	Length, Width, Depth, Perimeter, Volume, Area.
Attribute	Tolerance, Finish, Density, Mass, Material, composition.
Physical	Hole, Pocket, Skin, Core, PC Board, Cable, Spar, Wing.
Process	Drill, Lay, Weld, Machine, Form, Chemi-mill, SPF.
Assembly	Interconnect, Insert, Align, Engage, Attach.
Activity	Design Engineering, Structural Analysis, Quality assurance.

Table 1: Examples of features

Table 1 illustrates one level of feature definition; however, there are several levels of features definitions. For example, a feature of an aircraft could be a wing, yet this wing contains many parts, each of which consists of many lower level features. Therefore companies are also left to decide how to cope with the changing product definition and applying an appropriate feature based CER. Thus, the feature based costing approach is not yet fully established and the implications are not yet completely understood. Nonetheless, companies find the concept appealing. Other recent developments within the cost estimating community concern the use of artificial intelligence. The implications of which are discussed below.

4.4. Neural network based cost estimation

Neural networks (NNs) and fuzzy logic present the next generation in computerising the human thought processes [20]. Many researchers and practitioners are fast developing and investigating the use of artificial intelligence (AI) systems and applying them to cost estimating situations [21, 22, 23].

For cost estimating purposes, the basic idea of using NNs is to make a computer program learn the effect of product-related attributes to cost. That is, to provide data to a computer so that it can learn which product attributes mostly influence the final cost. This is achieved by training the system with data from past case examples. The NN then approximates the functional relationship between the attribute values and the cost during the training. Once trained, the attribute values of a product under development are supplied to the network, which applies the approximated function obtained from the training data and computes a prospective cost.

Recent work has demonstrated that neural networks produce better-cost predictions than conventional regression costing methods if a number of conditions are adhered to [21]. However, in cases where an appropriate CER can be identified, regression models have significant advantages in terms of accuracy, variability, model creation and model examination [22].

4.4.1.   Uses

The neural network does not decrease any of the difficulties associated with preliminary activities when using statistical parametric methods, nor does it create any new ones. The analyst is still left with a choice of cost drivers and must make a commitment to collecting specific cost data before analysis can begin.

Models can be developed and used for estimating all stages of a product life cycle provided the data is available for training. A great advantage that a neural network has compared to parametric costing is that it is able to detect hidden relationships among data. Therefore, the estimator does not need to provide or discern the assumptions of a product to cost relationship, which simplifies the process of developing the final equation [23].

4.4.2.   Issues

Neural networks require a large case base in order to be effective, which would not suit industries that produce limited product ranges. In addition, the case base needs to be comprised of similar products, and new products need to be of a similar nature, in order for the cost estimate to be effective. Thus, neural networks cannot cope easily with novelty or innovation. With regression analysis one can argue logically and audit trail the development of the cost estimate. This is because the analyst creates a CER equation that is based on common sense and logic. When considering neural networks, the resultant equation does not appear logical even if one were to extract it by examining the weights, architecture, and nodal transfer functions that were associated with the final trained model. The artificial neural network truly becomes a "black box" CER. This is no good if customers require a detailed list of the reasons and assumptions behind the cost estimate. The black box CER also limits the use of risk analysis tools, which as discussed below, is a prime benefit of parametric estimating.

A final estimating technique to discuss is the analogous method and more particularly that of case based reasoning.

4.5.   Case based reasoning

Analogy makes use of the similarity of products. The implicit assumption is that similar products have similar costs. By comparing products and adjusting for differences it is possible to achieve a valid and useable estimate. The method requires the means of both identifying the similarity and differences of items. This can be through the use of experience or databases of historical products. A more modern approach to the analogy method is case-based reasoning.

Case-Based Reasoning (CBR) can also be classed as a form of artificial intelligence since it can be used to model, store, and re-use historical data, and capture knowledge for problem-solving tasks. An important feature of CBR is the ability to learn from past cases/situations. A CBR system stores and organises past situations, then chooses situations similar to the problem at hand and adapts a solution based on the previous cases. An overview of the CBR process is illustrated in Figure 4.

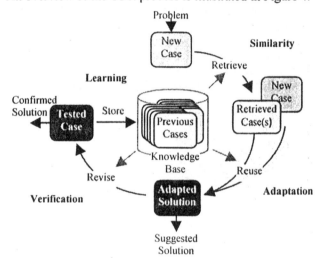

Figure 4: Case based reasoning process [24]

As with FBC, CBR relies on a feature description base. As previously explained this is not a straightforward task. Furthermore, CBR requires a number of past cases in order to be effective. In a highly innovative company past cases may not be available so will therefore reduce the effectiveness of the CBR system. Companies that use analogy estimates regularly may find CBR a more robust useful method.

Section four has described the main estimating techniques and their related issues and promotes a broad understanding of both traditional and state-of-the-art techniques. The next section describes and highlights the main techniques available for managing and reducing costs in order to achieve the target of an early estimate.

5. Cost management and cost reduction

5.1. Value analysis and value engineering

Although similar to each other these techniques serve different purposes. Value analysis (VA) is concerned with the analysis of a product with respect to reducing product/process costs. Typically, VA is a technique used on existing items/products in light of new processes, materials or assembly methods being available. Value engineering (VE) on the other hand is an approach that rigorously examines the relationship between a product function and cost and can be used during the concept stage. VE identifies the functions that are beneficial to the customer so that the value of a product is not just perceived as a low cost product, but rather one that satisfies the customer. This technique was used widely within the aerospace industry up until the 1970's. However, with the introduction of tighter defence budgets, a more stringent technique was required for ensuring cost targets were achieved and design to cost was introduced.

5.2. Design to cost

The objective with design to cost (DTC) is to make the design converge to an acceptable cost, rather than to let the cost converge to design. DTC activities, during the conceptual and early design stages, are one of determining the trade-offs between cost and performance for each of the concept alternatives. DTC can produce massive savings on product cost before production begins.

The general approach is to set a cost goal, then allocate the goal to the elements of the product. Designers must then confine their approaches to that set of alternatives that satisfy the cost constraint [25]. However, this is only possible once cost engineers have developed a tool set that designers can use to determine the impact of their decisions as they make them. Figure 5 illustrates an example of the types of input required for producing a DTC tool.

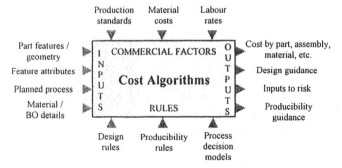

Figure 5: DTC model

It is the cost engineers who are responsible for bringing back, to the early stages of product development, enough information on cost that will enable the designer to use it for decision-making. They develop algorithms that designers can use to monitor the impact of their decisions as they proceed with their design [4]. In addition, they are responsible for updating and maintaining the validity of any algorithms used [26].

The tools to assist the designer in meeting and verifying cost goals are in most cases developed within the context of a specific industry or company [27]. One European aerospace manufacturer uses a computer tool called Cost Advantage™ [4]. However, it should be stated, the results produced from using such a tool are only as valid as the data that has been collated, normalised and input.

Both VA/VE and DTC help to manage the risk of failing to meet the required cost targets; however, they are not focused on risk as a main project objective. Therefore, the next section discusses risk management and its role within today's estimating community.

5.3. Risk Management

Because estimating is based on assumptions concerning the likely cost of an as yet unknown product outcome, there is an increasing trend to combine the statistical techniques of parametric cost analysis with statistical risk analysis methods. Parametric estimating, because of its statistical approach, offers the cost analyst the advantage of being able to quantify the risk of an estimate.

The introduction of risk assessment and risk analysis ensures that the consequences of risks are understood and taken into account throughout the project life cycle [28]. Furthermore, risk management ensures that the goals of the producer and consumer materialise and that they both benefit. It provides confidence concerning final costs and identifies actions needed to keep cost and schedule on target. There are five key steps to follow in the risk management process [29]. Figure 6 illustrates this process more clearly.

One of the most important benefits of using risk assessment is to generate a distribution/range of costs i.e. to move away from single point estimating, since a range of costs are much easier to estimate than a single cost [30]. Furthermore, once a risk analysis has been conducted the analyst can consider ways to reduce the risk e.g. by avoidance, deflection or contingency and then plan accordingly to control the reduction process.

Risk management along with VA/VE and DTC can be better utilised by combining them into a state-of-the-art cost management framework known as target costing.

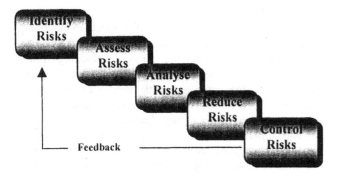

Figure 6: Risk Management Process

5.4. Target costing

Target costing (TC) is a cost management concept that is well suited for use within a concurrent engineering environment. It has mostly been used within the automotive industry as a means of strategically managing cost. TC provides a framework that places cost management issues into the forefront from the early phases of product development and can be used throughout all phases of a product life cycle. However, it is mostly practised during the design and development stages where most of the decisions that impact life cycle cost are made [31]. TC is a framework in which estimating becomes an integrated element. It combines the concepts from existing cost management and cost estimating/engineering tools e.g. VA/VE, DTC, risk management, and bases its philosophy on the logic and benefits of activity based costing.

5.3.1.   Unresolved TC Issues

TC is not suited to all industries. It is best used on new products, which characterise small incremental development changes from past similar products. The concept falls down when addressing the cost estimation of innovative products. Chiefly because, the process requires a breakdown of how the components of a product will effect the functionality of an, as yet, undefined product, and furthermore, what the cost of each product feature or component will cost in relation to whole product. This is not possible unless some sort of system has been developed that has the capability of producing a detailed product definition/breakdown during these early stages. Therefore, it is not as yet, widely used for companies that develop highly innovative products.

This concludes the discussion concerning available tools for estimating within a concurrent environment. Both the advantages and limitations of current and future estimating techniques have been summarised. The remainder of the paper discusses how industry uses the above techniques and also provides a matrix to assist with the planning and application of a particular technique throughout the product life cycle.

6.   State-of-the-art-practices

Seven high technology concurrent-engineering companies were interviewed as part of recent work carried out by Cranfield University [32]. The analysis was conducted within three main areas: the use of CER's, general costing, and the types of computing tools adopted.

Only a few of the companies had developed CERs for the manufacturing processes. These were either developed using computer tools or using the experience of highly skilled cost engineers. There seemed to be a lack of formal validation procedures for the CER's and no type of documentation seemed to be in place. There was generally no formal approach for costing the conceptual or detail design stage. Companies that did attempt these estimates seemed to rely mostly on expert knowledge with regards to past data, which is fraught with subjectivity. Non of the companies had CERs to predict their design activities.

For general costing analysis there was a tendency for companies to use a computer-based tool at the detailed manufacturing cost estimation level. The results produced from these analyses seemed to be fairly accurate. Most companies could validate this through feedback from production. Overall it was found cost benefit analysis was not being conducted. However, the companies did review their costing processes regularly, although there were no costing standards used as guidelines for this process.

A variety of costing software was used for both high level and detailed costing, and there was a mix of the level of integration with other business systems. Examples of the tools used included KAPES, PRICE (H), TIMSET and specifically developed in-house systems.

6.1   The challenges faced by European manufacturing industry

The snapshot view highlighted that the application of CERs within industry was not widely practised, and those that did use them did so badly. Companies could greatly enhance CER effectiveness and use by examining their procedures and methodologies for creating them. The application of CER's for the design process was one not even considered by the above companies. This was one of the underlying reasons that Cranfield University devised a methodology to take account of both the quantitative and qualitative issues of designing, and developed a CER methodology for costing the design process [10, 11].

The use of features, artificial intelligence and case based reasoning techniques were not used within any of the companies visited. And as mentioned earlier, few of them had adjusted their costing practices after the adoption of IPT or concurrent engineering practices. Few companies had completed a benefits analysis on the costing function.

In summary, there appeared to be a general lack of planning and order to the estimating process. In view of cost becoming an ever-increasing concern cost estimating and management needs a better focus. Companies considering the adoption of a concurrent engineering philosophy should use the opportunity to re-examine current practises and evaluate the possibility of adopting some of the more recent developments within the field of cost estimating and engineering. Benchmarking the leaders can also assist this process.

6.2. Benchmark the leaders

In cost estimating and cost engineering the USA leads the way in both practice and development [3, 33]. In Europe the European Space Agency (ESA) actively promotes the sharing of estimating best practices [34].

One leading European aerospace manufacturer is currently examining the feasibility of developing a seamless cost-estimating environment. Their early development plans and intentions are to adopt a feature based costing approach [4]. The company embraced the philosophy of an IPD approach and has demonstrated a strong commitment towards concurrent engineering. They have invested extensively into digital product assembly methods and information management systems, which are used to discharge information in line with their concurrent engineering process development. The emergence of the new IPD processes rendered their existing parametric estimating algorithms out of date, particularly for the design process. They seized this opportunity to embrace and integrate new estimating processes.

They recognised the potential of providing non-specialist cost estimators (design engineers) with a computer tool to inform them about the costs incurred with particular design approaches; in real time. This capability would empower non-cost specialists to make decisions related to cost improvements as they designed the product. This potential was realised due to the advent of 3D CAD modelling systems, which store information, related to features throughout the product hierarchy.

The idea of the process is to capture features from the CAD modelling tools, which can then be integrated to a design for manufacture (DFM) expert system that can price the cost of a design in real time. The DFM tool under evaluation is called cost advantage. It can accept part geometry directly from feature based modelling tools such as Pro-E and Unigraphics. It can be populated with design and manufacturing knowledge in the form of producibility algorithms so that it can evaluate a design based on the features, materials, and manufacturability. This then empowers the designer to make decisions related to cost as s/he worked.

Figure 7 illustrates a high level concept of the companies intent to integrate their cost modelling capabilities using a feature based approach throughout the concurrent engineering phases.

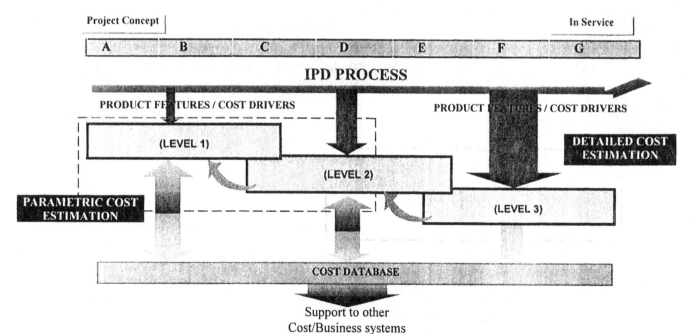

Figure 7: Integrated cost modelling

Companies wishing to use this approach would need a complete set of computerised tools that interface with each other. An obvious drawback for companies that may want to follow such an approach is the requirement for a comprehensive suit of expensive computerised tools. However, as computing power increases these tools become available and accessible for other industries to use. This development work may provide future estimators with an almost seamless system that can be used throughout the product lifecycle.

7. Matrix

Table 2 summarises where and when each of the techniques and processes discussed in this paper are best used throughout a product lifecycle.

The matrix shows that as a product moves through development the estimating processes need to change. The table suggests hard breaks between where one technique should be used against another. However, it should be borne in mind that parametric estimating (PE), neural networks (NN), and case based reasoning (CBR) could be used during later project phases, whereas ABC and detailed cost estimating cannot be used during the earlier product phases. Target costing (TC) is shown as useful throughout the product lifecycle, however, this is only possible when other estimating techniques and tools are integrated into the TC framework. Neural networks are not deemed suitable in the concept phase of innovative products since the estimates they produce are of a 'black box' nature. That is, they do not provide a facility to demonstrate the assumptions and reasoning behind the final estimate.

TOOLS AND PROCESSES USED WHEN:	PE	NN	CBR	ABC	Detailed Cost Estimation	Cost Management			
						VA	VE	DTC	TC
Concept design phase (innovation)	✓	✗	✓	✗	✗	✗	✓	✗	✗
Concept design (similar products)	✓	✓	✓	✗	✗	✗	✓	✗	✗
Feasibility Studies	✓	✓	✓	✗	✗	✗	✓	✗	✓
Project definition	✓	✓	✓	✗	✗	✗	✓	✓	✓
Full Scale development	✗	✗	✗	✓	✓	✓	✗	✓	✓
Production	✗	✗	✗	✓	✓	✓	✗	✓	✓

Table 2: Estimating process matrix

8. Summary and Conclusions

This paper describes how cost is an increasingly important factor of success within industry. And how cost estimating and cost management is essential to the survival of leading companies. Several state-of-the-art-techniques and processes, used to facilitate cost estimating, have been discussed with particular reference to their applicability within a concurrent engineering environment. This provided a broad overview of the strengths and weaknesses of each method. A snapshot view of several leading concurrent engineering companies was provided, which demonstrated the general lack of formal, organised processes to the estimating function. One leading European aerospace company was discussed with particular reference to their efforts at utilising and advancing the cost estimating process. And finally a matrix was provided that details where each of the discussed estimating processes should be used throughout the product life cycle.

In conclusion, there are a wide variety of emerging techniques available that companies can utilise to improve their cost estimating and management processes. Artificial intelligence will play an increasingly important role within the estimating communities. Because cost has become such an influential factor cost estimators and engineers should be aware of these technologies so that they can utilise them to improve their cost management processes. Although only a snapshot view of several companies was conducted a general observation was the lack of formal, disciplined approaches to the estimating process. Companies that want to continue succeeding and winning contracts will need to become more efficient and proficient at estimating their new developments. In a world of rapid change, increasing competition both local and global, the winners will be those that can confidently predict and successfully manage the cost of their developments. The tools are available lets start using them!

Acknowledgements

The authors would like to thank Geoff Tuer for his input and guidance to the work of this paper. This work has been performed within the research project 'The integration of quantitative and qualitative knowledge for cost modelling'. BAE SYSTEMS and EPSRC (Engineering and Physical Sciences Research Council) are joint sponsors. The work is jointly developed and supervised by BAE SYSTEMS and Cranfield University.

References

[1] HOULT D. P., MEADOR, C. L., DEYST, J., DENNIS, M. *Cost Awareness in Design: The Role of Data Commonality*, SAE Technical Paper, Number 960008, 1996.

[2] STEWART, R., WYSKIDSA, R., JOHANNES, J., Cost Estimator's Reference Manual, 2nd ed., Wiley Interscience, 1995.

[3] DEPARTMENT OF DEFENCE. *Parametric Estimating Handbook*, 2nd Ed., DoD, http://www.ispa-cost.org/PEIWeb/cover.htm, (1999).

[4] TAYLOR, I. M. *Cost engineering-a feature based approach.* In: 85th Meeting of the AGARD Structures and Material Panel, Aalborg, Denmark, October 13-14, 14:1-9, 1997.

[5] MILEHAM, R. A., CURRIE, C. G., MILES, A. W., BRADFORD, D. T. A Parametric Approach to Cost Estimating at the Conceptual Stage of Design. *Journal of Engineering Design,* 4(2): 117-125, 1993.

[6] PUGH, P. Working Top-Down: Cost Estimating Before Development Begins. *Journal of Aerospace Engineering,* Part G, Vol. 206, pp. 143-151, 1992.

[7] I.L., CROZIER, P. AND GUENOV, M. [M.D.]. *Concurrent conceptual design and cost estimating.* Transactions of 13th International Cost Engineering Congress, London 9-12 October, 1994.

[8] MEISL, C. Techniques for Cost Estimating in Early Program Phases. *Engineering Costs and Production Economics,* 14: 95-106, 1988.

[9] WESTPHAL, R., SCHOLZ, D. A Method for Predicting Direct Operating Costs During Aircraft System Design. *Cost Engineering,* Vol. 39, (No. 6): pp. 35 – 39, 1997.

[10] ROY R., BENDALL D., TAYLOR J.P., JONES P., MADARIAGA A. P., CROSSLAND J., HAMEL J., TAYLOR I. M. *Development of Airframe Engineering CERs for Military Aerostructures.* Second World Manufacturing Congress (WMC'99), Durham (UK), 27-30th Sep., 1999.

[11] ROY R., BENDALL D., TAYLOR J.P., JONES P., MADARIAGA A. P., CROSSLAND J., HAMEL J., TAYLOR I. M. *Identifying and Capturing the Qualitative Cost Drivers within a Concurrent Engineering Environment.* Advances in Concurrent Engineering, Chawdhry, P.K., Ghodous, P., Vandorpe, D. (Eds), Technomic Publishing Co. Inc., Pennsylvania (USA), pp. 39-50, 1999.

[12] DEAN, EDWIN B. *Activity Based Cost from the Perspective of Competitive Advantage.* NASA, http://mijuno.larc.nasa.gov/dfc/abc.html

[13] COKINS, GARY. ABC Can Spell a Simpler, Coherent View of Costs. *Computing Canada,* Sep 1, 1998.

[14] WIERDA, L. S. Linking design, process planning and cost information by feature-based modelling, *Journal of Engineering Design,* 2 (1), pp. 3-19, 1991.

[15] BRONSVOORT, W. F., JANSEN, F. W. *Multi-view feature modelling for design and assembly.* In: Advances in Feature Based Manufacturing, Ch.14, pp. 315-329, 1994.

[16] CATANIA, G. Form-features for mechanical design and manufacturing. *Journal of Engineering Design,* 2 (1), pp. 21-43, 1991.

[17] OU-YANG, C. AND LIN, T. S. Developing an Integrated Framework for Feature Based Early Manufacturing Cost Estimation. *The International Journal of Advanced Manufacturing Technology,* 13, pp. 618-629, 1997.

[18] BRIMSON, J. A. Feature costing: Beyond ABC. *Journal of Cost Management,* pp. 6-12, 1998.

[19] KEKRE, S., STARLING, S., THERANI, M. *Feature based cost estimation in design*

http://barney.sbe.csuhayward.edu/sstarling/starling/working2.htm (accessed 22nd February, 1999).

[20] VILLARREAL, J. A., LEA, R. N., SAVELY, R. T. *Fuzzy logic and neural network technologies.* In: 30th Aerospace Sciences Meeting and Exhibit, Houston, TX, January 6-9, 1992.

[21] BODE, J. (1998). Neural networks for cost estimation. *American Association of Cost Engineers,* 40(1), 25-30.

[22] SMITH, A. E., MASON, A. K. Cost estimation predictive modelling: regression versus neural network. *Engineering Economist,* 42 (2), pp. 137-162, 1997.

[23] HORNIK, K., STINCHCOMBE, M., WHITE, H. *Multilayer feed-forward networks are universal approximators. Neural Networks.* Vol. 2, pp. 359-366, Cited in: Smith, A. E., Mason, A. K. (1997). Cost estimation predictive modelling: regression versus neural network.. *Engineering Economist,* 42 (2), pp. 137-162, 1989.

[24] AAMODT, A., PLAZA, E. Case base reasoning: Foundational Issues, methodological variations, and system approaches. *Artificial Intelligence Communications,* IOS Press, Vol. 7: 1, pp. 39-59, 1994.

[25] MICHAEL, J., WOOD, W. *Design to cost.* Wiley Interscience, 1989.

[26] SIVALOGANATHAN, S., JEBB, A., EVBUOMWAN, N. F. O. *Design for cost within the taxonomy of design function deployment.* In: 2nd International Conference on Concurrent Engineering and Electronic Design Automation, Bournemouth, UK, pp.14-19, April 7-8, 1994.

[27] HEINMULLER, B., DILTS, D. M. *Automated design-to-cost: Application in the aerospace industry.* In: Annual Meeting of the Decision-Science-Institute, San Diego, CA, Vol. 1-3, ch.569, pp.1227-1229, November 22-25 1997.

[28] EDMONDS, R. J. *A case study illustrating the risk assessment and risk analysis process at the bid phase of a project.* British Aerospace Defence Limited, Dynamics division, Ch. 2.

[29] TURNER, R: J. *The handbook of project-based management.* McGraw-Hill International (UK) Limited, 1993.

[30] FORSBERG, S., KELVESJÖ, S., ROY, R., RUSH, C. *Risk analysis of parametric cost estimates within a concurrent engineering environment.* **Proceedings, 7th International Conference on Concurrent Engineering, University Lyon 1, France, July 17-20th, 2000.**

[31] HORVATH, P., NIEMAND, S., WOLBOLD, M. *Target Costing a State of the Art Review.* CAM I Research Project, Niemand, University of Stuttgart, 1993.

[32] ROY R., BENDALL D., TAYLOR J. P., JONES P., MADARIAGA A. P., CROSSLAND J., HAMEL J., TAYLOR I. *Development of Airframe engineering CERs for military aerostructures.* MSc group project, Cranfield University, UK, 1999.

[33] HERNER, A. E. *Joint Strike Fighter Manufacture Demonstrator,* RTO Workshop on Virtual Manufacture, Aalborg, Denmark, October, 1997.

[34] NOVARA, M. AND WNUK, G. *An ESA approach to linked cost-engineering databases: Preparing for the Future.* Vol. 7, no. 1, March, 1997.

Linking Engineers with the Voice of the Customer

Alan Dutson

Department of Mechanical Engineering, University of Texas at Austin, Austin, TX 78712

Robert Todd

Department of Mechanical Engineering, Brigham Young University, Provo, UT 84602

Spencer Magleby

Department of Mechanical Engineering, Brigham Young University, Provo, UT 84602

Brent Barnett

Marriot School of Management, Brigham Young University, Provo, UT 84602

Abstract

Many corporations today are experiencing intense competition in both foreign and domestic markets. The importance of gaining a true understanding of customer needs, and developing products that meet or surpass those needs, is more important to a company's survival today than ever before. Too often, however, corporations waste valuable time and resources developing products that fail to delight the customer. Better methods for understanding and internalizing the needs of the customer are often needed at all levels of a corporation, especially among the engineering staff that make critical design decisions during product development. This paper reviews guidelines presented in the literature for helping engineers better understand customer needs. An initial evaluation of the effectiveness of these guidelines is also presented.

1 Introduction

A recent effort by many corporations to restructure their organizations to better serve their customers emphasizes the increasing need for both technical and "marketing" expertise in today's industry. Fierce competition prevents companies from relying on technical expertise alone for success in the market. While technical excellence is important for the success of a product, understanding the *needs* of the customers can often be even more critical. A poignant example of this fact was found in a study of 224 product innovations in over 100 different electronic companies. The key factor affecting the success or failure of the products was found to be the degree to which the products were shaped to meet customer needs. In fact, "Technological lead and technical capability were named as reasons for success by less than 2 percent of the study sample" [1]. Technical excellence is a necessary, but no longer sufficient, condition for success in today's competitive market.

While the importance of properly identifying customer needs before embarking on the development of a new product may seem obvious, history shows that this critical step is often either poorly understood or poorly executed. Examples abound of technically successful products that fail to meet customer needs and, consequently, perform poorly in the market.

The Lockheed L-1011 jetliner is a good example of such a product. The L-1011 jetliner was an impressive technological achievement and is considered to be one of the best jetliners ever developed. It contained more backup systems than any other plane of its time, including several hydraulic control systems. It was well engineered and well built. The L-1011 jetliner, however, was over designed and over priced for most potential customers. Jetliner customers preferred the competing DC-10, and production of the L-1011 eventually had to be discontinued [1]. The L-1011, although technologically superior to its competition, proved to be a drain on company resources simply because it was not designed to meet customer needs. In short, Lockheed successfully built the wrong airplane.

Similar examples are found in the RCA videodisk, the Sony Beta cassette, and the Sinclair C5 electric car. Each product was technically innovative but commercially unsuccessful.

After conducting an extensive study of several companies, Walsh reported that, "Many of our observations - like the importance of good communications and an understanding of user needs - appear at first sight to be common sense, but it is surprising how many firms do not adopt apparently 'common sense' practices" [10]. There is an urgent need for companies to realize the importance of understanding customer needs and learn how to identify and incorporate those needs into the design process. (The *design process* here refers to those activities associated with identifying a specific need in the market and developing a product to fill that need.)

1.1 Engineers and the customer

The activities associated with identifying customer needs have traditionally been viewed as a responsibility of management or marketing. Even if an organization is structured so that marketing and management are the only groups that interface directly with customers, information on customer needs must be disseminated throughout the organization. While it may not be everybody's responsibility to interact directly with the customer, it *is* everybody's business to *understand* customer needs. Peter Marks explains that, "In a team-oriented organization everyone needs to think about customers and competition" [7].

As a fundamental part of product development teams, engineers have a special responsibility to understand customer needs. Engineers who have an accurate understanding of the needs and values of the customers will be more capable of making appropriate assumptions and design decisions than engineers who do not have such an understanding. Understanding and internalizing customer needs will enable engineers to design a product that will delight the customer and succeed in the market.

1.2 Engineering vs. marketing

The responsibilities of marketing personnel in identifying customer needs are different from those of engineering personnel. Marketing is often involved in such tasks as analyzing consumer buying habits, conducting market surveys, and identifying product opportunities in various market segments. To begin development of a product, marketing defines an opportunity within a particular market segment and gathers information on specific customer needs for the product. Engineers become involved in gathering information on specific customer needs after the product and target market have been identified. Engineers identify customer needs by working with internal sources, such as marketing, interacting directly with the customers, or both (see Figure 1). The methods for identifying customer needs that are discussed in this paper refer to the *engineering* role rather than the *marketing* role.

2 Guidelines from the literature

The fields of engineering design, marketing, human development, and needs assessment contain valuable principles that are applicable to identifying customer needs in product development. The fundamental principles from the literature are illustrated graphically in Figure 2. These principles include the following:

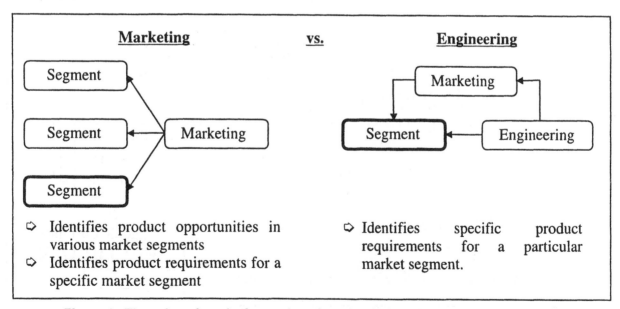

Figure 1. The roles of marketing and engineering in identifying customer needs.

- Establish a relationship of TRUST with the customer
- Determine WHAT is needed
- Determine WHY it's needed
- ANTICIPATE future and latent needs
- Establish internal COMMUNICATION about customer needs
- ORGANIZE needs information
- Receive ongoing customer FEEDBACK

Each principle is discussed briefly below.

2.1   Trust

A relationship of respect and trust must be established with customers in order to obtain honest input and feedback. This becomes especially critical when trying to understand customers' values and perspectives in order to identify latent and future needs.

Donald Gause and Gerald Weinberg claim that one reason customers turn away from the design process is that professional designers often treat them in a patronizing manner. Customer participation in the design process, they say, is much more likely if developers, "create an environment of shared expertise" [3].

Stephen Covey observed that, "...if there is little or no trust, there is no foundation for permanent success" [2]. A foundation of trust is essential for successful interactions with the customers throughout the entire development project.

2.2   What

Various techniques that can help engineers better understand what functions and attributes customers expect from a product are described in the literature. Such techniques include the following:

- Talk to customers or marketing directly (face to face).
- Conduct telephone or mail surveys with customers.
- Conduct focus groups with customers.
- Analyze competitors' successful products.
- Study consumer information journals.
- Talk to the product dealers and service personnel.
- Analyze service reports.
- Participate in fairs, exhibitions, conventions, etc.

Additional guidelines help developers implement the techniques described above more effectively. Each guideline is listed below with its accompanying reference.

- Identify required *functions* instead of product *features* [9].
- Gather customer requirements information from *several sources* [10].
- List the product requirements in the customers' own words. Translating the requirements into engineering parameters is done later in the design process [8].
- List customer requirements in *positive* terms, or what composes an ideal design [8].

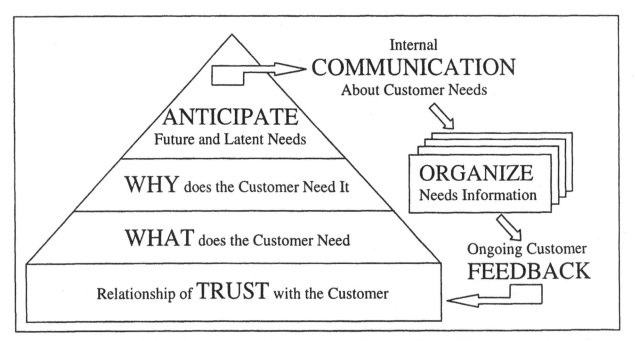

Figure 2. Principles for Identifying Customer Needs in Product Development.

2.3 Why

The step beyond understanding *what* customers want in a product is understanding *why* they want it. Understanding the values and perspectives of target customers enables engineers to better understand why customers want certain things in a product. The enhanced understanding that comes from knowing the *why's* of customer needs enables engineers to make sound design decisions and assumptions during product development.

Understanding why customers need certain things from a product requires familiarity with the customers themselves and with the *use environment* of the product. Griffin explains that a product developer should, "Live with your customers and watch them experience your products in their own environment" [5]. Hartley adds that, "...we want to understand the customer. We must use knowledge and insight, not data and information, to make our decisions. And knowledge...comes from being able to assume the perspective of the customer" [6].

2.4 Anticipate

Going one step beyond understanding the *why's* of customer needs requires anticipating future or latent needs. Hartley explains that companies that rely only on today's customers describing today's needs will likely lack the proactive approach necessary to develop successful new products [6]. Chris Galvin, CEO of Motorola, Inc, stated that, "...we must listen to our customer while also researching, risking, and anticipating the next innovative solution" [4].

The basic techniques suggested in the literature for anticipating future or latent needs include using *stretcher concepts*, working with *lead users* of the product, and *interacting personally* with customers.

Stretcher concepts "represent the range of technologically feasible products of the future" [6]. Using stretcher concepts allows product developers to explore the possibilities for future products with potential customers.

Lead users are defined as, "customers who experience needs months or years ahead of the majority of the marketplace and stand to benefit substantially from product innovations" [9]. By working with lead users, a development team may be able to, "identify needs which, although explicit for lead users, are still latent for the majority of the marketplace. Developing products to meet these latent needs allows a company to anticipate trends and to leapfrog competitive products" [9].

Interacting personally with customers involves the same principles outlined in the previous (Why) section.

2.5 Communicate

The next principle shown in Figure 2 refers to internal communication about customer needs. The primary suggestion from the literature for enhancing communication among different departments in a company is simply to have them work together whenever possible. For example, having both marketing and engineering representatives visit customers together, "...not only helped to assess and understand customer needs but also ensured good communication between marketing and technical staff" [10]. Information on customer needs must not be confined to select groups within an organization. Mechanisms must be established which facilitate communication about customer needs among the various disciplines involved in the development of a product.

2.6 Organize

Another common principle outlined in the literature involves organizing the customer needs information. Needs can be organized into groups, a hierarchy, and/or a ranking of importance. Organization is necessary to facilitate implementation of the customer needs into the design process and to ensure that the information is complete.

Gause presents a detailed method of organizing customer needs according to functions, attributes, constraints, preferences, and expectations [3]. Ullman presents an organization method in which needs are organized according to various *types* of needs, such as performance, appearance, cost, safety, etc. [8]. Ulrich presents a method in which customer needs are first placed in a hierarchy and then assigned a relative importance value or weighting. Customer needs are weighted either through a consensus of the design team or through additional interaction with customers [9].

2.7 Feedback

Obtaining customer feedback throughout the development cycle of a product provides refinement and clarification of customer needs. Many subtle requirements or expectations for a product can only be identified through customer input on physical mock-ups or prototypes.

Gause and Weinberg note that, "The easiest way to avoid user dissatisfaction is to measure user satisfaction along the way, as the design takes form" [3]. Walsh, reporting on a study of product development companies, noted that, "...firms with high turnover and profit growth were significantly more likely to subject prototypes to direct customer and user feedback than those that relied just on intuition or simple, indirect methods like asking sales staff for their comments" [10].

3 Evaluation of guidelines

Although many guidelines are outlined in the literature for identifying customer needs, no type of evaluation of the effectiveness of the various guidelines has been attempted. Since time constraints may prevent a development team from using all of the principles presented above, those methods which are most effective in providing the foundation for successful products should be identified and highlighted.

An initial evaluation of the effectiveness of the guidelines was conducted through a survey of engineers involved in product development. The survey contained questions on the degree to which the engineers felt that the guidelines outlined in this paper were employed during development of a product. The survey also contained questions about the perceived success of each product. A summary of the survey topics is shown in Table 1.

The purpose of the survey was to determine whether or not a correlation could be shown between the degree to which engineers incorporated the guidelines suggested for identifying customer needs and the perceived success of a product. The survey was sent to 25 product development companies. The companies were selected from those that have sponsored projects in the senior-level Capstone design course at Brigham Young University.

4 Survey results

A total of 35 individual responses representing 16 projects from 9 companies were collected. The average number of respondents per project was 2.19. The duration of the projects ranged anywhere from under 5 months to over 25 months. The majority of the respondents were mechanical or manufacturing engineers.

Simple and multiple regression analyses were performed on the survey data. The results of the simple regression analysis are shown in Table 2. The variable numbers shown in Table 2 correspond to the variable numbers and question topics shown in Table 1. The variables which are highlighted in Table 2 are those which were included in the multiple regression analysis.

The multiple regression analysis produced only one model for which all of the factors were statistically significant (t-Stat greater than 1.94). The model is shown in Table 3. The two factors in the model which show the strongest positive correlation with the success of a product are (1) the amount of face-to-face communication with the

Table 1. Summary of survey questions.

Variable	Question Topic
1	Amount of communication with customers about their needs.
2	Frequency of communication with customers about their needs.
3	Amount of engineering participation in conventions, benchmarking, etc.
4	Amount of face-to-face communication with customers about their needs.
5	Amount of visiting customers in their own environment.
6	Followed a formal method for identifying required functions, not features.
7	Perceived importance of identifying required functions, not features.
8	Frequency of communication among engineers about customer needs.
9	Frequency of communication with marketing about customer needs.
10	Frequency of communication with boss or supervisor about customer needs.
11	Followed a formal method for prioritizing or weighting customer needs.
12	Perceived importance of prioritizing or weighting customer needs.
13	Amount of customer feedback received during product development.
14	Followed a formal method for anticipating latent or future needs.
15	Perceived importance of anticipating latent or future needs.

Table 2. Summary of simple regression analysis.

Category	Variable	Slope	Std. Error	t-Stat	r^2 Value	y-Int
WHAT	1	0.290	0.148	1.95	0.214	2.64
	2	0.349	0.239	1.46	0.132	2.79
	3	0.247	0.190	1.30	0.107	2.64
WHY	4	0.261	0.159	1.64	0.161	2.78
	5	0.233	0.140	1.67	0.166	2.97
	6	-0.024	0.206	-0.116	0.001	3.65
	7	0.621	0.169	3.68	0.492	1.26
COMMUNICATE	8	0.244	0.294	0.830	0.047	2.60
	9	0.122	0.208	0.585	0.024	3.30
	10	-0.173	0.330	-0.524	0.019	4.27
ORGANIZE	11	0.046	0.214	0.216	0.003	3.51
	12	0.374	0.174	2.15	0.249	2.17
FEEDBACK	13	0.121	0.154	0.787	0.042	3.27
ANTICIPATE	14	0.044	0.224	0.199	0.003	3.52
	15	0.771	0.225	3.02	0.395	0.33

customers about their needs (variable 4) and (2) the frequency of communication about customer needs among participating engineers (variable 8). A negative correlation is shown between the frequency of communication about customer needs with the boss or supervisor and the success of the product (variable 10). The negative correlation is likely due to increased communication with the supervisor due to existing product difficulties (i.e. increased communication was the *result* of existing product difficulties, not the *cause*), although such cause-and-effect relationships can not be firmly established through this study alone.

Table 3. Best model from multiple regression analysis.

Equation	Variable	Coefficient	Std. Error	t-Stat	r2 Value
y = 2.01 + 0.373*x4 + 0.777*x8 - 0.716*x10	x4	0.374	0.145	2.57	0.459
	x8	0.777	0.314	2.47	
	x10	-0.716	0.340	-2.11	

References

[1] E. E. Bobrow, and D. W. Shafer, *Pioneering New Products; A Market Survival Guide*, Dow Jones-Irwin, 1987.

[2] S. R. Covey, *The Seven Habits of Highly Effective People*, Simon & Schuster, 1989.

[3] D. C. Gause, and G. M. Weinberg, *Exploring Requirements; Quality Before Design*, Dorset House Publishing, 1989.

[4] C. Galvin, Communication with Motorola Employees, 1999.

[5] A. Griffin, *Proceedings, Product Development: Best Practices for Defining Customer Needs*, The Product Development Roundtable, 1995, Section IV.

[6] J. Hartley, *Proceedings, Product Development: Best Practices for Defining Customer Needs*, The Product Development Roundtable, 1995, Section III..

[7] P. Marks, *Defining Great Products; How to Gain a Competitive Edge Through Knowledge of Customers and Competition*, Management Roundtable, 1991.

[8] D. G. Ullman, *The Mechanical Design Process*, McGraw-Hill, 1992.

[9] K. T. Ulrich, and S. D. Eppinger, *Product Design and Development*, McGraw-Hill, 1995.

[10] B. Walsh, R. Roy, M. Bruce, and S. Potter, *Winning By Design*, Blackwell, 1992.

Designing an Electronic Ordering System to Overcome Barriers to Internet-Based Electronic Commerce[1]

Charles V. Trappey, Professor
Department of Management Science, National Chiao Tung University
Amy Trappey[2], Professor
Rick S.-C. Wen, Graduate Student
Department of Industrial Engineering, National Tsing Hua University

Abstract

Product distributors must reduce retailers' ordering process times in order to successfully implement Quick Response (QR) in the chain of commercial activities. An electronic ordering system (EOS) is a time-saving approach for product ordering through the Internet and helps distributors increase the number of transactions. A current problem is that many store managers are adverse to new technology such as EOS and re-order stock by phone even though Internet based systems provide greater advantages. This research uses a survey to define the factors that inhibit the drug store managers' decision to order products via the Internet. After careful analysis of these barriers, a Web-based EOS framework is developed. The system is designed to automatically receive, process and store the ordering data in a distributor's database. The system offers a paperless and continuously available ordering service environment. The Architecture of Information System (ARIS) tool set is used to model and integrate the processes, data, functions, and organizations in regards to the barriers of adaptation. The end result is a technically advanced ordering process that is aligned with the end-users perceptions and needs.

Keywords: Electronic Commerce, Electronic Ordering System, Enterprise Modeling

1 Introduction

In a competitive business environment, effective communication plays a critical role and is largely responsible for the success of the distribution channel. The communication process can be divided into information flows within an organization and information flows between organizations. For retail organizations and for chain stores in particular, the internal information flow includes item prices, sales data, and a range of stock keeping information gathered and processed by bar-code readers and Point of Sales (POS) systems. Chain stores and international retail enterprises are leaders in the installation of high technology information systems that facilitate communication. For small and medium-size retail enterprises, the transmittal of messages internally and communication of data externally with suppliers is often manual and paper-based (i.e. by phone or by fax). Although the use of Value Added Networks (VANs) in large and medium-size enterprises has grown over the last twenty years, small and medium-size enterprises (SME) have found few advantages with their implementation. The usage fees and technology requirements of VANs tend to limit the widespread implementation by smaller companies. Only large companies that transmit greater amount of information within and between organizations tend to gain advantages using VAN-based Electronic Data Interchange (EDI) [1].

A wholesale supplier's goal is to quickly satisfy the demand of the customers and the retail stores in the supply chain. The daily routine of processing information (faxes, phone calls, computer based requests) is costly, time-consuming and error prone [2]. In order for information to

[1] This research is supported by ROC National Science Council Research Grant.

[2] Please send all correspondence to Prof. Amy Trappey at Department of Industrial Engineering, National Tsing Hua University, Hsinchu, Taiwan, ROC (after July 1999). E-mail addresses: trappey@ie.nthu.edu.tw and trappey@cc.nctu.edu.tw.

flow smoothly, all members of the channel have to have compatible tools of communication or bottlenecks will occur. Ballou [3] reports that the ordering process includes order preparation, order transmittal, order entry, order filling, and order status reporting. Retailers choose and maintain supplies depending on their ability to complete the activities of the ordering cycle.

A retail supply chain faces greater order cycle complexities as its markets become global and competition intensifies. Greater dependence is placed on information technology (IT) to facilitate communications and reduce costs in a global marketplace. VANs, the Internet and the World Wide Web (WWW) have replaced many of the costly, time-consuming and inefficient paper driven ordering processes. Electronic Ordering Systems (EOS) not only help retailers procure products through VAN or Internet, but also encourages the standardization of product data in the distribution channel. The continuing improvement of public information infrastructures and increasing linkages to the global Internet is causing a revolution in the ways businesses communicate. Business-to-business electronic commerce (B2B EC) via the Internet is emerging as a viable and effective means to gain competitive advantage.

Many countries are moving in this direction and Internet based B2B EC are planned as a major component of national information infrastructure [4]. However, many problems remain in implementing B2B EC globally. For example, some countries have a narrow network bandwidth, legal structures are inadequate [5] and the security of commerce on the Internet is questionable. A very significant barrier to Internet-based B2B EC adaptation is human acceptance. People tend to resist change from paper-based systems that they are familiar with to new electronic systems.

The Taiwan Pharmaceutical Strategic Alliance (PSA), a voluntary purchasing alliance of stores and franchises, began operations in 1995 [6]. PSA currently provides more than 2,000 types of pharmaceutical products to its members via an Internet-based EOS. However, less than 60% of the alliance members use EOS to order products. The majority of the drug store managers prefer to order goods by phone rather than use an Internet-based EOS. This research poses hypotheses that there are barriers (based on usage, value, perceived risk, and psychological perceptions) that slow a manager's adoption of an electronic ordering system. By evaluating and identifying these key factors and by properly engineering an Internet-based merchandise ordering system, it is further hypothesized that these barriers can be reduced or eliminated. In a global market place, efficient processing of information between retailers, distributors and manufacturers underlies profitability and competitive advantage. The automation of information processing faces not only technological barriers but barriers to adaptation by people as well.

1.1 Research objectives

Since 1980, chain stores have emerged around the world as generally accepted retail store formats. Suppliers have modified their operations to handle larger volume of requests in limited time frames and to process hundreds of customer requests for thousands of products in very short periods of time. Acceptable levels of service require quick response technologies and Electronic Data Interchange (EDI) to manage the information flow [7]. The traditional model of information exchange between retailers and suppliers (Figure 1) shows that a supplier accommodates the exchange standards of each of its customers. Likewise, small retailers comply with the ordering protocols of the suppliers. Thus, suppliers and retailers face a dilemma with an ever increasing number of data formats, business processes and procedures. As new communication channels are created, particularly channels with dedicated or specialized communication protocols and procedures, new costs are incurred.

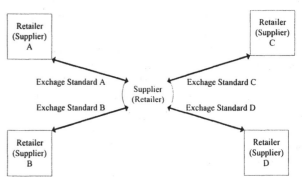

Figure 1. Traditional model of data exchange between a supplier and retailer(s).

Figure 2 shows channel communications using a standard protocol and established ordering procedures via a virtual warehouse. There are several benefits to this approach. First, a virtual warehouse offers around-the-clock services. Second, the Internet based warehouse offers global distribution. Third, the computer-based warehouse can automatically aggregate requests and transmit large purchase orders to manufacturers to minimize costs. Finally, a virtual warehouse provides standard electronic forms and ordering procedures that reduce unintentional errors and confusion.

Evan though these benefits are significant, Forrester Research [8] reports that only about 5% of America's Fortune 500 firms conduct business transactions via the

Internet although more than 80% of them have Web sites. A key to the success of electronic commerce relies heavily

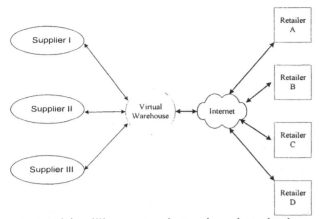

Figure 2. Transmitting documents in a standard format.

upon people's willingness to adopt and use the technology.

Pharmaceutical retailers are used as this paper's case example to illustrate and test barriers to adopting Internet-based electronic commerce. In order to capture the view of the pharmacists, a questionnaire measures factors critical to the adoption of EOSs. After the factors are measured, an EOS system model is constructed to counteract and limit barriers to adaptation. The EOS system model is developed using the Architecture of Information Systems (ARIS) concept and toolset that provides a holistic view of business processes [9].

2 Background

Bloch, Pigneur, and Segev [11] define electronic commerce (EC) as the buying and selling of information, products and service via computer networks. The authors extend the definition of EC by including "support for any kind of business transactions over a digital infrastructure." Electronic commerce is not only sharing business information, maintaining business relationships, and conducting business transactions by means of telecommunication networks, but also carrying out corporate processes that support the internal and external commerce and communication electronically among companies [12]. Ferguson [13] points out that EC systems directly connect buyers and sellers and offer paper-less and digital information exchange. The end result is a system that reduces time and location constraints, provides a dynamic and interactive user interface, and offers real-time updates to ensure information accuracy and currency.

The concept of EC has emerged with the inception of EDI and is rapidly changing as the Internet and the Web technologies are diffused globally [14, 15, 16]. Mak and Johnston [17] indicate that traditional VAN-based EDI

cannot be adopted by most SMEs because the private networks and proprietary technologies are expensive to setup, maintain and operate. Ferguson [13] indicates that two-thirds of all companies using EDI were forced to do so in order to comply with the requests of a powerful channel member.

The Internet makes the implementation of EDI faster, easier and cheaper for SMEs. Van-based and Internet EDIs are complementary, enabling companies to enhance their existing EC capabilities and expand the number of trading partners rapidly [18]. Overall electronic data exchange is moving from proprietary private networks to the open Internet. Flexibility for handling documents and data is greatly enhanced by the Internet EDI approach. A traditional EDI document is a text file conforming to UN/EDIFACT or ANSI X.12 standards, whereas an Internet EDI document can be output in EDI format (EDIFACT or X.12) or other widely accepted document standards (e.g., HTML, XML). Thus, an Internet EDI document can incorporate multimedia effects and the user-friendly interfaces of common document browsers.

Traditional non-EDI trading partners require a front-end application to translate the EDI document into a proprietary file for the existing software or translate the proprietary file into an EDI document for transmittal to the trading partners (Figure 3).

Figure 3. Traditional EDI model

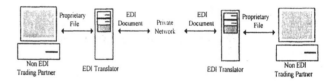

Internet technology (e.g., FTP, Email, Web Pages) provides many ways of transmitting messages. For example, an EDI-enabled trading partner can build form-based Web pages to transmit and receive business information from its non-EDI-enabled trading partners (Figure 4). Applications using this approach can interface with existing EDI systems and provide a facility to translate EDI documents into form-based Web pages. The resulting pages can then be accessed and updated by non-EDI-enabled trading partners using a standard Web browser. Further, the EDI-enabled trading partners can establish a private mailbox or FTP server for their non-EDI-enabled trading partners to retrieve or send standard Web-form documents. The only requirements for the non-EDI-enabled trading partner are Internet access and a Web browser (e.g., Netscape Navigator, Internet Explorer).

Figure 4. Direct exchange via Web pages

The transmittal of business documents using Web-based electronic commerce has gained acceptance in a very short period of time. Barber [1] notes that there are several advantages to Web-based EC. First, Web browser technology is inexpensive and the ISP charging rates are relatively inexpensive. Compared to the price of EDI software, Web browsers cost much less than dedicated EDI software. Second, Web sites can be updated and readily accessed by the public. Third, major firms have the ability to access the Internet and participants share the cost of the information infrastructure. Fourth, additional information or functions that may be valuable to the trading partners can be shared via the Web sites.

2.1 Barriers to adopting new products or services

Lind [19] indicates that there are many barriers to EDI technology adoption. These barriers are split into categories, including implementation issues, organizational issues, technical issues, work process issues, and market potential issues. Also, fear of changes and ignorance of the benefits creates barriers to EDI implementation. Beam and Segev [14] point out that public perceptions of the risk of internet fraud has generate significant barriers to the growth of electronic commerce. Porter and Cauffiel [20] note that aspects of culture and technology inhibit the expansion of electronic commerce and show that the primary inhibitors are the security and reliability of transactions, privacy concerns, and non-repudiation and authentication. Kang [21] proposes that complete access to the Internet is vital to EC adoption and that organization size, investment capability, IT expertise, and the nature of interactions with external organizations facilitates its spread through the organization. Mason [22] applies theories of organization decision making and organization learning to explain the reason for SME's EC technology adoption, linking its diffusion to perceptions of EC benefits and costs.

Wells and Prensky [23] divide products into continuous innovations, dynamically continuous innovations, and discontinuous innovations. These categories are further defined by the degree to which they embody technological and behavioral changes. A continuous innovation is the modification of an existing product that involves little technological innovation and little behavioral change. A dynamically continuous innovation involves a new product that provides some technological change or new benefit but requires little or no behavioral change. A discontinuous innovation involves a new product that provides some technological change or new benefit and requires new consumer purchase and usage behaviors.

An EOS is a discontinuous innovation because it introduces new technology and requires significantly different behavior on the part of the consumer. According to classical models, technology is adopted after a person moves through cognitive, affective, and behavioral stages [24]. During the cognitive stage, managers and executives are exposed to and consider the technology. After becoming aware of the technology's existence, the firm moves into the affective stage where managers develop feelings for the technology. If the feelings are favorable, then the firm moves into the behavioral stage and adopts the technology. Strong [25] proposes that the firm first becomes aware of the technology and awareness leads to interest. Interest in technology then creates a desire for the associated benefits. The desire for benefits accrued through the use of technology leads to action to adopt the technology.

The classic Innovation Adoption Model [26] demonstrates that the consumer purchase activities are linked to the adoption of a new product. According to the model, the adoption of a new product involves knowledge, persuasion, decision, implementation, and confirmation. Another classic model is the communications model [27]. In the awareness stage, the adopting firm is exposed to the technology and the exposure creates a perception. The perception leads to a cognitive or mental response. As the firm moves into the affective stage, an attitude about the product is formed. Finally, the attitude leads to action.

Ram and Sheth [28], as well as Ellen, Bearden and Sharma [29] enumerate four barriers to adoption. These barriers are usage, value, perceived risk, and psychological barrier. A product faces a "usage barrier" when consumers refuse to adopt a new product because it is incompatible with their existing (usage) behaviors. Consumers tend to resist adopting new products that do not offer better value than their existing alternatives. This resistance is known as the "value barrier." Further, consumer may be reluctant to adopt a product if they "perceive risk" about their purchase and usage of the product. Finally, consumers may hesitate to adopt a new product as a result of "psychological barriers" that stem from their prior attitudes and image.

3 A survey to measure barriers to EOS adoption

The four barriers to EOS adoption [28, 29] are used as a guideline for designing a questionnaire. The major constructs measured are barriers to usage, the value of EOS, perceived risks, and psychological barriers. These barriers to adoption are used because they are related to the actions of decision makers and may ultimately lead to EOS rejection. The difference between the barriers of adoption for (EOS and paper systems) is tested according to the following hypotheses:

H_1:   Overall an EOS will be harder to adopt than a paper system.

H_2:   Rated in terms of usage, managers will favor paper systems over an EOS.

H_3:   Rated in terms of value, managers will favor paper systems over an EOS.

H_4:   An EOS will have a higher perceived risk than a paper system.

H_5:   An EOS creates greater psychological barriers than paper systems.

Each barrier (x_i) is given a weight (w_i) and a multi-attribute model ($y = \sum w_i x_i$) is used to represent the overall score for the barriers to adoption. Forty managers were interviewed at drug stores across the northern part of Taiwan. In order to predict survey execution time and to standardize survey field procedures, a pretest of five surveys was conducted. The test method utilized face-to-face interviews to ensure the accuracy of the results. Each survey was conducted at a scheduled time (e.g. 7 p.m.) and required less than 30 minutes to complete.

3.1 Survey results

Table 1 shows that an EOS has a higher barrier to adoption than a paper system (H_1). Also, three sub-barriers of EOS (H_2, H_4, H_5) are rated less favorably than paper systems except for the value barrier (H_3, t=3.961 > $t_{(.05, 39)}$=1.648). Managers report that an Internet-based EOS offers higher accuracy, higher efficiency, and lower costs than a paper-based ordering system. Thus, managers recognize the value of an EOS but favor the usage, reduced risk and perceived benefits of a paper system.

Table 1. Paired-Samples T Test of the four barriers

Pair	Mean	Std. Deviation	Std. Error Mean	95% Confidence Interval of the Difference Lower	Upper	d.f	t-value
Usage1 – Usage2	−24.675	9.9341	1.5707	−27.852	−21.497	39	−15.709
Value1 – Value2	6.075	9.6990	1.5336	2.973	9.176	39	3.961
Risk1 – Risk2	−34.500	9.2847	1.4680	−31.530	−31.530	39	−23.501
Psych1 – Psych2	−18.300	7.4565	1.1790	−15.915	−15.915	39	−15.522
Total1 – Total2	−71.400	18.1966	2.8771	−65.580	−65.580	39	−24.816

(α=0.05)

USAGE1 represents the usage barrier to adoption of the EOS; USAGE2 represents the usage barrier to adoption of the paper-based system. Only the Value barrier is statistically non-significant.

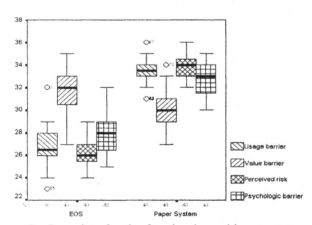

Figure 5. Box plots for the four barriers with respect to EOS and paper system.

Figure 5 shows that the usage barrier, the perceived risk, and the psychological barrier of adoption for an EOS receive lower evaluations than those of paper systems. Therefore, the adoption of the electronic ordering system can be improved by eliminating these barriers. The evaluation of the statistical means for the EOS survey data shows that managers rate the usage barrier and the perceived risk as the greatest factors limiting the adaptation of the technology. The usage barriers and perceived risk can be reduced through the design of the EOS.

Thus, the design objective of this research focuses on how to make the EOS easier to learns and easier to use, as well as how to create a system that is perceived as secure from fraud.

4 Overview of ARIS model

The Architecture of Integrated Information Systems (ARIS) concept is derived from a holistic view of business processes [10]. ARIS is used by ERP implementers to describe enterprises by means of a structured modeling technique that traces processes from the top strategic level (requirement definitions) down to the system design specifications and implementation. ARIS not only describes an enterprise and its business processes via models, but also uses the models for analysis and simulation in regards to process costs and times. The ARIS concept (and the tool used to apply the concept) offers a unique means to model and improve business processes.

ARIS uses descriptive views that divide business processes into the function view, the data view, the organization view, and the control view. When designing an integrated information system, all the elements such as processes, activities, events, conditions, users, organization units, and information technology resources, are taken into account. The first step for building the model requires defining functional processes, events, statuses, organization units, data and IT resources using individual descriptive views. Next, the control view is used to integrate the models and describe the interrelationships using common representation method, such as Process Chain Diagram (PCD) and Even-driven Process Chain (EPC) diagram.

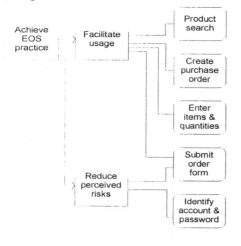

Figure 7. Function tree for the drugstore EOS.

4.1 Objective Diagram

An ARIS-based systems model (the objective diagram) provides the initial objectives and the related EOS functions incorporated to eliminate barriers of adaptation (Figure 6).

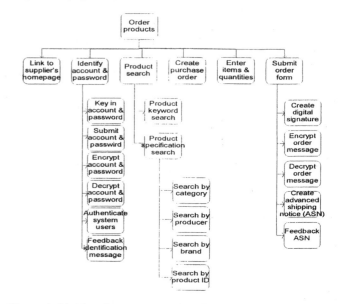

Figure 6. Linking the survey goal to the objective diagram.

4.2 Function view

In Figure 7, a complete set of EOS function elements are shown as a function tree in the function view. For instance, the function "identify account and password" contains the lower level sub-functions "key in account and password," "send account and password," "encrypt messages," "decrypt messages," "authenticate system users" and "send back identification message." The function "product search" can be divided into two sub-functions, including "product specification search" and "product keyword search."

4.3 Process chain diagram

A PCD for the Web-based EOS is shown in Figure 8. Procurement is the organization unit responsible for executing the functions of the ordering process. The events that occur during the order process are placed in the second column. The arrows link events in the sequence the functions are executed. The fourth column contains data objects that define the input or output.

4.4 Control view

The procedural sequence of the business process

functions are expressed in the form of process chains. By arranging a combination of events and functions in a sequential, event-driven process chains (EPCs). Figure 9 depicts the EPC for the Web-based EOS. The starting event is "Requirements are reported" and the end event is "Order form submitted." When the inventory of products is lower than the re-ordering quantity, drug store managers can link to the homepage with a client browser via the Internet. Then, the pharmacist's account and password are entered to ensure secured transactions. If the user identification fails, the user will be expected to key in again until the account ID and password are correctly keyed in. After the verification of the user's identity, the user is permitted to search for products using a search engine. Either a product specification search (e.g. brand name, producer and product number) or a keyword search

is used. Once products are selected, the order request is grouped into a purchase order with the specifications and demanded quantity inserted into the electronic order form. After submitting the order form, the data are stored in the database. Finally, an advanced shipping notice is written and the delivery time is sent to the client's browser.

In order to improve the security, non-repudiation and authentication, cryptography (encryption and decryption) and digital signature technology are applied to the process model as modeled in Figure 10. When sending important messages (e.g. the user's ID and password) or submitting ordering data, the documents are encrypted for transmission and decrypted at the time of receipt by the trading partner.

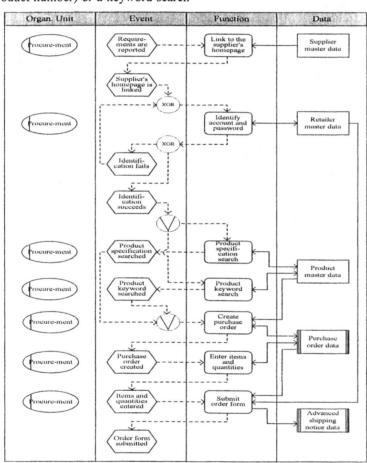

Figure 8.   Process Chain Diagram (PCD) for the Web-based drugstore EOS.

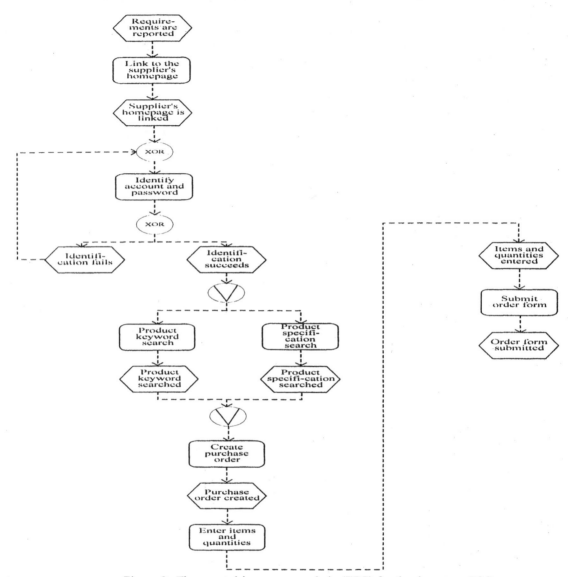

Figure 9. The event-driven process chain (EPC) for the drugstore EOS.

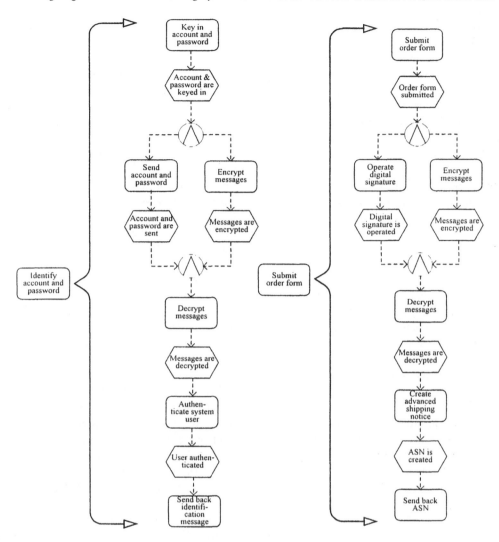

Figure 10. The detailed EPC for functions "identify account and password" and "submit order form"

5 System architecture

In the previous section, the EOS enterprise models are developed using the ARIS concept. The schematic architecture of the EOS is shown in Figure 11, where the drugstore manager orders products via the Internet EOS. The system is placed on the pharmaceutical distributor's NT server with IIS 3.0 to coordinate remote PC data transmission, collection, and ASP file execution from client browsers. On the NT platform, a Web server for NT and a database server for NT are required for the EOS environment.

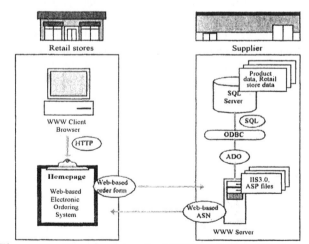

Figure 11. The pharmaceutical EOS system architecture.

From the Web environment, retailers enter ordering information into a form format using the Web pages. The Web server translates the ASP files into HTML pages. If the contents of the ASP file consist of SQL script and ActiveX Data Objects (ADO), it will interact with the SQL Database Server for proper database accesses. After processing ASP files, the results are translated into the Web server and received by the user via the server in HTML format. Therefore, a retail store only needs a network-enabled PC with a Web browser to send order forms, receive Advanced Shipping Notices (ASN) and conduct any other EC communications and transactions.

6 Conclusion

From the results of the survey, the perceived risk must be reduced to increase the likelihood that retail pharmacists will adopt the system. Therefore, the electronic ordering system is designed to enable and ensure authentication, non-repudiation and secured transactions. First, the managers key in the ID and the password of the store. Second, the documents are encrypted, transmitted, and then decrypted when received. Third, digital signature technology is employed to enable non-repudiation and authentication.

The system allows retailers to order products with an easy to use search engine in a secure environment. The managers receive an ASN document from the Web server after entering and submitting the standard order form. For suppliers, troublesome and error-prone order requests by telephone and fax can be minimized so that a supplier can save time and cost while increasing its trading opportunities.

The objective of the research is to plan and model a pharmaceutical merchandise ordering system for the Internet. This Web-based ordering system is designed to reduce the potential barriers for ordering products through the Internet. Finally, the Web-based EOS reduces error-prone order requests by phone voice and fax, saves expenditures in order handling, enlarges the client base and trading opportunities, offers around-the-clock customer services, and provides a user friendly interface for pharmacists.

References

[1]   Barber, N. F., The Strategic Shift to Electronic Commerce from EDI, *Journal of Electronic Commerce*, Vol. 11, No. 1, pp. 54-63 (1998).

[2]   Segev, A., Porra, J., and Roldan, M., Internet-Based Financial EDI: The Case of the Bank of American and Lawrence Livermore National Laboratory Pilot, The Fisher Center for Information Technology and Management, Institute of Management, Innovation and Organization, University of California, Berkeley, Working Paper 96-WP-1018 (1996).

[3]   Ballou, R. H., *Business Logistics Management* (Prentice-Hall, Englewood Cliffs, N.J. (1996).

[4]   Wong, P.-K., Leveraging the Global Information Revolution for Economic Development: Singapore's Evolving Information Industry Strategy, *Information Systems Research*, Vol. 9, No. 4, pp. 323-341 (1998).

[5]   Chen, G.C., Electronic Commerce on the Internet: Legal Developments in Taiwan, *Journal of Computer & Information Law*, Vol. 16, No. 1, pp. 77-123 (1997).

[6]   PSA, Interview with Peggy Zen, Manager of Information System, Pharmaceutical Strategic Alliance Headquarters, Taipei City (1997).

[7]   Ko, E. and Kincade, D. H., The Impact of Quick Response Technologies on Retail Store Attributes, *International Journal of Retail & Distribution Management*, Vol. 25, No. 2, pp. 90-98 (1997).

[8]   Bers, S. J., Dividing into Internet EDI, *Bank Systems & Technology*, Vol. 33, No. 3, pp.38-40 (1996).

[9]   Scheer, G., ARIS Toolset General Basics Manual (Springer-Verlag, Germany, 1997).

[10]   Scheer, G., ARIS Toolset Methods Manual (Springer-Verlag, Germany, 1997).

[11]   Bloch, M., Pigneur, Y., and Segev, A., On the Road of Electronic Commerce – a Business Value Framework, Gaining Competitive Advantage and Some Research Issues, The Fisher Center for Information Technology and Management, Institute of Management, Innovation and Organization, University of California, Berkeley, Working Paper 96-WP-1013 (1996).

[12]   Zwass, V., Structure and Macro-Level Impacts of Electronic Commerce: From Technological Infrastructure to Electronic Marketplaces, *International Journal of Electronic Commerce*, http://www.mhhe.com/business/mis/zwass/ecpaper.html (1998).

[13]   Ferguson, D. M., The Real Facts of EDI in 1997, *Journal of Electronic Commerce*, Vol. 11, No. 1, pp. 18-25 (1998).

[14]   Beam, C., and Segev, A., The Rise of Electronic Commerce: Contributions from Three Factors, The Fisher Center for Information Technology and Management, Institute of Management, Innovation and Organization, University of California, Berkeley, Working Paper 96-WP-1015 (1996).

[15]   Guay, D. and Ettwein, J., Internet Commerce Basics, *International Journal of Electronic Markets*, Vol. 8, No. 1, pp. 12-15 (1998).

[16]   Webber, D. R., Introducing XML/EDI Frameworks, *International Journal of Electronic Markets*, Vol. 8, No.

1, pp. 38-41 (1998).

[17]   Mak, H. C. and Johnston, R. B., Tools for Implementing EDI over the Internet, *Journal of Electronic Commerce*, Vol. 11, No. 1, pp. 44-53 (1996).

[18]   Tuten, D., Internet EDI: How the Internet is Shaping the Future of Electronic Commerce, *Journal of Electronic Commerce*, Vol. 10, No. 4, pp. 38-43 (1997).

[19]   Lind, M. R., Reducing the Barriers to Inter-organizational Electronic Data Interchange, *International Journal of Electronic Markets*, Vol. 8, No. 1, pp. 42-44 (1998).

[20]   Porter, A. and Cauffiel, D., Electronic Commerce Shows Positive Prospects, *Journal of Electronic Commerce*, Vol. 10, No. 1, pp. 64-67 (1997).

[21]   Kang, S., Adoption of the Internet and the Information Superhighway for Electronic Commerce, IEEE, Proceedings of the 31st Annual Hawaii International Conference on System Science, Vol. 4, pp. 12-21 (1998).

[22]   Mason, R. M., SME Adoption of Electronic Commerce Technologies: Implications for the Emerging National Information Infrastructure, IEEE, Proceedings of the 30th Hawaii International Conference on System Science,

Vol. 3, pp.495-504 (1997).

[23]   Wells, D. W. and Prensky, D., *Consumer Behavior* (John Wiley & Sons, Inc., Canada, 1996).

[24]   Williams, L. R. & Rao, K., Information Technology Adoption: Using Classical Adoption Models to Predict AEI Software Implementation, *Journal of Business Logistics*, Vol. 18, No. 2, pp. 10-26 (1997).

[25]   Strong, E. K., *The Psychology of Selling* (McGraw-Hill, New York, 1925).

[26]   Rogers, E. M., *Diffusion of Innovations*, pp. 79-86 (Free Press, New York, 1962).

[27]   Kotler, P., *Marketing Management: Analysis, Planning and Control*, Fifth Edition (Prentice-Hall, New York, 1984).

[28]   Ram, S., and Sheth, J., Consumer Resistance to Innovations: The Marketing Problem and Its Solutions, *Journal of Consumer Marketing*, Vol. 6, pp. 5-14 (1989).

[29]   Ellen, P. S., Bearden, W. O., and Sharma, S., Resistance to technological Innovation: An Examination of the Role of Self-Efficacy and Performance Satisfaction, *Journal of the Academy of Marketing Science*, Vol. 19, pp. 297-308 (1991).

Chao, Ping Yi and Su, Yu Ming
Department of Mechanical Engineering, National Sun Yat-Sen University, Kaohsiung, Taiwan
e-mail:chaopy@mail.nsysu.edu.tw

Abstract

Design change plays an important role in product development. The change can be a small modification, or a major reconstruction. The related components and data must be modified accordingly. The change of related components often causes the change of other components, which we call it the chain reaction, and this makes the design change difficult to handle. The data must be modified during the design change process and the related personnel must be informed to handle proper actions. In this research, we will analyze the relationship between each component first, and then derive a suitable guideline to proceed the chain reaction. This guideline can be used to reduce the scope of the change and minimize the number of components that need to be replaced.

To handle the related units and personnel in an enterprise, we model the process as well as the organization of the enterprise so that people in different sectors of the enterprise can follow and the corresponding documentation can be generated. In this paper, the ARIS modeling method is used as the modeling tool. ARIS is a modeling tool that models functions, organizations, and data, then integrates them under a control model. It can be used for analyze processes in an enterprise, which is the goal of the Business Process Reengineering, BPR. It is also the foundation of the Enterprise Resource Planning, ERP.

To verify the proposed guideline and models, we use a PC assembly as an example to carry out the design change process. Necessary documents can be generated from the process model. By using the documents, related divisions and personnel can be informed and proper actions can be taken. The proposed method demonstrates its potential in shortening the product development time, ensuring the product quality, and laying a firm basis for the future ERP for the enterprise.

1、Introduction

Design change is an important yet regular operation in product development. To ensure the quality of the product, a design will be modified several times before it is finalized. The modification process is one of the major tasks in the design change processes. The faster the design change process, the faster the product can be developed and the lower the cost it generates. When a product change is initiated, the design, the manufacturing, the purchasing, the marketing, and inventory divisions, etc. will all be affected. The change can be a small modification, or a major reconstruction. The change of related components may cause the change of other components. This causes the chain reaction of the change. All related components and data have to be analyzed and modified as well. To handle such complex change process and personnel involved, a product data management system, PDM, is a necessary tool to be used.

The design change process also involves many different divisions in a company such as the research and development division, the inventory division, the purchasing division, etc. Each division should cooperate with others to generate a new version of the product. The updating of changed data usually causes inconsistency of product data and thus, results in poor efficiency of the process. To coordinate people in different divisions, a standard process is mandatory.

In this research, we investigate the design change process and analyze related information, then propose a procedure to guide the change process and model the process to integrate necessary resources. The procedure can be used to reduce the scope of the change and minimize the number of components that need to be replaced. With the model we build, necessary resources are integrated and illustrated so that the Enterprise Resource Planning, ERP, can be made more easily. We will analyze the relationship between each component first, and then derive a suitable guideline to proceed the chain reaction. Then, a commercial modeling tool: the ARIS, which stands for the ARchitecture of Information integration System, will be used to model all related resources for the design change process.

2、Literature Review

Product data includes all data generated in its life cycle, such as design, planning, manufacturing, inventory, delivery, and recycle. A product data management system, PDMS, is used to manage all the related data and processes. It has to ensure the consistency and correctness of the data, and provides users in different stages with the data they need. These PDM functions are helpful in the product development. They are good not only for integrating product data, but also for integrating related departments and personnel. The design change process that relates to various data and people should be handled within a PDM environment.

However, there are other factors that complicate this issue, such as the standard data format, process model, and so on. This review covers some related issues.

Hardwick[2] proposes an architecture to manage changes and related processes concurrently. His system was based on a STEP/EXPRESS format and constructed based on the viewpoint of database. Data changes are the major activities of the system. The architecture is based on the concurrent engineering concept and use an object-oriented database -ROSE for implementations.

The configuration of a product includes its structure, the relationship among components, the geometric features and data. Yeh[9] offers an architecture of an automatic configuration design system based on computer network and the STEP standard. The system transfers the design specification from the product configuration to a detailed component table. All information on components are specified by using STEP. A designer can search for suitable components from the table.

Chao, et.al.[10] uses an object-oriented database to manage standard mechanical components. Comparing with relational databases, the object-oriented databases have the advantage of more flexibility and ease of maintenance. They also use the STEP standard in order to communicate with other systems.

XCON [7] is a rule-base system that is developed by Digital Corporation in late eighties for configuring their products. It is used to validate the technical correctness of customer orders and to guide the actual assembly of these orders. If new components are added into the system, the related rules will become complicated and the rule base becomes unmanageable. To solve the above-mentioned problem, the PC/CON system[8] uses a constraint-based method. The system is constructed with a relational database under a constraint-system shell—Saturn. The drawback of the system is that its database needs to be updated frequently.

Shaw[5] offers an architecture by using semantic data models. He describes product data using an object-oriented method, and presents the relationship of product data by using interactions among objects. Krause [4] describes four features for a good data model:(1) short lead-time, (2) high quantity, (3) low costs, and (4) a general concept for the whole product. Ranky[6] builds up his data model from the viewpoint of an enterprise and implements the concurrent engineering concept. He uses DFD (DataFlow Diagram) and the IDEF0 as his analysis method. Gu and Chan[1] build up a "Generic Product Modeling System" (GPM system) to integrated all necessary product-related information by using the STEP standard. In this GPM system, they organize product information into six libraries. These libraries can link with STEP data model and a GPM/CAD interface. Wang[3] also offers an integrated data model based on STEP. He uses an object-oriented database to store data and builds a mapping interface

between the STEP and an object-oriented database.

The past research shows that design change can be handled in a PDM environment. The product data model must be built to represent the necessary product information. The STEP and the configuration management concept are the future trend in representing product information. The configuration management concept provides designers with a better way in dealing with the relationship between components, so that product development can be more efficient. The modeling tool that integrates different aspects will be helpful in handling complex activities in an enterprise.

3、The Design Change Process

Product data include different type of data, such as the its geometry, its structure, the specification data, manufacturing processes, etc. The modification of data is the major task in the design change process. Apparently the scope of the design change is very large and complex. In this research, the change in an assembled product is discussed.

Design change process can be initiated for different reasons. It can be a major change that the overall structure of the product is changed. In this case, the components as well as the product structure must be changed. The change can also be a minor change that only a component is switched without affecting any other components. In this research, we focus on the change of one component in an assembly. The changed component will in turn affect other components as occurred in the chain reaction.

In such case, the key issue is the relationship between each component of a product. Once a component is changed, other related components in the product may be changed. The related components may cause other changes and thus, changes the whole product. In order to avoid possible divergence of such change, the scope of the change must be specified first.

Secondly, it can be found that the relationship between each component affects the complexity of the design change process. The relationship of a component refers to its parent components, its child components, and those with the same parent components. That makes three basic directions for verification. The change of each direction influences the complexity of the verification in other directions. The larger the changes are made, the more complex of the analysis. Because of the large amount of relations and the resulting chain reactions, a procedure that can handle such complex relationships will be helpful in bringing up the design efficiency and quality.

The relationship between each component is important for the design change. To find out such relationship, product structure is required. There are two methods to represent a product structure: the bill of material, BOM, and the part list. They both use top-down expansion strategy to show the hierarchy of a product. The difference between these two methods is that the part list individualizes each sub-part in the expansion, i.e. the part list expands the specific product without putting the same type of components at the same location in the structure.

As mentioned earlier, design change involves different personnel, divisions, and components. It may bring up a vast amount of information to be analyzed. In order to reduce the effort in making the change, the first rule is to minimize the scope of the change.

The path to trace the related components is complex when a change order is issued. In Figure 1, for example, the component D4 is to be changed. There are three main directions to be checked. The components C2, D3, D5, E3 and E4 have to be checked. We find that C2 component influences more components than D3 and D5 do, and D3 or D5 components influence more components than E3 and E4. The parent component relates more components than its child components. Therefore, we can set up the second rule for design change: to handle the component which has more influence first. So the sequence of the verification follows the direction: Dir1→Dir2→Dir3.

The procedure for the design change process is then organized as the following steps:

Step 1.   Find out the component to be changed from the product structure: the BOM or the parts list. Label this component as the initial node.

Step 2.   Find out a parent component to the initial node.

Step 3.   Check to see if a parent component needs to be changed. If it does, go to Step 2. Otherwise, go to Step 4, and label this component as the end node.

Step 4.   Find the data of child components. Check to see whether a component needs to be changed. If there are no child component, go to Step 8.

Step 5.   Search for alternative components from the product database for the component that needs to be changed. If found, go to Step 7.

Step 6.   Generate new versions of components if no alternative component exists.

Step 7.   Analyze the alternative components. If the component can be used, go to Step 4. Otherwise, go to Step 5.

Step 8.   Summarize all data changes, and update the database.

In this procedure, we need some rules and specifications to see if a component has to be changed or not, or to find suitable alternative components. After the analysis, we can find alternative components to replace unsuitable ones. If no suitable components are found, we have to generate new components to match with the specification of the required component. Generally, when we search for suitable alternatives from the product database, we can use the specification of its child components. Such specification can be used as the selection criteria for alternative components. This may reduce the selection effort for alternative components, and increase the efficiency for the design change process.

4、The Modeling Method

It is easy to make errors during the design change process if no standard procedure is followed. To build a standard procedure, we need to analyze the related resource and procedures involved. A good modeling tool is helpful in doing this job. Through the modeling analysis, the enterprise is easy to find possible bottlenecks in its processes, or find potential problems in its organization. Such problems can be made transparent to the managers and be solved earlier. The system can be integrated and its efficiency can be improved.

The ARIS modeling methods is based on an integrated view. Like the IDEF modeling method, ARIS also uses a graphical form to represent complex functions, relationships and procedures. Moreover, it analyzes the system process from an integrated viewpoint. The organization, the process, and the data can be integrated in the model. All separated models are related with each other. An integrated ARIS model integrates the functions, the organization, and the related data of an enterprise. By analyzing the information flow, the system process can be shown in detail. However, an enterprise model contains many entities, processes, and functions. We can imagine the size and complexity it may appear. To overcome this difficulty, ARIS divides the integrated model into three views - the Data View, the Organization View, and the Function Views.

It is much easier to build up separate views of a model than build an integrated model. With this concept, there are two steps to build an integrated model. The first is to build up separate views of the model, then to combine them with the so-called "Control View" in ARIS. The control view model can be regarded as an integrated model. Figure 2 shows the structure of an integrate model. From the analysis of an ARIS model, we can find potential problems within the organization or processes and make necessary modification for improvement. This makes it more attractive than IDEF in modeling the design change process for further integration within an enterprise. Considering the scope of ERP, or Business Process Reengineering, BPR, the ARIS modeling method has the potential for further implementation with other ERP systems.

5、Implementation

To demonstrate the proposed methodology, the design change process is carried out following the procedures discussed in section 3, and the process model is built by using the ARIS modeling method. In this research, we use documents from a personal computer manufacturer as a reference to build up design change process models, which include data models, function models, organization models, and process models. Then we use the assembly for a personal computer as an example to implement the proposed approach.

5.1 The modeling process

For an enterprise, the design change is related to its process, people, equipment, and data. We use ARIS to model the design change process. The first step is to build up its organization model, function models, and the data models, then we build up the process model to integrate these three models in the control view.

The "function tree" modeling method is used to describe the required function during the design change process. We firstly expand the function "design change" into several sub-functions. Each sub-function can be expanded into several sub-sub-functions if needed. The function should be expanded until its detail is clear enough to show the necessary information. The guideline discussed in section 3 is modeled, in which its functions are modeled. Both these two models use the process-oriented methods to expand.

Models in the data view are used to describe the required data for an enterprise. From the component viewpoint, the most important data is the related component data, such as the product definition data, the product structure data, etc. From the management viewpoint, there are also many related data. We build up four data models: the Product Structure Data Model, the Product Definition Data Model, the Approval Data Model, and the related actions of design change.

We use the eEPC methods to build up models for the change process. The method uses events and functions to represent a process. The functions in the control view models are linked from the functional models. The data and organization are linked from the data models and the organization model.

The first model shown in Figure 3 demonstrates the process from the management viewpoint. The model starts from the "change request" event, and then sends this request to the management division for approval. The "Deal with the design change application" function represents this action. The "Management Division" organization unit is linked to this function to demonstrate the division that executes this function. The "Request for design change" form is the input to this function. After the form is approved, the next thing to do is to handle and analyze the relationship between components.

This control model is expanded from the "Design change analysis" process interface in Figure 3, and it is built up based on the guideline proposed earlier. This model integrates the function models, the data models, and the organization model. When the design change process is initiated, we can follow the control

models to execute the proposed guideline, and generate data through the data model. All related divisions and personnel will be coordinated through the organization model.

The analysis is handled by the "change execution group" organization unit in the R and D division. After the analysis is completed, the related divisions have to be informed. The purchase division must be informed to purchase the necessary components to manufacture the new product. The applicant should fill out the "order form", and send it to the purchase division. At the same time, the manufacture control division has to be informed of the related manufacturing process change. The information division should also be informed to update the product data. After all these procedures are done, the next thing to do is to trace and evaluate the result of the design change. If the design is acceptable, the design change process is completed. Otherwise, the whole process continues until the product is satisfactory.

5.2 The design change for a PC assembly

After ARIS models are built, we can use them to carried out the design change process. The following paragraphs demonstrate an example by changing a component in a PC assembly.

The function and geometry of PC components have standard specifications. A PC assembly can be divided into three main parts: the case, the IC cards, and the peripherals. Generally, the case is seldom changed, but the IC cards and the peripherals are to be changed frequently due to the nature of the PC industry.

Before implementing the design change process for a PC assembly, its structure data has to be established first. The IC cards of PC components can be divided into several types: the mainboard, IO cards, sound cards, video cards, CPUs, and RAMs. The peripherals include the keyboard, the monitor, hard disks, floppy disks, CD-ROMs, MO disks, the mouse, speakers, etc. By assembling these components, a PC structure is established. Figure 4 shows a general structure of a PC.

The design change task starts from an existing part. We use a 486-level personal computer as the component that needs to be changed. Its specification is listed in Table 1.

Suppose that the product specification is to change the hard disk from a Seagate 540M to a Seagate ST32550W, which is a SCSI-3 ultra wide hard disk. The applicant applies for the design change from a management division by filling out the "Request for design change" form. This form is evaluated by the manager in the management division. After it passes the evaluation, the division informs the R&D division to form the "change execution group" to handle the related problems with the change. The group will analyze the whole product, and determine the part that needs to be changed. The relationship issue is handled in this group.

By carrying out the proposed guideline, we found that the CD-ROM and the hard disk have to be changed. The hard-disk is changed to an ST32550W SCSI hard-disk according the specification of the change. The only component to be handled is the CD-ROM drive. It can be connected to the IDE bus type only. There have two ways to solve it. One is to change the CD-ROM to a SCSI device, the other is to connect the CD-ROM with the mainboard IDE joint. The mainboard supports the IDE joint. The specification of the new PC is listed in Table 2. Figure 5 shows the new structure after the design change.

After the analysis of design change process, the related product data has to be updated. There are different divisions that are related to the changed data and must be informed. Two main divisions need to be informed in this example: the purchase and the manufacture control divisions. After starting manufacture the new product, the next step is to trace and evaluate the result of the design change. If everything is fine, the design change process is completed. Otherwise, the whole process continues until the product is satisfactory.

6、Discussion and Conclusion

Design change plays an important role in a product's life cycle. The goal of this research is to analyze the problems we may encounter in the design change process and build up models to analyze the process. To deal with the effect of chain reactions, we organize a guideline so that the change process can be conducted in a systematic way.

To model the design change process, we use the ARIS modeling methods to build up necessary models. The ARIS modeling method provides an integrated viewpoint that integrates the organization view, function view, data view, and the control view models. With such integrated viewpoint, complex process such as the design change can be modeled so that the related data, documentation, and personnel will be coordinated. The guidelines we proposed for the process are modeled in ARIS's control view model. It is integrated with an organization model and a data view model so that the related personnel and data of the design change can be coordinated. The proposed method and the models built provide an enterprise with a better way in handling the design change task.

We use a PC assembly as example to verify the proposed guideline. The result of the verification proves that this guideline can deal with the problems in the chain reaction caused in the design change process. However, this guideline is not a complete solution to this process. Some other factors such as assembly rules are needed. A product database is also helpful in selecting alternative components.

From the ARIS models we built, a process can be analyzed, and standardized procedures can be organized

to ensure the quality of the process as well as the product. The scheme is the so-called Business Process Reengineering, BPR, which is one of the original purposes of the ARIS. It is also a basis for the Enterprise Resource Planning, ERP.

Reference

1. Gu, P., and Chan, K., "Product modelling using STEP", *Computer-Aided Design.* Vol. 27 , No.3, March. 1995.
2. Hardwick, M.,and the RPI DICE Tean, "Managing Change Using STEP/EXPRESS", Rensselaer Design Research Center.
3. Wang, Jui-lu, "Integrated Product Modeling Using STEP and Object-Oriented Database Management System", *Automation*, 1996.(in Chinese)
4. Krause, F. L., Kimura, F., Kjellberg, T., and Lu, C.Y., "Product Modelling", *Annals of CIRP*, pp. 695-706, 1993.
5. Shaw, N.K., Susan, M.,Bloor, A., and Pennington, De, "Product Data Models", *Research in Engineering Design*, pp. 43-50, 1989.
6. Ranky, P.G., "Concurrent Engineering and Enterprise Modelling", *Assembly Automation*, vol. 14, No. 3, pp. 14-21, 1994.
7. Barker, V. E. and O'connor, D. E., "Expert Systems for Configuration At digital: XCON and Beyond", *Communications of the ACM*, Vol. 32, No. 3, pp. 298 - 318, March 1989.
8. Fohn, S. M., Liau J. S., Greef, A. R., Young, R. E., and O'Grady, P. J., "Configuration computer systems through constraint-based modeling and interactive constraint satisfaction", *Computer in Industry*, Vol. 27, pp. 3-21, 1995.
9. Yeh, S. C., "Automatic Configuration Dsign", Master's Thesis, Department of Mechanical Engineering, National Sun Yat-Sen University, Kaohsiung, Taiwan, 1977.
10. Chao, P. Y., Yeh, S. C., and, Chu, J. J., "Application of Open Object-Oriented Database in Standard Mechanical Components", Automation9', Taichung, Taiwan, 1997, pp. 873-880.(in Chinese)

Acknowledgement

This research is sponsored in part by the National Science Council of the Republic of China under the grant NSC86-2212-E110-016.

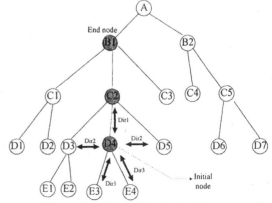

Figure 1. A product structure example.

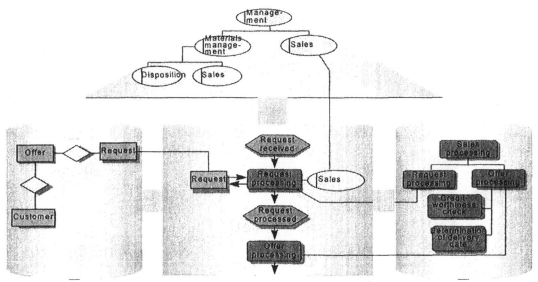

Figure 2. The ARIS structure of an integrated model

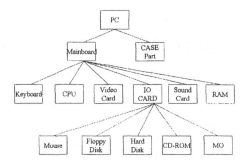

Figure4. A structure for a PC

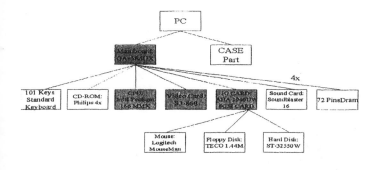

Figure 5. The structure after design change

Table 1. A specification list for a PC

Component	Specification	Qnt.
MainBoard	GA-486VS with 3 VL bus and 4 ISA bus	1
CPU	Intel 486 DXII 66	1
RAM	4M 72 pin DRAM	4
Video Card	ET-4000 Video Card	1
SoundCard	SoundBlaster 16 bits(ISA)	1
IO Card	ISA super IO card	1
F.D.	TECO 1.44 M Floppy-Disk	1
H.D.	Seagate 540 M Hard-Disk(IDE)	1
CD-ROM	PHILIPS 4x CD-ROM(IDE)	1
Keyboard	101 keys standard keyboard	1
Mouse	Logitech MouseMan/ 3 buttons	1
Monitor	ADI 4GP(15 inch)	1
Speaker	120W PC speaker	1

Table 2. The specification list after design change

Component	Specification	Qnt.
MainBoard	GA-586UX with 4 PCI bus and 3 ISA bus	1
CPU	Intel Pentium 166MMX	1
RAM	4M 72 pin DRAM	4
Video Card	S3-868 PCI Video Card	1
Sound card	SoundBlaster 16 bits(ISA)	1
IO Card	AHA-2940UW	1
F.D.	TECO 1.44 M Floppy-Disk	1
H.D.	Seagate ST32550W(SCSI-3 ultra wide)	1
CD-ROM	PHILIPS 4x CD-ROM(IDE)	1
Keyboard	101 keys standard keyboard	1
Mouse	Logitech MouseMan /3 buttons	1
Monitor	ADI 4GP(15 inch)	1
Speaker	120W PC speaker	1

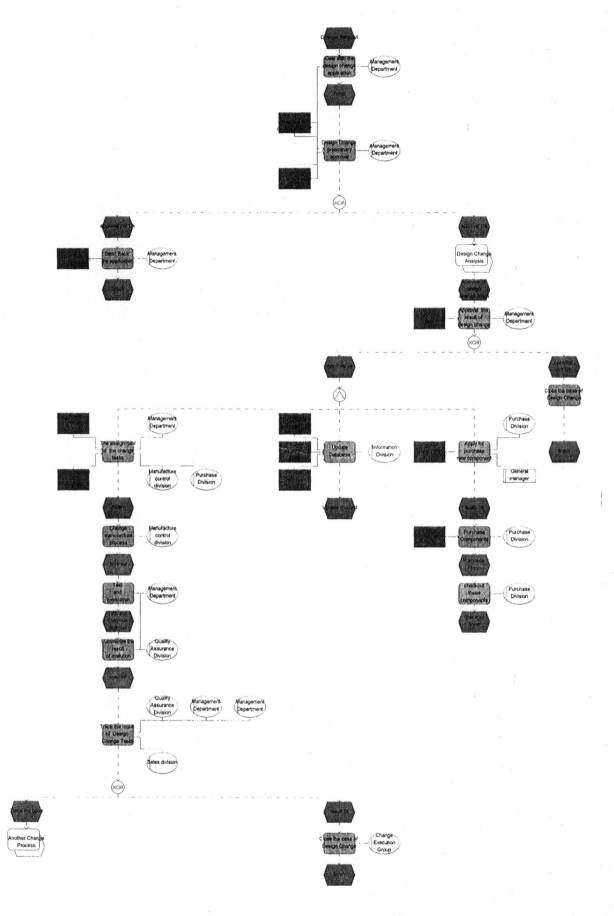

Figure 3. The ARIS model for the design change process

CHAPTER 2

Collaboration Technologies

A Model for Collaborative Search and Retrieval of Domain Specific Information on The Internet

M. Sun, N. Bakis

School of Construction and Property Management
University of Salford
Salford, M7 9NU, UK

Abstract

The exponential growth of the World Wide Web (WWW) poses new challenges to the on-line information search and retrieval task. The existing widely adopted solutions, such as information gateways and Internet search engines, have various limitations. The paper introduces a model for collaborative search of domain specific information based on software agent technology. It has two distinct features: (1) supporting multi-user collaboration; (2) applying domain specific knowledge and user profile knowledge to the information search. A prototype is implemented for the search of construction information. However, the model can be equally applied to other domains.

1 Introduction

The Internet and the World Wide Web (WWW) application will be remembered as one of the most important innovation of the twentieth century. The recent high profile merge between Time Warner - an entertainment contents provider, and American On Line (AOL) – an Internet service provider has given us a glimmer of the even more pivotal role Internet will play in the coming years. Today the WWW is already the most popular vehicle for disseminating information and delivering services on the Internet. It is estimated that the volume of information on the WWW is doubling every fifty days. The amazing growth of the WWW poses some new challenges in helping the users, especially the non-technical users to benefit from the information on the WWW. A number of characteristics of the WWW have compounded the difficulties of the on-line information search and retrieval task:

The WWW is expanding rapidly. Because the barrier for publishing information on the WWW is very low, organisations and individuals alike create new servers all the times. To set up a WWW site, one only needs a modem connecting a PC to one of the many free Internet Service Providers (ISP) and uses standard office management software to author HTML documents. Therefore the trend of the exponential growth of the WWW is set to continue.

The WWW is a disorganised network. There is no central ownership of the Internet and the main WWW application. To enforce any standards of quality control is almost impossible. For the time being, all information providers will provide information using their own style and information structure. The information and information service on the Internet are heterogeneous in nature.

The WWW is constantly changing. Documents are being added, modified and removed from the Internet all the time. Information available today may not be available tomorrow, *vice versa.*

Lack of user collaboration. At the moment information search and retrieval is very much a personal activity. Each user spends substantial amount of time on locating the right information sources. The found information is usually also useful to other users with similar interests. If the searching results can be shared between a community of users, everyone will benefit.

There are two common types of solutions to the information discovery and retrieval problem on the Internet. The first is a "yellow-pages-like information gateway" in which the on-line information is organised in logical categories. The user can locate the right information by browsing through the listings. The second type of solutions is the use of Internet search engines. A search engine is a special WWW server that gathers information about documents located on other WWW servers. The user can locate the required information by supplying keywords to the search engine which in turn finds the documents by matching these keywords. Both types of solutions enjoy a certain degree of success. However, they both have serious limitations.

Given the shear size of the WWW, it is impractical to build a single system that combines all the information resources. A more realistic approach is to build specialised information retrieval systems that provide access to information resources of a particular domain, i.e., construction or medicine. In the longer term, these domain specific information systems can collaborate by sending requests to each other to provide a unified service to the users. This paper describes a Collaborative Construction Information Network (CCIN) project which addresses the issue of collaborative information search and retrieval for construction related topics using the World Wide Web (WWW) on the Internet. Its key features include: (1) a one-stop-shop information search service for Internet users with construction related interests; (2) support for collaboration between these users; (3) application of construction domain knowledge and user profiles to improve the information search.

2   Limitations of the Existing Solutions

Today information is increasing at an accelerated pace. Information search and retrieval solutions are urgently needed to manage the "information overload". However this is by no means a new problem. When a collection of information reaches a certain size, there is always a need to find ways to organise the information and retrieve it when required. In the paper-based information age, card catalogues and classification systems like the Dewey Decimal System [1], are effective information retrieval tools. Today, the amount of digital information is growing exponentially and much of it is available through the Internet. The World Wide Web has become an expanding hypermedia database where information in various formats can be found on many related and unrelated topics. The traditional methods alone are no longer able to manage this information growth. A large number of new tools have emerged aimed at helping users to locate information on the WWW. These tools can be broadly divided into two types:

- Information gateways.
- Internet search engines.

2.1   Information Gateways

Information gateways use the same principle as the "Yellow Pages" business directory. Using the Yellow Pages, a user can locate correspondence details of a firm by looking through the classification hierarchy. At the initial stage of the WWW development, surfing and browsing were the main navigation methods. Yellow Pages type of information gateways were popular, ranging from personal hotlists to comprehensive lists of international services covering multiple publication-types and subjects [2]. Good examples of this type of information gateways include the WWW Virtual Library [3] and Yahoo [4]. There were also subject specific gateways, for example, the Construction Industry Gateway in the UK an initiative to provide a gateway service for construction specific information on the WWW [5].

There is no doubt about the usefulness of these systems. However they also have some major weaknesses. Because classifying the on-line information resources often involves human intervention which makes the expansion in scope difficult. As a result, each of the existing gateways only covers a very small fraction of the WWW resources, with Yahoo as an exception. There are no widely accepted standards for classification. Usually each information supplier decides the structure and content of its materials governed by chance, occasional decisions and staff responsible for the implementation. The consequence is a lack of consistency and reliability and a lack of independence from the individuals performing the task. Another weakness of information gateway solutions is the difficulty of keeping information updated. Due to the rapid changing nature of the WWW, many resources are quickly becoming non-usable. Because there are no effective automatic updating mechanisms it is very difficult for the gateways to follow the rapid changes of their contents, addresses, appearing and disappearing of documents on the sites they cover.

2.2   Internet Search Engines

The search engine solution is based on the principle of keyword matching. A user describes his/her information needs using a number of words. If these words are found in a document then there is a higher probability that this document is relevant to this particular user's information needs. Although in reality the keyword matching methods are much more complicated and sophisticated the principle remains the same. An Information Retrieval System (IRS) iterates through a document collection and builds an index of the most important words found in the collection. It associates each word with the locations of all the documents containing it. When a user specifies a query with a set of keywords, the system returns the documents containing all or some of the words in the query. Internet search engines are programs that roam the Internet (with flashy names like spider, worm or searchbot) to build up an index of meta-information about everything available on the net [6]. The gathered information, characterised by a number of keywords (references) and perhaps some supplementary information, is then put into a large database. The database is increasing all the time as more information becomes available on the net. Anyone who is searching for some kind of information on the Internet can then try to localise relevant documents by giving one or more query terms (keywords) to such a search engine. Since their first appearance in 1994, the number and diversity of Internet search engines have increased rapidly and continue to do so. The most popular existing Internet search engines include Infoseek, Excite, HotBot, Lycos, Alta Vista, etc. Each of these engines covers a percentage of the Internet, usually bigger than what an average information gateway covers. There are some overlaps between their coverage. There are now meta-search engines available, which pass the user's queries to multiple search engines in order to perform searching in a wider scope.

These search engines are still providing a useful service in the face of information overload on the WWW. However, they have many weaknesses which become more and more evident [7]. All these search engines rely on the user being able to formulate queries effectively. To use the advanced query, a user needs to apply sophisticate Boolean functions, e.g., "window AND (NOT (comput* OR Microsoft)) AND building". This is often beyond the capability of ordinary construction professionals. Keyword matching usually assumes that if a word occurs in a document more often the document must be more relevant to the search. However, this is not always the case. Furthermore, there is no good rating mechanism to rank the many hundreds or even tens of thousands of hits in response to each user's query. In addition, most search engines are domain independent and used in a single user mode.

3 WWW based Construction Information Services

In recent years, numerous construction specific WWW based information services have been piloted world-wide. Some examples are introduced very briefly below. Their WWW addresses are provided for readers who are interested in more details on each project.

3.1 Construction Industry Gateway

In 1996, the Department of Environment in the UK sponsored a scoping study for the establishment of a Construction Industry Gateway (CIG) aimed at improving information dissemination and sharing of knowledge in the construction industry [8]. The study has identified four possible designs of the Gateway:

1. Standard only option: In this option, there is no physical central server. The effort is on developing common standards for storing and retrieving information. Each information provider is encouraged to adopt these standards and share links.

2. An information signpost: In this option a gateway server is established that acts as a "signpost" for other information sources. It will provide users a single starting point for their information searches.

3. A value-added gateway: This option provides a gateway with indexing and searching functions. It also recommends the adoption of standard classification schemes to facilitate the interaction between the gateway and other information providers' servers.

4. A database host and gateway: In this approach the gateway acts as a host for some of the construction information databases. Because it provides a centralised service, tasks of ensuring the information quality and security are comparatively easy. However, keeping information update and keep pace with the information expansion will be more difficult.

A prototype with all the above design features has been developed to prove the concept and demonstrate the feasibility (http://archhive.ncl.ac.uk/cig/). However, so far the information on the server is very limited. It showed the difficulty if a system has to rely on other information providers.

3.2 M4I Knowledge Exchange

The Movement for Innovation (M4I) of the UK construction industry aims to develop a knowledge exchange as one of the central resources for the built environment industries to raise the flag about innovation (see http://www.m4i.org.uk/). The knowledge exchange is:

- A forum for sharing ideas and information
- A way of making your voice heard on important industry matters
- A vehicle for finding people and businesses that you can work with

To achieve this, organisations 'raise their flag' to state that they have information they wish to share on a particular topic. Currently, this relates mostly to demonstration projects identified in response to Sir John Egan's report on "Rethinking Construction". The Knowledge Exchange is expected to expand to include: technical papers; presentations; reports; industry news and views; process flow charts; and toolkits. As a result of the approach adopted the information on the Knowledge Exchange is usually of high quality and highly relevant to the construction users. On the other hand, it requires extensive human intervention in assessing the information sources. The scope of this system will always be limited.

3.3 CONNET

CONNET is a construction information service network developed by a consortium in Europe [9]. It aims at providing a one-stop-shop of technology transfer for the construction industry of Europe. It consists of a number of loosely coupled national nodes which are able to provide users with transparent access technical information, product data, newspaper service, etc. More details about the project can be found at the following WWW site (http://www.connet.org).

3.4 CARE-IS

Construction And Real Estate – Information Service (CARE-IS) is another Internet based "one-stop-shop" information service for the construction professionals [10]. The emphasis of this system is to collect information from other Internet sources, store the collected information in various databases and meet

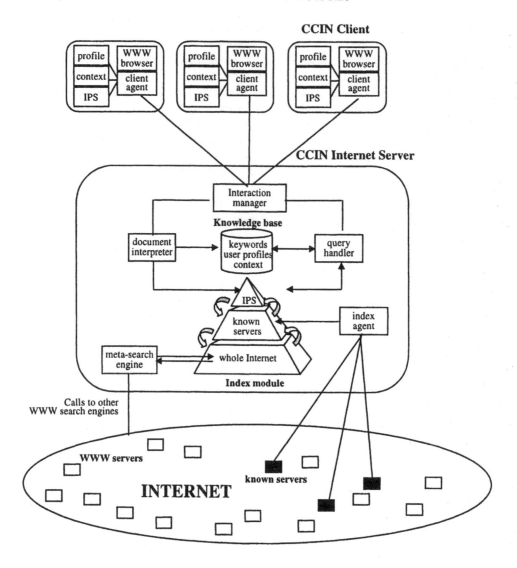

Figure 1 Architecture of Collaborative Construction Information Network

users' information requirements by providing local and Internet search services. This WWW site can be consulted for details of the project (http://care.bre.polyu.edu.hk).

CARE-IS and the systems mentioned above all require human intervention in gathering information from third party sources. This will seriously limit the speed and extent these services can expand. They also rely on, to various degrees, the co-operation of other information providers. Unfortunately these providers especially the commercial information providers in the construction domain usually see these systems as a threat to their own on-line information services. Their supports are not always forthcoming. Another weakness of these systems is the lack of support for user collaboration.

4 Collaborative Construction Information Network (CCIN)

4.1 CCIN Architecture

The Collaborative Construction Information Network (CCIN) is an on-going project carried out at the University of Salford. The aim of the project is to develop a value-added gateway to the construction information network, which facilitates the collaboration between users for the benefit of improved information search and retrieval on the WWW.

It is acknowledged that the WWW based information network will remain a distributed resource with information hosted on many servers. The gateway of this project seeks to provide a recognised starting point for the user to search for information and a repository for shared knowledge base that helps the information search. Figure 1 illustrates the architecture of the Collaborative Construction Information Network. It shows a typical three layer architecture with the CCIN servers playing the intermediary role between the users (clients) and information suppliers (WWW servers).

The system is developed using agent technology, consisting of a CCIN server and multiple collaborating client agents. The CCIN client agent is a plug-in package to standard WWW browsers, such as Microsoft Internet Explorer and Netscape, to provide additional functions. Once it is installed on a user's computer, the agent will facilitate the communication with the CCIN server and provide intelligent information search and retrieval services. To achieve this, the agent needs to gather information of user profile, information query context and information search history. Part of this information is communicated to the CCIN server and stored in the knowledge base on the server. This accumulated knowledge helps to improve the information search task of the user community as a whole. The CCIN server is in essence a construction domain specific WWW search engine. It does not host the original information documents. Its main function is to match information sources to users search queries.

The following explains how CCIN improves the search for construction information through the use of users profiles, user collaboration and construction domain knowledge.

4.2 User Profiles

User profile technology is originally developed for the purpose of information filtering. The goal of an information filtering system is to sort through large volume of dynamically generated information and present to the user those that are likely to satisfy his or her information requirement [11]. With the growth of the Internet and other networked information, research in automatic information filtering has been on the increase in recent years. Using a filtering system, a user does not need to search for information using queries. Instead, the user specifies his or her interests and the system monitors the information sources to inform the user when information of the required nature becomes available. To achieve this the system needs to know a profile of the user.

The CCIN server has a collection of built-in user profiles characterising the main type of users related to the construction industry, such as architect, building surveyor, project manager, cost consultant, academic researcher, etc. Each profile is associated with a set of interests which in turn include a set of keywords and information context. These profiles can be used as independent key for formulating queries. When a user connects to the server, he or she can select one or more standard profiles according to particular interests. The user's profile is further personalised during the process of querying. When a user downloads a document and gives a high rating to the document the client agent on the local computer will extract key concepts from the document and use them to improve the user profile. At end of each session, the updated user profiles will be communicated to the server, of course with the explicit permission of the user. The user can also explicitly modify the user profile by creating new interests, adding keywords and giving weight to existing keywords.

The server will update the user profile after collating and analysing inputs from a large number of clients. The update of profile means interests and keywords can be removed as well as added to reflect the user interaction with the server. Through such a learning process, the shared profile knowledge on the server is improving all the time. The server will progressively improve the understanding of the users' information requirements and the accuracy of the information search.

4.3 Layering Index Management

At the centre of the CCIN gateway server is an index module that consists of three hierarchical layers. The top layer stores information of documents that have been accessed by users and communicated to the server through the CCIN client agents. The data are stored in a meta-information format similar to Jasper's Intelligent Page Store (IPS) [12]. For each WWW page, the index module stores at least the following information gathered automatically by the client agents with a degree of user manual input:

- the document title
- a summary of the content
- a set of keywords
- a set of user profiles who recommended the page
- information type (product data, technical publication, news, project data, etc)
- users' annotations
- universal resource locator (URL)and,
- date and time of storage or update

The middle layer contains an index of documents located on known WWW servers which are more likely relevant to the construction users interests. These servers become known to the gateway server through two ways. (1) The information providers register their servers explicitly. (2) When the client agent submits a document to the top layer of the index module, the host server becomes known to the gateway implicitly. The CCIN server agent, a mobile information agent, visits all the known servers periodically and gathers information about documents hosted on these servers.

The bottom layer of the index module covers most of the accessible web hosts on the Internet. Given the constant growth of the Internet, the gateway server does not store any documents. Instead, it will rely on a Meta search engine to call upon other web search engines, e.g., Alta vista, Infoseek, etc.

The hierarchical layering implies that the amount of information accessible increases as one moves down from the top to the middle and the bottom layers, but the average potential relevance to the users' interests decreases. The advantage of the layering approach is that it allows a user to specify the scope and manner of a query. One can raise a query just for the IPS layer where more criteria can be applied apart from keywords. Alternatively, the user can make a general query using enhanced keywords supported by the server's knowledge base to a wider index in the middle layer or even the whole Internet.

A document's position in the index module is not fixed. There are migration paths through which a document can be moved from one layer to another as a result of users' search and retrieval actions. For example, when a document in the middle layer is retrieved by a user, the client agent will communicate that fact to the server. The server will upgrade the document to the top layer in the index module. On the other hand, if a document in the top layer has not been used by any user for a period of time (pre-defined threshold), the server will degrade it to the middle or even bottom layer. The purpose is to ensure the efficiency of the index system.

4.4   Construction Domain Knowledge

If the server knows more about a user's information needs, it can provide more accurate results for the user's search queries. The knowledge base of the CCIN server keeps information of three aspects that are construction domain specific.

Keywords: Keyword based searching remains an important method of information retrieval on the WWW. Instead of considering it as a simple syntax comparison, the CCIN server identifies a set of keywords and associates them with conceptual meanings. Relationships between these keywords are also analysed. As a result the server is able to support concept based searching. For example, when a user elect to search for information about "enclosing building element", he/she can choose to have extended the search to include sub-types "wall", "window", "door", "roof", or association-types "building material", "component product", etc. These keywords are identified based on recent data and processing modelling works of other research projects and the common classification standards used in the industry.

User profiles: A user profile articulates the features and the information needs of a distinctive type of users. The CCIN server has general profiles for the common types of construction information users, such as architects, engineers, project managers, researchers, academics, etc. The CCIN users can choose one or a combination of these profiles for their searching query. They can also derive personalised profiles based on these general ones.

Context: Context is a description of the purpose behind a user's search for information. It may be about stages of the construction process, design tasks or a research topic area. The CIG scoping study identified the information requirements of construction professionals during each stage of the construction project life cycle [8]. It can be used as a framework to organise the construction information query and retrieval.

The knowledge base is populated at the server set up stage. The server also has the ability to learn from its interactions with the client agents. The knowledge base is updated constantly.

4.5   User Collaboration

Various studies show that when people search for information, the first thing they usually do is to ask other colleagues who have done similar searching before [5]. The existing browsing and searching systems focus on technical information discovery methods while neglecting perhaps the single most important method of discovery that people rely on - other people. Collaborative browsing, sometimes also know as social resource discovery, assumes the existence of other users of similar interests who have located and evaluated relevant resources. The goal of collaborative search systems is to aggregate and share the fruits of the individual activity and knowledge of Internet information retrieval [13].

In the CCIN, when a user retrieves a document of interest, the client agent installed on the user's computer extracts some information about the document, such as the title and a set of keywords. This information is stored locally in an Intelligent Page Store (IPS) format. At the same time the client agent sends it to the CCIN server together with the user profile information and its context. The Document Interpreter on the server is responsible for processing this information and storing the document summary in the index module and updating the knowledge base if necessary. When another user with the same profile raises a query, the documents of the IPS layer will be searched before other information sources. However, to avoid the scope of search become progressively smaller the CCIN server will always return a proportion of the hits from the other two layers of the index module. This will ensure new relevant documents can be discovered by the user.

The other aspect of user collaboration is the update of the user profile as described in section 4.2. The advantage of this approach is that less skilled users can benefit from the more competent users.

4.6   Agent Based Solution

Agent technology is originated from the branch of Artificial Intelligence known as Distributed Artificial Intelligence. In recent years, it has grown into a fast expanding research area. As any other new research field, there are diverse definitions of the term agent. Here we want to quote a very generic definition given by Janca [14], "an agent is a software that knows how to do things that you could probably do yourself if you had time". There is a growing consensus in the Internet community that one of the most promising solutions to the problem of Internet information retrieval is the use of software agent technology.

The main feature of an agent based WWW is that information agents perform the role of managing, manipulating or collating information from many distributed sources [15]. Some information agents have intelligent features for example they can be mobile, can learn and can co-operate with other agents. Mobile agents are able to roam the WWW, interact with WWW servers, gather information on behalf of their owner and return home after performing duties set by their users. The learning ability of information agents refers to the fact that they can react and interact with external environment and improve their efficiency over time. While individual agent is able to work on its own, a number of agents can co-operate with each of them performing a role to achieve a collaborative goal.

Although there is still scepticism considering software agent is just another buzzword, the initial development in this area offers promising potentials. The benefits of software agents approach include:

- More intelligent information search
- More robust and flexible system
- Opportunity to incorporate domain knowledge
- Personalisation
- Better user collaboration.

The CCIN system adopted a multi-agent architecture. The CCIN client agent is implemented as a plug-in package to standard WWW browsers, such as Microsoft Internet Explorer and Netscape, to provide additional functions. It is available for free to the users. Its main purpose is to monitor the user's information retrieval activities and communicate with the CCIN server agent with the explicit permission of the user.

The CCIN also has an index agent which is a mobile information agent. Its function is to traverse the list of known WWW servers and build up an index for all the documents on these servers. It has the ability to handle different data formats, HTML document, postscript files, compressed files, databases, etc. In the case of interaction with databases on a third party server, minimum human involvement may be required for the interpretation of the data structures until industry wide data standards are universally adopted.

5   Conclusions and Future Work

As the WWW continues to expand the challenge for helping users to locate and retrieve the right information effectively will become more acute. This paper discussed the limitations of the two existing types of solutions, information gateways and search engines. Two areas are identified, user profile and application of domain knowledge, which will help to improve the information search on the WWW.

This paper reported an on-going CCIN project which seeks to improve the search for construction related information. It takes advantage of the emerging technologies in information filtering, collaborative browsing and information agents. The main contribution of the project is the user collaboration aspect and the inclusion of construction domain knowledge and user profiles. For example, when a user supplies a keyword "window" to the CCIN server, the server has the knowledge to qualify the query as "window" in the building context not in the computing context as "Windows 95". The principle behind CCIN should equally apply to any other domains.

At present, the system is still in its development stage. Several non-technical issues will need to be addressed such as the user privacy, information ownership and generation of critical mass of users. Like other collaborative systems, CCIN relies on the co-operation of its users. It needs to monitor and gather information about the user's WWW query activities. Ways need to be found to overcome the common fear of intrusion on user privacy. In the construction domain, a large amount of information is owned by the traditional information providers, many of them offer commercial on-line information services. While CCIN does not need to dispute the information ownership with these providers, it needs to get access or to make interface arrangements with their indexing services. To achieve the benefits of the CCIN system, a sizeable number of users are required. It is a challenge at the initial phase to generate sufficient interests amongst the targeted users.

References

[1]   A. Fowler, Dewey Decimal System, Children's Press Inc., USA, 1997

[2]   A. Brümmer, M. Day and et al, The role of classification schemes in Internet resource description and discovery, Research Report, NetLab, Lund University Library, Sweden, 1997 http://www.ub2.lu.se/desire/radar/reports/D3.2.3

[3]   G. Manning, About the Virtual Library, 1999, On-line WWW page: http://vlib.org/AboutVL.html

[4]   Yahoo: 1999,   On-line   WWW   Page: http://www.yahoo.com

[5]   S. R. Lockley, Amor R., "The Construction Information Gateway", in R. Amor (ed) Proceeding of ECPPM'98

Product and Process Modelling in the Building Industry, BRE, UK, 337-348, 1998

[6]   P. Gilster, Finding It On The Internet; The Internet Navigators Guide To Search Tools And Techniques, John Wiley & Sons, 1996

[7]   B. Hermans, Intelligent software agents on the Internet, Tilburg University, Tilberg, The Netherlands, 1996 http://www.hermans.org/agents

[8]   PE Consulting, Scoping Study for the Construction Industry Information Service, for Department of Environment, June 1996

[9]   Z. Turk and R. Amor, "CONNET – design issues", in Proceedings of INCITE 2000, Hong Kong, 17-18 January 2000, pp416-428

[10]   D. Scott, C.K. Pang, and K. Shen, "Internet-based information for construction professionals", in Proceedings of INCITE 2000, Hong Kong, 17-18 January 2000, pp869-886

[11]   S. C. Newell, "User models and filtering agents for improved Internet information retrieval", User Modeling And User-Adapted Interaction, Vol.7, No.4, pp239-256, 1997

[12]   J. Davies, R. Weeks and M. Revett: Jasper: Communicating information agents for WWW, technical report, BT laboratories, 1997, to access the paper through: http://www.labs.bt.com/projects/knowledge/jaspaper.htm

[13]   M. B. Twidale, D. M. Nichols, "Computer supported cooperative work in information search and retrieval", Annual Review of Information Science and Technology, Vol.33, pp259-319, 1998

[14]   P. Janca, Pragmatic application of information agents, BIS strategic decisions, Nowell, United States, 1995

[15]   H. S. Nwana, "Software agents: an overview", Knowledge Engineering Review, Vol..11 No..3 pp205-244, October/November 1996 http://www.labs.bt.com/project/agents/publish/papers/review1.htm

Synchronized "Design for X" Explorer on the World Wide Web[1]

G. Q. Huang, J. Shi, and K. L. Mak

Department of Industrial and Manufacturing Systems Engineering, the University of Hong Kong, Hong Kong, China
Email: GQHuang@hku.hk ; MakKL@hku.hk

Abstract

This paper is concerned with establishing a synchronized platform that can be used for a variety of DFX analysis on the Internet through the standard web browsers. Four associated facilities are provided in the platform. The first one is the Product Explorer, used to capture the product data. The second Process Explorer helps the user acquiring the process information. The third Measure Explorer provides measurement information of the interaction between process activities and product elements. The last Worksheet Explorer is an integrative worksheet displaying DFX analysis results. Within the DFX Explorer, the activities of multiple users must be synchronized and their conflicts must be resolved. Finally, a case study is introduced to show how the DFX Explorer can be used for Design for Disassembly (DFD) analysis.

1 Introduction

The aim of this paper is to present a web-based platform for Design for X (DFX) tools. This DFX platform is developed based on a conceptual DFX model that is called PARIX (Huang, 1996; Huang and Mak, 1997; Huang and Mak, 1998), as shown in Figure 1.

This model adopts ideas from the theory of the product realization process produced by Duffey and Dixon (1993), which considers the realization procedure as a triple composition (P, A, R) of Product competing with the market, Activities realizing products and Resources used for the realization. P, A, and R are interrelated and interacted with each other. These Interactions can be interpreted by that product elements consume activities and process activities consume resources. X in the term of DFX is just used to define the measurement of these interactions. It is a variable that is made up of two parts:

life cycle business process (x) and performance measures $(bility)$: $X = x + bility$.

The remainder of this paper discusses this platform in detail. First, the DFX project is briefly introduced and the overview of the resulting systems is presented. The DFX platform or Explorer is only one of the several deliverables from the project. Section 3 outlines the overview of the DFX platform or DFX Explorer. The implementation constructs and main components of the platform are discussed. Section 4 describes the detailed procedure of the development and application of the four main components. Section 5 is concerned with two issues regarding the collaboration among various users: synchronization control and conflict resolution. Section 6 introduces a case study to demonstrate how the DFX Explorer can be used for Design for Disassembly (DFD) analysis. Implications derived from the case study are analysed for further development and improvement.

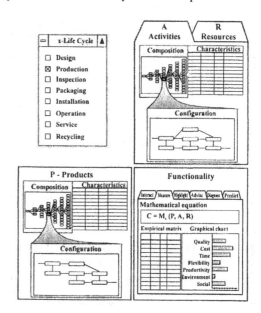

Figure 1 PARIX - A conceptual DFX model

---

[1] The prototype system is accessible at http://147.8.86.81/dfx/shell/.

2   The DFX Project

Since its initiation in 1998, the DFX project has proceeded into its final stage. Main outcomes include three web-based prototype systems and four supporting databases, as shown in Figure 2. The three systems are MetaDFX Explorer, DFX Explorer and Guideline Explorer. They are briefly summarized as follows:

- *MetaDFX Explorer.* It is a "mother" system that can be used by DFX tool developers to generate different DFX tools. It is basically a process-oriented performance measurement system. The MetaDFX Explorer provides a library of processes. The use of the MetaDFX Explorer involves five main steps. Processes are first selected for inclusion in the DFX tool. For each chosen process, appropriate performance indicators, measuring units, aggregating algorithms, and benchmarking method and values are identified. In-depth discussion is given separately (Shi *et al.*, 1999).
- *Guideline Explorer.* This is a web-based system of DFX guidelines. There are two modes of operation. The first mode is for editing or compiling DFX guidelines so that the DFX knowledge / data bases expand with applications and are updated with

developments. The other mode is for browsing and selecting DFX guidelines related to certain applications. DFX guidelines are defined by their patterns that are dependent on the characteristics of process activities and product elements. The patterns are used extensively for guideline search and retrieval. In addition to patterns, DFX guidelines have their contents that include diagnostic, advisory, measuring, benchmarking information. In-depth discussion on the Guideline Explorer is given separately (Huang *et al.*, 1999).

- *DFX Explorer.* This is a synchronized platform that can be used for a variety of DFX analysis on the Internet through the standard web browsers. Four associated facilities are provided in the platform. The first one is the Product Explorer, used to capture the product data. The second Process Explorer helps the user acquiring the process information. The third Measure Explorer provides measurement information of the interaction between process activities and product elements. The last Worksheet Explorer is an integrative worksheet displaying DFX analysis results. Within the DFX Explorer, the activities of multiple users must be synchronized and their conflicts must be resolved. This paper is primarily concerned with the DFX Explorer.

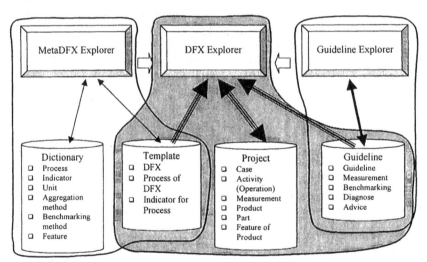

Figure 2 Framework of the DFX project

The three explorers are supported by four web databases. They are

- *Dictionary Database.* This is a process dictionary within which a collection of business processes, activities, and operations stored for the use by the

MetaDFX Explorer. All other explorers do not have the access to this database.

- *Template Database.* This maintains the definition information of individual DFX tools generated from the MetaDFX Explorer. The DFX Explorer retrieves the DFX definition from this DFX Template database. No changes can be made to the template database from the DFX Explorer. Instead, the MetaDFX is the only point where the DFX definition can be changed.

- *Project Database.* This is the working database exclusively used by the DFX Explorer for maintaining product, process and all the associated information related to DFX projects. Other explorers do not have the access to this database.
- *Guideline Database.* This is the knowledge / data base providing DFX guidelines. This database is primarily maintained by the Guideline Explorer

through which the DFX Explorer can retrieve relevant DFX guidelines during the process of a DFX analysis project. However, the DFX Explorer does not have the privilege to make any changes to the Guideline database.

The rest of this paper concentrates on the DFX Explorer and its supporting databases, as shaded in Figure 1.

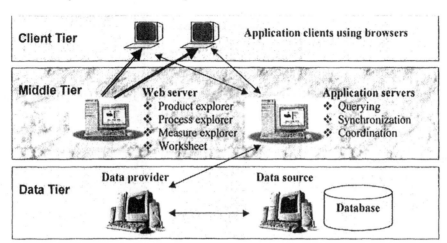

Figure 3 System architecture

3   Overview of the DFX Explorer

3.1   System Architecture

Figure 3 shows the architecture of the prototype system, which follows a typical three-tiered architecture:

- *Client tier* consists of a variety of application clients. An *application client* is a program running at a local computer within the open-standard web browser such as Internet Explorer. The client is not actually a part of the web-based system. It becomes one part of the system only when it visits the web server and gets data from the database.
- *Middle tier* includes the web server and application servers. *Web server* is a Windows NT Server computer running the web server software such as Internet Information Server (IIS). The operation interfaces of the DFX platform is provided in this tier in the format of web pages within which ActiveX components are embedded. *Application servers* are at

the center of the system. They provide the remote objects for application clients and thus control the information communication between various tiers. The objects provide three main facilities for the system: querying, synchronization and coordinating. Querying facility retrieves and updates the data in the remote database. Synchronization facility updates information synchronously without interruption. Coordinating facility treats the team decision making and the conflict resolution. The application clients and servers are deployed using DCOM (Distributed Component Object Model) standard.

- *Data tier* contains two parts: data source and data provider. *Data source (Database server)* provides the data for the system. It is a remote computer running the relational database management system such as SQL server. *Data provider*, which may be another remote computer, acts as the information broker between the data source and the application client using the Open Database Connectivity (ODBC) and ActiveX Data Objects (ADO).

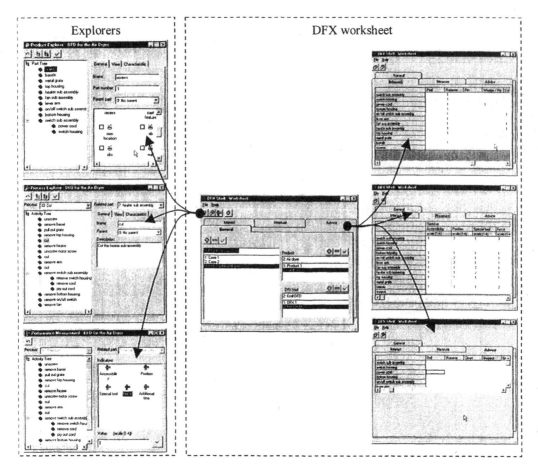

Figure 4 Main components of the system

3.2   Main Components

Figure 4 shows main components of the system. These four facilities reflect the front-end part, i.e. the application clients in the form of web pages provided by the web server. They share the same back-end application server through DCOM to be discussed in detail in the subsequent section. The four components are:

- *Product Explorer* helps the user to capture the product design data. Its left-hand side is a tree-view control that contains the product element/part information in the format of a hierarchical structure. The right-hand side is a multiple tab page control. Three tab pages are contained in it. The General tab page offers the user with some simple verbal description about current part in the left-hand side product tree. The below list-view in this tab page lists all product features. Those related to current selected part are checked out. The View tab page provides the

characteristic descriptions of current select part, while the facility for setting up the characteristic system is provided in the Characteristic tab page.

- *Process Explorer* helps the user to capture the process-related data. Main part of the interface is similar with that of the product explorer excepting that they are related to the process data. The left-hand side is the process tree. Here, a process means an operation for processing certain part of the product. Two combo-boxes in the topside show the process type/category and the corresponding part processed. The right-hand side contains three tab pages presenting the detailed information for the selected item. It can be concluded that, to some extent, this facility can be seen as a simple editor for process planning.

- *Measure Explorer* helps the user to carry out performance measurement of interactions between product elements and process activities. Its left-hand side of this facility is also a process tree although the data in it cannot be changed. The right-hand side is a list-view control reporting all measures in various indicators. The area below the list-view provides a facility to change the value of the measurement result.

- *Worksheet Explorer* provides a facility to display all data entered in previous three explorers in suitable form. Main element in it is a multi tab page control. Four tab pages are contained in it. The first tab page is just a simple project management facility. Every project is in relation to a DFX case study, which is a composition of one DFX tool and one product. Other three tab pages are in relation to main functions of a DFX tool. Each one containing a matrix stands for one certain viewpoint of the interaction measurement information between product and process.

4   Development of the DFX Platform

The four components have been first developed as ActiveX components and then attached to and/or embedded in HTML web pages. This section discusses the details of the four components in terms of their underlying methodologies.

4.1   Product Explorer

The product explorer is concerned with *P* (Product) in the PARIX model. Basically it models the product design decisions. This facility can help the user to collect, clarify and represent the product information, whether at the stages of concept design, detailed design production stage, or at the post-manufacturing stages. In the DFX platform, a pragmatic and easy-to-use product data model is employed. An exploded part list is usually given, as well as brief descriptions of individual parts. The objective of this kind of approach is just to minimize the amount of data inquiry work and prevent the information explosion in the DFX tools.

- *Hierarchical product analysis.* A product is often a complex assembly of a number of low-level components: subassemblies and elementary single-piece parts. Additionally, products are assorted into families, each of which consists of a number of similar products (Krause *et al.*, 1993). Owing to the multi-level nature, a product bill of materials (BOM) is often shown in the hierarchical format known as "indented explosion". Information in this kind of BOM includes: (1) part ID is a number that uniquely identifies a component; (2) a brief part description; (3) part number shows the quantity of part that is required to produce the assembly; and (4) parent part ID shows the relationship with other parts.

- *Characteristic-based part description.* Each component in a BOM, that may be an end product assembly, an intermediate subassembly, a purchased item, an elementary part, or a low-level feature, is characterized by a set of attributes and parameters. The key characteristics are incorporated with the BOM to offer the product elements in the BOM with the attributes and parameters. In general, key characteristics can be divided into several categories: geometry characteristics (shape, size, etc.), physical characteristics (weight, density, etc.), technological characteristics (tolerances, limits and fits, etc.), material properties (hardness, flexibility, etc.), and so on. Different DFX tools may require different sets of characteristics. For example, characteristics considered in Design for Assembly may include product structure, component forms and shapes, limits and fits, component orientations, component symmetry, weight and size, component rigidity, etc.

- *Features in part description.* A feature can be viewed as information set that refers to a variety of form and other attributes for the part, thus this set of information can be helpful for downstream business processes within the part life. The feature-based technology provides a fine approach for implementing DFX analysis by incorporating downstream considerations into the feature definition during the product design process. In our DFX project, these considerations are represented in a variety of DFX guidelines. These guidelines are manipulated and managed in the guideline system. A DFX guideline in the database is stored in patterns of feature and process. Thus the DFX platform can retrieve guidelines from the remote database by the types of feature and/or process.

4.2   Process Explorer

The process explorer is concerned with *I* in the PARIX model, which is used to model the interactions between product and process. In the DFX platform, this kind of interaction is represented by an activity, which is a composition of process and part. That is, an activity, or an operation, first belongs to certain type of business process, including the product development, purchasing, manufacturing and assembly, etc. Second, an activity is often related to a product element. In most cases, the product element is a part of the product.

- *Hierarchical process analysis.* In accordance with the hierarchical structure of the product, the process activities are also arranged into multiple levels. Take a normal gear reducer as example. The total assembly is always separated into a few levels. The first level may be the assembly of the whole gearbox, that is, assembling the spindles into the box. The second level may be the assembly of each spindle and gears along it. The process plan can be generated based on this approach, level by level, and end with the basic process activities. Types of the business of process that the activity belongs to are defined in MetaDFX according to the DFX tool. For example, the Kroll DFD employs about 10 elementary operations. Within the Hitachi AEM, assembly operations are categorized into about 20 elemental tasks. Lucas DFA tool also breaks assembly operations into a few fundamental tasks such as feeding, handling, fitting, gripping, etc. In conclusion, general description information for each process activity includes: (1) activity ID uniquely identifies the item; (2) a brief activity description; (3) part ID shows the product element subjected; (4) process ID shows the process type that the activity belongs to; and (5) parent ID shows the relationship with other activities.

- *Characteristic-based activity description.* Each activity, that may be an operation in a higher level or a lower level, can also be characterized in various attributes and parameters. For example, feed rate, cutting speed, cutting depth, number of cutting, and length of feed can be associated with machining (metal cutting) activities. This kind of activity-specific information is provided as the key characteristics of the process activity as the methodology in the product modelling.

4.3   Measure Explorer

The measure explorer studies X in the PARIX model. This facility can help the user to carry out the performance measurement of process activities. In the DFX platform, an activity is measured by a few (commonly 2-5) indicators, according to the process type it belongs to. This kind of process-oriented performance measurement has been defined in MetaDFX and saved in the form of DFX template, which is a triple composition of (DFX tool, process, and indicator). For each process of a DFX tool, various indicators can be selected during the DFX analysis, thus to measure the activities. The indicators may include process time, operation difficulty, environment consideration, service, safety, etc.

4.4   Worksheet Explorer

The DFX worksheet Explorer is used to manage the DFX analysis cases and presents all data provided in previous explorers in a suitable format. In the developed

system, the format is a matrix. One axis of the matrix represents product elements and the other process types. All analysis results are all indicated in corresponding cells, which can be categorized in three groups: *Interact*, *Measure* and *Advice*.

- *Interact* indicates the number of processing for certain product element. For example, if totally two activities with a common process type are applied on a product element, the cell related to the product element and the process type will be assigned a value of 2.
- *Measure* provides measurement information of the interaction with various performance indicators. The value is summarized of all interaction measures between the process indicator and the product element. It will be compared with the benchmarking value that is provided in the DFX tool. Those fail to satisfy the benchmarking condition will be highlighted with red color.
- *Advice* information is retrieved from the DFX guideline database. The guidelines associated with current process and features are retrieved from the database. The related advice information will be displayed below the matrix.

5   Brokering Collaboration

The application server is at the center of coordinating activities of multiple users who may use the different explorers or the same explorer at the same time.

5.1   Synchronized Web Application

The DFX Explorer has been developed as a synchronized web application. By synchronization it is meant that all the users of the same project share the same workspace, whether or not they are logged on to the system at a particular time. Synchronization ensures various project members to get the up-to-date and consistent information.

- how to select actions for synchronization and
- how to carry out the synchronization procedure.

As far as the first issue is concerned, basically any action that changes the status of the workspace should be synchronized. A variety of synchronization procedures have been developed for various actions in the system. Take the product explorer as an example. The synchronization actions include "*add new part*", "*delete part*", "*update part description*", etc. Therefore, when a part is added to the analysis by one member, all the other members can share this new part in subsequent actions.

However, some complication may occur when one member is trying to delete a part from the workspace while another member is updating the part information in the same workspace. This should be prevented. The application server of the DFX explorer employs a simple locking mechanism. That is, a decision is locked if any member is currently working on it. All the other members are not permitted to change the status of this decision until the first party (member) releases the decision. Five steps are contained in the synchronization procedure. They are:

1. The member commits the action to the team center.
2. Team center checks the lock flag of related resource for the action. If the resource has been locked, the team will reject the action. Otherwise, the action will be accepted and the related flag will be set to the locked status.
3. If the action is rejected, what the member can do is to try the action again and again until the team accepts the application or just to give it up. If the action is accepted, the member continues to ask for carrying out the action.
4. The team center carries out the action and sends effects to all related online members.
5. As requested by the team center, the member releases the lock on the team center. After this, another action can be committed to affect the related resources.

As for the system implementation, a variable is provided in the application server. This variable presents the locking status (locked or unlocked) for every resource that will be affected by the synchronized actions. Here the resource is in relation to the query result of the database, for example, all parts information of certain product.

When the user enters the working team, two objects will be created on the client machine: server object and client object. The server object will help the user to request information from the application server, for example, get query result, ask for action execution, etc. The client object has a reference linked to the application server so that the server can pass effects to the client during the synchronization procedure.

5.2   Conflict Explorer

In a collaborative working environment, individual contributions made by project members become the team decisions after certain form of consensus. Individual contributions may be contradictory to each other. Team members may disagree with each other's decisions. When a variety of users are involved in the working team, conflicts are to some extent unavoidable among them. Some conflicts are resulted from the relationship between members. This kind of conflict is related to the personal

feeling among members, which is obviously beyond the scope of this paper. More conflicts occur when there are different or disparate ideas about which action should be performed and how it should be accomplished.

The DFX Explorer provides several facilities for conflict resolution. One facility is the conflict explorer that tracks and represents the potential conflicts between members. The conflict should be clearly defined so that the whole team can understand them. A action-oriented methodology is introduced in the DFX platform. In this methodology, the conflict's representation is just equal to another problem, namely, the description of the action for synchronization. The information concerned with this kind of representation mainly includes a simple name of the action, the member proposing the action, related date and time, some parameters to describe the action, as well as a list of members related to the action. Take the action of "add a child part" as example. The parameters include ID numbers of the new part, the related product, and the parent part, and so on.

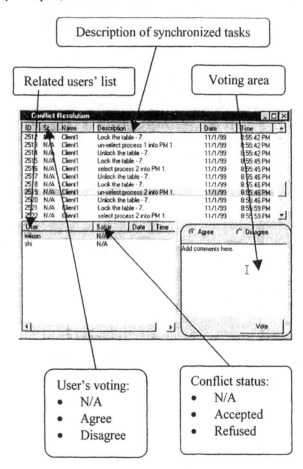

Figure 5 Conflict Explorer

When conflicts occur between project members, several strategies can be followed to resolve them. At present, three simple strategies are incorporated in the Resolution Explorer of the DFX Explorer. They are

- *Independent members.* Any idea from one member will be automatically considered as the team decision without considering if other clients agree or not.
- *Team leader.* The project manager or the team leader makes the finial decision.
- *Voting.* A voting group can be set up for each action. Each voter in the group will propose his or her idea about the action: agree or disagree. The final decision can be developed based on the ideas of all voters in the group.

In this prototype system, a simple facility for the conflict viewer is provided, which is just a simplified voting assistant, as shown in the Figure 5. The topside list view of the facility is to display information about conflicts or actions proposed by members. The conflict status reflects the team decision: "N/A" means the final decision is not achieved yet; "Accepted" means the team approves the action; and "Refused" means that the whole team considers it is not a good idea to carry out the action. The bottom side list view of the facility is to capture ideas from the voting group in relation to the conflict highlighted in the topside list view. The voting area helps the current member to commit his idea about the conflict action. The user can provide responses to the action carried out: agree or disagree, as well as some additional comments.

1, screws, 3;
2, barrels, 1;
3, metal grate, 1;
4, top housing, 1;
5, heater subassembly, 1;
6, motor screws, 2;
7, fan subassembly, 1;
8, lever arm, 1;
9, on/off switch subassembly, 1;
10, bottom housing, 1;
11, switch subassembly, 1 (12-13);
12, switch housing, 1;
13, power cord, 1.

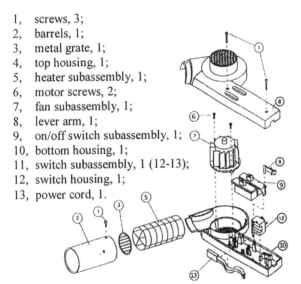

Figure 6 Exploded view of the hair dryer (Kroll, 1996)

6 Case Study

This section presents a case study to demonstrate a distributed and synchronized disassembly analysis procedure for a hair dryer shown in Figure 6. The purpose of the case study is to use the DFX Explorer discussed in previous sections in a more realistic environment to identify its potential strengths and/or limitations so that further improvements and developments can be made.

6.1 Background of the Case Study

This case study employs the Design for Disassembly (DFD) tool based on the methodology proposed by Kroll et al (1996). In our previous work (Shi *et al.*, 1999), the MetaDFX Explorer has been used to generate the web-based DFD tool. The resulting DFD tool includes eleven process types or categories. They are Pull/Push, Remove, Unscrew, Flip, Cut, Grip, Deform, Peel, Pry out, Drill, and the other kind. Each of them is related to and measured by several indicators such as accessibility, force, etc.

Client for DFX analyzer

Computer Configuration:
PII 300;
Memory: 96M
OS: Windows NT server

Web server,
Application server,
Database server

Computer Configuration:
PII 300;
Memory: 96M
OS: Windows NT server

Client for product designer

Computer Configuration:
PII 300;
Memory: 64M;
OS: Windows NT server

Client for process planner

Computer Configuration:
PI 166;
Memory: 32M;
OS: Windows NT workstation

Figure 6 Team members and their computer supports for the case study

6.2 Setup for the Case Study

The case study experiment has been configured as shown in Figure 7. The project team is assumed to include only three members. Product designer is responsible for using the product explorer to enter the description information of the hair dryer machine. Process planner is concerned with using the process explorer to provide the process-planning information for the product. The other member in the working team acts as both the team leader/manager and the DFX analyzer. The three members use three different computers connected to the Internet. In addition, another computer acts as the web server, application server and database server. Note these three servers can be physically deployed on different computers as long as they are connected to the Internet.

6.3 General Operations

All the three members must log on to the DFX Explorer by following the steps below:

1. The user uses the web browser and enters the URL (e.g. http://147.8.86.81/dfx/shell/docShell.vbd). Then, the application clients are downloaded and executed in the web browser.
2. The user provides the access information including IP addresses of the application server/client, user name and password.
3. The user clicks a button in the home page to trigger the system. The DFX worksheet is brought to the client, ready for using.
4. Different members will move to various facilities by clicking certain button in the toolbar of the DFX worksheet. The first three buttons can lead product designer, process planner and DFX analyzer to the product explorer, process explorer and measure explorer individually.

After the connection has been established, five main activities are involved. Firstly, the product designer enters product design data using the product explorer. Thirteen product elements or parts are contained in the hair dryer and will be inputted one after another. A simple textual name should be offered for each part, as well as the part number.

Secondly, the process planner carries out the process analysis. The seventeen process activities will be added using the process explorer. Activities of the same type on the same part will be considered in the same group in the Worksheet Explorer.

Thirdly, DFD analyst, together with the other two members, carries out the performance measurement. For every activity to disassemble the product, members provide measurement information in various indicators using the process explorer. Ideally, measurements can be obtained from the DFD guidelines through the Guideline Explorer.

The fourth step is to use the DFX Worksheet Explorer to display the analytical results centrally in a matrix format. Based on the information collected in the tab pages of interaction and measure, the team can get insights into the ease to disassemble the product.

The last step is concerned with the conflict resolutions. It should be pointed out that the operations about the conflict resolution could be carried out at any time when the member promotes an action. After the user clicks third button of the toolbar in the DFD worksheet, the system pops up the conflict viewer. The member provides the idea about the conflict. When a final decision is achieved through certain kind of mechanism, as discussed in the section 5, the team leader could promote a new action to fulfil the new decision or ask another member to do so.

6.4 Discussions on the Case Study

The case study has shown many good characteristics of the DFX platform. First, certain kind of DFX tool can be integrated and implemented in the web-based platform. With four facilities provided in the platform (product modelling, process modelling, interaction measurement and integrative worksheet), the platform can offer same fundamental functions as a standalone DFX tool. One advantage of the web-based version system is that there is no installation in the client machine. Second, the system introduces methodologies of synchronization and conflict resolution. Thus a team can be built up dynamically and team members can collaborate with each other to carry out the DFX analysis simultaneously. Third, although the band width problem might be encountered when the system is published in the Internet and accessed by remote client, the system gets high running speed in a university intranet environment. The application's downloading time is less than 20s. The time of opening a table is about 1s. Several synchronization actions are examined in execution time. The actions include adding a new part, deleting a part, and updating a part record. The results show that all the time is limited within 1-2s. It shows very good performance of the system as a web-based application. Final, using the ActiveX technology, high interactive interface and clear operation workflow are contained in the system. The user can become familiar with the system operation just after several experiments.

The case study also indicates that some modification or improvement work should be carried out. For example, in the voting facility for the conflict resolution, two choice of agree and disagree are often not enough for members to speak out their ideas. Other types of variables should be introduced and related aggregation algorithms for integrative decision could be developed. Take the process performance measurement value as an example. Various members could enter individual measures based on their own experience. The team might take the average value or something like this as the final decision.

In fact, the work on the synchronized DFX platform is still on going. Many aspects require further investigation. First, more case studies should be carried out to demonstrate and refine the prototype system. The areas of the case study may include design for recycling and disposal, design for testing and inspection, design for service and repairing, design for packaging and shipping. Second, when a variety of DFX tools are integrated into the system, how to find and select most appreciate DFX tool is another important issue that should be considered in the future. To solve the problem, a DFX tool repository model would be built up and methodologies of evaluation for both functionality and operability of DFX tools would be developed.

7 Conclusion

This paper has presented a synchronized web-based prototype system, the DFX Explorer, that can be used as a uniform platform for a variety of DFX analysis. The web-based approach offers the support for collaborative teamwork at different locations. The synchronization overcomes the limitation of ordinary web applications where decision consistence between team members is impossible to maintain. However, the design, implementation and operation of the synchronized web applications are much more complicated. Further research is needed for team-based decision-making including conflict resolution so that the strategies can be formalized and incorporated into the applications.

Acknowledgements

The authors are grateful to the Hong Kong SAR Research Grant Council (RGC) and the Committee on Research and Conference Grants (CRCG) of the University of Hong Kong for the financial supports for this project.

References

[1]. Duffey, M.R. and Dixon, J.R., "Managing the product realization process: A model for aggregate cost and time-to-market evaluation," Concurrent Engineering: Research and Applications, Vol. 1, pp. 51-59, 1993.

[2]. Huang, G.Q. (Ed.), "Design for X: Concurrent Engineering Imperatives," Chapman & Hall, London, 1996.

[3]. Huang, G.Q. and Mak K.L., "Developing a Generic Design for X Shell," Journal of Engineering Design, Vol. 8, No. 3, pp. 251-260, 1997.

[4]. Huang, G.Q. and Mak K.L., "The DFX Shell: A Generic Framework for Applying "Design for X" (DFX) Tools," International Journal of Computer Integrated Manufacturing, Vol. 11, No. 6, pp. 475-484, 1998.

[5]. Huang, G.Q., Shi, J. and Mak K.L., "Web-Based Design for X Guidelines," Proceedings of the 15th International Conference in Computer-Aided Production Engineering, April 19-21, 1999, Durham, pp. 65-71.

[6]. Krause, F.L., Kimura, F., Kjellberg, T., Lu, S.C.-Y., van der Wolf, A.C.H., Ating, L., ElMaraghy, H.A., Eversheim, W., Iwata, K., Suh, N.P., Tipnis, V.A. and Weck, M., "Product modelling," CIRP Annals, Vol. 42, No. 2, pp. 695-706, 1993.

[7]. Kroll, E., Beardsley, B. and Parulian, A., "A methodology to evaluate ease of disassembly for product recycling," IIE Transactions, Vol. 28, pp. 837-845, 1996.

[8]. Shi, J.; Huang, G.Q.; Mak, K.L. and Lee, W.S., "MetaDFX – A system for Developing Design for X Tools," Proceedings of the 4th International Conference on Industry Engineering Theory, Applications and Practice, November 17-20, 1999, San Antonio, Texas, USA.

A Product Model Enabling SMEs to Co-operate in a Distribute Engineering Environment

Franca Giannini, Marina Monti

Institute for the Applied Mathematics – CNR, via de Marini 6, 16149 Genova Italy

Tel. +39 010 64751   Fax +39 010 6475660   email: giannini(monti)@ima.ge.cnr.it

Domenico Biondi

Democenter, viale Virgilio 55, 41100 Modena Italy

Tel +39 059 848810   Fax +39 059 848630 email: d.biondi@democenter.it

Flavio Bonfatti, Paola Daniela Monari

Dept. of Engineering Sciences, University of Modena, via Campi 213/B, 41100 Modena Italy

Tel +39 059 376732   Fax +39 059 376799   email: bonfatti@unimo.it

Abstract

The paper illustrates a product model to support co-design activities developed within the Esprit Project EP25360 COWORK (COncurrent project development IT tools for small-medium enterprises netWORKs)

The idea of co-design between SMEs arises from the need of putting together the competencies that the participating nodes can offer in order to afford a design project that goes beyond the possibilities of the single node. Each partner develops a part of the whole project and exchanges with the other partners the only information of common interest, thus preserving its specific know-how.

The paper focuses the Product Manager module, specifically devoted to manage the knowledge that is strictly necessary to support the co-design activities of the single node within the network: hence, the Product Manager includes all and only those concepts pertaining to the description of the product whose design is in charge of the node.

1.   Introduction

Today SMEs operate in a market characterised by strong economic pressures and to short time-to-market and to maintain a high quality level since the beginning have became the main success factors

Generally, all manufacturing SMEs, and particularly enterprises working in the mechanical sector, need to frequently innovate themselves, both to create new products and to enhance the quality of the existing ones: the development of a new product can not only be reduced to find the appropriate configuration of standard components already existing on the market, but very often requires the development of new ones, possibly not only for the modification of the physical interfaces but also in functional terms. Since SMEs often do not have the complete knowledge for developing all the components needed for their products, they have to outsource the development of specific parts, starting in this way a co-operation with other enterprises working in the same industrial sector, in order to achieve a market objective that otherwise could be out of reach. Often the co-operation implies not a simple made-to-order development [1] but a real collaboration among the companies in the definition of the new product. It is then possible to say that these enterprises give rise to special type of Virtual Enterprise (VE) [2], in which each company maintains the greatest flexibility and business independence [3].

To be fast in the establishment of such VE is thus a key issue for being successful in the market, as well as to define the proper collaboration methodology: to achieve this aim it is fundamental to have proper IT tools supporting companies during all the phases of a VE's life cycle, starting from the identification of the most suitable partner, but also during the contract definition and the product specification.

Despite of the investment made in these last years, both in the research and the industrial field, currently the market still presents a lack of IT tools able to support SMEs in co-design activities in particular at a cost sustainable by small enterprises; there are some IT instruments whose aim is to support the organisation of engineering activities, partially targeted to shorten engineering time but their characteristics of use are absolutely incompatible with SME procedures.

In addition, most of PDM systems are mainly devoted to support process organisation and product data management internal to a specific company, possibly geographically distributed, but no real support for external negotiation is really offered.

This paper describes a software tool aimed at enabling SMEs, working in the mechanical sector, to co-operate in a distributed engineering environment.

The tool is the result of the Esprit Project 25360 COWORK (COncurrent project development IT tools for small medium enterprises netWORKs) [4]. In particular the focus will be on the Standardised Product Model and on the software module implementing the model itself, i.e. the Product Manager: it will be described how it handles the product specification and the negotiation on technical data.

The paper is organised as follows: in Section 2 the user requirements for a product manager gathered at the beginning of the COWORK project will be illustrated. Section 3 gives an overview of the COWORK methodology and system. Section 4 describes the developed Product Manager in details. Section 5 gives an overview of the process of validation of the developed tool. Finally, Section 6 ends the paper with the conclusions.

2. The user requirements framework

A good system analysis begins by capturing the requirements of an application [5] , therefore with the aim of developing a Product Manager specifically oriented to support users in their co-design activities, the first fundamental step has been to build a realistic user requirements framework; guided interviews have been proposed to users, i.e. SMEs from Italy, Spain and Germany, working in the mechanical sector: they had to explain the details about their current co-design activities, such as information exchange, constraints of that exchange, IT tools used in technical office and so on. Users were requested to specify the information on the basis of actual examples of the company, such as the current aspects of design activities, in particular for the paper topic, with regard to the management of the product data [6].

In a successive analysis phase, the end user requirements have been homogenised, and requisites for the different software modules have been established. Additionally, several discussions have been held between the end users, the intermediate users and innovators of each country to identify the most relevant needs concerning the design processes.

The user requirements collection has been carried out by national networks, due, among other reasons, to the ease of using the same language and the closeness (not only geographical closeness).

2.1 Requirements on Generic IT tools

First of all the framework of the IT tools has been analysed: the objective was to have a clear understanding about aspects such as the operating system in which the COWORK software would have to work, the kind of computer generated documents that COWORK would have to manage and other similar questions.

With regard to the hardware, it came out that the most part of the companies use Personal Computers (PCs) for design activities; the operating system used by the designers was mostly Windows NT, and moreover, all of them have expressed Windows NT as preferred operating system for the future.

One important aspect about the CAD programs used by the companies is that the end users have different types of CAD, and moreover they are often changing their CAD system. However, this heterogeneity is not so critical, because the end users have remarked that the information exchanged, concerning geometrical aspects, are usually 2D drawings in IGES or "dxf" formats. The users remarked that CAD systems are intended as a support tool to perform specific, often sophisticated, design activities: much knowledge on design aspects remains outside the CAD system, and it is very often on paper support.

2.2 The co-design scenario

The design programs currently used in the technical offices of the end users have also been analysed with the objective of knowing what kind of information they interchange during a design project.

As to the product codification, the enterprises use different modalities to codify their projects, purchased parts and parts that are co-designed. Thus, some enterprises apply a 'self explaining code' or a 'self meaning code' to codify the product resulting from the internal design activity, others use automatic means for producing incremental code number. From this information, it can be seen the necessity to communicate, receive and maintain codes related to co-designed parts and designs.

A further point regards information that is presently used to describe the product. The information about the product resulting from the internal design activity is often the following: CAD files (2D and 3D) and tables, paper 2D-drawings (generated by CAD but not only) and documents regarding technical information. As regards the paper drawings and sketches, information about their place and dimensional characteristics are kept. Always the R&D office (which produces the design) generates a bill-of-material and then communicates it to the commercial office. Further maintained information relative to a part could be its name, its amount in the assembly, short description, material, treatments and remarks. As regards the final product, the main characteristics are also described into the commercial catalogues.

The information about the purchased parts is often linked to the technical and functional characteristics reported in commercial catalogues. Sometimes, records about the purchased parts are stored; in those records the description of the part (type,dimensions,...) and information about the supplier are kept. The commercial and purchaser departments take track of the parts and their relative supplier as well.

The information about the parts co-designed by lower level co-designers are paper 2D drawings and sketches made by the enterprise itself as contractor and communicated to the lower level partners. These sketches are general designs in which specifications, tolerances and requirements are reported. Sometimes, files containing this information are exchanged and maintained. Further information regards the co-designer (information about the enterprise).

Some enterprises adopt their own conventions to handle the product data files but without efficient results. Generally, a proper computer-supported system to manage the product data is not used by any user.

An essential difference between a distributed design process and a centralised design process is that in the latter case the involved enterprises do not have a total visibility about the primary object of the design activity, that is, the detailed product model. They have visibility only about the information needed to each enterprise to execute at best its own co-design activities. This occurs because each enterprise feels proprietary of its own expertise, and therefore it does not intend to make its knowledge fully accessible to others. In fact, a typical situation is the one in which the same enterprise is able to take part in various projects of co-design, from time to time with different enterprise networks. It is unbelievable that each time an enterprise network arises, which generally will last only for a specific project, all the component designers transfer to the customer the total knowledge about its own part.

Since the co-design refers to a limited set of exchangeable information, it is important that such information is:

✓ Structured, and thus not ambiguous

✓ Exchanged in the right way and negotiable in time, until a shared version is reached

As the information visibility, we can focus on each co-design group. It is composed by an enterprise, its customer (that can be in turn a co-designer for an upper level customer) and the involved co-designers carrying out design activities assigned by the enterprise in discourse. The co-design group should manage

✓ the information of interest for the enterprise and for its customer

✓ the information of interest for the enterprise and each of its lower level partners

Considering the cases found at the user sites, some typical partnerships and network structures have been identified:

Customisation of an existing product : The first network has only two nodes: customer node and co-designer node. The co-designer has to make changes

in its old projects based on customer specifications. An example has been observed with one of the users: when a customisation is required, two parts are involved: the equipment that uses the crane jib (customer activity) and the crane jib itself, like shown in Figure 1.The relationship among the company and the customer is very simple and the times are not too long.

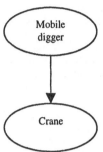

Figure 1 An example of product customization

Design of a new project : another case presents a contractor (Main Contractor), some co-designers, and several suppliers. This can occur when marketing analysis shows the advantage to design a new product family, see Figure 2. During the design, the co-designers and the contractor have a deep interaction.

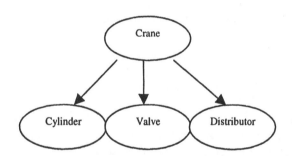

Figure 2 An example of design of a new product

Entrustment of the design and production of a product: sometimes, in a network it occurs that a co-designer entrusts the design and the production of a product to a sub-contractor. This can be due to lack of time or peculiarity aspects of the product that has to be produced. An example has been observed, with one company in the role of co-designer, which entrusts the production of a cylinder to sub-contractors, like illustrated in Figure 3.

churn. There is a trade-off, i.e., partial information exchange involving some level of acceptable change early in the process (churn) mitigates downstream risk through avoidance of design versions. Conversely, the higher the overlapping of activities and the lower the functional participation, the higher is the probability of design versions, but the lower the probability of churn. Increasing overlapping without any functional participation has the effect of increasing the probability of design versions only (Figure 9b).

The higher the degree of functional participation, the higher the probability of churn in the early phases, but the lower the probability of design versions in all phases (Figure 9a). Conversely, if there is little functional participation in upstream activities; then, there is a higher risk of design versions taking place. In the sequential process model, there is no churn when there is no functional participation, and there is a high probability of design versions in all phases except A. This is so, since it is assumed that once a new product idea is accepted, there is little probability of revisiting the preliminary activities involved in scoping out that idea. However, as functional participation increases with no overlap, churn increases to a limited degree, and design versions decrease (Figure 9c).

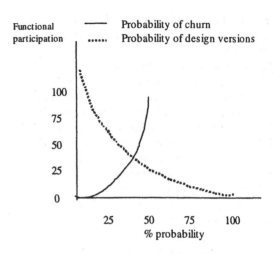

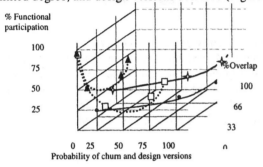

Legend:
—●— Churn curve for 25% functional participation
···□··· Design versions curve for 25% functional participation
—+— Churn curve for 50% functional participation
···▲··· Design versions curve for 50% functional participation

Figure 9a   A three-dimensional plot of the probabilities of churn and design versions versus functional participation and overlap

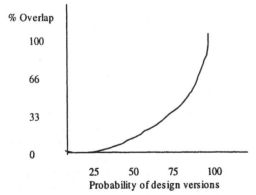

Figure 9b   Probability of design versions versus overlap

Figure 9c   Functional participation versus the probabilities of churn and design versions

3.2.6 Learning

Learning curves are used to capture the effect of knowledge accumulation during a given task which reduces effort during rework. Various learning curves are used as effort multipliers. A high effort multiplier is used for a sequential process, and smaller multipliers are used in accordance with increasing levels of functional participation. The higher the functional participation, the smaller the multiplier is, since it is assumed that more learning is taking place during activities through cross-functional interaction. As rework is initiated, the duration of activities is reduced and the probability of having to do further rework is also reduced. This occurs at each rework loop.

3.3.7 Actors

Actors are the resources that represent the functional roles required to perform the activities of a task. These resources process the production work and communicate and coordinate information. Resources are defined by skill type and allocable time. The model in this study assumes a limited resource capacity level, which realistically captures the effect of time delays due to resource constraints.

In the models, the activities are executed by the responsible resources, which are part of the supporting organizational structure. The various resource groups that participate in an NPD process have been grouped into three teams: the Business Unit, R&D and Operations. The Business Unit team consists of Marketing specialists. The R&D team consists of Design Engineers. The Operations team consists of Test Engineers and Manufacturing personnel. It is also assumed that an actor's skills match the task requirement; that is, each resource assigned to a task is qualified to perform that task. The costs for different resources are assumed to be equal.

3.4 Validation

The models were tested both internally and externally. Internal validation took place by running simulations and comparing results against predictions, while external validation was effected through a case study. Data for the case study was collected at the company site. This data included processes, subprocesses and their corresponding structures; activities, and their corresponding durations, complexity, and probabilities of rework; and resources which perform the tasks. This data was used in the model design and probability distributions for decisions. The results of the simulations were compared to actual CE-NPD process performance to validate the models.

4   Computer Modeling and Simulation

In this study, a process modeling and simulation approach was used; stochastic models of a generic NPD process, both sequential and CE, were built and simulated. A number of scenarios of NPD processes were simulated to study process parameters; the scenarios were: a strictly sequential process, a sequential process with varying levels of functional participation, a CE process with varying levels of overlap, and a CE process with varying levels of overlap combined with functional participation. The purpose of stochastic modeling was to simulate micro-level actions (functional participation, parallelism, etc.) as well as interactions among actors, and to measure the corresponding macro-level effects on sequential and CE processes through the performance indicators of project effort, project duration, coordination effort, and rework. Results showed the relationships between functional participation, and effort and span time (TTM), and between overlapping and effort and span time. For NPD, cost was considered equivalent to time since the major cost in a development process is worker wages.

The process modeler/simulator used was FirstSTEP™, a product of Interfacing Technologies Ltd. (Montreal). It is a simulation software which allows mapping of processes as a network of activities, dynamic simulation of the model, and quantitative analyses such as resource usage, bottlenecks, cost and duration of activities. FirstSTEP™ uses discrete-event simulation. FirstSTEP™ technology is based on object-oriented programming. Its simulation engine has the usual components: state variables, simulation clock, event list, counters, random-number generators, etc. The main program adopts the innate advantages of a process approach, i.e., it allows the user to create a model in terms of process building blocks, which are transform, transport, store/retrieve, and verify.. Probability distributions can be chosen for all activity durations in the model, including those of the activities and the decision points (which require durations as well, e.g., a decision may require a meeting to be held). These distributions are attributes of the activities in the process modeler.

The base CE-NPD model as described in this paper was composed of about 1000 activities.

5   Results

The variables of interest in the study are functional participation and overlapping of activities; their respective effects on effort and span time from simulations of the NPD models are shown in Figure 10 and 11. Specifically, the curves are the outcomes of increasing levels of functional participation and overlapping against effort and span time parameters.

In Figure 10, initially increasing functional participation from 0-25% slightly reduces overall effort, but further increases begin to show an upward trend. This is indicative of the fact that putting more people on a task will substantially increase churn. Note that span time significantly decreases as functional participation is increased from 0-25%, but that higher levels increase span time (Figure 11).

Increasing the level of overlapping has the effect of decreasing span time (Figure 11). This decrease in span time is accompanied by a decrease in effort with a 0-33% change in overlap, but a 33-66% change in overlap considerably increases effort (Figure 10). Thus, there is a trade-off between reduced span time and the level of effort.

As shown in Figure 12, there is a minimum point for effort versus span time for the amount of overlap in activities. As the amount of overlap starts to increase, there are efficiencies in reducing effort and span time; however, as overlap further increases in the process (33-66%), span time decreases, but effort increases because of the greater effect of rework due to design versions. The curve for functional participation in Figure 12 shows that for the modeled process there is a limit to the beneficial effects of greater teamwork. After a point (25%), further increases in functional participation lead to an increase in span time and a slight increase in effort.

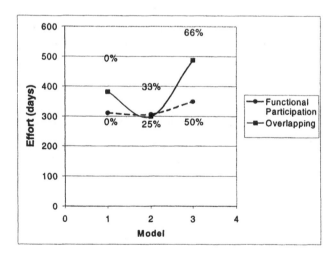

Figure 10 Effort versus different levels of functional participation and overlapping

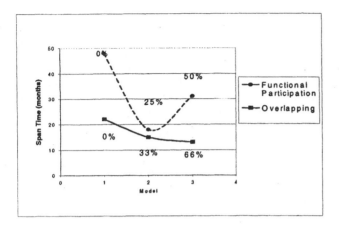

Figure 11   Span time versus different levels of functional participation and overlapping

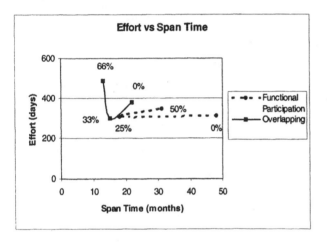

Figure 12   Effort versus span time for different levels of functional participation and overlap

6   Conclusions

The study showed that models of an NPD process can be built being composed of micro-models of concurrent engineering features to demonstrate macro-effects in the process.   Results have shown the tradeoff between reducing span time and the amount of effort spent during the process.   Knowing the precise shape of the effort versus span time curve for different variables is important since it would allow companies to design their development processes better.

With respect to managerial implications, the results of the study can be useful in terms of coordinating teams in a highly integrated process (identifying appropriate levels of functional participation), scheduling activities in the process (identifying appropriate levels of overlapping), evaluating alternative process structures (varying levels of functional participation, overlapping, learning, and risk), and estimating the completion time and cost of a project (obtained as results of the simulations).

Future research will be focused on understanding the interaction of the various parameters in the CE-NPD process in order to be able to optimize processes for span time and level of effort.   In addition, sensitivity analysis will be done to understand the importance of team learning and rework probabilities on reducing process effort and span time.   Finally, work needs to be done on how the completeness or certainty of information impacts the design of the NPD process to deliver shorter span times and reduced levels of effort.

References

[1] Haddad, C. J., "Operationalizing the Concept of Concurrent Engineering: A Case Study from the U.S. Auto Industry," <u>IEEE Transactions on Engineering Management</u>, Vol. 43, No. 2, May 1996, pp. 124-132.

[2] Handfield, R. B., "Effects of Concurrent Engineering on Make-to-Order," <u>IEEE Transactions on Engineering Management</u>, Vol. 41, No. 4, November 1994, pp. 384-393.

[3] Hauptman, O. and Hirji, K. K., "Influence of Process Concurrency on Project Outcomes in Product Development: An Empirical Study of Cross-Functional Team," <u>IEEE Transactions on Engineering Management</u>, Vol. 43, No. 2, May 1996, pp. 153-163.

[4] Prasad, B., <u>Concurrent Engineering Fundamentals-Integrated Product Development, Volume II</u>, Prentice Hall PTR, New Jersey, 1997.

[5] Shina, S. G., "New Rules for World-Class Companies," <u>IEEE Spectrum: A Special Report on Concurrent Engineering</u>, July 1991, p. 23.

[6] Krishnan, V., "A Model-Based Framework to Overlap Product Development Activities,"   November 1993, p. 3.

[7] Eppinger, S. D. and McCord, K. R., "Managing the Integration Problem in Concurrent Engineering," August 1993,

[8] Yassine, A. *et al*, "A Decision Analytic Framework for Evaluating Concurrent Engineering," *Working Paper, Wayne State University.*

[9] AitSalia, F. *et al*, "Is Concurrent Engineering Always a Sensible Proposition?", *IEEE Transactions on Engineering Management*, Vol. 42, No. 2, May 1995, pp. 166-170.

[10] Tian, H., Xu, W., Wend, H., Wu, Q., "A Review Upon Concurrent Engineering," *The 9th IFAC Symposium on Information Control and Manufacturing*, France, June 1998, Vol. II, p. 511-516.

[11] Vernadat, F. and Mhamedi, A., "The ACNOS Approach for Performance Evaluation of Enterprise Processes and Activities", IFAC, 1998, p.264.

A Generic Computer Support for Concurrent Design

Jacques Lonchamp
LORIA, BP 254, 54500 Vandœuvre-lès-Nancy, France
Jacques.Lonchamp@loria.fr

Abstract

Concurrent Design (CD) involves collaboration, coordination, and information-based co-decision making within a potentially distributed multifunctional team. This paper shows that a generic process-centered environment kernel, based on fine grain and decision-oriented task modeling, using customizable product models, providing capabilities for task model refinement at run time, and true collaboration support, is a good candidate for building dedicated computer aided CD environments. DOTS ('Decision-Oriented Task Support'), a Java prototype of such a generic environment kernel, is described in this paper and its usage in the CD application domain is discussed.

1 Introduction

The research described in this paper deals with computer support for *Concurrent Design* (CD), i.e., for the early phases of the Concurrent Engineering process. During CD, multifunctional teams, possibly distributed in time and space, work together for designing some product. In such a setting an efficient support for *collaboration, coordination, and information-based co-decision making* is needed.

Collaboration can be defined as a group process in which the group has common goals and produces one unanimous result (i.e., contributions are no longer attributes to group members, and the whole group takes responsibility for the result). Information sharing is the basic prerequisite for collaborative work: it implies common data models, shared data, and controlled access to them [1]. But data sharing is not sufficient for establishing and maintaining a true 'shared understanding' among the participants. Knowledge integration is also required, for instance through collective idea generation and discussion. In fact, a 'common information space' is negotiated by the actors involved [2].

Coordination is concerned both with the synchronization of activities (sometimes called 'activity-level coordination' [3]) and the synchronization of concurrent access to shared objects (called 'object-level coordination' [3]). Design processes are complex and intellectually demanding, and cannot be completely captured in a fixed process definition beforehand. To achieve flexible activity-level coordination, facilities are needed to support design process modeling, model execution, and (possibly collaborative) model refinement at run-time.

Co-decision making is central to CD [1]. It implies first allocation and sharing of responsibilities among the participants, and secondly, flexible support for various co-decision making processes.

The objective of this paper is to show that a generic process-centered kernel, based on *fine grain and decision-oriented task modeling*, using customizable product models, providing capabilities for task model refinement at run time, and true collaboration support, is a good candidate for building dedicated computer aided CD environments, because it takes in account explicitly the three aspects above-mentioned. The paper describes DOTS ('Decision-Oriented Task Support'), a Java prototype of such a generic kernel, and discusses its usage in the CD application domain.

We begin the paper by defining the general objectives of DOTS project. We then describe, in section 3, its conceptual meta model, with a particular emphasis on the argumentative reasoning aspect. In section 4, we discuss the architecture and usage of the current prototype, and we offer a small example of use in the CD domain. The paper closes by outlining further research directions.

2 The Objectives

This section summarizes the main objectives and requirements of DOTS project. They were elaborated with different application domains in mind.

(1) The system should support a small group of people (from two to less than ten) participating in a collaborative task, mainly distributed in time (asynchronous) and space; 'occasionally synchronous' work should also be considered. The coordination with other individual or collaborative tasks, i.e., the classical workflow aspect, is left outside DOTS prototyping effort because two other projects in the same research team focus on different aspects of Java-based workflow support [4, 5].

(2) The system should support a range of tasks, through a generic infrastructure, parameterized by a task model; this model should be often the customization of a basic library model, with some aspects that could remain unresolved until the execution (dynamic task model refinement).

(3) The system should provide an efficient assistance in three domains: guidance (i.e., task performance assistance and task model refinement assistance), argumentation and decision assistance, and group awareness (both asynchronous and synchronous).

(4) The project should provide a complete system for initial model development (editing, compiling, and verifying), system deployment support (installing and instantiating), and model execution and dynamical refinement. In simple cases, it should be possible to generate a fully operational customized system from the task model and from the standard kernel. In more complex cases, the environment designer should have to customize the generic product and tool types, and rarely should have to work at the generated code level.

(5) Both the entire infrastructure (client, server, development tools) and the generated code should be Java code, mainly for taking advantage of Java platform portability property.

(6) The project should provide a library of generic task models for brainstorming, collective review/ inspection, collective confrontation/merging of conceptual descriptions, free argumentation (i.e., emulation of an argumentation groupware), etc.

3   The conceptual meta model

In DOTS, a task model is described according to four perspectives: activity-oriented, decision-oriented, product-oriented, and organization-oriented, as shown in Fig. 1. Of course these perspectives are not independent and there are many relationships between them.

The next four subsections describe these perspectives and the last subsection emphasizes the argumentative reasoning aspect.

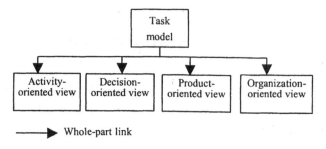

Figure 1 The overall conceptual organization of a task model.

3.1 The activity-oriented view

A collaborative task is structured into phases, in which the nature of the work (see section 3.2), the number and the identity of the participants can vary. During a phase (individual or collective) decisions are taken that modify the products under construction: operations are the elementary chunks of work, triggered by a decision. During a phase, participants can also freely access tools for performing activities not related to decisions (e.g. through query tools, server-side scripts, client-side external tools). The activity-oriented view of the task model mainly describes the phase types, the operation types, and the tool types (see Fig. 2).

When the task model is instantiated, a graph of phase instances is built, with phase precedence links. This instantiation can take place either statically (i.e., before execution) or dynamically (i.e., during execution).

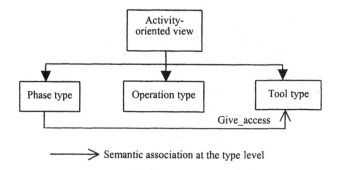

Figure 2 Main elements of the activity-oriented view.

Fig. 3(a) shows such an instantiated model describing how a design document is reviewed. First, the initial document is written down (here, we do not describe this task in details). Then, during the 'Review model refinement phase', a review model is chosen (individually or collectively) and dynamically instantiated. In the first refinement solution of Fig. 3(b), defects are first proposed individually and privately; then, during the 'Public defect evaluation phase', the proposed defects are collectively discussed and evaluated, i.e., accepted or rejected; finally, the document editor modifies the document in accordance with the review results. Then a new review can take place, whose model is one more time dynamically chosen and instantiated. A simpler review, without private phases can be sufficient at this stage as depicted by Fig 3(c).

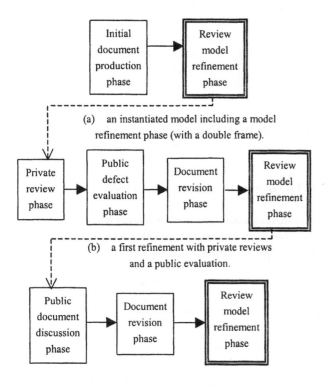

(a)   an instantiated model including a model refinement phase (with a double frame).

(b)   a first refinement with private reviews and a public evaluation.

(c) a second refinement with a public discussion.

Figure 3 A task model instance and two possible refinements.

3.2. The decision-oriented view

An issue is a problem that must be solved, generally concerning the products under construction. But the choice between different task refinements, as discussed in the previous section, is also an issue.

In most tasks, the different issue types are progressively taken into consideration. A phase is mainly defined by the subset of the task issue types taken in consideration at this stage.

Several option types specify how the issue type can be solved (e.g., AcceptDefectOption, and RejectDefect Option for Evaluate DefectIssue).

At the level of the task execution, i.e., at the level of the instances, users argue about the options of each issue instance. The decision takes (more or less) into account this argumentation in relation with the resolution mode of the issue type (see below).

Arguments are instances of a single Argument type and include a free textual rationale. Participants argue about the options and about the arguments themselves. They can also give qualitative preference constraints between the arguments (MoreImportantThan or >, LessImportantThan or <, EquallyImportantThan or =). Participants can also argue about the constraints: constraints as arguments are refutable. All the time, the system computes 'the best solution' in accordance with the current argumentation state (see section 3.5); but the actual decision is generally kept independent from the argumentation.

To each option type can be associated an operation type. The operation is triggered when the option is chosen. This operation can modify:

- a product or one of its components (e.g. add a defect to the list of proposed defects); this product evolution can in turn suggest new issues;
- the task content (e.g. dynamical creation of issues, options, phase instances, tools).

An option can also terminate a phase (see Fig. 4).

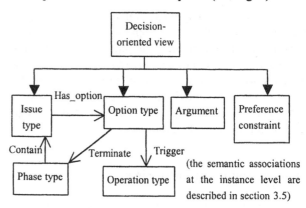

(the semantic associations at the instance level are described in section 3.5)

Figure 4 Main elements of the decision-oriented view.

The main characteristic of an issue type is its resolution mode:

- individual: the issue is solved by its creator;
- individualPrivate: similar to the previous mode but the issue and its consequences are only visible by the creator of the issue;

- collectiveDemocratic: the resolution is collective because the solution is necessarily 'the best solution' proposed by the system (see section 3.5) and at least two different users must take part in the argumentation;

- collectiveAutocraticWithoutJustification: the argumentation is collective but the choice is individual (autocratic); the choice is independent from the best solution proposed by the system and does not require any formal justification;

- collectiveAutocraticWithJustification: the choice is autocratic but requires a formal justification: an explanation step follows the argumentation in which only the decision-maker can argue in order to make the best solution equal to his/her own solution.

From a dynamical point of view, the life cycle of an issue is a sequence of interactions (expressed below as regular expressions):

- RaiseIssue: creates the issue instance (generally with parameters) and the corresponding option instances,

- (GiveArg | GiveConstr) $^+$: creates the argumentation tree,

- SolveIssue: solves the issue and triggers the operation associated to the chosen option (this operation generally makes use of the issue parameters).

There exist two important simplified cases. For an individual issue with a single option, only RaiseIssue is necessary (all the remaining is automatic): the issue is just an elementary action. For an individual issue with several options, only RaiseIssue and GiveArg are necessary: the issue is just an individual choice between several elementary actions, the argument can be understood as the rationale for the individual choice.

3.3. The product-oriented view

A product includes components at different levels of granularity. Products are currently specialized into textual product, list product, image product, and graph product (see Fig. 5). A parallel classification exists for tools (e.g. textual viewers, list viewers, image viewers, graph viewers). A minimum set of features is provided by these generic types (such as automatic graph layout methods); more specific features can be introduced through specialization or choice among predefined constraint verification rules, in the spirit of generic concept map editors such as [6]. And/or goal structures or design rationale descriptions are examples of specialized graphs

useful in CD task models, which can be provided by a customized kernel. Documents and tools can be instantiated either statically or dynamically.

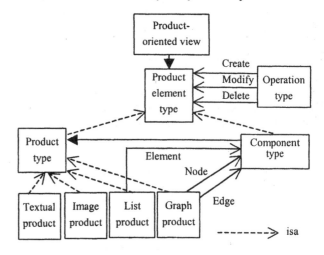

Figure 5 Main elements of the product-oriented view.

3.4. The organization-oriented view

Actors (currently restricted to human participants) play roles. Role types define what actors are allowed to do (see Fig. 6). Actors are instantiated statically or dynamically

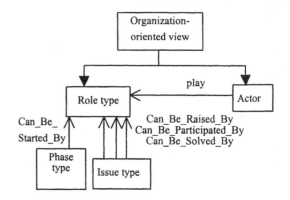

Figure 6 Main elements of the organization-oriented view.

3.5. The argumentative reasoning aspect

The system provides participants means of expressing their individual arguments and qualitative preferences, the aim being the selection of a certain solution. We discuss the evaluation procedure in two steps related to the absence (presence) of qualitative preference constraints.

Without preference constraints. The issue, the options, the arguments 'for' and 'against' the options, the arguments 'for' and 'against' the arguments form an argumentation tree. A score and a status (active, inactive) that derive from the score characterize each node of the tree. The score of a father node is the sum of the weights of its active child nodes that are 'for' their father minus the sum of the weights of its active child nodes that are 'against' their father. If the score is positive the node is active otherwise it is inactive. Only status propagates in the tree (because scores have no global meaning).

Without preference constraints, all nodes have the same weight (for instance 5, middle of the arbitrary interval 0-10 used in the next subsection, where 10 denotes the maximum importance). Leaves are always active. The preferred option (best solution of the issue - one or several) has the maximum score among all the options (see Fig. 7).

With preference constraints. Preference constraints are qualitative preferences between arguments of different options (global constraint) or between arguments of a same father argument (local constraint). One argument (source) is compared to the other (destination).

options is chosen (score strictly higher than the others) and if it is consistent with the other constraints (the evaluation is based on a path consistency algorithm). For instance, if arg1, arg2, and arg3 have the same father, and arg1 > arg2, arg2 > arg3, then arg3 > arg1 is inconsistent. This last constraint becomes (provisionally) inactive.

The weight of all arguments having the same issue as grand father (global constraint) or the same argument as father (local constraint) is computed by the following heuristics:

- all > relationships are defined by propagating them along the = relationships,
- for each argument arg involved in a constraint:
  . its max weight is computed, by subtracting 1 (starting from 10) for each argi such argi > arg;
  . its min weight is computed, by adding 1 (starting from 0) for each argument argj such arg > argj;
  . its final weight is computed as the average of its max and min weights.
  . the weight of an argument not involved in any constraint is kept to the average value (5);
- the rules of the previous item for computing the scores and the status are applied with these computed weights.

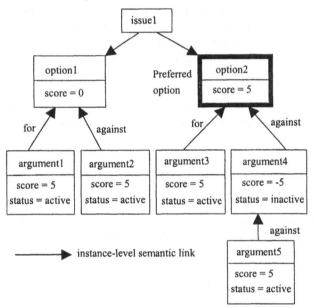

Figure 7 An argumentation tree without constraint.

Consistency is evaluated when the constraint is created and evaluated when another constraint becomes inactive.

To each constraint is associated a ConstraintIssue with three positions: MoreImportantThan (>), LessImportant Than (<), EquallyImportantThan (=). A constraint is active if both the arguments are active, if one of its

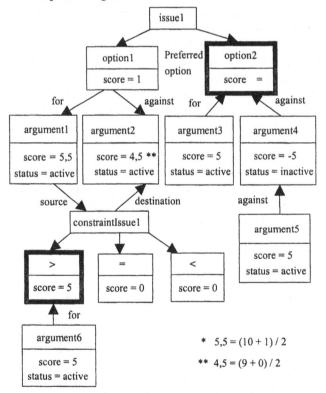

Figure 8 The argumentation tree of Figure 7 after introducing a constraint.

After each modification the whole tree is re-evaluated: for instance, inactivating an argument can re-activate a constraint that was inactive because it was inconsistent with the former constraint, which changes the status of an argument, which propagates on the upper level, and so on. As an illustration, in Fig. 8 a constraint is added to the argumentation tree depicted by Fig. 7. This argumentative reasoning technique is based on both Hermes and Zeno approaches [7,8].

4. DOTS PROTOTYPE

4.1. The system architecture

The system has a client/server architecture around an object database with a Java API. The database provides persistency, consistency, safety, and security. Communication and notification aspects are managed by a specific Java infrastructure. Persistency is conforming to the ODMG Java binding: persistent classes are declared statically and pre-processed before the Java compiler is called. This makes impossible dynamical schema evolutions. So, in the current prototype, we have chosen to generate (transparently) one separate database for each version of a task model. All these databases are accessed through a 'super base', and can be located on different machines. Each database contains the kernel, one task model version, and all the task instances conforming to this model. If a task model is changed, it is possible to run instances of these two different versions located in the two different databases.

The client is independent of the task model. It is written in Java and swing. The development and deployment environment includes three other tools, all written in Java and swing: a development tool (editor and compiler), an instantiation tool, and a static analyzer of instantiated models.

4.2. The main functionalities

The user enters the system with a registered user name (created with the instantiation tool and kept in the 'super base'), in one of the task instances in which he/she plays a role. The user can then act in accordance with the task model, the current task status, and his/her role. The user receives a threefold assistance: guidance (how to perform the task and how to refine the task model), argumentation and decision assistance, synchronous and asynchronous group awareness.

For task execution, the user can obtain the list of possible next interactions in accordance with the current task status and his/her role: issue types that can be raised, issue instances that can be participated in, and solved, phase instances that can be started, etc. Obviously only those possible interactions are accepted by the client. The user can also access to different textual and graphical views of the task model and of the task history (with colors highlighting for instance the active elements). Refining a task model is solving an issue that defines the different available solutions. As for each issue some static guidance is provided. Dedicated tools (e.g. query tools) can also provide dynamical information to make the choice easier.

At the argumentation level, the best option of each open issue is shown in color in the graphical view, as the active arguments and constraints; scores and weights can be displayed. The user can also list all open issues that are currently inconclusive (no option with a higher score than the others).

The main mechanism for asynchronous awareness shows what has evolved since the last connection of the same user in the same task (textual list, and specific color in all graphical views). For "occasionally synchronous" work, the user can obtain the list of all the connected users in the same task, can receive the notification of all constructive public actions from these other users in a notification window, and is warned when a document or a graphical representation becomes out of date (its background color changes).

Fig. 9 shows a client during the evaluation phase of a simple review, whose model is shown in window 1. Window 2 is the log window that contains the results of the interactions (here, a "what can I do? " request). Window 3 is the notification window: one can see that another user has logged in and has proposed a new defect. Window 4 is the NotYetEvaluatedDefect viewer tool and its content has become out of date after the creation of the new defect (the out of date marker is the dark background color). Window 5 shows graphically the current state of an issue; this description as the task description in window 1 is up to date (white background). Icons with a colored frame highlight active phases in window 1 and active nodes in window 5.

4.3. A CD example

We consider a multifunctional team of domain experts participating in a collaborative goal-directed acquisition task. The objective is to build collectively a goal-subgoal structure for a particular system (and/or graph).

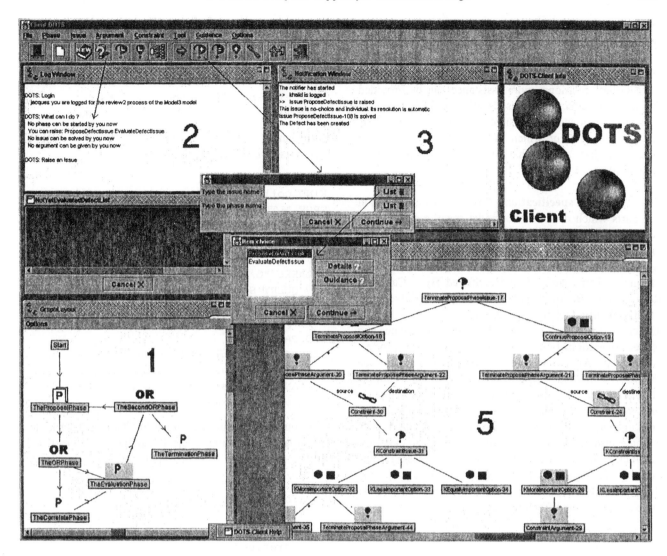

Figure 9 DOTS Client.

The task model organization. Classical strategies can apply: for instance, a private brainstorming phase for eliciting a maximum number of goals, followed by a public discussion phase for searching duplicate or irrelevant proposals, followed by an initial goal structure construction phase (for instance by the team leader), and terminated by an iterative review-revise cycle for improving the initial proposal.

First, DOTS provides the ability to manipulate graphical goal structures (graph management, graph layout, node expansion, etc.): the generic concept map library can be customized for this kind of graph, with customized presentation characteristics, and basic or specific properties verification.

Private brainstorming can be supported. In this mode, each participant cannot see the proposals of the others. Relaxed privacy is also possible (e.g. with a tool allowing to see a random choice of proposed goals), as public brainstorming. Proposing a new defect is just solving individually a ProposeGoalIssue, the argument being the rationale of the proposal.

In the public discussion phase, participants can raise issues for resolving duplicates, and for challenging irrelevant proposals. These issues are discussed, possibly with a very complex argumentation tree, and solved in accordance with some resolution mode (section III-B). Asynchronous and 'quasi synchronous' (i.e. through immediate notifications and out of date markers) working modes are available.

The initial construction and the iterative review-revise cycle are similar to the process discussed in section III-A.

By introducing in the goal structure notation objects which are processor for actions and the 'is responsible for' and 'wishes' relations, more assistance can be provided through goal reduction heuristics [9]. In DOTS, dedicated query tools can suggest possible goal reductions (such as the list all goals for which the responsibility is shared among several agents that are candidate for further reduction).

The task model specification. A task model is divided in two parts which describe the model specific entity types (specialization of phase, issue, option, role, document, tool, component, operation types) and the relationship types between them (Contain, Give_Access, Has_Option, Trigger, Terminate, Can_Be_Started_By, Can_Be_Raised, Can_Be_Participated_By, Can_Be_Solved_By, Create, Modify, Delete – see section 3).

The core part of each task model specifies issue types and related operation types that aim at changing product components. An issue type has a name, a resolution mode (KIND), a boolean saying if the issue is a simple alternative (in this case all the argumentation takes place on a single option, otherwise it is necessary to argue 'for' and 'against' all the options), two booleans saying if only a single instance or a single active instance can exist, the parameters of the issue (with the interaction messages and possibly OQL queries for generating list boxes), a textual description, and optionaly a static guidance on how to choose among the different option types:

```
<issue-type-name>
  KIND <mode>
  TRUE_ALTERNATIVE <boolean>
  UNIQUE_INSTANCE <boolean>
  UNIQUE_ACTIVE_INSTANCE <boolean>
  PARAMETERS
   (LABEL<message>
     [QUERY <OQL-query>]";" )*
  END_PARAMETERS
  DESCRIPTION <text>
  [GUIDANCE <text>]
```

The document types and the tool types are classified by content: text, image, list (giving the component type name), and graph (giving the node type name, and the edge type name). The component types can have attributes:

```
ATTRIBUTE_TYPES
  (<type-name> <attribute-name>
    ["=" <constant>] ";" )*
END_ATTRIBUTE_TYPES
```

The operation types are described through a list of elementary action specifications (of type CREATE, MODIFY, DELETE, EXECUTE - a script -, MAILTO):

```
ACTION
  (<action-specification> ";" )*
END_ACTION
```

For instance, the ChallengeGoalIssue type of the public discussion phase is specified by:

```
ChallengeGoalIssue
 KIND collectiveDemocratic
 TRUE_ALTERNATIVE true
 UNIQUE_INSTANCE false
 UNIQUE_ACTIVE_INSTANCE false
 PARAMETERS
   LABEL "Give the goal identifier: "
     QUERY SELECT * FROM Model.GoalExtent
       WHERE status = "proposed";
   // parameter 1 with a list box
 END_PARAMETERS
 DESCRIPTION "Collective evaluation of one
   of the individually proposed goal"
 GUIDANCE "Choose either to reject the
   goal or to keep it"
```

Two options types are associated to ChallengeGoalIssue: KeepGoalOption, and RejectGoalOption. RejectGoal Option can trigger an operation of type Invalidate GoalOperation:

```
InvalidateGoalOperation
  ACTION
    MODIFY Goal PARAM(1) WITH status =
    "refused";
    // PARAM(1) is a reference to the
    // parameter 1 of the associated issue
    // type
  END_ACTION
```

In more complex cases, the component can also be retrieved through an OQL query, possibly including references to the issue parameters. Moreover, several actions in the same operation can be related through local variables. The second example below shows the dynamical creation of a goal substructure, of the graph document associated to this component, and of its viewer tool:

```
CreateGoalSubstructure
  ACTION
    CREATE GoalStructure AS gstruct
      // local variable gsstruct
      WITH IName = PARAM(2)
      WITH enclosingGoalStructure =
```

```
      PARAM(1);
  CREATE GoalSubGraph AS gsgraph
    // local variable gsgraph
    WITH referent = gsstruct
    WITH referentAttribute = "enclosing"
    WITH componentIconPath =
      "Images/GoalGraph.GIF";
  CREATE GoalGraphViewer
    WITH TheGraph = gsgraph
    WITH IName = PARAM(3);
END_ACTION
```

The local variables gsstruct and gsgraph are useful for linking the three dynamically created components.

5. CONCLUSION

The generic infrastructure described in this paper aims at assisting participants of decision-oriented collaborative tasks. The approach is mainly based on fine-grain modeling of these tasks and the use of different assistance techniques: guidance, argumentative reasoning, group awareness.

To sum up, DOTS makes a synthesis of classical features of flexible generic process-centered environments, of argumentation and decision support systems, and of synchronous/asynchronous groupware systems.

Most of existing CD environments just act as a repository for and a controller to design artifacts (e.g. CASCADE [10], CoConut [11], Flecse [12], SHARE [13]). These systems do not provide activity-level coordination support (process support). On the opposite, Workflow Management Systems (WFMS) support predefined procedures and sometimes also ad-hoc processes, but do not have adequate support for synchronous or asynchronous collaborative and co-decision making activities. Therefore, some approaches aim at integrating more or less tightly WFMS and

collaboration tools (such as WoTel [14] or iDCSS [15] in the concurrent engineering domain). Only few systems truly integrate collaboration and coordination facilities. SCOPE [16] is the closest system from DOTS: it provides flexible support for specification, modification, monitoring, and execution of session-based collaborative processes. However, DOTS argumentation support has no counterpart in SCOPE.

From the concrete feasibility point of view, a previous mock-up system written in Smalltalk had already convinced us of the approach interest, in particular through a real size experiment [17]. The fundamental 'issue-argument-decision-operation' cycle seems easy to understand and use, even for inexperienced end users.

Our central claim is that building dedicated computer aided environments (in the CD application domain for instance) is made easier with DOTS. The main part of the work is to write the task model, possibly with model refinement alternatives. It is worth noting that most models will be constructed as a combination and customization of generic building blocks. Another part of the work is to tailor the generic product and tool types, such as OQL-based query tools for dynamic guidance. What is given for free are the client/server architecture, the client interface, the secure server storage, the process engine, the argumentation engine, the guidance and awareness capabilities.

In the next future, we plan to use DOTS for studying in depth and systematically several collaborative tasks that constitute the basic building blocks of many cooperative processes, such as concept graph co-authoring, and concept graph merging.

In a longer perspective we want to investigate other kinds of assistance, that could be plugged in DOTS kernel. For instance, the collaboration could take place not only between human participants, but could be assisted by software agents customized for participating to *issue instances production* and, possibly, to *issue instances evaluation and resolution.*

Acknowledgements

We would like to thank all members of the ECOO INRIA project for helpful discussions.

References

[1] R. Reddy, K. Srinivas, V. Jagannathan, R. Karinthi, "Computer Support for Concurrent Engineering", IEEE Computer, Vol 27, pp 12-16, 1993.

[2] K. Schmidt, L. Bannon, "Taking CSCW Seriously ", CSCW Int. Journal, vol 1, 1/2, Kluwer Academic Publisher, pp 7-40, 1992.

[3] C. Ellis, J. Wainer, "A Conceptual model of Groupware", in Proceedings of ACM CSCW'94, pp. 79-88, 1994.

[4] K. Benali, M. Munier, C. Godart, "Cooperation models in co-design", International Journal of Agile Manufacturing (IJAM), 2, 2, 1999.

[5] G. Canals, P. Molli, C. Godart, "Tuamotu: support for telecooperative engineering applications with replicated versions", IGROUP Workshop, WorkingPaper B-56, Oulu University Press, www.idi.ntru.no/~igroup/proceedings/canals.doc, 1998.

[6] R. Kremer, "Constraint Graphs: a concept map meta language", PhD Thesis, University Of Calgary, www. cpsc.ucalgary.ca/~kremer/dissertation/index.html, 1997.

[7] N. Karacapilidis, D. Papadias, "A group decision and negotiation support system for argumentation based reasoning", in Learning and reasoning with complex representations, LNAI 1266, Springer-Verlag, 1997.

[8] N. Karacapilidis, D. Papadias, T. Gordon, "An argumentation based framework for defeasible and qualitative reasoning", in Advances in Artificial Intelligence, LNAI 1159, Springer Verlag, pp 1-10, 1996.

[9] A. Dardenne, S. Fickas, A. van Lamsweerde, "Goal-directed concept acquisition in requirements elicitation", in Proceedings of 6th Int. Workshop on Software Specification and Design (IWSSD), pp 14-21, 1991.

[10] C. Branki, "The acts of Cooperative Design", CERAs, 3, 3, pp 237-245, 1995.

[11] U. Jasnoch, H. Kress, K. Schroeder, M. Ungerer, "CoConut: Computer-Support for Concurrent Design using STEP", in Proceedings WetIce, 1994.

[12] P. Dewan, J. Riedl, "Toward Computer-Supported Concurrent Software Engineering", IEEE Computer, Vol 27, pp 17-27, 1993.

[13] G. Toye, M. Cutkosky, L. Leifer, J. Tenenbaum, J. Glicksman, "SHARE: A Methodology and Environment for Collaborative Product Development", in Proceedings of IEEE Infrastructure for Collaborative Enterprises, 1993.

[14] M. Weber, G. Partsch, S. Hoeck, G. Schneider, A. Scheller-Houy, J.Schweitzer, "Integrating Synchronous Multimedia Collaboration into Workflow Management", in Proceedings of GROUP'97, pp 281-290, 1997.

[15] M. Klein, "iDCSS: Integrating Workflow, Conflict and Rationale-based Concurrent Engineering Coordination Technologies", CERAs, 3, 1, 1995.

[16] Y. Miao, J. Haake, Supporting Concurrent Design by Integrating Information Sharing and Activity Synchronization",

[17] J. Lonchamp, B. Denis, "Fine-grained process modelling for collaborative work support: experiences with CPCE", Journal of Decision Systems, 7, Hermès, pp.263-282, 1998.

Negotiation in collaborative assessment of design solutions: an empirical study on a Concurrent Engineering process

Géraldine Martin
AEROSPATIALE MATRA AIRBUS
BTE/SM/CAO M0101/9 316 route de Bayonne 31060 Toulouse cedex 03, France
Françoise Détienne
Eiffel Group "Cognition and Cooperation in Design" INRIA
Domaine de Voluceau, Rocquencourt, BP 105, 78153 Le Chesnay, France
Elisabeth Lavigne
AEROSPATIALE MATRA AIRBUS
BTE/SM/CAO M0101/9 316 route de bayonne 31060 Toulouse cedex 03, France

Abstract

In Concurrent engineering, design solutions are not only produced by individuals specialized in a given field. Due to the team nature of the design activity, solutions are negotiated. Our objective is to analyse the argumentation processes leading to these negotiated solutions. These processes take place in the meetings which group together specialists with a co-design aim.

We conducted cognitive ergonomics research work during the definition phase of an aeronautical design project in which the participants work in Concurrent Engineering. We recorded, retranscribed and analysed 7 multi-speciality meetings. These meetings were organised, as needed, to assess the integration of the solutions of each speciality into a global solution.

We found that there are three main design proposal assessment modes which can be combined in these meetings: (a) analytical assessment mode, (b) comparative assessment mode (c) analogical assessment mode. Within these assessment modes, different types of arguments are used. Furthermore we found a typical temporal negotiation process.

1 Introduction

In the face of growing competition, firms are increasingly having to reconsider their organisation and their processes. The development of methodologies supporting collective (or team) work is one key to success: Concurrent Engineering (CE), assumed to be one solution to achieve greater efficiency in the collective design process. The CE model prescribes various phases of design together with their temporal organisation. It consists in a "systematic, integrated and simultaneous" way of developing products and associated processes, in particular manufacturing processes, e.g., production [2]. Based on this model, methods and tools aim at guiding the organisation of the design, both at individual and team level.

Some companies have implemented CE in the course of re-engineering of their design processes. To do so, they extensively deploy tools for design support and for technical data management. In one aeronautical company, this change is being supported by research in Cognitive ergonomics. It is in this framework that a field study has been performed.

In Concurrent engineering, design solutions are not only produced by individuals specialized in a given field. Due to the team nature of the design activity, solutions are negotiated [1]. Our objective is to analyse the argumentation processes leading to these negotiated solutions. These processes take place in the meetings which group together specialists with a co-design aim.

We conducted cognitive ergonomics research work during the definition phase of an aeronautical design project in which the participants work in Concurrent Engineering. 10 different specialities are involved. We recorded, retranscribed and analysed 7 multi-speciality meetings. These meetings were organised, as needed, to assess the integration of the solutions of each speciality into a global solution.

After a brief presentation of our theoretical framework and hypotheses, we present an empirical study aimed at analysing the negotiation in an industrial Concurrent Engineering context. Our approach is strongly oriented by cognitive ergonomics work on the notion of constraint, and linguistics work on argumentation.

2 Theoretical framework and hypotheses

For design problems, the solutions are not unique and correct but various, and more or less satisfactory according to the constraints that are considered. The designers assess the solutions they develop according to their own specific constraints, which reflect their own specific points of view, in relation with the specificity of the tasks they perform and their personal preferences ([6],[7]). Also assessment modes may vary and involve more or less explicit constraints.

Constraints are cognitive invariants which intervene during the design process. The notion of constraints has been understood from different angles (1) according to their origin - prescribed constraints, constructed constraints, deduced constraints, (2) according to their level of abstraction, and (3) according to their importance – validity constraints and preference constraints ([3][6][8]).

Up to now, the assessment of design solutions has been mostly studied in the individual design process. In design activities, the assessment intervenes (1) to appreciate the suitability of partial solutions to the usual state of resolution of the problem, and (2) to select one of the solutions envisaged ([2][3][4]). The finality of this assessment is to make the decision to change one of its components, or to pursue the design if the assessment is positive[5].

Previous studies on individual design [4] have shown that various kinds of assessment modes may be involved. Bonnardel distinguishes between the following three assessment modes:
(a)   **analytical assessment mode**, i.e., systematic assessment according to constraints,
(b)   **comparative assessment mode**, i.e., systematic comparison between alternative proposed solutions and,
(c)   **analogical assessment mode**, i.e., transfer of knowledge acquired on a previous solution (accepted or not) in order to assess the current solution.

In collective design, we make the assumption that similar assessment modes may be found. With respect to linguistic work on argumentation ([9][10]), we will consider that these modes involve, in a meeting situation, the use of various types of arguments.

In the collective assessment, different specialities are going to be present, and they are going to have to justify their design choice so they are going to produce arguments. The purpose of these arguments is to provide information to convince the other people of the pertinence and veracity of the information provided in order to tend towards a conclusion that pushes them towards accepting the proposal [9]. When everyone has a joint will to reach agreement, we shall talk about negotiation. Negotiation does not force a person to accept a solution, dialogue makes it possible to go towards one conclusion rather than another, i.e., for example, the conclusion can be a compromise between what each person wants.

Linguists distinguish different kinds of arguments , argument by comparison, argument by analogy, argument of authority.

argument by comparison
Argument by comparison compares several objects in order to assess them in relation to each other. Comparisons can be made by opposition, by classification and quantitative classification.

argument by analogy
These are arguments that highlight a precedent, i.e., they enable the present case to be compared to a typical case proposed as a model.

We consider that the comparative assessment mode and the analogical assessment mode may involve what linguists call argument founded on an example, argument by comparison or argument by analogy. Most of these arguments can take the status of argument of authority depending on factors which give a particularly strong weight to the argument.

argument of authority is an indisputable argument which is built on a quotation of statements, so it is in no way a proof, even if it is presented as such. In general, the proposer's argument is the fact that it has been expressed by a particular authorized person, on whom he/she relies, or behind whom he/she hides.

Furthermore, due to the collective nature of the assessment process, we expect to observe combined assessment modes: in this case, each participant in an assessment meeting may use one or several assessment modes in order to convince the other participants.

Our research questions are :
(1) whether such combined assessment modes occur in assessment meetings;
(2) if so, whether there is a typical temporal organization of these assessment modes (or temporal negotiation patterns).

3   Methodology

3.1   Context

We conducted a field study on a design project in an aeronautical company. The project actors followed a CE methodology. The goal of the project was to design a new aircraft. The total duration of this project was three years. The study focused on the definition phase for the design of the aircraft centre section. This phase involved nine various fields of expertise for a total of approximately 400 actors. These actors use computer aided design (CAD) and product data management (PDM) to support design.

All the specialities work on the same part of the aircraft but each person according to his technical competence. "Informal" inter-speciality meetings are organized, as needed, to assess the integration of the solutions of each speciality into a global solution.

This research work involved seven of these "informal" meetings representing a representative sample of the meetings observed in the integrated design group. These meetings last between 15 minutes and an hour. We recorded two types of meetings :

-   **meetings between Design Office (D.O) specialities.** We have a configuration of meetings between designers in different fields, in which a D.O field (structure speciality) presents the same problem to five other D.O. fields (system installation speciality). We thus have an invariant in the structure solution proposed.

-   **meetings between D.O. specialities** (structure speciality) **and specialities which traditionally intervene later in design** (production and/or maintenance).

3.2   Collection of data

We took part in 7 of these meetings as observers. On the basis of audio recordings and notes taken during the meeting, we retranscribed the full content of the meetings. We also conducted interviews afterwards with the various participants to validate the coding we had made of them and make explicit a certain amount of information that was implicit in the meetings.

3.3   Coding scheme

The protocols resulting from the retranscriptions were broken down according to the change of locuters. Each individual participant utterances corresponds to a "turn". Each turn was coded according to the following coding scheme and broken down again as required to code finer units.

Our coding scheme comprises two levels :
  - **a functional level** : it highlights the way in which collective design is performed. Each unit is coded by a mode (request/assertion) an action (e.g., assess) and an object (e.g., solution n). At this level, a turn can be broken down into finer units according to whether there is a change in mode, activity or object.
  - **an argumentative level** : the aim here is to bring out the structure of the speech on the basis of a dialogue situation.

We coded the proposals for solutions made and the different types of arguments used by the speakers during the meetings.

4   Results

We found that there are three main design proposal assessment modes which can be combined in these meetings: (a) analytical assessment mode, (b) comparative assessment mode (c) analogical assessment mode. Within these assessment modes, arguments presented to defend a proposal for solution may take the status of " argument of authority ". Furthermore, we found a typical temporal negotiation pattern.

4.1   A general model of the assessment process

For the seven meetings analysed, whatever the problem involved, a solution is proposed by a speciality M1. This solution is called the initial solution. M1 will give arguments to support it in order to convince the other speciality, M2 (or the other specialities when more than two specialities are present). This solution may be accepted immediately by M2 who is convinced of the pertinence of the solution. On the other hand, M2 could refuse it, which is the most frequent case. Then follows a negotiation between the two specialities in order to reach a consensus. However, sometimes the negotiation fails and M1 and M2 must then find a compromise. An alternative solution is then proposed by M1 or M2 which will in turn be assessed. Often, several alternative solutions are proposed before a negotiated solution is reached. Finally, it sometimes happens that the meeting does not enable a result to be achieved. Each speciality must then work again before another meeting is convened.

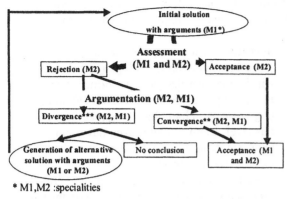

* M1,M2 :specialities

**Convergence : agreement ***Divergence : disagreement

Figure 1 : The assessment process

4.2   Combined assessment modes

The first type of result involves the way in which the proposals for solutions are assessed during these meetings. We have revealed the existence of analytical, comparative or analogical assessment modes in these meetings. This type of result is similar to the assessment modes analysed in individual design [4].

In addition, we have highlighted combined assessment modes, e.g. analytical/analogical. We present these modes and illustrated them graphically through examples.

4.2.1   analogical /analytical assessment

This mode combines analogical assessment and analytical assessment. In the framework of analogical reasoning, the current solution (the one which is proposed for evaluation) is called the target solution whereas the analogical solution (a previous solution which is brought up in the argumentation process) is called the source solution.

Figure 2 illustrates graphically such a combined assessment mode. In this example, specialists M1 use the analogical/analytical assessment to convince specialists M2 to accept the solution S1 proposed by M1.

Specialists M1 propose a solution, the target solution S1, which is rejected by specialists M2. In order to convince specialists M2 of the adequateness of S1, specialists M1 make reference to an analogical solution, the source solution S2. S2 is a solution which was accepted in a past context. In this context S2 was a solution negotiated between M1 and M3 : even if this solution was not so easy to use by specialists M3 (this solution was not ideal in terms of some constraints important for these specialists), they finally accepted it. In their argumentation, specialists M1 analyse the source solution S2

according to a set of constraints (analytical assessment). They make explicit positive arguments as well as negative arguments and defend the idea that the specialists M3 were able, in the past, to accept this evaluation and therefore the source solution S2. The conclusion of this negotiation process is the acceptance, by M2, of the target solution S1.

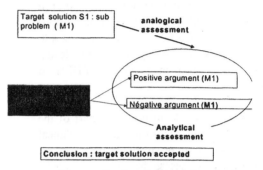

Figure 2 : Analogical /analytical assessment

4.2.2   comparative/analytical assessment

This mode combines comparative assessment and analytical assessment. The comparative assessment mode involves systematic comparison between the current solutions and one or several alternative proposed solutions. These solutions are alternative to the current proposed solution (the one originally to be assessed).

Figure 3 illustrates graphically such a combined assessment mode. In this example, each specialist will propose his own alternative solution. None of them accept the current proposed solution.

Specialists M1 propose an alternative solution Salt 1 whereas specialists M2 propose another alternative solution : Salt 2. Each alternative solution is then analytically analyzed by participants of both specialities. Specialists M1 positively assess Salt1 (their own proposed alternative solution) and negatively assess Salt2. Conversely, specialists M2 positively assess Salt2 (their own proposed alternative solution) and negatively assess Salt1. These analytical assessments allow each specialist to compare the suitability of the two alternative solutions according to various design constraints. In doing so, each speciality makes explicit the design constraints which are judged more important in his/her field. The conclusion of this negotiation process is that neither of the two proposed alternative solutions are accepted. Rather, a third alternative solution, which is a compromise between Salt1 and Salt2, is generated.

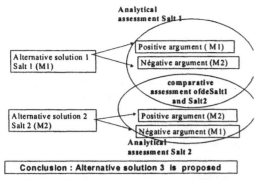

Figure 3 : Comparative/analytical assessment

4.2.3 Comparative/analogical assessment

This mode combines comparative assessment and analogical assessment. Figure 4 illustrates graphically such a combined assessment mode. In this example, specialists will propose an alternative solution (comparative assessment) and will defend this solution in reference to a previous source solution which was accepted in the past (analogical assessment).

Specialists M1 propose and defend the current solution S1. Specialists M2 propose an alternative solution Salt1. In order to defend this alternative solution, they make reference to a source solution, accepted in a past context, which is analogical to Salt1.

This source solution is then analogically assessed by the different specialists. This evaluation allows the specialists to compare the advantages (positive arguments) and drawbacks (negative arguments) of the current solution S1 and its alternative solution Salt1.

Specialists M1 give negative arguments toward Salt1 based on negative arguments toward the source solution ; this allows them to show, by comparison, the advantages of solution S1. Conversely, specialists M2 give positive arguments toward Salt1 based on positive arguments toward the source solution; this allows them to show, by comparison, the drawbacks of solution S1. The conclusion of this negotiation process is the absence of any negotiated solution or any consensus. In fact, due to the disagreement between the specialists on the source, a task is planned in order to verify information related to the source solution. The design rationale about the source solution has to be reconstructed for the next meeting.

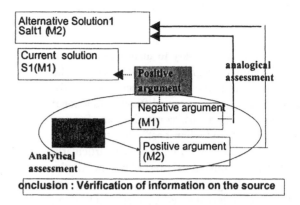

Figure 4 : Comparative/analogical assessment

4.3 Argumentation

Argumentation means provoking or increasing other people's support of the theories submitted to them for agreement. For argumentation to be effective, the designers, in the collective assessment use different types of arguments according to an order which seems to have become customary. Arguments used are of different nature. Due to the nature of the task, a design task, many arguments explicitly or implicitly make reference of design constraints. Furthermore, arguments can take the status of argument of authority depending on specific factors.

4.3.1 Use of Constraints

Arguments enabling a design solution proposal to be defended are often characterized by the use of constraints.

Constraints can be explicit or implicit in the argument as it is expressed by a speaker. The implicit or explicit nature can depend on the postulate of shared knowledge made by the speaker. Now this postulate is not always confirmed.

An argument often covers not only the explicit constraint but a hierarchical network of implicit constraints; this network can be broken down in order to convince the other participants.

We have observed that specialists firstly make explicit the constraints at the top of the hierarchy and, when it is necessary to bring other arguments to convince the other specialists, then the constraints lower in the hierarchy are made explicit.

We observed that the same constraint (the same terms are used by different speakers) can have different meanings according to the speakers, and more specifically, according to the speciality of the speaker.

In this case it is necessary to distinguish the two slopes of the sign, the signifier and the meaning. The meaning can have the same generic seme for different speakers but very different functional seme. For example, a cost constraint can, for one speciality, mean "production cost" and, for another speciality, mean "design cost". It seems particularly true for general constraints prescribed for all the actors of the design process (e.g., the cost) as opposed to constraints derived by a speciality (e.g., structure).

We also found that constraints can be weighted differently according to specialities. There is no absolute weighting except for certain constraints (1). The weighting is performed in a context according to the type of problem considered. One assumption is that the weighting is done not on the constraint as such but on its meanings. The context of the problem would make it possible to select a particular meaning.

4.3.2   Use of argument of authority

Any argument can take the status of authority depending on specific factors of the situation. This argument is presented as incontestable and therefore it has a particularly strong weight in the negotiation process.

We have found that an argument can take the status of argument of authority depending on :

- the status, recognised in the organisation, of the speciality that expresses it.

- the expertise of the proposer. The argument is going to make reference to a person recognized by all to be an expert in the speciality. It will be something like " It's Alphonse who said it would be more logical like that to pick up on these parts of the stringers".

- the "shared" nature of the knowledge to which it refers. This is typically the case in analogical assessment, when the participants in a meeting have shared knowledge about the source solution, e.g., everybody agrees that it works in this similar context. In some cases, we observed that participants do not share knowledge about the source (as in 4.2.3).

4.4   Temporal negotiation patterns

As explained before, we have found that combined assessment modes occur in assessment meetings. Another research question was whether there is a typical temporal organization of these assessment modes. We found that different assessment modes are used in the order shown in Figure 5:

- Step1: Analytical assessment mode of the current solution;
- Step 2: if step 1 has not led to a consensus, comparative or/and analogical assessment is involved;
- Step 3: if step 2 has not led to a consensus, one (or several) argument(s) of authority is(are) used.

Firstly the current solution is assessed. This is made using an analytical assessment mode. Arguments used by the two (or more) specialities may use more or less explicit design constraints. Specialists M1 use arguments to convince M2 and M2 does the same thing. Based on this analytical assessment, a consensus can be found and negotiation is finished.

If no consensus has been found, then either M1 or M2 (or more rarely both) use either an analogical assessment mode or a comparative assessment mode of the solution. The two types of assessment may also be combined. This can lead again to a consensus toward the initial solution or toward a proposed alternative solution.

If no consensus has been found, either M1 or M2 propose one or several arguments of authority. This generally leads to a consensus.

An example of a non converging negotiation process was illustrated in 4.2.3 (Figure 4). In this particular case, each specialist had different arguments related to the same source. The use of the source could have led to a consensus based on the shared knowledge concerning the source (argument of authority). In this particular case, this process was disrupted and no consensus could be found.

---

1 Each speciality has some specific strong constraints : for the structure specialists, for example, there are weight constraint and structure constraint.

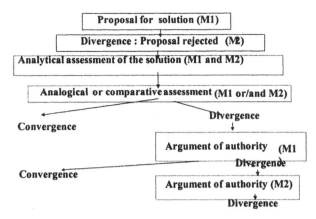

Figure 5 : The Argumentation process

5   Conclusion

To sum up, we have found that there are three main assessment modes which can be combined in design assessment meetings. Within these assessment modes, different types of arguments are used. Furthermore we found a typical temporal negotiation process.

Two courses of action are now being studied. The first is to improve design rationale traceability. Indeed, only a part of this design rationale is now absent from the minutes of meetings. The second is capitalization of the knowledge brought into play in the logical assessment and analogical /analytical assessment. This knowledge is associated with particular problems encountered in the past and procedural type general knowledge. This capitalization would be done for reutilization purposes.

Our long-term objective is to analyse and support the integration of points of view in multi-speciality design in order to improve the search for a compromise between designers in design reviews. Indeed, it is in assessment meetings that we can observe the confrontation of the points of view of the various participants in design. Owing to the collective nature of the activity, points of view are expressed, more or less explicitly, through argumentation (Plantin,1996).

6   References

[1] Béguin, P, (1997) « L'activité de travail : Facteur d'intégration durant le processus de Conception » In P.
[1] Bossard, C. Chanchevrier et P. Leclair (Eds) « Ingénierie concourante de la technique au social ». Economica. Paris

[2] Bonnardel N ; (1991) « L'évaluation de solutions dans la résolution de problèmes de conception et dans les systèmes experts critiques »In D ; Hein-Aime, R. Dieng, J.P.Regopuard (Eds). Knowledge modeling &Expertise Transfer. Frontiers in Artificial Intelligence

and Application(serie). Amsterdam Washington D.C, Tokyo I.O.S.Press.

[3] Bonnardel N. (1992) " Le rôle de l'évaluation dans les activités de conception " thèse de doctorat, Spécialité Psychologie Cognitive, Université Aix en Provence, France.

[4] Bonnardel N.(1999) « L'évaluation réflexive dans la dynamique de l'activité du concepteur » In J . Perrin (Ed.): pilotage et évaluation des processus de conception. L'Harmattan. 87-104

[5] Darses, F., (1994) « Gestion des contraintes dans la résolution de problèmes de Conception » thèse de doctorat, Spécialité Psychologie Cognitive, Université Paris 8, France.

[6] Eastman, C. M. (1969) Cognitive processes and ill-defined problems: a case study from design. In D.E. Walker and L. M. Norton (Eds): Proceedings of the First Joint International Conference on Artificial Intelligence. Bedford, MA: MITRE

[7] Falzon, P., Bisseret, A., Bonnardel, N., Darses, F., Détienne, F., & Visser, W. (1990) Les activités de conception: l'approche de l'ergonomie cognitive. Actes du Colloque Recherches sur le design. Incitations, implications, interactions, Compiègne, 17-19 octobre 1990.

[8] Janssen,P ;, Jégou,p, Nouguier, B ;, & Vilarem, M ;C ; (1989) Problèmes de conception : une approche basée sur la satisfaction de contraintes. Actes des neuvièmes journées Internationales : les systèmes experts et leur applications (PP 71-84), Avignon, France.

[9] Perelman C.; Olbrechts-Tyteca L. (1992) " Traité de l'argumentation" Ed de l'université de bruxelles

[10] Plantin C. (1996) " L'argumentation" Seuil

Realising Inter-enterprise Remotely Synchronous Collaborative Design and Planning within Multi-platform Environment

Z. Deng

Narvik Institute of Technology, 8505 Narvik, Norway

J. Pettersen, E. N. Jensen, B. Bang, and R. Davidrajuh

Narvik Institute of Technology, 8505 Narvik, Norway

Abstract

Enterprises are nowadays increasingly concerning with inter-enterprise collaboration. To obtain efficient inter-enterprise collaboration, one of the methodologies is to make use of Internet/Intranet communication and net-meeting technologies to support inter-enterprise remotely collaborative product design, production planning, manufacturing, supply-chain management, etc.

However, different enterprises often install different vendors' applications on different vendors' computing platforms. Thus, the collaborative net-meeting among enterprises has to be run in the multi-vendor and multi-platform environment.

For realising such kind of methodology, it is necessary to study: (1) how the inter-enterprise collaborative net-meeting can run in multi-platform environment, and (2) how the net-meeting can run much efficiently based on limited band-width capability of existing communication network. This paper has explored those two problems and the results are given.

1 Introduction

Manufacturing enterprises today are faced with challenges globally. These challenges cannot be effectively met by isolated effort within a single enterprise. Therefore, enterprises are nowadays increasingly concerning with *inter-enterprise collaboration*.

The architecture for inter-enterprise collaboration consists of a *communication layer*, a *collaboration layer*, and an *application layer* as shown in Figure 1 below[1].

The *communication layer* functions to enable multi-point inter-enterprise communication. Based on its support, the collaborative application tasks can be as follows (refer to upper part of Figure 1):

- Collaborative product design by means of computer aided design (CAD) *applications*,
- Collaborative process planning by means of computer aided process planning (CAPP) *applications*,
- Collaborative production planning and control (PP&C) by means of computer aided PP&C *applications*,
- Collaborative manufacturing by means of computer aided manufacturing (CAM) *applications*,
- Collaborative logistics management by means of computer aided logistics system (CALS) applications,
- Etc.

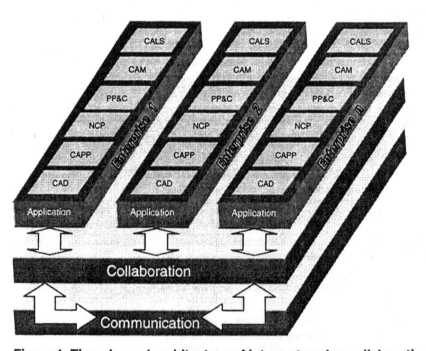

Figure 1 Three layers' architecture of inter-enterprise collaboration

To enable inter-enterprise collaborative tasks (design, planning, etc), we need not only the communication layer to provide multi-point communication service, but also the collaboration layer (shown in the middle of Figure 1) to provide multi-point collaboration tools to realise the collaboration within multi-vendor-product environment.

In practices, synchronous and asynchronous collaborations are encountered. *Synchronous collaboration* means that geographically dispersed partners are working in an *on-line* mode. They are working collaboratively at different locations but *at same time*. It is sometimes called a net-meeting or a net-conference. Thus, we *need tools to realise the synchronous multi-point collaboration.* By means of those tools, the partners can work together and exchange ideas via text chat or white board drawings and annotations on the screens, and share the applications (CAD, PP&C, etc) on the screens. Even more, they can make use of the multimedia means to see each other via video display on the screens, talk with each other via audio phones. *Asynchronous collaboration* means that the collaborative partners work in an *off-line* mode, i.e. the partners *do not work at the same time*[1]. This paper focuses only on the study of synchronous collaborations.

Often, different enterprises install different vendor's applications on different vendor's computing platforms such as PCs with Windows platform or with Macintosh OS platform; HP, SGI, SUN and DEC workstations with their various UNIX-like platforms. In reality, collaborative enterprises often work together in such a multi-vendor and multi-platform environment.

For the purpose of realising collaboration in the multi-platform environment, it is necessary to study: (1) how the inter-enterprise remotely synchronous collaboration (net-meeting) can be realised, and (2) how the net-meeting can run much efficiently based on limited band-width capability of existing communication network. Those problems will be studied in sections 2 and 3 of this paper respectively.

2 Realising Inter-enterprise Synchronous Collaborative Design and Planning in the Multi-platform Environment

2.1 Requirements for running an inter-enterprise net-meetings

Inter-enterprise collaboration possesses dynamically changeable feature. The individual enterprise in the collaboration is independent with each other. They are combined for a common product to market. Whenever market situation of the product has changed, the product must be changed, and the combination of enterprises should also be changed. Thus, it is not feasible to build up a dedicated communication network to permanently support a dynamically changeable inter-enterprise collaboration. Due to such kind of dynamic feature, reasonably, realising multi-point communication via the Internet/Intranet is the natural solution for the communication layer of the three layers' architecture (Figure 1).

As mentioned above, without the help of the collaboration layer of Figure 1, it is impossible to realise the multi-point inter-enterprise net-meetings within a multi-vendor and multi-platform environment. If an application is supported only multi-point communication service from the communication layer (Figure 1), users can only host a multi-point inter-enterprise conference only if every one in the net-meeting is using the exact same product. For example, if a user at enterprise *A* is running its CAD application on a HP workstation, and a collaborating user at distant enterprise *B*, who is running a PC, will be unable to see and participate in the design process at the enterprise *A* remotely, or say, unable to share the CAD application running at enterprise *A*. Thus, we need multi-vendor and multi-platform application sharing tool to realise the application sharing requirement.

In addition to the requirement of application sharing, furthermore in a net-meeting, naturally people want to exchange ideas by means of whiteboard drawings and annotations, to chat with each other by means of text, to exchange files by means of multi-point file transfer, to talk with each other by means of audio multimedia tool, to see each other by means of video multimedia tool. Thus, to run an inter-enterprise net-meeting, followings are basic requirements:

- Multi-point communication,
- Multi-point application sharing,
- Multi-point interoperable whiteboarding,
- Multi-point file transfer,
- Multi-point interoperable chat,
- Multi-point audio multimedia talking,
- Multi-point/multicast video multimedia seeing,

Where the first item, multi-point communication, is the task of communication layer of Figure 1; the other six items are the tasks of the collaboration layer of Figure 1.

2.2 Detailed architecture for inter-enterprise synchronous collaboration

From the discussion above, we can re-depict Figure 1 into a detailed one (Figure 2) where the multi-point communication service function is depicted in communication layer and the application sharing, whiteboarding, file transfer, chat, audio and video functions are depicted in collaboration layer.

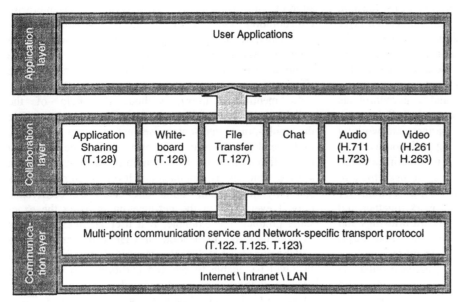

Figure 2   Detailed architecture of inter-enterprise collaboration

2.3   Necessity of standardisation and the T.120 and H.323 standards

As mentioned above, Inter-enterprise net-meeting involves in a multi-platform and multi-vendor environment. To realise such net-meetings, there are needs to standardise the protocols and services for the seven items, which are listed in section 2.1 and depicted in the communication layer and collaboration layer of Figure 2.

The International Telecommunication Union (ITU) has worked out the T.120 series standard in which its components of T.122, T.125 and T.123 define the multi-point communication service and the network-specific transport protocol (see lower part of Figure 2). As well, its components of T.128, T.126 and T.127 define the protocols and means for the application sharing, whiteboarding and file transfer respectively (see middle left of Figure 2)[2][3]. Also, the International Multimedia Teleconferencing Consortium (IMTC) has worked out the H.323 series standard in which its components H.261, H.263, H.711 and H.723 define the protocols for audio and video multimedia teleconferencing protocols (see middle right of Figure 2)[4].

If developers of net-meeting products, developed from multi-

vendors and supported by multi-platform, implement those standards into their products, then we can realise inter-enterprise remotely synchronous collaboration upon those standard-implemented multi-vendor net-meeting products in the multi-platform environment.

Over 100 key international vendors, including Apple, AT&T, British Telecom, Cisco Systems, Intel, MCI, Microsoft and PictureTel have committed to implement T.120-based products and services[2]. In next section, we shall have a survey on the state-of-the-art of some leading net-meeting tools. From this survey, it will give us a hint of how we can establish the inter-enterprise synchronous collaborations within a multi-vendor and multi-platform environment.

2.4   State-of-the-art of some existing leading net-meeting tools

2.4.1   Microsoft NetMeeting

Microsoft NetMeeting runs on PCs with Windows 95, Windows 98 or Windows NT 4.0. Architecture components of Microsoft NetMeeting are implemented upon T.120 and H.323 standards (shown in Figure 3). Its capabilities include[5][6][7]:

- Application sharing/Collaboration (T.128 standard),
- Electronic whiteboarding (T.126 standard),

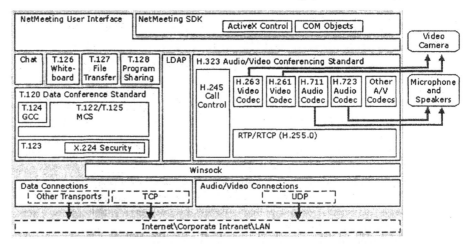

Figure 3   Architecture components of Microsoft NetMeeting

- Text-based chat,
- File transfer(T.127 standard),
- Audio & video capable (H.323 standard),
- Internet or intranet access,
- Multi-platform support.

2.4.2 HP VISUALIZE Conference

HP VISUALIZE Conference (HPVC) runs on HP VISUALIZE Workstations with only the HP-UX 10.20 ACE version is supported. It is implemented only with T.120 standard and does not support the H.323 standard. Its capabilities include[8][9]:

- Application sharing (T.128 standard),
- Whiteboarding (T.126 standard),
- Text-based chat,
- File transfer (T.127 standard).

2.4.3 SGImeeting

SGImeeting runs on SGI workstations with IRIX 6.5.2, 6.5.3 or 6.5.4. It is implemented upon T.120. Its capabilities include[10]:

- Application sharing (T.128 standard),
- Whiteboarding (T.126 standard),
- Text-based chat,
- File transfer (T.127 standard).

- Application sharing (T.128 standard),
- Whiteboarding (T.126 standard),
- Text-based chat,
- File transfer (T.127 standard),
- Audio and video conferencing (H.323 standard across standard TCP/IP networks).

2.4.5 DC-Share for UNIX

DC-Share for UNIX runs on UNIX workstation (Sun, SGI and HP). It is implemented upon T.120 and H.323 standards. Its capabilities include[12]:

- Application sharing (T.128 standard),
- Whiteboarding (T.126 standard),
- File transfer (T.127 standard),
- Text-based chat,
- Multipoint communications over TCP/IP,
- H.323 audio and video.

2.4.6 Timbuktu Conference

The Timbuktu Conference is a conferencing tool for Macintosh OS 8.0 or later. It is implemented upon T.120 standard. Its capabilities include[13]:

- Application sharing,
- Communications over TCP/IP.

Table 1   Summary of some existing leading net-meeting tools

	Microsoft NetMeeting	HP VISUALIZE Conference	SGImeeting	SunForum	DC-Share for UNIX	Timbuktu Conference
Application sharing (T.128)	X	X	X	X	X	X
Whiteboarding (T.126)	X	X	X	X	X	
File transfer (T.127)	X	X	X	X	X	
Text-based chat	X	X	X	X	X	
Audio \ video (H.323)	X			X	X	

In addition, users can transmit video input to the whiteboard through live video capture. Sharing video input enables users to position a video camera on a factory floor and transmit a potential line problem to a remote user, enabling distributed problem solving.

2.4.4 SunForum

SunForum runs on SUN workstations with Solaris 2.6 and Solaris 7. It is implemented upon T.120 and H.323 standards. Its capabilities include[11]:

2.4.7 Summary

Above surveys can be summarised and compared as shown in Table 1 where we find that Microsoft NetMeeting, SunForum and DC-Share for UNIX net-meeting tools can support all functions listed in Table 1 for multi-platform net-meetings. The HP VISUALIZE Conference and SGImeeting net-meeting tools can support most functions but without audio and video multimedia conferencing capabilities. As for the Timbuktu Conference net-meeting tool, it provides only multi-platform application sharing capability, but this is the most important capability for inter-enterprise synchronous collaboration.

2.5 An example of net-meeting in the multi-platform environment

Figure 4 gives an example of how the net-meetings can run upon the multi-platform environment. The screen shots shown in Figure 4 show a Windows NT system running NetMeeting 2.1 (on the left) and a UNIX system running DC-Share 2.1 (on the right) in a net-meeting[12].

The NetMeeting and DC-Share user interfaces are visible at the top right of the respective screens (Figure 4). The Windows user has shared Word (see bottom right of the left screen of Figure 4) into the net-meeting with the result that the UNIX user can view and control the shadow Word window (see bottom right of the right screen of Figure 4) as if it was running locally (even though it is actually running remotely on the hosting Windows system).

3 How to Run Net-meeting More Efficiently Based on Limited Bandwidth Capability of Existing Communication Network

3.1 Analysis on bandwidth requirement of various net-meeting functions

It is well known that the multimedia based net-meeting takes up a great amount of communication network bandwidth. This problem often remarkably delays the interaction among the multi-point net-meeting participants during the net-meeting, and makes the net-meeting's quality poor. In order to see the problem quantitatively, a calculation on bandwidth requirement of application sharing, video and audio conferencing functions are given below.

Microsoft NetMeeting 2.1 on Windows NT   **DC-Share 2.1 on UNIX workstation**

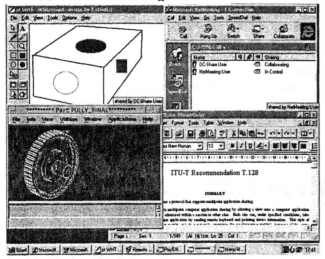 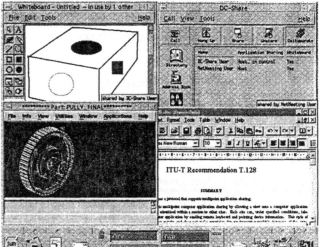

Figure 4   Example of net-meeting in the multi-platform environment

The UNIX user has shared CAD application, Pro/Engineer (see bottom left of right screen of Figure 4), into the net-meeting with the result that the Windows user can view and control the Pro/Engineer 3D model (see bottom left of right screen of Figure 4) as if it was running locally (even though it is actually running remotely on the hosting UNIX system).

The users have also started whiteboards and have loaded up a simple drawing with annotations (see top left of screens of Figure 4).

While the two screens are shown here side by side, in practice they might be very far apart (e.g. across the Atlantic), yet NetMeeting and DC-Share allow the users to collaborate irrespective of platform or location.

One experimental example is taken as calculation example shown in Figure 5[1]. The screen shots of Figure 5 show that two distantly collaborative product designers (the user of left screen and the user of right screen) are sharing the CAD application, ProEngineer, to design the part shown in the lower left of screens. They are using also the video and audio multimedia means to communicate with each other in the collaboration. By means of the video function and camera, they can see each other via movable photos as shown in the lower right of screens. By means of the audio function and microphones and speakers, they can talk with each other at the real time (see upper left of screens where microphone symbols and volume regulators are shown, and upper right of screens where speakers symbols and regulators are shown).

Thus, the calculation involves in the bandwidth requirements of CAD application sharing, video and audio multimedia communication.

3.1.1 Bandwidth requirement of CAD application sharing

Given following assumptions with respect to data requirements: a vertex is composed of 6 floating point values ((x, y, z) co-ordinate data and (nx, ny, nz) normal data) in a typical CAD application, primitives are composed of triangle strips of, on the average, seven triangles per strip (this strip contains 9 vertices which adds up to 54 floats as opposed to 126 floats for 7 individual triangles with 21 vertices)[14].

Assume that the CAD application sharing function transfers 30 triangle strips (i.e. 210 triangles) per second. Then, we get

54 floats/tri-strip = 1 620 floats/sec (30 tri-strips/sec) = 6 480 bytes/sec (4 bytes/float) = 51 840 bits/sec (8 bits/byte).

and 30 mm vertical. We use MPEG standard for motion video compression. Thus,

- Pixels for each photo (motion picture): [(40 x 30) / (310 x 235)] x (1024 x 768) = 12 954 pixels,
- Bits of data while using 24 bits of colour palette: 12 954 x 24 = 310 896 bits,
- Bit-rate for each motion picture at 25 frames per second, MPEG standard[15]: 310 896 x 25 = 7 772 400 bps (~7 772 Kbps),
- Bandwidth needed at compression rate, 130:1, MPEG standard[15]: 7 772 / 130 ≈ 60 Kbps,
- Then bandwidth needed for two movable pictures is **120 Kbps**.

3.1.3 Bandwidth requirement of audio multimedia communication

Assume that the audio system is configured as stereo (2 channels and 44.1 KHz sampling rate). Thus, the necessity of communication bandwidth can be calculated as follows (herein, we use MUSICAM standard for stereo audio compression):

Figure 5   Collaborative product design with application sharing, audio and video functions

It means that for transferring 30 triangle strips per second, **50 Kbps** bandwidth is required.

3.1.2 Bandwidth requirement of video multimedia communication

Assume that the resolution of full screen is configured as 1024 pixels horizontal and 768 pixels vertical with 24 bits of colour palette. The size of full screen is 310 mm horizontal and 235 mm vertical. The size of each photo (lower right of screens in Figures 5) is 40 mm horizontal

- Bit-rate of each channel while using 16 bits for each sampling: 44 100 x 16 = 705 600 bps,
- Bit-rate for 2 channels: 705 600 x 2 = 1 411 200 bps (1 411.2 Kbps),
- Then, bandwidth needed at compression rate, 7:1 (MUSICAM standard[15]): 1 411.2 / 7 ≈ **200 Kbps**.

3.1.4 Summary

To summary, in the example above, the total bandwidth needed is **370 Kbps** (50 + 120 + 200). If we take the **2 Mbps** bandwidth capability of outlet line from our Institute campus LAN to UNINETT (Norwegian university Internet) as example, collaborative design of Figure 5 has used up a rather large part of the line capability (18.5 %). Here we know why we often encounter remarkable delay in the video and audio conferencing.

3.2 Proposals on combination of existing net-meeting tools with telephone conference tool for efficient net-meeting

Notice that in the example of Figure 5, the size of the photo is so small as if a stamp. If we increase the size of photo three times of both width and height (i.e. 120 mm x 90 mm), then the total bandwidth needed will increase to 1.33 Mbps (50 + (120 x 9) + 200 = 1330 Kbps). Thus, 66.5 % capability of the outlet line of our Institute would be used by this example. Then, a terrible delay would make the net-meeting actually unable to be going on.

However, if we still use the small photo (40 x 30 mm) in the net-meeting, it is too small to well see each other. Using video you might want to look in the remote partners' eyes, but you can't, because the resolution is so poor you can't see their eyes. That isn't all. We have experienced, in best cases, the video have a one-second transmission delay. This makes a lot of confusion. Therefore, the small size of video is not quite helpful, but takes up a lot of bandwidth. So, one of our proposals is to shut down the video function while the net-meeting delay getting worse while a net-meeting is in the quick and large amount of inter-operation status. (At the beginning of a net-meeting, you may use it to introduce the participants each other to see whoever is attending.)

Audio communication is essential to ensure the net-meeting discussions. However, the audio multimedia communication also takes a lot of bandwidth and makes the application sharing to be delayed remarkably. So the other proposal we suggest is to use telephone meeting function in normal telephone system (e.g. ISDN telephone system) simultaneously with the net-meeting system. Thus, we can focally use bandwidth capability in the application sharing, whiteboarding, and test-based chat in case of limited bandwidth capability of the communication network.

4   Conclusions

This paper has explored of how to realise inter-enterprise remotely synchronous collaborative collaboration within multi-platform environment. After a survey of existing net-meeting standards and the state-of-the-art of some leading net-meeting tools, a conclusion is made that existing leading net-meeting tools are basically functioned at supporting the multi-platform inter-enterprise remote collaboration.

However, because of limited bandwidth capability of existing communication network, remarkable delay happens often in a net-meeting. Therefore, for efficient net-meeting, proposals on restrictive use of video and combinational use of net-meeting tools (T.120 based) with telephone conference tool are given by the authors.

For running net-meeting successfully, tool support is only one aspect, which we have discussed in this paper. However, you need also to well organise and manage the meeting process, and to pertinently use tools in different stage of the process. This is a organisation philosophy problem. The authors have done some preliminary research on this problem and further work will be done in the future.

References

[1]   Z. Deng, B. Bang, A. Laksa, and S. Nadarajah, "A Model of enterprise integration and collaboration tools and communication infrastructure for inter-enterprise collaboration", Globallization of Manufacturing in the Digital Communications Era of the 21" Century, Kluwer Academic Publishers, ISBN 0-412-83540-1, 1998.

[2]   "A Primer on the T.120 Series Standard", http://www.lotus.com/products/sametime.nsf/standards/8DDB25B6C08E70E5852566640072FCD2.

[3]   IMTC, "T.120 OVERVIEW, Multimedia Teleconferencing Standards", http://www.imtc.org/t120.htm.

[4]   IMTC, "H.323 Overview, Multimedia Teleconferencing Standards", http://www.imtc.org/h323.htm.

[5]   B. Summers, Official Microsoft NetMeeting Book, Microsoft Press, 1998.

[6]   Microsoft Co., Microsoft Windows NetMeeting 3 Resource Kit, http://www.microsoft.com, 1999.

[7]   Microsoft Co., "Windows NetMeeting," http://www.microsoft.com/windows/netmeeting/

[8]   Hewlett Packard, http://www.hp.com/visualize/collaboration/visconf/faqs.html.

[9]   Hewlett Packard, http://www.hp.com/visualize/fyi/bissue/july99/newsprods.htm.

[10] Silicon Graphics, Inc., "SGImeeting 1.1 Product Information", http://www.sgi.com/software/sgimeeting/datasheet.html.

[11] Sun microsystems, "Workgroup Collaboration Tools", http://www.sun.com/desktop/products/software/sunforum/.

[12] DATA connection, "DC-Share for UNIX", http://www.dataconnection.co.uk/conf/Dcshare.htm.

[13] Netopia, Inc., " Timbuktu Conference", http://www.netopia.com/software/tb2/mac/conference/index.html.

[14] Hewlett-Packard, "Hewlett-Packard's Large Screen Multi-Display Technology - An Exploration of the Architecture Behind HP's New Immersive Visualization Solutions", http://www.hp.com/visualize/products/immersive/whitepapers/3dsls/3d_sls.html

[15] W. Buchanan, Advanced Data Communications and Networks, Chapman & Hall, ISBN 0-412-80630-4, 1997.

In and Out : a New Methodology to Develop Team Effectiveness

M. Legris

C.E.R.S.O., Dauphine University, Paris, 75016

Abstract

The author presents a case study of a co-built methodology to organize team-work within a french cable tv distributor. In an organizational change context, the CEO's decided to develop a man centred engineering process. By enhancing participative management, giving more autonomy to team members, the aim was to integrate the different company's poles and to improve quality. Members of these « working groups », multifunctional and gathering operational and functional managers, had to undertake some projects, vital for the firm's development. To be efficient, the groups were to include the people concerned by the projects (i.e : the « actors »), whose working conditions, or job could be modified by the project implementation. A participative management was needed to carry this experience, as well as the creation of a new organization to spread information within the company.

Innovation is here seen as the global process of creation and organization surrounding the « work groups » and its impacts on the entire enterprise. Innovation consists also in the project results (the way the team strive to achieve its goal and find new abilities or skills, new ways of working and thinking) and their diffuse within the organization. A methodology in action was built by the author, the team members and the CEO's. It was based on few principles or rules, decided in common and on an opened learning process. If deadlines were fixable from the start, the objectives, the way of working and thinking, the rules when necessary, were appraised and reviewed to integrate any new idea, need or suggestion in the process. Team members auto-organized their work, consulting the CEO's whenever they felt it necessary. This methodology was based on two axis : integrating every concerned actor and learning from the experience itself. One of the principle motive was to try to assume a meta position, so that all project dimensions (technical, financial, human, operational , organisational, cognitive) could be integrated from the start of the project. The results show a great diversity and richness of the method in action.

1 Introduction

The problems of organizational change and complexity have loomed large in organizational theories as well as in management thought and practice. The growing requirement to study dynamic phenomena requires new theories and new methodologies. Project management usually implies a more effective horizontal cooperation within the firm, especially between conception and production departements, and the creation of specific project structures. Learning how to work on projects is learning new organizationnal design. It also means the modification of the collective actions' rules and some knowledge production from shared experience.

In this article, we intend to explore the first results of an experimental research on participative change through project teams work. The discussions held in the research group "knowledge and action" from the CERSO murtured the author's thought. This methdodology fits with concurrent enginnering organizational approach. First we will wonder why a new project team methodology would be needed. Then the methodology itself will be developed. The role of manangement and coordination will be highlighted. Finally the results are discussed in the last section.

2 Why would we need to seek for a new project team methodology ?

2.1 A new sociotechnical paradigm

Product diversification and product differentiation are both increasing very quickly while simultaneously, the product cycle is reducing. Product cycles overlap. Competitive success depends on the transformation of a company's key processes into strategic capacities that really provide superior value to customer (DRUCKER, 1992 [1] et STALK 1992 [2]). The last technical and economical evolutions renew the industrial systems' efficiency conditions. The search for « new production rationalization » (DE TERSSAC et DUBOIS 1992 [3]) shows the will to find new consistency between technical, economical and social spheres. As stated in sociotechnical paradigm : an open system adjusts itself while learning new informations. It enables actors to find equilibria between human and technical needs (LIU 1983 [4]). New information technologies are part of this process. They claim for new organizational modes and ways of working together.

The project based organization is sometimes seen as a response beyond others, to these stakes. The need for team work and for building new cooperation and coordination styles is aknowledged widely (MINTZBERG [5], SENGE [6], MORIN [7], REYNAUD [8]). Many authors have defined a new sociotechnique paradigm based on information and learning processes amongst the organization (NONAKA [9] ; CROZIER [10]). They stress on the organizational capacity to mobilize its employees at all levels to face change. Still there is no more « one best way » : each system has different effects on work organization. The main point is that the number of options increases. There have of course always been options : for example there was no absolut need to adopt the assembly lines at the end of XIX century, even if the cost of breaking away from a dominant pattern would probably have been high.

2.2 Cross fertilization

Managers and researchers need to search together for new solutions fitting local constraints. That is what the author intended to do with managers of a cable tv distributor. We co built a new team working methodology. From the research point of view, the interest was to experiment some hypothesis in real life conditions (participative research) and to be part of the change steering committee. From the company top management point of view, the situation was too fuzzy to handle it the usual way. They needed help to analyse the situation as well as to suggest new ways of working and managing inside the firm. They also wanted new ressources to be able to coordinate the change process. For the actors taking part to the teams, the motivation was to learn, to be associated to the decision process, to defend one's own intesrests within the company and to work together. So in one way, the researcher was involved in the action as well as the company's actors were implicated in the research. That's what I call cross fertilization, through participative action research (LIU, 1997 [11]).

The research lasted 2 years, fom 1997 to 1999. The data for this article were derived from multiple sources including archival data, participant observation, meetings, interviews, discussions, organizational and sociologne diagnosis[1] and periodical restitutions of the research results to the main actors.

2.3 The change process of the firm

The research took place in a french cable tv distributor, created in 1989. 130 people worked for this company when the study began. It was built on a holding design : one central office in a Paris suburb and 9 regional offices. The activity began included the wire cable network construction, the tv programming, the cable tv subscription sales, ending with the after sales service. In december 1998, the firm had 71 000 subscribers (turnover of 85 millions).

For ten years the firm had no competitor : a contract was signed with the town halls giving monopoly in cable tv distribution to the firm for at least 20 years. 9 french middle towns had signed for it. But in 1997 the satellite tv came up, intruding this market. Simultaneously, its rate of technical change (analogic technics, internet, pay per view, new broadcasts…) increased very quickly. The customers became more demanding.

The need for change was mostly exogenous but the CEO's also wanted to respond to an internal request for more responsability and autonomy. Indeed the middle management (i.e ; the 9 regional offices managers) had lost during the last years part of their power of decision, partly because the whole organization had been strongly centralized (creation of 5 central services such as marketing, financial reporting, accounting, HR). There was a strong demand for more responsability as well as implication in the decision process.

When one aims at driving an organizational change, one realises the importance of the principles of a systemic approach and of a constructivist perspective (LEMOIGNE [13], MORIN [7] , WATZLAWICK [14], ARGYRIS SCHON [15]) . Change is really a process of articulating one's vision of the world. In this case, the problem was to change mentalities as well as organizational structures. The aim identified was to move from a monopolistic culture to a customer service one, through participation and quality. It involved challenging the implicit

---

[1] Diagnosis : study of a company's organization, culture and main components as developped by Sainsaulieu [16].

assumptions[2] that have shaped the way people in an organization have historically looked at things. It also requires new communication techniques that actually get people to implicate themselves and experience new working practices.

3. A co built methodology

As we previously said, the change process was decided in a top down approach. When the research started, the change was rather directive. Nevertheless, the CEO's was conscient of the need to mobilize its employees. The definition of the target was not clearn, partly because incertainty on the business market was really high (mergers, new competitors, Technical changes). There was also internal pressure for change coming from shareholders. The firm as well as its competitors was far from rentability. The business plan provided equilibrium for year 2000. After 10 years, the firm was at the end of its first cycle : the birth (ADIZES[3]), trying to reach the childhood stage. The whole organizational design needed to be evaluated to find a new balance between centralization and managers responsabilisation, quantitative growth and qualitative development.

3.1. The team-work basics

The basic idea was to start a new dynamic inside the organization, to foster change capacities. The firms middle managers were asked to start working in teams to achieve some projects, vital to the firm success. After a few attempts to impose the team composition as well as the projects definition, the CEO's aggreed with us to try a more participative management. As they wanted to develop a man centered engineering process, they decided to trust their collaborators.

A new process began. During a two days sessions, the 14 managers were trained to classical project methodology. They also analysed the forces at stake and discussed the projects that had to be handled in priority. A schedule was set. 5 working groups were formed handling 10 projects. The issue was middle- term : one year at least to implement the changes decided.

Team composition	Projects handled	objectives	Results
« commercial work group » Marketing service manager, 3 agencies director	Collective habitations prospection	To enhance collective agreement subscriptions	New prospection methodology and tools
Value analyse work group Financial reporting manager, acounter, 2 agencies director	Monthly auto contrôle 3 projects	Contrôle results and salaries calculation	New auto contrôle procedure on agencies
« Personnel work group » Personnal service manager, 2 agencies director	-Risky missions - Technical service duty	- Work safely - Define the duty system	- Progressive plan to reach quality standards - New reward system and duty organization
« New Technics work group » Technical service director central and local, 1 agency director	- Maintenance - Local project implementation	- Improve service quality and network reliability - Improve response quality and rapidity	- definition of a maintenance scheme - local projects organization
« information work group » Informatic manager and 1 agency director	Customer service	Trace and Answer to customers demands	Customers consultation and new information organization

Figure 1. Work group composition and projects handled.

As one can see, the projects chosen implied cross functional team work and frame global innovations. By enhancing participative management, giving more autonomy to team members, the aim was to integrate the different company's poles and to improve quality. Members of these « working groups », multifunctional and gathering operational and functional managers, had to

---

[2] Schein [17] did define culture as « a set of shared taken for granted implicit assumptions that a group holds and that determines how it perceives, thinks about and reacts to its various environnement » 1992, p 236. Norms become a fairly visible manifestation of these assumptions.

[3] Adizes Les cycles de vie de l'entreprise. Diagnostic et thérapie.Organisations, 1991.

undertake some projects, vital for the firm's development. The problems at stake had long been relagated to « a time where we would have enough time to handle them » (as said by one manager). Some of them were of primordial importance (security, technical duty, maintenance) and lots of the firm's employees were waiting for a response, at last !

The main rules, decided by the team members and the CEO's, were power delegation from the CEO's to its managers, the back setting of the CEO from the project process, the participative aim to include all actors concerned by the projects, the role of the work group as proposition force (after study and discussion, they proposed solutions or actions to the CEO's), the Ceo's role was to give strategic vision, to inform the work group of some eventual modifications, to appraise their propositions and to take the final decisions (an allocation of some supplementary mean to develop the projects). The coordination role was the author's responsability as well as the process consistency. The author also provided methodological help, analysing the work group evolution and suggesting eventually some organizational evolutions first to the team members then, if agreed, to the CEO's.

Team members agreed on a shared definition of the concept of project inside the firm : a project developped by a work group is cross functional and shall be implemented in middle term (at least one year). It deals with innovations or/and stakes important for the whole organization. The work group use a participative methodology to allow every employee concerned by the actions envisaged to be informed, to give his opinion and eventually to take part in the project itself.

3.2 A cyclical approach

The team work methodology was conceived from the start as an open dynamic. There was no pre defined target to reached, for it was impossible for anyone to make projections in the long term (i. e. more than one year). The change process itself was perceived as a spirale like dynamic, made of progress and reverse, difficult to controle. The idea was to keep a smooth approach to allow learning on the way.

For example each project included the initial thought stage, the objectives setting and the definition of the necessary mean jointly by team members and top managers, the information gathering (internal and external), the presentation of a global diagnosis, the solutions proposed, the implementation and the following through of the actions decided, in agreement with operational services.

The main tools of projects conduct usually aim at controling time and actions. One shall not forget the

deadlines nor the cost limits. The complexity of project progress makes it difficult to find one simple and efficiente methodology. Planning change consists generally of searching for the best path from the initial situation to the one desired (logic networks, ordonnancement). Those linear approaches have shown their limits in rapid changing contexts (EMERY, TRIST [10], NONAKA [12], LEMOIGNE [13]).

In this case a pragmatic approach was tempted. The methodology built itself on the way. The work groups did find step by step the tools they needed. Two progress axis were highlighted. First, the need for concertation and information (inter and intra groups as well as inside the firm). Second, the insight that the classical sequential approach could not solve the problems. This is the reason why the change was seen as cyclical, allowing a permanent retroaction[4] from below phase to upstream phase so that new data and evolution could be taken into account in the project progress.

On the next figure the author try to caracterize the main differencies between linear sequential approach, and the cyclical one (one can also think of Argyris and Schön theories of single loop and double loop learning).

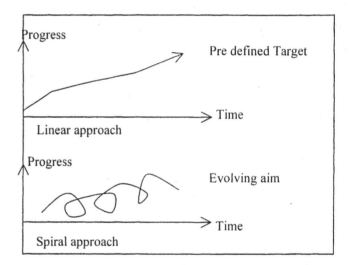

Figure 2 : linear and spiral like vision of the change processsus

Figure 2 is based on a metaphorical thought. One shall not read it as the description of reality. It shows two reality modelisation. Those modelisations have a great impact on one's action and of course on global organization. They can muture the implicit actions that drive people's action, even if are not able to express these « used theories » (Argyris et Schön [14b]).

The linear sequential approach often try to solve

---

4 See Herbst <u>Alternatives to hierarchy</u> 1976 p 64.

« visible problems » in short term, acting in a fragmented way on some of the system components without seeing the interrelationships. This logic « forgets » individual comportements and collective action while simultaneously seeking for the actors mobilization.

The cyclical approach is based on systemic modelling, aiming at a global and long lasting progress. The methodology is based on an integrated approch of sense making. Thus the exercice would be to operate a cultural pattern change[5] to break off usual thought and action frame that lead to « toujours plus de la même chose».

3.3   Main stages defined in the firm

5 main stages have been defined in this methodology : writing of a schedule of conditions, the diagnosis, the communication plan, the implementation, the following through and the appraisal.

One can figures the methodological progress developped by the work groups as follows :

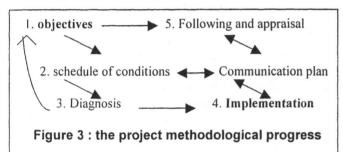

Figure 3 : the project methodological progress

Each stage had been defined previously in a general assembly so that nobody would be surprised. Inside thsi frame a lot of innovations was possible. The teams did not adopt the same working practices. Some did chose to have a leader, some did not. Some team members were expert on their fields and other had more generalist and pragmatic skills.

For example, the personnel service manager was member of the work group « personnel ». She was given the expert role inside the team. But she did not agree to this, and she asked for a common leadership. The other team members did realize they could also be seen as « experts » on some fields, usefull for the project achievement. The leadership became collegial.

In another work group, one of the member, an agency director, did not know anything about financial reporting, and he was part of the « value analyse » work group . He first felt very uncomfortable. Then he tried to withdraw

---

[5] Watzlawick clearly explained that 2 models of change coexist. The first leads to an evolution inside the same mental frame. The second one is a complete breakthrough one's mental marker, producing another frame. Also see Argyris et Schön.

from the team. He was not allowed to do so by top managers. Then he began to work with the others and discovered a new world. He did not became a perfect fiancial analyst but he did not feel such a repulsion for it. He found out he could help the team to stay simple and to explain in usual worlds their projects. He also realized how much financial report could help him in his manager role to get better results.

4. What about coordination and management ?

4.1   Leadership and organization

A specific organization was created to coordinate and manage the whole change process. This steering committee did meet once a month, more often, or less depending on the actuality. Its members were the 2 top managers of the firm and the author. The personnel service manager was also integrated to the thought.

Figure 4. Specific organization to manage the working groups

The steering committee was the main process regulator. It supported the work groups action, mainly by giving them clear direction on strategic issues and on global consistency. It defined any rules change requested by team members. It appraised the team effectiveness (according to the objectives and the schedule conditions). It also took any final decisions on what ought or could be done.

The working groups were not part of the pyramidal hierarchical structure of the company. Because they were cross functional and temporary (they existed the time to achieve the project), their team members did not get new official functions. The work groups did not appear on the charts. Nevertheless, team members got time and money, transport etc to work.

The company employee that did not take part directly in the work groups (the majority) were informed by a « letter » published monthly on the internal information network. Every work group was also responsible for its communication : explaining their ideas, asking questions, trying to understand and get information. Some of them actually developped consultation of employees concerned by their project, analysing the situation and its constraints with them. Some others did not, sometimes because they needed to find an agreement inside the group before, sometimes because they were freightened of people's possible reaction, occasionnaly because they did not find the ways to do it, and, at last, because they did not see it as important. One rule was always followed : if one team did seek for other employees efforts and time this group had to tell back what results they gathered, how they did use it and what they intended to do.

4.2 The project evolution on 2 years time

The work group had different projects, different in scope and difficulty. They did not evolve concurrently. Some work group spent more time on the initial thought stage, some on the data collection, depending on the project nature (creation, organization, rationalization) and on the team members skills.

Common deadlines were settled by the steering committee to enhance productivity and learning process through experience exchange, discussion, information gathering, results diffusion. Another postive force towards change was the others firm employees expectations. As everyone knew what projects were going on, whose team was in charge of it, and also because they could ask questions, make suggestions, they were waiting for concrete results. Managers involved in work groups wanted to prove their skills to the top manager but also to their subordinate, which were sometines taking part of that new experience. The team members were looking for legitimacy and recognition. They worked for the community benefits. A great pression to achieve concrete resuslts supported the change process.

From the beginning of this research, to its end, 3 major deadlines are to be noticed : 06/98 were the first propositions were argued, 09/98 with the first work group seminar and the decision to implemente some actions, and

02/99 were the first appraisal of the projects took place. The main dates are of course related to others firm events, like new strategies, new agencies etc... The work groups follow in this case the firm's global change cycle.

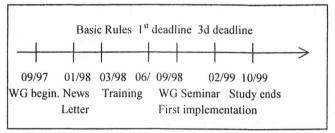

Figure 5. Major dates of the work group evolution

Figure 5 shows the global process. The top managers had decided to set back from the process from april to september 98, to show evidence of their will to empower the team members. They did not interfere with the project developpment, unless they were asked to do so by team members (none of them did). The coordination process relied on the author's care. The link between groups, the consistency watch was of my responsability. This choice comes from the will to give evidence of the reality of the team empowerment and aimed also to break off the usual working practice. Thus we introduced uncertainty and liberty in the project conduct.

In order to view the concurrent process at stake, see figure 6 which describes each work group activity rate, measured on its production (reports, information, procedures, data collection, proposition, results, group dynamic, meetings, number of people integrated to the team, collective learning).

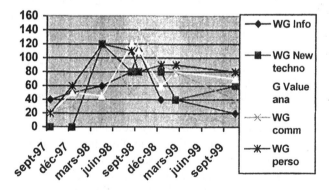

Figure 6 : concurrent work group productivity progession

Each work group follows a different path, but the majority of them meet on some peculiar moments (the deadlines, mostly). Another way to say that this methodology is adaptative and can fit to different kind of objectives and teams.

4.3 Management : the role of appraising

From this experience it seems that the major management role is to organize a fair appraisal system. In this case, the appraisal was collective and individual. The team was evaluated on the basis of its former objectives, compared to its results. The schedule of conditions was merely a contract between the team members themselves (who's doing what for when ?) and between the team members and the steering committee. When it was possible, the group dynamic and its learning capacities were also taken into account. Achieving a project meant for the individuals an increase in the chance to get a bonus.

From this point of view the definition of the objectives was crucial as a point of agreement and of construction of the group representations about itself and about its role. A negotiation process had begun, driving people to get involved and to argue about their points of view. The objectives were also used as a guideline, to help keep main ideas in mind whatever information, new perspectives etc could change the project nature itself. In one case, objectives had to be reviewed to integrate new aspects.

The work group introduced their objectives to the others in the news letter. In doing so, objectives also became part of the communication of the team and of its identity. Appraisal cristallizes several dynamics which concurrently gave to work groups a pattern of action. A fair competition between groups appeared, enhancing their productivity and innovation diffusion.

5. Discussion

5.1 What's participation ?

Discussions of participatory research usually fail to distinguish two important dimensions : the participatory intent of the research and the degrees of participation actually achieved by a particular project. One can't impose participation. Here the degree of participation achieved is the joint result of the character of the problems, the cultural context, the aims and capacities of the top management, the perception of the actors and the skills of the researcher.

This methodology is participative in the sense that it allowed the managers to take more responsabilities and autonomy. But it did not reached the level of global participation. The « official » team members were the managers. Other actors were consulted, questionned,

informed, trained but had no responsability in the project. They could ask questions in return, make suggestions, but a few did spontaneously.

The power structure did not really change. As the projects were implemented, a need for more decentralization was expressed. The CEO's did not take it into account. Among others, the General manager did refuse to allow work group to manage a dedicated budget. Their autonomy was then restricted.

5.2 Learning opportunities

Allowing every team member to embrace wider responsabilities and to learn from each other, the work group become a learning and mind opening center. Each work group followed the basic rules decided in common (see the upper section) and created its own « ad hoc » organization (MINTZBERG [5]). The challenge lay in the coordination of the skills usefull to achieve a project, avoiding a hierarchical power structure, so that a real dialogue can begin between managers and « basic employees». A winner / winner logic can provide an efficient dialogue frame.

Participation is not only a nice idea. It provides usefull barriers against major inconsistencies. By taking into account the actor's culture, representation and skills, it avoids unrealistic plans, failure to understand the stake in « real working life », impossible to know without being implicated in everyday work and practices. Participative management can also avoid the top down approach bias : actors won't feel forced in a change they were not informed of, don't understand the meaning and don't see the advantages nor the difficulties to find a solution (or why this solution was chosen). A project is more likely to be implemented with success if the solution fits its context while respecting global values like equity and justice amongst every salaried employee of the company.

Some projects really widened their application field. The project handling « risky missions » ended with a global security approach. To find the laws suitable to the cable tv distribution business, team members had to contact their competitors to build a « meta work group ». The firm became an innovator not only for itself, but also for its activity branch. This project raised also the question of the cultural habits around security. When techniciens have been working for years outlaws, there is a tremendous work to do to make them adopt new practices.

Another aspect that many team members highlignt is the enrichment they gain mixing functional and operational skills and points of view. Work groups were the first attempt to make agencies director and central services managers to cooperate. This was not so easy, a certain rivalry and mistrust at first could have prevented them from working together. Their strategic interests were differents. Finally they manage to work efficiently on common projects.

For example in one of the work group, a conflict appeared between two of its members : one central service manager, a kind of specialist, and an agency director, ancientmarketing man, very operational oriented. Because this tension threatened their capcities to work together, the team members asked for help. After discussion with the author, some peace could be find. My analyse was that the main opposition was not linked to a personnal matter but was the result of the confrontation of two contradictory world. The world of the written, very precise and rigourous but also monotonous, bureaucratic like faced the speech universe. More spontaneous, based on the world given, the latter was satisfied with good discussions. The first, on the contrary, thought that work started when things were written, planned, regsitered. None of them was satisfied for different reasons and accused the other of laziness. Once this analyse explained, both of them agreed. A modus vivendi was found : both took the commitment to regularly write small reports while allowing disgressions during discussion (thus enhancing creativity and group dynamic).

Some authors like BELBIN [15], argued that to enhance team effectiveness, the key point is the choice of the right personns composing it. It is evident that a team must be balanced and if possible, gather most of the competencies requested to achive its goal. From this case study, no evidence shows that CEO's ought to think and analyse in every details who sould be part of which team. The focus point seems to be more in the possibility to discuss freely the matter, from different point of views. The rules seem more crucial to enhance group effectiveness and innovation, if simple principles are aggred : team diversity, no hierarchical binds, equale repartition of speaking time, openess of the minds, and support provided to team members (training, encouragement, respect).

6. Conclusion

The projects' complexity was taken into account and became an opportunity for a change of vision, thanks to cooperation and open dialogue between most level concerned. Combining specialists and novice skills, functional and operational backgrounds, created a learning and comprehensive atmosphere which lead to new values : quality, equity, realism and curiosity. The firm evolved from a reproductive to an innovative dynamic, able to integrate , renew and enrich ways of working and thinking amongst the organization.

The cyclical vision of the change was validated as a change facilitator. First people stopped hopping for immediate results. Managers experimented the different times of change : they could no more stay focused on emergency work : to achieve their projects they had to step back and look at things differently. The confrontation between production and field support organizational and strategic logics was in that sense a rich source of emergent theories and ideas.

Besides, the author found out another cyclical process. Managers did not realize while acting that they were changing. They only became conscious of what they had learned (or of some of their learning) after the « doing time ». There is a gap between knowledge application into action and knowledge explicitly learned. The first manifestation is action, followed after a while (several month) by a construction of sense that allows people to talk about and explain what they learned. We join here the works on knowledge creation of NONAKA [9].

Multifunctionnal team need a methodology fitting to the firm context and giving enough autonomy for sense making activities to take place. The methodology developped int this article seems to fulfill concurrent engineering requirements. We don't know enough about this process to pretend driving it. The aim is more to not prevent it from happening. This is what this methodology intended to do. Results are nor definitive or universal. This research calls for more experimentation. This is not a ready to use methodology, if there can be some.

Thise case study highlight the necessity to complete the usual quality cost delivery logic, based on cost reduction, with a value management .

References

[1]   Drucker Peter « The New Society of Organizations » HBR, September october 1992.
[2]   Stalk George, Evans, Philip « Competing on Capabilities ; The New Rules of Corporate Strategy », HBR, march april 1992.
[3]   De Terssac et Dubois, Les Nouvelles Rationalisations de la Production. 1993
[4]   Liu, (Michel), Approche socio-technique de l'organisation, éditions d'organisation, 1983.
[5]   Mintzberg, (Henry), Le management. Voyage au Centre des Organisations, éditions d'organisation, 1998.
[6]   Senge, P. , La cinquième discipline, First, 1991.
[7]   Morin, E. La méthode, Seuil, 1977.
[8]   Reynaud, Jean-Daniel, Les Régulations dans les Organisations : Régulation de Contrôle et régulation Autonome » in Revue française de sociologie, vol 29, n°1, janv/mars 1988 , pp. 5-19.

[8] Nonaka, Ikujiro, The knowledege Creating Company, 1992.

[9] Crozier, (Michel), Sérieyx, (Hervé), Du Management Panique à l'Entreprise du XXI ème Siècle, Eska, 1993.

[10] Emery, (Fred), Thorsrud, (Eimar), Democracy at Work. The report of the Norvegian Industrial Democracy Program , Mennen Asten, 1976.

Trist et Bamforth, Some Social and Psychological Consequences of the Longwall Method of Coal Getting, Human Relations, vol 4 n°3, 1951.

[11] Liu Michel, Fondements et pratiques de la recherche-action, L'Harmattan, 1997.

[12] Nonaka, Ikujiro, A Dynamic Theory of Organizational Knowledge Creation in Organization Science, vol 5, n°1, février 1994, pp 14-37.

[13] Le Moigne, Jean-Louis, La Modélisation des Systèmes Complexes, Dunod, 1990.

[14] Watzlawick, Paul, Weakland, John, Fisch, Richard, Changements, paradoxes et psychothérapie, Seuil, 1975.

[15] a Argyris, Chris, Double Loop Learning in Organization, HBR, sept-oct 1977

b Argyris, Chris, Savoir pour agir. Surmonter les obstacles à l'apprentissage organisationnel, Intereditions, 1995.

[16] Belbin, Meredith, The Coming Shape of Organization, Butherford-Heinemann, 1996.

[17] Sainsaulieu R. Piotet F., Méthodes pour une Sociologie de l'Entreprise, Presses de la FNSP-ANACT, 1994.

[18] Schein, E, Culture : the Missing Concept in Organization Studies, Administrative Science Quaterly, vol 41, n°2, juin 1996, pp 229/240.

CONCURRENT ENGINEERING MODELS

AnandKumar Balakrishnan, M.Eng., Vince Thomson, Ph.D.
Department of Mechanical Engineering, McGill University
Montreal, Canada H3A 2K6

Abstract

In order to achieve their goals better, more and more corporations are using concurrent engineering (CE) as part of their new product introduction (NPI) process. An investigation of several companies was made to determine if the NPI-CE processes used by various companies were different. Company specific NPI-CE processes along with CE characteristics relative to corporate goals were investigated. The results showed that the NPI processes used by the companies were very similar, and that companies emphasized different parts of the process depending on their goals. Correspondingly, CE processes were structured differently according to process and corporate goals.

1   Introduction

With growing product complexity and more rapid new product introduction (NPI), the demand on companies' product development processes is increasing. As a result companies have started to use concurrent engineering (CE) approaches for new product development. However, the goals driving these companies during their use of CE differ with each organization. The question arises: do companies tailor their product development process according to their goals? Therefore, a research project was organized with the objective of investigating whether there are different or specific models for CE depending on the goals for NPI.

Many companies use new product development as a basis for successful competition. They use different strategies which include: innovative design, timeliness (correct time-to-market), reduced cost, and/or improved quality. Successful product development requires approaches that can organize the process, reduce waste, provide products to meet customers' needs and also respond to global competition effectively.

In general, the NPI process involves the following stages [1]:
1.   Concept generation
2.   Detailed specification
3.   Preliminary design
4.   Detailed design
5.   Product introduction
6.   Volume production

"Concurrent Engineering is a systematic approach to the integrated, simultaneous design of both products and their related processes, including manufacturing, test and support." [2] CE emphasizes the exchange of information early in the development process in order to reduce overall effort. To do this, communication is stressed and accomplished with the establishment of development teams, use of communication protocols (standards), and adoption of technology which aids information exchange [3-5]. In terms of time compression, processes are organized so that many activities can be executed in parallel.

With regard to NPI process design and the use of CE, several questions arise.
- For different NPI goals, do NPI-CE processes follow different models?
- If the NPI processes are different, how and in what way do they differ?
- If the NPI processes are not very different, does the structure of CE differ with different goals?

2   The Study

Companies were chosen which developed new products and which used CE for this development. Various project and CE characteristics were investigated. For the companies, Canadian Marconi Company, Newbridge Networks Corporation, and Nortel Networks, their general NPI process and CE policies were studied by the authors. From this information, hypotheses were developed with regard to the effect of goals on the NPI-CE process. The work by Swink et al. [6], where new product development projects for five companies were studied, was used to confirm and develop further the hypotheses developed for the NPI-CE processes relative to process goals.

Projects were characterized by primary goals (quality, cost, timeliness, innovation), complexity (low, medium, high), innovation (low, medium, high), technical risk (low, medium, high), and team structure. Team structure varied from a single product development team to the use of two types of teams, a core team and support teams.

152

Quality, cost, timeliness and innovation are all important goals; yet, it was found that organizations stress one more than another for all product development or that they stress different goals during the development of different products. Innovation is an important driver for product development and was considered during the research; however, while it was important to many of the companies studied, it was not the primary goal for any of the development projects.

For CE, it is possible to have different types of activities executed in parallel, i.e., activities related to different products, phases within a project, and design functions [6].

- Product parallelism is the simultaneous development of separate, but related products where coordination is required between the NPI teams and possibly between the different products in terms of common designs, manufacturing processes and/or shared development personnel and facilities.
- Project phase parallelism is the simultaneous development of market concepts, product designs, manufacturing processes, product support structures, etc., for the development of one product.
- Design parallelism is the overlap of various design functions, e.g., hardware, software, mechanical, electrical, etc., within a project.

Shown in Table 1 is a list of the companies studied along with the program or project characteristics of their NPI-CE processes. The general NPI-CE processes of Canadian Marconi, Newbridge and Nortel were studied as they applied to several projects, while the CE processes of Boeing Commercial Aircraft Division, Cummins Engine Company, Red Spot Paint and Varnish Company, Texas Instruments, and Thomson Consumer Electronics were studied for a single development project [6].

3 Process Environment

3.1 Boeing Commercial Aircraft Division

Boeing used CE for its 777 aircraft development. In order to achieve a high quality product, Boeing stressed customer and supplier participation in the design process, and attention to detail during the design process and into the test and build stages. Boeing's NPI process encouraged cross-functional integration and communication. Communication between product designers, key suppliers and customers was frequent permitting problems to be resolved quickly. Three dimensional modeling capabilities of the design system allowed the designers to test the fit of parts. The designers could access up-to-date designs for any of the 700,000 parts of the aircraft. Physical prototypes were lab

tested under severe environmental conditions and extensive testing was carried out throughout the project in order to avoid any defects. The product development teams were all co-led by design and manufacturing engineers in order to cut down development cost. Boeing improved coordination of the project by having several product development teams and by dividing the responsibility across multiple levels of their hierarchical organization. By doing so, the teams with the highest degree of interdependency were made to work close with each other.

3.2 Canadian Marconi Company

CMC's product development process is a stage-gate process. Two teams, namely, one core team and one support team are involved in the product development. In addition to these teams, two other teams, namely, a Bid Capture Team (BCT) and a Components Action Team (CAT) are also involved. The mandate of the BCT is to review a proposal from the customer, in full detail, and to ensure the complete understanding of technical and contractual requirements. The CAT includes members from Engineering, Components Engineering and Quality Assurance to build quality into the design of components. The 'voice of the customer' is given great consideration; suppliers also take part during the development. In order to imbue the spirit of leadership in its personnel, CMC emphasizes leadership 'rotation'. The members of the team are also specially trained to work in a CE environment.

As cost is the main goal for product development, there are frequent meetings to review product design and there is an emphasis on using computer based tools for design, test and manufacturing. Simulations are performed during the system design before full scale development and also during the simultaneous design of product and manufacturing processes.

3.3 Cummins Engine Company

Cummins used CE in the development of its heavy duty diesel (HDD) engine. Quality was the main goal for the development process. Cummins NPI team included internal representatives from design, manufacturing, etc., and also external suppliers. An important aspect of this project was to provide a high degree of product customization to the customers. This necessitated extensive experimentation and testing of various design alternatives which also helped to identify the specifications that provided good performance. Manufacturing engineers were involved in all phases of the product design. Prototypes were built using full

Company Project	Boeing 777	Cummins HDD	Marconi	Newbridge	Nortel	Red Spot TPO	Texas Inst. AVFLIR	Thomson DSS
Project Characteristics								
Primary goal	Quality	Quality	Cost	Time	Quality	Time	Cost	Time
Complexity	High	Medium	High	Medium	Medium	Low	Medium	Medium
Innovation	Medium	Medium	Medium	High	Medium	High	Low	High
Technical risk	Low	Low	Medium	Medium	Medium	High	Low	High
Team structure	Complex hierarchy	Core + support	Core + support	One team	Core + support	One team	Core + support	One team
Parallelism								
Product	Medium	Low	Medium	Medium	Medium	None	None	Medium
Project Phase	Medium	Medium	Medium	High	Medium	Medium	High	High
Design	Low	Low	High	High	High	None	Medium	High
Project Priorities								
Quality	High	High	Medium	Medium	High	Medium	Medium	Medium
Product cost	Medium	Medium	High	Low	Medium	Low	High	Medium
Timeliness	Medium	Medium	Medium	High	Medium	High	Medium	High
Innovation	Low	Low	Medium	Medium	Medium	Medium	Low	Medium

Table 1　Comparison of Development Projects with Respect to Project Priorities and Characteristics

scale production equipment whenever possible, thus bringing production issues to the surface early in the development process and spurring interaction among suppliers, designers and production personnel.

3.4 Newbridge Networks Corporation

Newbridge competes with new products in the telecommunications market, and thus, timeliness is the main goal for product development. A team is formed at project 'kick-off'. The members are dedicated throughout the entire project and are all aware of their functional responsibilities. The team does not have a formal leader as the organization emphasizes project oriented roles over organizational roles. The test engineers and hardware design engineers are co-located to improve communication to ensure a more testable design. Design, test and manufacturing use computer based tools. Simulations are performed during design which eliminates problems, thus resulting in fewer design changes. The early involvement of test and hardware engineers makes functional test code available before prototype production. The early involvement of test and hardware engineers contributes greatly to the timeliness of the development project.

3.5 Nortel Networks

Product development at Nortel emphasizes quality such that there is great deal of attention paid to detail throughout the NPI process. One core team and one support team are involved in the product development task and good communication between the members is stressed. The design and test engineers are co-located in order to ensure a good testable design before the prototype phase. Customers take part in the design and testing stages to voice their opinions. This ensures a good quality design for the customer. External agencies for standards and testing also take part in the product development task to check various standards and procedures that enable the product to meet market and industry requirements on time. Tools like DFM, DFA, DFMA and FMEA are used to facilitate design and to reduce the number of failures. Monthly meetings are conducted to review the progress of the project. Though formal training is not given to the members of the product development teams for a CE environment, Manufacturing and Operations are trained to understand each others' concerns.

3.6 Red Spot Paint and Varnish Company

Red Spot provides specialty paints and coatings, primarily to the auto industry. Red Spot's management realized that it was crucial for the company to develop a coating system that could compete effectively in the emerging area of Thermo-Plastic Olefin (TPO).

Timeliness for product development was important in order to satisfy Ford Motor Company, one of its major customers. To mitigate the risks of falling behind in the development of the new technology, Red Spot staff participated in capability discussions and shared information on-site with the original manufacturers of the TPO coatings. Red Spot's marketing and engineering representatives participated directly in defining the needs and uses of the product. Simultaneously, Red Spot engineers rapidly developed and tested numerous coating samples for TPO materials, thus saving a lot of time.

3.7 Texas Instruments

Texas Instruments used CE in its AVFLIR project (airborne vehicle forward-looking infrared system). The AVFLIR converts infrared radiation into visible images and supports video projection, guidance and data processing functions on aircraft. Since controlling product cost was the main goal of the development process, integration of manufacturing planning and design was a priority throughout the AVFLIR development process. Process design activities were started early and manufacturing representatives had a strong say in finalizing the design. Process engineers, NC programmers, and tool designers were co-located with design engineers to address manufacturing concerns. The engineers involved in the teams ensured that the product was affordable, producible, reliable, testable and easily maintainable.

3.8 Thomson Consumer Electronics

Thomson designs and manufactures televisions and peripheral equipment. They used CE in the development of the Digital Satellite System (DSS), a new product for television that ensured consumers a smaller receiving dish, clearer television reception, and a capacity to handle a larger number of channels than traditional home satellite systems. Rapid time to market was important for the development. A high level of communication between internal and external groups which performed parallel activities was formalized to ensure timely information exchanges. Communication was primarily between design engineers and technical experts from different vendors and partner firms. Manufacturing and design personnel were placed in separate teams with separate budgets, and integration of design and manufacturing issues occurred at top levels of management.

4 Results

Previous research into product development has shown

that product development processes are tied to organizational goals [1, 3, 6]. Despite this, the present study found that the NPI processes followed by the organizations in the study were more or less the same. Nevertheless, the emphasis on various stages of the NPI process was different depending on the process development goals (Figure 1). A summary of how different aspects of the NPI-CE process are stressed depending on process goal is given below.

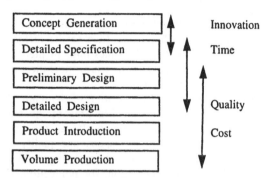

Figure 1 **NPI-CE process with indication of areas emphasized for different process goals**

- Organizations focusing on **innovation** emphasize the activities involved in concept generation and detailed specification. They use small, multidiscipline teams and stress methods for creativity.

- Organizations focusing on **time** emphasize the activities from detailed specification to detailed design. They structure teams to be dedicated, to have autonomy and to make decisions as early as possible in the NPI process. Finalizing design specifications early permits timely development of manufacturing and testing procedures. They stress a high level of communication between internal and external groups. They also ensure a manufacturable and testable design.

- Organizations focusing on **quality** emphasize equally the activities along the NPI process from preliminary design to volume production in order to design quality into the product and to effect quality during production. The companies stress cross-functional integration and efficient communication in design teams. They also use field proven technologies along with extensive experimentation and testing.

- Organizations focusing on **cost**, like quality, emphasize equally the activities from preliminary design to volume production in order to ensure the manufacturability of a design. They stress manufacturing and design integration within the product development team, and thus, foster intense interaction between design and manufacturing personnel as well as the use of design for manufacturability and assembly tools.

Goals Type of Parallelism	Innovation	Timeliness	Cost	Quality
Product	High	Medium	Medium	Medium
Project Phase	Medium	High	High	Medium
Design Function	High	High	Medium	Medium

Table 2   The Degree of Parallelism Used for Specific Goals

It was found that the different emphasis placed by companies during the NPI process (due to the different goals) modified the way that companies implemented CE. The main areas of difference were team integration and the amount of process parallelism.

For team integration, the composition of the team, the intensity of participation, and the life time of the team vary.   As has been discussed previously, as the goals change from innovation to timeliness, cost and quality, the area of emphasis in the NPI process changes (Figure 1).

With regard to team composition, for innovation, CE team members are mostly designers; as the process moves closer to manufacturing, team membership includes more manufacturing people.   This mitigates risk due to the effects of design change when product specifications are quite fluid. People are permanent or part time participants in the process depending on the duration of the project. The longer the project, participants such as, purchasing, NPI, testing, etc. tend to contribute to the project at specific points of the process instead of full time participation such as design.   For timeliness, diverse people are brought together for short periods of time for a concerted effort. The use of cross-functional teams during the design phase helps to zero in on a set of product specifications quickly. For cost and quality, participation in teams is spread throughout the NPI process with many players having specific roles at appropriate times.

The degree of parallelism in a project changes with goals and the type of parallelism used. The results from process analysis are shown in Table 2. The different types of parallelism listed in the table are: product - simultaneous product development; project phase – simultaneous development of marketing concepts, product designs, manufacturing processes, product support structures, etc.; and design functions – simultaneous execution of designs for hardware, software, mechanical, electrical, etc. In essence the more parallelism desired, the more interaction is required from team members. From a different perspective and possibly a more relevant one, the more integrated the team, i.e., the more interaction which already exists, the easier it is to implement parallelism in the NPI-CE process.   Thus, project goals influence the NPI process which in turn influences the degree of parallelism used.

5   Summary

An investigation of the NPI-CE processes used by several corporations revealed that
1.   the NPI processes used were quite similar,
2.   different product development goals resulted in different emphasis put on the various stages of the NPI process, and
3.   the way in which CE was implemented also depended on specific process goals.

In addition to designing the NPI-CE process to accomplish specific process goals, companies set NPI-CE process characteristics to help accomplish company goals, especially where it was necessary to manage a portfolio of simultaneous product development projects. The main characteristics involved in designing the NPI-CE process were the degree of team integration and the type and degree of process parallelism.

References

[1] J. Turino, "Managing Concurrent Engineering: Buying Time to Market", Van Nostrand Reinhold, NY, 1992, pp. 1-12.

[2] L. Trygg, "Concurrent Engineering Practices in Selected Swedish Companies: A Movement or an Activity of Few?", Journal of Product Innovation Management; Vol. 10, No. 5, 1993, pp. 403-415.

[3] R.B. Handfield, "Effects of Concurrent Engineering on Make-To-Order Products", IEEE Transactions on Engineering Management; Vol. 41, No. 4, 1994, pp. 384-393.

[4] V. Krishnan, "Managing the Simultaneous Execution of Coupled Phases in Concurrent Product Development", IEEE Transactions of Engineering Management; Vol. 43, No. 2, 1996, pp. 211.

[5] B. Prasad, "Concurrent Engineering Fundamentals – Vol II", Prentice Hall, NJ, 1996.

[6] M.L. Swink, J. C. Sandvig, V. A. Mabert, "Customizing Concurrent Engineering Processes: Five Case Studies", Journal of Product Innovation Management; Vol. 13, No. 3, 1996, pp. 229-244.

EVALUATING THE PERFORMANCE OF CONCURRENT PROCESSES

Nadia Bhuiyan, Vince Thomson
Mechanical Engineering, McGill University
Montreal, Canada H3A 2K6

Donald Gerwin
School of Business and Department of Systems & Computer Engineering, Carleton University
Ottawa, Canada K1S 5B6

Abstract

A stochastic computer model was built to simulate concurrent engineering processes in order to study how different process mechanisms contributed to new product development (NPD) performance. Micro-models of the phenomena: functional participation, overlapping, communication, decision-making, rework, and learning, were included, and their effects on the NPD process were related to time and cost. The effects studied were: the relationship between up-front, functional participation in the development process versus downstream work reduction; the trade-off between more up-front involvement versus product development time reduction; the cost of process coordination as up-front participation is increased; and the trade-off between reduced development time and more rework as parallelism is increased.

1   Introduction

Product innovation is increasingly being used as a competitive strategy. While market trends are forcing shorter product development times in order to meet time-to-market (TTM) targets, companies are trying to develop mechanisms to streamline their NPD processes. One approach that has provided much success towards achieving shorter TTM is concurrent engineering (CE) [1-5]. CE integrates inter-related functions at the outset of the product development process in order to minimize risk and reduce effort downstream in the process. Because it is a knowledge-based process, the coordination and communication of timely information exchange is a key factor. In manufacturing, departmental or functional organization has little effect on resultant productivity. It is process design as well as the conceptual framework (model) behind the process that is important, for example: transfer lines - subassemblies, part flows; Kanban - controlled, steady flow; and cellular manufacturing - group technology, cell design. The power of these process models is that they inherently provide coordination mechanisms as well as opportunities for building further coordination around them.

Previous researchers, such as Krishnan *et al* [6], Eppinger *et al* [7], and Yassine *et al*, [8] have studied different types of information dependencies in CE and have evaluated how to optimally overlap activities using mathematical and risk/decision analysis techniques. These authors have developed micro-models which describe the interaction and information exchange between detailed low-level activities, and analytically study their behavior. Thus, they view task interaction within a limited context and discuss the effects, but do not couple these results with the downstream payoffs for time compression. AitSalia *et al* [9] attempted to mathematically evaluate the costs of CE versus sequential processes; however, simplifying assumptions do not reflect many complex, real-life situations. The research described in this paper extends these efforts by studying a computer model of the entire NPD process which is composed of micro-models of coordination mechanisms in order to understand how the mechanisms contribute to overall process performance.

1.1   Research Objectives

The aim of the research was to study CE processes through stochastic computer models in order to demonstrate the dynamics of coordination versus work reduction and time compression. Insight was gained into optimum conditions for using CE by 1) studying the use of overlapping activities and functional participation in early phases of the process to reduce NPD span time, 2) understanding how coordination (design of process and team management practices) affects the CE process and contributes to span time compression, and 3) focusing on how the dynamic interaction of actors affects the process and contributes to downstream results. The study also considered the effect of the increased coordination as up-front functional participation and parallelism are increased to reduce span time.

2   NPD Process Definition

The NPD process can be defined as the undertaking of the activities and information gathering required to develop a product from its conceptual design up to its introduction to the market. A generalized process, consisting of organizing, managing and coordinating activities and the resources required to develop a product, is shown in Figure 1.

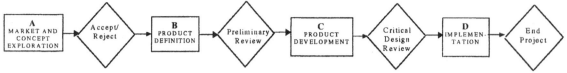

Figure 1   A schematic diagram for a general stage-gate NPD process

The traditional NPD process is sequential in nature where the phases of the process are performed one after the other. This process is highly, functionally segregated with little or no communication between the functions responsible for the different phases. In this process, all steps logically follow one another, and although there is little risk in terms of information transfers since all functions use information that is in its complete and finalized form, it nevertheless produces longer overall project duration and costs.

CE does not eliminate steps in the NPD process. The timing of activities and the coordination of work, however, do change, and the main benefits are from the increase in cross-functional communication, reduction of rework, and very importantly, improved learning processes. CE is concurrency, not only in the sense of parallel activities, but more importantly, concurrency of information. CE uses two main mechanisms to reduce the length of time for NPD: 1) increased information sharing from the start of a project to reduce rework and catalyze learning of the various functional responsibilities (functional participation), and 2) increased overlapping of activities. The relationship between product design and the design of its related manufacturing processes is mutually interdependent [10]. CE can be used as a tool to manage the interdependencies, so that the communication in a sequential process is converted into a bi-directional one. A high degree of coordination is needed in CE to manage interdependent activities. In contrast, there is little need for coordination in a sequential process, where there is only a one-way dependence between phases. While better management of interdependencies through CE does lead to shortened development time as compared to a sequential process, the price is higher cost of coordination.

3   Process Modeling Concepts

Process modeling is a technique used to study the operation of a process by considering it from its functional, operational, informational and/or resource viewpoints. In order to study the effects of CE, conceptual models of an NPD process were created, and then, simulated with a stochastic process simulator. The process models developed in this study included the modeling elements shown in Figure 2. The key conceptual components of the models are represented by process structure, activities, actors, e.g., designers, test engineers, and coordination structures, dependent relationships among activities, that govern communication and information processing.

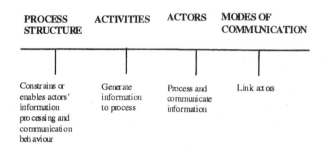

Figure 2   Four conceptual, modeling components

3.1   Process Structure

In this study, process models begin with an overall structure of the organization. This includes the departments consisting of various functions. Within an organization, tasks are required to support the structure and a set of activities is required to perform these tasks. The organizational structure is considered implicitly in that it has no tangible effect on the results; its purpose is to assist in modeling a real organization. The structure of the organization constrains the actors' information-processing and communication behavior.

The NPD process which is modeled consists of a task which must be performed, in this case, a project. This task contains four interdependent phases (A, B, C, D) which may be overlapping (Figure 1). Each phase consists of three sequential activities (A1, A2, A3). Each activity is composed of three units: *work*, which is made up of parallel sub-activities; *communicate*, which is the work needed to coordinate and exchange information; and *feedback*, which is an activity completion decision that indicates i) activity success and moving ahead or ii) activity failure and the need for rework of the *work* sub-activities. Figure 3 shows an overview of the top-level phase-activity model and Figure 4 shows a schematic of the second-level activities which are composed of work-communicate-feedback sub-models.

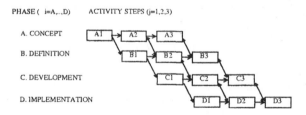

Figure 3   Overlapping of activities

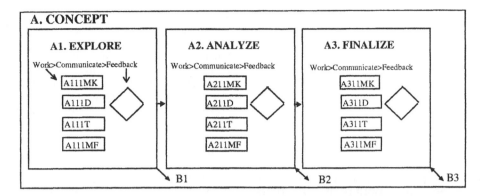

Figure 4   Activity-task model for phase A. The four parallel tasks are performed by personnel from marketing (MK), design (D), testing (T), and manufacturing (MF)

Work-communicate-feedback is the fundamental work loop for NPD teams in the model.

3.2   Activities

Activities are the main unit of work in the process. Activities are divided into sub-activities in the development processes model, but further sub-division is not done. Activities can be value-added production work, where a set of inputs is transformed into a set of outputs, or coordination work, which is required to coordinate activities through communication. Production work is based on specifications of the product, and it directly adds value to the process. Coordination work consists of communication and decision-making which facilitates production work, but it does not add value to the process. *Independent* activities refer to those which are capable of being completed without any communication. *Dependent* refers to activities requiring information from the other activities. Finally, *interdependent* refers to the case when tasks depend on receiving information from one another (arrows depict information transfers). Dependent (single-headed arrows) and interdependent (two-headed arrows) cases are illustrated in Figure 4; no arrow means independent.

Activities have the properties of duration, cost and complexity. Activity complexity defines how difficult activities are with respect to one another. These properties depend on a number of important variables: a) functional participation, b) overlapping, c) communication mechanisms reflecting the way cross-functional teams operate, d) decision-making processes which provide the move forward or rework outcomes depending on reviews and cooperative decisions, e) rework, and f) learning.

The process structure of the models is built on precedence requirements. Each process is a semi-structured process, wherein "all process steps can be identified, but only a partial order of the execution sequence is *a priori* known" [11]. In other words, alternative routings of the process are left open and decisions take place at run time depending on state variables. Therefore, it is partly deterministic, and partly stochastic. In the study, the activities are fixed in terms of initial execution time: this portion is deterministic. Randomness takes place in the decision-making process; alternatives exist at decision points, and each alternative is

tagged with a probability of execution. Randomness or uncertainty is reflected through potential decision alternatives and loopbacks. Initial activity durations were chosen to be approximately equal to average cycle times for general, complex NPD processes. Thus, model execution times for different processes are relative.

3.2.1 Functional participation

One of the two main drivers for achieving shorter cycle times using CE is functional participation. Functional participation is the collaboration of individuals from different functions on a team, where participation can be at different levels of effort for different activities during the NPD process. The percentage of time that an individual participates in a task is referred to as the percent functional participation.

Functional participation is modeled by assigning each member of the team who is presently in a non-traditional activity, i.e., an activity in which they would not normally participate in a sequential process, a percentage of the actual duration of the activity. For example, if a design activity has a duration of 5 days, and each non-traditional function dedicates 25% of their time, a designer would participate during the entire length of the activity (5 days), while the remaining functions would devote only one quarter of their time (1.25 days) to the activity (Figure 5). For modeling simplicity, it is assumed that all non-traditional functions participate in equal proportions throughout the process in a given model, and that all functions begin the activity simultaneously working full-time until their task time expires.

In the sequential process, it is assumed that each phase is executed by only one function: Marketing is responsible for the Concept phase; Design Engineering for the Definition phase; Test Engineering for the Development phase; and finally, Manufacturing for the Implementation phase. This represents the sequential process with no functional interaction between sub-process activities. In the CE model, a cross-functional team is created by having each resource participate in all activities. The percentage of their participation is explicitly defined for each scenario.

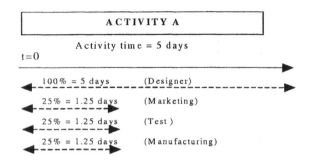

Figure 5   An example of an activity in which assist functions are participating at 25% work levels

3.2.2 Overlapping

Another important element of CE is the overlapping of product development activities. The sequential NPD process proceeds in a serial fashion, where information from one activity gets handed off to the next only after the completion of the initial activity. When activities are done concurrently, information is transferred at many points due to the execution of the two activities. There is a relatively high degree of risk and uncertainty in the early stages of a CE process during the transfer of incomplete and often highly complex information between functions, which tends to introduce more opportunity for rework. This makes overlapping difficult to manage; therefore, the potential risks associated with overlapping must be carefully examined to ensure that added effort and reduced quality are kept to a minimum [6].

Overlapping of activities describes how early a downstream activity B starts with respect to an upstream activity A (Figure 6, 7). Increased levels of overlapping guarantee an earlier message exchange/transfer between the two activities. The goal is to freeze acceptable specifications from activity A early in the process, which B can then use. However, there is a clear tradeoff between early message exchange and the corresponding risk it entails. The earlier the downstream activity B starts with respect to A, the more uncertain is the information B receives. Thus, whenever there is a possibility of a change within A, activity B risks a higher amount of rework.

Overlapping of tasks is modeled through precedence relations among activities, which are controlled by the decision-making process. When activities are independent or sequential, there are no mutual activity interdependencies, and therefore, little or no need for coordination. When activities are reciprocal or mutually interdependent, team members communicate with one another at many points in the process to decide if the partial information which they have exchanged is sufficient or acceptable to begin a downstream task. If members of the team decide that partial information is enough to begin downstream work early, then, overlapping takes place. Otherwise, there is a delay while they wait for more or better information to be available. They may continue to deem information insufficient to start downstream tasks, and in the extreme case, they will simply wait for the upstream activity to finish before beginning the downstream one; the process, then, becomes

sequential. There is a relationship between functional participation and overlapping of activities; in general, the higher the degree of functional participation, the greater is the potential for overlap. For the scenarios described in this paper, each of the decision alternatives has equiprobable outcomes, i.e., the alternatives of whether or not to proceed have equal probability. Depending on the situation, different probabilities of outcomes can be assigned to model various risk alternatives.

3.2.3 Communication

Communication mechanisms refer to the way in which information is transferred or exchanged between two actors in the NPD process. The communication media used is not considered; this assumption is expected to have a negligible effect on the outcome since differences in communication time are not expected to be substantial with respect to production time. Thus, this study assumes that the richness of communication, i.e., the availability of the appropriate communication media, is already embedded in the organization. Communication is facilitated by the process and coordination structure. These mechanisms reflect the type of activity interdependency that can take place. The communication between phases A and B describes the way that the two activities are executed relative to each other. In this study, two communication models are considered: sequential and mutually interdependent. In the first case, there is a one time, unidirectional information transfer from phase A to B, which occurs when A is finished. At the end of each phase is a 'gate' (stage-gate process) where an approval is required in order for the next phase to proceed. The coordination activity, i.e., decisions, at these gates reflect a high level of risk and a high probability of rework. The means of coordinating work between activities is through a simple notification process where A notifies B that it may begin, or through mutual interdependency, where teams in activities A and B work together and mutually depend on one another for information in order to proceed with their work. A and B are closely linked with a frequent exchange of messages. The coordination mechanisms are cross-functional teams (modeled through the functional participation in the work-communicate-feedback activity loops), activity partitioning (dividing phases into activities, which in turn are composed of work sub-activities), parallel activities (modeled through the parallel work sub-activities within activities), and frequent reviews of the development process (modeled through decision points throughout the process). Figures 6 and 7 show the communication patterns for sequential and mutually interdependent models; the arrows indicate the points in time that transfers take place and the direction of the messages. The case of independent, parallel phases was not considered in this study.

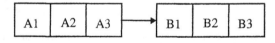

Figure 6   Sequential communication model

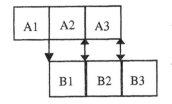

Figure 7 Mutually interdependent parallel activity model

3.2.4 Decision processing

A decision process begins with a plan of action consisting of alternative tactics. These alternatives must be formulated, evaluated, and appraised, and then, action is taken based on a decision. After appraisal, there may be a need for refinement or feedback. In this study, the decisions to be made are those of sufficiency and acceptance. Sufficiency refers to a decision made by a downstream function regarding whether or not information transferred from an upstream function is 'enough' for the downstream activity to begin. Acceptance refers to a decision made by the team leader regarding whether or not information exchanged within a team is satisfactory for one or more downstream activities to commence or continue their present work. If the information is acceptable, subsequent activities within the same phase and in the following phase can proceed, thus creating an overlap of activities. However, if the information is unacceptable, there is partial or complete rework for the activities involved. At each decision point, alternative tactics are available to the decision-maker, including the decision to move ahead, or do complete or partial rework for the current activity. Each of the alternatives is assigned a probability of success in the model. Once the alternatives are appraised, a decision is made. The CE-NPD model is structured from an information-processing point of view, where the decisions being made reflect the inherent risk of using incomplete information. So, if a given activity has a high risk of proceeding with partial and incomplete information, the outcome of the decision point at the end of the activity will exhibit a high probability of rework. Decisions are made at many points in the process. The user of the model can choose probability distributions across alternatives based on intuition and/or experience of an NPD process.

3.2.5 Rework

The two types of rework in the model are churn and design versions. Churn is the rework attributed to the informal changes to specifications as they are being determined. These changes take place prior to the formal adoption of the specifications, and therefore, they occur early in the process. Increased functional participation increases the amount of up-front effort, thus increasing the amount of churn in upstream activities, but reduces the probability of design versions since the quality of specifications being passed onto subsequent activities increases. This is significant, as downstream errors tend to be more consequential than those made upstream. Design versions reflect major rework due to decisions at

formal reviews at the end of a phase. They involve revisiting major work items. If at a review point, a new design version is created, entire sections of specifications need rework, although at a fraction of the original effort. However, as they affect entire sections of specifications, they contribute significantly to overall rework cost and process timeliness.

There are two critical drivers for rework: level of overlap and functional participation. As each of these variables is increased in the process, the level of effort and risk which occurs early in the process increases, thus increasing the amount of churn in the upstream activities and simultaneously reducing the probability of design versions downstream. Churn tends to involve a smaller amount of risk and rework since it means making changes early in the process when the design has not yet been solidified as compared to design versions where a great deal of change, and thus, time and cost, are likely to be necessary to rectify problems. For churn, rework decisions occur after the completion of an activity, while decisions for design version rework occur at the end of an entire phase (Figure 8). The specific reasons for churn or design versions are not of concern in this study, only the fact that the probability of occurrence depends on the amount of overlap and functional participation in the NPD process.

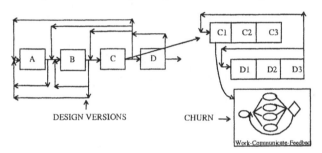

DESIGN VERSIONS   CHURN →

Figure 8 Rework decisions occur at the end of phases for design versions and at the end of the work-communicate-feedback cycles for churn

Activity uncertainty is modeled through the probability of reworking activities and/or phases in the process. Decision points are present at the end of each activity and phase. Decisions at the end of activities allow churn to take place in CE models, while decisions at the end of phases allow design versions to take place for all models. Churn and design versions are reflected in the randomness of the duration of a phase/activity. The risk inherent in one process model as compared to another is implied through the probabilities of churn and design versions occurring, and these are related to a) the timing of information exchange (overlapping) and b) to the degree of functional participation. The graphs shown below in Figures 9a, 9b, and 9c were built on assumptions based on the literature and experience. These distributions were used to populate the CE-NPD models.

Probabilities of churn and design versions versus functional participation are shown in Figure 9a. These graphs demonstrate that an increase in overlapping of activities requires a corresponding increase in functional participation to minimize the effect of rework due to increased design versions. However, this will create more churn. There is a trade-off, i.e., partial information exchange involving some level of acceptable change early in the process (churn) mitigates downstream risk through avoidance of design versions. Conversely, the higher the overlapping of activities and the lower the functional participation, the higher is the probability of design versions, but the lower the probability of churn. Increasing overlapping without any functional participation has the effect of increasing the probability of design versions only (Figure 9b).

The higher the degree of functional participation, the higher the probability of churn in the early phases, but the lower the probability of design versions in all phases (Figure 9a). Conversely, if there is little functional participation in upstream activities; then, there is a higher risk of design versions taking place. In the sequential process model, there is no churn when there is no functional participation, and there is a high probability of design versions in all phases except A. This is so, since it is assumed that once a new product idea is accepted, there is little probability of revisiting the preliminary activities involved in scoping out that idea. However, as functional participation increases with no overlap, churn increases to a limited degree, and design versions decrease (Figure 9c).

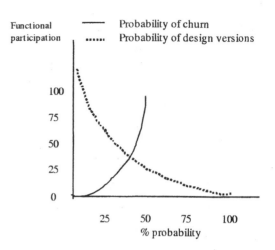

Figure 9c   Functional participation versus the probabilities of churn and design versions

3.2.6 Learning

Learning curves are used to capture the effect of knowledge accumulation during a given task which reduces effort during rework. Various learning curves are used as effort multipliers. A high effort multiplier is used for a sequential process, and smaller multipliers are used in accordance with increasing levels of functional participation. The higher the functional participation, the smaller the multiplier is, since it is assumed that more learning is taking place during activities through cross-functional interaction. As rework is initiated, the duration of activities is reduced and the probability of having to do further rework is also reduced. This occurs at each rework loop.

3.3.7 Actors

Actors are the resources that represent the functional roles required to perform the activities of a task. These resources process the production work and communicate and coordinate information. Resources are defined by skill type and allocable time. The model in this study assumes a limited resource capacity level, which realistically captures the effect of time delays due to resource constraints.

In the models, the activities are executed by the responsible resources, which are part of the supporting organizational structure. The various resource groups that participate in an NPD process have been grouped into three teams: the Business Unit, R&D and Operations. The Business Unit team consists of Marketing specialists. The R&D team consists of Design Engineers. The Operations team consists of Test Engineers and Manufacturing personnel. It is also assumed that an actor's skills match the task requirement; that is, each resource assigned to a task is qualified to perform that task. The costs for different resources are assumed to be equal.

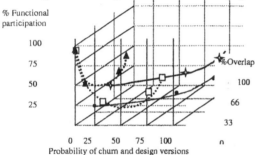

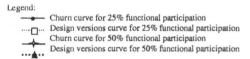

Figure 9a   A three-dimensional plot of the probabilities of churn and design versions versus functional participation and overlap

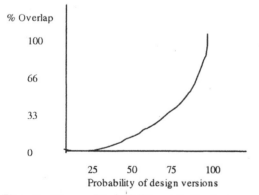

Figure 9b   Probability of design versions versus overlap

3.4 Validation

The models were tested both internally and externally. Internal validation took place by running simulations and comparing results against predictions, while external validation was effected through a case study. Data for the case study was collected at the company site. This data included processes, subprocesses and their corresponding structures; activities, and their corresponding durations, complexity, and probabilities of rework; and resources which perform the tasks. This data was used in the model design and probability distributions for decisions. The results of the simulations were compared to actual CE-NPD process performance to validate the models.

4 Computer Modeling and Simulation

In this study, a process modeling and simulation approach was used; stochastic models of a generic NPD process, both sequential and CE, were built and simulated. A number of scenarios of NPD processes were simulated to study process parameters; the scenarios were: a strictly sequential process, a sequential process with varying levels of functional participation, a CE process with varying levels of overlap, and a CE process with varying levels of overlap combined with functional participation. The purpose of stochastic modeling was to simulate micro-level actions (functional participation, parallelism, etc.) as well as interactions among actors, and to measure the corresponding macro-level effects on sequential and CE processes through the performance indicators of project effort, project duration, coordination effort, and rework. Results showed the relationships between functional participation, and effort and span time (TTM), and between overlapping and effort and span time. For NPD, cost was considered equivalent to time since the major cost in a development process is worker wages.

The process modeler/simulator used was FirstSTEP™, a product of Interfacing Technologies Ltd. (Montreal). It is a simulation software which allows mapping of processes as a network of activities, dynamic simulation of the model, and quantitative analyses such as resource usage, bottlenecks, cost and duration of activities. FirstSTEP™ uses discrete-event simulation. FirstSTEP™ technology is based on object-oriented programming. Its simulation engine has the usual components: state variables, simulation clock, event list, counters, random-number generators, etc. The main program adopts the innate advantages of a process approach, i.e., it allows the user to create a model in terms of process building blocks, which are transform, transport, store/retrieve, and verify.. Probability distributions can be chosen for all activity durations in the model, including those of the activities and the decision points (which require durations as well, e.g., a decision may require a meeting to be held). These distributions are attributes of the activities in the process modeler.

The base CE-NPD model as described in this paper was composed of about 1000 activities.

5 Results

The variables of interest in the study are functional participation and overlapping of activities; their respective effects on effort and span time from simulations of the NPD models are shown in Figure 10 and 11. Specifically, the curves are the outcomes of increasing levels of functional participation and overlapping against effort and span time parameters.

In Figure 10, initially increasing functional participation from 0-25% slightly reduces overall effort, but further increases begin to show an upward trend. This is indicative of the fact that putting more people on a task will substantially increase churn. Note that span time significantly decreases as functional participation is increased from 0-25%, but that higher levels increase span time (Figure 11).

Increasing the level of overlapping has the effect of decreasing span time (Figure 11). This decrease in span time is accompanied by a decrease in effort with a 0-33% change in overlap, but a 33-66% change in overlap considerably increases effort (Figure 10). Thus, there is a trade-off between reduced span time and the level of effort.

As shown in Figure 12, there is a minimum point for effort versus span time for the amount of overlap in activities. As the amount of overlap starts to increase, there are efficiencies in reducing effort and span time; however, as overlap further increases in the process (33-66%), span time decreases, but effort increases because of the greater effect of rework due to design versions. The curve for functional participation in Figure 12 shows that for the modeled process there is a limit to the beneficial effects of greater teamwork. After a point (25%), further increases in functional participation lead to an increase in span time and a slight increase in effort.

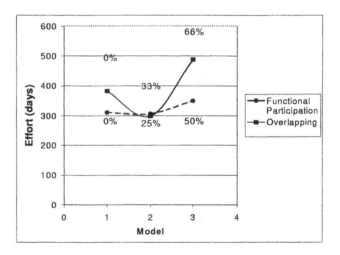

Figure 10 Effort versus different levels of functional participation and overlapping

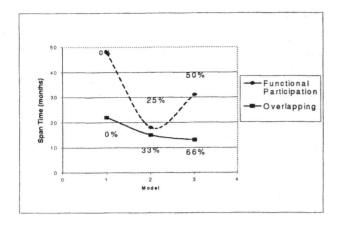

Figure 11   Span time versus different levels of functional participation and overlapping

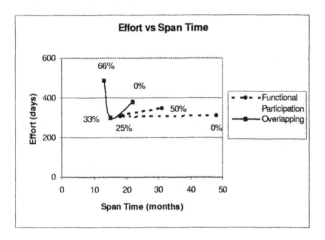

Figure 12   Effort versus span time for different levels of functional participation and overlap

6   Conclusions

The study showed that models of an NPD process can be built being composed of micro-models of concurrent engineering features to demonstrate macro-effects in the process. Results have shown the tradeoff between reducing span time and the amount of effort spent during the process. Knowing the precise shape of the effort versus span time curve for different variables is important since it would allow companies to design their development processes better.

With respect to managerial implications, the results of the study can be useful in terms of coordinating teams in a highly integrated process (identifying appropriate levels of functional participation), scheduling activities in the process (identifying appropriate levels of overlapping), evaluating alternative process structures (varying levels of functional participation, overlapping, learning, and risk), and estimating the completion time and cost of a project (obtained as results of the simulations).

Future research will be focused on understanding the interaction of the various parameters in the CE-NPD process in order to be able to optimize processes for span time and level of effort. In addition, sensitivity analysis will be done to understand the importance of team learning and rework probabilities on reducing process effort and span time. Finally, work needs to be done on how the completeness or certainty of information impacts the design of the NPD process to deliver shorter span times and reduced levels of effort.

References

[1]   Haddad, C. J., "Operationalizing the Concept of Concurrent Engineering: A Case Study from the U.S. Auto Industry," IEEE Transactions on Engineering Management, Vol. 43, No. 2, May 1996, pp. 124-132.

[2]   Handfield, R. B., "Effects of Concurrent Engineering on Make-to-Order," IEEE Transactions on Engineering Management, Vol. 41, No. 4, November 1994, pp. 384-393.

[3]   Hauptman, O. and Hirji, K. K., "Influence of Process Concurrency on Project Outcomes in Product Development: An Empirical Study of Cross-Functional Team," IEEE Transactions on Engineering Management, Vol. 43, No. 2, May 1996, pp. 153-163.

[4]   Prasad, B., Concurrent Engineering Fundamentals-Integrated Product Development, Volume II, Prentice Hall PTR, New Jersey, 1997.

[5]   Shina, S. G., "New Rules for World-Class Companies," IEEE Spectrum: A Special Report on Concurrent Engineering, July 1991, p. 23.

[6]   Krishnan, V., "A Model-Based Framework to Overlap Product Development Activities," November 1993, p. 3.

[7]   Eppinger, S. D. and McCord, K. R., "Managing the Integration Problem in Concurrent Engineering," August 1993,

[8]   Yassine, A. et al, "A Decision Analytic Framework for Evaluating Concurrent Engineering," Working Paper, Wayne State University.

[9]   AitSalia, F. et al, "Is Concurrent Engineering Always a Sensible Proposition?", IEEE Transactions on Engineering Management, Vol. 42, No. 2, May 1995, pp. 166-170.

[10] Tian, H., Xu, W., Wend, H., Wu, Q., "A Review Upon Concurrent Engineering", The 9th IFAC Symposium on Information Control and Manufacturing, France, June 1998, Vol. II, p. 511-516.

[11] Vernadat, F. and Mhamedi, A., "The ACNOS Approach for Performance Evaluation of Enterprise Processes and Activities", IFAC, 1998, p.264.

CHAPTER 3

Information Modeling

An Object-Oriented Database System to Support Concurrent Design

T. W. Carnduff[1], W. A. Gray[2], J. C. Miles[3], A. Al-khudair[2]

[1]: Information Systems and Computing Division, University of Wales Institute, Cardiff, UK
[2]: Department of Computer Science, Cardiff University, UK
[3]: Cardiff School of Engineering, Cardiff University, UK

Abstract

We have been investigating the use of object-oriented databases to support the activities of teams of designers working concurrently on large design projects. This paper describes the results of these investigations. We present an analysis of the requirement for object versioning and configuration management, data distribution with controlled access between individuals and groups of designers and the need for integration between database systems and CAD tools. We conclude that commercial object-oriented database systems are deficient in all of these areas and proceed to show how we have produced a federated distributed database system which supports design object versioning and configuration management within an integrated CAD tool environment. The system has been evaluated by practising designers to establish its validity.

1. Introduction

This paper describes an object database system which is a component of a design-engineering environment that supports concurrent work by a design team. The focus of the DESCRIBE[1] project, which is creating this design environment, is that of design reuse in concurrent engineering. In this project the focus of our experimental application is bridge design where bridges consist of components, such as foundations, abutments and bridge decks with their constituent beams and other elements.

Designers of complex artefacts, seldom design without reference to previous work, rather they adapt features of previous designs into current designs when they recognise familiar design features and constraints [1]. Thus they need to be able to recover designs and identify components they wish to reuse and modify, in their evolving design. This design environment is one in which

the design artefacts associated with a project are semantically and structurally complex, and the designers involved have different skills and concentrate on developing different aspects of the design. In this situation it has been claimed that concurrent working can bring benefits, such as a shorter design period [2, 3]. The designers work concurrently and are organised into groups. In this environment the need for inter-designer communication is paramount to inform other designers about changes in the evolving components. There is also an important requirement that each component can have several versions, which are in different states. This allows a group to evolve a design, and let other groups view it when they need to for any reason connected with their design task. A computer system to support such a design environment must support the evolutionary nature of the design process amongst designers, who are frequently distributed across the nodes of a local or even wide area computer network, in a flexible communication environment. An important component of such a design environment is the database system, which will control access to the different versions of the evolving design components by the members of the different groups. It should also be able to manage the evolving design configurations and the connections between components. In concurrent design, as well as the designer being involved, those who are to manufacture the product also have an input to ensure that the final design meets their needs.

2. Requirements for Object-Oriented Database Support of Concurrent Design

Although product design has been traditionally performed as a series of consecutive tasks in a process known as sequential engineering, the demand for shorter design times coupled with the need to create better quality products, has led engineers towards concurrent engineering. This is a design process whereby some or all

---

[1] This work was supported by EPSRC grants GR/K29241 and GR/K29234.

of the product development phases, including the design phases are executed in parallel, with the designers involved undertaking different aspects of the design task and using various methods of communication to make the other designers aware of their evolving design. It is claimed that substantial gains are being achieved using concurrent engineering [2, 3]. We have determined that a computer system that aims to support the concurrent design process should provide the following features [4]:

- *sharing of information and coordination of the engineering process* to provide for the sharing and management of all the available project resources (human expertise, computer tools, design data, etc.);
- *conflict identification and resolution* providing for the compromise of the possibly different views of engineers from different backgrounds or with conflicting priorities and the need to maintain the connections between the different components and ensure that there are no constraint violations at these connections;
- *versioning and configuration management* of design artefacts as a means of maintaining the project's history and design alternatives, as well as for representing structurally complex design artefacts. This involves storing all the information about a component including the design and the details produced by its structural analysis.

In addition, when we consider providing ease of use for the designer, the system should provide:

- *transparent integration with CAD tools* with which the designer is familiar, i.e. we should not be placing unacceptable learning curves on designer users due to the introduction of new tools.

Furthermore, as we intend that the design-engineering environment should support concurrent working, the computer system that supports it should also provide:

- *design autonomy* which will allow groups to work independently on components of the design or features of the complete design and communicate their evolving ideas to other members of the design team.

We seek to provide these system requirements within a computer design environment using an object-oriented database system to hold the drawings and structural details of the versions and configurations of the design components. We chose an object-oriented data model to support the storage of our design data as it offers strong

data modelling support for the persistent representation of engineering design artefacts [5], concurrent access and sharing of data is a mature feature of commercial database systems and object versioning and configuration management have been widely investigated in object database research for more than a decade [6].

2.1 Sharing of Information and Coordination of the Engineering Process

An environment supporting concurrent design must allow the contributing designers to share design components in some controlled manner and to communicate design information to other designers. One means of supporting this environment is to extend a versioning system to classify each object version as transient, working or released [6, 7, 8, 9, 10, 11]. These versions are stored in a hierarchically organised design database with private, subproject and project partitions, respectively. This has been widely suggested in the literature but as far as we know has not been tested in a working environment. We conducted an experiment in group design in concurrent engineering [12], with teams of students designing different components concurrently and noted the communication requirements between the teams. This led us to modify the above model (see section 4.1) as it led to a too rigid communication framework between the designers.

2.2 Conflict Identification and Resolution

Conflicts may arise between individual designers working on the same component or on different interconnected components of a design. It is normal in a hierarchically organised design database (see section 2.1), that conflicts at one level in the hierarchy are resolved at the next level up, so for example conflicts between two individual designer's private databases, would be solved at the subproject level. This is imposed by the hierarchical model and was found in our experiment to be too rigid, as the teams needed to communicate and resolve conflicts at the same level in the tree.

2.3 Versioning and Configuration Management

The evolution of designs through a series of changes can be recorded through the facilities of an object versioning system. Katz provides a survey of object versioning in [6] where he compares this feature in several early object-oriented systems. Later approaches to object versioning are described in [9, 10, 11]. Systems providing object versioning to support concurrent engineering design should provide the following facilities:

- A means of imposing version information on objects, such as version number, parent version number and a reference to the version graph to which they belong. An object containing versioning information is known as an object version.
- The capability to represent a version evolution graph which records the three types of version derivation operation, sequential refinement, alternative development and consolidation or merging. The nodes of the graph are either the object versions themselves or references to these versions. The objects which contain this graph and the operators required to manage the graph in terms of the three types of version derivation and access to the object versions, are know as generic objects. For any versioned object there will be only one generic object and potentially many object versions.
- The ability to store the design with its structural features so that component connections in versions can be checked for constraint satisfaction.

Configuration management is necessary in the context of a composite object which forms the root of a composition graph, each of whose nodes is a component of its parent node and has its own version evolution graph [6]. If, for example, a composite object has m components, each having n versions, then there may be up to m^n possible different configurations of the composite object. A configuration is a set of mutually consistent versions within a composition graph. One of the issues which has to be dealt with in a versioning scheme which has some configuration management capability, is to determine how the derivation of new versions of artefacts positioned at lower levels of a composition graph affect the versions of artefacts positioned at higher levels. Broadly speaking there are two approaches to handling this problem. In the first approach the propagation of the derivation of new versions at higher levels resulting from changes resulting in new versions at lower levels in a composition graph is left to the user of the system, with no system intervention [11, 13]. Systems employing the second approach utilise some form of change propagation, which is managed by the system with varying levels of intervention from the user [7, 8, 9, 10, 14, 15]. With either method it is important to have the structural information about the component available with the design so that connection constraint violations can be resolved.

2.4 Transparent Integration with CAD Tools

Having completed the concept design during which sketches will have been produced and initial calculations will have been undertaken, design usually progresses to a detailed drafting and artefact performance calculation stage using CAD drafting tools such as AutoCAD© [16] with unconnected finite element analysis and other routines being used to analyse the behaviour of the design. AutoCAD is a design tool for the creation, manipulation and visualisation of 2-dimensional and 3-dimensional representation and solid models. Solid models are 3-dimensional drawings, which include mass and material properties. Such tools are in wide use by designers in such diverse areas as civil, mechanical and electronic engineering and the designers are therefore familiar with their interfaces and features.

2.5 Design Autonomy

It is important that individual designers or groups working collaboratively on a design are able to work on their design in isolation from other versions of this design, being developed concurrently for different purposes, so that it can evolve to meet the design aims of these designers, without concern for alternative possibly conflicting goals. When such a design is stable it can be released for other designers to access and determine its effect on their project goals. This autonomy is important if design progress is to be made in a concurrent engineering environment.

3. Commercial Object Database Support for Concurrent Design

We have investigated the support offered by commercial object database systems to concurrent design through examining the degree to which they offered the five system requirements identified in the previous section. The prominent object database systems investigated were GemStone, ITASCA, O2, ObjectStore, POET and Versant.

✔: Feature supported   ✗ : Feature not supported   ? : No data

Criterion	GemStone	ITASCA	O2	ObjectStore	POET	Versant
1. Standards	OQL	?	Fully ODMG compliant ODMG data model OQL C++ binding	ODMG	ODMG OQL C++, Java Binding	ODMG C++ binding
2. CAD Application Support	✗	✗ Through STEP	✗ Through ODBC, CORBA or STEP	✗	✗	✗
3. Distributed DB Management	✔	✔	✔	✔	✔	✔
4. Transaction Management	✔	✔	✔	✔	✔	✔
4.1 Long-duration transactions	✗	✔	✔	✔	✔	✔
4.2 Shared (cooperative) Transactions	?	✔	?	✔	?	?
4.3 Distributed Transaction Management	✔	✔	?	✔	?	?
5. Version Management	✔	✔	✔	✔	?	✔
5.1 Version derivation hierarchy	✔	✔	✔	✔	?	✔
5.2 Version states (transient, working, released)	?	✔	?	?	?	✔
5.3 Current version	?	Latest or user defined	?	User selection	?	?
5.4 Version merging	?	✔	✔	✔	?	?
5.5 Version derivation (linear, non-linear)	?	✔	✔	✔	?	?
6. Configuration management	?	✔	✔	✔ composite object type	✔ via cross-database references	✔
6.1 Versioning of object configuration	?	✔	✔	?	?	?
6.2 Object sharing	✔	✔	✔	✔	?	✔
7. Workspace management	?	✔	✔	✔	✔	✔
7.1 Private/shared workspace	?	✔	?	?	?	✔
7.2 CheckIn/CheckOut of objects	?	✔	✔	✔	✔	✔
7.3 Private objects in shared workspace	?	✔	?	?	?	?
7.4 Workspace authorisation	?	✗	✔	?	?	?

Table 1: Comparison of Prominent ODBs

Information on these commercial systems is based on a combination of direct investigation of the software (where available), the technical manuals and the World Wide Web pages for each of these database system vendors. Table 1 shows the features investigated for each of the database systems and a coarse indication of the outcome of this investigation. Our analysis of these commercial systems was much more comprehensive than is indicated in this paper, but due to the focus of the paper and the space limitations we have chosen to limit the information reported here.

3.1 Sharing of Information and Coordination of the Engineering Process

Most of the database systems studied support the concepts of workspaces and versions at some level (see section 3.3). Workspace management may be used to provide contexts for both shared and private work. In ObjectStore, for example, when a user wishes to work on a shared version of a configuration residing in a particular workspace, he/she typically creates a child of that workspace, and then checks out the configuration's version from the parent workspace as a new version in the child workspace. When he/she finishes working on the new version, the user could check it into the parent workspace and make it available to other users. A parent-child hierarchy of workspaces created in this way provides many levels of privacy and sharing. This would provide a level of support for the hierarchical model described in section 2.1

ITASCA seems to offer the most comprehensive degree of support for the sharing of information and coordination of the engineering process, whereas we believe that neither GemStone nor Poet offers much in the way of direct support.

3.2 Conflict Representation and Resolution

None of the studied database systems appears to provide explicit mechanisms for managing conflicts between individual designer's design artefacts, except in the support that they offer for a hierarchical version model as discussed in the previous section.

3.3 Versioning and Configuration Management

All except one of the studied database systems offered support for object versioning. ObjectStore's versioning scheme was based on the concepts of configurations and workspaces. A configuration consisted of one or more objects that were to be versioned together. This definition is not in accord with that which we gave in section 2.3, where we presumed that it is the objects themselves that are versioned and then grouped into configurations. The versions of a configuration were structured into a single rooted tree, version derivation graph. ObjectStore did not store complete object versions, instead it provides a differencing mechanism by storing backward deltas as described in [17]. We contend that although deltas may be useful for versioning relatively simple objects, with the very complex objects found in typical engineering designs, a versioning system that supports complete object versions is more appropriate, especially in the context of a multiple designer environment where design objects must be capable of being accessed in their entirety without time-consuming object version reconstitution. ObjectStore started from the viewpoint of the configuration, and to version at a lower level of granularity, the user had to create subconfigurations. We contend that it is more intuitive for a versioning system to offer direct support for direct object versioning, with the means to group object versions within a composition graph, into configurations as this reflects how engineers design.

ITASCA offers the strongest support for versioning and seems to be in accord with the model proposed in the Orion project [7, 18, 19, 20]. This support extends through to a configuration management scheme which groups object versions within a composition graph.

To be versionable in O2 an object must belong to the class VERSION. As an instance of VERSION is made up of an arbitrary collection of O2 objects, this is quite similar to the ObjectStore configuration and thus suffered from the same shortcomings. Thus these versioning and configuration management sub-systems would not support our requirements in this area

3.4 Transparent Integration with CAD Tools

None of the database systems incorporate any features that directly offer transparent integration with CAD tools, although ITASCA and O2 support the means to integrate with CAD tools through the use of middleware. Furthermore, none of the database systems performs an analysis of the properties of design artefacts represented by CAD tools such as AutoCAD.

3.5 Design Autonomy

As none of the studied database systems, bar one, offered direct support for the sharing of information and coordination of the engineering process, they could not provide the degree of design autonomy that we propose. ITASCA, on the other hand, appears to offer good support within the context of the hierarchically organised design database described in Section 2.1. Unfortunately, we have been unable to put these facilities to the test as we do not posses a copy of the software at the time of writing.

4. The DESCRIBE Object Database System

4.1 Sharing of Information and Coordination of the Engineering Process

We have produced a data model, which supports concurrent design by individual designers working on different aspects/components of a design project [21]. This distributed design data model came about as a result of an experiment to determine the way in which designers work within a team [12]. We built on the notion of transient, working and released versions residing in various design project databases as described by [7, 13] to produce the architecture illustrated in Figure 1. Each designer has a local database partition managed by a DESCRIBE environment, consisting of a private and a public partition. The private partition consists of transient versions of design components while the public partition holds working versions and released versions. The latter two types of version cannot be amended but are available to other designers under conditions related to their status, where a released version is the current design version of the artefact. Version creation and classification is undertaken in the public partition and in a controlled manner between designers' private partitions through check-out and check-in operations. The three version classifications enable the different types of inter-designer communication needed to relay the current state of components and versions to other designers working on different aspects of the project concurrently. These different types of communication and how the versions support them were identified by the analysis undertaken in the experiment and are described fully in [4, 21].

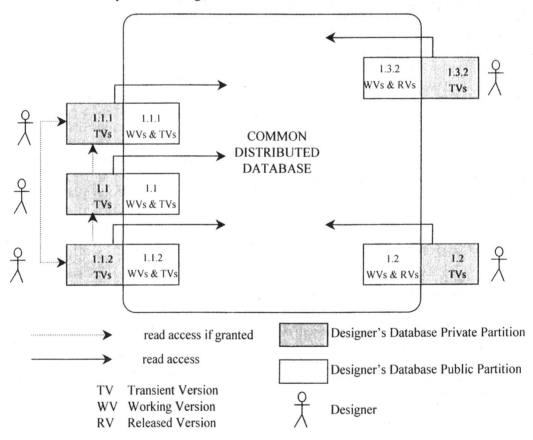

Figure 1: DESCRIBE Distributed Data Model

4.2 Conflict Resolution

This is undertaken by the designers, who detect conflicts and resolve them using the information provided by our system. The storing of versions of the CAD designs with their calculated behavioural properties is an important aid to designers in their work as it assists them to detect design conflicts and resolve them.

4.3　Versioning　and　Configuration Management

Object versioning in DESCRIBE has been described in [22], but in essence versioning capabilities are imparted to design artefacts by inheritance through the features presented in our two design classes VERSIONABLE and GENERIC, which when instantiated, produce versionable and generic objects, respectively. This version model provides for all three version-derivation operations referred to in the previous section, but of particular significance to design reuse, it includes the means to merge two object versions to produce a new object version [23]. The principle behind this approach is that, given a previous design artefact stored as an object in a database, we can merge it with another previous design or a developing design to incorporate those features of each that we want to embody in a new design. This will be further described in section 4.5.

A novel approach has been taken to configuration management in DESCRIBE, in that the means of dealing with the management of the versioned objects in a configuration is conceptually identical to the approach taken in managing the object versions in a version evolution graph. The two classes used to achieve this are called CONFIGURABLE and GENERIC_CONFIGURATION. Each of the object versions from a composition graph, which contributes to a configuration through inheritance, is an instance of the CONFIGURABLE class. There is a single generic configuration object – an instance of GENERIC_CONFIGURATION – for each configuration, and this object contains a graph whose nodes are references to the object versions comprising the configuration. Any object version can belong to more than one configuration, unless it is at the root of a configuration graph.

4.4 Transparent Integration with CAD Tools

Although in very widespread use, CAD drafting tools have some limitations when considered in the context of a design project. The reuse of existing design components is problematic, particularly as these tools do not have the means to provide the part associations between the drawings of component artefacts within a design artefact composition graph. It would be desirable for designers to retrieve clusters of design components within an artefact composition graph, that is subgraphs, and then adjust some of their features, for example dimensions, to suit the requirements of a new design. So for example, a road bridge designer might in considering the design of a bridge deck, recognise that the particular configuration of abutments (structures that support the bridge deck) and span is broadly similar to those encountered in some previous design. He/she would seek to incorporate in the new design a deck, which was largely similar to that of the previous design but with different dimensions. Bridge decks are themselves complex artefacts made up of many different components, many of which would have their own drawing. The designer's task would be eased if the CAD drafting tool employed, recognised and supported the artefact composition graphs underlying the design of structurally complex design objects.

A further limitation of CAD drafting tools is that they do not usually include the facility to carry out behavioural analyses of the designed artefacts. Performance-related calculations are normally undertaken separately. It is important that as part of the design cycle, comprehensive design calculations are carried out, for example finite element analysis, particularly if the outcome of calculations indicates deficiencies in the performance characteristics of the design artefact, then the design of that artefact must be reconsidered as a design conflict may result (see 4.2). This leads to an iterative design development cycle in which the designer must balance the various constraints to achieve an acceptable design.

DESCRIBE is a federated information system which integrates AutoCAD with the distributed data model shown in Figure 1. The architecture will be described with reference to Figure 2. The components of the architecture consist of AutoCAD, an extended toolset for AutoCAD, software to map AutoCAD files and utilise external calculation routines to analyse the designed artefacts into database objects and vice-versa, and a distributed designer database. These will be described in detail by considering the migration of information to and from the designer workstation interface.

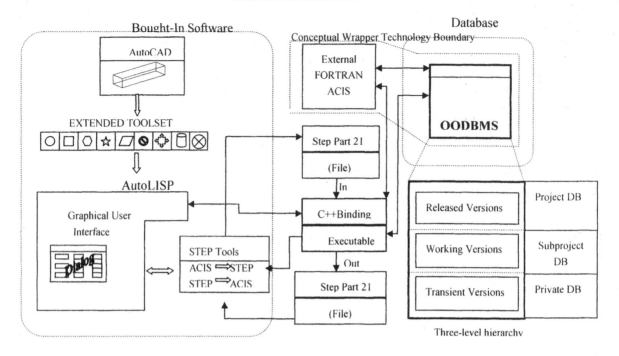

Figure 2: Software Architecture of DESCRIBE

4.4.1 Transferring Design Data from AutoCAD to the Object Database: The artefacts that we have created in AutoCAD are represented as solid models, that is, 3-dimensional models with mass and material properties. Using proprietary software (identified as *STEP Tools* in Figure 2) we are able to convert AutoCAD solid models to STEP files [24] through an intermediate graphical format. STEP (Standard for the Exchange of Product data) is a set of ISO standards (ISO10303) which support the exchange of engineering product model data. The standards are divided into several classes of parts, and can be grouped into infrastructure components and industry specific information models. The intermediate graphical format was necessary due to the limitations of the version of AutoCAD that we used. It is these files which are analysed by other software, e.g. finite element analysis software to determine the properties of the artefact. Through the C++ binding of our object database system O2 [25], we are able to wrap STEP files as attributes within database objects. These database objects also conceptually wrap behavioural analysis routines which are called when importing a STEP file into a newly created database object, to instantiate the properties of the artefact therein. For each of the component types in the design, there is a corresponding schema class with STEP file attributes, spatial parameters, strength-related data and associated methods that include the wrapped analysis routines.

Thus, checking-in an object to the database causes the behavioural analysis routines to run if the artefact is a new object, but this is also the case if it is a modified object, that is a new version of the object. Because bridge component objects are instances of component classes, they can be versioned through the versioning and configuration management scheme described earlier in section 4.3.

4.4.2 The User Interface: One important feature of the architecture described here is that the designer/user is presented with an interface within a familiar environment. This has been achieved by the production of an extended toolbar within AutoCAD which offers extended functionality to standard AutoCAD, namely it allows the storage and retrieval of designs in the database and the means to create, manipulate, retrieve and view design artefact versions. Significantly, these complex objects (in the object database sense) can be manipulated and viewed within AutoCAD at any level of granularity. The extended toolbar has been implemented using AutoLISP and C++. Figure 3 shows a functional abstraction of the DESCRIBE architecture.

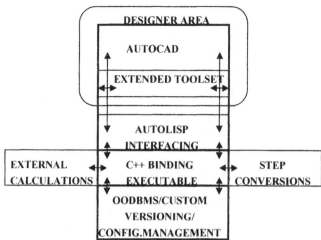

Figure 3: Functional architecture of DESCRIBE

4.4.3 Transferring Design Data from the Object Database to AutoCAD: When a user requests the retrieval of an artefact version from the database, using the AutoCAD extended toolbar, the retrieved AutoCAD drawing can be enhanced on command to include details of the properties of the design artefact such as the detail or summary of the finite element analysis, the object name, its type, its version number and a graphical representation of its version evolution graph. Upon retrieval from the database, each design object is checked for consistency, to determine whether the object has been moved in relation to the global co-ordinates.

4.5 Design Autonomy

Design autonomy is realised through the data model shown in Figure 1, and described in section 4.1. It is achieved by each designer's private partition of the local database, which holds the independent transient versions created and evolved by that designer. When a designer wishes to share a design version with other designers, the transient version is promoted to a working or released version in a public partition. This supports the required level of design autonomy.

5. Evaluation and Conclusions

5.1 Evaluation

The form of the system has partially been determined by an experiment using undergraduate students [12] and partially using experience of the authors, two of whom have worked as designers. However, it was considered that an independent evaluation by a practising designer was vital to ensure that what has been produced to date is relevant and of use. Engineering designers are used to using reasonably good quality software packages with user

friendly interfaces. Past experience has shown that if they are asked to evaluate the product of a research project, then this is only worthwhile towards the end of the project when the environment has been develop to a reasonable degree of robustness and possesses a passable user interface.

The evaluation was undertaken by a senior design engineer from Ove Arup and Partners. The form of the evaluation was a demonstration by the system developer who worked through the various aspects and features of the environment while the designer observed. At each stage the purposes of the various capabilities of the DESCRIBE environment were discussed. At the end of the demonstration the evaluator provided feedback. Her comments were positive. Interestingly she had fully comprehended the system and grasped the wider potential. She suggested possible extra domains to which it could be applied and gave strong encouragement for its continued development. Obviously, evaluation by one person is very subjective and further work is needed in this area.

5.2 Conclusions

In this paper we have identified and described how an object-oriented database system can be used to support a concurrent engineering environment. This component has been built and tested in a prototype implementation of our design environment. We intend focusing future work on constraint representation and management, and design reuse.

References

[1]	Pahl, G. and Beitz, W. *Engineering Design: A Systematic Approach*, 2nd ed., translated into English by A. Pomerons and K. Wallace, Design Council, London, 1988.

[2]	Sriram, R., Livezey, B. K. and Perkins, W. A. *A Distributed Shared Database System for Concurrent*

Engineering. Proc. of CE and CALS Conference, pages 29-43. Washington DC, 1992.

[3] Lawson, M. and Karandikar, H. M. *A Survey of Concurrent Engineering.* Concurrent Engineering: Research and Applications, 2(1), pages 1-6, March 1994, 1994.

[4] Santoyridis, I., Carnduff, T. W., Gray, W. A. and Miles, J. C. *Assessing a Versioning System for Collaborative Engineering Design with a Bridge Design Experiment.* The International Journal of Computer-Integrated Design and Construction (CIDAC), 1 (1), pages 17-28, 1999.

[5] Carnduff, T. *Supporting Engineering Design with Object-Oriented Databases,* PhD Thesis, University of Wales, Cardiff, UK, 1993.

[6] Katz, R. H. *Towards a Unifying Framework for Version Modeling in Engineering Databases.* ACM Computing Surveys, 22(4), pages 376-408, 1990.

[7] Chou, H. T. and Kim, W. *A Unifying Framework for Version Control in a CAD Environment.* In W. Chu, G. Gardarin, S. Oshuga and Y. Kambayashi eds., Proceedings of the 12th VLDB Conference, pages 336-344, Kyoto, Japan, 1986.

[8] Krishnamurthy, K. and Law, K. *A Data Management Model for Design Change Control.* Concurrent Engineering: Research and Applications, 3(4), pages 329-343, 1995

[9] Talens, G., Oussalah, C. and Colinas, M. *Versions of Simple and Composite Objects.* In R. Agrawal, S. Baker and D. Bell eds., Proceedings of the 19th VLDB Conference, pages 62-72, Dublin, Ireland, 1993.

[10] Ahmed, R. and Navathe, S. *Version Management of Composite Objects in CAD Databases.* In J. Clifford and R. King eds., Proceedings of the ACM SIGMOD International Conference on Management of Data, pages 218-227, Denver, USA, 1991.

[11] Park, H. J. and Yoo, S. *Implementation of a Version Manager in an Object-Oriented Database Management System.* In J. Murphy and B. Stones eds., OOIS '95 (Conference on Object-Oriented Information Systems), pages 323-336, Dublin, Ireland, Springer-Verlag, 1996.

[12] Miles, J. C., Moore, C. J., Carnduff, T. W., Gray, W. A. and Santoyridis, I. *Information Flows between Designers - an Experimental Study.* Proc. 3rd Workshop of the European Group of Structural Engineering Applications for Artificial Intelligence, University of Strathclyde, pages 13-25, 1996.

[13] Katz, R. H. and Chang, E. *Managing Change in a Computer-Aided Design Database.* In S. Zdonik and D. Maier eds., Readings in Object-Oriented Database Systems, pages 400-407. Morgan Kaufman, 1990.

[14] Kim, W., Ballou, N., Garza, J. and Woelk, D. *A Distributed Object-Oriented Database System Supporting Shared and Private Databases.* ACM Transactions on Information Systems, 9(1), pages 31-51, 1991

[15] Grundy, J. C., Mugridge, W., Hosking, J. and Amor, R. *Support for Collaborative Integrated Software Development.* In Proceedings of the IEEE Conference on Software Engineering Environment, pages 84-94, Noordwijkerhout, Netherlands, 1995.

[16] AutoCAD *AutoCAD Release 13 Command Reference,* Autodesk Inc., 1995.

[17] Eckland, D. J., Eckland, E. F., Eifrig, R. O. and Tonge, F. M. *DVSS: A Distributed Version Storage Server for CAD Applications,* Proceedings of the 13th VLDB Conference, pages 443-454, Brighton, UK, 1987.

[18] Chou, H. T. and Kim, W. *Versions and Change Notification in an Object-Oriented Database System.* Proceedings of 25th Design Automation Conference, ACM/IEEE, pages 275-281, 1988.

[19] Kim, H-J and Korth, H. F. *Schema Versions and DAG Rearrangement Views in Object-Oriented Databases,* University of Texas – Report Tr-88-05, Austin, Texas, 1988

[20] Kim, W. *Introduction to Object-Oriented Databases,* MIT Press, Cambridge, MA, 1990.

[21] Santoyridis, I., Carnduff, T. W., Gray, W. A. and Miles, J. C. *An Object Versioning System to Support Collaborative Design Within a Concurrent Engineering Context.* Proc. British National Conference on Database Systems (BNCOD15), pages 184-199, Springer, 1997.

[22] Kim, I., Carnduff, T. W., Gray, W. A. and Miles, J.C. *DESCRIBE: An Object-Oriented Design System to Support Concurrent Reuse of Data in Building and Engineering Design.* Proc. OOIS '95 (Conference on Object-Oriented Information Systems), pages 39-44. Springer-Verlag, 1995.

[23] Carnduff, T., Miles, J., Gray, A and Faulconbridge, A. *Case Adaption and Versioning in Concurrent Engineering.* In C. J. Anumba and N. F. O. Evbuomwan eds., Proc. 1st International Conference on Concurrent Engineering in Construction, Institution of Structural Engineers, pages 45-54, London, 1997.

[24] ISO (International Organisation for Standardisation) *Industrial Automation Systems and Integration - Product Data Representation and Exchange Part 1: Overview and Fundamental Principles,* ISO/IS 10303, 1st edition, December 1994, 1994.

[25] O2 Technology *The O2 System ODMG C++ Binding Reference Manual Release 4.6,* O2 Technology, 1996.

Multi-Aspect Description of a Design Process in a Relational Database Application

W. Marowski

Institute of Machine Design Fundamentals, Warsaw University of Technology, 02-524 Warsaw, Poland

J. Jusis

Institute of Machine Design Fundamentals, Warsaw University of Technology, 02-524 Warsaw, Poland

Abstract

Any design process can be described as a system of mutually dependent operations performed to develop the project. On the other hand, the hierarchy of project objects, which forms the logical structure of a product, as well as the hierarchy of project documents, is created in the project development process. Design processes can be managed and visualised by a relational database application. Data of nodes of all systems of structural dependencies mentioned above, together with relationships between nodes of the same and different structures, can be stored in the relational database. User interface of the database application should enable defining and manipulating those nodes, as well as visualising their relationships. Hierarchical structures can be presented in ActiveX controls of the TreeView type, while for the structure of operations a special presentation tool must be developed. Object oriented approach can be used to manipulate a node from the application code. Application of that type has been developed in Microsoft Access. Since only commonly used office software is required for that purpose, this solution can be convenient for small companies.

1   Introduction

During any design process large amount of data must be stored and accessed. To increase the efficiency of the design process and the product quality concurrent and collaborative design techniques should also be applied. Therefore design projects should be developed in integrated computer environment, which is able to perform various tasks by means of its software components (see e.g. [5]). The description of the design process itself is of particular importance for the functionality of the design environment. Any design process can be considered as the set of mutually dependent operations, in which the product structure (the hierarchy of project objects) is defined and modified. Results of operations are described in project documents. Thus, that model of a design process consists of definitions of operations, objects and documents together with description of relationships between its elements. This form of description of a design process can be stored in a relational database, which can be used as a kernel of an integrated computer design environment. That environment should perform its tasks by means of its software components. Object oriented programming language, as well as the automation technology can be used to integrate functions of particular components in one system (see e.g. [5]).

The use of a relational database application to visualise and manage design processes is discussed in the paper. Though a lot of advanced commercial software has been developed for that purpose, it mostly cannot be applied in small companies because of their very limited financial resources. On the other hand, those companies own commonly used office software, e.g. Microsoft Office. Therefore, that software together with free distributed ActiveX controls can be used to develop dedicated systems for particular companies. Those systems should be relational database applications and they should be developed by means of the programming tool owned by the given company. That tool should allow the use of object oriented programming language and the ActiveX technology. Solutions described in the paper have been developed in Microsoft Access using Visual Basic for Applications.

2   Model of the design process

Within the model of a design process the following components must be taken into consideration:

- project objects, i.e. elements of the logical structure of a product,
- project operations, i.e. steps in the project development process,
- project documents, in which results of project operations are described.

Components of each type can be considered as nodes of the corresponding structure. Types of those structures, as well as relationships between nodes of different structures, are discussed below.

Decomposition of a product into the multi-level hierarchical structure of project objects, known as object oriented product modelling (see e.g. [2], [9]), is a well-known approach, which gives an intuitive representation of the logical structure of a product. It is similar to the one applied in those modern CAD systems, which use feature-based solid modelling (see e.g. [1]). The product itself is a root node of the hierarchy of assemblies, sub-assemblies and parts. Except the root object, each node has exactly one parent node and an arbitrary number of child nodes.

The structure of project operations cannot be presented in the hierarchical form. Since some operations require results of another ones as input data, while, on the other hand, another operations are mutually independent, the project development process can be considered as a system of one-directional paths between nodes assigned to particular operations. Moreover, the sequence in which nodes should be visited during the design process need not be unique, for example if different start nodes can be selected. This concept is similar to the maze model of a design process presented e.g. in [7] and [8]. It is assumed that any operation node can be connected by means of one-directional paths (connections) with an arbitrary number of nodes, which precede or follow it in the structure.

The project development process is documented by creating or modifying various forms of project documents (technical drawings, calculations, simulation results, administrative documents, etc.). Any revision of a project document can be considered as a node of the hierarchical structure of project documents. The project contract can be suggested as the root object of that hierarchy. It is assumed that any revision of a project document has only one parent document (of course, except the root one). If any revision of document should be a child of more than one parent document (e.g. if the same detail is used in more than one assembly), multiple instances of that

revision of document should be created and assigned to the corresponding nodes of the hierarchy. To simplify the terminology, the term "project document" is used below instead the term "revision of a project document".

Each project object can be related to an arbitrary number of operations performed in the design process. In course of those operations the definition of that project object is created or modified. On the other hand, activities of any operation can be performed on an arbitrary number of project objects. Thus, the relationship between project objects and project operations can be qualified in terms of the relational databases theory as the "many-to-many" relationship. Since project documents describe modifications of project objects during particular operations of the design process and it cannot be excluded, that some documents can be modified in more than one operation, the relationship between project documents and project operations can also be of the "many-to-many" type.

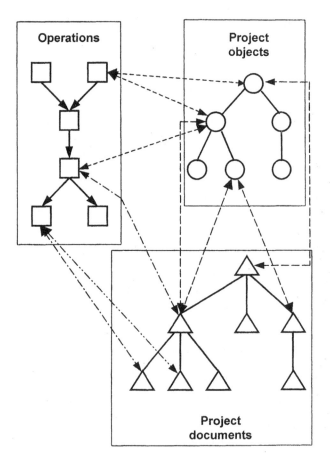

Figure 1   Model of a design process – components and relationships

It is obvious, that many project documents can be related to a given project object. On the other hand, a single document can describe more than one project object. Therefore the "many-to-many" relationship exists also between project objects and project documents.

The simple scheme of the model of a design process described above is shown in the Figure 1.

During the project development process some external documents, like standards, regulations, catalogues etc. must be also considered. Those documents can be assigned to particular project objects, documents or operations. They can also be ordered in hierarchical structures and they can be manipulated in similar way like project documents (except modification). This problem is out of scope of this paper.

All elements of the model of a design process, as well as relationships between them, should be stored in a database application used to manage the design processes. The suggested method is described in the next section of this paper.

According to the requirements of the quality management, each object, document or operation should have a responsible person assigned. Therefore it can be recommended to store in the database the organisational scheme of the design office and the hierarchical structure of employees. Those structures can then be visualised in the user interface of the database application, which enables the intuitive selection of the corresponding person.

3 Database representation of the model of a design process

Description of the considered design process in the relational database includes:

- descriptions of elements of the model (operations, project objects and project documents),
- descriptions of relations of elements of the same types,
- descriptions of relations between elements of different types.

Essential attributes of each project object, project document or operation should be included in its definition stored in one row of the corresponding database table. A separate table should be created for any type of component of the design process model. Moreover, an additional table is required to store data of revisions of any particular document. Set of essential attributes of a given entity (i.e. of an element of a given type) depends on the considered design office. It must contain a unique identifier of the entity as well as identifiers of a project, of a responsible person etc. Additional attributes can be selected from the predefined set and used to describe a particular object, document or operation, according to its character. Identifiers and values of those attributes are stored in separate tables together with the corresponding object identifiers.

To define the position of an object or document node in the corresponding hierarchical structure, an identifier of its parent node must be included in the set of essential attributes of that element. For the root node this attribute is not applicable, i.e. its field in the corresponding table record has a null value. To avoid the violation of data

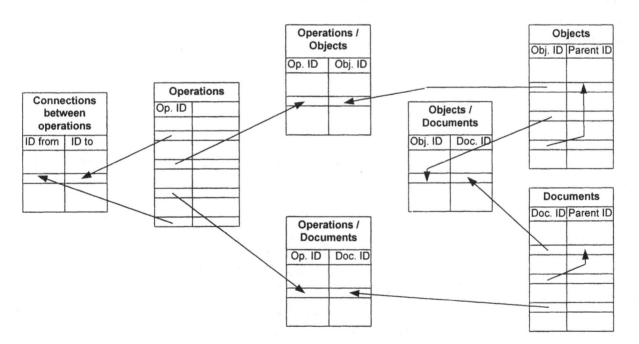

Figure 2 Database representation of the model of a design process – tables and relationships

integrity constraints the "one-to-many" relationship must be defined between the element identifier column and the parent element column of any table, in which descriptions of hierarchically ordered elements are stored. A detailed description of this problem can be found in [4].

The structure of operations of a design process can be described by means of the system of one-directional paths (connections) between two operations (the start and destination one). Descriptions of connections are stored in rows of a separate database table. An identifier of the connection, as well as identifiers of its start and destination node are indispensable attributes of the definition of connection, while another attributes can either be stored in the same table (if they are used to define most of connections) or selected for the particular connection from the predefined set and stored, together with their values, in a table created for that purpose.

A database table must also be defined to store data of project documents related to a given project object. The same approach is used to store data of project objects or documents defined or modified in a given operation. Thus, each "many-to-many" relationship is transformed into two "one-to-many" relationships between records of definitions of considered elements and an intermediate table.

The scheme of the structure of database tables used to store data of a design process model is shown in the Figure 2. To simplify the scheme, revisions of project documents are not shown in the figure.

4 Visualisation of hierarchical structures in the user interface of a database application

Project objects and project documents ordered in hierarchical structures can be visualised in the graphical user interface of the database application by means of application forms containing ActiveX controls of the TreeView type. That type of ActiveX control can be used to present hierarchical structures in the same form like the structure of drives and catalogues is presented in the Windows Explorer.

ActiveX controls are custom controls designed for the use in various object-oriented applications. Their events, properties and methods can be recognised and referenced by the object oriented application code if the corresponding control and its object library (a .OCX file) have been registered in Windows (see e.g. [6]). Numerous of ActiveX controls are free distributed with some software packages or they can be downloaded from the Web sites. One of those controls is also the Microsoft TreeView control (controls of that type are delivered also by another software companies, like e.g. Oracle).

All nodes presented in the Microsoft TreeView control are objects and members of the Nodes collection of the

control. Filling the control with nodes consists in defining elements of that collection in the application code written in an object-oriented programming language, like Visual Basic for Applications. To add a new node, a new instance of a Node class must be created and values of its properties must be defined to determine both the appearance of the node (e.g. its text, label or icon) and its position in the hierarchical structure. The reference to an item represented by that node should be also stored in one of its properties. The following approach can be suggested for that purpose:

- data of all items, which should be represented by nodes in the TreeView control must be queried from the database,
- if data of those items are stored in more than one table (as it is for revisions of project documents), the saved query, which contains all data required to define nodes of the considered TreeView control, must be created,
- identifiers of an item and its parent item must be stored in single fields of a record of a data source (table, query) of the TreeView control,
- to determine the position of a new node in the hierarchy presented in the TreeView control, its position relative to one of nodes existing already of the control (i.e. to one of elements of its Nodes collection defined by an index or by a key) must be defined,
- one of the not-displayed properties of a Node object of the character string type (e.g. the Key or Tag property) should be used to store an identifier of an item represented by the node.

If values of properties, which determine the appearance and position of the new node have been determined, that node can be added to the Nodes collection of the TreeView control. The scheme of defining of a new node is shown in the Figure 3.

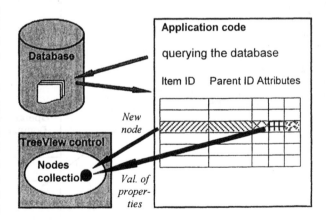

Figure 3 Defining nodes of the TreeView control in the application code

Two basic methods of filling the TreeView control with nodes can be suggested:

- defining all nodes of the hierarchical structure (or structures), which should be presented in the control in one process, e.g. when the corresponding form is opened or the contents of the control should be refreshed,
- defining only root nodes of all hierarchies presented in the control, another nodes are defined only if their parent node has been double-clicked by the user to display the corresponding lower level of the hierarchy.

nodes, which should be presented in the control. Two code-created collections are used for that purpose:

- the collection of node objects placed in the level of the hierarchy, which is currently filled by the code execution,
- the collection of nodes, which are parent nodes for nodes mentioned above.

When all nodes of the current hierarchy level have been defined already, the contents of the collection of parent nodes is replaced with nodes of the most recently filled level of the hierarchy, since they become parent nodes for the next hierarchy level.

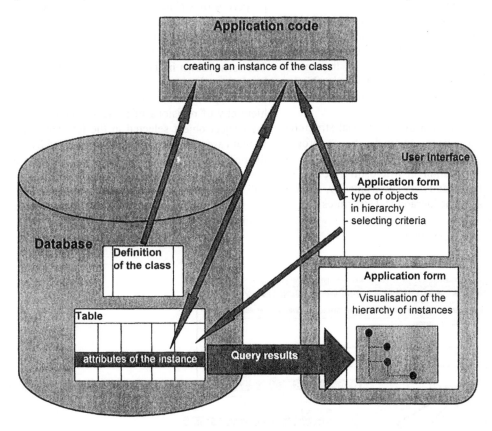

Figure 4   Using a code-defined class to fill the TreeView control

The advantage of the first approach is that all items of the hierarchy are represented by their nodes at once, if only the TreeView control is displayed. It is of particular importance, if the nodes are referenced by the application code, e.g. to visualise the required part of the tree. However, a considerable time can then be needed for filling the TreeView control for complex hierarchies.

The detailed description of an algorithm used for defining the complete set of nodes of the TreeView control is given in [4]. The method used consists in filling the subsequent levels of hierarchies, starting with all root

An alternative method can be the use of user-defined classes of nodes of the given hierarchy (e.g. nodes of the project objects hierarchy or nodes of the hierarchy of revisions of project documents). Instances of a given class are created by the code basing on results of a database query, in which the select criteria resulting from the current context of the session with the database application are used. Those objects are then added to one of the following collections:

- the collection of root nodes, i.e. nodes without parent node,

- the collection of nodes, for which their parent nodes are defined in the database.

Elements of the first collection are used to define start nodes for all hierarchies, which should be presented in the TreeView control. Then the second collection is searched to find objects with parent nodes defined already in the control. For any of those objects a new node of the corresponding hierarchy can be defined and then it can be removed from the collection. The considered group of statements is repeated for elements of the second collection until this collection becomes empty. Detailed description of this method can be found in [3] and its scheme is shown in the Figure 4.

The second approach is much simpler, since only child nodes of a node selected by the user should be defined. Dynamic database query, in which an identifier of the current parent object is applied as one of the search criteria, together with the Add method of the Nodes collection of the TreeView control should be used for that purpose.

The concept of visualisation of hierarchical structures described above has been applied in a project documentation management system developed as a relational database application in Microsoft Access '97 environment. The Microsoft TreeView Control v. 5.0 has been applied to visualise the hierarchical structures of:

- project objects,
- revisions of project documents,

- departments of the design office,
- employees of the design office.

Since the use of TreeView controls gives the clear overview of hierarchical structures and enables intuitive navigation through the hierarchy, the required item can be found much easier as e.g. by means of list or combo boxes. Therefore forms containing TreeView controls are applied not only to visualise hierarchical structures, but also as searching tools for some another forms of the application. For example, the parent object for the currently defined project object can be easily found this way (as it is shown in Figure 5), since values of properties of a Node object clicked by the user allow to identify an item represented by that node.

Forms containing two coupled TreeView controls can be used to visualise two related hierarchical structures. The form shown in the Figure 6 is applied to browse the hierarchy of project objects (the left TreeView control) and to present in the right TreeView control the hierarchy of revisions of project documents assigned to the project object selected in the left TreeView control. In that case each NodeClick event caused by the user for the primary (left) control results in refilling operation for the secondary (right) control, in order to display there only nodes, which satisfy searching criteria determined by values of properties of the node selected in the primary control.

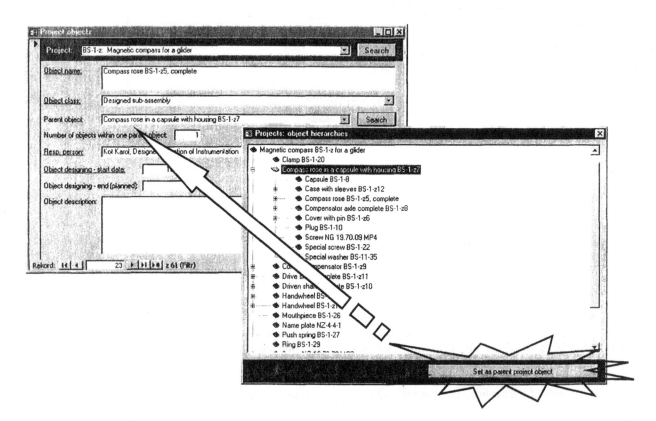

Figure 5   Defining the parent object by means of the TreeView control

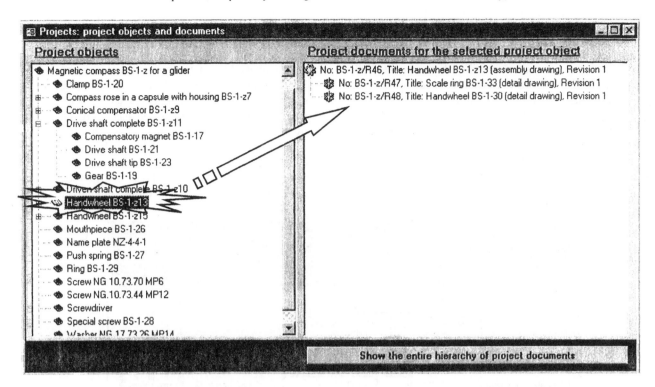

Figure 6   Two coupled TreeView controls

5   Tools for defining and visualisation of the structure of design operations

Since for any operation in the design process an arbitrary number of preceding and following operations can be defined, the TreeView control cannot be used to display the structure of operations of a design process. Therefore a special tool is required for that purpose. In the prototype of an integrated design environment developed by the authors of this paper the following approach has been applied:

- data of operations and of connections between operations are stored in database tables, as it is shown in the Fig. 2,
- operations and values of their basic attributes (i.e. attributes, which are obligatory for all operations, e.g. name or project identifier) are defined in the database form of fixed structure by means of text fields and combo boxes,
- one-directional connections between a selected operation and operations, which precede or follow it can be defined or presented in quasi-graphical mode by means of the form described below; the same form can also be used to inspect particular paths in the structure of operations and to calculate the sum

of values of the selected property of operations along the path,

- additional properties can be used to describe any operation or connection; the property sheet of any of those objects allow both to select properties, which should be used, and to set their values for that object,
- the set of additional properties (names and allowed values of properties, if necessary) is stored in the database and can be modified in the run time.

The form applied to define or examine connections of the selected operation is shown in the Figure 7. It consists of two pages. In the first page, shown in the figure, the scheme of connections of an operation displayed in the central field of the form is presented, while in the second page all connections between operations of a given design process, as well as connections "from" and "to" the selected operation are presented in alphanumeric form by means of list boxes. The first page of the considered form is divided into two areas: in the upper one connections, for which the selected operation is a destination point, are presented. In the lower area connections between the selected operation and operations, which follow it, are displayed. Thus, in any case the direction of both groups of connections is "from top to bottom". Up to five connections can be presented in each area. If the number of existing connections exceeds this limit, the contents of the corresponding part of the form can be scrolled horizontally by means of scroll buttons placed in the right part of the form near the borderline between the both areas. Buttons with arrows located outside (above or below) the fields of external points of connections can be used to move forwards (downward arrow) or backwards (upward arrow) along the selected path. After any click on one of those buttons, an operation displayed in the nearest text field is selected as the current operation, i.e. its name is moved to the central field of the form and its connections are displayed.

- Replacing the outer operation of the selected connection with an another one. The text field of the operation, which should be replaced, must be double clicked for that purpose. Then, a combo-box is displayed instead of it and another start or destination point of the connection can be selected.

- Deleting the connection, which has been selected by the right mouse button click while the <Alt> key is pressed down. Then, the "Delete" button should be used to delete the selected connection.

- Displaying the property sheet of the selected operation or connection. The mouse pointer and the

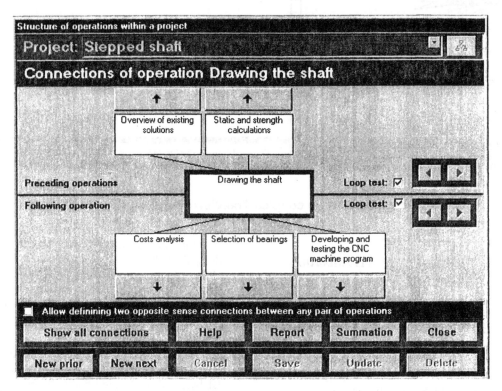

Figure 7   Visualisation of connections between operation in the user interface

The following operations on the set of connections of the given project can be performed by means of this form:

- Defining a new connection to or from the operation displayed in the central field of the form. The button "New prior" or "New next" should be clicked for that purpose, and then an empty template of connection is displayed in the corresponding area of the form. Instead of a text field, a combo box is displayed on the outer end of the connection line to make possible to select the name of the outer operation of the connection (neither the name of the current operation, nor names of operations already connected with the current one are displayed in the list).

right mouse button are used for that purpose. An example of the property sheet of an operation is shown in the Figure 8.

- Calculating the sum of a selected property along the path examined by means of arrow buttons. If the current operation is changed, its value of that property is added to the sum, if the downward arrow button has been clicked or it is subtracted from it, otherwise. The result of calculation is presented in the dialogue window, displayed if the "Summation" button has been clicked. In that window the name of the property should be also selected (see the Figure 9).

All operations described above are performed by the application code. To manipulate data of connections and operations Recordset objects and dynamic queries built at the runtime are used.

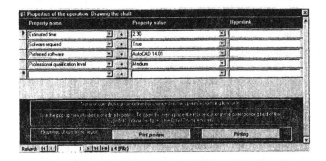

Figure 8 Property sheet of an operation

Defining of new connections can result in creating loops in the structure of paths between operations. Unintentionally created loop can cause problems in the further analysis of a design process, therefore the user should be informed, whether the planned or modified connection would close a loop in the structure. This test procedure can be time consuming, therefore it is executed only if the option "Loop test" has been selected for the corresponding area of the form (see Figure 7). An algorithm used for that purpose is based on the unidirectional character of existing connections. Therefore, if by means of those connections (without the planned or modified one) it is possible to reach the start point of the planned or modified connection starting in its destination point, then that connection would close the loop. An exception is a direct connection existing already between the destination and start point of a planned or modified connection, since it has been assumed that a pair of opposite sense connections between two operations is not considered as a loop.

In the course of testing of the existing system of connections the structure of paths examined already is written to the records of the temporary database table created by the application code. Each record of that table corresponds to one path, which is described by means of the following attributes:

- the origin of the path, i.e. an operation in which that path is starting,
- the end of the path, i.e. an operation with no connections to another operations (this attribute is equal to null until it is possible to move forwards along the path),
- the operation reached in the last step along the path,
- the former start point, i.e. the start point of the recent segment of the path.

During the test procedure dynamic queries are build in the code to return from the database data of connections with an origin in the current operation. Recordset objects are applied to analyse query results in the code. If an operation reached along at least one path is the one, in which the planned or modified connection should have its start point, the warning message is displayed and the testing routine is terminated.

Single step of the testing routine can be described as follows:

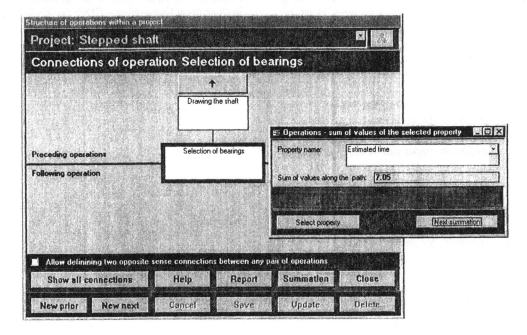

Figure 9 Total estimated time calculated along the path

- operations reached along all not completed paths are examined to find in the database data of all connections having their origins there,
- if any of those connections leads to the previous operations along the given path, it is removed from the set of new connections (to avoid analysing the pair of opposite sense connections between two operations),
- if some paths bifurcate in those operations (i.e. if more than one connection has its origin in a given operation), additional rows are inserted into the temporary table,
- if a given operation can be reached along more than one path, redundant rows are deleted from the temporary table,
- if no another operation can be reached from a given one, the path end field in the corresponding record of the temporary table is filled with an identifier of that operation.

Steps described above are repeated until in all records of the temporary table the path end fields have not null values or until a loop is detected.

6   Summary and conclusions

Some concepts concerning models of design processes and their database representation have been presented in the paper. The use of that approach makes possible to develop a database application, which can be a kernel of an integrated design environment. Some components of a model of a design process can be considered as ordered sets of elements. Data of elements of those sets, including their positions in the corresponding structures must be stored in the database. Ordered structures should be also presented in the user interface of a database application in an intuitive and easy to browse form. The use of ActiveX controls of the TreeView type to visualise hierarchical structures is discussed in the paper. Also the special tool for presentation of general structure of project operations is described. Practical realisations of those concepts have been developed in Microsoft Access.

Solutions presented in the paper can be applied to build an integrated design environment, which does not require any special and expensive software components. Only the popular and commonly used office software, like Microsoft Access, as well as free distributed ActiveX controls, should be used for that purpose. This approach can be especially interesting for small companies.

Acknowledgement

The Polish State Committee of Science has supported this research with grant No. 7T07C00812.

References

[1] "Getting Started with Solid Edge. Version 4", Intergraph Corporation, Huntsville, 1997

[2] P. Ghodous, D. Vantorpe, "Actual Product Representation Techniques", Advances in Concurrent Engineering – CE99, pp. 269-276, Technomic Publ. Co, Lancaster, PA, 1999

[3] W. Marowski, "Object Approach to the Description of a Design Process in a Database Application", Proc. of a 3rd Workshop on Computer Aided Design, Manufacturing and Exploitation, pp. 85-92, Helion, Gliwice, 1999 (in Polish)

[4] W. Marowski, "Object Oriented Description of a Design Process in a Database Application with ActiveX Controls", Computer Integrated Manufacturing, Proc. of the International Conference CIM '99, Vol. II, pp. 15-24, WNT, Warszawa, 1999

[5] W. Marowski, J. Wróbel, "The Process Approach to the Computer Aided Machine Design", Przegląd Mechaniczny, No. 17-18/98, pp. 13-18 and 23-24, 1998 (in Polish)

[6] T. M. O'Brien, S. J. Pogge, G. E. White, "Microsoft Access '97 Developer's Handbook", Microsoft Press, Redmond, 1997

[7] J. Pokojski "Product Model Transformations in Maze Model of Design Process", Computer Integrated Manufacturing, Proc. of the International Conference CIM '99, Vol. II, pp. 121-128, WNT, Warszawa, 1999

[8] J. Pokojski "Knowledge Based Support of Machine Dynamics Analysis", Advances in Concurrent Engineering – CE99, pp. 336-341, Technomic Publ. Co, Lancaster, PA, 1999

[9] T.Taura, Y. Aoki, H. Takada, K. Kawashima, S. Komeda, H. Ikeda, J. Numata, "A Proposal of the Activity Chain Model and Its Application to Global Design", Advances in Concurrent Engineering – CE99, pp. 29-38, Technomic Publ. Co, Lancaster, PA, 1999

Model Based Management of Product's Design Specifications

Pekka Savolainen

Embedded Software, VTT Electronics, P.O. Box 1100, FIN-90571Oulu, Finland

http://www.vtt.fi/

Abstract

Contemporary PDM and CM systems do not provide any specification management for product variants at a detailed level, i.e., at the level where the variation really occurs. This is likely to result in unnecessary burden due to extra maintenance work, while also hiding the real product model from the designers of the product. This paper presents an approach to presenting the feature model of a product in the form of structured text, and using the thus formed knowledge base in sharing the common design knowledge over a distributed development environment. The approach also demonstrates how the detailed design specifications of a family of sophisticated electronics products, with product variants in several dimensions, can be brought into an automatically customisable format.

Keywords: concurrent engineering, domain modelling, feature model, specification management, mass-customisation, reuse, software development, product documentation, design data management, structured text, SGML, XML

Introduction

Shortening the time-to-market of electronics products requires that the various parties of the manufacturing company, SW design, HW design, marketing, user manual production, and their management, work in parallel when producing products and bringing them onto market. The HW components of the implementation platform should be implemented when SW is ready for the integration test, and user manuals should be ready for translations when mass-production starts. This kind of parallel work towards a common goal, i.e. a product on the market, requires an information system that supports and enhances the efficiency of communication. The various disciplines of a manufacturing company need a common ground where the design knowledge can be shared.

Good candidates for such a medium of communication are PDM systems (Product Data Management Systems, considered here from the viewpoint of ERP, Enterprise Resource Planning) [1], and VRP (Virtual Reality Prototyping) systems [2][3]. Yet, the contemporary commercial systems still fail to meet the challenges set by the required accuracy of product variant management in the design process of embedded products (i.e. an electronics product with embedded software), which is essential for the production of mass-customised products. UML based CASE tools have also been used in specification modelling [4]. Though UML explicitly expresses generalisation and aggregation relationships, nor are the use case model or the CASE tools designed the network of interrelated product variations in mind.

Models are the horn of plenty in software development: they help in understanding the domain, communicating design problems, and solving them. The feature modelling method has been applied with success in reuse-oriented systems engineering. The next chapter reviews the feature modelling approach from the point of view of modern electronics product design. The third chapter introduces embedded product's design process and the prospects that we see for sophisticated specification management therein. Also our model to formal management of product specifications is described. The fourth chapter demonstrates how the model is maintained during product design, and how it can be used to support the design process. The presentation concludes with a discussion of the benefits of the approach and presents future challenges for the integration of design data management.

The principle employed in our study can be applied in the further development of several kinds of sophisticated electronics products. When a company develops a product based on a new core technology, the company is often interested in bringing the product to market as quickly as possible, and product differentiation is of secondary concern. However, as competitors enter the market, the product may need to start competing along user and aesthetic dimensions. On the other hand, incorporating the preferred new features into the product is usually a trade-off between product's quality and price: an electronics device – or an aeroplane – can be made lighter, but this action will probably increase manufacturing costs. One of the most difficult aspects of product development is recognising, understanding, and managing such trade-offs in a way that maximises the success of the product. This raises the need for product variants management. The different market sections are usually satisfied by different design compromises, and all the thus formed various products must be supported and developed further with minimum extra costs. This evolution can already be seen in hand-held telecommunications products, and will probably be present in the future personal devices of ubiquitous computing [5][6].

Feature modelling in systems design

Feature modelling is a method of modelling sophisticated electronic products in the design process. It has its origin in mechanical engineering [7][8], where the structures of mechanical systems are presented as features. A feature was originally considered as a portion of part geometry that would correspond to specific machining operations [9]. In comparison with geometric models, feature models represent parts in terms of higher-level, functionally significant entities, such as holes, slots, or pockets. (By contrast, ordinary geometric models represent parts in terms of low-level geometric entities, such as faces, edges, and vertices.) For some time, feature modelling has been used for modelling software components [10][11][12], and lately [4][13][14][15] have applied feature modelling to object oriented system analysis and design.

The purpose of software feature modelling is:
– to identify customer needs in family-based products;
– to create a basis for the design of software components for product families;
– to support customised software production and delivery.

The knowledge that is presented in the form of a feature model can be used to increase the effectiveness of application design though reuse in mass-customisation. The principle of mass-customisation is to use the same designs or components, slightly varied as necessary, over several products or applications [16]. In software development, mass-customisation is implemented by constructing a platform of generic components, and deriving the product specific code from that platform [17][18]. Features – in this context – are the instrument used for modelling, designing, and developing this platform. Kang & al. [14] use the term FORM, i.e. feature-oriented reuse method, emphasising the principle of feature modelling.

An intuitive notion of features is fairly clear: features are distinctive aspects or characteristics of some system or entity that is the subject of inquiry [19]. For example in a familiar, consumer-reports style comparison chart, features are the columns of attributes by which a set of similar systems or products are compared, contrasted and evaluated. Kang & al. define [14] the feature as "an attribute or characteristic of a system that is meaningful to, or directly affects, the user, developer, or other entity that interacts with a system". In the domain of telecommunications, for example, Call-Forwarding, Call-Handling, and Call-Answering are features, and Binary-Search and Depth-First-Search algorithms are more technical features. In other words, features are abstractions of characteristics within an application domain visible to the user or the developer.

In software development, a set of products derived from a common platform is called a product family. The feature model of a family of products defines all the features in the platform and the relationships between them. When the product development focus is laid on the common platform – instead of the individual products – from which the design of individual products can be derived, all the products in the product family can be developed further at once. This is likely to bring in savings of time and total development cost.

The foundation of software platform development is comprised of the detection of domain entities and the analysis of their variability and commonality. In feature modelling, the model is constructed on the basis of a technical perspective (the information sources including expert users, manuals, analysts and design documents) and used for designing of SW architecture and the components therein. Kang & al. [14] present a formal method of modelling the application environment as features and their relations, designed for detecting commonalities and deriving variants from the model.

Kang's fine-tuned method of comprehensive feature detection and classification is primarily designed for building a product platform anew. Based on our experiences in electronics industry, the development of products happens incrementally. Perunka & al. [12] have already presented, see Figure 1, an iterative approach for an incremental development of the model. A tentative model is derived from the existing sources of domain knowledge, and the features of the product family are described using the entities of the model, refining the model as necessary. The software (SW) platform, and the components therein, are then enhanced to contain the entities of the model and to produce product variations automatically. The procedure can be repeated for new product releases, where new features are modelled as new entities of the model. Both models [12][14] emphasise the SW designer's role in both constructing and using the model, though Kang & al.[14] mention system users as a source of information and Perunka & al. see marketing (sales) as a natural group of feature model's users.

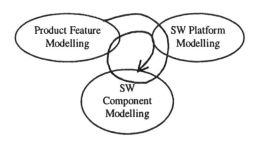

Figure 1 Development of feature model

Acquisition cycle wide feature modelling

Kalaoja [13] proposes a more comprehensive utilisation of feature model, integrating the design, implementation and configuration knowledge in the same model, which will also be used by sales managers and customers. In our recent studies we have worked on this subject and widened the scope of feature modelling to presenting design features in a format usable also in other activities of industrial design and product development process [20], such as concept design and testing of the prototypes, see Figure 2.

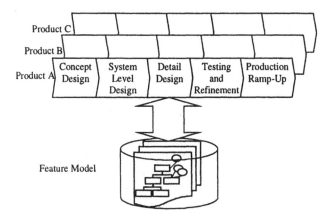

Figure 2 Use of the feature model in design of the products

Software design, however, still retains its central role in our framework. Thanks to this improvement, the new feature modelling approach provides support, as will be shown in the next chapter, to a concurrent advancement of various design activities, the results of which are put together in the assembly of end products.

Feature model

Our definition for features follows the common rules of configuration knowledge description [18]. The features describe the behavioural objects – and technical objects behind them to the needed extent – available in the application domain, and also the features describe the relations among object instances.

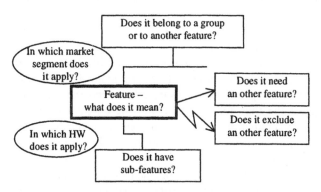

Figure 3 What is a feature

In Figure 3 the abstract term of feature is put into more specific terms. As already defined, features are abstractions of characteristics within an application domain visible to the user or the developer. In our approach, defining a feature means answering to the type of questions presented in Figure 3. The formal structure in which the answers are authored is the underlying instrument of our model-based management of product specifications. The structure is described in the following sections in detail.

Together the feature specifications form a model that describes the product family in the form of detailed requirements. The underlying purpose of explicit feature definitions is to make the decision process explicit, allowing everyone in the team to understand the decision rationale and reducing the possibility of moving forward with unsupported decisions. The feature model is released, thus creating a record of the decision making process for future reference and for educating newcomers. The feature model also helps to locate reusable assets in the development domain.

Kang & al. [14] manage four classes of features according to the types of information they present: application capabilities (services), operating environments, domain technologies, and implementation techniques. While sophisticated analysis of features is vital for detailed software design, in our framework for feature specification management, all the features are modelled with the same basic structure. This is attributable to the higher abstraction level of our modelling scheme, and to the optional attributes (labels) that are included in the model and that can be used for specifying the nature of a feature.

Structured modelling method

A formal presentation method is required for automatic processing of feature specification documents. During the last decade, the use of Standard Generalised Markup Language, SGML (ISO 8859)[21], has become general in technical documentation [22]. We have found the SGML format appropriate for manipulation of the feature specifications.

The most common way to use SGML language in technical documentation is to author manuals in SGML format, using an SGML aware editor, and thereafter to adjust the thus formed structured presentation in batch for dissemination over various kinds of media. The adjustments include page formatting, in the case of paper publishing, and generation of hyper-linked indexes and cross-references when the documents are to be published electronically. Automatically produced card-based documents, used for supporting training and market presentations [23], in addition to the traditional scroll-based documentation, are one example of the benefits of structured technical documentation.

The vantage of SGML is that the information that is presented in a common, readable format is at the same time in a format formal enough to be automatically processed. As a matter of fact, SGML-based authoring allows setting up a knowledge base of product's feature specifications, from which information can be queried – optionally in combination with other information sources – to satisfy all the information requirements of the feature processing to be described [24].

Elements of feature structure

The use of SGML requires a document structure to be specified. The structure must correspond to the actions required for producing the intended result document [25], which in this case is the compiled product specification. Thus, the features and their properties that form the basis of configuration have to be modelled as an SGML document type definition, DTD. In the modelling, the existing specification documents, together with the knowledge of the most experienced specifiers, have proved to be invaluable for finding the structure's building blocks for the structure. Special care has to be

taken when modelling the interrelationships and product platform dependencies needed in the compilation of the end-products. The document structure dictates what kinds of configurations are possible: feature details not authored at the beginning cannot be subsequently derived either.

In our model, the DTD is common to all features, meaning that all feature documents obey the same document structure. Besides a body that describes the behaviour of a feature, the structure contains common document header and footer information, with change histories, glossaries and indexes, as presented in Figure 4.

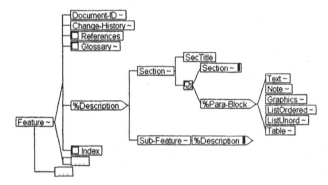

Figure 4 SGML structure of features

Named boxes in the structure tree are the elements of the feature. A real-life feature structure is likely to contain several dozens of elements. The structure is simplified here to focus on the elementary idea of presenting the feature as structured text. (The extra elements would express various details, e.g. the change history typically consists of a list of change entries, each entry comprised of the name of author, description of the change, and dates of committing and accepting the change.)

Tilde after element's name in the illustration means that the element has attached attributes (that the tool, where the illustration is captured from, does not show together with the element tree). The attributes are used for several purposes in product specification management. However, they are not directly used for formatting the document layout. The formatting of SGML documents is specified in a separate style-sheet file and is based on the semantic structure of the document, which also the attributes are describing for their part.

There are property attributes, which are needed in case of interrelated features, along the hierarchy. The property attribute of the related element (the feature or part of it, e.g. section, paragraph, or graphics) indicates the other feature that the element is related to. For example, if a part of the feature must be omitted in the case of omission of the other (referred) feature, or a selected sub-feature or section excludes another feature, the property attributes convey this information to the configuration solver. The property attributes, together with the possible sub-feature structures, are exported into the rule base of the configuration solver for the solver to use them in building and checking up the proposed product configuration. There are also attributes that express specialisation of features in the appropriate parts of features. The specialisation could occur for example in case of telecommunication devices per system or in case of handheld devices per keys in the keypad. The specialisation attributes are treated in configuration in the same way as property attributes, though they do not refer to another feature but a possible argument of the configuration procedure.

Beside the described feature structure no specific variation or extension points are needed: the fixed and variant parts are presented in the same structure. Typically, the body of the feature contains recursive structures, which are shown in the example DTD, also. With recursive structures the depth of sectioning (the exact number of "subs" in sub-sections / -features) does not have to be known in time of constructing the model.

The details of the specified products are in the format of blocks of descriptive texts, or highly parameterised atomic textual or graphics items, at the leaf level of the feature tree. Examples of leaf level items are gadget's voice signals or texts to be presented on the screen. The parameters describing the items are attached to them as attributes, to be used in automatic product customisation

to substitute the contents with the actual, product specific values.

Model's support for the design process

During the design the feature model is updated with new features and subsequently utilised to benefit from the detailed information in the feature specifications. After the specification database has been created, the designers can produce the views they need to the data to contribute their inherent knowledge towards further development of the product family.

Update of the model

As described above, the entire functionality of a product family is divided into a set of features that, when put together and adjusted (the adjustment will be described in detail below), form a comprehensive specification of a product. After its initial formation, the knowledge base is developed incrementally.

When a new feature is thought up, its contribution is judged against the existing specification knowledge, represented by the latest specification release. As a new feature is accepted, the knowledge base is augmented with this new feature and its premises. The augmentation involves describing the function of the feature as a feature specification, and fine-tuning the existing features in relation to the added functionality. This activity is iterative and contains the formation of both new and already existing features.

As the specification evolves, modernised releases of

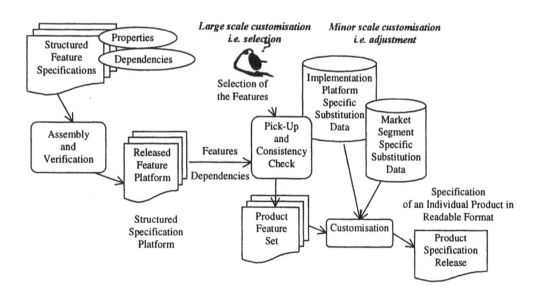

Figure 5 Automatic generation of the specification release for an individual product

the platform are introduced, where new features are added, and some old ones may be dismissed.

Utilisation of the model

During the DTD development process it had to be ensured that the required customisations could be established as selection and substitution of elements included in the product feature structure. In structured documentation, the structure routinely reflects the layout to be produced, i.e. the titles, subtitles, paragraphs and lists can be located in the structure [25]. Above and beyond, the structure should resemble the cognitive logic comprising a feature, and furthermore also the model comprising the generic platform and the individual product specifications.

In order for the knowledge base to be really useful, it must produce accurate specifications for the design of end products. The extraction of product specification for a single product comprises the following steps, see figure 5: assembly of the feature specifications to form a general product release, picking the features to be included in the product, and subsequent adjustment, i.e. customisation, of each selected feature.

The generic release assembly collects the feature specifications accepted for a specific release into a general release platform. The required version and status information is included in the specifications. At this point, the feature base is verified against references to non-existent features. This kind of inconsistencies could be checked up during authoring, but in case of excess database it can be too laborious to be executed during interactive processing.

After assembly, the designer (or even a marketing person) can start picking up the features for the end-product to be built. The selection of the features is based on the unique names of features and utilises the information provided by the selection tool of the relations between the features: some features are required, while others are optional; some features require the existence of another feature, while others may reject each other. This knowledge is recorded in the knowledge base, in the feature specifications, and exported to the selection tool prior to the selection process. The knowledge of feature relations, like all the content in the knowledge base, is specific to the current release of generic feature platform.

The structural levels of customisation are likely to divide into two categories: minor scale customisations, i.e. substitutions, and large-scale customisations, i.e. selections. Minor scale changes are substitutions of general terms with their product-specific values,

substitutions of variables with their values set in the customisation arguments, or substitutions of display items with their localised equivalents. Large-scale customisations occur in two phases, as direct selections of features and sub-features (to be explained shortly), and as indirect selections of property sections based on the selection of another feature.

The property sections are marked parts (from a paragraph to a section in size) of a feature that are meaningful only with another feature. They can be used for describing functions produced by two or more features together, when the combined effect would be too small for a full-scale feature.

In some occasions it may be reasonable to leave the joint effect out even if both features are selected. In this case, the property section is named with unique name during authoring and included as an individual feature into the feature set – which turns the property section into a sub-feature.

In the case of collaborating features, the feature-picking tool needs to know the requirement for the linked features. These requirements are expressed as attributes of the features or, in case of sub-features, of the sub-features themselves.

It is evident that this compilation of product specification is closely connected with the authoring of the features. Besides the interrelationships of the features, there are properties in the product that depend on the pre-set aesthetic or hardware dimensions. These can include the keys in keypad, the size and resolution of the display, the type of the display (black and white, greyscale, or colour), and the system platform of the product. All these properties are described, and marked up, during authoring and traversed during product compilation. Furthermore, the translation of interaction examples – in case of customisable graphics – and substitution of variables with product-specific values are carried out during the product compilation process.

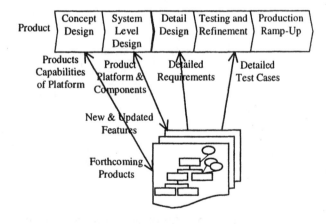

Figure 6 Model's support for the design process

Figure 6 presents a sum-up of the benefits of the model for the development of embedded products. The described process can be executed partially or as a whole. The complete execution is required to produce an up-to-date specification of an end-product, to be used in detailed design of the product. For example the need of translation, with detailed information on the display item occurrences provided during authoring of feature specification, could be traced by further querying this document.

Execution of the first part of the process provides an updated release of the generic platform. This can be compared to a previous platform release to see the intended differences between successive product families, which would be useful information in developing the software platform, for example. Furthermore, if the configuration of software platform resembles closely enough the specification platform (it is likely that the software configuration is far more complicated), the relations extracted from the generic feature specifications could be used in mass-customisation of the software, also.

Conclusions

This paper presents a structured text application of feature modelling. The presented feature model contains detailed requirement specifications of an embedded product, including configuration data for product variations. According to earlier studies, the main application area for feature modelling is to be found in software architecture design. Based on this study, feature modelling can be extended to support also the other parts of product development cycle, from concept generation to testing of the product.

The general advantage of the approach is that designers are able to produce human-readable design information that is, as such, formal enough for automatic processing, also. These specifications act as a common ground of the several disciplines involved in the embedded product design. Feature specifications, or selected parts of them, can be easily disseminated, via Intranet, up-to-date to readers in remote locations, which is essential in a distributed concurrent engineering environment.

Both the down-stream and up-stream parts of the development cycle benefit from structured feature specifications. It is possible, for example, for concept designers learn the options of contemporary products by studying related product specifications, their differences and similarities. The results of their work can even be included in the specifications of incoming products, e.g.

as VRP-simulations that can be run – in the electronic version of documentation – with a click of the mouse.

The marketing and user documentation, for their part, benefit from the specifications as they can get detailed descriptions of product's behaviour, with accurate (product specific) examples of e.g. displays and keypads, at an early phase of product development. This will also facilitate the thinkable translation and localisation processes, encouraging concurrency to the tasks that otherwise tend to be sequential. These kinds of options of interactive communication between the various product development parties are a major step towards incremental feature development, as addressed to by [12].

A technical benefit arises from basing the specification management on standard SGML, which makes also partial development of the specification environment possible. For example, the editor or the dissemination strategy can be changed without any need to change the customisation process. Furthermore, though our studies are based on SGML, the processing can be implemented by – or fairly easily converted into – XML [26] that is likely to offer an opportunity for using more economical tools in the future.

The described method of producing re-usable product specifications can be seen as an expedient for integrating the now separate islands of electronic product design. Further challenges include to study the integration of the described system and the design data management systems (ERP, Enterprise Resource Planning Systems, or SCM, Software Configuration Management Systems) used for managing other parts of product's design information.

Acknowledgements

I would like to thank all the members of our research group, Embedded Product Data Management at VTT Electronics, for the long-term co-operation that made writing this paper possible. I am especially grateful to my colleagues Mr. Jarmo Kalaoja and Dr. Eila Niemelä for sharing their profound notions about the software implementation of feature-based modelling and Mr. Mikko Kerttula for the discussions on the virtual reality prototyping.

References

[1] Pikosz, P.; Malmqvist, J.: Possibilities and Limitations When Introducing PDM Systems to Support the Product Development Process. Proceedings of NordDesign'96, Espoo, Finland 1996. p. 165–175.

[2] Loh, P. K: K.; Hura, G. S.; Khoon, C. C.: Virtual Prototyping of Cellular Phones. Software – Practice and Experience. Vol. 29, No. 10, 1999. p. 897–929.

[3] Kerttula, M.: Virtual Design and Virtual Reality Prototyping. MET Publications No. 13 1999, Metalliteollisuuden Keskusliitto, MET, Finland (http:/www.met.fi/kustannus).

[4] Vici, A. D.; Argentieri, N.; Mansour, A.; d'Alessandro, M.; Favaro, J.: FODAcom: An Experience with Domain Analysis in the Italian Telecom industry. International Conference on Software Reuse, 2-5 Jun 1998. p. 166-175.

[5] Thimbleby, H.: Design Probes for Handheld and Ubiquitous Computing. In: Gellersen, H.-W. (ed.) Proceedings of Handheld and Ubiquitous Computing, First International Symposium, HUC'99, Karlsruhe, Germany, Sep 1999. Springer-Verlag, Berlin. p. 1-19. (392 p.) ISBN 3-540-66550-1. ISSN 0302-9743.

[6] Time Special Report: The Communications Revolution. TIME, Vol. 154, No. 15, 11 Oct 1999.

[7] Laakko, T.; Mantyla, M.; Mantyla, R.; Nieminen, J.; Sulonen, R.; Tuomi, J. 1990. Feature Models for Design and Manufacturing. Proceedings of the Hawaii International Conference on System Science, Vol. 2. 2–5 Jan 1990. p. 445–454.

[8] Liu, T–H.; Fisher, G. W.: An aproach for PDES/STEP compatible concurrent engineering applications. Concurrent engineering: Tools and technologies for mechanical system design. In Haug, E. J. (ed.) NATO ASI Series, Berlin; Spring–Verlag, p. 433–464. ISBN 0-397-56532-9.

[9] Laakko, T.: Incremental Feature Modelling: Methodology for Integrating Features and Solid Models. Acta Polytechnica Scandinavica, Mathematics and Computer Science Series No. 63, Helsinki University of Technology; Helsinki, Finland, 1993.

[10] Kang, K. C.; Cohen, S. G.; Hess, J. A.; Novak, W. E.; Peterson, A. S.: Feature-Oriented Domain Analysis (FODA) Feasibility Study. Techical Report CMU/SEI90-TR-21, ESD-90-TR-21, Nov 1990. Software Engineering Institute, Carnegie-Mellon University, Pittsburgh, PA.

[11] Ihme, T. 1991. Reuse-oriented structured analysis for embedded systems. VTT, Helsinki, 1991, Publications / Technical Research Centre of Finland : 85. ISBN 951-38-4063-8. 142 p. (In Finnish)

[12] Perunka, H.; Niemelä, E.; Kalaoja, J.: Feature-Oriented approach to design reusable software architectures and components in embedded systems ESI, European Reuse Workshop '97 Position papers and presentations. Brussels, 26–27 Nov 1997. p. 74–77.

[13] Kalaoja, J.; Niemelä, E.; Perunka, H.: Feature modelling of component-based embedded software. STEP 1997: Proceedings Eighth IEEE International Workshop on Software Technology and Engineering Practice incorporating Computer Aided Software Engineering. London, UK, 14–18 July 1997, p. 444–451.

[14] Kang, K. C.; Kim, S.; Lee, J.; Lee, K.: Feature-oriented engineering of PBX software for adaptability and reusability. Software – Practice and Experience, Vol. 29, No.10, 1999. p. 875–896.

[15] Niemelä, E.: A Component Framework of a distributed control systems family. VTT Publications 402, Technical Research Centre of Finland, Espoo, Finland, 1999.

[16] Ross, A.: Selling uniqueness. Manufacturing Engineer, Vol. 75, No. 6, Dec 1996, p. 260–263.

[17] Freuder, E. C.: The role of configuration knowledge in the business process. IEEE Intelligent Systems & their applications, Vol. 13, No. 4, July–August 1998, p. 29–31.

[18] Sabin, D. & Weigel, R.: Product Configuration Frameworks – A Survey. IEEE Intelligent Systems & Their Applications, Vol 13, No. 4, July–August 1998, p. 42–49.

[19] Simos, M.; Anthony, J.: Weaving the Model Web: A Multi-Modeling Approach to Concepts and Features in Domain Engineering. International Conference on Software Reuse, Jun 2-5 1998, IEEE Comp Soc, p. 94-102.

[20] Ulrich, K. T.; Eppinger, S. D.: Product design and development. McGraw-Hill, New York, 1995. ISBN: 0-07-065811-0; 0-07-113742-4. 289 p.

[21] Goldfarb, C. F.; Rubinsky, Y.: The SGML Handbook. Clarendon Press, Oxford. 1990. ISBN: 0-19-853737-9. 663 p.

[22] Rath, H. H.; Wiedling, H.-P.: Making SGML Work: Introducing SGML into an Enterprise and Using Its Possibilities in Advanced Applications. Computer Standards & Interfaces, 18, 1 Jan 1996. Elsevier Science B.V. p. 37–53.

[23] Hsu, L. H.; Liu, P.; Dawidowsky, T.: Multimedia authoring-in-the-large environment to support complex product documentation. Multimedia Tools and Applications, Vol. 8, No. 1, 1999. Kluwer Academic Publishers, p. 11–64.

[24] Savolainen P.; Konttinen, H.: A Framework for Management of Sophisticated User Interface's Variants in Design Process: A Case Study. In: J. Vanderdonckt, A. Puerta (eds.), Computer-Aided Design of User Interfaces II, Proceedings of the 3rd International Conference on Computer-Aided Design of User Interfaces CADUI'99 (Louvain-la-Neuve, 21-23 October 1999). Kluwer Academics, Dordrecht. p. 205-215. (355 p.) ISBN 0-7923-6078-8.

[25] Maler, Eve; Andaloussi, Jeanne: Developing SGML DTDs from text to model to markup, Prentice Hall PTR, New Jersey, 1996. 532p. ISBN 0-13-309881-8.

[26] Harold, E. R.: XML: Extensible Markup Language. IDG Books Worldwide Inc., Foster City, CA. 1998. ISBN 0-7645-3199-9. 426 p.

A Relational Model for Analyzing Subclass Structures of Manufacturing Product Models

Stephen C. F. Chan & John W. T. Lee

Department of Computing, The Hong Kong Polytechnic University, Hong Kong

Email: {csschan, csjlee}@comp.polyu.edu.hk

Abstract

Data Models of manufacturing data, such as models of products, are often represented as subclass structures, which can be considered sets of classes constrained by specialization, intersection, union, and disjoint relations, some of which are explicitly specified but many of which are implicit. Here a real world object is an instance that may belong to multiple classes. In a subclass structure of even modest complexity, it is often difficult to visualize and analyze the relations between classes. This paper describes a relation-based model of the subclass structure and associated inference rules to uncover implicit relations. It provides well-formed and easy-to-understand answers to common questions such as whether two arbitrarily chosen classes are mutually exclusive, or whether the subclass structure is inconsistent, rendering it impossible for a particular class to have any valid instances, etc. The model has been tested on selected data models specified in the data modeling language EXPRESS, for the manufacturing product data exchange standard STEP.

1 Introduction

The object-oriented paradigm is increasingly popular in the modeling of manufacturing systems. In many projects, such as the modeling of computer-integrated manufacturing systems proposed by Yong & King [1], the emphasis is on modularity - the "object" as encapsulation of data and operations defined on it. In projects such as STEP/EXPRESS [2, 3], the emphasis is on the use of subclass inheritance to model generalization and specialization relationships. In the STEP/EXPRESS project as well as others, difficulties are often encountered in visualizing and analyzing large and complex subclass structures.

Models of manufacturing data typically involve large numbers of entity classes and complex relations between them. In some data models such as EXPRESS, subclasses of a common superclass may be mutually exclusive or overlapping. Given such a data model of even moderate complexity, it is not easy to determine whether the inheritance hierarchy is consistent, i.e., whether there are entity classes defined that cannot have any instances because of conflicting constraints. On the other hand, it is also not easy to determine whether a certain complex entity instance - an instance that belongs to a certain combination of entity classes - is allowed by the subclass hierarchy.

A complicated algorithm was developed for determining allowed instantiations of complex entity classes for data models defined in the language EXPRESS [3], which was defined by the International Standards Organization as part of the product data exchange standard STEP. This algorithm generates the complete set of instantiable complex entity classes through iterative traversals of the inheritance structure. It is difficult to describe and visualize, and has not yet been proven to be valid.

Chan & Lee [4] briefly introduced a relational model for generic subtyping structures and a set of rules for the derivation of more explicit relations between entity types in the structure. This papers expands on that model and the rules, and discusses how the model can be used to determine answers to questions such as consistency of the structure and allowable instances of complex entity classes. The theoretical foundations for the model, and the soundness and completeness of the set of rules are discussed in more detail in Lee & Chan [5].

2. The Relation Model

Consider a class structure defined by four kinds of relations:

- a class is the **specialization** of another class;
- a class is the **intersection** of two or more classes;
- a class is the **union** of two or more classes; and
- two classes are **disjoint**.

These relations together define a set of membership constraints among the classes, forming a class structure. In the following we define a model of the class structure using relations.

2.1 Modelling of Class Structure as Relations

A class structure specification $\mathcal{K}$, in a class universe T of classes, is defined as a tuple

$$(T, S, J, G, D)$$

where

- S, J, G, and D are binary relations in T, i.e., subsets of the Cartesian product T×T; and
- are represented as Boolean membership matrices.

In general a homogeneous relation $R \subseteq U \times U$ of a set of objects U can be represented by an n×n Boolean matrix [6] where

- $n = |U|$;
- $R(x, y)$ denotes the (x, y) element of the membership matrix R; and
- $R(x, y) = 1$ if $(x, y) \in R$, and the element (x, y) is said to be in R.

S is the specialization specification, where (x, y) is in S if x is specified as a superclass of y $(x \geq y)$. The superclass / subclass relationship is given a set inclusion semantics, i.e., x is a superclass of y if and only if at all times, {objects of class x} $\supseteq$ {objects of class y}. Some models distinguish between the intentional and extensional perspectives to permit greater flexibility in modelling. In this paper, however, we are only concerned with membership constraint, and will regard class inclusion and superclass as synonymous and use these terms interchangeably.

EXPRESS example:

```
ENTITY odd_number SUBTYPE OF (integer_number); …
                  ⇒
          integer_number ≥ odd_number
       S (integer_number, odd_number) = 1
```

In this example odd_number is specified as a subclass of integer_number, perhaps distinguished by having additional attributes or constraints on its attributes.

J is the intersection specification, where (x, y) is in J if x is specified as the intersection of a set of classes, and x is then referred to as an intersection class. It does not have an equivalent in the EXPRESS model. The concept in EXPRESS that is closest to it is multiple inheritance - a subtype that inherits from multiple supertypes.

EXPRESS example:

```
ENTITY edge_loop SUBTYPE OF (path, loop); …
                  ⇒
         Iset(edge_loop) = {path,loop}
J(edge_loop,path) = 1 & J(edge_loop,loop) = 1
```

The EXPRESS semantics for multiple inheritance specifies that every instance of edge_loop is an instance of path, as well as an instance of loop. It does not follow that every object that is an instance of path as well as an instance of loop is an instance of edge_loop, For the purpose of testing the model proposed in this paper, however, we shall assume that an EXPRESS subtype that inherits from multiple supertypes is the intersection of the supertypes.

It is possible that some classes in an intersection are themselves intersection classes. A canonical form of J can be defined as

$$J_C =_{DEF} J^+ \sqcap -(J^\circ L)^T$$

where

- J^+ is the transitive closure of J,
- L is the universal relation where $L(i, j) = 1$ for all i, j,
- $R^\circ S$ (or simply RS) is the relation $\{ (x, y) \mid \exists z \in T: (x, z) \in R \land (z, y) \in S \}$,
- the transpose or converse of R, R^T, is the relation $\{(x, y) \mid (y, x) \in R\}$,
- the complement of R, $-R$, is the relation $\{ (x, y) \mid (x, y) \in T \times T \land (x, y) \notin R \}$, and
- the intersection $R \sqcap S$ is the relation $\{ (x, y) \mid (x, y) \in R \land (x, y) \in S \}$.

This corresponds to repeatedly substituting intersection classes in an intersection set by their constituent classes. Thus J_C is the transitive closure of J with all intersection classes removed from the set of classes whose intersection makes up an intersection class. In the following example,

$$T = (a, b, c, d, e, f, g)$$
$$ISet (c) = \{e, f\}$$
$$ISet (a) = \{c, d\}$$
$$ISet (g) = \{a, b\}$$

as illustrated in Figure 1.

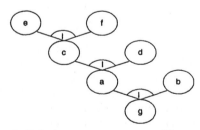

Figure 1. Relation among a group of intersection classes.

Then

$$
J= \begin{matrix} 0 & 0 & 1 & 1 & 0 & 0 & 0 \\ 0 & 0 & 0 & 0 & 0 & 0 & 0 \\ 0 & 0 & 0 & 0 & 1 & 1 & 0 \\ 0 & 0 & 0 & 0 & 0 & 0 & 0 \\ 0 & 0 & 0 & 0 & 0 & 0 & 0 \\ 0 & 0 & 0 & 0 & 0 & 0 & 0 \\ 1 & 1 & 0 & 0 & 0 & 0 & 0 \end{matrix}
\quad
J^+= \begin{matrix} 0 & 0 & 1 & 1 & 1 & 1 & 0 \\ 0 & 0 & 0 & 0 & 0 & 0 & 0 \\ 0 & 0 & 0 & 0 & 1 & 1 & 0 \\ 0 & 0 & 0 & 0 & 0 & 0 & 0 \\ 0 & 0 & 0 & 0 & 0 & 0 & 0 \\ 0 & 0 & 0 & 0 & 0 & 0 & 0 \\ 1 & 1 & 1 & 1 & 1 & 1 & 0 \end{matrix}
\quad
J_C= \begin{matrix} 0 & 0 & 0 & 1 & 1 & 1 & 0 \\ 0 & 0 & 0 & 0 & 0 & 0 & 0 \\ 0 & 0 & 0 & 0 & 1 & 1 & 0 \\ 0 & 0 & 0 & 0 & 0 & 0 & 0 \\ 0 & 0 & 0 & 0 & 0 & 0 & 0 \\ 0 & 0 & 0 & 0 & 0 & 0 & 0 \\ 0 & 1 & 0 & 1 & 1 & 1 & 0 \end{matrix}
$$

Note that if there are additional constraints on the intersection class so defined, then the model has to be refined. For example, if the edge_loop class above had additional constraints imposed on it, e.g., in the form of additional attributes or constraints on its attributes, then edge_loop is only a subclass of the intersection class of path and loop (path&loop), i.e.,

$$
\text{Iset(path\&loop)} = \{\text{path, loop}\}
$$
$$
J(\text{path\&loop,path}) = 1 \ \& \ J(\text{path\&loop,loop}) = 1
$$
$$
\text{and}
$$
$$
\text{path\&loop} \geq \text{edge\_loop}
$$

as illustrated in Figure 2. In this case edge_loop is not a proper intersection class of path and loop.

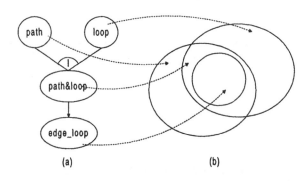

Figure 2. (a) The intersection class path&loop and its subclass edge_loop; (b) the relation as represented in set diagrams.

G is the generalization specification, where (x, y) is in G if x is specified as the generalization of a set of types, and y is in the component set of x.

EXPRESS example:

```
TYPE boolean_operand = SELECT (solid_model,
csg_primitive); ...
                  ⇒
       USet(boolean_operand) =
       {solid_model,csg_primitive}
    G(boolean_operand,solid_model) = 1 &
    G(boolean_operand,csg_primitive) = 1
```

Analogous to the intersection specification, we define a cannonical representation, G_c, for generalization classes in terms of non-generalization classes.

$$
G_c =_{DEF} G^+ \sqcap -(GL)^T
$$

D is the disjoint specification, where (x, y) is in D if x and y are specified to be disjoint (x ‖ y). Two classses x and y are disjoint if and only if at all times, {objects of class x} ∩ {objects of class y} = ∅.

EXPRESS example:

```
ENTITY curve SUPERTYPE OF (ONEOF(line, conic));
                  ...
                  ⇒
           line ‖ conic
        D(line,conic) = 1
```

2.2 Inference Rules for Derived Relations

From the class structure ? , we can derive a number of relations which fully describe the subclass and disjoint relations of a class and all other classes in T, based on a set of sound and complete inference rules.

(IR1) Inclusion Reflexivity: Every class a in T is a superclass of itself.

IR1 can be represented by the operation

$$
M_s = S \sqcup I
$$

where I is the identity matrix, i.e., $I(x, x) = 1, \forall x$,
the union $R \sqcup S$ is the relation $\{ (x, y) \mid (x, y) \in R \wedge (x, y) \in S \}$; and
M_s is the explicit subclass relation.

(IIR1) Intersection Inclusion: If y is in the intersection set of x, then y is a superclass of x.

IIR1 can be represented by the operation

$$
M_s = M_s \sqcup J^T
$$

For example, in a class universe where
$$
T = \{\text{edge\_loop, edge, path}\},
$$
and edge_loop is specified as the intersection of edge and path, i.e.,
$$
\text{ISet (edge\_loop)} = \{\text{edge, path}\}
$$
then by IIR1,

$$
J = \begin{matrix} 0 & 1 & 1 \\ 0 & 0 & 0 \\ 0 & 0 & 0 \end{matrix}
\quad
M_s \sqcup J^T = \begin{matrix} 1 & 0 & 0 \\ 0 & 1 & 0 \\ 0 & 0 & 1 \end{matrix} \sqcup \begin{matrix} 0 & 0 & 0 \\ 1 & 0 & 0 \\ 1 & 0 & 0 \end{matrix} = \begin{matrix} 1 & 0 & 0 \\ 1 & 1 & 0 \\ 1 & 0 & 1 \end{matrix}
$$

where M_s (edge, edge_loop) = 1 and
M_s (path, edge_loop) = 1 indicate that edge and path are superclasses of edge_loop.

(IUR1) Generalization Inclusion: If y is in the component set of x, then x is a superclass of y.

IIR1 can be represented by the operation

$$M_s = M_s \sqcup G$$

(IR2) Inclusion Transitivity: If x is a superclass of y and y is a superclass of z, then x is a superclass of z.

IR2 can be represented by the operation

$$M_\geq = M_\geq \circ M_\geq$$

where $M_\geq$ represents the superclass / subclass relation between classes in T,
$M_\geq = M_s$ initially, and
R°S is the relation {(x, y) | $\exists z \in$ T: (x, z)$\in$R $\land$ (z, y)$\in$S}.

A more explicit representation of $M_\geq$ can be derived by applying IR2. For example, in a class universe where

$$T = \{geometry, curve, conic\},$$
$$geometry \geq curve, \text{ and }$$
$$curve \geq conic;$$

then by IR2,

$$M_\geq = M_s = \begin{matrix}1 & 1 & 0 \\ 0 & 1 & 1 \\ 0 & 0 & 1\end{matrix} \text{ and } M_\geq \circ M_\geq = \begin{matrix}1 & 1 & 0 \\ 0 & 1 & 1 \\ 0 & 0 & 1\end{matrix} \circ \begin{matrix}1 & 1 & 0 \\ 0 & 1 & 1 \\ 0 & 0 & 1\end{matrix} = \begin{matrix}1 & 1 & 1 \\ 0 & 1 & 1 \\ 0 & 0 & 1\end{matrix}$$

where $M_\geq \circ M_\geq(1,3) = 1$ indicates that geometry is a superclass of conic.

(IIR2) Implicit Intersection Inclusion: If all types in the intersection set of x are superclasses of y, then x is a superclass of y.

IIR2 can be represented by the operation

$$J' = J \sqcup (I \sqcap \neg JL)$$
$$M_\geq = (\neg J') \lozenge M_\geq$$

where
- J' is a modification to J where all non-intersection classes are treated as intersections of themselves and
- R◊S is the relation {(x,y) | $\exists$ z$\in$U: (x,z)$\in$R $\lor$ (z,y)$\in$S}..

For example, in a class universe where
$$T = \{edge\_loop, X, edge, path\},$$
$$ISet (edge\_loop) = \{edge, path\},$$
$$edge \geq X, \text{ and }$$
$$path \geq X;$$

as illustrated in Figure 3.

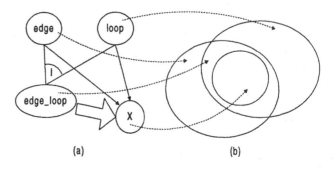

(a)   (b)

Figure 3. An example of implicit intersection inclusion between edge_loop and X.

Then by IIR2,

$$J = \begin{matrix}0 & 0 & 1 & 1 \\ 0 & 0 & 0 & 0 \\ 0 & 0 & 0 & 0 \\ 0 & 0 & 0 & 0\end{matrix}$$

$$J' = \begin{matrix}0 & 0 & 1 & 1 \\ 0 & 1 & 0 & 0 \\ 0 & 0 & 1 & 0 \\ 0 & 0 & 0 & 1\end{matrix}$$

$$M_\geq = (\neg J') \lozenge M_\geq = \begin{matrix}1 & 1 & 0 & 0 \\ 1 & 0 & 1 & 1 \\ 1 & 1 & 0 & 1 \\ 1 & 1 & 1 & 0\end{matrix} \lozenge \begin{matrix}1 & 0 & 0 & 0 \\ 0 & 1 & 0 & 0 \\ 1 & 1 & 1 & 0 \\ 1 & 1 & 0 & 1\end{matrix} = \begin{matrix}1 & 1 & 0 & 0 \\ 0 & 1 & 0 & 0 \\ 1 & 1 & 1 & 0 \\ 1 & 1 & 0 & 1\end{matrix}$$

where $M_\geq(edge\_loop, X) = 1$ indicate that edge_loop $\geq$ X.

(IUR2) Implicit Generalization Inclusion: If all types in the component set of x are subclasses of y, then x is a subclass of y.

IIR2 can be represented by the operation

$$G' = G \sqcup (I \sqcap \neg GL)$$
$$M_\geq = M_\geq \lozenge \neg G'^T$$

For example, consider the following class universe where
$$T = \{boolean\_operand, X, solid\_model,$$
$$csg\_primitive\},$$
$$USet (boolean\_operand) = \{ solid\_model,$$
$$csg\_primitive \},$$
$$X \geq solid\_model, \text{ and }$$
$$X \geq csg\_primitive;$$
as illustrated in Figure 4.

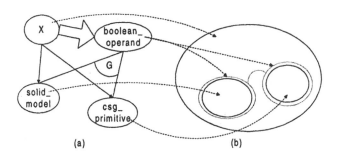

Figure 4. An example of implicit generalization inclusion: (a) the class structure; (b) the corresponding set diagram.

Then by IIR2,

$$G = \begin{array}{cccc} 0 & 0 & 1 & 1 \\ 0 & 0 & 0 & 0 \\ 0 & 0 & 0 & 0 \\ 0 & 0 & 0 & 0 \end{array}$$

$$G' = \begin{array}{cccc} 0 & 0 & 1 & 1 \\ 0 & 1 & 0 & 0 \\ 0 & 0 & 1 & 0 \\ 0 & 0 & 0 & 1 \end{array}$$

$$M_\geq = M_\geq \lozenge (-G')^T = \begin{array}{cccc} 1 & 0 & 1 & 1 \\ 0 & 1 & 1 & 1 \\ 0 & 0 & 1 & 0 \\ 0 & 0 & 0 & 1 \end{array} \lozenge \begin{array}{cccc} 1 & 1 & 1 & 1 \\ 1 & 0 & 1 & 1 \\ 0 & 1 & 0 & 1 \\ 0 & 1 & 1 & 0 \end{array} = \begin{array}{cccc} 1 & 0 & 1 & 1 \\ \mathbf{1} & 1 & 1 & 1 \\ 0 & 0 & 1 & 0 \\ 0 & 0 & 0 & 1 \end{array}$$

where $M_\geq(X, \text{boolean\_operand}) = \mathbf{1}$ indicate that $X \geq$ boolean_operand.

(DR1) **Disjoint Symmetry:** If x is disjoint with y, then y is disjoint with x.

DR1 can be represented as
$$M_\emptyset = D \sqcup D^T$$

where $M_\emptyset$ represents the disjoint relation between classes in U.

(DR2) **Disjoint Inheritance:** If x is disjoint with y, and y is a superclass of z, then x is disjoint with z.

DR2 can be represented as

$$M_\emptyset = M_\emptyset M_\geq$$

(DR3) **Implicit Disjoint:** If all types in the component set of x are disjoint with y, then x is disjoint with y.

DR3 can be represented as

$$M_\emptyset = (-G') \lozenge M_\emptyset$$

(DR4) **Null Class:** If x is disjoint with x, then x is disjoint with all y.

2.3 Derived Relations and Their Application

Combining IR1, IIR1 and IUR1, we define the Explicit Superclass Relation M_s as

$$M_s =_{DEF} S \sqcup J^T \sqcup G \sqcup I$$

By repeated application of IR2, IIR2, and IUR2, all the implicit superclass relations can be represented explicitly. Using fixed point semantics, we define **the Superclass Relation $M_\geq$** as

$$M_\geq =_{FP} M_\geq M_\geq \sqcup ((-J') \lozenge M_\geq) \sqcup (M_\geq \lozenge (-G')^T)$$

where $M_\geq := M_s$ initially. The right hand side of the equation can be shown to be monotonic. As the set of classes is finite, the equation will have a fixed point.

Similarly, by repeated applications of DR1, DR2, and DR3, all the implicit disjoint constraints can be made explicit. We define the **Class Disjoint Relation $M_\emptyset$** as

$$M_\emptyset =_{FP} M_\emptyset^T \sqcup M_\emptyset M_\geq \sqcup ((-G') \lozenge M_\emptyset)$$

where $M_\emptyset = D$ initially.

Inconsistency occurs in a class structure when constraints are in conflict with one another, resulting in a class is defined that cannot have any elements. This occurs when a class is disjoint with itself, i.e., when any of the elements (x, x) is in $M_\emptyset$.

A class structure defines allowable instantiations of complex classes, instances that belong to multiple classes. A complex class is not allowed by the class structure if any pair (x, y) of classes that it contains is in $M_\emptyset$. Hence this relation model can be used to determine allowed complex entity type instantiations.

Annex B of the EXPRESS Language Reference Manual [3] listed an algorithm which generates the complete set of complex entity classes (called complex entity types in EXPRESS) instantiable from the class structure. In contrast, a simple algorithm to determine the legitimacy of a complex class based on our relation model can be as follows:

Let V_c be an vector representing the complex entity class C, where

S_c = set of classes included in the complex class,

$|V_c| = |T|$,

$V_c(x) = 0$ if $x \notin S_c$, and

$V_c(x) = 1$ if $x \in S_c$.

If $(M_\varnothing {}^\circ V_c) \wedge V_c <> 0$, then C is illegal.

For example, in a in a class universe where
$T = \{a, b, c, d, e\}$,
$a \geq b$,
$a \geq c$,
$b \geq d$,
$b \geq e$, and
$b \parallel c$;
as illustrated in Figure 5.

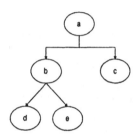

Figure 5. Checking the validity of complex instance $C = \{c, e\}$ in a sample class universe.

For $C = \{c, e\}$

$$M_\varnothing {}^\circ V_C = \begin{array}{ccccc} 0 & 0 & 0 & 0 & 0 \\ 0 & 0 & 1 & 0 & 0 \\ 0 & 1 & 0 & 1 & 1 \\ 0 & 0 & 1 & 0 & 0 \\ 0 & 0 & 1 & 0 & 0 \end{array} {}^\circ \begin{array}{c} 0 \\ 0 \\ 1 \\ 0 \\ 1 \end{array} = \begin{array}{c} 0 \\ 1 \\ 1 \\ 1 \\ 1 \end{array} \quad \& \quad (M_\varnothing {}^\circ V_C) \wedge V_C = \begin{array}{c} 0 \\ 0 \\ 1 \\ 0 \\ 1 \end{array}$$

where $(M_\varnothing {}^\circ V_C)(c) = 1$ indicates that c is mutually exclusive with some element of C, and $(M_\varnothing {}^\circ V_C) \wedge V_C(c) = 1$ indicates that c is also in C. Hence some element in C is mutually exclusive with some other element in C, and C is thus illegal.

3. A Practical Example

As a practical example of the use of the derived superclass and disjoint relations, consider the following much simplified EXPRESS model on geometry, as illustrated in Figure 6.

```
ENTITY geometric_item;  SUPERTYPE OF
   (ONEOF(point, curve, surface));
END_ENTITY;
ENTITY point; SUBTYPE OF geometric_item;
   END_ENTITY;
ENTITY curve; SUBTYPE OF geometric_item;
   END_ENTITY;
```

```
ENTITY line; SUBTYPE OF curve; END_ENTITY;
ENTITY conic; SUBTYPE OF curve; END_ENTITY;
ENTITY surface; SUBTYPE OF geometric_item;
   END_ENTITY;
ENTITY x; SUBTYPE OF (conic, surface);
   END_ENTITY;
```

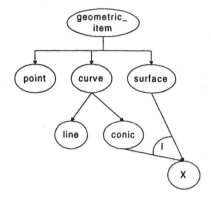

Figure 6. An example class structure to illustrate the derivation of implicit relations.

The relation model for this class hierarchy is derived in the following, where T = {geometric_item, point, curve, surface, line, conic, x}.

```
S= 0111000   J= 0000000   G= 0000000   D= 0000000
   0000000      0000000      0000000      0011000
   0000110      0000000      0000000      0101000
   0000000      0000000      0000000      0110000
   0000000      0000000      0000000      0000000
   0000000      0000000      0000000      0000000
   0000000      0001010      0000000      0000000
```

By applying the inference rules discussed above, we arrive at the following explicit relations:

```
M≥ =    1111111   M∅ =  0000000
        0100000         0011111
        0010111         0101001
        0001001         0110111
        0000100         0101001
        0000011         0101001
        0000001         0111111
```

Examining $M_\geq$ and $M_\varnothing$ we discover a number of constraints among entity types that were not specified explicitly in the original model. For example,

$M_\geq(1,5) = 1 \Rightarrow$ geometric_item $\geq$ line
$M_\geq(1,6) = 1 \Rightarrow$ geometric_item $\geq$ conic
$M_\geq(1,7) = 1 \Rightarrow$ geometric_item $\geq$ x
$M_\geq(3,7) = 1 \Rightarrow$ curve $\geq$ x
$M_\geq(4,7) = 1 \Rightarrow$ surface $\geq$ x
$M_\geq(6,7) = 1 \Rightarrow$ conic $\geq$ x
$M_\varnothing(2,5) = 1 \Rightarrow$ point $\parallel$ line
$M_\varnothing(2,6) \;\; = 1 \Rightarrow$ point $\parallel$ conic
$M_\varnothing(2,7) = 1 \Rightarrow$ point $\parallel$ x
$M_\varnothing(3,7) = 1 \Rightarrow$ curve $\parallel$ x
$M_\varnothing(4,5) = 1 \Rightarrow$ surface $\parallel$ line

```
M∅(4,6) = 1 ⇒ surface ‖ conic
M∅(4,7) = 1 ⇒ surface ‖ x
M∅(5,7) = 1 ⇒ line ‖ x
M∅(6,7) = 1 ⇒ conic ‖ x
M∅(7,7) = 1 ⇒ x ‖ x
```

The last constraint indicates that x is disjoint with itself, rendering the class structure inconsistent.

On the other hand, any complex entity including the combination of point-line is disallowed because $M_\emptyset(2,5) = 1$, but one including line-conic may be allowed because $M_\emptyset(5,6) = 0$.

4. In Conclusion

We described a relational model which allows the subclass structure to be represented and analyzed in a matrix notation, producing answers to questions such as the legality of complex entities and consistency of the subclass structure. Hence the model can be very useful in the evaluation of the quality of data models developed for various aspects of manufacturing, and in uncovering potential ambiguities and problems in the applications of these data models. In the implementation of the relational model, the matrices involved may become very large and the related operations can be computationally intensive. However, the matrices are generally quite sparse and optimization techniques may be used to speed up such operations.

5. Acknowledgement

The work described in this paper was partially supported by the Hong Kong University Grants Committee Research Grants Council, Project HKP51/94E.

6. References

[1]   Yoon, D.H.H. & King, L.S. "An Object-Oriented Approach to Computer Integrated Systems", *Journal of Systems Integration*, 6, p.159-179, 1996.

[2]   *ISO/IS 10303-1 Product Data Representation and Exchange - Part 1: Overview and Fundamental Principles*, International Standards Organization, 1994.

[3]   *ISO/IS 10303-11 Product Data Representation and Exchange - Part 11: The EXPRESS Language Reference Manual*, ISO, 1994.

[4]   Chan, S. & Lee, J. "Declaration and Modeling of the Type Structure in EXPRESS", *Proceedings, Express User Group International Conference*, October 1996, Toronto, Canada.

[5]   Lee, J. & Chan, S. "A Relation Approach to Consistency Checking in Class Structures", working paper.

[6]   Schmidt, G. & Strohlein, T. *Relations and Graphs*, Springer-Verlag, 1993.

FORMALISATION OF KNOW-HOW RULES FROM CONCEPTUAL STRUCTURES MODELING. APPLICATION TO CYLINDER HEAD MACHINING

A. Lefebvre [1] [3], J. Renaud [2]
L. Sabourin [3], G. Gogu [3], J.-F. Plusquellec [1]

[1] RENAULT, DM-DITS/PCIM
Rueil-Malmaison, F-92 562

[2] LRGSI, ENSGSI/INPL
Nancy, F-54 010 Nancy cedex

[3] LaRAMA, IFMA/Université Blaise Pascal - Clermont II
Aubière, F-63 175 cedex

Abstract

In the manufacturing industry the quick evolution of products seems to affect particularly geometric and functional parameters. Each company wants to take this evolution into account from the engineering design so as to use previous knowledge for new products. In a concurrent engineering context, the setting up of a knowledge base easily useable by everyone into a project will allow to save money and to decrease time to market .

Our research area concerns the definition of a methodology and the characterization of manufacturing constraints for machining process planning of prismatic parts. The first step of the work deals with the formalisation of expert knowledge related to machining process. The use of conceptual models enables us to precisely view several functional conditions to respect for a product. Moreover conceptual models will also be used to determine know-how rules so as to re-use previous knowledge and to limit technical risks. The application field of the work is the RENAULT Powertrain Division.

1 Introduction

The presented work is supported by the Research Division of RENAULT. This article points out the modeling of process planning for prismatic parts and we focus on the cylinder head part. The RENAULT Powertrain Division has been set up flexible lines since four or five years. These production lines are able to support high variations of volume and products mix (five to seven head cylinder types). The process plan is the result of several computed constraints (manufacturing constraints, economic constraints etc.). Our main goal is to identify these constraints and to represent them using conceptual models. Then they will be identified as input of a decision system which will generate alternatives of process plans (cf. figure 1).

This article presents a modeling approach based on conceptual graphs. We attempt to identify and formalize the expert knowledge as regards manufacturing constraints for the elaboration of process plans for head cylinder parts. We also attempt to create know-how maps from previous experiences and parameters chosen by experts.

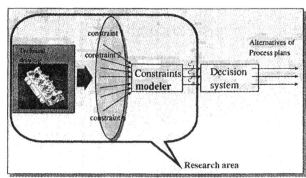

Figure 1 Research area

2 Approach and objectives

The work in progress related to the modeling of expert knowledge is composed of five steps. Steps 1 to 4 are achieved and step 5 will come in the near future.

i.   identification of manufacturing features (2.1)
ii. decomposition of manufacturing features according to their attributes and toleranced geometric items from the *product* point of view (2.2.)
iii. collection of the expert knowledge: formalization of manufacturing constraints (2.3)
iv. modeling expert rules using conceptual graphs (2.4)
v.   determination of know-how rules

Case study: camshaft lines on a cylinder head

Camshaft lines machining is a hard process to make: the parameters choice lies in expert knowledge because of the use of long cutting tools (e.g.: reamer) and very small tolerance values. Today the camshaft lines process plan depends a lot on each process engineer. We use the first step of the Design Rules methodology [10] to describe the operations sequence: for each product (G9, F4, F9), we set up a list of the activities sequence with resources and machined features as shown in figure 2.

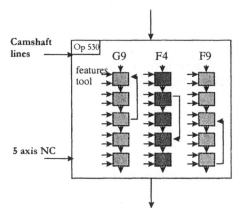

Figure 2 Activities sequences of camshaft lines machining for G9, F4 and F9 products

2.1 Manufacturing feature [6] [9]

The approach of knowledge expert analyze is based on the manufacturing feature concept which is detailed in this paragraph.

Manufacturing features enable to understand the overall piece by both the designer and the manufacturer and also by everyone who is involved in the project. Their are composed of product and process data. The following definition is close to the one of the GAMA group [2]: a manufacturing feature is characterized by a set a geometric surfaces associated with attributes (roughness, tolerancing etc.) and a sequential order of operations (machining, assembly, heat treatment etc.). Figure 3 shows that our work is focused on the identification and the formalisation of product/process constraints and figure 4 gives some details of the product part of a feature.

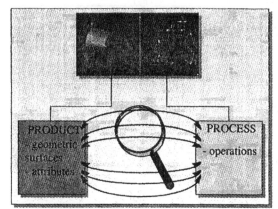

Figure 3 Manufacturing feature

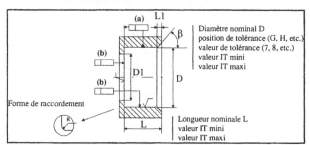

Figure 4 Details of the product part of a feature

2.2 Manufacturing features splitting up

Once we have identified features which make up camshaft lines, we spilt them up from the product point of view into their attributes (diameter, diameter tolerances, roughness) and their toleranced geometric elements (axis, face, etc.). Information given by figure 5 come from process oriented plans.

FEATURE	ITEM	TYPE	VALUE	REFERENCE
Housing	diameter	Øρ		
	tolerance	Hρ		
	roughness	Rρ		
	real axis	coax	Ød1	M-N
	bearing δ com. face	loca	η	ref
		encadre	φ	ref1
	F300 com. face	dim	x + 0.2	ref1

FEATURE	ITEM	TYPE	VALUE	REFERENCE
Bearing α	diameter	Øα		
	tolerance	Eα		
	roughness	Rα		
	reference axis M	loca	Ød2	ref

FEATURE	ITEM	TYPE	VALUE	REFERENCE
Bearing β	diameter	Øβ		
	tolerance	Eβ		
	roughness	Rβ		
	reference axis N	loca	Ød3	ref

Bearing γ	diameter	γ		
	tolerance	Eγ		
	roughness	Rγ		
	axis	cyl	T	common zone
		rect	Ød4	common zone
		para	Ød5	ref common zone
		para	Ød6	restriction ref

Figure 5 Manufacturing features: attributes and toleranced geometric surfaces

2.3 Expert knowledge collection and know-how rules

The aim of that step is to represent the process engineer reasoning which comes to sequences described by figure 2: activities order, choice of features to be machined, cutting tools selection etc.. The bi-directional links formalization between product and process parameters and the determination of know-how rules can help a lot for the making of an expert knowledge base [8].

Expert rules (*if then else* form) that have been developed are of three types that we call f1, f2 and f3. Figure 6 gives an illustration for each one from the machining of a joint housing coaxial to a camshaft line bearing.

✓ **f1**: if {parameter}$_{product}$ → then {parameters}$_{process}$

if $\varnothing_{housing} = \varnothing\rho$ if $H = H\rho$ if $R = R\rho$ if $loca_{bearing} = loca\ \varnothing d2$ if *no restriction*	
	then **passes_nbr = 2**

✓ **f2**: if {parameters}$_{process}$ → then {parameters}$_{product}$

if tool_type = *multidiam* if machining = *finition* if jig_number = *1*	
	then **coax Ød1 /M-N Ok**

✓ **f3**: if {parameters}$_{product/process}$ → then {par.}$_{process}$

if cycle time = *tcy* if *coax Ød1/M-N* if *M-N is loca$_{bearing}$ Ød2*	then **multidiam tool for finition**

Figure 6 Example of expert knowledge rules

2.4 Conceptual graphs modeling

2.4.1 Definition

A conceptual graph is an abstract representation for logic with nodes called concepts and conceptual relations linked together by arcs [ISO/IEC 10646-1]. A more formal definition has been given by M.-L. Mugnier [7]:

A conceptual graph is related to a support S which defines the basic vocabulary useful to represent manipulated knowledge in an application domain. A support is a five-tuple $S = (T_C, T_R, \sigma, I, \tau)$ where:

i. T_C is a set of concept types structured in a lattice representing an AKO -is a kind of- hierarchy and allowing multiple inheritance

ii. T_R, is a set of relation types

iii. σ is of star graphs showing for every relation type what kind of concept types it can link

iv. I is a set markers for concept vertices: one generic marker and individual markers which allow to distinguish and name distinct entities

v. τ is a conformity relation from I to T_C which associates a marker m with a concept type t -"there is an individual m that IS-A t"-.

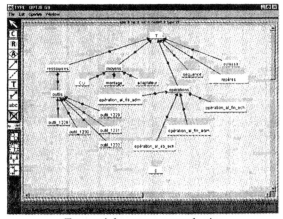

T_C : partial concept types lattice

I = {alésage, 1231, M4, ébauche}

τ = {(alésage, entité), (1231, outil), (M4, montage), ...}

Figure 2 Partial definition of a support S

An S-graph related to a support S is a multigraph[1]
$G = (R, C, U, lab)$ where:

i. *R* et *C* are two classes of relation and concepts vertices ($C \neq 0$)

ii. *U* is a set of edges; the set of edges adjacent to a vertex *r* is totally ordered

iii. *lab*: every vertex has a label defined by a mapping *lab* which obeys to the following:

if r $\in$ R, the lab(r) = type (r) $\in T_R$

if c $\in$ C, then lab(c) = (type(c), ref(c))

2.4.2 Conceptual graphs and CharGer© editor interests

We choose CharGer© editor [1] for constraints modeling. This conceptual graphs editor is developed by the University of Alabama and is still a beta-version.

CharGer© allows:

i. to build data models animated when the user changes concept instances

ii. static and "dynamic" data modeling using simple conceptual relations (localization, property etc.) and "dynamic" conceptual relations (functions computing, databases queries etc.). Graphs can also be linked together by using co-reference links from one context to another one. One important point is that CharGer© editor can simulate processes when instances change [4] [5].

Figure 3 shows that a cutting tool has three characteristics. When the user changes the tool number, the editor points at the DBOutils.txt database through a "dynamic" operator and reads information related to the attributes. The DBOutils.txt database belongs to a *text* type database and contains all the required fields specified in figure 3 (*outil, repères, diamètre, type*).

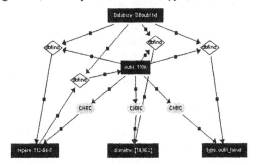

Figure 8 Simple conceptual graph

2.4.3 Modeling objectives

The main objective is to describe the process engineer reasoning when he chooses process parameters. It means that we have to describe a set of product and process constraints. From the process tolerances of the piece, we focus on parameters which determine the manufacturing process like: cutting tools selection, manufacturing sequences, number of machining passes by feature, type of jig etc..

2.4.3.a Product/process constraints modeling

In that paragraph we show an example of product/process modeling using a conceptual graph.

Among constraints related to process planning generation, we have a look at manufacturing constraints which link product and process data. We take as instance a *boring* feature machining whose design parameters are well known (cf. figure 9).

The conceptual graph of the *boring* feature is based on the concept feature (cf. 2.1) and is also composed of product generic concepts (as *caractéristiques*) and process generic ones (as *opération, outil*). Relations between concepts are either simple {POSS, CHRCED, SUP, INF}, or mathematical {fonction} or database queries {dbfind}. CharGer© editor allows also to simulate processes when the user changes concept markers (e.g.: tool number). Finally, all databases used in CharGer© are of *text* type and must contain specified concepts in the graph that stand for fields in the database.

The problem the process engineer wants to solve is to find the main tool parameters (length, diameter) for correct machining. We can see that the depth and the boring diameter tolerance are dependent on the length and the tool diameter which are read through a query in DBCOI.txt database [3]. We apply the same to the roughness and boring quality which are dependent on the machining sequence. In that case, the query is based on an heuristic.

---

[1] Graph which allows more than one edge between two vertices, here between a concept vertex and a relation vertex

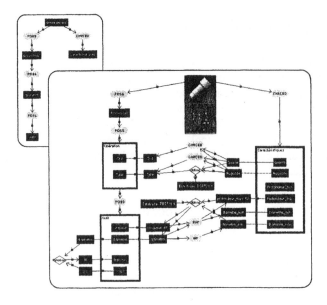

Figure 9   Product/process constraints

2.4.3.b   Expert   knowledge   rules modeling

The conceptual graph of figure 10 shows the three types of constraints f1, f2 and f3 defined in 2.3. Heuristics must now be found that is we have to determinate dynamic relations f1, f2, f3 which allow the expert knowledge to be re-usable. Our working axis for that point moves towards 2D graphical representations of parameters. The number of parameters is up to five; so some of them shall be bring together according to experts.

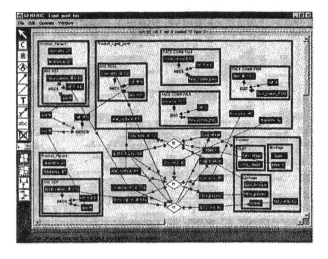

Figure 10   Constraints graph from expert rules

3   Conclusion and further works

Considering an innovation context, the identification of expert domains related to a set of product characteristics will allow a project team to analyze whether a new product can be machined on the production line. Parameters which are out of the expert domain are then detected at an early stage of the industrial process.

The context of the overall project is to develop an integrated tool (CAD) of process plans generation for prismatic parts on flexible lines as mentioned in figure 1; further works will deal with the problem of integrating formalized constraints into a solver within the CAD modeler. This forthcoming process planning modeler which will integrate product (numerical definition of parts) and process data (cutting tools, jigs etc.) shall be of great help to limit technical risk.

Bibliography

[1] CharGer 2000 v2.2b 1999, Copyright 1998-1999 by Harry S. Delugach

[2] Groupe GAMA, textes réunis par P. Bourdet et F. Villeneuve, *La gamme automatique en usinage*, Ed. Hermès, ISBN 2-86601-255-0, Paris, 1990

[3] A. Lefebvre, L. Sabourin, J.-F. Plusquellec, *Entités de réalisation, référentiel commun entre concepteur et fabriquant. Modélisation par graphes conceptuels*, Colloque CPI'99, pp. 108-115,Tanger, Maroc, 1999

[4] G.W. Mineau, *Constraints on Processes: Essential Elements for the Validation and Execution of Processes*, Dept. of Computer Science, Université Laval, Québec City, 1998

[5] G.W. Mineau, *The Representation of Semantic Constraints in Conceptual Graph Systems*, Dept. of Computer Science, Université Laval, Québec City, 1998

[6] J. Quichon, *Programmation Logique sous Contraintes: Application à la génération automatique de gammes d'usinage en fraisage*, Thèse de Doctorat de l'Université Blaise Pascal, Clermont II, 1999

[7] M.-L. Mugnier, M. Chein, *Représenter des connaissances et raisonner avec des graphes*, R.I.A., vol. 10, n°1, pp. 7-56, 1996

[8] J. Renaud, *Démarche de capitalisation de connaissances-métier*, Thèse de Troisième Cycle, ENSGSI Nancy, 1994

[9] L. Sabourin, *Machining features: integration of the machining function in the modelling of parts for automated process planning*, IDMME, pp. 103-112, 1997

[10] J.-M. Voirpin, J. Renaud, M. Dufour, B. Mutel, C. Guidat, *Design Rules: a method for the modeling of technical data in the context of innovation*, INCOM'98, Nancy, 1998

Taxonomy of Information and Knowledge Management in a Concurrent Engineering Context

M. GARDONI

GILCO laboratory, ENSGI, 46 avenue Félix Viallet, 38031 Grenoble Cedex1, France
mail : gardoni@gilco.inpg.fr

E. BLANCO

Soils, Solids, Structures laboratory, Domaine Universitaire, BP 53, 38041 GRENOBLE Cedex 9, France
mail : blanco@hmg.inpg.fr

Abstract

Concurrent Engineering approaches heavily rely on reliable and efficient shared information among people involved in the design, engineering, industrialisation and even manufacturing of products shared information. In a context of integrated design methodology, the nature of information exchanged by the design team changes and enterprises tend to support this information through co-operative technology (CSCW). So, Non-Structured-Information becomes increasingly important within Integrated-Team. Pieces of information flowing in this kind of team are heterogeneous (verbal sketching, writings, drawings etc...) and seem hardly controlled. Although informational aspects become more and more strategic, they are barely controlled. Indeed, it could be difficult to structure, share and access pieces of information to enhance Integrated-Team working and to capitalise knowledge and know-how to learn from past experiences and to avoid doing the same mistakes twice.

In this paper, we propose taxonomies of information flowing in an Integrated-Team starting from the analysis of the design activities. We focus both on the characteristics of information and the materiality and the use of the intermediary objects, which support this information. Then we discuss the need that information system has to meet in Concurrent Engineering context.

Moreover, tracked information are needed to supply raw materials to knowledge management. Ability to capitalise depends on tracked information. In this context, we propose to highlight limits of knowledge management because of the lack of both linguistic and graphical information support.

1 Evolution of Engineering Information Requirements

Several organisational strategies have been deployed by many enterprises to adapt to evolving economical constraints over the century (standardised production, economy of scale, economy of scope). Traditionally, engineering activities are performed in a sequential order. Over the last ten years, strong market pressures have forced manufacturing companies to drastically reduce the time-to-market of their products. This is why companies tend to apply the Concurrent Engineering approach (CE) [1] [2], which is opposed to the sequential engineering approach.

Concurrent Engineering (CE) takes into account the well-known QCD (Quality, Cost and Delay) objectives. This is why companies tend to apply the Concurrent Engineering approach, which is opposed to the sequential engineering approach as summarised in Figure 1. Figure 1 introduces two fundamental way of working starting from receiving results or from exchanges of information. This is why many companies tend to apply the CE approach to parallelise as much as possible their activities [3] [4].

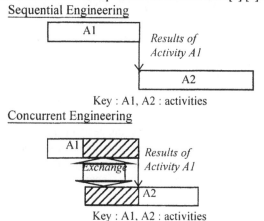

Figure 1 Comparison between Sequential Engineering and Concurrent Engineering

Most of the information exchanged in sequential engineering are results of the different tasks. Actors who are responsible of the next task can trust it. In CE, actors of the design have to exchange conjectures [5] that allow the others to work simultaneously. The actors have to discuss ideas, drafts of solutions etc... These information are partially true, and have to be updated often. The nature of information is then different and the role of product data management is therefore increasing [6] [7]. The actual systems have to support this co-operation however they had been mostly designed for sequential design process. That's why, some enterprises promote the exchanges by design teams, then CE relies on so-called Integrated Teams made of (Figure 2):

- a Project Leader, who is responsible of the project for achieving the quality, cost and delay objectives,
- Department Representatives, who are still member of their originating department (with which they keep a hierarchical link). They have a functional relationship with the Project Leader.

"control" can be characterised in terms of four criteria: structuring, sharing, access and capitalisation [8].

The concept of information is complex. Indeed, among various taxonomy of information, we distinguish three of them, associated respectively with

- the levels with knowledge,
- the structuration of information,
- the object that support information.

2 Taxonomy of information related to knowledge

In reference to a European experiment over a period of nearly fifteen years [9], four levels of information were identified: rough information, organised information, treated information and advanced information.

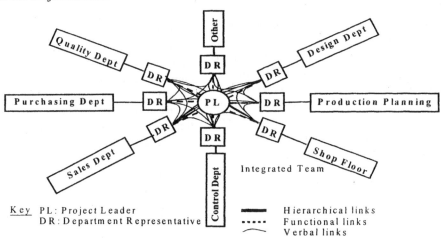

Figure 2 Possible representation of an Integrated Team

The operations in Integrated Teams facilitate the exchange of information between the various services. This brings more flexibility and more reactivity to the project. This results in reduction of cycle times, development costs and product costs. Then in the design process two information flows exist simultaneously:

- The information relative to the product that are controlled on the PDM [6]. Those product data are mainly results of the team decisions.
- Another information flow deals with all the communications within the design team.

Those communications are mainly performed verbally in face-to-face talks or on the phone. Unfortunately, this information is therefore poorly controlled. This term

In order to be closer to the industrial context, we consider that these four levels of information are vectors of communication representing the four levels of knowledge which characterise the knowledge and know-how [10] [11]:

We thus associate

- rough information with the representation of data,
- organised information with the representation of information,
- treated information with the representation of knowledge,
- advanced information with the representation of the theory /expertise.

2.1 Data

Data are materialised by rough information or basic facts. They are made of symbols and figures and they reflect parts of the concrete world. With each event or thing can be associated data, it is then possible to collect and store it in the shape of "white matter" (documentation in shape of paper or electronic). What makes these data strategic, it is in particular:

- Proliferation of the networks (interns to the companies with the intranet/extranet and external with Internet), as well as the continual fall in the costs of communication which facilitates the search for information,

- Need to circulate information in the companies to accelerate their reactivity. Indeed, to be reactive, it is necessary that flows which cross companies are fast : financial flows, matter flows but also decision or data flows.

- The power of the data-processing machines now allows the realisation of virtual models which tend to supplant the physical models because of the low costs of transport, the easy management of the various versions of the prototypes, etc.

- The "revolution" of the multi-media incites people of the company to communicate in a different way such as incorporating in their numerical documents, other documents, images (fixed or animated) and sound. The use of multi-media risks to take time because of our significant paper culture.

All the data of a fiel could not be the "white matter" because everything can not be retranscribed in the shape of symbols or of figures (example: a delicate manual operation).

2.2 Information

Information are the base of the communication. In our context, they are represented by organised information. They are the result of a significant selection activity and aggregation of useful data. The nature of the selection depends on several factors and several combinations of factors, which are related to the individual who treats the data. This one is influenced by his social system, its culture, its department in the company, etc. Generally, information are conceived with an aim of making a decision. They can be constituted of a multidisciplinary structuration of rough information, such as:

- The cluster of data concerning a product (example: a plan of a part with its cartouche accompanied by a Quality documents, etc).

- " Intelligent packs " of information: this concept, borned in Great Britain [9], is the result of efforts made by the large libraries and the resource centres to bring out answers adapted to the needs (listed by investigations) of the users.

Organised information is a sum of rough information, they are truly useful only after processing by an individual who is able to understand them.

2.3 Knowledge

Knowledge is materialised by treated data. It could be considered as a refining of information. During this refining process, information which are supposed to be re-used is often synthesised and systematised. This is why, the life cycle of a knowledge is longer than those of information. There are two types of knowledge [12]:

- Explicit, which refers to knowledge made up of formal elements (data, functions, algorithms, models, information, etc.) which can be represented either in the shape of models, or of words, graphs, etc.

- Tacit, which cannot be expressed in a language (or all other symbols) because it is made of more abstract elements (experiment, skill, etc).

Tacit knowledge would require a refining in which knowledge itself would be directly generated without taking the form of processed information. Tacit knowledge is very significant for the company.

Rough and organised information "are not worth anything" if nobody is able to treat them and to exploit them. This processing or cycles of knowledge acquisition (Figure 3) is based on other sources of information and validations with experts of the field. Rough and organised information becomes elaborated information then and thus acquires added value.

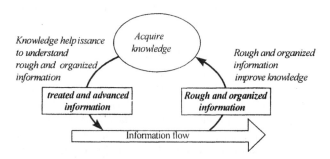

Figure 3: Cycle of knowledge acquisition

This cycle of knowledge acquisition is dynamic. Knowledge is thus not static but dynamic, it is in constant evolution (what is "true" at the "t" time will perhaps not be it anymore at the "t+1" time). Elaborate information is a synthesis with levels of different processing, including more or less disciplines and offering a more or less global solution of the treated topic. A state of the art can take the shape of:

- Theoretical collections based on methods, which make it possible to rationalise, systematise and structure.
- Presentations of the conditions of technical implementation where the opinion of the expert, are the most important. This one is based on the common sense, the intuition and the experiment.

This synthesis is often not very operational and thus not easily usable directly by the decision-makers. Thus appear advanced information.

2.4 Theories / expertise

The theories / expertise are represented by advanced information. The theory is a generalisation of knowledge, that affects our experiments and the activity which consists in giving a significance to a thing. The expertise is to find the relevant piece of information in a significant flow of data. These tasks of generalisation and search will make it possible to emphasise information. They require at the same time a capacity of extraction starting from documents and a capacity of development of information advanced in contexts of analysis which can be complex. This concept of advanced information appears because of the increase of two parameters:

- Complexity: competition becoming world wild, the factors likely to influence the life of the company become large too. Moreover, these factors become more and more diversified: macro-economic, technological, scientific, financial, political, sociological, demographic, etc. However, they remain accessible via a growing number of sources from information. So the volume of data to be located, treated, then to integrated, does not cease increasing: "too many information kills information?". Moreover, there are many interrelationships between these data.
- Speed: the life cycle of the products shortens, which imposes more flexible means of production, increased logistic performances etc. Thus, the rhythm of evolution of much of these data accelerates. The decision-makers have to treat on the one hand volumes of information increasingly significant and varied and on the other hand, they have to reduce the cycle information-decision.

To meet the requirements of these decision-makers, it s not only a question of providing good information, to the right person, at the good time what constitutes in fact a real problem, it should be done with the level of " added value" suitable for an immediate use.

- Against complexity, it is necessary to incorporate in a coherent way varied data in very dispersed origins, compared to the treated topic,
- Against speed, it is necessary to provide an adequate added value answer.

Advanced information are more than simple synthesis, they have to take into account the potentials of the company and its environment. They present information directly usable. They meet the needs of the decision-makers who seek to find operational solutions to solve daily practical problems. They also meet the strategic needs of the company by proposing elements of decision on the long term: development, diversification, launching of new products, attacks of new markets, etc.

The management of these four types of information (rough, organised, treated and advanced) could be connected with the management of the industrial intellectual capital. Indeed, the transformation of a data into information then into knowledge and theory/expertise represents a cost. When a person solves a problem, he acquires knowledge, and even theory/expertise. If this problem occurs again, it would be judicious to re-use knowledge or theory/expertise previously produced rather than starting again the cycle of acquisition of knowledge. That's why, companies have a growing interest for the "Capitalisation of the knowledge and know-how" in other words "Knowledge Management".

3 Taxonomy of information related to structuration

In order to meet the needs of rigour of the companies without going down on a too fine level of granularity, we chose an instructional design of the significance of information. We thus consider that the construction of a sentence corresponds to a whole of instructions formulated in term of variables, which provide a sense to the statement. Exchanged information is then an abstracted entity, a theoretical object which consists of linguistic components and components rhetoric [13]:

- The linguistic components build the significance of information starting from instructions. They are characterised by the clearness of their formalism more or less structuring which leaves the possibility of having or not various possible interpretations. For instance, industrial designs formally described the geometrical shape of a part and leave only few little possible interpretations. On contrary, talks can be interpreted in several ways according to the receivers.

- The rhetoric components bring a sense to information by addition of contextual information. This construction is characterised by the facility to identify information context. For instance, an industrial design associated with a cartouche not have to, in theory, require the possession of other information for its understanding.

According to instructional semantic description (Figure 4), we agree that an Information System is structuring when information has one and only one significance (with clear linguistic components) and a well defined sense (with accurate rhetoric component).

The properties of this structuring enable us to define *Structured-Information* (*SI*), *Semi-Structured-Information* (*SSI*), and *Non-Structured-Information* (*NSI*).

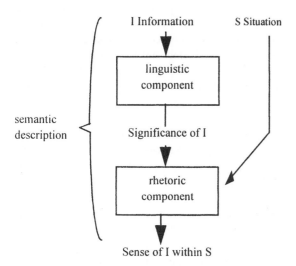

Figure 4: Instructional design of the significance of information

Characteristics of the *SI*:

- Linguistic components of the *SI* are generally imposed. The employed formalisms are accurate and logic. They leave little place to interpretation (example: an industrial design, etc),
- Rhetoric components of the *SI* are also imposed. They are clearly defined and have to be indicated to validate information. These *SI* have contractual values, this fact is accentuated by the installation of

the standards like ISO (example: an industrial design is associated with the name of the designer, the product, the part, the date, etc.).

The emission of a *SI* is, in general, an integral part of the work of the transmitter. He has not the choice of the *SI* type (design, text, etc). He has to respect the rules concerning the container, the content and the circulation of the *SI*. These constraints were elaborated so that all the receiver have, thanks to the *SI*, necessary and sufficient information to carry out their tasks (either information are contained in the *SI*, or the *SI* indicate the other *SI* which contain them). Thus sending *SI* does not presuppose to know the person but the knowledge and know-how of the receiver. This generates uniformity of the *SI* within a group (service, company or group of companies).

Characteristics of *SSI* :

- Linguistic components of the *SSI* are little formalised. They can take the shape of texts, tables, graphs, etc. They could be hardly understandable by all but more easily for direct receivers (for instance: graphs without legend, design without industrial formalism, etc).
- Rhetoric components could be parsimonious. Indeed, the transmitter knows the receiver and adapts the level of granularity of the rhetoric components according to the supposed knowledge of the context that the receiver has (example: meeting reports are not always easily interpretable by a person who did not take part to the meeting).

The *SSI* are stored less longer than the *IS* because the context is not always associated with information, they can thus quickly become not useful.

Characteristics of *NSI* :

- The *NSI* are very little formalised. The formalisation of the linguistic components employed depends on the degree of complicity between transmitters and the receivers, which can leave place to a multitude of interpretations.
- The rhetoric components can be very light if they ensure a sufficient degree of relevance for the comprehension of information by the receiver (example of a verbal message which can be relevant according to the transmitter and of the receiver: 'we do as we talked').

The *NSI* are essentially volatile, because even if it is possible to preserve a piece of information of a talks (' we do as we talked'), it is sometimes more difficult to remind the context.

The appreciation criteria of structuring an Information System are:

- The rigour, if the Information System incites the transmitter to use clear linguistic components and to associate with information accurate rhetoric component,
- The facility to associate linguistic and rhetoric components with information. In this case, the facility is connected with the intuitivity,
- The possibility to structure enough information to be able to differentiate each piece of information,
- The unicity of the piece of information in order to answer to the preoccupation of coherence and an integrity.

4 Taxonomy based on the objects supporting information.

Previous accounts of design work point to various ways in which design works depends on communicative activity. [14] Design work is related to the production and the use of information. But the information produced by the actors of the design is quite heterogeneous, it could be as well digital models, drafts, tables of data, as plans or prototypes, etc…This communicative activity is supported by a lot of artefacts all along the design process. The notion of information is poorly related to the use and production of those artefacts. For example, a prototype is generally not managed in information flow as well as most of the sketches used by designers. The notion of information focuses on the communication aspects of the design. Those artefacts are also representation of the future product. Our hypothesis is that it is important to understand both the network of designers and the content of the design work. In order to perform this analysis we propose to follow the artefacts that support the design information.

We use the notion of intermediary objects [15] [16] to describe all the objects or documents that appear or are used in the process, whatever their form, their origin or their destination: schedules, minutes, functional graphs, calculation results, drafts, 2D plans or 3D models, prototypes, etc.

Those intermediary objects can be seen as resulting from the design work but also as supporting and highlighting it. This term of intermediary object serves as a generic designation that is useful in its globality. It enables us to raise general questions about the how the design processes under study actually function.

The ability of these objects to help us understand better both the network of designers and the content of the design work is due to their hybrid character: modelling the future product they act as communication vectors between the product designers. These two aspects are so connected in the reality of the process that we cannot isolate one from the other without deforming their nature.

As a vector of communication, the objects structure the design network, like models of the future product, they highlight its evolution.

All the intermediary objects do not have the same characteristics in the design. Those characteristics depend on the properties of the object itself but also of the situation of action in which it is committed.

Those characteristics can be decline along two axes [16].

4.1 Commissioning / mediating

This first axis qualifies the ability of the objects to transmit the intention of the transmitter. A commissioning object is a transparent object that makes only transmit one intention, an idea. It does not modify of anything the idea or the intention of its producer. When we analyse information without taking into account the medium or the object, we privilege this direction.

On the opposite direction of this axis we met the mediating objects. Each object corresponds to a particular way of objectivizing the idea or the intention by inscribing it in a specific organised matter, whose structures and rules it has to come to terms with. Thus, for example, a drawing is not simply a faithful representation of the mental idea or of the specifications. It is a translation, i.e. a realisation and a transformation which has been carried out according to its specific constraints (including those of the material used, which may vary depending on whether it is on paper or on screen) and its own rules and conventions. Thus, object modifies the idea, the initial intention, from its existence and its use as support of transmission. In fact many objects are mediators because they offer to their user different capacities of action. They introduce new constraints in the process. For example many Studies that focus on the role of the sketches refer to the phenomenon of reinterpretation and emergence [17]. Designers have to take into account that the shift from one form to another changes the contents. The prototype is not merely a new formulation of what is represented in the drawing: it is a new version of the final object, specifying some aspects and modifying others

4.2 Open / closed

The second axis Open / closed qualifies the margin which is left to the user. A closed object transmits a strong regulation whereas an open object is a support of negotiation. The position of an object on this axis depends on the statute of the information given by the actor, and on the object itself.

In a Sequential model of design process, closed objects support most of the information. For example the formal drawings are distributed from designers to the various downstream user groups. They give instructions for the fabrication stages. On the opposite sketches are "group thinking" tools for the designers and serve to the negotiation or problem solving. Sketches are generally open objects that can play the role of conjecture while official drawings transmit decisions from a group to another.

5 conclusion

To be competitive, companies have to decrease the time to market of their products. To this aim, they tend to apply the concept of Integrated Engineering. This mode of organisation is based on the parallelisation of the activities of design, preparation, manufacture, etc. This overlapping of the activities cannot be effective without exchanges of *SSI* and *NSI*. The importance and flows of these types of information are increasing. The companies have to manage a larger mass of information of different typologies (*SI*, *SSI* and *NSI*) and their supports have to adapt with these new constraints.

In the organisation of design teams work, it is important to take into account the characteristic of the objects. For example a CAD model that needs 200 hours of work had lost its ability to be an open object. This model couldn't help the negotiation between manufacturers and designers because of the cost of any modification. The implementation of CE then failed.

The figure 5 proposes to map the taxonomy of information related to structuration and the point of view referring to object characteristics.

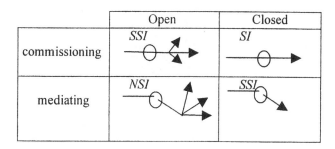

Figure 5: Mapping of taxonomies

For example sketches are Mediating/Open objects. We had shown that their sense emerges during the interactions [5]. They are built as local conventions that allow the group to build a common understanding of the situation. They are sometimes unusable out of the situation of action where they were built. The linguistics components are partially defined (using technical drawing rules to sketch a part for example) but the designers play with the ambiguity of the representations to exchange conjectures of solutions. Sketches correspond to *NSI* and as verbal exchanges are not controlled in industrial design process.

On the opposite we consider that *SI* tend to correspond to Commissioning / Closed objects. The producers of this type of objects tend to define the rhetoric components that allow a unique understanding of the information. It is possible because the situation of use is known. It is supported by conventions of high level that are shared by the actors of the design process. Those objects play a role of transmitters of information whereas open objects are support for negotiation.

The characteristics of objects of course are not fully fixed. In the course of action actors can open a closed object using it to support negotiation process. For instance an official drawing can be used as a starting point for a new research of solution, designers could sketch on it.

The figure 6 highlights the different needs that the information system has to meet in a CE context. It shows the characteristics of the objects needs to support the designer's work. During co-operation phases actors needs mainly open objects whereas closed objects are required to transmit formal design decisions.

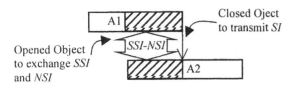

Keys

SI: Structured-Information
SSI: Semi-Structured-Information
INS: Non-Structured-Information
A1, A2: Activities 1 and 2

Figure 6: The different information needs in CE context

This is why, MICA approach [18] [19] (a specific interactive messaging system) has been developed and experimented in order to harness of *NSI*. Indeed, a GroupWare tool called also MICA, was created through an Intranet. It has been implemented and put into operation in an engineering team of twenty people since May 1998. Some return on experience has already been extracted about improvement of the efficiency of the Integrated Team.

Moreover, MICA tracked parts of the interactions and information exchanges between members of Integrated Teams in solving an engineering problem. Then a method to capitalise relevant knowledge deployed during a product design project has been tested. This way, data-mining approach seems to be useful in the context of *NSI* synthesis. Indeed, we achieve to extract some IF/THEN rules from the magma of *NSI* tracked during the experimentation. These exchange treated as semantic networks are used to capitalise knowledge about problem solved during the project. This knowledge could be re-used in further studies. This treatment allows to upgrade relevant information in knowledge as mentioned in the cycle of knowledge acquisition in section 2.

Nevertheless, this kind of capitalisation from linguistic data is limited because of the graphical *NSI* lack. Indeed Henderson emphases on visual culture of the engineers [20]. She says that sketches are "the real heart of visual communication". Sketching takes a large place in the group understanding of technical problems. Sketches could be considered as graphical *NSI*. They are both models of the product and communication vectors. Here is one of the actual the limit of the MICA approach and more generaly of Knowledge Management system. It only takes into account linguistic data and does not include management of graphical *NSI*. Generally GroupWare

solutions for interactive sketching have to be improved. Furthermore, how could we treat graphical *NSI* to capitalise knowledge in design process? It is still an open question for research?

References

[1]   G. Sohlenius, " Concurrent Engineering, Keynote Paper" Annals of the CIRP, 41/2,pp. 645-655, 1999

[2]   B. Prasad, Concurrent Engineering Fundamentals - Integrated product and process organisation, Vol. 1, Prentice Hall, Englewood Cliffs, NJ, 1996

[3]   D. Brissaud, O. Garro, " An approach to concurrent engineering using distributed design methodology" CERA journal, V4, N°3. Pp303-311, sept 1996

[4]   S. Tichkiewitch, L. Roucoules, " Methodology for innovative design" "International CIRP Design Seminar ", Integration of process knowledge into design support systems pp 81-90, March 24-26 1999.

[5]   E. Blanco, L'émergence du produit dans la conception distribuée, PhD thesis of industrial engineering INPG 1998.

[6]   J.C. Bocquet, Ingénierie Simultanée, conception intégrée, Conception de produits mécaniques (méthodes, modèles et outils), coord. M. Tollenaere, Editions Hermès, Paris, 1998

[7]   F.B. Vernadat, Enterprise Modeling and Integration: Principles and Applications, Chapman & Hall, London, 1996

[8]   M. Gardoni, M. Spadoni, F. Vernadat, " Information and Knowledge Support in Concurrent Engineering Environments", 3rd International Conference on Engineering Design and Automation, EDA'99, Vancouver, B.C., Canada, August 1-4, 1999

[9]   P. Degoul, Le pouvoir de l'information avancée face au règne de la complexité, Annales des mines, Avril 1993

[10]   S. Tsuchiya, " Commensurability, A Key Concept of Business Re-engineering", 3rd International Synposium on the Manangement of Information and Corporate Knowledge (ISMICK'95), Institut International pour l'Intelligence Artificielle, Compiègne, France, October 23-24, pp.81-89, 1995

[11]   T.V. Van Engers, H. Mathies, J. Leget, C.C.Dekker, " Knowledge Management in the Dutch Tax and Customs Administration : professionalisation within a knwoledge intensive organisation", 3rd International Synposium on the Manangement of Information and Corporate Knowledge (ISMICK'95), Institut International pour l'Intelligence Artificielle, Compiègne, France, pp. 71-80, October 23-24, 1995

[12] J.P. Barthes, M. Grundstein, "Discussion Summary", 3rd International Symposium on the Management of Information and Corporate Knowledge (ISMICK'95), Institut International pour l'Intelligence Artificielle, Compiègne, France, October 23-24 1995

[13] Moeschler, J. Modélisation du dialogue (représentation de l'inférence argumentative), Editions Hermès, Paris, 1989

[14] L.L. Bucciarelli Designing engineers MIT Press 1994.

[15] D. Vinck and A. Jeantet, Mediating and commissioning objects in the sociotechnical process of product design: a conceptual approach. in proceedings of management and new technologies: COSTA3 workshop Designs, Networks and Strategies, European Community, 1995

[16] E. Blanco, O. Garro, D. Brissaud, A. Jeantet, Intermediary objects in the context of distributed design proceedings of CESA, Lille, July 1996

[17] A.T. Purcell and J.S. Gero, "Drawings and the design process" Design studies 19, 1998, 389-430

[18] M. Gardoni, M. Spadoni, F. Vernadat, Requirements analysis for better information support in Concurrent Engineering Environments, International conference on industrial engineering and production management, (IEPM'99), Glasgow, July 12-15 1999

[19] M. Gardoni, Maîtrise de l'information non structurée et capitalisation du savoir et du savoir-faire en Ingénierie Intégrée - Cas d'étude Aérospatiale Matra, PhD thesis of Metz University 1999

[20] K. Henderson On line and On paper visual representations, visual culture and computer graphics in design engineering, MIT press 1999

Reuse and patterns for viewpoints integration of Automated Production Systems

Djibril NDIAYE, Michel BIGAND, Didier CORBEEL, Jean-Pierre BOUREY

Laboratoire d'Automatique et d'Informatique Industrielle de Lille

LAIL, CNRS UPRESA 8021

Ecole Centrale de Lille, BP 48

59651 Villeneuve d'Ascq France

email: [Djibril.Ndiaye, Michel.Bigand, Didier.Corbeel, Jean-Pierre.Bourey]@ec-lille.fr

Abstract

The improvement of productivity must pass by the co-operation of various designers and experts who use varied modelling tools. The exchange and information sharing are then necessary. There are standardised solutions (CIM-OSA, BASE-PTA) proposing a top-down approach based on the preliminary definition of a reference model. In this paper we perform a bottom-up approach, which consists in gradually integrating the users' viewpoints in a repository, after meta-modelling them. We created several patterns based on graph theory, and made an application on the CASPAIM project developed in our laboratory.

1. Introduction

Nowadays, businesses live trough a highly volatile environment, in a world vigorously competitive market, filled with risks resulting from constant change and ever-increasing complexity. So to improve their productivity, businesses must reduce time-to-market with a proactive approach. This is done by the co-operation between various designers and experts. Since the 90's the Computer Integrated Manufacturing (CIM) concept aims to cover all activities related to the manufacturing business. CIM is then a multi-disciplinary subject, which is mainly concerned with information management and processing.

The intervention of several experts with particular viewpoints and modelling tools needs integration tools and methods in order to ensure the consistency of the whole approach.

There are standardised solutions proposing an approach based on the preliminary definition of a general model, which uses reference architecture. By specialisation, this reference model is adaptable to particular systems: it is a top-down approach. For example this is the case of the CIM-OSA, which proposes a multi-viewpoint integration of the enterprises.

In the same way, the BASE-PTA model constitutes a reference framework used as a basis for the design of new systems.

This integration approach based on a reference model (top-down approach) is well adapted to the design of a new production system. But the integration of existing applications with regard to these reference models remains difficult, because this can only be done by identification and adaptation of their characteristics, in relation to those proposed in the reference models.

In view of this situation, we defined a bottom-up approach, which consists in gradually integrating the users' viewpoints, after meta-modelling based on the Unified Modelling Language (UML) [25]. Thus our approach requires a thorough study of what exists. So the UML models are closely related to the viewpoint of each expert. To obtain a generic repository we must allow a large reuse of the obtained models, based on patterns [1].

After justifying our bottom-up approach, we briefly present CASPAIM project, and develop our approach based on meta-modelling and reuse. Then we present some patterns based on graph theory, and finally show an application on the control viewpoint of the CASPAIM method.

2. Context

2.1. The repository

The repository ensures, to the various users, the exchange and the sharing of relevant information (data, processing, documents, resources, etc.) relating to a product (process, part, etc.) during its life cycle.

The objective is to obtain a consistent global model by data integration, therefore by resolving conflicts such as [20]: the polysemy, the redundancy, the data's inaccessibility, the risks of inconsistency related to a modification.

A first approach consists in using a reference model.

It is the case for example of the CIM-OSA model [26] that proposes a multi-viewpoint integration of the enterprises. Various models are proposed and each of them is dedicated to a specific field of the enterprise (organisation, resources, information, and functions).

The derivation models the description of the passage from the expression of the company's needs to the design specifications, then to the implementation. Since the reference architecture is consistent, the particular architecture, which results from it by instantiation, is also consistent. This type of integration is well adapted to the design of a new flexible production system (FPS), but the consideration of existing means remains difficult.

In the same way, the BASE-PTA model [2] constitutes a reference framework intended to be used as a basis for the design of new Computer Aided Design systems, by supporting the exchange and the transfer of data with the existing ones. A data conceptual model based on the entity/association formalism describes the reference model.

Integration between the tools is done using a neutral exchange file. This file is built starting from the reference BASE-PTA model. The construction mechanism of the neutral file is based on the extraction of an Applied Conceptual Sub-Diagram (ACSD). This ACSD is obtained by successive specialisation and restrictions phases that seek to identify the significant and common elements handled by the tools to integrate.

As shown in [7], the integration of existing applications with regard to these reference models remains difficult, because it can be done only by identification and adaptation of their characteristics in relation to these proposed in the reference models.

However the objectives of the Automatic Engineering, are:
➢ on the one hand, the development of the ASP (Automated Systems of Production) in a continuous and consistent way,
➢ on the other hand to take into account some independence of the design activities so as not to impose a development method.

In view of this situation, we decided to study and develop an ascending approach that consists in gradually integrating the users viewpoints after modelling them. Indeed, the various users' models are worked out, and constitute many viewpoints that we have to integrate using the repository.

2.2. CASPAIM

The LAIL has developed an assisted design method for the Flexible Production Systems (FPS). This method allows the assisted generation, of the control part of an integrated production system, and its architecture.

The work concerns several aspects (figure 1).
➢ Planning / Scheduling [18], generates a production plan that distributes the various tasks between the production means, by taking their capacity and their availability into account.
➢ Control [3], defines the operational sequences according to the concurrent use of the production resources.
➢ Supervision [9], defines the control that must perform the production plan under the best conditions. The supervision consists in several views such as:
- Maintenance [19], which manages the preventive or curative actions on the production system;
- Monitoring [12] [23], which detects the errors and automatically diagnoses their causes.

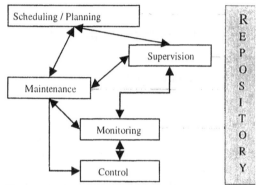

Figure 1: the various viewpoints of a FPS

Thus, various models and tools (Petri nets, Statechart, bar chart, specific formalisms, etc.) were simultaneously developed and used, in the various points of view. We will see how to ensure the consistency of these various models by defining and using the repository.

3. Global approach

3.1. Principle of meta-modelling

Our main objective in this paper is to represent in a common formalism the various designers' models, and to federate the data, which they handle. According to [21], when the act of modelling applies to models, the approach is called meta-modelling. This approach is particularly appropriate for the definition of the data structure, because it allows a better comprehension of the models. A neutral model can be used as dialogue technique between the various designers. Thus they handle relevant data of a FPS with their own models, which are stable and validated, but their specialisation in a given field does not facilitate the exchange and the sharing of information. To remove the partitions (figure 2) between these models while keeping their independence, we chose to translate them according to the UML (presented on section 3.3) object formalism.

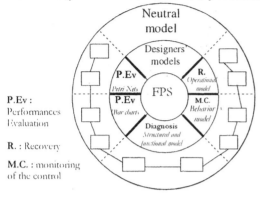

P.Ev : Performances Evaluation

R. : Recovery

M.C. : monitoring of the control

Figure 2: meta-modelling approach

3.2. Bottom-up approach

The repository is not built with an a priori knowledge of developed models. From the moment where the repository is built for the existing viewpoints, the repository must envisage the possibility of integrating new ones.

Thus, in our approach, we do not want to influence neither the viewpoint to be defined, nor the choice of the designers.

Thus we perform an ascending and progressive integration. The models produced by the various designers are the starting points of this approach.

Starting from the standard architecture of an IRDS (Information Resource System Dictionary) [20] we defined our repository architecture, which is based on three levels (figure 3).

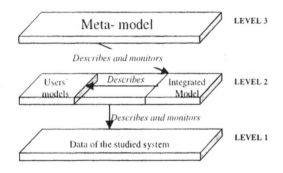

Figure 3: three levels of the repository

➤ The level 1 corresponds to the data handled by users.

➤ The level 2 constitutes the dictionary of the level 1 objects. At this level the intersections between users' models are detected, in order to build the integrated model and preserve the consistency of the whole set of models.

➤ The level 3 allows the description and the control of the obtained UML models. It corresponds to the package "foundation" of UML meta-model defined in UML Semantics v 1.3 [25].

3.3. Object oriented approach: UML

The Unified Modelling Language [25] is a general-purpose object formalism, which is extensible, broadly applicable, tool-supported and standardised by the Object Management Group. UML is the result of the convergence of the three main object oriented approaches [22], [8] and [17]. The object-oriented paradigm focuses on constructing reusable units and encompasses the conceptualisation and specification principles of abstraction, encapsulation, inheritance and polymorphism. Reuse mechanism can be performed at different levels, so UML proposes packages for organising elements (classes, packages, diagrams etc.) into semantically related groups. Packages are excellent mean of model structuring and show immediately the general organisation of the system.

In our approach, UML is used as mean for communication in order to support the users' models integration. Given that we build an information system, we mainly use class diagrams to show a group of structured elements. For correctly translating the links between these elements, we enriched the class diagrams with the Object Constraint Language (OCL) [25] and additional constraints (exclusive and existential) [4], [5], [6]. OCL is used to express in a precise way the constraints, which cannot be specified within UML diagrams. These constraints can be related on the modelled field or the structure of the objects. The OCL language allows us:

➤ to specify the invariant on the classes and the types,

➤ to describe the pre and post conditions on the operations and methods,

➤ to specify the constraints on the operations.

The keyword **context** defines the context of OCL expression. The keywords **inv**, **pre** and **post** indicate the stereotypes "invariant", "precondition", and "postcondition". **Self** is used to refer to the contextual instance. The value of an object property defined in a class diagram is specified by a dot followed by the name of the property.

context atype **inv:**
Self.property

4. Reuse mechanisms

4.1. Structure of the repository at level 2

Level 2 of the repository consists of a group of three packages: "Design Patterns", "Basis Models" and "Integrated Model". These packages constitute a first layer.

Each one of these packages contains class diagrams grouped into other packages constituting a second layer. In accordance with the layers architecture models, the packages of a level will use only packages of the same level or lower level (here packages of layer 2 cannot use packages of layer 1).

➤ **"Design Patterns":**
This package contains the design patterns, which are defined after study of the concepts used in the various viewpoints. The elements of these packages are not directly usable; they are generic elements and must be instantiated. We developed several design patterns to capture standard structures such as directed graph, bipartite directed graph, tree structure, and composite…
In the next sections we only present two of them.

➤ **"Basis models" : Domain Patterns**
In the FPS, there are models corresponding to the standard viewpoints. For example, the objects of a FPS are often studied according to these following aspects: structural, functional, behavioural, operational, etc.
We use "domain patterns" to capitalise these recurring models which are used as a basis to the experts intervening in a project. A domain pattern defines a field, which can be used in several viewpoints.
The elements of this package are built by instantiating (①) some design patterns previously defined.

➤ **"Integrated models"**
This package contains the various viewpoints of a project. The division into viewpoints depends on the project, and takes place rather early in the development of the project. The designer of the repository does not have consequently any influence on this division, and guarantees thus a perfect autonomy to the various experts. The models used in the viewpoints are often derived from the basic models, hence the use relation (②) between the two packages;
Models of viewpoints can also be obtained by instantiating design patterns. For this reason, we envisage also to use (③) the package " design patterns" if the basis models are not appropriated.
Some viewpoints consist of several other viewpoints. Then we can be brought to make a first integration, represented by the relation (④) before obtaining the final integrated model.
For example, for CASPAIM project, it is necessary to integrate the supervision and control viewpoints. However the supervision is the result of integration of several viewpoints: diagnosis, management of the modes, recovery, monitoring, etc.

We use the traditional methods to perform the viewpoints integration, by resolution of conflicts (synonymous, homonymous, polysemy, problems of multiplicity, etc.) resulting from semantic recoveries of the various elements [27].

The objective is to build UML models of each viewpoint by using the elements of packages "Basis Models" and " Design Patterns" as well as possible then integrate them.

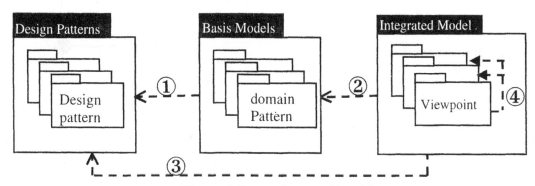

Figure 4: the packages of level 2

4.2. Reuse method

The level 2 of the repository enable us to practice reuse; however, enabling reuse does not ensure that reuse will occur. We developed on figure 5 a cohesive reuse strategy. This method optimises all possible ways to reuse existing patterns.

The initial data are the following ones:

➢   The object of the study (machine, cell of production etc.)
➢   Various points of view;
➢   Design Patterns resulting from former knowledge;
➢   Domain Patterns resulting from former knowledge.

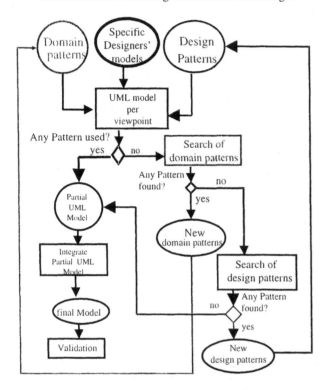

Figure 5: reuse approach

The first situation consists in using the available basis models or instantiating design patterns in order to build UML models of each viewpoint, then integrate them.

If the available models (Domain patterns and Design Patterns) are however not appropriated, we search for new invariant, and try to define a new Domain Pattern. This one reuses, if possible, a design pattern and enriches the package "Basis Models".

At the end, if no Domain Pattern is found, we search for Design Patterns that enrich the package "Design Patterns". If there is no solution, we build the UML model without any reuse mechanism.

It is an iterative and incremental approach, which is based on the capitalisation of the expert's know-how.

5.   Design Patterns and Domain Pattern

A design pattern gives a name, isolates and identifies the fundamental principles of a general structure in order to make of it a useful means for the development of a reusable oriented object design [13].

More particularly we have developed structural design patterns. They show how objects merge to constitute greater structures.

The models developed by the experts within the framework of CASPAIM project generally manage composite objects. A composite is the aggregation of components [15] [16]. The composite objects have a very rich semantics [10] [11].

The graph theory [14] is frequently used in the various viewpoints to apprehend the composite objects. We thus studied these structural objects to build "structural design patterns" [13].

5.1. Design   pattern: "Bipartite   Directed Graph" (BDG)

This pattern is used to represent a second alternative of composite objects, which is presented in the form of bipartite directed graph [14]. They are frequently used in the various points of view. Petri nets are typical example of BDG.

A directed graph G=[X, U] is known as bipartite if the set of nodes X can be partitioned in two disjoint subsets X1 and X2 so that, for any arc u(i, j) • U:

➢ $i \cdot X1 \Rightarrow j \cdot X2$
➢ $i \cdot X2 \Rightarrow j \cdot X1$

There are thus two types of nodes in a bipartite directed graph, as indicated towards figure 6.

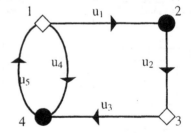

Figure 6: Bipartite directed graph

The class diagram on figure 7 describes the structure of the pattern "bipartite directed graph". We obtain now two distinct classes Node_1 and Node_2 for the two types of nodes.

We defined one OCL constraint to avoid that a node stays without any link, because of the zero minimal multiplicity.

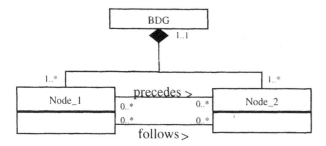

Figure 7: structure of the design pattern "BDG"

5.2. Design pattern: "Tree structure"

The composite objects can be also presented in the form of tree structure (figure 8).

A tree structure is a directed graph where each node has one and only one predecessor, except one: the root.

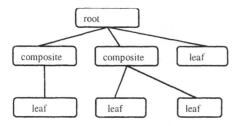

Figure 8: composite

The structure of this pattern is represented on the figure 9. It is used to represent the model of composite objects, which has a tree structure. These objects represent composite/component hierarchies.

According to the level of granularity wanted, we can continue the decomposition of a tree. Thus a component considered as leaf is likely to become composite. For this reason, we propose the class diagram on figure 9.

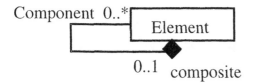

Figure 9: structure of the design pattern "tree structure"

An element is made up of several other elements. An element can belong to one and only one other element.

To prevent that an element is neither component nor composite, (because of the zero minimal multiplicity); we add the following OCL constraint.

> **context** Element inv:
> Element.**Allinstances** → **count** >1
> **implies**
> (**self**.component →
> union(**self**.composite)) → **notEmpty**

5.3. Domain pattern: "structural Model"

To represent information necessary to the various points of view, several models are developed. They are complementary and make it possible to treat specific aspects of a system (a system of production or one of its elements). These models (domain patterns) which are elements of the "Basis model" package are established by instantiating the design patterns previously defined.

The first aspect studied in a system, is naturally its physical aspect through the description of its structure. Any element of a production system consists of components; this is a tree structure aspect. The figure 10 shows the example of the decomposition of a machining centre [24].

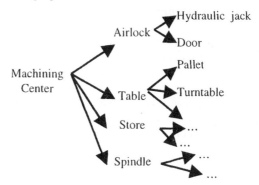

Figure 10: structural model of a machining centre

We saw that any object being presented in this form can be represented by the design pattern "tree structure", hence the representation of the structural model on the class diagram on the figure 11. A physical system consists of several elements that in their turn can be broken down into other elements. As we can see it on this example, a domain pattern can be based on both an instantiation of a design pattern and a domain specific class or structure

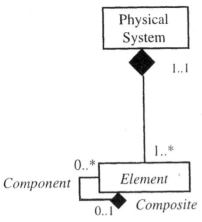

Figure 11: domain pattern "structural model"

For the need of this article we will limit ourselves to two Design Patterns and one Domain Pattern. In the next section we apply our approach on the control viewpoint of the CASPAIM project.

6. CASPAIM: Generation of the control model

6.1. Control viewpoint

The designer of the control viewpoint uses logical sequence based on Petri nets model (cf. figure 12). A place models the state of the product and a transition models a function that modifies the product state or creates a new product from existing ones. The logical sequence of a product describes the operation sequencing (order constraint). Thus it defines the manufacturing process which allows the achievement of the final product(s) from the raw one(s).

So a logical sequence is a functional description of the manufacturing process, without taking the resources and the manufacturing technique into account.

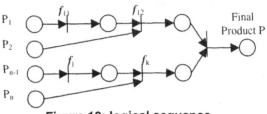

Figure 12: logical sequence

The objective here is to define operating sequence from logical sequence. A determinist sequence of operations is established (for example we have to choose which operation come first when their chronology is irrelevant), and the production means are chosen according to their accessibility and concurrent uses. This is done by three steps (cf. figure 13):

➤ **Initialisation:**

For each operation of the logical sequence, the resources that are able to perform the operation under consideration are selected. A final place is added for each logical sequence to represent the final area.

➤ **Integrating the manufacturing resources:**

Several resources may perform the same operation; this alternate structure is due to the flexibility in resources choice and flexibility in operations chronology.

➤ **Obtaining basis operations:**

On operating sequence we have only basic operations, so transfer and processing operations are separated.

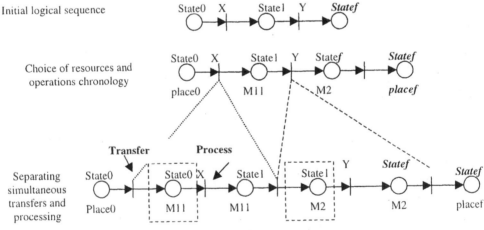

Figure 13: operating sequence

6.2. UML model of control viewpoint

> **Logical sequence**

A logical sequence is based on a Petri nets model. This model is a bipartite directed graph. Thus the pattern " Bipartite Directed Graph " can be instantiated. But by using directly this pattern, a place (product state) belongs to one and only one Petri net while one place of a logical sequence can belong to another logical sequence. For example, the final state of the logical sequence of a part can be the initial state of another logical sequence. Then we have to adapt the design pattern "Bipartite Directed Graph" to this situation by removing the composition relation between the classes "Logical sequence" and "Place" (figure 14).

In order to do the product nomenclature, we reuse the design pattern "tree structure".

Since a place of a logical sequence models a product in a specific state, an association links the two instantiated patterns.

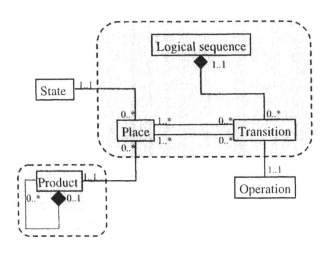

Figure 14: UML model of a logical sequence

> **Production system architecture (figure 15)**

A production system is composed of several resources (machines, robots, and conveyor...). The products move through the production system from a physical zone to another one. The designer of control viewpoint uses the notion of accessibility between physical zone, which compose the resources. The products' moving is performed by positional operations. We reuse the domain pattern "structural model" to describe the production system. This pattern is adapted because of the physical zones, which are not decomposable.

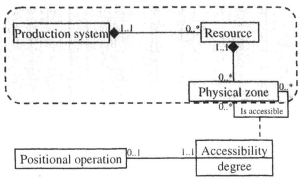

Figure 15: UML model for the structure

> **Operating sequence (figure 16)**

The operating sequence is generated from the logical sequence and production system architecture. An operating sequence is also based on Petri nets model. We reuse the design pattern "Bipartite Directed Graph" on the same way than for the logical sequence. From one logical sequence it can be generated several operating sequences. So the places and transitions of an operating sequence (OS_Place, OS_transition) are derived from logical sequence ones (LS_Place, LS_transition). Now we distinguish two kinds of operation: the positional operations for the transfers and the functional ones for product transformations.

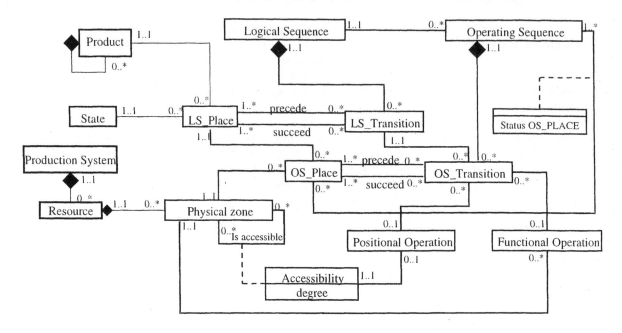

Figure 16: Final class diagram for control viewpoint

7. Conclusion

The difficulty in managing information of a production system designed in the CIM concept is mainly due to:

➢ the multi-disciplinary subject,
➢ the independence of designers,
➢ the specialisation of design tools which are characterised by a very high evolutivity,
➢ the simultaneity and multiplicity of information sources,
➢ the progressive and iterative nature of design process.

The developed bottom-up approach aims to guarantee the consistency of a set of models resulting from various independent viewpoints.

The principal advantage of this approach lies in the fact that it starts from what exists: expert models, standard structures... The various contributors in a project do not undergo any constraint, contrary to the top-down approach that proposes a model of reference.

Meta-modelling is used as dialogue technique between the various designers; it enabled us to define the intersections between the various models in order to build integrated model.

This meta-modelling is carried out using the visual language of modelling UML, which is an OMG standard.

The use of OCL language allows a precise expression of the constraints.

The reuse performed by using design patterns and domain patterns, guarantees a generic repository.

The difficulty consists in finding the significant invariant on the various used concepts and models.

The division into packages allows to gradually enriching the repository. It also allows us to control more easily the impact of a modification through the relations between the elements of packages.

We created software based on ORACLE to test each design pattern, domain pattern. An implementation of the presented class diagram was carried out. It allowed an easy development of the algorithm for the creation of the operating sequences from the logical sequences.

The integration of the other viewpoints (supervision, performance analysis...) developed in the CASPAIM project are currently under development.

References

[1] ALEXANDER C., ISHIKAWA S., SILVERSTEIN M., JACOBSON M., FISKDAHL-KING I., ANGEL S., A pattern language, Oxford University Press, 1977.

[2] AFNOR Z 68-901, Représentation des systèmes de contrôle et de commande des Systèmes Automatisés de Production, Génie Automatique, AFNOR standard, 1992.

[3] AMAR S., Systèmes automatisés et flexibles de production manufacturière : méthode de conception du système de coordination par prototypage orienté objet de la partie procédé, PhD thesis Lille1 university, 1994.

[4] BIGAND M., BOUREY J.P., CORBEEL D., MAIK J.P., A generalised object approach for the design of flexible manufacturing systems, CESA'96 IMACS Multiconference, Lille, 1996, pp352-357.

[5] BIGAND M., CORBEEL D., BOUREY J.P., The design of Flexible Manufacturing Systems by using a knowledge based system, 15th IMACS World Congress, Berlin, 1997, pp541-546.

[6] BIGAND M., CORBEEL D., NDIAYE D., BOUREY J.P., Extensions of object formalism for representing the dynamics: application to the integration of viewpoints in the design of a production system, IDMME'98, Compiègne, 1998, pp1169-1178.

[7] BOITARD L., Contribution à l'Intégration d'Outils de Génie Automatique autour d'un Système d'Information Unifié. Prototypage d'un Atelier Intégré de Génie Automatique, PhD thesis Nancy1university, 1998.

[8] BOOCH G., Object Oriented Design with Applications, The Benjamin/Cimmings publishing Company Inc, 1991.

[9] CAMUS H., OHL H., KORBAA O., GENTINA J.C., Cyclic Schedules in Flexible Manufacturing Systems with Flexibilities in operating sequences, First International Workshop on Manufacturing and Petri Nets, 17th International Conference on Application and Theory of Petri Nets, Osaka, 1996, pp.97-116.

[10] COUFFIN F, Modèle de données de référence et processus de spécialisation pour l'intégration des activités de conception en Génie Automatique, PhD thesis E.N.S of Cachan, 1997.

[11] DJERABA C., NGUYEN G.T., RIEU D., Objets composites et liens de dépendance dans les systèmes d'information, INFORSID 93, Lille, 1993.

[12] EL KHATTABI S., CRAYE E., GENTINA J.C., Supervision by the behavior modelling, IEEE International Conference on Systems, Man and Cybernetics, SMC'95, Vancouver, 1995, vol.2, pp.1416-1422.

[13] GAMMA E., HELM R., JOHNSON R., VLISSIDES J., Design Patterns, Elements of Reusable Object-Oriented Software, Addison-Wesley, 1994.

[14] GONDRAN M., MINOUX M., Graphes et algorithmes, EYROLLES edition, 1995

[15] GUARINO N., PRIBBENOW S., VIEU L., Parts and Wholes: Conceptual Part-Whole Relations and Formal Mereology, ECAI'94 Workshop, Amsterdam, 1994.

[16] HALPER M., GELLER J., PERL Y., Part Relations for Object-Oriented Databases, Proceedings of the 11th International Conference on the Entity Relationship Approach, Karlsruhe, Germany, 1992, pp. 406-422.

[17] JACOBSON I., CHRISTERSON M., JONSSON P, OVERGAARD G., Object Oriented Software Engineering, Addison-Wesley, ACM Press, 1992.

[18] KORBAA O., CAMUS H., GENTINA J.C., FMS Cyclic Scheduling with Overlapping production cycles, 2nd International Workshop on Manufacturing and Petri Nets, 18th International Conference on Application and Theory of Petri Nets, 1997.

[19] LY F., TOGUYENI A.K.A., CRAYE E., Predictive Maintenance and Monitoring in Flexible Manufacturing Systems, 15th IMACS Congress, Berlin, 1997, vol.5, pp.415-420.

[20] MARTIN D., La norme de référentiel IRDS : objectif et fonctions, Génie Logiciel & Systèmes Experts, Nr 22, 1991.

[21] OUSSALAH C. et al, Ingénierie objet, Concepts et techniques, InterEditions, 1997.

[22] RUMBAUGH J., BLAHA M., PREMERLANI W. EDDY F., LORENSEN W., Object oriented Modelling and design, Prentice-Hall, 1991.

[23] THAM K.D., CIM-OSA : Enterprise Modelling, Enterprise Integration Laboratory, University of Toronto, 1992.

[24] TOGUYENI A.K.A., ELKHATTABI S., CRAYE E, Functional and/or structural approach for the supervision of flexible manufacturing systems, CESA'96 IMACS, Symposium on discrete event and manufacturing systems, 1996, pp 716-721.

[25] Unified Modelling Language, Semantics and notations, available on http://www.rational.com

[26] VERNADAT F., Enterprise Modelling and Enterprise Integration using a Process-based Approach, Information Infrastructure Systems for Manufacturing, Amsterdam, 1993, pp.65-84.

[27] WON K., CHOI I., GALA S., SCHEEVEL M., On resolving Schematic Heterogeneity in Multidatabase systems, in Modern database Systems, WON Kim ED., ACM Press, 1995.

An Approach for Building Product Models by Reuse of Patterns

L. Gzara

GILCO Laboratory, INPG, Grenoble, France - Lilia.Gzara@gilco.inpg.fr

D. Rieu

LSR Laboratory, IMAG, Grenoble, France - Dominique.Rieu@imag.fr

M. Tollenaere

GILCO Laboratory, INPG, Grenoble, France - Michel.Tollenaere@gilco.inpg.fr

Abstract

This paper deals with an approach for Product Information Systems (PIS) engineering by reuse of patterns. The pattern approach provide an engineering guide to model data by organizing hierarchically and functionally modeling problems and the manner to resolve them. This would contribute to accelerate building and implementing product and process models during PIS engineering. A special interest is given to identify and specify different patterns for product modeling. However, a pattern-based approach can be developed only for disciplines which acquired a certain maturity, i.e. those for which there is both a consensus around a finite set of problems and a variety of known solutions for solving these problems. There is no universal agreement on the knowledge needed in product information systems, let alone on the representation of this knowledge. The first step consisted thus of a field analysis providing a common terminology and a semantic of the principal concepts managed in PIS and proposing various models to fix these concepts. It forms a basis for exploring the problems frequently occurring during PIS specification. A pattern catalogue is then defined to resolve the identified problems.

1 Introduction

Industrial companies must today obtain a rigorous control of their product information system (PIS) in order to increase their reactivity to the different changes involved in the product development process or later during the product life. PIS manage all types of information used to define, manufacture, and support products. This may include part definitions, specifications, CAD drawings, project plans, feasibility reports, engineering change orders, etc.

Since the mid 1980's, software industries have developed a new class of software packages, called Product Data Management Systems (PDMs) to support management of all product engineering data. PDMs constitute for industrial companies the core of Product Information Systems, such it is the case of DBMS for Management Information Systems. PDMs are tool platforms adapted to characterization of items, bills of material, documents, procedures, etc. by providing class libraries dedicated to PIS.

However, PIS engineering raise many difficulties:

- At analysis stage: lack of 'formal' specification models for users requirements formulation which can be easily understood by the users and lack of approaches able to draw up clear and unambiguous functional specifications,
- At design stage: discontinuity in specification when evolving from analysis stage (functional specifications) to design stage (technical specifications),
- At implementation: complexity in PDMs implementation due to the complexity and consistency of such tools and also due to a poor documentation of the proposed components.

Furthermore, we should highlight that the development of such systems is always accompanied with organizational changes (e.g. new objectives, new information flows, evolution of quality procedures) and regular evolutions of needs, that the PIS must take under consideration and anticipate. Industrial companies are therefore obliged to constantly evolve their PIS, within reduced times in order to avoid the delivery of obsolete systems, unfitted to new organizations.

In conclusion, Neither specification nor implementation of PIS in organizations are currently dealt with methodically, thus generating considerable costs and malfunctions. The objective of this research is to propose a methodological framework for PIS engineering, supporting specification "by deviation" both as regards capitalization and reuse of knowledge already captured in past experiences and consideration of the available software resources (components and systems). This should accelerate the PIS development process.

The paper is organized as follows. Section 2 outlines the need for a reuse approach to resolve the among problems and focuses on reuse of patterns. Section 3 presents the patterns engineering process conducted to identify and specify various patterns for product modeling. An overview of the developed patterns is given.

2. The reuse approach

2.1 The need for reuse in PIS development

The reuse approach, already effective in software engineering, is a key factor to our PIS development methodological approach, both for specification and implementation of PIS and for their evolutive maintenance. Reuse is a new development approach by which a system can be built from existing components already described, carried out, tested and accepted in past experiences. The aim is to avoid remaking all with each new application or when changing a software. Thus, we deliberately are in the framework of a capitalization and reuse approach of *functional*, *organizational* and *software* components, at all stages of the PIS development process. Today the reuse approach is extensively used in software engineering, and different forms of reusable software components have already been proposed. Three major components can be identified: toolkits [1], frameworks [2,3] and patterns [4,5]. A special interest is given to patterns.

2.2 Patterns

The Pattern concept was initially proposed in architecture field by C. Alexander [4]. Then, the pattern concept has been adopted into software engineering by K. Beck and W. Cunninghan [6] to resolve development problems. More recently, P. Coad [7], E. Gamma [5], etc proposed patterns dedicated to Information Systems engineering. Several pattern definitions are presented in the literature. It comes out from these definitions that a pattern constitutes *a know and know-how base to resolve a problem frequently occurring in a particular field*. This know-how allows to identify the problem to resolve, to propose a correct solution to take it into account and finally to give indications to adapt this solution in a particular context. Patterns form then "know-how" oriented components, while other forms of

reusable components are only "know" oriented components. These latter provide only solutions to problems whereas "know-how" oriented components provide solutions but also the manner to construct these solutions. Furthermore, reuse of patterns is, in our opinion, the most suitable form of reuse for PIS engineering, as it can be used in all stages of the PIS development process (analysis, design, implementation). The objective is thus *to adapt the pattern approach to a particular field, that is of PIS, according to a target technology, the PDMs*.

2.3 Patterns in PIS

The proposed methodological framework is based on a consistent set of models of various abstraction levels that can be developed by reuse of patterns. Each level proposed must enable a problem to be solved (of a functional, organizational, technical nature, etc.) specific to PIS development. Just as for management information systems engineering, two main aspects must be taken into consideration in PIS engineering. First, an organizational aspect specifying the organizational information system (OIS) and with which "business" patterns will be associated. Second, a technical aspect (software) specifying the part of this OIS that will result in computerization. This is the computer-based information system (CIS) which will result in specifying and reusing "software" patterns.

- At organizational level, **business patterns** are important in the stages of analysis and design. They provide solutions for application field problems. They must be able so to take into account a set of information needs associated both with product description (structure of the product, documentation and various representations) and PIS processes management (engineering change process management, configuration process management, etc.). Two forms of modeling are thus essential: *product* modeling and *process* modeling. The aim is to use and adapt product and process modeling solutions taken from industrial engineering work (design modeling [8], enterprise modeling and integration [9], techniques of BPR [10]) as well as work carried out in information systems engineering (engineering process modeling [11,12]).

- At technical level, definition of **software patterns** is strongly linked to the PDM system which is at the basis of PIS implementation. The generic nature of modeling based on software patterns stems from its independence from a PDMs. This modeling is the expression of a technical solution that takes two major problems into account: implementation of product and process models and communication of the PIS with other systems. With respect to implementation of models, PDM systems use database models (relational or object) and workflow models. In both cases the models proposed in tools are extensively used and can thus be integrated in the proposed methodological framework.

Definition of software patterns associated with PIS is based on functions proposed by most PDMs. With respect to business patterns, the aim is to identify them from the field's analysis. We give interest in the present paper to *business patterns* and specially those for *product modeling*.

2.4 Formalism

The formalism retained to describe a pattern is an adaptation of the Gamma formalism [5] by retaining only the five headings that we judge essential :
- **Name**: name of the pattern,
- **Classification**: relating to *product* or *process*,
- **Intention**: problem to which the pattern addresses,
- **Motivation**: a scenario of application of the pattern describing particular problems,
- **Semi-formal description**: solution suggested by the pattern, expressed using UML[1] diagrams [13] and describing in particular the participants in the pattern and their collaborations.

3. Patterns Engineering Process

Introducing patterns in PIS engineering process requires first to construct a patterns library or catalogue wich will be then reused in various PIS development projects. Two new complementary processes[2] are then needed in PIS development (fig. 1):
- one process dedicated to *patterns engineering For reuse*, i.e. identification and specification of various patterns to use during specification of PIS. This process aims to explore recurrent problems and then to formalize associated solutions in various patterns.
- one other process dedicated to *PIS engineering By reuse of patterns*, i.e. specification of PIS using patterns identified in the previous process. This

process, based on user's requirements, identifies the problems to resolve, select from the patterns catalogue the patterns which resolve these problems and then considers the proposed solutions.

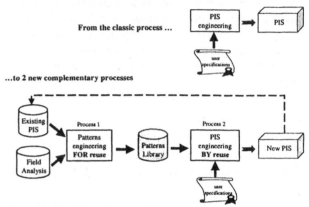

Fig. 1. New development process

This new process is recurrent, i.e. models obtained in the new PIS, as a result of the By reuse process, supply the For reuse process in order to constantly take into account new problems.

In the next, we are focusing on the first process, that is of *patterns engineering For reuse*. Highlighting business patterns requires first, identification and comprehension of the problems common to PIS (related to product definition, representation and processing) and second, specification of the associated solutions. The first step of problems identification is based on the study of existent PIS models. However, due to the absence of a consensus on the concepts managed in PIS, we focused our efforts on a field analysis, providing a *generic framework* for the field. This framework provides a common terminology and semantic of the principal concepts and proposes a reference model to fix these concepts. The model thus obtained forms, with the existent PIS models, a basis for exploring the problems frequently occurring in the field. Once problems identified, definition of patterns associated to these problems is based partially on design patterns catalogues available in software engineering. Indeed, some of PIS problems can be brought into general modeling problems raised in these design catalogues.

In the sequel, we present the Generic Framework resulting from the field analysis. Then the developed business patterns catalogue is described.

3.1 Generic Framework

The aim of the generic framework is to fix a terminology and a semantic for the concepts managed in

---

1 the choice for UML notation was dictated by its capacity to support dialog between users and designers of the system and the variety of abstraction levels of diagrams it offers so to ensure a continuum specification.

2 The concepts of "Design FOR reuse" and "Design BY reuse" were introduced by many authors. We quote mainly the FODA method [14].

PIS and then to model the variety of relationships existing between these concepts. As we specified in section1., PIS are centered on product. They manage all product engineering information used in various related-processes. Thus, concepts managed in PIS are articulated on the one hand around the product and on the other hand around the processes acting on the product. We report particularly on the product axis.

3.1.1 Product abstraction levels

The term product is a general concept whose employment and significance differ from the context. It takes various aspects according to the business considered. Furthermore, many terms referring to the product are used, such as exemplary, generic product, product family, specific product, basic product, etc. Thus, the term product is referred sometimes to virtual objects and sometimes to physical objects. To relieve this ambiguity, we define three abstraction levels of the product concept:

- *exemplary-product* which is the product delivered to the customer (a physical object resulting from manufacturing process, such as *my peugeot 206 S16*),
- *product-type* which is the product model according to which exemplars are manufactured (such as *peugeot 206 S16*),
- *generic-product* which is a product model including all possible options and design variants[3] according to which definition of a particular product-type is done by a particular choice of options and variants (such as the line of cars *peugeot 206*).

The passage from one generic-product to product-types and then from one product-type to product-exemplars is done according to design variants and optional items. A generic-product has at least two non-interchangeable[4] design variants of the same component in its composition. A product-type doesn't present any choice between design variants or is composed only of interchangeable variants. Thus:

- the passage from the generic-product to product-types is made by fixing some variants for each variant item. In the particular case of non-interchangeable design variants, only one design variant must be chosen. Furthermore, the choice of one variant of a particular variant item constrains the choice between variants of an other variant item. For example, the choice of the variant "4cylinders" for "engine" implies the choice of

the variant "tank" for "container" whereas the choice of the variant "electric" for "engine" implies the choice of the variant "battery" for "container".

- the passage from the product-type to product-exemplars is made by fixing one design variant through a list of interchangeable variants. For example, the choice of one of the variants "green frame" or "blue frame" for "frame" doesn't trouble the composition of the remainder. All exemplars of Peugeot 206-S16 are composed of a "4cylinder engine", "2 liters cylinder", "16 valves", etc. whatever of their frame color.

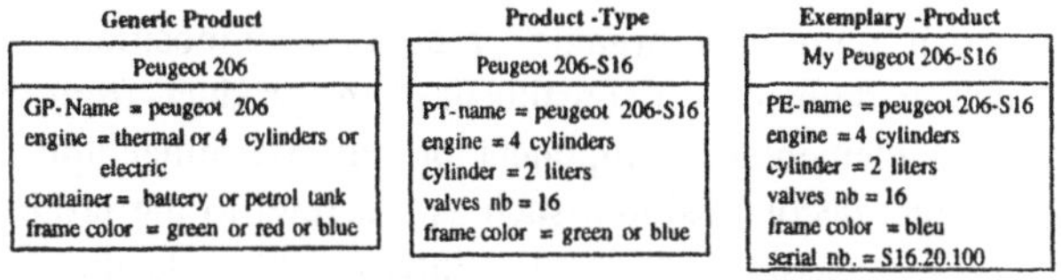

Generic Product	Product -Type	Exemplary -Product
Peugeot 206	Peugeot 206-S16	My Peugeot 206-S16
GP- Name = peugeot 206 engine = thermal or 4 cylinders or electric container = battery or petrol tank frame color = green or red or blue	PT-name = peugeot 206-S16 engine = 4 cylinders cylinder = 2 liters valves nb = 16 frame color = green or blue	PE-name = peugeot 206-S16 engine = 4 cylinders cylinder = 2 liters valves nb = 16 frame color = bleu serial nb. = S16.20.100

Fig. 2. Three levels of peugeot 206

This distinction between abstraction levels is made not simply to raise a terminological ambiguity but especially because the knowledge associated with a product doesn't relate to the "same" product but to particular levels of this product.

We should note furthermore that each product level has a life cycle through which it evolves.

3.1.2 Representations of knowledge

During the whole product life cycle and depending on the state of the product, various knowledge is associated with this one to describe it. The product state is function of the knowledge associated to it. This knowledge is supported by various representations forming thus the product model. Two types of representations are distinguished: *documents* and *Bills Of Material* (BOM).

- A document is defined as a container for knowledge aiming at describing an object and seized on a support (paper, tape, cassette, disc, microfilm, etc). According to its finality, a document can be a *Model* (aiming at defining the product such as CAD drawings or controlling the way in which the associated processes must be carried out such as process routings) or a *Record* (reporting the activity results such as control sheets). It can be therefore composite or elementary. A folder is thus a composed document gathering in some way a set of documents in order to treat easily various data at operational status. Like products, a document has its own life cycle (created, valid, void, etc.).
- A Bill Of Material is defined as a graph structure composed of nodes with same nature and linked by

---

3 a coordinated set of alternatives in the design which produce a different product. Design variants represent sets of variations which evolve in versions consistent with the rest of the product [17].

4 Two objects A and A' are interchangeable if A' can be substituted for A without involving evolution of any assembly using A [16].

composition relationships. It describes the product, at a particular level, from several points of view, such as its finality (functional BOM), its definition (engineering BOM), its manufacturing process (Manufacturing BOM), its maintenance process (logistic BOM), etc. A BOM is thus a recursive composition of technical objects whereas a technical object is a component of the product. It can be an item (case of engineering or manufacturing BOM), a function (functional BOM), a feature (geometric BOM), etc.

The most used technical object in BOMs is *item* which represents organic product decomposition. An item is defined as a constituent of the product, from the most elementary component or part (terminal unit in product decomposition, such as a screw) to the product itself, passing by all intermediate layers of decomposition. Many terms are used in the literature to express organic decomposition. Some of them correspond to logic criterion of decomposition (such as component, system, and sub-system), some others to material criterion (such as part, mechanism). The item concept describes the product from a management point of view and designates all the terms introduced above to express organic decomposition. Furthermore, an item can be perceived at two abstraction levels. One can deal with virtual item and physical item. It can be also described by various representations (documents, BOMs for composite-items). Thus, virtual products are composed of virtual items and physical products are composed of physical items.

Fig.3 gives examples of product BOMs and illustrates the virtual and physic aspects of the item.

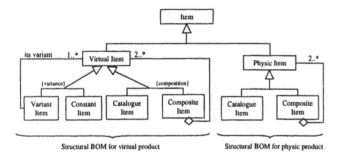

Fig. 3. Example of Product BOMs

The whole representations associated to the product forms then the product model. Several views can be defined to express a selective perception of the product,

by extracting some information from the model. It serves a particular community in the enterprise.

As for evolution, a product, an item or a representation can evolve. We distinguish three mechanisms supporting these evolutions:
- *version* and *revision* for products, items and representations,
- *correction* for representations.

We define below each of these mechanisms:
- **Revision**: definition or semantic change, which ensure the interchangeability of the initial object and the modified one, thus without impact on the whole employment cases of the object.
- **Version**: definition or semantic change that doesn't ensure the interchangeability of the initial object and the modified one, thus impacting at least one employment case of the object.
- **Correction**: editorial change (format) without impact on the representation dependant objects (such as correction of a spelling mistake).

3.1.3 Reference Model

Based on the partial models built throughout the field analysis to fix each concept, a global model is obtained by linking these sub-models (fig.4). Therefore some of the above-described concepts have knowledge in common, we built a super-class to these objects to which we associate this knowledge.

Description of the reference model : It comes out from the descriptions given in § 3.1.1 and §3.1.2 that a product whatever of its level is described by various representations. By representation, we mean documents and Bills Of Materials.
- Each product level is documented by various documents. "Product-Exemplary" level is essentially documented with "record-documents". "Generic-Product" and "Product-Type" levels are rather documented with "model-documents". In the same way, an "item" some is its nature (virtual or physic) is documented by various documents (such as a plan or a CAD model for a mechanical item). Then, the class "document" can be associated for both items and product levels. We built so a super-class to *Generic-product*, *Product-Type*, *Product-Exemplary*, *virtual item* and *physic item*, named "documented object" which is relied to the class "document" and so the associated sub-model.
- On the other hand, each product level is described by various structures or BOMs. A Bill Of Material is a recursive composition of technical objects, which can be items, functions, features, etc. Thus, a BOM is not

managed in the PIS as an object with an own existence; it concretizes the composition relationships between composite objects and related components. It is materialized in the product model by an association linking the structured object, i.e. the object to whom a BOM is associated such as product, to the root of the corresponding BOM graph which is a composite technical object, hence:

- Structural BOMs describing "Generic-Product" or "product-type" are based on recursive composition of virtual items. They are modeled by an association between the product level class and the "composite virtual item" class, corresponding to the top level of the associated BOM. Since several BOMs based on items are existing, each one is modeled in the item recursive composition sub-model by a particular composition link stereotyped according to its nature ("engineering BOM", "manufacturing BOM", etc.).
- Structural BOMs describing "Product-Exemplary" are based on physical items. They are modeled by an association between the "Product-exemplary" class and the "composite physic item" class.

- Functional BOM is modeled by an association between the "Generic-Product" class and the "composite external function" class.

Analysis of the reference model : this reference model was built, basing on the hypothesis of three levels of product. It relies then functional BOM or virtual structural BOMs to only "generic-product", for example. We highlighted the existence of three product levels in the enterprise but some organizations may not have generic products. In this case, functional BOM or virtual structural BOMs is rather associated with "Product-Type". Some other organizations may not manage product-exemplary level. In this case, there is no reason to distinguish "physical items". Furthermore, the number and type and BOMs managed is widely varying with organizations (functional BOM and hence "functions" may not be managed whereas geometric BOM exists for example). Type of "documents" is also varying. Then, fig. 4 can be perceived as a possible product model for PIS based on 3 levels of product, three BOMs (functional, engineering, manufacturing) and record & model documents.

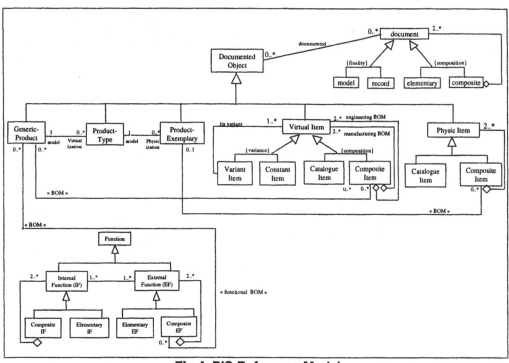

Fig.4. PIS Reference Model

The reference model so obtained is then both too general and too specific. It is general because it doesn't express specific properties to organizations (for example product property "engine" for car industry). Thus it is not easily used by organizations. On the other hand, it is specific in the mean where it expresses a particular case of three product levels, three BOM types, etc.

The aim to build a reference model, raising all differences due to the luck of a consensus about PIS concepts, is very ambitious. We can only express the variability about these concepts. The Key remaining challenge is so to keep the reference model general, even more general and more complete than the current model, and give a method to express variability.

- One solution consists of proposing first a *meta-model* too general and then declining various models from this meta-model by instantiating it according to various specificity points [18]. This solution remains restrictive since it fixes in advance the specificity.
- An alternative solution is to propose first a general model and then to give a method to adapt this model, basing on various patterns allowing each one to specify one particular variability point. This approach seems to be more suitable since it is more flexible.

The retained approach to offer a method for PIS engineering is so to propose a general reference model and then to offer a pattern catalogue. That consists of proposing a set of patterns, which allow adapting this reference model according to various variability points, so to fit it into one organization specificity.

4. The Pattern Catalogue

4.1. Overview

The retained approach for PIS engineering is made possible by reuse of patterns. It is based on the use of a succession of patterns resolving various problems of PIS specification and providing a progressive refinement of the reference model according to the organization specificity. In this section, an overview of the developed business pattern catalogue for product modeling is presented. We should note that a pattern catalogue is a set of patterns relied with a set of links [19]: *use, extension/refinement, alternative*, etc.. We only define here the two links used in the present pattern catalogue:

- *Use link:* a pattern B "uses" a pattern A when the solution of pattern A constitutes completely or partially the intention of pattern B, i.e. the problem treated in B is a sub-problem of that in pattern A. For example pattern A is for engineering change process management and pattern B is for engineering change request.
- *Refinement link:* a pattern B "refines" a pattern A when the intention of pattern B is a special case of the intention in pattern A, i.e. the problem treated in pattern B can be resolved also by the solution in pattern A. For example pattern A is for BOM

construction and pattern B is for functional BOM construction.

The proposed catalogue is thus a succession of business patterns relied by these links. We can not develop here the whole pattern catalogue. We just present an overview of these patterns and the links between them (fig.5). We only provide the *Intention* and partially the *Solution* of major patterns. However, in order to illustrate the concept of pattern, one particular pattern is developed at the end of this section.

As stated before, there is variance between organizations about some points related to product levels, kind of BOMs, document, etc. The first step in the PIS specification project is so to fix specificity of the organization according to these variance points, which constitutes the entry point in the specification approach.

The first **pattern "Fixing Variance Points"** has as intention to identify all variance points in the organization in order to fix them. The solution consists of directing, by *use* link, to various patterns fixing each one, a particular variance point.

The immediate patterns have hence the intention to provide a guide to fix each variance point. Then:

- the *pattern "Product Levels"* allows identifying the number and type of product levels in one organization. The solution suggests using three patterns to represent one, two or three levels. For example the suggested *pattern "Two Levels"* proposes a solution to represent the product concept in 2 abstraction levels and shows how to integrate in the model a new product in one level. This pattern will be particularly developed at the end of this section.

- the *pattern "Applied BOMs"* allows to fix the type of BOMs managed in the organization (functional BOM, generic engineering BOM, logistic BOM, etc.) and suggests to use various patterns to construct these BOMs. The solution suggests to use the *pattern "Construct Basic BOM"* to construct standard BOMs[5], such as Product-exemplary structural BOMs based on physic items and it directs to others patterns for constructing specific BOMs such as:

  - *"Construct BOM With Variants"* pattern for BOMs based on variant components such as generic geometric BOM,

---

5 A standard BOM is a recursive composition of objects

- *"Construct BOM With Options" pattern* for BOMs based on optional components such as generic functional BOM,

Each of these two patterns uses the pattern "Construct Basic BOM". Furthermore, others patterns are defined to complete some characteristics of the above BOMs, according to the semantic of the composition link in the considered BOM. Then, *"BOM With Exclusive Composition" pattern* allows to specify BOMs where the components existence depends totally on composite

existence such the case of structural exemplary BOM based on physic items (if one compound is removed, all its components are also removed).

- Once BOMs are built, they are associated to the appropriate product levels by linking the root of each BOM to the associated product level class. That's the intention of the *pattern "Associate BOMs To Product"*

- Others patterns are defined to identify applied documents, to associate documents to the objects they represent, etc.

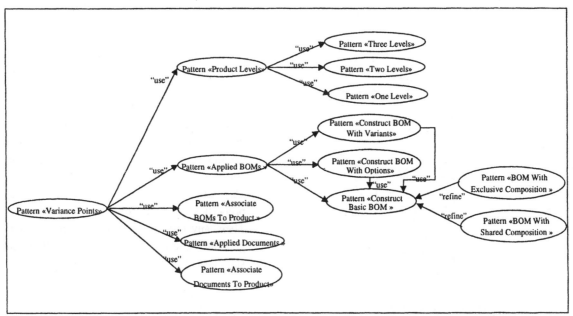

Fig.5. Overview of the developed pattern catalogue for product modeling

4.2. Pattern "Two Levels"

Name: Two Levels

Classification: Product pattern

Intention: This pattern represents the product concept in 2 abstraction levels. This allows first, to partition knowledge between these levels (common properties in higher level and specific properties in lower level) and second, to define relations between the levels so to allow propagation of knowledge between levels. This pattern can be applied for representing Generic-product and Product-type levels as well as Product-Type and Product-Exemplary levels.

Motivation: Let us consider the example of *Car* product. We illustrate the particular case of two levels: Product-Type and Product-Exemplary. We suppose here that no product lines exist. For example there is only the type peugeot 206.

Product level	Peugeot 206 (type)	My peugeot 206 (exemplary)
Charact -eristics	**PT-Name** = peugeot 206 **Engine** = 4 cylinders **Color** = green or blue	**PE-Name** = peugeot 206 **Engine** = 4 cylinders **Color** = blue **Serial n°** = 206-10-50

Fig. 6. Two levels of product

Knowledge associated for these entities is of two types:

- **entity properties** (an entity takes a value for each one of its properties), such as the property *engine = 4 cylinders* for the product-type *Peugeot 206*.

- **constraints** expressed at a certain abstraction level and bearing on values of lower level properties, such as the constraint *color = green or blue*, in *Peugeot 206*, constraining the value of the property *color* of *my Peugeot 206 (color = blue)*.

The class diagram in fig.7 organizes the knowledge associated to the product levels.

- The class "car" holds properties of all car types. *Peugeot 206* is an instance of that class.
- The class "product-exemplary" holds properties of all car exemplars.

For each car type, for example the type *peugeot 206*, we define a sub-class of "product-exemplary" class (for example *"peugeot206-exemplary"*) corresponding to all exemplars of that type (exemplars of *peugeot 206*) and which holds specific properties of these exemplars. Each exemplary of one car type (for example *my peugeot 206*) is an instance of that sub-class.

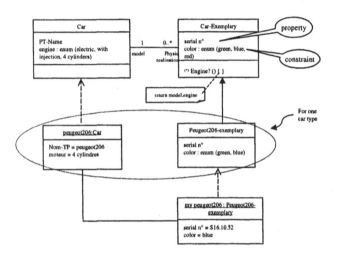

Fig. 7. Properties and constraints

peugeot206 is an instance of "Car" class and *my peugeot206* is an instance of "Peugeot 206-exemplary" class. Associations between levels allow relying each exemplary to its car type.

The properties of a level remain visible at the lower level by defining consultation methods allowing to propagate a property value through associations (see the method *Engine? ( )* in fig.7).

Solution:

1. A company organizes its activities around two product abstraction levels. There exist two abstract classes (fig.8):
- Product higher level "PHL", holding common properties to lower level products deriving from it and containing a set of possible values for some properties
- Product lower level "PLL", holding specific properties of lower level products associated to one

higher level product and containing fixed values for some properties.

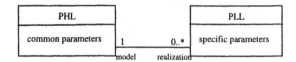

Fig. 8. Two abstraction levels of product

Two scenarios can exist: "PHL is a generic-product and PLL is a product-type" or "PHL is a product-type and PLL is a product-exemplary".

2. When the company decides to create a new "Product Higher Level" (for example the new product-type *peugeot 206*), let's *phl1*. That results in the object *phl1: PHL*.

3. With the creation of the new set product *phl1*, there will be probably various lower level products deriving from the new set *phl1* (for example various exemplars of peugeot 206). One then creates a concrete class for all lower products resulting from *phl1*, let's "PLL (phl1)". This class contains specific properties to *phl1* lower level products (fig.9). Then, any creation of a new higher level product generates one instance of the class "PHL" and one sub-class of the class "PLL".

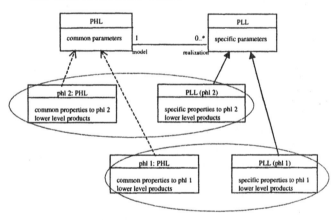

Fig. 9. Creation of two Higher Level products

In the proposed solution, we deal with a partition of the class "PLL". In fact, for each instance of "PHL" class (for example the product-type *peugeot 206*), there exist a sub-class of "PLL" associated to this instance to represent all lower level products raising from that instance of higher level product ("peugeot 206-exemplary", for all exemplars of the product-type instance: *peugeot 206*). Thus the population of "PLL" class is partitioned into sub-classes. This partition is controlled by an inheritance link. This concept was largely discussed by Dahchour [20].

We should finally note that the pattern "Two Levels" can be perceived as refinement of more general patterns known in the literature as "item-description" [7] or "Materialization" [20] patterns.

5. Conclusion

The approach by reuse of pattern provides an engineering guide for building various product and process models which should accelerate the process of PIS specification. A pattern catalogue for product modeling is developed. Basing on this catalogue, a PIS designer aiming at specifying product description in the PIS parts with the first pattern "Variance Points" to fix the specificity of its organization. It fixes then the levels of product managed, thanks to the pattern "Product Levels". Once the sub-model representing the product levels is built (with "One, Two or Three levels" patterns), the designer can fix the number and type of BOMs associated for each product level. According to the characteristics of these BOMs, the sub models representing product BOMs are built by using one or more of the patterns "Construct BOM With Options", "Construct BOM With Variants", "Construct Basic BOM", "BOM With Exclusive Composition", etc. These BOMs sub-models are then linked to the product levels sub-model to associate each BOM to the appropriate product level (pattern "Associate BOMs To Product").

An analogue procedure is applied to documents associated to the product : applied documents are fixed and modeled (pattern "Applied Documents") and then associated to the product levels sub-model (pattern "Associate Documents To Product").

The product pattern catalogue so presented is actually under test and validation by our industrial partner. Extensions of the current catalogue for a complete description of the PDM domain including versioning mechanisms, passage between product levels mechanisms, etc. is under progress.

References

[1] Poulin J.S., « Populating Software Repositories: Incentives and Domain Specific Software ». Journal of Systems Software, 30, pp. 187-199, 1995.

[2] Wilson D.A., Rosenstein L.S., Shafer D., « *Programming with MacApp* ». Reading, Massachusetts, Addison-Wesley, 1990.

[3] Fukanaga A., Pree W., Kimura T.. *Functions as data objects in a data flow based visual language.* ACM Computer Science Conference, Indianapolis, 1993.

[4] Alexander C., Ishikawa S., Silverstein M., Jacobson M., Friksdhl-King I., and Angel S., "A pattern Language", Oxford University Press, New York, 1977.

[5] Gamma E., Helm R., Johnson R., and Vlissides J., "Design Patterns: Elements of reusable Object Oriented Software", Addison-Wesley Publishing company, 1995.

[6] Beck K., Cunningham W., "Using Pattern Languages for Object-Oriented Programs", Norman MEYROWITZ, ed. OOPSLA'87, 1987.

[7] Coad P., "Object-Oriented Patterns", communications of the ACM, 35(9), September 1992.

[8] Xue D., Yadav S., Norrie D.H., "*Knowledge base and database representation for intelligent concurrent design*", CAD journal, vol 31, n°2, Feb. 1999.

[9] Vernadat F., "Enterprise Modelling and Integration: Principles and Applications", Chapman & Hall Ed, 1996.

[10] Jacobson I., Ericsson M., Jacobson A., "*The object advantage: Business Process Reegineering with object technology*", Addison Wesley, 1995.

[11] Rolland C., *Modeling the Requirements Engineering Process.* Proc. Fino-Japanese Seminar on Conceptual Modeling, 1993.

[12] Jarke M., Mylopoulos J., Schmidt J.W., Vassiliou Y., *DAIDA - An environment for evolving information systems.* ACM TOIS, vol. 10, n°1, 1992.

[13] Rumbaugh J., Jacobson I., and Booch G., "Unified Modeling Language Reference Manual", ISBN: 0-201-30998-X, Addison Wesley, December 1997.

[14] Kang K.C., Cohen S.G., Hess J.A., Novak W.E., Spencer Peterson A., "Feature-Oriented Domain Analysis (FODA) Feasibility Study", Carnegie-Mellon University - Software Engineering Institute - Technical Report AD-A235 785, 1990

[15] OMG, "Product Data Management Enablers–Request For Proposal", Object Management Group document, 1997.

[16] Maurino M., "La gestion des donnees techniques – Technologie du concurrent engineering", Ed. Masson, 1993. *(in French)*

[17] CIMdata Inc., « Product Data Management: the definition. An introduction to concepts, Benefits, and Terminology», (http://www.cimdata.com/), 1997.

[18] Gzara L., "Towards a Generic Framework for Product Information Systems", Technical Report RR-1999-11, GILCO Laboratory, INP-Grenoble, France, 1999.

[19] Rieu D., Giraudin J.P., Saint-Marcel C., Front-Conte A., " Des operations et des relations pour les patrons de conception ", In Proc. of Inforsid 1999, 1- 4 Juin 1999, La Garde, France. *(in french)*

[20] Dahchour M., "Formalizing Materialization Using a Metaclass Approach", In proc. Of CASSE'98, 1998.

Approach for Technical Data Integration in Manufacturing Engineering

Rebiha BACHA

Research Department, Technocentre Renault, TCR RUC 4 47, Guyancourt 78288

Bernard YANNOU

P/L Laboratory , ECP, Gde Voie des Vignes, Chatenay-Malanbry, 92295

Abstract

The technical data management (TDM) problems faced by methods units (particularly in the automotive industry) result in part from the vertical design of process engineering software (CAPE): one application – one database, which puts up partitions between the databases used in different fields of activity. In addition, the changes in process data scope are no longer handled by conventional Information Systems (IS). Our objective here was to overcome those difficulties which are real concurrent engineering bottlenecks, by encouraging systematic structuring of product-process technical information in order to provide a universal PDM technology-oriented model. We have identified different design approaches for these models, which are derived both from systemic concepts and from software engineering. The approach that we prefer is based on an Information System Architecture (ISA) known as the Zachman Framework. It attempts to reconcile existing trends by combining the Top-down and Bottom-up approaches in IS design. Our work used standard specification diagrams to formalize TDM specifications and the process activity, with special emphasis on the data handled during the engineering phase.

1. Introduction

Probably one of the key factors in the success of concurrent engineering at industrial companies is the deployment of information technologies, particularly Engineering Data Management Systems (EDM)[1]. The first generation of these systems was limited to documentation and administration of bill of Materials. The scope grew progressively to encompass process management with the use of Workflow systems, now bundled into software packages[15][2]. Today's systems include more and more integrated functions [16] that can be broken down into two main categories.

Internal specifications: Modeling, centering, archiving, revision tracking, sharing, and data security[3]

External specifications: The links with Engineering Computer Systems(CAD/CAM/CAPP) eliminate redundant data entry.

Functions such as data classification (with the advent of Group Technologies), display (with graphic viewers), configuration management (to handle product-process diversity and multiple variants). See Figure 1.

The methodological approaches to corporate implementation of EDM systems are covered by several research efforts [1][7] and reported in publications such as [11]. There are many of them and they are sometimes interrelated because based on two major currents of thought.

Systemic approach: a Top-down approach, which evaluates evolution of corporate IS based on its organizational structure and strategic decisions[4] (See Figure 2). One of the most popular of these techniques is Business Process Reengineering (BPR)[5] [10].

Software Engineering Workshops (SEW) Approach: a Bottom-up approach that uses functional breakdown principles[6], to structure the system lifecycle. With the emergence of object modeling methods in the US, several projects have been initiated [6].

Structuring technical information (the subject of our work) is an activity that is a prerequisite for implementation. It reflects the specifications and the internal TDM specifications: work that is often specific to the industrial activity involved. The proposed approach is characterized by:

- Satisfaction of engineering data requirements (data is often unstructured and/or stored on differing media).

---

[1] Also known variously as Product Data Management (PDM), Engineering Management System (EMS), CAD/CAM Database Management and others.

[2] In addition to the CRISTAL project (Cooperative Repositories & Information System for Tracking Assembly Lifecycle) exposed in [15], there is a large number of projects covering the same ground. Examples include: SIMNET, EP 25584, INTEREST (Integrated Environment for Durability & Reliability Design Support Tools),

EP 26892, Rapid PDM (Rapid Implementation of Product Data Management)

[3] The data can be in various formats: documents, drawings, geometric digitalization or tests

[4] The theory is discussed by Tardieu, H. & Guthmann, B. "Le triangle stratégique: stratégie, structure et technologie de l'information" - Édition d'Organisation - Paris - 1991

[5] A concept propounded in: Hammer, M. & Champy, J., "Reengineering the Corporation, a Manifest for Business Evolution" - Harper Business - NY - 1993

[6] Example of modeling by Computer Integrated Manufacturing (CIM) specialists

- Recovery of existing databases (designed for automation of functions and causing redundancies and multiple entry of the same information).

- The combination of the Top-down and Bottom-up approaches enables better integration of the objects modeled, and shrinks design times.

We will start with a discussion of the architecture framework for the IS project on which we have based our model. We will go on to discuss our design approach for the new diagram, illustrating the principal modeling milestones with concrete examples germane to the automotive industry[7].

2. IS Architecture Framework

The Zachman Framework[8] is an outgrowth of existing architecture standards [20]. Most frameworks, however, have two shortcomings:

- Inadequate integration of the different views of the company (function, data, process and others) and/or the level of abstraction of modeling (project lifecycle: contextual, conceptual, logical, and physical). Example: IDEF methods.

- Incomplete representation of framework concepts by software tools. Examples: FirstStep and PrimeObjects derived from the CIMOSA architecture.

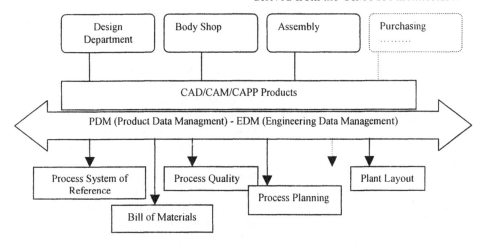

Figure 1: EDM within a Manufacturing Company, Main Functions

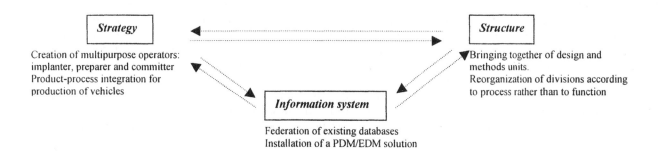

Figure 2: The Strategic Triangle, Typical Interactivity between the Three Poles

---

[7]   For reasons of confidentiality, we will limit the number and content of the examples

[8]   J. Zachman is the inventor of the ISA conceptual framework. The architecture is illustrated by a 36-cell square matrix. It covers the six axes of corporate modeling: Who, What, Where, When, Why an How, which are subdivided into six levels of abstraction (representing the lifecycle), starting from the top level (contextual level of the company) and working down (technical design details). The Zachman Institute (www.zifa.com) is working with several users and suppliers (principally Popkin Software) of toolkits to promote rational implementation of its concepts. For more information on the semantics and structure of the framework, see the documents cited here and their websites.

We have studied BPR and IS design tools to select a solution that fits our specifications [3]. The results of our study prompted us to select System Architect 2001 from Popkin Software[9]. It has the following principal characteristics:

- Complete, contiguous integration of diagrams for the principal information system modeling methods (company modeling, business process reengineering, object oriented methods, data modeling, SA/SD and real time methods);

- Zachman Framework modeling: the flexibility of this metamodel has been demonstrated since it was selected to develop industrial[10] and academic[11] applications. It is better known for design and construction of data-warehouses[12];

- Support for Catalyst methodology: a method of specifying a corporate information system (BPR concept); the concept is developed, marketed and distributed principally by Computer Sciences Corporation[13];

- Coverage of the IS project lifecycle (queries, analysis, design, coding: code generation and reversal).

3. Principal Modeling Milestones

The projected IS is the outgrowth of the integration of models developed in stages I and II. Stage I involved modeling of the processes in our scope, i.e., the vehicle industrialization process. Stage II involved modeling technical process data. Since the stages are not interrelated, we are working on both models concurrently.

[5] has highlighted the importance of studying interactions between the models in both directions. It then becomes possible to identify for each player, the process, the milestone, the information systems involved, and the way to develop or improve them to modify the current situation. Conversely, for each existing or projected IS, we can determine all the strategic advantages that can be reaped without being restricted to operational or local use.

The English-speaking community is dedicated to IS developments based on process analysis, thereby transcending the functional modeling methods used previously [8].

3.1. Stage I: Process Modeling

Level 1: Classification into Who, What, Where, When, Why and How
In order to cope with the complexity of the field to be modeled, we suggest initially classifying the different perspectives of the company by asking the following questions (See Figure 3):

What: Addresses all existing data stored in relational databases or on media relative to CAPE Prodcuts (simulation, layout design, ...) or office automation.

How: covers the functions, processes, and activities in Manufacturing Engineering Departments.

When: To determine the sequence of activities through corporate milestones: professional branch and vehicle project milestones.

Where: Reflects the place where the process is running: company departments and units and also networks and production.

Who: Determines the players involved in the industrialization process, classified into agents (operator), participant in the industrialization project, and external partners.

Why: Clarifies the motivations and objectives of the corporation through activities.

Level 2:
The work then involves formalizing MED activities using standard specification diagrams inspired by the Catalyst method[14]. This is used to structure the above breakdown without sacrificing the links between the views and by entering the details for each. The principal types of diagrams used are discussed below:

Process Chart: One of the diagrams commonly used in BPR. It illustrates the process sequence (activities, functions, or actions) in the company, showing the time sequencing. They are triggered by "events" and produce "results" and are connected by mandatory or optional sequence. Figure 4 shows the sequence of operations (EBP) to obtain an assembly cascade[15] (result) from a part assembly (event).

Process Decomposition: It is used to break down the industrial process hierarchically and to identify the parent-child relationships. There are therefore primary, secondary, and Elementary processes (See Figure 5).

Process Map: The objective of this model is to show the path of a process through different organizational units. The workstation preparation process, illustrated roughly

[9] System Architect 2001: a complete tool to model the information system for a company; this upper case tool has been available on the international market for over 11 years. It was developed by Popkin Software (www.popkin.com)

[10] A typical application (aerospace industry) is dicussed in Karlin, J. and Welch, D., A framework of Mission Operations System Models - 1996 - www.esoc.esa.de/external/mso/spaceOps/1-10/1-10.htm

[11] Examples include a course program based on the Zachman Framework: Van Vliet, P. J. A. Information Systems Architecture & Organization Syllabus - Information Systems & Quantitative Analysis Department, College Of Information Science and Technology, University of Nebraska at Omaha, 1997 – See Whitten, J. L. and Bentley, L. D., Systems Analysis and Design Methods - 4th Ed. - Irwin McGraw-Hill - 1998

[12] See Inmon, W. H., Zachman, J. and Geiger, J. - Data Stores, Data Warehousing and the Zachman Framework - McGraw-Hill - 1997

[13] French CSC website: www.csc.com/france/solutions/solutions domaines.htm

[14] We have only used a subset of diagrams available in Catalyst.

[15] One part of the body work engineer's job. From a part assembly, he must determine the best way to distribute assembly operations (with the appropriate techniques and necessary resources) to existing or future work stations

in Figure 4, involves the body work engineer's units (choice of the process within the cascade) and layout engineers (calculation or work area needed), automate (robotic feasibility), as well as the production and cost control branches (to evaluate tool costs).

Functional Hierarchy[16]: As is evident from the name, this diagram helps us analyze the process according to its functions and to arrange them into a family tree. The function-process link results in one or more functions that contribute to completion of an EBP. We then see that a single function can contribute to completion of several EBP.

Other diagrams have been used in our modeling work. They are in the following categories:

Organization Model: Used to establish the logical and hierarchical structure of organizational units within the scope of the process, to identify the role of players, the associated skills, and the intra- and extra–departmental decision process.

Application Model: Used to establish the functional characteristics required from the new application, the processes supported, and the functions that the application must perform.

Who ?	What?	Where ?	When ?	How ?	Why ?
. Operator . Process Engineer . Manager . Layout Designer	. Existing Manufacturing DB . Identification of CAPE Products	. Factory Site . Network . Organizational Units	. Project Milestones . Other Events Significant to the Business	. Assembly . Flow Simulation . Costs, Layout . Conditionning	. Updated documents . process complexity considerably ......

Figure 3: Corporate Objects Classification

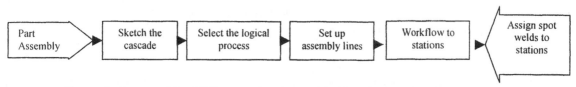

Figure 4: Process Chart " Sequencing Assembly Line Preparation Operations"

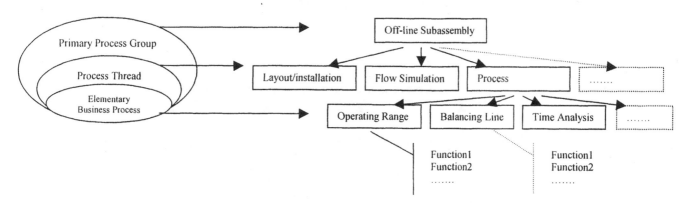

Figure 5: Engineering Manufacturing Process Decomposition

---

[16]   This type of diagram is widely used in the field of systems functional analysis

3.2. Stage II: Data and Basic Concept Modeling

Level 1: Classification into products, process, resources and configurations of the project/factory

This stage was used to convert information into data by initially classifying them as follows:

Products: All objects relative to the vehicle: assemblies, generic parts, variant, function.

Process: This contains the types of operations (manual or automatic), their description (assembly, procurement, or bonding operation) as well as parameters (operating time, acquisition position).

Resources: Covers all layout, human resource (operator, architect), machine and tools assigned to the station.

Configurations of the project/factory: This category contains all objects differentiating one factory site from another. It involves industrial hypotheses of the project (volume/mix, production rate), the characteristics of the plant (surface area, labor, and location) as well as strategic options for the project.

Level 2: Identifying Similarities and Determining Data Requirements

The principle sources of information are:
- The stated requirements of future users of the application;
- The model of business engineering determined from Stage I;
- Documents produced during performance of these processes: records of collaborative work;
- Existing relational databases that can be read to generate the ERD diagram using the reverse reengineering function; we can thus review data redundancies in different databases and identify shared business objects and their relations.

Identification of similarities is used to generate the classes of global objects from potential objects (See Figure 6).

Level 3: Object-Oriented Data Modeling[17]

There are two types of diagram: static and dynamic diagrams[18]. We will only discuss the static portion of modeling, particularly Class diagrams.

There are two different paths to reach the Class diagram: [19]
- The Use Case[19] diagram;
- Relational databases (we start from the physical model and convert it to a logical model; the logical model is then used to generate the skeleton of a class diagram. In Figure 7, we have boldfaced the paths that we took during modeling.

Figure 8 illustrates a simple example of a class diagram that illustrates the different objects handled and the relations between objects[20].

Product	Process	Resource	Configurations of factory site
assembly/part Graphic 3D Functional grouping...	Parameters Feasability Ergonomics	Position on site Investment Human	Area Volume/mix Degree of automation ....

Product	Process	Resource	Configurations of factory site
Variant/option Part Supplier part	Operation dictionary Operating Range	Equipment Cell/ Lay out Objects	Area Grouping Number of workers Foreign factory ....

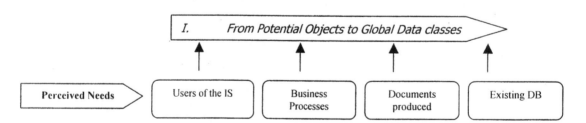

Figure 6: Requirements and similarity analysis of objects

---

17 The object approach selected is the UML concept
18 Static: Class diagram, Use Case diagram
   Dynamic: Activity model, sequence and collaboration diagram, etc.
19 Some authors such [9] explain that Use Cases can be the starting point to identify processes and that object-oriented modeling can after all cover all aspects and views of a company. It is therefore sometimes classified as a BPR technique.
20 There is abundant literature on UML syntax, and in particular [19]: this was our reference

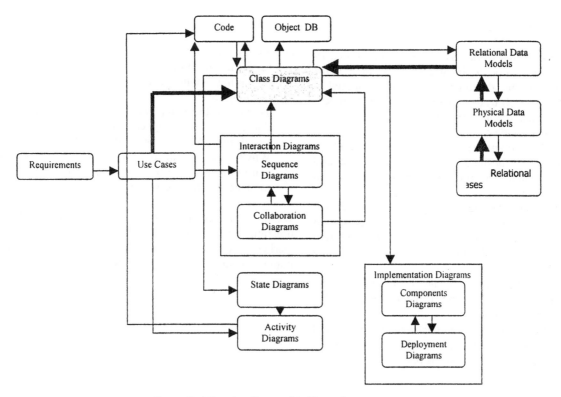

Figure 7: Ways leading to the Class diagram

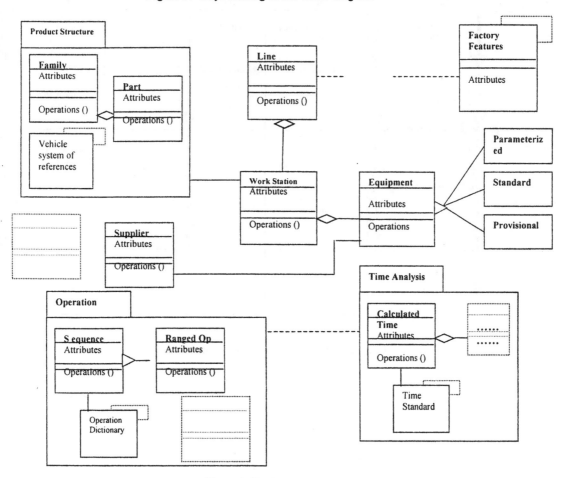

Figure 8: Basic concepts in data modeling

241

4. Summary

The objective of this work was to construct a unified IS for the process scope to be implemented in a PDM technology. In this document, we have proposed a systematic approach to development of product-process IS that is summarized in Figure 9.

The approach provides:

- Real integration of the different views and levels of detail of the business process through standard diagrams and specifications known in system analysis area.
- A bottom-up (starting from existing database) and top-down (by analyzing the operational processes of the domain) approach, calling into question sequential approaches that do not start with existing data;
- Better rationalization of the IS product-process design thanks to the 'Catalyst' methodology and the Zachman metamodel;
- Use of the object approach both for model construction (coding), and for model conceptualization and analysis phases.

In addition, the work allowed us to:

- Capitalize on functional properties of the TDM, promoting maintenance and upgrading of the resulting IS;
- To master the complexity of the industrialization domain, by proposing classification and modeling based on (What? How? Who? Where? When? Why?) and (products, processes, resources and configurations of project/plant) axes corresponding to Stages I and II, respectively.

We are currently completing the first version of TDM functional specifications of the We are preparing a market survey of PDM in order to select one for implementation of the resulting IS.

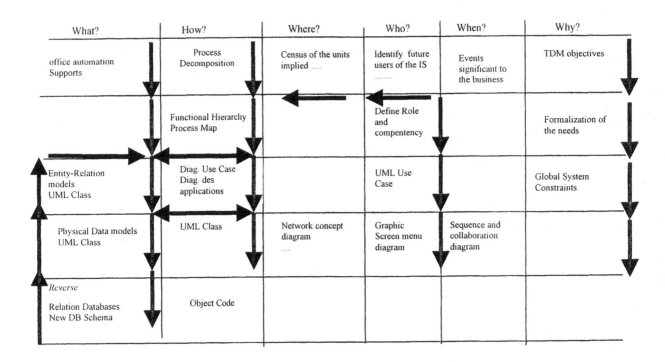

Figure 9: Principles modeling milestones

References

[1] AIT for European Manufacturing Industry, "Integration Platform, Specification 1.0", Consortium distribution, December 1998, www.ait.org.uk/projects

[2] E. Alsene, "The Computer Integration of The Enterprise", IEEE Transactions on Engineering Management, Vol. 46(1), 1999.

[3] R. Bacha, "Évaluation des outils BPR et de conception de SI, Choix d'une solution", Annual progress report, Research Department, Renault Technocentre, 1999

[4] J.E. Bailey, R.H. Rucker, "Automated Management of Design Data : Product, Process and Resource structures" , International Journal of Industrial Engineering , Vol. 5(1) , 7-16 , 1998.

[5] J-A. Bartoli, J-L. Le Moigne, "Organisation Intelligente et Système d'Information Stratégique", Economica, 1996.

[6] C. Cauvet, D. Rieu, B. Espinasse, J.P. Giraudin, M. Tollenaere, "Ingénierie des Systèmes d'Information Produit : une approche méthodologique centrée réutilisation de patrons", Inforsid, 1998.

[7] Y.M. Chen, T.H. Tsao, "A Structured Methodology for Implementing Engineering Data Management Robotics and Computer Integrated Manufacturing" Vol. 14, 275-296, 1998.

[8] M. A. Cook, "Building Enterprise Information Architecture Reengineering Information Systems", Hewlett-Packard Professional Books, 1996.

[9] I. Jacobson, E. Ericsson, A. Jacobson, "The Object Advantage : Business Process Reengineering with Object Technology", Addison Wesley Publishing Company, 1995.

[10] B. Kalpic, A. Polajnar, "Model of the holistic Information Integration of an enterprise", Strojarstvo, Vol. 39(6), 275-280, 1997.

[11] K. G. McIntosh, "Engineering Data Management , A Guide to Successful Implementation", McGraw-Hill Book Company, 1995.

[12] R. Martin, E. Robertson, "Formalization of Multilevel Zachman Frameworks", Technical Report n° 522, Computer Science Department, Indiana University, 1999.

[13] B. Morand, "Trois arguments et trois propositions pour concevoir des systèmes d'information organisationnels", Ingénierie systémique : de la conception orientée objet à la conception orientée projet, GRASCE, URA CNRS 935, 1994.

[14] C. Oussalah, "Génie Objet", Hermès, 1999.

[15] Z. Kovacs, J-M. Le Goff, R. McClatchey, "Support for product data from design to production", Computer Integrated Manufacturing Systems, Vol. 11(4) , 285-290, 1998.

[16] J-M. Randoing, "Les SGDT", Hermès, 1995.

[17] C. Roland, "L'ingénierie des processus de développement de systèmes : un cadre de référence Ingénierie des Systèmes d'information", Vol. 4 (6) , 705-744, 1996.

[18] J.F. Sowa, J.A. Zachman, "Extending and Formalizing the Framework for Information Systems Architecture", IBM Sytems Journal , Vol. 31 n° 3, 1992.

[19] Modeling Systems with UML , A Popkin Software White Paper , Version 1.1 , 1998

[20] F. Vernadat , "Techniques de modélisation en entreprise, Application aux processus opérationnels", Economica, Paris, 1999.

[21] J.A. Zachman, "Building an Enterprise Architecture", The Popkin Process, Draft version 0.93, www.popkin.com, 1999.

[22] W.J. Zhang, Q. Li, "Information Modelling for Made-to-Order Virtual Enterprise Manufacturing Sytems", Computer Aided Design (CAD) , Vol. 31, 611-619, 1999.

Component-based Software Engineering
on the Basis of Active Network

G. V. Smerdina

Russian Research Institute for Artificial Intelligence,
SB RAS, Institute of Informatics Systems,
Lavrent'jeva Av 6, Novosibirsk, 630090, Russia

Abstract

The active network as multiagent system composed from cognitive components with an active code is considered. Such decomposition is a natural projection of the cognitive process of thinking about knowledge representation and knowledge processing. The software engineering is proposed that consists in the specification of cognitive components: the agents-objects and agents-relations, specification of a model, i.e. the cognitive network, and specification of control mode by the network as the script that consists in activation rules of network links. The specification of the model of problem area uses hybrid knowledge representation and processing medium, such as objects, constraints, rules and imperative programming. In proposed examples the universality of such a framework for the real tasks from areas of scheduling, planning and resources allocation, etc is shown. The advantage of such representation consists in a simplicity of programs, which have a linear structure with two-level control.

1 Introduction

The active network as multiagent system composed from cognitive components with an active code is considered.

The cognitive approach considers thinking of a person as a mechanism of knowledge handling. It seems natural to find the most simple way of the knowledge representation for complicated practical problems that would considerably facilitate their representation and handling process. The representation idea as cognitive network composed from a collection of the cognitive software components: intelligent agents-objects connected with intelligent agents-relations is a natural projection of the cognitive process of thinking about knowledge representation and knowledge processing in the head of investigator. The concept of a cognitive network was introduced in paper [1].

Agent-object is such a structure that maps an independent essence of the exterior world. This essence is specified by a set of attributes, restrictions of the attribute values and possible calculations or actions that can be performed under them.

Agents-relations are used to describe interinfluence of agents. They determine restrictions on interacting agents and (or) rules of their coordinated behaviour.

The control of such network is a plan that consists of rules governing the activation of network links. The rules are executed after checking conditions.

Such representation is universal for many practical tasks that naturally reflect some projections of interactions of the real world essences. As the prototype of such representation the language of knowledge representation on the basis of objects and constraints [2,3,4] has been served: the control is performed through productional rules by which the pattern matching in a semantic network takes place.

To describe these components such customary abstractions as data types, classes, objects, inheriting, constraints, rules, procedures are used. In this sense the environment based on cognitive network is object-oriented. Taking into consideration that the intelligent objects and relations are created from primitive objects (agents is endowed with knowledge of their role and behaviour in the context of the real world), it is possible to name such environment an agent-oriented one. Basic property of cognitive network is its configurableness, i.e. a possibility to include and to delete separate links of the network.

The *constraints* mechanism helps to adapt dynamically the behaviour of an agent according modifications of the environment. The specification of constraints on the basis of subdefinite computing model [5,6,7,8] represents a natural way to describe restrictions.

2 Cognitive network as the basis of component-based software engineering

2.1 Cognitive network

Cognitive network is a collection of cognitive software components: intelligent agents-objects connected with intelligent agents-relations.

2.2 Agent-object (O)

The knowledge base of the agent-object can be represented by the triple:

$O = < A, C, R>$,

where $A = \{a_1,.., a_k\}$ are attributes described by a name and a type. For attributes the following types can be defined: *integer, real, atom, subdefinite integer*, given by values as an interval, for example: *start: integer (10.. 20)*. Except for simple types one can use structured types such as *set* and *tuple* to describe attributes.

This part creates dynamic components of agent knowledge and maps symbolical representation of facts both about the exterior world and the system interior condition.

$C = F (A)$ are functional dependencies that specify constraints of attributes, for example, *finish = start + duration*. Constraints represent the static part of agent knowledge.

$R = \{r_1,..,r_k\}$ are serially ordered set of rules as r_i: *Cond -> Act*,

where *Cond* are conditions of attributes of the agent, *Act* are evaluations over attributes or operations on output of attribute values. Each rule is executed once, if the operator *repeat* which causes the cycling of the rule is not specially used. Operator *non-execute* interrupts execution of a sequence. This part contains knowledge about the behaviour of the agent.

Characteristic attribute for the agents can be one with the help of which possible states of an agent are described. Any part of an agent description may be absent so the agent turns either to object or controlling component.

2.3 Agent-relation (L)

The agent-relation $L(O_1,.., On)$ can install associations for two or several agent-objects. It allows to link by functional dependencies attributes of connected agents-objects and also to describe an coordinated behaviour of agents-objects with the help of rules of their interaction.

The knowledge base of the agent-relation can be represented by the triple:

$L = < A, C, R>$,

where $A = \{a_1,..,a_k\}$ are attributes of cognitive relations. These attributes are set similarly to the agent-object by a name and a type.

$C = F (a_1,..,a_k \in O_1,.. ,a_1,..,a_k \in On)$ are functional dependencies on attributes of agents-objects connected with this relation.

$R = \{r_1,..,r_k\}$ are serially ordered set of rules as r_i: *Cond -> Act*,

where *Cond* are conditions of attributes of agents connected with this relation, *Act* are evaluations over attributes of the relation or over attributes of the linked agents-objects or the output of attribute values. Each rule is executed once, if the operator *repeat* which causes the cycling of the rule is not specially used. The operator *activate* used in *Act* allows to transmit control to other agents-objects or agents-relations.

For the agents we distinguish two kinds of communications: the output of information about specified values of attributes and refinement or modification of controlled attributes.

2.4 Operations over the cognitive network

For the cognitive network the following operations are defined:

create – to create the current set of agents,
add – to add a new agent,
delete – to delete an agent from a network,
edit – to edit attributes for the given agent,
activate – to activate an agent.

2. 5 Properties of the cognitive network

The cognitive network has properties of universality, autonomy, activity and dynamic adaptability.

*Universality of the cognitive network

Any problem area can be represented as some interacting really existing or abstract essences connected by physical laws or spatially-temporarily or more complicated cognitive laws which can be represented by mathematical assotiations. Thus, any problem area can be described with the help of independent (autonomous) cognitive essences, namely cognitive (intellectual) objects, i.e. agents-objects and cognitive (intellectual) relations between them, i.e. agents- relations.

*Autonomy

The autonomy is realized in description of an individual essence of problem area with the help of a unique set of attributes. The description of an agent contains knowledge of attributes, admissible discrete or continuous values of the attributes, restrictions on values of the attributes, rules that define behaviour of the agent. The agent-relation incapsulates shared knowledge and behaviour of the agents connected by a relation.

*Activity

Ability of an agent to execute some calculations above available attributes and to output information messages about the state of the agent. Possibilities of calculations over global variables-messages that are used for coordinated control of the network. Except for the agents the property of activity is applied to separate attributes of the agents connected by functional dependencies given in constraints parts. For real applications in many cases a value of attributes can be defined by intervals of admissible values. On creating a cognitive network the uniform computational network is generated. It consists from local computational networks of autonomic agents. On modifying the value of an attribute the automatic calculation of attributes connected with him takes place.

Both an agent-object and agent-relation have properties of activity and access to attributes of the connected agents-objects.

*Dynamic adaptability

The dynamic adaptability property concerns to cognitive network as a whole and is based on a possibility of dynamic creation of a network, dynamic inclusion and deletion of links of a network, dynamic refinement of agent attribute values specified by intervals and further constraint propagation on a network through the agents-relations.

2.6 Control by the cognitive network

The calculations are produced after creation of a current set of the agents-objects and agents-relations, i.e. creation of the cognitive network. On creating the cognitive network the uniform computational network composed of attributes of the agents connected by constraints is generated.

2.6.1 The computational network (CmN)

The computational model is based on the method of subdefinite calculations [5] that can be considered as one of variants of constraint propagation tools applied for solving the computing tasks, and briefly can be circumscribed as follows.

The algebraic expressions of an initial mathematical model are divided on binary and unary relations with the help of introductions of additional variables. For obtained set of relations is created the computational network. This network is represented as the graph, which tops are the initial and additional variables, and arcs are relations connecting these variables. Values of variables are subsets of area of admissible values which are represented as intervals. Initial values of variable are either a set in an initial system, or infinite intervals.

The computing process represents a queued sequential evaluation relations. At the result of this calculation the new values of variables are determined as intersection of the previous value with calculated one (in common case they are interval). Thus, for a variable sequence of not extending intervals is obtained, certainly containing all solution of the task.

The computational network may be considered as a bidirectional oriented graph with variables as its nodes and relations between them as its arcs.

Every time on creating a new agent-object in a network the local computational network (LCmN) is generated on the basis of dependencies described in the part *constraints*. When a new agent–relation is created CmN can link attributes of the various agents and in this sense they can have a global character for the cognitive network.

The CmN puts into practice the subdefinite computing model whose theoretical background is given in works [5,6,7,8]. When an attribute is modified, calculations based on a data-flow principle for attributes with subdefinite values are performed in the CmN.

2.6.2 The script of control and controlling operators

The script of control contains a sequence of rules, i.e. a plan. To control activation of the agents in a network an operator *activate* is used.

The calculation in cognitive network is performed in the following order:

1. On creating agents in the cognitive network the LCmN are generated from attributes of the agents connected with constraints and are executed. If any attribute in the CmN is modified on a data-flow principle these calculations are executed: for the attributes with subdefinite values when the change of an attribute value takes place the values of other attributes connected with it by functional assotiations also change.

2. The sequence of rules of the plan is executed. In the right part of the rules the operators of manipulation of the cognitive network are executed. The operator *activate* activates the agents-objects or the agents-relations.

3. For an activated agent the rules in the part *reaction* are sequentially executed. If a rule contains the operator *repeat*, the rule is fulfilled until an inconsistency in the CmN appears. In this case the rollback on the previous step of calculations takes place and the next rule is performed. If the operator *non_execute* is used then the section *reaction* is not executed after it.

2.6.3 *Role of messages in control of the cognitive network*

The agents have access to the common information in the CN through the mechanism of messages. The messages are determined at the global level for all network and are described as global variables of *integer, atom* types. Any agent in the network can send a message, i.e. to set a value of a global variable, or to receive the message, which then can be used for choicing the strategy of its behaviour.

2.7 Principles of component-based programming and types of agents

The component-based programming assumes:

*Decomposition of problem domain into autonomous cognitive component.

*Description of classes of cognitive components, i.e. agents-objects and agents- relations.

*Creation of a dynamic framework, i.e. the cognitive network from concrete copies of the agents.

*Creation of the script, i.e. plan of activation of the agents in the cognitive network.

On executing the script the specific agents are activated, and the active code of the agent is started. The active code is rules connected to the modification of attribute values. The execution of the rules happens in the correspondence to a type of the agent. There are the following types of the agents:

– sequential execution;

– direct execution.

For the sequential agents rules are executed one by one, and each rule is executed once, if the operator *repeat* which causes cycling a rule is not specially used. In this case a rule is executed until an inconsistency in the LCmN appears.

For direct execution agents it is typical the execution of only one rule in the part *reaction*. This rule is chosen from the group of active rules on a condition and usual is finished by an operator *non_execute*. The used rule can be then deleted.

3 Examples of component-based software engineering

The agents are described with the help of classes using the mechanism of inheriting. The numerous examples of using the decomposition as a cognitive network are proposed in [9,10]. In this paper there are 3 examples that demonstrate the technology suggested.

3.1 Example: a moving agent

Example of a moving agent description in MARS language [9]:

Let an object represents a square that autonomously moves in parallel to axis to the right until it reaches the fixed wall.

Figure 1 The moving agent is a square

```
structure POINT
  x: integer (0..100);
  y: integer (0..100);
end;
class ACTIVE_SQUARE
    centre: POINT;
    side: integer (10..20) ;
    state: (move, stop);
constraints
  centre.x <= 75;
reaction
  state = move => centre.x := centre.x + 1;
        Vis_square (centre, side); repeat;
                    //on repeat executing of the rule
  => state := stop;
end;
```

In this example the class *ACTIVE_SQUARE* is described, for which the attributes *centre* and *side* are defined. In the part *constraints* the restriction which inspects transition of square is given. In the part *reaction* there are 2 rules that determine behaviour of the square.

Creation of the cognitive network and control of the network:

```
create new Active_square:
        ACTIVE_SQUARE (side = 50,  centre = 0,0);
plan
  read (cl) = Enter => Active_square.state :=move;
                activate (Active_square);
end;
```

The current set of agents is created by the operator *create* and contains one agent *Active_square* of the class of the agents *ACTIVE_SQUARE*. The plan determines a model of control and contains one rule, in which by keystroke *Enter* the square is activated. The first rule in *reaction* is executed until inconsistencies appear, i.e. until the square reaches the wall. When inconsistency arises the next rule is executed.

The current set of the agents is created by the operator *create* and contains one agent *Active_square* of the class of agents *ACTIVE_SQUARE*. Plan determines the model of control and contains one rule, in which by keystroke *Enter* the square is activated. The first rule in the *reaction* part is executed. Each time when a new value of the coordinate *centre.x* is to be calculated the CmN is started to execute. The CmN is generated on the basis of assotiations described in *constraints* part. For the given example execution of the rule is similar to execution of a cycle operator in traditional programming languages.

3.2 Scheduling a complicated task

The task consists of 5 subtasks, it is required to define maximum time necessary for realization of the task. It is known, that the subtasks 2,3 and 4 should begin after a termination of the subtask number 1, the subtask 4 should begin after realization of the subtask 2, and subtask 5 should begin after termination of the subtasks 3 and 4. Besides, it is known, that a duration of realization of the tasks are inexact and are determined as intervals. In this example duration of task are defined in conditional units. It is required to estimate execution time of the task in whole.

Cognitive network consists of 5 agents-objects of a class TASK and one agent-relation of a class LIST_TASK:

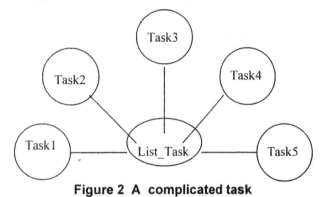

Figure 2 A complicated task

type TASKS (Task1, Task2, Task3, Task4, Task5);
class TASK
  name: TASKS;
  start: integer;
  finish: integer;
  duration: integer;
  constraints
  finish = start + duration;
end;
relation LIST_TASK (task1, task2, task3, task4, task5: TASK);
  start: integer;
  finish: integer;
  constraints

task1.start = start;
task2.start >= task1.finish;
task3.start >= task1.finish;
task4.start >= task2.finish;
task4.start >= task1.finish;
task5.start >= task4.finish;
task5.start >= task3.finish;
finish >= task5.finish;

reaction
= > print (" time necessary for realisation of the task
           equal ", finish);
end;
create Task1: TASK (duration = (10..12));
    Task2: TASK (duration (5..30));
    Task3: TASK (duration (10..20));
    Task4: TASK (duration (3..40));
    Task5: TASK (duration(10..20));
    List_Task: LIST_TASK (Task1, Task2, Task3,
                          Task4, Task5:TASK);
plan
  activate (List_Task (start := 0));
end;

On creating the CN the computational network on the basis of constraints is generated and the automatic refinement of the subdefinite values of the attributes takes place. It is only necessary to print the new calculated value of the attribute *finish*.

3.3 Compiling the schedule for the processor of conveyor type on the example of fast Fourier transformation algorithm

This example demonstrates the use of the cognitive network for more complicated problem area. The figure below represents the oriented graph FFT (Fast Fourier transformations algorithm) for case N = 8 [11]. Each arc is an image of transformation as

$$x^r_{i+1} = x^r_i + W^r_i * x^r_k + W^j_i * x^j_k;$$
$$x^j_{i+1} = x^j_i + W^j_i * x^r_k + W^r_i * x^j_k;$$

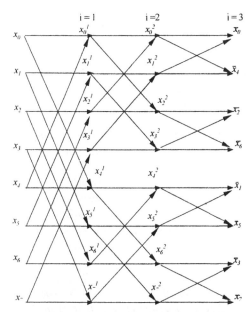

Figure 3 The oriented graph Fast Fourier transformations algorithm

It is required to describe the FFT algorithm for the universal computing device of conveyor type that performs operations of addition, division and multiplication and has the depth equal to 10. The stream of calculations enters on the input of the pipeline. On 10 steps the result of the input calculation can be obtained from the output and then be used as an operand for the following calculations. It is necessary to supply the maximum loading of the pipeline. The outcome of the task should be the stream of calculations that can then be used as the input of an actual processor of the conveyor type. This task belongs to the class of the resource allocation and consists in compiling the schedule that support the stream of calculations. This example demonstrates distributed algorithm description with the help of decomposition of the task as a cognitive network.

The principles of solving the task are based on modelling the process of calculations. The cognitive network represents complex variables (class X) connected with transformation relations (*TRANSFORM_R*, *TRANSFORM_J*). As soon as transformations in a separate relation are completed, the relation deletes itself from the network. Complex variables in the cognitive network have the states (*READY, UNREADY*), that reflect their readiness for consequent transformations. The relations of transformation *TRANSFORM* have 5 steps of transformations, which result in calculation of new complex variables. Being used the rule is removed from agent. The class *SCHEDULE* is used to create the calculation stream intended for feeding the input of the conveyor computing device.

```
type J : real (-100.0..100.0);
type R: real (-100.0..100.0);
class X
  x_r: R;
  x_j: J;
  R1, R2, R3: integer;
  w_r: R;
  w_j: J;
  state: (READY, UNREADY);
end;
relation TRANSFORM_R (arg1:X, arg2:X, res:X)
counter: integer (0..10);
reaction
  P1:  arg1.state  =  UNREADY  /  arg2.state  =
UNREADY => non_execute;
    P2:  => CELL :=
          pack (res.R1, arg2.x_r, arg1.w_r, MLTP);
          activate (SCHEDULE);
    P3:  => CELL :=
          pack (res.R2, arg2.x_j, arg1.w_j, MLTP );
          activate (SCHEDULE);          '
          counter := COUNT; delete P1, P2, P3;
  P4: COUNT – counter < 10  => non_execute;

  P5:  => CELL :=
          pack(res.R3, res.R1, res.R2, PLUS );
          activate (SCHEDULE);
          counter := COUNT; delete P5;
  P6: => CELL := pack(res.x_r, arg1.x_r, R3, PLUS);
          activate (SCHEDULE);
          counter := COUNT; delete P6;
  P7:  =>  res.state = READY; delete ();
          CNT_DEL:= CNT_DEL + 1;
end;
relation TRANSFORM_J (arg1:X, arg2:X, res: X)
    counter: integer (0..10);
reaction
  P1:  arg1.state  =  UNREADY  /  arg2.state  =
UNREADY => non_execute;
    P2:  => CELL :=
          pack (res.R1, arg2.x_j, arg1.w_r, MLTP);
          activate (SCHEDULE);
          counter := COUNT;
    P3:  => CELL :=
          pack (res.R2, arg2.x_j, arg1.w_r, MLTP );
          activate (SCHEDULE);
          counter := COUNT; delete P1,P2,P3;
  P4: COUNT – counter < 10  => non_execute;
  P5:  => CELL :=
          pack(res.R3, res.R1, res.R2, PLUS );
          activate (SCHEDULE);
          counter := COUNT; delete P5;
```

```
P6: => CELL := pack(res.x_j, arg1.x_j, R3, PLUS);
          activate (SCHEDULE);
          counter := COUNT; delete P6;
P7: => res.state := READY, delete ();
          CNT_DEL := CNT_DEL + 1;
end;
class SCHEDULE
   schedule: tuple of integer [1000];
reaction
   => schedule (COUNT) := CELL;
      COUNT := COUNT+1;  MES_EXE := YES;
end;
pack (a_result, a_operand1, a_operand2, cod_op)
         (32-25)a_result, (24-16)a_operand1,
            (15-8)a_operand2, (7-0)cod_op;
end;
class CONTROL
 reaction
    CNT_DEL = 48 => stop;
    => CELL := 0; MES_EXE := NO;
    => activate (TRANSFORM_R, TRANSFORM_J);
    MES_EXE = NO => activate (SCHEDULE);
end;

create
  new x0: X (x_r: 0.0, x_j:0.0, w_r:1.0, w_j:0.0,
             state: READY);
  new x4: X (x_r: 0.0, x_j:0.0, w_r:1.0, w_j:0.0,
             state: READY);
  new x01: X (state:UNREADY);
  new x41: X (state:UNREADY);
  new x0_x4_r: TRANSFORM_R (arg1: x0, arg2: x4,
                            res: x01);
  new x0_x4_j: TRANSFORM_J (arg1: x0, arg2: x4,
                            res: x01);
  new x0_x4_r: TRANSFORM_R (arg1: x0, arg2: x4,
                            res: x41);
  new x0_x4_j: TRANSFORM_J (arg1: x0, arg2: x4,
                            res: x41);

  new x1: X (x_r: 1.4142131, x_j:0.70710655,
             w_r:0.70710655, w_j:-0.70710655,
             state: READY);
  new x5: X (x_r: -1.4142131, x_j:-0.70710655,
             w_r:1.0, w_j:0.0, state:READY);
  new x11: X (state:UNREADY);
  new x51: X (state:UNREADY);
  new x1_x5_r: TRANSFORM_R (arg1: x1, arg2: x5,
                            res: x11);
  new x1_x5_j: TRANSFORM_J (arg1: x1, arg2: x5,
                            res: x11);
  new x5_x1_r: TRANSFORM_R (arg1: x1, arg2: x5,
                            res: x51);
```

```
  new x5_x1_j: TRANSFORM_J (arg1: x1, arg2: x5,
                            res: x51);

  new x2: X (x_r: 2.0, x_j:1.0, w_r:0.0, w_j:-1.0);
  new x6: X (x_r: -2.0, x_j:-1.0, w_r:1.0, w_j:0.0);
  new x21: X (state:UNREADY);
  new x61: X (state:UNREADY);
  new x2_x6_r: TRANSFORM_R (arg1: x2, arg2: x6,
                            res: x21);
  new x2_x6_j: TRANSFORM_J (arg1: x2, arg2: x6,
                            res: x21);
  new x6_x2_r: TRANSFORM_R (arg1: x2, arg2: x6,
                            res: x61);
  new x6_x2_j: TRANSFORM_J (arg1: x2, arg2: x6,
                            res: x61);

  new x3: X (x_r: 1.4142131, x_j:0.70710655,
             w_r:-0.70710655, w_j:-0.70710655);
  new x7: X (x_r: -1.4142131, x_j:-0.70710655,
             w_r:1.0, w_j:0.0);
  new x31: X (state:UNREADY);
  new x71: X (state:UNREADY);
  new x3_x7_r: TRANSFORM_R (arg1: x3, arg2: x7,
                            res: x31);
  new x3_x7_j: TRANSFORM_J (arg1: x3, arg2: x7,
                            res: x31);
  new x7_x3_r: TRANSFORM_R (arg1: x3, arg2: x7,
                            res: x71);
  new x7_x3_j: TRANSFORM_J (arg1: x3, arg2: x7,
                            res: x71);

  new x02: X (state:UNREADY);
  new x12: X (state:UNREADY);
  new x22: X (state:UNREADY);
  new x32: X (state:UNREADY);
  new x42: X (state:UNREADY);
  new x52: X (state:UNREADY);
  new x62: X (state:UNREADY);
  new x72: X (state:UNREADY);
  new x03: X (state:UNREADY);
  new x13: X (state:UNREADY);
  new x23: X (state:UNREADY);
  new x33: X (state:UNREADY);
  new x43: X (state:UNREADY);
  new x53: X (state:UNREADY);
  new x63: X (state:UNREADY);
  new x73: X (state:UNREADY);

  new x01_x21_r: TRANSFORM_R (arg1: x01,
                              arg2: x21, res: x02);
  new x01_x21_j: TRANSFORM_J (arg1: x01,
                              arg2: x21, res: x02);
  new x21_x01_r: TRANSFORM_R (arg1: x01,
                              arg2: x21, res: x22);
```

new x21_x01_j: TRANSFORM_J (arg1: x01,
arg2: x21, res: x22);

new x11_x31_r: TRANSFORM_R (arg1: x11,
arg2: x31, res: x12);
new x11_x31_j: TRANSFORM_J (arg1: x11,
arg2: x31, res: x12);
new x31_x11_r: TRANSFORM_R (arg1: x11,
arg2: x31, res: x32);
new x31_x11_j: TRANSFORM_J (arg1: x11,
arg2: x31, res: x32);

new x41_x61_r: TRANSFORM_R (arg1: x41,
arg2: x61, res: x42);
new x41_x61_j: TRANSFORM_J (arg1: x41,
arg2: x61, res: x42);
new x61_x41_r: TRANSFORM_R (arg1: x41,
arg2: x61, res: x62);
new x61_x41_j: TRANSFORM_J (arg1: x41,
arg2: x61, res: x62);

new x51_x71_r: TRANSFORM_R (arg1: x51,
arg2: x71, res: x72);
new x51_x71_j: TRANSFORM_J (arg1: x51,
arg2: x71, res: x72);
new x71_x51_r: TRANSFORM_R (arg1: x51,
arg2: x71, res: x52);
new x71_x51_j: TRANSFORM_J (arg1: x51,
arg2: x71, res: x52);

new x02_x12_r: TRANSFORM_R (arg1: x02,
arg2: x12, res: x03);
new x02_x12_j: TRANSFORM_J (arg1: x02,
arg2: x12, res: x03);
new x02_x12_r: TRANSFORM_R (arg1: x02,
arg2: x12, res: x13);
new x02_x12_j: TRANSFORM_J (arg1: x02,
arg2: x12, res: x13);

new x22_x32_r: TRANSFORM_R (arg1: x22,
arg2: x32, res: x23);
new x22_x32_j: TRANSFORM_J (arg1: x22,
arg2: x32, res: x23);
new x32_x22_r: TRANSFORM_R (arg1: x22,
arg2: x32res: x33);
new x32_x22_j: TRANSFORM_J (arg1: x22,
arg2: x32, res: x33);
new x42_x52_r: TRANSFORM_R (arg1: x42,
arg2: x52, res: x43);
new x42_x52_j: TRANSFORM_J (arg1: x42,
arg2: x52, res: x43);
new x52_x42_r: TRANSFORM_R (arg1: x42,
arg2: x52, res: x53);

new x52_x42_j: TRANSFORM_J (arg1: x42,
arg2: x52, res: x53);

new x62_x72_r: TRANSFORM_R (arg1: x62,
arg2: x72, res: x63);
new x62_x72_j: TRANSFORM_J (arg1: x62,
arg2: x72, res: x63);
new x72_x62_r: TRANSFORM_R (arg1: x62,
arg2: x72, res: x73);
new x72_x62_j: TRANSFORM_J (arg1: x62,
arg2: x72, res: x73);

new SCHEDULE;
new CONTROL;

MES_EXE: (YES, NO);
// control message: YES: the CELL was packed,
NO: the CELL was empty
COUNT: integer;
// message for counter of schedule item

CNT_DEL: integer;
// message for counter deleted network
relation (TRANSFORM)
plan
=> COUNT:= 0; CNT_DEL:= 0;
=> activate (CONTROL); repeat;
end;

4 Conclusion

The suggested decomposition of problem area on the basis of software components: agents-objects and agents-relations represents two-level method to handle knowledge, for which the complicated algorithms become linear ones and it facilitates the process of representation of the task. The solving the task is reduced to specification of the task as the cognitive network and control of that network. Component-based software engineering is universal since it represents the real world in a natural way, and a distributed algorithm represents a mode of programming by means of modelling.

The offered model of knowledge representation, i.e. the cognitive network, reflects wider understanding of multiagent interaction than conventional, taking into account both static and dynamic models of interaction of agents, and thus expands the range of the tasks considered from the point of view of multiagent programming.

The use of customary abstraction such as classes, objects, rules, procedures creates comfort for the programmers who have been brought up on languages like *C* and *Pascal* types.

This model can be applied to different fields of knowledge: from real time planning, scheduling, and resource allocation up to the expert systems and problems of adaptive control in conditions of varying environment, for example, an automatic control of the automobiles on the given route [10].

As a toolkit for construction cognitive networks the system MARS-V is projected. The system MARS-V is an open environment for incremental development of applications and in its base variant assumes the use of the *C++* compiler, libraries of *C++*, a set of software components for visual description of a framework of problem domain and knowledge components specification.

References

[1]   G.Smerdina Cognitive Network is a New Metaphor for Multiagent Programming. // Proc. of The 1st International Workshop of Central and Eastern Europe on Multi_Agent systems. CEEMAS'99.June 1-4, 1999. St.Petersburg, Russia P.249-255.

[2]   Yu. A. Zagorulko, I.G. Popov. A Software Environment based on an Integrated Knowledge Representation Model // Perspectives of System Informatics (Proc. Of Andrei Ershov Second International Conference PSI'96). - Novosibirsk, June 25-28, 1996. -P.300-304.

[3]   Yu. A. Zagorulko, I.G. Popov. Knowledge Representation Language with Objects and Constraints // Proc. Of 6th East-West International Conference on Human-Computer Interaction - Human Aspects of Business Computing (EWHCI'96). -Moscow, Russia, 12-16 August, 1996. - P.56-66.

[4]   Yu.A. Zagorulko, I.G. Popov. Object-Oriented Language for Knowledge Representation Using Dynamic Set of Constraints // Proc. Of the Third Joint Conference on Knowledge-Based Software Engineering in Smolenice, Slovakia, 1998, -P.124-131.

[5]   Narin`yani A.S. Subdefinitness in Systems of Knowledge Representation and Handling. In: Inf. AS USSR, Engineering Cybernetics. 1986.-N5.- .3-28 (In Russian).

[6]   Vitaly Telerman, Dmitry Ushakov. Data Types in Subdefinite Models. In: Jacques Calmet and others (eds)., Art. Intell. And Symbolic Mathematical Computation, Lecture Notes in Computer Science; Vol. 1138, Springer, (1996). -P.305-319.

[7]   Igor Shvetsov, Vitaly Telerman, Dmitry Ushakov. NeMo+: Object-Oriented Constraint Programming Environment Based on Subdefinite Models. In: Gert Smolka (eds)., Principles and Practice of Constraint Programming - CP97, Lecture Notes in Computer Science; Vol. 1330, Springer, (1997). -P.534-548.

[8]   D.M. Ushakov. Some Formal Aspects of Subdefinite Models. –Preprint. Russian Acad. of Sci. Siberian Division. A.P.Ershov Iinstitute of Informatics Systems.Vol.49. -Novosibirsk, 1998. P.24.

[9]   G.Smerdina. MARS: Multiagent Active-Reactive System. Sixth National Conference with International Participation "CAI`98". -Pushchino. Russia, 5-11 October. -P. 59-65.(In Russian).

[10] G. Smerdina. Cognitive Network and its Using for Dynamic Adaptive Problems. Complex systems: Control and Modelling Problems. //Proc. of the International Conference, -Samara, Russia, June,15-17, 1999. P. 178-183 (In Russian).

[11] Henri J. Nussbaumer. Fast Fourier Transformation and Convolution Algorithms. Springer Series in Information Sciences. By Springer-Verlag Berlin Heidelberg 1981 and 1982 (in Russian).

CHAPTER 4

Management and Organization

RE-ENGINEERING THE ENGINEERING CHANGE MANAGEMENT PROCESS

G.Q. Huang, W.Y. Yee, K.L. Mak

Department of Industrial and Manufacturing Systems Engineering, The University of Hong Kong,
Email: GQHuang@hku.hk ; MakKL@hku.hk

Abstract

The Engineering Change Management (ECM) process is a special business process that is concerned with managing changes and modifications in forms, functions, materials, etc. in product and process design. Recent investigations indicate that the ECM process affects the product development and rationalisation significantly in terms of time to market and costs. This paper proposes to apply the Business Process Re-engineering (BPR) concept to rationalise the ECM process. A framework is proposed by embedding a micro ECM procedure as a single step in a macro BPR procedure. If an ECM system does not exist in a particular company, the framework can be used to introduce and implement an appropriate ECM system. If an ECM system already exists but does not operate as effectively and efficiently as expected, the framework can be used to improve and rationalise it. The primary role of ECM is to provide a clear focus and vision necessary for BPR. The primary role of BPR is to institutionalise a good ECM system that is able to make substantial improvement in the business.

1. Introduction

Changes and modifications on forms, fits, materials, dimensions, functions, etc., of a product or component are usually referred to as Engineering Changes (ECs). ECs can be as simple as documentary amendments, or as complicated as the entire redesign of the products and the manufacturing processes. Business, organisational, and operational changes such as schedule changes, order changes, etc., are not directly related to product design, and are therefore not usually termed as ECs. However, they may be the consequences of ECs, or they may become the dominating cause for ECs.

The management of engineering changes (ECs) has serious resource implications in any manufacturing company because it involves almost all the functions across the entire organisation. These functions may not only be the sources, but may also be the victims of ECs. Manufacturing companies have to cater for these ECs by constantly adjusting their activities. Indeed, an inefficient and ineffective management of ECs cripples the robustness of manufacturing, irrespective of the advances in manufacturing technologies.

A number of surveys on Engineering Change Management (ECM) have been conducted recently in manufacturing industries (Maull et al, 1992; Boznak, 1993; Huang & Mak, 1998). These surveys have revealed two significant phenomena. On the one hand, these surveys have exposed that ECs are a noticeable problem in many companies. For example, Huang and Mak (1998) have found an average of about 65 active ECs in the surveyed companies. This finding is consistent with that reported by Maull et al. (1992) and Boznak (1993). Boznak (1993) also has reported the annual EC administrative processing cost in surveyed companies ranged from US$3.4 million to US$7.7 million. Maull et al. (1992) have found out that ECs may incur a cost up to 10% of annual turnover. It has been found that ECs have adverse effects on delivery time, production schedules, scrap and rework, etc. On the other hand, these surveys have revealed that only one third of the companies have formal ECM systems and procedures to deal with ECs while the rest majority deal with ECs informally and on an ad hoc basis. The significance of EC problems and the inefficiency of the current ECM systems trigger the need for a systematic ECM re-engineering framework.

Whether dealt with formally or informally, the average throughput time of an EC from its identification through processing to implementation is unreasonably long. For instance, Watts (1984) discovered a "40-40-40" pattern of time distribution in ECM. That is 40 days are, on average, for identifying an EC, 40 days for processing paperwork and approval, and 40 days for implementing it. The "middle 40 days" for paper processing is of particular interest here. Although the term of BPR was not widely used, Watts attempted similar concepts for re-engineering the entire ECM system and process in order to reduce the "middle 40 days". Remarkable improvements were achieved, including 82.5% reduction in middle processing time and 50% reduction in amount of paperwork. These achievements have clearly indicated the potentials of BPR in re-engineering the ECM process.

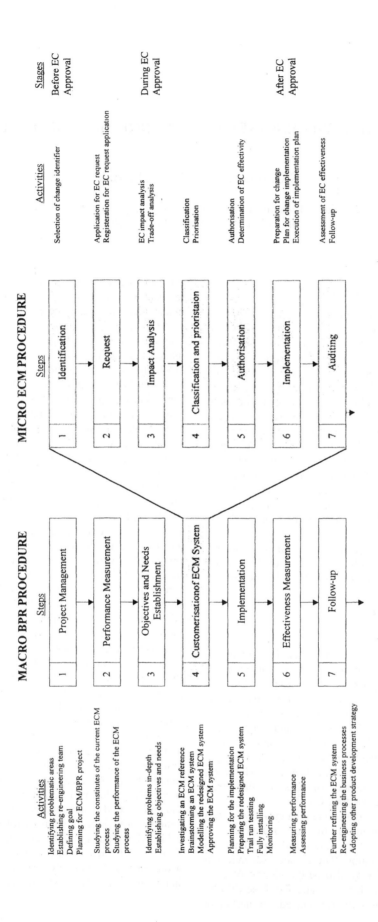

Figure 1. The ECM/BPR framework

The aim of this paper is to propose a generic and systematic framework that can be tailored or extended for re-engineering and implementing the ECM process in manufacturing industries. The framework is based on the key concept of BPR in the context of ECM implementation. The ultimate objective is to reduce the problems and improve the efficiency and effectiveness of ECM.

2. The ECM/BPR Framework

Business Process Re-engineering (BPR) has become one of the most popular concepts of the business management in 1990s. Hammer & Champy (1993) have defined BPR as "the fundamental rethinking and radical redesign of business process to achieve dramatic improvements in the critical, contemporary measurements of performance, such as cost, quality, service, and speed". Although there are several variations of BPR definitions (Davenport *et al.*, 1990; Manganelli and Klein, 1994), the following four characteristics are common:

- BPR focuses on an existing business process for re-engineering. One advantage to reengineer the entire process, instead of individual task of functions, is that it can be tried to solve and minimise all the problems incurred from the process and all the relevant impacts on business completely.
- BPR requires fundamental and critical rethinking on the current practice of the existing business process through thoughtful analysis. The fundamental rethinking is beneficial because it is able to indicate why the process is undergone re-engineering and what the expectations of this process are.
- BPR recommends redesigning the process radically in order to generate a new process that is totally departed from the existing process. The radical redesign is advantageous because the expected performance can be delivered from the redesigned process. But, it involves a large amount of time and efforts in practice.
- BPR is able to deliver dramatic and remarkable improvement through implementing the redesigned process in the practice within a short period. In addition, the improvement can make a breakthrough to gain competitive advantage to the business. Very often, the improvement can be measured quantitatively by the speed and cost and qualitatively by the service and quality.

The major objective of BPR is to rationalise a chosen business process in order to improve the performance of a process and thus the entire business. Remenyi & Whittaker (1994) have suggested two approaches to implementing a BPR project. They are radical approach and incremental (or creeping) approach. On the one hand, radical approach is usually addressed to an urgent problem that is very critical to the survival of the business. This approach is adopted to the project that aims to solve the problem through a great transformation across the organisation. Although high risk and huge cost are involved in the transformation, re-engineering the process can give substantial benefits to the process with far-reaching impacts on the business.

On the other hand, the incremental approach focuses on dealing with single issue or smaller process cautiously. It is accompanied by performing a series of BPR projects of which each project is processed step-by-step. Unlike the radical approach, the project performed by the incremental approach is usually monitored easily and costs reasonably; however, it is incapable of delivering substantial improvement from only one project.

In practice, which approach is selected for re-engineering a business process depends on the urgency of the problem, organisational strategy, and available resources of the organisation. Quite often, the practice may lie between the radical and incremental approaches. It is also possible to use the radical BPR approach to "kick start" a re-engineering project at the beginning, and afterwards, the incremental BPR approach is adopted to support and continue the project forward (Hess & Oesterle, 1996).

With the reference to the characteristics and the approaches of BPR, researchers and practitioners, such as Davenport (1993), Jacobson et al (1995), Keily (1995), have proposed many models to implement BPR. Of particular interest to this research is a recent work by Chan & Choi (1997). They have proposed a so-called RADIE (Recognition, fundamental Analysis, radical reDesign, Implementation and Evaluation) model for BPR. The model does not only address to the characteristics and approaches of BPR, but also takes the success and failure factors of performing a BPR project into account. In addition, the model incorporates the BPR guidelines suggested by Hammer and Champy (1993) and Romney (1994). The RADIE model provides a clear guidance to carry out a BPR project successfully by showing how to start and continue the BPR project properly.

The RADIE model is simply divided into two phases: BPR study phase and BPR implementation and evaluation phase. The object of the BPR study phase is to prepare a new process model to replace the existing process in the practice. This phase starts with recognising a value-added business process for re-engineering from various processes, then analysing the recognised process in detail, and lastly redesigning the process radically to suit the business. Within this phase, loops can be cycled from the redesign step back to analysis or recognition step.

The second phase, BPR implementation and evaluation, is primarily concerned with the institutionalisation of the new process model in practice. The redesigned process proposed from the previous phase is implemented in the real environment and then evaluated to seek further improvement. After the evaluation, it is possible to seek performance improvement of the redesigned process and to trigger another BPR project to improve the business performance continually.

The ECM process is here the subject matter for reengineering. It involves almost all the functions across the entire organisation. It is also closely coupled with another key business process, i.e. the product development and realisation process. Because of these characteristics, the ECM process must be rationalised so that other related business processes are not adversely affected by ECs. In a separate work, a systematic ECM framework has been established, including procedure, organisational and documentary aspects (Yee et al, 1998). This paper focuses on establishing another framework for implanting and institutionalising the ECM system in manufacturing organisations. The concept of the BPR approach will be used in establishing such a framework. An overview of the resulting framework, BPR/ECM is shown in Figure 1. The left-hand side of the figure presents the macro BPR framework, highlighting the main steps and activities involved in implementing an ECM system. The right-hand of the figure shows an overview of the micro ECM procedure. As can be seen from the figure, the micro ECM procedure is only a single step in the macro BPR framework. The primary role of ECM is to provide a clear focus and vision necessary for BPR. The primary role of BPR is to institutionalise a good ECM system that is able to make substantial improvement in the business. The micro ECM procedure has been discussed elsewhere (Yee et al., 1998). This paper will focus on discussing the macro BPR procedure.

3.   Macro BPR Procedure

This section presents the macro BPR procedure which follows the RADIE model (Choi and Chan, 1997) with modifications. The seven steps in the macro procedure characterise the BPR properties of the RADIE procedure, for re-engineering a business process on the whole, and particularly specialise in re-engineering the ECM process. These steps are capable of identifying and analysing the problems in the current ECM process, redesigning a new model for ECM process, and implementing and evaluating the redesigned model. This methodical approach is advantageous to develop and improve the ECM process effectively and efficiently.

There are two major modifications between the RADIE procedure and the macro BPR procedure. (1) In the "Recognition" of the RADIE procedure, it explores as many re-engineering opportunities as possible among various business processes, and then selects the most valuable process out for re-engineering through a series of analysing and prioritising work. However, in the "Project Management" of the macro procedure, the ECM process is already assumed as the unique process for re-engineering.

(2) For the sake of easy controlling and monitoring a process, it is not unusual to break down the big activity into smaller activities. Therefore, "Fundamental Analysis" and "Evaluation" in the RADIE procedure are broken down into "Performance Measurement" and "Objectives and Needs Establishment", as well as "Effectiveness Measurement" and "Follow-up" in the macro procedure respectively. Although the "Fundamental Analysis" and "Evaluation" in the RADIE procedure are broken down, their BPR characteristics are still retained in the smaller activities of the macro procedure.

Step 1 - Project Management

Project Management is the initial step in the macro BPR procedure. The objective of this step is to get the company ready for re-engineering the ECM process. ECM/BPR project is used as the name to term for re-engineering the ECM process. The outcome of this step is a project plan that guides the company, in particular the re-engineering team, to pursuit the goal for re-engineering the ECM process. The following major activities are involved.

☐   *Identifying problematic areas.* To start with the ECM/BPR project, it is the most important to identify what the problematic areas exist with the current ECM process. The identification can be performed by comparing the ECM performance internally against the historical ECM record, and/or externally with the other manufacturing companies. Some questions that can be asked during the comparison are suggested as below:

- Is the ECM process costly?
- Is the ECM process time-consuming?
- Is there a problem with preparing ECM documents?
- Is there a problem with flowing EC-related information?
- Is there a problem with using EC techniques, such as EC impact analysis, EC classification and prioritisation, EC effectivity determination, etc.?
- Is there a problem with the procedure for processing ECs?
- Is there a problem with the responsibility to manage ECs?

The questions is used to manifest the potential needs for re-engineering the ECM process, and therefore prompt the project to be carried out forward. Simple questions are usually sufficient to identify the problematic areas in the ECM process. Answers of the questions will be used later to investigate the problems in-depth and set goal for the ECM/BPR project.

☐ *Establishing re-engineering team.* After the ECM/BPR project is triggered, substantial and subtle work is sequentially followed to deal with. A re-engineering team can be formed to take the work forward. The re-engineering team is made of two parties: core group and extended group. The members in the extended group are responsible for contributing their opinions, knowledge, ideas, and comments to the project, and giving support to the core group to process the re-engineering activities. They may be IT specialists, ECM consultants, BPR practitioners, and staff concerned with managing ECs. On the other hand, the core group is responsible for performing the necessary activities of re-engineering the ECM process in great efforts. It may involve the representatives from various disciplines, such as design, manufacturing, marketing and etc., within the company. Moreover, it is advantageous to assign a person among the core group as team leader. He/She is especially responsible for facilitating teamworking and overseeing the progress of the project.

☐ *Defining goal.* The goal of the ECM/BPR project is the ultimate achievement expected after the completion of the project. The importance of defining the goal is that the goal can guide the re-engineering team with direction and clarify the ultimate target of the project. To define the goal, guidelines can be obtained from the identified problematic areas in the ECM process and the business strategy. The goal may be very general at this moment, but should be relevant to the ECM process. Refinement is possible later throughout the project.

☐ *Planning for ECM/BPR project.* An action plan is required to guide the project approaching the defined goal through careful consideration. Necessary actions and available resources are brought into consideration so that an adequate and practical plan can be generated to show clearly about that who should do what and when. Revising the plan is essential at later steps in order to ensure that proper action is done at the right moment.

Table 1. Examples of quantitative metrics.

Transaction of change

Volume of paper for requesting an engineering change
Volume of paper for evaluating an engineering change
Volume of paper for approving an engineering change
Volume of paper for auditing an engineering change

Time

Amount of time for requesting an engineering change
Amount of time for evaluating an engineering change
Amount of time for implementing an engineering change
Amount of unscheduled idle time for processing an engineering change
Amount of net time for processing an engineering change

Cost

Net cost for making an engineering change
Administration cost of an engineering change
Cost of rework and scrap due to engineering change
Unestimated cost for processing an engineering change

Table 2 Guidelines for brainstorming an ECM system.

ECM Documentation

- Is it required an EC request form to request a change need?
- Is it required an EC evaluation form to analyse the potential impacts of a requested EC?
- Is it required an EC notice form to inform the approval of an EC?
- Is it required an EC auditing form to check the effectiveness of EC implementation?
- What items are included in the form?
- How can the forms be transmitted among the disciplines across the company?

ECM Procedure

- Is it required an ECM procedure to guide the ECs for being managed?
- How can an EC be requested, processed, implemented, and audited?
- What are the tools and techniques used to manage ECs?

ECM Organisational Structure

- Is it required an EC co-ordinator to co-ordinate and monitor the EC activities?
- Is it required an ECM Board (ECMB) to authorise the requested ECs? Is it required a leader in the ECMB?
- Is it required a Workforce to support both EC co-ordinator and members in ECMB in handling EC activities?
- Who will be the EC co-ordinator?
- Who are the members in the ECMB? Who is the leader in the ECMB?
- Who are the members in the Workforce?

Others

- Is it necessary to have an EC record to keep the relevant data of ECs? If necessary, which kind of record, paper-based or computerised record?
- Is it necessary EC meeting to authorise the ECs? If necessary, how often and how long?
- Is it necessary to have an ECM software package to aid the managing ECs?
- Is it necessary to have other supporting software package, such as PDM, MRP, etc. to enhance the effectiveness of the ECM system?

Step 2 - Performance Measurement

The intention of this step is to study how severe of the problematic areas of the ECM process through understanding the fundamental components and the performance of the existing ECM process. The output from this step is a set of performance data that provides valuable insights to express the expectations of the ECM process in the next step. The following activities are involved in this step.

❑ *Studying the constitutes of the current ECM process.* It is necessary and important to study the current ECM process before determining what performance is expected to deliver in the ECM process. A model can be set up for studying the existing ECM process. The model should include the main elements involved in managing ECs, such as the ECM documents, procedures, technique, responsible parties, etc. Based on the developed model, studying the operations among the elements is followed. Simple studies is enough because it is worthless to spend a great amount of time and efforts in detailed understanding the operations without moving forward to actually do in re-engineering. Flowchart and simulation software packages are commonly used to develop a model for studying the ECM system.

☐ *Studying the performance of ECM process.* The performance of the existing ECM system can be measured by two approaches. One is quantitative measurement and the other is qualitative measurement. In most occasions, quantitative measurement is highly desirable because it can quantify the performance of ECM system. There are many quantitative metrics that can be used. Table 1 shows some examples of the quantitative metrics, which are categorised into three groups, namely, transactions of change, time, and cost. On the other hand, qualitative measurement is possible and often used to measure the performance, particularly when the information is uncertain and lack. In practice, which approach is most appropriate for measuring the performance depends highly on the availability and the integrity of the existing performance data of the ECM process. It is also possible to measure the performance by both approaches. Relevant performance data can be collected from the EC record. The collected data is then processed and presented into desired format, as the input required by the next step.

Step 3 - Objectives and Needs Establishment

The object of this step is concerned with improving the performance of the ECM process through analysing the problems highlighted and performance data received of the existing ECM process from the previous steps. The output from this step is the objectives and needs of the ECM process expected in re-engineering the ECM process. The objectives and needs is able to give the focus for redesigning the ECM process in subsequent step. The following two activities are involved in this step:

☐ *Identifying problems in-depth.* Having been highlighted the problems and obtained the performance data of the existing ECM process from the previous step, further analysis is necessarily followed to trace back the roots for causing problems and evaluate the impacts affecting the business. Both roots and impacts can give valuable insights to redesign a new ECM system that is effective to solve and minimise the identified problems and their impacts. Cause-effect diagram may be one of the common tools to investigate the roots and impacts.

☐ *Establishing objectives and needs.* Without the establishment of objectives and needs for the redesigned ECM system, it is very hard to reveal what kind of ECM system is desirable. The objectives are the performance that is expected to achieve in the ECM process. The needs are the performance that is expected to deliver from the ECM system, for

example, the duration for processing an EC. The objectives and needs can be refined in subsequent steps, especially when new idea is popped up or unexpected problem is discovered in the subsequent steps.

Step 4 - Customisation of ECM System

Customisation of the ECM system is the core step in the macro BPR procedure. The aim of this step is to explore as many as alternatives for redesigning an ECM system in order to make a breakthrough of the performance of the ECM process. The outcome from this step is a custom ECM system that is able to overcome the problems and improve the performance of the existing ECM system. The following four activities are involved in this step.

☐ *Investigating an ECM reference.* With the reference to the established goal, objectives, and needs, it is known what the redesigned ECM system is desirable to have and do. This information seems sufficient to reengineer the ECM system already. However, the redesigned ECM system may be insufficient to deliver dramatic performance improvement. To overcome this, investigation of an ECM reference is a useful way. Through the investigation, innovative ideas for re-engineering the ECM system are usually popped up. The investigation can be benchmarking with the best ECM practice of other companies and studying the reference model, such as IDEF0 (CAMI-i, 1979) and ECM framework (Yee, *et al.*, 1998).

☐ *Brainstorming an ECM system.* Brainstorming an ECM system is a key activity in this step. It is a big and difficult task, and therefore, much effort is highly required to identify the opportunities to redesign the ECM system. For the ease of brainstorming, it is worth considering dividing the ECM system into four parts so that more attention can be paid to brainstorm each part. The four parts and some examples of the corresponding items are shown in table 2 as guidelines.

☐ *Modelling the redesigned ECM system.* Having been brainstormed a new ECM system, it is quite often unable to build up and operate as brainstormed. Modelling the ECM system is necessary and advantageous to make the system as realistic as possible and to weep out the susceptible potential errors and risks from the system. The model should not only show the elements in the ECM system, but also include how the ECM system works as it was designed to and what the performance is expected. More important, the performance resulted in the model should be examined whether the performance fulfils the goal, objectives, and needs established in previous steps before the redesigned ECM system is finalised. The model can be created by the commercial simulation software packages.

- *Approving the ECM system.* Having been established a model that can perform what are expected, it is of need to demonstrate to the top management for approval. The demonstration may show how the new ECM system is designed, how it works to process ECs, and what the performance can be delivered. The demonstration is not only to get approval, but also to obtain support from the management to implement the redesigned ECM system, if it is approved.

Step 5 - Implementation

This step is primarily concerned with putting the redesigned ECM system institutionalised from the preceding step of customisation of ECM system into practice. The output from this step is that the redesigned ECM system can be utilised to process ECs in the real environment. It involves the following main activities.

- *Planning for the implementation.* In order to organise the activities followed by the approval of the redesigned ECM system, an implementation plan may be commenced to schedule the necessary activities with thoughtful consideration on the available resources. Gantt Chart can be used to present the plan.
- *Preparing the redesigned ECM system.* To carry out the implementation plan, three main areas should be paid more attention to. First, staff is very often rarely happy with and unwilling to be involved in the changes arisen from re-engineering the ECM process because of the human nature. Perhaps, they are short-sighted that they do not foresee the long-termed benefits from re-engineering. It is necessary for the managerial level to clarify clearly the goal and long-termed benefits of the ECM/BPR project to every level in the organisation. Next, the redesigned ECM system may introduce various changes to the techniques and workflow in managing ECs. Concerned staff may be unsure about how to use the techniques and how to be involved in the workflow. To solve such situation, training may be provided for the concerned staff in order to get them familiar with the redesigned ECM system in advance. Lastly, new components, such as computer hardware, ECM software package, etc., of the redesigned ECM system may be obtained from outsourcing. Therefore, the availability of new components should be ensured before the start of trial run test.
- *Trial run testing.* It is hard to envisage that the redesigned ECM system which can replace the current system completely without any neither disruption nor problem. The trial run test may be employed in order to minimise the incurred disruption and problem. It can be performed in three approaches. The first approach is to select a group of staff from concerned disciplines to use the redesigned ECM system for processing ECs. One advantage is that every concerned discipline can experience the redesigned system to certain degree. It is disadvantageous that the selected staff may be unwilling to be involved in using a new system which is usually troublesome. The last approach is to choose a product to test the redesigned system. The ECs originated from such selected product are handled by the redesigned system. It is advantageous that every company, whether it is small or large, may easily find a product out for the purpose of testing. The disadvantage is that no EC is originated from the selected product during the testing period, especially when the testing period is short, and consequently, the system cannot be tested. The last approach is very similar to the second approach that a family of products, instead of a product in the second approach, is selected out for testing the redesigned system. It is advantageous that enough volume of ECs can be experienced the redesigned system in the normal occasion. One disadvantage is that a large amount of ECs may be originated from various products within the family. Great amount of downstream work is often incurred from the originated ECs, but the redesigned system is immature to deal with such heavy downstream work. In practice, what approach is selected for trial run test depends on the business strategy and the practical situation.

- *Fully installing.* Upon the approval of trial run test, the redesigned ECM system is installed to take the replace of the current system. Typically, the earlier the implementation of a new system and the longer the installation time, the higher the cost required. Therefore, in order to minimise the cost, it is suggested to select the most appropriate and economical time to install the redesigned ECM system and to keep the installation time as short as possible.
- *Monitoring.* At to this moment, the implementation plan and project plan are nearly completed. But, it is usual problems may be encountered during implementing the plans and the progress of implementing the plans may behind the schedule. It is of necessity to reviews the plans in order to get everything properly and on schedule and to facilitate the company moving forward towards the completion.

Step 6 - Effectiveness Measurement

Although the redesigned ECM system is already put into practice, it does not mean the redesigned system is working properly and is able to deliver the performance as desired. Very often, measuring the effectiveness of the redesigned system is in need to ensure the desired results are achieved and sustained. This step aims to measure the performance resulted in the redesigned ECM system. The outcome from this step is a set of performance data that gives valuable insights to improve the redesigned ECM system in the subsequent step. The following activities are involved in this step.

☐ *Measuring performance.* The performance delivered from the redesigned ECM system can be evaluated in quantitative and qualitative approaches. For quantitative measurement, time and cost conducted in the redesigned system can be measured. Some examples of quantitative metrics are suggested, as shown in table 1. On the other hand, qualitative measurement can be taken from the feedback of concerned staff. The feedback of the concerned staff can provide the comments on the service delivered from and the satisfaction with the redesigned system. The qualitative measurement can be accomplished by conducting interview and survey to the concerned staff.

☐ *Assessing performance.* The improvement delivered from the redesigned ECM system is assessed through comparing the measured performance of the redesigned system with the performance of the original ECM system. There are two approaches to assess the performance: quantitative assessment and qualitative assessment. The selection of the approach depends on what kind of approach is used to measure the performance at advance. For quantitative assessment, the measured performance is minus the original performance so that a net value is obtained. The net value is then compared with the goal, objectives, and needs previously established in order to check whether the desired performance is delivered as expected. On the other extreme, the qualitative assessment is to compare the performance from comments given by the concerned staff with the goal, objectives, and needs previously established to check whether the expected performance is achieved already. Both assessments can indicates a gap between what is achieved and what is expected, and meanwhile, it also highlights a path for further improvement in the subsequent step.

Step 7 - Follow-up

In the proceeding step, it is known what have been improved and are needed to improve. The intention of this step is to seek the means for the improvement in order to increase the gain for the business ultimately. It includes the following activities.

☐ *Further refining the ECM system.* Refinement for the redesigned ECM system is necessary to improve the performance, if it is not yet achieved as desired; or to enhance the performance, if it is already achieved as

expected. The refinement may focus on solving the problems embedded in the ECM system.

☐ *Re-engineering other business processes.* ECM process is closely associated with other business processes, for instance, material supply process, customer request process and customer service process. Improvement can be accomplished by re-engineering another business process in the effort of BPR in order to solve the problems of the corresponding process.

☐ *Adopting other product development strategy.* Three strategies are widely used in product development. They are (1) avoid design changes as far as possible; (2) make design changes as early as possible; and (3) optimise the necessary design changes. A number of techniques and tools, such as Design For X, Quality Function Deployment, etc. have been developed as the aids to improve the product design, based on these strategies. Regarding to the ECM process, it adopts the third strategy that the necessary changes of the product design are made at the optimal situation. Adopting the other strategies of product development is also an effective means to improve the business.

4. Concluding Discussions

This paper has proposed an ECM/BPR framework for implementing an ECM system in manufacturing industry if it does not yet exist in companies, or re-engineering the ECM system if it already exists but does not operate effectively and efficiently. The proposed framework integrates the BPR concept into the ECM process. Within this framework, the ECM provides a sharp focus and a clear vision for the BPR while the BPR gives a good ECM practice and makes significant improvement in the business.

It is necessary and advantageous for the framework to integrate BPR and ECM procedures. The necessities and advantages are concluded as below:

- From the micro point of view, an ECM procedure deals with the ECs systematically step-by-step through a series of ECM activities, from the identification throughout the processing to the auditing. From the macro point of view, these ECM activities are condensed into one single step in the BPR procedure.
- The ECM procedure provides the macro procedure with a sharp focus for the re-engineering team to concentrate on important and relevant issues of the ECM process. It also provides the macro procedure a clear vision for radical changes on the activities, processes, procedures, and organisational structures of the ECM process.
- The BPR procedure provides a methodical mechanism for implementing radical changes and

sustaining the benefits as suggested by ECM. These changes and benefits are followed by repercussions throughout the business.

- The step of customisation of ECM system can ensure the ECs are processed thoroughly from being identified to obsolesce without neither being ignored nor missing important steps.

The ECM/BPR framework has been developed as being generic. Neither the ECM procedure nor the BPR procedure should be followed rigidly. Instead, they are flexible enough to be extended or tailored to suit the particular circumstances.

References

[1] Boznak, R.G. (1993) Competitive Product Development, Milwaukee, WI: Business One Irwin/Quality Press

[2] Choi, F.C., Chan, L. (1997) "Business Process Re-engineering : Evocation, Elucidation and Exploration", Business Process Management Journal, Vol. 3, 39-63

[3] Davenport, T.H., et al. (1990) "The New Industrial Engineering: Information Technology and Business Process Redesign", Sloan Management Review, Vol. 31, No. 4 summer, 11-27

[4] Davenport, T.H. (1993) process Innovative: Re-engineering Work through Information Technology, Harvard Business School Press, Boston, MA

[5] Hammer, M. and Champy, J. (1993) Re-engineering the Corporation: A Manifesto for Business Revolution, Haper Wiley Chichester.

[6] Hess, T., Oesterle, H. (1996) "Methods for Business Process Redesign: Current State and Development Perspectives", Business Change and Re-engineering, Vol. 3, No. 2, 73-83

[7] Kiely, T.J. (1995) "Managing Change: Why Re-engineering Projects Fail", Harvard Business Review, Vol. 73, No. 2, March/April, 15

[8] Huang, G.Q., Mak, K.L. (1998) "Engineering Change Management : A Survey within UK Manufacturing Industries", submitted to International Journal of Production and Operations Management

[9] Jacobson, I, Ericsson, M, Jacobean, A. (1995) The Object Advantage: Business Process Re-engineering with Object Technology, Addison-Wesley as ACM Press Books

[10] Maull, R., Hughes, D., Bennett, J. (1992) "The Role of the Bill-of-Materials as a CAD/CAPM Interface and the Key Importance of Engineering Change Control", Computing & Control Engineering Journal, March 1992, 63-70

[11] Manganelli, R.L., Klein, M.M.(1994) "Your Re-engineering Toolkit", American Management Association, Vol. 83 No. 8, August, 26-30

[12] Remenyi, D., and Whittaker, L. (1994) "The Cost and Benefits of BPR", Business Change and Re-engineering, Vol. 2, No. 2, 51-65

[13] Romney, M. (1994) "Business Process Re-engineering", The CPA Journal, Vol. 64, No. 10, October, 30-33

[14] The ICAM Definition Method, IDEFO, The Architect's Manual, CAMI-i, 1979

[15] Watts, F. (1984) "Engineering Changes : A Case Study", Production and Inventory Management Journal, Vol. 25, Part 4, 55-62

[16] Yee, W.Y., Huang, G.Q., Mak, K.L. (1998) "Towards a Reference Engineering Change Management Framework", Technical Report, Department of Industrial and Manufacturing Systems, University of Hong Kong

Organisational Integration of Meeting Results

C. J. Costa

Departamento de Ciências e Tecnologias de Informação, ISCTE, Lisboa.

P. A. Antunes

Departamento de Informática, Faculdade de Ciências, Universidade de Lisboa, Lisboa

J. F. Dias

Departmento de Ciências de Gestão, ISCTE, Lisboa

Abstract

This paper discusses the problem of organisational integration of outputs generated by Electronic Meeting Systems (EMS). The level of organisational integration supported by actual systems is still very weak, especially in what concerns the post-meeting phase. In order to tackle this problem, we developed a framework, based in the concept of genre, specially tailored for meetings supported by EMS. The paper illustrates the application of the model to a specific organisation, based on a sample of 214 decisions taken in meeting sessions of a management team. The incorporation of communication genre in a software system that supports organisational integration of meeting outcomes is also analysed in this paper.

1 Introduction

Meetings are the most widespread and – possibly – the most expensive type of coordinating teams of people in organisations. We have seen in the literature that a meeting may cost up to US$1000 per hour in salary costs; and that there are more than three billion meetings per year in the United States [21]. Because of this huge potential market, Electronic Meeting Systems (EMS) have been viewed since the beginning of the 1980's as the Holy Grail to improve meeting processes and outcomes [16]

The role of EMS can be broadly defined as facilitating two fundamental aspects of group work: content and process [20]EMS change the static contents of traditional meetings (e.g. data in a flip chart) into dynamic contents, which people can easily manipulate, model and share. EMS have also the potential to change traditional meeting processes, either increasing participation, stimulating collaboration, guiding individual and group tasks to assure coherent results or avoiding conflicts.

Unfortunately, the success of EMS seems to depend on too many factors. For instance, Dickson et al. [15] found out that some kinds of process support decrease group effectiveness (in particular, inflexible kinds of process support); Miranda and Bostrom [20]also found out that some kinds of content support have a negative impact on meeting outcomes while others have positive impact (e.g. anonymity).

To complicate these matters, the role of EMS may not be confined to support meetings. They can extend their support to meeting preparation ([6][7]) for instance with the purpose of defining an agenda or clarifying preliminary positions that people may want to bring to meetings. EMS can also extend their support to the post-meeting phase, for instance to support evaluating the outcomes or increasing commitment.

Arrived to this point, we should stress that, besides few notable exceptions, which will be described later, there is not much research work done in the subject of post-meeting support. Our objective, reported in this paper, is to tackle this issue using a divide and conquer strategy. Since EMS functionality can be divided in content and process support, we have started by addressing the process facet.

But how to characterise post-meeting processes? They primarily deal with organisational integration and effectiveness. In fact, independently of the quality of the outcomes produced by a meeting, they must flow to the organisation and induce the production of goods and services or influence people's opinions. If a decision is a consequence of a question or request, a response must be sent to the ones who made the request. If it was decided that somebody would execute a task, so this person must be informed and instructed. During the meeting, participants may notice that there is not enough information to take a decision. In this situation, information must be requested to other internal or external entities. At the same time, all these events must be organised, orchestrated and monitored. This paper describes such a framework

1.1   Related Work

Few researchers have discussed EMS support to the post-meeting phase. One of them is Milan Aiken, who has for some time been experimenting the integration of expert systems with EMS ([3],[1],[2],[12],[4] and [5]). Among the proposed systems, there is an Expert Session Analyser (ESA) imposing structure to meeting outcomes such that they can be used as inputs to other systems. Later, a data retrieval agent ([2],[12]) and a natural language translation agent [2], were also proposed. These new systems organise the results of brainstorming sessions ([1]). Another tool, designated idea consolidator, was proposed to automate the process of organising ideas ([1]). This tool condenses text, by identifying key words and matching them with users' comments.

Cire [25] is a collaborative information retrieval system dedicated to support cooperative information seeking and retrieving. Although the major purpose is to support the meeting process, this system constructs a shared memory that may be used across and outside meetings, thus falling in the post-meeting phase.

Raikundalia and Rees [24] proposed a system named LoganWeb, which is an electronic meeting document manager for the World Wide Web. LoganWeb tools provide meeting transcripts with information in various readable and navigable forms. LoganWeb also provides a *secretarius* that moderates contributions to meetings.

2   Framework for studying post-meeting processes

Post-meeting processes primarily deal with organisational integration and effectiveness. In fact, independently of the quality of the outcomes produced by a meeting, they must flow to the organisation and induce the production of goods and services or influence people towards positive emotions and constructive relations. If a decision is a consequence of a question or request, a response must be sent to the ones who made the request. If it was decided that somebody would execute a task, so this person must be informed and instructed. During the meeting, participants may notice that there is not enough information to take a decision. In this situation, information must be requested to other internal or external entities. At the same time, all these events must be organised, orchestrated and monitored.

According to Herbert Simon [26]), this level of organisational integration and effectiveness requires accomplishing at least three steps:

(1) developing a plan of behaviour for all implicated members of the organisation;
(2) the communication of the relevant portions of the plan to each member; and
(3) willingness of these members to follow the plan.

Considering that developing such a plan is a necessary component of meeting outcomes, and willingness can only be achieved through psychological means, we should focus on the matters of communication.

One recent research attempt to characterise organisational communication in concrete terms is based on the concept of genre. The concept of genre was imported from the literature [28], but was generalised to the organisational context (see, for example[13]). A genre of organisational communication is an institutionalised communicative action (e.g. memo, report, resume, enquiry, letter, meeting, announcement, expense form and training seminar).

Genres are characterised by their purpose and form. The purpose is not a private motive, since the community members must socially recognise it. In a empirical study examining the communication exchanged by a group of workers that relied on electronic mail for coordination, Orlikowski and Yates [22] identified the following purposes: informational message; comment on group process or use of medium (meta comment); proposed rule, feature or convention (proposal); request for information, clarification or elaboration (question); reply to previous message or messages (report); and residual category (e.g. thanks, apologies, ballots). The form of the genre refers to observable aspects of the communication, such as medium, structural features and linguistic features. In the same study [22], several forms were also identified: embedded message, graphical element, heading, opening, sign-off, sub-heading, subject line and word or phrase emphasis.

Genres may be linked together in a way that constitutes a communicative process. This circumstance creates a genre system; with interdependent genres that are enacted in some typical sequence. Orlikowski and Yates [23]presented such a genre system intended to characterise meetings. According to these researchers, a meeting is a composition of four genres: meeting logistics; meeting agenda; the meeting itself; and the meeting report.

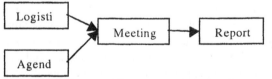

Figure 1- The meeting as a system of genres (adapted from [23])

Of course, this system of genres is so broad that is hardly useful to describe post-meeting processes in detail. However, it provides a starting point for studying the issue. One way to proceed in this subject is either by decomposing or specialising the genre system (like in [18]). Through decomposition, we can "divide" the system in a set of components. The meeting may be divided in a set of issues (or decisions). The agenda may be decomposed in agenda topics and the minutes may also be decomposed in communication statements. Typically, each agenda topic has a direct relation with a meeting issue and communication statement.

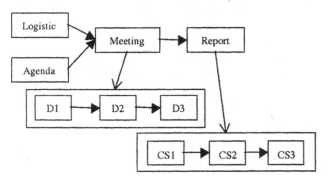

Figure 2 - Decomposition

We should note however that decomposition rapidly reaches a point where the notion of genre is lost, because social recognition is lost. So, we may try to specialise the system of genres.

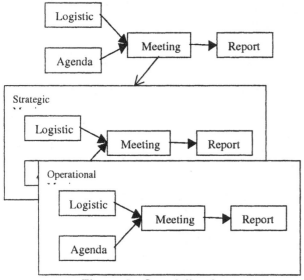

Figure 3 - Specialisation

Specific logistics, agenda, meeting and minutes genres (or reports) may come together to form specialised meeting genres (systems), such as strategic meetings, operational meetings, brainstorm meetings, etc. Contrary to the decomposition approach, specialisation preserves social recognition.

We thus came to a framework for analysing post-meeting processes based on the communication of socially recognisable communicative acts – genres – which, combined together assemble communicative processes – genre systems. Genre systems may then be decomposed and specialised. Finally, the post-meeting process comprises agenda topics, meeting issues, communicative statements, all combined in different manners to form specific meeting genres. In the next section we will show how this framework can be used in practice

5 Using the framework

Genre analysis is not an abstract categorisation exercise, but closely tied to the situated activities in the community using them. It is what the community members (or at least the most skilled ones) recognise as genres, that counts. The situated nature of genres makes it difficult to develop post-meeting support without analysing how the communities of people do their meetings and transfer outcomes to the organisation.

As an attempt to understand these matters and, at the same time, assess the proposed framework, we analysed a long collection of outcomes produced by meetings of directive members of a public organisation. The number of sessions was 30 and those sessions have taken place during a period of 4 year (from 1996 to 1999). The process employed had the following steps:

- Identification of logistic genre;
- Division of each meeting in decisions;
- Identification of decision genres;
- Identification of agenda topic genres;
- Identification of meeting minutes genres (also called report genres or communication statement genres).

This process had the support of a group of people related to the organisation, or to other organisations working in similar departments, who helped in the identification and analyse of genres.

According to the people involved in the process, there is just one logistic genre. Those types of decisions have a high level of formalisation and date and time are established in the regiment of this department.

The 30 meeting minutes were decomposed in decisions. The number of decisions was very changeable, from 3 to 16 decisions in each meeting. In the final, a list composed by 214 decisions was produced.

Those decisions were grouped in the following meeting genres (or decision genres): "decide actions", "decide unitary plan", "postpone decision" and "decide continuos plan".

"Decide action" is a genre that groups all decision that are oriented to a specific action to be performed by a specific actor in the organisation or outside the organisation.

"Decide unitary plan" is a genre of decisions that have as result the planning of one activity or group of activities but that don't have a continuous application in the time.

"Postpone decision" is a genre that is used where there is no information or there in a specially reason not to decide in the present moment.

"Decide continuos plan" is a genre that has as result the production of politics, rules and regulations that have a certain level of continuity.

Now, According to the process, were identified the agenda topic genres. The following genres of agenda topic were identified: occasional requests and repetitive requests. Occasional requests are those topics that aren't foreseen with exactitude. Repetitive requests are those requests that periodically happen, like the annual budget approval.

After identifying logistic genres, agenda genres and decision genres, were identified minutes genres. For simplicity purpose those genres were named communication statement genres.

Nine main genres were identified this way. Figure 1 lists the obtained genres.

The response is a genre which main purpose is answering to some formal proposal, offer or request. Its form is typically a yes or no answer. The receiver is the one who asked for that request. The request may be a complaint about some service that the organisation provides or a contracting proposal presented by a department.

Instruction is a genre that has the objective of instructing to do something in a given moment or in a given way. The receiver of this communication is the one who will be responsible for the fulfilment of the task.

Document approval is a communication to "everybody" stating that something was approved. For instance, the budget approval is the responsibility of this organisational structure, but other department or service makes the budget. After being approved, it is available to everybody that needs it.

"Rule, regulations or explanation" is a communication to every department or external entity related to the organisation. The objective is to create new ways of doing work in the future. Those rules, regulations and explanations are produced by this organisational structure during the meeting as a result of the meeting decision process. If other department has produced a rule or

regulation, and it is only approved during the meeting, the genre used is the one described before.

Document transfer is a communication saying that some document may be given to a specific department, person or external entity.

Information request is a genre with the main purpose of asking for information to a specific department, person or external entity. For instance, it may be decided that, to take a decision about a contract, some opinion must be requested to a consultant.

Delegation is the genre used to nominate a person or group to a position. It can also nominate a commission to take charge of a problem. This genre must define the problem and objectives involved, as well as the people that will deal with it.

Information note is a genre which main purpose is to inform a certain department, person or external entity. Typically, every genre has as purpose of giving information, but there is one communication genre where the main purpose is this one.

Genres	Purpose	Form	Receiver	Examples
Response	Respond to some formal request	Decision value: Y/N	Who requested the response: - Typical internal departments (e.g. financial) - Workers - External entities - Other services	Expense acceptance
Instruction	Order/Instruct formally some action	Task definition	Explicitly defined (Individual workers or departments)	"Pay employee X"
Document approval	Approve some document	Decision value: Y/N	Not explicitly defined (Only by content)	Budget approval
Agenda	Schedule for later Postpone decision	Agenda	Meeting	Decide to decide next meeting
Rule, regulation, explanation	Define an organisational rule		Not explicitly defined (Only by content)	
Document transfer	Transfer document (Approve transfer)	Attached document	Explicitly defined	Publishing the annual report
Information request	Ask information		Explicitly defined	Ask information to law consultant about contract
Delegation	Delegate power to a person or a commission	Must indicate the mission as well as the name of the commission or person	People involved	Creation of a committee to evaluate a proposal
Information note	Inform /clarify		Explicitly defined	

Figure 4 – Communication Genres

After identifying the main genres used in this organisational context, people were asked to categorise each of the 214 decisions that were produced, using the communication genres that have been proposed.

Genre	Percentage of communications that use this genre	Percentage of decisions that use this genre
Response	7%	13%
Instruction	33%	63%
Document approval	2%	5%
Agenda	1%	3%
Rule, regulation, explanation	11%	21%
Request for information	3%	6%
Document transfer	0%	1%
Information note	33%	62%
Delegation	8%	15%

Figure 5 – Communication genres observed

After analysing each genre in what can be called a meeting genre repertoire, genres were linked. The Figure 6 shows the observed genres. The only exception is the lack of reference to the logistic genre. This is not explicitly reference in the figure because the genre is the same whatever the other genres.

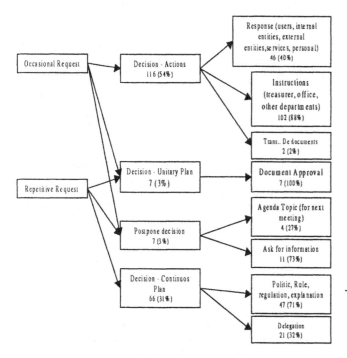

Figure 6 – Genre meeting system

This process of classification and the final results obtained showed us some characteristics of genres:

- It is possible to create a structure that is a system of meeting genres and also a repertoire of meeting genres, with a meaningful quantity of genres.
- The concept like specialisation and decomposition give an important tool for operationalisation of genre analysis.
- There is a need of a relatively small number of genres; for example, in the meeting minutes genre, a large percentage of decisions correspond to instruction given by the group to other departments, such as accounting and administrative, equipment, secretary or parking services.
- Some decisions may produce more than one genre. For instance, there are several decisions that have as consequence an instruction to a service and information to others.
- Each genre can be decomposed in sub-genres with slight differences (different receivers, different form). For instance, responses can be sent either to users, employees or external entities.
- According to the opinion of several users, the generic genres (logistics, agenda, meting and report) are insufficient to describe a meeting genre system. They suggested that the description of a meeting genre system might include other generic genre. This genre could be a "context genre" or a "document genre". This genre would include documents being analysed in the meeting session, documents used to support decisions or documents used to give context to the meeting and make decisions understandable in the future. It was also suggested that this genre could be called Context or Documents.

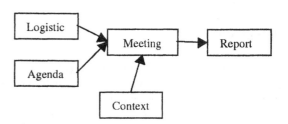

Figure 7- A "new" meeting system of genres

- Some people involved in the analysis process suggested the creation of a software that could help in the identification and documentation of communication genre and genre system. This information could be helpful in a future meeting. Consequently, this genre system could be incorporated not only in a methodology but also in an analysis and design tool.

6 Discussion

The framework that is proposed in this paper is useful because it allows meeting participants to have a general environment where everybody involved shares concepts. By definition, genres result from the interaction of the community of users and, consequently, are understood by involved people. With those genres, meeting participants do not need to identify or define specific components of genres, because they have already been identified before.

The use of genres may have a great advantage by identifying communications genres actually used for meeting dissemination, evaluate this system eventually redesign it if necessary.

By creating or making explicit a genre repertoire, the process of close up and dissemination of meeting results is significantly improved.

The use of genres may also be useful for the production of software that supports the organisational integration of meeting outcomes.

This software would have two modules: an analysis module and a management module.

The analysis module is where communication genres and genre systems used in the organisation are introduced. Its purpose it is to simplify and help the analysis process broadly illustrated in the last section. The final result would be a repertoire of meeting genre systems and communication genres.

The management module is used during the meeting process by the facilitator that introduces additional data. The system may also capture some log files produced by the EMS.

Having a repertoire of genres, the system would be capable of suggesting the more adequate or more used genre for a specific situation.

7 Conclusion

The problem analysed in this paper is the organisational integration of meeting results produced by EMS. In order to solve the problem, we use a framework where the concept of genre is a central one. In the implementation level, genres may be explored in detail, allowing the creation of templates that improve the productivity of the decision making process.

The great advantage of this framework is that it is supported in the concept of genres. In fact, because users understand genres, this approach allows an adequate process of choice during the meeting process. So, during the meeting, participants choose a concrete type of communication, defined by them, instead of discussing the philosophy of communication. On the other hand, those genres must be adequately and clearly defined, in order to correctly establish the communication with the receiver. We found out that a relatively reduced number of genres are necessary to describe meeting outputs in organisations.

This framework also proved to be useful in the analysis of lack of integration from the decisions taken in meeting sessions into the organisation. For instance, it was observed that in some situations, requests did not have responses because such response was not institutionalised for a number of situations.

The incorporation of communication genre in a software system that supports organisational integration of meeting outcomes is one opportunity being considered.

References

[1] M. Aiken, J. Carlisle, "An automated idea consolidation tool for computer supported cooperative work"; Information and Management; Vol. 23; pp. 373-382. 1992;

[2] M. Aiken, C. Govindarajulu, "Knowledge-Based Information Retrieval for Group Decision Support Systems"; Journal of Database Management; Vol. 5; No. 1 Winter. 1994

[3] M. Aiken, L Motiwalla, O. Sheng, J. Nunamaker jr, "ESP: An Expert System for Pre-Session Group Decision Support Systems Planning"; Proceedings of the Twenty-Third Hawaii International Conference on System Sciences, Hawai, January 2-5; pp279-286, 1990

[4] M. Aiken, A. Shirani, T. Singleton, "A Group Decision Support System for Multicultural and Multilingual Communication"; Decision Support Systems; Vol. 12; No. 2, September; pp. 93-96, 1994.

[5] M. Aiken, M. Vanjani "An Automated GDSS Facilitator" SWDSI 1998 Conference Dallas, Texas, .1998

[6] P. Antunes, T. Ho. "Facilitation Tool - A Tool to Assist Facilitators Managing Group Decision Support Systems". Nineth Workshop on Information Technologies and Systems (WITS '99). Charlotte, North Carolina. December 1999.

[7] P. Antunes, T. Ho, L. Carriço. "A GDSS Agenda Builder for Inexperienced Facilitators". Proceedings of the 10th EuroGDSS Workshop. Copenhagen, Denmark. June 1999.

[8] M. Bergquist, J. Ljungberg. "Genres in Action: Negotiating Genres in Practice"; Proceedings of the 32nd Hawaii International Conference on System Sciences. Maui; Hawaii; January 5-8, 1999

[9] R. Bostrom, R. Anson, V, Clawson, "Group facilitation and group support systems". Jessup and Valacich (editors), Group Support Systems: New Perspectives. Macmillan. 1993

[10] A. S Butler.; Team Think; McGraw-Hill, 1996

[11] M. D. Cohen, J.G. March, J. P.; Olsen; "Garbage Can Model of Organizational Choice"; Administrative Science Quarterly; Vol. 21; pp246-275. , 1972

[12] S. Colon, M. Aiken, B. Reithel, A. Shirani,. "A Natural Language Processing Based Group Decision Support

System"; Decision Support System; Vol.12 pp.181-188, 1994

[13] K. Crowston , M. Williams; "The Effects of Linking on Genres of Web Documents"; Proceedings of the 32nd Hawaii International Conference on System Sciences. . Maui; Hawaii; January 5-8, 1999

[14] T. H. Davemport, J. Short, "The new Industrial Engeneering: Information Technology and Business Process Redesign"; Sloan Management Review; Summer; p. 11-27, 1990

[15] G. Dickson, J. Partridge, L. Robinson, "Exploring Modules of Facilitative Support for GDSS Technology"; MIS Quarterly 17(2); pp. 173-194, 1996.

[16] J. Fjermestad, S. Hiltz, "An assessment of group support systems experimental research: Methodology and results". Journal of Management Information Systems, 15(3), 7-149, 1999

[17] C.E. Lindblom, "The Science of Muddling Through" Public Administration Review; Vol. 19 pp. 79-88, 1959

[18] T. Malone, K. Crowston, J. Lee, B. Pentland, C. Dellarocas, G. Wyner, J. Quimby, C. Osborne, A. Bernstein, "Tools for inventing organizations: toward a handbook of organizational processes"; CCS WP 198, 1997

[19] H. Minzberg, D. Raisinghani, A. Theoret, "The Structure of Unstructured decision making"; Administrative Science Quarterly; Vol. 17; pp 1-25; 1976

[20] S. Miranda, R. Bostrom, "Meeting facilitation: Process versus content interventions". Journal of Management Information Systems, 15(4), 89-114, 1999.

[21] J. Nunamaker, R. Briggs, D. Mittleman, D. Vogel, P. Balthazard, "Lessons from a dozen years of group support systems research: A discussion of lab and field findings" Journal of Management Information Systems, 13(3). 1997.

[22] W. Orlikowski, J. Yates, "Genre Repertoire: The Structuring of Communicative Practice in Organizations"; Administrative Science Quarterly 39; pp. 547-574, 1994

[23] W. Orlikowski, J. Yates, "Genre Systems: Structuring Interaction through Communicative Norms" CCS WP 205; Sloan MIT WP 4030; July, 1998

[24] G. Raikundalia, M. Rees, "Scenario of Web User Interface Tools for Electronic Meeting Document Generation and Presentation" QCHI95 Symposium; Bond University; 21 August, 1995

[25] N. Romano, J. Nunamaker, D. Roussinov, H. Chen, "Collaborative Information Retrieval Environment: Integration of Information Retrieval with Group Support Systems"; Proceedings of the 32nd Hawaii International Conference on System Sciences. . Maui; Hawaii; January 5-8, 1999.

[26] H. Simon,. Administrative behaviour: a study of decision-making processes in administrative organizations (4 th ed.). Simon & Schuster Inc, 1997

[27] The 3M Meeting Management Team Mastering Meetings. McGraw-Hill, Inc, 1994

[28] J. Yates, W. Orlikowski, "Genre of organizational communication: A structurational approach to studying communication and media"; Academy of Management Review; 17 pp- 299-326, 1992.

Control of operational industrial project needs

M. Barouh

Université de la Méditerranée, Institut Universitaire de Technologie
Avenue Gaston Berger, Aix en Provence F13647

J. M. Linares

Laboratory Mécasurf
CER ENSAM, 2 cours des arts et métiers, Aix en Provence F13617

J. M. Sprauel

Laboratory Mécasurf
CER ENSAM, 2 cours des arts et métiers, Aix en Provence F13617

Abstract

Functional group concept is the result of the systemic approach. In many situations we can model and highlight internal flows of the studied system. We can also suggest several control strategies.

In the description carried out, a 4-P type-classification is done (process, proceeding, procedure and product). This makes possible the modelling approach of any processes according to the level of desired detail. The properties of the functional group confer to it a real power of description. Modelling can be done according to a multidirectional approach. (Transversal, top-down and bottom-up).

We propose in this contribution to show how the functional group is used in the concurrent engineering application field.

We focus our presentation on this top-down and transversal approach (from the need towards the detail, a process to be implemented to reach an objective).

The topic is the project work methodology. We show a technical project work can be described according to the typology (limited to the operational system) by this modelling tool. Finally we explain that information-flows and material-flow of the industrial project require at least 4 control levels.

1   Introduction

Nowadays companies have to face up to the change, the complexity, the diversity of the market and the industrial competition. This is why they have to finely control products, process and methods. The control of time, uncertainties and complexity are the actual priorities since the Taylor's economy.

Quality concept imposes now to place the customers in the heart of the industry where the cost, quality, delay are the main performance criteria. Concurrent engineering aims to make work different services, different functions, different work groups in the same time in order to break the compartments of the system. The concept imposes a very important communication between each groups and services. The work done to produce the product is more and more constrained by the delivery time [1].

Now the product is studied for its whole lifetime, this is why methods of management as systemic methods have to be implemented.

Time to market is a second goal for which the project work methodology implements the concurrent engineering.

Principles and systemic characteristics of the project work methodology are presented. We suggest an original modelling of the project work methodology by the functional group concept [2]. This first step of study provides the specifications of needs. The transversal characteristic of the functional group concept is clearly employed in this modelling.

2   Project and project methodology

2.1   Project

Project is defined as a set of actions done to reach a goal in a specific mission clearly defined. The work project is different than the product project. The work project is named engineering project or customer project. The product project is named development project or market project. Its goal is to develop and produce a product for a specific market. The market project is specially studied in this paper.

The systemic approach describes the project as an active objet with finalities which progresses in an environment.

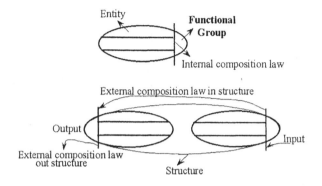

Figure 1 Functional Group

3 Functional group concept

The functional group concept is the result of the systemic approach. The method allows to describe transversal systems as well as top down systems. The systems concerned can be a product or a process. This is a set of entities that realise a function in order to satisfy a specific need. The terms function and need are defined in the general case. The functional group concept is not based on a rigid formalism or computation rules to keep its systemic characteristics. An entity that is a simple or complex element can be divided without losing its integrity. It is a whole indivisible. It can be described according to different points of view without changing its attributes.

A functional group is created by applying an internal composition law on entities. Overabundant data are firstly classified according to an internal hierarchy. That provides the internal behaviour of the functional group concerned. Also the functional group has its own characteristics that can hide the properties of few entities.

External composition law are done to associate functional group together. Each functional group is placed in a net and they have an input and an output. An element can constitute a functional group.

Several functional group can be included in a structure. In this case they are managed by an intra-structure internal composition law. This operation creates a new entity studied in its global functioning. The output flow is the result of the operation between the input flow and the functional group.

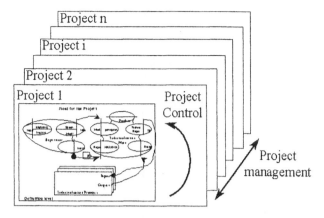

Figure 2 Management project

4 Application for the operational industrial project

4.1 The industrial process

Usually industry has to lead several project in the same time. Each project is considered as a set of subprojects in interaction. Roots of a project are a command from the principal that transcribes a real need.

The project management is a topic widely studied in the research bibliography [3]. [4] The cognitive activity network model is based on an oriented hierarchical structure. [5] [6] Its goal is to control the project planning. The figure 2 illustrates that each project is composed with several tasks. In our modelling we suggest to group all the activities realised at the same level into an only work deck. The work deck allows to satisfy the three characteristics : Time, place, action. This is favourable to implement the concurrent engineering. This method is very used in automotive industry. It is also possible to group all the work decks in a global work deck.

4.2 Definition level

Input data are issued from the direction at the higher level. The output of this level is constituted by the product. The detected or created need will be transform as a specific product. The creation of the functional requirements specifications is a very important step.

The main participants of the project take place in this phase. There are the supervisor and the responsible of industrialisation.

Now we emphasise that there is a discontinuity between the detected need and the idea that just is a transcription. The supervisor realises a first modelling. The marketing service is an entity which contributes to

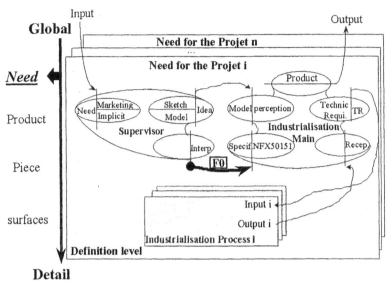

Figure 3 Project definition level

define the functional group of "Need" (FG_{need}). Drafts, drawing, notes represent the entities of the functional group of the "Idea" (FG_{idea}).

The responsible of industrialisation perceives a modelling of Idea.

The responsible carries out a functional requirements specification (FG_{requi}) in which all the functions are described by their designation, success criteria and degree of flexibility. The described loop can be employed several times before the writing of the final technical requirements ($FG_{Tech\ Req}$). This phase deals with the juridical aspects in relation with the contract negotiation [GIM96].

The supervisor creates a new technical specification of need that is a new interpretation thus is a further possible discontinuities generator. The functional group attributes represent the input flow for the design process at the second level. This defines the functional group of receiving FG_{receiv} employed by the responsible of industrialisation. The product is the output of the system. The F0 function validates the functional specifications. It constitutes a reference for the responsible of industrialisation in order to establish the technical specifications of need (FG_{TR}).

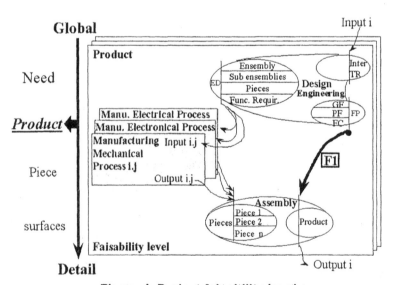

Figure 4 Project faisability level

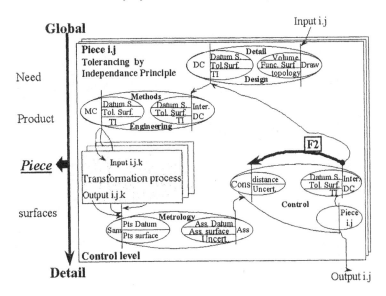

Figure 5 Project control level

4.3 Feasibility level.

The product is generated by several manufacturing processes. We focus our study on the description of the mechanical processes. The system input is the interpretation of the technical specification. The designer realises the design drawing from this information. The designer work constitutes the input of the different manufacturing processes (electronics, mechanics…).

Manufactured parts are assembled to constitute the product. The F1 function goal is to verify if technical specifications provided by the designer are respected by the functional group of the product ($FG_{product}$). Presently CAD software propose a virtual assembly module to guaranty the feasibility of this function. In the following we focus our investigations on the manufacturing mechanical process. The deck work should be composed at least with the supervisor, the industrialisation responsible and the industrialisation process. These 3 participants can introduce specialist members from their staff.

4.4 Control level

The control level can be described by two different ways (independence principle or requirement E, M, L). When the independence principle is used, the behaviour of the control level is the following (figure 5).

The functional group of the product drawing owns attributes as volumes, functional surfaces and brut surfaces. The tolerancing of the piece is done by an internal composition law (FG_{DC}).

It has to respect the functional constraints defined by the precedent work deck. The manufacturing methods technician transcribes data from design service. Data is represented by the functional group "$FG_{interDC}$". This functional group is composed by functional surfaces, reference surfaces, nominal surfaces and tolerances. This is the result of an interpretation of the experience of work. $FG_{interDC}$ provides the functional group of manufacturing condition (FG_{MC}) by using an intra-structure external composition law. This data constitutes an input for the transformation process.

After the manufacturing task the metrology service samples surfaces expressed in terms of reference axes (CMM) (FG_{sam}). An internal composition law transposes all measured points in terms of reference surfaces axes. An external composition law generates the functional group of associated surfaces (FG_{ass}). Associated reference surfaces, associated surfaces and their uncertainties are included in this functional group.

The control operation starts by a set of geometric calculations (usually distances computation). The functional group of construction (FG_{cons}) tries to respect the properties required by design tolerancing. It propagates uncertainties of the associated functional group in the geometric computation. The controller compares the FG_{cons} to FG_{DC}. Piece is right if comparisons are correct. Results are given with α is equal to 5%. The control of uncertainties is usually very difficult. This slows down the level progression.

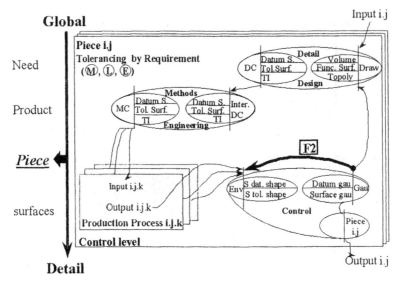

Figure 6 Project control level

The F2 control function is different while the designer uses the requirement (L, M, E). In this case (figure 6), the structures of metrology and control are grouped in the same structure. The control test is based on material gauges or virtual gauges. This operation is closer to the functional surfaces than the preceding one.

At this level (deck work) the three functions of manufacturing, design and control are present.

4.5   Operational level

The figure 7 illustrates the transformation process (figure 7) (Cutting, milling, grinding…). Interpretation of

the FG_{MC} allows to operator to use statistical process control. Also the operator studies the manufacturing process capabilities. The production structure manufactures the real surfaces according to the control board.

The metrology structure is used as a feedback to control the production. When the F3 function is satisfied, the real surfaces become the output functional group.

At this deck work the three functions of manufacturing methods, production and metrology are present. The output represented the actual surfaces of the piece.

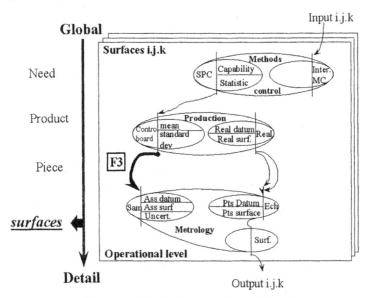

Figure 7 Project operational level

5 Conclusions

This paper brings to the fore the potentialities of the functional group concept. Our methodology is a top-down and bottom-up analysis from the creation of the project to manufacturing of the surfaces. Many investigations deal with the global aspect without taking into account the practical aspect of the work. Our study emphasises characteristics of the practical experience.

This paper focuses only on the operational part of the project. This modelling allows to control the quality of the product.

The four control levels have been presented in the four work decks. This organisation helps the concurrent engineering development.

The control functions guarantees the desired quality of the product. The overabundance of control slows down the project process.

Now the modelling of the project management will be investigated. This means to define and create the different links between all the part of the project (figure 1).

References

[1] C. Marty, J.M. Linares, Industrialisation des produits mécaniques, Tome I Conception et industrialisation, Hermes science, 1999.

[2] J.M. Linares, M. Barouh, J.M. Sprauel, "Pilotage du processus d'industrialisation d'un produit mécanique, Congrès International de Génie Mécanique, 26-28 May, Montréal (Canada), Vol.3, pp1661-1671, 1999.

[3] K.K Yang, C.C. Sum, "An evaluation of the due date, resource allocation, project release a,d activity schedulling rules in a multiproject environment", European Journal of Operational Reseach. Vol.103, n°1, pp.139-154,1997.

[4] T.Thamhain, D.Wilemon, "Building high performing enginerring projet teams", IEEE Transactions on Engineering Management, n°34, Vol.3,pp.130-137,1991.

[5] O. Grunder, P.Baptiste, O. Barakat,"An approach to model and evaluate innovation and concurrent engineering projects", Fifth international workshop on project management and scheduling, April 11-13, Poznan (Poland), Vol1,pp.94-97,1996

[6] O. Barakat, O Gunder, P. Baptiste, "Vers un multi modele projet produit process pour la description et la planification cognitive des projets d'innovation", Congrès International de Génie Mécanique, 26-28 May, Montréal (Canada), Vol.3, pp1487-1493, 1999.

Method of reorganization of SME in concurrent engineering

Pr. O. Garro, Dr. E. Ostrosi, Dr. J-P. Micaëlli

University of Technology of Belfort and Montbéliard (UTBM)
Mecatronique3M
F 90010 – Belfort Cedex
Tel: 33 3 84 58 31 36 / Fax: 33 3 84 58 31 46
olivier.garro@utbm.fr

Abstract

This paper describes a method for supporting the concurrent engineering in SME. The objectives of this work include improvement of design quality, reduction of design time, retention of designer expertise. The method is based on the action-research. It is the observer-actor of the process design, who takes information on the design activities and the product to be designed using the concept of intermediary object of design,. The matrix of analysis is used to analyse the design process. The proposed solution is a new organisation based on the Design Assistance Tool, which affect the design process. The reorganisation of design process based on this method has shown a good effectiveness in terms of objective realisation

1 Introduction

Concurrent Engineering is often seen as gathering together designers, manufacturing engineers, process monitors, marketing personnel to work in teams at the pre-design and the design stages [1]. Today indeed, the majority of the large companies are structured in concurrent engineering. On the other hand most of SME remain on sequential diagrams. The objectives of this paper consist in proposing a method of reorganisation of the SME in concurrent engineering organisations. We take the example of a work completed in a French SME. This work was completed starting from a six months study (April 99 - September 99) in the ABC company. In the beginning, the objective was to propose solutions to reduce the design delays. Quickly, work was directed towards an action of analysis. The result was a proposal for a reorganisation of the company in the form of Concurrent Engineering. The work was undertaken in the form of action-research. ABC has engaged an engineer from our university, for a six months period, to carry out this analysis. This engineer had observed and analysed the design process. In parallel, the problematic points of this

process were gradually elucidated, and solutions were proposed in order to reduce the deadlines and to improve the design process. The second section represents the method of reorganization of the SME in concurrent engineering. The third section presents the observation stage. Here, it is shown the double role of observer – actor. The intermediary objects of design are observed such as to find the best one for communication and cooperation. The fourth section analyses the product and design process. The matrix of analysis: *"Requirement x Function x Feature"* is used. The fifth section present the new propositions for the firm reorganization.

2 Method of reorganization

The method used is based on three different stages according to figure 1.

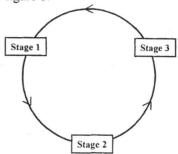

Figure 1 Reorganization Method

Stage 1: Design process observation. The actor engaged in this phase takes part in the process through a design task (for example designer, project leader, associated quality...) [2]. This point seems important to us for the following reasons. As an actor, he finds his legitimacy compared to the other actors. He will be accepted and accommodated better for his observations. Insofar as he takes part in the common objective, he will have more easily access to the information. From the moment the observer acts, he will be in relation with the dynamics of the project. Because of his action, he will be informed for all action. Furthermore, he will be able to collect information, which could escape to him. Lastly, from the moment the observer is engaged in technical task, he will necessarily be in a better position to understand the project. The observation in the ABC firm was practiced according to two points of view: the design activities point of view and the designed product point of view. The design activities are observed through the actors organization and their relations inside and outside of the firm. The product is observed through its intermediate representations.

Stage 2: Design process analysis. In this stage the product and the activities are analyzed. The origin of problems and the ways to avoid them are searched [9]. The matrix of analysis is used as diagnostic tool. The following problems are searched: first the problems related to the product design, second the problems related to the design activities and then the problems related to the couple product - activities.

Stage 3: Proposition of a new organization. From the design process analysis, a solution is proposed. This solution is double facetted. First, it is a new tool and a new intermediary object. Second, it is affects the design process organisation. The new tool and the new design process organisation are obviously strongly dependent. The new design process organization is centered on the new proposed design tool. In fact the design tool is in the same time a key actor in the design process. It has a good impact on the information flow. Pascal Laureillard shows for example how, starting from new intermediate objects, the organisation of the design in a large company evolves [3].

Loops: Implementation. The proposed solution must be implemented. The evolution of the design process is again necessary to be observed (stage 1) and analysed (stage 2). It is important to see the changes in order to detect the undesirable effects of them and if necessary to correct them (stage 3). In the study presented in this paper, the work has been limited to the first 3 stages.

3 Design process observation

Observation has focused in one hand on the product to be design and on the other hand on the different design activities.

3.1 Design product

The tools, which are used to analyse the design product are based on the concept of intermediate objects of design [4,5]. An intermediate object is an object produced during the design activities. These objects are very significant. They indicate how the various actors of the process organise themselves in order to design. From these objects, one can for example deduce the real nature of the process [Hansen 99]. The intermediate objects that have been founded are the following:

— Plans,
— CAD models realised with Pro-Engineer CAD system,
— Devise,
— Manufacturer catalogues,
— Nomenclatures,
— Calculations and formulas useful for the construction and the dimensioning of the machine.

The intermediate objects that we followed the trace are the plans, CAD models and the nomenclatures.

Plans. The plans are drawings of the product responding to customer's requirements. The plan represents the relative techniques part to the customer or to subcontractor. It is a privileged support of the communication between the customer, the ABC firm and subcontractor. It makes several returns between different actors, with modifications brought by one or the other of concerned parts. This procedure is very time consuming.

CAD models. It is considered often that the CAD system is the most suitable to the co-operative work. It represents the part model in relationship with other parts models that compose the system. It is easy to modify the CAD model and its edition on the computer network makes it accessible in real time. In the context of repetitive design tasks, parametric CAD models can be realised. The development of feature modelling strategies can contribute to give a new dimension to co-operative work. Both aspects are insufficiently developed at ABC entreprise. On the other hand, the CAD systems require a good control. The technical aspect can limit the creativity. For the innovative product, the traditional CAD system must be coupled with the other innovative tools.

Nomenclatures. ABC used a nomenclatures management system, created for its own needs, which links the nomenclatures with the references. This is used to estimate a product during the design process. A new system related to the design process must be connected and interfaced with the nomenclatures management system.

3.2 Design activities

The activities of design were observed through 3 points of views:
—The organisation of a project,
— The relations between firms,
—The role of each actor in the company,

3.2.1   Actors involved in the design process at ABC

ABC is a small company of 10 people, of which 6 work within the design office,
—Mechanical design (4 persons),
—The design of the part controls (2 persons),
—Business management, purchase and links with the suppliers (3 persons).

In fact the organisation of ABC is not also simple because certain people engaged in the technical part are engaged in the follow-up of the design, the implementation of the equipment, etc. As in any SME, the division of the labour is not thorough. By detailing the role of each various actor of the design at ABC, we have four different actors, which are:
—The mechanical design office: three people work on drawing boards, one on console CAD. Each project relates to two or three mechanical engineers: one for the overall plans and one or two for the detailed plans.
—The electrical design office works in margin of mechanical design board. It deals with the part controls machines (automation), electric device, as well as power supply of the machines.
—The direction has a role of supervisor of the projects and is implied in the negotiation of the contracts (management of business).
—The purchase service which is in relation with all the subcontractors.

The various members of the company are divided on various projects (electric, automatism, management, purchases, secretariat, and mechanics) according to their skill. Each one works its part on its side, then the team meetings are organised according to the encountered difficulties. The mechanical engineers draw only the mechanical machine elements. The electric parts are not represented on the mechanical levels. Only the supports of the electric parts are drawn. There is not an explicit design method. In addition, new technologies are not yet completely introduced, as for example advanced CAD systems.

3.2.2   Organisation of project at ABC

ABC manages two types of projects: routine project; innovative project. This second type of project is riskier. The factors of risk come from the new technologies, which can not be tested. Many problems can thus emerge and increase the costs of the study. Normally, these two types of studies must follow different processes. It is not the case at ABC. All the project follows a very general course as it is shown on figure 2.

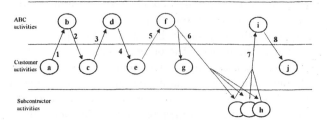

Figure 2 Development of a project

On the previous figure, all the letters represent activities and all number represent intermediary object exchanged between the different actors.

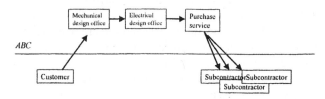

Figure 3 Activity f

The ABC' activities are resumed on the table 1.

Activities		Actors
a	Call for Tender	customer
b	Estimation	ABC : direction
c	Choice + requirement chart	Customer
d	Project	ABC : direction + mechanical design office
e	Choice + new requirement chart	Customer
f	Plans of the machine	ABC : mechanical design office + electrical design office + purchase service
g	Validation	Customer
h	Production	Subcontractors
i	Implementation	ABC : electrical design office
j	Utilisation	customer

Table 1 The different activities and their corresponding actors

The development of a project is relatively linear. ABC is directly implied in four stages (figure 2). But at the same time as ABC is the project manager, the company remains generally responsible for the manufacture part.

The two preliminary activities (**b** and **d**) correspond to the bargaining between the company and a potential customer. ABC responds to many call for tenders. Approximately 1 to 2 % of these responses are materialised by a contract. The ABC firm sells between five to ten machines per year. One of the objectives of the reorganisation of the design can be to increasing of positive responses percentage to calls for tenders. In the main design activity (*f*), three different actors work. As shown in figure 3, the work organisation of these actors is very linear. For manufacturing, ABC chooses the subcontractors, if those are not imposed by the contract. The subcontractors manufacture the machine (*h*) and, if necessary, they carry out the civil engineering. After delivery of a machine, it is proceeded to its starting-up (*i*). This phase continues with a service of after-sales.

Intermediary objects exchanged between the activities are shown on table 2.

Intermediary objects exchanged between activities	
1	Call for tenders
2	Estimation + Prefeasability
3	Requirement chart
4	Project
5	New requirement chart
6	Plans of the machine
7	Validation
8	Realisation
9	Implementation
10	Use

Table 2: Intermediary objects

4 Design process analysis

The encountered problems can be classified on the following:
1. *Problems related to the product design.* These problems concern the routine products and the innovative ones. The matrix of actions is used for the product design diagnostic.
2. *Problems related to the design activities.* The graph representation is used to analyze these problems in the case of design process.

4.1 Problems related to the product design

The problems related to the product design analysis are raised based on the features based product analysis. We view features as generic or specific forms of a part or an assembly to which engineers associate certain attributes and engineering knowledge used in the different phases from product design to their manufacturing [6]. It is believed that features implement a set of functions so as to satisfy a set of requirements. The *Requirement x Function x Feature* matrix (figure 4) is build to perform this analysis:

	Requirement	Function	Feature
Requirement	R_{11}	R_{12}	R_{13}
Function		R_{22}	R_{23}
Feature			R_{33}

Figure 4: Requirement x Function x Feature Matrix

The matrix of action shows the following relationships:
1. Requirement - Requirement Relationship, noted R_{11};
2. Requirement - Function Relationship, noted R_{12};
3. Requirements - Feature Relationship, noted R_{13};
4. Function - Function Relationship, noted R_{22};
5. Function - Feature Relationship, noted R_{23};
6. Feature - Feature relationship, noted R_{33};

Requirement–Requirement Relationship. A product (part or an assembly) is the result of design requirements' satisfaction. Design requirements are multilevel demands and wishes. In the context of the ABC' activities, the space of the requirements is considered as limited. It is not unusual that some requirements could have an implicit or explicit relationship between them, or a requirement can involve a set of sub requirements. The requirement-requirement relationship decomposition permits to cluster into groups the requirement relationships, which can be related.

Requirement-Function Relationship. In the first phase of the process design, the designer needs to transform the product requirements into functions. This transformation can be represented by the Requirement – Function Relationship. Considering the case of ABC firm, the space of requirements is limited, so it is possible to construct this relationship. For example, in the design activity, in the case of a shaft, the requirements were expressed in the following forms: "transfer the force X". For that requirement, the designers used the following functions:

Case 1: shaft positioning and shaft fixing,
Case 2: positioning-fixing.

In the first case, the designer considers two requirements to be satisfied, in the second case, he considers "one" requirement. A requirement expressed as "positioning-fixing" has often produced a "fixing" solution rather then a "positioning - fixing" solution. The grouping of Requirement - Function on the same class can permit to avoid the previous errors.

Function-Features Relationship. A part or an assembly is considered as a set of features. Each feature is believed to implement a specific or some specific functions. The function – features mapping is done using the help of an expert. On one hand, for one function, some concurrent solution is found, on the other, for some functions a feature or some features can be implemented to satisfy the function [7].

From this analysis some problems are raised:

1- The actual solutions are not always optimal in terms of the function – features relation implementation. The principal reason is that the space of function – features relationship is not completely examined;

2- Some functions can be implemented by some kind of features satisfying better the manufacturing process requirements;

Function-Function Relationship. Functions are implemented by features, which are related to form the part as a whole. It is clear, the part is represented as interrelated features, which must satisfy interrelated functions. From the function-function relationship, the following conclusion are drawn:

1- Two or more functions can generate one or more supplier functions. For example, additional geometric tolerances can be generated to satisfy some functions' cohabitation.

2- There is conflict between functions. The features based solutions are not relevant.

Feature–Feature Relationship. This relationship permits to consider the similarities between the different products. In fact, each product is decomposed on a set of relevant connected features. The features net can represent this decomposition. The comparison between the parts nets has inferred:

1- There are some class characterized by a strong product nets similarities;

2- There are some exceptional product nets.

The first inferred conclusion shows the products (parts and assemblies) are similar in the term of features. This kind of products represents similarities in the term of the previous described relationships. These are called the routine products. The second inferred conclusion shows the products are exceptional. They are called innovative products.

Requirements-Features Relationship. This relationship can be inferred from the **Requirement–Function Relationship** and **Function–Features Relationship** by transitivity. This relationship permits to give a rapid idea on the product solution based on the requirements.

4.3   Problems related to the design activities

The analyze of the design activities presented in the figure2 highlight two main problems:

First, there is no links between activities (*b*), (*d*) and (*f*) which are specific activities within ABC firm. The realized work in the activity (*b*) is not used in activities (*d*) and (*f*). In the same way, the realized work in activity (*d*) is not used in activity (*f*). The first consequence is a knowledge impoverishment on the product design. The second consequence is an important lost of project time development.

Second, the peculiar activity (*f*) is very linear as shown in the figure 3. By knowing that it is the most important activity in the project development, it is obvious that the project delay development can be reduce by the concurrent work of each actor.

5   Proposition of a new solution

From the precedent analysis two interrelated propositions are done. The first proposition is the development of the Design Assistance Tool. The second proposition, related to the first one, is the new organisation of design process.

5.1   Design Assistance Tool.

The objective of this stage is to study the feasibility of certain design tasks automation. For that, the principles of a Design Assistance Tool are defined. A Design Assistance Tool defines a set of data, algorithms, procedures and interfaces coupled to a CAD systems in order to assist design process. According to the conclusions of design process analysis, the development of the Design Assistance Tool is addressed in priority to the routine product design.

Global Architecture of the Design Assistance Tool. The tool receives input data (A) and provides the output results (D). There is an interaction with the actors taking part in the design. The Design Assistance Tool permits to exchange the information between the actors and its modules. The actors can ask for the information (C), then the Design Assistance Tool can propose the solutions, or propose an interactive procedure to modify this solution or to establish a new one (B).

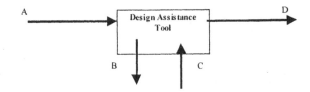

Figure 5 DAT Global Architecture

This architecture supposes the interaction between the different actors. In fact, considering the triple

Requirement x Function x Feature, the Design Assistance Tool can propose some solutions, for example by identifying the client requirements. The modifications on the requirements has a good impact on solutions, consequently the Design Assistance Tool represent an open environment for the consensual solution finding between the customer, the firm and the subcontractors. This permits to reduce the time to market.

Local Architecture of the Design Assistance Tool. The Design Assistance Tool is composed by a set interrelated modules which supports the design process. The architecture of each module can be given on the following figure.

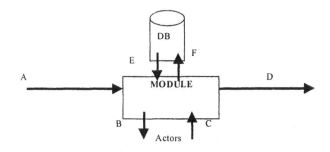

Figure 6 DAT Local Architecture

The input (A) for a module are data provided from the different actors or from the other modules. The module provides the outputs (B) which can be intermediates results or the final results. The module informs the actor on the inferred result (B) and it can take the orders (C) to transmit or not the result to the modules or to the other actors. Moreover, the enrichment of database is considered for the unexpected cases.

Modules actions. The process design requires a certain number of actions are to be realised [8]. These actions can be seen through exchanges between the Design Assistance Tool and its environment. The following actions are considered:

- Exterior data using;
- Exterior data validation;
- Project progress report visualisation;
- Details project visualisation;
- Results project validation related to a project phase;
- Solution propositions;
- Solution choosing and validation;
- Multilevels (customer, manufacturing,…etc) information providing;
- Proposition of parametric solutions;
- Product based components visualisation;
- New information providing (plans, heuristic, new features, etc.).

Design process with the Design Assistance Tool. The design process with the Design Assistance Tool can be divided in the following phases:

Phase 1: Preliminary solutions. The goal of this phase is the preliminary solution configuration, which must satisfy the requirements of customer. The transformation of requirements into features (Requirement-Function, Function-Feature and Feature-Feature modules) permits to give a preliminary solution or so called the minimum technological solution. The solution is requirement sensitive, that is, it can change by adding and removing the requirements. In the case of alternative features, the convergence on a minimum technological solution can be done by interactively features selection. This is the pre feasibility solution.

Phase 2: Final Solution. The goal of this phase is to refine the preliminary solution. The final features selection must satisfy the compatibility between features. The compatibility between functions is checked too. The geometric constraints related to a feature or the dimension between features are defined by replacing the parametric values. The features modifications such as they can satisfy the subcontractor requirements (manufacturing, assembling…etc) is another action in final solution searching.

5.2 New organisation in concurrent engineering.

The implementation of Design Assistance Tool has affected the organisation of design process in ABC (figure 7). Three design cases are distinguished:

Routine design. In this case, the Design Assistance Tool can almost as completely respond to the customer or to the subcontractor requirements. The Design Assistance Tool plays a double role: it is a tool and in the same times an actor in the design process. It is the protocol of Design Assistance Tool, which permits co-operation to the actors. It is important to underline that an actor must be specialised in communication with the Design Assistance Tool, and consequently he will be the actor specialised in the routine design. Its role is to follow the routine design project, to ensure the communication between actors through the Design Assistance Tool, to check the errors and to propose the intervention for Design Assistance Tool improvement.

Semi routine design. In this case, the Design Assistance Tool can not completely respond to the customer or to subcontractor requirements. This case must be underlined and must be better understood because a bad interpretation of Design Assistance Tool capacities can produce its non-usability. The solution represented by Design Assistance Tool is not completely or it could have the errors. It is a solution where the routine and the innovative design must cohabit. In term of the actors, here the co-operation between the specialised routine design actor and innovative design actors is necessary. The capitalisation of the new solution for the re-use must be considered.

Innovative design. In this case, the Design Assistance Tool can not respond to the customer or to subcontractor requirements. It is the case of the innovative product. The project management must satisfy the criterions of an innovative product. The innovative design process must be treated separately from the routine design process because it demands the innovative method and tools. However, the Design Assistance Tool can be consulted for particular aspects, too. In the end, the innovative product must be transformed in a routine product. So the co-operation between the actors of the routine design and innovative design is necessary.

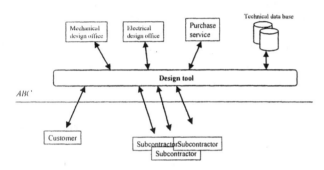

Figure 7 CE organization

6 Summary and Conclusions

A five stages method for concurrent engineering in SME is presented. The first stage concerns the observation phase. The observer – actor in the design process is used. The second stage realises the design process analysis. The requirement x functions x features matrix analysis is used. From that analysis two interrelated propositions are done in the stage three:

1- The first proposition is the development of the Design Assistance Tool.

2- The second proposition, related to the first one, is the new organisation of design process.

The Design Assistance Tool is double facetted: it is a tool and in the same time an actor in the design process. Three cases have been distinguished. The first case is the routine design, where the Design Assistance Tool gives the almost as completely solutions. The second case is the semi routine design, characterised by a not completely solution given by the Design assistance Tool. The third case is the innovative design process.

The results of the method show it is possible to execute a great number of orders in shorten delivery times. Before the method application the design times were 9 months, today it is reduced in less than 2 months. The errors of the design are reduced. The knowledge capitalisation and the new competencies acquisition are some other estimated profits. However, the method implementation requires the investment on CAD systems with specific tools such as the Design Assistance Tool. Finally, the reorganisation of the process design must be followed in order to correct the errors during the implementation phase.

References

1. Blanco E., Garro O., Brissaud D., Jeantet A. Intermediary object in the context of distributed design, CESA IEE-SMC, Lille 96.
2. Hess D. The new ethnography and anthropology of science and technology in knowledge and society : the antropology of science and technology, London JAI press 92.
3. Laureillard P., Boujut J.F., Jeantet A. Conception intégrée et entités de coopération, *in* Les objets de conception, Paris, ed. Hermes 97
4. Vink D., Jeantet A. Mediating and commissioning objects in the sociotechnical process of product design : a conceptual approach, pp 111-129 in management and new technology, Cost A3, Vol 2, Bruxelle 95
5. E. Blanco, *L'émergence du produit dans la conception distribuée*, phD Thesis, University of Grenoble, 1998.
6. E. Ostrosi, A. Coulibaly, B. Mutel, Ph. Lutz, *A Plex Grammars Approach for manufacturing Features Recognition*, International Journal of Robotics and Automation, Vol. 13, Nr. 3, pp. 33-42.
7. A.Coulibaly, E. Ostrosi, B. Mutel, *An Integrated Mechanical Design Approach based on the product FST Specifications Models*, Proceedings of 2nd International Conference on Integrated Design and Manufacturing in Mechanical Engineering, IDMME'98 PRIMECA, May 27-29, Compiegne, France, 1998
8. O. Garro, I. Salaü, P. Martin. Distributed Design Theory and Methodology. Int. J. of Concurrent Engineering: research and applications, vol 3, number 1 March 95, pp 43-54.
9. Hansen C. T. Identification of design work patterns by retrospective analysis of work sheets, ICED99, Munich 1999.

CHAPTER 5

Collaborative Decision Making

A Total Design Process Framework & Knowledge Management Methodology for an Engineering Product Design Process

I.R. Reid

School of Engineering, Sheffield Hallam University, Sheffield, S1 1WB, UK
Tel: (44)(0114) 225 3091, Email: I.Reid@shu.ac.uk

C. Pickford

School of Engineering, Sheffield Hallam University, Sheffield, S1 1WB, UK
Tel: (44)(0114) 225 3403, Email: C.Pickford@shu.ac.uk

Abstract

Knowledge Management (KM) is increasingly becoming a source of improving competitive advantage. This paper will consider the development of Knowledge Transfer (KT) in the context of the engineering design activity, focusing on Make-to-Order (MTO) product markets. The 'Action Research' study was carried out in UK based multinational company. The study entailed looking at the design process of their core product, which took place in April 1999. In conjunction with this action research a mailed questionnaire was also distributed to similar engineered product manufacturers, so that comparisons could be draw . In conducting these research activities a number of issues during the design process were examined including the tools and techniques used. Following on from this the paper will discuss the potential of a conceptual KM framework to support the design process.

1 Introduction

The success of management to respond to increased global competition and their ability to recognise the value of internal and external information has led to the knowledge management process becoming a key activity in sustaining corporate growth [1]. A critical success factor in today's market is the firm's ability to understand the customers' needs and to provide a satisfying and preferably customised solution to every customer in a way that exceeds the competition. This goal, although crucial for the long–term survival of the firm can create conflicting objectives. Even when the performance of departments is optimised independently to achieve their respective best operating level, it may lead to sub-optimal performance of the firm as a whole [2].

It is widely accepted that the interface between Marketing and Design has a direct impact on the customer and can significantly influence the firm's competitiveness in the market place [3][4]. Clearly organisations will need new practices, methods and tools in order to live up to new requirements [5].

The paper will look at Knowledge Management (KM) within the process of New Product Design (NPD). The paper will focus on the case study that was carried out in a company that produces centrifugal pumps for the oil and gas industry. In carrying out this Action Research a number of issues within the company which effect NPD were examined. These included supply chain, market analysis, product modularisation, knowledge transfer and the tools and techniques used.

2 Methodology for Investigation

2.1 Questionnaire

At the early stage of the research, a questionnaire was mailed to 150 UK based multinational companies to gauge the extent that best working practices within the design activity have been adopted. 23 companies responded making the take up rate 15.3%. The questionnaire was targeted towards a selection of companies working within the market sectors of multi-functional products i.e. gas turbines, locomotives, power generators etc.

Specifically the survey aimed to discover the methods and techniques being used to support the transfer of data and information from the customer to the finished product. The study also considered the operational environment where these methods are being applied e.g. Concurrent Engineering (CE).

Accordingly an element of the survey questioned the organisation of the technical departments from which the designs were produced, and considered the design system. For the purpose of the questionnaire the design activity was considered to follow the structure illustrated in fig 1.

Figure 1.   Product Design Activity

2.2 Action Research

The action research provided a unique perspective into all of the Organisation: Design, Marketing, Technology and Manufacturing for company X. Certainly the case of company X cannot be generalised to all MTO firms however there will be parallels. Of particular interest was the use of soft technologies such as: organisational structure and communication channels, as well as the more traditional analysis of design tools.

There were four principle methods of analysis utilised:
- Structured interviews with key people within the firm.
- Data collection from company information sources.
- Process mapping with 3 members of the organisation.
- Collection of benchmarking information from industry, and case material.

The research has enabled a detailed theoretical model of the design activity to be constructed. The success of this activity was underpinned by the accumulation of both qualitative and quantitative data from Company X , see fig 3.

This data can be broken down into four areas:
- **Awareness** (document any existing work related to the theory of product and process design)
- **Usage** (study reported design projects with reference to MTO product creation)
- **Experiences and outcomes** (study reported engineering design projects with reference to customised engineered products)
- **Implementation** (study the methods of Action Research and case material for executing change within the design process)

3   FINDINGS

3.1   Questionnaire Analysis

3.2   Product Design Activity:

It became apparent from the survey that 43% of the respondents were using dedicated teams and 57% part time teams. In these companies it was discovered that dedicated teams were used when urgent projects had to be rushed through and the drain on human resources within other areas could be dealt with. These companies reported dedicated teams worked very well for this type of project but the managerial and resource problems prevented them from being used continuously. The survey was unable to gain on exact measurement of how much quicker a dedicated team was compared to a part time team. However the focus towards dedicated design teams is encouraged throughout the NPD process.

3.3   Design Tools

After the management environment in which design process was investigated the questionnaire went on to evaluate the tools used in this process. These can be split into three groups:

- 17% of the respondents used 'Human Resource' based e.g. Brainstorming and Taguchi methods.

- 83% of the respondents used 'Computer' based tools e.g. CAD and CADCAM.

- 78% of the respondents used combined 'human resource' based and 'computer' based tools

3.4   Design systems and Design Tool Awareness

The next part of the survey looked at the current working practices within the design process and focused on New Product Design (NPD) see Fig. 2.

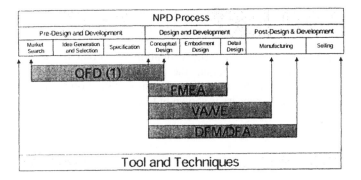

Figure 2. New Product Design (NPD) Process

Of the respondents, 60% used the following techniques, quality function deployment (QFD), failure mode and effect analysis (FMEA), design for manufacture and assembly (DFMA), and value analysis (VA). These techniques were considered effective in identifying opportunities to commence engineering design only, or avoiding unnecessary design change [6]. For example, QFD helped to translate customer requirements into engineering requirements, and to set the common objectives which different business units must strive to achieve. DFMA helped to locate the possible sources of difficulties to downstream activities at the early stages of product design.

1. Quality Function Deployment
Surprisingly this technique was only used by 20% of respondents. From the analysis of the raw data it became apparent that most of the companies had heard of the technique but found it too complex to use as their expertise was limited. Those who did use the technique found it very effective especially in a team-based environment.

2. Failure Mode and Effects Analysis
This technique was used by 17% of the respondents. A selection of the respondents felt that FMEA was effective and it helped identify problem areas within the product or process function. Unlike DFMA it was a very well understood technique which was well practiced in the transport and auto industries. It was also observed that this technique was most effective especially if used in a team-based environment.

3. DFM/DFMA
The technique was used by 33% of the companies surveyed and was of the opinion by those who used it to be a valuable tool.

4. Value Analysis Value Engineering
This technique was used by 23% of the companies surveyed. This is the general method used for evaluating functional costs.

5. OTHER Techniques
Of the respondents, only 7% used other techniques that supported the design process. These techniques included:

- Brain Storming
- Benchmarking
- Simulation
- Rapid Prototyping

3.5 Questionnaire Synopsis

Over 80 per cent of the respondents agreed that design changes should be encouraged as early as possible; changes at the early stages of design and development cost considerably less than after full-scale production.

The questionnaire also highlighted the factors that influence the design process, also supported by the literature survey. The two most significant barriers highlighted by respondents (over 80 per cent) were "poor communications", and "problems are discovered too late, resulting in panics and leading to quick fix solutions". Poor communication among the parties involved in product development was seen to be a major cause for design change. Once again, "the later the stage for the problem to be identified, the more expensive and time consuming to solve it" had been highlighted.

3.6 Action Research

3.7 Characteristics of Make-to-Order (MTO)

Identification of the products functional requirements and subsequent co-ordination with the customer needs contributes to its overall success. Traditionally this has been done on a product by product basis [5]. This is a typical of MTO. MTO firms have few standard products, and volatile difficult-to-predict demand. A product order usually stems from a successful bid, which represents the MTO firm's terms (price and lead-time) to satisfy a prospective customer's stated requirements. The customer may decide to accept, reject or modify these terms [7][8]. Many of the operational decisions rely on the intellectual capital of the decision-making process. To fully understand this complex environment action research was the chosen methodology.

3.8   Company X

Company X is based in the UK, it is the principle plant for the groups' core pump systems product range. The company is a market leader, has an excellent reputation, and can be considered to be successful when compared to its sister companies and competitors.   One of the companies strengths is its readiness to review its' operations and receive external inputs, hence its' involvement with this research.

The pump systems have evolved over the last 15 years, although excellent from an engineering perspective they are not yet fully optimised for competitiveness. These factors have increased the need to review NPD to maximise competitiveness, and improve the organisation's sustainability within the marketplace. Company X's business process for pumping systems is:

'market–sell–engineer–make–install–maintain/operate'

This model is quite different to the traditional 'standard' consumer product design sequence of: 'market-design–manufacture–make-sell'. Therefore for Company X the following criteria are in play [4].

- They have one opportunity to optimise the design.
- They are therefore compelled to over engineer.
- They need to establish exact costings very early in the process.

Following a two month review of the design process, a post project exercise was conducted, this was designed to capture significant aspects of the decision making process when developing products. This would either identify aspects of best practice or to highlight new opportunities for improvement within the design process. The Action Research methodology used has resulted in a model based framework of company X's existing design process for pre-manufacturing. The framework presents current design process within company X and describes the main constraints and complexities of the engineering design process in MTO product environment studied.

3.9   Pre-Manufacturing Process

Customer requirements are determined and analysed by the sales engineer to establish which technical and commercial solutions may meet the customer's needs. The tendering process may be repeated a number of times until the customer requirements are satisfied. Significantly it is only when the customer decides to turn the proposal into a

official contract order, that technical and commercial data is verified by the engineering department for any inconsistencies regarding the completeness and feasibility of the order. Only when the validation is correct is the complete part list generated, the production plan is then fixed in detail and the manufacturing production process is started [9]. This process is shown in Fig. 3

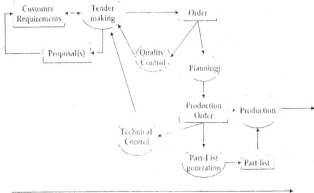

Tendering Process Time 7-8 weeks

Figure 3. The Pre-Manufacturing Process

3.10 Characteristics of the Pre-Manufacturing Process

1. Sequential Process.
There was a general acknowledgement that the pre-design and development stage was a linear process. This was recognised to be the result of the sales department being preoccupied with customer negotiations rather than interacting with the other business functions involved with satisfying customer requirements.

2. Incomplete knowledge of customer requirements.
This was evidenced by frequent modifications to the product throughout the business process. The causal factors were the use of incomplete or inappropriate historical data, and a sub-optimal software tendering package.

3. Functional Organisational Structure.
Clearly this very traditional approach to the provision of products and services from receipt of order leads to the well documented problems of lack of ownership, fragmented understanding of customer requirements throughout the organisation, long lead times and little evidence of cross functional team working.

4. Product starts the manufacturing process prior to completion of engineering stage.
The need for this is clearly to attempt to satisfy customer delivery dates (previously negotiated by the sales

engineer). The significance of this in this environment is that, due to the sequential nature of the whole process, manufacturing is more prone to making mistakes and incurring unnecessary costs.

5. Limited use of Design Tools other than CAD/FEA i.e. 'hard technologies'.

Although there was an awareness in the organisation of other tools and techniques i.e. QFD, FMEA, DFM, CE, there was no evidence of a deep understanding of these techniques or their application [10].

3.8 Recommendations:

Design is a translation of information and knowledge about customer requirements into products and services. The information flow usually begins with a loosely defined requirement and should end with an exact specification for a product and associated process that exceeds customer requirements. Information about developing the product is generally not archived. This information can be viewed as a result of a series of engineering and business decisions, each of which is one step in the transformation of the initial requirement into an exact specification. Each decision is based upon previously developed information and knowledge of requirements. Each decision has an impact on the final product specification; accordingly a **knowledge transfer support system** will enhance the process of NPD. Specifically the following recommendations have been made to company X:

- Review lost/won contracts to establish failure/success criteria and categorise the criteria identifying whether the contract was lost/won for performance or process reasons.
- Benchmark the design process.
- Develop a comprehensive information and knowledge transfer framework.
- Investigate restructuring the organisation to implement a more concurrent process.
- Review the appropriateness of a wide range of design tools.

4 Discussion

The results of our study are supported by the recent literature in areas of NPD, market-design interface and Engineered Product Design (EPD). Insufficient product and process knowledge reduces the product's overall performance within highly competitive markets. Individual organisations focus on different areas to conduct their business. This reflects their strengths, the nature of their

business and the inclinations of their personnel. In practice most organisations pursue one or more of the following Knowledge Management (KM) strategies such as creation, capture, transformation, and use. The managerial challenge is therefore to improve the processes of knowledge acquisition, integration, and utilisation, but any improvements must stem from the organisation's ability to capture and harness knowledge [11].

An increasing number of researchers and commentators have been recently tuning their attention to KM and particular the role of KM and innovation [12]. The focus of future will be the integration of the business processes within the design activity via a KM system.

4.1 Knowledge Transfer (KT) Infrastructure

The requirement from the business process to the design process is the elimination of all unnecessary information that may result in the design process losing focus, this requires the creation of a dynamic information system [13]. Infrastructure issues at the business process level should be identified. The responsiveness of the process lies in the integration of three major resources i.e. organisations with a networking management structure, people with special skills and knowledge, and flexible and intelligent technologies into a co-ordinated independent and synergistic system [14].

- Tacit knowledge
- Dynamics, networking
- Values and trust
- Motivation and commitment
- Rewarding knowledge sharing
- Focus on processes

4.2 KT Process within NPD

The paper has already highlighted that NPD is a highly complex process, involving the assimilation and co-ordination of a great wealth of data. One legitimate explanation for the high failure-rates in NPD is the ineffective exploitation of expert knowledge, both internally and externally. From an internal perspective pressure to embrace vast quantities of knowledge within an ever-shortening time frame entails few opportunities for organisations to adequately transfer knowledge. Figure 4 illustrates the relationship between the higher-level business process and the lower-level design process. Linking these activities, is the horizontal and vertical transfer of information, knowledge and know-how that are required to improve the decision making process encompassed within NPD.

4.3 A Knowledge Management (KM) Strategy

KM has strategic and operational perspectives. At an operational level KM is more detailed and focuses on facilitating and managing knowledge related activities, Its function is to plan, implement, operate and monitor all the knowledge-related activities and programmes required for effective Intellectual Capital Management (ICM). The most frequently used differentiation of Knowledge is into tacit and explicit knowledge [15].

The questionnaire and action research indicates that there is a proliferation of design tools. An obvious concern is the low utilisation and understanding of existing tools and techniques. This continuing programme of research will focus upon the design activity in MTO, organisation and KM methodology to support this process, see Fig 4.

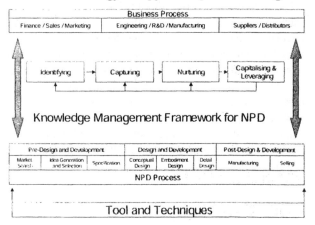

Figure 4 Knowledge Management (KM) Framework

- **Identifying Knowledge**: focusing on the grouping and identification of the inputs from the external environment (customer's specification, supplier chain) into categories of importance and relevancy to NPD. The 'Know-What' of NPD.

- **Capturing Knowledge**: Focuses on the codification the information and data (explicit knowledge) during the commencement of NPD design process.

- **Nurturing Knowledge**: Focuses on the reliance of the individual's mindset and 'know-how' (tacit knowledge) to support the decision making process and combing the reuse of existing knowledge to optimise the knowledge transfer process. The 'Know-How' of NPD.

- **Capitalising and Leveraging Knowledge**: focuses preventing the disappearance of this new knowledge and transfer the knowledge across functions via training programmes, tutorials, KBSs, and IT networks.

4.4 The KM Approach to managing the business process of NPD

The literature, questionnaire and action research highlighted the limited transfer of knowledge during the design process. Clearly successful product design is a vulnerable process, knowledge transfer creates the opportunity to increase the value of the knowledge through wider application. How much value depends upon how well the knowledge is transferred and maintained i.e. communication, storage and updating? To optimise the opportunities for successful knowledge transfer within the design process changes to the organisation structure and culture will probably be required. The KM framework activities must also fit the IDEF0 model, see Fig.5

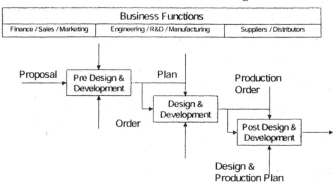

Figure 5. IDEF0 Model for NPD

4.5 Competitive Pressures

The framework will have to accommodate the rapid introduction of new products into changing markets with different order winning criteria. Cost, delivery and quality will still be major factor in winning orders, however ensuring that the customer unique 'added value' order winning criteria are also satisfied and surpassed is crucial.

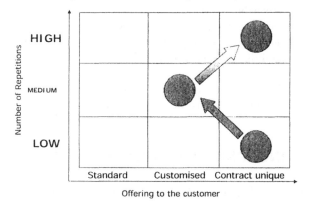

Competitive Pressures are forcing organisations to change their position

Figure 6.   Competitive Pressures Business Process (modified model)

Ingersoll Engineers 1993 [4].

These new goals are illustrated in fig 6, MTO firms' have already moved from 'unique-low repetition' to 'customised-medium repetition'. The objective for such organisations is to achieve increasing repetitions whilst still providing unique solutions to customers. It would seem unlikely that current design mechanisms with their focus upon explicit knowledge and specification driven information systems will achieve this.

5   Results

5.1   Conceptual Infrastructure Model for KT

Management theory has accepted that hidden assets knowledge of employees, customers and suppliers play a vital role for the sustainability of the company [15].

The framework illustrated in Fig 7 recognises that the development of an information and Knowledge Transfer Support System requires a multi-faceted, integrated solution. It is not solely dependent upon I.T. nor can it be addressed by simply re-structuring the organisation or investing in people. All three aspects of the model, Technology, Organisation and People, need to be included in the development of the system.

- **Technology**: consists of explicit knowledge (information) that supports the product development activity, data, documentation, etc.
- **Organisation**: consists of the organisation structure, functions and culture.
- **People**: consists of the capability of the employees as well as values, attitudes and beliefs (tacit knowledge).

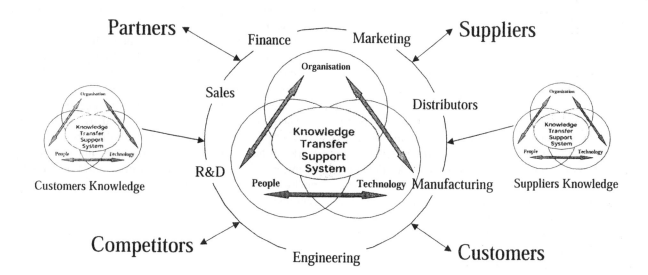

Figure 7. Infrastructure Model for KT

Current research has already identified some of the characteristics of a knowledge management support system. These can be summarised as follows:

Technology Characteristics
- Robust
- Accessible
- Friendly
- Temporal
- Indicate relevance or weight
- Compatible
- Flexible

Organisation Characteristics
- Inclusive i.e. suppliers, customers experts and partners.
- Culture of sharing rather than retaining knowledge.
- Reward system that recognises a wide range of contributions, rather than sole performance of individual, within well defined work assignments
- Responsive
- Encourage collaboration
- Values team work
- Rewards innovation

People characteristics
- Capacity and motivation to learn
- Highly developed interpersonal skills
- Unique expertise
- Shared values
- Ownership

5.2 Application of the infrastructure model for KT to the NPD of M.T.O companies

The complexity of the KM process would appear to almost overwhelming for many organisations, perhaps even more so for those operating in mature markets with a history of organisational functionality and task orientated employees. However our research into the design process of MTO firms, although observational and therefore by its nature at the stage generalised, leads us to believe that such organisations have some 'natural assets' that will facilitate the development of a knowledge management support system. These can be summarised as follows:

1. They are highly conversant with technology and are able to evaluate technological solutions to complex problems
2. They recognise the abundance of tacit knowledge within their broader organisation (supply chain, customer chain). Whilst simultaneously recognising that they do not fully harness this knowledge to provide competitive leverage.
3. They have experts with unique knowledge.
4. They have isolated pockets of good practice in multi-disciplinary team working

Equally there are significant gaps in their practice particularly in the areas of organisational culture and values.

6 CONCLUSIONS & FURTHER WORK

The initial findings of the first qualitative study stage of this work i.e. the study of the design process within company X and the initial sampling of a broader range of organisations is complete. The initial conclusions indicate that the design process of MTO firms' is sequential, has incomplete knowledge of customer requirements, a functional structure, commits to products and costs at an early stage and utilises few of the wide range of design tools and methodologies available.

An examination of current generic design framework within company X revealed weaknesses in certain areas, such as sequencing, monitoring, controlling, and displaying the process.

To create a design process based upon the integration of natural assets and technology within the organisation via a KT support system will require the adoption of a human-centred approach to the design process, rather than focus upon 'hard' technologies.

At this moment in time there has been little contemporary research within this field. Most of the current research has been directed towards to the further development to design tools rather than methodologies for the implementation of successful engineering design.

There are a number of issues raised by this research which have given the need for a further programme of work within company X to develop the Knowledge Transfer Support System within MTO product environments. The findings of our second study will be reported subsequently.

References

[1] Gilbert .M and Cordey-Hayes .M, 'Understanding the process of Knowledge Transfer to achieve successful technological innovation'. Technovision; Vol 16, No 6, pp. 301-316.

[2] Lancaster G, 'Marketing and engineering revisited' Journal of Business & Industrial Marketing; 10:1; pp. 6-15; ISSN: 0885-8624, 1995

[3] Mukhopadhyay S.K. Gupta A.V. 'Interfaces for resolving marketing, manufacturing and design conflicts', European Journal of Marketing, Vol. 32 No1/2, pp.101-124, 1998.

[4] Hill A, 'Competitiveness and processes: integrating engineering in the capital goods industry' World Class Design to Manufacture; 02: 5 1995; pp. 27-31, ISSN: 1352-3074.

[5] Zamirowski E.J. and Otto K.V. 'Identifying product portfolio architecture, modularityusing function and variety heuristics'. Proceedings of the 11th International Conference on Design Theory and Methodology, 1999 ASME Design Engineering Technical Conference, 1999.

[6] G.Q. Huang, Design for X, Chapman & Hall, 1996

[7] Easton F. 'Pricing and lead time decisions for make-to-order firms with contingent orders', European Journal of Operational Research, No11 6 pp. 305-318 1999.

[8] Hendry et al 'Production planning systems and their applicability to make-to-order companies', European Journal of Operational Research. No 40 1989, pp. 60-71.

[9] Vanwelkenhuysen J, 'The tender support system', Knowledge-Based Systems, Vol 11, Issue 5-6, pp 363-372, 1998.

[10] Spring et al 'The use of quality tools and techniques in product introduction an assessment methodology', The TQM magazine Vol 10 No 1 1998, pp. 45-50.

[11] Jordon J., Jones P. 'Assessing your Company's Knowledge Management Style' Long Range Planning Journal, Vol.30 No 3, pp.392-398, 1998.

[12] Wiig K M, 'Integrating Intellectual Capital and Knowledge Management', Long Range Planning Journal, Vol 30, No 3, pp 399-405, 1997.

[13] Court A. 'Issues for integrating knowledge in new product development: reflections from an empirical study'. Knowledge Based Systems, No 11 1998 pp. 391-398.

[14] Haken H. 'Advanced Synergetics', Springer-Verlag, New York/Berlin, 1983.

[15] Coombs R, Hill R, 'Knowledge Management Practices and path dependency in innovation', Research Policy Journal No. 27 pp.237-253, 1998

A Collaborative Approach to Path Guidance System

Teruaki Ito

Department of Mechanical Engineering, University of Tokushima, Tokushima 770-8506, Japan

S.M. Mousavi Jahan Abadi

Graduate School of Engineering, University of Tokushima, Tokushima 770-8506, Japan

Abstract

Today's Automatic-Guided-Vehicles (AGV) are used to transport materials, work-in-process, and finished goods in factories and warehouses. A Path Guidance System (PGS) is necessary to guide AGVs to find their fastest paths toward their destinations. At present, there are two mechanisms for implementation of PGS: Centralized PGS and In-Vehicle PGS. This paper proposes third mechanism: Collaborative PGS, which is according to the concept of Multi-Agent System (MAS). In collaborative PGS, AGVs are guided by the Intersection Agents (IA), which are standing on each intersection. IAs figure out the fastest path for each AGV through the collaboration with each other. For evaluation of the suggested PGS, three application models with different assumptions are introduced and the methods of implementing of collaborative PGS in them are elaborated. The first application model is static travel time model with no limitation of roads capacity. The second application model is the extension of the first model with dynamic structure of roads. The third application model is static travel time model with limitation for roads capacity.

1 Introduction

Today's Automatic-Guided-Vehicles (AGV) are used to transport materials, work-in-process, and finished goods in factories and warehouses. For transportation of loads, AGVs have to move in the paths which lead them to their destinations within minimum time. These kinds of paths are called the fastest paths.

For determination of fastest path, AGVs need a Path Guidance System (PGS). At present, there are two mechanisms for implementation of PGS: Centralized PGS and In-Vehicle PGS. In the centralized PGS, AGVs are guided centrally. This center keeps the data of streets and traffic situation of AGVs in real time, and with processing the data, it guides the AGVs to their fastest paths. In the in-vehicle PGS, all AGVs are autonomous and they choose their paths toward their destinations by themselves [1].

This paper proposes third mechanism: Collaborative PGS, which is according to the concept of Multi-Agent System (MAS). In collaborative PGS, AGVs are guided by the Intersection Agents (IA), which are standing on each intersection. IAs figure out the fastest path for each AGV through the collaboration with each other. The collaboration mechanisms of IAs for defining the fastest paths of AGVs are elaborated in this paper.

In the section 2, path guidance problem is defined and its relation with AGV-load assignment system is defined. In the section 3, different mechanisms for implementing the AGV's PGS are presented and the advantages and disadvantages of each of them are explained. In the section 4, the suggested PGS of the this paper with the name of collaborative PGS is presented and explained. For evaluation of the suggested PGS, three application models with different assumptions are introduced and the methods of implementing of collaborative PGS in them are elaborated in section 5. The first application model is static travel time model with no limitation of roads capacity. The second application model is the extension of first model with dynamic structure of roads. The third application model is static travel time model with limitation for roads capacity. In section 6, the collaborative PGS and the results of applying it in the three mentioned models are evaluated.

2 Path Guidance Problem

Specific number of AGVs have the responsibility of transportation of materials, work-in-process, and finished goods in factory and warehouses. According to the demand for transportation of loads, number and distribution of AGVs, AGV-load assignment system specifies the kind of the load, starting point, destination, and the transportation time of each AGV. This paper does not discuss about the internal mechanism of AGV-load assignment system. It only uses the results and outputs of this system (Figure 1).

After the assignment of load to AGV, it is going to choose the path which takes the shortest travel time to the destination. This path is called the fastest path for the AGV. The AGVs are guided to find the fastest path by the Path Guidance System (PGS). Dynamic features such as dynamic travel time, limited capacity of the streets,

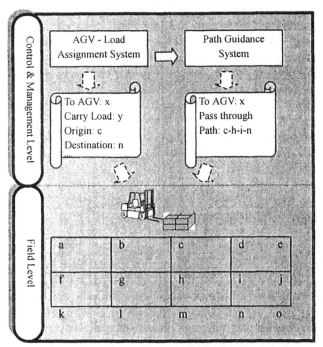

Figure 1 **Path guidance problem.**

changes in the roads' structure, and etc. make it difficult to find a solution for path guidance problem. In this paper, different mechanisms for implementing the PGS are presented and proper conditions for using them are suggested.

3   Mechanisms for Implementing Path Guidance System

At present, one of the following mechanisms are used to guide AGVs to find their fastest paths (Figure 2):
  –   Centralized path guidance system; and
  –   In-vehicle path guidance system.

In the centralized PGS, AGVs are guided centrally in the path guidance center. This center keeps the data of streets and traffic situation of AGVs in real time, and with processing the data, it guides the AGVs to their fastest paths.

If the network of streets is not so expanded and there are not so many crossroads, centralized PGS can be properly used, because the required data and processing in the path guidance center is logical. But if there are many crossroads in the network, the required data and processing will be increased. Also, the reliability of system will be decreased, because as a result of the appearance of any problem in the path guidance center, the control of AGVs will be difficult and even AGVs may be needed to be stopped. Furthermore, the required flexibility of the system to control the AGVs locally will be decreased.

In the in-vehicle PGS, all AGVs are autonomous and they choose their paths toward their destinations by themselves. If the environment of AGVs (network of roads) is static, the in-vehicle PGS has the ideal condition. In this case, AGVs should have a proper model of their static environment (network of roads) and choose their fastest path according to their states. But, as it was mentioned before, in the real situations, there are dynamic features, such as travel time, capacity of streets, structure of network of roads, and so on. In these conditions, it will be difficult for the AGVs to trace the dynamic changes in the environment with their sensors.

With regards to the two above mentioned PGS, and their advantages and disadvantages, this paper presents the third kind of PGS under the title of collaborative PGS. Structure of this PGS is on the basis of the concept of Multi-Agent System (MAS). In the next section, collaborative PGS is elaborated.

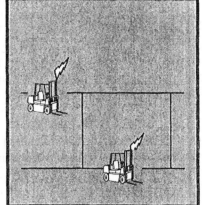
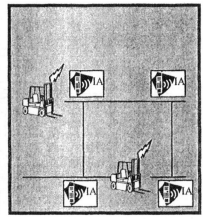

(a) Centralized Path Guidance System   (b) In-Vehicle Path Guidance System   (c) Collaborative Path Guidance System

Figure 2 Mechanisms for implementing the path guidance system.

4   Collaborative Path Guidance System

Collaborative PGS is a PGS which applies the distributed database and control structure to guide AGVs toward their destinations through their fastest paths. Guidance of AGVs by collaborative PGS is done through the collaboration among the distributed guidance centers. Structure of collaborative PGS is based on the concept of Multi-Agent System (MAS). MAS is an artificial system composed of a population of autonomous agents which co-operate with each other to reach common goals, while each agent pursues its individual objectives. In the other word, MAS is a loosely coupled network of problem solvers that work together to solve common problems that are beyond the individual capabilities or knowledge of each problem solver. The overall performance of MAS is not globally planned, but develops through the dynamic interactions of agents in real time [2].

Performance of the collaborative PGS is on the basis of following three principles:
- Data must be kept locally (distributed database).
- Decisions must be taken locally.
- Local decisions must be taken collaboratively.

In collaborative PGS, Intersection Agents (IA) are applied as distributed path guidance centers (Figure 3). An IA is an agent which stands on a specific intersection and keeps the local data of the intersection and its out-going branching streets. In response to the order of the arrived AGV, the IA offers the data of the fastest path from itself to the AGV's destination. In other words, IA can be simply assumed as a data board standing on an intersection, which receives the data about AGV's destination and shows the fastest path toward the destination.

For determining the fastest path from an IA to a

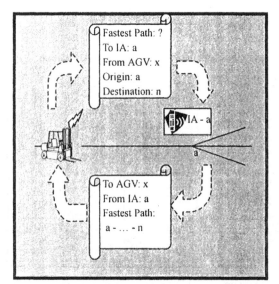

Figure 3 Collaborative path guidance system.

destination, usually IA needs to have the data of other intersections and streets. But as it was explained before, each IA keeps only the local data of its intersection and its out-going branching streets. For solving this problem, IAs must have collaboration together. Mechanism of collaboration among IAs for determining the fastest paths, differs according to the conditions of path guidance problem. On the section 5, three application models for implementing collaborative PGS and the mechanisms of collaboration among the IAs for determining the fastest path are presented.

5   Three Application Models for Collaborative PGS

For evaluating the collaborative PGS, three application models for implementing collaborative PGS are explained in this section. In the first application model, it is assumed that the travel times are static and the streets capacity is unlimited. The second application model is the extension of the first model with dynamic structure of roads. In the third application model, it is assumed that the travel times are static and the streets have limited capacity for the number of AGVs moving through them.

5.1   Static Travel Time with Unlimited Capacity (Application Model 1)

5.1.1   Assumptions of Application Model 1

- Travel times are static; that is, the duration of a trip from an intersection to an adjacent intersection does not depend on departure time.
- Travel times can be deterministic or stochastic. Although the travel time of a street is stochastic and it follows a statistical distribution, it can be considered as deterministic travel time, if the expected of statistical distribution is regarded as the travel time. In this case, the fastest path is determined with calculating the path with the fastest expected total travel time [3].
- The capacity of streets is unlimited. In other words, the possible number of AGVs in a street is unlimited. As it was mentioned in the first assumption, it is assumed that the travel times are static and traffic of AGVs in a street has no impact on the travel time of that street.
- Structure of roads is static and it does not change among the time.

5.1.2   Goal of Application Model 1

As it was mentioned before, the goal of path guidance problem is finding the fastest path of each AGV toward its destination. According to the assumptions of model 1, the fastest path from a specific starting point to a specific destination is a fixed path, which is time-independent. Travel time of this path is fixed too.

5.1.3   Solution of Application Model 1

While the centralized PGS is implemented, path guidance center distinctly finds fastest path for each AGV. In this regard, usually Dijkstra's shortest path algorithm is used, where the costs of each street are their travel times [4].

While collaborative PGS is implemented, the mechanisms must be used which can determine the fastest path according to the collaboration of distributed IAs.

As it was explained before, in the first model, the fastest path from a specific starting point to a specific destination is a fixed and time-independent path. In other words, the fastest path is static, not dynamic. Therefore, if an AGV can be informed of its fastest path by the closest IA, it can reach its destination in the minimum time. Figure 4 shows the internal performance mechanism of AGVs.

IAs determine the fastest path toward specific destination through the collaboration with each other. The internal performance mechanism of IAs is shown in figure 5.

In figure 5, an IA receives the request for the fastest path from itself to a defined destination. This request can be asked by an AGV or an adjacent IA. Then the IA checks itself to know whether or not it has the required data for finding the fastest path from itself to the destination. If it has the complete data, it calculates the fastest path and its travel time and send the information to the relevant AGV or adjacent IA. On the other hand, if the IA does not have the complete data, it needs the collaboration of other IAs. Therefore, the IA sends messages to related adjacent IAs to offer their suggestions with a deadline for the fastest paths from each of them to the destination and the travel times. When the applicant IA receives the required information, it calculates the fastest path from itself to the destination and its travel time and sends the final information to the relevant AGV or the adjacent IA. Calculation of fastest path and its travel time is according to the following formulas:

$$T_{iz} = \text{Min}_j \{ t_{ij} + T_{jz} \}$$

$$P_{iz} = (i \overset{*}{\rightarrow}) \ P_{jz}$$

T_{iz} = Travel time of the fastest path from the intersection i to intersection j

j = an adjacent intersection to intersection i

t_{ij} = Travel time for passing through the street from intersection i to adjacent intersection j

P_{iz} = Fastest path from intersection i to intersection z

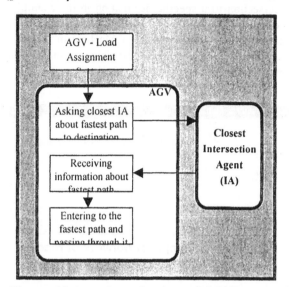

Figure 4 Internal mechanism of AGV in model 1

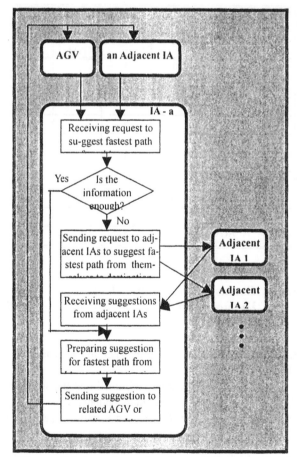

Figure 5 Internal mechanism of IA in model 1.

j^* = The adjacent intersection which makes the minimum amount of T_{iz}

Table 1 shows a complete process of collaboration among AGV and IAs in order to find the fastest path from an intersection to a specific destination in the example of figure 6. This process is called "Just-In-Time (JIT) Communication for Collaboration Mechanism", which is on the basis of the concept of distributedly implemented dynamic programming.

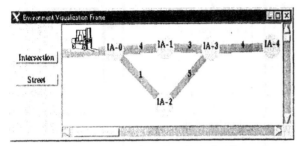

Figure 6 An example of network of roads.

Table 1 An example of JIT communication for collaboration mechanism.

Message From	Message To	Message Contents			
		Origin	Destination	Fastest Path	Travel Time
AGV	IA-0	0	4	?	?
IA-0	IA-1	1	4	?	?
IA-0	IA-2	2	4	?	?
IA-1	IA-3	3	4	?	?
IA-2	IA-3	3	4	?	?
IA-3	IA-1	3	4	3-4	4
IA-3	IA-2	3	4	3-4	4
IA-1	IA-0	1	4	1-3-4	7
IA-2	IA-0	2	4	2-3-4	9
IA-0	AGV	0	4	0-2-3-4	10

5.2 Extension of Model 1, In The Case of Dynamic Structure of Roads (Application Model 2)

5.2.1 Definition of Dynamic Structure of Roads

In the first model, it was assumed that the structure of streets is static and time-independent. But in a real situation it is not always true and usually there are some changes in the network of roads. These changes can be caused by opening or closing a street. The changes in the structure of roads can be intentional or unintentional, for example AGVs accident or falling down of a AGV's loads.

5.2.2 Managing the Dynamic Structure of Roads in PGS

For managing the changes in the structure of roads and for guiding the AGVs, the two following functions must be done by PGS (figure 7):

1. *Updating the model of PGS for the network of the roads:* PGS has to receive the data of the changes in the network and update its model. Through this process, transference of the data of the changes to the PGS has a great importance, specially when the changes in the network of roads are unintentional and neglecting them may cause danger or problem for the system.

2. *Guiding the AGVs according to the new structure of streets:* PGS has to do path planning again for the AGVs which their paths are related to the closed or opened streets, and they need to be re-guided.

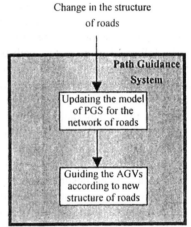

Figure 7 Managing the dynamic structure of roads in PGS.

5.2.3 Managing the Dynamic Structure of Roads in the Collaborative PGS

Same as all kinds of PGSs, the collaborative PGS, for managing the changes in the structure of roads, has to update its model of the network of roads and then guide the AGVs according to the new structure.

Since in collaborative PGS, the data of the network of the roads are kept locally and distributedly in the IAs, updating the collaborative PGS' s model of the network of roads can be done quickly and easy. If a street is opened or closed, the related IA updates its local database of its out-going branching streets.

Calculation of fastest path in collaborative PGS is done locally by the IAs. Therefore, the IAs can find the fastest path for each AGV which faces changes in the structure of network of roads. This mechanism is on the basis of JIT-communication for collaboration mechanism (it has been explained in the model 1).

In this regard, figure 8 shows an example. In this example, the AGV, that explained in the example of model 1 (figure 6), is moving toward its destination through its fastest path 0-2-3-4. The new path 3-5-4 is open for this AGV. In this case, IA-3 calculates the new fastest path for the AGV on the basis of JIT-communication for collaboration mechanism. In this example, the fastest path of AGV has changed from 3-4 path to 3-5-4 path.

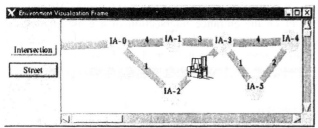

Figure 8 An example of change in the structure of roads in the model 2.

Table 2 An example of JIT-Communication for collaboration mechanism in model 2.

Message From	Message To	Message Contents			
		Origin	Destination	Fastest Path	Travel Time
IA-3	IA-5	5	4	?	?
IA-5	IA-3	5	4	5-4	2
IA-3	AGV	3	4	3-5-4	3

5.3 Static Travel Time with Limited Capacity (Application Model 3)

5.3.1 Assumptions of Application Model 3

- Travel times are static.
- Travel times are deterministic
- Streets have limited capacity. If a street full and an AGV wants to enter this street, it has to wait on the intersection till it finds a place to enter the street. In this case, it is assumed that the AGV is on the intersection, not in a street. Therefore, the travel times of the streets are static and if it is needed for the AGV to wait for passing through the street, waiting time as well as travel time must be calculated [5].
- Structure of roads is static and it does not change among the time.

5.3.2 Goal of Application Model 3

Similar to the first model's, the third model's goal is finding the fastest path of each AGV for reaching its destination. On the contrary, in the third model the fastest path from a specific starting point to a specific destination is dynamic and time-dependent. It is due to the limited capacity of the streets and the relevant waiting times.

5.3.3 Solution of Application Model 3

In collaborative PGS the following mechanisms are defined for finding the fastest paths in model 3:
- Internal performance mechanism of AGV
- Internal performance mechanism of IA for finding fastest path
- Internal performance mechanism of IA for updating its local data
- Just-In-Time (JIT) communication for collaboration mechanism
- Fastest path registration mechanism

The difference of first and third models is the limitation of streets capacity. In this regard, each IA has to keep the local data of AGVs passing through its out-going branching streets. Therefore, the IA can calculate the remained capacity of these streets. Furthermore, each AGV has to announce its path to all IAs which are standing on its path toward the destination.

Internal mechanism of AGVs in the third model is similar to the one in the first model. The only difference is when an AGV is going to take the fastest path to its destination (according to the suggestion of its closest IA), it announces its path to all the relevant IAs on the basis of fastest path registration mechanism (Figure 9).

Internal mechanism of IAs for finding fastest path in the third model is similar to the first model (Figure 5). The only difference is in the formulas of calculating fastest path and its travel time. In the second model these formulas are changed in order to cover the waiting times as well:

$$T_{iz} = Min_j \{ t_{ij} + w_{ij} + T_{jz} \}$$

$$P_{iz} = (i \rightarrow^*) P_{jz}$$

w_{ij} = Required waiting time for AGV to enter to the street from intersection i to adjacent intersection j

Figure 10 shows the internal mechanism of an IA for updating the local data of its out-going branching streets capacity and the traffic of AGVs.

In the third model, the Just-In-Time (JIT) communication for collaboration mechanism for calculation of fastest path is the same as first model (Figure 6).

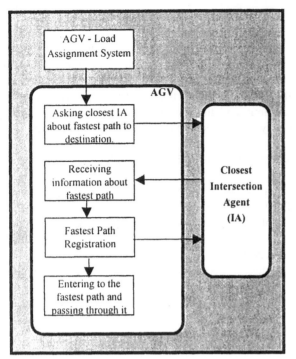

Figure 9 Internal mechanism of AGV in model 3.

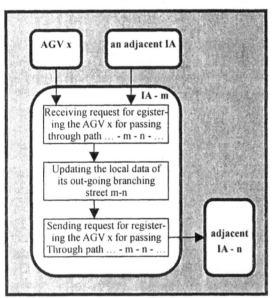

Figure 10 Internal performance mechanism of IA for updating its local data in model 3.

As it was mentioned before, each AGV has to announce its path to all the IAs which are standing on its path toward the destination. This function is done according to the mechanism of fastest path registration.

5.3.4 An Example of Application Model 3

Figure 11 shows a network of roads and the travelling times of its streets. It is assumed that all streets have limited capacity and only one AGV can go through each street in a time. Four AGVs have entered the intersection 0 of the network of roads at $t = 0$, $t = 2$, $t = 5$, and $t = 8$. They are going to reach their destinations (intersection 4) through their own fastest paths. Figure 12 shows the time schedule of AGVs passing through the streets and table 3 shows the fastest paths and waiting times of AGVs. Table 4 is an example of fastest path registration for AGV-3.

As it is was explained in application model 3, due to the limited capacity of streets, some times AGVs may need wait on the entrance intersection of a crowded road and also their fastest path may be changed among the time. This matter is shown in figure 12 and table 3.

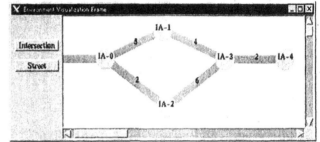

Figure 11 Network of roads for example of application model 3.

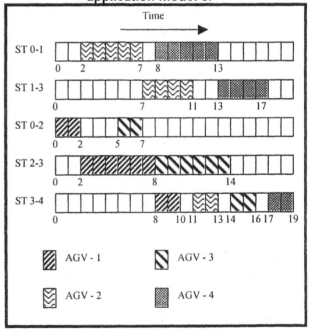

Figure 12 Time schedule of AGVs passing through the streets.

Table 3 Results for example of application model 3.

	Fastest Path	Traveling Time	Waiting Time
AGV - 1	0-2-3-4	10	0
AGV - 2	0-1-3-4	11	0
AGV - 3	0-2-3-4	10	1
AGV - 4	0-1-3-4	11	0

Table 4 An example of fastest path registration mechanism for AGV-3.

Message From	Message To	Message Contents				
		Fastest Path	Registration For Street	Waiting Time	Entrance Time	Leaving Time
AGV-3	IA-0	0-2-3-4	0-2	0	5	7
IA-0	IA-2	0-2-3-4	2-3	1	8	14
IA-2	IA-3	0-2-3-4	3-4	0	14	16

6   Evaluation of Collaborative PGS

Evaluation of collaborative PGS in the three mentioned models shows that the collaborative PGS can be used properly in order to guide AGVs for finding their own fastest paths. Implementation of distributed data base and control system in collaborative PGS increases the reliability of this system in comparison with centralized PGS. Because temporary break down of one or few IAs does not make problem for the whole collaborative PGS, and the rest of the IAs can manage the PGS. Furthermore, required flexibility for managing PGS locally is provided in collaborative PGS. For example, each IA can easily manage the changes caused by opening a new street or closing a street with no need for making changes in the states of other IAs (as explained in model 2). In collaborative PGS, the required data and calculation for each IA is less and collaborative PGS can manage complex and expanded network of roads in the basis of the collaboration among IAs.

The main problem for collaborative PGS, is the great number of exchanged messages among the IAs. This problem is much noticeable when the number of AGVs increase or the network of roads is expanded. For solving this problem in a real situation, a combination of centralized PGS and collaborative PGS can be used. If the network of roads is so expanded, it can be divided into different zones. In each zone, the IAs collaborate with each other according to the collaborative PGS. Each zone has its coordinator which has contact with all IAs within the zone

and all other coordinators. The AGVs in an expanded network of roads can be easily managed according to the collaboration among coordinators of zones.

7   Conclusion

Collaborative PGS for guiding AGVs in finding their fastest paths to their own destinations has been explained in this paper. For implementing the collaborative PGS, IAs are introduced. An IA is an agent which stands in a specific intersection and keeps the local data of the intersection and its outgoing branching streets. Collaborative PGS determines fastest path for each AGV on the basis of collaboration of IAs.

For evaluating the collaborative PGS, its implementation has been studied in three models with different conditions. The first model is a static travel time model with unlimited capacity of roads. The second model is the extension of the first model with dynamic structure of roads and the third model is a static travel time with limited capacity of roads.

References

[1] Kim, H., Choi, K. and Ahn, B., "Evaluation of estimated shortest travel time using traffic network simulation," IEEE International Conference on Systems, Man, and Cybernetics, Tokyo, Japan, pp. IV 586-588, Oct. 12-15, 1999.

[2] Ito, T. and Mousavi Jahan Abadi, S. M., "A multi-agent approach to job-based communication," Design and Systems Conference, Osaka, Japan, pp. 195-198, Nov. 28-Dec. 1, 1999.

[3] Wellman, M. P., Ford, M. and Larson, K., "Path planning under time-dependent uncertainty," Eleventh Conference on Uncertainty in Artificial Intelligence (UAI-95), Montreal, Quebec, Canada, Aug. 18-20, 1995.

[4] Dijkstra, E. W., "A note on two problems in connection with graphs," Number. Math. 1, pp. 269-271, 1959.

[5] Hattori, Y., Hashimoto, T. and Inoue, S., "A study for the traffic flow control considering the capacity of the road by cellular automation method," IEEE International Conference on Systems, Man, and Cybernetics, Tokyo, Japan, pp. IV 569-573, Oct. 12-15, 1999.

A System Performance Scale for Engineering Team Decision-Making

Oded Maimon
Department of Industrial Engineering
Tel-Aviv 69978, Israel Email: maimon@post.tau.ac.il

Eran Reuveni
Department of Industrial Engineering
Tel-Aviv 69978, Israel ,Email: reuveni@post.tau.ac.il

Abstract

This paper presents a method to conduct collaborative decision process during system conceptual design stage. The method leads team engineers to define System profit function (SPF), The SPF then may be used as an evaluation criteria for different system concepts. Once a concept is selected, the SPF continues to support the design process, providing the means for easily prioritizing all proposed design changes to selected concept. The proposed method has been implemented in a case study, which demonstrated the improvement of collaborative decision process.

Introduction:

Concurrent Engineering (CE) philosophy calls for involvement of personnel from all required disciplines: engineers & nonengineers. In CE, all major parties involved in getting the product to market contribute to the development process. Prasad [5] states that CE approach requires a parallel, interactive, and cooperative multidisciplinary teamwork. Just as manufacturing leveraged people's skills CE is about leveraging teamwork (economy of cooperation) to handle information and make informed decisions and utilizing the team intellectual power. One of the known facts about product development is the immense amount of information gathered during the process. This fact puts a heave load over the capability to carry a quick effective decision process. In this paper we suggest a way to perform this collaborative decision process based on unified target function-

System Profit Function (*SPF*)

System Profit Function is an explicit mathematical expression for system cost-effectiveness. The *SPF* function is based on a similar target function suggested by Butler [1] as an alternative to decision-processes in regard to initial provisioning.

The *System Profit Function*, as business profit function, is expressed as the difference between system yield value for the customer and the overall expenses required to establish, operate and maintain that system, throughout *the* system's life-cycle, i.e. *Life Cycle Cost (LCC)*.

This *SPF* is formulated as follows:

$$(1) \quad \Pi = E * VPC - LCC$$

Where:

    E stands for system effectiveness.
    VPC - *Value of Performance* Capacity (expressed in \$ units).
    LCC - Life Cycle Cost.

System effectiveness shall be defined as weighted sum of the expected effectiveness for each one of the missions performed using the system, i.e.:

$$(2) \quad E = \sum \omega_m * E_m$$

Where:

    E_m - denotes System Effectiveness when performing mission m.

    ω_m - denotes weight number allocated for mission m.

And where:

$$(3) \qquad \sum \omega_m = 1$$

Weights may be generated using Saaty's [7] AHP method. Yet, it should be noticed that implementing the *SPF* approach, there is a need to allocate weights only to **system missions** (usually a small number), as opposed to decision methods that requires to allocate weight to all system parameters (which some times gets to tens or hundreds).

Eisner [2] Klain [4], describe number of definitions for System effectiveness. We shall use the term, established by WSEIAC [9] for System Effectiveness:

$$(4) \qquad E_m = A_m * D_m * P_m$$

Where, for mission *m*:

A_m - stands for *System Availability*.

D_m - stands for *System Dependability*.

P_m - stands for *System Performance*.

Based on the System *Profit Function*, a prioritization factor may be defined as follows:

$$(5) \quad l_i = \frac{\Pi_i - \Pi_o}{c_i}$$

Where:

l_i : Prioritization factor for Engineering Configuration Proposal (*ECPi*).

Π_i Value of the *System Profit Function* after implementation of *ECP i*.

Π_o: Value of the *System Profit Function* at basic system configuration.

Calculating expression (5) for each Engineering Configuration Proposal (*ECP*) reveals the most cost-effective configuration proposals. The working team can now clearly identify which engineering configurations are more attractive for further engineering effort. Doing so, the team has got a common set of consistent goals. Clear & supporting goals provide constancy of purpose. They allow every one to set aside frivolous issues and focus on what is really important.

Case Study

In order to demonstrate the capabilities of the proposed method, a case study was conducted. The goal of the case study was to demonstrate the use of *SPF* in the conceptual design of an upgrade configuration for an attack helicopter.

The program team was composed in accordance with CE philosophy and included operational personnel, system engineers, system specialists (aeronautic, avionics and propulsion engineers) and maintenance engineers. Each team member also acted as a liaison with the original manufacturer's engineering teams in order to receive additional engineering and cost data as required for the decision process. The role of the team was to identify and assess operational gaps; identify technological solutions and alternatives; and to define new system architecture, which could be implemented within a framework of technological risks and budget and time constraints.

Functional gaps:

Analyzing the attack helicopter capabilities at battle field, focused the team on 5 functional gapes:

1. Since the helicopter has undergone numerous modifications in the course of its operational lifetime, the aircraft's overall weight has increased significantly while its weapon load limit has decreased proportionally.

2. Since engine performance is related to air temperature, maximum take-off weight decreases on hot days. As a result, given a fixed load or amount of ammunition, more attack sorties are necessary on a hot day in comparison to a colder one.

3.   At the current avionic configuration, each avionic system is operated by a unique control unit. As a result, pilot workload increases, affecting both aircraft safety and mission success.

4.   It takes precious time to transmit mission information between helicopters that are assigned to the same mission. Transmitting mission information using a skeletal map would shorten mission preparation time.

5.   Since different lubricants (oil, water) accumulate in the helicopter's belly, the lower fuselage antenna suffers from a continuously wet environment that degrades the conductivity of coax cables and connector pins. Also, in antennas, the inner impedance matching circuits tend to break. Poor antenna fuselage connection leads to a poor antenna radiation curve.

Technological Alternatives

Technological innovations offer several system upgrade alternatives:

A. Implementation of a four-blade rotor, instead of current two-blade rotors, to achieve better take-off load capability.
B. Implementation of better engine materials and processes for hot parts to create better engine thrust.
C. Implementation of a common control unit for all avionics system.
D. Implementation of CRT display & data link for graphically transmitting a skeletal battle scenario.
E. Implementation of special antennas with sealed installation housings.

The Collaborative Decision Process

This case study confronts the team with the challenge of conceptually designing a new system configuration, based on **different**

engineering proposals. Each engineering proposal has different performance domain. *SPF* shows itself to be a powerful tool for the collaborative decision process, especially in multidisciplinary cases such as this one. Once an appropriate performance function is defined, the *SPF* can be used for measuring and evaluating different system configurations.

In order to define the *SPF* function, engineering team should define *A, D, P, LCC* terms, that best represents the system on hand.

Step 1: System availability may be calculated using the common expression:

$$(6) \quad A_o = \frac{MTBMA}{MTBMA + MTTR}$$

Where:
  MTBMA –Mean time between Maintenance Action.
  MTTR - Mean Time to Restore.
Any other, definitions may be appropriate as well.

Step 2: Dependability function definition.

Dependability function is defined as the probability to conduct the mission without any interrupting incident of any kind (human, technical or other). Hence, the dependability function may take the general form:

$$(7) \quad D_m = 1 - \prod_j R_{mj}$$

Where: *Rmj* represents the probability of risk *j* to occur during performance of mission *m*.

Step 3: System Performance function definition

The System Performance function should be defined as the probability to achieve mission goals within mission constraints.

Any function may be used as system performance function even boot-strap simulation based on measured data or simulated data. In case there is no such data, the performance function should have the form of expression (8):

$$(8) \quad P_m = \int_D a_m(x) f_m(x) dx$$

Where: $a_m(x)$ is the acceptance function, which defines the acceptability level for design parameter x.

In case of more than one design parameter, the performance function should have the form of expression (9):

$$(9) \quad P_m = \int_D U_m(x_1.x_i.x_n) f(x_1.x_i.x_n) dx_1.dx_i.dx_n$$

In case of statistical independent parameters, expression (9) takes the form:

$$(10) \quad P_m = \int_D U_m(x_1.x_i.x_n) f(x_1).f(x_i).f(x_n) dx_1.dx_i.dx_n$$

In order to keep the performance function simple as possible, it is recommended to implement an hierarchical approach and differentiate between first order mission capabilities, sub capabilities, sub-sub capabilities and so on. Doing so, system performance function may include first order capabilities only, where as for each first order capability, a specific performance function shall be defined.

In the current case study of an attack helicopter, five mission types were identified. For each mission, mission first order capabilities were listed. For example:

Mission A: antitank mission
  Capability1: To perform mission preparation.
  Capability2: To Fly to target area.
  Capability3: To identify targets.
  Capability4: To identify targets to each team
      member.
  Capability5: To perform target assault.
  Capability6: To perform Target kill
      assessment.
  Capability 7: To perform next target assault.
  Capability 8: To perform fly back to refuel
      point.
  Capability 9: Refuel, rearm, repair (if needed).

The total mission performance function was defined as:

$$(11)$$
$$P_m = \frac{Ca}{Tg} * \iiint_K \frac{MT}{\sum_{i=1:K} T_i} f(T_1)...f(T_i)..f(T_K) dT_1..dT_i..dT_K$$

Where:

Ca- denotes maximum ammunition capacity.
MT- denotes maximum mission time.
Tg – denotes total ammunition delivery goal.

Term (11) may be easily calculated using any calculating tool implementing monte-carlo simulation technique.

Engineering proposal	Sys. parameter	Symbol
Engine improvement	Take off load	Ca
Four blade rotor installation	Take off load	Ca
Unified Control Unit	Mission preparation time	$f(T_1)$
CRT display & data link	Identifying targets to team members	$f(T_3)$
CRT display & data link	Target kill assessment	$f(T_6)$
Antenna installation	System Availability	Am

Step **4:** Life Cycle Cost definition:

The last term to be defined is the System Life Cycle Cost calculation. General form of Life Cycle cost may be defined in term (12):

$$(12) \quad LCC = \sum_{P=1}^{5} (NRE_P + RC_P + LSC_P + OPC_P)$$

where:

NRE: Represents *Non Recurrent Expenses*.
RC: Represents *Recurrent Cost*.
LSC : Represents *Life Support Cost*.
OPC: Represents *Operating Cost*.

Where the *P* subscript represents typical operating phases in system life cycle:
  Phase 1: System development.
  Phase 2: Initial Capability.

Phase 3: Full scale Capability.
Phase 4: Mid life upgrade period.
Phase 5: System draw down.

The *Value of Performance Capacity (VPC)* should be defined by program management as derived from customer willingness to pay for improved system performance. See also Gale [3] approach for managing customer value.

Once, the *SPF* is completely defined, the team may evaluate each engineering proposal using term (3) and then calculating prioritization factor (l_i) using expression (5). The team-work results for the attack helicopter mission A is presented in Table 2:

Engineering proposal	Cost (M$)	$\Delta\Pi_i$	*l* factor
Engine improvement	120	178.75	149%
Four-blade rotor installation	150	49.76	33%
Unified Control Unit	10	3.6	36%
Airborne data link	10	17.65	177%
Antenna installation	1	2.66	266%

Table 2: Prioritization results

Based on the results it is easy to select the most cost-effective ECPs to work on. It should be noted that the *SPF* evaluating approach grades all engineering configurations in reference to customer effectiveness goals, where as other methods as QFD or Pugh's [6] method, grade configurations in reference to selected benchmark. This difference may be very important, especially in case that all proposed system configurations fails to satisfy customer goals. In such a case the alternative

of rejection of all proposals should be considered. Therefor it is important that decision method would support such a decision.

In case of several alternatives for each functional requirement, working team may use Zwicky [10] approach to form new configurations. In other case, working team may use the genetic algorithm to form different combinations of ECPs. It may be noted that for such cases it would be useful to have a software program that automatically computes the *SPF.*

Summary and Conclusions

In this paper a method for evaluating concept alternatives was presented. The method is based on *System* Profit *Function (SPF),* which measures both system effectiveness and system life cycle cost. A test case of a concept design for an attack helicopter, implementing the proposed method was described.

It is important to note that the *SPF* function is suitable for use as a target function for all product development phases (requirement definition, product definition, process definition, delivery and support). At each one of these stages, more decisions have to be taken. Most of new decisions are generated as further breakdown of the system components and system performance is discussed. Other decisions are required as new *information* emerges that contradicts previous assumptions. Either way, the team may continue to use *SPF* as the basis for decision making process. Should the team wish to do so, it may be required, that all data from former stages would be available and would be organized in such a way that it may

support the process of automatic computation of the *SPF.*

References

[1] Butler, Robert A., An Alternative to Readiness-Based Decision Making ,System Exchange, August 1993.

[2] Eisner, howard, Computer Aided System Engineering. Prentice Hall 1987

[3] Gale,Bradley T. Managing Customer Value, The Free Press, 1994.

[4] Kline, Melvin B., Concept of system effectiveness, Naval Postgraduate school, Monterey, California.

[5] Prasad, Biren, Concurrent Engineering Fundamentals, Volume 1,. Prentice Hall 1996

[6] Pugh, S., Concept Selection- A Method That Works, *Proceeding of the international Conference on Engineering design,* (ICED) Rome, 1981.

[7] Saaty, T.L, "The Analytic Hierarchy Process", McGrow-Hill New York, 1980

[8] Susman. Integrated Design & Manufacturing for competitive Advantages.

[9] WSEIAC-weapon System Effectiveness Industry Advisory Committee report-1963

[10] Zwicky F., "The Morphological Approach to Discovery, Invention, Research & Constraction", in the collection: New Methods of Thought & Procedure, *Springer Verlag, 1967.*

An Agent-Based Decision-Supporting Framework for Taguchi Experiment Planning

J. W. Lee

Div. of Mechanical Aerospace and Automation, Eng. Inha University, Inchon, Korea 402-751

S-J Cho

Departement of Automation, Eng. Inha University, Inchon, Korea 402-751

Abstract

In this paper, we present an agent-based decision-supporting framework for Taguchi experiment planning. A Taguchi experiment has four distinct phases: planning the experiment, designing the experiment, conducting the experiment, and analyzing the experiment. Among these, the planning phase includes the most important decision-making tasks such as determination of experiment objectives, quality characteristic, and control factors. The planning phase, however, has not been paid proper attention by experimenters. In many cases, experimenters tend to rush into the designing phase, not spending much time in planning tasks. We were unable to find any software system that supports the planning of a Taguchi experiment. In this paper, we suggest a decision-supporting framework for Taguchi experiment planning. The framework is composed of two agent-based mechanisms. The first one employs Internet agent that collect the domain knowledge from knowledge providers, who may be distributed in remote places. Another agent then visualizes the collected knowledge and reports the results to the experimenter. Engineers who normally would have difficulties in collaborating because of limitations on their time or because they are in different places can easily work together on the same experiment team and brainstorm to make good decisions. The second agent-based mechanism offers context-sensitive advice generated by another intelligent agent during the experiment planning process. For example, if an experimenter chooses an undesirable quality characteristic, the agent will warn against potential problems. By using the proposed framework, unnecessary spending will be minimized, and the feasibility of the experiment will be increased. It is also expected that effective implementation of this framework will encourage the application of Taguchi experiments as an effective method of concurrent engineering.

1. Introduction

Engineering design can be viewed as the definition of a set of design variables, called 'parameters' and the assignment of appropriate values to them. To determine these values, engineers have made use of various theories, mathematical formulae, and heuristics. They have also utilized experimental methods that are either actual ones or computer simulations. The conventional experiments usually consider one variable at a time based on trial and error, and require much time and money to finish the design, often preventing optimization of the product. Eventually the manufacturer is forced to provide imperfect and expensive goods that cannot complete in the market. The Taguchi method, however, is a 'robust design', and as a systematic and efficient design optimization method, enables the study of many design variables in a small number of experiments. Using a mathematical tool called 'orthogonal array' and a new quality concept, 'S/N ratio (Signal-To-Noise ratio)' that are contained in Taguchi method, one can maintain the required performance in various circumstances. Since the early 1980s, when the Taguchi method was introduced in the United States, it has been applied by a number of large companies. And today it is widely used as one of the most useful quality tools in the world [1, 2, 3, 4, 5].

In opposition to these widespread applications, the attempts to enhance the efficiency of the Taguchi using computer has not been tried many times. There are only a few researchers and software developers who tried to computerize the method. WinRobust by Dermiggio, commercial software ANOVA-TM and STN are examples [1, 6, 7]. With these tools, matrix experiments can be designed and tedious calculation for analysis of the result can be avoided. We found an artificial intelligence application. Lee et al. developed an expert system for designing matrix experiment. The system, developed in Prolog language, obtains the information about control factors and their levels to choose a fitting orthogonal array and design matrix experiment [8]. All these previous works put their focus on designing matrix experiment and calculating the result of the experiment, but not on the initial stage of planning the experiment.

In this research, we propose an agent-based decision-supporting framework for experiment planning in the Taguchi method. The planning phase contains the most fundamental decision-making issues such as experimental objective, quality characteristic, and control factors. Experiment designers are required to hold both domain specific knowledge and the method-related knowledge. To present both kinds of knowledge to the experiment designer, the proposed framework has two distinct agent-based decision-supporting mechanisms. The first is to collect information from the knowledge providers, who are involved in the experiment, and visualize it to generate an easy-to-understand report. The agent approaches to the knowledge providers via Internet to ask them a question and collect their answers. Another agent visualizes the collected information and makes the report to support the decision-making of an experiment designer. The second agent-based decision mechanism is to give helpful advice that is generated by an agent during the experiment planning. For example, if an experiment designer chooses an undesirable quality characteristic, the agent will warn against potential problems.

The rest of this paper is organized as follows. Section 2 provides a research background and previous works. The description of the architecture of the proposed framework, each agent and their functionalities are presented in Section 3. Section 4 will describe the prototype implementation and evaluation. Finally we present conclusions and future research issues in Section 5.

2. Conceptual Background

In this section, we will briefly introduce the conceptual background involved in this research.

2.1. Taguchi Method: The Significance of Experimental Planning

A few researchers suggest their own procedures for performing the Taguchi experiment that are slightly different from others [2, 3, 4]. The procedure proposed by Peace is composed of following four distinct phases [2].

- Planning the experiment: The key issues of the experiment are determined – forming an experiment team, determining experiment objective, quality characteristic, measurement planning, selecting control/noise factors and their levels, and determining experiment strategy.
- Designing the experiment: Matrix experiment is designed. Calculating DOF (Degree Of Freedom), selecting orthogonal array, and assigning factors are included in this phase.
- Conducting the experiment: The actual run and measurement of experiment is carried out.
- Analyzing the experiment: Performs tabular and graphical analysis, determining optimum parameter set, and confirmation run.

It is agreed that the Taguchi method is easy to use. Beginners, however, frequently make a mistake that invalidates the power of Taguchi method and fail to get any useful information through the experiment. From time to time, shortage of information and haste in making decisions are possible pitfalls even for the experienced. Undesirable decisions in the planning phase can affect the entire experiment and cause unwanted costs and resource spending. In the worst case, the experiment may be all in vain. Peace put an emphasis on the significance of experiment planning in the four phases he suggested. He also explained that errors could be minimized if the following procedure is observed. First, an experiment team should be organized. The team holds continuous brainstorming meetings to obtain a variety of ideas. The team makes use of quality tools such as Pareto chart, cause-and-effect diagram, and FMEA (Failure Mode and Effect Analysis) to make careful planning [2]. This guideline of Peace has inspired the key idea of this research.

2.2. The Software Agent

Software agent technology is becoming the most important area of research and various applications are being challenged in most fields of computer science. An agent is a software program that performs a user-delegated task. An agent commonly has the following characteristics [9]:

- Delegation: An agent performs a certain task on behalf of a user.
- Autonomy: Agents work without direct instruction.
- Communication skills: Agents interact with user or other agents.
- Monitoring: Agents should monitor circumstances in order to perform a task autonomously.
- Intelligence: Agents should identify the given conditions and determine what to do.

An agent may have other various characteristics – mobility, security, and personality [9, 10].

We found a few works in regard to this research. Wang developed a decision-supporting system using data mining and agent technology [11]. Bui and Lee proposed an agent-based framework for building decision-supporting systems [12]. They also have given us an important research idea. Compared with the work of Wang, whose information source is a database, the information source of our system is human knowledge providers.

2.3.   Internet-Based Survey System

The questionnaire has been widely used to gather the opinions of large numbers of people. Currently this traditional tool is applied on the Internet. There are some commercialized examples. The Questionnaire & Survey System (QSS) is a program that can make a survey on the Internet and Intranet. This program generates a HTML questionnaire and shows the result whenever it is necessary [13]. Pop-Up is another example of a web-based survey system. It randomly pop-ups and invites the user to participate in the survey as many people visit the web page [14].

We employed the concept of these systems, with a few small modifications. First, our system has an agent-based architecture. The purpose of our system is to collect technical information, while that of the example systems is to obtain information for business, marketing, or politics. And unlike the range of these systems, at least as they are usually designed, the range of our survey is predetermined.

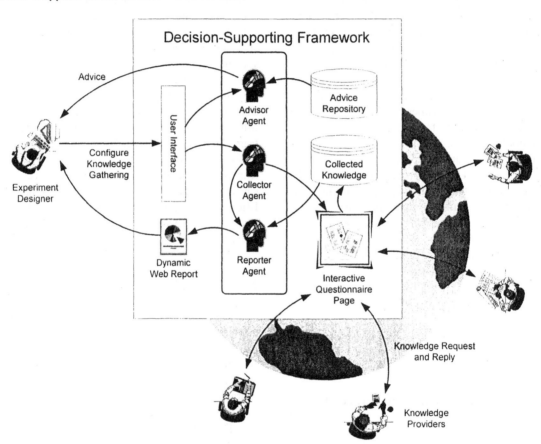

Figure 1 The overall architecture of agent-based decision-supporting framework for Taguchi experiment planning

3. Agent-Based Decision-Supporting Framework

The overall architecture of decision-supporting framework for Taguchi experiment planning is shown in Figure 1. Here, experiment designer is a person who leads the experimentation team and makes all decisions in the experiment. Knowledge providers are ones who are organized in an experiment team in order to provide domain knowledge. A knowledge provider group may include engineers, researchers, operators, and all other people who might help the experiment designer. In order to cooperate in the framework, they should be able to freely use e-mail and World Wide Web.

As mentioned in Section 1, planning the Taguchi experiment requires both domain knowledge and Taguchi method knowledge. To provide both kinds of knowledge to the experiment designer, the proposed framework employs two distinct decision-supporting mechanisms. In Section 3.1, we will describe the domain knowledge providing mechanism, and in Section 3.2, the context-sensitive advice mechanism.

3.1. Domain Knowledge Providing Mechanism

The main purpose of domain knowledge providing mechanism is to collect information from distributed knowledge providers and to report it with visualized format so that the experiment designer might understand it quickly and easily. The mechanism is composed of two agents: the 'collector agent' that gathers the information from the knowledge providers and the 'reporter agent' that visualizes the collected information and reports it to the experiment designer (and to the knowledge providers if needed). Figure 2 and Figure 3 illustrates the architectures of each agent.

To obtain the required information, the experiment designer poses the question and configures various options about answering manner and visualization. For example, the experiment designer can ask the question "Select 5 items that are thought to have strong effect on the quality and order them descending manner" that is 'select-and-order type'. Figure 4 shows a part of the user interface for question configuration. Table 1 presents the types of questions and options available in this research. The collector agent generates an interactive web page and requests that the knowledge providers answer the question via e-mail. The knowledge providers can easily access the interactive web page through a hyperlink in the e-mail if they use mail client program such as Outlook Express. If they answer the question and submit it, the

agent adds it to the knowledge storage. Meanwhile, the agent monitors the number of knowledge providers who answer the question. If the predefined condition is fulfilled (if 80 % of knowledge providers answered the question for example), the collector agent launches another agent 'reporter'. The reporter agent retrieves the collected information from the knowledge storage and generates a web report in a predefined format. Then the reporter agent announces by e-mail that the report for the question has been prepared. Finally, the experiment designer has access to the report by web browser to use it in decision-making. Employing this mechanism, engineers who are widely distributed and who have difficulties in getting together can be organized in the same experiment team and provide knowledge to each other easily, since the knowledge collecting tasks are carried out in the Internet environment.

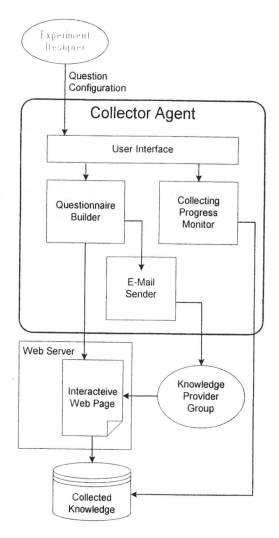

Figure 2 The architecture of the collector agent

3.2. Context-Sensitive Advice Mechanism

The researchers on the Taguchi method have described some desirable situations in the experimental process, as well as undesirable ones [1,2,3,4,5].

The context-sensitive advice mechanism should support experiment designer when these situations occur. An agent called 'advisor' helps him reduce erroneous decision-making by giving him proper advice. Figure 5 illustrates the architecture of the advisor agent.

Table 1 Questions types and available options

Question types	Answer option
Multiple-choice	Allow multiple selection Allow adding a new item
Fill-in	Word or phrase Description
Ordering	Give weights to each item Select and order

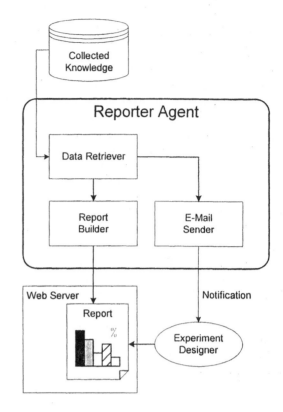

Figure 3 The architecture of reporter agent

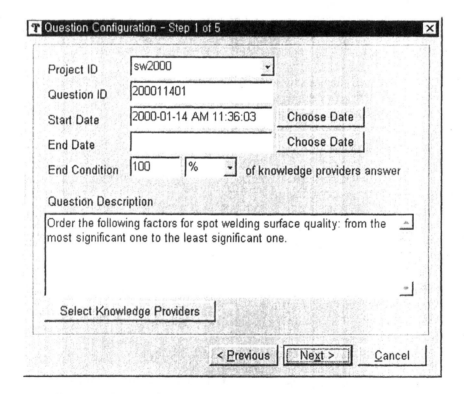

Figure 4 A part of user interface for question configuration

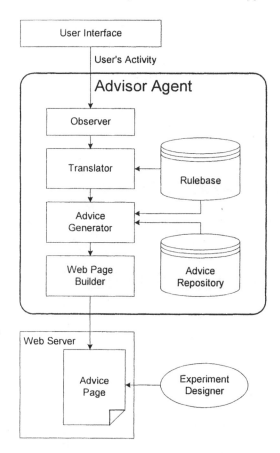

Figure 5 The architecture of advisor agent

The advisor agent observes the experiment designer's activities in performing the experiment planning and updates the record of the status of the experiment designer continuously. If the current status is thought to require help, the advisor agent generates a recommendation that fits the situation. The rule base contains the rules that describe the relationship between the current status and available advice in the advice repository. For example, if the experiment designer chooses the type of quality characteristic 'yes or no type' and does not fill in the text box for 'available measuring equipment', the advisor agent will pop up an window to ask him more questions about the situations such as "Does the measurement depends fully on the inspector's sense?", "Is there a clear boundary between yes and no?", and "Do you have enough man power to inspect the result?". Then the advisor agent will give warnings about the potential problems or give him a recommendation such as "You should provide a visual aids to facilitate the inspection task" or "You had better choose an alternative quality characteristic".

On the other hand, the experiment designer can activate the advisor agent intentionally if needed. In this case, advisor gives advice that can be applied to the current situation.

4. Prototype Implementation and Evaluation

4.1. Prototype Implementation

The prototype system is developed using the following tools and platforms:

- Operating system: MS Windows NT 4.0.
- Web Server: MS Internet Information Server 4.0.
- Web browser: MS Internet Explorer 5.0.
- Agent implementation: MS Visual Basic 6.0, MS Visual C++ 6.0.
- Communication between agents: COM/DCOM.
- Dynamic generation of interactive web page and report page: MS Active Server Page (ASP) 2.0.
- User interface: MS Visual InterDev 6.0 and MS Visual Basic 6.0.
- Knowledge storage: MS SQL Server 7.0.

4.2. Prototype evaluation

It is not easy to quantitatively evaluate the performance of a decision-supporting system. In this research, we conducted a simple test of knowledge collecting example with the aid of colleagues. Figure 6 shows a generated report for the test question "Order the following factors for spot welding surface quality: from the most significant one to the least significant one". We think this report would be helpful in choosing control factors among the given factors.

Discussing the process and the result, we obtained the following comments:

- One correct bit of opinion about a particular problem may well be accompanied by two bits of opinion about the same problem that are wrong. To avoid this problem, the answer of knowledge providers should have different weights according to such factors as their experience, education, and area of expertise.
- The framework will show its usefulness when the number of knowledge providers is greater than a certain number. So it is desirable that middle/large size companies or laboratories employ the

framework. In case of small companies, several companies that are interested in similar technology can organize a community for knowledge gathering. It is possible that the knowledge providers may not participate in the knowledge collecting session steadily since it is not a face-to-face meeting and they, of course, have other work to attend to. In this case, some incentives to increase the rate of participation should be considered. Support from upper-level management is also desirable to make the knowledge providers have more interest. Question types and answer types are restricted to a few formats. A wider variety of formats should be designed and implemented in order to have clear conversations between the experiment designer and the k nowledge providers.

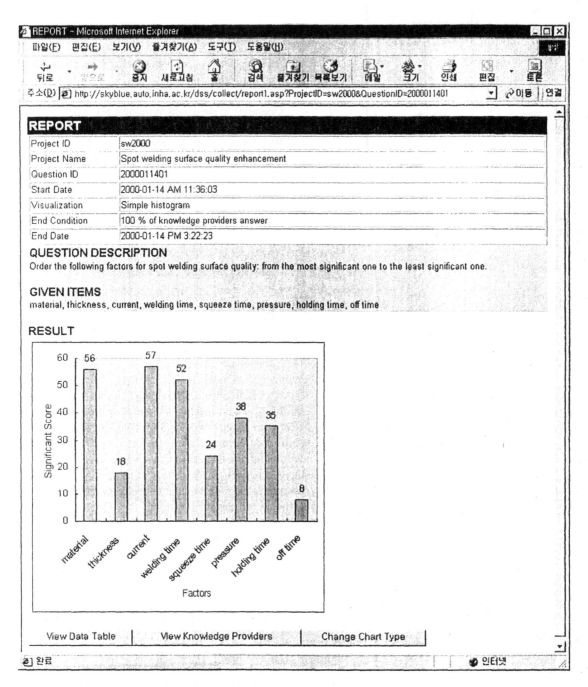

Figure 6 An example report generated by domain knowledge providing mechanism

5. Conclusions and Future Work

In this research, we proposed an agent-based decision-supporting framework for Taguchi experiment planning. Summarized below are the characteristics of the framework and their significance.

The framework provides both domain knowledge and Taguchi experiment knowledge. It has a domain knowledge collecting mechanism that is composed of collector agent and reporter agent. Since the mechanism works on Internet environment, it is expected that the problems of arranging a conventional face-to-face meeting can be overcome to a certain degree. It also has a context-sensitive advice mechanism to prevent improper decision-making proactively.

The framework supports the decision-making in the planning phase of Taguchi experiment. The planning phase has not been paid proper attention so far.

We expect that using this framework will increase the feasibility of experiments, so that unnecessary expenditure of time and resources due to erroneous decision-making might be minimized. It is also expected that effective implementation of this framework will encourage the application of Taguchi experiments as an effective method of concurrent engineering.

In the future, we have a plan to extend the area of research to support a total Taguchi experiment process. The comments obtained in the prototype evaluation (in Section 4.2) should be also considered.

References

[1]   W. Y. Fowlkes and C. M. Creveling, Engineering Methods for Robust Product Design: Using Taguchi Methods in Technology and Product Development, Addison-Wesley, 1995.

[2]   G. S. Peace, Taguchi Methods: A Hand-On Approach, Addison-Wesley, 1993.

[3]   M. S. Phadke, Quality Engineering using Robust Design, Prentice Hall, 1989.

[4]   P. J. Ross, Taguchi Techniques for Quality Engineering 2nd Edition, McGraw-Hill, 1996.

[5]   G. Taguchi, Introduction to Quality Engineering, Asian Productivity Organization, 1987.

[6]   Advanced Systems & Designs Inc. Users Guide: ANOVA-TM for Windows, Advanced Systems & Designs Inc. 1986.

[7]   InfoMate, STN for Windows: Quality Engineering Software, InfoMate, 1989.

[8]   N. S. Lee, M. S. Phadke, and R. S. Keny, "An Expert System for Experimental Design in Off-Line Control", Expert Systems, Vol. 6, No. 4, pp. 234-249, 1989.

[9]   A. Caglayan and C. Harrison, Agent Sourcebook, John Wiley & Sons, 1997.

[10] M. Knapik and J Johnson, Developing Intelligent Agents for Distributed Systems: Exploring Architecture, Technologies, and Applications, McGraw-Hill, 1998.

[11] H. Wang, "Intelligent Agent-Assisted Decision-Support Systems: Integration of Knowledge Discovery, Knowledge Analysis, and Group Decision Support", Expert Systems with Applications, Vol. 12, No. 3, pp. 323-335, 1997.

[12] T. Bui and J. Lee, "An Agent-Based Framework for Building Decision Support Systems", Decision Support Systems, Vol. 25, pp. 225-237, 1999.

[13] http://www.survey.co.kr.

[14] http://www.surveysite.com/docs/index-showapplet.html.

Alignment of Goals between Agents in a Collaborative Decision Process

R.T.Sreeram and P.K.Chawdhry
Department of Mechanical Engineering
University of Bath, Bath BA2 7AY
United Kingdom.
Tel: +44-1225-826826 ext. 5376
Fax: +44-1225-826928
Email: enprts, enspkc@bath.ac.uk

Abstract

This paper views collaborative product development as a multi-agent decision process where the agents have their own preferences towards a common goal. These preferences often lead to conflicts between agents on certain shared design parameters. For the resolution of such parametric conflicts, the authors use a combination of game theory and dependency based reasoning encapsulated in a negotiation scheme. The scheme also presents certain additional criteria for isolating a single negotiated solution. A case study of the design and manufacture of a Geneva mechanism is used as a test vehicle to investigate conflicts between: (a) software- software agents, (b) software - human agents and (c) human - human agents. The results obtained from this study indicate that the combination of game theory and dependency based techniques assists in maintaining rationality and alignment of preferences between agents in a decision process.

1 Introduction

A collaborative product development process involves a heterogeneous set of human and software agents [5]. In a product development process, these heterogeneous agents have individual preferences on certain shared design attributes such as strength or cost. The existence of such preferences often leads to conflicts that need to be resolved in a timely manner. For the resolution of such conflicts researchers have identified techniques based on game theory and constraint relaxation [3]-[2].

The game theoretic approaches evaluate a single point negotiated solution but the effect of dependen-cies between design parameters is generally ignored. This can lead to localised solutions as stated by Ku-siak *et al.*, [2]. In order to maintain global consistency in the negotiated solutions, the use of qualitative and quantitative reasoning assist in propagating the effect of parametric variation uniformly throughout the domain. In addition, the use of aspiration levels to mark preferences relates well to the real world scenario where decision makers change their preferences in the midst of a negotiation process. The unification of the two approaches and a quantitative comparison with that of previous research was presented in [3], [4].

This paper examines the feasibility of a conflict negotiation scheme,[3] [4], in the context of a design and manufacture case study of the Geneva mechanism. In particular, the conflicts are presented as cases of bilateral negotiation between (a) software-software agents, (b) software - human agents and (c) human - human agents. In a given conflict situation, these agents are assumed to have differing preferences towards a shared design attribute. The alignment of such preferences is carried out via negotiation schemes. In addition, the rationality in the negotiated solutions is maintained using dependency based reasoning techniques [2]. Both qualitative and quantitative results are presented as a means to evaluate the extent of rationality and alignment of decisions between agents.

The rest of the paper is organised as follows. Section 2 reviews the key aspects of the unified negotiation scheme developed by the authors for negotiation in a preference-based process [3], [4]. Section 3 applies the scheme to a design and manufacture case study of the Geneva mechanism. The results obtained from the case study are presented in Section 4 followed by a detailed analysis in Section 5. Section 6 presents the conclusions of this research.

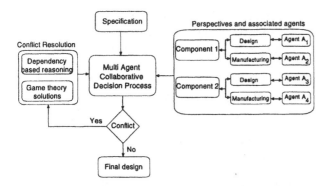

Figure 1: A generic multi-agent decision process [4]

2 A unified scheme for conflict negotiation

A comparison of game theory and constraint relaxation approaches in [3] and [4] emphasised the need for single-point consistent design solutions. Therefore a negotiation scheme with the advantages of the two approaches could be considered beneficial. Such a scheme captures the following key attributes:

- agents' preferences
- a single point solution
- consistency of design solutions

An abstract negotiation process which captures these attributes in terms of formal procedures is shown in Figure 2. Agent's preferences need to be explicitly stated for negotiation to be realistic. This is carried out by specifying certain utility functions for each agent involved in a conflict. A single-shot game-theoretic technique would yield a negotiated settlement based on techniques such as Nash and Kalai-Somordinsky [1]. Based on a global constraint network the effect of the negotiated solution is propagated throughout the domain to ensure that the obtained solutions are consistent.

2.1 Identification of conflicts

Kusiak and Wang [2] have developed a negotiation formalism which included the ideas of perspectives, design variables and constraints between them. We retain the notation used in [2].

In a collaborative process ($\mathbf{P_c}$), expert agents (A_i, $i = 1....n, n \in \Re$) belonging to several prespectives (P_j, $j = 1....m, m \in \Re$) such that ($P_j \subset \mathbf{P_c}$). Each perspective P_j could correspond to product development phases such as design and manufacturing. Expert agents associated with each perspective ($A_i \equiv P_j$)

are responsible for task processing. Each agent associated with a perspective consists of several design variables $\{V\}$ such that $\{V\} \subset (A_i \equiv P_j) \subset P_c$.

- The design specification is usually captured as a set of design variables $\{V_S\} \in \{V\}$ which are assigned values at the start of the design process. This set of variables contains certain design requirements common to all agents.

- Each agent (A_i) in the decision process is assumed to have a set of preferences U^{A_i} over a set of shared variables $\{V_{SH}\} \in \{V\}$ such that $U^{A_i} \neq \{\phi\}, \{V_{SH}\} \neq \{V_s\}$.

- The evaluation of unknown variables including the decision [$\{V_D\} \in \{V\}$] and performance variables $\{V_P\} \in \{V\}$ are carried out and a initial design is formed.

- Each agent (A_i) in a decision process controls certain key decision variables that are also directly related to the performance variables.

- As the agents interact to examine the acceptance of the initial design, a conflict can be detected due to differences in preferences over shared variables, i.e., $A_i(\{V_{SH}\}) \neq A_j(\{V_{SH}\}), i \neq j$.

2.2 Reasoning on a design network

A design network graph ($\mathbf{G}$) is formed based on the analytical formulae [of the form $f(V)$] which describe the domain. Such a network is useful in propagating any changes throughout the domain. This network provides quantitative as well as qualitative dependency between any two design variables. The Figure 2 shows a directed graph $\mathbf{G}$ represented as a four tuple $\mathbf{G} = (\{V^e\}, \{E\}, \Omega, \Psi); \{V\} \in V^e, \delta \in \Omega$ and $\psi \in \Psi$, where:

> $\{V^e\} \in \{V\}$ are the vertices of $\mathbf{G}$.
> $\{E\}$ is the set of directed edges.
> δ is the qualitative dependency of each edge (E).
> ψ is the quantitative dependency of each edge (E).

When a full network is specified it has all the edges, vertices and dependencies. Of immediate interest are the decision and performance variables and the link between them. Thus the aim is to establish effective dependencies between the decision and performance variables. A general reasoning procedure is given in [3].

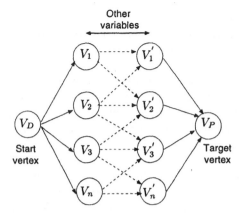

Figure 2: Graph (G) showing the start (V_D) and the target (V_P) vertices

2.3 Trade-off and a mechanism to enforce decisions

Based on the utility functions, game-theoretic solutions such as Nash [N(S)], Kalai-Somordinsky [K(S)], Egalitarian [E(S)] and Utilitarian methods [U(S)], a one-shot compromise is reached resulting in four different game-theoretic solutions in general. However, only one solution out of the four can be chosen as a final negotiated settlement. The need to isolate a single solution arises for two reasons: (a) The pay-off for each agent obtained by a given scheme are not always identical. Therefore, an agent may wish to choose a solution which yields highest possible utility to itself but not necessarily to other agents; (b) The need to prevent secondary conflicts which could delay the negotiation process. In the case of human conflicts, Nash (1950) suggested the use of a generic mechanism for breaking deadlocks in bargaining. Certain IF-THEN rules have been developed in [3] with the aim of obtaining a single point solution from a set of game-theoretic outcomes.

3 Example: A collaborative environment for the Geneva mechanism

3.1 Objective

The design and manufacture of a Geneva mechanism has been chosen as a test vehicle to validate the alignment of preferences between agents (both software and human) in a collaborative process for a product of medium complexity. Specifically the Geneva mechanism has been chosen as the representative product for the following reasons:

- A Geneva mechanism represents a product of reasonable complexity with respect to its geometry and closely interacting geometric features.

  - Large number of design variables.
  - The geometrical features strongly interact.
  - Tightly coupled design parameters.
  - A multi-part product with kinematic functions.

- The mould making process would require high quality tools and precision. This associated with the shrinkage allowance for the plastic makes the design of the mould intricate. Such a design would also involve designing the ejection mechanism for the removal of the product from the mould.

- To present a case for testing negotiation schemes involving a variety of agents. In particular, three cases of negotiation behaviour are studied: (a) software agent - software agent (b) software agent - human agent and (c) human agent - human agent.

In addition, there were several practical considerations, including available facilities and expertise of human agents, which served as the criteria for this choice.

3.2 Experimental methodology

The experimental methodology consisted of the identification of the design problem which is suitable for representing conflict negotiation situations. The Geneva mechanism case provided a problem which allows representation of agents' preferences on certain shared design attributes. Within the life-cycle process, of interest to conflict situations and closely interacting phases are design and manufacture of the wheel, driver and mould ejection mechanisms. Before carrying out the actual experiment, it was decided that the design related decisions of the Geneva mechanism should be assigned to software agents whereas manufacture related decisions should be assigned to the human agents. In order to facilitate task solving, certain product-specific software agents were built (e.g., wheel and driver agents). The choice of human agents was based on their expertise in mould design and manufacturing. The product development framework also included manufacturing machines (e.g., milling and injection moulding machines) and software tools to support collaborative work [5]. In particular, the reasoning and negotiation agents [3] were present in the framework to enable the resolution of conflicts between agents. These agents, software tools and machines were configured in a collaborative product development process.

3.3 The product development process

The Figure 3 shows how the agents and other tools were configured in the collaborative design and manufacture of the Geneva mechanism. The human agents were responsible for ensuring the correct working order of both software and hardware tools. For example, the human designer (HA1) was responsible for testing the software agents (for reasoning, negotiation and design). Figure 3 captured the product development process for which the quantitative analysis and relevant product information is presented in detail.

3.4 Design specification

The human agents (HA1, HA2 and HA6) were responsible for identifying the specification for the design problem. The specification of the design consisted of the following key stages:

- the initial specification

- formation of design network and evaluation of dependencies

- formation of utilities

The design of Geneva mechanism is specified by assigning initial values to the design variables shown in Table 1.

4 Negotiation process applied to the Geneva mechanism

4.1 Construction of dependency graph

With this set of design specifications, a dependency network graph, as shown in Figure 4, was created to determine the qualitative and quantitative dependencies between design variables. The human designer (HA1 in Figure 3) was responsible for specifying this network graph (by inputting the edges and their corresponding qualitative dependencies) to the reasoning agent [Table 2]. This was done in order to enable the reasoning agent to evaluate the qualitative and quantitative dependencies between any two nodes in the network graph.

Table 1: Design specification and constraints

SPECIFICATIONS		
Number of slots, N_s	-	6
Speed of rotation, ω	rpm	1000
Inertial load, I_l	$Kg - m^2$	1.1298e-06
Material, P_{mat}	-	Polypropylene
OTHER VARIABLES		
Wheel diameter, D	m	0.06
Wheel tip width, w_{tip}	m	0.005
CONSTRAINTS		
$0.003 \geq w_{tip} \leq 0.007$; $C \leq 0.03m$;		$N_s \leq 8$
$0.05 \leq D \leq 0.07$; $w_t \leq 0.01m$		

4.2 Evaluation of dependencies

In the present case study of Geneva design, the key decision variables are N_s, D, w_{tip}, r_p and S_w. The influence of these decision variables on the final performance parameters such as τ_C, σ_R and σ_T was considered important in this study. In particular, emphasis is placed on slot width (S_w), which is a common variable for both wheel and driver agents.

To evaluate the dependencies between design variables, the qualitative and quantitative influence on critical parameters was determined. An effective dependency relationship was thus established between the key decision and other performance variables. The effect of change in slot width (S_w) on shear stress (τ_C) was one of the deciding factors for choosing preferences for agents. The contact shear stress (τ_C) between the pin and the wheel slot is of considerable importance in Geneva mechanism design; life of the Geneva mechanism is critically dependent on τ_C [Lee (1985)]. The following observations can be made in Table 2.

- increasing the wheel diameter (D) increases the shear stress. For example, the quantitative dependency relating D and τ_C is $\psi_{D,\tau_C} = 2.56811$; increasing wheel diameter by 10% increases shear stress by approximately 26%. Thus the increase

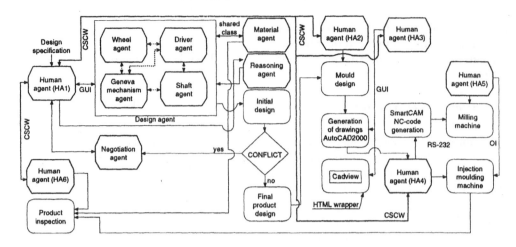

Figure 3: Configuration of the product development process

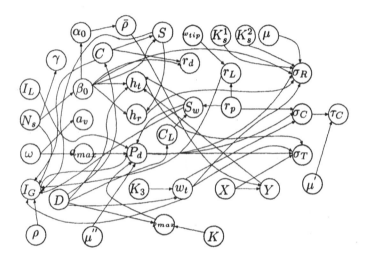

Figure 4: Design network for the Geneva mechanism

in D was not considered desirable due to (a) limitations on the size of the Geneva wheel (manufacturing constraints) and (b) material breakage at wheel root due to high root and shear stress.

- The pin radius (r_p) has a negative influence on the shear stress values. However, r_p cannot be randomly increased without affecting the wheel size; in particular, the wheel diameter (D). Hence any large increase in the value of r_p would not be welcome.

- The influence of centre distance (C) was found to positively influence the state of shear stress. C is one design parameter which was found to be very sensitive to perturbations. Even a minor change in the value of C infuences the state of shear stress by a large extent. Such a choice would

pose a disadvantage for negotiation though centre distance is of common interest to both wheel and driver agents.

- Increase in the value of wheel tip width (w_{tip}) also shows marked increase in the value of τ_C. This study found that increasing w_{tip} minimises root stress (σ_R) and tip stress (σ_T). However, such an increase could be made possible without affecting the overall wheel size only up to a certain limit imposed on the wheel diameter. Moreover, since w_{tip} is governed by the wheel agent alone, this choice is not in the common interest of both agents.

- Of all the variables considered in Table 2, the influence of wheel slot width (S_w) on τ_C was found to be minimal. In addition, any increase in S_w has only a negative influence on the state of stress. This is one common design parameter that can be explictly approached based on mathematical relations from both wheel and driver design perspectives.

4.3 Identification of conflict situations

During the product development process three distinct conflicts were identified:

- Based on the initial design specification for the Geneva mechanism, the wheel and driver agents proposed two different values for the slot width which evaluated to be $0.0085m$ and $0.0094m$ respectively. This presented a conflict situation between the wheel and the driver agents.

Table 2: Effective parametric dependencies between design variables

Effective parameter dependencies								
$D \to \tau_C$			$D \to \sigma_R$			$D \to \sigma_T$		
NP:	δ_{D,τ_C}	ψ_{D,τ_C}	NP:	δ_{D,σ_R}	ψ_{D,σ_R}	NP:	δ_{D,σ_R}	ψ_{D,σ_R}
6	+	2.56811	7	+	24.4438	8	-	-2.14665
$r_p \to \tau_C$			$r_p \to \sigma_R$			$r_p \to \sigma_T$		
NP:	δ_{r_p,τ_C}	ψ_{r_p,τ_C}	NP:	δ_{r_p,σ_R}	ψ_{r_p,σ_R}	NP:	δ_{r_p,σ_R}	ψ_{r_p,σ_R}
3	-	-0.35064	2	-	-0.00120	3	-	-0.88780
$S_w \to \tau_C$			$S_w \to \sigma_R$			$S_w \to \sigma_T$		
NP:	δ_{S_w,τ_C}	ψ_{S_w,τ_C}	NP:	δ_{S_w,σ_R}	ψ_{S_w,σ_R}	NP:	δ_{S_w,σ_R}	ψ_{S_w,σ_R}
2	+	0.12838	2	-	-0.00135	3	-	-0.99350
$C \to \tau_C$			$C \to \sigma_R$			$C \to \sigma_T$		
NP:	δ_{C,τ_C}	ψ_{C,τ_C}	NP:	δ_{C,σ_R}	ψ_{C,σ_R}	NP:	δ_{C,σ_R}	ψ_{C,σ_R}
2	+	2.17024	3	+	24.4384	2	-	-3.0
$w_{tip} \to \tau_C$			$w_{tip} \to \sigma_R$			$w_{tip} \to \sigma_T$		
NP:	δ_{w_{tip},τ_C}	ψ_{w_{tip},τ_C}	NP:	$\delta_{w_{tip},\sigma_R}$	ψ_{w_{tip},σ_R}	NP:	$\delta_{w_{tip},\sigma_R}$	ψ_{w_{tip},σ_R}
1	+	2.37219	1	-	-0.02498	1	-	-3.27936

[NP: number of paths]

- The wheel thickness (w_t) evaluated by the wheel agent turned out to be $0.0084m$. This was found to be in conflict with human agent coalition (HA3 & HA4) which proposed a value, $w_t \leq 0.0064m$.

- During the mould design, a conflict was identified between two human agent coalitions. The HA1 and HA2 suggested a six-pin based ejction mechanism whereas the HA3 and HA4 coalition preferred the use of two ejector pins.

Having presented the conflict situations, their resolution was carried out using the unified scheme in [3].

5 Conflict resolution

5.1 Conflict on slot width: between software agents

The human designer agent specified the utilities shown in Figure 5 to the wheel and driver agents. Based on these utility functions (specified by the wheel and driver agents), the negotiation agent evaluated four different game theoretic solutions. A summary of the results of negotiation is presented in Table 3. The Kalai-Somordinsky solution is taken as the final negotiated settlement due its property of equal utilities to both the agents. For the present case the negotiated solution evaluated to a slot width of $0.009m$. The change in S_w need to be propagated locally as in Kanappan and Marshek (1993) or globally. Depending on the problem and the number of independent

Table 3: Game theory solutions for conflict on slot width

Agents	Conflict variable	Preferred values	Game theory solutions				Final game solution	Negotiated value for S_w
			N(S)	K(S)	E(S)	U(S)		
Wheel agent	S_w	0.0085	0.6795	0.7566	0.7566	0.6795	0.7566	0.009
Driver agent		0.0094	1.0	0.7566	0.7566	1.0	0.7566	

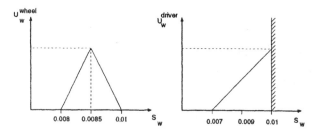

Figure 5: Utility functions for slot width

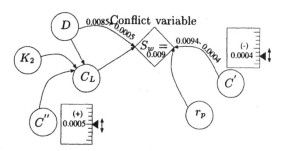

Figure 6: Design network pertaining to slot width

variables associated with the conflict variable, the effect of negotiated solution can be included in the design. For example, Figure 6 shows the graph of immediate variables that affect S_w. Of these variables, C' and C'' are the two constants that are used to represent clearance and are varied from both wheel and driver design perspectives to suit the negotiated solution; the values for C' and C'' are changed by +0.0005 and −0.0004 respectively[1]. The human designer agents (HA1 and HA2) prepared the part drawings using AutoCAD based on the initial design. The drawings of Geneva wheel and driver thus generated were then analysed for manufacturabiity.

5.2 Conflict on wheel thickness: between software-human agents

The human agents examined the part drawings based on dimensions and limits set by the design specification. Though all geometrical and material constraints were identified prior to the initial design, there were some resource restrictions on machine tools such

---

[1]The idea of clearances presented in this study is based on the limits and fits for cylindrical parts. In particular an approximate estimate is obtained from running and sliding fits in the nieghbhourhood of RC2 to RC7 [9].

as availability of milling cutters, etc. The machinist agents suggested a reduction in the wheel thickness. The need for thickness reduction is due to a possibility of incomplete edge formation in the manufactured product. As shown in Figure 7, a drill tool with diameter greater than $2mm$ would result in rounded edges affecting the tip width and pin entry. A drill tool with

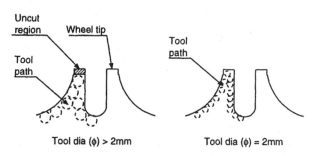

Figure 7: Effect of drill tool on wheel tip width

less than 2mm diameter was, however, not preferred by the machinist agents due to a high depth of cut required. As a result a conflict was detected between the wheel design agent and the human agents on the value of wheel thickness (w_t). As the first step towards conflict resolution, the human agents specified

Table 4: Game theory solutions for conflict on wheel thickness

Agents	Conflict variable	Preferred values	Game theory solutions				Final game solution	Negotiated value for w_t
			N(S)	K(S)	E(S)	U(S)		
Human agents	w_t	0.0084	0.6884	0.5899	0.5899	1.0	0.5899	0.0070
Wheel agent		0.0060	0.5161	0.5899	0.5899	0.29	0.5899	

a utility description for w_t [2]. Another utility description representing the choice of wheel design agent was also specified [Figure 8]. These utility functions were input to the negotiation agent which evaluated a negotiated settlement. Table 4 indicates that the Kalai-

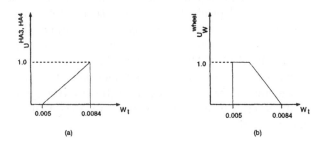

Figure 8: Utility variations of agents on wheel thickness

Somordinsky value which yielded a utility of 0.59 to both agents was chosen as the final settlement. The finalised value for the wheel thickness was found to be $0.007m$. However, this negotiated value for wheel thickness had to be propagated throughout the design network. As shown in Figure 9, a network of all variables relating to wheel thickness was formed. The change in w_t is propagated to K_3. Since K_3 [3] is an independent constant and is also directly proportional to w_t, this was an apt choice. The corrected value of K_3 was found to be 0.1167. Based on the changes which occured in the values of S_w and w_t, an updated design was evaluated. Only the performance parameters (such as stress and load) have been altered and not the key geometric parameters such as the wheel diameter and wheel tip width.

---

[2] The utility description formed here represents a valid group coalition function of HA2, HA3 and HA4.

[3] The range of K_3 is between 0.05 to 0.14.

5.3 Conflict on the number of ejector pins: between human agents

Based on the initial design, the human designers (HA1 and HA5) created the drawings of the mould using AutoCAD. The mould drawings for wheel and driver without the positioning of the gates were generated with a $0.5mm$ allowance for shrinkage [4] as shown in Figure 10. This design satisfied the general re-

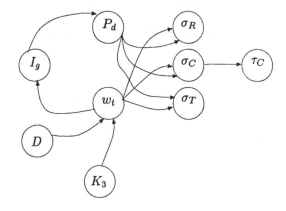

Figure 9: Design network pertaining to wheel thickness

quirements of a good mould stated above and was thus finalised. However, there was still a need for an ejection mechanism. Based on the brainstorming discussions between the machinists (HA3 and HA4) and designers (HA1 and HA2), it was decided that ejector pins will need to be arranged in the base of the mould. A conflict was then detected on the number of pins required. The machinist agents suggested the use of two pins in opposite corners of the mould as shown in the Figure 11. The designers, however, preferred

---

[4] this allowance for polypropylene was assumed based on the information provided by GE Plastics, Inc. USA.

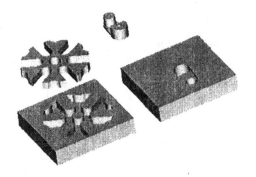

Figure 10: The Geneva wheel and driver moulds

six pins arranged in symmetric locations. As a means

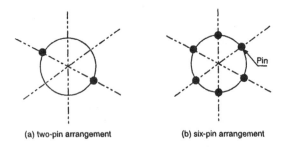

(a) two-pin arrangement   (b) six-pin arrangement

Figure 11: Pin configuration for ejection of the product

of solving this conflict between human agents, certain rational criteria were identified to proceed with negotiation:

- shrinkage allowance

- nature of plastic material

- effect on the mould

- effect on the product

- machining time

The process of negotiation between the human agents is qualitatively represented in Table 5. The machinists argued that due to the shrinkage of plastic upon cooling, the moulded product would seperate from the walls of the mould. Therefore, only a sufficient shrinkage allowance and two ejector pins would be adequate for proper ejection. The effect of pin force on the product is unsymmetrical and may lead to twisting of the wheel upon removal from the mould. As Table 5 indicates the quality of decisions for HA3 and HA4 coalition varies between medium (M) and high (H) except for the case of the effect on product. The quality of decisions made by HA1 and HA2 differs from that of

HA3 and HA4. As Table 5 shows, the quality of decision considering machining time and effect on mould were rated between low (L) and medium (M). Since the HA1-HA2 coalition has two low ratings, the negotiated settlement was to use two pins initially and add pins later, if necessary. In addition, a final assessment of the product would be possible only after manufacturing trials with the injection moulding machine. But based on the available information prior to manufacturing, the settlement was considered rational.

The moulding process was carried out after making the necessary changes in the settings for the injection moulding machine. The moulded product was checked for its dimensional accuracy. Interestingly, the cooling patterns indicated a higher rate of shrinkage in the thickness direction ($w_t \approx 6.4mm$). The decision to use six-pin based ejection mechanism for the wheel was found to be effective. However, for the driver mould, the two-pin arrangement was found sufficient. The manufactured product was then assembled on to an aluminium base with a specified centre distance [Figure 12]. The assembled mechanism was functionally tested (by HA1, HA2 and HA6 in Figure 3) for its slot entry, exit and other prescribed motions. The product exhibited fair shrinkage characteristic throughout and the clearances on slot width, etc, were found to be preserved.

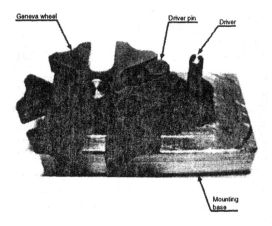

Figure 12: The assembled Geneva mechanism

6   Discussion

This case study has identified the goal-level integration in product development as a key issue. The multi-agent collabarative design and manufacturing of the Geneva mechanism with multiple goals led to the need

Table 5: Negotiation results for conflict between human agents

Agents	Conflict variable	Preferred value	Quality of decisions					Solution
			Shrinkage	Plastic type	Mould	Product	Time	
HA3, HA4	Number of ejector pins	2	M-H	M-H	M-H	L-M	M-H	2 pins
HA1, HA2		6	M-H	M-H	L-M	M-H	L-M	

for a negotiation mechanism to ensure the alignment of goals in a preference-led decision process. Such a conflict resolution mechanism needs to be flexible enough to deal with conflicts between heterogenous agents (both human and software).

The case study investigated the conflicts arising due to varied individual preferences of agents and their resolution via a novel scheme for negotiation. The negotiation scheme (developed as a combination of game theory and dependency based reasoning) has been found to offer consistent design solutions while preserving the preferences of agents. The conflict resolution between agents has been approached as a two-player game with specified utility functions. With respect to the chosen domain (which is parametric design) where dependencies exist, the present results have validated the combination of dependency and game theory approaches. However, this claim excludes possible situations where only fuzzy relations (between design parameters) are present.

Based on the quantitative results presented in Section 5, the Geneva mechanism presented a problem of higher complexity which has involved several more variables as compared to cases presented in Kanappan and Marshak [1], Kusiak and Wang [2] and Berker and Brown [6]. The unified approach has been validated in the context of parametric design problems with well defined analytical relations. The game theory based approaches need the specification of objective or utility functions. Therefore, the genericity of the unified approach can be seen at par with that of Kusiak and Wang [2] and Kanappan and Marshek [1]. Klien [10] presented generic schemes (mainly rule-based) for conflict resolution that is independent of the domain and hence the genericity of Klien's approach is high compared to that of the authors' approach [5].

The importance of consistency of design solutions has been demonstrated via the qualitative and quantitative reasoning. The reasoning based approach ensured that the solutions are consistent and globally updated for any change. In the present study, this is evident in instances such as the effect of the slot width (S_w) and stress levels (at wheel root and tip). With respect to design consistency, the authors' approach is superior to that of Kanappan and Marshek [1] or Berker and Brown [6]. This is due to the fact that the latter did not consider the consistency of solutions in their negotiation schemes.

Several of the previous investigations have focussed on applying standard techniques based on game theory to evaluate a set of negotiated solutions [for example, see Kanappan and Marshek [1]]. A weak approach of contraction of feasible design space was adopted by Kusiak and Wang [2] for obtaining a single point solution for a specific case[6]. Hence most of the previous research did not focus on obtaining a single point solution. The unified scheme of Sreeram and Chawdhry [3] offers additional rules for isolating a single solution from a set of game-theoretic outcomes.

With respect to a scheme for conflict resolution, the author's approach has developed, implemented and validated a scheme. Several of the previous investigations, including that of Berker and Brown [6], Kanappan and Marshek [1], did not systematically analyse their negotiation protocol. For example, Kanappan and Marshek obtained negotiated solutions using game-theoretic schemes but did not consider the effect of parametric dependencies or analyse the quality of negotiated solutions. Kusiak and Wang [2], however, presented a formal dependency based procedure for

---

[5]However, Klien [10] did not quantify the results obtained from his generic schemes applied to a case study. This is not surprising since the schemes developed in his study were independent of the domain which bears a large proportion of the quantitative information

[6]The approach of contraction generally cannot ensure convergence of feasible design space to a single point solution when preferences between agents exist [see Sreeram and Chawdhry [3]]

reasoning in a design network.

As seen in real-world scenarios, the proposed research takes a view that conflict always occurs due to differing preferences with a realistic assumption that total cooperation between agents may not always exist. Such a view was not adopted in Berker and Brown [6] and Kusiak and Wang [2] where total cooperation was assumed. Finally, the proposed approach and other approaches including that of Kanappan and Marshek have been applied to conflict situations involving two agents or two coalitions. The conflicts between n-agents or n-coalitions, though an interesting research issue in game theory, has not been investigated by the authors since this does not fall within the identified scope of this research.

The following key points emerge from this discussion:

- In a collaborative decision process, there is a need for explicit representation of agents' preferences (both software and human).

- These preferences are usually contradictory.

- For conflict resolution between agents in parametric design, it is necessary to combine the game theory and dependency based reasoning techniques. Such a combination has been shown to yield consistent single point solutions within a feasible design space.

7   Conclusions

This paper examined a multiple-goal decision process and ensuing conflicts between agents in a case study of a collaborative design and manufacture of a Geneva mechanism.

This research viewed the goal level integration in product development as a key issue. Furthermore, it investigated the conflicts due to preferences that occured between software and human agents through the development of novel schemes for negotiation. The negotiation schemes (developed as a combination of game theory and dependency based reasoning) have been found to offer consistent design solutions while preserving the preferences of agents. The conflict resolution between agents has been approached as a two-player game with specified utility functions. With respect to the chosen domain (i.e., parametric design) where dependencies exist, the present results have validated the combination of dependency and game theory approach to negotiation.

The proposed research has improved upon certain methodological limitations suffered by the previous research with respect to aligning goals between agents in a collaborative framework consisting of software and human agents, CSCW[7] tools and legacy non-networkable facilities.

Acknowledgements

The first author would wish to acknowledge the Postgraduate Research Studentship award from the University of Bath and the Overseas Research Scheme (ORS) award from the Council of Vice Chancellors and Principals (CVCP), London, for this research.

References

[1] S. M. Kanappan and K. M. Marshek, "Engineering Design Methodologies: A new perspective. Chapter in Concurrent Engineering Automation , Tools and Techniques", (ed.) Kusiak, A., John Wiley & Sons, New York, 1993.

[2] A. Kusiak, J. Wang, and D. W. He, "Negotiation in Constraint Based Design," *ASME Journal of Mechanical Design*, Vol. 118, 9, pp. 470-477, 1996.

[3] R. T. Sreeram and P. K. Chawdhry, "A Unified Scheme for Conflict Negotiation in a Multi-Agent Decision Process," *Proceedings of the sixth ISPE Conference on Concurrent Engineering: Research and Applications*, September 1-3, Bath, UK., 1999a.

[4] R. T. Sreeram and P. K. Chawdhry, "Towards the Unification of Game-Theoretic and Constraint Relaxation Techniques for Conflict Negotiation," *Proceedings of the sixth ISPE Conference on Concurrent Engineering: Research and Applications*, September 1-3, Bath, UK., 1999b.

[5] R. T. Sreeram and P. K. Chawdhry, "Human-centred Integration of a Stand-Alone Manufacturing Facility in a Networked Product Development Environment," *International Journal of Computer Integrated Manufacturing*, Vol.12, 4, pp. 338-360. 1999c.

[6] I. Berker, and D. C. Brown, "Conflicts and Negotiation in Single Function Agent Based Design Systems", *CERA Journal*, Vol.4, 1, pp. 17-33. 1996.

---

[7]CSCW = Computer Supported Cooperative Work

[7] J. Nash, "The Bargaining Problem", *Econometrica*, Vol.18, 2, pp. 155-162. 1950.

[8] M. P. Wellman, *Formulation of Tradeoffs in Planning Under Uncertainity*, Pitman Publishing, New York, 1990.

[9] J. E. Shigley and L. D. Mitchell, *Mechanical Engineering Design*, McGraw-Hill, New York, 1983.

[10] M. Klien, "Supporting Conflict Management in Co-operative Design Teams", *Group Decision and Negotiation*, 2, pp. 259-278. 1989.

Appendix 1

Nomenclature of symbols used for the Geneva Mechanism

β_0 = Angle of Geneva wheel (deg.)
R = Radius of Geneva wheel (m)
D = Diameter of Geneva wheel (m)
r_L = Locking drum radius (m)
a_{max} = Maximum acceleration of geneva wheel (rad/sec^2)
S_w = Wheel slot width (m)
w_{tip} = Wheel tip width (m)
w_t = Wheel thickness (m)

N_s = Number of slots
C = Distance between centres of wheel and driver (m)
S = Wheel slot distance (m)
I_L = Load inertia $(kg - m^2)$
I_G = Moment of inertia of geneva wheel $(kg - m^2)$
P_{mat} = Material of the wheel and pin
ρ_{wheel} = Density of wheel material (kg/m^3)
ν_{wheel} = Poisson's ratio of wheel material
E_{wheel} = Modulus of elasticity of pin material (Pa)
σ_T = Tip stress (Pa)
σ_R = Root stress (Pa)
τ_C = Contact shear stress (Pa)
C_L = Clearance between wheel and slot (m)
K_1, K_2, K_3, K_4 and K_5 = design constants
k_1^s, k_2^s = Stress concentration factors
μ = Coefficient of friction
α_0 = Angle of driver (deg.)
r_P = Pin radius (m)
r_D = Driver radius (m)
σ_C = Pin contact stress (Pa)
ρ_{pin} = Density of wheel material (kg/m^3)
ν_{pin} = Poisson's ratio of wheel material
E_{pin} = Modulus of elasticity of pin material (Pa)

CHAPTER 6

Standards in CE

A framework for adoption of Standards for Data Exchange

Ricardo Jardim-Gonçalves   and   Adolfo Steiger-Garção

UNINOVA – Instituto de Desenvolvimento de Novas Tecnologias
Dep. de Eng. Electrotécnica da Fac. de Ciências e Tecnologia da Univ. Nova de Lisboa
Campus FCT/UNL, Quinta da Torre, P2825-114 Caparica – Portugal
Email: rg@uninova.pt, asg@uninova.pt

Abstract

One of the main problems usually found not contributing for an easy adoption of standards for data exchange, is the complexity and effort required to develop the interfaces to adopt it, acting as converters between the Application's data model and the standard's. In environments where models for data exchange are very often updated, this problem is more significant requiring flexibility in its interfaces.

In order to give some support for the development and update of these interfaces for standard-based data exchange, standards include Implementation Methods (IM) together with Standardised Data Access Interfaces (SDAI). This is the case for example of 20s Series of the ISO 10303 STEP (STandard for the Exchange of Product model data).

The paper starts presenting in detail the scope and architecture of SDA, based on the proposed by Part22 of STEP, to be the basis for the understanding of the presented research work in next sections, providing the background for those not aware of SDAI.

After, it describes the requirements for the development and linkage of converters for applications willing to adopt a standard for data exchange.

Following, it presents results of the research at UNINOVA on automatic code generator toolkits, acting as facilitators for the development of these converters. The architecture to support extensions on top of these facilitators, to support data rules validation and knowledge representation are presented.

Also, it describes a strategy for the reuse of already existent standard data models, like STEP's Application Protocols (APs) in its 200s series. This strategy is based on a Multi-level reuse of APs to cover the data model requirements for data integration of Applications, when the Application's data scope embraces several of existent APs. Therefore, it shows how facilitators can contribute to support the implementation if this Multi-level APs.

The paper concludes with an overview on the projects where this research was developed.

1. Bridging applications to Neutral Format Platforms

For one application to adopt a Standard for Data Exchange, it should implement the interface that bridges the Application's data represented usually in proprietary format, translating such own data format to the Standard's.

In order to make easier the development of these translators, an Application Programming Interface (API) is normally defined by the Standard, providing a description of the required basic mechanisms to support the implementation of the translators for data described according to the Standard model [18].

Defined as a Standard, this API works as the normative layer establishing the communication between the data objects themselves and those objects implemented by the application, and it is very often called Standard Data Access Interface (SDAI).

Figure 1 depicts the role of SDAI in the development of a converter between an Application internal data format and a standard Application Protocol.

To those interested to adopt a standard, the importance to have this APIs tested and ready to be used and linked by applications is enormous, since it will facilitate its adoption in a convenient level of abstraction without the requirements in terms of details of implementation for those developing the translators for import and export of data from the beginning.

Analysing Figure 1, for Application1 to adopt a standard data model (Application Protocol) for data exchange, it needs to develop a translator for data import and export. Such converter can be developed on the top of a SDAI that provides the basic mechanisms to access and manage the data in Standard format. Therefore, the required development is thus the Application Data Model to SDAI converter (DM2SDAI), which links with the SDAI to have direct access to the Application1's data.

333

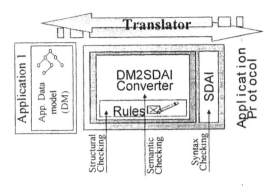

Figure 1. The role of SDAI in the development of a converter

Another important feature a SDAI should provide is the possibility to support *early* and *late* bindings of the Application Protocol for the programming language of implementation.

Early binding means that SDAI will adopt a static structure dedicated exclusively to manage the specific AP for what it was developed.

Late binding is based in a general implementation of SDAI, assisted by a dictionary of data ready to support any application model, and able to be dynamically managed and handled in run-time.

While the early binding approach is static, although with an easier and faster interface for implementation purposes, the late binding is very flexible allowing run-time model changes and updates.

Examples of these APIs are the STEP's SDAI, described in detail in next section, and the XML's DOM.

The Document Object Model (DOM) [4] is an API to access and manage data represented in XML (eXtensible Markup Language) format [5]. This API understands the XML data described as a tree-based representation, and defines the mechanisms required to navigate and manage the data across such tree in width and depth.

These mechanisms enable access and handle of its elements and attribute values as tree data nodes, allowing insert and delete of such nodes, and the conversion of the tree structure back into XML data format.

The root of the tree is the XML document. Each root's child represents the top-level instances of XML data Elements. Each Element can have related attributes and other Elements as children nodes, representing the data content and its sub-elements, which may have also children, and so on.

The DOM is thus a useful interface to handle and manage XML data format files, viewing its structure mapped as a DOM tree. Modification or production of new XML documents as output, or construction of a new DOM tree

by beginning and after convert it to XML, can be done using this API.

This mechanism provide a very flexible way of access and produce XML data format output, usually easier than simply writing or reading directly to a file in that format.

2. STEP's Standard Data Access Interface

The ISO10303 STEP [3][6][8] describes its API for standard format data access in its part #22 - Standard Data Access Interface (SDAI) [7], specifying its functionalities in a general and neutral way independently of a programming language.

Others parts in the same 20s series of this standard (e.g., Part#23, Part#24), provides a description of SDAI binding for specific programming languages (e.g., C, C++, Java), that when implemented are ready to be linked with the converters for the Application using such programming language.

This API enables the management of the data structures for the standard model, and provides the functionalities and means to handle and instantiate its objects, where objects and data dictionary are kept in an SDAI-repository.

The export of data in standard format is done using specific SDAI's procedures, and reflects the data stored in the SDAI's repository. The import of data using the SDAI specific functions allows reading data in standard format and populating the SDAI's repository to be accessed afterwards through several SDAI mechanisms.

This section gives a detailed explanation of the aim and architecture of SDAI, with the intent to be a support for the understanding of the research work presented in following sections, to those not aware of SDAI.

2.1. Architecture of SDAI

STEP standard uses the EXPRESS language [9] to describe data models. The architecture of SDAI is designed to support the implementation of interfaces for the creation and manipulation of instances of EXPRESS entities, independently of the storage technology.

The fundamental principle of SDAI is to make easier one Application to develop the required interfaces to import and export data in Neutral Format [10] following the specifications, rules and data structure of the set of Application data models (known as Schemas) to be adopted.

From the user point of view, the architecture of SDAI (Figure 2) is based on the set of operations that establishes the front-end with the application adopting a Standard protocol.

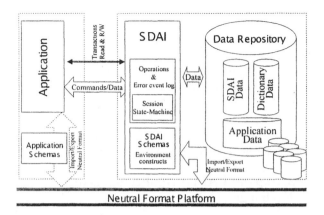

Figure 2. Architecture of SDAI

The Application communicates with SDAI using SDAI's commands, and an Error Event log can be generated to keep track on SDAI's usage and modus operandi.
Operations on SDAI run inside one SDAI Session. When the application intends to use SDAI, to Open one Session must be the first command to execute.
After, and until the execution of the Close command, all SDAI commands are available controlled by the Session's State Machine.
The Session State-Machine, implemented within SDAI, controls the sequence of available SDAI's commands in one point of time within one Session. This mechanism controls the appropriate sequence of use of these commands, obligating to be executed in a consistent and controlled way.
A Data repository supports the SDAI operations. Although very often just one Data repository is used, several can be adopted and used together.
The role of the Data repository is to store and keep updated the public and private data for exchange and control referent to the Application data, Schemas dictionary and SDAI related information. The Data Repository acts as the support for SDAI operations, manage of data structures and manipulation of instances of Application data schemas.
Schemas are the entities that define the structures of data inside repositories.
The SDAI schemas describe the data structures for data exchange environment constructs. They are fixed and are part of the SDAI specifications.
The Application data is store based in the data structure provided by the Application as Application Schemas. During its compilation process, the information referent to these schemas is stored in the Dictionary Data working as a support for the Application Data management.

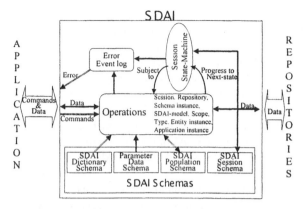

Figure 3. Internal structure of SDAI

The integration of one Application with a Neutral Format Platform using SDAI is established through these Applications Schemas, describing the data structures of the information to be exchanged as import/export mechanisms between the application and the external platform.
For implementation, SDAI provides the means to implement these mechanisms, putting them available as a set of operations to be used as an API.
Figure 3 describes the internal structure of SDAI.

2.2. The commands of SDAI

To use SDAI, the Application sends commands through the SDAI's API, together with the data to be exchanged between the Neutral Format Platform and the SDAI's repository. An Error Event log is available by SDAI to be used by the Application, if wanted, for check of reliability of the executed procedures.
All commands executed using SDAI must be done within one SDAI Session. All other SDAI commands will be executed under the scope of one session and are available controlled by the Session State Machine until the Session is closed.
The SDAI commands are grouped as:

- Session commands – Set of operations to handle and manage sessions, including the event recording tasks. These operations actuate directly on the Session Schema SDAI-model.
- Repository commands – Manages the access to SDAI's data repositories, supporting transaction capabilities. A transaction consists of a series of operations whose effect may be saved or undone as a unit. A session can initiate a transaction where a sequence of operations accessing to the repository can be executed in Read/Write or Read Only mode. When a transaction is closed,

modifications in repository can be Committed or Aborted, confirming or discarding all updates done in the meantime in repository.

- SDAI-model commands – A SDAI-model consists in a group of related entity instances based on a Schema or set of inter-related Schemas, where the Data Dictionary describes its entities implicitly or explicitly. Operations of this group manage SDAI-models and actuate based on the information stored in the SDAI's data dictionary.
- Schema instance commands – A Schema instance represents a logical set of related SDAI-models, and acts as the domain for validation rules and for the inter-schema referencing mechanism between external entities. Its operations comprehend management of SDAI-models and verification of its rules.
- Scope commands – The scope concept make available the possibility to have inside an SDAI-model a set of entities grouped and nested within an entity instance. This possibility enables the system to act on this scoped group as a unit.
- Type commands – Actuating directly on the data provided by the Dictionary, this set of commands allows getting information about the Schema entities, including its attributes and relationship (e.g., is_a) with other entities and schemas.
- Entity instance commands – These commands provide access to the SDAI-model instances in the repository, and includes the required mechanisms to handle, manage and validate instances for conformity with the Schema model rules from where its data structures and rules are defined. A sub-set of the Entity instance commands is defined as Application instance, describing the set of commands only applicable to instances entity data types defined in SDAI-models created by Application Schemas.

2.3. Execution of SDAI commands

One SDAI command is executed receiving the input parameters required for its execution, returning its results and updating accordingly the SDAI's Error Event log. When one operation is called, and before to be executed, the Session State Machine checks if the operation can be executed, consulting the Session status kept in the Repository's Session Data.

If positive, the operation is executed and the State Machine progresses for the next state. If not, the Error Event log is updated and the command operation returns without modifying the parameters.

For the handling of Application Data using the SDAI's repository, the Dictionary Data provides the information referent to the Schema Data Structures for the data to be exchanged, while the Population data keeps up to date the logical organization of Application instances inside the repository.

The Parameter's Data Types are also checked against the SDAI Parameter Data when instance data is passed as parameters through the SDAI's API.

2.4. Set-up of SDAI for Application Data Exchange

The integration of one Application with a Neutral Format Platform is made through a set of Application Schemas describing the data structure and rules of the data to be exchanged between the parties.

After describing these Schemas using a Standard Language, e.g. EXPRESS in case of the adoption of the standard STEP, they should be compiled to prepare the SDAI with the required information to support the data exchange based on these data models.

Resulting from this process of compilation, two kinds of binding to SDAI can be considered as described before: The Early binding and the Late binding.

In the Late binding process, the Application Schema is compiled and the Dictionary SDAI-model based on the SDAI's Data Dictionary is populated with the information describing the application model structure and rules. This allows the Application and SDAI to have access later to this meta-data information, and enable the SDAI operation to run accordingly.

The result of the Application Schema compilation for an Early binding process is a set of new SDAI commands to be added and linked to the set of the already described native and general ones. These new commands are Application Schema dependent, and implement a direct access to the SDAI-model entities and its related data (e.g., attribute values) without need to access first to the Dictionary of Data.

In the Early binding approach the Dictionary of data is not mandatory to exist in SDAI, since the required information about the Application Schema is put available directly through the new SDAI commands. This generates a "compiled preset" approach where the commands for Application Schema data is determined in an early phase during the compilation process, what cannot be updated during run-time.

This Early binding approach offers a facilitated access and use of SDAI using the Schema-based SDAI commands, although it is static during the run time.

The Late binding approach provides flexibility since, during runtime, access to the Dictionary SDAI-model is possible, modifying and updating the description of the Application Schemas if wanted.

Because the SDAI commands in the Late binding approach are based on the information stored in the Dictionary, the use of SDAI implies a direct handling of Schemas meta-data, making its usage more complex. Some times a hybrid approach is used, intending to bring the benefits of both approaches.

Figure 4 gives an overview of the SDAI and Application data flow, for one Application to be integrated in a Neutral Format Platform.

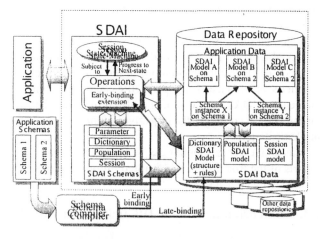

Figure 4. SDAI and Application data flow

After the compilation process, independently of the Early or Late binding approach, an Application is ready to use the SDAI commands to exchange data in Neutral Format based on the Application Schemas compiled.

As explained before, the SDAI commands are available subject to the SDAI's Session State Machine regulations, and run supported by the SDAI-models created and handled in the SDAI Data as part of the SDAI repository, as described by the SDAI Schemas.

The Application data is organized in a set Application SDAI-models, each one storing instances of entities belonging to one Schema.

The Schema instances provide the general data about one Application Schema, and its role is to behave as a logical aggregator of all SDAI-models belonging to that Application Schema.

Inter-Schema references between instances of entities is allowed if two or more SDAI-models are associated to a same Schema-instance, as is the case of the SDAI-models A and B or the B and C, in Figure 4. In addition, for this to be allowed at least one pair of Entity types have to be declared as domain equivalent in the SDAI data dictionary between the involved schemas.

References between SDAI-model A and C are not thus permitted, as presented in Figure 4.

3. The role of facilitators in the adoption of Standards for Data Exchange

For one Application to adopt a Standard for Data Exchange, the development of a converter on top of SDAI includes the development of an Implementable Application Protocol Interface (I-API), described in a general way in Figure 1 as a DM2SDAI converter. This interface is responsible for the translation between the Application's internal data, structured in an internal format (and most of the times proprietary), to the Neutral Format, structured as described by the Application Schemas (e.g., Application Protocols – APs).

3.1. Motivations to have facilitators

After transform the Declarative representation of an Application Schema e.g., described in EXPRESS, in an Implementable form as the result of the compilation of the schema, the I-API for the Application's converter for the data exchange can be developed on top of SDAI.

The implementers of the converters, especially those not experienced with the standard being adopted, recognize very often that the set of commands put available by SDAI is of very low-level and difficult to use and understand.

This is true even when the Early binding approach of SDAI has been adopted, adding to SDAI some more schema-dependent facilities.

When the Application developers are small companies with a reduced number of programmers, this problem is still more relevant since they do not have most of the times the required manpower to study and understand all the requisites of the Standard and of use of SDAI to enable them to implement the converters.

To put available toolkits acting as facilitators for the development of these converters, making easier the implementation of these converters, will stimulate the development of the converters and thus the adoption of standards.

Because models differ from case to case, facilitators should be flexible supporting any model described in the modeling language.

This is the case of the toolkit called Genesis, developed by UNINOVA in the scope of several European projects, aiming to be a facilitator for the development of Application data to SDAI interfaces I-API, when schemas are described using the EXPRESS language.

3.2.   The use of facilitators to develop I-APIs

The facilitators to develop I-APIs are toolkits that generates automatically code for a target programming language, based on the meta-data corresponding to an Application Schema. The generated code will act as an Implementable AP High Level Interface.

The generated code implements an Abstract Data Type (ADT) of the declarative Application Schemas described in EXPRESS and compiled to the SDAI's Dictionary SDAI-model.

Figure 5 depicts a framework to integrate Applications using High-level Interface facilitators.

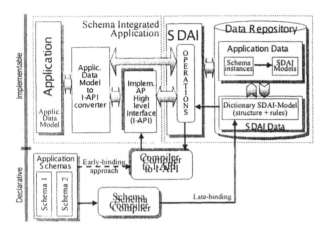

Figure 5. Framework to integrate Applications using High-level Interface facilitators

The facilitator toolkit (i.e., Genesis) works as a Compiler to I-API, generating automatically code based on the information stored in the Dictionary SDAI-model after compiled by an EXPRESS compiler using a SDAI Late-binding approach.

The interface generated automatically provides a set of methods of access of higher level than those provided by SDAI. Although not offering all the operations SDAI can provide, it encapsulated these operations and gives the most the functionalities required for the development of the translators for the integration process.

If sought, the integrator can use SDAI directly for those operations not supported by the generated interface, and to use the generated code facilities for those the facilitator was prepared, making easier the development work on top of SDAI.

Because Application Schemas are described in EXPRESS, that is a descriptive object-oriented language, Genesis generates C++ code, transforming the abstract description in an implementable one, represented as an Abstract Data Type (ADT) of such Schemas.

The generated code is thus a set of C++ classes representative of the ADT of the compiled Data Schemas. The methods associated to each class have consult, assignment and import and export mechanisms for instances and its attributes following the rules described by the standard.

The generated code should be linked with a library of classes provided together with toolkit, and developed to support some of the basic functionalities of its classes' methods. It also virtualizes some of the basic mechanisms of SDAI, like those to manage and handle SDAI sessions, access the data repository or manage the system's memory.

Though that different SDAI platforms can be adopted to be linked with the generated code, allowing an immediate adoption of other implementations of the standard or even to other standards (see [18]), a specific class supporting the general SDAI functionalities is also provided to be instanciated accordingly the objectives of the implementation.

Details about Genesis toolkit can be found in several reports of the RoadRobot and funStep European project and in [12][21][22][17].

3.3.   Extension to provide DB and KB facilities

The integration of one Application with a Neutral Format Platform can get an add-value if the Application can also have available Database (DB) and Knowledge base (KB) facilities.

To support this aim, an extension for the High-level I-API should be put available including a direct link between this library and a DataBase Management System (DBMS) and Expert System Shell (ESS).

This extension implies the code generated automatically should provide methods to handle and manage the data and knowledge stored in the DB and KB, as well as a set of mechanisms to query data, trigger rules and start the reasoning mechanisms.

Both, DB and KB, should support object-oriented construction to be more suitable for the mapping with the High Level I-API. Though Relational DataBases are of most common use nowadays, an extension layer to support Object-oriented feature should be considered on top of it. Nowadays most of the known RDB already provide it automatically.

Also an Expert System Shell supporting Object-Oriented capabilities for data and knowledge representation should be selected and integrated for use.

Figure 6 depicts this extended framework for integration, highlighting its relationship with the I-API compiler, SDAI and Application Data Model to I-API converter.

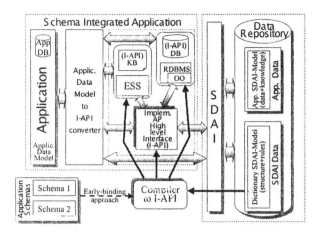

Figure 6. Extended framework for integration

To provide this extension the Compiler to I-API should generate automatically the equivalent Application structures and functionalities in the KB and in the DB, based in the information stored in the Dictionary SDAI-model, including both data structure and rules. Also, the interfaces in the High-level I-API to access to the data and knowledge of the DB and KB should be generated.

In the case of the RDB or the OODB, it will be generated the structures for data storage mapped with the structure of the Application Schema, providing facilities in conformance with the Application Schemas.

Also, the Schemas rules can be loaded in the DataBase, after translated to its format of representation by the compiler, and be prepared to trigger to assure data conformity with the model.

In a similar way, data structure and rules can be generated in the KB format in order to make possible to use the ESS to produce an Expert System based in the Application data structure and rules.

Having the Application Schema been developed to be prepared to represent knowledge, it can be exchanged between different applications and executed by different ESS, though the compiler is able to convert the data structures from EXPRESS representation to the several ESS structures.

This possibility is very important thinking in conformance testing and interoperability checking of data exchange among different applications, where the rules flowing between the systems in neutral format can include knowledge and rules able to check the data and aid to assure the reliability of the integrated systems.

4. Multi-level integration of protocols

One of the typical problems found in organizations is when they intend to integrate their applications with a Neutral Format Platform, and the scope of its applications cover more than one existent Application Protocol.

An example of that is what is taking place in the funStep project, which addresses the integration of Applications associated with furniture product data representation and furnishing decoration projects.

The funStep project developed one model following the STEP methodology called the funStep AP. The model was submitted to ISO TC184/SC4 to proceed with the official procedures to become an ISO International Standard, and in this moment it has its New Work Item approved by this community and it is registered as ISO10303 Part 236 (i.e., STEP AP236).

In the same community, other teams are addressing other scope of implementation and they are seeking harmonization between its own Application Protocol and other APs that should be referenced by them.

A real example of this scenario is related with the Furniture and the Building and Construction Industries. STEP provides one Application Protocol, e.g., AP225, that describes, among others, the representation for the space of a room. On the other hand, STEP is also providing an AP for furniture product data representation. Thinking in one Application devoted for decoration of houses, this application needs data representation from the room and from the furniture to be placed inside the room for decoration. A similar situation happens with the Shipbuilding Applications where furniture should be placed inside the ship to furnish it.

One possibility to face this need and integrate such application for decoration with a neutral format platform is to create a new schema or set of schemas devoted for this application, based on the model structure of the AP for Building & Construction, taking the part relative to room description, and on the AP for furniture, taking the part required from the furniture representation.

Another approach is to consider the immediate reuse of the already existent APs, and to integrate and harmonize them using a new high-level Schema responsible for this integration task. This new Schema will act as a meta-AP responsible for the links between the required parts resulting from the APs it refers, building a multi-level integration of APs (Figure 7).

This multi-level approach brings some advantages in terms of reuse of the existent APs.

One immediate advantage is the complete reuse of the existent APs. The second one is related with the conformance testing and interoperability checking tasks on the new AP.

If already exists developed libraries implementing the interfaces for these APs, when reusing them following this Multi-level approach, all these interfaces can be immediately reused.

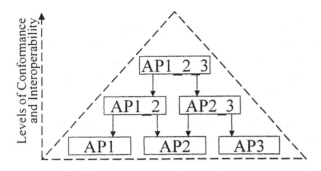

Figure 7. Multi-level integration of APs

Therefore, just a new layer should be developed to implement the top-level interface, supporting the inter-relationship among the different entities from the different schemas in the low-level.

Though the interfaces for the low-level schema interface implementation are already verified for conformance and interoperability, now the special concern should be focused in the verification tasks of the new Top-level Schema, in order to assure compliance of the global multi-level APs.

In this Multi-level approach, the policy for adoption of rules for verification can be two fold.

One is to keep all the rules of the lower level, and complement them defining in the top-level the rules required to cover the complete model requirements.

Other is to define in the top-level all the rules required in the model, with the possibility to consider, or not, those rules already defined in the low-level.

Although in the first option the reusability takes and important role, sometimes it is difficult to define in the top-level the all required rules for the complete model without enable/disable some of the rules already existent in the lower levels.

4.1. The use of facilitators to support Multi-level APs implementations

For the implementation of interfaces in a scenario where a unique complete new schema is created, the procedure is the same as described in the previous section, using the I-API compiler for the development of the High-level interface.

In the case of reusing the already existent APs following the Multi-level approach, several High-level I-APIs should be generated, one for each AP adopted, and linked all together according to the Multi-level APs structure. Figure 8 depicts the framework to integrate Applications adopting Multi-level APs structure, using I-API High Level Interfaces.

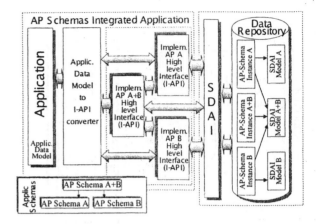

Figure 8. Framework to integrate Applications adopting Multi-level APs structure

5. Conformance requirements

Validation procedures should be performed after developing the interfaces to assure conformity of the application with the adopted standard. Also, to assure a complete interoperability with this application with all other parties integrated in the global system, verification and validation should be performed with all parties in order to assure a reliable exchange of information conforming with the standard, as syntactically as semantically.

Conformance and interoperability testing are procedures that should be performed to validate and assure the quality of the global integrated system, as a monitoring procedure or when a new party is plugged in. This issue is even more sensitive when different standards (e.g., Application Protocols) are concerned, and semantics and harmonisation of concepts and structures have to be realised.

Validation is directly related with the fact that erroneous assumptions in the early phase of development (e.g., data modelling) will cause correction work in later stages and consequently worst quality. To guarantee the syntactical correctness of a data model with a standard is not the main problem because it comprehends the complete formal description of its methods and grammar, and parsers are usually available.

The main difficulty is typically related when dealing with semantics. The more semantics are included the more complex the conceptual data model is. Thus, standard models have been including little formal semantics to avoid not realistic implementations due to the complexity of its representation. Together with difficulties related with the extension of the model not giving easily a complete overview of the global model, this causes too much freedom in the model interpretation by the user. From the architecture of Figure 1, the Syntactical validation is performed by SDAI, and provided as a library ready to be used and adopted. The DM2SDAI implementer will assure the structural validation (e.g., check of array bounds, attribute value out of type range) and verification of semantics.

6. Scope and source of contributions of the presented work

The work presented in this paper results mainly from the bibliographic compilation, research and developments done by the authors contributing for the results of the European projects presented in summary below. Details are available from their public reports and deliverables.

6.1. CIMTOFI project

The basic aim of the BRITE/EURAM CIMTOFI (CIM sysTems with improved capabilities fOr Furniture Industry) project was to reduce the "Idea-to-Production" cycle for furniture product development, by integration of tools selected in the market [11]. Tools and methods developed and implemented in this project intended to contribute for the introduction of the CIM philosophy in a step-by-step way, not only in furniture industry, but also in other industrial sectors. The tools integrated were modelled using EXPRESS, and the integration tasks supported by a Standard-based Integration Platform called SIP.

6.2. RoadRobot project

The main goal of the ESPRIT III RoadRobot (Operator assisted Mobile Road Robot for Heavy Duty Civil Engineering Applications) project was the development of a generic architecture to be used to design and develop various modules of an automated road construction site [19][20]. A road paver and an excavator working on the field were used to test the integrated system functionality. RoadRobot's information models were described in EXPRESS and the information integration platform was based in such developed for CIMTOFI project.

6.3. funStep project

The ESPRIT IV funStep (Development of a STEP-based Environment for the Manufacturer-Customer Integration in the Furniture Industry) project aims to develop a general integrated STEP based environment for the manufacturer-customer integration and for the support of

the management information flows in the factory, applied to the furniture industry [1][13][14][15][16]. The project have been developed a STEP based specification of information models regarding product development, customer orders processing, process planning, and customer's project management. While the modelling work developed during the CIMTOFI project addressed the manufacturer process (internal to the factory) of the furniture industry, the funStep project addresses the business one (external) where several applications have been adapted and integrated using a STEP based integration platform.

6.4. ECOS project

The ESPRIT IV ECOS (Lite e-commerce operative scaleable solution for SMEs) is a 30 months project started in late 1998. ECOS aims to develop a set of services and software tools to implement standard-based data exchange to assist in electronic business activities among industrial collaborators in the furniture sector chain, i.e. material suppliers, furniture manufacturers and commerce [2]. This project has considered existing standardisation initiatives like ISO/STEP for product data modelling, and UN/Edifact and XML for business related data exchange. The available technology at the SMEs and the existent internet-related technology including data security and privacy have been used and adapted when required. ECOS has a vertical approach (furniture sector oriented), although its architecture based on that developed during the previous presented projects, can be applied to other sectors, resulting so in a set of services and software tools to be exploited regardless industrial sector.

7. References

[1]   http://www.funstep.org/funstep/
[2]   http://www.funstep.org/funstep/ecos/ecos.htm
[3]   http://www.mel.nist.gov/sc5/soap/
[4]   http://www.w3.org/TR/REC-DOM-Level-1
[5]   http://www.xml.com/pub/98/10/guide0.html
[6]   Introducing STEP – The foundation for product data exchange in the aerospace and defence sectors, Government of Canada, 1999, ISBN 0-662-64382-8.
[7]   ISO TC184/SC4, "IS - ISO 10303, Part22 – Standard Data Access Interface
[8]   ISO TC184/SC4, IS - ISO 10303, Part1 - Overview and Fundamentals Principles, 1994
[9]   ISO TC184/SC4, IS - ISO 10303, Part11 – EXPRESS Language reference manual
[10] ISO TC184/SC4, IS - ISO 10303, Part21 – Clear text encoding of the exchange structure
[11] Jardim-Gonçalves, R., Barata, M., Steiger-Garção, A., CIMTOFI Project - Brite / EURAM Project BE-3653, Deliverables and reports, 1990/94.

[12] Jardim-Gonçalves, R., Sousa, P., Pimentão, J.P., Steiger-Garção, A., Furniture commerce electronically assisted by way of standard-based integrated environment – the ESPRIT funStep project proposal, 1999, pp. 129-136, ICE'99 - 5th International Conference on Concurrent Enterprising, The Hague, Netherlands, CCE-DMEOM, UK, ISBN 0-9519759-8-6.

[13] Jardim-Gonçalves, R., Sousa, P., Pimentão, J.P., Steiger-Garção, A., Integration of furniture manufacturing systems - the funStep project approach, Eight International Manufacturing Conference - IMCC'98, 1998, Singapore, ISBN 981-04-0209-0.

[14] Jardim-Gonçalves, R., Sousa, P., Pimentão, J.P., Steiger-Garção, Furniture commerce electronically assisted by way of a standard-based integrated environment - the ESPRIT funStep project proposal, ICE'99 – International Conference on Concurrent Enterprising, Hague, Netherlands, 1999, ISBN 0 9519759 8 6

[15] Jardim-Gonçalves, R., Sousa, P., Pimentão, J.P., Steiger-Garção, Borras, M., Gresa, I., An integrated architecture to promote furniture business. The funStep project and established industrial initiatives, Conference in Product Data Technology Europe - PDT99, Stavangar, Norway, 1999, QMS, UK, ISBN 1 901782 03 4

[16] Jardim-Gonçalves, R., Sousa, P., Pimentão, J.P., Steiger-Garção, ESPRIT #22056: the funStep project Integration of product and business data for furniture industry, 1998, ECPPM'98 – Product and Process modelling in the Building Industry, BRE, ISBN 1 86081 249 X

[17] Jardim-Gonçalves, R., Sousa,P., Pimentão, J.P., Steiger-Garção, A., Integrating manufacturing systems using ISO 10303 (STEP): An overview of UNINOVA projects, 1999, International Journal of Computer Applications in Technology, pp. 39-45, Vol.12, No 1,[17] ISSN 0-952-8091

[18] Jardim-Gonçalves, R., Steiger-Garção, A., to be published in book: Agile Manufacturing: 21st Century Manufacturing Strategy, Chapter 48: "Putting the pieces together" using standards. Elsevier Science Publishers.

[19] Pimentão, J.P., Jardim-Gonçalves, et al., The RoadRobot project - from theory to practice, 1996, Lisbon, Basys'96: Balanced Auomation Systems II - Implementation challenges for anthropocentric manufacturing", pp.126-133, Chapman & Hall, London 1996, ISBN 0-412-78890-X.

[20] Pimentao,J.P.; Azinhal,R.; Goncalves,T.; Steiger-Garcao,A., RoadRobot project (ESPRIT III- 6660) - Deliverables and reports, 1993-1994

[21] Sousa, P., Pimentão, J.P., Jardim-Gonçalves, R., Steiger-Garção, A., Towards compatibility between product data libraries using the Genesis' environment, 13th International Conference on Systems for Automation of Engineering and research (SAER'99), 1999, September , Varna-Bulgaria

[22] Sousa, P., Pimentão, J.P., Jardim-Gonçalves, R., Steiger-Garção, ISO 10303 Application Interfaces supported by Genesis environment – The funStep ESPRIT project experience, 3rd IMACS International Multiconference on Circuits, Systems, Communications and Computers (CSCC'99/IEEE), 1999, Athens-Greece

SDAI as the Common Access Interface for Object-Oriented Database Management Systems[1]

Thu-Hua Liu, Professor
Department of Industrial Design, Chang Gung University
Amy J.C. Trappey[2], Professor
Chii-Shi Lin, Graduate Student
Department of Industrial Engineering & Engineering Management
National Tsing Hua University, Hsinchu, Taiwan, R.O.C.

Abstract

This research describes the specifications for the ISO10303 Standard Data Access Interface (SDAI) and depicts the SDAI role in the integrated software development environment. Primarily, the research documents the process of implementing the SDAI on a specific object-oriented database management system (OODBMS), called Web-based Object Oriented-DataBase (WOO-DB). WOO-DB is developed by the Institute for Information Industry and uses a proprietary object definition language (ODL) and object manipulation language (OML) to describe and manipulate data. The implementation is achieved by mapping the functions and schemata between the SDAI data model and WOO-DB ODL and OML. Because C language binding is available in SDAI, the research focuses on the C-API of WOO-DB and implements the SDAI C language for the standard application interface. Since the data access interface is independent of data storage technologies, the integration of applications can be achieved under a common data accessing protocol.

Keywords: ISO10303 (STEP), SDAI, OODBMS, data access interface

1 Introduction

Since the first release of ISO10303 (STandard for the Exchange of Product model data - STEP) in 1994, STEP has gradually been adopted as a neutral product data interchange standard by applications ranging from CAD, CAM to product data management (PDM). However, in most engineering applications, information and data are stored and managed electronically by specific methods in their own formats. STEP specifies the implementation methods, in the Part 20 series, for manipulating the data defined in STEP-based data models. Currently, STEP data accessing methods are still at the level of file implementation, i.e., using the STEP physical file format in ISO10303 Part 21. Although ISO TC184/SC4 is developing a data accessing standard interface at the database level, the concept and method are still not widely adopted in engineering applications.

In modern enterprises, Database Management Systems (DBMS) play an important role as data repositories. A new access method that can achieve the STEP data manipulation at the database implementation level is developed as the Standard Data Access Interface (SDAI). SDAI is a set of Application Programming Interface (API) functions that can manipulate the information models defined by EXPRESS language. A number of computer language bindings for the STEP implementation methods are under development to make the SDAI operations available in different computer language environments. An important reason for developing the SDAI is the capability of porting the applications to various platforms. Most of the commercial DBMSs provide their own APIs for writing applications. Thus, the application software will be developed only under the specific API environment. The situation makes the concurrent engineering practice impossible unless their existing access interfaces are designed to support the requirement of universal access, such as SDAI. If all DBMSs support the SDAI standard, the STEP-compliant programmers can write the applications regardless of specific features of each access interface. Hence, the barrier free communication for concurrent engineering can be achieved by allowing applications to access data in various databases.

The main objective of this research is to construct a data access implementation at the database level. As a degree of effectiveness, the database level implementation is more important than the file level

---

[1] This research is supported by ROC National Science Council Research Grant.
[2] Please send all correspondence to Prof. Amy Trappey; E-mail: trappey@ie.nthu.edu.tw.

implementation. Loffredo [7] presented three types of SDAI access architectures based on the quantity of data and the required time span of data transfer. According to Loffredo, the SDAI access architectures are divided into three categories, i.e., (1) file upload/download SDAI binding, (2) cached SDAI binding, and (3) direct SDAI binding. Our work here is to apply the direct SDAI binding to an Object-Oriented Database Management System (OODBMS). The direct SDAI binding allows applications to access, manipulate and update data in databases without any intermediate.

Although the STEP standard provides a set of protocols to make the data exchange with application domains possible, the shared engineering database is not yet widely used in STEP environment. This research develops a standard data accessing methodology to an OODBMS applying SDAI. In the implementation, a commercial OODBMS product called Webbased Object Oriented-DataBase system (WOO-DB) is chosen to be our target DBMS. WOO-DB, an Object Oriented Database Management System (OODBMS) developed by the Institute for Information Industry (III), provides two sets of API for developing applications, which are C and C++ interface respectively. Among them, the C++ interface is an ODMG-93-compliant programming environment. Because the SDAI C++ language binding (Part 23) is not yet available, this research focuses on the C-API of WOO-DB and implements the SDAI C language binding (Part 24) on this interface. This research produces a library of functions for the SDAI-based applications to access data in WOO-DB.

2 Background

In 1994, Goh *et al.* [3] implement the SDAI specification on an object-oriented database called Ontos. This commercial database has its own API for programmers to develop applications of specific purposes. The SDAI library routines are divided into six parts, including instance identifiers, query, data creation, modification, meta-data queries and validation. Most of the implementation issues are product-related. This means issues presented in implementing SDAI function routines are achieved using the specific API provided by Ontos. Herbst [4] proposes an EXPRESS-modeled database that handles data manipulations via the SDAI. The research defines relevant data models for the Scientific and Statistical Database (SSDB) using EXPRESS language and constructs an EXPRESS-modeled database prototype on a commercial object-oriented database, called ObjectStore. Nink [6] builds a SSDB using the STEP standard. Specifically, the author applies STEP on the Flexible Image Transport System (FITS), which is a scientific data exchange standard. Nink describes the FITS data model using EXPRESS and makes it be the schema inside the system. This system uses a code generator called E2C++ to generated

C++ classes and initializes the SDAI data dictionary schema. The data access mechanism in this system is built by applying the SDAI concept.

Li *et al.* [8] develop a Product Modeling System on STEP (PMSS). PMSS is a manufacturing-feature-oriented modeling system and uses a mechanism, called STEP preprocessor, to translate the feature abstraction file into a STEP file. The STEP neutral file can be accessed by STEP file platform that is relatively independent from other modules in PMSS through SDAI. Colyer *et al.* [2] also introduce a project of the Magnet Integrated Design and Analysis System (MIDAS). The main objective of the MIDAS is to create a STEP compatible and an open environment for engineering designs. All the engineering data is stored in a specific database called DEVA, which is developed by Rutherford Appleton Laboratory. This system adopts a client/server environment and the SDAI as its data access architecture.

Rando and McCabe [9] discuss the problems encountered in designing the C++ binding to the SDAI. Although the C++ programming language possesses object-oriented features, there are differences between the data description method of C++ and the EXPRESS language. The problem of the semantics of inheritance is highlighted in designing phase of the C++ binding and the corresponding solution is discussed in the paper.

Botting and Godwin [1] give a formal definition of SDAI operations and its EXPRESS schemata. The formal method used in the paper is to create a formal model by a language called VDM-SL. VDM-SL can describe and analyze the structures and the semantics of SDAI operations and models. In this paper, several points of ambiguity in the SDAI are presented and discussed.

3 Standard data access interface

In today's software environment, most of application software is developed using specific data access methods, i.e., APIs provided by DBMS vendors, to manipulate the data in specific databases. When the existing application software requires manipulating a different database whose access interface is different from the original one, the application program must be modified to incorporate the new data repository. Thus, programmers need a universal data access method for efficiently accessing a variety of databases. The universal data access method has to be independent of all database systems. Hence, programmers can write the application software regardless of the specific data storage technology. Under such standardized data access environment, concurrent engineering can be achieved more easily. ISO TC184/SC4 proposes a solution that can solve the problem of various access methods, called Standard Data Access Interface (SDAI). The SDAI is defined in Part 22 of ISO 10303 series standards and provides a standard API to the STEP data that is defined using the EXPRESS language. The SDAI also provides a consistent data access environment to develop application software. Thus, the application software can be developed regardless of specific data storage technologies.

3.1 Binding styles of the SDAI

All the language bindings of the SDAI can be categorized into two groups, which are early and late binding, respectively. The use of the binding styles depends on the characteristics of a programming language. The programming language of an early binding system, such as SDAI C++ binding, can provide some specific data structures (e.g. class definitions of C++) for programmers to describe the application schema. Early binding systems utilize these data structures provided by certain programming languages to directly define the specific schema in applications instead of consulting EXPRESS data dictionary. Using early binding systems, the type check of data structures that derive from the application schema can be executed when compiling the application [7].

Unlike early binding systems, late binding systems access data values by consulting the data dictionary. Usually, this type of systems provides some kinds of mechanism to consult the data dictionary at run-time. The schema defined in EXPRESS model may be translated into a data dictionary that can be manipulated by the late binding system. Programmers do not need to describe the data model in applications specifically. All the references to the data dictionary can be done via parameters passed to the SDAI operations. Because all operations of referring to the data dictionary are executing at run-time, this type of systems is called the late binding system [7].

3.2 Part 22: standard data access interface specification [5]

3.2.1 SDAI schemata:
The SDAI data schemata consist of four major subdivisions, which are the dictionary, session, population, and parameter schema, respectively. By definition, the objective of the SDAI dictionary schema is to provide formal definitions for the structure of a data dictionary. Dictionary model includes two major parts, i.e., type definition and entity definition. In the SDAI specification, there are seven types and thirty-eight entity definitions in the dictionary schema. According to ISO 10303-22, the objective of the SDAI session schema is to provide a structure of data for describing the current state of a SDAI session. This schema aims to define the states such as sessions, access modes, transactions and error logs, etc.

The third subdivision in SDAI schemata is called population schema. This SDAI schema defines a structure of data that can be used to describe the

organization, creation and management of the instances of the EXPRESS entity type manipulated in the SDAI. The organizational objects created by the SDAI application during the SDAI session are specified in this schema. If an implementation does not access to the data dictionary, the population schema shall be used to describe the instances manipulated by the SDAI.

The last subdivision of SDAI schemata is the parameter data schema. This schema can provide conceptual descriptions for the data passed to SDAI operations through the parameters. All the EXPRESS declarations are defined to provide description of the SDAI operations and the definitions of the SDAI environment in which entity instances exist.

3.2.2 SDAI operations:
The SDAI service to the data repository can be achieved by a series of functions. Those functions are also the main body of the SDAI specification. Part 22 describes the general behavior of each SDAI operation and the minimal operational capability that every SDAI language binding shall provide. However, Part 22 does not specify how those functions are implemented in different programming language environments. The functional specification of each programming language is described in other parts, such as Part 23 (C++ language binding), Part 24 (C language binding), and Part 26 (Interface Description Language-IDL binding).

In general, the SDAI operations can be classified into seventeen categories according to the different operational properties. These categories include environment operations, session operations, repository operations, schema instance operations, SDAI model operations, scope operations, type operations, entity instance operations, application instance operations, entity instance aggregate operations, application instance aggregate operations, application instance unordered collection operations, entity instance order collection operations, application instance order collection operations, entity instance array operations, application instance array operations and application instance list operations. Each category includes one or more definitions of operations.

4 Research method

This research selects a commercial OODBMS as the target database system, called WOO-DB. WOO-DB stands for Webbased Object Oriented-DataBase and is developed by a quasi-official software organization called Institute for Information Industry (III). WOO-DB currently provides two sets of API, which make applications available in two programming language environments, i.e., C and C++ respectively. This research selects the WOO-DB's C programming interface as the target system for implementing the SDAI.

4.1 Communications between the SDAI and WOO-DB's C-API

There are three types of SDAI implementation methods that can be used. The methods are the file up/download SDAI binding, cached SDAI binding and direct SDAI binding. The file upload/download SDAI binding uses some kinds of intermediate file, usually the STEP file (defined in Part 21), as the temporary file. When a file upload/download binding is running, the model will be extracted from the database and written to the temporary file. Then, SDAI applications manipulate the temporary file. When the SDAI service is processed, the updated file is loaded back into the database. A similar architecture to the file upload/download binding is adopted by another SDAI implementation method called cached SDAI binding. The different part is that the cached SDAI binding utilize the main memory instead of intermediate files. The extracted model from database is transferred into the main memory of the computer system and manipulated by the SDAI application. After the SDAI service is completed, the cached SDAI binding loads the updated data back into the database. Both the above SDAI implementation methods use the intermediate mechanisms to implement SDAI applications. Finally, the direct SDAI binding works without any intermediate mechanisms and updates the data inside the database immediately. Each SDAI operation can be achieved by calling the native function(s) of the API provided by the database system.

The direct SDAI binding possesses the advantages such as lower latency and concurrent access. However, the direct SDAI binding is the most difficult to implement among the three architectures. Complex algorithms and considerable coding efforts may be needed to build a direct binding system. Using direct binding architecture, the key of the implementation work is the communication between the API of the target database and SDAI specifications. The characteristic and design of an API will significantly influence the effect of an implementation.

In this research, direct SDAI binding is adopted to implement the SDAI specification. A number of functions defined in SDAI C binding will be implemented by calling WOO-DB's C-API. Many SDAI C functions make up an intermediate library of functions and the concept is illustrated in Figure 1. Once the intermediate library is complete, it can be linked to the SDAI applications. This procedure makes the SDAI functions available under WOO-DB's environment.

4.2 Example schema

For executing SDAI services in WOO-DB, this project utilizes a number of SDAI applications based on the example specified in the appendix of the SDAI C library reference manual published by STEP Tools Inc

(1996). The example includes several SDAI services, such as environment operations, session operations, creating model, creating instance, updating the data, and retrieving data values. It uses a set of EXPRESS definitions as the application schema. The schema contains four entities including a point, a circular, a line and a string of characters. The applications can access the attributes of these entities via the SDAI methods. Figure 2 illustrates the instances created in WOO-DB by the application with the EXPRESS definition of the application schema shown as follows.

```
SCHEMA example;
    ENTITY Point;
        x : REAL;
        y : REAL;
    END_ENTITY;
    ENTITY Line;
        enda : Point;
        endb : Point;
    END_ENTITY;
    ENTITY Circle;
        radius : REAL;
        center : Point;
    END_ENTITY;
    ENTITY Text;
        label : STRING;
        center : Point;
    END_ENTITY;
END_SCHEMA;
```

5 Implementation

5.1 Environment operations and session operations

The execution flow of WOO-DB's environment operations and session operations is slightly different form the SDAI C binding. Figure 3 shows the flow after executing the two types of operations in WOO-DB and SDAI C binding. The sequences of opening the repository and starting the transaction between WOO-DB and SDAI C binding are reversed. Because the two operations can not be changed in WOO-DB, a modification must be made for these two processes in SDAI applications to coordinate the sequence of WOO-DB. The codes of the SDAI application shown in Figure 4 exhibit the modified sequences of these two operations.

The SDAI operations *sdaiOpenSession* and *sdaiStartTrx* can be implemented because WOO-DB provides the same functions as the two SDAI operations. The argument *mode* of the *sdaiStartTrx* conveys two access modes of the transaction, which are *sdaiRW* (for read-write) and *sdaiRO* (for read-only) modes, respectively. However, WOO-DB only provides the read-write mode for C-API users in this version so the *sdaiRO* mode is unavailable in this implementation. In

Table 1. Necessary WOO-DB's functions for achieving corresponding SDAI operations

Process	SDAI operations	Necessary WOO-DB's functions
Open session	sdaiOpenSession()	omInitialize()
Open repository	sdaiOpenRepositoryBN()	omCreateOB() omConnectOB() omSetDefaultObjectBase()
Start transaction	sdaiStartTrx()	omTransact()
End transaction	sdaiEndTrx()	omTransact()
Close repository	sdaiCloseRepository()	omDisconnectOB()
Close session	sdaiCloseSession()	omFinalize()

the *sdaiOpenRepositoryBN* function, the implementation has a simple algorithm to determine whether the specified repository exists. If it does not exist, system will create a repository first and then connect to this repository. Otherwise, the system will connect to the target repository directly. Except *sdaiOpenRepositoryBN* function, most of the environment operations and session operations achieved in this research are one-to-one implementations. Table 1 is the mapping that specifies the necessary WOO-DB's functions for achieving the SDAI environment operations and session operations.

5.2 Creating model

In WOO-DB, all the definitions of the data dictionary are written in applications and created by calling specific functions provided by WOO-DB. This makes implementors hard to design an EXPRESS compiler to automatically transform the EXPRESS models into WOO-DB. The solution adopted in this implementation is to create the data dictionary into WOO-DB by writing the creation routine of the application schema. The coding of a SDAI C function called *sdaiCreateModelBN* can put such idea into practice. Figure 5 shows the entity declarations of the data dictionary in WOO-DB and the implementation of *sdaiCreateModelBN* for creating schemata.

The creation of WOO-DB's schema requires two major steps. The first step is to describe entities of the specified schema using the data structure declaration function *omAttrSpec*. The second step is to create these entities into database by calling WOO-DB's functions. In the second step, there are three necessary procedures to create an entity (class) in WOO-DB. The first procedure is to create a dynamic array as a temporary area, and then put the entities into the array, which have been described in *omAttrSpec*. The second procedure is to create the specified entity using the function called *omCreateClass*; meanwhile, the dynamic array that stores the data structure *omAttrSpec* is also passed into the *omCreateClass* function through the first argument. Finally, the dynamic array created in first procedure should be deleted for releasing the RAM. The entire process to create a WOO-DB entity can be illustrated in

Figure 6.

The application schema can be completed when the creation of all necessary entities is finished. Figure 7 shows the schema using the Graphical User Interface (GUI) provided by WOO-DB. These entities and the corresponding attributes are created by the SDAI method.

5.3 Creating data into database

In the SDAI C binding, the creation of data involves two necessary procedures, which are to create instances and to put corresponding data values into instances. The first function called *sdaiCreateInstanceBN* creates an instance named *point1*, and returns the identifier of this instance. The second function called *sdaiPutAttrBN* puts data values of attributes into the specified instance using the instance identifier returned from *sdaiCreateInstanceBN*. In order to simulate these SDAI operations, we utilize several procedures of WOO-DB to construct the two SDAI function routines. The implementation of *sdaiCreateInstanceBN* is shown in Figure 8.

WOO-DB has a function called *omCreateObject* that can be used to create new instances into database; however, it needs to work with the corresponding entity identifier conveyed by the *PointerToClass* argument. The *omFindObject* function can get such identifier from database and pass it to the *omCreateObject* via the *PointerToInstance*. Then, the identifier of the new instance is passed through the *PointerToInstance* and returned to the next procedure of the SDAI application, *sdaiPutAttrBN*. This operation of creating an instance is achieved using the entity name. However, in this case, another SDAI function called *sdaiCreateInstance* will be easier to be constructed because this function directly uses the entity identifier rather than the entity name to create a new instance. Figures 9 illustrates the data input interface which triggers SDAI functions to update WOO-DB data instances.

5.4 Updating data values

This section introduces the SDAI operations of

updating data values implemented in this research. Figure 10 shows the SDAI operations for retrieving and updating data values of all *point* instances in WOO-DB. The function called *sdaiGetEntityExtentBN* can retrieve all of the instances of the *Point* entity by the entity name and return an aggregate that stores identifiers of these instances. The *sdaiCreateIterator* function creates an iterator for referring members of the specified aggregate. After initializing the iterator, the following *while* statements utilize the identifier retrieved from the aggregate by calling the *sdaiGetAggrByIterator* function to put the new data values into each point instance.

Figure 11 is the implementation of the *sdaiGetEntityExtentBN*, a WOO-DB function called *omSelectObjects* can be used to achieve the functionality of retrieving all instances of a specified entity and returns a dynamic array that stores identifiers of these instances. The dynamic array can exactly be the aggregate and its identifier can be returned to the SDAI application. However, the *omSelectObjects* function works with the *Class* argument that indicates the entity of the specified instance. The *omFindObject* obtains such entity identifier from the system and returns it through the *Class* argument.

The iterator is a mechanism that can be used to refer to members of the specified aggregate. However, WOO-DB doesn't provide such type of mechanisms for accessing aggregates. In order to construct the iterator in WOO-DB, we create a structure type that has two members in header file, called *Iterator*. The two members of *Iterator* are (1) a pointer to the specified aggregate called *Array* and (2) a variable for positioning the member of the aggregate called *IteratorCounter*. Using the structure, the implementation of the iterator and its related function can be shown in Figure 12.

In the *sdaiCreateIterator* function, the implementation declares a pointer (*NewIterator*) to the structure *iterator* and set its member (*Array*) to the specified aggregate. In the *sdaiBeginning* function, the member *IteratorCounter* is a counter for moving the reference to the next element and is positioned at the beginning of the array by allocating the value. The SDAI function *sdaiNext* firstly adds one to the *IteratorCounter* member and then tests whether the current member is null. If the current member is available, the program gets into the *while* statements and updates the data value. Otherwise, the program will break the loop and proceed to next statements. Further, the function *sdaiGetAttrBN* in the SDAI application retrieves values form system, which includes primitive values, instances and ADB data type.

5.5 Retrieving (query) data and deleting Instances

Programmers can utilize the function routines discussed in above sections to access the data values stored in WOO-DB. For instance, *sdaiGetAttrBN* function is defined to retrieve the value of an attribute for a point instance. Similarly, programmers can also use iterators or other operations to navigate the values of specified instances that programmers attempt to query. The *sdaiDeleteInstance* is a routine to delete an instance. In this function, the code design of deleting instance is quite simple because WOO-DB provides an exactly identical function that can be used to implement the *sdaiDeleteInstance*, called *omDeleteObject*. This proves that the degree of complex in implementing the SDAI function routines depends on the design of the specific API provided by the target database system.

6 Discussion

Based on the experiences of implementing the SDAI on WOO-DB, we conclude a number of basic features that an OODBMS should support when constructing the SDAI access mechanism. The analysis can be listed as follows:

1. A comprehensive, well-design identifier recording mechanism for indicating each object stored in the database is necessary. In WOO-DB, each object in the database has a unique identifier provided by WOO-DB and the system can manipulate these objects according to their identifiers. However, WOO-DB can only manipulate one schema in each database so WOO-DB does not assign any identifier for schemata. This makes model-related functions of the SDAI difficult to be completely implemented.

2. The data dictionary-driven system should provide a good capacity to access the data dictionary. Usually, this type of systems utilizes a set of functions to consult the data dictionary when manipulating the data of the database. Nowadays, many database systems are in C environment, and some of them may use the programming language as the ODL, such as WOO-DB. Most of WOO-DB's operations can be successfully performed by consulting the data-dictionary; however, the input data operations must be done through specific structure data types instead of calling put functions.

3. The target OODBMS should possesses an ODL that can strongly perform mappings between EXPRESS models and database schemata. SDAI is an API that aims to manipulate the EXPRESS-defined data. Any OODBMS attempts to deal with the EXPRESS data models should have such feature; also, this is a basic feature for implementing the SDAI. Some C API systems directly use the programming languages as the ODL. Due to the lack of full object-oriented characteristics, these systems are possible to produce problems when translating EXPRESS definitions into the database schema. An OODBMS that has a powerful ODL or object-oriented programming

environment, such as C++, for building the schema is a better choice to perform the data model mappings.

4. Database systems should provide the capability to support the rule management or constrain validation. However, many database systems do not provide this functionality to meet such requirement. The ease of the validation implementation depends on whether the target system provide such feature.

7 Conclusion

The research presented in this paper investigates the issues including the OODBMS, access interface standards (SDAI) and SDAI implementation on an OODBMS (WOO-DB). WOO-DB provides object-oriented features, such as inheritance and reference relationships, for implementers to develop the SDAI implementation. The implementation methods of STEP are categorized into four types, which are level one - the file exchange, level two - the working form exchange, level three - the shared database, and level four - the shared knowledge base. The data exchange methods of current STEP-compliant systems are commonly at level one and executed via translators provided by vendors. This type of system cannot satisfy the requirements of managing and sharing the product data through a single interface. Database-driven applications are the trend in today's software development. This research focuses on the level three (the shared database) and constructs a direct binding architecture on an object-oriented database instead of the traditional file exchange. In Figure 13, we can observe the differences between the shared database environment and the file exchange environment.

At the database level, the SDAI is a good solution to the problem of integrating different access interfaces of various database systems. Once the SDAI is supported, the concept of the neutral application can be utilized in the environment of current product data exchange. As shown in Figure 14, the neutral SDAI application can access different database-driven platforms and manipulate the STEP-defined data in a shared database environment. Thus, improvement in the reusability and portability of software and data repositories is achieved.

References

[1] Botting, R.M., and Godwin, A.N., 1995, "Analysis of the STEP standard data access interface using formal methods," *Computer Standard & Interfaces*, Vol. 17, pp. 437-455.

[2] Colyer, B., Simkin, J., Trowbridge, C.W, Barberies, U., Picco, E., Gutierrez, T., Longo, A., Greenough, C., Thomas, D., Alotto, P., Molfino, P., Molinari, G., Jared, G., and Sormaz, N., 1997, "Project MIDAS: Magnet Integrated Design and Analysis System," *IEEE Transactions on Magnetics*, Vol. 33, No. 2, pp. 1143-1148.

[3] Goh, A., Hui, S.C., Song, B., and Wang, F.Y., 1994, "A study of SDAI implementation on object-oriented databases," *Computer Standards & Interfaces*, Vol. 16, No. 1, pp. 33-43.

[4] Herbst, A., 1994, "Long-term database support for EXPRESS data," *Proceedings of the 7th International Working Conference on Scientific and Statistical Database Management*, Charlottesville, VA, USA, pp. 207-216.

[5] ISO 10303-22, 1996, *Industrial Automation Systems and Integration - Product Data Representation and Exchange-Part 22: Implementation Methods: Standard Data Access Interface*, Subcommittee 4 of Technical Committee 184, International Standard Organization, Geneve, Switzerland.

[6] Nink, U., Hansen, D., and Ioannidis, Y., 1997, "Using the STEP standard and databases in science," *Proceedings of the 1997 9th International Conference on Scientific and Statistical Database Management*, Olympia, WA, USA, pp. 196–207.

[7] Loffredo, D.T., 1998, "*Efficient Database Implementation of EXPRESS Information Models*," Ph.D. Thesis, Department of Computer Science, Rensselaer Polytechnic Institute, Troy, NY, USA.

[8] Li, B., Meng, M., Li, J., Sun, Y., Yang, L., and Li, Z., 1996, "Research on the development of framework based application interface," *Man and Cybernetics Proceedings of the 1996 IEEE International Conference on Systems, Man and Cybernetics*, Beijing, China, Vol. 2, pp. 1559–1562.

[9] Rando, T., and McCabe, L., 1994, "Issues in implementing the C++ binding to SDAI," *Computer Standards & Interfaces*, Vol. 16, No. 4, pp. 331-340.

SDAI Function Library

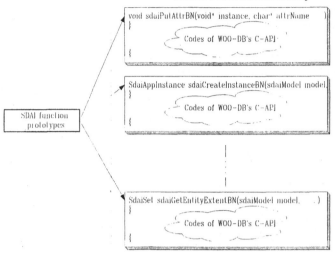

Figure 1. The methodology of SDAI implementation in this research.

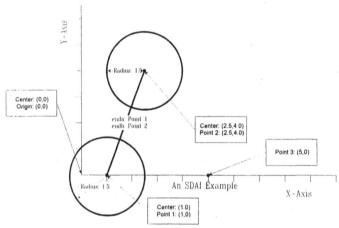

Figure 2. Instances created in WOO-DB by this implementation.

```
/* start session */
mySession = sdaiOpenSession();

/* start transaction */
sdaiStartTrx(myRepository, sdaiRW);

/* open repository */
myRepository = (SdaiTransactionRepository)sdaiOpenReposito

    { SDAI operations }

/* commit transaction */
sdaiEndTrx(myRepository, sdaiCOMMIT);

/* close repository */
sdaiCloseRepository(myRepository);

/* close session
sdaiCloseSession(mySession);
```

Figure 3. Environment operations and session operations between two systems.

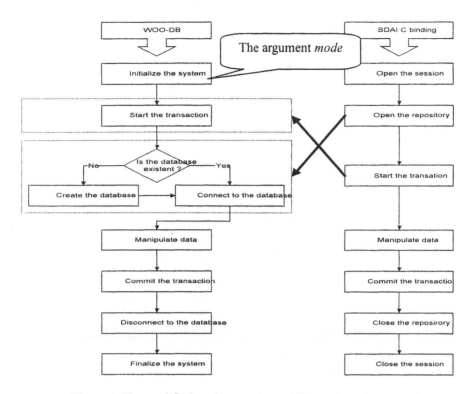

Figure 4. The modified environment operations and session operations.

```
omAttrSpec AttrsInLine[] = {
        {"enda", OM_NULL, "Point", 1, 0, 0, OM_NULL},
        {"endb", OM_NULL, "Point", 1, 0, 0, OM_NULL},
};

omAttrSpec AttrsInPoint[]={
        {"x", OM_NULL, "omDouble", 1, 0, 0, OM_NULL},
        {"y", OM_NULL, "omDouble", 1, 0, 0, OM_NULL},
};

SdaiModel sdaiCreateModelBN (SdaiRepository repo, SdaiString string, SdaiSche
{
        omDyStr  Attrs=OM_NULL;
        /* Create the Schema inside WOO-DB */
        // Line

        Error=omCreateDyStr(&Attrs,sizeof(omAttrSpec)*2,AttrsInLine);
        Error=omCreateClass(ObjbaseObj,"Line",OM_NULL,Attrs,OM_NULL,0,
                &LineClass);
        Error=omDeleteDyStr(&Attrs);

        // Point

        Error=omCreateDyStr(&Attrs,sizeof(omAttrSpec)*2,AttrsInPoint);
        Error=omCreateClass(ObjbaseObj,"Point",OM_NULL,Attrs,OM_NULL,0,
                &PointClass);
        Error=omDeleteDyStr(&Attrs);
```

Figure 5. The implementation of creating schema (extracted).

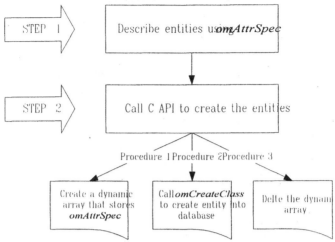

Figure 6. The process to create a WOO-DB entity

Figure 7. The entity definitions created in WOO-DB.

```
SdaiAppInstance sdaiCreateInstanceBN (SdaiModel model, SdaiStrin
{
        omObject PointerToInstance;
        omObject PointerToClass;

        omFindObject(OM_NULL, EntityName, &PointerToClass);
        Error=omCreateObject(OM_NULL, PointerToClass, &Pointe
        return PointerToInstance;
}
```

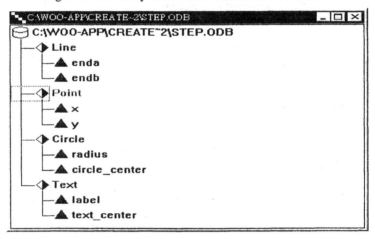

Figure 8. The implementation of the *sdaiCreateInstanceBN* function.

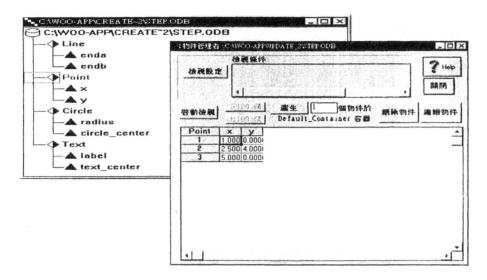

Figure 9. Three *Point* instances created in WOO-DB.

```
/* update instance */

itor = sdaiCreateIterator(sdaiGetEntityExtentBN(myModel, "Point")
sdaiBeginning(itor);
while(sdaiNext(itor))
{
        (void) sdaiGetAggrByIterator(itor, sdaiINSTANCE, &Add_
        (void) sdaiGetAttrBN(Add_Point, "x", sdaiREAL, &x);

        sdaiPutAttrBN(Add_Point, "X", sdaiREAL, (x+10.0));
}
```

Figure 10. SDAI operations for updating all *Point* instances

```
SdaiSet sdaiGetEntityExtentBN(SdaiModel model, SdaiString Name)
{
        omObject Class;
        omDyStr Result;
        omU4B Size;

        Error=omFindObject(OM_NULL, Name, &Class);
        Error=omSelectObjects(Class, OM_NULL, OM_FALSE, OM_RE
                        OM_NULL, &Result);
        Error=omGetDyStrSize(&Result, &Size);

        return Result;
}
```

Figure 11. The implementation of the *sdaiGetEntityExtentBN* function.

```
Sdailterator sdaiCreateIterator ( void *aggregate )
{
        Sdailterator NewIterator = (Sdailterator)malloc(sizeof(struct Iterator))
        NewIterator->Array = aggregate;
        return NewIterator;
}
```

Declare a pointer to *struct Iterator*
{
    omDyStr Array;
    int IteratorCounter;
};

```
void sdaiBeginning (Sdailterator Iter)
{
        Iter->IteratorCounter=-1;
}
```

```
SdaiBoolean sdaiNext ( Sdailterator Iter )
{
        Iter->IteratorCounter++;

        if(((omObject*)Iter->Array)[Iter->IteratorCounter]!=OM_NULL)
        {
                return sdaiTRUE;
        }
        else
        {
                return sdaiFALSE;
        }
}
```

```
void sdaiGetAggrByIterator( Sdailterator Iter, SdaiPrimitiveType valueType, vo
{
        //retrieve the value from the aggregate
        (*(void **)identifier)=((omObject*)Iter->Array)[Iter->IteratorCounter];
}
```

Figure 12. The implementation of the iterator mechanism.

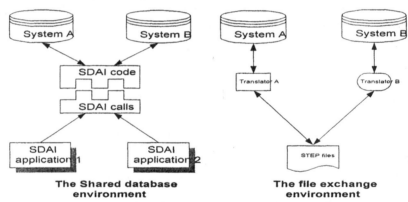

The Shared database environment   **The file exchange environment**

Figure 13. The comparison between two data exchange environments.

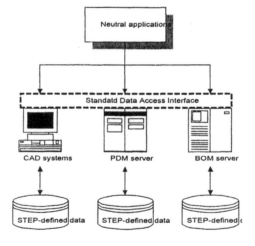

Figure 14. The concept of neutral applications.

354

Supporting product development in distributed heterogeneous environments

Luis Alberto Bertolero

Dipartimento di Automatica e Informatica, Politecnico di Torino, Corso Duca degli Abruzzi 24, 10129 Torino, Italy
and FiatAvio S.p.A., Via Nizza 312, 10127 Torino, Italy. lbert@cclinf.polito.it

Silvano Rivoira

Dipartimento di Automatica e Informatica, Politecnico di Torino, Corso Duca degli Abruzzi 24, 10129 Torino, Italy

Abstract

This paper describes how an integrated information base compliant to STEP models can be constructed, and how this can be shared among different applications in a distributed environment by means of CORBA.

A prototype system, currently under development in our Department, is then introduced.

The system is based on the ISO/DIS 10303-203 Config Control Design and 10303-214 Application Protocol: Core data for automotive mechanical design processes.

1 Introduction

The increasing outsourcing of technological development to suppliers is going to demand strong integration of distributed development sites in order to support joint and concurrent development processes.

Two major problems are then encountered:

- how to capture product information in information models able to serve as a reference throughout the product life-cycle
- how to integrate heterogeneous CAD/PDM tools from different companies to form a completely interoperable system.

The main activity for developing standardised data models for industrial products is today concentrated in the evolution of the standard ISO 10303 – STEP.

Most of the work insofar done in the definition of the standard has been finalised to guarantee the exchange of product data between different application systems [9], [10]. But the purpose of STEP includes storing, transferring, accessing and archiving product data in order to allow industrial co-operation during the whole life cycle of a product.

Interoperability among different hardware and software components has been the primary objective of the Object Management Group (OMG) in the definition of the Common Object Request Broker Architecture (CORBA).

The result is an application integration technology that allows applications to communicate with one another, regardless the platform they are running on and where they are executing.

2 Standard product information models

The standard ISO 10303 – STEP [1] defines semantically complete data models for various application contexts *(Application Protocols)* and specifies a standardised access to the data *(Standard Data Access Interface)*, which assures a sharable and evolving software layer to manage data objects [2].

Each Application Protocol *(AP)* represents a "view" of a data subset relevant for some applications (mechanical or electronic CAD, CAM, CAE, Bill of Material,...) and coherent with the associated processes.

Examples of actual Application Protocols are:

- AP 203 – Configuration Controlled 3D Designs of Mechanical Parts and Assemblies
- AP 207 – Sheet metal die planning and design
- AP 209 – Composite and metallic structural analysis and related design
- AP 210 – Electronic assembly, interconnect and packaging design
- AP 212 – Electrotechnical Design and Installation
- AP 213 – Numerical control process plans for machined parts
- AP 214 – Core Data for Automotive Mechanical Design Processes
- AP 218 – Ship structures
- AP 221 – Functional data and their schematic representation for process plant
- AP 225 – Building elements using explicit shape representation.

Every Application Protocol is based on two models: the *Application Reference Model (ARM)*, which is a user data model and provides a logical view of the data, and *the Application Interpreted Model (AIM)*, which is a standard oriented model.

The semantic consistence among different APs is obtained by deriving every AIM from a common set of definitions *(Integrated Resources)* independent from the application contexts: each AP refines and constrains the use of the Integrated Resources according to the information requirements of its application area formally stated in the ARM model.

The possibility of viewing the same data through different APs is currently offered by some *Application Interpreted Constructs (AIC)*, which are subsets of data models common to several APs *(Geometrically bounded wireframe, Topologically bounded surface, Mechanical design shaded presentation, Constructive solid geometry, Associative draughting elements, ...)*.

The representation of physical data in a neutral format [3], and the definition of semantically complete models for various application contexts *(Application Protocols)*, formally described in a standard language [4], ensure that the data generated in a given context by an application system can be correctly interpreted and used by a receiving application in the same context.

The initial use and success of the standard, in fact, has been largely limited to data exchange between CAD systems.

In 1998 the Automotive Industry Action Group (AIAG) performed a comparison of commercially available STEP and IGES translators, and the STEP implementations resulted to be superior to their counterparts in every measured way: on the average, IGES conveyed significantly less surface area, but required larger file sizes.

The received information, however, will usually need to be integrated with industrial data from other sources and then stored and subsequently used by other applications, according to a complex network of data exchanges.

Therefore, the integration of product data into long term archives, concurrently shared by multiple users and applications in the different phases of the life-cycle, becomes a much more strategic goal to be achieved than the direct exchange of physical data files.

The expected benefits range from re-use of enterprise knowledge to improvement of customer services, from flexibility in enterprise organisation *(outsourcing, business process re-engineering)* to exploitation of new techniques for designing and manufacturing products *(virtual enterprise, concurrent engineering [8])*.

The architecture of STEP allows today the construction of databases shared by different application processes, with the limitation that the data model is described by a single Application Protocol or it includes pre-defined subsets of data models common to several APs *(Application Interpreted Constructs)*.

A more extended interoperability among APs, which allows one to select and combine parts from different APs in order to obtain an "Enterprise STEP Protocol", is in the scope of the standard but currently not yet available: reusable collections of scope statement, information requirements, mappings and module interpreted models (STEP Application Modules) will support specific usage of product data across multiple

application contexts, and will be used together in Application Protocols [5].

3   Storing and accessing product data

Figure 1 shows how it is possible to create consistent, STEP compliant archives, by using development tools currently available on the market.

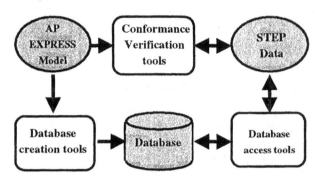

Figure 1 Creating a STEP compliant database

STEP data files produced by various applications (CAD, Bill of Material, ...), and represented in accordance to the ISO 10303-21 part [3], can be checked for conformance to an EXPRESS data model (usually an Application Protocol). The test checks the syntactic correctness of the data, the coherence of explicit and derived attributes, and the satisfaction of constraints on values, while allowing visualisation and editing of the data file.

The AP EXPRESS model have to be transformed into a corresponding conceptual schema of database, represented in a given DDL *(Data Definition Language)*, by means of the database creation tools.

The STEP data conforming to the model are then inserted into the database by primitive functions activated through the SDAI library [2]; similar functions will allow data retrieval and modification.

4   Data   sharing   in   distributed environments

Two main concepts are the basis of many current applications:
*   Object oriented programming
*   Distributed architecture.

The resulting systems allow many users to interact with many objects dispersed on different machines in a client-server model.

Of course, not all client-server models using object-oriented techniques necessarily offer all the OO power, elegance and facilities one could expect. Many of them have used OO technologies only at the client side and there is no reason why not to do it in the server side [11], [14].

In this sense, the OMG's CORBA *(Common Object Request Broker Architecture)* is an OO middleware that specifies a full-fledged distribution of objects across networks with interfaces accessible from anywhere. CORBA hides the complexity involved in programming distributed applications in heterogeneous environments.

The CORBA specifications are developed on the basis of the OMA *(Object Management Architecture)* which defines an Object model and a Reference model [7], [12], [13], [15]..

The Object model specifies how distributed objects must be described in a heterogeneous environment.

The Reference model defines the interactions between these distributed objects.

In the Reference model, one part defines the ORB and the different object interface categories, while the other part defines the concept of Object Frameworks which are the descendants of object-oriented class libraries.

The ORB is the core of the architecture and the main responsible for establishing communication between the different interfaces, known as:

- Object Services
- Common Facilities
- Domain Interfaces
- Application Interfaces.

An Object Framework is composed by several components; each component will support at least one of the interfaces mentioned above. Object Frameworks are domain-specific.

Interfaces, which specify operations and types an object supports, are defined by means of the OMG Interface Definition Language (IDL).

The importance of the IDL language is that it is just a declaration language, and therefore independent from the programming languages used in the implementations which is very convenient in heterogeneous environments.

In CORBA the interoperability is achieved by means of the *General Inter-ORB Protocol* (GIOP). Particularly used is the IIOP *(Internet Inter-ORB Protocol)* which specifies GIOP running on the TCP/IP transport layer.

5 A Web based STEP database server

A pilot project, aiming to build a Web based database server compliant to the STEP *Application Protocol 214: Core Data for Automotive Mechanical Design Processes* [6], is currently under development by our Department and FiatAvio S.p.A..The purpose is to experiment the integration and sharing of data among remote industrial processes so far poorly interacting or directly exchanging physical data.

The AP214 defines the context, scope, and information requirements for various development stages during the design of a vehicle, covering the mechanical aspects only.

Even if the Application Protocol has been developed in the framework of automotive industry, it can usefully support the design of any complex mechanical product.

The following data, in fact, are within the scope of the AP214:

- products of manufacturers and suppliers that include parts, assemblies of parts, tools, and assemblies of tools. The tools are used by various manufacturing technologies, such as shaping, transforming, separating, coating, or fitting;
- process plan information to manage the relationships among parts and tools used to manufacture them and to manage the relationships between intermediate stages of parts or tools;
- product definition data and configuration control data pertaining to the design phase;
- changes of a design, including tracking of the versions of a product and data related to the documentation of the change process;
- management of alternate representations of parts and tools during the design phase;
- identification of standard parts and library parts;
- release and approval data for products, versions of products and representations of products;
- data that identify the supplier of a product and any related contract information;
- seven types of representation of the shape of a part or tool: 2D-wireframe, 3D-wireframe, surface, faceted-boundary, boundary, compound-boundary, constructive solid geometry;
- data that pertain to the presentation of the shape of the product;
- representation of portions of the shape of a part or a tool by form features.
- product documentation represented by explicit and associative draughting;simulation data for the description of kinematics structures;
- analysis data for linear static FEA;
- properties of parts or tools;
- surface conditions;
- tolerance data.

5.1 The Prototype

Our first prototype under development based on the AP203 (for later extension into the AP214) has been realised using currently the next available commercial software products:

- ECCO Toolkit shareware versions
- STEP Tools ST- Developer 7.0
- STEP Tools ST- Oracle 7.0
- Oracle 8
- OrbixWeb 3.1

- Java Development Kit 1.1.8
- MS Visual C++ 5.0

Figure 2 shows the sequence of steps followed to obtain all the modules required to construct the database server, starting from an EXPRESS model.

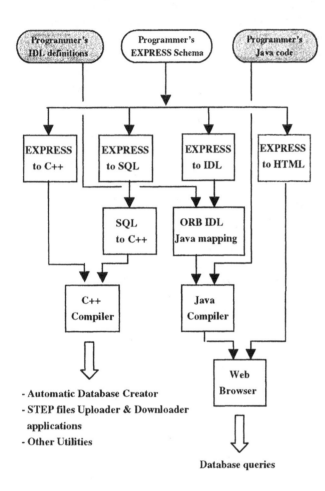

**Figure 2   From an EXPRESS
schema to application programs**

Firstly, the EXPRESS schema is translated into several different representations:

- C++, for exploiting the power of this language when creating application tools;
- SQL, for querying the relational database;
- IDL, for designing the CORBA distributed application;
- HTML, for having Web pages with all the entities, types, constants and so on present in the EXPRESS schema.

After that, some mappings of SQL into C++ and of IDL to Java is performed.

Finally, by compiling the C++ modules we obtain applications for the automatic creation of the database and for the uploading/downloading of STEP data files, while from the Java compiler we get the classes

necessary for querying the database by means of a Web browser.

5.2 Translations of an EXPRESS schema

It is well known that the definition of a complex and ambitious standard such as the ISO 10303 – STEP has required a lot of efforts during several years.

It is not surprising, therefore, that the firstly published documents, namely the EXPRESS models of the Integrated Resources and the AP203, contained many bugs that have made practical implementations more difficult

To solve the so-called "mysteries of AP203" seemed to need too much work with respect to the stability and robustness of the result.

Some years have passed since then, and it can be said that the progress done leads now to a good possibility of practical implementations. The adolescence of the ISO 10303 is ending, compilers are more reliable and compliant with the last revisions of the part 11 - EXPRESS language.

Even if our project is based on the AP214, we have worked also with the AP203 in order to have a larger variety of step data files available.

The first phase of the work was the review of the parts AP203 and AP214 and the removing of syntactic errors. In this work two different EXPRESS compilers have been used (STEP Tools and ECCO Shareware versions).

It is worth noticing that many errors were located in the entities constraints (WR#) of the EXPRESS schema and the lists or errors detected by the compilers were often quite different.

During the tests, the results and diagnoses generated by the different compilers have been compared in order to obtain more reliable corrections and take advantage of the different products performances and facilities.

As an example, four pages of errors were reduced to half of a page with one compiler, and the remaining errors did not prevent the other compiler from generating code for the database creation.

The debugging tools have proven to be very useful for discovering errors not exactly detected in the compilation phase.

5.3   Database Creation

Once the EXPRESS schema has been compiled and translated, the creation of the relational database is a straightforward process. For this we have used:

- A STEP to RDBMS Interface;
- an SQL to C/C++ translator;
- a C++ compiler;
- a relational Database.

The STEP to RDBMS Interface is a tool to generate an RDB schema from an EXPRESS model and to upload and download STEP files in the relational database. For this, the next code is generated:

- DDL statements for the corresponding EXPRESS Schema;
- Code to load/unload the STEP files.

The next information is STEP Tools' "ST-Oracle" specific. This interface supports entities, aggregates, selects and all the primitives (dynamic SQL is not supported). For being able to unload and reconstruct STEP files the tool creates the database converting the EXPRESS schema into SQL normalised tables definitions; in this way inherited attributes of an object are normalised into many different tables. For keeping track of this, base tables are also used in order to be able to identify the object and any information necessary about the original STEP file.

There are four base tables, three of them just to hold information about the original STEP file; the remaining one, the most important, called "oid_mapping" table, as shown here, will hold identity information about the STEP object in the database.

```
CREATE TABLE oid_mapping (
  Oid          CHAR((42)    NOT NULL,
  oid_key      INTEGER      NOT NULL PRIMARY
                            KEY,
  design_id    INTEGER      NOT NULL,
  entity_type  VARCHAR(80)
);
```

In this table the "oid_key" field is used as a foreign key for tying together all of the tables into which the inherited attributes of an object have been normalised out.

For the primitives types the STEP to RDBMS mapping is as follows:

EXPRESS	SQL
Integer	integer
Real	double
Boolean	integer
Logical	integer
Enumeration	varchar
String	varchar
Binary	number

With respect to an entity, as shown below, this is mapped into a single relation where the attributes are represented by columns.

```
      ENTITY entity_a;
      primitiv_attrib1  : [EXPRESS primitive type];
      primitiv_attrib2  : [EXPRESS primitive type];
      END_ENTITY;
```

```
ENTITY entity_b;
primitiv_attrib3  : [EXPRESS primitive type];
object_attrib1    : entity_a;
END_ENTITY;
```

```
CREATE TABLE entity_a (
entity_a_id    INTEGER NOT NULL
               REFERENCES
               OID_MAPPING(
               OID_KEY),
primitiv_attrib1 [SQL equivalent type],
primitiv_attrib2 [SQL equivalent type]
);
```

```
CREATE TABLE entity_b (
entity_b_id    INTEGER NOT NULL
               REFERENCES
               OID_MAPPING (
               OID_KEY),
primitiv_attrib3    [SQL equivalent type],
object_attrib1_id  INTEGER
);
```

As it can be seen each relation table have an identifier key field corresponding to the "oid_key" key in the "oid_mapping table that will serve for unique identification of an object in the relation. EXPRESS primitives attributes are mapped into SQL equivalent types as mentioned above whereas object attributes use the "oid_key" foreign key from the corresponding "oid_mapping" table.

The tables created have columns only for the immediate attributes, so inherited attributes are stored in their the respective supertype table representation. For example in the next table

```
ENTITY entity_c  SUBTYPE OF (entity_a);
primitiv_attrib4        : [EXPRESS primitive type];
END_ENTITY;
```

the mapping will be

```
CREATE TABLE entity_c (
entity_c_id    INTEGER NOT NULL
               REFERENCES
               OID_MAPPING (
               OID_KEY),
primitiv_attrib4  [SQL equivalent type]
);
```

So, an EXPRESS instance of type "entity_c" will be stored in the SQL tables entity_c and entity_a. with their tuples   tie together by the same "oid_key" numeric value.

The mapping for aggregate types is realised using two tables as shown here

```
ENTITY entity_d;
aggregat_attrib1  :LIST OF entity_a;
END_ENTITY;


CREATE TABLE entity_d  (
entity_d_id          INTEGER NOT NULL
                     REFERENCES
                     OID_MAPPING(
                     OID_KEY),
aggregat_attrib1_id  INTEGER
);


CREATE TABLE listofentity_a  (
listofentity_d_id    INTEGER NOT NULL
                     REFERENCES
                     OID_MAPPING (
                     OID_KEY),
entity_a_id          INTEGER,
entity_a_index       INTEGER
);
```

The elements of the aggregate are represented by tuples in the table. The identifier "aggregat_attrib1_id" references an identifier from the "listofentity_a" column that will point to all the tuples in the listofentity_a belonging to the aggregate type. The elements can be also ordered using the "entity_a_index" index.

For more information about the EXPRESS to RDB mapping consult the STEP Tools documentation.

Due to the long names of some entities or attributes and to the restrictions in the length of the identifiers in the databases, in order to avoid name collisions, an abbreviation generator is available. The created abbreviation files are then provided to the program when the step files are loaded/unloaded on/from the database.

The SQL to C/C++ translator, is a precompiler that accepts SQL statements embedded in a C/C++ program and translates them into a standard C code before compilation with the C++ compiler.

The database creation usually requires a large amount of table space.

5.4 CORBA implementation

Once the STEP files have been uploaded in the database, what remains is to make the database accessible from the Web by means of CORBA [16].

Putting together STEP and OMG's CORBA seems quite natural, since both are implementations based on the Object Technology.

The software involved is :
- JDK (Java Developer Kit);
- JDBC driver;
- an ORB OMG compliant (Java mapping).

Figure 3 shows the flow of data from a client node to Ine database.

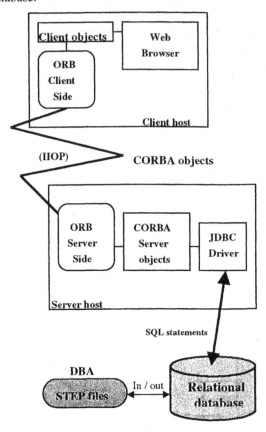

Figure 3   CORBA database connection

Clients applets will submit their requests using the ORB. The Server application receives the requests and redirects them to the database by means of the JDBC driver.

In the figure 4 shown below, we have an applet that illustrates some of the work already implemented.

6   Summary and Conclusions

In the present paper we have examined the basis and elements involved in the implementation of a virtual enterprise [8].

The goal of the first prototype is to make feasible the sharing of industrial data product in distributed and heterogeneous environments.

Full-fledged product descriptions can be stored in relational database systems that support the entire product life cycle.

Mainly two standards have been used:
ISO 10303 – STEP, and particularly the application protocols AP-203 and AP-214, for product data management;

OMG-CORBA for developing full object-oriented applications in distributed heterogeneous environments.

The information system is automatically developed starting from the EXPRESS model.

Communications in distributed heterogeneous environments are performed by means of applications implemented on a Java ORB.

Commercial software and tools have been used for creating all the components of the system.

Much work has to be done in the field of database access in order to have full standard object adapters abl to simplify the interfacing of object-oriented and relational systems

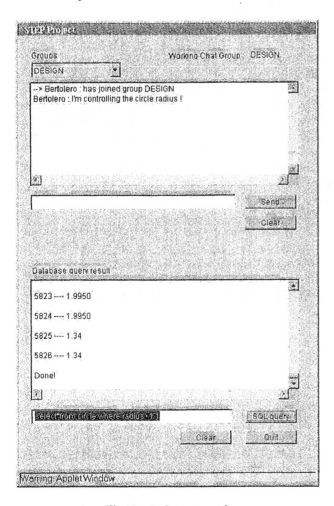

Figure 4 An example

In the next future we plan to improve the entire path from the information models to the database implementation.

References

[1]   ISO, "IS-10303-1, Product Data Representation and Exchange. Overview and Fundamental Principles", ISO, 1995.

[2]   ISO, "DIS-10303-22, Product Data Representation and Exchange, Implementation methods: Standard data access interface specification", ISO, 1996.

[3]   ISO, "IS-10303-21, Product Data Representation and Exchange, Implementation methods: Clear text encoding of the exchange structure", ISO, 1994.

[4]   ISO, "IS-10303-11 Description methods: The EXPRESS language reference manual", ISO, 1994.

[5]   ISO, "TC184/SC4/WG10 N219, Industrial Framework Model", ISO, 1998.

[6]   ISO, "DIS-10303-214, Product Data Representation and Exchange, Application protocol: Core Data for Automotive Mechanical Design Processes", ISO, 1999.

[7]   S. Vinoski, "CORBA: Integrating Diverse Applications Within Distributed Heterogeneous Environments", IEEE Communications Magazine, February 1997, pp. 46-55.

[8]   M. Hardwick, D. L. Spooner, T. Rando, K. C. Morris, "Sharing Manufacturing Information in Virtual Enterprises", Communications of the ACM, February 1996.

[9]   J. Fowler, "STEP for Data Management Exchange and Sharing", Technology Appraisals 1995.

[10] M. Susan Bloor and Jon Owen, "Product data exchange", UCL Press 1995.

[11] Paul Haggerty and Krishnan Seetharaman, "The benefits of CORBA-Based Network Management", Communications of the ACM, October 1998, Vol.41 No.10.

[12] John Siegel, "OMG Overview: CORBA and the OMA in Enterprise Computing", Communications of the ACM, October 1998, Vol.41 No.10.

[13] Michael Guttman and Rob Appelbaum, "The Next Generation of CORBA", Component Strategies August 1998.

[14] Amjad Umad, "Object-Oriented Client/Server Internet Envirinments", Prentice Hall, 1997.

[15] Alain Pope,"CORBA Reference Guide", Addison Wesley 1997.

[16] Geib,Gransart,Merle, "CORBA, des concepts a la pratique", Laboratoire d'informatique Fondamentale de Lille.

Extracting E-R models from IDEF0 diagrams

*Manabu Kamimura, **Satoshi Kumagai, *Kiyoshi Itoh

*Department of Mechanical Engineering, Sophia University
7-1 Kioi-cho, Chiyoda-ku, Tokyo, 102-8554, Japan
**Research & Development Headquarters, Yamatake Co., Ltd.,
1-12-2,Kawana, Fujisawa, Kanagawa, 251-8522, Japan

Abstract

We propose a method of extracting an E-R model from IDEF0 diagrams in designing entities and relationships. Extracting E-R models from IDEF0 diagrams enables us to capture important information from diagrams. IDEF0 defines activities to represent tasks performed by a system and defines ICOM (Input, Control, Output, and Mechanism) both to represent things in a system and to connect activities to form a diagram. To make the extraction of E-R models easier, we limit our method to the well-disciplined IDEF0 diagrams defined by Kumagai, Itoh and Inoue. The well-disciplined IDEF0 is an IDEF0 with specialized role and use of ICOM to enable more detailed description and analysis. In well-disciplined IDEF0 diagrams, control has special role to trigger activities. We define entities and attributes using ICOM, and define relationships using activities. We exclude control from ICOM before the extraction to save the designer from the task of excluding the information manually.

1. Introduction

E-R models [Hawryszkiewycz 1990] [Korth, Silberschatz 1986] are widely used in designing databases for commercial and engineering applications. An E-R model consists of entities, relationships and attributes. Specifying entities and relationships is usually done manually, and no standard and systematic method exists to guide such specifications. The process of design, therefore, relies on designer's intuition and often causes inadequate formalization of complex requirements. On the other hand, IDEF0 [Marca MacGowan 1988], a method to systematically analyze and to design functions of complex systems, is standardized [FIPS 1993] and is widely used. We propose a method of extracting E-R models from IDEF0 diagrams. This process enables us to capture information that the specialists think important and also

enables to use the discipline of IDEF0.

In section 2, we first introduce IDEF0. Next, we will talk about well-disciplined IDEF0 [Kumagai 1998][Kumagai 2000][Inoue 1999]. IDEF0 is very flexible so sometimes the model gets ambiguous. We introduce a disciplined to make the model very clear and precise.

In section 3, we introduce E-R models, which are widely used in designing databases, and the problems they have.

In section 4, we propose a guide of extracting E-R models from IDEF0 diagrams. We propose a notion of mapping entity.

In section 5, we demonstrate the guide by applying it to a practical example of well-disciplined version of IDEF0 modified from the diagram of "Build Special Part"[Marca, MacGowan 1988].

We conclude the paper by evaluating our guide and discussing issues of applying the extracted E-R models to next phase of design.

2. IDEF0

The U.S. Air Force Program for Integrated Computer Aided Manufacturing (ICAM) developed IDEF (Integration Definition method) as a systematic method of analyzing system requirement and modeling and as communication. IDEF is a series of techniques to analyze the system and includes function modeling, information modeling, object-oriented modeling, process, ontology, and etc.

IDEF0 is one of them, and produces a "function model". A function model is a structured representation of the functions, activities or processes within the modeled system or subjected area. IDEF0 is based on SADT™ (Structured Analysis and Design Technique™), developed by Douglas T. Ross and SofTech, Inc and it is standardized [FIPS 1993]. IDEF0 includes both a definition of a graphical modeling language (syntax and semantics) and a description of a comprehensive methodology for developing models.

2.1 Syntax and usage of IDEF0

In this section, we introduce the syntax and usage of IDEF0 [Marca MacGowan 1988].

(1) Activity and ICOM

IDEF0 models business processes as a collection of *activities*, and *things* which activities use. Those things are classified into *ICOM* (Input, Control, Output, and Mechanism). IDEF0 diagrams are described as in Fig. 1.

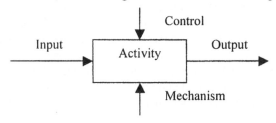

Fig. 1 IDEF0 diagram

An activity uses *mechanism* to convert *input* to *output* under the constraint of *control*.

ICOM includes anything that can be described as a thing in a business process such as data, material, information, organization, personnel, rule, product, etc. We write the definition of the ICOM on Table 1 in detail.

Table 1. The definition of ICOM

Input	Things to be converted by activity
Control	Thing which cannot be changed by an activity and show how and when the activity is to be executed
Output	Thing produced or created at the result of execution by an activity
Mechanism	Personnel, equipment, machine of other organizations, which are used by the activity

(2) Forking and Joining ICOM

An arrow rarely represents one thing by itself. Generally, it stands for a collection of things. For example, in Fig. 2, the arrow name "Selected tools" contains "Machine tools" and "Hand tools". Because arrows are collection of things, they can have multiple sources and multiple destinations. Therefore, arrows fork and join.

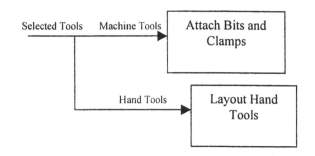

Fig. 2 forking arrows (ICOM)

(3) Layer of the diagrams

IDEF0 gradually introduces greater and greater levels of detail through the diagram structure comprising model.

The top diagram is called A-0 diagram and only has one activity. This diagram shows the system boundary and the interfaces to function outside the system. The A-0 activity drawn on A-0 diagram is decomposed into A0 diagram, which describes the whole process of the system. Since A0 diagram describes the whole system, activities and ICOM are written in general. Activities in the A0 diagram are named A1, A2, etc. from the left most and upper most. These activities are described in general and are supposed to be described in detail in other diagrams. Lower diagram has the same name of the activities drawn in the upper diagrams. For example a diagram that describes activity A1 in the A0 diagram is called A1 diagram.

The activities that represents the system as a single module is detailed in another diagram with activities connected by interface arrows.

The number of activities in the diagram should be no fewer than three to make the diagram not too general but no more than six box to keep it simple.

We show the structure of the IDEF0 diagram in Fig 3.

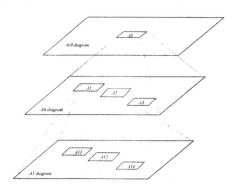

Fig. 3 IDEF0 layer

2.2 Well-disciplined IDEF0

IDEF0 is a flexible method and sometimes this flexibility causes the model to get ambiguous. We introduce more strict discipline to make the model precise, comprehensive, and consistent. This discipline is based on the concept that many people, machines and etc. collaborate in business process. This concept takes notice to the moment these collaborators meets, which enables us to capture important scenes. We call this version of IDEF0, *well-disciplined* IDEF0 [Kumagai 1998][Kumagai 2000][Inoue 1999].

We use well-disciplined IDEF0 diagrams when we extract the E-R model. Well-disciplined IDEF0 diagrams satisfy the following conditions.

(1) Viewpoint is controlled.

The designer stands in a place, person, organization, or thing to view and to model the system. This place is called *viewpoint* in IDEF0 and should not be moved during the analysis, or the system gets ambiguous and large unnecessary.

We show an example of the difference of viewpoint in Fig. 4 and in Fig 5.

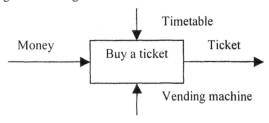

Fig. 4. Model from the viewpoint of the buyer

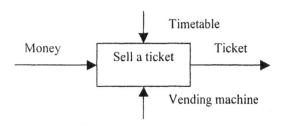

Fig. 5. Model from the viewpoint of the seller

In Fig.4, the diagram is described from the viewpoint of the person who buys the ticket and the activity is named "buy the ticket". In Fig. 5, the diagram is described from the viewpoint of the person who sells the ticket.

It is hard to keep the viewpoint controlled for the author because of the steps in the design process. The designer first lists up the ICOM, and then names the activity from the connected ICOM. The designer tends to name the activity in the viewpoint of the performer of each activity, but viewpoint of the whole system is different from the viewpoint of the performer of each activity. This causes the model to have multiple viewpoints, while IDEF0 should only have one viewpoint.

In this example, there are two viewpoints. One is the person who buys the ticket, and the other is the person who sells the ticket. If this model describes "how to get on the train", the viewpoint has to be in the person who gets on the train and this activity also has to be written from that viewpoint. The designer cannot name the activity when they only think about the particular activity alone.

(2) Materials and information are classified clearly.

Materials and information are sometimes very ambiguous. For example, when we look the "Chart" as material, we take importance to what it is made of. It may be made of paper or a plastic. People working for paper companies take interest in it. When we look it as information, we take importance to what is written on the chart and do not care whether it is written on a paper or on an electronic file.

(3) Input and control are classified clearly.

Input and control are usually classified by the intuition of the designer and sometimes the classification gets ambiguous. IDEF0 only says that the difference between input and control is that input is changeable by the activity but control is not. For example in Fig.4 or Fig. 5, we can regard the timetable as control because the timetable is already written and cannot be changed by the person who buys the ticket. We can also regard the timetable as input when we assume that the person looks into the timetable and determine the train he or she would take. We think that the information in the timetable is converted.

This provokes the discussion we will describe in (5).

We propose to add the following conditions for well-disciplined IDEF0. These conditions are also based on the concept of collaboration.

(4) Forking/Joining arrows are tagged.

About forking and joining arrows, Marca defines the labeling of branch as in bellow.

1.Branches which are not labeled are assumed to contain all things indicated by the label before the branch (i.e., all things go to the branches). [Marca,MacGowan 1988]

Fig. 6 Interpretation of untagged ICOM

2. Branches which are labeled after the branch point are assumed to contain either all or some of the things indicated by the aggregate label before the branch (i.e., each branch label makes explicit what it carries). [Marca,MacGowan 1988]

Fig. 7 Tagged ICOM

This definition is clear in the case described in Fig.6 and Fig.7. We agree with these cases, but next we talk about an another case that makes the diagram ambiguous. There are some diagrams described as in Fig. 8.

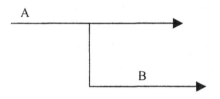

Fig. 8 Partly untagged ICOM

In Fig.8, one of the branching arrows is labeled but the other is not. FIPS suggests an interpretation as in Fig.9.

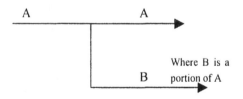

Fig.9 Interpretation of Fig.8 [FIPS 1993]

FIPS says that one branch carries *B* that is a portion of *A* and the other untagged branch carries *A* that means the same thing before the arrow branches. This interpretation is natural when we think A as information because duplication is very easy.

On the other hand, when we regard A as a material in Fig. 8, the untagged arrow is assumed to be the rest part of A because making copies is not easy. Then, it is natural to think that the untagged part in Fig. 8 is "A-B" (the other half of B).

We propose not to allow such untagged ICOM as in Fig. 8 and to tag each of them because these cause the diagram to get ambiguous. We should not rely on the FIPS interpretation because the interpretation is unnatural when we regard the concerning ICOM as material. If the forking arrows are completely the duplication of the part before the forking point as in Fig.6, we agree with the interpretation to avoid redundancy.

(5) Special Role of Control

Because of the ambiguous definition of the role of input and control we described in (3), we may have ICOM that describes many meaning that causes the diagram to get ambiguous. We show an example of this case.

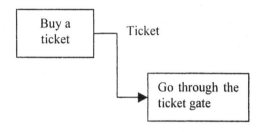

Fig. 10 Ambiguous classification of input and control

In Fig.10, the ICOM "ticket" can be material when we concern about the paper itself. We can also regard this ICOM as information that triggers the next activity "Go through the ticket gate". If this ICOM include the information, which only triggers the activity to start, the information is not used after the activity started. The information is not converted to the output by the activity so it should not be included in the input.

To solve this problem, we suggest that input should be a material or information, which exists in the system for a long time, and control should be a trigger of the activity. The model in Fig. 10 is supposed to be drawn as in Fig. 11.

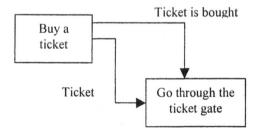

Fig. 11 Classification of input and control

In Fig. 11, the input "Ticket" is material and exists in the system for a long time. The control "Ticket is bought" triggers the activity "Go through the ticket gate". When the control comes in, the activity starts its process. After the activity started, the information doesn't have to be kept. This classification also enables us to describe the situation that the arrival of the input is different from the starting of the plugged activity.

We can write Fig.4 as Fig. 12.

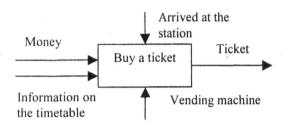

Fig. 12 Clear classification of input and control

Fig. 8 shows an example of clear classification of input and control. When you arrive at a station and you are to ride a train, you have to buy a ticket so the activity "buy a ticket" is triggered. The activity converts the money and the information on the timetable to a ticket by using the vending machine. When we think the timetable as input, we regard it as information of many trains and the train is going to be selected by the activity. On the other hand, when we think as control, timetable is information that says the particular train the person is going to take and is already selected before the activity starts.

3. E-R models

The E-R (Entity-Relationship) model [Hawryszkiewycz 1990] [Korth, Silberschatz 1986] is based on a perception of a real world, which consists of a set of basic objects called entities and relationships among these objects. It is developed to facilitate database design.

3.1 Entities and Entity Sets

An *entity* is an object that exists and is distinguishable from other objects. An entity may be either concrete, such as a person or a book, or it may be abstract, such as a holiday or a concept.

An *entity set* is a set of entities of the same type. The set of all persons having an account at a bank, for example, can be defined as the entity set customer. Similarly, the entity set account might represent the set of all accounts in a particular bank.

An entity is represented by a set of *attributes*. Possible attributes of the customer entity set are name, social security, street, and city.

3.2 Relationships and Relationship Sets

A *relationship* is an association among several entities. For example, we may define a relationship that associates customer "Harris" with account 401. This specifies that Harris is a customer with bank account number 401.

A *relationship set* is a set of relationships of the same type.

3.3 E-R Diagram

E-R models are described graphically by an E-R diagram, which consists of the following components:

1. Rectangles, which represent entity sets.
2. Ellipses, which represent attributes.
3. Diamonds, which represent relationship sets.
4. Lines, which link attributes to entity sets and entity sets to relationship sets.

We show an example in Fig.13.

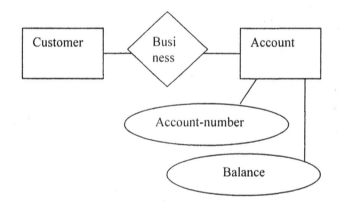

Fig. 13 E-R diagram

4. Process of extracting E-R models from IDEF0 diagram

We propose a guiding process for extracting an E-R model from well-disciplined IDEF0 diagrams. The process is not automatic and it needs to have the designer's decisions as to which items are made entities and which are attributes. This is because designing entities and relationships using only syntactic structure of IDEF0 will result in unnatural decisions in many cases and the designer need to consider the meaning of activities and connected ICOM for more natural output. Due to use of disciplined IDEF0 diagram, we define entities using ICOM except controls. Let IOM denote ICOM except control.

(Step 0) Generally, we will map IOM to entities and attributes, and activities to relationships. IOM connected by forking and joining is mapped to the same entity or attributes of the same entity. For each IOM a, the entity is defined as mapped entity of a and denoted as $E(a)$. Each activity x defines relationships $R(x)$ among mapped entities of IOM connected to the activity.

(Step 1) The process starts at the top level of IDEF0 diagram, i.e., A-0. First create an entity for each input or mechanism a of IOM of the diagram and make a an attribute of the created entity. The mapped entity of such IOM is the entity defined. Then, for output b, we need to consider its meaning. First, we need to decide if the information on the output is included in the E-R model or not.

1A) If we decide not to include the information, $E(b)$ is *NIL*.

1B) If we decide to include that, we then decide the following.
1B-1) If b is very closely tied to one of the input a, and can be attached to a, define $E(b)$ to be equal to $E(a)$ and b is made an attribute of $E(a)$.

1B-2) If b is very closely tied to one of the mechanism c, and can be attached to c, define $E(b)$ to be equal to $E(c)$ and b is made an attribute of $E(c)$.

1B-3) If b is very closely tied to both of the input a and the mechanism c, and can be attached to both a and c, define $E(b)$ to be equal to $E(a)$ not to $E(c)$ and b is made an attribute of $E(a)$.

1B-4) If this is not possible, a new entity is created and $E(b)$ is defined to be that entity.

1C) If there is at least one output entity created in this way, the activity of A-0 is made relationship. The relationship relates all mapped entities defined here.

(Step 2) The process proceeds to lower diagrams, i.e., A0, A1, etc in a similar way. When we consider IOM in each diagram, we can assume that all IOM that have connections to externals are already mapped to entities. Now we need to define $E(a)$ for internal IOM a. Again we need to consider the meaning of a as before. We start from the upper most and left most activity and proceed to lower and right activities in similar manner. When we consider an activity, we can assume that its input and mechanism are already mapped to entities. If there is an input or mechanism d that come from outside with no specific origin, we need to treat d as an exception and introduce new entity as the mapped entity of d and make d an attribute of the entity. Now for output b of an activity, we consider the following. As before, we need to decide if the information on b is included in the E-R model or not.

2A) If we decide not to include the information, $E(b)$ is *NIL*.

2B) If the b is used as a control for some activity, $E(b)$ is *NIL*.

2C) If we decide to include that, we then decide the following.
2C-1) If b is very closely tied to one of the input a, and can be attached to a, define $E(b)$ to be equal to $E(a)$ and b is made an attribute of $E(a)$.
2C-2) If b is very closely tied to one of the mechanism c, and can be attached to c, define $E(b)$ to be equal to $E(c)$ and b is made an attribute of $E(c)$.

2C-3) If b is very closely tied to both of the input a and the mechanism c, and can be attached to both a and c, define $E(b)$ to be equal to $E(a)$ not to $E(c)$ and b is made an attribute of $E(a)$.

2C-4) If this is not possible, a new entity is created and $E(b)$ is defined to be that entity.

2D) If there is at least one output or mechanism created in this way, the activity of IOM connected is made relationship. The relationship relates all mapped entities defined here.

(Step 3) If relationship exists both on upper level diagram and on lower level diagram, choose the relationship on the lower level diagram. Lower diagrams describe the system in further detail than upper diagrams.

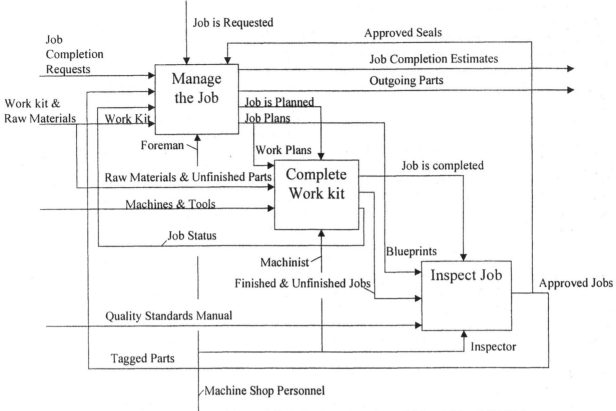

Fig. 14 A0 diagram of the well disciplined version of "Build Special Part" (IDEF0)

Well-disciplined IDEF0 diagram guarantees that Step 2B) and 2C) will not cause conflict. Note that steps 1A), 1B), 2A), and 2C) require judgment by the designer using semantic information of IOM. This means that the process cannot be done automatically. To automate the process, we need additional information.

5. An Example of the extraction

We demonstrate the method by applying it to an example of well-disciplined version of IDEF0 modified from the diagram of "Build Special Part". This IDEF0 model has 9 diagrams.

We show the A0 diagram in Fig. 14 to capture the overall structure and omit the other diagrams because of the space of this paper.

Next in Fig. 15, we show the E-R model we extracted from the IDEF0 diagram we showed above.

6. Conclusion

We conclude this paper by evaluating our method.

Extracting E-R models from IDEF0 diagrams enables us to design E-R models by using the information in the IDEF0 diagrams and not to rely on the designer completely. This also enables us to design database by using IDEF0 diagrams.

But the information on the IDEF0 diagrams is not complete and the extracted models have the following problems.

1   Extracting attributes are not enough. Especially key attributes are not defined automatically.
2   The diagram is not complete and is not normalized. It requires steps such as redundancy elimination for refinement.
3   Transactions are not extracted.
4   Detailed relationships such as the number of entities with which another entity can be associated (i.e. "mapping cardinalities") are not extracted.

We believe that such information cannot automatically be extracted and the designer has to add the information. The information is not explicit in the IDEF0 diagrams.

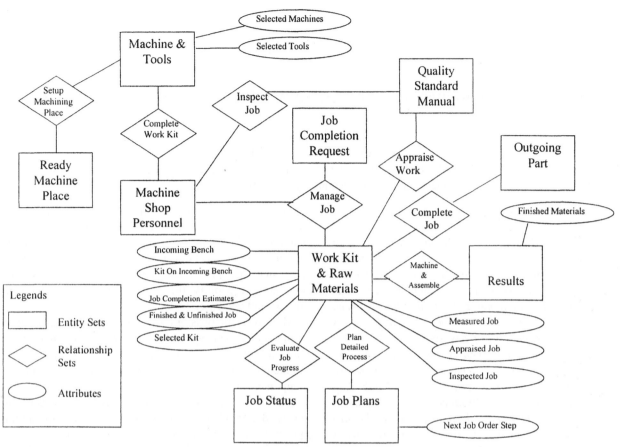

Fig. 15 Extracted E-R model from "Build Special Part" IDEF0 diagram

We plan to apply our method to a number of larger examples to ensure its effectiveness.

Reference

[FIPS 1993] Draft Federal Information Processing Standards Publication 183 "Announcing the Standard for INTEGRATION DEFINITION FOR FUNCTION MODELING (IDEF0)" (1993)

[Hawryszkiewycz 1990] Igor T.Hawryszkiewycz Relational Database Design An Introduction, Prentice Hall (1990)

[Inoue 1999] Kiyofumi Inoue, Satoshi Kumagai, Ryo Kawabata, Kiyoshi Itoh: Enhancement of Collaboration Process by Bi-directional Converter between IDEF0 and Multi-Context Map, FOSE (Foundation of Software Engineering) '99 p.260-267 (1999) (in Japanese)

[Korth, Silberschatz 1986] Henry F.Korth, Abraham Silberschatz: Database System Concepts, McGraw-Hill Advanced Computer Science Series (1986)

[Kumagai 1998] Satoshi Kumagai, Kiyoshi Itoh: Designing Collaborative Work in IDEF0 using Interface Model, CERA Journal, Vol.6, No.4, p.333-343, (December 1998)

[Kumagai 2000] Satoshi Kumagai: A Study on Analysis Methods in Collaboration Domain, Doctoral Thesis, Sophia Univ., (2000).

[Marca, MacGowan 1988] D.A. Marca C.L. MacGowan, IDEF0/SADT Business Process and Enterprise Modeling Eclectic Solutions Corp. (1988)

Integration of component descriptions in product data management systems

M. El-Hadj Mimoune

Y. Ait-Ameur, G. Pierra and J.C. Potier

LISI/ENSMA, BP 109, Téléport 2, F-86960 Futuroscope cedex, France

e-mails: {mimoune, yamine, pierra, potier} @ensma.fr

Abstract

This paper addresses two fundamental aspects of concurrent engineering processes. It suggests the integration of component and product data. On the one hand, the Parts Library standard, denoted P-Lib, has been described for exchanging part libraries between CAD systems, which use a big amount of heterogeneous data related to components or parts. Its data model allows the capability to exchange complex data between library management systems : data schemas, queries, constraints, domains and methods are described in this P-Lib data model. On the other hand, Product Data Management systems, denoted PDM, have been defined, in the last decade, in order to manage thousands of heterogeneous data (CAD/CAM data, BoM etc.) related to products. Technical data management permits the management of the all-informational patrimony of a product starting from its design until its discharge.

Products are essentially composed of pre-existing components stored in component or part libraries. This paper presents the approach we developed to integrate component data and product data. This approach has been applied with the particular standardised P-Lib data model for representing components and a particular PDM system namely SmarTeam. It will allow the automatically storage of component data in a PDM and the easy referencing of these data in the products during their whole lifecycle.

Keywords: Parts Library (PLib), PDM, data integration, meta-data, tracability, EXPRESS language.

1 Introduction

The growth, the heterogeneity, the security of data motivates the creation of information processing systems to facilitate heterogeneous products data management during their whole lifecycle (CAD-CAM Files, Quality folders, BoM etc.). These systems are named PDM (Product Data Management Systems). In a number of industrial fields, products are constituted with pre-existent components stored in component or parts libraries.

The capability to refer to component data from a PDM, in order to allow the use of PDM to represent and to manage product data in parallel with component data, permits to represent both product and component data in the same design environment. As a result of this referencing capability, the development and management costs would decrease.

Moreover, the possibility to refer to and to manipulate component knowledge and data as well as the reason for choosing this component in a given product facilitate the maintenance of those products when, for example failing components are to be replaced.

Two main approaches can be used for describing several data models. The first one, called multi-modelling approach, consists in describing several data models. Integration of these data models requires a translation from a data model to another. The second one is the meta-modelling approach where each data model is a particular instance of a data meta model.

The multi-modelling approach may be used for representing product data models in PDM (it is the one followed in the STEP standard for example in the Application protocols). The meta model approach has been followed by P-Lib standard for representing components (or parts library) data model.

Therefore, the integration of both approaches is a crucial issue when addressing both product data and component data. So, the goal of this paper is to present an approach to integrate both component and product data and knowledge in a common framework.

In the context of our approach, component data models are defined by means of instances of the P-LIB data model, developed in the Parts Library ISO-13584 standard. The P-LIB model is a formal data model allowing to represent and to exchange part libraries.

Product data are represented in a particular PDM. For implementation purposes, we have chosen the *SmarTeam* PDM, developed by the Smart Solutions Company. So, the goal of our approach is the representation of P-LIB data in a PDM.

The integration of these models requires the description of a common language allowing to represent them. For this purpose, we have used the EXPRESS data modelling language and its graphical representation. This choice was motivated by the fact that this language allows the representation of all the knowledge categories related to both components and products.

So, our approach uses the EXPRESS language and its graphical representation as specification language for representing all the data models presented in this paper.

The approach we developed is a mixed (or hybrid) approach. Indeed it does not totally use a meta modelling nor a multi modelling approach. It uses the capabilities of both approaches and their ways of representation. It consists in representing the P-Lib data components in two different representation levels: the meta level representation for component families descriptions (component classes) and the hard encoded (or the direct representation) for representing data stored in tables. The first represents component families while the second represents the components themselves. So, some P-Lib descriptors are represented by the PDM objects (representation at the meta level) and the other entities representing tables are represented by the PDM tables (direct representation). This choice allows to use the capabilities of both the LMS (Library Management System) for selecting components and PDM for managing product data. Therefore, the PDM becomes capable to refer directly to the components data.

This paper is structured as follows. Next section presents an overview of the formal data modelling language EXPRESS and its associated graphical representation : EXPRESS-G. Section 3 gives an overview of P-Lib data models and its specifications. The overview of a PDM systems and description of a particular PDM (SmarTeam) is presented in section 4. Finally, last section presents the integration approach of the P-Lib data model in the selected PDM system.

2 Overview of the EXPRESS language

Knowledge modelling and automatic data processing have a great importance in different kinds of engineering knowledge. Several formal representations are used to formalise these knowledge models among them, we can cite the NIAM method, OMT[1], UML[2] etc. These formalisms were developed to allow a real world modelling (a universe of discourse) by semantic entities.

The use of CAD/CAM systems in engineering domains has begot an earnest problem, which concerns data exchange and sharing between contractors and subcontractors. This situation is due to their use of heterogeneous systems. Moreover, several formats were developed to allow this exchange. Among the different developed standards we can quote the STEP (STandard

for the Exchange of Product data model) international standard which has developed the EXPRESS formal data modelling language (ISO 10303-11). The goal of this language definition is to allow data models description in order to exchange and to share data compliant with these models [3]. Moreover, it defines an exchange format for any data model specified in EXPRESS.

In the following, we will focus on the EXPRESS language constructs necessary to understand this paper. More details on the EXPRESS language can be found in [3] and [14].

2.1 General structure and EXPRESS concepts

EXPRESS is an object oriented data modelling language handling the important characteristics of object oriented language like abstraction, encapsulation, modularity and hierarchy [4].

In EXPRESS a data model is represented by a set of modules, called SCHEMAs. Each SCHEMA describes a set of entities, which represent the objects we want to model. Each entity is defined by a set of characteristics called attributes. Finally, each attribute has a data type where it takes its values. A SCHEMA may also contain a set of functions and procedures, which allow data coercion (integrity constraint) and derived attributes expressions (behavioural knowledge).

Entities represent, in the real world, objects that have physical existence (screw, bearing) or conceptual existences (address, course). In EXPRESS the entity represent a class of objects, which share common properties and behaviours. Inheritance between entities, representing structural knowledge, is allowed.

Attributes represent the properties which characterise an entity. They allow descriptive knowledge modelling. An attribute is defined by an identifier and a data type which can be simple (integer, real, character), structured (lists, sets, bags, arrays which are rigorously defined in EXPRESS), or user defined like in programming languages (C++; ADA etc.). Attributes can refer to entities (aggregation); i.e. attribute value can be an instance of this entity.

Three kinds of attributes are distinguished: Explicit attributes, Derived attributes and Inverse attributes.

Constraints: the EXPRESS language supplies the capability to describe constraint expressions. Constraints on data models are introduced by two different rules types: local rules which are applied on each entity instance separately , and global rules which are globally applied on several entity instances at the same time.

Local rules shall be checked against for each instance of a constrained entity. They are declared, at the same time as the attributes. Local rules are applied on the value domains of set of attributes.

Inheritance: An EXPRESS data model is composed of a great number of entities, which are connected by father/son relationship (*is_a* relationship). If an entity A is a subtype of another entity B then A contains the properties of A more its own properties. A inherits all its parent properties entity.

In EXPRESS the inheritance can be simple, when an entity inherit only one entity, or multiple when an entity inherits several entities at the same time.

2.2 EXPRESS-G

EXPRESS-G is the graphical representation of textual EXPRESS data models. It allows synthetic presentation of textual EXPRESS data models, which are hard to read by human being. The graphical representation permits to make a partial representation of the full data model. Moreover, this formalism may be used to design a data model in preliminary modelling stages. EXPRESS-G allows structural and descriptive data representation with a graphical annotation. This format helps to have a global view of a data model. It should be noted that a graphical model increases readability and comprehensibility.

We have used the EXPRESS-G to model both the meta-model of P-Lib and the multi-representation model of PDM (SmarTeam) to confront the integration of their different approaches. So, the EXPRESS-G representation allows us to make a sort of interface between the two systems that will allow us to use components data (P-Lib) in PDM systems. We have chosen this formalism instead presenting program sources that we have developed.

Example

To illustrate the EXPRESS representation we will present, in the following, an example of a simple EXPRESS data model. We take the example of the geometry. Circle and point are both geometric entities. The entity Point has the attributes X, Y, Z representing co-ordinates and the inverse attribute *is_centre_of*, which allows expressing the following constraint: a point can be a centre to the maximum of two circles. The entity Circle has the following attributes: centre, radius and perimeter. The last one is a derived attribute.

The textual data model of this example can be represented as follow:

```
ENTITY Geometric_entities
SYPERTYPE OF (point, circle);
END_ENTITY;

ENTITY POINT
SUBTYPE OF (Geometric_entities);
X, Y, Z : REAL;
   INVERSE
Is_centre_of: SET [1:2] of circle for centre;
END_ENTITY;
```

```
ENTITY Circle;
SUBTYPE OF (Geometric_entities);
  Centre : Point;
  Radius : REAL;
  DERIVE
  Perimeter : PI*2.0*(SELF.Raduis)
END_ENTITY;
```

The representation of this example in EXPRESS-G is given on figure 1.

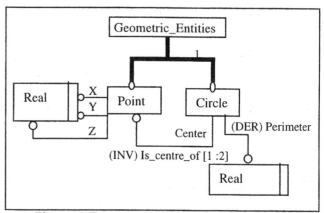

Figure 1 Example of EXPRESS-G model

Legend :

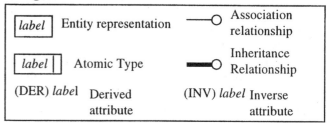

3   P-Lib : the parts libraries data model

The lifecycle of a product is a complex track. It requires different knowledge and involves several experts that work on the same product but with different perspectives and points of views.

The resulting process includes specification, design, manufacturing, and maintenance etc. [5]. Those activities are essentially informational processes and need to be handled by models, documents, computers etc.

The Computer Aided Engineering (CAE) area studies computer systems that allow to support these processes [6]. These computer systems supply efficient tools and mechanisms to represent different knowledge categories and to make it available for each expert in the suited format.

Ideally, these systems should allow each expert to work on his own perspective and should ensure co-ordination and interaction between the narrows perspectives.

Often, products to be designed are made of pre-existent technical objects [7]. It is the case for several domains of engineering such as electronic, mechanic etc. Hence, in such domain, a very significant part of design knowledge is related to components design. This knowledge

data and store them in their own libraries. The data contain at the same time the component descriptions (general models) ant their representations (functional models).

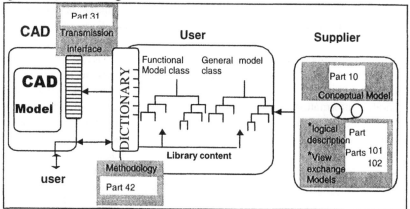

Figure 2 general Architecture of parts library

corresponds to the capability to select a component, to evaluate its behaviours and to create different representations related to each discipline.

The goal of P-LIB-based digital catalogues is to allow exchange of component knowledge between suppliers and designers. These catalogues should be able to transmit both component characteristics and their representations. It also should include the various information element included in catalogues like figures, documents, tables etc.. Notice that, in practice, these catalogues are used by product designers to choose the components to be inserted in products.

In order to represent catalogues and component knowledge, the P-Lib data model has been described and standardised. The next sections give an overview of this data model. More details can be found in [8], [9], [10], [17]

Library data are structured into classes according to the object-oriented paradigm. Three kinds of classes are considered in P-Lib:

General model classes enable library data suppliers to provide the definition of parts represented by a hierarchy of part family classes.

Functional model classes enable library data suppliers to provide various representations (e.g. Geometric, schematic, procurement data etc.) for these collections of parts.

Functional view classes enable the specification of the kind of representation provided in different functional model classes.

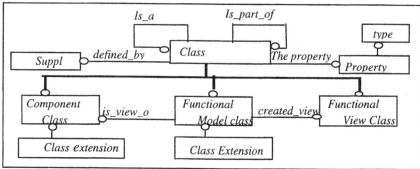

Figure 3: class hierarchy in P-Lib

3.1 P-Lib architecture

The general architecture of P-Lib is shown on figure 2. Suppliers describe components they supply within library (suppliers library) according to the formal P-Lib data model defined in EXPRESS and documented in the ISO 13584-24 and ISO13584-42 [8], [9]. Users recover these

The figure 3 summarises the description given above. It gives an overview of the whole P-Lib data model.

These classes shall be defined as a dictionary. It consists of a set of entries associated with a human-readable and computer-sensible representation of the meaning associated with each entity. This dictionary provides a referencing mechanism between library data obtained from different suppliers and enables the user to obtain an understandable view of parts held in the library.

3.2 Fundamental principles

The international standard P-Lib is based on different principles [10]:
- It separates the representation of information held in a parts library from the implementation methods used in data exchange.
- It separates information about the structure of a parts library from the information relating to the different representations of each part or part family which belongs to the parts library.
- It uses a formal data specification language, EXPRESS to specify information about the structure of a library.
- It permits the information about the different representations of each part or family of parts within the library to be specified by different standards. The information is referenced within the information specifying the structure of the library.

3.3 P-Lib specification

The P-Lib standard was developed to allow designers to recover component data together with their different representations. A designer can, thus, choose component based on the properties supplied in P-Lib. The selection process consists of the following stages:
- the first stage is to choose a particular component family (component category), which corresponds to the desired functionality ,
- the object of the second stage is to choose, in the family, one or several accurate component instances which are adapted to the requirements expressed by the insertion context. Notice that this insertion context is usually a PDM or a CAD system.
- the end of this process is to choose a component well defined which we can precisely identify. This method of selection enables us to define a design problem to be solved. This process allows interchangeability in a maintenance context for example.

The users can access to suppliers' libraries across the interfaces, which exist between each system and P-Lib models. In this manner designers can use component data in their CAD or PDM systems.

As it was stated previously P-Lib uses the EXPRESS language to model and formalise component data. These data are stored in EXPRESS exchange format it provides and are encoded in ASCII files.

3.4 Conclusion

In conclusion we can say that the P-Lib standard has provided an approach and information models to facilitate components data exchange and their sharing between suppliers and designers systems. It allows exchanging digital component catalogues which integrate the supplier knowledge on the components he/she supplies (intelligent catalogues).

The P-Lib data model is a meta-model, which allows intentional and extensional description of component catalogues for their exchange between heterogeneous systems. This model characters both the structure and the characteristics of components (structural and descriptive knowledge), and the mathematical relations existing between different characteristics (procedural knowledge).

Use of P-Lib component data in CAD systems is possible using the interface which have been defined between P-Lib data and CAD systems. But such interfaces do not exist yet between P-Lib and PDM systems. This is the main motivation of our work.

Components are largely used in industry to design and manufacture new products. Companies use PDM systems (Product Data Management) to manage data related to the products they manufacture (CAD/CAM files, quality folders etc.). So, it is crucial to offer some capability to use component data defined in P-Lib in PDM systems.

4 PDM systems description

With the use of computers in more and more activities, a company needs to manage efficiently thousands of heterogeneous data created each year. These data are related to CAD/CAM files, Specification files, Numerical Command (NC) programs etc. Because of the big volume of information, of the various actors using this information and of the diversity of data processing tools, controlling this information generates several problems. Among these problems, we can cite:
- incoherence between various document and file versions used by different actors in development phases,
- delivery of incomplete manufacturing data for production,
- delay between product manufacturing and the delivery technical document.

To address these problems and subject to control the various technical information, involved in product development, in production, in marketing and in logistical support of industrial product, a new data management method was developed: the Product Data Management. This field is subject to an active standardisation work: STEP (STandard for the Exchange of Product model data); CALS-CE that is an initiative of the American Department of Defence aiming at defining methods and tools to manage and exchange technical information

between all computer systems used inside the company. The following figure shows existing interaction between PDM and various other systems.

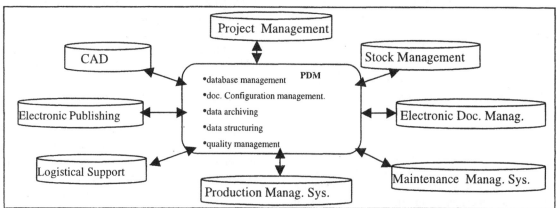

Figure 4: PDM systems interfacing with computer company systems.

related to weapon manufacturing.

4.1 Technical data

Technical data are data concerning product and process definition, which are used during the whole of product life. They include all the data that allow product description throughout its lifecycle. These data can be structured or not structured. They can have either electronic or hardcopy support. These data are necessary to identify and to describe state and configuration of a product, to manufacture a product, to control product evolution, to use and maintain the product.

These data are generated by different computer tools: CAD/CAM systems (Computer Aided Design/ manufacturing), CAE (Computer Aided Engineering), QFD (Quality Function Deployment) etc. and they are related to design, to engineering, to the manufacturing, to quality management, to the logistical support etc.

Technical data are associated with product during its whole lifecycle (starting from its design until its discharge). These data have particular characteristics. Each datum is associated to a particular computer system, which allows to interpret its content.

Lifecycle encompasses several phases. It includes the following: requirement definition, concept design, production, operation, maintenance and discharge.

4.2 PDM definition

A PDM is a system which allows to organise and to manage all the heterogeneous data described in the previous section. It also allows to give the right data to the right person and, to provide to everyone his/her own view of the project. A PDM manages technical data access, modification and sharing. It permits multiple access at the same datum and/or at the same time. It ensures interfacing

Figure 4 shows how a PDM manages various data relating to a product. This management is done at two levels [12]. each set of product data (content) is recorded in a container associated with meta data. A PDM stores the content and manages the container.

Container management: supports the BoM (Bills of Material). It encompasses two functions: description of element components (Object ID, attributes etc.), and composition links that exist between different elements (component/composite relation). The composition link, between the component object and the composite object, is defined as another object.

Content storage: is a storage system with a great capacity: **the vault.** Vault content represents for example the drafting files, pictures digitised by a scanner, ASCII files etc. Visualisation of such object is not be possible without the engines that generated them and without PDM viewer. Without these engines we can't interpret their content. These objects are called BLOB (Binary Large Object). Objects, which are put in the vault, are controlled by the PDM. Any creation or modification is submitted to the validation by an ECO number (Engineering Change Order).

4.3 SmarTeam: description and structure

For our experimentation, we used a particular PDM system supplied by Smart solution called SmarTeam.

SmarTeam is founded on the principle of object-oriented databases.

We used this tool for study any feasibility of integration of P-Lib in a PDM. In the following, we will outline the structure of SmarTeam this will help us to present the solution we propose to achieve this integration.

The structure of SmarTeam database is based on the concept of a project that represents the main class of the database. This global structure is constituted by a set of parallel class hierarchies where the first hierarchy represents the main classes. This structure permits to access data easily because they are generally bound to the instances of a main class and one can reach them through objects of this class (by the Browser or by the associated Application Programming Interface API).

4.3.1 Classes

In SmarTeam data are regrouped in several hierarchies of classes that represent categories of objects and then support simple inheritance. These classes are hierarchised to permit simple inheritance. In SmarTeam classes are classified in the following manner:

a) The main class: as stated above, the structure of data is organised around a hierarchy of main classes that plays a particular role in the database. It represents described product(s). Instances of the main class are displayed when SmarTeam is launched. This class has links with the other classes. Via these links, the user can reach the other objects directly.

Example: we can put the class *Projects* in the main class. Instances of this class can be motors, or gearbox for example. Others objects, such as parts of motors or gearboxes, will have a link with the instances of the class *Projects*.

b) Super-classes: they are the higher level classes in the hierarchies of which the first described is the main class. They are used to regroup classes in a specialisation hierarchy, and they constitute a separate tree. To every Super-class, it is necessary to associate indexes that allows the system to verify the uniqueness of objects in the class and that play the primary key role in a database.

Super-classes permit the construction of link classes between all classes that are in the hierarchy. The link between two classes is an object of the link class that binds their super-classes.

c) Leaf classes: Leaf Classes are classes of the lower level in all parallel hierarchies. In SmarTeam objects are only instances of leaf classes. The other classes are not instanciable (super-classes and intermediate classes).

d) Intermediate classes: intermediate classes are those classes that are between super-classes and leaf classes in the parallel class hierarchies. They permit common attribute factorisation. Both intermediate classes and super-classes are abstract classes.

e) Internal classes: In SmarTeam, predefined classes are used by the system to manage the database. For example the identification of users (their names, passwords etc.) is recorded in the internal class USERS. The other classes are used for the management of documents (identification of applications that are applied to each type of file to permit its visualisation for example).

f) Link classes: we distinguish two types of links in SmarTeam: aggregation link and association link.

Aggregation (or composition) link classes: aggregation is an association between a parent object and one or several son objects. This type of link is used typically to bind an object with its components. For example a wheel is composed of a tire, of an air room and of a rim. We can associate to the wheel these components through the use of the composition link. The composition link classes are created automatically for every super-class. These classes permit to stock aggregation links between the different objects (the link composite-component between two objects is an object of the composition link class).

Association (or logical) link classes: SmarTeam permits to bind between two classes (an instance of these classes) without taking into account their place in the hierarchy. We can make this link by the association link classes. For example, we can bind a screw CHc10 that is an instance of *screws* class with a document describing this screw.

g) Lookup table classes: they contain a list of strings and they are used only to define an enumerated type of data for an attribute that can only take its values in this list.

The general structure of the described classes is illustrated by the EXPRESS-G diagram of figure 5.

The first one allows documentation management (in a file-managed class, each object is associated with a

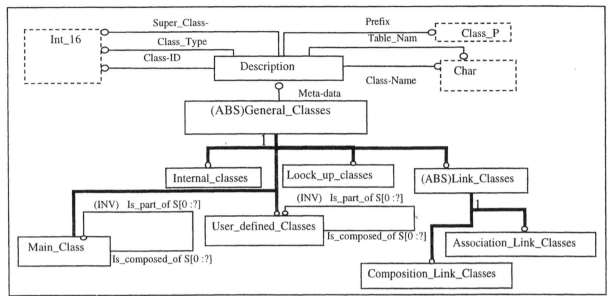

Figure 5 : general structure of classes in SmarTeam

All classes have some predefined common attributes called class attribute. They allow the definition and the characterisation of classes (class_name, class_ID etc.). They take the same values for all class instances.

4.3.2   attributes

In object oriented paradigm, a class has a number of attributes that permit to distinguish objects in one class. In SmarTeam, these attributes, themselves, have a certain number of properties that distinguishes them from some others attributes. Figure 6 illustrates properties of attributes in SmarTeam.

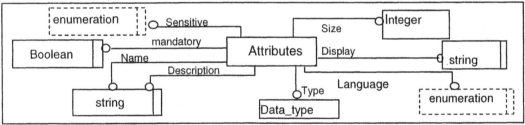

Figure 6: attribute properties

SmarTeam distinguishes four types of attributes
-   class Attributes : are attributes that allows the definition of classes (CLASS_NAME, ID etc.) and they take the same values for every class instances.
-   obligatory attributes: are predefined attributes assigned with each class and their values are managed by the system.
-   mechanism Attributes : there are two mechanisms in SmarTeam: file management and revision management.

document and has file management attributes), the last one allows revision management and verification (register, checkin, checkout, approval etc.)
-   user defined Attributes.

4.3.3   objects

Objects are instances of classes. They differ from each other by their object identification id (OID) and by their attribute values. For example, for classes Screw and Drawing, a Screw CHc10 will be an object of class Screw, its representation would be an object "Draw_screw", instance of the Drawing class, and link between these two objects would be an object of the corresponding link class.

Objects have attributes of the class to which they belong (predefined attributes and attributes defined by the user). Values of predefined attributes are assigned directly by the system. Object_ID value is assigned by the system and it takes an unique value used by the system to identify the object in the database.

4.4   PDM specifications

PDM systems are used to manage heterogeneous data, which describe enterprise products (CAD files, Production

Management System (PMS) data, BoM etc.). These data represent the static characteristics of products. Bill of Material (BoM) is the decomposition of a product in standard components purchased or in components manufactured in the enterprise from raw materials. It is well-known in mechanics and it is the point of departure of the PDM. In the PDM, at the physical components, the maps, the technical draws, the quality folders which allow the better description of the products, were appended. This decomposition is called METADATA. In the PDM the link between product and its components is an object of composition link class. So, both the OID of parent object (composite) and the child object (component) are stored in the link class. The same approach is used for the association links, which are used typically to link the physical components with their technical draws.

4.5   Conclusion

PDM systems are software developed in the last years to allow management of heterogeneous data, which describe products. These systems support products lifecycles including: requirement description, concept design, production, operation, logistical support, maintenance and discharge. They also permit quality management. This is very important to have ISO 9000 certificate to support concurrent engineering process. It is well known that using pre-existing components to design and to manufacture a new product is very useful. It allows to considerably decreasing prices of products. Pre-existing component data are recorded in a P-Lib's models manager. So, it is useful to reference directly from PDM component data. We will present the approach we developed to integrate P-Lib in a PDM system in the rest clause.

5   Integration approaches

5.1   Convergence and difference between PLib and PDMs

P-Lib describes component families by meta models. So, the data models, which hold the component (Physical files) are instances of the meta models. It has been defined so as to be easily extensible and to be very general. PDM systems describe products and manage BoM and documents which are related to these products. The modelling approach used in PDM is called the multi-modelling approach, because a specific model is developed for each BoM. Each software editor chooses a model which fits with its needs. In STEP a generic PDM data model has been defined from which everyone can

derive its own specific data. So, the main difference between both approaches is due to the representation levels. Meanwhile, both approaches are intended to describe technical components and products. This is the reason which lead us to study integration of P-Lib in a PDM.

5.2   Aims of integration

The aim of integration of P-Lib in PDMs is specially to allow designers manufacturers and users to benefit from the suppliers' knowledge on components, as it is embedded in electronic catalogues. This knowledge, described by suppliers on components, must be available for use by designers, manufacturers and users and workable by constructors. So, this integration should allow storage of P-Lib component data in a PDM and should provide referencing those data in product data, therefore an efficient maintenance of products.

5.3   Integration approaches

a) Direct representation: in the PDM, data are stored in tables where each line represents an object of the class represented by this table and each column represents values of an attribute. One needs model not only table content but also table schema. Instead, in P-Lib another approach has been developed to represent tables because this approach requires that the table schema are defined before population such table. This approach consists in representing them by a list of lists representing each attribute in a column of a table. So, tables are represented by a number of EXPRESS entities. Moreover, in P-lib models, the classes representing families of components and their properties are also represented separately (meta-representation).

So, the direct representation consists in extracting component families from the meta model and in representing them by PDM objects. As stated before, component class properties are represented by separated entities in P-Lib. These entities contain a set of attributes that are common to all the instances. In SmarTeam class attributes are predefined, so to avoid the redundancy of data we have opted to represent the common attributes which describe a class by a description classes and to link the two classes. For example, for a Screw_family class we associate directly the object attributes: diameter, length etc. and then we create a class Screw_family_description that is associated with the Screw_family class. In SmarTeam dynamic creation of classes is not allowed, so this approach does not enable us to make an automated integration.

b) **Meta levels representation approach:** this approach consists in representing each P-Lib entity, such as it is defined in the EXPRESS model and represented in an exchange file, by PDM objects. Every P-Lib entity will be represented as a class in the PDM. For example, each of the entities *supplier, class, component class, functional view class,* etc. will be represented by PDM objects. The weakness of this approach is the impossibility to represent, in SmarTeam, aggregate data types (used to encode P-Lib tables). It is also the complexity to browse and to query such a meta model.

c) **Hybrid or mixed representation approach:** this approach consists in using both approaches in parallel. It allows us to benefit of the advantages of both approaches. We separate the representation of tables and the representation of classes. On the one hand, we represent tables that contain component attribute values by relational tables provided by the PDM (direct representation). On the other hand, classes such as *supplier, class, component class, functional view class,* and so on are represented by PDM objects (meta-representation). In the meantime link classes will be created to link objects with tables that contain their attribute values.

This approach has been implemented on the Smarteam PDM system and several examples have been processed.

6 Summary and Conclusions

In this paper we presented the approaches we developed to integrate P-Lib defined component catalogues in to PDM system. We have discussed the possible approaches and we have proposed to merge two approaches: the direct representation and the meta level representation. The resulting approach is named hybrid or mixed approach. It allows us to represent in the same model P-Lib descriptors, which characterise component families and suppliers, and P-Lib entities representing tables by PDM relational tables. The advantage of this approach is the possibility to make a completely automated integration of P-Lib defined catalogue and to keep the specificity of P-Lib data model, which contains a complete description of component families. Notice that no changes on the PDM system are required. As a result, this integration will allow the use of a PDM systems to manage at the same time component and product data. It allows the easy recovery of any component data as well as the knowledge on these components from its reference. It will also allow the integration of LMS's (Library Management System) and PDM's functionalities. This integration leads to increasing the productivity and to simplification of the component selection and their

referencing within products. It makes also easier products maintenance by making possible the automatic exchange of component referenced when some components become e.g. obsoletes.

7 References

[1] J.Runbaugh, M.Blaha, W.Premerlani, F.Eddy, W.Lorensen, Object oriented modelling and design, Prentice-Hall International edition, 1991.

[2] I. Jacobson, G. Booch and J.Runbaugh, The unified Software development process, Addision-Wesley Eds, 1999.

[3] M. Bouazza, Le langage EXPRESS, Editions Hermès, 1995.

[4] G. Booch, Object Oriented Design, Redwood City, Calif: Bejamin/Cummings, 1991.

[5] G. Pierra, Intelligent electronic component catalogues for engineering and manufacturing, International symposium on global engineering networking Antwerp, Belgium, pp. 331-352, 1997.

[6] G. Pierra, Modelling classes of prexisting components in a CIM perspective: The ISO 13584/ENV 400014 approach, revue internationale de CFAO et d'Infographie, vol 9, pp. 435-454, 1994.

[7] J.M Moranne, conception assistée par ordinateur d'ensembles mécanique avec recherche d'une bonne solution: le logiciel SICAM, Proceedings of MICAD'86, Paris, Hermès, pp. 41-71, 1986.

[8] G. Pierra, Y. AIT-Ameur and E. Saedet, Parts library: Logical resource: Logical model of supplier library, ISO document: ISO/IS 13584-24,1999.

[9] G. Pierra, H. U. Wiedmer, description: methodology for structuring parts families, ISO document, ISO/IS 13584-42, 1997.

[10] P. Harrow, M. West, Parts library: overview and fundamental principles, ISO document, ISO/DIS 13584-1, 1997.

[11] T. Schreuber, B. Wielinga, J. Breuker, KADS: A Principled Approach to Knowledge-based System Development, Acadimic Press, London, Forthocoming, 1992.

[12] J.M Randoing, Les SGDT, Edition Hermès, 1995.

[13] M Maurino, La gestion des données techniques, Edition Masson, 1993.

[14] ISO 10303-11, Industrial automation systems and integration -- Product data representation and exchange -- Part 11: Description methods: The EXPRESS language reference manual, 1994.

[15] F. Feru, C. Viel, Echanger avec le protocole d'application 203 de STEP: Echange et partage de données CAO et GDT, Aerospatiale & Goset, 1998.

[16] CIMdata, Product Data Management, http://CIMdata.com, December 11, 1998.

[17] E. Sardet, G. Pierra, Y. Ait-Ameur, Formal Specification, Modelling and Exchange of components according to P-Lib, A case study, International symposium on global engineering networking Antwerp, Belgium, pp. 179-200, 1997.

A Standardised Data Model for Manufacturing Management : the ISO 15531 MANDATE Standard

A. F. CUTTING-DECELLE

Université de Savoie/ESIGEC/LGCH, Domaine de Savoie Technolac, 73376 LE BOURGET DU LAC, France

J. J. MICHEL

CETIM, Département Informatique, 52 avenue Felix Louat, 60304 SENLIS, France

Abstract

The aim of this paper is to present the work undertaken within the International Standardisation Organisation, by the ISO TC 184/SC4 « Industrial Automation Systems and Integration / Industrial data » Group in the field of industrial manufacturing management data, resulting in the ISO 15531 « MANDATE » (MANufacturing management DATa Exchange) standard.

After an overall description of the standard, of its main features (scope, basic principles) and a presentation of the global standardisation environment in the field of industrial data, this paper will analyse the benefits that can be expected from the use of standards in the domain of the management and exchange of manufacturing information.

1   Introduction

Manufacturing systems have become increasingly dependent on the adequate supply of information related to their control [23], [24]. To address this problem, many software applications have been developed to support manufacturing process control. However, too little attention has been paid to the flow of information between these applications. This has lead to a situation whereby manufacturing software applications have been described as « islands of information » [10] without inter-relations. The work presented here is aimed at providing a possible answer in terms of a standardised communication approach among production management systems.

In a first part of this paper, we will present the industrial needs in the domain of the manufacturing management information, through the description of the new environment of manufacturing, through an information model of a manufacturing system, then through the presentation of a generic structure of manufacturing management systems.

The second part will present the standardisation context, mainly at the international level, in the domain of « industrial data », then, the third part will describe the

ISO 15531 MANDATE standard, with its scope, main features and structure.

The last part will provide some elements related to the impact on the manufacturing industry of the use of standards for managing and exchanging manufacturing data, first through a description of the available (or under development) standards, then through an analysis of the potentialities of these standards.

2   Industrial needs for manufacturing management

2.1 The new environment for manufacturing

Recent developments, particularly the success of Japan in world markets and the feeling that this success has been derived, to a significant degree, from superior manufacturing systems, have changed people's perception of the role and importance of the manufacturing function in the industrial firm [10]. Business managers and business authors, such as Hayes and Wheelright [19] now regard manufacturing as a competitive weapon in the market-place and recommend that each company includes in their business plans specific goals in the area of achieving manufacturing excellence. Manufacturing is now an equal partner at the corporate boardroom. The focus in manufacturing is competitiveness, and the emphasis is on enterprise integration, application integration as well as on the factory of the future as the means of achieving it.

As a result, manufacturing firms find themselves in a totally changed environment. This change is not confined to any one industry, and evidence of it can be seen in such varied industries as the automotive industry, consumer goods, electronics and white goods. Management faced with rapid changes must devise new strategies to deal with the competitive nature of this new environment. The old strategy of mass production derived from notions of economics of scale is no longer seen as valid and is being discarded in favor of a strategy that facilitates *flexibility*, reduced *design cycle time*, reduced *time to market cycle* for new products and reduced *order cycle time* to the customer for existing products.

Some important characteristics of this new environment are :

- Increased product diversity ;
- Drastically reduced product life cycles ;
- Increased awareness and understanding of the environmental impact of manufacturing systems and their products ;
- Changing cost patterns ;
- Great difficulty in estimating the costs and benefits of integration technology ;
- Changing social expectations.

To assist managers with such requirements, manufacturing systems must be designed with the necessary level of flexibility in order to increase the speed of implementation of the required responses. This revolves around the **variability handling capability** of a system [8]. This property can be provided by incorporating the appropriate level of resource flexibility in the system by design ; it can also be provided through the development and incorporation of rules, procedures and guidelines during systems operation which can reduce the consequences of change during manufacture at the point of impact [15].

We present below a generic answer that can be provided to managers, in terms of the overall structure of the manufacturing system (process integration). Both levels are necessary to improve the overall manufacturing context.

2.2 Generic structure of manufacturing management systems

The Fig. 1 below presents a schematic description of the manufacturing management environment [22] :

- **Product design** : set of actions, requirements involving the use of a model of manufacturing operations to estimate the performance of the manufacturing system for different levels of demand. A key-input of the model is the process plans that specify the job steps, including resource requirements, to make the product.
- **Capacity requirements planning and analysis** : seeks to determine whether manufacturing operations can process the shop orders released in a timely manner. First, a finite capacity analysis determines the level of resources required to meet current demand.
- **Scheduling** : when capacity levels are set, detailed scheduling can be accomplished by using the model

to allocate available resources at specified start times to the actual jobs included in the shop orders. If the model contains the detailed process plans of job steps, the start and completion time of each operation can be established : these schedules can be distributed for schedule management which provides an ability for the user to easily view and modify the schedule. The outputs of schedule management are dispatch lists for each resource detailing the scheduled time to perform each job.

- **Status presentations/statistics** : Data and statistics on operational status are fed back to scheduling in order to determine the frequency with which new schedules need to be prepared. The display of this status information provides a basis for on-going decision making. Through this feedback link, continuous improvements in manufacturing operations can be made and information is gathered for future design assessments and new scheduling algorithms.

The feasibility of integrated planning, scheduling and execution relies heavily on the ability to build on existing data, but also on the models created to handle these data. The use of a common modelling language to build models across the functional tasks makes the integrated problem solving using the feedback control loops described above plausible.

An other advantage of the use of models is that they contain information about manufacturing processes, and by using and updating models continually, manufacturing processes can be better understood. It is worth stressing that understanding leads to improved manufacturing operations, it also enables the identification of information necessary for improving design.

This schema focuses on the manufacturing process (and the information related to its management) in itself, rather than on the products/materials/components used to make the manufactured product. Information about materials, components and products used in the manufacturing process mainly appears in the arrows « material », « purchase orders ».

The aim of this presentation of the ISO 15531 MANDATE standard is to provide elements of answer in terms of manufacturing management data modelling.

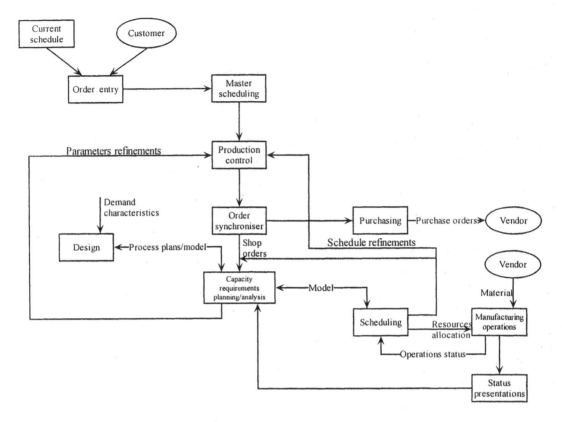

Fig. 1 A generic representation of the manufacturing management environment [22]

3   The standardisation context applied to industrial data

Various international standardisation committees, subcommittees and working groups of expert, address different industrial or business concerns. Some of them are rather specialised or focused on specific technical area, whereas the others are related to more general purpose.

The standardisation committee presented here is the ISO TC 184. The scope of the ISO TC 184 Committee (Industrial Automation Systems and Integration) is : « *standardisation in the field of industrial automation and integration concerning discrete part manufacturing and encompassing the application of multiple technologies, i.e. information systems machines and equipment, and telecommunications* ».

The committee is composed of four sub-committees :

- SC1: *Numerical control devices*
- SC2 : *Robots*
- SC4: *Industrial data*
- SC5: *Systems architecture and communications*

The mission of the SC4 is to develop and promulgate standards for the representation of scientific, technical and industrial data, to develop methods for assessing conformance to these standards and to provide technical support to other organisations seeking to deploy such standards in industry.

The standardisation efforts for *Industrial Data Representation* appear through the four areas of scope, which are : representation of Product Definition Data (ISO 10303 STEP standard) [10303-1], Structure of Part Libraries (ISO 13584 P-LIB standard), Industrial Manufacturing Management Data (ISO 15531 MANDATE standard) [13], [14] and representation of data for Oil and Gas (ISO 15926) [20].

Consistency of those standards in between them will strongly benefit from the unique development framework. ISO TC184/SC4 covers the whole domain of industrial data and has developed a set of rules, guidelines and tools to ensure quality and consistency of the data models developed in its various working groups [HBK99]. The EXPRESS language and its variants (EXPRESS_G,...) are some of them.

MANDATE appears by the way as a core element of the set of standards involved in industrial application integration.

4 Presentation of the ISO 15531 MANDATE standard

4.1 Scope and basic principles

The information generated about the manufacturing process of an industrial product is very important for the life cycle of this product, notably in a context of sustainable development [2]. Manufacturing may be defined as the transformation of raw material or semi-finished components leading to production of goods, products or intermediate products. Manufacturing management is the function of directing or regulating the flows of goods (and information related to it) through the entire production cycle from requisitioning of raw materials to the delivery of the finished product, including the impact on resources management.

A manufacturing management system manages the flow of materials and products through the whole production chain, from suppliers, through manufacturers, assemblers, to distributors and sometimes customers. The relations among those partners may be identified and structured in an electronic form with a view to facilitate electronic exchanges. Then, information handled during these exchanges have to be identified, modelled and represented in such a way that they may be shared by a maximum of partners through the usage of standards for product and manufacturing data modelling.

The production planning functions within the supplier plants are assumed to have strong relationships with people in charge of the master production scheduling of the main plant, who share with them information on the likely pattern of the future demands to allow suppliers to plan in turn their production. On a day-to-day basis, the operational planning system of the main plant sends orders to the suppliers to ensure the availability of components, sub-assemblies and others such as resources needed to its manufacturing and assembly process.

From this approach, three main categories of data related to manufacturing management may be distinguished :

- **information related to the external exchanges**, e.g., between plants and suppliers;
- **information related to the management of the resources** used during the manufacturing processes;
- **information related to the management of the manufacturing flows**.

MANDATE is an International Standard for the computer-interpretable representation and exchange of industrial manufacturing management data. The objective is to provide a neutral mechanism capable of describing industrial manufacturing management data throughout the production process within the same industrial company and with its external environment, independent from any particular system. The nature of this description makes it suitable not only for neutral file exchange, but also as a basis for implementing and sharing manufacturing management databases and archiving. The standard is focused on discrete manufacturing, but not limited to it. Then any modification or extensions to industrial that do not belong to discrete part manufacturing have always been under consideration when they did not imply any contradiction or inconsistency with the initial objective of the standard.

The standard addresses these three types of data. It is does not standardise the model of the manufacturing process. The aim is to provide standardised data models for those three types of manufacturing management data. The purpose of the standard development is to facilitate the integration between the numerous industrial applications by means of software able to provide standardised representations of these three sets of data.

The aim is to provide standardised data models for the three types of manufacturing management data above (usually complex, strongly time dependent in a very complex way, with close relationships among them) in order to facilitate the integration among the numerous industrial applications by means of common models able to represent these three sets of data shared, exchanged during the entire production lifecycle, but also belonging to the core of the manufacturing process.

Manufacturing management information usually includes these three main types of data sometimes described through different modelling methods since their concerns and usage are different. They have nevertheless to be consistent between them and with the other data exchanged in the production process, such as product data, component data, cutting tool data ... .

4.2 Structure of the standard

MANDATE is divided into three separated series of parts, strongly related and developed in close co-operation in order to preserve the consistency of the whole standard. Furthermore, since the standard is developed within the SC4 committee, it addresses operations dealing with product manufacturing, and makes use of component description, it will also have to be consistent with the ISO 10303 STEP and ISO 13584 P-LIB standards :

- **Parts 15531-2x series** : Production data for external exchanges : referring to the representation of product information to be exchanged with the external environment (customers and suppliers), with the aim to model the main production information exchanged between industrial companies, in order to improve exchanges and integration through the use of EDI protocols.

- **Parts 15531-3x series**: Manufacturing resource usage management data: referring to the way of managing the use of resources, such as resource configuration and capabilities, operation management of manufacturing devices, installations and facilities.

- **Parts 15531-4x series**: Manufacturing flow management data: refering to the material flow control and aiming at the representation of data and elements supporting the control and monitoring of the flow of materials in manufacturing processes

- **Specificity of the standard**: the specificity of the data model as provided by MANDATE is to completely separate the model describing the resources used during the manufacturing process from the time representation and from the model of flow management. This point provides an important, and original feature of the standard.

According to the fact that the Part 2x series will be based on the BSR work (ISO 16668) and limited to the proposal of some specific Semantic Unit we will here focus on the 3x and 4x series, since they are the most advanced in the standardisation process.

4.3   ISO 15531 3x series : Manufacturing Resource Usage Management Data

Manufacturing resources form the basis and long term potential of any production company. The efficient usage of resources is one of the main goals in production management. Comprehensive information about available manufacturing resources is required to enable an efficient resource usage [3]. Since various enterprise functions are dealing with resources, with the consequence for resource data to be processed by different Information Technology (IT)-systems, a common, standardised model to represent, share and exchange resource usage management data is required. This format is based on a generic information model covering all aspects of resource usage management [EBD97], e.g. : determination of resource demands, development, determination of resource inventory, ordering, adjustment of resource capacity and maintenance.

A complete representation of manufacturing resources (e.g. shape aspects), is not within the scope of the standard [4]. Only data relevant to decision making concerning the use of resources (e.g. within process planning or job scheduling) are considered. In order to meet all the tasks, the information model is structured in

a modular way (Fig. 2). The entity[*] *resource* forms the central element within the schema. Each further description classifying or detailing a resource characteristics is related to *resource*. The schema entities can be clustered into logical units representing : *resource hierarchy, structure of resource characteristics, resource status, definition of resource views, definition of resource characteristics and resource configuration*, as shown on the Fig. 2 below :

Considered as extensions of the instantiation of cutting tool data, all types of manufacturing resources can be represented by the generic information model.

Note : for a global approach on manufacturing management data, as dealt with by a standard such as MANDATE, the concept of « *resource* » cannot be limited to the mere tooling, facilities, machining-equipment and other material resources : examples have been added in the standard about the use of this schema for other types of resources, such as human resources, data (e.g. provided by NC machine-tools control) and software.

4.4   ISO 15531 4x series : Manufacturing flow management data

The manufacturing management services needed to achieve the integration of the operations within industrial companies include resource management, operation management and time management. These information and functions, necessary to support the manufacturing management activity require the definition of a common semantics for the description of material flows, but also of the related flows of information needed to control them.

The 4x series of parts is focused on the classification and the semantics of information and functionalities required to support scheduling and controlling of flows of materials (especially for discrete manufacturing environments), through a time model, a conceptual model for flow monitoring and control and the related set of building blocks from which standardised data models and representations for planning, scheduling, controlling, monitoring and exchanges (with internal and external partners) of material flows are specified.

The dissociation between the representation of the time from the other manufacturing management concepts (such as resources management, data and control flows, etc.) is a specificity of MANDATE and contributes to extend the application range of the standard, notably to

---

[*] « *entity* » is taken here with the meaning of the ENV 12204 and not the meaning of the ISO 10303 (STEP) or ISO 15331 (MANDATE) standards. For further explanation see ISO DIS 15531-1 (MANDATE overview) clause 3.6.12 enterprise entity (Note).

continuous processes, through the possibility to express continuous intervals of time.

To be open to existing and future concepts, this work is based on a general approach for modelling industrial process planning and controlling activities, through a conceptual model of industrial process monitoring and control.

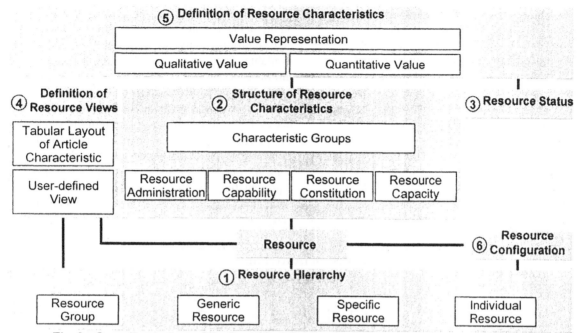

Fig. 2 : Overall structure of the Part 32 of the standard (Resource Information Model)

This conceptual model must be able to [5] :

- perform the tasks of managing an industrial process contained within the monitoring, and management information system (representation of industrial processes, of the management activities needed for monitoring and controlling the industrial process) ;
- define the interfaces enabling a communication between those two representations (transmission of control orders to the industrial process and feedback from actual information to the control information system).

These two functions will be achieved through :

- the modelling of elements and processes to be monitored : workers, materials, machines, information ;
- the modelling of elements and processes for managing the monitoring and manufacturing management activities : data.

The modelling methodology defines concepts and constructs enabling a graphical representation of the main features of an industrial company.

The main concepts defined within the two conceptual models of the 4x series are the following :

- **Operations and elements** : an operation is defined as a transformation of specific elements from one state to another (from at least one input element into one or more output elements). An element is a static

representation of anything which can be characterised by a behavior and attributes (context dependent and able to change over time). Only elements requiring management decisions need to be specified within the model.

- **Time** : Modelling a dynamic behavior requires the consideration of a time measure, defined in this context as a sequence of repeatable events. A *point_in_time* is the representation of a possible *event_occurrence* within any *time_domain*.

In order to enable the observation and comparison of relevant events in the system to be modelled a time domain has to be specified. The particular time domain to be chosen depends on the modelling objective and on the dynamic behavior of the elements and operations to be modelled. To compare events assigned to different time domains, it may be necessary to define transformation rules (in the extreme case for each combination). : this feature is a specificity of the time model of MANDATE, and contributes to the power of the standard.

- **Paths for elements and data transfer**, among which :
  - ♦ **element path** : element_in_state_categories characterise states of elements where decisions are pending. Operation categories represent state transitions transforming input elements from one state into output elements in another state. Establishing links between the category_nodes -

symbolised by directed lines between the corresponding category_nodes - allows for the description of a possible flow of elements from one category to another (flow with the meaning of assigning elements to categories, not with the meaning of a physical movement). The resulting model is a directed graph of connected element_in_state_categories and operation_ categories describing the transformation of input elements into output elements. The system behavior is represented by the flow of the elements through the graph. Elements entering an operation are consumed, by the end of the operation new element_in_states are generated. Along the defined paths elements are assigned by local monitoring and control activities to the succeeding element_in_state_categories. An element can only play one role in an operation.

♦ **control information path** : in addition to the flow of elements, an exchange of information between the categories is often required for the execution of the process monitoring and control activities. This kind of information is called global control information and comprises additional constraints relevant for the monitoring and controlling activities which are not locally available. To mark the possible flow of control information an additional path is established, symbolised by a dashed line. Unlike the element path there is no restriction in the sequence of the nodes. Fig. 3 shows the aggregation of the single components to a compound conceptual information model for process monitoring and control. Let us mention the 3rd axis of this representation schema, provided by the time axis sequencing the element_in_state or operation categories.

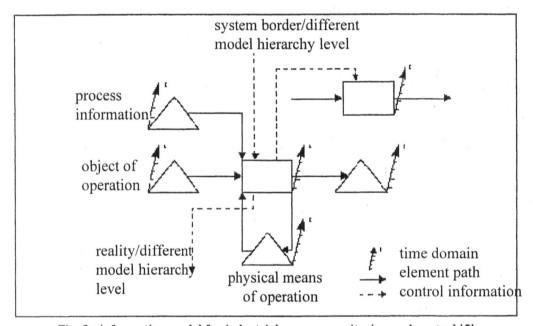

Fig. 3 : information model for industrial process monitoring and control [5]

♦ **data for data exchange path** : when data for the control and monitoring of flows of materials need to be exchanged within the model, a data exchange relationship between the particular nodes is required. These relationships are defined for element and control information paths. Data exchange relations, as well as all data for data exchange need to be described. However, the objective is not to define the transmitted data, but to provide the modelling resources needed to

describe the data exchange. Data for qualifying other data can also be needed (e.g. « order » or « this is a stock of the node xy of the model zx »). In order to be able to describe the data defining the way of qualifying data, it is necessary to classify processes in which these data are transmitted, even if these processes do not themselves belong to the standard.

The Fig. 4 provides an example of a model of a manufacturing and material flow process.

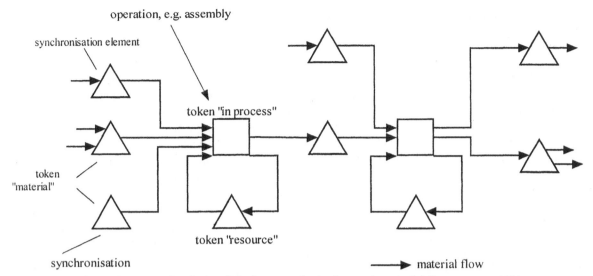

Figure 4 : example of a model of a manufacturing and material flow process [11]

- The ISO 15531-42 part : Time model
Given the current development of the parts, it is interesting to present the part 42 of the MANDATE standard, the « time model » :
The subject of the time schema is the definition of concepts related to the time representation, necessary for modelling flow management data [6]. It allows representation of time domains, points in time and time units.
Concerning the management of industrial manufacturing systems an assignment of a time domain is necessary to enable the observation and comparison of relevant points in time of the system to be modelled.
Time is defined by a domain containing a sequence of points in time. A point in time is defined by a selected location on the time axis. A time unit is the unit which is used to measure the duration in the related time domain.
It is important to notice that, in the Time model, the representation of elements of time is completely independent of any event. Time and time concepts exist prior to any use by a model to order/trigger activities.
The potential of this schema is to enable a generic representation of the time, applicable whatever the cases.
Then, once applied to a specific use, it is possible to adapt this representation either to a calendar (discrete set of points in time), or else to a duration (dense and continuous set of values of points in time) -- or a combination of both cases.
The EXPRESS-G representation of the Time Model is provided on the Fig. 5 next pages.

- The ISO 15531-43 part : Conceptual model for flow monitoring and manufacturing data exchange
The part ISO 15531-43 [7], « Conceptual model for flow monitoring and manufacturing data exchange » related to the representation of data for manufacturing

flow management is currently under development. This part will provide a set of building blocks enabling representations of planning, scheduling, controlling and monitoring of information and material flows. This conceptual model can be decomposed into three main models, which are :
• conceptual model for modelling elements ;
• conceptual model for flow modelling ;
• conceptual model for data exchange.

- Conceptual model for modelling elements :
This schema provides for the structural aspects of manufacturing processes in the context of planning, scheduling, monitoring and control. The EXPRESS [1] constructs of the schema allow for the representation of constraints being relevant for the previous functions. Resources defined in the 3x series can be seen as kinds of elements whose behavior may be represented in this model.

- Conceptual model for flow modelling :
This schema provides for the dynamic aspects of manufacturing processes in the context of planning, scheduling, monitoring and control. The EXPRESS constructs of the schema allow for the representation of the flows of elements within a manufacturing system and the corresponding information.

- Conceptual model for data exchange :
While the conceptual model of industrial process monitoring and control provides the basis to define the data for monitoring and control of the flow of material in industrial processes, a more detailed addition of the conceptual model for the description and control of the flow of data is also needed. For this reason, the description of all flows and negotiations that are to be executed within a manufacturing model will be

analysed in more detail in this part, in order to provide bases for the definition of data describing data exchanges and negotiations between the elements of the model.

4.5   Use of MANDATE in industry

Once combined together, the joint use of the different parts of the MANDATE standard provides a precise, and standardised representation of the data necessary to the management of manufacturing information, based on an approach in terms of resource, time and flow management.

These features of the standard make it useful to deal with an important category of management problems met in manufacturing systems, notably the problems related to the hierarchical levels at which the decisions must be taken, with the corresponding conflicts [9]. We may distinguish at least the following three levels :

- production planning : planification of the production without specifying every detail, with a relatively long time horizon and an aggregate view of the manufacturing system. At this level, it is possible to consider the production capacity as a decision variable, whereas at the lower levels the capacity constraints are not negotiable ;
- production scheduling : at this level, we take a closer look at the manufacturing facility, with a shorter time horizon. We have a set of tasks to be carried out with a set of resources ; we must specify exactly what happens, where and when ;
- production control : this is the lowest level in the hierarchy, it is also a task carried out in real time. At this level, we must ensure proper implementation of production plans/schedules despite the occurrence of random events. Another task is to monitor the production activities in order to provide the upper levels with statistics about the execution (e.g. about processing times) and up-to-date information necessary to revise the long term plans or schedules.

Actually, the decision levels can be extended both upward and downward : going upward, we may find facility location problems and manufacturing facility design problems ; going downward leads to issues related to numerical control of single machine.

Separating the concept of time (and all the related features) from other concepts more specifically related to manufacturing management provides MANDATE with the capacity to deal with the three previous levels, but also to integrate them into a common approach of *manufacturing management*. From this point of view, MANDATE can be considered as an enabling integration technology.

More generally, the expected benefits of the use of MANDATE in industry, are :

- non-ambiguous agreement on the identification of

the different processes used in manufacturing engineering, with the corresponding modelling ;
- identification of the flows handled during the manufacturing process : control, resources, data ;
- contribution to a product and production data-based integration approach, together with the ISO 10303 STEP and ISO 13584 P-LIB standards ;
- better identification of the functions to be assumed during a manufacturing process ;
- use of methodologies such as MRP, MRPII, where this type of information is already identified, recognized and agreed on ;
- clear distinction between external/internal information, as well as information coming from data warehouses, or other kinds of information dealing with resources.

5   Conclusion

Within production management systems, functions of planning and scheduling can be considered as core-processes, actions leading to their improvement will have consequences on the whole manufacturing process.

Current trends in planning and scheduling are towards greater coordination across manufacturing operations [22] ; that is, scheduling areas and plants, and the integration of planning and scheduling systems as part of manufacturing execution systems. These trends will be supported by improved data identification and collection, improved shop floor control systems, advanced database management systems and improved software programs for developing graphical interfaces. Through the use of detailed models of the manufacturing system, the barriers between engineering design, planning, scheduling and execution can be decreased significantly [12].

Another trend will be that manufacturing management functions, such as planning and scheduling can support the sales process by providing realistic shipment dates for sales representatives to promise to customers. These systems will support the purchasing of material with respect to the availability of resources while meeting promised ship dates.

The operation of the production system will be coordinated with downstream logistics requirements to produce combined manufacturing and distribution cost savings. Lastly, the data from shop floor collection systems and from manufacturing functions such as planning and scheduling will be organised to support the decisions of managers and executives (through the use of enterprise modelling standards).

The standardisation work presented here aims at providing a contribution to industrial integration, it is also important to notice that the MANDATE standard fully supports Concurrent Engineering features.

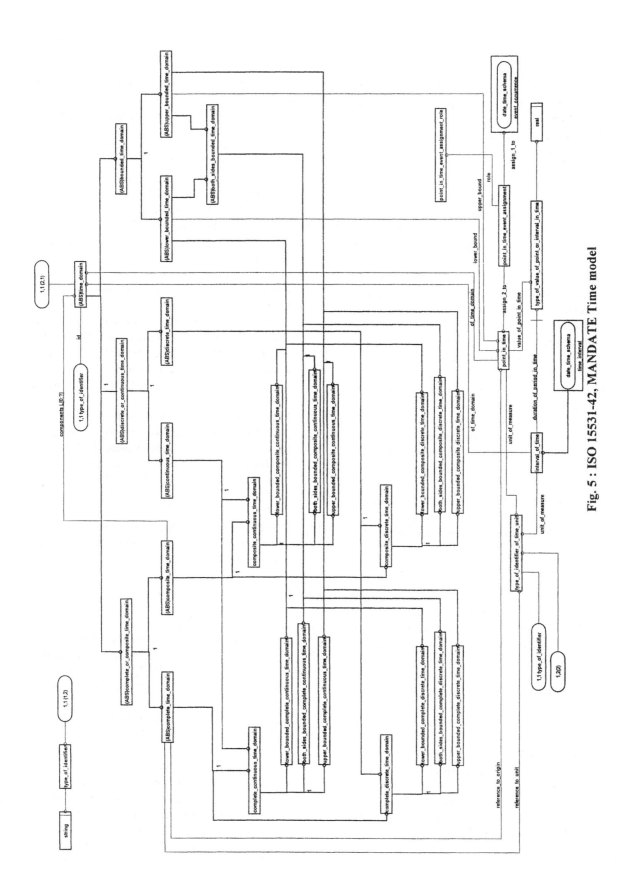

Fig. 5 : ISO 15531-42, MANDATE Time model

389

References

[1] Industrial automation systems and integration - Product data representation and exchange, ISO IS 10303-11, Part 11 : Description methods : the EXPRESS language reference manual, 1994

[2] Manufacturing management data exchange : overview and fundamental principles, ISO/CD 15531-1, ISO TC184/SC4/WG8 N147, 1997

[3] Jochem, R., ISO/CD 15531-31: Industrial Automation Systems and Integration - Manufacturing Management Data - Part 31: Resource Usage Management Data - Overview and Fundamental Principles. Committee Draft. 1997.

[4] Deuse, J., Ciesla, M.: ISO/WD 15531-32: Industrial Automation Systems and Integration - Manufacturing Management Data - Part 32 : Information Model for Resource Usage Management Data, Working Draft. 1997.

[5] Manufacturing management data exchange :manufacturing flow manag-ement data, structure of the series of parts and basic concepts, ISO/CD 15531-41, ISO TC184/SC4/WG8 N151, 1997

[6] Manufacturing management data exchange : manufacturing flow manag- ement data, time model, ISO/WD 15531-42, ISO TC184/SC4/WG8 N191, 1998

[7] Manufacturing management data exchange : manufacturing flow management data, Conceptual model for flow monitoring and control, ISO/WD 15531-43, ISO TC184/SC4/WG8, 2000

[8] S.K. Banerjee, Methodology for integrated manufacturing planning and control systems design, in «The planning and scheduling of production systems, methodologies and applications», A. Artiba and S.E. Elmaghraby Eds, Chapman & Hall, 1997

[9] P. Brandimarte, A. Villa, Advanced models for manufacturing systems management, CRC Press, 1995

[10] J. Browne, J. Harhen, J. Shivnan, Production management systems, Addison-Wesley, 2nd Edition, 1996

[11] AF Cutting-Decelle, standardisation in product and process data modelling : the ISO STEP and MANDATE standards contribution to the integration of the life cycle of buildings, CIB W78-95, Stanford University, 1995

[12] AF Cutting-Decelle, The use of industrial management methods and tools in the construction industry - application to the construction process, Concurrent Engineering in Construction, London, 1997

[13] AF Cutting-Decelle, J Deuse, JJ Michel, Standardisation of Industrial Manufacturing Management Data : the MANDATE (ISO 15531) Approach, PDT Days 1998, March 1998, London, UK, in proceedings of the European Conference, Product Data Technology Days 1998, ISBN 1 9011782 02 6

[14] A.F. Cutting-Decelle, J.J. Michel, MANDATE (ISO 15531) : a standardised way towards a dynamic modelling of manufacturing data, ISATA, Wien, 1999

[15] R.P. Duimering, R. Safayeni and L. Purdy, Future factories and today's organisation, proceedings of 8th Int ; conf on CAD/CAM, Robotics and Factories of the Future, Metz, 1992

[16] Advanced manufacturing technology, systems architecture - Enterprise model execution and integration services, V. 3.3, CEN TC310 WG1, 1998

[17] Eversheim, W.; Deuse, J.: Gestaltung der NC-Verfahrenskette - Integration marktgängiger DV-Systeme auf der Basis eines feature- basierten Produktdaten modells. In: VDI-Berichte 1322. VDI-Verlag. Düsseldorf. 1997, S. 195-214.

[18] SC4 Organisation handbook, ISO TC184/SC4, 1999

[19] R. Hayes and S. Wheelright, Restoring our competitive edge : Competing through Manufacturing, John Wiley and Sons, New York, 1984

[20] Open Information Interchange, http ://www2.echo.lu/oii/

[21] B. Prasad, Concurrent Engineering fundamentals, Vol. 1, Prentice Hall, 1996

[22] A.A.B. Pritsker, K. Snyder, Production scheduling using FACTOR, in «The planning and scheduling of production systems, methodologies and applications», A. Artiba and S.E. Elmaghraby Eds, Chapman & Hall, 1997

[23] R Riddick, A Loreau, Moels for integrating scheduling and shop floor data for data collection systems, NIST, 1997

[24] P Timmermans, Modular design of information systems for shopfloor control, PhD thesis, Eindhoven University of Technology, 1993

[25] R. Wild, The techniques of production management, Holt, Rinehart and Winston, London, 1971

A necessary response for achieving concurrency and integration in engineering: The standard ISO / STEP

D Molin

Association GOSET, Nanterre, 92000

Abstract

For a long time, industrials have been knowing that one of the mandatory ways to achieve concurrency and integration in engineering is to have a solution for exchanging and sharing product data quickly, easily, safely, and over the entire life-cycle. To solve this problem, the development of the standard for of the exchange of product data is underway for several years inside the ISO TC184/SC4 community. STEP is based on the principle that the integration of the processes requires the integration of the product data involved in each process. This unique set of data can then be managed during the entire life-cycle of the product. This paper outlines the different STEP related works achieved by entreprises.

Then, it presents some examples of STEP implementations, where the integration of the product data inside a same model enables the industrial processes. It particularly shows that STEP is now, enabling the management of digital mock-up and the co-operative design of product in domains such as mechatronics, automotive and aeronautics.

1 Introduction

Among the techniques experimented by the industrials to develop the simultaneous engineering in the enterprise, some of them are based either on a product and process data.

The idea on which these techniques are based sums up thus: to achieve the integration of the processes through the data integration.

For more than ten years, international experts have been working on these subjects that are in the scope of the project ISO 10303, called STEP project for "Standard for the Exchange of Product data".

Today STEP is sufficiently advanced so that the industrial world can experiment and use it.

In some domains, actions of industrial deployment of STEP have already followed these experiments.

This article exposes some fundamental ideas that are the foundations of this project.

It makes a report on the results of some experiments of STEP in the domains of the aeronautic, automotive, and the mechatronics.

These examples illustrate how concurrent engineering can be achieved in the enterprise by the use of the results issued from the STEP project.

2 Fundamentals of ISO 10303

2.1 Purpose of STEP

The purpose of ISO 10303[1] is to specify a form for the unambiguous representation and exchange of computer-interpretable product throughout the life of a product[2]. This form, independent of any computer system, enables consistent implementations across multiple applications and systems.

ISO 10303 separates the techniques of representation of product information from the implementation methods used for data exchange.

The representation techniques provide a single representation of product information common to many applications. This common representation can be tailored to meet the needs of specific application.

ISO 10303 defines a formal data specification language, *EXPRESS* [3], that is used to specify the representation of product information. The use of a formal language provides unambiguous and consistent representation and facilitates development of implementations.

2.2 Integrated resources

The integrated resources constitute a single conceptual model for product data. The constructs within the integrated resources are the basic semantic elements used for the description of any product at any stage of the product life-cycle.

The specification of the representation of product information is provided by a set of integrated resources. Each integrated resource comprises a set of product descriptions, written in *EXPRESS* known as resource constructs. One set may be dependent of other sets for its definition. Similar information for different applications is represented by a single resource construct.

The integrated resources are divided into two groups: generic resources and application resources. The generic resources are independent of application and can reference each other.

NOTE 1 – The aspects included in these generic resources are dealing with the fundamentals of product description and support, the geometric and topological representation, the representation structures, the product structure configuration, the material, the visual presentation, the shape variation tolerances, and the process structure and properties.

The application resources can reference the generic resources and add other resource constructs for use by a group of similar applications. Application resources do not reference other application resources, but can reference generic resources.

NOTE 2 – The aspects included in these applicative resources are dealing for example, with draughting, finite element analysis, or kinematics.

2.3   Application Protocols

2.3.1   Purpose

An Application protocol (AP) specifies the representation of product information for one or more applications.

NOTE – The different application protocols that presently exist, are dealing with the following subjects:

- Explicit and associative drawing;
- Configuration controlled design,
- Mechanical design using boundary and surface representation;
- Sheet metal die planning and design;
- Life-cycle management;
- Change process;
- Composite and metallic structural analysis and related design;
- Electronic assembly Interconnect, and packaging design;
- Electrotechnical design and installation;
- Numerical process plans for machined parts;
- Core data for automotive mechanical design processes;
- Ship arrangements, moulded forms; piping, structures;
- Functional data and their schematic representation for process plant;
- Exchange of product data for composite structures;
- Exchange of design and manufacturing product information for casting parts;
- Mechanical product definition for process plans using machining features;
- Building elements using explicit shape representation;
- Ship mechanical systems;
- Plant spatial configuration;
- Exchange of design and manufacturing product information for forged parts;
- Process engineering data: process design and process specification of major equipment;
- Technical data packaging core information and exchange.

2.3.2   Elements of an Application Protocol

The elements of an AP are the following:

2.3.2.1   Definition of the scope

An AP includes the definitions of scope, context, and information requirement that are established from these industry requirements. An activity model (AAM) that describes the processes, information flows, and functional requirements of the application supports the statement of scope.

NOTE 1 – The activity model is generally formalised in IDEF0.

2.3.2.2   Expression   of   the   application   requirements

The information requirements and constraints for the application context are defined by means of a set of units of functionality and application objects using application-based terminology. These requirements are gathered to constitute an application reference model (ARM). An ARM is a formal information model that is documented in an informative annex in the Application Protocol.

NOTE 2 – The different graphical representations of an Application Reference Model may be EXPRESS-G[4], IDEF1X[5], or NIAM[6].

2.3.2.3   Interpretation process

The resource constructs for representing the information requirements of the application are specified in an interpreted model (AIM) using the *EXPRESS* language.
The AIM is assembled from resource constructs specified by the integrated resources through the interpretation process that consists in mapping the application requirements defined within the defined context and scope of the Application Protocol onto the resource constructs.

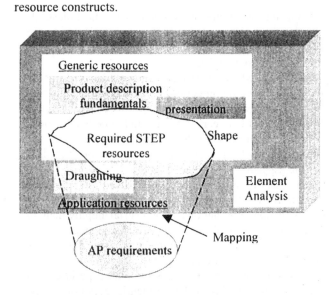

Figure 1 The interpretation process during the elaboration of a STEP AP.

This mapping defines the use within the AIM of resource constructs from the integrated resources to represent the information requirements of the application. The result of this interpretation process is an integrated model.

3 Exchange versus sharing data

Exchanging product data among the software systems used in the entreprises is the first use of STEP. However, STEP also enables to share product data among software systems. This concept seems to be more relevant to concurrent engineering. Product data sharing and data exchange can be defined as follows.

3.1 Data exchange in the context of STEP

Data exchange is the transfer of information from one software system to another via a medium that represents the state of information at a single point in time. The data exchange characteristics can be summarised as follows:

- Initiated by data originator;
- Transformed into a neutral file;
- Content determined by discrete even in time;
- Redundant copy of data created.

In the exchange model, the software system maintains a master copy of the data internally and exports a snapshot of the data for the others to use. This use is without explicit controls on the changes on the internal data.

3.2 Data sharing in the context of STEP

Data sharing provides a single logical information source to which multiple software systems have access. Controls over access to the information, updates to the information, ownership of the information, etc., are typically provided in implementing and administering the information source. The data sharing characteristics can be summarised as follows:

- Initiated by data originator;
- Data on a demand of receiver;
- Data access levels embedded in protocol;
- Appears as a single data source;
- Read (real-time) and update capabilities.

In the sharing model, there is a centralised control of the ownership and there is a known copy of the data, which can be accessed and revised under controlled circumstances.

The mechanism used for a product representation is a digitally encoded file (with a neutral or proprietary format).

Software systems interfacing with a PDM system still face the issues of translation and interpretation of the data files. Consequently, the fundamentals of STEP remain applicable.

4 Related standards

Two other standards that are developed inside the ISO/TC184/SC4 community are the standards ISO 15531 MANDATE, ISO 13584 PLib, and ISO 15926 Oil & Gas. These standards may be considered as related to the ISO 10303 standard, because of these reasons:

- Like STEP for the product data exchange or sharing, MANDATE is dealing with manufacturing management data, PLib is dealing with the provisions to transfer part library data, and Oil & Gas is dealing with oil and gas production facilities life-cycle information.
- They contribute each one in its own area, to improve the simultaneous engineering by a similar way as the one followed when using STEP.

These standards are summarised hereafter.

4.1 ISO 15531 MANDATE

This standard is dealing with manufacturing[1] management data. And the problematics raised in this standard may be summarised as follows:

The information generated about the manufacturing process of an industrial product is important for the life-cycle of this product, particularly in a context of sustainable development. A manufacturing management system manages the flow of materials and products through the whole production chain, from suppliers, through manufacturers, assemblers, to distributors and customers.

The relationships among the partners may be identified and structured in an electronic form with a view to facilitate electronic exchanges. Then, information handled during the exchange has to be identified, modelled and represented in a way that it may be shared by a maximum of partners through the usage of standards for product and manufacturing data modelling.

The production planning functions within the supplier plants are assumed to have strong relationships with the master production scheduling people of the main plant, who share with them information on the likely pattern of the future demands to allow suppliers to plan in turn their production.

From this approach, three main categories of data related to manufacturing management may be distinguished as follows:

- Information related to the external exchanges,

---

[1] Manufacturing may be defined as the transformation of raw material or half-finished components leading to goods production.

e.g., between main plant and suppliers;

- Information related to the management of the resources used during the manufacturing processes;

- Information related to the management of the manufacturing flows.

Then, the objective of this standard is to provide a neutral mechanism capable of describing industrial manufacturing management data throughout the production process within the same industrial company and with its external environment, independent from any particular system.

4.2   ISO 13584 PLib

The ISO 13584 enables a representation of parts library information together with the necessary mechanisms and definitions. These representation and mechanisms enable parts library data to be exchanged, used and updated.
The exchange may be between different computer systems and environments associated with the complete life-cycle of the products where the library parts are used, including product design, manufacture, utilisation, maintenance, and disposal.
The standard provides a generalised structure for parts library system.
Bridges have been clearly established between a product data described with a STEP Application Protocol and the PLib compliant references of this product data.

4.3   ISO 15926 Oil and Gas

The ISO 15926 is a standard for the representation of oil and gas production facility life-cycle information. The approach chosen for modelling this application domain is different from the STEP approach even though the enabling technologies are the same (EXPRESS language, EXPRESS-G representation, physical file format ...).

- The representation is specified by a generic conceptual data model that is suitable for being the basis for implementation in a shared database. This approach is known as "Data Warehouse".

- The data model is designed to be used in conjunction with reference data, i.e. standard instances that represent information common to a number of users, production facilities, or both. These references are available inside a Reference Data Library (RDL) and are to be standardised.

- The support for a specific life-cycle activity depends on the use of appropriate reference data in conjunction with the data model.

NOTE – this new approach, initiated with the EPISTLE framework[7], has been experimented particularly with the ESPRIT III project PROCESSBASE, and ESPRIT IV project PIPPIN. Now, this new approach is underway and will depend on the capability of ISO 15926 representatives by the industries of the Oil & Gas sector to build a complete and consistent Reference Data Library.

5   STEP implementation experiments and industrial deployments

This last clause presents three cases of effective use of STEP:

- The first one consists in the elaboration of the entire data model of the Product Data Management system of an enterprise.

- The second one reports on an experiment of the capability of STEP AP 203 to be used for transferring data from CAD systems to another software system in aeronautics and automotive sectors. This project named " Great Innovating Project (GPI)" is the first project about the STEP standard organised in France.

- The third one reports on the decision of the Boeing company to use STEP as a data exchange support with its partners. It shows, thus, that STEP standard is now an industrial reality.

5.1   Integration of data in a PDM using a STEP data model

5.1.1   Purpose of the activity

An automotive enterprise was to choose a Product Data Management system (PDM) to improve the management of all the technical data and associated documents produced by from its various departments.

The purpose of the activity led by this enterprise and described hereafter was, then, to elaborate the data model to be implemented inside this PDM system.

This data model includes electrical, mechanical, and management data that are built with resources exclusively defined within:

- The STEP Application Protocol 214
  "Core data for automotive mechanical design processes"
  (see the information requirements §5.1.2) and,

- The STEP Application Protocol 212
  "Electrotechnical design and installation"
  (see the information requirements §5.1.3).

Moreover, the STEP post-processors of CAD and CAM software used by the entreprise feed the PDM System.

5.1.2 Scope and information requirements of AP 214

The global objectives of the application protocol ISO 10303-214 may be summarised as follows:

- continuous management of the information all along the life-cycle of the products;
- description of the products without redundancy;
- independence of the information from the implementations;
- better quality of the information and of the data transfers thanks to non ambiguous and internationally validated definitions;
- capability of archiving.

For some companies involved in the development of the protocol, the protocol also provides a reference data model for the conception of new information systems.

Beyond the data types it specifies, AP214 can also be seen as an enabler of various functionalities. Limiting our list to the data usually dealt with PDM systems, AP214 allows:

1. The identification of parts and tools under design.
2. The description of bills of materials and of explicit relationships between parts.
3. The description of manufactured objects.
4. The identification of the functions to be achieved by the products and of the generic organic structure common to a range of products.
5. The specification of families of commercial products
6. The description of variational bills of materials.
7. The description of Activities and projects
8. The description of Process-planning (to describe process-plans for manufacturing and control, to identify the inputs and outputs of a process operation, to identify the necessary resources to achieve a process operation, to assign process related characteristics to a process operation).
9. To describe the structure and the characteristics of Documents.

In addition to these specific functionalities, AP214 provides generic capabilities that can be applied to almost any kind of data. It allows:

10. To associate external documents with product data; to associate a physical mockup to a part; to associate an external document, computerized or no, to refer in a geometric model to a CAD file; to refer in a drawing to an external digital picture
11. To manage the Effectivity data (dates, temporal intervals) and characteristic events.
12. Classifications
13. To manage characteristic properties

14. To manage the description of the changes (modifications of property values; modifications in a geometric model or in a drawing) and to relate these modifications with the concerned product data.

5.1.3 Scope and information requirements of AP 212

The STEP application Protocol ISO 10303-212 deals with Conception and realisation of electrical installations. This standardisation work is made within ISO TC184/SC4, in narrow collaboration with the IEC (International Electrotechnical Commission).

The protocol STEP AP212 enables to manage and to exchange data of conception, installation and partially of electrical system maintenance. It is organised following a modular structure that covers functional, equipment and geographical aspects, as well as the configuration and the documentation of installations. Its main covered domains are the following:

Functions and equipment:

- complete functional description of systems (functions and connections),
- equipment and physical connections,
- possibility of classification of items,
- structure of product (harmonised with that of AP214).

Installation:

- complete definition of the cabling,
- definition of the spatial positioning of equipment,
- reference to the 3D geometrical description of equipment.

Properties of equipment:

- predefined and specific components, Standard components described according to standards ISO 13584 (PLib) and IEC 61360.

It is in order to prolong works, realised more than ten years ago, in Germany and in France (VNS and SET standards), and to insure a narrow link with standards of the mechanical sector, that the development of the protocol AP212 of STEP has been undertaken from 1992.

5.1.4 Results of the activity

The resulting data model constitutes a consistent subset of resources that have been kept from the AIMs of the both APs.

This data model, illustrated Figure 2, will be implemented inside the database of the PDM System chosen by this entreprise. This data model is illustrated in the figure below.

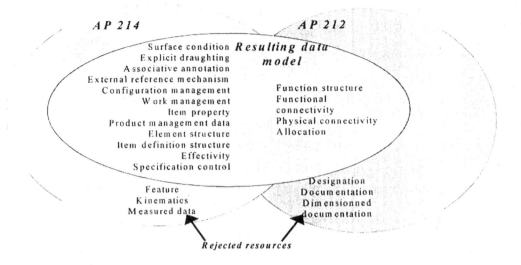

Figure 2 Resources included in the data model coming from the both STEP APs.

Note: These resources are grouped in Units of Functionality[2].

5.2 Mechanical data : an experiment of the AP 203 in an industrial environment

5.2.1 Goals and context of the experiment

French manufacturers summoned up their strengths to experiment the STEP 203 application protocol.

The industrial experimentation project of the AP203 was gathering four industrial partners around association GOSET: Aerospatiale, Dassault Aviation, PSA Peugeot-Citroën and Valeo.

The goals of the project named "Great Innovating Project (GPI)" were:

- To compare the AP203 model with the needs expressed by companies,
- To experiment the data exchanges between CAD-CAM systems,   i.e. 3D geometric model exchanges,
- To check the ability of exchanging or sharing data covering the two AP 203 application fields and thus to support conventional industrial process of modification management and shared products conception[8].

5.2.2 Scenarios and demonstration

Two scenarios, one in the aeronautic domain, the other in the automotive sector were set up. For the first time, demonstrations of some software systems were realised in real time.

The aeronautic scenario consisted in simulating the addition of a pump within a plane by a part supplier. The concerned part of the digital mock-up of a plane had to be extracted and transferred to the part supplier so that he inserts the pump and sends back the corresponding information to the plane manufacturer.

The automotive scenario demonstrated the need to define work process and methods for exchanging. The provider, Valeo, had to design an headlight with its electrical corrector from requirements defined by the end-user, PSA Peugeot-Citroën, then to send back the complete design for integration within the digital mock-up of the vehicle.

In the two cases, exchanges between the CAD and PDM heterogeneous systems were materialised by AP203 STEP files, both including geometry and configuration data.

5.2.3 Results of the GPI

Industrial partners have gained an understanding of industrial processes and where the protocol could be used and its benefits for CAD-CAM and PDM.

These two experiments have demonstrated the STEP benefits in a scope of configuration information and exchange between CAD/CAM systems of industrial partners.

The GPI has proved that the STEP protocol 203 can be used to exchange data within a shared conception, for scenarios covering both PDM and CAD-CAM field[8].

---

[2] A Unit of Functionality is a collection of concepts of the application and their relationships that are defined in the A.R.M.

5.3   Example of industrial deployment of STEP in the aeronautics

5.3.1   industrial partners involved

Boeing Commercial Airplane group with Pratt & Whitney, Rolls Royce and GE Aircraft Engines decided in 1998 to use STEP as data exchange support as part of the 777 and 767-400 ER production programs.

5.3.2   Context of the d industrial deployment

Boeing and its engine suppliers exchange product data during " digital pre-assembly", to check form and part assembly that fit together the engines' and plane. Previously, solid model assemblies were exchanged in CATIA format developed by Dassault Systems.

5.3.3   A new exchange process

First, Boeing, Pratt & Whitney, Rolls Royce and GE Aircraft Engines implemented a new exchange process based on STEP ISO 10303-203.

It was perfected in the PowerSTEP project and relied on the technology developed by the AEROSTEP experimentation project led by PDES Inc.

These two projects are the results of sustained effort between Boeing and its engine suppliers with their CAD-CAM system suppliers: Dassault Systems (CATIA), EDS (Unigraphics) and Computervision (CADDS).

5.3.4   The first results

Today more than 500 industrial models have been exchanged between Boeing and the three engine suppliers with STEP.

Exchanges have been executed by CAD-CAM users and not by the STEP experts. Approximately 90% of models are correctly transferred without any intervention. Appropriate measures are taken to improve this degree of success in order to no longer use old direct interfaces, or any manual operation.

STEP exchanges are realised with interfaces marketed by software suppliers. Exchanged data cover:

- the solid model geometry,
- the product identifications,
- component/assembly relationships and,
- administrative data.

6   Conclusions

The fundamentals of STEP and of the related standards that are based on the same concepts have been outlined. Some examples of the actual use of this standard in the industrial context have been given.

These examples demonstrate the real contribution to the simultaneous engineering concept brought by the technologies concerning the data integration developed within the STEP project.

However, the industrial deployment of concurrent engineering will not satisfactorily spread out without a re-engineering of the processes inside the entreprise and of its relationships with its partners.

Valeo and PSA Peugeot Citroën have performed a first stage in that direction during the GPI project[9]. But this process should advance more quickly, taking advantage of a rigorous modelling, as Dr. Ghodous works[10] show.

References

[1] S. Arbouy, A. Bezos, A.-F. Cutting-Decelle, P. Diakonoff, P. Germain-Lacour, J.-P. Letouzey, C. Viel. *"STEP : Concepts fondamentaux"* – AFNOR.

[2] Standard ISO/IS 10303-1 *"Industrial Automation Systems and integration – Product Data Representation and exchange - Overview and fundamental principles"* – ISO, Geneva 1994.

[3] Standard ISO/IS 10303-11 *"The EXPRESS Language Reference Manual"* – ISO, Geneva 1994.

[4] Standard ISO/IS 10303-11, annex D: *"The EXPRESS Language Reference Manual, graphical representation."* – ISO, Geneva 1994.

[5] *"IDEF1X (ICAM definition Language 1 eXtended)"*, federal information processing standard 183, Integration Definition for Information Modelling (IDF1X), FIPS PUB 184 National Institute of Standards and Technology, December 1983.

[6] G. M. NIJSSEN, and T. A. HALPIN: *"Conceptual Schema and Relational Database Design: A Fact Oriented Approach"*.

[7] Chris ANGUS, Peter DZIULKA: *" EPISTLE framework"*, version 2.0, issue 1.2.1 April 1998.

[8] Frédéric Féru (Aérospatiale-Matra), Christophe Viel (GOSET) *"Echanger avec le protocole d'application 203 de STEP"* – GOSET.

[9] Dr. Frédéric Chambolle (PSA) *"Un modèle de données piloté par des processus d'élaboration : aplication au secteur automobile dans l'environnment STEP"* PHD Thesis – Ecole Centrale Paris – France April 1999.

[10] Dr. Parisa Ghodous (CAD/CAM and Modelling Lab. University of Lyon I – France) *"An Integrated Product and Process data model "* September 1996.

CHAPTER 7

CE in Virtual Environment

Tolerance Assignment Adequate for Manufacturing and Assembly

Pavel G. Ikonomov

3D Inc. Yokohama
Dept. of Production, Information and System Engineering
Tokyo Metropolitan Institute of Technology
6-6, Asahigaoka, Hino, Tokyo, 191-0065, Japan
Tel: (81-42) 42- 585-8600, Ext.4212
E-mail: pavel@ddd.co.jp

Shuichi Fukuda

Dept. of Production, Information and System Engineering
Tokyo Metropolitan Institute of Technology
6-6, Asahigaoka, Hino, Tokyo, 191-0065, Japan
E-mail: fukuda@exmgfkta.tmit.ac.jp

Abstract

Concurrent Engineering technology reduces the length of the design-manufacturing cycle while at the same time has a higher requirement for quality and tolerances. Most of the inadequacies of tolerance assignment at the design stage are discovered late, at the production or assembly stage. We propose a computer system to be used by the designer for appropriate tolerance assignment to meet manufacturability and assemblability requirements. In order to match tolerance requirements for real manufacturing and assembly process we propose usage of virtual gauge and introduction of a statistical tolerance model at the design stage. Evaluation of manufacturing tolerances early in design stage will examine the validity of design tolerances and define whether re-design or adjustment of the design is needed. Similarly evaluation of assembly tolerances, which will define assemblability of the proposed design, will be done.

1 Introduction

The purpose of this paper, tolerance adequate for manufacturing and assembly, is to develop a system that assists the designer to assign proper tolerances for the real parts. The aim is to improve manufacturability and assemblability work at an early design stage and to reduce production problems that lead to expensive or erroneous production cycles, even impossible production. The ideal design engineer shall have thorough manufacturing experience and adequate assembly skills. Unfortunately this is not the case in the modern design-manufacturing system. This is the reason we proposed a system for design tolerance suitable for manufacturing and assembly in concurrent engineering.

By providing manufacturability we can help the designer to modify the design, if necessary to balance the needs for efficient machining against the needs for quality product. The process plan precedence influences greatly how we achieve machining tolerance requirements. Fixture problems and change of the designated datum also influence accuracy. There are numerous papers dealing with design, process planning and scheduling problems. For the purpose of simplicity, we will assume that the process planning problem is solved regarding the sequence optimization, machine tool and fixture selection. An automatic way to generate and evaluate alternative operation plans for a given design was proposed by Gupta[1].

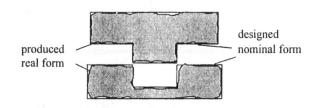

Figure 1 Assembly view of designed nominal and

Let define the problems we are dealing at design and manufacturing stages. The designer assumption is that the part is produced at nominal shape applying maximum material principle to assure feasible assembly between parts, see Figure.1. In practice the produced real part form defer considerably not only on the in its shape, but also the way different surfaces are related between and to datum's. Moreover the datum itself is assumed plane that

401

can be ill defined as well. In the manufacture and assembly stage those difference are discovered and decision has to be made to redesign or continue with production while apply some statistical tolerance method to work out the non-perfect produce sequence of parts to be assembled.

Even though the planning process is integrated, the main problem to be solved is how to reach the prescribed accuracy and assemblablity of the designed part. If we can prove that the designer's decision is correct, the green light can be given for the transformation from design stage to production stage. In most cases the design model is assumed to have flaws, so redesign will be necessary. We have emphasized here the importance of tolerance checking and assistance to the designer in choosing correct and possible to manufacture tolerances. Similar problems arise with assemblablity.

2 Design and manufacturing systems

system for tolerance for manufacturing and assembly works in an interactive way, see Figure.2. For easy perception we divide the working space into Real world and Computer interface. The Real world is the way designers see the part and its assembly on the monitor screen. The computer interface displays data and is used for interaction between the designer and data. The Data Base contains manufacturing and assembly data and is built with a specific data structure.

3 Tolerance analysis at design stage

Concurrent engineering places a great significance on the role of the designer. Small errors in the design phase will have a significant impact on the manufacturing and assembly phases. It is desirable to clear all possible obstacles affecting tolerance requirements. There are many considerations regarding achievement of the described accuracy of the machine part. Here we will specify some

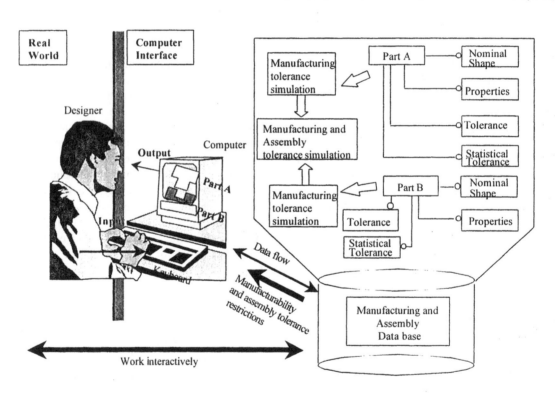

Figure 2. Simulation system for manufacturing and assembly

The designer works interactively with the computer to design parts in accordance with the worst-case tolerance. When those tolerances cannot be met statistical manufacturing tolerances and assembly statistical tolerances are applied instead. This ensures that the designed parts can be machined and assembled at the actual manufacturing process to follow. The proposed

general requirements, solutions and the needs of tolerance evaluation.

There is not possible to predict all variation of manufacturing process, but we shall be able to help designer with tolerance setting based on existing variation (mathematical) or statistical tolerance practices in manufacturing and assembly. We suggest using of combined method to assist design practice.

3.1 Existing tolerance practice

At present, tolerances are assigned, based on ISO tolerance standard. When the part is manufactured and assembled the tolerance is based on those standards. The implementation of those tolerances is a choice between worst-case tolerancing (ISO tolerance standard) and statistical tolerancing. In case of worst-case tolerancing everything is straightforward, it gives a 100% guarantee in satisfying ISO tolerance. On the other hand the economical advantage of statistical methods has forced many companies to produce parts that do not comply

according to function requirements. He checks the designed product for conformance with worst-case tolerance. The designed product model is submitted for manufacturing. Here manufacturability check is performed on the product model against manufacturing tolerance. Next the part is produced following worst-case tolerance or statistical tolerances. Statistical tolerances are used when costs for achieving worst-case tolerance is very high or it is not possible to produce all the parts with the prescribed tolerance. In some cases when it is not possible to produce the part, the design model is returned for redesign. After manufacturing the product, it is checked for manufacturing tolerance and then sent to the assembly. At assembly phase part is checked against worst-case tolerance or statistical tolerance. Statistical tolerance is

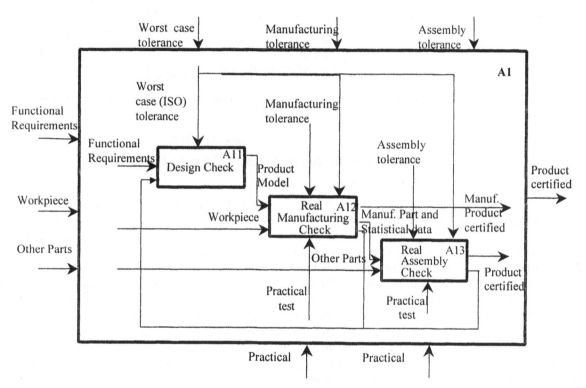

Figure 3. Design and real practice

completely with tolerances, see Figure 3. Here is the dilemma, part is designed with worst-case tolerance, but is produced with less than 100% guarantee of the tolerances. There is a need for designer to consider the real manufacturing and assembly practices. Figure 4. shows how at design, manufacturing and assembly stages tolerances are checked. The designer develops the part

used when costs for achieving worst-case tolerance is very high or it is not possible to assemble all the parts with prescribed tolerance. In some cases when it is not possible to assemble the parts, the design model is returned for redesign. Assembly checking is performed on the parts and if they satisfy assembly tolerance then the products are certified.

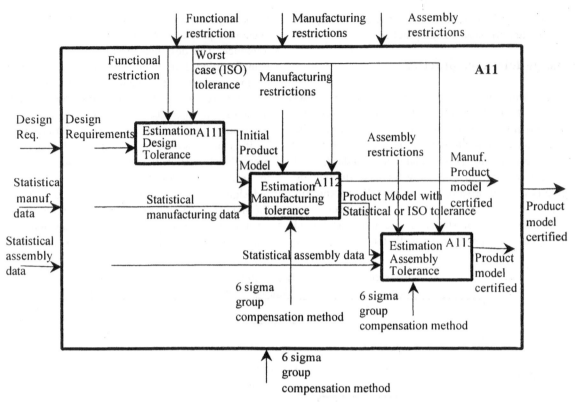

Figure 4. Proposed Statistical Tolerance Usage

3.2 Advantages of the statistical tolerance

Statistical tolerance gives also high process guarantee up to 99.73% (6σ) that the part will meet the tolerance, but the cost of production is considerably cheaper. Usually production can throw away remaining small percentage of parts with errors, rather than trying to develop perfect process to meet the 100% objective. Therefore the economical benefit is well understood so statistical tolerance is widely applied to manufacture and assembly, yet it is not used in design. This is the reason we propose application of the statistical tolerance in the design phase. Mathematical interpretation of the statistical tolerance that is well known shall be introduced to the designer as an alternative to worst-case tolerancing. Further an ISO statistical tolerance standard shall be developed.

Here we propose usage of worst-case tolerance, variation method for accuracy calculation and introduction of the statistical tolerance in design practice. Figure 4. shows the new design process that includes worst-case tolerances, manufacturing and assembly tolerances. The designer develops initial product model according to functional requirements and worst-case tolerances. This model is

checked against statistical manufacturing data. If those requirements cannot be met by manufacturing process, tolerance is increased using some statistical method (6 sigma, group or compensation) to estimate the manufacturing tolerance. Subsequently the new product model that was tested against manufacturing tolerance is the one used for assemblability testing. Again, this product model is used for checking against statistical assembly data, and if those requirements cannot be met by the assembly process, tolerance is increased using some statistical method (6 sigma, group or compensation) to estimate the new assembly tolerance. Successfully tested product model is certified for the real production stage to follow. It is clear that the new proposed design process that includes statistical, manufacturing and assembly tolerance estimation is based on the real design-production process, shown in Figure 3. This will ensure proper tolerance assignment at the design stage and production process without design flaws, as the consideration regarding manufacturing and assembly requirements has been done at the design stage

From the above it follows that the designation of the adequate tolerance based on the existing manufacturing and assembly practices will ensure manufacturability and assemblability of the designed products. On the other hand, such a system gives the designer greater flexibility to assign bigger tolerance; so cheaper products can be produced.

4. Variation method for accuracy calculation

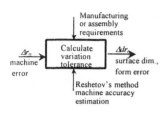

Variation method for accuracy calculation of machine surfaces is based on Reshetov - Portman's method for machine tools accuracy estimation [3]. It is based on the mathematical model of the main error of the mechanical system.

The errors of position of the machine units and elements are input parameters, and the errors of dimension position and form of machined surface are output parameters.

The model uses form shaping system code to represent different machining processes. This code is further used in calculation of the accuracy of machined surfaces. The models of form shaping system of cutting tool for single point, linear tool and surface tool are constructed and also machine layout can be achieved. From those models calculation of form-shape layout output is done. Positional error vector and nominal radius vector of the surfaces are calculated using the following equations.

$$\Delta r_0 = A_{0,l} r_l \tag{1}$$

$$f_j(q_1, \Lambda, q_{n+m}) = 0, \quad j = 1, \Lambda, L. \tag{2}$$

After some substitutions.

$$\Delta r_0 = \sum_{i-1}^{l} \sum_{j=1}^{6} \left(A_{0,l} D^j A_{i,l} r_l \right) \delta q_i^j \tag{3}$$

Here D^j are matrices, δq_i^j is the positional error of the unit for j^{th} coordinate.

Let consider now output accuracy estimation.
The equation of the base surface is.

$$\Delta r_b = e_b r_0 + d r_0 \tag{4}$$

$$d r_0 = \sum_{i=1}^{m} \left(\partial \frac{\partial r_0}{\partial q_{0i}} \right) \Delta q_{0i} \tag{5}$$

Where Δr_b is defined as a sum of vector of dimensional and positional errors.

e_b is the matrix (4x4) of the positional error of the coordinate system, which is coupled with base surface relative to the main system [3]. The elements Δq are used to estimate dimensional and positional errors. Form errors can be estimated as standard deviation from base surfaces. Second method is by constructiion of new surfaces that includes distortion of the form of the nominal surfaces. For details refer to the origin source [3]. This method for evaluation of accuracy can be applied to calculation of tolerances from the designer when machining specifications are known.

5. Worst-case (standard) tolerance

When solving tolerancing problems, one must choose between worst-case tolerancing and statistical tolerancing. Worst-case tolerance is the standard tolerance that guarantees 100% inter-exchangeability of the design parts. Worst-case tolerancing is the safer approach. If the inputs are within their respective tolerances, the output is guaranteed to be within its worst-case tolerance. This is especially important for products like heart values or critical components on airplanes. However, this guarantee comes at a high cost [5]. Worst-case tolerancing guards against a worst-case scenario that is highly unlikely if not impossible in most situations. Worst-case tolerancing guards against all inputs simultaneously being at the same extreme. In order to reduce production cost the designer shall consider statistical tolerance as an alternative tolerancing method.

5.1 Test of the designed part for assemblability

Existing assembly models inside the CAD systems have done a great deal to assist engineers toward assembly checking for interference and dimensional tolerance checking. This works in case of top-down assembly approach, when the same CAD system is used for all members of the enterprise. In case of concurrent engineering usage of the same type of CAD software is impossible, so we have to find some common solution for each participant. The needs of the designers we will explain with the aid of some examples.

We specify two levels of checking for assemblability regarding tolerances. [2]

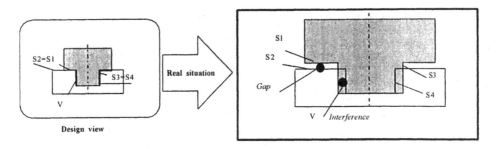

Figure 5. Dimensional restriction

- The first level is the ordinary dimension restriction, related to size and form, usually defined on the part drawing (Figure 5).
- The second level check is for geometrical restriction derived from relations between mating parts or from geometrical tolerance requirements (Figure 6). Simply stated this is to ensure that geometrical restriction from one part will not interfere with physical dimension of the other part. That means to meet geometrical tolerance requirements.

Having geometrical restriction as a reference, the designers can decide by themselves to make new design changes or request design changes on the other mating part. It is assumed that in most of the cases it will be possible to make a design and test it against requirements of all three restrictions without consulting the other partner.

5.2 General assemblability problems to be considered at design stage

5.2.1 Dimension (nominal shape) restriction

Dimension restrictions are based on nominal shape of the designed part. These restrictions are alignment and

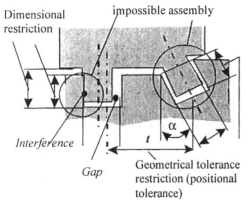

Figure 6. Geometrical tolerance restriction

interference checking and have multiple engineering solutions even in CAD software, so we will consider this problem solved on the basic level.

Lets analyze the problem of the Virtual assembly depicted in Figure 5. Suppose we have a simple assembly, requiring mating surface S1 with S2 and S3 with S4. As can be seen alignment of the two parts are correct, but surfaces S1 and S2 leave a gap and S3 and S4 interfere. The designers specify assembly, use visualization tools and verify visually the correctness of the assembly, without considering tolerances.

5.2.2 Geometrical tolerance restriction

Geometrical restrictions are more complicated in their matter. They are derived from functional requirement of the parts. Ordinarily they are represented with geometrical tolerances and aim to give unconditional geometrical assembly between parts. Their nature is static, but the complicated three dimensional spatial relationships between related parts is an intricate problem. Lets analyze the problem of the Virtual assembly depicted in Figure 6. Suppose we have a simple assembly, requiring mating surfaces of the upper and lower part. It is possible to find a solution for dimensional and geometrical restriction, but still both parts cannot be assembled. The designers shall have a mechanism to prevent such mistakes. The only way to do that is to consider the mating parts dimensions, geometry and tolerances. We had proposed Virtual gauge base Virtual assembly solution that is based on geometrical tolerance. An extended Virtual assembly model using STEP standard Express-G model schema shown in earlier publication. [2]

5.3 Geometrical tolerances testing

ISO geometrical tolerance is intended to guarantee the assembly. This is the reason why in our proposed assembly model we rely on geometrical tolerances. In our previous work [3], we have proposed Virtual gauge as a base for evaluation of the measurement with Coordinate Measuring Machine. The concept of Virtual gauge is applied here for representation of the relationship between features and datum's when we have geometrical tolerance. The spatial relationships between feature and datum are represented with constraints. Figure 7., shows an example of the Geometrical constraint perpendicular to datum.

This constraint model use vectorial representation and its mathematical formulation and comply with ISO tolerance specifications.

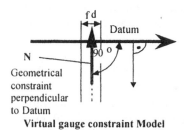

Figure 7. Constraint Model

The required spatial relationship can be given with geometrical constraint applied to the geometrical element during the evaluation. This example indicates the necessity of evaluation procedure with geometrical constraint that represents explicitly the spatial relationship between feature and datum.

The spatial relationship between geometrical element and datum is carried out from geometrical constraints. Geometrical constraint can be positional and directional. The geometrical element is plane, cylinder, cone, etc., and is related to axis2_placement_3d, that define it position and direction. This axis2_placement_3d has separate position constraint and direction constraint.

5.3.1 Virtual gauge representation

Virtual gauges are defined by position, direction vector and sizes. For example virtual cylinder is defined by position vector P, direction vector N of the axis and radius R of the size.

The presented model is based on the application of constraint to the position and direction vectors of the virtual gauge in order to obtain results satisfying ISO geometric tolerance requirements. To define and apply constraints representing the relationships between features and datum the Geometrical constraint model was developed. As shown before, virtual gauge is represented with position and direction vectors and sizes. We propose the mathematical representation of the constraint applied to the virtual gauge. Constraint applied to the position vector *P(PxPyPy)* is given by Equation (6) and constraint

applied to the direction vector *N(NxNyNy)* is given by Equation (7).

$$A \cdot P_X + B \cdot P_Y + C \cdot P_Z + D = 0 \qquad (6)$$
$$E \cdot N_X + F \cdot P_Y + G \cdot P_Z = 0 \qquad (7)$$

These constrain equations can be combined. One or more constraints for position and direction vectors can be used to define different geometrical tolerances. Application of combined or separate constraints gives the following cases: for the position - 4 cases, and for direction - 3 cases, or all together 12 cases.

In Figure 8. and 9, the virtual gauge cylinder, defined with its position vector P, direction vector N, and size R, and the Constraint plane for position vector are shown. In Figure 8., the physical meaning is that the position vector is constraint to stay on the plane given by Equation (6).

In Figure 9., the physical meaning is that the direction vector is constraint to be perpendicular to the Constraint line that is given by Equation (7).

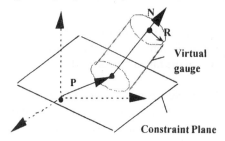

Figure 8. Representation of the position constraint for cylinder

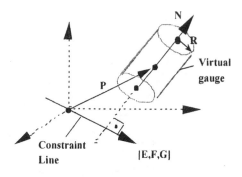

Figure 9. Representation of the direction constraint for cylinder

Proposed constraint model has been successfully applied with SDS (Torsor) mathematical method for geometrical tolerances evaluation and can be applied for assemblability purpose.

5.3.2 Testing procedure based on the Virtual gauge and SDS (small displacement screw)

5.3.2.1 SDS method

A SDS is used for testing of the assemblability of the assembled parts.

The small displacement of each point in Euclidian space can be described by two vectors: $\vec{D}_M$ and $\vec{\theta}$. Vector $\vec{D}_M$ corresponds to three small translations, vector $\vec{\theta}$ to three small rotations. These two vectors define SDS model for torsor $T_{M,\theta}$:

$$\vec{\theta} = [\alpha\beta\gamma]^T, \qquad \vec{D}_M = [uvw]^T$$

$$T_{M,\theta} = \begin{bmatrix} \vec{D}_M \\ \vec{\theta} \end{bmatrix} = [uvw\,\alpha\beta\gamma]^T \qquad (8)$$

Tolerance related to TTRS can be represented as a tolerance torsor representing the small displacement, possible within the tolerance zone [7].

5.3.2.2 Testing for assembly contact condition

We have a contact between surfaces of two parts when they are opposite and apart. Between the two surfaces we have only avoid. The contact condition can be fixed (weld, bolts, glue) or free in other cases. M1 and M2 are two nominal points belonging to each opposite contact surface. The contact offset between and M2 is the scalar product between $\overrightarrow{M_1M_2}$ and $\vec{n}$, where $\vec{n}$ is t external to the material of the surface S1.

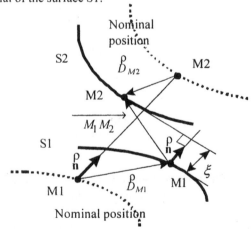

Figure 10. Contact conditions

We have free contacts when assembly constraints are linked and there is no interpenetrating between surfaces. This can be expressed by linear relations of the components of the torsion vectors related to opposite points: contact offsets are positive or null. When we have fixed contact, contact offsets are null. Assembly is warranted when we have no penetration, for each contact. MGDEp of the different parts are fixed in nominal position of the assembly. Each contact is tested separately. We want to obtain the minimal value of contact offsets. To have a non is no interpenetrating of surfaces, the modification of the nominal position or the nominal size is needed. In such a way, nominal stacks can be evaluated. This is called Uncoupling of tolerancing and nominal dimensions.

To ensure assembly of two surfaces at their contact surfaces, it is only necessary to modify the nominal position (relatively to its MGPEp) or the nominal dimension (relatively to its MGPEs) of one of the two surfaces. The value of this modification is equal to the value of is no interpenetrating.

The Virtual gauge represents the functional requirements of tolerance model. We have proposed an extension of the SDS (torsor) method to represent relationships between datum's and toleranced surfaces [7].

The designed part with defined tolerance can be checked for assemblability using outlined method, if error occurs immediately new version of he design model can be specified an checked again thus providing us with possibility to establish concurrent design process.

6. Statistical tolerances

6.1 Statistical tolerances method

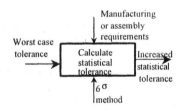

Statistical tolerance methods differ from worst-case tolerance in that they allow tolerances to be increased which lead to a more economical process.

Statistical tolerances are determined by a target, t_x, and maximum standard deviation, σ_x. They are denoted by $t_x \leq 3\sigma_x$. They are more commonly denoted by $t_x = \pm\Delta_x$. When using the <ST> symbol, it is most common to have

$\Delta_x = 3\sigma_x$. However, this relationship is not formalized in the standards and the six sigma approach uses $\Delta_x = 6\sigma_x$. [4]

$$\sigma_i^2 = \sum_{i=1}^{m-1} \sigma_i^2 \qquad (9)$$

If we denote with t risk coefficient for the sizes out of tolerance, then.

$$t_i = \frac{\delta_i}{\sigma_i} \qquad (10)$$

Taking into account Equation (6).

$$\frac{\delta_\Delta^2}{t_\Delta^2} = \sum_{i=1}^{m-1} \lambda'_{mean} \delta_i^2 \qquad (11)$$

Here δ_Δ is the closing link tolerance, t_Δ risk coefficient for percentage of sizes that are out of tolerance δ_Δ.

$$\delta_{mean} = \frac{\delta_\Delta}{t_\Delta \sqrt{\lambda'_{mean}(m-1)}} \qquad (12)$$

$\lambda'_i = 1/3$ for equality distribution law, $\lambda'_i = 1/6$ for Simpson law, $\lambda'_i = 1/9$ for normal distribution law. More accurate calculation gives.

$$\lambda'_i = \frac{k_i}{t^2} \qquad (13)$$

Tolerance increasing can be calculated.

$$R = \frac{1}{t_\Delta} \sqrt{\frac{m-1}{\lambda'_{mean}}} \qquad (14)$$

The basic advantage of the statistical method is the possibility to increase tolerances, comparatively to the worst-case tolerance method. That makes possible for the designer to designate higher tolerances for design parts that have tolerances that cannot be met by the traditional worst-case tolerance method. It also makes sure that parts can be produced with existed machines and assembly can be carried out. When there is an additional increase of the tolerance or special requirements for manufacturing and assembly, the group method and the compensation method can be used.

Table 1. Relative increasing R of the medium value of the tolerance related to risk percentage λ'_{mean}.

Risk %	λ'_{mean}	R
0.27	1/3	1.41
0.27	1/6	2
0.27	1/9	2.45
1	1/3	1.65
1	1/6	2.35
1	1/9	2.85

6.2 Group (selective) method and method of compensation (regulation) link

6.2.1 Group method

The group method guarantees assembly inside for each group distribution.

Tolerance mean values increases n times relatively to the worst-case tolerance method.

$$\delta'_{mean} = n \delta_{mean} \qquad (15)$$

The group method allows an increase in accuracy for high-precision group products, like bearing, engines, machine cutting tools, etc. [5]

6.2.2 Method of compensation (regulation)

Assemblability is guaranteed by changing the size (by machining or adjusting) of one chosen link. Compensation is calculated from Equation (16)[5].

$$\delta_k = \delta'_\Delta - \delta_\Delta = \sum_{i+1}^{m-1} \delta_i - \delta_\Delta \qquad (16)$$

Here δ_Δ is the tolerance of closing link, defined by functional requirements or restrictions, δ'_Δ is the economically achievable tolerance, m is the number of the dimensional links.

Δ_k changes to the mean value as a result increased tolerance of δ' of the compensation link [5]. If we have symmetrical value of the tolerances δ' the equation is simplified to:

$$\Delta_k = \frac{\delta_{max}}{2} \qquad (17)$$

From the above it is clear that the designer shall pay attention not only to the standard tolerance practice based on ISO standard, but common factory and industry practice in order to develop products that meet manufacturability and assemblability requirements.

7. Conclusions

We proposed a model for design tolerance suitable for manufacturing and assembly in concurrent. In order to carry on tolerance requirement we propose introduction of statistical tolerance model to design stage. For checking of machining and assembly tolerance we proposed usage of worst-case tolerance when 100% inter-exchangeability of the design parts is required. To reduce tolerance requirements in order to design products according to the actual manufacturing and assembly practice, usage of statistical tolerance, 6 sigma, group method and method of compensation (regulation) is proposed. Necessary mathematical equations and considerations are provided.

Using statistical tolerance will improve designers work, reduce tolerance requirements and needs for redesign in concurrent engineering environmental that place great importance on the design.

References

[1] S. K. Gupta, Automated Manufacturability Analysis of Machined Parts, doctor thesis, Univ. of Maryland, 1994.

[2] P. Ikonomov, et al, Navigation System for Virtual Assembly, Proceeding of Advances in Concurrent Engineering pp.277-284, 1997.

[3] D. N. Reshetov and V. T. Portman, Accuracy of machine tools, ASME Press, pp.21-125, 1988.

[4] W. A. Taylor, Process Tolerancing: A solution to the dilemma of worst-case statistical tolerancing, http://www.variation.com/var-soft.html, 1997.

[5] B. C. Balakshin, Theory and practice of machine building, Vol.2, pp.65-88, 1982.

[6] Q. Bjorke, Computer-Aided Tolerancing, pp. 6.4-6.9, 1978.

[7] P. G. Ikonomov, T. Kishinami, F. Tanaka, Virtual Assembly Using STEP Data, Proceedings of the conference: Advances in Concurrent Engineering CE96, Toronto, Canada, 26-28 Aug 96. p.363-369

Virtual Reality Assembly Process Simulation

Pavel G. Ikonomov

3D Inc., Yokohama
Dept. of Production, Information and System Engineering
Tokyo Metropolitan Institute of Technology
6-6, Asahigaoka, Hino, Tokyo, 191-0065, Japan
E-mail: pavel@ddd.co.jp

Shuichi Fukuda

Dept. of Production, Information and System Engineering
E-mail: fukuda@exmgfkta.tmit.ac.jp

Yutaka Kanou
3 D Inc., Yokohama
E-mail: kanou@ddd.co.jp

Abstract

Concurrent Engineering technology reduces the length of the design-manufacturing cycle while at the same time has a higher requirement to the design work regarding manufacturability and assemblability. Virtual assembly assist designer to improve product manufacturability and assemblability and allow him to predict and understand possible production problem. Virtual reality technology allows instant examination of the accuracy of the design concept early in the development cycle before even the physical model is produced. Further this VR system can be used for training of the personal of line before the real process is delivered. In virtual assembly simulation, all parts, machines and tools placed in their corresponding place and user can manipulate them in ordinary way. It is implemented based on real assembly practice while achieving high-speed frame rate on ordinary personal computer.

1 Introduction

In this paper we show development and implementation of one Virtual Assembly system based on the real product and assembly process.

Virtual Assembly for Concurrent Engineering presents a tool in hands of designer for correct and efficient design modeling. Most of the inadequacies of design assignment are discovered late, at the production or assembly stage. A virtual computer model to be used by the designer to deal with manufacturing and assembly problems at the design stage is needed. We will show the capability of the virtual technology to assist designer and help him with knowledge that are not relevant to his work, experience. The proposed virtual model leads designer trough assembly processes for the purpose of assembly planning and assemblability, to achieve one successful design that doesn't need to be redesign, changed or rearranged. Virtual manufacture and assembly idea is to synthesize relevant requirement of design, manufacturing and assembly. Shifting of manufacturing and assembly requirements to the design stage will have remarkable impact on the design/production development cycle from conceptual stage to realization and verification.

Value of the propose system for virtual manufacturing and assembly:

The system being developed has several important differences from usually developed VR based simulation. It is implemented based on the real assembly practice, it achieves high speed frame rate on ordinary personal computer. Other very important issue is the user interface with VR environmental in the natural manner. No special knowledge or computer skills are required as the work within the immerse space VR assembly environmental with gloves, HMD and tracking sensors offer an easy non interrupting style of work.

A virtual assembly/disassembly simulation for the Japanese manufacturing company was developed and tested. The goal of the project was developing of the VR assembly process simulation so designers can test their concept and if it proves correct to use it later for training of the personal. The factory settings have all parts, machines and tools placed in their corresponding place and user can manipulate them in ordinary way. The software and peripheries provides immerse feeling, 3 dimensional interface and workspace that ensures undisruptive usage of all elements taking part in the simulation.

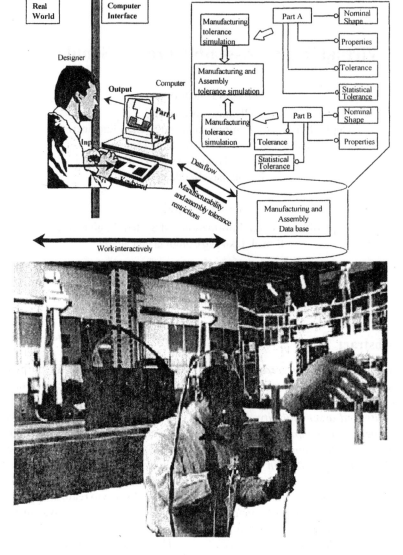

Figure 1. From interactive PC work to immerse environment

2   Concurrent Engineering

Engineering is a cooperative process in which work is coordinated over an extended period of time. In the course of development effort, individual aspects of the product may be worked on concurrently by several engineers or teams. The situation is called Concurrent engineering (CE). There are three types of cooperative work in which data needs to be shared: work group of engineers, across the product life cycle and between enterprises involved in the joint development. [1].

3   Virtual manufacturing and assembly

3.1 Simulation system for virtual manufacturing and assembly

At present the designer works interactively with the computer to design parts in accordance with manufacture and assembly restriction and tolerance requirements. When those requirements are met redesign in accordance with manufacturing and assembly tolerances are applied instead. This ensures that the designed parts can be machined and assembled at the actual manufacturing

data, and if the assembly process can not meet those

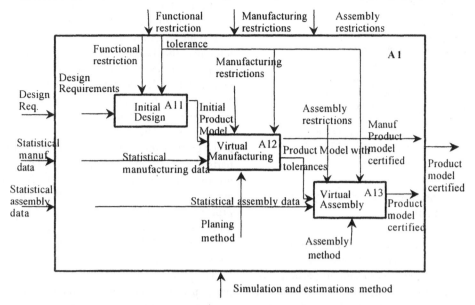

Figure 2. Manufacturability and assemblability at design stage

process to follow. The proposed evaluation system for tolerance for manufacturing and assembly works in an interactive way, see Figure.1. For easy perception we divide the working space into Real world and Computer interface. The Real world is the way designers see the part and its assembly on the monitor screen. The computer interface displays data and is used for interaction between the designer and data. The database contains manufacturing and assembly data and is built with a specific data structure.

3.2 Virtual manufacturing and assembly process

Concurrent engineering places a great significance on the role of the designer. Small errors in the design phase will have a significant impact on the manufacturing and assembly phases.

Figure 2. shows the new design process that includes manufacturing and assembly restrictions and tolerances. The designer develops initial product model according to functional restriction. This model is checked against statistical manufacturing data. If those requirements can not be met by manufacturing process, tolerance is increased using some statistical method or empirical method to meet the manufacturing requirements and tolerances. Subsequently the new product model that was tested against manufacturing restriction and tolerances is the one used for assemblability testing. Again this product model is used for checking against statistical assembly

restrictions, tolerance is increased using some statistical method or empirical method to meet the new assembly tolerance. Successfully tested product model is certified for the real production stage to follow. It is clear that the new proposed design process that includes manufacturing and assembly restrictions and tolerance estimations is based on the real design-production process. This will ensure considering of the manufacturing and assembly restrictions at the design stage and production process without design flaws.

From the above it follows that the designation of the adequate tolerance and following constraints of the based on the existing manufacturing and assembly practices will ensure manufacturability and assemblability of the designed products.

4 Virtual assembly requirements

4.1 Virtual assembly problems an requirements

4.2.1 Test of the designed part for assemblability

Existing assembly model inside the CAD systems have done a great deal to assist engineers toward assembly checking for interference and dimensional tolerance checking. This works in case of top-down assembly approach, when same CAD system is used for all members of the enterprise. In case of concurrent engineering usage of same type CAD software is impossible, so we have to

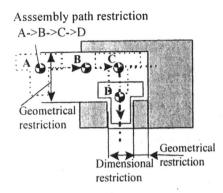

Asssembly path restriction
A->B->C->D

Geometrical restriction

Geometrical restriction

Dimensional restriction

Figure 3. Assembly (machining) path restriction

find some common solution for each participant. We propose this solution as a virtual assembly as a solution. The aim of this virtual assembly is to support participants in concurrent engineering, suppliers and contractor, during their design process. What are the needs of the designer we will explain with the aid of some examples.

We specify three level of checking for assemblability.

- The first level is the ordinary dimension restriction; related to size, form usually defined on the part drawing.
- The second level check is for geometrical restriction derived from relations between matching parts or from geometrical tolerance requirements (Figure 6). Simply stated this is to ensure that geometrical restriction from one part will not interfere with physical dimension of the other part. That means to meet geometrical tolerance requirements.
- Third level, based on assembly path restriction is more complex (Figure 3). Designer shall have initial assembly path description in order to check for obstacles on the assembly path between designed by him part, including size of assembly holder (robot or human hand with tool), and the matching one. This path is defined in the initial stage of the design, and passed to him together with overall design requirements or is defined by designer based on the matching part size. This path is simplified assembly process path develop

by manual, semi-automatic or automatic manner. At present level we are considering that only manual and semi-automatic method is applicable.

Having assembly path and geometrical restriction as a reference, the designer can decide by himself to make new design changes or request design changes on the other matching part. It is assumed that in most of the cases it will be possible to make a design and test it against requirements of all three restrictions without consulting other partner. The Virtual assembly is an effective tool to assist participants of concurrent engineering for designing parts that are accurate from assembly point of view and do not require redesign or new design. It is transferring the assembly process testing at the front end of design process without requiring special kind of software or knowledge to participant. As designed and assembly data are standard STEP data and information file, they are hardware and software independent.

The proposed Virtual assembly shall be applied at each participant site, as well as at the contractor site to ensure consistent design and concurrency at all stages of the concurrent engineering design process.

4.2.2. Assembly (machining) path restriction

New challenge is the requirements deriving from assembly path restriction. Similar problems exist in machining processes for accessibility and size restrictions.

The assembly (machining) path to be considered is so called basic motion planing problem (Figure 7.), as it deals with geometric issue only. It is used to check geometrical feasibility of the planed operation. It is assumed that only restricted subset off the workplace is considered. For example lets consider the problem of adding a part to assembly with holder (or machining with cutting tool) (either robot arm or manually by hand or tool). We can brake it down into five problems:

1. Move the holder position near the part
2. Grasp the part (tool)
3. Transfer the part to the current assembly (move the tool to starting point)
4. Mate the part with the subassembly (machine the part)
5. Retract the holder (cutting tool)

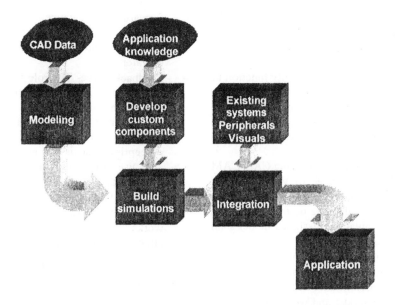

Figure 4 Creating a virtual reality application

As the motions are with small amplitude, we can restrict our attention to the workplace as a small volume around the assembly. If one wants to be sure that the generated path is safe in 4, the motion of the arm with the holder can be checked for possible collision. If such collision occurs the planer shall have another solution for state 4.

From the above mentioned it is clear that designer shall have some guidance system to deal with the complexity of all Virtual assembly (machining) problems.

As concurrent engineering is a dynamic process we propose using of the Virtual manufacturing and assembly. We had point that assembly (manufacturing) path of the part with holder (tools) is the one that shall be consider from designer. At present there are numerous manual and semi-automatic method to solve assembly (machining) path using robot assembly path planing. The assembly path requirements considered here is only basic motion planing model that is relatively easy solved. Moreover we are interested of the assembly restriction of the part only on some points on the assembly path, that additional simplifies assembly path planing. For details refer to Latombe [5].

5 Application of Virtual Assembly

The system was designed and implemented in 3 D Inc. in collaboration with TMIT. Once the mechanical parts system is designed in CAD software the geometrical data is imported to VR package EON Reality. For the complicated geometry with polygon reduction of the irrelevant parts is done before importing in order to keep performance of the simulation enough high. Geometrical relationship and location of the local axis between parts is preserved during. Depending of the assembly requirements additional tools and machinery like screw driver, wrenches, torque wrenches, push tools, press machine, crane machine and controllers are developed and set. Further room interior: walls, floor, working tables is created and imported to VR package. New 3D GUI user-friendly interface is designed to be comfortable for ordinary user, not accustomed to PC environment. Later setups, behavior model, and the sequence flow and control functions of the assembly system is programmed.

The VRAD supports a choice of VR peripheral devices. The system consist of:

5.1 Virtual assembly principle using visualization virtual reality software

In this section of this paper we will describe what are requirement for VR based assembly.

Usually one expects all assembly and visualization to occurred using CAD software. While this sounds an attractive idea, several problems arise immediately. First problem is that CAD data is very big, so perforce is very slow, unless one uses supper fast computer. Frame rate is so slow that even without assembly interaction with an object, user has to wait. The other problem is that CAD software are designed for other type of usage, mainly drafting, but when it come to interface, working space and immerse filling it cannot fulfill those needs.

Other issue is that that VR peripherals are not designed for and does not work with CAD systems. Certainly some CAD companies have tried to include assembly function in the software, but virtual reality presents can not be achieved in most of the cases. The design shall be evaluated in Tree-dimensional environmental to achieve an immerse filing of the of the model, workspace and user movements.

After completing the design using and existing CAD/CAM system CAD data shall be transferred to VR system. In most of the case it requires considerable data polygon reduction in order to achieve adequate performance, high frame refresh rate of the simulation. Designed model shall be modified for virtual assembly:

- Data reduction - As the virtual assembly simulation is dynamic process, using motion, animations, etc., huge data model will cause performance degradation.
- Immerse feeling of the model and workspace -Virtual working space shall be designed in the matter that is close to the real situation to carry out an immerse feeling
- Peripherials-3d tracing sensors, gloves for new task used in virtual assembly such as grasping and caring objects, tracking the movement of the assembled parts and the user during assembly process
- HDM, Stereo Glasses-use of Head Mounted Display or stereo Glasses to achieve an immerse filing in Tree-dimensional environmental.
- 3D Space User Interface (SUI), user position set, forward button, cancel, redo etc. - New easy to use SUI shall be designed, including gesture mechanisms closely resembling real gesture movement of the operator.
- Feedback Force- The role for the feedback force is to implement touch feeling and physical quantities such as friction, dumping, and collision detection between the user and virtual objects. Without such physical quantities, simulation is basically orientation to visual perception, that is not enough for person to control and use VR simulation, moreover without such feelings user can get disoriented and lost in VR space.
- Keeping VA version update and synchronization in virtual enterprise -As the virtual assembly design works in condition of the concurrency, special attention shall be paid to VA data version update and synchronization.

5.2 Virtual assembly example-machine assembly schema

Assembly ordered tree constrain-it is a common practice that the assembly process sequence follows assembly tree hierarchy. Assembly precedence order constraint shall be implemented in the designed assembly simulation structure to prevent confusion or inaccurate assembly result.

Physical constraint simulated by assembly constraint – Assembly between parts shall met constraint such as axis alignment, surface matting and alignment. Further those constraint shall be implemented in the assembly simulation to secure and help the assembly operation during the

Figure 5. Virtual assembly system hardware

assembly, as the is no possible for him to fulfill those constraints in the real assembly

Collision Detection – Assembled parts are visualized, but collision detection shall prevent any possibility that a part pass trough or interfere with other part. This is relatively easy done for convex collision detection, but concave collision detection shall be implemented as well. On the other hand collision detection is computer expensive, so close attention shall be paid not to slow the simulation frame rate.

Using Tolerance for VR assembly- It is insufficient as the virtual parts are already simplified there are no mean to check the real tolerance situation. Another problem is that ISO tolerance is aimed at two-dimensional testing, and there is no well-established standard for 3D representation of the tolerance. STEP standard gives adequate tolerance representation, so geometry data tolerance checking shall be implemented before virtual assembly. Recently there are several commercial CAD/CAM softwares that support tolerance checking of the CAD data.

5.3 Virtual assembly example

Here we present a simulation assembly and disassembly. Up to now assembly process and virtual reality graphics required high end powerful machine such as Silicon Graphics -Onyx, Octane, O2. One of the benefits of using of the ordinary business PC reduces the cost, bust more important is the possibility to use available peripherals to set up a complete system available even for medium size research laboratory. The system we use to

(also Sony Glasstron-Stereo HMD, Polhemus Fastrack Tracking System, Fake Space.

We run EON Virtual reality software that supports most of the peripherals. Key parts of this VR software are: EON Base the main engine, Data Exchange –Most geometry was imported using data exchange from CAD software IGES file format. The rest of the geometry was develop in Multigen, 3DSMax and Autocad r.14 and imported.

Collision Detection –was used to prevent passing interference of the part and as a sensor when part assemble or operator take the part. Assembly mode – allows straightforward process of the assembly operation and possibility to include hints, help messages to support user during the Virtual assembly.

Rigid Body Dynamics are applied to simulated dynamic force of the moving parts.

Force Feedback, Gravity Forces – gives feeling of touch and reality behavior of the assembled part and the sense of realism that add the visual simulation immerse feeling.

Peripherals – HMD V8, instead of the flat screen are used to simulate immerse visual feeling of the diving in the virtual reality environmental. The other HMD we used, Sony Glasstron, provides stereo image with high resolution SVGA 800x600 and is good for long lasting complex simulation, as was in our case. Gloves used with FakeSpace and Fastrack were used to proceed the VR simulation in natural, manner. Each part was assembled to other using usual hands to capture and move them. Constraint imposed to the sequence of the assembly does not allow improper assembly; one could not skip assembly stage.

Figure 6. Virtual assembly system implementation

develop and run the simulation, consist of PC Pentium 266, 64 RAM. For complete immersion HMD Virtual research V8

SUI-For the purpose of the VR simulation special 3D-SUI was developed. As a result system work using common human gestures, no special computer skills are required. It and can be applied to train new staff to do assembly before actual assembly process is applied, to optimize assembly sequence set operation at design stage.

This system allows user without prior knowledge of the assembly sequence to make one assembly trial run and improve quality of the intended assembly. Looking from the other side, from the designer position, such application flexible systems can be utilized easily for developing of assembly system without prior knowledge of computer languages and compiling code.

The assembly system represents hands and body, head (camera) motion for reality feeling. The geometry of virtual hands is visible, so that user hands motion are exactly reproduced in real time in the VR simulation. Left and right hands griping is supported. During the difficulties for the user we designed grabbing ability to be possible when hands approach close vicinity of the corresponding part. In order to use available equipment we decide to we simplified type pinch gloves widely available at low-cost performance ration. We manage to compensate the sensible feeling of the force feedback with adequate visual and sound effects, while keeping performance high (frame rate) and low overall cost of the system.

When hand collides with corresponding part to be assembled, hands changed its color that is a sign that it is possible that this part can be grabbed. This check assures unique assembly sequence. To achieve realistic grip we design attachment of the hand to the part to be in the most comfortable and easy for the user way. Each hand has motion sensor attached for accurate motion tracking. The user interface with VR environmental in the natural manner. No special knowledge or computer skills are required as the work within the immersive VR assembly environmental with gloves, HMD and tracking sensors offer an easy non interrupting style of work.

Conclusions

An implementation of the proposed Virtual assembly simulation using existing software and outline of a virtual reality realization schema was presented. Virtual Reality system will add the designer for analysis of the actual assembly and manufacturing on the early stages of the design process and will save valuable resources for testing of the real process. The system being developed has several important differences from usually developed VR based simulation. At first it is implemented from real assembly practice achieving high-speed frame rate on ordinary personal computer. Other very important issue is the user interface with VR environmental in the natural manner.

References

[1] Morris K.C., Dabrovski K., Fong E., Database Managment Systems in Engineering, NIST, Gaithersburg, Marylant 20899.

[2] Hardwick M., Downie B.L., Kutcher M., Spooler D.L., Concurrent Engineering with Delta Files., IEEE Computer Graphics and Applications, January 1995.

[3] S. K. Gupta, Automated Manufacturability Analysis of Machined Parts, doctor thesis, Univ. of Maryland, 1994.

[4] P. Ikonomov, et. al, Navigation System for Virtual Assembly, Advances in Concurrent Engineering 97. pp.277-284

[5] Latombe Jean-Claude, Robot motion planing, pp.5-43, 1991.

Towards Functional Virtual Prototyping : Virtual World Interaction through Automatic Specification Assembly Joints

Prasad Wimalaratne, Dr. Terrence Fernando and Kevin Tan
t.fernando@salford.ac.uk, {g.d.s.wimalaratne, k.t.w.tan}@pgr.salford.ac.uk
Centre for Virtual Environtments, University of Salford, Salford M5 4WT, United Kingdom.

Abstract

The use of Virtual Environment Technology to support product design process has received greater attention in the recent years. However very little research has been carried out in simulating interactive manual assembly/disassembly tasks with in virtual environments. This paper presents an approach to carry out assembly/disassembly tasks on a virtual prototype through an intuitive interface and to simulate physical realism within a virtual environment. The research aims to simulate assembly and maintenance tasks of products within an immersive virtual environment.

1 Introduction

The Concurrent Engineering (CE) approach has been recognized as one of the key engineering philosophies for lead time compression and product development cost reduction. In the CE approach, communication between representatives from design, engineering, testing, manufacturing, operations and maintenance helps to resolve problems at an early stage of the design by sharing their specific areas of expertise. This process requires advanced computer tools to assist in the product design and manufacturing processes. The drive to achieve increased concurrency and to allow product development issues to be dealt with very early in the design stages has led to rapid-prototyping. The ultimate development of rapid prototyping in terms of speed, cost and flexibility is virtual prototyping - the use of a 3D digital model of the product to explore the problems and to perform operations and analysis as if on a physical prototype. The virtual prototype (VP) needs to have a degree of functional realism that is comparable to that of a physical prototype. Virtual prototypes with capabilities for interactively manipulating the model can be categorized as functional virtual prototypes.

Virtual environment (VE) is a relatively new technology, which is beginning to influence engineering applications. Current virtual prototyping systems aim at using CAD data, with VE tools to replace, at least to some extents, physical prototypes. Most design evaluations are currently done using digital models. However, physical prototypes are still built in most cases, mainly due to its better spatial presence. VE can help provide greater presence and allow realistic manipulation of the model. Furthermore, use of VEs can allow design changes to be quickly incorporated into the virtual prototype and the designs can be optimized through a larger number of "virtual" design alternatives.

Typical CAD systems are not designed for real-time modelling. Product development teams are accustomed to traditional Computer-Aided Design (CAD) and Computer-Aided Manufacturing (CAM) systems. These systems have allowed them to accelerate the product development process. However these CAD/CAM tools still fall short of providing support for downstream processes such as product maintenance assessment. In a maintenance simulation scenario, it is necessary to allow the user to interactively carry out assembly and disassembly tasks. Therefore, the motivation for this work emerged from the increased interest by the manufacturing companies to consider downstream processes such as maintenance during product design.

Very little research has been carried out to develop software environments for assessing maintainability issues during design. Therefore, better software environments are necessary for supporting the assessment of the maintenance processes during the early stages of the design life cycle. A research programme referred to as Interactive Product Simulation Environment for assessing Assembly and Maintainability (IPSEAM) has been established with fully immersive VE facilities at the Centre for Virtual Environments at Salford University. A prototype system has been designed and implemented to import CAD models, visualize and carry out assembly and disassembly operations. This research explores VE as a potential useful tool for maintenance simulation, which involves interactive assembly and disassembly operations being performed on a virtual prototype.

A typical assembly process consists of a succession of tasks, each of which consists of joining assembly parts to form the final assembly. Parts are considered joined when the relevant contacts and alignments between parts are established. These contacts and alignments called assembly relationships can be described in terms of geometric constraints. Specification of an assembly relationship can be considered as a constraint specification and satisfaction problem. The disassembly operations involve breaking the previously defined constraints. The proposed approach presented in this paper, combines geometric constraint management techniques and precise collision detection techniques to support building and dismantling of virtual assembly models in a dynamic, efficient and intuitive manner.

The VE technology has allowed the user-interface to expand beyond the realism of the mouse and keyboard paradigm, by providing capabilities for interacting with 3D models with 3D interfaces. Today, most VEs use 3D menus floating in the scene to interact with the system as a natural extension to 2D menus. However, in applications such as an interactive assembly simulation system, conventional 2D or 3D menu-based systems for specifying required assembly joints and mating conditions are a cumbersome and time-consuming process. Therefore, as an alternative approach to menu-based systems to specify assembly joints and mating conditions, a more intuitive and efficient approach is being proposed for carrying out assembly and disassembly tasks on a VP. In the proposed approach, user is assisted with an intuitive user interface for carrying out assembly and disassembly operations through direct manipulations. The proposed approach uses precise collision detection techniques and geometric constraint management techniques to simplify the task of accurately positioning and assembling parts with in the VE. When the user manipulates the parts of the model, the system automatically identifies the surface mating conditions and provides feedback to the user. If the user provides confirmation, the relevant geometric constraints are satisfied. Once the assembly joints are defined through direct manipulation, kinematic simulation can be carried out. The collision detection and response handling techniques ensures

parts do not penetrate one another. The constraint deduction techniques assist with the real-time detection of possible constraint conditions between the colliding objects and provide feedback to the user on possible constraints. The initial prototype system has been implemented and tested using case studies provided by industry.

The scope of this paper is a review of the state-of-the-art in virtual prototyping technology, collision detection techniques and the current state of the IPSEAM. A description of the software architecture of the system and how the constraints, collision detection and user interaction are handled within the virtual world is presented. Finally, the results of the initial prototype system, identified system issues that are currently being addressed and future work are presented.

2 Related Work

This section briefly reviews the related work including, virtual prototyping, collision detection, and constraint-based modelling.

2.1 Virtual Prototyping

Traditional approach to assess maintenance issues is to use a physical prototype. Physical prototypes can be expensive to build and difficult to modify specially in the late stage of the design process. VP, which addresses the difficulties associated with physical prototypes, is capable of reducing the costs and expediting the design process.

Types of Virtual Prototypes : In industrial practice, prototypes can be categorized into three types [6] depending on the function they perform.
- *Geometrical Prototypes* - this type of prototypes represent the geometrical and visual properties (such as colour and texture etc). These are useful in discussions where the product alternatives are to be presented to non-engineers.
- *Functional Prototypes* - this type is mainly used for engineering analysis, tests, presentation and discussions about designs. Usually the prototype can be manipulated to demonstrate its main functions.
- *Technical Prototype* - this type of prototypes are used in the latter phases of the product development process. Usually Geometrical and Functional prototypes do not present the method of

manufacturing a product. These requirements are met by the technical prototypes. The prototype is used to demonstrate the manufacturability of a product and is used in production planning.

Application Areas of Virtual Prototyping:

The following are some of the major application areas in virtual prototyping [6].

- *Visual Design Review* - CAD models of the product designs are used to create the virtual prototype and used for evaluation. Examples are automotive designs, aerospace designs, power plant designs and architectural designs. The traditional approach for automotive design is to create clay models to evaluate properties such as colour, texture and reflection. The cost of building clay models during vehicle design process is very high and often there are several iterations of the design reviews.

- *Functional Simulation* - This involves the analysis of product functions like kinematics, dynamics and deformations.

- *Analysis of Simulation Results* - Virtual prototypes can be used for analysis of simulation results specially if the simulation results are associated with spatial data. For an example automotive companies such as General Motors, Volkswagen and BMW [6], uses virtual prototypes to analyze the effects of collision on new models of cars.

- *Packaging and Assembly Simulation* - Aerospace and automotive industry uses VEs for assembly simulation [6]. This involves assembly planning operations and clash detection.

- *Ergonomic Analysis* - This involves the use of virtual prototypes in an immersive VE for analyzing whether the man machine interface of the product is user friendly or not. For an example, in automotive industry, virtual seating bucks are built to investigate the designs for accessibility and functionality of the instrument buttons and displays.

- *Training* - Virtual prototypes can be used for equipment training. For example, Software Technology Branch of NASA's Lyndon B. Johnsons Space Centre has investigated the use of virtual prototypes for training astronauts to repair equipment in the weightlessness in the space with realistic dynamics [21].

- *Production Planning* - This involves the use of virtual prototypes for simulation of assembly and manufacturing process.

- *Product Presentation* - Virtual prototypes can be used for presenting products to the customer. The Virtual model can be changed according to the requirements of the customer. Matsushita's Virtual

Kitchen in Japan, uses VEs to help customers choosing appliances and furnishing for kitchens [25]. Customers bring their architectural plans to the store and a virtual prototype is created for selecting the appliances.

2.2 Collision Detection

Collision handling techniques are used to enforce solidness of objects in a VE. In general a collision phase may consists of three elements [18] : (1) collision detection, (2) contact area determination (3) collision response. The first element detects whether the objects would penetrate each other while the second uses the output of the first and determines the contact area, the third prevents the penetration if there is any collisions. The collision response may involve making the objects bouncing off each other, stopping and providing feed back via colour changes or aiudio signals etc.

A fast collision detection component is a necessary component in an interactive VE[5]. Due to length restriction the authors would like the readers to refer to [13,11,17] for more details on Collision detection techniques. IPSEAM currently uses public domain precise collision detection software library V-Collide[12] to detect collisions among the parts in the virtual enverionment.

2.3 Simulating Physical Realism

Two main approaches are being pursued by researchers to support constraint-based interaction between assembly parts: *Physically-based Modeling* and *Constraint-based Geometric Modeling.*

In physically based modeling [2,22,4], physical forces acting upon objects and their motion equations are integrated and solved at each time step, using standard numerical methods. During the simulation, collisions are detected [4,25] at each time step and forces arising from such impacts [2,22] are calculated and new initial conditions are passed to the dynamic integrator to continue the simulation. Unfortunately, this approach is computationaly intensive proecsses and therefore real-time simulation and interaction are only possible for a small number of components. Furthermore numerical instability can be a problem in this approach.

In constraint-based geometric modeling, objects are accurately positioned in terms of geometric constraints. In this domain, much research has been conducted to develop efficient geometric constraint solvers by exploiting the geometric domain knowledge together with degrees of freedom of objects.

The current constraint-based approaches can be divided into two main categories: *Equation-based* and *Geometric Constructive.*

In the equation-based approach, the constraints are described as a set of simultaneous equations and solved using numeric, symbolic or graph-based techniques. Numeric technique [19,20], solves equations using iterative methods such as Newton-Raphson. Symbolic technique [14], solves the equations through symbolic algebraic methods, such as Grobner bases [3]. Although the numeric and symbolic techniques are quite general, they can have convergence problems and are also computationally expensive, making them unsuitable for supporting interactive constraint-based operations in virtual environments. Graph-based technique first maintains constraints (equations) and variables in an undirected bipartite graph. This graph is then directed to give a sequence of constraint satisfaction. Examples of this technique can be found in [27,26].

In the Geometric Constructive approach, constraints are not translated into a unique system of equations as in the equation-based approach. Instead, a set of constructive steps is provided which place geometric elements relative to each other through rigid body transformations, according to the degrees of freedom (DOF) of the geometric entities. In this approach, DOF of geometric objects are considered as resources that are consumed by moving an object to satisfy a given constraint relative to a fixed geometry. Each constraint, upon being satisfied, reduces the DOF of an object and hence reduces the allowable rigid body motion of the object. Examples of this approach can be found in [15, 23,8]. The use of geometric knowledge, DOF of objects and graph-based techniques result in efficient constraint satisfaction algorithms in the Geometric Constructive approach. Therefore a constraint solver based on Geometric Constructive approach has been employed in this research to support constraint management within our Interactive Product Simulation Environment.

3. IPSEAM System

This section presents the system architecture of the IPSEAM, which is being developed at the Centre for Virtual Environments at the University of Salford. The system provides capabilities to import CAD models into the VE, visualise and interact with the models. The IPSEAM system architecture integrates different technologies such as constraint-based modelling, collision detection, virtual environments and user interface technology. This section presents the main system components and interfaces and explains how constraint-based modelling and virtual environment technology have been integrated together to support assembly/disassembly tasks on a VP.

As shown in Figure-1, there are two main components to this system architecture: *baseline virtual environment and the constraint manager.*

3.1 Baseline Virtual Environment

The baseline virtual environment is developed using the OpenGL Optimizer™[28]. The OpenGL Optimizer has been chosen as the graphics engine for the virtual environment due to its CAD capabilities. However, OpenGL Optimizer lacks CAD interfaces to import data into its scenegraph. Therefore a CAD interface was developed as a part of the IPSEAM project for importing CAD data into the OpenGL Optimizer scenegraph. This CAD interface is capable of importing Parasolid format CAD models into the scenegraph while preserving the integrity of the CAD data.

Once the assembly parts are loaded into the scenegraph via the CAD interface, the virtual environment interface allows the user to grab and manipulate objects in the 3D space. The system has been integrated with a SpaceOrb™ [15] six degree of freedom motion controller and a CyberGlove®[30] to support glove based interaction. The visualisation can performed via different display modes such as workstation desktop, RealityRoom™[29] : rear-projected stereo curved wall or ReaCTor™[29] : multi-walled rear-projected stereo room.

3.2 Constraint Manager

The constraint manager comprises of three main modules: constraint solver, assembly relationship graph manager and constraint management interface . Constraint management interface is the interface between the constraint manager and the baseline virtual environment. It processes the requests and directs the action to the constraint solver or the assembly graph.

Relationship Graph (RG): The Relationship Graph (RG) maintains assembly relationships between the mating surfaces of assembly parts. The RG is an undirected graph where each node represents either a geometric entity (mating surface) or a constraint. The nodes representing geometric entities are connected to constraint nodes using arcs to represent their assembly relationships.

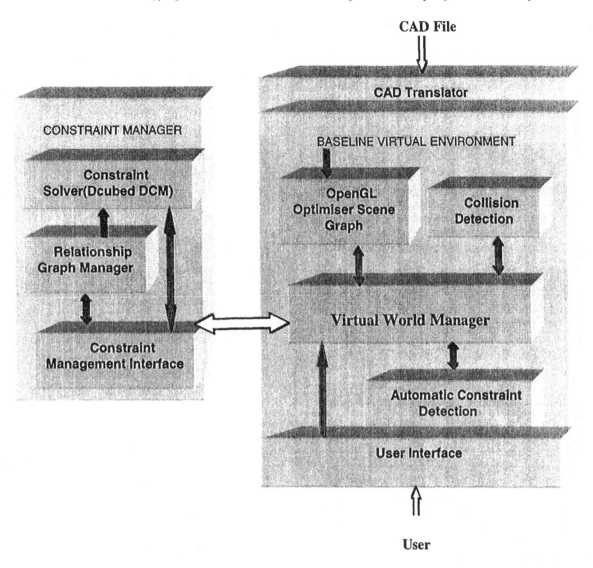

Figure-1 : Software Architecture of the IPSEAM System

Constraint Solver: The task of the constraint solver is to satisfy the specified constraints specified by the system in response to user interaction. The constraint solver used in this research is the 3D dimensional constraint manager (3D DCM) from D-Cubed Ltd, which is based on the geometric constructive approach. The constraint solver satisfies a given set of constraints and produces relative rigid body motion for assembly relationships.

3.3 Constraint Detection

A key contribution of this paper is the novel contarint detection approach developed to support assembly/disassembly tasks within the VE. This approch allows the user to precisely position the assembly parts to carry out assembly tasks within the VE without menu interaction. Most of the currently available commercial CAD systems provides limited facilities to carry out assembly tasks on digital product models using a 2D menu based approch. However this requires a large number of menu interactions to specify the geometric constraints between the parts even for a fairly simple assembly model. The use of 3D tracking devices, stereo projection video and advanced rendering hardware in VEs enable useres to visualize and interact with the models with a greater desgree of realism. However, inorder to carry out assembly and disassembly operations, user needs to specify the assembly/disassembly tasks by some means. The typical solution would be to use 3D menus within the VE to allow the user to specify assembly mating conditions. As Dai [6] points out, 3D menus within an immersive VE is not an easy means to interact with the VE system. The approch porposed in this paper uses direct object manipulation sequence to infer the assembly mating conditions automatically. Once the parts are assembled, the user is able to simulate kinematic motion of the parts through rigid body motion of the constrainted parts.

In proposed approach, the constraint detection module monitors user manipulations. While an object is being manipulated, the position of the moving object is sampled to identify new constraints between the manipulated object and the colliding objects. This is performed in four steps:

- *Step 1:* Detect precise collisions between the manipulated object and the surrounding objects and providing feedback.

- *Step 2:* Collision detection triggers constraint detection for detecting possible assembly contacts between the surfaces of the collided objects. A constraint is recognised if the geometric surface elements of the collided objects satisfy conditions of a particular constraint type. The current implementation recognises surface mating conditions such as against, concentric, cylindrical fit and spherical fit etc.

- *Step 3:* When a constraint is recognised, feedback is provided to the user by highlighting the mating surfaces. This newly identified constraint is ignored if the user continues to move the object to invalidate the condition for the constraint.

- *Step 4:* If the user decides to accept the newly identified constraint, surface description of the mating faces and the type of constraint to be satisfied are sent to the constraint manager. The constraint manager satisfies the recognised constraints. The constraint manager determines the precise position of the assembly part and its allowable rigid body motions. This information is used by the scenegraph to define the precise position of the assembly part.

Constrained Rigid Body Motion: The constrained rigid body motions of the assembly parts are used to support realistic manipulations of assemblies without breaking the existing assembly constraints. A particular manipulation of an assembly part(s) is not allowed if it is not supported by its allowable rigid body motion.

Functional Flow of the System: The following shows the functional flow of the system.

i.   User Interaction
         ⇓
ii. Detect Collisions
         ⇓
iii. Provide Feedback(Collision Response)
         ⇓
iv. Detect Geometric Constraints
         ⇓
v.   Provide Feedback(Detected Alignments)
         ⇓
vi. Accept User Confirmation
         ⇓
vii. Satisfy Constraints

4. Results

The basic framework of the IPSEAM has been implemented and tested using a series of assemblies. The current system runs on both Unix and Windows NT platforms. All the program development has been written using C++ language.

Figure 2 illustrates one of the test model, which have has successfully assembled using the IPSEAM system. Once assembled, the kinematic simulations of the assemblies can be performed interactively.

Figure 2: Test Model- Digger Mechanism

5.Conclusions and Future Work

This paper presented an approach to support the assembly/disassembly tasks on a VP through an intuitive and efficient user interface. The development of such a sophisticated environment involves bringing together many technologies such as virtual environments, constraint-based modelling, assembly modelling, CAD data representation and 3D direct manipulation techniques. This paper presets a software framework, which is being developed using constraint-based assembly/disassembly operations to support maintenance simulation on VPs. A number of case studies have been used to evaluate and demonstrate the constraint-based manipulation of assembly parts.

The future plans include further development of the system to perform maintainability assessment of several industrial case studies. During this research, further work will be carried on constraint management capabilities necessary for supporting complex maintenance tasks. A glove-based interface is being integrated with the constraint-based interaction techniques to develop an intuitive interface for supporting maintenance operations. This work is being carried out in collaboration with Rolls-Royce PLC, British Aerospace, EDS Parasolid and D-Cubed Ltd.

Acknowledgements

Thanks go to Rolls-Royce plc, British Aerospace, EDS Parasolid and D-Cubed Ltd. for providing case study material and support for this project.

References

[1]   Anderel, R. and Mendgen, R., 1996, "Modelling with Constraints: theoretical foundation and application", Computer Aided Design, vol 28, pp 155-168.

[2] Baraff, D. 1995, "Interactive Simulation of Solid Rigid Bodies", IEEE Computer Graphics and Applications, pp 63-74.

[3] Becher, T and Weispfenning, V, 1993, "Grobner Bases – A Computational Approach to Commutative Algebra", Graduate Texts in Mathematics, Springer-Verlag, New York

[4] Bouma, W.J. and Vanecek, G., 1991 ,"Collision Detection and Analysis in a Physically-based Simulation", Proceedings. Eurographics Workshop on Animation and Simulation, pp. 191-203.

[5] Cohen, J. and Lin, M. and Manocha, D. and Ponamgi, K.,1995, "I-COLLIDE: An Interactive and Exact Collision Detection System for Large-Scaled Environments", ACM Intl. 3D Graphics Conference, pp 189-196.

[6] Dai, F., 1998, "Virtual Prototyping - Principles, Problems and Solutions", Tutorial Notes : IEEE Virtual Reality Annual International Symposium (VRAIS).

[7] Fa, M., Fernando, T., and Dew, P.M., 1993 "Interactive Constraint-based Solid Modeling using Allowable Motion", ACM/SIGGRAPH Symposium on Solid Modeling and Applications, pp. 243-252

[8]   Fernando, T., Fa, M., Dew, P.M. and Munlin M., 1995, "Constraint-based 3D Manipulation Techniques for Virtual Environments", Proceedings, International State of the Art Conference (BCS) on Virtual Reality Applications.

[9] Gao, F. 1995, "Improving the Performance of Interactive Constraint-based Solid Modelling Systems", PhD Thesis, School of Computing, University of Leeds.

[10] Gao, S. 1998, "Constraint-based Solid Modelling in a Virtual Environment", ASME Computers in Engineering Conference(DETEC98/CIE 5544)

[11] Hubbard, P.M., 1995, "Collision Detection for Interactive Graphics Applications", PhD Thesis, Department of Computer Science, Brown University.

[12] Hudson, T., Lin, M, Cohen, J., Gottschalk, S., and Monaka, D., 1997, "V-Collide: Accelerated Collision Detection for VRML", VRML 97 conference.

[13] Klosowski, J. T., 1998, "Efficient Collision Detection for Interactive 3D Graphics and Virtual Environments", PhD Thesis, Department of Applied Mathematics and Statistics, State University of New York at Stony Brook.

[14] Kondo, K., 1992, "Algebraic Method for Manipulation of Dimensional Relationships in Geometric Models", Computer-Aided Design, Vol. 24, pp.141-147.

[15] Kramer, G. A., 1992, "A Geometric Constraint Engine", Artificial Intelligence, Vol. 58, pp. 327-360.

[16] Labtec 2000, "SpaceOrb-360 Six Degree Motion Controller, Labtec® ", on WWW: http://www.spacetec.com (Feb 2000)

[17] Lin, M.C. and Gottschalk, S., 1998, "Collision Detection Between Geometric Models: A Survey" ,IMA Conference on Mathematics of Surfaces , on WWW :http://www.cs.unc.edu/~dm.

[18] Lin, M.C. and Monaka, D. and Cohen, J. and Gottschalk, S. , 1996, "Collision Detection : Algorithms and Applications ", Algorithms for Robotics Motion and Manipulation , pp 129-142.

[19] Lin, V.C., Gossard, D.C. and Light, R.A., 1981, "Variational Geometry in Computer Aided Design", ACM Computer Graphics (SIGGRAPH'81), Vol 15, pp. 171-175.

[20] Light, R. and Gossard, D., 1982, "Modification of Geometric Models through Variational Geometry", Computer-Aided Design , Vol 14, pp. 209-214.

[21] Loftin, R.B. and Kenny, P.J.,1994 , "The Use of Virtual Environments for Training the Hubble Space Telescope Flight Team", Interservice/Industry Training Systems and Education , NASA, Johnson Space Center, on WWW :http://www.vetl.uh.edu/proj.

[22] Mirtich, B. and Canny, J., 1995, "Impulse-based Simulation of Rigid Bodies", Procedings. Simposium on Interactive 3D Graphics.

[23] Owen, J.C., 1991, "Algebraic Solution for Geometry from Dimensional Constraints", ACM/SIGGRAPH Symposium on Solid Modeling Foundations and CAD/CAM Applications, pp. 397-407.

[24] Ponamgi, M.K., Monocha, D., Lin, M.C., 1997, "Incremental Algorithms for Collision Detection Between Polygonal Models", IEEE Transaction on Visualization and Computer Graphics, Vol.3, No.1, pp. 51-64.

[25] Ressler, S., 1994, "Applying Virtual Environments to Manufacturing", National Institute of Standards and Technology, Technical Report -NISTIR 534.

[26] Sannella, M., 1993, "The SkyBlue Constraint Solver and its Applications", First Principles and Practice of Constraint Programming Workshop (PPCP'93), Newport, RI.

[27] Serrano, D. and Gossard, D,. 1992,. "Tools and Techniques for Conceptual Design", Artificial Intelligence in Engineering Design, Vol. I, C. Tong and D. Sriram (Eds), pp. 71-116.

[28] SGI, 1999, "OpenGL Optimizer™ Software", on WWW: http://www.sgi.com/software/optimizer/ (Nov 1999).

[29] Trimension, 2000, " Reality Cubic Tracked Environment(ReaCTor), RealityRoom, Trimension Inc", on WWW : http://www.trimension-in.com (Feb 2000)

[30] VirTec, 2000, "CyberGlove, Virtual Technologies-inc", on WWW: http://www.virtex.com (Feb 2000)

Virtual Construction Materials As A Decision-Making Support For Material Routing

J.L. Yang and L. Mahdjoubi
School of Engineer and Built Environment of University of Wolverhampton
Wolverhampton, WV1 1SB, UK
Tel. +44 (0) 1902 321000

Abstract

Materials movement is an important consideration for an effective project management planning in construction sites. However, conventional planing tools are often ineffective in aiding decision taking for materials routing in construction sites. Despite the growing use of Information technology in construction, this new technology has not yet affected this field of construction management.

This paper is to describe an ongoing research at the University of Wolverhampton, which attempts to develop a decision-support system to help construction sites managers to make informed decision about materials movement on sites.

This first stage of this work involves the integration of computer-aided design (CAD), Geographical Information Systems (GIS) and Fuzzy-logic (FZ). The software ability allows completely transparent (to the user) links to external commercial applications including AutoCAD, AutoMAP, and MS Access. This work demonstrated that the availability of these tools could greatly enhance the reasoning and decision taking process on material routing issues.

1. Introduction

Planning access routing for routing within a construction site is an important consideration in the development of effective material movement. Yet this field of study has received a very little attention.

Despite recent efforts primarily aimed at helping site managers to make informed decisions, this field of study is still largely underdeveloped.

However, an emerging body of research has attempted to address this issue. These studies are primarily aimed at improving materials management in construction sites. Nevertheless, little work has yet attempted to improve the decision taking process of materials routing in construction sites.

This paper reports on current progress to develop a decision support for materials routing in construction sites. Despite it is still in its early stages, this system has proven very beneficial for site managers.

2. Decision support systems and materials management

Decision-making is a processing of choosing among alternative course of action for the purpose of attaining a goal or goals (Simon 1977. Decision making tools are defined as interactive computer-based systems, which help decision-makers utilise data and model to solve unstructured problem (Robbins, 1991).

A great deal of work was carried out in the development of decision-making systems. Research focussed on decision-making tools to aid manager in construction project management processing. This requires a combination of the intellectual resources of individual with the capabilities of the computer to improve the quality of decisions (Kenn and Raiffa, 1978). Decision making tools intends to support managerial decision-makers in a semi-structured decision situation. These systems are designed to extend the capabilities of decision-makers but not to replace their judgement.

The development of decision-making systems is by no means new. Some systems have been widely used in decision support system such as Knowledge-Based Expert systems, Neural Artificial Intelligence systems and so on. These systems were designed to deal with specific practical problems in industry.

Fuzzy logic and expert system are now widely used in the decision-making processing in the practice. The construction industry has attempted to apply these modern information technologies to such as its project management and cost control. Using the fuzzy logic to select suitable design approaches was developed at the University of Nebraskz NE (Peak, etc. 1992). It suggested that a multicriterion decision-making methodology could be applied for selecting the best design/build proposal under uncertainty which relate the high technical factors and low construction cost

Expert systems were used in intelligent decision making processing and optimisation in construction management for many years. Amirhanian and Baker (1992) reported that their expert system acts as an equipment selector for choosing proper earth moving equipment in order to meet construction project budget and schedule requirements. Furthermore, an integrated simulation with expert system to choose and simulate the construction equipment earth-moving under a given set of job condition has been developed at Northeastern university, USA (Touran, 1990). The expert system acts as a selector and as an interface between the user and simulation software package.

However, materials management in construction sites has been lagging behind in this field. This industry is still characterised by the lack of advanced techniques.

Jeljeli and Russell (1995) suggested that use of decision analysis approach is to cope with uncertainty in construction industry. The approach has implemented a model to analyse the uncertainty in decision making for construction clear up in sites. In the context of construction project management, site managers is tend to identify a few key determining factors or more widely certain factors to control project budget by using neural network technique to decision making processing (Chua, Kog, Loh and Jaselskis, 1997). WorkPlan (Choo, Tommelein, Ballard and Zabelle, 1998) has been created to systematically develop a week work plan with using a database program. It considers possible factors and uncertain reasoning to choose the best scheduling to construction work. It uses database program during the decision-making processing.

Other researchers (Wu and Hadipriono, 1992) has presented a new method to estimate the duration of construction project activities called duration decision support system by using the fuzzy modus technique to evaluate the impact of different factors on activities duration. It uses linguistic values of these factors to represent in fuzzy set models, and allows a scheduler to partially match evidence with rules and select the suitable duration of construction project.

In recent years, neuro-fuzzy approach has dominated research in computerised construction management. Yu and Skibniewski developed (1999) a multi-criterion decision model for quantitative constructability analysis, which is based on a neuro-fuzzy knowledge. They suggest that with this system, the constructability can be quantified, measured and improved. It also incorporates construction manager's subjective preference information. They have used neuro-fuzzy network-based approach in providing a mechanism to trace back factors causing unsatisfactory construction performance and the necessary feedback to construction engineers for technology innovation (Yu and Skibniewski, 1999).

Most work in this area has been conducted at the University of Michigan. A series of tools have been developed to aid materials handling and site layout control. MoveSchedule was designed as a planning tool for scheduling space use on construction sites. It was suggested that this system solves constrains dynamic layout problem. Its objective is to minimise resource transportation and relocation costs (Zouein, 1996). MovePlan system was developed to assist planners and managers in the creation of sequences of layouts corresponding to an activity schedule. This system informs suppliers about material unloading locations and workers about material storage in construction sites (Tommelein, 1994).

Hazem and Bell (1995) suggested that material management systems should be integrated with computer systems that are used for design and scheduling. They proposed an Object Oriented Methodology (OOM) data structure for a materials-management system.

Some systems used fuzzy-expert systems in the implementation of decision taking process in construction project management. The resource allocation in construction project management (CPM) is a key issue, which is related to project scheduling. A fuzzy expert system has been used in CPM to determine the most effective factors, and generate the best selection of allocation in CPM (Chang, Ibbs and Crandall, 1990).

Despite these developments the issue of materials movements in construction sites has not received adequate attention. Varghese (1992) pioneered one of the few studies in this area, which involved routing large vehicles in construction sites. Using proprietary software in Computer Aided Design (CAD), Geographical Information Systems (GIS) and a Spreadsheet, this system has developed links to other custom programs to evaluate access scenarios of large vehicles to construction sites. Varghese et al. (1995) found that "access considerations have a strong influence on the layout of temporary facilities, location of assembly and storage yards, layout of roads, construction sequence, extent of assembly, and design of overhead and underground elements crossing the roads." Material storage location is allocated for short or long term use according to the duration of construction activities, which affects the routing decision.

Cheng and O'Conner (1996) developed an automated site layout system for construction temporary facilities. This system called ArcSite uses Geographic Information Systems (GIS) integrated with Database Management Systems (DBMS) to assist designers in identify suitable areas to locate the facilities in construction site.

The research described in this paper aims to develop a system, primarily aimed at the selection of the best routes for materials movements in construction sites. The "Virtual Construction Materials Router" combines fuzzy logic techniques with CAD and GIS to choose the best routes for materials to help planners to make a sound decision before materials are shifted from one location to another. In short VCMR is involved in the planning, executing, and controlling of routing of materials in sites.

3. Virtual Construction Materials Router

The VCMR is a novel tool to aid site managers and planners to select the best movement of materials in construction site. It uses advanced GIS-fuzzyTECH system to determine certainties and uncertainties during the selection of materials routing. It takes into consideration site layout, site access, temporary storage and new building blocks to choose the suitable materials' routing during construction work.

3.1 System Architecture of VCMR

Figure 1 System Architecture

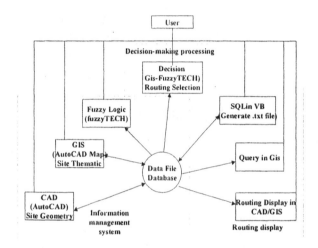

Figure 1 shows the main system structure of VCMR. It includes information management system, decision-making system and routing display system.

3.2 Information management

Information management system is a data representation. First, site managers or planner draw or digitise the site layout in CAD system (AutoCAD). This spatial data is then made usage in a GIS package (AutoCAD Map). This includes linking attributes to existing digital files. Subsequently the site thematic and geometry data is exported to database management system (MS access). Using Visual Basic (VB) programming, this information is then restructured and reorganised, ready for decision processing. All these steps are performed automatically between the various applications.

3.3 Fuzzy decision processing

Decision-making process is an integrated program to process the data and generate the final decision. The input variables are from a database management system.

This module enables users to perform decision analyses. By weighing various variables such as location of storage of materials and loading points, destination, site configuration and the like, the system is capable of generating various alternatives and suggesting the preferred route.

3.4 Graphical routing display

The resulting information from fuzzy-logic is appropriately displayed in a graphical format GIS/CAD package. This involves the uses Structure Query Language (SQL) to extract information about the final decisions, as suggested by the system. This information is displayed in a 2D format by connecting nodes between the relevant nodes in the CAD/GIS system.

4.   An Application of VCMR

4.1   User Interface

This section provides a practical application of VCMR. First, the site information was retrieved from data management system. The input variables were then exported to fuzzyTECH for decision analysis. The information is processed by the fuzzy logic systems and a final decision is suggested as a result.

Figure2 illustrates a user interface of VCMR, which includes site information management and data, ranges control bas and decision query command. Site thematic and geometry function can be edited and updated. The result of the analysis can be queried using SQL.

Figure3 & 4 show the information of site thematic and geometry in sub-data management system.

Figure 2

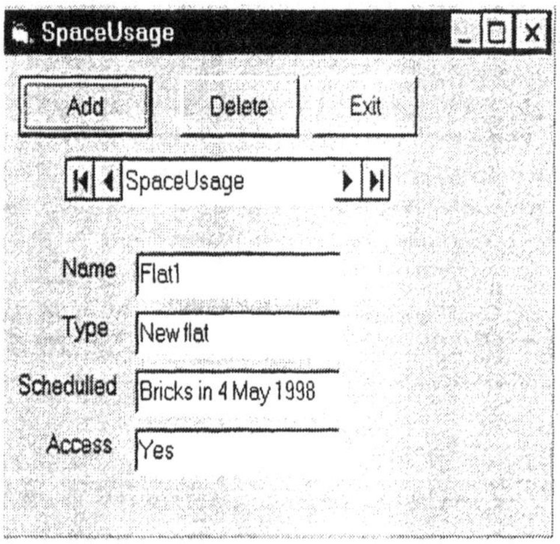

Figure 3

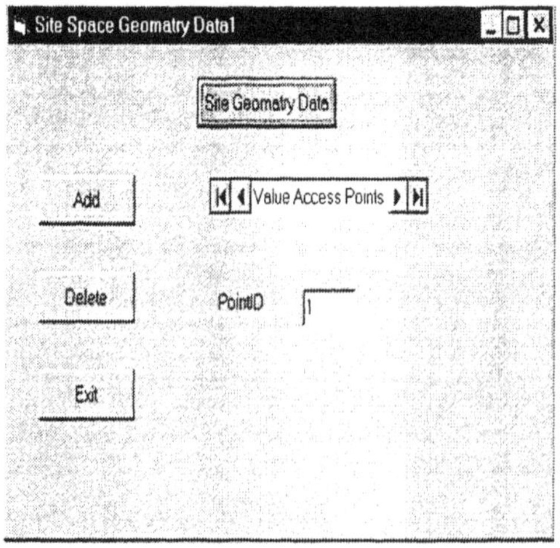

Figure 4

4.2 Decision Making Process

The decision-making process takes into consideration materials storage, loading and destination.

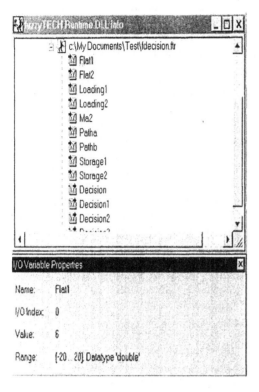

Figure 5

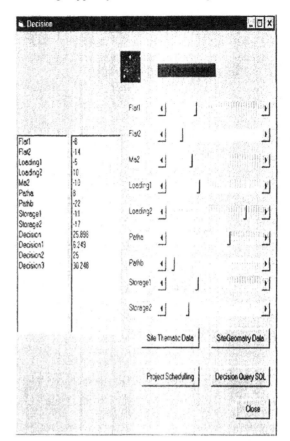

Figure 6

Figure 6 illustrates how the decision is modified according to the input variables. The output of the decision is stored in a file, which can be accessed using SQL.

The visual result of the decision is displayed in CAD/GIS package, simulating the suggested route for shifting the materials. This enables the manager to make the appropriate decision.

4.2.3 Routing Display

Figure 7 shows site layout and possible access network (in Red), storage (11 &12), loading (3, 4, 5, 6, 9 &10), new building block (1,2,7 &8) and the other usage (13 &14) in the site.

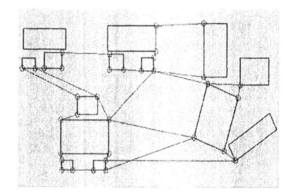

Figure7

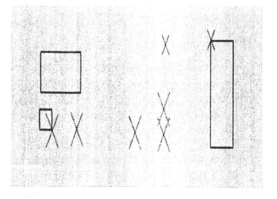

Figure 8

Figure 9 is to link these elements and display the final routing of materials. After whole the process, user can see the new building (block 1) need materials,
and know that the best is that where materials are from storage1. The displayed access is suitable routing to new building (Block1) for the materials movement.

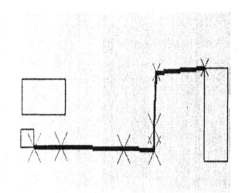

Figure 9

This example demonstrates that the VCMR is new method in construction materials management. It combines advanced GIS/CAD and fuzzy logic system to predict the future routing for materials movement in site.

5. Conclusion

Materials routing has suffered from a lack of interest for years. Current decisions in construction are primarily based on intuition and experience of the site manager.

The research has demonstrated that an appropriate decision support system can reduce uncertainty and help managers to make informed decisions to shift materials in construction sites. Although, this research is still at its early stages, its preliminary results are very promising.

Reference:

1. S.N. Amirkhanian and N.J. Baker, Expert system for equipment selection for earth-moving operations, Journal of Construction Engineering and Management, 118(2), pp318-331, 1992.
2. T.C. Chang, C. William and K.C. Crandall, Network allocation with support of fuzzy expert system, Journal of Construction Engineering and Management, 116(2), pp239-259, 1990.
3. M.Y. Cheng and J.T. O'Oconnor, ArcSite enhanced GIS for construction site layout, Journal of Construction Engineering and Management, 121(4), pp438-445, 1995.
4. D.K.H. Chua, Y.C. Kog, P.K. Loh and E.J.Jasflskis, Model for construction budget performance-Neural network approach, Journal of Construction Engineering and Management, 123(3), pp214-222, 1997,
5. E. M. Hazem and L. C. Bell, "Object-oriented methodology for material management system," Journal of Construction Engineering and Management, 121(4), pp438-445, 1995.
6. N.Jeljeli and J.Russell, Coping with uncertainty in environemnt construction: Decision analysis approach, Journal of Construction Engineering and Management, 121(4), pp370-380, 1995.
7. O. F. Olusegun, O. O Jacob and J. L. Dennis, "Interactions betweenconstruction planning and influence factors," Journal of Construction Engineering and Management, 124(4), pp245-256, 1998.
8. J.H. Peak, Yong W. Lee, Selection of Design/Building Proposal Using Fuzzy-Logic System, Journal of Construction Engineering and Management, 118(2), pp303-317, 1992.
9. S. R. Robbins, "Management," 3rd ed. Englewood Cliffs, NJ: Prentice-Hall, 1991.
10. H. J. E. Rodda S. Demuth, and U. Shankar, "The application of a GIS-based decision support system to predict nitrate leaching to groundwater in southern Germany." Hydrological sciences journal-Journal des sciences hydrologiques, vol.44, No.2, pp221-236, 1999.
11. H. Simon, the New Science of Management Decisions. Rev. ed, Englewood, Cliffs, NJ:Prentice-Hall, 1977.
12. I. D. Tommelein, R. E. Levitt and B. H. Roth, "Site-layout modeling: how can artificial intelligence help?" Journal of Construction Engineering and Management, 118(3), pp594-610, 1992.
13. A. Touran, Integration of simulation with expert system, Journal of Construction Engineering and Management, 116(3), pp480-493, 1990.
14. K. Varghese, " Automated route planning for larger vehicles on industry construction sites," Faculty of the Graduate School of University of Texas at Austin, 1992.
15. P. P, Zouein, "MoveScheduling: a planning tool for scheduling space use on

construction sites. Ph.D. Dissertation, Civil and Environment Engineering of University of Michigan, 1996.

16. Ribarsky, W et al., "Visualisation and analysis using virtual reality. IEEE Computer Graphics and Applications, January, pp10-12, 1994.

17. W. D. Yu and M. J. Skibniewski, "A neuro-fuzzy computational approach to constructability knowledge acquisition for construction technology evaluation." Automation in construction, Vol.8 pp539-552, 1999.

18. W. D. Yu and M. J. Skibniewski, "Quantitative constructability analysis with a neuro-fuzzy knowledge-based nulti-criterion decision support system." Automation in construction, Vol.8 pp533-565, 1999.

CHAPTER 8

Multi-Agents Architectures

Collaborative Design among Different Fields in Mobile-Agent Environments

Masataka Yoshimura

Graduate School of Engineering, Kyoto University, Kyoto, Japan 606-8501

Kouji Takahashi

Graduate School of Engineering, Kyoto University, Kyoto, Japan 606-8501

Abstract

Today's product design problems are increasingly large-scaled due to the diversification of engineering technologies and consumer needs. Designers must meet these needs rapidly, evaluate wide-ranging engineering requirements at the product design stage, and develop designs that distinguish their product from others in the field. This type of work often involves multi-disciplinary large-scaled optimization problems. When more than one discipline is involved in product design decision making, technical experts in different fields or divisions can benefit from sharing specific knowledge as they cooperatively and concurrently solve such problems. In this study, methodologies using mobile-agent systems to support collaborative product design activities are constructed.

1 Introduction

At present, the scale and complexity of design problems are increasing, as industrial technologies become more complicated and consumer needs more diversified. Designers must develop efficient and effective design strategies that quickly respond to the needs of sophisticated users. In order to realize optimum product designs, the performance and marketability of a potential product must be examined and evaluated at the earliest possible stage, by modeling of a variety of ideas and possible solutions. To meet such needs requires design systems that can draw upon widely dispersed resources during the evaluation stage, while maintaining an up-to-date awareness of competing products. To achieve optimum design solutions, these design activities should be treated as a large-scale multiobjective decision-making problem.

In such design problems, specialized areas belonging to different groups become mutually involved, as a number of engineers having different areas of expertise participate in designing the best possible product. These individuals often have unique technical opinions and working strategies, which affect the ease with which optimum design solutions can be achieved. Design problems can become further complicated when the number of experts increases, each of whom brings a different level of expertise and unique personal approach to their field. In such cases, collaborations should be carried out so that engineers can utilize their specialized knowledge cooperatively, in parallel, simultaneously, and with the greatest possible efficiency. This can be more easily achieved when information networking technology is available.

The difficulty of group decision making in such situations requires new information infrastructures that can assist and accelerate effective design activities while supporting multiple designer input and judgment.

As research related to the subject of this paper has shown, collaborative and cooperative design problems [1-6] have enjoyed increasingly important status as research subjects, with the common goal of more effectively realizing the benefits of concurrent engineering [7,8]. Collaboration among different divisions, different enterprises, or groups of engineers possessing diverse knowledge is considered to be one of the most promising methods for stimulating product design and manufacturing activities. Collaborative decision-making among designers and manufacturers has been discussed [9], where it was shown that this method can achieve more preferable design solutions. The advantage of exploiting synergy effects that arise during cooperative endeavors with a group of experts having different knowledge, via computer networks, is another important concurrent engineering subject [10]. In addition, methodologies to support group decision-making are also attractive subjects in the development of effective collaboration strategies among experts across different technological fields [11].

At present, networked computer systems are becoming increasingly powerful, and widespread. In order transfer the burden of routine and even semi-intelligent tasks from human being to computers, as much as possible, agent technologies deployed in information network environments have been actively studied [12]. Simple, but intelligent agent systems have been developed for use in such cooperative working environments [13,14].

Technological environments for realizing collaborative designs via computer networks have been the subject of research for considerable time, and are almost ready for actual use in the field. Truly practical methodologies and systems for assisting the solution of collaborative design and manufacturing problems are eagerly awaited.

This paper further develops the concept of concurrent engineering and a design support system that achieves more preferable designs by utilizing "agent technology", a next-generation data communication technology. This study examines methodologies using mobile-agent systems to support collaborative product design activities, where such mobile-agents are software programs operating across a computer network. In this study, the agent systems used in the optimization processes and information systems are implemented in the Java programming language.

Articulated robot design problems are employed to demonstrate the collaborative design processes taking place among different fields in mobile-agent environments. Specifically, a user of the product, together with experts in structural and control design, collaboratively engage in the design of robotic products, using mobile-agents in a networked environment.

2   Construction of mobile-agent systems

The agents used here are deployed in software and support human-type activities such as decision making, information searching, negotiation, and work on behalf of another person, all within networked information spaces. Mobile-agents are software programs that move within network spaces, and perform the work of agents at distant network nodes. Fig.1 shows a conceptual illustration of collaborative designs using mobile-agents, where mobile-agents operate across distant network nodes to perform a variety of agent-type tasks. Here, two designers are conducting collaborative work that is greatly assisted by mobile-agents.

In conventional client-server systems, the software that does the actual work exists on the server (the distant node when viewed from the client side), but the work is controlled at the client side node. During processing, the communication line between these nodes is kept busy.

In the proposed mobile-agent method, however, the agent software itself is transferred to distant nodes. The server node can have multiple agents, each operating autonomously on behalf of the individuals involved in the design problem. Software agents placed on client side nodes are transferred to the server side node, where they communicate with server software that orchestrates the agents' operations. After a software agent has been transferred to another node, communication with the originating node can stop while the agent performs its work autonomously. The communication line between the client and server nodes need only operate when the agents are transferred. Communication costs can therefore be kept low, since communication lines are idle except during these transfers.

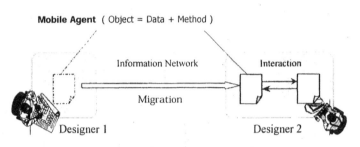

Figure 1   Conceptual illustration of collaborative designs using mobile-agents

3   Systems using mobile-agents to support collaborative product design activities

Fig.2 shows a conceptual illustration of a multi-agent system supporting collaborative product design. The process begins when a member of the product design group makes an outline or sketch that incorporates users' requests for a product. These requests can be essentially devoid of design specifics at the outset, but the inclusion of some specifics is beneficial. Experts in each engineering division of the design group then use their specialized knowledge to cooperatively construct concrete design solutions that attempt to realize the user's requests. Using the multi-agent system, individual members of the design group can acquire useful and concrete design information beyond their areas of expertise or engineering knowledge. Each expert obtains provisional optimum solutions by performing optimization for decision variables and evaluative factors belonging to his or her special engineering field, according to the user's ideas and requests for the product under consideration. Then, coordinated by mobile-agents, the provisional designs obtained by each expert are gradually integrated into overall product design solutions for the design team as a whole.

The multi-agent system constructed in this study is composed of mobile-agents, an optimizer agent for structural analysis, and an optimizer agent for control analysis, as shown in Fig.3.

Design proposals offered by other experts are checked and evaluated according to the recipient's criteria. If the proposals satisfy the criteria, the proposals are accepted. If they do not, the recipient returns new proposals to the partner. An outline of the negotiation processes is shown in the flowchart in Fig.4. By repetition of the negotiation processes, provisional designs obtained by each expert are gradually integrated into overall product design candidate solutions. In the applied examples explained later, genetic algorithms are used to perform each optimization.

To explain the design coordination processes in more detail, examples of simple robot designs are given in section 4. A structural model of the robot is shown in Fig.5.

The following three members are engaged in the robot link design problem:

      1. user
      2. expert structural design engineer
      3. expert control design engineer

The user is assumed not to have any technical knowledge concerning the product design, only a general idea of what is needed.

Each of the above experts only has knowledge in his or her specific engineering field. Table 1 shows the design variables for which each "optimizer", i.e. combination of a person and a software agent, is involved. The expert structural design engineer selects a motor from the standpoint of structural characteristics, while the expert control design engineer selects a motor from standpoint of control performance factors.

A multi-agent system must be constructed and then adjusted to accurately reflect the demands of the user who will buy the product. In the applied example, the user's strongest concern is the product's size, so this parameter is incorporated into the initial stage of the optimizing process, which affects subsequent processing. For this example, the user's desired size is set as an initial design variable and used during the optimization process, which reduces the chance that the final design solution will be unattractive to the user.

Based on user demands, the structural system expert and his software agent calculate the structural design specification. After the first iteration, the solution does not provide for robot link movement, since an evaluation for link control was not carried out, although the structural evaluation was. In other words, the structural system optimizer outputs a provisional design solution only in terms of a structural evaluation.

Mobile-agents receive and retain design specification data calculated by the optimizer when they are transferred to other optimizer agents on other nodes. After arriving at a different node, or optimizer location, the data are

transmitted and the next type of optimizer process is started. During the design collaboration, the transmission and processing of data by mobile-agents is of great importance. For example, design specification data handled by the mobile-agents are chronologically accumulated in agent databases (Vector Objects), enabling a grasp of the design solution progress among the optimizers during the crucial negotiation process. The agents transmit this type of data to the structural optimizers during the iterative process of the design optimization, until both optimizers agree upon a solution.

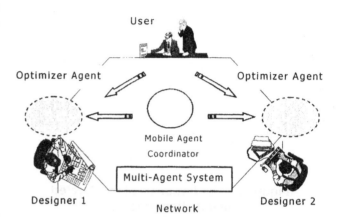

Figure 2 Conceptual illustration of a multi-agent system to support collaborative product design activities

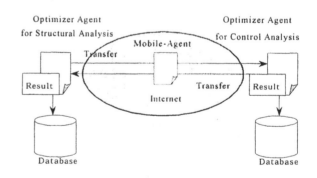

Figure 3 Mobile-agents and stationary agents in collaborative design environments

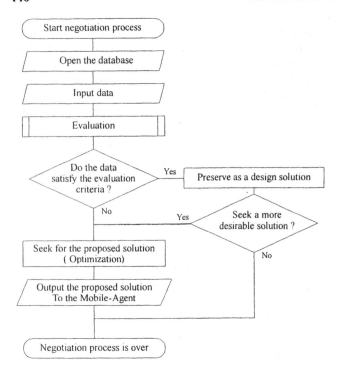

Figure 4   The flowchart of the iterated negotiation process

Table 1   Design variables in which each optimizer is involved

Structural optimizer	Design Variables	Control optimizer
Design variables involving structural optimization	Length of link1	
	Length of link2	
	Code no. of Motor	Design Variables involving Control optimization
	Code no. of Motor	
	Cross-sectional shape code no. of link1	
	Cross-sectional shape code no. of link1	
	Material code no. of link1	
	Material code no of link1	
	Feedback gain kp1	Design Variables involving Control optimization
	Feedback gain kp2	
	Feedback gain kd1	
	Feedback gain kd2	

Applied examples

Design problems of articulated robots such as shown in Fig.5 are used to demonstrate collaborative design processes among different fields in mobile-agent environments. Three different people, the user of the product, a structural design expert, and a control design expert, are collaboratively engaged in the product designs across a networked space, as shown in Fig.2. Fig.6 shows the constructed multi-agent system. In this study, collaborative product design activities using a multi-agent system were realized on a LAN (Local Area Network).

The criteria for the structural system optimizer are the manufacturing cost and deflection at the end-effector point, while the criteria for the control system optimizer are the operational accuracy and the electric power requirement.

The structural design engineer determines the material and dimensions of the robot links, while the control design engineer determines the feedback gains. Both of the expert engineers are engaged in the selection of motors. The structural design engineer selects a motor based on attributes such as mass and moment of inertia from among the motor attributes, while the control design engineer selects a motor based on attributes such as the rated torque, maximum torque, torque constant, induced resistance, moment of inertia, mass, and gear ratios. Eight types of motors (Code Nos. 0 to 7) are considered as alternatives. The motor attributes are the rated torque, weight, cost, maximum torque, torque constant, induced voltage, armature resistance, moment of inertia, and gear ratio.

Alternative cross-sectional shapes for the links include

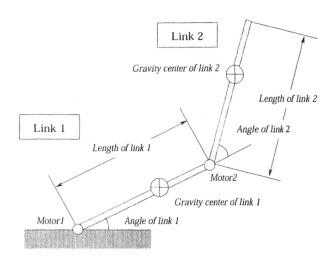

Figure 5   Diagram of a 2-degrees of freedom robot

those that are solid circular (Code No.0), hollow circular (Code No.1), solid rectangular (Code No.2), and hollow rectangular (Code No.4). Alternative link materials include steel (Code No.0), cast iron (Code No.1), aluminum alloy (Code No.2), and magnesium alloy (Code No.3). The attributes considered for each material include Young's modulus, the shearing modulus, material density, and the material unit price. For controlling link motions, feedback gains are included in the design variables.

The product manufacturing cost is obtained by summing the material purchase cost, link manufacturing processing cost, and motor purchase cost. The operational accuracy is evaluated by monitoring the maximum deviation from the specified path of the end-effector point during the operating time. The power consumption is calculated by summing the electric power consumed by each motor.

The design specifications are defined according to user's requirements for the product, as shown in Table 2.

Negotiations are started between the structural design and control design experts via their respective mobile-agents and stationary agents. Table 3 shows results of the negotiation process. As can be seen in Table 3, a possibly satisfactory solution was not achieved until the sixth iteration.

The negotiation histories are stored in mobile-agent memories. In addition, candidate design variables obtained during each transfer of information among the mobile-agents are stored in mobile-agent databases. This information may be provided to each expert and/or presented to the user, and is used in determining the final product design specifications.

The control expert conducted optimization for feedback gains for the motors proposed by the structural optimizer operating on behalf of the structural expert, and the design solutions (design solution 1) shown in Table 4 were obtained. Criteria values for design solution 1 are shown in Table 5.

The design solutions were obtained by optimizing the feedback gain design variables provided by the control expert according to the motor type recommended by the structural design expert. Thus, the type of motor was not subject to negotiation by the mobile-agents.

The design solution 1 results were obtained based on the acceptance of the motor type chosen by the structural optimizer. In order to obtain more preferable design solutions from the standpoint of control performance, optimization including motor design variables was conducted. This solution is shown in Table 6, as design solution 2. Criteria values for design solution 2 are shown in Table 7.

By comparing design solutions 1 and 2, it can be seen that the operational accuracy variable of the control

performance is greatly improved for design solution 2, although the manufacturing cost and deflection are less desirable than in design solution 1. However, the manufacturing cost and the other product characteristics satisfy the user's requirements, so design solution 2 is a candidate solution.

Based on the design solutions proposed by the control design expert, optimization was conducted for the structural design variables of the links by the structural optimizer. These design solutions are shown in Table 8, as design solution 3. In addition, feedback gain variables were optimized based on design solution 3, and the results (design solution 4) are shown in Table 9. Table 10 shows criteria values for design solution 4. Similar information exchange processes can be repeated to obtain better design results, with diminishing degrees of improvement. Here, design solution 4 was adopted as the final robot design, the outcome of a coordinated effort among all of the experts and the user.

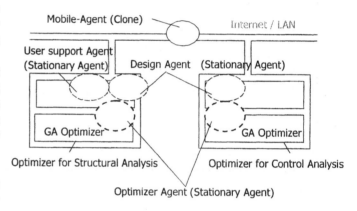

Figure 6   Construction of a multi-agent system

Table 2   User's requested values for the robot

User's requested values	
Length of link 1 [m]	1.0
Length of link 2 [m]	1.0
Manufacturing cost [yen]	580,000

Table 3 Results of the negotiation processes

	Number of Negotiation	Arm 1 length	Arm 2 length	Motor 1 Code no.	Motor 2 Code no.	Arm 1 Cross section Code no.	Arm 2 Cross section Code no.	Arm 1 Material Code no.	Arm 2 Material Code no.	Feed back Gain kp 1	Feed back Gain kp 2	Feed back Gain kd 1	Feed back Gain kd 2	Other expert's Evaluation	Stage of User Compromise
Structure	0	1.025	1.025	0	0	1	2	1	1					Control　　No	0
Control	0	1.025	1.050	7	7	1	2	1	1	181	64.0	1.41	1.00	Structure & User　No	0
Structure	1	1.150	1.250	0	6	0	0	0	2					Control　　No	1
Control	1	1.150	1.250	7	7	0	0	0	2	128	128	1.00	1.00	Structure & User　No	1
Structure	2	1.350	1.325	5	6	3	3	0	2					Control　　No	2
Control	2	1.350	1.325	7	7	3	3	0	2	128	181	2.00	1.00	Structure & User　No	2
Structure	3	1.025	1.000	2	0	0	1	2	0					Control　　No	3
Control	3	1.025	1.000	7	4	0	1	2	0	181	181	1.41	1.00	Structure & User　No	3
Structure	4	1.000	1.125	4	2	3	3	1	1					Control　　No	4
Control	4	1.000	1.125	7	6	3	3	1	1	181	181	1.00	1.41	Structure & User　No	4
Structure	5	1.350	1.275	4	0	0	2	3	2					Control　　No	5
Control	5	1.350	1.275	7	5	0	2	3	2	181	181	1.00	1.41	Structure & User　No	5
Structure	6	1.050	1.050	6	5	1	1	2	0					Control　　Yes	6

Table 4 Design solution 1 for the robot

Design Variables	Solution Value or Code No.	
Length of link 1 [m]	1.050	
Length of link 2 [m]	1.050	
Cross-sectional shape 1	1	Partial design solutions recommended by structual optimizer
Cross-sectional shape 2	1	
Material of link 1	2	
Material of link 2	0	
Code No. of Motor 1	6	
Code No. of Motor 2	5	
Feedback gain kp1	1.414	Design solutions proposed based on control optimization
Feedback gain kp2	45.25	
Feedback gain kd1	45.25	
Feedback gain kd2	2.000	

Table 5 Criteria values for design solution 1

Criteria values	
Manufacturing cost (yen)	522,280
Deflection (10^{-4}m)	2.08
Operation accuracy (10^{-4}m)	2.00
Power consumption (W)	2551.052

442

Table 6   Design solution 2 for the robot

Design Variables	Solution Value or Code No.	
Length of link 1 [m]	1.050	Partial design solutions recommended by structual optimizer
Length of link 2 [m]	1.050	
Cross-sectional shape 1	1	
Cross-sectional shape 2	1	
Material of link 1	2	
Material of link 2	0	
Code No. of Motor 1	7	Design solutions proposed based on control optimization
Code No. of Motor 2	4	
Feedback gain kp1	181.0	
Feedback gain kp2	181.0	
Feedback gain kd1	1.0	
Feedback gain kd2	1.0	

Table 7   Criteria values for design solution 2

Criteria values	
Manufacturing cost (yen)	522,280
Deflection (10^{-4}m)	2.08
Operation accuracy (10^{-6}m)	2.94
Power consumption (W)	3382.022

Table 8   Design solution 3 for the robot

Design Variables	Solution Value or Code No.	
Length of link 1 [m]	1.025	Partial design solutions recommended by structual optimizer
Length of link 2 [m]	1.025	
Cross-sectional shape 1	2	
Cross-sectional shape 2	0	
Material of link 1	1	
Material of link 2	3	
Code No. of Motor 1	7	Design solutions proposed based on control optimization
Code No. of Motor 2	4	

Table 9   Design solution 4 for the robot

Design Variables	Solution Value or Code No.	
Length of link 1 [m]	1.025	Partial design solutions recommended by structual optimizer
Length of link 2 [m]	1.025	
Cross-sectional shape 1	2	
Cross-sectional shape 2	0	
Material of link 1	1	
Material of link 2	3	
Code No. of Motor 1	7	
Code No. of Motor 2	4	
Feedback gain kp1	90.51	Design solutions proposed based on control optimization
Feedback gain kp2	181.0	
Feedback gain kd1	16.00	
Feedback gain kd2	1.000	

Table 10   Criteria values for design solution 4

Criteria values	
Manufacturing cost (yen)	554,917
Deflection (10^{-4}m)	2.08
Operation accuracy (10^{-4}m)	1.98
Power consumption (W)	4999.805

5   Concluding remarks

In this paper, product design methodologies utilizing mobile-agents operating in a networked computing environment were proposed to allow the collaborative integration of experts in different engineering fields, with different knowledge and technology, who are also physically dispersed. Here, beyond the simple aggregation of collective knowledge, the integration of expert knowledge across different fields was realized, in which coordination and negotiation between engineers having different expert knowledge were key factors. Experts in each engineering division, using their specialized knowledge, cooperatively constructed concrete design solutions that attempted to realize the user's requests. The engineering collaboration methods discussed in this paper are one of the most promising features of product design and manufacturing in the age of information networking.

References
[1]   A. B. Baskin, G. Kovacs, and G. Jacucci (Edit), Cooperative Knowledge Processing for Engineering Design, Kluwer Academic Publishers, 1999.
[2]   W. Mitchell and K. Singh, Survival of businesses using collaborative relationships to commercialize complex goods, Strategic Management Journal, Vol.17, pp. 169-195, 1996.
[3] A. G. Rechard, Virtual vertical integration: Partnership realized, ASQC 50th Annual Quality Congress Proceedings, pp. 757-758, 1996.
[4]   R. T. Sreeram and P. K. Chawdhry, A Single Function Agent Framework for Task Decomposition and Conflict Negotiation, 1998ASME Proceeding of DETC'98/DFM-5748, 1998.
[5]   V. Ravindra and J. Tappeta, and E. Renaud, Multiobjevtive Collaborative Optimization, 1997ASME Proceeding of DETC'97/DAC-3772, 1997.
[6]   K. Shibao and Y. Naka, Optimization of the Distributed System by Autonomous Cooperation, Distributed Automation System, pp. 29-40, 1994.
[7] B. Prasad, Concurrent Engineering Fundamentals Vol.I -- Integrated Product and Process Organization --, Prectice

Hall PTR, pp. 216-276, 1996.

[8] B. Prasad, Concurrent Engineering Fundamentals Vol.II -- Integrated Product Development --, Prectice Hall PTR, pp. 298-301. 1997.

[9] M. Yoshimura and H. Kondo, Product Design Based on Concurrent Processing of Design and Manufacturing Information by Utility Analysis, Concurrent Engineering: Research and Applications, Vol.4, No.4, pp. 379-388. 1996.

[10] M. Yoshimura and K. Yoshikawa, Synergy Effects of Sharing Knowledge During Cooperative Product Design, Concurrent Engineering: Research and Applications, Vol.6, No.1, pp. 7-14, 1998.

[11] M. Yoshimura and H. Kondo: Group Decision Making in Product Design and Manufacturing, Proceedings of 1997 ASME Design Engineering Technical Conferences, Sacramento, California, pp. 1-7, 1997.

[12] M. Bradshow (Edit), Software Agents, The MIT Press, 1997.

[13] J. D'Ambrosio and T. Darr, and W. Birmingham, Hierarchical Concurrent Engineering in Multiagent Framework, Concurrent Engineering: Research and Applications, Vol.4, No.1, pp. 47-56, 1996.

[14] Yan Jin Weihua Zhou, Agent-Based Knowledge Management for Collaborative Engineering, 1999 ASME Proceeding of DETC'99/EIM-9022, 1999.

A Multi–Agent System for Distributed Factory Decentralized Emergent Control

C. Bournez
G. Beslon
J. Favrel

Laboratoire PRISMa
Institut National des Sciences Appliquées de Lyon
69621 Villeurbanne Cedex, France

Abstract

We propose a Multi–Agent Architecture for the control of a distributed manufacturing system. Such a production facility reflects geographically distributed physical systems, encountered for example in a virtual factory context. Thus, it is characterized by a highly dynamic structure, and therefore a need for flexibility and agility.

The use of an agent–based framework allows hot–reconfiguration so as to achieve a "light–weight" integration, since our architecture is completely heterarchical and decentralized.

To take advantage from the distributed features of the physical system, we use mechanisms that rely on the implicit cooperation between the agents and their coordination within a competitive marketplace. These mechanisms involve dynamic reasoning of each agent, using both individual preferences, acquired while playing economic roles in the market, and partially pre–designed behavioral rules. The agents construct communication strategies as well as negotiation skills and local task scheduling management, thanks to a unified and synchronized learning system.

keywords: Decentralized systems, Multi-agent learning, Agent models and architectures, Agent Market.

1 Introduction

In Computer–Aided Manufacturing Production Management, Multi-Agent Systems can operate at different control levels. *Advanced Scheduling*, sometimes called planning, advanced planning, predictive planning or scheduling, or even just scheduling, is a long or medium term prevision of calendars for all resources of the production system.

Our work also deals with *reactive task scheduling*, which we call *dispatching*. It is an additional scheduling feature that is able to react to the production system's disturbances and evolutions. When distributed, the production system can change its layout in short delays: these delays may even be shorter than the predictive schedule horizon. Therefore this dispatching functionality is essential to reactivity (to disturbances) and flexibility (to changes in the production system layout). Multi–Agent Systems are also useful in technical data management.

This paper presents the Multi–Agent System architecture designed for OCEAN (Organization and Control Emergence with an Agent Network). It is organized as follows: We first describe the coordination mechanism, which relies on a Contract Net. Collective cooperation emerges from individual competitive behaviors of the agents. We briefly justify our special interest in emergent control. In a second part, we explore further inside the agent. Starting by categorizing its activities, we explain the construction of a reasoning system adapted to contract negotiation, but precisely in our distributed production management scope. Before concluding, we try to demonstrate the need for synchronized learning in the various activities of the agents and propose an implementation of such a learning system over Case-Based Reasoning techniques.

2   Decentralized Control Architecture

2.1   Background

Usual Architectures in multi-agent systems for Control, especially for distributed facilities, are:

- hierarchical, with a central agent, or a agents unit, that coordinates the entire system. Some restricted autonomy can be let to agents in sub-levels (considered as "slaves" by the top-level master). This architecture is a tree-like control system.

- heterarchical, completely decentralized, without masters nor slaves, that is to say no levels or sub-levels at all. All the agents are not necessary equivalent in functionalities, but have no control over each other.

- hybrid architectures combine these two approaches, with levels of distributed autonomous agents.

In order to deal with the physical distribution of the production facility, we propose a heterarchical architecture. We will give further justifications of our choice in section 2.3, from a production control point of view.

In advanced scheduling, some market-based Multi-Agent Systems rely on heterarchical architectures where agents are associated to workstations. This is the case in Shaw's work (see [6]), where the agents sell intermediate products in a push method. The same architecture is used by A. D. Baker [2], but with backward propagation of offers from final demand, that is to say pull control policy. Time constraints in production (for the schedules) are local problems of the individual agents. In another view, Maley and Duffie considered agents as pieces of information, like routings. In both cases, negotiation rely on production or transportation steps. An agent can exclusively manage one production or transport kind of task, which corresponds to a piece of the necessary routings of a product order.

In a dispatching scope, Parunak proposed a hierarchical-like architecture based on an organizational model : a factory is constituted of workshops, which are constituted of cells that are groups of machines. The Multi-Agent Marketplace is a nodular structure. The nodes between the hierarchical levels are the agents of the system. Parunak criticized this system, by telling it was suffering "severe communication bottlenecks", because of the important centralization.

2.2   A Multi-Agent Contract Net

Our heterarchical architecture is strongly inspired by the original Contract Net Protocol (CNP), first described by Smith in [9].

The Contract Net Protocol is a negotiation mechanism that can be described in 4 steps :

1. Auction: An agent takes the role of a contract manager. It sends an auction for a contract, to some or all agents of the system.

2. Bidding Phase: Each agent that receives this auction evaluates it and can send a bid to the manager or simply ignore its offer, depending on what makes sense for itself, i.e. whether it can earn money from this contract.

3. Evaluation and Awarding the contract: When the manager of the auction is supposed to have received all bids that potential bidders could send, it chooses the best bid, by means of calculations from a reasoning system. Then it sends a message awarding the contract to the winner.

4. Execution: The winner of the contract executes it and sends the result back to the manager. It is *engaged* to execute the contract, if it cannot execute it, it has to pay a penalty. This is called *commitment*.

The bidding phase itself can follow various auctioning systems:

- In an ascending bid, an agent has to offer more than the concurrent agent. The bidders make the pricing offers. The first price (lower limit, or reserve price) is initially given by the manager. The more the prices increase, the less bidders stay in the auction. The auction terminates when there is only one agent left.

- In a descending bid, the manager announce the higher price limit (reserve price) and decreases by steps. The first agent which agree with the announced price wins the auction.

- Sealed Bid are offers sent in a private message to the manager, and the higher price wins. We use this method, which accelerates the negotiation, since there is only one round.

- The Vickrey Auction is a variant in which the second higher price wins.

All along the negotiation process, the agents use a currency unit to evaluate the contracts. Prices, penalty rates, time, and so on are parameters of the Contract Net. They have influences on the agents' strategies. Obviously, the history of the negotiations in the Agent Network should play an important role in the way the agents behave during the auctions. These strategical aspects of the agents suppose that they acquire preferences, based on information gathered in contracting processes.

Figure 1 shows the Multi-Agent Network we use in OCEAN. The white stars represent customers demands, converted in basic production orders. They are randomly dispatched to some agents of the system. The agents are the little round tokens. Some of the agents have resources to manage. The resources are either machines, transport systems, storage places, human resources. These agents have in their individual environment the necessary technical data for the exploitation of their resource when they have one. The resources and associated knowledge are distributed. Note that not all the agents have a resource to manage and they never have more than one resource. The allocation of the physical production system resources to the agents is kept random, like the allocation of the customers orders. When an agent receives an order, it may be able to perform a part of the required work. Not all the agents involved in a contract negotiation will perform a task for this order final achievement. The number of schedule actors is inferior to the number of negotiation participants. The agent that has received the order may not be able to perform the entire order. Then it sends one (or several) auction(s) in order to have the other part of the work done by one (or several) other agent(s). The number of auctions it sends depends on how it fragmented the rest of the work in subtasks.

The agents situated in the "Auction field" will receive the proposition. After the negotiation, the subtasks are awarded to the best agents. In fact, an agent may have no resource, but take a task in order to subcontract for it. It sends an auction, and it allows an agent outside the auction field of the first one to get the task. The intermediate agent gets a benefit from its role of middleman.

2.3 Emergent control

The need for agility and flexibility, particularly in Virtual Factories, led us to draw some essential concepts[3]: At the contrary of the architectures we described previously by Parunak, Shaw, Baker, Ma-

ley or Duffie, the architecture in OCEAN has no correspondence between agents and physical system resources. Agents are merely virtual entities that negotiate and execute. From a similar idea, there is no correspondence between any organizational model of the company or factory and the Multi-Agent System. The interest of this absence of links is to let a greater part to play to emergence. Indeed, the Multi-Agent System is a Marketplace, with a kind of economy, illustrated by currency and product flows. Each Agent is involved in the Contract Net, but it is not Auctioneer or Bidder in particular. It plays all the available roles on a contract net : manager (=auctioneer), bidder, middleman.

Agent point of view

The individual behavior of the agents of the system depends on several decisions just dealing with contracting.

- In the Auction phase: The auction field (or "addressee choice", according to Okho et al. terminology in [5]) offers at least two options. One is to broadcast the offer to all the agent's environment (not all agent marketplace, but the part of the marketplace that the agent knows). A second possibility is to address the offer to some specified agents, depending on historical knowledge of the agent.

- Announcement of the winner: The Auction Manager can broadcast the result, so that the agents who lose know that they have lost, or it can address only an award message to the winner.

- Re-emission of offers: Let us imagine that no bid is received within the time limit specified in the offer. The Auction manager can restart emitting its offer, either with another addressing method or with the same method but in a later period.

- The commitment level: A bidder can be considered as *engaged* with the task since it has sent a bid, or only later, if the contract is awarded to him.

These decisions are not parameters of the Contract Net Protocol, at the contrary of what can be found in some studies around CNP ([7], [8]). Each agent makes its own choices about the way it has to behave, according to its contracting history and the preferences it has built over this information. We will explain in section 3 how it achieves such a knowledge acquisition and information treatment.

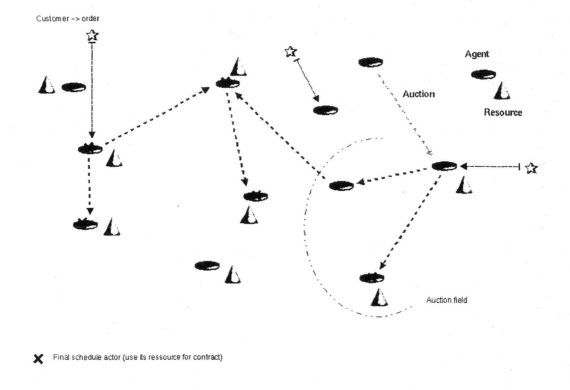

Figure 1: Architecture of OCEAN Multi-Agent Market

The ability of agents to fragment and make tasks clusters may lead to a great simplification in local scheduling on a given resource (monitored by one agent). This infers benefits (by scale savings), but there is another challenge here : the fragmenting and clustering actions must offer a large possibility of emergent control policies. A frozen policy like Kanban can be classically modified by reconfiguration of the routings. But in OCEAN, the process of distributing work or factorizing work is an implementation of such a reconfiguration, thanks to the emergence of control. An agent can decide to produce more than required by a task and store the excedent production until the next similar task. This occurs when there is a repetitive order contracted by the same agent. We conclude that the choice of a control policy (for each product or family) between 1- produce on order or 2- produce in big regular batches and store is achieved thanks to emergent individual behavior.

Production management point of view

As we said previously, we would like to use the same system for achieving scheduling (advanced schedul-

ing), dispatching (reactive scheduling), and planning execution. We argue that a mere hierarchy does not allow this structure. Decentralization is the key for increasing agility, which is essential to reach this direct interaction between scheduling (and control) horizons (short/medium/long term). Moreover, a heterarchical system avoids propagation of disturbances between physical level and control level.

A second and even more essential goal is to let a control policy emerge for each product or product family: In traditional manufacturing control systems, the user chooses a control policy when he adds a product or product family in the Technical Databases. For example he will choose Kanban or MRP. This choice is difficult and frozen. We aim at letting the system decide. When the context of demands or production resources will change, the system would automatically adapt its policy. At the beginning, this Multi–Agent System will use a "degraded mode" control policy (during adaption phase), and then a better, adapted, control policy will emerge from the inner multi–agent system's behavior. That is to say, the cooperative result of individual competitive behavior of all agents of the system. The emergence of adaptive control policies for each product relies on our Multi–Agent System self-organization.

3 Agent Design

3.1 Agent activities in the market

Each agent adapts its behaviour according to its state in the marketplace. It can play three roles regarding to a given contract.

`Auction manager` (Auctioneer): It has to deal with the problem of addressing its auction: broadcast (to all its environment) or addressed (to which agents) ?

`Bidder` (not for an auction that it manages, of course): Its strategy of answering or not to an auction mainly relies on a utility function. It calculates what benefits it can expect from this contract. Of course, the resource that the contract will use has to be free at a suitable period of time. Thus, its resource schedule is important as well.

If the agent has no resource to make the job, it may sub–contract the task, which may be hazardous. Depending on the amount of the penalty if it cannot achieve the task and its past experience in sub-contracting this kind of tasks, the agent's strategy will vary. In this case it has to play the "secondary role" on the market. It can act as a `retailer or wholesaler`, by fragmenting and clustering the tasks. In production management terms, its benefits equals to scale savings .

Our study of these roles in the global CNP organization allowed us to identify activities required from the agents. As far as we know, each of them may jointly use pre–cabled behavior rules and acquired behavior features. Thus, we may refer to them with the word `skill` as well as `activity`. However, this alternate use of the two words, which are in fact slightly different in meaning, may not be too much confusing. Let us now present what these "primitive" skills are.

Communication

Communication is the support for negotiation. We designed a small communication protocol, with a compact abstract syntax and pragmas, so as to keep the use of the communication system simple. (Note: In the implementation, the communication system is independent of the language, as we see later in this paper).

The communication activities of an agent are :

1. Discovering its environment (presence of other agents).[1]

2. Interacting with other agents, using the Contract Net protocol (the 4 steps described earlier) and a communication underlying protocol that we have written.

3. Getting technical information about products (routings, nomenclature links. This information is distributed in the marketplace).

Agent's preferences

The agent's use of the currency unit on the marketplace is coupled with the following:

4. The agent needs to evaluate prices and costs of tasks, and eventually products and resources use. The utility function, which enables to make a bid for an auction, uses the pricing system of the agent.

5. The choice of contracts to bid for. When the agent plays the role of a bidder, it can not answer to all auctions. The process of choosing interesting auctions is one of its essential activities.

6. The choice of a winner. When the agent plays the role of auction manager, it has to select the bidder to whom it will award the contract.

7. Clustering and Fragmenting tasks. We already discussed the importance of this process in control policy emergence. This activity needs much learning, so at the beginning the agents will tend to use direct contracting without clustering/fragmenting. The emerging control policy will then be a well-known (frozen) policy. Then learning will fully enable this agent activity and different policies will emerge.

Local scheduling

Local scheduling is the real problem solving process:

8. Each agent has to schedule its own activities, in order to send auctions, bids, and contract awards in reasonable times. This can be a repeated sequence, so there is no particular difficulty there.

---

[1] We call `market` (marketplace) the entire multi-agent system, and `Environment` of an agent the part of the market that it knows, in fact: what it has explored. **There is no model of their environment in the agents.**

9. It also has to schedule tasks for a resource if it owns one (machine, transport resource or storage place). Remember that not all the agents are resource-managers.

Local scheduling can use one of the best known methods: linear early (scheduling the task in the earliest available time period), linear least dense (scheduling the task in the least dense period of the calendar), or hierarchical, in a possible combination with constraint chains method.

3.2 OCEAN agent global design

The general conception schema of an agent in OCEAN is shown in figure 2. The agent internal structure is composed of the following elements:

A Transceiving System is connected to the network through a connection wrapper. Other agents are connected, not necessarily to the same wrapper, and a message transport mechanism assure the transmission between wrappers. The transceiving system uses FIFO (First–In First–Out) queues, with encapsulated messages, to add/remove extra communication parameters, as transmission security, reliability, etc.

A Message Handler constructs internal message structure into contract–protocol–messages, according to an abstract syntax. Then it passes these messages to the transceiver (that will encapsulate it). In the other way, it receives protocol messages and passes the information toward the learning system. Further details are given in section 3.4

The Case–Base handler relies on reasoning rules that are partially cabled (identical for all the agents), to deal with the contract net protocol, and acquired rules, according to the 4R learning cycle [1]: Retrieve a similar case (or several similar cases), Reuse knowledge and information stored in this (these) case(s), Revise the solution after it has been applied and evaluated, Retain a new case (or modify old ones). The preference-related rules are strategies learned during negotiations. We will present in section 3.4 the relations between strategies learning and information learning.

Synchronized activity handlers are sub–systems that give structure to the activities presented in 3.1, plus a communication chooser, that does not handle protocol–message internal information but message passing itself (we will give a brief overview in 3.3). The three boxes near the agent frontier symbolize the information learned about the environment. Planning and timer give necessary information to the scheduler and contract evaluator. Technical data is essential to the contract evaluation, but it is related to acquaintances too (the agent knows about its acquaintances' production capabilities). Finally, the environment, or internal market image, is a sort of acquaintances base, that includes the market frontier data (customer orders...).

3.3 Brief overview on communication strategies

The Contract Net Protocol itself will be influenced by some extra-protocol individual strategies. OCEAN allows us to study five categories of these strategies:

- Intentional message leakage
- Randomized address errors
- Communications eavesdropping
- Fake instances of protocol messages
- Identity usurpation

The classical Contract Net Protocol describes how contract negotiation takes place over the network. The agents can develop strategies for their communication learning task, which is crucial for environment discovery and preferences construction. The messages in our Contract Net Protocol implementation include an encapsulated compact–syntax (the encapsulated messages are the real CNP messages) and additional information for the communication mechanism: confidentiality (against eavesdropping), reliability (against leakage and address errors), signature (against identity usurpation), and eventually acquaintance trust level. Some choices infer an increased communication cost, or even an increased message processing cost. Thus, decision relies on a synchronized learning between communication strategies and the use of these strategies in negotiation. Each part of the decision process requires learning. Next section will discuss further about synchronization of learning, not only in the communication scope.

3.4 Features for a synchronized learning

The Contract Net Protocol implementation used by A. Baker in [2] involves no learning capabilities at all

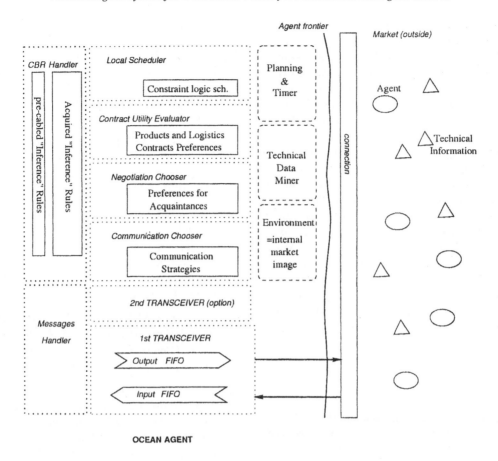

Figure 2: Design of an agent in OCEAN

in the agents activities. The role of each agent is notably forced by the fact that all agents are situated. We argue that our agents, being independent from physical structure, are more adequate to allow control emergence, required for a Virtual Factory distributed facility for example. Thus, we use the same basic Contract Net Protocol, the agents can easily play different roles in the negotiation process that takes place in the network, whereas situated agents would generally assure a predefined double–sided manager-contractor role tightly related to its position in a "supply–chain-like" production facility. The drawback of the choice of the open agents lies in agents' dependance on their learning system.

For each of their possible roles, the agents have to learn from past experiences in order to improve their strategies. Moreover, within each role, several skills are involved, each of them needing learning. From an agent internal point of view, a negotiation is a sequence of messages, which are processed using the knowledge about 1- the other agents

(acquaintances–related preferences), 2- the contracts' subjects (product/logistic–related preferences), and 3- the current state of schedule (density,...).

The Case–Based Reasoning handler allows to use a given message according these three ways, each of which is managed by its own sub–system: In fact, each sub–system (local scheduler, contract evaluator and negotiation chooser) has its own representation of Cases. The CBR handler takes in charge the conversion of messages into Case Representations, as a traditional CBR system. But it extracts several representations of the same message, to enable the learning sub–systems synchronous exploitation of the information.

The communication strategies are also learned this way, that's why it is shown as a fourth sub–system in figure 2 although it is not a basic CNP activity.

At the contrary to CBR, a Reinforcement Learning (RL) system, which uses a trial–and–error mechanism, would not allow such a unified and synchronous learning. In addition, RL may not be very efficient in such a dynamic environment. In OCEAN it could be useful for the Evaluator activity. A standard universal contract utility formula, could be found by successive adjustments according to the rewards received. In the same way, the agents could try to compute values for communication choices (reliability, confidentiality, signature...). The main difficulty encountered in RL is the important number of criteria makes it nearly impossible to get a relevant formula. Non–existent values or irrelevant ones (for example, wealth is irrelevant at the beginning of simulation) infer strong disturbances in the learning process. This problem would be the same in CBR, if we would not have designed subsystems close to each other: in cases, reasoning about actions is influenced by some criterias or others. Multiple combinations of criterias mean multiple case representations. We believe that a meta–representation with derived reprensentations for sub–systems resolve this without dissociating learning processes nor introducing asynchrony.

4   Conclusion

In the field of Manufacturing Control, Multi–Agent Systems have proved their adequacy for *advanced scheduling* and *dispatching*. We can classify their applications according these two scopes, and within these scopes, identify architectures. With OCEAN architecture, we aim at integrating predictive and reactive scheduling. A hierarchical control architecture should not be used, because the system could not be as flexible and agile as we would like, whereas a Marketplace is a reactive and adaptive architecture. Moreover, a distributed manufacturing control naturally infer a desire of parallelism. OCEAN heterarchical architecture is constructed on the idea of a marketplace that follows the Contract Net Protocol basic principles. Therefore it enables such a parallelism, thanks to efficient decentralization with emergent coordination.

The agents are free, autonomous, individually rational and self–interested. They aim at maximizing their own profit. Thanks to their individual learning capabilities, we expect cooperative control policies to emerge from individual behaviors. The independance of the agents versus physical and organizational enterprise structure is meant to allow production management policies to emerge, differing from existing ones when relevant. An interesting feature of the agent network may be that it intrinsically performs autonomous reaction to disturbances, as it self–(re)organizes in case of hot–reconfiguration of the distributed factory.

References

[1] Agnar AAMODT and Enric PLAZA. Case–Based Reasoning : Foundational issues, methodological variations, and system approaches. *Artificial Intelligence Communications Journal*, 7(1):39–59, 1994.

[2] Albert D. BAKER. *Manufacturing Control with a Market–Driven Contract Net*. Ph.D. dissertation, Rensselaer Polytechnic Institute, Troy, NY, USA, May 1991. 303p.

[3] Carine BOURNEZ, Guillaume BESLON, and Joël FAVREL. A multi-agent market-driven contract net for emergent manufacturing control. In *Proc. MultiConf. on Systemics, Cybernetics and Informatics (SCI'99)*, volume 3, pages 296–303, Orlando, Florida, USA, July 31–Aug 4 1999.

[4] G. M. O'HARE and N. R. JENNINGS, editors. *Foundations of Distributed Artificial Intelligence*. Sixth Generation Computer Technology. Wiley-Interscience, New York, 1996. 576p.

[5] Takuya OHKO, Kazuo HIRAKI, and Yuichiro ANZAI. Reducing communication load on Contract Net by Case–Based Reasoning — Extension with directed contract and forgetting —. In *Proc. of the 2nd Int. Conf. on Multi–Agent Systems (ICMAS'96)*, pages 244–251, Kyoto, Japan, 10–13 December 1996.

[6] H. V. D. PARUNAK. *Applications of Distributed Artificial Intelligence in Industry*, chapter 4, pages 139–164. In O'HARE and JENNINGS [4], 1996. 576p.

[7] Tuomas W. SANDHOLM. Contract types for satisficing task allocation : I theoretical results. In *Proc of AAAI Spring Symposium : Satisficing Models*, Stanford University, California, USA, 23–25 March 1998. 8p.

[8] Tuomas W. SANDHOLM and Victor R. LESSER. Advantages of a leveled commitment contracting protocol. In *Proc. of the National Conference on Artificial Intelligence*, pages 126–133, Portland, USA, 1996. Extended Version 1995, Technical Report TR-95-72.

[9] Reid G. SMITH. The Contract Net Protocol : High-level communication and control in a distributed problem solver. *IEEE Transactions on Systems, Man and Cybernetics*, C.29(12):1104–1113, 1980.

A Multiagent System for a Cooperative and Distributed Vision System

Edouard Duchesnay   Jean-Jacques Montois   Yann Jacquelet

LTSI/GRAID

IUT de Rennes-Saint-Malo, Université de Rennes 1

35409 Saint-Malo France

Abstract

In a production line context, it is sometimes necessary that several vision system have to cooperate. Each image recognition system is highly context dependent which implies the development of high cost/time-effective specific systems.

The multiagent approach naturally allows cooperations between vision systems. But it also can make a rapid adaptation of the vision system to a new configuration of the production line easier.

This paper presents a method of machine vision architecture based on a hybrid and fine granularity multiagent system that encourages incremental design via modular and hierarchical structuring of knowledge and pattern recognition mechanisms.

This cooperative work relies on a multiagent architecture (OSAgent) developed to support an adjustable granularity of agents, a physical distribution of computation and several transparent message passing mechanisms to allow local communication through memory between agents of the same vision system and distant communication through TCP/IP between agents of different vision systems.

1 Introduction

In a production line context, it is sometimes necessary that several vision system cooperate, i.e. quality control of the same production good from different angles (fig 1), or quality control of a production good at different steps of the production process. The multiagent distributed aspect provides a good prerequisite to implement this feature. The underlying architecture of the multiagent system should support various kind of message passing mechanisms: communication via a network or via memory. Figure 1 shows the agents'topology among the production line and the

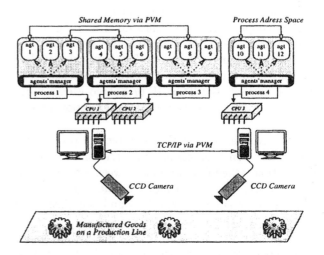

Figure 1: Agents'topology among a production line

corresponding used message passing mechanisms. [1]

We propose a multiagent approach for a machine vision system that aims at facilitating the networking and distribution in a concurrent engineering aspect. In this approach, the segmentation and interpretation processes are unified in a agents'network structured as a growing pyramid.

In this article we will focus on a single vision system which is itself composed of many agents. But the long term goal is to extend our architecture to several cooperating agent-based vision systems.

All the knowledge of the system relies on an other agents'network composed of detectable patterns. Each agent of the pyramid receives some specific behaviour from an agent in the knowledge network. The inherited behaviour allows to detect more complex patterns taken into account by a new agent in the next level of the pyramid.

Basic pattern detection consists of homogeneous region detection or segment/curve detection. Therefore

---

[1] Details on terms used in this figure are given in section 6.

453

the segmentation problem can be seen as a pattern detection problem and then be unified to the interpretation.

However complex pattern detection is not treated in this paper so that the evaluation (section 7) can be compared to a segmentation process.

The problematic section justifies why the unification of the segmentation and interpretation processes allows an earlier integration of the information that comes from interpretation in order to guide the agents fusion efficiently.

The multiagent aspect allows the use of simple entities, easy to create, on which learning mechanisms can be applied. The fusion of the behaviour of those entities provides robustness and genericity more adapted to interact in a complex environment.

This cooperative work relies on a multiagent architecture whose major characteristics are: (1)An adjustable granularity of agent; (2)a physical distribution of computation; (3) a transparent message passing system that uses TCP/IP or Shared Memory depending on the respective physical situation of two agents [2].

Low level services of this architecture provide facilities for: synchronization, resources sharing, communication, scheduling of and between agents. High level services provide facilities to manage: the acquaintances, competences, beliefs and negotiations of and between agents.

To implement this work, Java has been chosen, for its good compromise between portability, security, speed and facility to develop symbolic reasoning systems, as well as numerical mechanisms.

2   Problematic

The usual conception of architectures for artificial vision compels the conceptor to place the whole system intelligence on a few units of "monolithic" treatment. This induces an over-specialization that handicaps such a system which is merged into a complex world. By shifting the complexity from the treatment to a structure composed of simple, independent and communicating entities, it is hoped that, through the merging of their behaviour, a robustness and a genericity to interact with a complex environment will appear. Classical images processing architectures are usually split into three "monolithic" units treatment: the preprocessing layer, the segmentation layer and

---

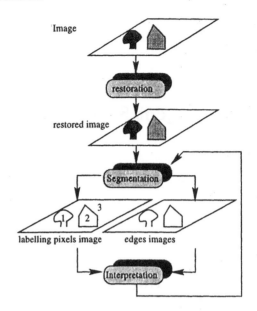

Figure 2: Traditional conception of pattern recognition (in image) architectures

the interpretation layer. As we will see, this splitting poses many problems.

2.1   Problems posed to segmentation

(1) The global treatment (centralized) is poorly adapted to the nature of the images, which present various textures. (2) A lack of flexibility appears in the use of the operators. Indeed, a great number of operators is available, the problem rather lies in knowing which ones to use. This choice is to be hard coded in the algorithm (the central one). If we wish to add new operators, or to deal with new textures, it is necessary to "open the system", to insert the operator and the choices that lead to it, and to take the risk of making the algorithm ineffective on textures which it treated correctly before.

2.2   Problems posed to interpretation

The interpretation layer works on a map of extracted primitives(labelling pixel image) or/and an edges image. It is going to try to put into relation an observation (Obs) with a model (Mod) where usually $dim Mod \ll dim Obs$. This operation is realized with two functions: (1)*reduced()* that allows to pass from Obs space to Mod space. This problem is known as the feature generation problem [18]: how are features generated? What is the best number of features to use? (2)*recognized()* that matches with the best vector of Mod. This is the classifier design problem[18].

---

[2]If agent are on the same vision system memory can be used to exchange message otherwise TCP/IP should be used

Mathematics approaches (called "transformed" techniques) are often used to implement *reduced*() and *recognized*(). Let's mention the Fourier descriptor technique [20], the Hough transform [19] or techniques based on the extraction of particular edge points combined with mathematics applied on shape geometry. One can already notice that such techniques are extremely specialized on a specific image type and will not tolerate important variations around an expected observation. Other methods issued from AI (symbolic approaches) faced the same problem where rules could not reflect all the possible observations. Although having brought a kind of flexibility, the use of meta rules has not solved the problem. The connexionist methods, however being more adapted to a difficult formalizable problem and being able to learn, have showed their limits; because a neural network keeps specialized on the treatment of an information type (for example a full-face). Three reasons can be diagnosed for these limitations : (1)**Non locality :** The same functions *reduced*() and *recognized*() are not necessarily adapted to the whole image, because in the same image the nature of the observations can be extremely varied. (2)**Non progressiveness :** *reduced*() and *recognized*() do not look up from an information source, which is quantitatively rich but qualitatively poor (labelling pixel image). They must jump over an important semantic ditch at once. It forces the conceptor of these two functions to have a very precise idea of all the possible observations. (3) **Centralization :** *Mod* is almost impossible to build because it must reflect all the possible varieties of observations. It also has to record all the necessary attributes for a good discrimination of the observations.

2.3 Problems posed to the interaction between segmentation and interpretation

This non progressiveness also finds expression in the interaction between segmentation and interpretation. Indeed, if the segmentation unit is represented as a "monolithic" block that, after treatment, submits its results to another "monolithic" block which is the interpretation unit, the interaction between the last two blocks will be late, rough and brutal. Indeed, interpretation can only correct segmentation after the last has been executed : this correction will be based on the whole image (here appears the non locality aspect) and will express something like "restart from the beginning using these new parameters". As a conclusion to this first part, we do not in any way pretend to

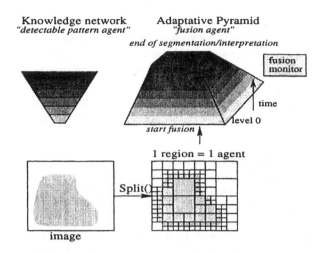

Figure 3: The knowledge network (pattern agent network)

put all these techniques aside; but we simply propose a new way of using them with the help of multiagent system. Works have already been done in order to evaluate the pertinence of using distributed artificial intelligence in the computer vision domain; let's note some architectures based on blackboards [1] or on multiagent architectures [2], [3], [4], [5], [6].

3 System architecture

We propose a multiagent approach organized according to the architecture of figure 3. The architecture (named **SmaVision**) is composed of two agents networks:

1. The pyramid of "fusion agent": This one is progressively built level by level during the segmentation/interpretation phase. This process is controlled by a monitor agent with contract net protocol.

2. The "pattern agent network": This one represents the knowledge of the system, all of those agents possess one or several pattern detector mechanism, those mechanisms are used by the "fusion agent".

Characteristics of the architecture:
These agents act through a behavioural fusion of treatment techniques and they allow:

1. **A local adaptation** of *reduced*() and *recognized*() to the image local context.

2. **A progressive application** of simple *reduced*() and *recognized*() allows: (a) to progressively reduce the dimension of the observation space; (b) to make a recognition in the same progressive way; (c) to adapt *reduced*() and *recognized*() to the current semantic level. For example, symbolic techniques, seem well adapted to a high semantic level.

3. **A fine, subtle and progressive retroaction feedback loop** between segmentation and interpretation. This is done by merging the interpretation phase with the segmentation.

4. **A scattered and local conception of** *Mod* in the form of an agent pattern networks, one of the major trump of which is the capacity to make use of learning mechanisms.

4   Irregular Adaptative Pyramid of agents

4.1   Initialization of the Pyramid: Split

A Split [17] algorithm is performed on the image. A "fusion agent" is affected on each region. Those agents, that form the first level of the pyramid, are then concentrated on high frequency parts of the image (see agents' density on fig 8).

Algorithm 1 Initialization of the system
---
regions ← recursive-split(image)
create the fusion monitor agent
for all *region_i* ∈ regions **do**
    createNewAgent(Level 0; inherit behaviour from *defaultPattern*; initialize with *region_i*)
    initialize the acquaintances of the agent
end for
send start fusion **to** fusion monitor

4.2   The construction of the pyramid: Merge

The fusion agent network (fig 3 right part & fig 4) presents itself as a structure with a pyramidal aspect composed of agents. This multiagent network is inspired by Irregular Adaptative Pyramid [16] [15]. The distribution of agents in the plan formed by the horizontal and the depth shows their spatial distribution in the image. The vertical axis represents a semantic enrichment. The nearer the base we are, the more

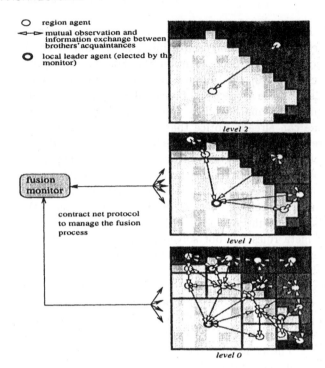

Figure 4: The construction of the fusion agent pyramid

agents there are whose semantics is low levelled ; the farer we go, the less agents there are whose semantics has grown richer.

The director scheme of the agents' cooperative work is based on contract net protocol [9]. The fusion monitor agent (the contractor) broadcasts a task announcement message to its acquaintances that are in the current level (fig 3 right part & fig 4) (algorithms 2 & 3) of the fusion pyramid. The task consists in making a fusion between agents where a local pattern has been detected. The eligibility is done on a satisfaction coefficient that reflects the quality of the detected pattern.

Each fusion agent uses the knowledge **inherited from a specific pattern agent** to:

1. collect and propagate informations by reflexes;

2. enrich the beliefs by deduction;

3. detect some new patterns and then fusion with acquaintances if elected by the fusion monitor.

Those mechanisms inherited from a specific pattern agent are detailed in section 5. The fusion algorithm (algo 3) always manipulates a "Generic_Pattern class" that provides a standard and generic interface [3] to call

---
[3]for example `dettectPattern(Beliefs, Acquaintances)`

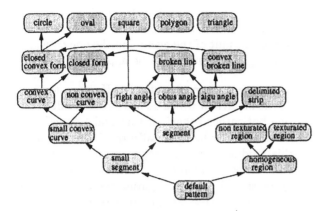

Figure 5: The knowledge network (pattern agent network)

the previous mechanisms. In fact, the called function will be the one of the current inherited pattern.

Algorithm 2 fusion monitors'behaviour

---

repeat
  for all active $agent_i \in$ currentLevel **do** **send**
  init & make proposition **to** $agent_i$
  repeat
    when received proposition **do**
      class the proposition
      if all agents answer **then send** elected **to** the
      agent that made the best proposition
    end
  until $\exists$ an active agent $\in$ current level
  currentLevel $\leftarrow$ currentLevel+1
until $\exists$ an active agent

---

5 The agents of the knowledge network

5.1 The knowledge network architecture

This agents network represents the distributed model of patterns that we will try to recognize in the image. The knowledge network (fig 3 left part & 5) presents itself as an agents network. The latter does not evolve during the segmentation but only during the learning phases. The vertical axis on this network represents a semantic enrichment. Two natures coexist within the agents:(1)A nature of the "edges" type that represents "high frequency" components, the structure of which is more or less complex (simple segment, long segment, curve, right angle, …). (2)A nature of the

---

Algorithm 3 fusion agent behaviour

/*this code is asynchronously executed by each agent*/
when received init **do**
  $Beliefs \leftarrow$ calculateLocalCharacteristics($Beliefs$)
  $Beliefs \leftarrow$ observAcquaintances($Acq_{brothers}$,
  $listOfRequiredData$)
  $Beliefs \leftarrow$ deduction($Beliefs$)
end

when received make proposition **or** update proposition **do**
  $Acqs \leftarrow$ selectAcquaintances(from $Acq_{brothers}$,
  where $activ = true$)
  $patterns \leftarrow$ detectPatterns($Beliefs$, using acquaintances: $Acqs$)
  $patterns = (\ldots (pattern_i, satisfactionCoef_i,$
  $involvedAcquaintanceList_i), \ldots)$
  $proposition \leftarrow (pattern_i, satisfactionCoef_i)$
  such $satisfactionCoef_i > satisfactionCoef_j \forall j$
  send $proposition$ **to** fusionMonitor
end

when received elected **do**
  /*the agent becomes a local leader*/
  father$\leftarrow$createNewAgent(currentLevel+1; inherit
  behaviour from $pattern_i$)
  for all $acc_i \in involvedAcquaintanceList_i \cup$
  $mySelf$ **do**
    send become a child of father **to** acc_i
    $affectedAgents \leftarrow affectedAgents \cup$ **receive**
    $affectedAgents_i$
  end for
  for all $agt_i \in affectedAgents$ **do**
    /* some acquaintances may have to update
    their detected pattern because the current agent
    is now involved in a fusion*/
    send update proposition **to** agt_i
  end for
end

when received become a child of father **from** leader **do**
  $Acq_{father} \leftarrow$ father
  $activ \leftarrow false$
  send $Acq_{brothers}$ **to** leader
end

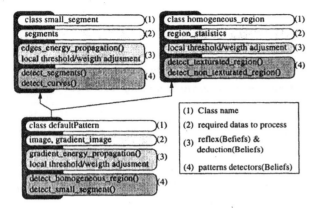

Figure 6: Some Pattern agents

"region" type that represents stationary components of the image. From the association of the last two natures results a new nature. For example, an association "is-the-border-of" between a "border" nature and a "region" nature emerges on a "form" nature.

We add a third semantic dimension to the two dimensions of the image. One of the originalities of our approach deals with this work on semantic deepness where the treatment is done by an agent which relies on semantically enriched and reduced informations by the agents of the lower levels. This provide progressiness in the pattern recognition process. It also makes easier the conception of the used pattern recognition mechanisms, and should allow the addition of learning procedures mentioned below.

Let's precise that those agents do not make the pattern recognition themselves. They have recognition unit and other mechanisms which they bequeath to the fusion agents.

5.2 The Pattern agent architecture

Each Pattern agent contains several mechanisms that allow:

1. to calculate the local characteristics `calculateLocalCharacteristics()` from the children characteristics;

2. to make some deduction `deduction()`. This unit is made of rules that are going to enrich the beliefs. For example, local threshold may be adjusted and then adapted to the local situation. A similar action is applied on weights allocated to statistical or edges characteristics. The strategy can then be tuned on the local context;

3. to detect local patterns `detectPatterns()`. This unit has to detect potentially interesting patterns

on the beliefs and to calculate a satisfaction coefficient for each detected pattern. This is the most important unit of the agent and the one on which learning procedures must be applied to;

4. to implement specific reflexes. Here we regroup all the behaviour which do not have a result intention but rather behaviour that make "laws" that we consider as being **locally** true. Consequently, this module contains simple treatments that we wish generic and robust. The developments in progress deal with propagation algorithms of contour energy waves aiming at detecting forms by concentrating energy on the zones where there exists a line or a curve. Thus, if the cognitive aspect of our agents fail to bring out semantically richer forms, the informations that have merged from these local interactions will be susceptible of restarting and orienting our treatment. This unit implements the reactive aspect of our agent which is important research axe on multiagent system [7].

The previous mechanisms are working on the agent beliefs that are taking into account some informations that have been previously asked to his acquaintances (children[4] and brothers [5]).

For agents with poor semantics, those mechanisms are purely procedural, they are working on numeric values such as the average, the variance, the gradient or the number of common pixels (also called affinity) between two agents. For agents with richer semantics, symbolic facts and predicates are calculated. So the previous mechanisms are implemented with coded rules in a Rete network [10]. The use of those rules finds two justifications :(1)The need of efficiently detecting potentially interesting patterns on the beliefs, what allows to limit the number of times we calculate the agent satisfaction function. (2)The need of being able to define dynamically new rules of data extraction. Indeed, if we dynamically add new pattern agents, it will be necessary to precise on which data relies the satisfaction function and how to extract those data. The interpreted aspect of the rules allows this dynamic definition of new behaviour easily.

Learning perspectives : Even if the current architecture has a static knowledge, it was realized to support, in the future, learning mechanisms. The basic component of the knowledge network must be "hand built" (fig 3 right part). But the purpose is

---

[4]lower level acquaintances that fusion to form the current agent

[5]same level acquaintances

to propose a hybrid approach (using the symbolics as well as the numeric) and a distributed approach, supporting the learning mechanisms.

Those adaptation characteristics seem, indeed, essential to us, when a machine is spelled what it is to do according to a certain situation, we are bound to fail if the machine is confronted to a complex world. In a world of incomplete information, an unforeseen situation can always happen. Moreover, as D.H.Wolpert and K.Tumer [11] point it out, "the multiagent systems which do not integrate any learning mechanisms need an important manual configuration work and often are not as supple and strong as we would like". It then seems necessary to pass on to the "meta" level, that is to say to tell the system how to learn to do. This current of thought is inspired by the "dynamic approach" of the cognition which considers that the cognition has to be physically incarnated to be developed thanks to a perception action loop. To allow this learning, the agents are structured in a network, the links between agents (in this network) and their behaviour according to a situation are doomed to evolve thanks to the use of various techniques such as neuronal networks, the Bayesian classifier, the "Case Based Reasoning", genetic algorithms [14] or reinforcement learning techniques [12]. More precisely, it is the link Situation$\rightleftharpoons$Action which will undergo, within each agent, whether a reinforcement or a weakening in accordance to rewards. A learning mechanism through rewards, inspired of Q-learning [12] [13] or "Bucket Brigade" from the classifiers systems[14], will allow a propagation of the rewards and, in consequence, a learning in cascade.

6 OSAgent: A tool to develop a multiagent system

The tools available to develop multiagent system did not answer our needs because of problems such as the mode of communication, i.e., in TCP/IP exclusively with no shared memory, or the excess of granularity. As an agent is connected to a process, we initiated the development of "OSAgent" which aims at providing a set of services rendering multiagent system development easier, since its information processing constraints are close to ours.

6.1 Features

Agent/computer process decoupling (fig 1): The concepts of computer process and agent are often mixed. They indeed share numerous notions, such

respective positions of two agents	message passing mechanisms	scheduling mechanisms
Same process, many threads	process address space	OS scheduler
Different processes, same computer	Shared memory *(PVM)*	OS scheduler
remote computers	*TCP/IP (PVM)*	OSs schedulers

Table 1: Message passing & scheduling mechanisms

as synchronization, mailboxes, synchronous and asynchronous communications, semaphores, etc. Actually, a process can carry one or a multitude of agents, which, besides, is often the case when, for instance, the agents'granularity is too fine for a whole process to be used. So a single Thread [23] is allocated to each agent. Threads are "lightweight" which means that the context switching overhead is low because all threads share the same address space.

Adjustable granularity: One should be able to insert very light (reactive) agents as well as heavier, more cognitive agents. There should be a potential of a high number of agents, i.e., tens of thousands.

Physical distribution of computation (fig 1): Given the great number of computations to make, we want to take advantage of the intrinsically-distributed feature of agents to share out the computation. Then the threads'computation can be distributed among multiprocessors (SMP) architecture or even among several different computers in a network.

Adequate mechanisms to send messages: Using the respective physical situation (table 1) of two agents (and the two threads that execute them) which want to communicate, a specific mechanism to send messages is required. The mechanism actually used should be transparent for the agent. PVM (Parallel Virtual Machine) [24] is used to facilitate interconnection between the processes.

High portability: To implement this work, the Java language has been chosen for its good compromise between portability, security, speed and facility to develop symbolic reasoning systems, as well as numerical mechanisms.

6.2 Service Layers

The agent manager: An agent manager is located in each (computer) process; it **schedules** the execution of the agents which are present in the process. Each agent is allotted a thread. This thread can be

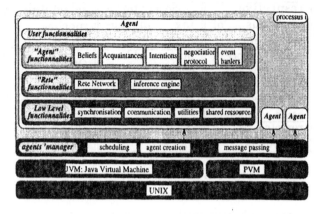

Figure 7: Generic agent architecture

level	number of agents	time in second	
		2 proc.	1 proc.
0	1833	24	39
1	298	11	12
2	48	0.6	0.72
3	10	0.1	0.16
4	2	0.003	0.003
total	2191	35,7	51,88
		gain: 1.45	

Table 2: Number of agents by level and required time to process with a biprocessor SMP (2 proc.) and single processor (1 proc.)

scheduled on the level of the manager or delegated to the operating system. The other role of the manager is to provide a **message passing system** as in operating system micro kernel architecture [22]. The message passing system routes messages among the agents by using the adequate mechanism to send messages. This, sometimes, involves the use of PVM. The messages take the form of objects or hierarchies of objects which, if needed [6], are serialized then deserialized on reception. Agents can thus send and receive complex-structured objects in a way which is totally transparent for us. The message structure also provides facility to implement, in the future, the KQML [8] message format.

Low level functionalities: supply an interface to send messages, **synchronization** and management services of **shared resources** among agents, a model of execution in asynchronous environment, as well as a set of tools enabling the filtering of incoming messages for instance.

"Rete" functionalities include an inference engine of the first order, using Rete algorithm, along with an engine of the same order working in backward chaining. Apart from the specific AI features given by such a tool, the data-driven aspect led us to include the tool in the agent by default.

"Agent" Functionalities include a set of services specific to agents but relatively generic in so far as they are not peculiar to our application. This layer, even though it is incomplete, already offers a set of services allowing the management of the **acquaintances** and **competences** possessed by an agent. A generic data base permits storing the agent's **beliefs**. Moreover a mechanism automatically collects the informations observable by the agent -in its environment

---

[6]if use of Shared memory or TCP/IP

and acquaintances- and automatically triggers reactive procedures if the environment changes. Protocols, i.e., contract net, enable the agents to start **negotiations**.

7 Evaluation

Our architecture was tested on the image "original image" of figure 8. This is a 192x192 image extracted from another image taken from the Brodatz album.

The characteristic used to merge or not agents are the following:

- average of grey level

- variances of grey level

- frontier length & frontier gradient. Those characteristics represent the affinity between two agents.

All those characteristics are weighted according to the local context.

7.1 Feasibility and technical aspect

The evaluation has been done on a bi Pentium® II (450MHz) PC computer, running Linux as operating system. The Linux kernel had to be modified to support up to 4090 threads.

The **Execution time** (table 2) shows that the monitor acts as a bottleneck. In fact, the gain in using two processors in only 1.45. In the future, a complete distributed architecture (with no monitor) is planned. But this ratio still shows an ability to really distribute the computation.

The **memory footprint:** is about 129 megabyte for 2192 agents; then each agent requires 59 kilobyte of memory. This is reasonable for a non embedded system.

7.2 Comments on results

The result of the segmentation (on the top figure of 8) shows a similar quality compared to a classic Adaptative Pyramid segmentation ([21] p. 342).

For each levels of the pyramid, an arbitrary color is given to all agents present in this level, the region represented by each agent is painted with this color.

In Levels 3, some agents[7] have disappeared (in black) because they no longer are interested in merging with other agents The same thing happens with the triangle in level 4.

But those agents, of course, appear in the "result image".

8 Conclusions and future work

In this paper, we have presented a multiagent approach for a machine vision architecture where the segmentation process is considered as a low level pattern recognition problem. All recognition is gradually processed in a growing pyramid of agents taking into account the knowledge (reflexes, pattern recognition mechanisms etc ...) contained in the network of pattern agents.

The tests realized on SMP architecture show an ability to really distribute the computation although the "fusion monitor" (which controls the merging process through a contract net protocol) acts as a bottleneck. The monitor should be replaced by a decentralized collaboration protocol that solves the problem created when the same agent is involved in several pattern detections. Indeed, patterns are locally detected in parallel by agents. The problem is to select among the detected patterns the one that will be in the next level of the pyramid without forgetting the non selected patterns.

Low level patterns should be improved in order to segment more difficult images. Development on higher level is in progress : work is done on the edge geometry to allow the recognition of simple shapes.

The key factor of a pattern recognition system is the way knowledge is represented to enable the observations to be matched with this knowledge representation efficiently. An essential feature of this representation should be its ability to learn. We consider that the agent philosophy provides an efficient method of representing knowledge and its learning ability.

Our concern is to have a network of agents composed of heterogeneous classification mechanisms operating on heterogeneous observations, cooperating for

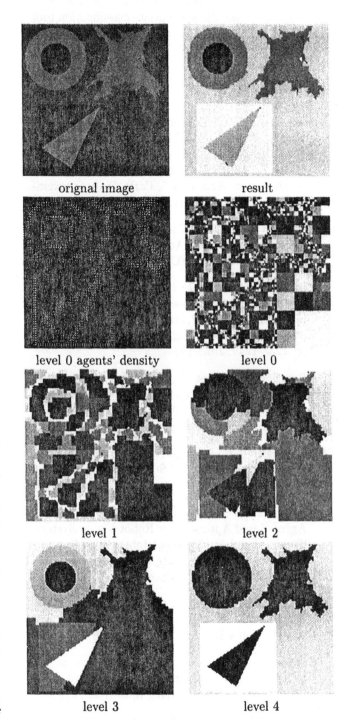

orignal image result

level 0 agents' density level 0

level 1 level 2

level 3 level 4

Figure 8:

---

[7]the small disc and the irregular shape

recognition as well as learning. Thus meta-knowledge must be extracted from the classification mechanisms and coupled with data-analysis mechanisms in order to automatically select the appropriate classification tool associated with discriminating features drawn from observations.

References

[1] Garnesson P., Giraudon G., Montesinos P.: Messie: Un Système Multi Spécialistes en Vision, Application l'Interprétation en Imagerie Aérienne. AFCET: "Reconnaissance des Formes et Intelligence Artificielle", Paris, France, november 29th - december 1st, 1989.

[2] Baujard O. Garbay C.: KISS: Un Système de Vision Multi-agents. Actes du VIIème Congrès "RFIA", pp 89-98, AFCET/INRIA, 1989.

[3] Baujard O Garbay C.: KISS: a Multi-Agent Segmentation System. Special Issue "From Numeric to Symbolic Image Processing : Systems and Applications". Optical Engineering, Vol. 32. No. 6. pp 1235-1249, June 1993.

[4] Boucher A., Ronot X., Garbay C.: Segmentation de séquences d'images cytologiques par un système multi-agents. JFIADSMA'96. (J.P. Mller & J. Quinqueton), eds, p. 125-135. Hermès, 1996.

[5] Gallimore R.J., Jennings N.R., Lamba H. S.: Co-operating Agents for 3-D Scientific Data Interpretation. IEEE SMC, vol 29, no 1, February 1999

[6] Cheng T. K. Kichten L. Liu Z. Cooper J.: An Agent-Based Approach for Robot Vision System. Technical Report 95/34, October 1995, 7 pages. Computer Vision and Machine Intelligence Laboratoy, University of Melbourne

[7] Brooks R. A.: A Robust Layered Control System For a Mobile Robot. IEEE J. Robotics and Automation v RA-2 nr 1, March 1986

[8] Finin T. & all: DRAFT, Specification of the KQML Agent-Communication Language. The DARPA Knowledge Sharing Initiative, 1993

[9] Davis R., Smith R.G.: Negociation as a Metaphor for Distributed Problem Solving. Artificial Intelligence, vol.20(1), p. 63-109

[10] Forgy C. L.: Rete: A Fast Algorithm for the Many Pattern/Many Object Pattern Match Problem. in Artificial Intelligence journal, vol. 19, p. 17-37, 1982

[11] Wolpert David H., Tumer Kagan: An Introduction to Collective Intelligence. Tech Rep. NASA-ARC-IC-1999-63.

[12] Sutton R.S., Barto A.G.: Reinforcement Learning An Introduction. MIT Press, 1998

[13] Watkins C., Dayan P.: Q-learning. Machine Learning journal, 1992

[14] Goldberg: Genetic Algorithms. Addison-Wesley, 1989

[15] Meer P.: Stochastic Image Pyramids. CVGIP, 45, p. 269-294, 1989

[16] Jolion J.M., Montanvert A.: The Adaptive Pyramid : A Framework for 2D Image Analysis. CVGIP:IU, 55, p. 339-348, 1992

[17] Horowitz, S.L., Pavlidis, T.: Picture Segmentation by a Directed Split and Merge Procedure. ICPR74, p. 424-433, 1974

[18] Theodoridis S., Koutroumbas K.: Pattern Recognition. Academic Press, 1999

[19] Risse T.: Hough Transform for Line Recognition: Complexity of Evidence Accumulation and Cluster Detection. ECCCV94, p. 95-100, 1994

[20] Yoo M.Y., Huangs T.S.: Pattern Recognition by Fourier Shape descriptor. Purdue Report Nov.1976-Jan.1977

[21] Cocquerez J.-P., Philipp S.: Analyse d'image: filtrage et segmentation. Masson edition, 1995

[22] Tanenbaum A.: Modern Operating Systems. Prentice Hall, 1992

[23] Bacon J.: Concurrent Systems: An Integrated Approach to Operating Systems, Database and Distributed Systems. Addison-Wesley, 1993

[24] Al Geist A., Beguelin A., Dongarra J.: PVM: Parallel Virtual Machine A Users' Guide and Tutorial for Networked Parallel Computing. MIT Press, 1994

CHAPTER 9

Design Technologies

Determination of the parameters of cutting tools in integrated design of products

K. Mawussi and V.-H. Duong

GRPI, Université Paris 13 (IUT de Saint Denis), Place du 8 mai 1945, 93206 Saint Denis Cedex

K. Kassegne

Département Génie Mécanique, Ecole Nationale Supérieure
d'Ingénieurs (Université du Bénin), B.P. 1515 Lomé - TOGO

Abstract

Concurrent engineering recommends a production process whereby the conceptual development of the products as well as the preparation of their manufacturing are carried out in parallel. The research works presented in this paper fall under this approach and more particularly in the integrated design, i.e. the satisfaction of the most significant manufacturing constraints in the design stage of the products. The method for the specification of cutting tools and the determination of their parameters proposed is based on the application of the constraints to each machining feature identified beforehand on the part. It leads to the creation of the minimal machining configurations.

1 Introduction

Concurrent engineering uses two main mechanisms to reduce the industrialization time of new products: on the one hand, the increase in the sharing of information since the beginning of the project to reduce the redundancy of the work carried out within the various services and functions of a company (functional participation) and on the other hand the extension of the overlapping of the activities (parallelism) [Bhu99]. The combination of these two mechanisms which the DFX (Design for X) proposes makes it possible to reduce considerably the problems encountered during the integration of the design and manufacturing. This integration consists in ensuring that the part has an intrinsic possibility to be manufactured [Bri98].

The DFX is an emergent philosophy making it possible to work out decisions related to the design of products and their production process simultaneously by examining their links [Hua99]. This improvement passes by the development of constraints relating to the products and process like their integration in engineering tools. There are several constraints related on the various features defined on the parts, on the relations feature-process, the machine tools, the cutting tools and the costs and lead times of the concurrent development of products [Gay99]. Just like the DFX, the integrated design express the need for taking into account immediately in the design stage of the product the constraints related to the manner of manufacturing it. Currently, there are not CAD/CAM tools allowing this taking into account.

The work presented in this paper falls under the step of integrated design. The purpose of it is principally the development of a integrated design tool: V-VIEW. More precisely, we expose here the method of determination of the cutting tools parameters which come to supplement their specification. The determination of the parameters, just like the specification of the cutting tools are based on voluntarily simplified models which we also detail. They lead to the generation of constraints whose satisfaction or not makes it possible to inform the designer about the validity of the produced model considered.

2 Basic models

During the design and manufacturing process of products, several constraints are defined. They affect the parameters related on the one hand to the definition model of the product and on the other hand to the means used in manufacturing. In order to facilitate the development of our integrated design tool, it is necessary to specify models with which the parameters are associated. Within the research work presented, two models were retained: the model of the part based on machining features and the simplified model of cutting tool. The minimal machining configurations which we seek to determine during the specification of the cutting tools are defined starting from an association of the elements of the two models selected.

2.1 Simplified model of cutting tool

The cutting tools represented are revolving tools used in milling of prismatic shapes. The generic model having been used as a basis for our work is that whose 54 parameters are defined in [Com92]. In practice, the tools often used for this type of shapes are the drills, the 2 or 3 sizes cylindrical mills and the tee slot mills or dovetails. The machining of surfaces corresponds in general to two principal machining modes: end milling and flank milling with a side or axial plunge. A complementary mode can be obtained by the combination of the two principal modes. These types of tools and machining modes lead to a simplified model comprising an axis with which we associate flat ends (a main end and possibly a secondary end) and/or a generatrix. The schematic representation of this simplified model is the following one:

- Axis = half right-hand line.
- Flat = segment of line.
- Generator = segment of line.

Functional surfaces of the tool which can be with three, we indicate them in the order by BGB (main end noted B1, noted generatrix G and secondary end noted B2). In the designation of the types of cutting tools, the absence of each of the three functional surfaces is announced by the letter X.

is quite obvious that for the choice of cutting tools in the development stage of machining process, the parameters (numbers some very reduced here) must correspond to those defined in the catalogues provided by the cutting tools manufacturers. It is for this reason that the simplified model has to integrate constraints defined explicitly or not by the cutting tools manufacturers.

2.2 Machining feature

In order to precisely locate the possible problems involved in the non satisfaction of constraints and to facilitate the processing, the part is broken up into machining features. Each one of these features is composed of a set E_k of faces f_i defined by their design data (dimensional and geometrical specifications, surface finish, surface treatment, etc...) and having to satisfy the following conditions:

- $\forall$ k, Card $(E_k) \geq 2$
- $\forall$ k and i, $f_i \in E_k \Rightarrow \exists f_j \in E_k$ / *material angle* $(f_i, f_j) \geq 180°$
- $\forall f_u$ and f_v, *material angle* $(f_u, f_v) \geq 180° \Rightarrow \exists k$ / $\{f_u, f_v\} \subset E_k$

The identification of the machining features on a part express the point of view of an expert in process planning who must define the shapes machined starting from the design data. As an example, the various classes of the features defined in our research work are given below (see figure 2).

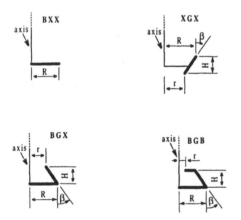

Figure 1   Cutting-tools classes of the simplified model

This simplified model which counts 4 types of tools (see figure 1) is used for the specification of the cutting tools. It

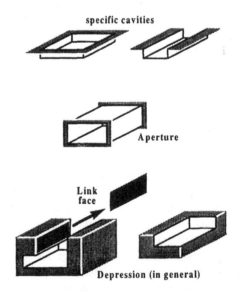

Figure 2   Classes of machining feature.

3 Determination of the parameters of cutting tools

The determination of the parameters of cutting tools requires their identification as a preliminary. In the integrated design tool that we develop, this identification is made in the specification stage of the cutting tools. In this specification stage, we primarily seek to determine the different minimal machining configurations (MMC) necessary to the realization of each machining feature identified on the part considered. The concept of MMC is clarified hereafter. The existence or not of the required MMC allows, in the context of integrated design, to inform the designer about the machinability of the part from a cutting tool point of view. For example the inexistence of MMC for a feature of the part makes it possible to locate the difficulties of choice of cutting tool. In the same way the fact that a feature can be machined only with one MMC whose cutting tool is relatively expensive makes it possible to draw the attention of the designer to the portions of a part which increase the total machining cost.

3.1 Minimal machining configuration

It is characterized by:
- on the one hand, the type of cutting tool (BXX, XGX, BGX or BGB), the position of the axis of the cutting tool (direction and orientation defined compared to the characteristics of the feature) and the parameters R, R, H and β;
- in addition, faces of the feature machined by the cutting tool.

According to the type of cutting tool, one can thus distinguish four categories of minimal machining configurations. The simple fact of specifying that a MMC is of type BGB is enough to inform about the type of cutting tool which characterizes it. Thus the standard information of cutting tool is implicitly related to the identity of each category of MMC.

3.2 Suggested method

Our approach for the specification of the cutting tools and then the determination of their parameters is based on the expertise of the process planner on the milling of the various classes of features. This expertise made it possible to set up constraints and their procedures of application for each type of MMC except for that of type BXX which is intended for the surfacing of the isolated faces coming from the envelope of the part. Let us note that in the

worked out procedures, the directed satisfaction of the constraints suitable for a type of MMC makes it possible to reduce the processing time and end to create one of its instances. On the whole, 26 constraints were defined (9 for the XGX, 9 for the BGX and 8 for the BGB) in [Duo99]. Their presentation in this communication would be too tiresome. We will thus limit ourselves to the presentation of the constraints defined for the MMC of the type BGB.

3.3 Constraints and procedures: case of the MMC of the type BGB

3.3.1. Definition of the constraints

The type of cutting tool BGB corresponds to the cutters 3 sizes mills and slot mills which are used to machine 3 faces at the same time. Let us note that theoretically, these cutting tools make it possible to machine at least two faces and with more the five faces. To identify a MMC of this type, let us start from two sets of faces B={f_i, f_j} and F={g_k / k=1 to n}. The faces f_i and f_j are machined respectively (except indication) by the ends B1 and B2 of the cutting tool. As for the faces g_k of the set F, they are machined by the generatrix G of the cutting tool.

Let's assume P_i a point of the face f_i and P'_i the image of P_i by orthogonal projection on the face f_j. The 3 following constraints express the specificity of the faces of B, the fact that these two faces are in opposite and the position of the faces of the set F (see figure 3a).

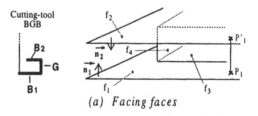

(a) Facing faces

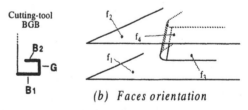

(b) Faces orientation

Figure 3　General layout of the faces of the feature.

- *Constraint C1*: f_i and f_j are plane faces.
- *Constraint C2*: $\overrightarrow{P_i P'_i} \cdot \overrightarrow{n_i} > 0$ and $\overrightarrow{n_i} = -\overrightarrow{n_j}$.
- *Constraint C3*:   All the faces of the set F have a concave link with the faces f_i and f_j.

In the example of the figure 3b, the faces f_1 and f_3 have a material angle (angle measured inside the material ranging between two faces and/or generatrix) equal to 90° whereas that of the faces f_1 and f_4 is different from 90°. In this case it is impossible to in addition consider the machining of these faces with only one MMC of the type BGB. The fact that the axis of the conical blending face is not perpendicular to the face f_1 makes impossible its machining with a MMC of the type BGB. The 2 following constraints allow to validate all these situations.

- *Constraint C4*:   All the faces of the set F have the same *out-of-material* angle (in opposition to the material angle) with the face f_i.
- *Constraint C5*:   The axis of any cylindrical or conical face belonging to the set F is perpendicular to the face f_i.

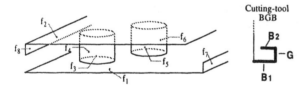

Figure 4   Division of the set F in islands

In the case of figure 4, the set F is composed of two islands of cylindrical faces. The machining of these two islands by a MMC of the type BGB is impossible, at least with regard to the space which separates them. Moreover, the face f_7 concave linked to the face f_1 makes it possible to consider the machining of the latter by surface B1 of the cutting tool provided dimensions and position of the face f_2 give us an accessibility. According to this observation, the simultaneous presence of the faces f_7 and f_8 make impossible the machining of the feature by a MMC of the type BGB. The constraints C6 and C7 presented below allow to validate all these situations. As for the last constraint C8, it expresses the fact that the use of such a MMC can be justified only if the set F is not empty.

- *Constraint C6*:   Whatever the couple of faces of the set F considered, there exists always a path making it possible to go from the one to the other by gradually traversing the faces of the set F dependent between them.
- *Constraint C7*:   At least one of the faces of B is not concave linked to faces not belonging to F
- *Constraint C8*:   The set F is not empty.

3.3.2. *Procedure of application of the constraints*

To create an instance of a MMC of the type BGB, all the constraints defined above must be satisfied without exception. The principal role of the procedure is to direct the application of the constraints in order to reduce the execution time of the algorithm during the implementation. In the event of satisfaction or not of the constraints, the procedure also specifies the actions to be carried out or the conclusions to be drawn. The procedure that we worked out comprises the five following stages.

- *Stage 1*:   Among the faces of the feature considered, we seek the faces f_i and f_j of the set B. At this occasion, we apply the constraints C1 and C2. If no set B is identified, the procedure stops.
- *Stage 2*:   For any set B identified at stage 1, we apply the constraint C3 to the others faces of the feature (not belonging to the set B identified). All the faces satisfying this constraint C3 constitute the set F. Finally, by application of the constraint C8 we eliminate any set B for which the constraint C3 is never satisfied.
- *Stage 3*:   We apply the constraint C7 to the faces of the set B to check if f_i and f_j are well *machinable*. If this constraint is not satisfied, we eliminate the set B considered. If only a face of B is not concave linked to faces not belonging to F, it is it which is machined by surface B1 of the cutting tool.
- *Stage 4*:   We apply the constraints C5 and C6 to check the validity of the faces of the set F. If any of these two constraints is not

satisfied, we eliminate the set B considered.

- •Stage 5: If necessary, we break up the set F into subsets whose faces satisfy the constraint C4. With each subset (or the set F if it is not broken up) we specify the angle β starting from the material angle who characterizes it.

3.3.3. Determination of the parameters of cutting tool

A the resulting one from the application of the constraints, the parameters of cutting tools must be evaluated to define the MMC of the type BGB corresponding to the sets B and F identified. Let us note that only the determination of the parameter β forms integral part of the procedure of application of the constraints. The limit value H_{max} of the parameter H corresponds to the distance between the faces f_i and f_j. Its determination leads to the specification of an interval of values $[H_{min}, H_{max}]$ for which H_{min} is a constant fixed by the manufacturers of cutting tools.

To determine the parameter R, we consider the face f_i machined by the end B1 of the cutting tool and limited by its common edges with the other faces of the part. It is significant to announce that a distinction is made between the edges which connect the face f_i to the other faces of the part (concave link) and the other edges (convex link). Initially, we treat only the edges which connect the face f_i to the other faces of the machining feature. The procedure of determination of the radius R comprises the four following stages.

- •Stage 1: We search for all the triplets of rectilinear edges (a_i, a_j, a_k) related one to the other by concave vertex. For each triplet of edges found, we calculate the diameter $\delta_{i.j.k}$ of the tangent circle with the three edges (see figure 5a).
- •Stage 2: We search for the rectilinear couples of edges (a_i, a_j) bound by a common convex vertex. For each couple of edges found, we calculate the distances from the vertex compared to all the other edges (see figure 5b). So that a distance thus calculated either considered validates, it is necessary that the point of minimal

distance is located on the edge. When it is not the case, it is the smallest distance compared to the vertex of the edge which is retained (see figure 5c). The machining diameter of $\Delta_{i.j}$ associated with the couple of edges considered is the smallest of all the calculated distances.

- •Stage 3: We associate a machining diameter of Γ_i with each non rectilinear concave edge (arc of circle). This diameter is simply that of the basic circle (see figure 5d).
- •Stage 4: We consider all the edges not linked to two other edges. In the same way that at stage 2, the isolated limiting points (not common runs with several edges) of these edges are regarded as convex vertex (see figure 5e). After processing of the isolated limiting points, we associate a machining diameter of ϕ_i with each edge considered.

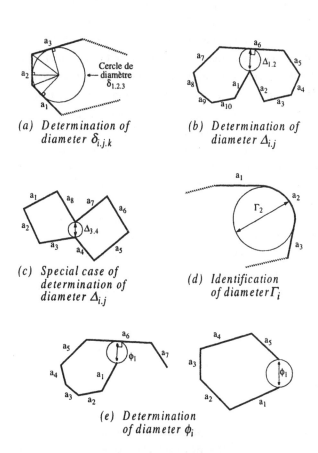

(a) Determination of diameter $\delta_{i.j.k}$

(b) Determination of diameter $\Delta_{i.j}$

(c) Special case of determination of diameter $\Delta_{i.j}$

(d) Identification of diameter Γ_i

(e) Determination of diameter ϕ_i

Figure 5   Determination of the parameter R for

the concave link edges

In the procedure of determination of R, the non rectilinear convex edges (arcs of circles) are replaced in an iterative way by curvilinear edges (see figure 6). These last being convex also between them, one associates to them at stage 2 a machining diameter of $\Delta_{i.j}$. The method of approximation of the arcs of circles was already validated in [Maw98] for the decomposition of ruled surfaces.

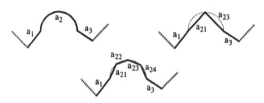

Figure 6   Approximation of a convex arc of circle by rectilinear edges

At the end of the stages, let us note D the smallest of the $\delta_{i.j.k}$, $\Delta_{i.j}$, Γ_i, and ϕ_i given previously. The maximum value of R (R_{max}) is equal to half of D. When the required final diameter corresponds to a $\delta_{i.j.k}$, machining leads to the disappearance of the edge a_j to the profit of an arc of circle. At this level, one can define a coefficient of recovery C_r (lower than one) whose multiplication with $\delta_{i.j.k}$ makes it possible to avoid the disappearance of the edge a_j.

For the edges b_k which connect the face f_i to faces of the part not belonging to the machining feature (convex link), the procedure of determination of the radius R appears simpler. Indeed, the minimal value of the radius R (R_{min}) is largest of $\Delta_{i.j}$. As the examples of figure 7 show it, $\Delta_{i.j}$ represents the distance between the vertex common to the edges b_i and b_j (i=0 for a vertex common to two faces a_u and b_i) and one of the edges a_k. If the point associated with the distance is not located on the edge a_k, the distance is calculated compared to the limit points of the edges a_k.

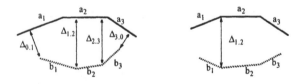

Figure 7   Determination of the parameter R for the convex link edges

In all the procedure of determination of the minimal value of the radius R, all the arcs of circles (convex and concave) are replaced by rectilinear edges according to the method of approximation presented previously.

For the face machined by the end B2 of the cutting tool, we determine in the same way the radius R' (R' max and R'_{min}) adapted to its machining. The maximum and minimal values of the radius r related to those of all the other parameters have as an expression:

$$r_{max} = R_{max} - (H_{max}.\tan\beta + R'_{min})$$
$$r_{min} = R_{min} - (H_{max}.\tan\beta + R'_{max})$$

A negative value for the parameter r corresponds to a problem of accessibility. In a general way, the study of the accessibility of the forms by the cutting tools cannot be dissociated from the method of determination of the parameters. The continuation of the work currently undertaken on this subject [Maw97] will make it possible to supplement the method presented in this communication.

4   Application to the integrated design: project V-view

Project V-view (Virtual VIEW) under development within the GRPI (Group of Search in Integrated Production) of the University Paris 13 aims at installing an integrated tool for the design of mechanical parts. Currently, this tool is articulated around the 4 following modules:

- topological analysis
- virtual specification of the means of production
- assistance to the development of machining process.
- development of machining process.

The first two modules are used to evaluate the model of the part in progress or at the end of the design. The evaluation results mainly in the identification of the machining features and the satisfaction of constraints associated with the means for production, in particular the cutting tools and the machine tools. These two modules thus provide information necessary to the third module. At the time of a request, the module of assistance to the design goes up information resulting from the first two modules in the form of difficulties and cost of machining and assists possibly the designer in the search for solutions of replacement. Let us note that the difficulties and cost of machining are given compared to problems such as: the incomplete cover of the part by the features, the non satisfaction of constraints, the use of specific means of production, etc...

Method of specification of the cutting tools and determination of their parameters presented here formed integral part of the second module. In its application to the integrated design we added to the already quoted constraints and procedures two new requirements. To express these two requirements, let us consider the examples of figure 8. On the one hand, the presence of two concave faces of the set F dependent (see figure 8a) led to a not machined zone which requires a blending surface. In addition, when the blending surface exists, its radius must be sufficiently large to cover the not machined zone (see figure 8b). The requirements thus make it possible to announce these problems of connection to the designer.

- Requirement 1: No plane face of the set F is concave linked to another plane face different of f_i and f_j.

- Requirement 2: The radius of any cylindrical or conical face of the set F is not smaller than those of the MMC considered.

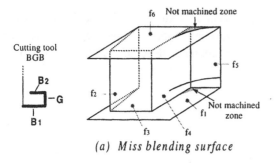

(a) Miss blending surface

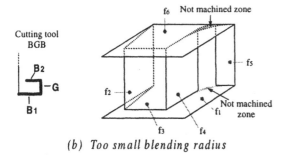

(b) Too small blending radius

Figure 8   Problems associated to the connection of the faces of the set F

The non satisfaction of the constraints related to the parameters of the cutting tools must also go up systematically (in suitable forms) with the designer. These constraints are defined starting from the recommendations of manufacturers of cutting tools. Lastly, certain information as for example the coefficient of recovery must be given by the designer.

With the exit of the application of the constraints to a machining feature, several cutting tools can be given. In this case, the cutting tools evaluated according to two criteria:

- Tool Cost: it is the minimal cost of the cutting tool (according to the manufacturers data base) corresponding to the most economic combination of its parameters (defined in the form of intervals).
- The number of blending surfaces between the faces of the machining feature induced by machining according to the minimal machining configuration (MMC) considered.

At this stage, the information provided to the designer relates to on the one hand the MMC having the weakest tool cost with the number of associated blending surfaces, and on the other hand the MMC having the lowest number of blending surfaces with its associated tool cost. In both cases, it can be the same MMC. Let us note in addition that this information is useful for the fourth module of V-view (development of machining process) on the level of which a best choice of cutting tools is essential.

Conclusion

The specification of the resources is a significant stage in the manufacturing preparation. In integrated design, the constraints on which it is based mainly determine the machinability of the shapes of the parts. Consequently, it is necessary to integrate, in the design tools, the constraints resulting from the specification of the resources.

In this paper, we presented part of the constraints related on the specification of the cutting tools and the determination of their parameters. The specification of the cutting tools integrates the minimal machining configurations which associate the characteristics of the cutting tools with the various faces of the machining feature considered. The presentation of the constraints is followed of that of a procedure whose role is to direct their application in order to reduce the execution time of the module concerned in V-view. The first implementation model which has been developed enabled us to validate the constraints. The research work presented here is in the course of implementation in V-view. It will be supplemented by other constraints which better define complex tools configurations.

Bibliography.

[Bhu99]   Bhuiyan N. and Thomson V. *Simulation of concurrent engineering process*. Proceedings of the international conference on industrial engineering and production management FUCAM 99, Glasgow, July 12-15, 1999, pp. 211 - 221.

[Bri98]   Brissaud D., Martin P. *Process planning: from automation to integration*. IFAC Symposium on Information Control Problems in Manufacturing, Nancy-Metz, june 1998, pp. 231 - 236.

[Cai98]   Caillaud E., Lamothe J. and Lacoste G. *Concurrent engineering and cooperative design: sharing the risk and integration of the know-how*. Proceedings of IDMME'98, Compiègne - France, May 27-29, 1998, pp. 1061-1068.

[Com92]   Computervision Corporation. *CADDS 5 Release 2, CVNC M3 User Guide*, Copyright June 1992.

[Duo99]   Duong V.-H. *Spécification des outils de coupe en conception intégrée*. Mémoire de Recherche, DEA de Production Automatisée, Juillet 1999.

[Gay99]   Gayretli A., Abdalla H. S. *An object-oriented constraints-based system for concurrent product development*. Robotics and Computer-Integrated Manufacturing, Vol. 15 (1999), pp. 133 - 144.

[Hua99]   Huang G. Q., Lee S. W., Mak K. L. *Web-based product and process data modelling in concurrent "design for X"*. Robotics and Computer-Integrated Manufacturing, Vol. 15 (1999), pp. 53 - 63.

[Maw97]   Mawussi K. *Détermination de l'accessibilité de formes prismatiques en conception intégrée*. 2ème Colloque National de Productique, Casablanca Maroc, Novembre 1997.

[Maw98]   Mawussi K., Bernard A. *Three-dimensional cutting-tool-path restriction. Application to ruled surfaces approximated by plane bifacets*. Computers in Industry 35 (1998) p. 247 - 259.

[Tia98]   Tian H., Xu W., Wend H.-D. and Wu Q. *A review upon concurrent engineering*. Conference IFAC'98, 1998, pp. 511-516.

An Intuitive Approach For The Segmentation Of Parametric Curves And Cross-Sections

A. Hatna[*] and R. J. Grieve

Systems Engineering Department
Brunel University, Uxbridge, Middlesex
UB8 3PH, UK
[*]Email: empgamh@brunel.ac.uk

Abstract

In general, displaying, plotting or generating a tool path along a curve goes through the discretisation of this curve into a set of successive points. We present a method achieving the discretisation of parametric curves with two features: it is error-control based and direct since it allows the discretisation within the specified tolerance and without using any iterative scheme. The method is extended to resolve the overspecified problem of parametric surface/plane intersection.

Keywords: *Parametric Curve, Parametric Surface, Discretisation, Intersection*

1. INTRODUCTION

Parametric functions are omnipresent in Computer Aided Design and Manufacturing (CAD/CAM) systems and are used for the modelling of objects due to their numerous mathematical features. Typically, the physical representation (as a drawing or tool path) of a parametric function is usually reduced (by interpolation) to a simpler model before its subsequent processing into a form suitable for use by, for example, CNC machines. Two main methods of interpolation are used: linear and circular interpolation. In such cases, the CAD/CAM software approximates the parametric curve into a set of lines and/or arcs and an internal program associated with the printing device or machining converts them into smaller linear motions within real time. In other words the linear interpolation of a curve (also called *segmentation, discretisation* or *polygonisation*) means that it may be represented by a set of linear segments (P_0P_1), .., $(P_{i-1}P_i)$, .., $(P_{q-1}P_q)$ within a specified tolerance. By tolerance, segmentation error, or chord error we mean how closely the line segments follow the curve. Circular interpolation is the representation of a curve by a set of arcs of circle. In the following, we are only concerned with *linear interpolation*. Details about circular interpolation can be found elsewhere [1, 2]. Recently, spline interpolation has been integrated into many CNC

units and many researchers have shown interest in exploiting this method [3, 4].

In the following we present two methods related to the segmentation of parametric curves. The first one deals with the linear interpolation of curves within a specified tolerance. An intuitive approach is used to establish an incremental algorithm of segmentation and some examples are discussed and compared to a standard method. In the third section, this approach is expanded to consider the resolution of the problem of the Parametric Surface and Plane Intersection (PSPI). This is important because the PSPI (usually used for the computation of tool paths, visualisation, and surface and solid manipulations) is considered as an *overspecified problem*, where a non-linear equation in two unknowns is to be resolved. The algorithm proposed bypasses this problem by using dimensional criteria rather than the conventional intersection equation.

2. SEGMENTATION OF PARAMETRIC CURVES

To perform a linear interpolation of a parametric curve $P(u)$, $0 \le u \le 1$, the simple and direct method commonly used is the *uniform method*. It consists to increase the real parameter u by a constant amount Δu (that is why the term *uniform* is used) such that

$$u_0 = 0$$
$$u_{i+1} = u_i + \Delta u$$

So a set of points $P_0 = P(u_0)$, $P_1 = P(u_1)$, $P_2 = P(u_2)$, .., P_q ($u_q = 1$) approximating $P(u)$ is generated. To make this method faster, the *forward difference* technique can be used [5].

These methods, though they are simple, do not take into account the shape of the curve and its variation, and most importantly, the accuracy of the segmentation is not known in advance. Also the variation of the distance between the successive points of interpolation is unknown, while from a CNC metal cutting perspective it can result in an unknown and undefined variation of the feedrate (consequently quality machining problems) along the curve [6]. This makes it difficult to use these methods for machining purposes. Some methods have been proposed to overcome this problem. *Dichotomy* approaches can be used and are based upon successive division, in the middle, of the parameter interval of the curve wherever the chord error exceeds the specified maximum chord error. This method allows control over the segmentation error but it is time consuming; it may also fail in same cases of curve symmetry.

The method given in [6] allows the calculation of the next parametric value u_{i+1} on the basis of the current value u_i and the first partial derivatives $\dot{x}_i$ and $\dot{y}_i$ of the current position such that

$$u_{i+1} = u_i + \frac{VT}{\sqrt{\dot{x}_i + \dot{y}_i}}$$

where V is the feedrate and T is a sampling period.

Kiritsis [7] has found that the next value of the parameter u corresponding to the next point on the curve can be obtained as

$$u_{i+1} = u_i - \frac{\sum\limits_{i=1}^{3} (X_i - x_i)\dot{x}_i}{\sum\limits_{i=1}^{3} (X_i - x_i)\ddot{x}_i - \sum\limits_{i=1}^{3} \dot{x}_i^2}$$

where (X_1, X_2, X_3) are the co-ordinates of the interpolating point and (x_1, x_2, x_3) are the co-ordinates of the corresponding point on the curve. This formula is given from the *normality condition*: $\mathbf{n}.\mathbf{t} = 0$ where $\mathbf{n} = (X_1-x_1, Y_2-y_2, Z_3-z_3)^{\mathrm{T}}$ and $\mathbf{t}$ is the tangent vector to the curve at the point (x_1, x_2, x_3). A similar approach is used in [8] for a "quick" evaluation of the segmentation error δ.
In [9] an approach based on the use of genetic algorithms is used to perform a polygonisation of curves.

A comparison, from the point of view of memory size, feedrate fluctuation and CPU time, between two methods of linear segmentation has been conducted in [10]: the first method uses an increment on the cartesian co-ordinates on the curve whilst the second increments the value of the parameter of the curve.

The methods described above do not take into account the error of segmentation (except for the dichotomy method but this is too slow) which is the main criterion used to evaluate the quality of an approximation in, say, a machining work. The approach proposed below aims to overcome this disadvantage and to provide a fast and chord - error - based algorithm.

2.1 Chord - error - based segmentation curve method

Let us assume that a point $P(u)$ on the parametric curve (P) is the current point and we want to determine the next point $P(u+\Delta u) = (x(u+\Delta u), y(u+\Delta u), z(u+\Delta u))^{\mathrm{T}}$ which verifies the condition

$$\delta = \left\| P(u + \frac{\Delta u}{2}) - \frac{P(u + \Delta u) + P(u)}{\beta} \right\| < \delta_0 \quad (1)$$

where

$$\left\| P(u + \frac{\Delta u}{2}) - \frac{P(u + \Delta u) + P(u)}{\beta} \right\| =$$

$$\left[\sum_{i=1}^{3} \left[x_i(u + \frac{\Delta u}{2}) - \frac{x_i(u + \Delta u) + x_i(u)}{\beta} \right]^2 \right]^{1/2}$$

$$\quad (2)$$

- δ is the maximum deviation between the curve and the chord between successive points and δ_0 is the allowed chord error.
- β is a real and positive number corresponding to the maximum chord error δ of segmentation. In practice β is taken as the projection of the point $P(u+\Delta u/2)$ onto the segment $P(u)P(u+\Delta u)$. Resulting is β being generally close or equal to 2.

Using the Taylor expansion around Δu we can define $P(u+\Delta u)$ as

$$P(u + \Delta u) = P(u) + \Delta u \dot{P}(u) + \frac{\Delta u^2}{2!} \ddot{P}(u)$$

$$+ \frac{\Delta u^3}{3!} P^{(3)}(u) + ... + \frac{\Delta u^m}{m!} P^{(m)}(u) \quad (3)$$

$$+ \zeta(\Delta u^{m+1})$$

where Δu represents a small variation of u; $\dot{P}(u)$, $\ddot{P}(u)$, .., $P^{(m)}(u)$ are the partial derivatives of the first, second until m^{th} order of $P(u)$ respectively, and $\zeta(\Delta u^{m+1})$ is the approximation error.

In order to simplify the writing we note $P(u) = P$, $\dot{P}(u) = \dot{P}$ and so on.
Considering that
- the third and higher order terms can be neglected,
- the chord error δ is maximum (the worst case) and
- β is **not equal** to 2 (this assumption has been included in order to avoid the trivial case in equations (7) and (8)).

Therefore we can rewrite the inequality (1) as

$$\left\| \frac{\beta - 2}{\beta}(P + \frac{\Delta u}{2}\dot{P}) + \frac{\beta - 4}{8\beta}\Delta u^2 \ddot{P} \right\| \approx \delta_0 \quad (4)$$

Since the relationships

$$\sum_{i=1}^{3}\left[\frac{\beta - 2}{\beta}(x_i + \frac{\Delta u}{2}\dot{x}_i) + \frac{\beta - 4}{8\beta}\Delta u^2 \ddot{x}_i\right]^2 \geq 0$$

$$(5a)$$

and

$$\delta_0 \geq 0 \quad (5b)$$

are always true so we can state

$$\sum_{i=1}^{3}\left[\frac{\beta - 2}{\beta}(x_i + \frac{\Delta u}{2}\dot{x}_i) + \frac{\beta - 4}{8\beta}\Delta u^2 \ddot{x}_i\right]^2 \approx \delta_0^2$$

$$(6)$$

Further developments of the above equation give the following:

$$\left[\frac{a^2}{4}\sum_{i=1}^{3}(\dot{x}_i^2 + a \cdot c \cdot x_i \dot{x}_i)\right]\Delta u^2$$

$$+ a^2 \sum_{i=1}^{3} x_i \dot{x}_i \Delta u + a^2 \sum_{i=1}^{3} x_i^2 \quad (7)$$

$$- \delta_0^2 = 0$$

where $a = \dfrac{\beta - 2}{\beta}$ and $c = \dfrac{\beta - 4}{4\beta}$

(7) is a quadratic polynomial equation, thus, analytical solutions can be obtained.

If the third order terms of Δu are kept we obtain

$$(a(\frac{1}{24} - \frac{1}{3\beta})PP^{(3)} + \frac{a}{2}(\frac{1}{4} - \frac{1}{\beta})\dot{P}\ddot{P})\Delta u^3$$

$$+ (\frac{a^2}{4}\dot{P}^2 + (\frac{1}{4} - \frac{1}{\beta})aP\ddot{P})\Delta u^2 + (a^2 P\dot{P})\Delta u \quad (8)$$

$$+ a^2 P^2 - \delta_0^2 \approx 0$$

where $P^{(i)}P^{(j)} = x^{(i)}x^{(j)} + y^{(i)}y^{(j)} + z^{(i)}z^{(j)}$

The segmentation produced by (8) is much better than that one produced by (7) but we limit our investigation to the second order partial derivatives (equation 7) because:

• CAD/CAM systems rarely use curves of degree greater then 3, so higher partials may not exist.

• In general it is not possible to resolve analytically polynomial equations of degree greater than 2. Using

a numerical scheme may result in a slower segmentation algorithm because in this case all roots must be found numerically.

If the discriminant of the equation (7) is equal to or greater than zero, we will have double or two different zeros. If there are two different zeros the value to hold is the one that gives the closest deviation to the allowable tolerance but without exceeding it. However and in order to avoid an additional computing time it is preferable to take the smallest value of the two zeros.

If the curve being processed is a tool path where the feedrate along it must be kept constant for machining considerations [6], the method developed above can be used to perform the segmentation in respect of that requirement as follows.

Let L be the length of a linear segment between $P(u)$ and $P(u+\Delta u)$ that the cutting tool must follow within a time T with a feedrate V such that $L = VT$. So

$$\|P(u + \Delta u) - P(u)\| = L \quad (9)$$

Such an equation can be written as

$$[x(u + \Delta u) - x(u)]^2 + [y(u + \Delta u) - y(u)]^2 = L^2$$

$$(10)$$

Substituting the expansion, until the second order, of $P(u+\Delta u)$ in equation (10) gives

$$(\Delta u\dot{x} + \frac{1}{2}\Delta u^2 \ddot{x})^2 + (\Delta u\dot{y} + \frac{1}{2}\Delta u^2 \ddot{y})^2 = L^2 \quad (11)$$

By neglecting the terms of Δu with powers greater than 2 we get

$$\Delta u^2(\dot{x}^2 + \dot{y}^2) \approx L^2 \quad (12a)$$

or

$$\Delta u = L / \sqrt{\dot{x}^2 + \dot{y}^2} \quad (12b)$$

Since that $\Delta u = u_{i+1} - u_i$ and $L = VT$, so

$$u_{i+1} = u_i + VT / \sqrt{\dot{x}^2 + \dot{y}^2} \quad (13)$$

which is equivalent to the *equation 23* given in reference [6].

2.2 Application

To evaluate the performances of the proposed method, hundreds of parametric curves have been segmented by the proposed method and compared to the uniform method presented in the beginning of this section. For lack of space, we present the two examples of Bézier curves having the control polygons shown in the appendix.

• The first example is a Bézier curve of degree 6 with no singularities. The co-ordinates of its vertices are given in *Appendix* as curve 1. The curve is segmented using constant parameter steps (Δu) of 1/200, 1/300 and 1/533. The last value corresponds to the uniform parameter step that guarantees the segmentation of the curve within a chord error of 1μm. This error is also used as the maximum allowable error for the proposed approach. The parameter β has been taken as $\beta = 1 - \delta_0 / 20$. The results of this comparison are shown in *Fig. 1*.

• The second example is a Bézier curve of degree 9 with an inflection point (control polygon is given as curve 2 in the appendix). The method proposed is used to linearly interpolate the curve within a tolerance of 0.01 mm. Constant parameter steps of 1/100, 1/150 and 1/215 are used for the uniform method. It is the step $\Delta u = 1/215$ which generates the segmentation within the tolerance of 0.01 mm. The results are shown in *Fig. 2*.

Figures 1 and 2 show the clear superiority of the proposed approach. Meanwhile CPU time tests have shown that for equivalent segmentation precision the computation times for the suggested method are very short. The tests have been conducted for precision of

values as small as 0.1 μm, and polynomials of degree as high as 30.

2.3 Discussion

The main drawback of the uniform method and the methods given in [5, 6, 7] is that the segmentation error δ could not be known in advance. In contrast, the method proposed in this paper is based on the condition that the segmentation error must be less than the specified tolerance. This aspect is very important in form analysis and machining.

Another aspect of the proposed approach is the sensible improvement of the number of segments needed to perform the discretisation in the case of small tolerances. For example, a reduction of at least 30% in the number of segments is achieved with curve 2 when the segmentation error of 0.1μm is imposed. Such a feature is important in CNC machining when memory and size of G-code files are to be considered.

During the experimentation of the proposed method, we have realised that if the error segmentation is excessively small or β is not enough close to 2, the method doesn't guaranty the precision required. It has been found that the interval [$1-\delta_0/10$, $1-\delta_0/20$] is a suitable range for the choice of β to get precision going until 0.1 μm.

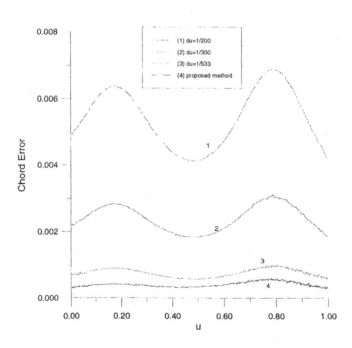

Fig. 1. Comparison between the uniform segmentation and the method proposed for curve 1 of appendix.

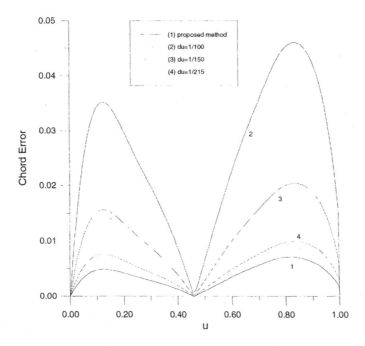

Fig. 2. Comparison between the uniform segmentation and the proposed method for curve 2 of appendix.

3. SEGMENTATION OF CROSS-SECTION CURVES

The problem of intersection between a surface and a plane is very important. It is used to plan tool paths [11, 12, 13], to obtain cross-sections and sliced objects used to show the hidden details of an object or to get a new object. The intersection may produce open curves and /or closed curves.

Assume we have a biparametric surface (P) defined as

$$(P) = \{ \ P(u, v) = (\ x(u, v), \ y(u, v), \ z(u, v)\ )^{\mathrm{T}} \\ \text{for} \ \ u \times v = [0, 1] \times [0, 1] \ \} \qquad (14)$$

and a plane π to be intersected. The implicit equation of the plane π may be expressed as

$$a\,x + b\,y + c\,z + d = 0 \qquad (15)$$

By substituting the co-ordinates of the surface into the equation of the plane we get

$$a\,x(u, v) + b\,y(u, v) + c\,z(u, v) + d = 0 \qquad (16)$$

which is the equation of a 3D curve. To make this curve solvable (because it is in fact an equation in two unknowns u and v) one must have a method that allows single points on it to be determined. The method commonly used consists of three steps [12, 13]:

- detection of an Initial Point of Intersection (IPI),

- determination of a *step of progress* along the curve and
- determination of the *next point* of intersection.

To perform the first and second steps, Choi [13] has suggested a method that involves resolving two linear equations in order to get an approximate position for the next intersection point. The *real* intersection point is then refined by intersecting a 3D curve and the plane π.

In [12] it is suggested the use of the step length L defined as $L^2 = 4\delta(2\rho - \delta)$, where δ is the maximum deviation between the true intersection curve and the chord between successive points, measured normal to the chord and ρ is the radius of curvature of the surface. Once L is obtained, the next point of intersection must be determined by resolving a non-linear equation.

In this section we present our development of an algorithm for the Parametric Surface/Plane Intersection (PSPI). The approach is based, as for the curve segmentation, on the intuitive definition of the linear interpolation error. The approach consists of 3 steps:

- segmentation of the *drive curve* (DC),
- determination of the *next point* and
- termination condition.

In the following we assume that the surface will be sliced perpendicularly to the Z axis. However, the approach remains the same whatever the slicing axis is.

3.1 Segmentation of drive curve (DC)

We define a drive curve $S(t)$ as an isoparametric curve of the surface $P(u, v)$ that contains the initial points of the PSPI curves. Ideally, the highest and lowest Z co-ordinates occur on the DC. Otherwise more than one curve may be used for the same surface. In the case of open curves, it is generally the case that one of the edges of the surface is the DC. Thus, the DC is defined as

$$S(t) = \{\{P(u, v = 0) \text{ or } P(u, v = 1), 0 \leq u \leq 1, t = u\} \\ \text{or } \{P(u = 0, v) \text{ or } P(u = 1, v), 0 \leq v \leq 1, t = v\}\} \tag{17}$$

In the case of close curves, the DC is taken as

$$S(t) = \{\{P(u, v = v^*), 0 \leq u \leq 1, t = u, v^* = \text{constant}\} \\ \text{or } \{P(u = u^*, v) \ 0 \leq v \leq 1, t = v, u^* = \text{constant}\}\} \tag{18}$$

The initial point of intersection (IPI) for the kth slicing plane may be determined by resolving the equation
$$k \, h = S_z(t) \tag{19}$$

where h is the distance between the successive planes of slicing (the height step of slicing) and S_z is the z co-ordinate of S.

A fast and reasonably accurate method can be deduced as follows.

$$S_z(t) - S_z(t + \Delta t) = h \tag{20}$$

Dividing both sides by $t - (t + \Delta t)$ gives

$$\frac{S_z(t) - S_z(t + \Delta t)}{t - (t + \Delta t)} = -\frac{h}{\Delta t} \tag{21}$$

Δt is very small thus

$$\Delta t = \frac{h}{\dfrac{\partial S_z(t)}{\partial u}} \tag{22}$$

3.2 Determination of the next point

The parameter steps Δu and Δv of two parameters u and v for a given surface (P) are to be found from the current point $P(u, v)$. These parameter steps must verify the conditions

$$\begin{cases} \left\| P(u + \dfrac{\Delta u}{2}, v + \dfrac{\Delta v}{2}) \dfrac{P(u + \Delta u, v + \Delta v) + P(u,v)}{\beta} \right\| \\ < \delta \\ z(u + \Delta u, v + \Delta v) = z(u,v) \end{cases} \tag{23}$$

where β has the same significance as stated in §2.1 and δ is the chord error in the XY plane.

The Taylor expansion around Δu and Δv of the biparametric function $P(u+\Delta u, v+\Delta v)$ can be written as

$$P(u + \Delta u, v + \Delta v) = P + \Delta u P_u + \Delta v P_v + \frac{1}{2!}\Delta u^2 P_{uu}$$
$$+ \frac{1}{2!}\Delta v^2 P_{vv} + \frac{1}{2!}\Delta u \Delta v P_{uv} + \zeta(\Delta u^3, \Delta v^3) \tag{24}$$

where $P = P(u, v) = (x(u, v), y(u, v), z(u, v))^T = (x, y, z)^T$, $P_s = \dfrac{\partial P}{\partial s}$ and $P_{ts} = \dfrac{\partial^2 P}{\partial t \partial s}$.

Limiting the expansion to only the first order derivatives and neglecting the second and higher order terms give the following analytical solution

$$\Delta v = \frac{(x^2 + y^2 - a)z_u}{z_v(xx_u + yy_u) - z_u(xx_v + yy_v)} \tag{25}$$

$$\Delta u = -\Delta v \frac{z_v}{z_u}$$

where $\quad a = (\dfrac{\beta \delta}{2 - \beta})^2$

The implementation of this solution may occasionally give unacceptable values (more precisely it may not guaranty the suitable segmentation accuracy). Using the expansion with the second derivatives but neglecting the third and higher order terms yields

$$4a^2(x^2 + y^2) - \delta^2 + c_1 \Delta u + c_2 \Delta v + c_3 \Delta u^2 + c_4 \Delta v^2 + c_5 \Delta u \Delta v = 0 \tag{26a}$$

$$z_u \Delta u + z_v \Delta v + (z_{uu}\Delta u^2 + z_{vv}\Delta v^2 + z_{uv}\Delta u \Delta v)/2 = 0 \tag{26b}$$

where
$a = (2 - \beta)/2\beta$
$c = (4 - \beta)/4\beta$
$c_1 = 4a^2(xx_u + yy_u)$
$c_2 = 4a^2(xx_v + yy_v)$
$c_3 = a^2(x_u^2 + y_u^2) + 2ac(xx_{uu} + yy_{uu})$
$c_4 = a^2(x_v^2 + y_v^2) + 2ac(xx_{vv} + yy_{vv})$
$c_5 = 2a^2(x_u x_v + y_u y_v) + 2ac(xx_{uv} + yy_{uv})$

These two equations are not linear. They can be resolved by Newton - Raphson method. We notice that the position P and partials $\dot{P}$ and $\ddot{P}$ are to be computed only once for each step and they are not to be re-evaluated at each iteration of the Newton-Raphson process. The computation of P and its partials may consume significant CPU time. The algorithms of de Casteljau and Cox – de Boor are easily extensible to surfaces as shown in [14]. Other algorithms for that purpose can be found in [15,16,17,18]. An example of slicing a Bézier surface is presented in Fig. 3.

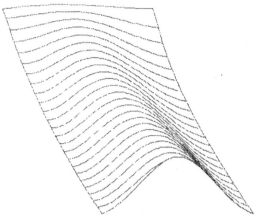

Fig. 3. Slicing of a Bézier surface by the proposed method: a case of open curve where the error in the plane *XY* is 0.01mm and the depth of slicing is 10mm

3.4 Termination condition

For an open curve, the process stops when u or v goes out the interval [0,1]. Then the curve will be completed by searching a point of intersection between the current slicing plane and an edge curve. In the case of closed curves we suggest the use of the accumulation of angles of the segmentation polygon (*AA*) rather than the distance test between the IPIs and the current position as given in [13]. The accumulation angle test is

$$AA = \left| \sum_{i=2} \theta_i \right| = \left| \sum_{i=2} (\angle P_{i-2}, P_{i-1}, P_i) \right| \geq 2\pi$$
(27)

where
$P_i = P(u, v)$ is the current point on the curve of PSPI,
$P_{i-1} = P(u - \Delta u_{i-1}, v - \Delta v_{i-1})$
and so on

In other words the termination condition is verified if the absolute value of the sum of the angles between the successive segments of the approximated curve is greater than or equal to 2π.

4. CONCLUSION

In this paper, two incremental segmentation algorithms are presented. The first one allows the

linear interpolation within the specified tolerance of parametric curves. It is direct and fast enough, with sufficient accuracy, to suggest it as an algorithm for interactive curve manipulation and toolpath generation. The second algorithm deals with the computation of the curve of intersection of a parametric surface and a plane. Its implementation and test over several examples has revealed that it is a suitable tool for the pre-processing of free forms roughing.

REFERENCES

[1] Y. Koren and O. Masory, "Reference - word circular interpolators for CNC systems", *Trans. ASME J. Eng. Indust.*, vol. 104, pp. 400-405, 1982.

[2] G. W. Vickers and C. Bradley, "Curved surface machining through circular arc interpolation", *Computers in Industry*, vol. 19, pp. 329-337, 1992.

[3] M. K. Yeung and D. J. Walton, "Curve fitting with arc splines for NC toolpath generation", *Computer Aided Design*, 26(11), pp. 845-849, 1994.

[4] D. S. Meek and D. J. Walton, "Approximating smooth planar curves by arc splines", *J. of Computational and Applied Mathematics*, vol. 59 pp. 221-231, 1995.

[5] I. Zeid, *CAD/CAM theory and practice*, McGraw-Hill Inc., 1991.

[6] M. Shpitalni, Y. Koren and C. C. Lo, "Real time curve interpolators", *Computer Aided Design*, 26(11), pp. 832-838, 1994.

[7] N. Kiritsis, "High precision interpolation algorithm for 3D parametric curve generation", *Computer Aided Design*, 26(11), pp. 850-856, 1994.

[8] D. Qiulin D and B. J. Davies, *Surface engineering geometry for computer aided design and manufacture*, Ellis Horwood, Cheichester, 1987.

[9] S. C. Huang and Y. N. Sun, "Polygonal approximation using genetic algorithm", Proc. of IEEE Int. Conf. on Evolutionary Computation (ICEC'96), pp. 469-474, 1996.

[10] D. C. H. Yang and T. Kong, "Parametric interpolator versus linear interpolator for precision CNC machining", *Computer Aided Design*, 26(3), pp. 225-234, 1994.

[11] J. C. Leon, *Modélisation et construction de surfaces pour la CFAO*, Hermès, Paris, 1991.

[12] I. D. Faux and M. J. Pratt, *Computational geometry for design and manufacture*, Ellis Horwood, Chichester, 1987.

[13] B. K. Choi, *Surface modeling for CAD/CAM*, Elsevier, Amestrdam, 1991.

[14] K. Spitzmuller, "Partial derivatives of Bézier surfaces", *Computer Aided Design*, 28(1), pp. 67-72, 1996.

[15] T. W. Sederberg, "Point and tangent computation of tensor product rational Bézier surfaces", *Computer Aided Geo. Design*, vol. 12, pp. 103-106, 1995.

[16] S. Mann and T. DeRose, "Computing values and derivatives of Bézier and B-spline tensor products", *Computer Aided Geo. Design*, 12(1), pp. 107-110, 1995.

[17] T. Saito, G. J. Wang and T. S. Sederberg "Hodographs and normals of rational curves and surfaces", *Computer Aided Geo. Design*, 12(4), pp. 417-430, 1995.

[18] G. I. Wang, T. W. Sederberg and T. Saito, "Partial derivatives of rational Bézier surfaces", *Computer Aided Geo. Design*, 14(4), pp. 377-782, 1997.

APPENDIX

- Control points of Bézier curve 1:
  (45, 165), (0, 112), (5, 37), (121, 0), (257, 24), (244, 108), (195, 165).

- Control points of Bézier curve 2:
  (0, 100), (40, 150), (80, 200), (180, 150), (250, 120), (200, 80), (300, 0), (400, 80), (350, 150), (300, 200).

Geometrical Modeling Based on Design Intention in 3-Dimensional CAD System (Concurrence between the consideration of specification and geometrical modeling)

H. Ishikawa   H. Yuki   K. Nishikata

Dept. of Mechanical Engineering and Intelligent Systems
The University of Electro-Communications
Chofu, Tokyo 182-8585, Japan

Abstract

Although conventional three-dimensional CAD systems are useful for modeling the geometry, those systems have to be operated by the view point of feature. In the present study, a strategy of the modeling based on design intention, that is named the intention-driven modeling (IDM), is proposed to achieve the concurrency between each process of the consideration for specification and the geometrical modeling in the mechanical design. The concept of IDM is described and its prototype system is developed with the object-oriented approach to demonstrate examples of the modeling. Since the system represents the component of the model as the object that has qualitative information, the designer can operate the system by the communication based on design intention from the stage of initial design. The system also brings such benefit that the designer can concentrate the attention on the specification in the design.

1 Introduction

Concurrent engineering (CE) aims the cooperative work by engineers who participate in design and manufacturing of industrial products [1]-[6]. To realize the concept of CE, a common work space, where cooperative workers can interactively define, recognize, modify and evaluate the product model, should be provided [7]. It is also required that the design intention, that is the purpose for generating the shape of products, is shared among the cooperative workers uniformly in order to communicate smoothly with each other.

Therefore, the design environment where anyone can operate the system centering on the design intention is effective in improving concurrence.

Three-dimensional (3-D) solid modelers are recognized as useful tools to provide such environment because they can represent geometrical data and the other information on the solid. Some kinds of computer system based on 3-D CAD solid-modeling have been studied and developed for the realization of concurrent engineering. In these studies, from a view point of CAD system, geometrical modeling is mainly concerned. Geometrical modeling for the product is usually carried out after the process of consideration for the realization of specification. In the mechanical design such as the design of mechanism and the structural design, the consideration for the specification is also very important. Designers spend much time for thinking on the consideration. 3-D CAD system should support not only for the geometrical modeling, but also the thinking on the specification by designers.

In the present study, in order to achieve the concurrence between each process of the consideration for specifications and the geometrical modeling in the mechanical design, a strategy of the geometrical and functional modeling based on design intention is proposed. The concept of the proposed method is formulated by object-oriented approach. A prototype system is developed. The system simultaneously allows the designer to consider the specification and to model geometrical shape by inputting design intention in 3-D CAD system.

2 Concurrence between the consideration of specification and geometrical modeling

2.1 Characteristics of concurrence of 3-dimensional CAD

Recently, 3-dimensional CAD system has been spread in the industrial world. However, the purpose of usage of 3-dimensional CAD is basically the same as ordinary 2-dimensional CAD, that is the geometrical modeling of the product, based on the design solution by designers. In the case of 2-dimensional CAD system, the geometry of the products is drawn by third angle projection method, so that after the determination of the value for design parameters, which are obtained by the consideration of specification, the drawing for the draft starts. The data of geometry in 2-dimensional CAD system does not directly correspond to real solid of the product and shows the projection of real solid, like shadow. Thus, the consideration of specification is completely separated from the geometrical design.

On the other hands, the data structure for solid modeling in 3-dimensional CAD system corresponds to real parts and product. The assembling state of product, including, for example, parts joined by contact with friction, can be realized. Using digital mockup of the product, the simulation for the motion of mechanism and for mechanical vibration can be performed. Also, the stress analysis for the solid model can be easily done by finite element analysis. These show that there is a possibility of simultaneous consideration of specification and geometrical modeling. The concurrent consideration makes the contribution to the shortening of lead time for product development and brings a benefit of easy consideration of specification from a variety of design aspects by using the function of graphical visualization for solids.

In the present study, the method for realizing the concurrent consideration is investigated.

2.2 Intention-driven modeling

As an example of the concurrent consideration of specification and geometrical modeling, based on the characteristics of 3-dimensional CAD system, the method of feature base modeling is adopted in the solid modeler of 3-dimensional CAD. However, the application of feature base modeling is limited to the partial shape that has geometry-based meaning. Then, since the true meaning of feature base modeling seems to be still the contribution to the modeling of shape

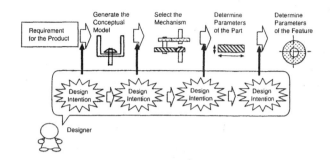

Figure 1: Role of design intention in the detailization of the model

of product, it is not sufficient for the concurrent consideration.

On the other hands, generally speaking, in the field of mechanical design, the initial design is realized by abstract modeling that corresponds to the stage of draft for conceptual design as drawing. In the conceptual design, the objective function (purpose) of the product is decomposed to some partial functions. The model is detailized through the realization of the specification for the product. The detailization is carried out through the process of conceptual modeling, selection of mechanism, the determination of parameters of the part, and the determination of parameters of geometry. Through the process, the shape of conceptual model is detailized to real shape. Though each part of conceptual model has the partial function, mechanism, parts and shape have also each intrinsic function. In the present study, the functions of mechanism, parts, and geometry are defined as design intention. The method, by which the above process from conceptual design to shape is proceeded, is proposed, based on the design intention. Figure 1 shows the process schematically. The idea in the present study is concerned on the support system, based on the design intention, for the process of the development of function from conceptual design stage, in other words, for the detailization process to detailed geometry from draft for conceptual design.

In the present study, each intrinsic function of mechanism, part and shape is expressed by the knowledge unit with the information of geometry in 3-dimensional CAD system. As the knowledge unit, object-oriented knowledge expression with the data of design intention and the method for the realization of the specification corresponding to the design intention is adopted. The message communication between objects is performed through the design intention. By the message communication, the detailed design can be achieved in the top-down scheme. Figure 2 shows

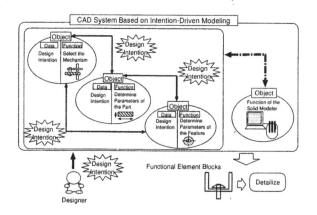

Figure 2: CAD system based on the intention-driven modeling

the idea for the detailization of conceptual design progressed by object-oriented modeling. The prototype for the proposed system for the concurrent modeling based on the design intention (intention-driven modeling: IDM) is developed. The designer can operate the system from the initial design. The system also bring such benefit that the designer can concentrate the attention on the specification for the design with the support of visualization function of solid modeling.

3 Development of conceptual design model to detailed design model

3.1 Design intention and functional element block

In the present study, a prototype system based on the solid modeler of 3-dimensional CAD for modeling the product, connecting with the consideration of its specification, is developed. In the system, the visualization function of CAD is utilized as an interface of intention-driven model (IDM). IDM starts when a designer draws simple solid models with arbitrary shape and size, like cube or cylinder, as figure for conceptual design. The solids are the object to consider the details of the geometry and functions and are called "functional element block". As a result of conceptual design, each of functional element block in the figure has a topological position and a connection with another functional element block.

The function of the functional element block is expressed as object-oriented knowledge in the system. Constructing the dependent relationship (*has-a* relationship and *kind-of-a* relationship) between the object for functional element block and those for the

previous mechanism, parts, and feature model, or that between the objects for mechanism, parts, and feature model, the functional element block is derived to the solid for the detailed design model. The dependent relationship corresponds to the semantics of mechanism, parts and feature models in a product.

IDM starts when a designer inputs design intention to functional element block. The design intention in functional element block is an object of the corresponding object-oriented expression and should include information necessary for developing it to comparatively more detailed model. In the present study, the following four items are included as design intention,

(i) function: purposive motion or state being satisfied by the functional element block under consideration,

(ii) input block: functional element block affecting the functional element block under consideration,

(iii) output block: functional element block affected by the function of functional element block under consideration,

(iv) element type: type of machine element or structure element to realize the function.

Examples for function (i) are "transfer of motion", "realization of motion" and "realization of state". Examples for element type (iv) are gear, shaft, motor and so on. Input block and output block are determined by the layout relationship between functional element blocks drawn in 3-dimensional CAD system.

Design intention is input as attribute information [8] of solid model for any cases of functional element block, mechanism, part and shape. Besides design intention, if there is information necessary for data or method of object-oriented expression and the information can not be obtained from the other object-oriented expression, the system request the user to input the information.

In the system, a class object is prepared in advance as knowledge expression for functional element block. Drawing functional element block on the screen and inputting design intention for that by a designer, it just means that an instance object which has the data of four information in design intention and the method for deriving the mechanism corresponded to the design intention is automatically produced.

3.2 Derivation of mechanism

The derivation of mechanism from the instance object is a process to actualize the decomposed function of the original function in the conceptual stage. In general, since there are some mechanisms, by which the partial and decomposed function could be actualized, the system selects a fitted function that serves its object under the constrain conditions of input block, output block and element type that the functional element block has. The system shows a result of the selection to the user. If the constrain condition presented by the functional element block is not sufficient, the system get the geometrical or topological information through the information on the connection with the other functional element blocks, or the necessary information by sending message to the other functional element block, or the necessary information by requesting a user.

For example, let's consider a case of "power transmission by the motion of rotation" and "shaft" as the function and element type of functional element block, respectively. To derive mechanism, the information on the direction of transmission and the conversion of torque is necessary, in addition to the information of function and element type. The former information is obtained from the topology of the functional element blocks for input block and output block. The latter information is obtained by the reply of the user to the system.

In order to derive mechanism, knowledge base system on the relationship between the constrain condition and mechanism should be prepared.

3.3 Derivation of mechanism, part and shape

In IDM, the relation between mechanism and part, and that between part and shape are denoted by object oriented expression with *has-a* relation. When a designer determines a mechanism for a functional element block, the block sends the information of given conditions concerned on the mechanism as a message. Let's consider the same example as the previous section, that is, a case of "power transmission by the motion of rotation" and "shaft" as the function and element type of functional element block, respectively. The information (numerical value and contents) of reduction rate of velocity, transmission direction, transmitted torque, distance between related shafts and so on is sent to an object embodying the mechanism. The object of mechanism holds the information as its own data.

For the object of part, it is necessary to determine the numerical value of parameters that characterize the geometrical shape of the part itself. For the determination, based on design intention, the object refers the information of the object of mechanism in the

higher rank or that of object of part that has a connection with itself, or requests the designer to determine it.

Similarly, it is also necessary for the object of shape to determine the size of the shape. For the determination, based on design intention, the object refers the information of the object in the higher rank or requests the designer to determine it.

When the objects of part and shape obtain the values of the parameters, characterizing themselves, they generate their shapes on the display of computer by using the visualization function of 3-dimensional CAD.

Through the *class-instance* relation and *has-a* relation between objects, the solid model generated by IDM has the information of design intention, in addition to the information of geometrical shape. This means that part and shape exist in the space of the relation of connection based on the constraints of mechanism and geometry. Then, for example, if a dimension of a part is changed, the dimension of the other part that has the relation of connection with it can be parametrically changed, satisfying the constrained conditions. This is a unique characteristics of IDM.

The process to derive mechanism, part and shape is carried out for each functional element block one by one. This means that there is a possibility that semantic consistence could not be hold for topological relationship between the detailized shape and the other functional element block that has not been detailized or between detailized shapes. If the topological relationship is added to the information of element type of design intention for the related functional element blocks, the problem might be solved.

4 Examples of modeling

4.1 Modeling of power transmission mechanism

By using a commercial software of 3-dimensional solid modeler (Ricoh Co.,Ltd., DESIGNBASE), we make a prototype system for IDM. Let's suppose a design modeling for washing machine as an example of execution of the system. A designer draws functional element blocks, shown in Figure 3. In the figure, block 1 is shaft, block 2 is motor, block 3 is rotational wing and block 4 is water tank. In general 3-dimensional CAD system, the concept of "frame" [9] that is the other kind of the attribute information can be introduced at any point in a solid model. Frame has also attribute information defined reflexively, including the information of the local coordinate. Frame defined

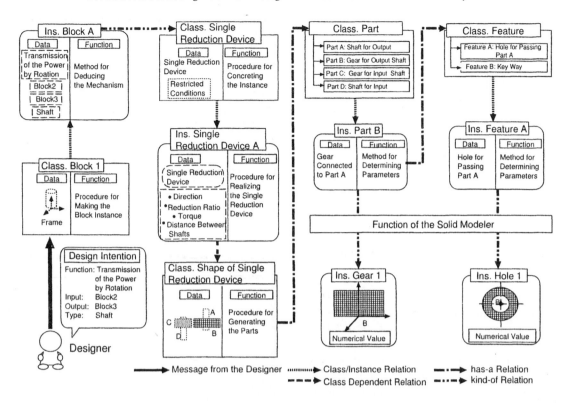

Figure 4: Schematic diagram of the modeling process

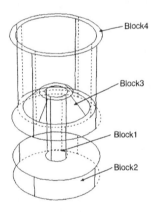

Figure 3: Functional element blocks those correspond to the electric washing machine

for functional element block is called block frame. A surface of the block coincides with the plane of local coordinate defined as the attribute information of block frame. Design intention is also inputted as the attribute information of the block frame. Then, for the block frame of block 1, design intention (shaft as element type, power transmission by rotation of mo-

tion as function, block frame ID of block 2 as input element, and block frame ID of block 3 as output element) is given.

4.1.1 Development of functional element block to detailed solid model

As an example of the development of functional element block, the process of the development of functional element block 1 is shown in Figure 4. Sending a message of design intention from a designer to class object "Block 1", an instance object "Block A" is generated. The instance object selects candidates for mechanism, based on the constrained conditions of the reduction rate of rotational velocity, the distance between shafts, transmission torque and the direction of transmission of power, determined by the information that is received from block 2, block 3 and a designer. After that, the object shows the candidates to a designer. If a designer determines single reduction device among the candidates, the instance object "Block A" sends a message to class object "Single reduction device". The class object yields an instance object "Single reduction device A".

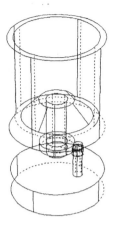

Figure 5: Detailized geometry of the shaft

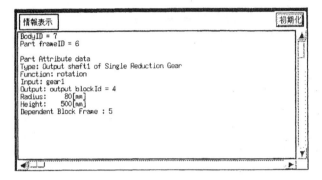

Figure 6: Example of referring the attribute information

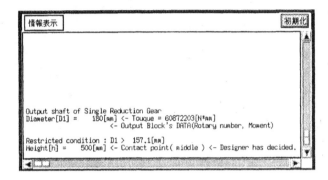

Figure 7: Example of referring the restricted condition

Class object "Shape of single reduction device" that is subordinate to instance object "Single reduction device A" sends a message to class object "Part" that has the information on composed parts, and produces an instance object for a part under consideration. For example, by the output data of design intention in instance object "Block A", part A is produced as instance object "Shaft connected with Block 2" and part B as instance object "Gear connected with part A". The dimension and position for each part are determined by referring the data of instance object "Single reduction device", and based on these information, the shape is modeled by the visualization function of solid modeler. If the characteristic parameters are necessary for the expression of the shape under consideration, such as gear, instance object "Part B" sends a message to class object "Feature", in which there are some kinds of feature modeling. Receiving the message, class object "Feature" produces instance object "Feature A". The geometry for "Feature A" is modeled by solid modeler. As a result of the whole process mentioned above, the shaft developed to detailed geometry is shown in Figure 5.

4.1.2   Reference of design intention

In IDM system, since solid model is treated as object, a designer can know the design intention and dimension of part from its shape on the display of the system. In addition to them, the constrained conditions for the part can be also referred. Figure 6 shows an example of the attribute information for output shaft of single reduction device (gear). Figure 7 shows an example of the constrain conditions for the dimension on output shaft of single reduction device (gear).

4.1.3   Design modification

The geometry of the shape of solid model produced by IDM is constrained by some conditions yielded by design intention. The fact means that a designer can parametrically modify the geometry holding semantic consistency based on the constrained conditions. Figure 8 shows an example that by a change of the diameter of output shaft of single reduction device (gear), the diameter of a hole of gear connected with the shaft is modified. Corresponding to the change of the diameter that becomes greater, the diameter of a hole of gear is modified to be greater consistently.

4.2   Modeling of table structure for supporting a heavy object

As a second example, modeling of cantilever type table for supporting a heavy object is considered. At first, the designer draws three functional element blocks for casing, narrow table and heavy object, as shown in Figure 9. Design intention for the block of table is given as follows: support as function, flat plate as element type, block frame connecting with heavy

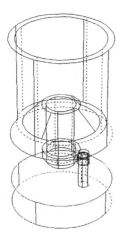

Figure 8: Example of design modification

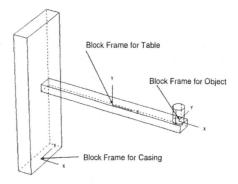

Figure 9: Functional element blocks those correspond to the supporting table

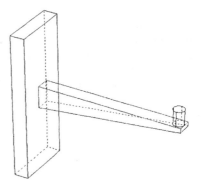

Figure 10: Detailized geometry of the table

subject as input and block frame connecting with casing plate as output. The information is inputted to the block frame of table as attribute information.

In order to start IDM for the model, the system needs the information on the idea how to support a table, that is given by the relationship between narrow table and casing plate, as the additional attribute information. Then, the system requests the designer to input the idea. Let's suppose that the designer inputs an idea of cantilever type. The system obtains the information on the weight of heavy object from the attribute information of input block. The system also obtains the information on the distance between the fixed point and the point for supporting the object on the table from the information on the distance between frames of input block and output block. Based on the information, the system shows the designer some candidates for supporting system by cantilever type. Let's consider the case that the designer selects the beam of uniform strength. The system requests the designer the information on the width of the table and that on the maximum deflection at the end of cantilever beam.

After getting the information, the system shows the allowable condition for the value of thickness of the table, based on the values of width and deflection. By determining the thickness to satisfy the condition, the system produces detailized geometry of the table, shown in Figure 10.

5   Concluding remarks

In order to achieve the concurrence between each process of the consideration for specifications and the geometrical modeling in mechanical design, a concept of the geometrical and functional modeling based on design intention, named intention-driven modeling (IDM) is proposed. 3-dimensional CAD system is defined as a common platform for these processes. A prototype system is made, based on object-oriented approach. In the present system,

(1) since a designer operate the system through design intention, the designer can pay attention to the consideration of the function of product, not to geometrical modeling,

(2) the consideration of the specification can be realized from various points of view,

(3) solid model with design intention can be produced,

(4) the system can be used at conceptual design.

References

[1] J. V. Harrington, H. Soltan, and M. Forskitt, "Frame-work for knowledge based support in a

concurrent engineering environment," *Knowledge-Based Systems*, Vol.9, No.3, pp.207–215, 1996.

[2] A. Retik and B. Kumar, "Computer-aided Integration of Multidisciplinary Design Information," *Advanced in Engineering Software*, Vol.25, pp.111–126, 1996.

[3] P. Dewan and J. Riedl, "Toward Computer-Supported Concurrent Software Engineering," *IEEE Computer*, Vol.26, No.1, pp.17–27, 1993.

[4] M. R. Cutkosky, R. S. Engelmore, R. E. Fikes, M. R. Genesereth, T. R. Gruber, W. S. Mark, J. M. Tenenbaum, and J. C. Weber, "PACT: An Experiment in Integrating Concurrent Engineering Ststems," *IEEE Computer*, Vol.26, No.1, pp.28–38, 1993.

[5] D. R. Brown, M. R. Cutkosky, and J. M. Tenebaum, "Next-Cut: A Second Generation Framework for Concurrent Engineering," *Proceeding of MIT-JSME Workshop: Computer-aided Cooperative Product development*, pp.8–25, 1989.

[6] F. Londono, K. J. Cleetus, and Y. V. Reddy, "A Blackboard Scheme for Cooperative Problem-Solving by Human Experts," *Proceeding of MIT-JSME Workshop: Computer-aided Cooperative Product development*, pp.25–50, 1989.

[7] K. Iwata, M. Onosato, K. Teramoto, and M.-S. Seo, "An Architecture for Form Feature Modeling for Concurrent, Cooperative and Consistent Product Design," *JSME International Journal, Series C*, Vol.40, No.3, pp.533–539, 1997.

[8] H. Ishikawa, H. Yuki, and S. Miyazaki, "Design Information Model in 3-D CAD System for Concurrent Engineering and its Application to Evaluation of Assemblability / Disassemblablity in Conceptual Design," *Advances in Concurrent Engineering*, pp.287–292, 1999.

[9] H. Chiyokura, *Function of graphics, Introduction of 3-D Modeling*, Sanngyou Tosho, pp.3–4, 1994. (in Japanese)

Processor for Maze Model of Design Process

J. Pokojski

Institute of Machine Design Fundamentals, Warsaw University of Technology, 02-524 Warsaw, Narbutta 84, Poland

Abstract

The paper presents a concept and implementation of multi-dimensional computer environment based on the maze model of the design process. The following aspects are considered: maze model – knowledge based systems, maze model – case-based reasoning.

1 Introduction

An important aspect of the computer aided design process is its formal model [3, 4, 5, 8, 11, 12, 13,16, 17]. Formal models decide about what is possible to do during the design process, what systems can be used, in what order, what the sequence of steps looks like.

The maze model [4, 6, 11, 12, 13, 14] has an open structure, this means the design process can be initiated from different initial nodes. It can go via feasible paths and it can finish in different nodes. The maze model is quite useful in conceptual design, where the designer concentrates on some important aspects of a problem which is directly solved by him. The missing data have to be generated by the designer or by the system on a reasonable way.

No matter whether the designer has longer or shorter professional experience, the main problem is how to create efficiently an environment which allows him many different operations and at the same time supports his operations with tools which help him to analyze the new solutions. Conceptual design is connected with actions whose goal is to improve the performances of new solutions.

Many applications, made in the past, the so called integrated environments, work according to a linear or a hierarchic model of the design process [1, 4, 7, 9, 10]. If we try to use this somehow valuable software for conceptual design analysis, it is not easy to operate with it in a flexible way.

The paper presents the concept and implementation of computer processor supporting design process which is based on the maze model. The developed concept tries to accept the structure of the so called integrated environment, which is based on the hierarchic model, but also offers new possibilities of functioning especially in the case of the conceptual design.

2 Maze model of the design process

Product and process modeling is a routine activity for persons who create environments for supporting the design process. Assumptions made in the model have significant consequences for that what can be done later with the implemented environments.

The rich literature about product and process modeling is based on the observations of human designers [3, 4, 5, 16, 17].

The maze model [4, 13, 14] is one of the concepts of models of the design process. In the maze model the designer can start the process of designing from some set of nodes. His way of moving in the maze can be created individually and he can finish designing every time in different nodes. The maze model was used in different applications, especially with conceptual design [4, 6, 13, 14].

The maze model which is a reasonable but idealistic concept always needs some kind of knowledge support which generates the missing data and solutions on the basis of the existing knowledge. The process of moving from one node to the other in the maze requires testing if all needed data really exist. If not, a special procedure is initiated and the missing data are created. Tools of artificial intelligence can fulfill the function of the data generators. In [4, 6] expert systems were used to support this process. The expert systems contain knowledge which provides the missing data while the knowledge bases contain domain knowledge.

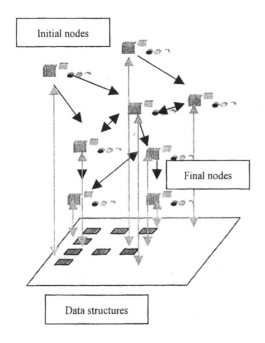

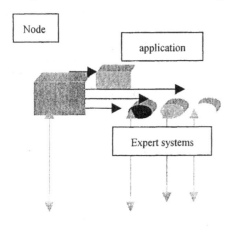

Figure 1 The maze model of design process

Figure 2 The single node of the maze model of the design process

3   Case based reasoning and maze model of a design process

Case-based reasoning is one of the artificial intelligence technologies [8, 17]. The cases are stored solutions of the problems solved in the past. Case–based reasoning means the solving of new problems on the basis of similar solutions from the past. Case-based reasoning is becoming more and more popular in design. There are some important issues of case-based reasoning which have to be mentioned. The first one is the problem of representation this means what information about earlier solutions can be stored? Design cases can consist of: problem formulation (with history of its re-formulations), result of designing (final project), all made decisions and analysis (with its history and background assumptions and knowledge).

The indexing and the way of storing are the next important aspects. The main difficulty is to solve the problem of indexing and storing adequately to the way how normally particular problems are solved. It means when we solve a new problem and look for some past solutions we should have the possibility to formulate our claims in a simple way. And we should get past solutions in an acceptable time.

Adaptation is the next stage of the case-based reasoning procedure. In the existing systems we can observe the tendency to automatic adaptations of solutions or cases with the knowledge "what is allowed to do with them". To leave the adaptation for a human and to use interactive edition of a case are alternative approaches to this problem (when there is more background knowledge, history).

4   Scenario of the designer's conceptual work

In [4, 5, 6, 8, 16, 17] we can find concepts how human designers work when they solve conceptual problems. Often they parallel solve alternative projects. When they come with some ideas to the dead end, they return and repeat some actions. Altogether it looks like "playing with the problem".

To realize these ideas in our concept of environment means that the designer can move in the maze, change data generators from expert system to case-based reasoning.

5   Processor for the maze model of the design process

In the chapter the implementation of a processor for the maze model of the design process is presented. The processor was built in CLIPS environment [6, 11, 12].

It is assumed that the processor will be universal and be used for the integration of different applications for supporting conceptual design of particular products.

The processor has as input the following data:

-   list of used applications in the design process,
-   list of connections between applications,

- lists of design variables connected with particular applications,
- modules supporting particular applications:
  - knowledge based system which supports missing data generation,
  - knowledge based system testing input data quality,
  - knowledge based system supporting the designer in the next step of the design process selection.

The data mentioned above decides about the kind of model of processor functioning. The environment created by the processor is based on this model.

The models of processors functioning are specific, because they depend on particular software, products, the designer and his way of solving problems.

The maze layer is the most important component of the processor. The maze layer consists of nodes. The node of the maze is implemented as a template with the following list of slots:

- name of node,
- list of connections with input nodes,
- list of connections with output nodes,
- name of application which is activated in this node,
- list of input data for the application,
- name of knowledge based system which supports the missing data generation,
- name of knowledge based system which tests input data quality,
- name of knowledge based system which supports the designer with the next step of the design process selection,
- list of paths (stored in the case based reasoning library) which go through this node.

The next objects of the whole system are paths, which are stored descriptions of plans earlier realised in the environment. The paths are treated like cases in case based reasoning. They are stored as templates with the following slots:

- name of path,
- list of steps realised in earlier process,
- list of addresses of data states of nodes earlier realised plans,
- notes - comments.

The system also stores structures which keep data states of nodes of earlier processes.

The system is controlled by commands. It is possible to indicate any node from which we want to start the whole

procedure. We can give the name of the realised project, we can test input data, we can correct these data, we can look for the paths, passing through this node, analyse the data states of the node and select the missing input data (cases can be examined from the point of structure, function or behaviour). We can also activate the application, which is connected with the node. After that we can analyse the result and select the next step in the maze model. Then the user can add his associations and comments to the system. These are special data structures which contain:

- verbal description of association,
- list of nodes which are connected with a particular association,
- name of path with the association.

The knowledge based systems which are connected with a particular node can work in two ways: 1) with dialogue (user has to make some number of decisions), 2) automatically (without dialogue with the user). In the actual version the knowledge based systems connected with different nodes work separately.

The user can open several projects at the same time.

6 Exemplary problem

As a test environment we use a system which supports the design process of a piping system (which was developed by the author as an integrated application [6, 9, 10, 15]). The specialized system consists of two subsystems: one models the geometry of the piping system and is connected to a second one which analyzes the fluid flow dynamics problems of fluid in the piping system. The whole system exists in several configurations.

For modeling the geometry of the piping systems the EASYPLOT system [7, 9] is used. The second system, which is used in an especially created complex system, is the system INES for hydraulic networks dynamics analysis [1]. This system exploits cell formalism. For the modeling of the specific hydraulic system it is necessary to describe a system of nodes, their connections, their parameters and their boundary conditions.

The modeling of the hydraulic model means a lot of effort for the designer. Therefore a module was made

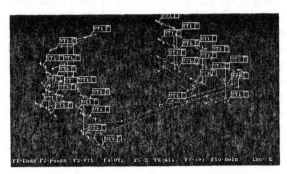

Fluid dynamics problem modeller

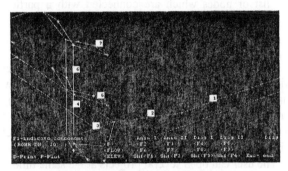

Graphic postprocessor

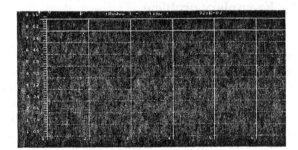

Diagram from postprocessor

Figure 3 The system for fluid flow dynamic analysis

which supports the designer with a fluid flow dynamics model generation [9, 10]. The generator of the fluid flow dynamics model is a knowledge based system. This system on the basis of piping system geometric data (from CAD system) and some physical data, and the user's class of model specification (for instance water installation or chemical installation etc.) creates a dynamic model of fluid in the piping system. This system exploits rule technology. After the simulation the results come back from the simulator to the CAD system. There they are presented to the user together with a geometric model as a system of diagrams and animations (fig.3).

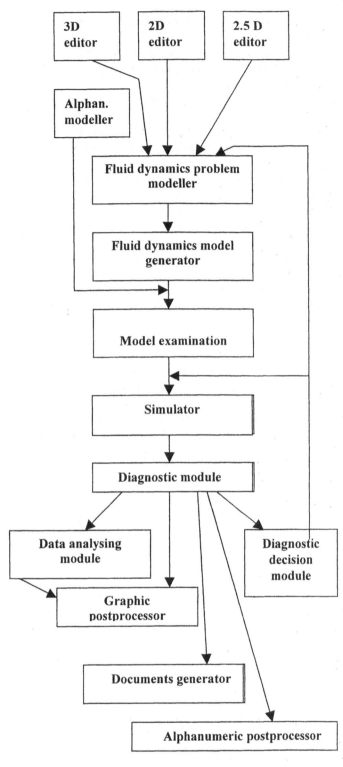

Figure 4 The applications of an exemplary system in the maze

Normally the designer doesn't stop after having solved one simulation problem. Often he has to create a sequence of simulation problems. The results of every simulation case can be fruitful and give him impulses for the next simulation. It means that the designer has to analyse the results of the simulations many times. He has "to fish" for expected and not expected phenomena.

If we look at the problem from the point of view of its dimensions we notice that the system can be used for the following range of problems: several hundreds of nodes, several state variables for every node, and about 500 time steps. It can give huge amount of data for analysing, also in the form of diagrams and animations.

Every expert in fluid flow dynamics knows phenomena which he understands as results of simulations and for which he looks. In principle he looks for phenomena similar to those which he knows from well considered cases in the past.

The system works in different modes: 1) with automatic fluid flow model generation, 2) with the hand made fluid flow model generation, 3) with knowledge based fluid flow model generation. But not every model is possible to build and not every analysis is possible to make in every mode. The modes differ in their range of possibilities and the way how easy it is to operate in the system.

At the present moment the processor is at the testing stage on the above exemplary problem.

In the version actually tested data integration is made with processor storing patterns of all data sources and with the help of this knowledge it can make data transfer to knowledge based systems.

7 Conclusions

The paper presents the concept of an environment based on the maze model of the design process.

The proposed concepts were developed on the basis of experience with some so called integrated design environments, and attempts of using them on the stage of conceptual design.

References

[1] Babala D.: A Brief Description of The Computer Program INES. AB ASEA - ATOM, Masteras, Sweden ,1989.

[2] Blackboard Technology Group, Inc. , GBB manuals, 1998.

[3] Borkowski A.(Ed.) :Artificial Intelligence in Structural Engineering. Proceedings of the 6th EG-SEA-AI Workshop Wierzba 99, WNT, Warszawa, 1999.

[4] Boulanger S., Gelle E., Smith I.: Taking Advantage of Design Process Models. IABSE COLLOQUIUM, Bergamo, pp. 87-96, 1995.

[5] Chawdhry P.K., Ghodous P., Vandorpe D. (Ed.): Advances in Concurrent Engineering CE99, Sixth ISPE International Conference on Concurrent Engineering: Research and Applications, Bath, UK 1999. Technomic Publishing Co. Inc., Lancaster, USA, 1999.

[6] Cichocki P., Gil M., Pokojski J.: Heating System Design Support. In [16], pp. 60-68.

[7] Maetz J.: Programm ISOM. Ein Programm zur Erstellung von Isometrien und Stucklisten. Manual: Kerntechnik –Entwicklung -Dynamik, Rodenbach, Western Germany,1990.

[8] Maher M., Pu P.: Issues and Applications of Case-Based Reasoning in Design. Lawrence Erlbaum Associates, Publishers, 1997.

[9] Pokojski J. : Manual for system "Pressure Drops". Kerntechnik-Entwicklung-Dynamik, Rodenbach, Western Germany, 1990.

[10] Pokojski J.: An Integrated Intelligent Design Environment on the Basis of System for Flow Dynamics Analysis. International Conference on Engineering Design, 1995, Praga, pp. 1333-1338.

[11] Pokojski J.: Product Model Transformations in Maze Model of Design Process. Computer Integrated Manufacturing, Proceedings of Int. Conf. CIM 99, Zakopane 9-12.03.1999, WNT, Warsaw, pp. 121-128, 1999.

[12] Pokojski J.: Blackboard Integration of Design Tools.. Computer Integrated Manufacturing, Proceedings of Int. Conf. CIM 99, Zakopane 9-12.03.1999, WNT, Warsaw, pp. 112-120, 1999.

[13] Pokojski J. : Knowledge Based Support of Machine Dynamics Analysis. In [5], pp. 336-344.

[14] Pokojski J., Cichocki P., Gil M.: Intelligent Personal Assistant for Machine Dynamics Problems. In [3], pp. 159-164.

[15] Pokojski J., Wróbel J.: An Intelligent Design Environment for Machine Design. IABSE Colloquium on Knowledge Support Systems in Civil Engineering, Bergamo 1995, Poster Session, pp. 3.

[16] Smith I. (Ed.) : Artificial Intelligence in Structural Engineering. Information Technology, Collaboration, Maintenance, and Monitoring. Springer-Verlag, Lecture Notes in Artificial Intelligence 1454, 1998.

[17] Sriram R.: Intelligent Systems for Engineering. Springer-Verlag, 1997.

A Descriptive Model of Collaborative Concept Selection Processes in Engineering Design

Maurice Girod

Department of Mechanical Engineering, Loughborough University
Loughborough, LE11 3TU, UK

Amanda C. Elliott

Department of Mechanical Engineering, Loughborough University
Loughborough, LE11 3TU, UK

Ian C. Wright

School of Engineering, Coventry University
Coventry, CV1 5FB, UK

Neil D. Burns

Department of Manufacturing Engineering, Loughborough University
Loughborough, LE11 3TU, UK

Abstract

The aims of this paper are (i) to introduce a newly developed set of descriptive concept selection process models and (ii) to discuss a number of observations that we made whilst generating these models. Two main process variants were found: formal and informal. We observed a number of characteristics for the way the process was undertaken depending on which variant was chosen. These are discussed at length in this paper. The applied research methodology was based on pattern coding of transcripts that were the outcome of 3 recorded workshops involving groups. The benefit of our research is to provide new insight in how collaborative concept selection processes actually take place.

1 Introduction

The engineering design process can be seen as a series of interrelated activities that is driven by decisions [1]. Thus, an effective design process relies heavily upon effective decision-making. As a consequence supporting decision-making can be a significant means for achieving design process improvements.

We are carrying out a research project that aims at improving the engineering design process by evolving decision-making support. Our particular interest is decision-making in selection problem situations during the conceptual design phase, i.e. design concept selection. After having shown that a variety of methods are available, which may support the resolution of such problem situations, we argued in [2] that further work should be dedicated towards empirically identifying designers' requirements for such methods. We believe a descriptive theory is needed that explains the behaviour of observed concept selection processes in design and implies requirements for support methods. Our project aims at establishing such a theory through the empirical generation of concept selection process models.

We organised 3 group workshops. The groups were observed whilst being engaged in collaborative concept selection processes. Two of these groups consisted of final year students and 1 group consisted of professional engineers. They were all given the same task, to evaluate a number of conceptual design solutions and to select, as a group, the most effective one. The design solutions were not developed during our workshops, but had been produced in a number of preceding brainstorming sessions, which involved our workshop participants. We also provided the groups with some evaluation criteria. The workshops were recorded and transcribed.

In [3] we discuss our research methodology and introduce our initial research results. We identified a set of activities carried out by the groups observed. We also quantified how much time our groups spent on these activities. This allowed for some preliminary conclusions about the characteristics of concept selection processes. However, the main benefit of the identified set of activities is that we can use them for the generation of descriptive process models. We see such models as the structured representation of relationships between process elements, which are the identified activities.

This paper will concentrate on (i) the development of our descriptive process models, which we have generated by exploring relationships between the identified activities and (ii) on a number of observations that we made. The benefits of the models and the observations are (i) to provide new insight in how collaborative concept selection processes actually take place and (ii) that they may be used for future work that aims at establishing the needed descriptive theory.

2 Activities in concept selection processes

Within this section we will briefly introduce the set of activities, including any sub-activities, that we previously identified in [3]. Basically, these activities represent particular parts of the multi-facetted discussions by the groups during the workshops. Each of these parts addresses a certain aspect of the concept selection process in a particular way. Table 1 lists the entire set of activities and delivers some explanations on their meaning. A more detailed discussion on these activities can be found in [3].

To identify the activities we applied a content analysis approach to our transcripts. The essence of such analysis is generally to define categories of interest and then assign the syntactic data to them; the many words in the gathered data (the transcripts) are thus transferred into many fewer categories of meaning (the activities) [4]. As it was a cross case content analysis using a replication strategy [5], the identified set of activities is generally applicable for all of our cases, i.e. all 3 workshops.

ACTIVITIES Sub-activities		MEANING
DISCUSSING THE PROCESS APPROACH		The group determines how to undertake the task of evaluating a number of design concepts and selecting one.
	Discussing the general process approach	Context independent approach: could be applied to any set of design concepts and any set of evaluation criteria.
	Discussing the specific process approach	Context dependent approach: supplements general process approach and is only relevant for the concepts and evaluation criteria at hand.
IDENTIFYING CRITERIA		The group identifies additional evaluation criteria.
DEFINING CRITERIA		The group tries to find a consistent understanding of the evaluation criteria's meaning and relevance.
WEIGHTING CRITERIA		The group expresses the level of importance for evaluation criteria.
	Weighting criteria informally	Using linguistic quantifiers, such as 'quite important'.
	Weighting criteria formally	Using a formal, pre-defined scale.
CLARIFYING CONCEPT WORKING PRINCIPLES		The group makes sure that they understand the working principles of the design concepts, being evaluated.
CLARIFYING THE PRODUCT ENVIRONMENT		The group makes sure that they understand what the environmental conditions are that the final product has to operate in.
	Determining the product environment	Identifying what the environmental conditions are.
	Making assumptions on the product environment	If the environmental conditions cannot be identified, assumptions are made.
DELIBERATING SUB-ISSUES		The group identifies how a concept behaves for a particular sub-issue.
	Discussing sub-issues	Generate and analyse solutions for sub-issue (sub-problem).
	Accepting assumptions about sub-issue solutions	Explicitly accept a generated solution for a sub-issue and assume it to be part of one of the design concepts.
GAINING EXTERNAL INFORMATION		The group gathers information that cannot be generated by them.
RAISING EVIDENCE		The group attempts to justify different types of statements.
	Raising evidence on restricted performance of concepts	Justifying a concept's performance with respect to a particular evaluation criterion (restricted performance).
	Raising evidence on comprehensive performance of concepts	Justifying a concept's overall performance. That is, with respect to the entire set of evaluation criteria (comprehensive performance).
	Raising evidence on criteria weights	Justifying a weight for an evaluation criterion.

Activities Sub-activities	Meaning
Determining or evaluating performances	The group finds out how a concept performs and evaluates the performance.
Determining or evaluating restricted performances informally	Using informal statements to address a concept's restricted performance. A perceived value may or may not be communicated.
Evaluating comprehensive performances informally	Using informal statements to address a concept's comprehensive performance. A perceived value is communicated.
Evaluating restricted performances formally	Using a formal, pre-defined, scale to address a concept's restricted performance. A perceived value is communicated.
Evaluating comprehensive performances formally	Using a formal, pre-defined, scale to address a concept's comprehensive performance. A perceived value is communicated.
Mapping intuition onto ranking	The group expresses whether the calculated ranking matches an intuitive ranking.
Controlling the process	The group manages (steers) the process without adding content.

Table 1: Explanations on meaning of identified activities and sub-activities.

3 Process models

After identifying the set of activities that were carried out during the observed concept selection processes, we then generated process models.

Methodology

Our methodology for the model generation was based on pattern codes. The above set of activities was found by general coding, which is a research device for summarising segments of data. Pattern coding then grouped these summaries into a smaller number of constructs. It identified emergent themes or configurations by pulling together material into units of meaning and can be seen as a sort of meta-code [6]. Units of meaning were found by going through the transcripts, marking off units that cohere because they deal with the same topic and then dividing them into topics and subtopics at different levels of analysis [7].

The first level of our pattern coding analysis revealed the concept selection process model as shown in figure 1. This is a general model showing the steps of the 3 observed concept selection processes including the variants identified. These steps are associated with tasks that the groups tried to resolve. The next level of analysis then concentrated on patterns *within* some of these tasks. That is, we identified patterns of activities that the groups carried out to resolve these specific tasks. The tasks for which we could identify activity patterns are highlighted in figure 1. The patterns themselves are graphically shown as specific process models in figures 2, 3 and 4.

General concept selection process model

The general concept selection process model in figure

1 shows that our groups went through 3 main process steps, which are: step 1, 'prepare the process'; step 2, 'explore criteria and concepts' and step 3, 'conclude the process'.

Step 1 consisted of a discussion on the process approach. The groups identified subsequent tasks and prepared to resolve them. Step 2 consisted of investigating relevant evaluation criteria (involving the exploration of their structure and meaning) and evaluating the alternative design concepts (involving the exploration of the concepts' performances). Step 3 consisted of ranking the concepts' overall performances and finally analysing the ranking. The groups condensed information that was gathered during the previous steps in order to conclude the concept selection process.

We identified 2 variants of how step 2 was approached. We called variant 2A 'comprehensive approach' and variant 2B 'criteria based approach'.

The comprehensive approach (2A) was adopted by 1 of our groups. It comprised step 2A.1 'evaluate concept comprehensively' and step 2A.2 'support comprehensive evaluation'. In step 2A.1 a concept was first evaluated with respect to its comprehensive (overall) performance. Then, in step 2A.2 the preceding comprehensive evaluation was supported by gathering particular advantages and disadvantages. This sequence was repeated for each concept. Step 2A.2 will be subject to further model development in a later section.

The criteria based approach (2B) was adopted by 2 of our groups. It comprised step 2B.1 'investigate criteria' and step 2B.2 'evaluate concepts restrictedly'. The evaluation of concepts in step 2B.2 was based on the criteria structure developed in step 2B.1. It was evaluated how the concepts performed with respect to (restricted to)

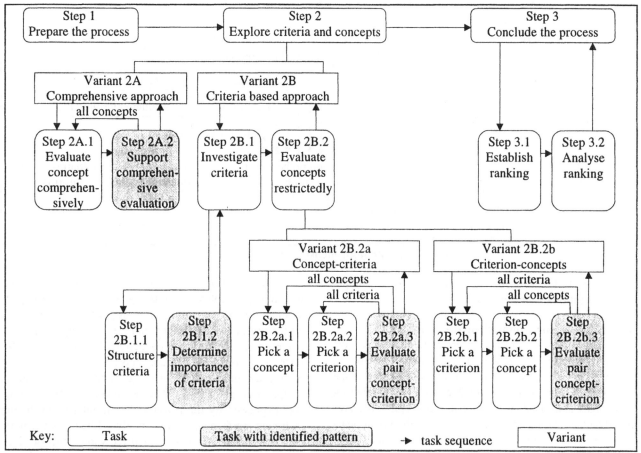

Figure 1: General concept selection process model.

each criterion individually. We called such evaluations 'restricted evaluations'.

In any case, whether the concepts were evaluated using the comprehensive approach or the criteria based approach, all observed groups established a ranking order in step 3.1 that prioritised the concepts' comprehensive performances. If the criteria based approach had been applied, for each concept a comprehensive performance needed to be determined first, before a ranking could be established. This was done by aggregating the results of the restricted evaluations as generated in step 2B.2. Once a ranking order was established the groups proceeded by analysing the ranking in step 3.2. This analysis either concentrated on deliberating various sub-issues with reference to the highest ranked concept or on mapping group members' individual, intuitive rankings onto the one established by the entire group. The former type of analysis seemed to have the purpose of thoroughly checking the highest ranked concept's feasibility. The latter type of analysis seemed to have the purpose of checking whether all individual group members were 'happy' with the ranking as established by the entire group.

For the 2 steps comprising the criteria based approach we identified some further sub-steps.

Step 2B.1, 'investigate criteria' consisted of the steps 2B.1.1, 'structure criteria' and 2B.1.2, 'determine importance of criteria'. Within the former step the decomposition of the given evaluation criteria into sub-criteria was discussed and evolved. Also, additional relevant evaluation criteria were identified and the groups tried to clarify the meaning of all criteria. Within the latter step the importance of evaluation criteria, as perceived by the entire group, was determined. Step 2B.1.2 will be subject to further model development in a later section.

Step 2B.2, 'evaluate concepts restrictedly' consisted of 3 steps, 'pick a concept', 'pick a criterion' and 'evaluate pair concept-criterion'. These steps were sequenced by 2 different variants. In variant 2B.2a, which we called 'concept-criteria', one *concept* is picked and then evaluated with respect to all criteria before the next concept is picked. At variant 2B.2b, which we called 'criterion-concepts' one *criterion* is picked and then used for the evaluation of all concepts before the next criterion is picked. Steps 2B.2a.3 and 2B.2b.3, which are basically

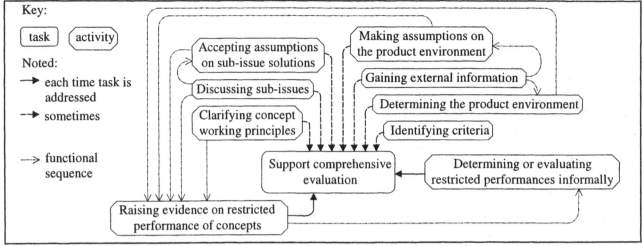

Figure 2: Support comprehensive evaluation (step 2A.2).

the same task, will be subject to further model development in a later section.

Step 2A.2: support comprehensive evaluation

The pattern of activities that were carried out to resolve the task 'support comprehensive evaluation' is shown in figure 2. The overall aim of this task was to support some concept's comprehensive evaluation by gathering advantages and disadvantages of this concept. Whenever an advantage or disadvantage was brought up at least some of the activities in figure 2 were carried out. These were carried out without any obvious *time* sequence. The activities 'raising evidence on restricted performance of concepts' and 'determining or evaluating restricted performances informally' were always carried out when the task 'support comprehensive evaluation' was addressed. The other activities were only carried out sometimes.

We perceived an apparent *functional* sequence of the activities for the task 'support comprehensive evaluation': 'Gaining external information' aided 'determining the product environment'. If this environment could not be determined, due to lack of information, the groups were 'making assumptions on the product environment'. 'Discussing sub-issues' lead to 'accepting assumptions on

sub-issue solutions'. The activities 'clarifying concept working principles', 'discussing sub-issues', 'accepting assumptions on sub-issue solutions', 'making assumptions on the product environment' and 'determining the product environment' all informed the activity 'raising evidence on restricted performance of concepts'. This activity, in turn, delivered a justification for an advantage or disadvantage of the concept at hand. The advantage or disadvantage was expressed by the activity 'determining or evaluating restricted performances informally'. 'Identifying criteria' was not carried out with any perceivable purpose. New criteria simply emerged in the course of resolving the task.

Step 2B.1.2: determine importance of criteria

The pattern of activities that were carried out to resolve the task 'determine importance of criteria' is shown in figure 3. The overall aim of this task was to express the importance of evaluation criteria by establishing formal weightings for each of them. Whenever the importance of a criterion was expressed at least some of the activities in figure 3 were carried out. As in the case discussed above, there was no obvious *time* sequence of activities. An exception is 'controlling the process' which most often initiated tackling the task. This was done by a control

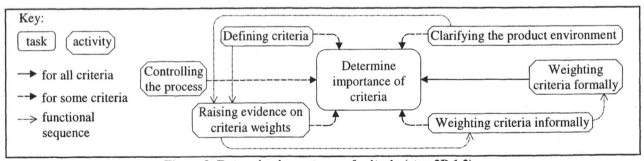

Figure 3: Determine importance of criteria (step 2B.1.2).

remark, e.g. *"Let's look at the importance of ...".* The activity 'weighting criteria formally' was always carried out when a criterion's importance was formally determined. The other activities in figure 3 where only carried out sometimes.

We perceived an apparent *functional* sequence of activities for the task 'determine importance of criteria': 'defining criteria' and 'clarifying the product environment' informed the activity 'raising evidence on criteria weights', which generated a justification for the outcome of the activity 'weighting criteria informally'. This outcome, i.e. an informal weighting, was then mapped onto a formal scale by the 'weighting criteria formally'.

Steps 2B.2a/b.3: evaluate pair concept-criterion

The pattern of activities that were carried out to resolve the task 'evaluate pair concept-criterion' is shown in figure 4. The overall aim of this task was to evaluate a concept with respect to an individual criterion (restricted evaluation). Two of our 3 groups aimed at establishing a formal evaluation for each concept-criterion pair. Apart from the activity 'evaluating restricted performances formally', which produces the required formal evaluation, the pattern of activities for the task 'evaluate pair concept-criterion' is very similar to that of the task 'support comprehensive evaluation of concepts'. This can be seen by comparing figures 4 and 2. Again, the groups carried out at least *some* of the activities in figure 4 whenever the task 'evaluate pair concept-criterion' was addressed. Also, there was no obvious *time* sequence regarding these activities.

We perceived an apparent *functional* sequence of activities for the task 'evaluate pair concept-criterion' which was very similar to the one for the task 'support comprehensive evaluation of concepts'. This can also be

seen by comparing figures 4 and 2. However, there were some differences. The activities 'controlling the process', 'defining criteria' and 'evaluating restricted performances formally' were not carried out during the task 'support comprehensive evaluation of concepts'. 'Controlling the process' most often initiated tackling the task 'evaluate pair concept-criterion'. This was through a control remark from some group member. 'Defining criteria' was an additional contribution towards informing 'raising evidence on restricted performance of concepts'. The activity 'evaluating restricted performances formally' mapped informal evaluations, as the outcome of 'determining or evaluating restricted performances informally', onto a formal scale.

4 Observations

Whilst we were analysing our transcripts to generate the above process models, we made a number of observations. These observations address particular characteristics of the 3 groups' individual processes.

Observation 1:
Consistency regarding the application of evaluation criteria was a main difference between the 2 variants for approaching the resolution of step 2 .

The comprehensive approach (variant 2A) started with an overall (comprehensive) evaluation of some concept. Then, this evaluation was supported by gathering advantages and disadvantages of the particular concept. These advantages and disadvantages referred to evaluation criteria. We noted that, in the case the particular concept had been comprehensively evaluated as 'positive', only those criteria where used for the support for which the concept performed advantageously, and

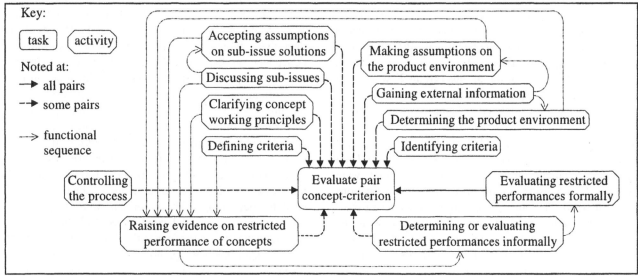

Figure 4: Evaluate pair concept-criterion (step 2B.2a/b.3).

vice versa. This meant that the comprehensive evaluation of a particular concept was based on a sub-set of the evaluation criteria only. Also, the comprehensive evaluations of other concepts were not based on the same sub-set of evaluation criteria. This was in contrast to the criteria based approach during which all concepts were evaluated consistently on the basis of all criteria.

The comprehensive approach was adopted by the 1 group that did not use any formal evaluation methods, whereas the criteria based approach was mainly adopted by the 2 groups that both used formal evaluation methods (evaluation matrices).

Observation 2:

None of the groups deliberated the evaluation criteria in a thorough and consistent manner.

We noted that only those 2 groups that used the criteria based approach spent some time on discussing a decomposition of the relevant evaluation criteria. Only 1 of these 2 groups also discussed and determined importance weightings for the criteria. Yet, some of these importance weightings were not explicitly justified. This is why they seemed to have a somewhat intuitive character. For the group that used the comprehensive approach no deliberations on the evaluation criteria's structure or their importance were noted.

Observation 3:

The groups that chose to use formal evaluation methods (evaluation matrices) had a tendency to evaluate the concepts' performances on a relative basis rather than absolute.

There were 2 variants for the resolution of step 2B.2: 'variant a', concept-criteria and 'variant b', criterion-concepts. Picking one *concept* and then evaluating it with respect to all *criteria* before the next concept is picked, as in 'variant a', seemed to suggest evaluations on an absolute basis. That means, the concepts' performances may be evaluated on an independent scale and not relative to each other by comparisons. This was done consistently, using 'variant a', by 1 of the groups (group 2) when they evaluated the concepts' performances with respect to those criteria that they defined as being constraints. These were binary evaluations: 'satisfaction' or 'no satisfaction'. The same group used 'variant b' to evaluate those concepts that they defined as objectives. In this case they did not attempt to evaluate on an absolute basis. So they were consistent with using 'variant b' for relative evaluations (comparisons). The other group (group 3) that also used a formal evaluation method applied 'variant a' with the stated intention to compare each concept to a datum only. Yet, they gradually involved more concepts to compare performances across them. When they did so, they leaned towards 'variant b', picking a *criterion* and then using it

for the evaluation of all *concepts* before the next criterion is picked. The same group then applied 'variant b' with the stated intention to evaluate the concepts' performances on an *absolute* scale, rather than relative. Yet again, despite their stated intention we noted a strong tendency towards performance *comparisons across* concepts. So they were inconsistent by having an unintentional tendency towards relative evaluations.

Observation 4:

The groups that chose to use formal evaluation methods (evaluation matrices) did not explicitly communicate a justification for the evaluation of each concept-criterion pair.

The task 'evaluate pair concept-criterion' was carried out during step 2B.2 of the criteria based approach (see figures 1 and 4). A justification for the evaluation of particular concept-criterion pairs was produced by the activity 'raising evidence on restricted performance of concepts' (see figure 4). However, we noted that this activity was not carried out consistently for all concept-criterion pairs by either of the 2 groups that used the criteria based approach.

Observation 5:

During the activities 'weighting criteria informally' and 'determining or evaluating performances informally' we noted information formats that did not agree with the formats required by the evaluation methods used.

Previously, we discussed the tasks 'determine the importance of criteria' (step 2B.1.2) and 'evaluate pair concept-criterion' (step 2B.2a/b.3). It was mentioned that, to complete these tasks the groups mapped the outcomes of the activities 'weighting criteria informally' and 'determining or evaluating performances informally' onto formal scales. Such mappings were in fact the transformation of informal information into formal scores. We considered any information that had a different format than these scores as informal. The scores' format was dependent on the particular formal evaluation method used by the group. A transformation of information from one format into another format may change the scope of the information. For example, by mapping quantitative performance information onto ordinal scales the scope of the original information is reduced because ordinal scales cannot express quantitative performance differences.

We noted especially 2 information formats that could not be expressed by the scales of the evaluation methods used by our groups. These formats were linguistically vague quantifications and expressions of likelihood.

Linguistically vague quantifications were statements that included linguistic quantifiers. A wide range of these quantifiers was used. For example: Concept A may be marginally/slightly/a little bit/a fair bit/much/a lot/a great deal better than concept B. When vague, linguistic quantifications were mapped onto a qualitative scale all inherent quantitative content was lost. When they were mapped onto a precise quantitative scale a level of precision was pretended that had not actually been inherent in the original, vague information.

Information that contained expressions of likelihood conveyed an indication of how probable it was that the content is actually true. Such likelihood was never expressed numerically in percent, but always by using linguistic terms. Most often terms were used that can be associated with a likelihood of 100%. For example: Concept A must be/is definitely better than concept B. Yet, we also noted terms that can be associated with a likelihood of less than 100%. For example: Concept A may be/could be/ is possibly/is probably better than concept B. Whenever information that contained expressions of likelihood was mapped onto the scales used by our groups these expressions were lost.

Observation 6:

None of the groups produced thorough records of their concept selection processes.

Those 2 groups that used formal evaluation methods produced some records by completing their evaluation matrices. However, these matrices only contained formal evaluation statements (scores), but no underlying justifications, assumptions, deliberated sub-issues, etc. The group that did not use any formal evaluation methods did not produce any records at all.

Observation 7:

The evaluation of concepts had an iterative nature on some occasions.

We noted that all 3 groups occasionally repeated the evaluation of concepts that had been evaluated before. This happened regardless of whether the criteria based approach or the comprehensive approach was used.

Observation 8:

Not all groups used all evaluation criteria for establishing a ranking of the concepts' comprehensive performances.

Those groups that had applied formal evaluation methods during step 2 (see figure 1) calculated a comprehensive performance for each concept in step 3.1. These calculations consistently aggregated all restricted evaluation results for each concept, as written down in the evaluation matrix, into a comprehensive evaluation statement for each concept. The calculation results for all concepts implied a ranking order.

The group that had not applied any formal evaluation methods during step 2 *informally* established a ranking in step 3.1. To do so, some of the different concepts'

advantages and disadvantages (based on some criteria - see observation 1), as gathered in step 2A.2, were compared to each other. We noted that not all advantages and disadvantages that had been gathered were applied again for establishing the final ranking in step 3.1.

Observation 9:

Sub-issues that had already been discussed during evaluations of concepts in step 2 were discussed again during an analysis of the final rankings in step 3.

Once a ranking order was established during step 3, the group that did not use any formal evaluation methods concentrated on the highest ranked concept and discussed again various sub-issues that they had already addressed during step 2. This appeared to have the purpose of reassuring that the concept would actually be feasible.

The other 2 groups, those that used formal evaluation methods, analysed the established ranking order with the apparent intention to understand and accept a particular concept's position within this order. During this analysis in step 3 the groups very frequently discussed sub-issues that had been already discussed during the concepts' restricted evaluations in step 2.

Observation 10:

The usage/non-usage of formal evaluation methods influenced the selection as well as the time taken for the process.

Those 2 groups, who had chosen to use a formal evaluation method, selected independently of each other the same concept. However, the group that had chosen not to use any formal methods, selected a different concept. Also, the groups using formal evaluation methods took considerably more time for their concept selection processes than the group that did not use any formal evaluation methods.

Observation 11:

None of our groups had apparently produced information that (i) could help answering the question on whether or not new, more effective concepts should be generated or (ii) that could direct a search for new, more effective concepts.

At the end of each workshop we asked whether the group considered the selected concept as the most effective one among the set of concepts being evaluated or whether it was considered sufficiently effective on an absolute basis. All 3 groups replied that their selections are only seen as the most effective among the set of concepts being evaluated. This was reflected by our earlier observation that there was a tendency to evaluate on a relative basis, by comparisons. None of the groups stated exactly what features should be addressed by efforts that aim at improving the concepts.

5 Discussion

Within this section we will discuss our observations by relating them to other researchers' results. We could not identify any work that had exactly the same research focus as our project. Yet, a number of studies have addressed the design process and, at least touched our specific aspect of design concept selection.

Ehrlenspiel and Dylla [7] have empirically investigated design procedures. According to them, a thorough analysis of demands (indicating evaluation criteria) and solution properties (i.e. performances of concepts) as well as defining the importance of demands are qualities of successful designers. We have observed that in particular those groups who used formal evaluation methods devoted some effort towards analysing the demands (criteria), including identifying their importance. Ehrlenspiel and Dylla [7] also state that a characteristic for less successful designers is being less concrete and rather emotional regarding demands and solution properties. We observed a higher level of rigour for those groups that applied formal methods. This rigour can be seen (i) by their consistency regarding the inclusion of *all* criteria in the evaluation of concepts and (ii) by agreeing as a group upon expressing concepts' performances on a *formal* scale. This was in contrast to the group that used the informal, comprehensive approach. They were less rigorous as they first *informally* evaluated the comprehensive performance of a concept and then *informally* supported this evaluation on the basis of only a *few* criteria. Therefore, those groups that used formal evaluation methods apparently showed characteristics of successful designers, whereas the group that did not used any formal methods apparently showed characteristics of less successful designers.

In agreement with Ehrlenspiel and Dylla [7], Feldy [8], speaking from years of practical experience, argues that efforts addressing criteria are a sign of effective decision-making because it helps all participants really understand the actual goals of the process. He also states that good decisions are accepted and supported and that there is some agreement of how to judge the effectiveness of their outcomes. We observed that at least the groups that used the formal methods intentionally tried to find general agreement and support for the established ranking. However, none of our groups made any statements on how to judge the general effectiveness of the concept selection process outcome. Feldy [8] claims this is where decision processes often fail.

Despite realising that formal decision approaches are time consuming (which became evident in our study too) Feldy [8] sees a number of benefits from using them. Among these benefits are the need for documentation and the provision of defensible reasons for a decision. This was only partly observed in our study. That is, although 2 of our groups used formal methods including the completion of evaluation matrices they still did not produce justifying reasons for all of their weightings and evaluations. Also, justifications were not documented.

Ullman et al. [9] have empirically researched design and established a mechanical design process model. They observed that when designers evaluated multiple concepts, they compared their performances and usually based these comparisons on only a few criteria. When 3 concepts were compared, 1 was taken as the focus and the other 2 were only compared to it rather than to each other. We did not generally observe this behaviour in our study. The informal group considered concepts individually and never explicitly used one concept as a focus, i.e. datum. The groups using formal evaluation methods did evaluate by comparing performances, but we found that the use of a datum did not seem practical. In fact, 1 of our groups intended to evaluate their concepts by exclusively comparing each one to a datum. But they gradually involved more concepts in these comparisons.

Our groups' general tendency to performance comparisons, rather than absolute evaluations, and to creating rankings rather than ratings was also observed in an empirical study by Ehrlenspiel and Lenk [10]. Apart from this, they noted that generally relatively little written notes were taken during the concept selection process; not all criteria were used for the evaluations; the importance of criteria were not explicitly stated and that designers see evaluations as an accumulation of advantages and disadvantages regarding the alternative solutions rather than establishing 'hard' scores. These are all characteristics that we have observed for the group that did not use any formal evaluation methods.

Ehrlenspiel and Lenk [10] have noted that evaluations are usually vague and not precise. We also observed this.

Dwarakanath and Wallace [11] conducted a descriptive study on general decision-making in engineering design. They noted that before decisions are made, designers explore sub-issues and sub-alternatives. They also noted that designers frequently forgot previously identified issues. Similar observations were made in our study. We observed that a number of sub-issues that had been discussed, without taking notes, during the evaluations of concepts in step 2 were discussed again in step 3 when the final ranking orders were analysed.

The above discussion highlights that a number of our observations have been addressed by previous research. By having a very specific focus on concept selection processes, including the usage of formal evaluation methods, our study had a different perspective to the above referenced research publications. This may be the reason why some of our observations do not entirely agree with the observations of others.

6 Conclusions

In this paper we developed a concept selection process model and discussed some observations we made. The model as well as the observations emerged from applying a pattern coding research methodology to the transcripts of 3 workshops. In these workshops we recorded 3 groups of decision-makers solving the overall task of evaluating a number of design concepts and selecting the most effective one.

Our general concept selection process model consists of various steps and sub-steps that the 3 groups carried out. These steps represented tasks, which were resolved by certain sequences. The 3 main tasks, observed for all groups, were: step 1 'prepare the process'; step 2 'explore criteria and concepts' and step 3 'conclude the process'. With respect to step 2 'explore criteria and concepts' we observed 2 variants of approaches to resolve the task: comprehensive approach and criteria-based approach. The former approach was informal and was characterised by little consistency regarding evaluation criteria. It was applied by the 1 group that did not use any formal evaluation methods. The latter approach was formal and was characterised by a high level of structure and consistency regarding evaluation criteria. Yet, more time was consumed by this approach. It was applied by the 2 groups that both used formal evaluation methods.

The tasks, associated with process steps, were resolved by activities that the groups carried out. For some of these tasks we could observe particular patterns of activities that contributed towards resolving the tasks. The patterns that we observed referred to the tasks: step 2A.2 'support comprehensive evaluation of concept'; step 2B.1.2 'determine the importance of criteria' and step 2B.2a/b.3 'evaluate pair concept-criterion'.

Whilst developing our models we made a number of observations regarding some characteristics of the 3 groups' processes. These observations addressed:

- Consistency regarding use of criteria;
- Thoroughness of deliberations on criteria;
- Tendencies towards relative evaluations;
- Justifications for evaluations;
- Disagreeing information formats;
- Records of process;
- Iterative nature of evaluations;
- Usage of criteria for establishing ranking;
- Repeated discussions on sub-issues;
- Influence of using methods on selection and time;
- Information on effectiveness of selection.

A concluding discussion highlighted that a number of our observations have been addressed by previous research. Some of our observations do not entirely agree with these previous observations. This may be caused by obvious differences regarding the underlying research foci. Also, previously stated expectations regarding the positive effects of using formal evaluation methods could not be entirely confirmed by our study.

The findings discussed in this paper will be the basis for our further research activities. These aim at establishing a theory that will explain our observations as well as indicate requirements for concept selection support methods.

References

[1] Midland, T., 1997, 'A Decision Directed Design Approach', *Engineering Designer*, May/June, pp.4-7.

[2] Girod, M., Elliott, A.C. and Wright, I.C., 2000, 'Decision-making and design concept selection', *Submission for Engineering Design Conference (EDC2000)*, Brunel, UK.

[3] Girod, M., Elliott, A.C., Wright, I.C. and Burns, N.D., 2000, 'Activities in Collaborative Concept Selection Processes for Engineering Design', *Submission for ASME2000 Conference on Design Theory and Methodology*, Baltimore, USA.

[4] Stauffer, L.A, Diteman, M. and Hyde, R., 1991, 'Eliciting and Analysing Subjective Data about Engineering Design', *Journal of Engineering Design*, Vol.2, No.4, pp.351-366.

[5] Yin, R.K., 1994, *'Case study research: Design and methods'*, (Applied Social Research Methods Series, Vol.5), Beverly Hills, CA: Sage Publications.

[6] Miles, M.B. and Huberman, A.M, 1994, *'Qualitative Data Analysis: an Expanded Sourcebook'*, Sage Publications,

[7] Ehrlenspiel, K. and Dylla, N., 1993 'Experimental Investigation of Designers' Thinking Methods and Design Procedures', *Journal of Engineering Design*, Vol.4, No.3, pp.201-211.

[8] Feldy, E.C., 1997, 'Introduction to Decision Making in Design Engineering', *Proceedings of the National Design Engineering Conference at National Manufacturing Week*, Chicago, USA, pp.153-158.

[9] Ullman, D.G, Dietterich, T.G, Stauffer, L.A., 1988, 'A Model of the Mechanical Design Process Based on Empirical Data', *Journal of Artificial Intelligence in Engineering Design and Manufacturing (AI EDAM)*, Vol.2, No.1, pp.33-52.

[10] Ehrlenspiel, K. and Lenk, E., 1993, 'Einflusse auf den Bewertungsprozess beim Konstruieren', *Proceedings of the International Conference on Engineering Design ICED'93*, The Hague, NL, pp.449-456.

[11] Dwarakanath, S. and Wallace, K.M., 1995, 'Decision-making in Engineering Design: Observations from Design Experiments', *Journal of Engineering Design*, Vol.6, No.3, pp.191-206.

m.girod@lboro.ac.uk

Mass Customisation: A Methodology and Support Tools for Low Risk Implementation in Small and Medium Enterprises

J. E. Mooney

Manufacturing Engineering & Industrial Management, University of Liverpool, Liverpool, L69 3BX

H. S. Ismail

Manufacturing Engineering & Industrial Management, University of Liverpool, Liverpool, L69 3BX

S.M. M. Shahidipour

Manufacturing Engineering & Industrial Management, University of Liverpool, Liverpool, L69 3BX

Abstract

Mass production can no longer satisfy the demands of increasingly turbulent competitive environments. New paradigms of agility, responsiveness and mass customisation have emerged. Mass customisation provides product variety and customisation at prices comparable to a mass produced equivalent. However, this introduces new demands on firms. Whilst larger organisations can afford the risk of making mistakes, small to medium enterprises (SME's) are typically more vulnerable and need a structured lower risk approach. It is argued that successful implementation starts with the design of product families that maximise the reuse of components. An approach for optimising the design of product families that can be configured for the needs of SME's is introduced. A measure of product similarity and component reuse in product families is also presented. The approach is applied through a case study in a local SME.

1 Introduction

Driven by complex social, political, geographic and technological factors, the past decade has seen dramatic changes in the global market environments. Manufacturing companies have been under tremendous pressure to meet apparently conflicting goals of efficiency and consumer choice. On the one hand customers are demanding that their orders be met faster and at lower cost. On the other hand, they are demanding highly customised products and variety. This has led a growing number of economists and scholars to declare that the paradigm of mass production is no longer able to satisfy such demands. New paradigms of agility, responsiveness and mass customisation have emerged.

Mass customisation is the application of technology and new management methods to provide product variety and customisation through flexibility and quick responsiveness at prices comparable to mass-produced products [1]. However, mass customisation in itself introduces new demands on firms. These include improved product development processes, flexible manufacturing planning and control systems, and closer supply chain management. Whilst larger organisations by their nature can afford the risk of making mistakes, small to medium enterprises (SME's) are typically more vulnerable, and hence need a structured low risk approach.

The authors argue that the successful implementation of mass customisation starts with design of product families that maximise the reuse of components. The paper presents a review of existing methodologies for the design of mass customised products and introduces an approach for optimising design that can be configured to the needs of SME's. The process of identifying and designing measures of product similarity and component reuse is described. These measures are used to assess the degree of flexibility of a product line, and to set boundaries within which new or customised designs should operate. This enables designers at the early stages of the design process to select cost effective configurations of components and to assess the implication of product changes on production costs.

The approach has been applied through an initial case study carried out on an existing and redesigned product group in a local SME to demonstrate the application of the measures developed.

2   Background

2.1   Mass Production: the legacy

The fundamental principle of mass production is to produce high volumes of standardised products at the lowest possible cost. Low costs are achieved through economies of scale (high production volumes that yield lower unit costs) and the division of labour (job specialisation and narrowly defined repetitive tasks). Productivity is maximised through sequential production flow with long product runs, inventory buffers for maximum capacity utilisation, and vertical integration for reduced supply and demand variations. However, with competitive environments experiencing increasing turbulence, these traditional means of maintaining an advantage no longer apply [2].

2.2   Mass Customisation: a new paradigm

In his book 'Mass Customisation: The New Frontier in Business Competition', Pine [1] states: *"Mass customisation is the new frontier in business competition for both manufacturing and service industries. At its core is a tremendous increase in variety and customisation without a corresponding increase in cost. At its limit, it is the mass production of individually customised goods and services. At its best, it provides strategic advantage and economic value."* On the surface, this notion of increased variety with no increased cost appears to be an oxymoron [3]. However, many companies have developed successful design and manufacturing strategies, based on the principles of mass customisation. Examples can be found in a diverse number of industries including computers [4, 5, 6, 7], electronics [8], chemical [9], and consumer goods [10, 11].

2.3   Mass Customisation versus Variety

Some companies have taken to trying to meet the demands of consumers simply by offering more variety. Products are manufactured using traditional techniques from a mass production heritage. Usually this entails add-on activities to existing processes. For example, some companies provide customisation through 'engineering change orders' (ECO) or via custom engineering departments. The principle of low cost and thus mass customisation is thereby sacrificed [12].

Companies must understand the distinction between customer focused *external* variety and inefficient *internal* variety. External variety gives a company a competitive advantage through product differentiation and customisation to satisfy the individual requirements of the customer. However, it should be noted that excessive external variety could be detrimental to a company. Customers often find too much variety confusing or consider it transparent. Internal variety is usually found within inefficient processes. Symptoms of internal variety include part proliferation, excess inventory and work in progress, and complicated manufacturing processes.

2.4   Implementing Mass Customisation

In any product development process it is important to translate the requirements of the customer into a product definition. For the development of mass customised products this becomes more complex. The requirements of the customer must be translated into a list of customisable attributes that represent a family of products (i.e. capture the *external* variety).

A number of techniques can be found in the literature. Typically, these are formal and based on a systematic methodology. Anderson [12] proposes *'QFD for Mass Customisation'* an approach based on 'Quality Function Deployment (QFD)' [13]. Tseng & Jiao propose *'Design for Mass Customisation (DFMC)'* [14]. The essence of the approach is to identify optimal 'product family architectures' formed from composite building blocks. The building blocks are defined by identifying the 'functional requirements (FRs)' for a family of products [15]. A density analysis of the FRs is the applied to group similar designs into product families. Based on their ability to fulfil the FR's, design parameters can then be selected.

In practice however, many SME's do not have the resources required to employ such formal techniques. Therefore it is more likely that less formal approaches based on subjective intuition and market knowledge will be used.

2.5   Modular Product Structures & Reuse

To minimise internal variety and the associated costs, whilst retaining the external variety that will meet the demands of the customer, *modular product structures* can be developed [16]. The essence of a modular product structures are modules and components that can be selected and combined in order to configure different customised products within a product family. Economies of scale are achieved through the reuse of internal modules and components, rather than finished products of the mass production paradigm. The potential benefits to manufacturing [17] include:

- simplified production planning and scheduling
- lower setup and holding costs
- lower safety stock
- reduction of vendor lead time uncertainty
- order quantity economies

Modular product structures also enable the task of differentiating a product for a specific customer to be postponed until the latest possible point in the supply chain [4]. The postponement of product variety (as illustrated in Figure 1) results in risk pooling and consequently reduces overall manufacturing, distribution, and inventory costs [18].

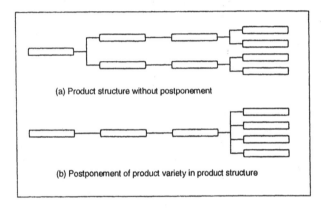

(a) Product structure without postponement

(b) Postponement of product variety in product structure

Figure 1: Postponement of product variety [18]

3 Similarity

A key enabler to mass customisation is the ability to identify the most economic modules and maximise their reusability and application. This section examines how internal variety can be reduced. It identifies the need to measure the similarity of products within a product family in order to optimise the reuse of common components.

To successfully develop modular product structures, companies need to measure the similarity between products and the 'reusability' of components in a product family. The measurements should be used to assess the degree of flexibility of a product family, and to set boundaries within which new or customised designs will have to operate. The measurements should also be used to optimise the use of common components across the product family.

One measure that is commonly used is the *'Degree of Commonality Index (DCI)'*. This is based on the average

number of parent items per distinct component, for a particular product structure [17]. The DCI is defined as:

$$DCI = \frac{\sum_{j=1}^{d} \phi_j}{d}$$

where: d = the total number of distinct components;

j = (1, ..., d)

ϕ_j = the number of immediate parents component j has over a set of end items or product structure level(s)

However the DCI does not measure the similarity between different product structures. Therefore, where a product family contains customisable products with differing structures, the DCI cannot be used. The relative commonalities are incomparable.

3.1 The Need for a Measurement of Similarity

To enable companies to develop flexible product families whilst optimising the reuse of common components and modules, a measurement of the similarity between customisable products within a product family is needed.

The objective of reuse is to minimise the total number of parts needed to build the maximum number of customised products. However, this can lead to a conflict. The demands of the customer for increased customisation can lead to an increase in the number of components (often of low-reuse) used within a product family. This conflict cannot always be resolved. Often a product will contain low-reuse components to differentiate it from other products within the family, or to differentiate it from competitors' products in the market (i.e. external variety). In this situation a compromise should be made. Companies should ensure that low-reuse components are necessary, and can be justified by adding to a product's differentiating attributes.

The measurement of similarity between products cannot be based on the reusability of components alone. This is too simplistic and does not account for low-reuse components that add value to a product in terms of differentiation etc. A more accurate measurement of similarity should also consider factors, such as:

- the cost of the components used to make a product
- the sales volume of each product in a family
- the contribution of each product in a family

Figure 2 shows the typical information relevant to an example product family. The product family needs six components (*B1* to *B6*), to build five finished products (*A1* to *A5*). The table includes each component 'cost', 'product volume' and 'product contribution'. A matrix is used to show where each component is used in each product.

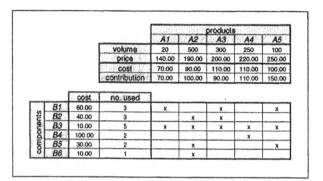

		products				
		A1	A2	A3	A4	A5
volume		20	500	300	250	100
price		140.00	190.00	200.00	220.00	250.00
cost		70.00	90.00	110.00	110.00	100.00
contribution		70.00	100.00	90.00	110.00	150.00

		cost	no. used	A1	A2	A3	A4	A5
components	B1	60.00	3	x		x		x
	B2	40.00	3		x	x		
	B3	10.00	5	x	x	x	x	x
	B4	100.00	2				x	
	B5	30.00	2		x			x
	B6	10.00	1		x			

Figure 2: Example product family

3.2 Similarity of Product Structures

At a basic level, the similarity between customisable products within a product family should be measured. This 'similarity coefficient' is based only on the product structures and reusability of components. The following factors must be considered:

- the number of components in the component group (i.e. the number of components needed to build every product in the product family)
- the number of distinct components used to build a particular product
- the number of other products in the family using the same components

A similarity coefficient should be calculated for each product with respect to every other product in the family. This should indicate those products with high-similarity and low-similarity to others in the product family. Likewise, a component reuse coefficient should be calculated for each component in the group. This should indicate those components with high-reuse and low-reuse.

In the example product family shown in Figure 2 above, product *A1* (made from components *B1* and *B3*) is considered to have a high similarity value. All of the components used to make *A1* are common to at least three products in the family (i.e. component *B1* is common to products *A1*, *A3* and *A5*). In contrast, product *A4* (made from components *B3* and *B4*) is considered to have a lower similarity value. This is because component *B4* is

not common to any other product. In the example above, it can be seen that component *B3* has maximum reuse value (i.e. it is common to every product in the family).

3.3 Cost

The cost of the components and modules used to make a product should be considered when similarity is measured. In an ideal world products would be made only from inexpensive components. In practice most products contain a number of components that are more expensive than the rest. Where possible these expensive components should only be used were reuse is high. It is undesirable to have a product family that contains expensive components with a low reuse value. Any measure of similarity should consider this.

The effect of component cost can be seen in the example above. Product *A4* (made from components *B3* and *B4*) is considered to have a low similarity value. As already discussed, this is because component *B4* is not common to any other product. In addition, component *B4* is the most expensive component in the group. Therefore the similarity value of product *A4* should be reduced.

3.4 Volume

The sales volume of each product in a family should also be considered, when similarity is measured. Most product families will contain some products that have low similarity values. This may be for a number of reasons that might include the completeness of a product family, and customer demand. If a product is in high demand and sells well, the similarity value becomes less important. In these circumstances the sales volume should negate the effects of low similarity. Any measure of similarity should consider this.

The effect of sales volume can be seen in the example above. Product *A2* is made from components *B2*, *B3*, *B5* and *B6*. The similarity value of product *A2* is lowered by component *B6* (because *B6* is not common to any other product). However, product *A2* has the highest sales volume in the product family. Therefore, the similarity value should reflect this high sales volume, and negate the effects of low similarity based on component reuse.

3.5 Contribution

Finally, the contribution of each product in a family should be considered when similarity is measured. As discussed, most product families will contain some products that have low similarity values. If a product has a high contribution, the similarity value becomes less important. In this circumstance the contribution should negate the effects of low similarity. Likewise, any measure of similarity should consider this.

The effect of contribution can be seen in the example above. Product A5 is made from components B1, B4 and B5. Product A5 has the highest contribution in the product family. Therefore, the similarity value should reflect this, and negate the effects of low similarity based on component reuse.

4   Measuring Product Similarity

This section presents five coefficients to measure similarity. They consider the factors of product structure, costs, volume and contribution. The coefficients are as follows:

- product structure similarity coefficient (R_n)
- product cost similarity coefficient (R_c)
- product volume similarity coefficient (R_v)
- product contribution similarity coefficient (R_t)
- aggregate product similarity coefficient (R)

4.1   The Similarity Matrix

A product family is defined with N distinct components (B_1 to B_i) needed to build M finished products (A_1 to A_j) within the product family.

i.e. N number of distinct components needed to build the product family
B_i component ($i = 1 \rightarrow N$)
M number of products in the product family
A_j product ($j = 1 \rightarrow M$)

A product-component matrix U_{ij} is used to represent the product family structure. Where:

$$U_{ij} = 1 \rightarrow B_i \in A_j$$
$$U_{ij} = 0 \rightarrow B_i \notin A_j$$

An example product-component matrix is shown in Figure 3 to represent a product family structure that uses seven components ($B1$ to $B7$), to build seven finished products ($A1$ to $A7$).

	products						
	A1	A2	A3	A4	A5	A6	A7
B1	1	1	1	1	1		1
B2	1	1	1		1		1
B3	1		1	1	1		1
B4	1		1		1		
B5	1	1		1	1	1	1
B6		1	1			1	
B7		1			1		

Figure 3: Product-component matrix U_{ij}

4.2   Rationalisation of the Component Group

The '*similarity analysis*' can be performed on a number of levels. At the lowest level, this would be for every component in the group. However, for most product families the analysis would be over complex. Therefore it is often necessary to rationalise the component group. Typically, this would usually limit the analysis to those components that collectively contribute to over eighty percent of the product family costs. A '*pareto*' sort is one technique that can be used for this type of rationalisation. At a higher level, an analysis can be performed on any sub-set of the component group. Alternatively, the reuse of particular module or component type can be analysed. For instance, this might be for a component critical to a particular process.

4.3   Product Structure Similarity Coefficient (R_n)

The '*product structure similarity coefficient*' R_n identifies the similarity of a product with respect to the other products within the product family (i.e. reuse of components). It is based only on the components used in the product structures. The following factors are considered:

- The number of components needed to build the products in the product family, N
- The number of distinct components used to build the particular product, n_j
- The number of other products in the family, using the same components, m_i

The number of unique components used to build product Aj is defined as:

$$n_j = \sum_{i=1}^{N} U_{ij}$$

The number of products using component B_i is defined as:

$$m_i = \sum_{j=1}^{M} U_{ij}$$

The 'product structure similarity coefficient' R_n for product A_j is defined to be:

$$R_{nj} = \frac{\sum_{i=1}^{N} U_{ij}(m_i - 1)}{(M-1)n_j}$$

It is not possible to specify what the minimum level of similarity or variation of similarity within a product family should be. This is dependent on the properties of the product family and the component group analysed. Therefore, each company must establish the level that their own product families should attain.

4.4   Product Cost Similarity Coefficient (R_c)

The 'product cost similarity coefficient' R_c identifies the similarity of a product with respect to the other products within the family, based on the costs of the components used in the product structures. Where:

c_i is the cost of component B_i

c_{max} is the maximum cost of all components used in the product family

The 'weighted' cost of component B_i is defined as:

$$w_{ci} = \frac{c_i}{c_{max}}$$

The 'product cost similarity coefficient' R_c, for a product A_j is defined to be:

$$R_{cj} = \frac{\sum_{i=1}^{N} U_{ij}(m_i - 1)w_{ci}}{(M-1)\sum_{i=1}^{N} U_{ij}w_{ci}}$$

The 'product cost similarity coefficient' R_c is an improved measurement of similarity because it introduces the factor of cost. The coefficient highlights the effect of costly components that are not reused widely with the product family.

4.5   Product Volume Similarity Coefficient (R_v)

The 'product volume similarity coefficient' R_v identifies the similarity of a product with respect to the other products within the family, based on the 'sales volumes' of the components used in the product structures. Where:

V_j is the volume of sales for product A_j

The 'weighted' volume for product A_j is defined as:

$$w_{vj} = \frac{V_j}{\sum_{j=1}^{M} V_j}$$

Therefore, the 'weighted' volume of component B_i is defined as:

$$w_{vi} = \frac{\sum_{j=1}^{M} U_{ij}w_{vj}}{m_i}$$

The 'product volume similarity coefficient' R_v for a product A_j is defined to be:

$$R_{vj} = \frac{U_{ij}w_{vi}}{n_j}$$

The *'product volume similarity coefficient'* R_v introduces the factor of volume to the measurement of similarity. Components that are used in products that have a high volume carry more weight. However, as the sales volume increases, R_v has a tendency to distort the measure of similarity. It is possible to achieve high R_v values for products, with a low product structure similarity (i.e. the product is built from 'low-reuse' components). Therefore, R_v should not be used in isolation.

4.6   Product Contribution Similarity Coefficient (R_t)

The *'product contribution similarity coefficient'* R_t identifies the similarity of a product with respect to the other products within the family, based on the 'contributions' of the components used in the product structures. Where:

P_j is the selling price of product A_j

The total cost of a product A_j is defined as:

$$C_j = \sum_{i=1}^{N} U_{ij} c_i$$

Therefore, the contribution of product A_j is defined as:

$$T_j = P_j - C_j$$

The 'weighted' contribution of product A_j is defined as:

$$w_{ij} = \frac{T_j}{\sum_{j=1}^{M} T_j}$$

Therefore, the 'weighted' contribution of component B_i is defined as:

$$w_{ti} = \frac{\sum_{j=1}^{M} U_{ij} w_{ij}}{m_i}$$

The 'product contribution similarity coefficient' R_t for a product A_j is defined to be

$$R_{ij} = \frac{U_{ij} w_{ti}}{n_j}$$

The *product contribution similarity coefficient* R_t introduces the factor of contribution to the measure of similarity. Components that are used in products that have a high contribution carry more weight. The *product contribution similarity coefficient* R_t exhibits similar behaviour to R_v. However, as contribution increases, R_t has a tendency to distort the measure of similarity. It is possible to achieve high R_t values for products, with a low product structure similarity (i.e. the product is built from 'low-reuse' components). Therefore, R_t should not be used in isolation either.

4.7   Aggregate Product Similarity Coefficient (R)

The 'aggregate product similarity coefficient' R combines the four similarity coefficients: R_n, R_c, R_v and R_t. Each coefficient can be assigned a 'weight' that corresponds to the influence it has on the measure of similarity. Where

w_{rn} is the weight assigned to the 'product structure similarity coefficient' R_n

w_{rc} is the weight assigned to the 'product cost similarity coefficient' R_c

w_{rv} is the weight assigned to the 'product volume similarity coefficient' R_v

w_{rt} is the weight assigned to the 'product contribution similarity coefficient' R_t

The 'product contribution similarity coefficient' R for a product A_j is defined to be:

$$R_j = \frac{R_{nj} w_{rn} + R_{cj} w_{rc} + R_{vj} w_{rv} + R_{tj} w_{rt}}{w_{rj} + w_{rc} + w_{rv} + w_{rt}}$$

At this stage of development, the weightings that should be applied to the similarity coefficients R_n, R_c, R_v and R_t. (i.e. w_{rn}, w_{rc}, w_{rv} and w_{rt}) have not be established These are depend on the properties of the product family, component types, sales pattern, profit margins etc. It is likely that they will also be unique to each product family. For initial measurements of similarity the 'weights' have been set to unity (i.e. $w_{rn} = w_{rc} = w_{rv} = w_{rt} = 1$).

5   Case Study

This section summarises the results of a case study to demonstrate the application of the similarity measures. The study was carried out in a local SME that design and manufacture shower enclosures and bathscreens. The similarities of an existing product family and a new product family being considered as a potential replacement for the company existing products were measured. The design was based on modular product structures. The purpose of this study was to identify if the internal variety within the new family was reduced.

5.1   The Existing & New Component Groups

To simplify the analysis, a sub-set of the product family component group was analysed. It was decided that this should be the components with the longest lead-time; the aluminium extrusions used to build the enclosure frame. The component group used to build the existing product families consisted of twenty-three unique extrusion-profiles. This number tripled if the finish was also considered (i.e. each profile is available in three different colours). As can be appreciated, the internal variety within the component group was high. The main reason for this is historical. As each style was introduced, extra extrusions were added. In the past the company had not considered the impacts of reusability and part proliferation.

As a result of a redesign exercise, the company reduced the number unique extrusion-profiles to twelve. Similarly, this number triples if the finish is also considered. It should be noted that the company did not employ a formal technique to establish the optimum modular product structures. The redesign exercise was based on subjective intuition and the experience of company's designers.

5.2   Similarity Analysis

The *similarity coefficients* were used to measure the similarity of products in the family and the reuse of components in extrusion group. Only four coefficients were applied (R_n, R_c, R_v and R). The information required to apply the 'product contribution similarity coefficient (R_t)' (i.e. the total cost of the components used to build the products) was not readily available at the time of the study.

In total *102* products within the existing product family were analysed (this still excluded an number of customisable attributes such as 'glass finish' and 'door hand' options). The identical number of products were analysed for the replacement product family. However, due to the nature of the design, the company did expect to offer more products than they do at present.

Similarity Analysis of Existing Product Family

Due to the number of components and products analysed, the size and complexity of the similarity matrix was considerable (i.e. 23 components by 102 products). Therefore, only the results of the analysis is summarised

The mean, maximum and minimum values for R_n, R_c and R_v can be seen in Table 1 below:

	Mean	Maximum	Minimum
R_n	9.0%	12.4%	0.0%
R_c	8.9%	11.9%	0.0%
R_v	9.4%	21.1%	1.0%

Table 1: R_n, R_c and R_v for the existing product family

The mean, maximum and minimum values for R can be seen in Table 2:

	Mean	Maximum	Minimum
R	9.1%	15.0%	0.3%

Table 2: R for the existing product family

Similarity Analysis of the Redesigned Product Family

With the number of components and products also being high (i.e. 12 components, 102 products), the size and complexity of the similarity matrix was considerable (i.e. 12x102). This section summarises the results of the analysis and highlights the important features.

The mean, maximum and minimum values for R_n, R_c and R_v are shown in Table 3.

	Mean	Maximum	Minimum
R_n	27.4%	38.6%	6.9%
R_c	27.3%	38.6%	6.1%
R_v	30.1%	49.5%	5.6%

Table 3: R_n, R_c and R_v for the redesign product family

The mean, maximum and minimum values for R can be seen in Table 4 below:

	Mean	Maximum	Minimum
R_v	28.2%	42.2%	6.2%

Table 4: R for the redesigned product family

5.2   Results of the Analysis

The average (or mean) *aggregate similarity coefficient* (R) for the existing product family was 9.1%; and 28.2% for the redesigned product family. This is an improvement of over 300%. Therefore, the analysis showed that the redesigned extrusion profiles significantly improved product similarity and the reuse of components used to build the products. The redesign products should therefore reduce the internal variety within the product family, and in turn lead to cost savings.

5.3 Discussion

It is argued that the successful implementation of mass customisation starts with the design of product families that maximise the reuse of components. The five *similarity coefficients* presented in Section Four measure such reuse by considering the factors of product structure, costs, volume, and contribution. However, in isolation the measures are abstract. The results merely indicate the similarity of products in a product family. This does not realise the benefits of reduced internal variety.

The foundation of mass customisation is agile manufacturing, and only by implementing such strategies will the benefits of reduced internal variety be realised.

The SME in which the case study was carried out has introduced such a strategy. This is based on a cellular manufacture system that employs *kanban* stock replenishment using *'Just In Time (JIT)'*. As a result of the similarity analysis, the company has set up a trial cell to manufacture the redesigned product family. The full benefits of reduced internal variety has not been fully realised at present. However, the improvements identified so far include:

- increased external variety and customisation
- a flexible rather than focussed manufacturing cell
- reduced numbers of components held in the cell
- economies of increased component order quantities

6. Summary

The paper examines how the reuse of common components in product families can reduce internal variety. Five coefficients that measure the similarity of products within a family are presented. These consider the factors of product structure, cost, volume and contribution. The *'product-component matrix'* is introduced to represent product families and calculate the similarity coefficients.

The similarity coefficients were used in a case study carried out in a local SME to identify if a redesigned product family actually minimised internal variety. The reuse of components within the existing 'vertical extrusions' component group was measured. This was compared to the measured reuse of components within the redesigned 'vertical extrusions' component group. The analysis showed that there was a significant increase in both the similarity of products in the family, and the reuse of the components. Therefore internal variety was minimised.

This analysis also demonstrated the versatility of the similarity coefficients. It showed that an analysis of individual component groups (and not just complete component families) could also be performed, and provide useful results.

References

[1] Pine J.B., Mass customisation: the new frontier in business competition. Harvard Business School Press, 1993.

[2] Singletary E.P., Winchester S.C., Beyond mass production: analysis of the emerging manufacturing transformation in the US textile industry. Journal of the Textile Industry, 87(2), pp.97-116, 1996.

[3] Dwyer J., Why tailoring is no longer on the shelf. Works Management, pp.18-21, Jan 1999.

[4] Feitzinger E., Lee H.L., Mass customisation at Hewlett-Packard: the power of postponement. Harvard Business Review, 75(1), pp.116-121, Jan 1997.

[5] Bowman I., Customise and make a profit. Manufacturing Computer Solutions, pp.21-22, Nov 1994.

[6] Fitzgerald B., Mass customisation - at a profit. World Class Design to Manufacture, 2(1), pp.43-46, 1995.

[7] Beaty R.T., Mass customisation. Manufacturing Engineer, 75(5), pp.217-220, Oct 1996.

[8] Eastwood M.A., Implementing mass customisation. Computers in Industry, 30(3), pp.171-174, Oct 1996.

[9] Gilmore J.H., Pine B.J., The four faces of mass customisation. Harvard Business Review, 75(1), pp.90-101, Jan 1997.

[10] Kotha S., From mass production to mass customisation: the case of the National Industrial Bicycle Company of Japan. European Management Journal, 14(5), pp.442-450, Oct 1996.

[11] Kotha S., Mass customisation: implementing the emerging paradigm for competitive advantage. Strategic Management Journal, 16(Special Issue), pp.21-42, 1995.

[12] Anderson D.M., Agile product development for mass customisation: how to develop and deliver for mass customisation, niche markets, JIT, build-to-order, and flexible manufacturing. Irwin Professional Publishing, 1997.

[13] Hauser J., Clausing D., The house of quality. Harvard Business Review, 66(3), pp.63-73, 1988.

[14] Tseng M.M., Jiao J., Design for mass customisation. Annals CIRP, 45(1), pp.153-156, 1996.

[15] Suh N.P., The principle of design. Oxford series on Advanced Manufacturing, 1990.

[16] Kohlhase N., Birkhofer H., Development of modular structures: the prerequisite for successful modular products. Journal of Engineering Design, 7(3), pp.279-291, Sep 1996.

[17] Sheu C., Wacker J.G., The effects of purchased parts commonality on manufacturing lead time. International Journal of Operations & Production Management, 17(8), pp.725-745, 1997.

[18] Tseng M.M., Lei M., Su C., A collaborative control system for mass customisation manufacturing. Annals CIRP, 46(1), pp.373-376, 1997.

Analysis of creative problem solving methods

G. Bertoluci

Laboratoire LOGIL, ENSAM, Paris

M. Le Coq

Laboratoire CPNI, ENSAM, Paris

Abstract

For several years laboratory CPNI of the ENSAM has undertaken work on the formalisation of the cycle of product design. Within the search framework engaged on the integration of TRIZ within this cycle, we initially endeavoured to include/understand the mechanisms and tools which it implements. As a second step, we developed an implementation process of this method. We will present the results of this study.

1 Introduction

Pierre FALZON defines problems of conception like problems where "...final state must be build and where initial state is bad defined. The task of the designer is to define the problem and its solution...". To achieve this goal, various tools have been developed in the design area. These tools aim at supporting the designer when characterising the product's environment and attributes, or managing the overall project.

```
1/  Project planning

2/  Identification and translation of
    the need

3/  Creative  search  for  solutions
    concepts     meeting     customer
    requirements

4/  Preliminary design

5/  Detailed design

6/  Evaluation  and  selection  of
    solutions
```

Figure 1 Classical design model

In addition, one can also employ such tools to develop project team creativity in order to raise the number of solutions concepts than can be applied. However, in order to benefit from the potential gains that may result from their utilisation, this group of tools has to be used at the appropriate step when processing a global design process like the one defined in a classical design model. (Fig.1) [17][2].

However, if "...we can use design methodologies, ... there is no step-by-step procedure leading to the solution: the problem definition is not the prior step to the design of the solution. They are build at the same time and interact with each other [7]." This analysis breaks out the sequential model of the process, as per which the utilisation of a solving problem tool is located in the phase 3 of the design process. Indeed, the analysis of cognitive processes associated to product design [6,25] has demonstrated than the theoretical model of refinement of the solution and transformation of the perspectives of representations of the object, developed during three successive phases of study – functional, structural, physical – is not followed by the actual behaviour of designers. This can be explained by the following reasons: Firstly – at the step of functional definition – the designers naturally tend to think of physical based on their own acquired knowledge.

Secondly, the design process systematically implies the breakdown of the system into smaller pieces which are treated separately down to the finest details of design. Consequently, incompatibilities and contradictions show up when putting these different local solutions back together. Unfortunately, the usual way to deal with such new issues consists of making compromises at the product finalisation stage.

Actually, one can think of using a problem solving method at any step of the design process provided that it allows to integrate the real nature of the problem to solve. However, the effectiveness of this method will depend on its ability to break out the designer's tendency to solve problems by transferring already known solutions applied to previous situations. [25, 4].

In the laboratory CPNI's context of research on design methodology, we have decided to conduct a study on the Theory of Problem Solving (TRIZ). This method - created to solve technological problems - is used to improve either existing systems or systems still in the course of definition.

Various publications have demonstrated the interest of this method. Nevertheless, a certain confusion reigns on the nature of TRIZ: is it a science, a method or a tool? Regarding each one of its elements: what is the related field of application and at what step of the design process can it be used? The information research carried out did not give satisfactory answers. Then we decided to work on the formalisation of the implementation of this method according to the formalism proposed by LE COQ [17] for the characterisation of a design method. In order to do this we identified the different elements that compose the structure of TRIZ: its basic concepts, steps and tools. We'll clarify the mode of analysis we have carried out on these tools through cards tools created by VADCARD [24]. Finally, we will explain how these tools can be included within the overall process of TRIZ implementation.

1 Underlying principles of the Problem Solving Theory

We can define the goal of TRIZ's creator, G.Altshuller, as the will to provide designers with a database of ways of solutions for a certain category of design problems. Based on his personal experiment and on the observation of the work of his colleagues, G. Altshuller came up with two major conclusions [1, 23, 13]:

-   Large majority of concepts of solutions suggested in patents originated in personal experience and technical and scientific knowledge of the designers working on the resolution of the problem. As a result, very broad fields of universal knowledge remain unexplored although they could potentially offer better solutions.
-   All the methods and tools aimed at developing creativity :
    -   Either by revealing the existing potential within each individual by counteracting the inhibitions that limit him.
    -   Or by taking advantage of the atmosphere happening in successful workshops (Brainstorming).

Nevertheless none of the available tools provided the designers with tangible elements enabling them to have access to ways of solutions requiring [8]:

-       To implement technologies or sciences they do not know anything about.
-       To transpose to their own issues the solutions already implemented for similar problems in other industries.

Therefore, the analysis of the patents revealed that similar principles of solutions are employed in industrial fields very distant from each other. In addition to this, several decades may pass before a solution concept already known in one field is discovered in another.

However, in order to be able to perceive these similarities it is necessary to consider them on a conceptual level. It is thus possible to be freed from the specificity inherent to a given vocabulary or technology. Without this ability of exploring what was realised in other fields, each speciality must " reinvent " its solutions. Years of search thus could be lost. A significant work was undertaken on a great number of patents (2 000 000 were analysed) which leads

	Importance of the solution obtained	Requested knowledge	Distribution / 100 patents
Lev 1	Conventional solution already known to the speciality.	Industry	32%
Lev 2	Minor invention belonging to the same paradigm. Improvement of an existing solution.	Company	45%
Lev 3	Real technological innovation constantly referring to the field of expertise. Existing system.	Industry	18%
Lev 4	Invention using external expertise. Resorts to science and not technology.	Society	4%
Lev 5	Major discovery thanks to a new science.	Universe	1%

Table 1 Inventiveness levels

to a classification of the degrees of inventiveness (Table 1) of the solutions suggested in the patents [19,20,21,26].
TRIZ is used in order to facilitate the emergence of solutions at level 2 and especially 3 or 4. These levels 3 and 4 have been identified like inventive. This concept of inventive problem is very important as far as TRIZ is concerned; it can be defined like this:
" A technological system originates an inventive problem when there is necessity to eliminate a drawback function and the method to do this is unknown or the knowledge or the means usually used to perform this generates a new problem [5]".

The timeframe to solve these problems can be reduced and especially the solution suggested does not initiate compromise. In order to realise this, the designer must be able to rely on tools that allow him to select some solutions concept already used for similar problem in foreign fields.
The difficulty was precisely to elaborate such tools.
The analysis of patents has demonstrated the following :
- There is a strong recurrence of creative problems in technical systems and of natural evolution of these systems.
- Few patents suggested inventive solutions

Based on this it has been decided to identify - among these inventive solutions - the functions and characteristics regularly in conflicts and their solutions. A work of conceptualisation of these data has provided standards for the characterisation of problems and their solutions.

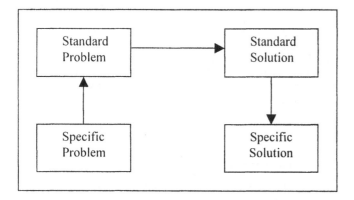

Figure 2 Abstraction principle of TRIZ

This work had to be realised at a conceptual level in order to extract solutions from the specificity (like vocabulary or technology used to implement the solution) of the area they come from. Then tools had to be created to help the designer find out the standard problem linked to his specific problem (Figure 2).

3 Solving tools of creative problems and implementation process

3.1 Tools

We have decided to focus our interest on tools specifically developed by G. Altshuller and the researchers who have followed in his footsteps in order to exploit the process of analogy.
We will not also describe in this article such tools like "little men" or "operators S(ize) T(ime) C(ost)". Indeed for us the objectives of these tools can be assimilated to those of classical tools used to increase creativity by their effect on the psychological inhibitions.

1/ Definition of the tool.
2/ Target.
Objective of the implementation of the tool.
3/ Level of intervention.
Explicative level, corrective level or preventive level.
4/ Principles/concepts
Theoretical assumptions underlying the tool.
5/ Procedures
Pragmatic approach to the tool
6/ Design tool category
There are seven categories: multi-field characterisation of the need, creativity, solution definition, solution materialisation, solution analysis, project management and quality tool.
7/"Space of representation"
Categories of information given by the tool.
8/ Utilisation
Fields of relevant theoretical application and traditional Industrial fields of application.
9/ Usage Cost
Time necessary to its implementation, appropriate human resources in term of quantity & quality (knowledge is required), hardware, and thus consequently the financial cost of use of the tool. (Possibility of an individual use of the multi-field tool).
10/ Note(s)
Additional comments on the tool. Advantages and Drawbacks identified by the practical experience of the tool. Other similar tools may be mentioned.
11/ Bibliography.

Figure 3 Rubrics of tool cards [VADCARD]

Moreover we will not mention the specific use of the laws ruling the evolution of technical systems (defined by G. Altshuller) that can be done in order to predict the product evolution. As far as we are concerned, we only use it through some suggestions they provide concerning the principles (Altshuller's matrix) and Solutions Standards (Data associated to S-Field).

In order to realise and formalise the analysis we have carried out on TRIZ's tools we have relied on the results of the work conducted by P.VADCARD [24]. He suggested building the analysis of creativity tools in design through the implementation of tool cards. These tool cards included 11 rubrics (Figure 3).

In order to adapt these cards to our needs, we have modified them because they initially were created to analyse all of the tools used in the product design area [10]. Therefore, we have removed the rubric on the "Space of representation". This concept does not make sense in this study. On the other hand, we have added two rubrics:

- 1 - Anglo-Saxon terms used to indicate the tool
- 2 - French Synonyms and foreigners used to indicate the tool.

Tool cards of this type have been created for:

- Ideation questionnaire
- Ideation problem Formulator
- Altshuller matrix
- S-Field models
- Separation principles

The result of this formalisation combined with the experiment of these tools on industrial problems has been the elaboration of the process described in figure 4.

3.2 Implementation process of Classical TRIZ

The implementation of the whole process, as we have just defined it, consists of fourteen stages (Figure 4). That does not mean that any problem requires the realisation of the totality of these stages. Indeed, these fourteen stages lead to the successive use of three of the principal tools of TRIZ : the technical contradictions, the S-Field models and the physical contradictions (Figure 5). The order of use of these tools as indicated here reflects the order in which they were created. It also reflects an increasing need of knowledge about the environment of the problem analysed in order to lead into interesting solutions.
The use of this particular tool enables the group to gradually widen one's knowledge of the environment. In addition to this, if the solutions resulting from the use of

1/ Description and analysis of the problem

2/ Formulation of technical contradiction

3/ Use of matrix of contradictions.

4/ Brainstorming of the principles retained

5/ Evaluation and choice of a solution

6/ Construction of the S-Field model

7/ Use of process algorithms

8/ Brainstorming of the standard solutions retained

9/ Evaluation and choice of a solution

10/ Refinement of the "time-space" analysis

11/ Formulation of physical contradiction

12/ Application of the principles of separation

13/ Brainstorming of the principles of separation

14/ Evaluation and choice of a solution.

Figure 4 ENSAM's implementation process of TRIZ

technical contradictions or the S-field models give satisfaction it is not necessary to continue the analysis with the following tool. Indeed, the implementation of each one of these tools consumes time and must thus be carried out only if it is considered to be essential. Moreover, when a solution considered to be satisfactory was identified one notes that it is difficult for the designer to keep his mind open to other solutions. His mind remains focused on the previous solution. Actually, if he continues his search he will generally work by complementation, i.e. that he will use the new suggestions to try to enrich his original solution. This approach is interesting since it increases the possibility of improving the initial solution. On the other hand, the drawback of this behaviour is a tendency " to complicate " the original idea without appreciable counterpart.

1.   Description and analysis of the problem.

This first stage is not formalised in the " traditional " TRIZ approach for the independent use of the various tools. On the other hand, the ARIZ method (algorithm of processing of TRIZ) [1,12,26,27] is intended to complete this work. However, due to the fact that the implementation of ARIZ is very time consuming, its use is limited to the resolution of very complicated problems. We have therefore decided to rely on the "problem first analysis" questionnaire developed by the Ideation

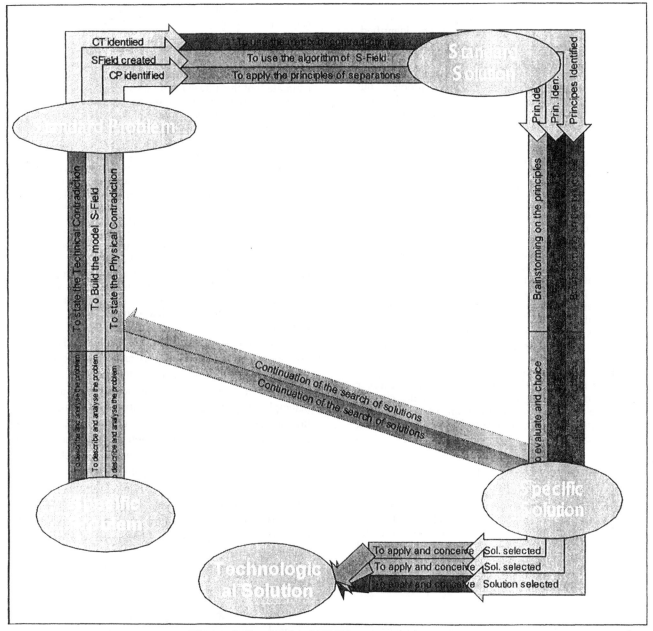

Figure 5 Position of TRIZ's tools in the process

company to formalise the search for information. We use its formalism as the basic working tool when dealing with the people raising the problems. The answers brought by the latter make it possible to define:

- the limits of the system
- the nature of the "super system" it belongs to and the "sub systems" it comprises
- the main function to perform
- the "downgraded" function (s)
- the on-going objective
- the tests already carried out in order to solve the problem
- the resources available
- the criteria which will make it possible to define the quality of the solutions suggested

2. Formulation of technical contradiction.

In addition to the census of information performed through the use of the questionnaire it also enables the user to define the two major points vital to the continuation of the process :

- The Ideal Final Result, i.e. the targeted system we wish to end up with

- The technical contradiction which defines the issue to solve. If this does not happen, it may be explained by the following reasons :

- The analysis carried out previously was insufficient and the problem not identified, then it must be done all over again.

- The analysis carried out revealed that it was not really about an inventive problem but rather about a problem badly analysed so far. In this case, solutions will show up when filling in the questionnaire.

If it turns out there is a real contradiction, it is recommended to state this contradiction with the most general terms possible. The terms used here " are then translated " thanks to the 39 input parameters on the matrix of Altshuller. Actually it is about converting a specific problem into a standard problem. One can note the heterogeneity of the parameters suggested in this matrix. Some of them seem to characterise an element (mass of the mobile object, volume of the motionless object...). Others seem to be intended for the characterisation of a function (precision of measuring, reliability...). In addition, the terminology employed is such as certain parameters are " neutral " (power, maintainability...) while others (complexity of the object, loss of substance...) already carry already a connotation of judgement on the element or the function considered. So far we are missing information about how this method has been made up to help us understand the reasons of this heterogeneity. However one can wonder about their influence on the choices which have to be carried out at the time of the modeling phase when a specific problem is converted into a standard problem.

3. Use of the matrix of contradictions

Once the input parameters selected, the use of the matrix consists in identifying the principles suggested for the resolution of contradictions which are related to the interaction of these parameters. This stage is the one that allows to move from the definition of the standard problem to the definition of the standard solutions.

4. Brainstorming on the principles

Finally the statement made during our experiments was that whatever the order in which the principles are considered the solution will always build gradually. As a result, it is difficult or even impossible to define afterwards what principle the solution is extracted from.

5. Evaluation and choice of a solution.

Nothing is proposed to evaluate the relevancy of the developed solutions in the traditional tools of TRIZ. The only means available is to bring together the solution considered and the Ideal Final Result which was defined at the beginning of the study. However this comparative presents only a qualitative character which cannot be sufficient within the framework of an industrial project.

At this stage of the process, it is no longer possible to rely on the assistance provided by the " traditional " TRIZ method since the idea is to ensure the transfer between the standard solution and the specific solution. It seems to us that the creative potential is a key factor at this point. As per the process we have developed, we estimate that this phase must be carried out in two steps. First, a preliminary work of analysis and search for solutions is carried out by the TRIZ project leader. Then a second brainstorming session involving the people raising the problem takes place. It gives the opportunity to present the principles resulting from the use of the matrix. Actually, the experiments undertaken so far encourage us to think that the effort of conceptualisation required to understand the meaning of these principles is difficult to realise for people unfamiliar with the use of the principles. On the other hand, because this stage aims at finding out a specific solution, it is generally necessary to have a good knowledge of the material, technical and scientific environment of the system. To be able to propose the specialists new ideas of solutions resulting from the analysis of the principles makes it easier for them to understand the meaning that can be given to these principles.

Moreover the first brainstorming being realised by a non-specialist, one limits significantly the risk to suffer from psychological inertia inherent to the experiment.

Finally it is generally recommended to classify the principles suggested by the matrix according to the number of times when they show up. However, it does not seem to have any correlation between the number of ideas raised during the brainstorming which follows this identification and the frequency of appearance of the principles in the proposed solutions. Whilst conducting a search for a solution, one also conducted a full review of all principles prior to the creation of the matrix [8]. This shows once again how embarrassing it was to be unaware of the rules and methods that were applied to analyse and synthesise patents.

For this reason we propose at this stage to use the traditional matrices of quotation of solutions according to the criteria defined at the time of the questionnaire. This phase of work is thus to be carried out with the people raising the problem. According to the results obtained it will be decided either to continue the search for solution by engaging step 6 - construction of the S-Field model - or to select one or more solutions in order to move to the stage of development of this (these) solutions. In this case, the following part of the project will come out of the competence of the TRIZ project team to be handed over to the manufacturer.

4   Conclusions

The methodical analysis of the means proposed within the Theory of Resolution of Inventive Problems made it possible to emphasise the possibility of distinguishing the attributes of a method. The existence of basic concepts legitimates tools that can be used within a structured implementation process. Such a formalisation of the method seems essential to us in order to allow its introduction and use in industrial circles and to fully benefit from its potential. The experts of the method generally state that a great deal of experiment is necessary to be able to fully benefit from TRIZ (two hundreds hours of training and a hundred studies would be required to acquire an expertise). We are actually convinced that the experiment makes it possible to improve the expertise. Indeed, part of the dynamics provided by TRIZ in the creativity area is based on the fact that it brings to tackle the problems and their solutions according to a mechanism completely different from the one we are accustomed to. To some extent using successively each of the tools makes us " to enter into the systems " and then to leave our usual position of observers. This new positioning seems to play a significant role as far as the direction that one will give to the inductive principles of the matrix of the contradiction, the suggestions of the standard solutions and the principles of separation are concerned. It is therefore necessary for us " to relearn " to think and it is conceivable that this stage will be long.

However, the opportunity to rely on a structured process is an essential asset to facilitate and thus shorten the training period. In addition, this structured approach will not constitute a barrier to the creativity. One could easily imagine that there is a contradiction between a structured thought process on one hand and on the other hand the need for leaving a large field of freedom to the designer so that he can give free course to its creativity. However, in the present case, this freedom remains intact at any stage of the process.

Finally the power and the distinctive feature of this method is to rely on a database of a great richness and robustness due to the significant number of analysed patents which are at its origin. We have described through the process we have presented a possible use and access mode of this database. However, there are today on the market available products - like TRIZ inspired softwares such as Tech Optimizer and Innovation Work Bench – that propose more or less different access modes. They have been elaborated in order to facilitate the work of the designers in the light of the experiments conducted by the experts having contributed to the development of the method. In this mode of use, one may think that the contribution of the TRIZ methodology directly related to the problem approach is modified compared to what we presented. On the other hand, the advantage of these products is to offer assistance in the search of solutions which require little knowledge and experience of TRIZ.

References

[1]   ALTSHULLER Genrich. (1984). « And suddenly the inventor appeared. » 2ème édition. Worcester MA. : Technical Innovation Center,. 173 p.

[2]   BASSEREAU Jean François. (1995). Cahier des charges qualitatif design par lemécanisme des sens . 190 pages. Thèse de doctorat : Génie Industriel : ENSAM Paris.

[3]   BECKER KAREN. (Page consultée le 24/08/99). NLP+TRIZ = accelerated creativity for product designers. www.triz-journal.com/archives/98aug/98aug_article1/98aug-article1.htm..

[4]   BURKARDT J.M. & DETIENNE F. (1995). La réutilisation de solutions en conception de programmes informatiques. Psychologie Française, 40 (1). Pages 85-98.

[5]   CLARKE Dana. (1995). TRIZ : through the eyes of an american TRIZ specialist., Ideation International Inc.

[6]   DARKE J. (1984). The primary generator and the design process. Developments in Design Methodology. Ed : Nigel Cross. New York : Wiley &Sons.

[7]   DARSE Françoise (1997). L'ingénierie concourante : un modèle en meilleure adéquation avec le processus cognitif de conception. Ingénierie Concourante : de la technique au social . Ed : Economica Gestion. Paris. 166 pages. ISBN : 2717834079.

[8]   DOMB Ellen. (1997) TRIZ : An approach to systematic Innovation. GOAL/QPC Ressearch Commitee, Research Report 1.1.

[9]   DOMB Ellen. (Page consultée le 24/08/98). Psychological inertia : two kinds in one stone story. www.triz-journal.com/archives/98aug/98aug_article5/98aug-article5.htm.

[10]   DUPONT Guillaume, VERVEN Ronan. (1999). Analyse et mise en œuvre de la méthode de conception TRIZ. Rapport PFE. ENSAM Paris. 56 pages + annexes.

[11]   FEY Victor. (Page consultée le 06/04/99). Dilemma of a radical innovation. www.triz-journal.com/archives/99apr/99apr_article4/99apr_article4.htm.

[12]   Ideation TRIZ Methodology. (1995) 4ème édition. Newport Beach. Ideation International Inc.

[13]   KAPLAN Stan. (1996). An introduction to TRIZ., Ideation International Inc.

[14]   KOWALICK Jame. (Page consultée le 24/08/98) .Psychological inertia. www.triz-journal.com/archives/98aug/98aug_article3/98aug-article3.htms.

[15]   KOWALICK James. (Page consultée le 06/07/98). Human functions, languages and creativity. www.triz-journal.com/archives/98may/98may-article5/98may_article5.htm.

[16] KOWALICK James. (Page consultée le 15/07/98). TRIZ and business survival. www.triz-journal.com/archives/96nov/article3/article3.htm.

[17] LE COQ Marc. (1992). Approche intégrative en conception de produits. 211 pages. Thèse de doctorat : Génie Industriel : ENSAM Paris.

[18] MANN Darrell. (Page consultée le 24/08/98).Digging your way out of the psychological inertia hole. www.triz-journal.com/archives/98aug/98aug_article2/98aug-article2.htm.

[19] RANTANEN Kalevi. (Page consultée le 15 juillet 1998). Levels of solutions.www.triz-journal.com/archives/97dec/dec-article4.htm.

[20] RANTANEN Kalevi. (Page consultée le 15/07/98). Brain, computer and the ideal final result. www.triz-journal.com/archives/97nov/nov-article1.htm.

[21] RATANEN Kalevi. (Page consultée le 15/07/98). Polysystem approach to TRIZ. www.triz-journal.com/archives/97sep/article1/paper1.htm.

[22] TERNINKO John, ZUSMAN Alla, ZLOTIN Boris (1996). Step by step : TRIZ : Creating Innovative Solution Concepts. 3ème édition. Nottingham. Responsable Management Inc.228 p.

[23] TERNINKO John. (Page consultée le 01/12/98).TRIZ.www.mv.com/ipusers/rm/TRIZ.htm.

[24] VADCARD Philippe. (1997). Aide à la programmation de l'utilisation des outils en conception de produits . 167 pages. Thèse de doctorat : Génie Industriel : ENSAM Paris.

[25] VISSER W. (1992). Use of analogical relationships between design problem-solution representations : exploration at the action-execution and action-management levels af the activiy. Studia Psychologica, 34 (4-5). Pages 351-358.

[26] ZLOTIN Boris, ZUSMAN Alla. (Page consultée le 06/04/99). Managing Innovation Knowledge. www.triz-journal.com/archives/99apr/99apr_article2.htm.

[27] ZUSMAN A., ZAINEV G., CLARKE D (1999). TRIZ in progress. Ideation Research Group. Southfield, MI, USA. 247 p.

Product Data Reduction:
Geometric models' Parametrization schemes

G. Wahu, A. Bouras
LIGIM, Université Claude Bernard Lyon 1, France

J.M. Brun
ESIL, Université Aix Marseille 2, France

Abstract

One of the main problems in Product Data exchange is to communicate the product geometry efficiently. Neutral files and normalized models intend to do so, and succeed well when the geometric models are compatible.

When the geometric models are not compatible, their conversion becomes necessary down to any detail of the modeling systems that use these geometric models.

In case of parametric surface models, data reduction is frequently needed in such conversions, and corresponds globally to a number of poles reduction.

A previous analysis has shown that there are two main criteria for control points reduction: the parametrization and the extremity conditions. Curvature extrema and inflexions are less critical and might impose unwanted limits to pole number reduction. The possibility to define an optimum optimorum solution for the approximations used in data reduction is noted. The relation between this optimum optimorum and the existence of an optimal parametrization leads towards a new approach of the curves' and surfaces' approximation, presented in this paper. This approach has the advantage of modifying the parameter setting in a transparent way, while matching easily the extremity conditions.

Introduction

Automatic conversions between different product representation schemes is a continuing challenge. The main differences that can occur are:

Systems do not use the same modeling schemes,

Models are of same nature but their mathematical formalisms are different (as it is the case for surface models of Bézier and Spline types for example),

Formalisms are identical but systems are different in parameters, such as surface degree or precision factors.

Apparently, the communication between systems based on *similar modeling schemes* can be done on a term to term translation basis. Unfortunately the sensibility to the design precision can lead to practical incompatibilities of models that are theoretically identical.

In the specific case of surface conversion, surface systems have no need for the topologic information found in solid models. Meanwhile, the models creation in highly constrained environments results into curves and surfaces, either over-segmented or of dangerously high degree, eventually both. Curves and surfaces, defined by number of poles larger than necessary, induce severe problems in further use. Data exchange with other systems can be impossible, if the degree exceeds what is allowed in the receiving system, or untractable if the number of curves or surfaces generated is too large. Data *quality* can be very poor, curves and surfaces which are defined by an extremely large number of poles have difficulties in behaving 'nicely'. In many cases, the conversion has to be customized, and some problems, due to the models' nature and complexity, have to be treated on a case-by-case basis.

The communication between these systems can thus be limited to the conversion of models of a B-Spline or Bézier types. Conversion of B-Spline into Bézier is apparently easier than the converse, since one can convert a B-Spline surface into a set of simple Bézier patches whose degrees are generally accepted by systems using the Bézier formalism. This conversion is also used to facilitate intersection and visualization calculations. Unfortunately this can lead to a proliferation of patches that the receiving system cannot accommodate. In such a case one has to replace this huge set of low degree Bézier patches by a smaller one of higher degree. Sometimes both can combine and it is necessary to reduce at the same time the number of patches and their degree, obviously at the cost of the model precision, but at a cost as low as possible. The classical approaches for Bézier and B-Spline models degree reduction are presented in section 2. In section 3, we recall the main criteria needed for degree reduction.

Precision problems often appear when one wants to limit surfaces' degrees or to make approximations. An optimum solution can be defined as one which obtains the minimum number of surfaces poles while maintaining a given precision. Unfortunately, this process needs the definition of new curves and surfaces parametrization. Such parametrization is proposed in section 4.

Bézier and B-Splines poles' number reduction

For Bézier curves and surfaces, the poles' number reduction can be seen as the reverse process of degree elevation [1,7,15]. In this case, if n is the curve's original degree, the reduced degree is $m=n-1$ so that to obtain an approximation of a lower degree, the reduction scheme must be used recursively.

Degree elevation is obtained step by step by a De Casteljau process of poles creation [1]. Each pole P_i is replaced by a barycentric combination:

$$P_i' = \frac{i}{n+1} * P_{i-1} + \frac{n-i+1}{n+1} * P_i .$$

Adding the extremities $P_0' = P_0$ and $P_{n-1}' = P_n$ provides the poles of a curve of degree $n+1$ identical to the original degree n curve, this process can be done also from P_n to P_0. Unfortunately, on curves of effective degree higher than n, the downgrading process gives two different results depending on the way it is processed (from P_0 to P_n or P_n to P_0).

The idea of a blend between these two elementary processes, with blending coefficients depending on the rank in each elementary process, was obviously used since the beginning.

It was an important insight in this process that enabled Eck [5,6] to find the optimal blending and to prove that it is the best componentwise approximation. However this optimal solution, aside its componentwise limitation assumes that the curve parametrization stay unchanged (Fig. 1). Improvements over the Eck's solution can be expected from a global optimization and parametrization's modifications. Some similar works were suggested by Bogacki and Brunnett [3,4].

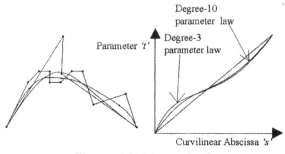

Degree-10 parameter law

Degree-3 parameter law

Parameter 't'

Curvilinear Abscissa 's'

Figure 1 Eck's method

Remark : To visualize the parameter setting of a curve, a graph of the *'parameter law'* $t = f(s)$ is used (where t is the parameter value and s the curvilinear abscissa). The linear parameter setting is used as reference.

The poles' number reduction is a crucial problem when converting from B-Splines to Bézier. An exact convertion is obviously possible since B-Splines are fundamentally sets of Bézier curves or patches of low degree (typically 3 with C2 continuity). Doing so produces huge sets of low degree Bézier curves or patches, which are sometimes untractable by Bézier based CAD systems.

For B-Splines, the parametrization law is often seen as a nodes sequence setting, the choice of critical points or zones, in which the nodes sequence must tighten, seems to be, at least, as important as the parametrization law. Nodes sequence and parametrization are of same nature anyway and the influence of curves' characteristic points on the quality of an approximation is useful to split the overall curves or surfaces in consistent area, according to the B-Spline degree.

For Bézier as well as for B-Splines, or more generally for NURBS, the pole number reduction can also be seen as an approximation problem, which involves a parameter setting problrm as well as a node sequence choice for B-Splines.

Critical criteria for curves' and surfaces' data reduction

The curves and surfaces modeled in a sending system are often an *approximation* of the canonical surfaces originally defined by the designer [9]. Effectively the surfaces found in such a model are more than often canonical surfaces (like planes, revolution cylinders or cones, spheres or Torii), which are defined by a limited set of parameters like axes and radii. Data reduction in this case consists in retrieving these few parameters from the approximated models to redefine their initial reduced definition. These parameters are hidden in rational formulations (like Nurbs or Rational Bézier) and approximated in polynomial ones (like B-Spline or Bézier). Retrieval of canonical parameters, out of surfaces' equations, needs first to recognize the type of surface then to compute the parameter values. A better solution is to find these parameters from some geometric properties as the *normals at the corners* of control nets of polar formalisms.

Curvature control is also considered as an important criterion for data reduction, but there is no study of its level of importance; moreover all the approximation theory is based on the minimization of some curves *distances* that derive directly from points distances, without any reference to curvatures. The quality of a data reduction is thus directly related to the distance between points of a same parameter value, this presents a drawback: points can be far away while laying on the same curve or surface, when they correspond through different parametrizations. The choice of a good parametrization of curves and surfaces is crucial when defining them by discrete points [7,8,10,12,13].

In a precedent study [14], it was expected to find two main criteria aside the extremity constraints: parameter's setting and curvature extrema. But it was incredibly easy to improve over our more subtle approach, only by interactively moving the curves poles. Then, when using the parametrization law of these interactive curves, the results were nearly independent of the points chosen to build the curve.

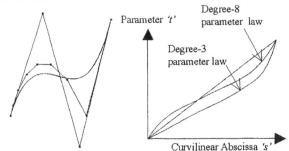

Figure 2 Parametrization variation

In Figure 2, the differences between the two parametrizations show that the geometrical characteristics of the curve cannot determine the parameter setting, independently of the degree of the curve. The parametrization law is the only critical criterion. Aside the extremity conditions while moving poles, creates directly the best possible approximation as well as the optimal parametrization.

Parametrization based reduction approach

Parametrization schemes

Our first study of parametrization [2] followed the Farin's statement that there is no intrinsically better parametrization in the classical schemes. He found thus of good practice to test chordal parametrization first, then Lee's centripetal scheme and ultimately the Foley's tangent variations one. Such parametrizations come from kinematics analogies where one travels on the curve at constant speed or slows down on curves depending on centrifugal forces or the speed to turn the steering wheel.

Another method, suggested by Hoschek [10], deserves a little attention. Hoschek proposed a scheme relating

parameter's modifications to tangential errors. This comes probably from the observation that, in the mean squares minimization process, the distances taken into account are between points of same parameter's value and not effective distances between points on the two curves.

In our previous study [2], we proposed parametrization laws defined after the geometric properties at both curves' ends. The position of extremities induces a linear parametrization where the parameter value t is proportional to the curvilinear abscissa s ($t=a*s$). Adding tangents, then curvatures at extremities produces parametrizations that are respectively parabolic ($t=a*s+b*s^2$) and cubic ($t=a*s+b*s^2+c*s^3$). These parametrizations were found convenient for curves depending essentially on these end conditions. For more complex curves a segmentation of the curves at critical points, with parameter continuity, was envisioned.

The optimal parametrization searched here can be defined for the approximation of curves or surfaces known everywhere, or nearly everywhere. It is the parametrization for which the maximum of the distances between a point on a curve and the approximated curve is minimal, which implies that these distances are perpendicular to the curve. It can be defined also as the parametrization for which distances are perpendicular and minimal in the mean squares sense.

Some heuristics to reproduce the designer's actions were used with some success, but were increasingly difficult to turn into systematic algorithms when the curve degree increases. In fact we were only able to improve over degree 3, which corresponds to the C_2 extremity conditions already used in the cubic law, to degree-4.

An analysis of these heuristics, to define an algorithm independent of the degree of the approximating curve, was undertaken successfully:

the optimal solution is found when the sum of the weighted maximal errors produces null vectors for each pole of the approximating curve.

This corresponds also to the fact that all errors are taken by projection of the approximating parametric curve on the given curve, the tangential error is thus cancelled.

The parametrization defined on the given curve cannot change from this solution without increasing the errors between the curves, it is then the optimal parametrization for the given degree of approximation.

If and when, such a parametrization of the curve is defined, the optimal solution is defined by the condition:

$$\sum_{j=0}^{m}\sum_{i=0}^{n}\left(P_iB_i^n(t_j)-P(t_j)\right)B_k^n(t_j)=0$$

where j corresponds to the maximal error values and k to the pole number, there is $n-1$ conditions for the poles P_1 to P_{n-1} since the poles P_0 and P_n are defined as the curves extremities. For G_1 continuity conditions, the poles P_1

and P_{n-1} are also constrained to lay on the extremity tangents of the given curve.

Independently of the extremity conditions, this condition corresponds to the minimization of the square of the maximum errors when the parameterization is optimized. It is thus somewhat related to a least square method and to a Tchebychev minimax condition [11].
The least square approach can be formalized as follows:

$$\frac{\partial E}{\partial P_k}=0$$

where

- $E=\sum_{j=1}^{m}\left(f(t_j)-P(t_j)\right)^2$

- $P(t_j)$ is a point of the given curve

- $f(t_i)=\sum_{i=0}^{n}P_iB_i^n(t_i)$ is a point of the approximating curve

then

$$\frac{\partial E}{\partial P_k}=2\sum_{j=1}^{m}\left(f(t_j)-P(t_j)\right)\frac{\partial}{\partial P_k}\left(\sum_{j=1}^{m}f(t_j)-P(t_j)\right)$$

$$\frac{\partial E}{\partial P_k}=2\left(\sum_{j=1}^{m}f(t_j)-P(t_j)\right)B_j^n(t_j)$$

which is the optimizing condition stated above.

If one replaces the m maximal errors $j=1,m$ by a regular sampling $k=1,N$ in the condition stated above, one finds a classical mean squares minimization:

$$\sum_{k=1}^{N}\sum_{i=0}^{n}\left(P_iB_i^n(t_j)-P(t_j)\right)B_k^n(t_j)=0$$

Error function

The Bézier curve degree reduction can be defined as follow:

Let $C(t)=\sum_{i=0}^{n}b_iB_i^n(t),t\in[0,1]$

be a n-degree Bézier curve defined by a pole set $\{b_i\}_{i=0}^{n}$
Then find

$$\overline{C}(t)=\sum_{i=0}^{m}\overline{b_i}B_i^m(t),t\in[0,1]$$

the lower m-degree (m<n) Bézier curve defined by a pole set $\{\overline{b_i}\}_{i=0}^{m}$ *that minimizes an error function* $d(C,\overline{C})$.

Therefore the approximating curve $\overline{C}$ depends on the chosen error function $d(C,\overline{C})$. Common error functions are:

- $d(C,\overline{C})=\max\left\{\|C(t)-\overline{C}(t)\|:t\in[0,1]\right\}$ which measures the maximal Euclidean distance between points of a same parameter value.

- $d(C,\overline{C})=\sqrt{\int_0^1\|C(t)-\overline{C}(t)\|^2\,dt}:t\in[0,1]$ which measures the least square Euclidean distance between points of a same parameter value.

Using these error functions assumes that the parametrization does not change within the curve degree reduction. We submit some new error functions that allows to take into account the parametrization variation:

- $d_{nm}(C,\overline{C})=\max\left\{\sqrt{\|P_i-\overline{C}(t_i)\|^2}:t_i\in[0,1],P_i\in C,i\in[1,n_i]\right\}$ which measures the maximum perpendicular Euclidean distance.

- $d_{ns}(C,\overline{C})=\sum_{i=1}^{i=ni}\sqrt{\|P_i-\overline{C}(t_i)\|^2}:t_i\in[0,1],P_i\in C,i\in[1,n_i]$ which measures the least square perpendicular Euclidean distance.

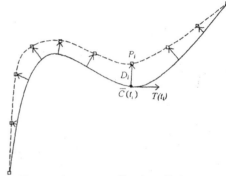

Figure 3 perpendicular distances

Remark: the P_i point's parameter value is not used, the errors between the two curves are measured on a set of points $\{P_i\}$ chosen on the reference curve. These points are projected on the approximating curve to compute the distances. Thus, one obtains distances between the curves independent of the parametrization.

Minimization process

To minimize the errors, an iterative method based on the poles displacements is used. For a Bézier curve, a pole displacement alters the totality of the curve, even tough an alteration of the pole b_i is maximum for the parameter value $t_i=\frac{i}{n}$ (where n is the curve's degree). Thus a point of parameter t_j is modified by all the poles with a relative influence corresponding to the Bernstein coefficient $B_i^n(t_j)$, which reaches his maximum when $t_i=\frac{i}{n}$.

To reduce the total error between the two curves, one has to minimize the distances d_{nm} defined above. For the first distance the displacement of each pole b_i will be:

$$\partial b_i = \sum \max(\partial b_{ij}) = \sum [\max(V_j) * B_i^n(t_j)]$$

where V_j is the error.

For the second distance the displacement of each pole b_i will be:

$$\partial b_i = \sum \partial b_{ij} = \sum [V_j * B_i^n(t_j)]$$

On the approximating curve the parameter t_j associated with an error V_j is automatically modified at each iteration, so the parameter setting evolves in a continuous way.

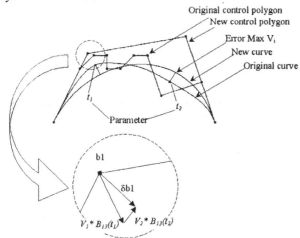

Figure 4 The minimization process

One can remark that a pole modification can have an influence perpendicular to the curve for some points and tangential for others. Tangential modifications modify weakly the perpendicular distance, but strongly the parameter value. For each pole, the weighted sum of error vectors defines the vector that the pole displacement has to cancel. When these weighted sums are all zero it is impossible to diminish an error without increasing other ones in such a way that the curves distance is increased.

Extremity conditions are at least G_0, so b_0 and b_1 are fixed. G_1 constraints must be respected quite often, if not always for CAD models, and it is easy to impose the poles b_1 and b_{n-1} displacements to stay on the extremities' tangents to verify the G_1 constraint (one has only to project the correction vector on the extremity tangents). However such a constraint has adverse effects on the final parametrization and to the curves' distance as well.

Classically, the search of the minimal degree is realized in a downward way i.e. the degree is reduced successively and the reduction process is used recursively. While proceeding, there is generally no guarantee that the target distance cannot be achieved at lower degrees so the data reduction is not optimal.

We propose here an upward approach, which guaranty to find the lower degree matching the target distance; such an approach adapts easily to our minimization process. Computations are initialized with a curve of the smallest possible degree. This degree can be defined from the curve's geometric characteristics, that is to say that it depends on the number of *"significant inflexions"* (inflexions impossible to smooth at the target precision). To match G_1 extremity conditions with some freedom left, degree-3 at least is needed. Then, if the optimal approximation (at the chosen degree) cannot match the needed precision, the degree is increased and a new minimization process started. Moreover, the former result is reused through a degree elevation, which avoids restarting from the beginning.

The extension of the basic algorithm to Bézier surfaces rests on the same principles. The displacement of each pole is:

$$\partial P_{lk} = \sum \varepsilon_j * B_l^n(u_i) * B_k^m(v_j)$$

Surfaces are computationally much more expensive than curves, but the possible data reduction is also much more effective. So the initialization of the process is very important in the case of surfaces. For that, we build a Bézier/Coons surface starting from the four curves obtained by employing our algorithm on the surface edges. Thus we obtain an initialisation surface suitably respecting the G_0 constraints which are necessary in CAD

Results

Curve results

The following example shows a three-dimensional Bézier curve of degree 9 (Fig. 5).

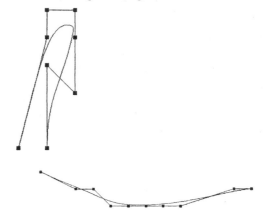

Figure 5 Original curve

The Bézier curve (Fig. 6) is a G1 approximation of (Fig. 5), with an average relative error of 0,00245

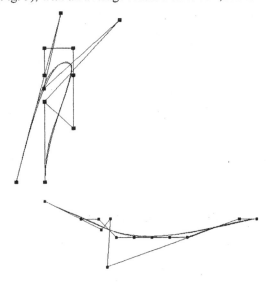

Figure 6 Approximation curve

The curve (Fig. 7) represents the error's evolution between the two curves. One can notice that this curve undulates almost regularly and that the maximum error is close to the average error (0,0042/0,0025). In fact the irregularities we can see are related essentially to the respect of the G_1 condition.

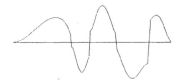

Figure 7 Error graph

Surface Result

Consider for example a B-Spline surface defined by a network of 23*23 poles (Fig. 8). Since it is a 3*3 degree B-Spline surface, the number of poles would be 16*22*22, if converted to a set of 3*3 degree Bézier patches.

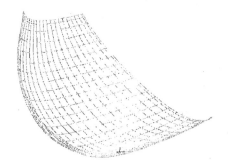

Figure 8 Original surface

It is possible to convert it to a 5*5 degree Bézier surface of 36 poles only and a relative error limited to 0.0012. Incidentally, one can note that this "real" surface is a difficult one since it looks like a three sides patch. This comes from the fact that two sides of the patch are tangent, which causes the isoparametric lines to be ill defined in the vicinity of the corner where these sides are tangent.

    a   **b**

Figure 9 Coons / Bézier surface

We use the degree-5*5 Coons/Bézier surface (Fig. 9a) as initial surface. The figure (Fig. 9b) represents the evolution of the error along the surface. At this stage of the process, the approximation's relative error is 0.0475. Then the minimization algorithm is applied to the surface (Fig. 9a) and one obtains the surface (Fig. 10a) whose relative error is 0.00126 only

    a   **b**

Figure 10 Approximation Bézier surface

The second surface is a typical CAD data transfer B-Spline (Fig. 11) of degree 3*3 defined by 22*22 poles.

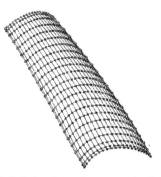

Figure 11 Original surface control polygon

For initialization, we use a degree 5*5 Bézier/Coons surface (Fig. 12) obtained starting from the four approximations curves of the original surface edges.

Figure 12 Initialization surface control polygon

To improve the precision of the approximation, the surface degree was increase during the processing. The final approximation was a degre 6*6 Bézier surface and the precision obtained is much better since the relative error is 0.0005 only.

Figure 13 Approximation surface control polygon

Conclusion

The number of pole reduction is an important problem for curve and surface models conversion, which is needed in CAD data exchange. It can also be useful to reduce unwanted undulations and to improve the quality of curves and surfaces in taking into account the *"significant inflexions"* only.

The critical criteria in this domain are the parametrization law and the extremity conditions (points and tangencies). Curvature extrema and inflexions are less critical and might impose unwanted limits to pole number reduction.

We can note that there was an optimal parameter setting making it possible to compute the best approximation for a given degree. This optimal parametrization creates perpendicular distances between the points of the approximation and the target curve. The minimization of these distances provides an optimum optimorum for the approximation, this minimization is a side result of the optimal parametrization achieved for Bézier approximating curves.

Some improvements were then proposed to take account of constraints suitable for the data exchange such as search of the minimal degree and the respect of the continuity constraints.

The transposition of this coupled approximation-parametrization scheme to surfaces is easy but the closeness of the surface, initiating the process, to the original surface is critical for the process performances, since the computation for surfaces are much heavy than for curves. Finally, the process has also interesting properties for other modeling area such as curves and surfaces reverse engineering, which starts from a scatter set of points needing a parameter setting. More generally any approximation of unspecified surfaces CAD surfaces can take advantage of the proposed process.

References

[1] P. Bézier, Courbes et Surfaces, Hermès, 1987.

[2] J.M. Brun, S. Foufou, A. Bouras, P. Arthaud, "In search of an optimal parametrization of curves", Proceedings of the fourth Int'l Conf. On computational graphics and visualisation techniques, Lisbon, Portugal, Dec 11-15 1995.

[3] P. Bogacki, E.W. Stanley, X. Yuesheng, "Degree reduction of Bézier curves by uniform approximation with endpoint interpolation", CAD, Vol. 27(9), pp. 651-661, 1995

[4]   G. Brunnett, T. Schreiber, J. Braun, "The geometry of optimal degree reduction of Bezier curves", CAGD, Vol. 13, pp. 773-788, 1996.

[5]   M. Eck, "Degree reduction of Bezier curves", CAGD, Vol. 10, pp. 237-251, 1993

[6]   M. Eck, "Least square degree reduction of Bezier curves", CAD, Vol. 27(11), pp. 845-851, 1995

[7]   G. Farin, Curves and surfaces for CAGD a practical guide, Academic Press, 1997

[8]   B. Hamann, J-L Chen, "Data point selection for piecewise linear curve approximation", CAGD, 289-301, 1994.

[9]   K. Harada, E. Nakamae, "Application of the Bezier curve to data interpolation", CAD, Vol. 14(1), pp. 55-59, 1982.

[10]   J. Hoscheck, "Intrinsic parametrization for approximation", CAGD, Vol. 5, pp. 27-31, 1988.

[11]   M. Lachance, "Chebyshev economization for parametric surfaces", CAGD, Vol. 5, pp. 195-208, 1988

[12]   E. Lee, "Choosing nodes in parametric curve interpolation", CAD, Vol. 21(6), pp. 363-370, 1984.

[13]   B. Sakar, C-H Menq, "Parameter optimization in approximating curves and surfaces to measurement data", CAGD, Vol. 8, pp. 267-290, 1991.

[14]   G. Wahu, "Critical criteria for data reduction in curve and surface models conversion", Proceedings of the first Swiss conf of CAD/CAM, Neuchâtel, Switzerland, Feb 22-24 1999.

[15]   M.A. Watkins, A.J. Worsey, "Degree reduction of Bézier curves", CAD, Vol. 20(7), pp. 398-405, 1988

Designing a new automotive seat function

BAUDU Samuel,
FAURECIA, route d'Etampes -ZI de Brières-les Scellés B.P. 91152 ETAMPES - France

JUDIC Jean-Marc,
FAURECIA, route d'Etampes -ZI de Brières-les Scellés B.P. 91152 ETAMPES - France

LE COQ Marc,
ENSAM, Paris 151, boulevard de l'Hôpital 75013 Paris - France

Abstract

Comfort and safety are two main fields of researches in the automotive industry. These researches are often led in a distinct manner. This article concerns a study whose aim is to improve both safety and comfort offered by the automotive seats. Allowing drivers to modify their seats positions while driving is recommended by physiological studies. This seat modification is also justified by the evolutions of driving conditions and the evolutions of driver physiological state. Designing a new seat function to be used while driving is different from the current seat adjusters' design process. The design method that we followed is described, in particular the use of CAD 2D human model.

1 Introduction

Improving safety and comfort for automotive occupants is a daily preoccupation for car manufacturers and suppliers. Today, everybody agrees on the fact that safety and comfort are linked together, however researches on those fields are often led in a distinct manner.

As a consequence each new equipment added to a vehicle gives either real benefits in terms of safety, or in terms of comfort, seldom for both of them.

This research fits in with a more ambitious approach, whose aim is to find ways to improve both safety and comfort. More precisely, we are aiming at improving primary safety, i.e. avoiding road accidents. Today, we think that postural comfort can be a bridge between comfort and primary safety. In order to do this, we want to integrate in automotive seats certain elements stemming from postural comfort considerations.

This article does not aim at describing the experimental analysis we use to quantify and qualify effects on primary safety, but rather deals with improving the driver's postural comfort, and highlighting the mechanical aspects present in our study.

In the first part, we will examine the previous work done on driving activity and human posture which led us to improve postural comfort by studying a mobility function to be used while driving.

The second part brings up the main differences between current seat adjusters and the mobility function that we want to design. In particular, we shall see that current seat adjusters can't replace the mobility function.

The last part compares current seat adjusters design process and the mobility function design process.

2 Generalities

An Automotive seat is composed of 2 main elements in contact with the occupant : the seat back and the seat cushion. Each of these 2 parts include a metal structure on which a foam cushion and covering materials are placed.

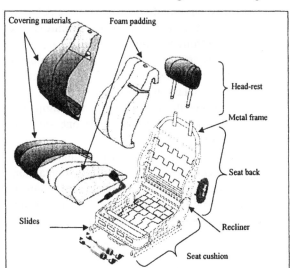

Figure 1 : automotive seat.

Some mechanical systems called seat adjusters allow the user to adjust and to modify the seat position, in particular :

- The longitudinal position,
- The height position,
- The cushion seat angle,
- The seat back angle.

3   Improving postural comfort

Before focusing on design aspects of our research, it's necessary to present and to explain the baseline of our study. Hence, this part concerns the existing knowledge in the field of the driving activity and human posture.

3-1   The driving activity

Driving a car demands a mental and a physical activity from the driver, even during a leisure ride.

The elementary Human-Vehicle-Environment system presented by D.Lechner et P.Van Eslande [1] gives us a general view of the driving activity.

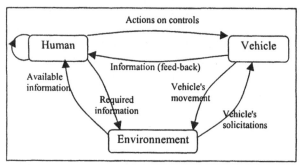

Figure   2   :   elementary   Human-Vehicle-Environment (D.Lechner et P.Van Eslande)

The links between the driver and the vehicle are composed of human actions on vehicle controls and information coming from the vehicle.

The relations between the vehicle and the environment consist of vehicle movements on the road and solicitations by the environment on the vehicle.

The relation between the driver and the environment, is an exchange of information, available on one side and required on the other.

We can notice that the driver has to regulate and to adjust himself to conditions resulting from interactions between the components of Human-Vehicle-Environment system. We shall now explain more precisely the mental and the physical components of the driving activity.

3-1-1 The mental activity

The mental activity while driving can be described with the Rasmussen model [2]. Three levels are presented : a tactical one, an operational one and a strategic one. The tactical one includes detection and research of information. The operational and the strategic one deal with the information processing within the brain. While driving, the most predominant and precious source of information for the driver is visual. It comes from external and internal environments.

3-1-2 The physical activity

The physical component of the driving activity consists in acting on the vehicle devices to control and to regulate the speed and the direction of the vehicle. This activity depends on the architectural design of the automotive interior. On current vehicles, driving controls are mainly the steering wheel, the speed pedal and to a lesser extent the brake, the clutch pedal and the gear level.

We have presented separately the two components of driving activity, but in fact both of them affect the posture. The driver's posture results from the manner, in which he adapts himself to requirements from mental and physical activities.

The visual activity is complex, due to the large panel of information, coming from different sources. The driver must prepare himself to receive that information.

The driver's eyes are bound to his visual constraints, the position of his head is ruled by the eyes, and the organisation of the other body parts results from physical activity requirements and seat design.

In consequence the driver's posture is relatively rigid.

Of course, some devices decrease physical exercises, by reducing the necessary driving actions, in particular the actions of the legs. For example, some vehicles fitted with an automatic gear box are equipped with only 2 pedals, some vehicles are also equipped with a speed regulator, which allows the driver not to keep his foot on speed pedal permanently. However, this reduction of activity requirements offers only a limited mobility. In fact, the driver must not keep his foot far from the pedals in order to be able to brake at any time. Thus, today, even with more sophisticated driving controls, the driver's posture is quite unchanging.

3-2   The driver's posture

Driving   conditions   sometimes   intensify   the phenomenon described above. For example, driving on a highway is a monotonous activity with few significant turns and rare gear changes.

From physiologists' and ergonomists' point of view, this situation, is undesirable. Indeed, a certain number of recommendations underline the consequences and the misdeeds of the unchanging body position.

SPÉRANDIO [3], 1981, makes the following recommendations: " a sufficient liberty of movement must be left to the operator so that he can move in various directions. The operator must also be able to change his posture as often as possible. "

In addition, in an article of LEUDER [4], 1983, we can read that comfort represents "no optimum or ideal fixed set of physiological conditions but rather...optimally varying physiological levels".

Researchers in the automotive industry, WEICHENRIEDER and HALDENWANGER [5], 1985, make the following remark : "even if it felt comfortable for a short time, an unchanging body position imposes a strain on the organism...although the body must be held by the seat, a degree of liberty must remain. If this is lacking, muscular tension must...be varied by slight frequent adjustments of the seat structure."

Other researchers GRIECO, 1986, and WINKEL, 1987, [6] pointed out that there might be a risk in postural fixity and that, to a certain extent, some body movements was required.

Activity analysis (physical and mental) and the consequences of maintaining a fixed posture led us to study a mobility function to be used while driving.

According to SCHOBERTH, 1978, [4] movements provide the following physiological benefits :

1-Muscle movement serves as a pump to improve blood circulation.

2-Afferent nerves send impulses to the central nervous system to maintain alertness.

3-The spine receives nutrients solely by passive diffusion occurring from changes in pressure caused by movement.

4-Different pressures acting on the spine and on the tissues are continuously redistributed.

All the benefits mentioned above, show the necessity of mobility during the driving activity.

This mobility is also justified by the very definition of posture.

Posture is the spatial arrangement of human corporal segments of a human during activity. It depends on :

1-Human anatomical and physiological characteristics

2-Physical attributes of the work station (here the driving space),

3-The demands of the task.

However, if we take a closer look at this activity, we notice that while driving, the state of the operator evolves, the requirements of the activity too, driving downtown for example is different from driving on a highway.

Previously, we explained that the driver's posture is relatively fixed due to the requirements of his activity. These requirements evolve during driving, so naturally, posture must evolve too. Moreover, even with constant driving conditions, moving is also a human need.

3-3   To move is a human need.

Moving is not only a theoretical recommendation justified by experts, but also a human need. Of course, everybody agrees with this affirmation, because we can observe easily that even people who work in a seated position move frequently, and not only to execute their daily tasks. For the automobile driver, some experimental results confirm our hypothesis.

BERTIN [7] (1996) carried out an experiment of simulated driving during which each subject had to drive 3 times 30' with 8 minutes of pause after each driving period (on the whole 1 hour 30' of driving)

This experiment was conducted on 34 subjects. Each subject did the test twice, i.e. drove for 3 hours. From the video recordings, the postural adjustments (movements of the conductor in his seat) were noted.

Among the 34 drivers, 23 of them moved their bodies while seated.

On average, there were 5 adjustments for the 3 hours of simulated driving. The lowest frequency was 1 adjustment for cumulated 3 hours of driving, and the highest was 25. For the subjects not having adjusted their posture, we can assume that they did not feel the need (the 30 minute periods of driving followed by rest periods of 8 minutes were not sufficiently long). This study also revealed another significant fact : the drivers consider that their need to move is accompanied by a drop in the level of alertness, which can be considered as an evolution of the drivers state.

An unchanging posture is not convenient, so drivers move in their seat. Is this situation acceptable? Should mobility while driving be offered by seat mechanisms?

First of all, movement while seated is a limited movement, because we can not move up or down. To obtain a more inclined back, we have to move our legs forward. So, moving in the seat is a residual mobility of a human behind the steering wheel.

Second of all, after moving, the thighs or the back are probably not totally in contact with the seat. So, this posture can not be maintained for a long time, because the distribution of pressure between the seat and the body is not homogeneous. Moreover, the devices designed to protect automobile passengers are totally efficient only if they are correctly placed in their seats. For example, the safety belt must be sufficiently tightened, and the head should be close to the head rest.

Faced with this situation, we think that it's not enough to allow mobility, but indeed we must encourage mobility. Mobility while driving must express itself through a new function. Thus, we have decided to design a "mobility function", in order to allow driver to change his seat position while driving. This function (called "mobility) function " presents 2 main interests :

• To allow the driver to adapt his posture to evolving activity requirements.

• To allow the driver to move while driving.

In the next part we will present the different stages of the design process. In particular, we will discuss the specific difficulties concerning the innovating characteristic of our function.

4　Can current adjusters be the mobility function?

Automobile are used by people (both male and female) with various body proportions. For example, among the French driver population (mixed), considering a 90% interval, the length between buttocks and knees is comprise between 53 cm to 63.1 cm, and the length of the trunk is comprise between 80.5 cm and 93.8 cm [8].

So, the first function of an automotive seat is to allow the driver to find correct support in order to position himself in the vehicle, and to drive under suitable conditions.

This is called the postural support function of the seat.

In order to assure this function, current seats are equipped with adjusters allowing the driver to adjust the seat in order to find a correct posture.

Currently, seat adjusters allow drivers to adjust, and to modify the longitudinal position and height of the seat, back and cushion angle, with various cinematic laws, coupled or not,

As a consequence, automobile seats have already mechanisms that provide a certain degree of liberty to the seat in the car. Thus, the driver also has a certain degree of mobility in the driving space.

This may raise a question : if such mechanisms already exist why study a new mobility function?

In the next part, we will specify why current adjusters do not seem adapted to offer mobility to the driver and why it is necessary to design another generation of adjustments.

The first opposition concerns the functional aspect of the adjusters. Indeed, seat adjusters are essentially intended to allow drivers with different body dimensions to drive in convenient postures.

Mobility functions are intended to allow a driver, with his own body proportions, to change his position. A well designed mobility function will also encourage the driver to use it more often.

Moreover, current adjusters are supposed to be used in the vehicle before starting. The adjustments are thus not thought nor conceived to be used while driving. As a consequence they are not adapted to be used while driving. Some situations illustrate these considerations.

Discontinuous adjusters, used on the majority of manual slides, are not convenient for the desired use. To be in an unlocked seat, even for only 2 seconds, is a risky situation for the driver, that can not be ignored by car manufacturers.

Another element which must be taken into account is that, a modification on a seat height adjuster may require a re-adjustment on the slides. An actions and corrections loop should not occur while driving.

These arguments in favour of the design of a new mobility function can appear theoretical. In reality drivers may use current seat adjusters while driving. But, do current adjusters allow mobility while driving? In other words, do the drivers modify the position of their seat during a long monotonous journey on a motorway?

To answer this question, we carried out a survey on 49 drivers. This survey took place on a service area of a French highway (A10-11). The results are clear : among the 49 questioned drivers, 45 hadn't modified the position of their seats during the journey that they had just made. The 4 other drivers modified their position because they had forgotten to adjust their seats before driving.

In conclusion, the adjusters are neither designed nor used as a function that enables the drivers to change their posture while driving.

5 -Design process

This part concerns the differences between current seat adjusters' design process and mobility function design process.

5-1 Aren't the adjusters' design method sufficient for the mobility function?

A seat is used by drivers with different body dimensions. Drivers of small height as well as tall drivers must be able to drive with a suitable body posture.

Generally, in automotive industry, a set of human models is used, in order to take into account the body dimensions diversity. In particular, 3 models are more often used : a 5 percentile model which represents a driver of small height, a 50 percentile model which represents a medium driver and a 95 percentile model which represents a tall driver.

The designer uses a set of human models, and each human model is placed in an optimised posture. As a consequence, on the design sketches of an automobile, in particular on the driving space, a set of postures and a set of different hip points are drawn.

In order to allow the seat to reach these different hip points (belonging to the different human models), current seat adjusters are designed.

According to theses design sketches, each human model seems to have only one posture, wrongly considered as ideal.

JUDIC [9], 1993, shows the existence of a mobility space, which is a group of postures that a given driver can adopt for driving.

The model used by JUDIC, in order to define the mobility space, takes into account :

1. natural human mobility and limits,
2. postural constraints,
3. driving space constraints.

JUDIC defines the mobility space, as a zone of least discomfort. In particular, a zone of least discomfort represented at the hip point, defines and visualises for a human model under consideration, the area for the hip points in the x and z plane where the previously formulated postural constraints are respected.

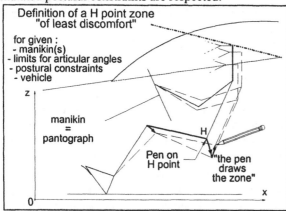

Figure 3 : definition of a Hip point least discomfort zone (JUDIC)

In reality, for a given 2D human model, if his hip joint is in the zone of least discomfort, then we can say that its posture is in a correct position for driving.

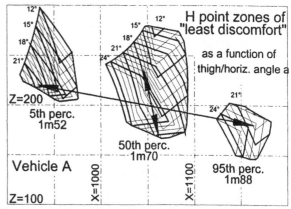

Figure 4 : Hip point of least discomfort (JUDIC)

These zones show that a given driver can adopt different postures, but they don't help us define the movement of the mobility function.

This representation (zone of least discomfort) only permits us to describe the exterior borders of the area where movements are compatible with the driving activity.

Moreover, this human model is not sufficient to design the mobility function, because it is intended to be used while driving. Therefore, a geometrical behaviour model is also necessary.

5-2 The mobility function design process

Since we attempt to offer drivers the possibility of changing their postures while driving, why not asking them with which seat position they would like to drive?

That seems to be a good idea, but in this case we have to face new difficulties.

Firstly, we must consider the difficulties due to the use of an old product, to define the characteristics of a new one.

In our case, we would use a conventional seat with 4 adjusters (slides, lifter, recliner and cushion angle adjuster). Then, we would ask the driver to choose a seat position different from the seat position he would have normally chosen.

This experimentation requires certain ability from the subject. The driver is supposed to have a precise idea of the position that he wants to occupy. During the experiment he must also be able to appreciate his current seat position, in order to operate adjustments to converge towards the target position. Moreover, this supposes a good comprehension of the possibilities, in terms of movements offered by the seat adjusters.

We carried out such an experiment to obtain a set of different postures (3 postures per subject) taken by a given driver while driving. This experiment was only conducted on 16 subjects. Thus, our results can not be considered as general facts, nevertheless, they give us certain indications.

1. Finding a new seat position with 4 adjusters (slides, lifter, recliner and cushion angle adjuster) was difficult for our subjects. In order to simplify this experiment, we prohibited the use of the cushion angle adjuster.

2. A diagram that showed the different stages of adjustment to find another posture was necessary.

3. On average, for the 16 subjects, the gap observed between extreme seat positions adopted is :

- 45 mm for the slides
- 37 mm for the lifter
- 4° for the angular position of the back.

Second of all, asking a person to define a driving position different from the one he/she usually adopts, is also a bit illusory. Indeed, previously we explained the reasons which led us to study an innovative mobility function. These reasons consisted mainly in specialists' consideration, and not as an expressed human need. Moreover, our study conducted on a French highway shows that the main factor that preventing the postural changes is to believe in the existence of an unchanging ideal posture. Thus, after a driver has chosen his seat position, he will not want to modify it because this change will be perceived like a bad adjustment.

This result isn't astonishing since the current seats are designed to be used in a vehicle before starting. As a consequence, the idea of possessing an unchanging ideal posture which must not be changed is reinforced.

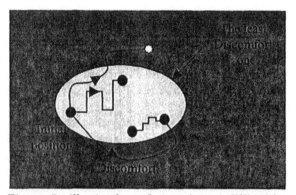

Figure 5 : illustration of experiment difficulties

The driver doesn't seem to be in a position to help the designer, but the latter has to choose a seat motion.

We are going to present the 2 main stages of the design process that we have followed. The first stage consists in using a driver representation which allows to represent the human mobility. The second stage is the choice of the final seat position of the mobility function.

Faced with the difficulties encountered by drivers to find a seat position different from their usual, the designer has used a CAD 2D biomechanical human model. Nevertheless, real drivers are not excluded from this design process, because all the hypothesis and choices will be issued considering a human driver.

With this CAD 2D human model, the theoretical mobility of human while driving has been analysed. The ARTOBOLEVSKY's formula allows us to calculate the degree of liberty of our model.

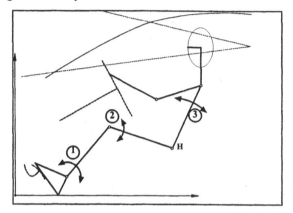

Figure 6 : 2 D human model.

The model is composed of 7 body parts (the foot, the lower leg, the thigh, the trunk, the upper arm, the fore arm and the head). These different parts are articulated by pivot joint. In order to determine the human mobility while driving, we have considered the foot to be fixed, the head to be vertical, and the hand to be fixed on the steering wheel.

At the beginning, our "mechanism" is a 2D model, thus it has 21 (7*3) degrees of liberty. The 7 pivot joints reduce the mobility to 7 (21-(7*2)). Moreover, the foot is fixed (-3), the head is vertical (-1), thus, we obtain at the end 3 degrees of liberty. In other words, the position of this human model can be describe with 3 parameters. Available parameters are the angular positions of the 5 free body parts (the lower leg, the thigh, the trunk, the upper arm, and the fore arm), and the Cartesian co-ordinates of the 4 free pivot joints (knee, hip, shoulder and elbow). For example, to set the angular positions of the lower leg, the thigh and the trunk would be equivalent to defining the human models' position.

For the designer, choosing the final seat position of the mobility function means describing the human model position with 3 parameters. Although several possible triplets exist (84), parameters selected must be linked with the reality and must be adapted to the movement philosophy.

Previously, the bibliographical analysis showed us that postural changes while driving can be justified in 2 situations : evolution of activity requirements, and constant driving conditions (to provide changes while driving). We will now present you a choice of mobility function concerning the adaptation to an evolution of driving requirements.

The table below presents the differences between driving downtown and driving on a highway according to the selected criteria (which influence the driver's posture).

	Downtown	Highway
Action on pedals	A lot of action on the clutch pedal	Essentially, keeping speed pedal at the same position.
Action on the steering wheel	large amplitudes actions	little corrections
Speed of the vehicle	Lower and various	Higher and uniform
Visual information	A lot of information, from all directions	Essentially, frontal and rear during overtaking
Body and mind of the driver	Stressed	Relax

Table 1 : comparison between downtown driving and highway driving.

We decided to design a function for downtown driving. More precisely, to design a function allowing the driver to adapt his seat to the downtown driving conditions after having occupied a position for another traffic condition.

Our analysis of the downtown driving characteristics, lead us to make the assumption that the driver's position adapted to downtown driving is :

- a position which permits the driver to be closer to the steering wheel for easiest actions
- a higher position in the driving space for a better view of the external environment.

The designer has to move the CAD 2D human model, to obtain the desired final position. Of course, several movements can offer the same qualitative modifications. In our case, we decided to move all the body parts forward, but the head much more than the rest of the body, to give a real sensation of a closer view of the road.

To do this, we have chosen to move the 2D human model with a rotation around the ankle point.

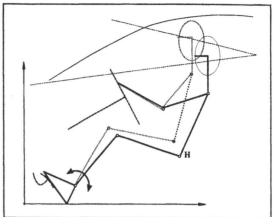

Figure 7 : the 2D human model movement for downtown driving.

The next stage concerns the seat mechanism design which reproduces a rotation movement around the ankle point, but this work is more usual for the designer.

We can notice that the movement is defining with human parameters and not with seat parameters. Ben COWAN and Marek KMICIKIEWICZ (CKE Technologies inc.) had found similar results for office work-seat, and maybe, that is a constant characteristic of mobility function

6-Conclusion and perspectives

A seat equipped with this mobility function is manufactured. An experiment will be carry out, in order to validate the interest of our new function. Although we have integrated human consideration throughout the design process, we don't know how the mobility function will be perceived. The users are the only ones who can judge this new mobility function.

Choosing the seat movement is maybe the most difficult stage for the designer. As RABARDEL [10] says, the designer invent a product, but at the end, users invent the use. Here, we have to search the manner to create a new use. Today, the only solution is to try and to experiment. In order to facilitate these iterative researches, FAURECIA is developing an experimental computerised seat powered with servomotors. This equipment allows the seat to follow motions laws without having to manufacture a new mechanism.

Although we only have to program a motion law and to . test it, several questions about the mobility function stay :

- What do drivers want to change?
- How many mobility function, and seat adjusters do we have to design?
- Which kind of controls do we have to offer?

Nevertheless, one thing is sure, the seat movement must be synchronised to the main element : the human.

References

[1]   D.LECHNER, P.Van ESLANDE, "Comportement du conducteur en situation d'accident", ingénieurs de l'automobile, n°717, pp.64-71, 1997.

[2]   F.NATHAN, "Apport des systèmes d'aide au conducteur", actes du congrès sécurité automobile-Rouen, Société des ingénieurs de l'automobile, 1997.

[3]   J.C.SPERANDIO, "l'ergonomie du travail mental", éditions masson, Paris, 1981.

[4]   R.K.LUEDER, "Seat Comfort : a review of the construct in the office environment", Human Factors n°25(6), 701-711, 1983.

[5]   A.WEICHENRIEDER, H.G.HALDENWANGER, "The best function for the seat of passenger car", SAE Technical Paper Series, n°850484, Detroit, 1985.

[6]   P.D.MICHEL, "Biomechanics of sitting postures", departement of industrial engineering state university of New York at Buffalo, 1991

[7]   C.PHILIPPS-BERTIN, "Apport d'une activité cognitive dans le maintien de la vigilance des conducteurs automobiles", thèse de doctorat en psychologie, Lyon, 1996.

[8]   R.REBIFFE,J.GUILLIEN, "Enquête anthropométrique sur les conducteurs français", laboratoire de physiologie et de biomécanique de l'association Peugeot -Renault, 1982.

[9]   J.M.JUDIC, "Contribution à l'élaboration d'outils d'aide à la conception des moyens de réglages posturaux des sièges d'automobile, thèse de doctorat en génie industriel, Paris, 1993.

[10]   P.RABARDEL, "Approche cognitive des instruments contemporains", Armand Colin, Paris, 1995.

CHAPTER 10

CE in Construction Industry

RISKCIM - Addressing Risk Analysis in the UK Construction Industry

Alan Tracey

Information Systems Institute, University of Salford, The Crescent, M5 4WT,
Salford/Manchester, UK
Tel: 44 161 295 4280 – Fax: 44 161 745 8169

Yacine Rezgui

Information Systems Institute, University of Salford, The Crescent, M5 4WT,
Salford/Manchester, UK
Tel: 44 161 295 3443 – Fax: 44 161 745 8169

Abstract

The paper proposes a model-based approach to risk analysis and prediction in the construction industry, using proven Object Oriented methodologies. First, the paper describes the risk management process. This is followed by a presentation of the RiskCIM model that captures all the artifacts involved in risk analysis. Then, the paper proposes IT-based methods of capturing the tacit knowledge and expertise of key construction actors, which is considered in decision making process, during qualitative risk assessment. This is achieved through the RiskCIM prototype, a comprehensive description of which is given in the paper. A global dynamic knowledge base of construction related risks is in the process of being populated. This is intended to be available across communication networks using web-based middleware Finally, the paper concludes with an overview of future development on the RiskCIM prototype.

1 Background

A large number of studies have been conducted by various researchers in order to highlight the benefits of information and communication technologies (ICTs) in the construction industry [1],[4]. These technologies include standards for data exchange and product modeling [10],[8],[5],[9] and integration techniques through the use of object-oriented technologies [13].

Most of the research in the area of Computer Integrated Construction aims at improving existing business processes within construction companies and projects using the latest developments in ICTs. Similarities have been drawn with the manufacturing industry leading onto the adoption of a process-driven view of construction projects [8].

The accurate specification of project processes is increasingly becoming a major contributing factor to the success of construction projects. Project processes are, however, fairly complex and in most cases difficult to predict, particularly for activities which are at lower levels of the project activity breakdown. This is largely due to the multi-disciplinary nature of construction projects that involve many actors with complementary or even sometimes-conflicting roles. In addition, construction teams are frequently brought together for projects and broken apart again upon completion.

Reality usually differs from the scenarios planned and drafted in engineering companies by architects, quantity surveyors or project managers. It is also frequently much more complex. For example, differences between planned and real tender figures have long been a source of complaint by clients and have resulted in most cases in contractual litigation between the parties involved [3]. Further analysis [3] reveals that a great deal of delays within construction projects are due to either exceptionally adverse weather conditions, amendments made to the project information base, or problems of information consistency [2]. Labour shortages, strikes, unavailable materials or late delivery also reveal to be important factors causing delays.

In fact, the decisions made during the lifecycle of a Construction project are multi-dimensional, combining together factors which range from the highly subjective to the perfectly objective. The decisions are made by many, often non co-located, actors belonging to different disciplines, throughout the construction project lifecycle. These decisions are made over very long periods of time in an iterative manner and are commonly revisited weeks, months and even years after they were originally taken. There is considerable potential for misunderstandings, inappropriate changes, and decisions which are not notified to all interested parties. In this context, the construction industry is often referred as a high-risk industry. Therefore, it is crucial for any project manager to be able to assess the risks associated with his project. Specifically, the Project Manager needs to be able to identify which of the risks can have a critical affect on the quality, cost and timescale of the project. An awareness of the critical risks will enable the project manager to take steps to avoid the identified potential risks or positively manage them. However, current practices in the industry reveal that construction risks can only be viewed

scenario that highlight the usefulness and the short-term benefits of the approach.

2 The Risk Management Process

Risk analysis forms a fundamental part of the early stages of a construction project. Risk analysis is the process whereby careful consideration is given to identifying risks that may arise throughout the lifecycle of

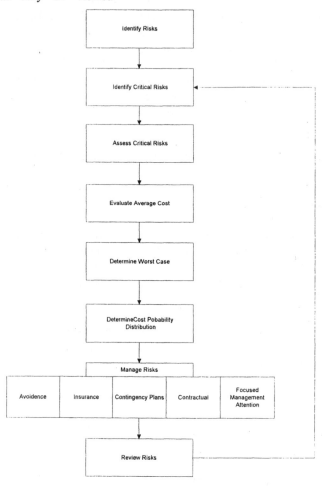

Figure 1 – The Risk management Process

subjectively due to a lack of historical facts and data, drawn from completed projects, describing objectively successful and failed experiences. Moreover, for most projects, risks probabilities are selected subjectively based upon the project teams tacit knowledge and experience.

The paper describes a model-based approach to risk assessment. The RiskCIM prototype that implements the proposed model is described, along with a business

a construction project.

In the construction industry, the main purpose of a risk analysis/management system is to support the Project Manager whilst making decisions pertinent to the project risks being assessed. Consequently, the project manager will gain a deeper understanding of a risky situation, and more important, to provide him/her with a deeper insight to what will be the likely impact of the risk. Figure 1 depicts a framework comprising the fundamental stages of the risk management process that has been adopted for use in the present research. A brief description of the main stages of this framework is given below.

2.1 Identify Critical Risks

Whilst performing the risk identification activities, typically, a project plan or Work Breakdown Structure (WBS) is used as an aid to the initial input to the analysis process. The WBS and the project plan are used to map out the project, classifying the various tasks and their related activities. Included in the WBS or project plan, generally, will be the duration of each task together with the duration of each separate activity related to that task.

The WBS and project plan can also be represented graphically in the form of a Gantt chart. The latter, will present the project manager with a graphical representation of how the project life cycle, together with, the activities and tasks involved within a particular activity. Furthermore, it also depicts how these activities and tasks relate with each other.

Anticipated risks pertinent to the construction project are then systematically tracked, identified and classified and associated to their particular activities and tasks. This process is generally employed by 'stepping' through the project plan stage by stage in the order of the project execution. Assumptions concerning any probable risky situations can be made explicit during this identification stage. For example, during the substructure stage of a building project, the activity of laying the foundations for a particular building will encapsulate a number of related tasks, e.g. excavating for the footings, shuttering and concreting and building the footings. Each of these separate tasks can have a degree of risk associated with it. For example, during the excavation stage, there is a risk of discovering human remains, or another example, might be the risk of bad weather during the excavation stage. The impact of the latter risk, i.e. flooding or the adverse affects that frost can have on the concrete, coupled with adverse ground conditions throughout the site. Consequently, causing a delay to the project, which will affect plans for completing the project to the specified completion dates.

To be effective, the risk identification stage must be carried out meticulously. A number of methods or (combination of these methods) may be employed throughout the risk identification stage. The phrase *'flows of experience'*, was adopted [7] in the context of capturing knowledge from individuals with past experience in project work. This is carried out with a view to using this knowledge and experience to supplement the risk identification stage. Some of the methods employed, involve the use of:

- *Brainstorming sessions:* Generally conducted with key actors involved in different levels of the project. However, to effectively employ this technique the choice of the actors must be carefully considered, ensuring a balance of expertise together with management experience. Also, generally, within a discussion group, personality traits must also be considered, since the person with the stronger personality often tends to be domineering and may inadvertently, sway the outcome of the discussion.

- *Structured interviewing:* This technique involves the use of pre-prepared material and questions, pertinent to the subject matter being discussed, for example, questionnaires and other related discussion material.

- *Standard questionnaires:* This technique is an effective method of stimulating a person's memory.

- *Checklists:* This technique is generally employed on large construction projects. The data on the checklists, generally, comprises historical data, gathered from previous projects. This technique serves as an effective thought stimulating method to employ during the risk identification stage. However, when this method is employed, careful consideration is needed, due to the nature of most construction projects being different. Therefore, it is difficult to cover every possible eventuality, consequently, this could constrain the risk identification process. Therefore, the person(s) using these lists, as mentioned earlier, should do so to initiate new ideas, and prompt thoughts on possibilities of other potential risks, by observing the risks on the checklist, and making comparisons in context. Subsequently, any risks that have been identified will be recorded for further analysis.

Research [15] identified that, the majority of organisations, tend to use list of standard risks for input and comparison during the process of risk identification. The identification of risks during this stage is critical, in that any risks that may not have been discovered, will inadvertently, have been accepted by the company. Since project plans will typically be interlocked (meaning that they affect other plans or are affected by other plans) the risk identification activities are usually wide in scope. The aim at this particular stage of the risk analysis process is primarily to seek out and identify all possible risks. Although some of the examples used in this section relate to activities at the operational level (construction), it is worth mentioning here, that there are different phases in a construction project where decisions will be made by different levels of management. For example, the management at the strategic level, will be interested in the risks that may affect the organisation in the long term; the middle management will be interested in risk analysis during the procurement and planning stages; and the operational management will be interested in the risks associated with the actual processes on site.

2.2 Assess Critical Risks

Having completed the identification stage of the risk analysis process, an assessment of the risks can be achieved by ranking each risk into their respective probability of occurrence. However, at this point, attention must be brought to the availability of historical data from previous construction projects. The availability of this historical data, will, in general, allow a reasonable estimate of the occurrence to be made, for some of the more mundane risks in the project. Although, because of the complex nature of construction projects, for most of the risks, the probability of occurrence will be purely subjective. Generally, the method used for classifying risks is to rank all the risks identified in the previous stage into classes or ranges of probabilities, this is an estimate of the likelihood of the risk occurring. For example, >75% (very high), *50 - 75%* (high), 25 - 50% (low) and < 10% (very low). Once the risks have been ranked into their various classes, each risk will be further assessed pertinent to the impact it will have on timescales, cost and quality in relation to the project involved. Project managers who have attained a high level of experience within construction projects will typically carry out this stage. The latter will be complemented, where possible, with historical data from similar construction projects. Subsequently, risks that fall into the high or very high-risk categories are classified as critical risks.

2.3 Manage Identified Risks

Once the risks have been identified and the impact they may have on the project has been assessed, there are several strategies that can be taken to manage and mitigate these risks. These strategies include:

- Avoidance
- Transfer
- Contingency Plans

The Avoidance strategy is directed mainly at removing the cause of the risk and can take on a number of forms, for example:

(i) **further research** - of the activity within which the risk lies, to determine the elements that may lead to the cause of the risk with a view to further reducing the uncertainty, a good example here would be the sample boring of the ground to reduce uncertainty pertinent to ground conditions on a site.

(ii) **Changing contract strategy** - for example, '*not putting all your eggs in one basket*' avoidance. In this context can be achieved by reducing the size of an workpackage assigned to a single contractor and allocating smaller workpackages to several contractors, thus reducing the dependency on any one organisation depending on a single contractor for services.

(iii) **Transfer strategy** - The transfer strategy basically means that the risk is passed onto some other organisation or subcontractor who is more competent in the particular work area or process subject to risk. Consequently, the organisation or subcontractor will be more capable of identifying and mitigating the particular risks involved. However, to employ the transfer strategy successfully, other factors relating to the recipient of the transfer must be considered. For example, will the contractor be financially stable enough to carry the particular risk? Is there a chance of it forcing them into liquidation? If the latter is the case then the risk will be returned back to the client. Another example of risk transfer being that, consideration must be given to the fact that if a subcontractor accepts responsibility (ownership) of the risk, he is likely to incur extra cost onto the client to cover the extra cost of carrying the risk. So is it worth transferring the risk in the first place? In order to make the transfer strategy work effectively, it should be ascertained that any risks transferred to a subcontractor would be within their control, coupled with some form of incentive to encourage effective performance on the project.

The development of effective contingency plans will allow the project manager to cater for particular circumstances which may arise should a related risk materialise. An example of this might be the relocation of human resources between project activities. The cause of this may be adverse weather conditions, whereby the ground-workers are unable to carry out their normal duties. For example, working outdoors, excavating for the drains and road layouts etc., they will be relocated, temporarily to another job inside a building, for example, preparation cleaning for internal tradesmen, until the normal weather conditions resume, then they will return to their normal duties.

2.4 Review risks

The preceding stages of the risk management process will generally be an ongoing process. Specific project team members will be responsible for monitoring activities and related risks throughout the lifecycle of the construction project. As each of the tasks and related activities progress, then some of the risks will cease to exist, although, other risks will materialise. Therefore, it is necessary for the project team to review the risks regularly, to ensure that they are aware of any change in the status of any risks pertaining to their particular construction project. In the event that any of the risk profiles do change, the review stage of the risk management process will capture this occurrence. Subsequently, the project team can implement the procedures necessary to amend the situation.

3. The RiskCIM Object Model

The, UML (the Unified Modeling Language) [13] was chosen for the development of the RiskCIM model in the Rational Rose CASE tool. This language is rapidly

oriented modeling, and is well supported by the Rational Rose CASE Tool. UML is developed primarily from two of the most popular modeling formalisms for object-oriented modeling, [6], and has been adopted as an international standard within the OMG (Object Management Group) that develops the Common Object Request Broker Architecture. The models shown in this paper are expressed in the UML. An overview of the RiskCIM model is given in Figure 2. Projects are described in terms of their activities. Each activity is implemented through a set of tasks. The model supports activity dependency tracking, as well as the notion of activity precedence. Activities are under the responsibility of precise roles that are carried out by specific actors. An example of a role is: project management. Each activity will have a set of elements of risk associated to it. These risks will be owned by specific roles. Risks are of two types: qualitative risks and quantitative risks. Each risk has potential risk impacts that can result into other risks. Risks might also involve mitigation techniques, captured through the proposed model.

4. The RiskCIM prototype

RiskCim, designed primarily to cater for the needs of the construction Project Manager, who had proficient

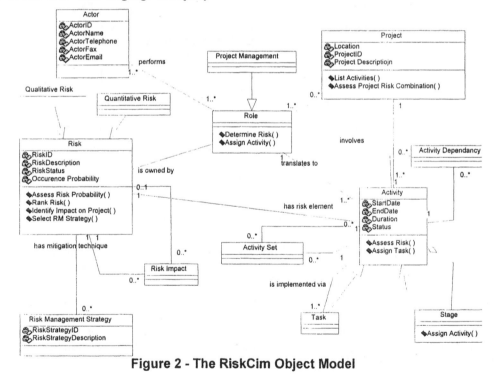

Figure 2 - The RiskCim Object Model

being established as a de facto standard for object-

expertise in the field construction risk management.

RiskCim can be used effectively, as a decision support system (DSS) for the PM, from inauguration of a

RiskCim was also designed to be a global dynamic data store, for storing, retrieving and manipulating risks,

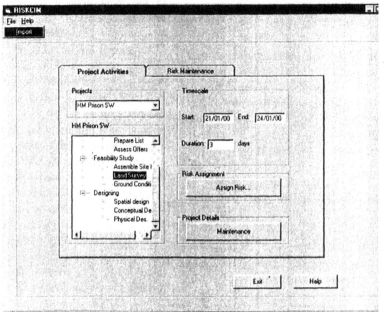

Figure 4 - Work Breakdown Structure

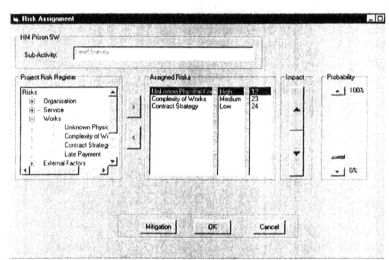

Figure 3 - Risk Assignment

construction through to completion, inclusive of the post project maintenance period. This process, inherently, will involve the processing of vast quantities of data, from a very large number of different stages and activities. One can probably imagine how tedious the task would be, to manually process such a vast amount of data. For example, imagine, manually assigning all the relevant risks, taken from a standard risklist, to the individual stages of a construction project.

pertinent to the construction industry. Global in that the design facilitates the use of a networking environment, enabling the database to be accessed from remote locations, facilitating for on site queries. Dynamic in that it can be continuously updated with new risks as they arise. The latter can be implemented through a standard WEB browser, utilising middleware technologies.

4.1 Populating the RiskCim Database

The risks are captured from historical data existing within the particular organisation. For example, existing risk lists, employee experience, senior management experience. (For more in depth discussion on data for project risk analysis, see [7]. Subsequently, new risks will be appended to the database as and when they are discovered.

As mentioned earlier the design of RiskCim facilitates the mapulation of information from a range of project planning applications. Essentially, construction project data from a construction planning package is imported into RiskCim where it is displayed in a Hierarchical Work package/activity structure. Subsequently, the functionality of RiskCim allows the person responsible for risk analysis/management to navigate through the tasks and activities previously imported from the construction project plan, and systematically enter (attach) activity related risks and associated details as and where necessary. The tasks, related activities and details are then stored in the database, the additional details that are stored are as follows:

- *Project stage details* - The stage of the project the task and related activity are related to.
- *Risk description details* - A brief description of the risk in question.
- *Duration of activity* - This is presented in three forms i.e. Startdate, enddate and duration.
- *Probability value details* - A numerical value between 0 - 100, representing the probability of the risk occurring. This decision is made from the intuition of the person responsible for entering the data, and may be supplemented by the use of historical data, checklists, and project teams experience.
- Probability function details - i.e. the appropriate type of function, for example, Normal, Beta, Pareto, Brlang etc. (See [9], [14] for a more in-depth discussion on the theoretical aspects of these functions)
- *Risk ownership details* - Details of the person responsible for the risk. E.g., the client, contractor, project manager etc.
- *Date of entry details* - The date and time the risk was entered into the database.
- *Data type details* - The data type of the risk, i.e. subjective or objective.

- *Risk type details* - The type of risk, i.e. fixed risk or variable risk.
- *Mitigation details* - Here the person responsible for risk analysis, will enter mitigation information, pertinent to the risk currently being processed.
- *Risk relation details* - Information pertinent to any other risks, related to the particular risk being processed. These could be parallel risks (see chapter three) i.e. risks whose occurrence is triggered by the occurrence of a particular risk.

Once the details have been entered for the particular construction project currently being processed, calculations are performed on the numerical data i.e. Monte Carlo Simulation, then the information is stored in the database. Subsequently, the data is exported back into the project planning application, and the amended data is compared against the baseline of the original project plan. The functionality of RiskCim allows this process to be carried out repeatedly, providing the project manager with a "what if' facility. Thus, enabling hypothetical scenarios to be input for comparison and other "what if' related queries.

4.2 Benefits of RiskCim

Although, initially the process of inputting risks into RiskCim will be tedious and time consuming. The dynamic nature of RiskCim will eventually accumulate into a vast repository of construction risks, corporate knowledge and experience. The benefits that can be derived from the system offer a competitive advantage to all the levels of management within the organisation. For example:

- Senior management decision making will be improved by virtue of the high visibility of risk exposure (and possibly opportunity) in potential construction projects.
- The organisations corporate image will inherently reflect in the eyes of their customers and clients, because of the professional approach they have adopted toward risk analysis.
- The likelihood of over pricing a project is reduced by being aware of the potential risks and their likelihood of occurrence and mitigation strategies.
- Improves understanding of the construction project by identifying risks and proper consideration of mitigation strategies.
- Reinforces the decision making process, thus enabling decisions to be more systematic and less subjective.

- The framework will encourage a deeper insight to a situation whilst searching for ways to mitigate Risks.

5. Conclusion

This paper has described construction risks in context and has presented the general stages of a Risk Management framework, which form an essential aspect of project management. The purpose of this process is to ensure that the person responsible for a project, generally the project manager, is made fully aware of any risks associated with the tasks and activities relating to the project. Subsequently, to identify the critical risks, with a view to minimising the affect they will have on the project. Furthermore it will supplement decision making by the provision of a formal structure for risk quantification.

A global dynamic knowledge base of construction related risks are in the process of being populated. This is intended to be available across communication networks using web-based middleware. As regards future work, the team is currently investigating the use of Case-Based Reasoning techniques to retrieve past project learnt lessons on the basis of risk similarity.

References

[1] Anumba C.J. (1998). Industry Uptake of Construction IT Innovation – Key Elements of a Proactive Strategy, *In: proceedings of the CIB W78 conference on The Life-Cycle of Construction IT Innovations*, Stockholm.

[2] Atkin, B. (1995). Information management of construction projects, "Integrated construction Information", E&FN, Spon, London, UK, 1995.

[3] BCIS (1988), Tender sum/final account study. BCIS News, (25), London, Building Cost Information Service, Royal Institution of Chartred Surveyors.

[4] Betts, M., *Information Technology for Construction, In: proceedings of the CIB W78 conference on IT Support for Construction Process Re-engineering*, Cairns (1997).

[5] Bjork B-C. (1997). INFOMATE: A Framework for Discussing Information Technology Applications in Construction, *In: proceedings of the CIB W78 conference on IT Support for Construction Process Re-engineering*, Cairns.

[6] Booch, G. (1994) Object-Oriented Analysis and Design with Applications, 2/e. ISBN 0-8053-5340-2. Addison-Wesley.

[7] (Bowers.J 1994), Data for Project Risk Analysis, International Journal of Project management, 1994 12 (1) 9-16.

[8] Cooper, R., Hinks, J., Aouad, G., Kagioglou, M., Sheath, D. and Sexton. M. (1998). The Development of a Generic Design and Construction Process, In: Proceedings of the European Conference on Product Data Technology, Building Research Establishment, Garston, Watford, UK, pp205-214.

[9] Flanagan (1994), *Risk Management in Construction*, Chapter 3, Page 46.

[10] Levitt, R. E. & Kartam, N. (1990). A, *Expert Systems in Construction Engineering and Management: State of the Art*, The Knowledge Engineering Review, Vol. 5.

[11] IAI, Industry Foundation Classes, Version 1.5.(1997) http://www.interoperability.com/

[12] ISO/TC184/SC4, *Part 1: Overview and fundamental principles*, International Standard, ISO, Geneva, (11) (1994).

[13] OMG (1995, *The Common Object Request Broker Services, OMG*.

[14] Pallaside. @Risk *Risk Analysis and Modelling*,1990,2000.

[15] (Semster, 1994), Usage and Benefits of Project Risk Analysis and Management, International Journal of Project Management, 12 (1) 5-8.

[16] UML (1997) UML Document Set, Version 1.0, 13 January, 1997 (Rational Software corporation) .http://www.rational.com/uml/index.html.

Development of Component Based Integrated Software System for the Design of Building Foundations

Bedilu Habte* and Udo F. Meißner[†]
Institute for Numerical Methods and Informatics in Civil Engineering
Technical University of Darmstadt
Petersenstrasse 13
64287 Darmstadt
Germany

Abstract

The wide ranges of software tools available for the structural design of building foundations offer, in general, partial and disconnected solutions. One has to solve separate portions of the design using the different software available and needs to manually transfer the data and/or results of one software to the other; thus, requiring a lot of time and effort. Attempts to develop software that could integrate the various design steps are increasing from time to time. In this study a software system is being modeled and developed for the design and analysis of building foundations based on component programming. The Java-Beans^{TM} component model has been used to develop the components required for the system.

1 Introduction

What do available software-systems for foundation analysis and design lack? Accordingly, what type of integration in foundation analysis and design is sought for? Considerations to be made in integrating software systems for the civil engineering field have been raised, and different approaches have been suggested at several occasions by various researchers in the field [1, 3, 4, 7, 8, 9]. This section points out some specific shortcomings of existing systems for foundation analysis and design and suggests an improved integration as applied in the system being developed.

1.1 Types of foundations

Most software system for this purpose support only the analysis and design of a particular or limited types of foundations. The majority of them are usable only for the analysis of a given foundation. Some systems assume that different foundation types need to be analysed using different software, thus provide disconnected systems for the analysis of foundations of a single project that may include different foundation types. The system developed here incorporates alternative foundation types and their design and analysis requirements.

1.2 A given foundation is an integral part of the whole building

A given foundation design problem is not an isolated task, rather, it is commonly an integral part of a building whose foundation elements may interfere with one another. In the software system developed here, the foundation design project is considered as one task for the whole building surrounded within a given boundary line describing the project limits. This could further enable the direct transfer of the effects of the substructure to the superstructure and vice versa.

1.3 Code of practices differ in place and/or time

Design and analysis needs to be performed based on different code of practice. Thus, the requirements of code of practices need to be separately modeled. This is important since code of practices could vary from time to time, and differ in different countries. At the moment two code of practices have been integrated in the system: the Eurocode and the Ethiopian Building Code Standards (EBCS). In addition, the necessary documentation of each code of practice could be referred to on-line, which will save a lot of the engineer's time that would otherwise be spent consulting the printed documents.

---

*MSc., Graduate student
[†]Prof. Dr.-Ing., Director of the Institute

1.4 A design problem does not possess a unique solution

Design problems do not have unique solutions, rather they may have satisfactory alternative solutions out of which one may select the best or optimal solution based on certain criteria. The foundation analysis and design system developed here incorporates the consideration of multiple alternatives and evaluates each alternative according to given criteria. Based on these criteria and using fuzzy decision making, it would be possible to select the optimal alternative.

1.5 Object Serialization and data formats for communication

Foundation analysis and design systems shall preferably be able to store the data they have been working on and retrieve them at any future time. In addition, the ability of a system to read the data (or objects) generated by another system is also an important issue for communication between different systems. For these reasons, it has been aimed to use standardized or widely used formats for input-output in the software system. Thus, it is possible to serialize the current state of a project as Java^{TM}-objects and reconstruct them later on. Furthermore, the system supports reading and saving all project data in XML-format (an emerging standard format) and saving project drawings in a DXF-format (a widely applicable format for drawing exchange).

2 Component Based Programming

A component is a reusable software building block that may be combined with other components. Components can be independently developed from one another. Then, they could be connected or integrated into a larger system (application). The main advantages of component oriented programming, among others, are:

- components could be development independent from one another,

- they could be combined and/or replaced as required,

- a component is ready for integration, i.e., no source codes or recompiling is necessary,

- maintenance and upgrade of components is easy and this does not affect the application user,

- a component could be executed from any location on a network.

2.1 Components for Foundation Analysis and Design

To completely express the foundation design problem, it will be necessary to supply information about the super-structural elements and also the underlying soil. Objects used to represent all these information are included in the **footing** component of the system. Figure 1 shows the partial class hierarchy of the footing component [5, 6]. All possible types of foundations are modeled as parts of the footing component.

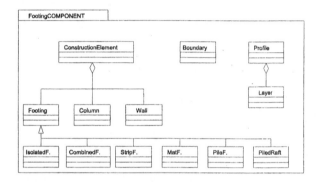

Figure 1: Partial Class diagram of the **footing** component

Another design information is organized in the **code** component. A user may have a project in a country where a different code of practice is used. Thus, the software system he uses for design shall be ready to exchange the code component. Separate implementations of the requirements of codes of practices have been included in the current system. Objects in this component provide the necessary code requirements needed for the design of reinforced concrete foundations. It also includes the respective code-of-practices publications for online reference.

Objects needed to make such calculations as the determination of the bearing capacity, settlements, stresses in soil, the necessary foundation dimensions [2, 12], necessary matrix computations, etc., have been organized into the **calculator** component. This component also includes objects that are needed for the various steps of the analysis and design of different foundation types according to a selected code of practices.

A component termed **guis** is provided to enable a graphical interaction with the system. Menus have been provided to be used for the various steps of the design process. The user can instantiate a project, design, analyse , and make any decisions about the it using the main window of this graphical user interface.

The **controller** component coordinates the whole application. Foundation design project is instantiated, alternative design solutions are investigated and saved for future reference, and the optimal design is selected with the help of objects of this component in cooperation with all the other components.

3   Realizing Integrated Systems

System integration can be realized connecting the components by hand coding. It is also possible to use a RAD- (Rapid Application Development) environment to visually integrate the components. In this study, the foundation design process has been divided into smaller and manageable parts and then appropriate software components are being developed for each of these parts. The public classes and methods of a component which may be used for integration are made visible in a RAD - Environment or they could be consulted in the API - (Application Programming Interface) documentation that usually accompanies a component.

3.1   Distributed Java-Beans^{TM} Components

Java-Beans components could be located at different places and dynamically instantiated from anywhere on the internet, and on any operating system that supports the Java^{TM}. This is one side of distributed computing, a very useful application in the prevailing network based computing environment [13]. Figure 2 shows how the internet could be considered as a large pool of software components and facilitate the development, integration and the dynamic instantiation of a component based integrated software system.

3.2   Online System Application

The system could be activated in a Java Virtual Machine (JVM)^{TM} by making a call to the main class of the **guis** component. This loads the other components needed for foundation design from their respective locations and makes them ready for use. The user can then open a foundation design project (or a design problem) using a data stored in a text file.

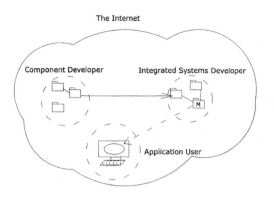

Figure 2: Distributed Components

Further editing or viewing the data graphically is possible by selecting the corresponding menu. After specify the code-of-practice to be used, the user can select the project to be designed or analysed. The user has the option to printout the design data directly from the guis or export these data into a particular data format for further manipulation with an external program. The application of the system at real time is partially shown in the Appendix.

4   Alternative Design Solutions

A structural design problem does not possess a unique solution, but a satisfactory one to some extent. To optimize the design or to select the "best" foundation system, some criteria must be set. Because it is not possible to compare design alternatives manually, fuzzification of some of the design criteria would be necessary. Then, out of given alternative solutions, the optimum foundation system is selected by considering factors such as:

- the least volume of material used or least construction cost

- the least maximum and differential settlements produced,

- the least stress induced at a critical depth in the soil,

- the least stress in the concrete, and/or steel,

- the least deformation and/or crack-width of the foundation element,

- the highest safety factor against sliding and over-turning, etc.

Fuzzy-logic principles are applied with optional weighting factors to select the best solution out of given alternatives. Accordingly, the optimum solution will be the one which maximizes the membership function for all the criteria considered [10, 11, 14]. As an example a project whose four alternative solutions is shown in Table 1. The corresponding membership functions for four equally important criteria considered (c1 to c4) is also given. Using the *min-max* method Alternative II is found to be the optimal solution.

Alternative	c1	c2	c3	c4	min
I	1.00	0.75	*0.60*	0.90	0.60
II	0.80	*0.65*	0.77	0.75	0.65
III	*0.55*	0.90	1.00	0.80	0.55
IV	0.80	*0.50*	0.67	0.90	0.50

Table 1: Design alternatives and membership values for four criteria, an example

5 Summary

Software components are independently developed, tested and then joined together with other related components to attain an integrated application. Component developers prepare general purpose, reusable and exchangeable components; integrated system developers on the other hand use components of their own or those developed by others. This enables in general a simpler and faster software development process than trying to model the whole system as one unit.

Unlike existing software for the analysis and design of building foundation, the system developed here assists the engineer in selecting a foundation considering the whole building as one unit. Such an integrated system would reduce the time that may otherwise be spent using disconnected software, and makes it easier to try out and compare alternative solutions.

Acknowledgement

This research work is being financed by the German Academic Exchange Service (DAAD).

Appendix

Real-time application of the foundation analysis and design system

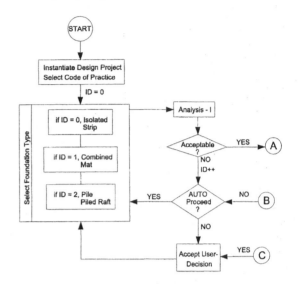

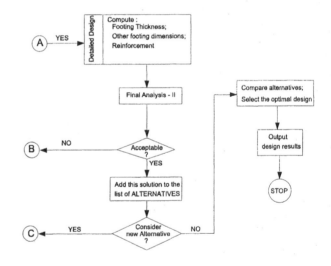

References

[1] Godfried Augenbroe, *COMBINE2 - Computer Models for the Building Industry in Europe - Final Report*, Available at URL: http://erg.ucd.ie/combine.html, November 1995.

[2] Bowels, J. E., *Foundation Analysis and Design, McGraw Hill*, New York, 1996.

[3] Fenves, S. J., *Successes and further challenges in computer-aided structural engineering*, Proceedings of the Sixth International Conference On Computing in Civil and Building Engineering, Berlin, 12- 15 July 1995.

[4] Fenves, S. J., *Computing in Civil and Building Engineering: Changes and Opportunities*, Proceedings of the Seventh International Conference On Computing in Civil and Building Engineering, Seoul, 19- 21 August 1997.

[5] Habte, B., *Entwicklung eines Komponentenmodels zur Bemessung von Stahlbeton-Fundamenten im Hochbau*, VDI Reihe 4 Nr. 147, VDI Verlag, 1998.

[6] Habte, B., *An Integrated Software System For The Design Of Building Foundations Using Software Components*, VDI Reihe 4 Nr. 156, VDI Verlag, 1999.

[7] R. Katzenbach and G. Festag, *The Geotechnical Information System as a New Object-Oriented Tool within Geotechnical Construction and Design*, Proceedings of the Seventh International Conference on Computing in Civil and Building Engineering, Seoul, 19-21 August 1997.

[8] U. Kolender and D. Hartmann, *Object oriented structural analysis, design and construction of industrial structural systems*, Proceedings of the Sixth International Conference on Computing in Civil and Building Engineering, Berlin, 12-15 July 1995.

[9] U. Meißner, F. Peters, and U. Rüppel, *Graphically interactive, object-oriented product modeling of structures*, Proceedings of the Sixth International Conference on Computing in Civil and Building Engineering, Berlin, 12-15 July 1995.

[10] S. S. Rao, K. Sundararaju, B. G. Prakash, and C. Balakrishna, *Fuzzy Goal Programming Approach for Structural Optimization*, AIAA Journal, Vol. 30, No. 5, May 1992.

[11] C. J. Shih and C. J. Chang, *Pareto Optimization of Alternative Global Criterion Method for Fuzzy Structural Design*, Computers & Structures, Vol. 54, No. 3, August 1993.

[12] Ulrich Smoltczyk, *Grundbautaschenbuch*, Ernst & Sohn Verlag, Berlin, 1997.

[13] Sun Microsystems, *Java and Java-Beans online documentation*, Available at URL: "http://java.sun.com/".

[14] Lotfi A. Zadeh, King-Sun Fu, Kokichi Tanaka and Masamichi Shimura, *Fuzzy Sets and Their Application to Cognitive and Decision Processes*, Academic Press, Inc., New York 1975.

An Information System in a cooperative building production context

Stéphane Lasserre

UMR CNRS n°694 MAP-GAMSAU, Ecole d'Architecture de Marseille. 13288, Marseille - Cedex 9

Farid Ameziane

UMR CNRS n°694 MAP-GAMSAU, Ecole d'Architecture de Marseille. 13288, Marseille - Cedex 9

Abstract

Our work is concerned with the communication of information in building construction activity. Our proposal is to allow partners of a construction project to share all the technical data produced and manipulated during the building process through the use of a system linking distributed databases enabling remote manipulation through the Internet. To control the evolution of the information during this production process is the main problem to resolve in the design of our system.

1 Introduction

The building construction activity encounters problems in information management and communication all along the following stages which carracterise its life cycle : design, engineering, construction and maintenance.

The architectural project tends to be complex because it must integrate a growing number of constraints and regulation needs :

- user conveniences (accessibility, acoustic and thermic, etc.),
- constructive (parasismic properties, complex cost management, etc.),
- realization (products prescription fields, technical laws, quality controls, etc.),
- reduced time management,
- management of dynamic information generated by multiple point of views of a growing number of partners with distinct competences.

With the help of new computer technology and tools, this domain area wants to increase the productivity of the building production by facilitating the circulation of manipulated information both among the partners and through the conception, realization, and maintenance stages of this singular industrial product. Acording to a specific request, the goal is to share and extract datas from the building description.

To reach these objectives, it is necessary to build an open structure of building description that would be able to assimilate specific changing professional information along the building life cycle.

In addition, it means that a normative aproach must exist in order to make the access and the re-use of technical solutions easier. We'll talk about re-engineering or knowledge capitalization first initiated by industrial areas.

2 Data Exchange in the building context

The building production is a non-linear complex activity. It is sequenced by distinct stages in which many actors enrich the project description with their own vocabulary and competences.

This team is linked by a common goal of realization achievement. Each partner of the team helps the others by sharing graphical and textual datas he produces. These communication medias are the most efficient way to manage a good information circulation.

Many research programs focus on building information management through its life cycle. The goal is to answer most of the actors in this process who want to go towards a concurrent engineering and more generally towards a cooperative work during the production processes.

The problem of data exchange between heterogenous CAD systems produced little or no successfull file exchange formats : DXF, IGES, SET, VDA, CALS, STEP and more recently Industry Fondation Classes (IFC) from the last research program, IAI. International Alliance for Interoperability. It must be the most spectacular one, because it links whole of the CAD system firms, many users and institutionnal partners.

From the following French projects (SUC, MOB, GSD, SIGMA, etc.) and international projects (COMBINE, ATLAS, RATAS, IAI, etc.) that make an analogy between architectural field and industrial field remain two tendencies in the way to describe a building :

- research programs focus on the building construction works, and
- research programs focus on the process description that lead the construction of the building.

3 The « Communication and CAD Tools » research project

Our work is inherited from the results of previous projects we've deal with that are part of « product/process model » research programs.

The aim is to contribute to the realization of a cooperative management informative sytem for building activity [1]. It would be initiated during the engineering stage, enriched through the execution stage, and maintained by the owner when the building is in use. Each actor would be able to get the building representation he needs. That's what the figure 1 proposes to suggest.

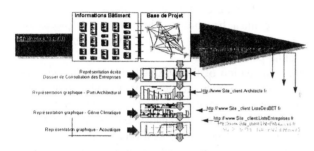

Figure 1 Schematic representation of the « Communication and CAD Tools » project

This project produced yet a conceptual schema of data according to the knowledge fields of architectural design. This schema allows us to build a coherent group of entities in a Data Base Management System in order to give particular representations of these complex assemblies, and to manipulate them by remote control. The following hypothesis guided our work :

- the knowledge universe of the building construction area is distributed, but we can access it through a network,
- a building may be described by a collection of construction works and the spaces contained. This being the construction economist partner's point of view,
- this description may grows step by step, and every partner can access it,
- we consider that a 3D CAD model was elaborated just before the engineering stage. It's a support to detail the architectural project.

To answer to these hypothesis, we produced a conceptual data diagram that describe the building with the construction works it is composed of.

Under a generic class of « entity-of_builing » we organise all the objects concerned with the construction of a building according to a hierarchic set of sub-classes.

Each of them are composed of attributes (product specifications) and methods (knowledge relative to a specific product – for exemple, a HVAC calculation depends on the volume and the capacity of a room).

In that perspective, we represent the set of objects manipulated in the description, in a computerized environment by coupling any ODBC data source (witch specifications map our building data schema) with solid and parametric CAD system (Mechanical Desktop by Autodesk Inc.) and the Internet. This environment allows us to maintain the building description through its data diagram that needs to be frequently updated because no real normative project are elaborated yet.

According to that, we attach a database to each architectural project, in order to complete its morphologic description with textual data that usualy caracterizes a construction work.

4 Help system for construction works choices

Many times, data manipulated by partners who colaborate to the raise of the building are introduced into professional oriented systems and cannot be re-used by the others. Data are transformed through these systems and the semantic added is misunderstandable.

Furthermore, at the end of the construction, the owner of the building may legaly request whole of the informations enriched during each stage. So, through the exploitation stage of the building (the longest one), information research is harder when we have to maintain or to find responsabilities for specific problems.

Our project try to find an answer to this problem by sharing a unique data model with the partners involved in the building construction processes.

At least in France, the following figure 2 describes what guide a decision relative to a construction work choice during engineering stage.

We also illustrate the role of each actor in this process, and the knowledge fields shared between each other.

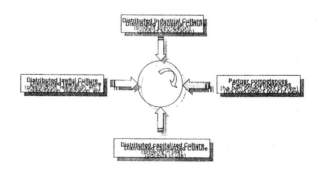

Figure 2 Culture circle for construction works decisions

According to its own building point of view and knowledge, each actor may want to make a proposition for a construction works. We found that they all have common legal constraints and they need to respect product functionnal features.

Going on the building life cycle, legal aspects may be considered as the leading decision parameter :

- As we said, during engineering stages in order to satisfy product specifications and use,
- during the realization, it is necessary to verify the conformity of works in order to improve quality process,
- when the building is in use, in order to improve its maintainability and find responsabilities for bad implementations by tracking design history of the building.

Consequently, by taking into account information continuity problems at every stage, we have refined our building description schema with detailed lawful specifications of construction works. We now consider that legal aspects are part of the most important points that lead building elaboration processes and decision processes.

The following picture mainly shows the idea of « radio buttons » that must be checked in order to verify the choices ability. At this stage, we need to stress on we don't really want to avoid final confrontation between partners, but only solve unnecessary exchanges.

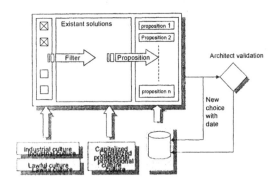

Figure 3 Three point of view for helping decision

Furthermore, we also prepare these confrontations by first filtering particular expectations :

- Functionnal expecting or features (mass, acoustic specification, etc.),
- Costs constraints,
- Design expectations (geometry, color, etc.),
- Constructive aspects (legality in a particular configuration)

Each point of view allows to reduce the amount of data picked up among networks.

5   professional approaches

We've dealed about the necessity of taking into account partner choices and decisions by integrating their knowledge during elaboration processes. We now need to allow them to make specific query into building database and to get personal representation of this building.

5.1   Professional representation of data

In France, by making specific folders (Cahier des Clauses Techniques Particulières, Cahier des Clauses Administratives Particulières, Dossier des Ouvrages Exécutés, etc.) we give a simple representation of building for each partner. The content of these documents depends on the partner's needs.

So, we must be able to filter the entire pieces of informations in order to deliver a specific answer to a particular partner request.

By anticipation of the products descriptions we can access today through the network, the objective is to demonstrate the specifications of a construction work may be extracted from an exhaustive set of informations which the use depends on professional needs.

This goal is very important, because answering to the multiple-expertise problem in a building context means we must be able to give a « real time » representation of the building for an actor. This representation is strictly made up of the informations he needs to achieve his expert's report.

5.2   Professional profiles

The lack of normative data classifications and the speed of their advancement during production process confers to informations dynamic aspects. Consequently, successive representations of these data are rapidly out of date.

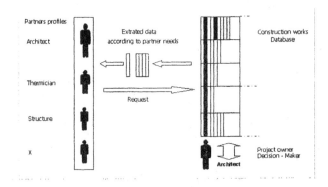

Figure 4 The multi-actors aspect : specific informations for each actor [1]

Often specific, it seems to be important to let the database owner to maintain and set database during the stages, while he must be able to give updated views.

Figure 4 shows the construction works description mecanisms based on existant databases and the different professional profiles associated to.

Furthermore, these pieces of information are not really confidential for other partners. Our professional approach is only focused on the possibility to make them more understandable.

In that way, we're sure to improve the quality and the efficiency of actor's expertise with the help of refine searchs.

We have worked on electronic representations of the folders elaborated during the design and engineering stages (figure 5). It comes with a personnal information associated to a professional profile. Each actor is able to customize his default profile. He can select data he wants to screen on next request, as well as graphic representations. Each space comes as a dynamic 3D VRML file (server side generated) that depends on the selected graphical entities. This network 3D representation uses the Virtual Reality Modeling Language (VRML).

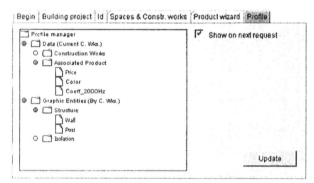

Figure 5 Customize your profile

6 Query and manipulation principles

To answer to those concepts, we suggest an open request system for a set of experts whose searchs sequences are based on different point of views. We are focused on actors experiences during the engineering stage because thise project step provides most graphical and textual medias, exchanges and communications between partners.

The figure 6 shows us that the structure of our query system may be seen as an incremental information research. It is sequenced by the following steps :
1. query database indentification,
2. partner login,
3. space choice,
4. selection of a construction work associated to the space,
5. information filtering mechanisms.

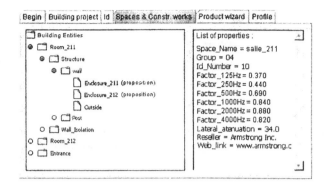

Figure 6 Incremental query scenario

7 Technology and tools

To answer a cooperative context, we've based this sub-system on a web environment that allow us to manage good conccurent access. We choosed the Oriented Object Language JAVA that offers good opportunities in :
Network implementation mecanisms for client/server applications,
Good security management,
Cross platform applications,
Database access through Java Data Base Connectivity and a three-party architecture that is independant of specific DBMS.
To maintain and enrich easily our information system, we found a good response in the powerfull Lotus Notes client from Lotus Inc.
About graphic files we manipulate, we'll remain mainly :
DWF (Drawing Web Format) from AutoDesk Inc. in order to publish vector drawings
VRML (Virtual Reality Modeling Language) as a 3D graphic interface support

8 Graphical interface of the experimental system

A query sequence is step by 2D or 3D interactive graphics. In the building communication process, graphic drawings are the most used media because it's the only media building partners have in common. We want to increase interactivity by using them as interfaces to point out the right spaces or construction works.

The following figure 7 shows the system state at the end of a query. We can see a list of construction works the selected space is composed of, attributes of a particular one (ceiling), and a 3D space representation for an architect profile.

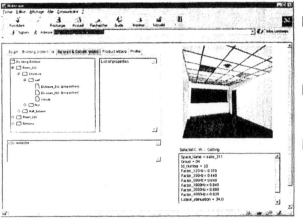

Figure 7 System state at the end of a query

9   Conclusions and perspectives

Our work is mainly directed by the improvment of :

- Data circulation between distant partners,
- Share of a common description diagram in a DBMS environment,
- Answers to manipulation of heterogenous documents (texts, drawings, multimedia files, etc.),
- Professional approaches on a specific industrial product,
- Help system for construction works decision.

Those results are important in our research center context because this project is since January 2000 part of a CAD software leader research : Nemetscheck A.G. (Allplan, Allfa, palladio X, etc.). This firm want to improve the links between CAD systems, Data share through the networks and DBMS for non graphic informations.

We want to extend this work that belongs to information systems in a building production context to the realization stage by taking into account processes models. In addition, we work on a data repository for electronic building documentation. This kind of data warehouse could fill a gap between spreaded informations among the network and their accessibility.

References

[1]   Ameziane F. 1998, "Structuration et représentation d'informations dans un contexte coopératif de production du bâtiment", Thèse de l'Université d'Aix-Marseille III, Faculté des Sciences et Techniques de Saint-Jérôme, N° 98AIX30012.

[2]   Armand J. et Raffestin, Y. 1993, «Guide de la construction», J. Armand et Y. Raffestin (Editions Le Moniteur).

[3]   Celnik O., Coste E. et Vincent P., 1997, "Internet, BTP et architecture", Editions Eyrolles.

[4]   Chan K. and Gu P., 1993, "A STEP-based generic product model for concurrent engineering", in P. Gu and A. Kusiak : Concurrent Engineering : methodology and applications.

[5]   Chen C. S. and Wu J., 1993, "Product modeling and data exchange", in P. Gu and A. Kusiak : Concurrent Engineering : methodology and applications.

[6]   Darses F., 1997, "L'ingénierie concourante : un modèle en meilleure adéquation avec les processus cognitifs de conception", in P. Bossard, C. Chanchevrier et P. Leclair (eds. Economica) : Ingénierie concourante, de la technique au social.

[7]   Harold E. R., 1997, «Programmation réseau avec JAVA», E. R. Harold, Editions O'Reilly.

Integration of Internet-oriented technologies for co-operative applications in construction dynamic virtual environments

Y. Rezgui

Information Systems Institute, University of Salford, Salford, M5 4WT, UK

A. Zarli

CSTB, 290, route des lucioles, B.P. 209, 06904 Sophia Antipolis cedex, France

Abstract

The paper introduces a comprehensive approach proposed within the frame of the OSMOS project, to support effective information sharing and smooth co-operation between non co-located teams, and the co-ordination of their concurrent activities in an environment that promotes trust and social cohesion in the construction industry. First, the paper gives a general overview of the construction industry process and organisational settings along with their associated limitations, and introduces the aims and objectives of the OSMOS project. Following this introduction, current as well as past research in the area tackled by the OSMOS project are presented. An emphasis is put on results and findings that can be potentially exploited. The OSMOS solution is then described via a generic system architecture addressing the information sharing and process control requirements of the project. The next chapter focuses on the role that XML can play to support information sharing and exchange within a Virtual Enterprise (VE). Finally, the paper presents the anticipated results of the project along with the technical, process, and business ingredients for a successful implementation and take-up of the resulting internet-based service prototypes. You did remember to send in your abstract, didn't you? This is the abstract of my paper. It must fit within the size allowed, which is about 3 inches, including section title, which is 11 point bold font.

1 Introduction

Advances in personal computer technology along with the rapid evolution of networking and communications have had a substantial impact on industry business processes. The emergence of Client / Server applications, at the end of the 80s, have offered a first promising answer to the problems of flexibility, scalability, and extensibility of modern businesses. Software applications were being downsized from expensive mainframes to networked personal computers and workstations that are often more user-friendly and cost effective. The introduction of the Internet along with advances in 3-Tier architectures and middleware technologies have brought new challenges and competitive advantages that the industry is now trying to comprehend and exploit.

The Building and Construction domain has always been referred to as a traditional industry, despite the fact that it has adopted for decades the modus operandi of the so-called Virtual Enterprise (VE). In fact, buildings have long been designed and constructed by non co-located teams of separate firms who come together for a specific project and may never work together again. Organisations and individuals participating in a team bring their own unique skills, knowledge and resources, which include proprietary and commercial software applications (Fig.1). Recent surveys [14, 15] reveal that current technology solutions in use in the building and construction domain present one or more of the following characteristics:

- Extensibility: despite recent evolutions, mainly due to the impact of the Internet, existing solutions are still often fixed and not open, with a lack of support for legacy, as well as new, upcoming systems in terms of hardware, software, databases, and networks.

- High Entry Level: IT solutions are still often expensive to buy for SMEs, as reported by the OSMOS end-users. More entry levels should be provided, e.g. from personal (low cost) to enterprise (high cost) editions.

- Lack of Scalability: most available proprietary and commercial solutions offer limited growth path in terms of hardware and software.

- Application Centric and lack of support for business processes: there is often a requirement to organise the enterprise around the adopted IT solution.

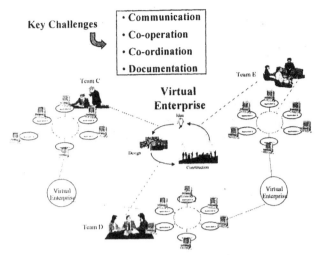

Figure 1 Construction Industry Context

Due to this context, various problems have been identified. These include the following:

- IT support for the fragmentation imposed by the very nature of the industry in terms of communication and information exchange still needs improving.
- Interactions between actors are still not well co-ordinated, especially because of the inherent dynamic business relationships taking place in the construction industry. Commercial workflow systems didn't have the expected impact on projects, and within companies, mainly because of the characteristics mentioned above.
- IT support for Information and document management varies from one company to another, but overall is still done in a traditional and ad hoc way.
- Project documents present a great deal of redundancy and often lack structuring.

This overall context has often resulted in information inconsistencies, business process inefficiencies, and change control and regulatory compliance problems. The overall aim of the OSMOS project is to enhance the capabilities of construction enterprises, including SMEs, to act and collaborate effectively on projects by setting up and promoting value-added Internet-based flexible services that support team work in the dynamic networks of the European construction industry. This translates into the Scientific and Technological measurable objectives described below:

- Specify Internet-based services for collaboration between dissimilar construction applications and semantic cross-referencing between the information they manipulate.

- Specify Internet-based services allowing the co-ordination of interactions between individuals and teams in a dynamic construction virtual enterprise.
- Specify a model-based environment where the release of, and access to, any shared information (including documents) produced by actors participating in projects is secure, tracked, and managed transparently (in real time whenever possible, otherwise asynchronously).
- Provide low entry level tools (cheap and user-friendly) to small enterprises to act and participate in construction virtual enterprises.
- Allow end-users to use their proprietary and commercial applications on projects, by implementing the specified services (e.g. via plug-ins), and allow them to transparently participate to collaborative work in dynamic virtual enterprises.
- Implement the model-based environment providing a distributed information management support for the virtual enterprise.
- Set up two OSMOS Internet-based team work service providers prototypes for the purpose of the project, and ensure their take-up, as commercial offers, after the completion of the project.
- Define the migration path to using the OSMOS approach.
- Analyse the likely benefits of adopting the OSMOS approach.

The paper gives a comprehensive overview of the OSMOS proposed solution. A generic system architecture is presented along with its technology constituents in terms of services, tools, and middleware supporting the construction virtual enterprise. This architecture is in the process of being deployed and interpreted within two of the three end-users involved in the project, to set up internet-based team work services.

2 State of the art research in the area of computer integrated construction

A large number of studies have been conducted by various researchers in order to highlight the benefits of information and communication technologies in the construction industry [1, 2]. These technologies include standards for data exchange and product modeling [4, 11] and integration techniques through the use of object-oriented technologies [6].

State of the art research in the application of IT in construction reveals that integration has been achieved, mostly, on static models that define the structure of shared information in the form of files or databases. The OSMOS research team advocates that integration should be made through frameworks which define semantic relationships between the interfaces of separate distributed components. This is an area where Construction needs advances. Such frameworks are already under development, especially within the Object Management Group through business object facilities, based on the Common Object Request Broker Architecture [6]. On the other hand, the World Wide Web has now emerged as a result of the growth of the Internet. It was, until the advent of HTML, mainly used in academia. However, HTML (which was derived from SGML), is mainly used to describe and exchange information contents as opposed to semantics. The recent XML, and the related DOM standards combined with semantic object models describing a building (e.g. STEP / IAI-IFCs) offer a unique opportunity to promote effective information sharing in the VE.

Back to the concern of "semantic frameworks", it is worth mentioning ongoing efforts on RDF, that should be a quite interesting fount of inspiration within OSMOS. RDF [8] deals with metadata in the sense of "information about information": e.g. acting with a document that is an information, one can think about getting its author or editor, that provides with information on the document. Most systems define their own combination of metadata and their own facilities for storing and managing it, without any facilities for sharing or interchanging metadata. As currently under development within the W3C, RDF provides with a simple common model for describing these medatata in terms of relationships between an object (any resource described by a URI - Uniform Resource Identifier, e.g. a URL on the Web, a specific section within a document, etc.), a PropertyType that identifies a resource (i.e. some piece of information) that has a name and can be used as a property, and the link (called a property) between the object and the PropertyType through a given value identifying an instance of the PropertyType.

However, the success of collaborative work relies not only on the ability to provide solutions to the problems of multi-criteria information representation, information sharing and exchange, information life-cycle support; but, also, the support of the various interactions taking place between individuals / groups / or corporations as well as the management of their authorities and rights over information in accordance with their precise roles in the VE. These important issues are tackled within the field of computer support for co-operative work (CSCW). CSCW is more generally concerned with the introduction and use of groupware systems to enable and support team work.

Groupware solutions include traditionally a subset of the following system components: Workflow (task scheduling), Multimedia Document Management, E-mail, Conferencing, and shared schedule of appointments. A recent survey of groupware constituent technologies [9] reveal a lack of homogeneity, and a diversity of applicable de facto standards and APIs from the leading Groupware vendors. In fact, the last decade has seen a tremendous activity in new specifications and developments of standards and architectures for CSCW and enterprise application integration. While these developments seem to offer a challenging opportunity for the VE, they do hide complex architectural problems in relation to the selection of the right tools, toolboxes and infrastructures.

Therefore, a suitable team work IT-based solution requires a broad methodological approach, a deep understanding of the information / process requirements, and also the understanding of the dynamics and the specificity of the context in which the support for team work is required, namely the Construction industry.

3 The OSMOS System Architecture

Recent and continuous investigations on the use of advanced computer-based technologies, especially in ESPRIT funded European projects, including VEGA [13], CONDOR [10] and GENIAL [7], have shown promising results. The today difficulties lie in identifying the right reference marks and methodology to relate user requirements and needs to the adequate technology. This chapter proposes a first investigation towards an approach to solving these issues, by establishing a common foundation to better take advantage of the benefits of distribution and Internet-oriented technologies.

Figure 2 introduces a generic functional architecture of the OSMOS project supporting co-operation between a variety of commonly used applications in a Construction VE. In particular, the OSMOS Integration services will provide means by which proprietary and commercial building and construction applications can inter-work by enabling the sharing and cross-referencing of information. This will be based on existing standards, including CORBA services, the (XML) DOM object API (W3C), and existing leading groupware vendors APIs. However, it is worth pointing out that the above APIs and services will not all be implemented. The OSMOS project will, rather, identify, extend whenever appropriate, and specify a subset of the above that are suited and adapted to the specific business and information requirements of the VE in the Building and Construction domain. In addition, an OSMOS VE manager will be developed. The latter will provide advanced functionality for team work service providers to manage the VE.

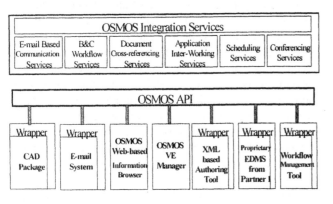

Figure 2 The OSMOS Functional Architecture

In fact, an effective implementation and deployment of the above functional architecture requires the following:

- Systems based on standards for data modelling and exchange. In addition to the media used for communication between applications, those applications need to have a common understanding of information (e.g. its semantics). Such a common understanding will be more and more developed for each vertical industrial domain: standardised vertical models will ease the interactions between systems in a business domain.

- Inter-operation between heterogeneous applications and systems (i.e. sharing and exchange of information). Typically, existing as well as future applications (including in the area of e-commerce) will have to get information from various corporate information systems, built on top of modules and components that have been most of the time developed with different technologies.

- Candidate technologies should adhere to the plug-and-play paradigm. These should offer easy integration through the provision of adequate gateways, in order to allow the selection of "best-of-breed" modules supplying the best solution to end users needs.

The aspects of data modelling have been tackled through various efforts including ISO STEP/10303 [11] and the IFCs developed by the IAI [4] which has a formal liaison with STEP. As regards applications inter-operation, there has been quite a long evolution from centralised architectures, client/server model, to the now well accepted 3-Tier based architecture (making a clear separation between client desktop, middle-tier application

server and DBMS/persistent storage), that has to be set up along with some middleware technology (e.g. CORBA, DCOM and MOM which supply asynchronous mechanisms for the routing and formatting "on-the-fly" of messages). From a business point of view, such an approach can be synthesised as in the following figure 3:

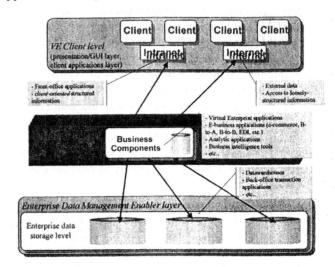

Figure 3 A Business Oriented View of the OSMOS Infrastructure

Translated into a more process oriented view, such a figure can transmute into Figure 4, that not only still introduces the three layers to be managed, but also two other intermediate levels so as to "disconnect" as much as possible the three fundamental layers. In order to target any type of client, especially over the Internet, a loose coupling, that deals with the enterprise business logic independently of any presentation and client desktop issues, should be provided. Note that the "Business Components" layer only manages what refers to the business of the enterprise (e.g. a banking transaction, a digital simulation of the behaviour of construction components based on a mathematical model, a notification module in a document management system, etc.): the specific ways in which information is requested, serviced and visualised are disconnected from the business process. A similar model of loose coupling can be envisaged between the intermediate layer and corporate information warehouses, once again in order not to mix the application logic with the specific rules governing the access to a given database.

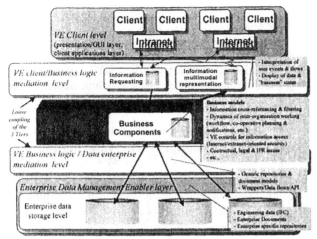

Figure 4 A Process Oriented View of the OSMOS Infrastructure

Besides the specification of various data semantics at nearly all levels, figure 4 induces the necessity to act with different technologies, targeting issues like communication architecture, integration with the Internet, transactional mechanisms, and the potential use of components:

- Software communication: The latest developments in this area provide a lot of opportunities, e.g. the CORBA specification (the new 3.0 specification has been released) and its family of various commercial or free implementations, Microsoft's COM+/Active X model, MOMs like IBM's MQ-Series or Microsoft's MSMQ, and so on.

- Integration with the Internet: A plethora of tools and technologies has emerged these last years, including data-oriented languages such as HTML, DHTML, VRML and now the XML family; scripting languages such as Perl, Javascript; server-side programming like CGI or ASP; and of course the various Java-based technologies (see below).

- Transactional mechanisms: These ensure that a set of operations embedded in a transaction are safely and coherently executed. Currently solutions are provided with tools implementing the CORBA OTS specification, the Microsoft's MTS, or the Java Transaction Service - JTS.

- Components: Current efforts include Microsoft's Active X and Sun's Java Beans. It is also worth mentioning, mainly on the server side, the EJB and its OMG generalisation, and the CORBA Components Model [3].

Eventually, some technologies clearly appear to deal with various issues indeed and can be found at various levels of figure 4: these are XML (described in section 4) and Java technologies. Java is today a well known technology. Initially a language developed so as to be light enough for small programs on embedded systems, it has grown to a set of programming techniques and tools that have invaded the Web. The main Java technologies can be found in Java2 Enterprise Edition - J2EE [5], with a Java Web server incorporating XML parsing, Java servlets, JSP, EJB, and JTS (described below):

- Java Servlets are programs running on the server side, acting in similar way than CGI scripts and using the HTTP protocol, but getting benefits of Java technology (including multi-threading).

- JSP, the Java Server Pages, is built on top of servlets. It can be viewed as a standard on the server side for dynamic generation of HTML code, with Web pages embedding Java code compiled in reusable servlets. Especially, JSP allows developers to separate functions for creating content from those for presentation.

- EJB is the components model developed by SUN, and that can be viewed as a generalisation of the Java Beans, but on the server side, which means with all the underlying infrastructure to deal with transactions access to data bases, etc.. EJB aims at supporting business components on the server.

- JTS is the Java Transaction Service, that deals with transactions in Java-based environments.

Likewise, it is worth mentioning that Java2 has its own CORBA implementation, and includes JMS, the Java Messaging Service that provides asynchronous messaging (i.e. a Java MOM). Among all these efforts, some are standards (e.g. developed by ISO, the W3C, or industrial consortia like the OMG), others are open technologies (this is the case of most Java developments), and others are still less open like Microsoft-based technologies. In this context, selecting the right set of technologies is difficult, as this has to take into account a variety of factors, including:

- The business application use cases, as well as the context for the future deployment of the selected technologies. For instance, in the case of data publication on the Internet, it becomes necessary to deal with clients desktops having different data management capacities (e.g. a simple Web Browser, or a browser powered with XML tools, or a CORBA-compliant client, or a mobile phone, etc.). Another issue is to identify whether all data can be managed by the client application, or if it is processed on the server.

- The nature and the capabilities of the legacy systems that exist in the different companies forming the VE and that have to be connected. These systems are more or less easily integrated, depending on whether they provide just some rough gateways, or that they publish a full API (e.g. through IDL interfaces).

- Time scale for deploying Extranet or VE network with the required services. Typically, in the construction industry, nearly each project is a one-of-a-kind project, and each time a new network has to be built with new partners. In such a context, it is hardly acceptable that the only elaboration of the VE infrastructure requires several months before its achievement: ways of realising this infrastructure in few days or weeks must be found.

It is one of the major tasks of the OSMOS project to review recent or emerging information and communication technologies as well as team work services, related to communication (including middleware), co-operation (including standardised shared information repositories), co-ordination (including task synchronisation, and access control), and documentation (including document routing, and version control). Available commercial offerings from the main vendors will be analysed, along with their underlying technology. An assessment of the potential risks regarding the construction industry acceptance and take-up of the identified Information and Communication Technologies will be conducted as well, in conjunction with the OSMOS end users.

4   XML: a promising solution for the VE

The origin of XML has to be found within SGML [12], a language for describing and inserting in a neutral way tags within any type of documents. XML has been developed by the W3C originally for the exchange of structured documents within Intranets or over the Internet, in a simpler way than when using SGML. Thus, XML plays a role similar to SGML, but in open and standardised networks. XML is text-based, self-describing, and, above all, gaining acceptance as a global standard. The XML language is indeed a meta-language, allowing the creation of any (XML-based) new language, and especially what can be called "data presentation language" : this leads as well to the definition of new file formats that can be instantly parsed by any XML-compliant application. Thus, for a particular domain, it is quite possible to create a new presentation (or exchange) semantics, which is however different from the semantics

of the data themselves (i.e. the content of the XML message).

Considering the various aspects of enterprise application integration (EAI) and the 3-tier architectures as previously exhibited, the XML model can be considered from 2 different view-points, depending of the fact one targets the client side or the server side.

4.1   XML for cataloguing and further distributing/recovering information

The main key-points involved are concerned with using XML-compliant browsers for searching XML documents, designing DTDs for various Web documents, and controlling those documents along with DTDs and style sheets for representation (making use of XSL: eXtensible Style-sheet Language):

- XML can be used for managing data as messages content: thanks to the various tags surrounding each data, new advanced and more powerful search engines can be envisaged through cross-search inside a structured and vast content, while today HTML only allows a "full-text" search.

- XSL allows to decorate the well-formed trees based on the underlying model of XML, in order to generate, for instance, one or several HTML pages from an XML message. It is worth noticing that this translation/formatting mechanism is not limited to HTML, and it is likely to interpret the structure of the XML message so as to produce any kind of protocol that could be further analysed by a VRML plug-in, or for CD-ROMs or a WebTV, etc.. XSL can also be used as a device to convert tags from a DTD into tags of another DTD, and several XSL-based transformation engines already exist that carry these types of conversion on the fly when fuelled with conversion specifications.

Some of the benefits that can be expected include:

- A clear separation between content and presentation, with a mapping to various visual representations (including HTML views), and it is quite conceivable to mix data originating from distinct sources in a single Web page on the client side: XML is a technology for flexible, dynamic document content information.

- The XML DOM (Document Object Model) standard object API (specified by the W3C) aims at giving developers programmatic control of XML document content, structure, and formats. The DOM is the XML underlying object model, enabling to manipulate XML documents through an object-oriented approach (with C++, Java, etc.). XML allows to display the data, as well as to find and extract information (by programming) within an XML data set on the client (and this is even true on the server side).

- On the client side, the data can be "naturally" re-introduced in a client local database. Moreover, it should be possible to ship the semantical meta-data as contained in the DB sources when data are streamlined, then allowing a better exploitation of these data.

4.2 XML for enterprise-level services

XML can be viewed as a standard way of passing data between many heterogeneous distributed application servers, as well as across multiple operating systems, therefore as a basic model for data exchange at level of middleware layer. It can be considered as a protocol offering in some cases an alternative to COM and CORBA protocols, especially in the Internet context (one can keep on using classical HTTP firewalls, without having to open those firewalls to DCOM or IIOP through proxies), thus facilitating distributed computing on the Internet. Even more, it could be considered as a link for communication between a COM domain and a CORBA domain, therefore using XML as a way of bridging protocols.

XML can be used to define the container for a message content, for any type of data provided by a repository. For example, considering two MOMs that are used within a project, it is possible to establish an XML grammar to encapsulate any message, thereby allowing to distribute the same message through both middleware environments, provided that there is a process written for each environment that is able to extract the route and content of the message. Thus, XML supports data portability as a platform-neutral document description meta-language that offers means for data serialisation. It is worth noticing, at that point, that it can be quite beneficial to associate XML with Java, which offers code portability, as supporting the development of platform-neutral applications. Some of the expected benefits are:

- XML appears to promptly become a standard, with a potential role as a universally accepted format for the exchange of information between heterogeneous applications. XML is expected as one of the primary means for developers to design multi-tier applications in heterogeneous environments.
- XML seems to be appropriate in co-operative applications because dealing with documents and means to convey business knowledge. An interesting perspective is to define and use meta-data (this

currently is few or not existing in most applications) that can be exposed through XML messages.

- XML supplies cost-effectiveness for implementing Internet and distributed applications based on XML software tools and components (both on client and server sides). Regarding the software market, a lot of actors (IBM, Oracle, Sun, etc..) integrate the parsing and generation of XML documents within their platforms. A lot of application servers use the XML format for sending information, and databases integrate as well an XML parser. In addition, a bunch of freeware tools are accessible through the Web. Thus, as soon as an application is powered with XML, this should lead to minimal effort to exchange information.
- XML provides a way of tagging data and objects as they are called for on a network. Extending this feature already identified on the client side, allows an automatic way of populating databases on the fly with XML (especially local databases for specific applications connected to Intra/Extranet). Besides, in order to deal with the semantics of data that compose the content of XML messages, current efforts are undertaken, among others in the W3C with DCD and XML schemas, to define more semantics attached to an XML document content, including element names and rich data types.

4.3 Comparison between XML and existing middleware solutions for OSMOS

One of the key challenge that the industry is facing is not only internal communication within the enterprise systems, but the ability to communicate seamlessly with the outside world. The question can be formulated as "how to quickly and easily exchange information with business partners independently of platforms, software applications and networks ?". This question is even more crucial in dynamic virtual environments in the Construction industry, where partners are almost never the same from one project to the next one. From that point of view, XML seems to offer promising perspectives as XML messages between applications can be realised through HTTP, thus dealing with firewalls. This is probably one of the main advantages of XML against technologies like CORBA, DCOM or even MOMs. Points in favour of XML include:

- XML provides explicit messaging between applications, whilst CORBA deals with implicit messaging (i.e. which is transparent and managed at level of the ORB): either this messaging is dedicated to the ORB in use, and only applications explicitly connected to this ORB can inter-operate, or the standard IIOP is used, but even in that case, management of IIOP messages is intrusive at level of applications codes as it need to be explicit, i.e. the application must say if it deals with IIOP messages or internal ORB's messages.

- XML is better adapted to manipulate large volumes of data, as it is oftentimes the case in Construction, while probably CORBA best fits with application accessing few remote data. IIOP is not configured for dealing with large messages, it has been developed following a method-based access model, and deals with messages containing methods calls with their typed parameters.

- Issues related to information persistency can be more easily managed with XML. At least, a basic persistence level is ensured with XML documents that are stream of bytes that can be directly stored in files, with storing of the appropriate DTD, while persistency has to be effectively managed within a CORBA-based implementation.

- Eventually, taking into account that most of Construction process models relate to an asynchronous communication type, XML (for standardised messaging between applications) and MOM-like middleware appear as essential technologies to tackle applications integration and workflow-based co-operation issues within the Construction industry.

On the other hand, distributed object computing technologies, because of their strong object typology, reveal to be more effective. For instance, if the requesting application needs to check or validate the type of information it gets, CORBA ensures the well typing of the returned result, through a fully object-oriented type checking, at least at level of all the types as manipulated in IDL. XML only deals with strings, as DTDs define a hierarchical structure for composition of strings. Consequently, data serialisation with XML files leads to provide the target application with string-based information, where the semantics (as defined in the source application) of the serialised data is lost. Another point is permanent data consistency versus management of inconsistencies. CORBA, as the information is accessed and modified remotely, provides with only one point of conservation of a valid information (on the server). Note that this also leads to powerful transactional mechanisms.

An initial conclusion is that presumably CORBA-like technologies are well adapted to local (even large) Intranets, but not easily customisable to the Internet, and probably, it will reveal oftentimes as an infrastructure offering too powerful mechanisms for client desktops (e.g. CORBA is not required so as to only deal with presentation concerns, where HTML or DHTML fit well). Java applets are an alternative, but have not been considered as fully satisfactory so far, due to some trouble with code portability. XML then emerges as a solution for conveying data in a structured way, still relying on http at level of transport for accessing any client platform. A future step could be to access distributed objects from a Web browser using XML for sending requests and receiving answers, as this is currently investigated with XML-RPC (XML Remote Procedure Call), for instance. However, in spite of the various advantages associated to the family of XML-based languages, this does not mean that everything must be managed via XML technology, but the integration of technologies must be finely tuned: such a tuning in the context of the Construction industry will be a major objective in OSMOS.

5   Conclusion

The paper presented the European OSMOS project. The latter involves leading research and academic institutions, along with key industrial players, in the building and construction domain. The OSMOS system architecture was described along with a set of base technology that are potential candidates for OSMOS. The services that OSMOS will provide are expected to enable construction industry software to be integrated with traditional Groupware software components, and, on the other hand, to accommodate intra-company and inter-company communication. It is expected that the project will advance the state of the art in the application of CSCW in the construction domain by:

- Providing construction specific, and scalable solutions, that take into account the particular organisational settings of each construction enterprise participating in the VE, including SMEs.

- Providing IT and organisational solutions that promote trust and social cohesion among the partners of a construction VE.

- Providing effective, model-based solutions, to support Communication, Co-operation, and Co-ordination between individuals and groups collaborating in a construction VE, based on the specificity and information / process requirements of the Construction domain.

- Providing models for business processes, working methods, organisation, contracts, and legal responsibilities related to CSCW in a VE.

One of the fundamental tangible objective of the OSMOS project is the ability to deploy a flexible adapted team work solution in a limited amount of time, e.g. in days or few weeks as opposed to months, as it is the case today as regards the deployment of Electronic Document Management (EDM) and Product Data Management (PDM) systems in construction companies. The project end-users are presently involved with the deployment and interpretation of the proposed system architecture within their organisation, to set up internet-based team work services. The OSMOS consortium is in the process of setting up four user interest groups in Finland, France, Sweden and the UK. These groups are expected to provide ways of translating the results to other industrial sectors across Europe. Finally, the consortium would like to acknowledge the support of the European Commission under the IST program (IST-1999-10491).

References

[1] C.J. Anumba, "Industry Uptake of Construction IT Innovation – Key Elements of a Proactive Strategy", In: proceedings of the CIB W78 conference on The Life-Cycle of Construction IT Innovations, Stockholm, 1998.

[2] Elsewise, The European Large Scale Enginnering Wide Integration Support Effort, ESPRIT 20876 project, Web site: http://www.lboro.ac.uk/elsewise/.

[3] D. Frankel, "CORBA Components – alive and well", Java Report, Vol. 4, No 10, p. 70-77, October 1999.

[4] IAI, Industry Foundation Classes, Version 1.5. (1997). Web site: http://www.interoperability.com/

[5] J2EE, The JAVA 2 Platform Enterprise Edition. Web site: http://java.sun.com/j2ee/.

[6] OMG, Common Facilities RFP-4: Common Business Objects and Business Object Facility, OMG TC Document Number 96-01-04, 1996. Web site: http://www.omg.org/public-doclist.html.

[7] E. Radeke et al., "GENIAL: Final report, ESPRIT 22 28, 1999", 51 p. Web site: http://wwwgen.uni-paderborn.de/GENIAL/index.html

[8] RDF, The Resource Description Framework, http://www.w3c.org/RDF/.

[9] Y. Rezgui and G. Cooper, "A Proposed Open Infrastructure for Construction Project Documents Sharing", Electronic Journal of Information Technology in Construction, Vol.3, 1998.

[10] Y. Rezgui et al., CONDOR: Final Report, ESPRIT 23 104, University of Salford, 1999.

[11] STEP - ISO/TC184/SC4, Part 1: Overview and fundamental principles, International Standard, ISO, Geneva, (11),1994.

[12] SGML, Information-processing -- Text and office system -- Standard Generalized Markup Language (SGML). - ISO 8879 document, 1986.

[13] J. Stephens, A. Zarli, M. Marache, J. Rangnes, J., R. Steinmann and H. vandeBelt, "VEGA Public Final Report", August 1999, PFR-01 report, ESPRIT 20408 VEGA. Web site: http://cic.sop.cstb.fr/ilc/ecprojec/vega/home.htm

[14] M. Vakola Business Process Re-engineering: Organisational Change Evaluation of Implementation Strategies, PhD doctoral thesis, University of Salford, 1999.

[15] A. Zarli, O. Richaud, and E. Buckley, "Requirements and Trends in Advanced Technologies for the Large Scale Engineering Uptake", In Proceedings of CIB W78 Conference on The Lifecycle of Construction IT Innovations, Stockholm, pp 445-456, 1998. ISBN 91-7170-281-4.

From Manufacturing to Construction : Towards a Common Representation of Process Information : the PSL Approach

A. F. Cutting-Decelle

University of Savoie/ESIGEC/LGCH, Domaine de Savoie-Technolac, F 73376 LE BOURGET DU LAC, France

J. J. Michel

CETIM, BP 80067, F 60304 SENLIS, France

C. Schlenoff

NIST, 100 Bureau Drive, Stop 8260, Gaithersburg, MD 20899, USA,

Abstract

In all types of communication, the ability to share information is often hindered because the meaning of information can be drastically affected by the context in which it is viewed and interpreted. This is true in manufacturing, because of the growing complexity of manufacturing information and the increasing need to exchange this information among various software applications. This is particularly true in construction, because of the number of the actors involved in the construction process and the diversity of the information handled during the construction stages (design, construction, operation of the building). A solution to this problem is the development of a common language enabling all the actors of the construction process to share the same semantic concepts intrinsic to the capture and exchange of information related to the construction process. The aim of this paper is to present the PSL language, its main features and the underlying ontology, and to analyse its " *applicability* " to the construction sector.

1 Introduction

In the past decades, and despite the great strides taken in information technology, the goal of integrated engineering and control systems (mainly for manufacturing, but the phenomenon has also started in construction for several years) has remained elusive [17]. Integration of systems requires at least compatibility of data representations, communication paradigms and system architectures. Advances have been made in each of these areas, such as product or component data exchange (ISO 10303 STEP [1], ISO 13584 P-LIB [3]), communication protocols (TCP/IP, OSI) and architectures (CORBA). Although it is the information that must be shared between these systems, it is the representation and the language that provide the mechanism to allow the sharing to take place.

One area of data representation which has received relatively little attention, when compared to product data, is process data. This is particularly interesting when one considers that every aspect of a manufacturing (or construction) enterprise involves some form of process. Just as significant is that each manufacturing function (or each construction function) dealing with process typically has its own representational approach. In manufacturing, a production scheduling system typically has its own means of entering, manipulating and representing a sequence of actions, a process planning system has another, a workflow management system yet another. This is particularly true in construction, where the different process functions, such as design, (cost and site) planning, engineering, construction, site logistics, monitoring and control, facilities management, etc. are most of the time carried out by different enterprises, each of them with its own vocabulary, way of working and scheduling of the operations.

Clearly there are benefits to be gained from a representation common to all of these applications, as shown in Fig. 1 [17].

Motivated by this growing need to share information process in the manufacturing environment, the PSL project is aimed at providing a generic *Process Specification Language* for describing process, building on existing methodologies.

Originally developed for a manufacturing environment, and given the similarities between this sector and the construction sector, it seems interesting to try to follow the same approach in a construction environment, thus to try to use the language in construction.

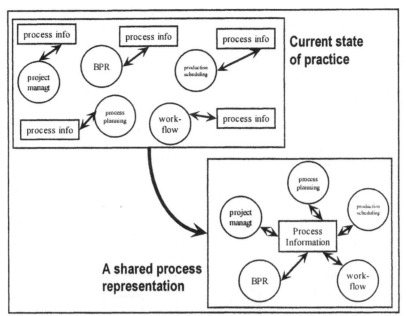

Fig. 1 : Motivation for project [17]

2 Manufacturing and construction industries : features, specificities

The criteria used for the analysis are mainly based on an approach in terms of communication of these two industrial domains. The analysis is done from the point of view of *information management*, we will here focus on the kind of information specifically relating to process, either for manufacturing industry, or else for construction.

Processes and process information are fundamental to manufacturing -- but also to construction. In the last decade, there has been an increase in the number and types of software applications which attempt to capture the essence of process. These range from tools that simply portray processes graphically to tools that enable simulation, analysis and/or control of processes. As industrial (manufacturing, or construction) companies move toward an increasing integration, there is a growing need to share process information: for example, project management software will use process data from workflow applications. All this is leading to the conclusion that as more and more processes are modeled, analysed, monitored and controlled, the probability is increasing of having these processes expressed in different and incompatible ways. On the other hand, as more and more of these automated applications are implemented and integrated, the need is increasing of a robust, standard method for representing processes in an unified way [17].

2.1 Information models for manufacturing systems

The variability of a manufacturing system can be expressed through the features of the information model describing the system. This model will have to include the functional intent of product and processes, encompassing product definitions, process definitions (process planning, execution and production engineering) and production operation and control (scheduling and building instructions). This is shown in Fig. 2 below. On this schema, the four main sets of data structured according to the model of the manufacturing system give everyone access to current data and share the up-dated information instantaneously as the product evolves from one stage to an other. As such, it encourages the use of common databases, eliminating potentials for wasted or duplicated efforts in traditional data re-creation and re-formatting. This model also provides a precise description of the activities, data connectivity and communication network of an enterprise. Through the acts of flowcharting, the model captures the information flows through the units involved in the design and development of the product, since it reflects the information needs of both the manual and automated units of an enterprise [7].

Described in terms of data « reservoirs », the model consists of four elements : *engineering data, manufacturing engineering data, manufacturing control data* and *master schedule data*. All these « reservoirs », although not directly filled with data coming from product, manufacturing or component representations can however be considered as usage categories, or usage classes of data. Applied to the management of a manufacturing system, a joint use of these multiple representations dealing with specific and different types of data (e.g. data related to product, to manufacturing, to manufacturing management, etc.) may lead to lacks of consistency among the different representations of these data, if the problem of their overlap is not carefully examined. Within the model, seven major activities run in parallel : product definitions, process definitions, production operations and control, manufacturing and assembly, quality control, vendor and supplier interface and product support. The benefit

of this model is to allow planning of additional investments in information technology needed to effectively manage complex configuration controls.

Note : since the schema focuses on the flows of information within the manufacturing system, input/output information is not represented.

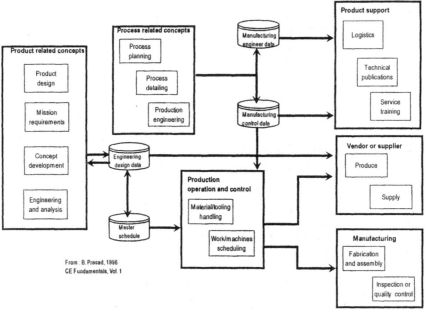

From : B. Prasad, 1996
CE Fundamentals, Vol. 1

Fig. 2 Common product and process model [15]

A benefit of this schema is to enable an easy location of concepts related to product, process, product support, manufacturing and production operation and control, even though, in common industrial practice, things are not always so simple.

2.2   Information models in construction

Information handled during the construction process can, from different points of view, be divided into several categories [6] :

- According to the first point of view, information can first of all state facts. This category of information is what design documents, which are the results of design decisions, primarily concern. Information that needs to be transferred between computing systems in the construction process is mostly of this type (ex. : « *the color of the surface x of wall y is green* »). This context of the information is not specific to construction. Secondly, information has to define goals and requirements which a particular project must fulfil, such as « *the building may not have more than 5 floors* ». This kind of information could be called requirement, or constraint. The term « requirement » seems more natural when discussing construction practice, whereas constraint has a clear meaning in database theory. Performance based systems are very much concerned with this type of

information. Constraints are typically defined using operators which restrict attribute values of building description objects within certain limits or to certain pre-defined values. The third category of information state rules which restrict facts, but which apply in general and are not tied to a particular project, such as the following « rule » : « *a beam, which is directly or indirectly structurally supported by a column cannot be erected before the column* ». These three categories fo information can be called « facts », « constraints » and « knowledge ». From a programming language point of view, facts can be constructed using simple assignment statements, requirements are mainly represented by inequality operators (or simple algorithms) and knowledge can be formalised using knowledge based systems.

- The second semantic point of view divides information into project-specific and more general information. Facts can be both project specific and general. Constraints are mainly project-specific and knowledge is usually general in nature.

- The third point of view concerns presentation and categorises the types of documents used to present the information for human interpretation. Some typical presentation formats used in construction are : drawings, schemas, realistic visualisations, written specifications, calculation results, bills of materials, contracts, orders and various tendering documents.

We will here limit our study to project-specific information, focusing on the semantics of the information. The reason of this choice is due to our primary concern to study information management within construction projects. The information which needs to be communicated to other parties in the construction process consists mostly of factual information and not of the constraints and knowledge used in the process of determining this factual information. Clearly constraints are very important in the early briefing stages of projects and in quality assurance applications. Knowledge mainly resides in application programs and its effect on the actual transfer of data between project participants will need to be examined further. Among the different formal modelling techniques enabling a description of the information management and the physical processes of a construction project, some of them focus on modelling the structure of the information describing the products, processes, resources and other elements of the construction process. We will here limit our study to activity modelling, this activity being mainly described in terms of processes and resources.

Several process models for Construction are available today, among which the MOPO model [9], covering the whole construction life cycle. Other models mainly focus on the design stage as the model described in [5]. Some process models introduce Concurrent engineering features, such as the model of [4].

A generic representation of the construction process, as provided by [6], consists in three main categories, which are : *activities, results, resources*. An activity uses resources to produce results. Traditional construction classification systems often tend to equate results to buildings and their parts. This is due to a desire to distribute total construction costs over building parts, which is useful for cost analysis purposes. It is, however, evident that information (mostly delivered as documents) and services are other important sub-types of results.

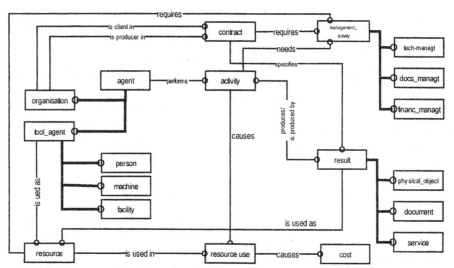

Fig. 3 : **EXPRESS-G diagram of a generic representation of the construction process (adapted from [6])**

The schema of the Fig. 3 (adapted from [6]) provides an EXPRESS-G [2] representation of some objects of a generic construction process. Among them :

- **Activity** : very kernel of the model, with relationships with most of the other objects of the model : an activity may have relationships with the result it produces, the resources it uses and the agents performing it.
- **Result** : example of an entity type needed for classification purposes, but which is intermediate in nature since most of the relevant information about results will be defined in the class descriptions of the sub-types of results.
- **Physical _object** : any physical object with shape and location. Both characteristics may be dynamic.
- **Service** : results of activities which are not physical

objects or documents (information), such as « guarding the site ».

- **Agent** : any organisation, person, machine, or facility which participates in the activities of the project. An agent performs some activity in the construction process. A fundamental aspect of an agent (distinguishing him from a product) is that he/it has an existence outside the project and usually participates in several projects.
- **Resource use and cost** : each activity in the project demands a number of inputs in the form of resources which are « consumed » or used. The actual use may be measured in manhours, tons, squaremeters, etc. A clear distinction is made between resource use and the resources entity in itself, which may be documents, materials,

machines or persons. The use of any resource involves a cost, which in most cases can be measured by the amount of the resource consumed, but in some cases by the opportunity cost of that resource, that is the cost the customer accepts to pay for the resource in question.

- **Management_activity** : super-type of the different types of management related to a construction project, such as technical management (planning, logistics, QS activity), document management (drawings, bills of quantities, calculation notes) and financial management. This entity is also fundamental to introduce CE concepts in the model.

2.3   Similarities, differences between manufacturing industry and construction

A number of lessons can be learned from the manufacturing sector [8] with regards to the implementation and practical use of « process view » within the construction industry [9]. The area within the manufacturing sector that relates to construction and building works is often called « new product development » (NPD). It concentrates on the development of an idea, need or client requirement to the final commercialisation of the product (building, or car). A number of similarities can be found between the two industries with regards to the activities used for developing new products. For example, they include :

- The start of a project can be initiated internally or by direct and/or indirect contact with the customers.
- The development of the product requires the participation of a number of specialists and functions such as designers, surveyors, marketing, stress analysts, etc.
- The successful construction or manufacture of a building or product can only be achieved if all external (suppliers and consultants) and internal resources are utilised and coordinated effectively.
- The building or product is handed over to the customer/client and provisions are made for future support.

However, there are a number of distinct differences, the most important of which is that in the manufacturing industry all NPD activities are coordinated, managed and controlled using a common framework which is the NPD process. The construction industry mainly uses *ad hoc* methods for achieving the latter and therefore reducing repeatability of process execution, resulting in the same mistakes occurring time after time.

The shift into the establishment of a consistent process for the construction industry requires a new way of thinking entailing a change of culture and working practices. Furthermore, it will require :

- A good understanding of current practices and

future trends.
- Effective communication mechanisms of such processes, such as modelling.
- Agreement of participating parties.

3   Motivation for the development of the PSL language

The goal of this project [16] is to identify or create a process specification language (PSL) that can be common to all manufacturing applications, generic enough to be decoupled from any given application, and robust enough to be able to represent the necessary process information for any given application. Additionally, the PSL should be sufficiently well-defined to ensure complete and correct exchange of process information among established applications. This PSL would facilitate communication between the various applications because they would all « speak the same language », either as their « native » language or a « second » language, for exchange.

Unlike many computerised languages in use today, it was the hope of the development team that this language would benefit from a formally described, notation-independent information schema underlying all of the possible notations which might arise. By formalizing the structure of the language in this way it becomes possible to use multiple alternative notations to convey the same information, therefore enabling multiple « views ».

The approach taken for the project has been to break it into distinct phases, namely : requirements gathering, existing process representation analysis, schema definition, language grammar/syntax development, language notation(s) development, pilot implementation and validation, and finally, submission as a candidate standard. The second phase, the representation analysis, has been designed to determine how well existing representations would support the requirements found in phase 1. This analysis has provided an objective basis on which to identify the representation or combination of representations that provides the best coverage of the requirements and to identify gaps in existing representations' abilities to address process specification requirements. The language schema, grammar, syntax, and notation have been developed as a result of this analysis.

Suitable scenarios or group of scenarios have also been identified for prototype implementations to ensure the completeness and ease-of-use of the specification language. It is this validated, documented language that is being submitted to ISO organisation as a candidate standard. Feedback and consensus from the process community have been and are still aggressively pursued during all phases of the project.

4   Presentation of the PSL

The Process Specification Language (PSL) project, at the National Institute of Standards and Technology (NIST) is aimed at creating a neutral, standard language for process specification to serve as a neutral representation to integrate multiple process-related applications throughout the manufacturing life cycle.

4.1 Scope

To keep this work feasible, the scope of study is limited to the realm of discrete processes related to manufacturing, including all processes in the design/ manufacturing life cycle. This includes, but is not limited to applications in process planning, scheduling, simulation, workflow, project management and business process re-engineering. Business processes and manufacturing engineering processes are included in this work both to ascertain common aspects for process specification and to acknowledge the current and future integration of business and engineering functions. In addition, the goal of this project is to create a « *process specification language* », not a « *process characterization language* ». Our definition of a *process specification language* is a language used to specify a process or a flow of processes, including supporting parameters and settings. This may be done for prescriptive or descriptive purposes and is composed of an ontology and one or more presentations. This is different from *a process characterization language,* which we define as a language describing the behaviors and capabilities of a process independent of any specific application. For example, the dynamic or kinematic properties of a process (e.g., tool chatter, a numerical model capturing the dynamic behavior of a process or limits on the process's performance or applicability), independent of a specific process, would be included in a characterization language [16].

4.2 The PSL language

A language is a lexicon (a set of symbols) and a grammar (a specification of how these symbols can be combined to make well-formed formulas). The lexicon consists of logical symbols (such as boolean connectives and quantifiers) and nonlogical symbols. For PSL, the nonlogical part of the lexicon consists of expressions (constants, function symbols, and predicates) chosen to represent the basic concepts in the PSL ontology. Notably, these will include the 1-place predicates 'activity', 'activity-occurrence', 'object', and 'timepoint' for the four primary kinds of entity in the basic PSL ontology, the function symbols `beginof`

and `endof` that return the timepoints at which an activity begins and ends, respectively, and the 2-place predicates `is-occurring-at`, `occurrence-of`, `exists-at`, `before`, and `participates-in`, which express important relations between various elements of the ontology [16].

The underlying grammar used for PSL is based roughly on the grammar of KIF [12] (Knowledge Interchange Format), formal language based on first-order logic developed for the exchange of knowledge among different computer programs with disparate representations. KIF provides the level of rigor necessary to define concepts in the ontology unambiguously, a necessary characteristic to exchange manufacturing process information using the PSL Ontology. Like KIF, PSL provides a rigorous BNF (Backus-Naur form) specification, enabling a rigorous and precise recursive definition of the class of grammatically correct expressions of the PSL language. In addition to the simple clarity, the BNF definition makes it possible to develop computational tools for the transfer of process information, one of PSL's central goals. In particular, by fixing the definition of the language precisely (and *only* by so fixing its definition), it is possible to develop translators between PSL and other, similarly well-defined representation languages

4.3 The PSL ontology

The foundation of the process specification language is the PSL ontology, which provides rigorous and unambiguous definitions of the concepts necessary for specifying manufacturing processes to enable the exchange of process information.

The PSL ontology is essentially two-tiered. The foundation of the ontology (the first tier) is a set of process-related concepts that are common to ALL manufacturing applications. These concepts constitute the core of the PSL ontology and include concepts such as objects, activities, activity occurrences, and time points. However, these concepts alone would only allow for the exchange of very simple process specifications. Therefore, this ontology includes a mechanism to allow for extensions to these core concepts (the second tier) to ensure the robustness of the ontology.

4.4 Theoretical foundations of the PSL

They are of two kinds : Model theory and Proof theory.
- **Model Theory** : the model theory of PSL provides a rigorous, abstract mathematical characterization of the semantics, or meaning, of the language of PSL. This representation is typically a set with some additional structure (e.g., a partial ordering, lattice, or vector space). The model theory then defines meanings for the terminology and a notion of truth for sentences of the language in terms of this model. The objective is to

identify each concept in the language with an element of some mathematical structure, such as lattices, linear orderings, and vector spaces.

Given a model theory, the underlying theory of the mathematical structures used in the theory then becomes available as a basis for reasoning about the concepts intended by the terms of the PSL language and their logical relationships, so that the set of models constitutes the formal semantics of the ontology.

- **Proof Theory** : the proof theory consists of three components : PSL Core, one or more foundational theories, and PSL extensions :

- **PSL Core** : set of axioms written in the basic language of PSL. The PSL Core axioms provide a syntactic representation of the PSL model theory, in that they are sound and complete with regard to the model theory. That is to say, every axiom is true in every model of the language of the theory, and every sentence of the language of PSL that is true in every model of PSL can be derived from the axioms. Because of this tight connection between the Core axioms and the model theory for PSL, the Core itself can be said to provide a *semantics* for the terms in the PSL language.

- **Foundational Theories** : the purpose of PSL Core is to axiomatize a set of intuitive semantic primitives that is adequate for describing basic processes. Consequently, its characterization of them does not make many assumptions about their nature beyond what is needed for describing those processes. The advantage of this is that the account of processes implicit in PSL Core is relatively straightforward and uncontroversial. However, a corresponding liability is that the Core is rather weak in terms of pure logical strength. In particular, the theory is not strong enough to provide definitions of the many auxiliary notions that become needed to describe an increasingly broader range of processes in increasingly finer detail. (Auxiliary notions are axiomatized in PSL *extensions*, discussed next.) For this reason, PSL includes one or more *foundational theories*. A foundational theory is a theory whose expressive power is sufficient for giving precise definitions of, or axiomatizations for, the primitive concepts of PSL, thus greatly enhancing the precision of semantic translations between different schemes. Moreover, in a foundational theory, one can define a substantial number of auxiliary terms, and prove important metatheoretical properties of the core and its extensions.

There are several good foundational theories. Of these, set theory is perhaps the most familiar, and perhaps, all in all, the most powerful. Set theory's foundational capabilities are well known. It is, in particular, capable of serving as a foundation for all of classical mathematics, in the <u>sense</u> that all

notions of classical mathematics – integers, real numbers, topological spaces, etc. – can be defined as sets of a certain sort and, under those definitions, their classical properties derived as theorems of set theory.

For PSL's purposes, however, a more suitable foundation is a modified and extended variation of the *situation calculus*. The reason for this is that the situation calculus's own primitives – *situation*, *action*, *fluent* (roughly, *proposition*) – are already highly compatible with the primitives of PSL; indeed, it is very natural to identify PSL primitives with, or define them in terms of, the primitives of the situation calculus. In addition, the situation calculus is also strong enough to define a wide variety of auxiliary notions and, with the addition of some set theory, it can be used as a basis for proving basic metatheoretic results about the Core and its extensions as well.

- **Extensions** : third component of PSL, a PSL extension gives one the resources to express information involving concepts that are not part of PSL Core. Given the importance of the concept, it is developed in the section 3.6 below.

The three components of the PSL architecture and their relations are illustrated in Fig. 4 : the solid arrows indicate the definability relation. The dashed lines indicate partial definability, i.e., the case where some, but not all the additional linguistic items in the language of an extension are definable. Two or more solid arrows pointing to the same oval indicate the possibility that more than one given theory might jointly be used to define a new extension. Therefore, we might have connected PSL Core to foundational theories, but this would not sufficiently distinguish the central role of the Core from the more auxiliary roles of extensions. Hence, we picture PSL Core as sitting directly upon the foundational theories. "(+ Foundational Theory) " in the PSL Core box indicates that PSL Core together with a foundational theory are typically used to formulate definitional extensions.

4.5 PSL core

PSL Core is based upon a precise, mathematical, first-order theory, i.e., a formal language, a precise mathematical semantics for the language, and a set of axioms that express the semantics in the language. There are four primitive classes, two primitive functions, and three primitive relations in the ontology of PSL Core. The classes are OBJECT, ACTIVITY, ACTIVITY_OCCURRENCE and TIMEPOINT. The four relations are PARTICIPATES-IN, BEFORE, and OCCURRENCE-OF. The two functions are BEGINOF, and ENDOF. ACTIVITIES, ACTIVITY_

OCCURRENCES, TIMEPOINTs (or POINTs for short), and OBJECTs are known collectively as entities, or things. These classes are all pairwise disjoint.

Intuitively, an OBJECT is a concrete or abstract thing that can participate in an ACTIVITY. The most typical examples of OBJECTs are ordinary, tangible things, such as people, chairs, car bodies, NC-machines, though abstract objects, such as numbers, are not excluded.

OBJECTs can come into existence (e.g., be created) and go out of existence (e.g., be « used up » as a resource) at certain points in time. In such cases, an OBJECT has a begin and/or end point. Some OBJECTs, e.g., numbers, do not have finite begin and end points. In some contexts it may be useful to model certain ordinary OBJECTs as having no such points either.

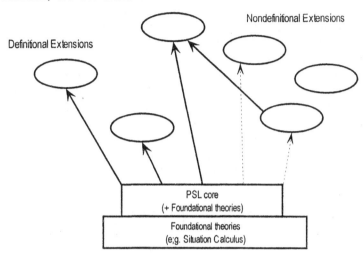

Fig. 4 : the PSL semantic architecture [16]

An ACTIVITY-OCCURRENCE is a limited, temporally extended piece of the world, such as the first mountain stage of the 1997 Tour de France or the eruption of Mt. St. Helen. Any ACTIVITY-OCCURRENCE is simply taken to be characterised chiefly by two things : its temporal extent, as determined by its begin and end POINTs (possibly at infinity), and the set of OBJECTs that participate in that ACTIVITY at some point between its begin and end POINTs.

TIMEPOINTs are ordered by the BEFORE relation. This relation is transitive, non-reflexive, total ordering. In PSL Core, that time is not dense (i.e., between any two distinct TIMEPOINTs there is a third TIMEPOINT), though it is assumed that time is infinite. POINTs at infinity (INF+ and INF-) are assumed for convenience. (Denseness, of course, could easily be added by a user as an additional postulate.) Time intervals are not included among the primitives of PSL Core, as intervals can be defined with respect to TIMEPOINTs and ACTIVITIES.

4.6 Extensions

Extensions give PSL a clean, modular character. PSL Core is a relatively simple theory that is adequate for expressing a wide range of basic processes. However, more complex processes require expressive resources that exceed those of PSL Core. Rather than clutter PSL Core itself with every conceivable concept that might

prove useful in describing one process or another, a variety of separate, modular extensions have been (and continue to be) developed that can be added to PSL Core as needed. In this way a user can tailor PSL precisely to suit his or her expressive needs.

To define an extension, new constants and/or predicates are added to the basic PSL language, and, for each new linguistic item, one or more axioms are given that constrain its interpretation. In this way one provides a « semantics » for the new linguistic items. A good

example of such an extension is the theory of timedurations. PSL Core itself does not provide the resources to express information about timedurations. However, in many contexts, such a notion might be useful or even essential. Consequently, a theory of timedurations has been developed which can be added as to PSL Core, thus providing the user with the desired expressive power.

4.7 Extensions in PSL Version 1.0

The set of extensions in PSL 1.0 fall roughly into three « families » : PSL Outer Core, Generic Activities, Schedules.

- PSL Outer Core : There is a small set of extensions that are so generic and pervasive in their applicability that we set them apart by calling them the PSL Outer Core. These three extensions are : Subactivity Extension, Activity-Occurrence Extension, States Extension. The Subactivity Extension describes how activities can be aggregated and decomposed. It also defines the concept of primitive activity, which can not be decomposed into any further activities. The Activity-Occurrence Extension defines relations that allow the description of how activity-occurrences relate to one another with respect to the time at which they start and end. The State Extension introduces the concept of state (before an activity-occurrence) and post-state (after an activity-occurrence).

- **Generic activities and ordering relations :** Figure 5 illustrates the modules in PSL that are required to define the terminology for generic classes of activities and their ordering relations. There are nine relevant extensions to PSL Core, four dealing with generic process modeling concepts and five dealing with schedules. The five focusing on schedules will be discussed in the next section. The four dealing with generic process modeling concepts are : Ordering Relations, Nondeterministic Activities, Complex Sequence Ordering Relations, Junctions.

- **PSL extensions for schedule :** These extensions were motivated by the applications in the PSL pilot implementation, in particular ILOG Scheduler 4.3 [http://www.mel.nist.gov/psl]. At the beginning of the pilot implementation of PSL, there were no extensions capable of completely defining concepts such as temporal constraints. It was therefore necessary to design new extensions containing terminology whose definitions correctly and completely captured the intuitive meaning of the ILOG Schedule concepts. Scheduling can be characterized intuitively as the assignment of resources to activities such that the temporal constraints are satisfied. Temporal constraints include the duration of activities and the temporal ordering of activity-occurrences. These intuitions lead to the introduction of five extensions within PSL 1.0, shown in Fig. 5 : Durations, Activities and Duration, Temporal Ordering Relations, Reasoning about State, Interval Activities.

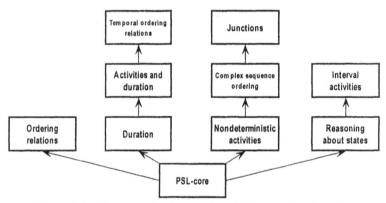

Fig. 5 : PSL modules for generic classes of activities and their ordering relations

4.8 Approach for developing extensions

From the above list of extensions, one may see that certain representational areas within PSL have been thoroughly worked out and some have not been addressed yet. For example, the area relating to « ordering of activities » has been well addressed within the extensions of « Ordering Relations for Complex Sequence Actions », « Ordering Relations over Activities » and « Temporal Ordering ».
However, other representational areas such as « Process Intent » have not yet been addressed.
The development of PSL has proceeded on an as-needed basis. The initial PSL ontology was developed using a single scenario, the EDAPS (Electromechanical Design and Planning System) scenario developed by Steve Smith at the University of Maryland [10]. The concepts introduced were defined and modelled within PSL and later extended as other scenarios were explored. The PSL ontology was then further expanded to incorporate the concepts introduced in various manufacturing software applications when PSL was

used to exchange process information among these packages. As more software applications become « PSL-compliant », PSL will be continually expanded to ensure that ALL process-related concepts are capable of being represented within the language.

5   Use of the PSL in construction

The possibility of using PSL in construction will be analysed through an example, provided by the Data Flow diagrams of a construction project, as described in [11].
In this study, a « construction company operation » is described, through a DFD (Data Flow Diagram) representation, in terms of :
- 1 : pre-tender procedures,
- 2 : pre-construction procedures,
- 3 : construction procedures.
Within the constuction procedures diagram, we find the following stages :
- 3.1 : initiating project controls,
- 3.2 : site management,

- 3.3 : external valuation procedures,
- 3.4 : cost value comparison procedures,
- 3.5 : resource management,
- 3.6 : project information procedures,
- 3.7 : external progress meeting,
- 3.8 : internal site review meeting.

We will focus here on the stage 3.2 « *site management* », developed in the Fig. 6 : this entity is a sub-type of the object « tech_managt » of the generic schema of the Fig. 3. The reason of this choice is that the concepts provided on this diagram, related to management information are close to concepts commonly dealt with in manufacturing, and thus in PSL.

The main concepts emerging from this view are the following :

- **coordinate site team**
- **determine site strategy** : having been selected by the company directors the site team and associated construction managers and surveyors have to determine the site organisation in terms of office layouts. The individual responsibilities are normally laid down in the company policy, however, a measure of flexibility should exist to account for the differing personalities and skills of the staff.
- **forecasting resource requirements** : process, role of a site team in forecasting resource requirement is seen to be the early consideration of all factors that may lead to delay on the contract. The need to place orders in good time and to raise requisitions for materials and plant long in advance of them being required on site is obvious but it can easily be overlooked in the heat of day to day problem solving, coordinating, planning, etc. The site agent is principally responsible for ensuring that resources and design information are available when required although the construction managers and surveyors who visit site regularly are able to take a more subjective view and hence have a major role to play in this aspect of site management.
- **site info management** : effective communication between all parties is an essential attributed on any project if it is to be succesful. Typically, there are four areas of concern : between site and regional office, between staff and workforce, between the company and its sub-contractors and suppliers, between the company and the client's consultants.
- **site performance monitoring** : process, typically the engineer working under the supervision of the site agent is responsible for setting out the works and providing all levels and benchmarks necessary for the work to procee accurately and smoothly. He is also generally responsible for checking the works as completed for accuracy and tolerance. The site

agent and foremen jointly coordinate and supervise the work and also inspect the works together to ensure it meets the required specification, quality of finish, etc.

- **site personnel management** : process dealing with : bonuses, progress target, performance reports, foreman's inputs. The site team should place great emphasis on encouraging and motivating the workforce and in developing and maintaining a high morale.
- **site scheduling** : process dealing with professionals' inputs, managers' inputs, info requirement, resource requirement, one week programme, eight week programme.

We will limit the presentation to a first level of identification between objects of the PSL ontology and elements of the management process of a construction project. We have to define an ontology of the construction process (here limited to the management features), if possible in the terms used within PSL : core, outer core, and extensions vocabulary. If the existing extensions reveal inadequate for construction, it may be necessary to develop new ones. However, as far as possible, we will try to deal with the existing ones.

- Method of work :

The idea consists in developing the scenario provided by the « site management » process of the example represented Fig. 6, extracted from the set of diagrams of [11]. This diagram is then represented in IDEF3 [13], in order to provide a process-centered view, since this method facilitates the capture of the description of what a system actually does. Applied to an object of the diagram (Fig. 6) as an example, we get :

```
(define-process site-management-proc
        :documentation « site management »
        :components (determine-site-strategy
        site-scheduling site-info-management
        coordinate-site-team
        site-performance-monitoring
        forecasting-resource-requirements
        site-personnel-management)
        :constraints nil)
```

The same textual version has to be provided for the other objects of the same view and of the other views, together with the constraints, rules and other requirements if any (any information or requirement related to a process and needed by it, including durations of the tasks, junctions, etc.).

Then, the set of textual elements must be translated into PSL concepts, through their expression in KIF. Today, however, this translation will have to be done by hand, but work is on-going to develop automatic translators.

Applied to the example above, we get :

```
(and  (doc      site-management-proc      « site
management »)
```

```
(subactivity determine-site-strategy)
(subactivity site-scheduling)
(subactivity site-info-management)
(subactivity coordinate-site-team)
(subactivity site-performance-monitoring)
(subactivity forecasting-resource-requirements)
(subactivity site-personnel-management)
(idef-process site-management-proc))
```

All the process entities need to be translated, but also all the relating concepts, such as : time, duration, objects, occurrences (as found in the PSL core), and the corresponding relations. Of course ordering activities, sub-typing, scheduling concepts also need to be translated into PSL elements : this work is on-going.

6   Conclusion

In a construction scenario, expressed here according to two different representations (EXPRESS-G for the generic level, DFD for the site management level), the PSL can be used to facilitate the interoperability of construction knowledge, shared among the numerous actors of a construction project, each of them with his own practice, and semantics of the language he uses. This same knowledge is very often exchanged with a lot of difficulties, leading to misunderstandings, thus to delays, cost increases and bad quality of the end-product, the building. In terms of Concurrent engineering, PSL could offer the possibility to integrate CE features, through rules, in the process model.

References

[1]   Industrial automation systems and integration - Product data representation and exchange, ISO IS 10303-1, Part 1 : Overview and fundamental principles, 1994

[2]   Industrial automation systems and integration - Product data representation and exchange, ISO IS 10303-11, Part 11 : Description methods : EXPRESS language reference manual, 1994

[3]   Industrial automation systems and integration - Parts library, ISO IS 13584-1, Part 1 : Overview and fundamental principles

[4]   C.J. Anumba, N.F.O. Evbuomwan, A concurrent engineering process model for computer-integrated design and construction, in Information Processing in civil and structural engineering design, 1996

[5]   S. Austin, A. Baldwin, A. Newton, A data flow model to plan and manage the building design process, Journal of Engineering Design, Vol 7, N° 1, 1996

[6]   B.C. Björk, A unified approach for modelling construction information, Building and Environment, Vol 27, N° 2, 1992

[7]   A.F. Cutting-Decelle, J.J. Michel, C. Schlenoff, ISO 15531 MANDATE : a standardised data model for manufacturing management, IJ-CAT Journal, special issue on « Applications in industry of product and process modelling using standards », to be published, 2000

[8]   AF Cutting-Decelle, The use of industrial management methods and tools in the construction industry - application to the construction process, Concurrent Engineering in Construction, London, 1997

[9]   R. Cooper, M. Kagioglou, G. Aouad, J. Hinks, M. Sexton, D. Sheath, The development of a generic design and construction process, European PDT Days, 1998

[10]   http://www.isr.umd.edu/Labs/ CIM/cimcontent.html

[11]   N. Fisher, S.L. Yin, Information management in a contractor, T Telford, 1992

[12]   M. Genesereth, R. Fikes, Knowledge interchange format (Version 3.0) – Reference manual, computer science dept, Stanford University, Stanford, CA, 1992

[13]   R. J. Mayer, C.P. Menzel, M.K. Painter, P.S. de Witte, T. Blinn, B. Perakath, Information integration for concurrent engineering (IICE) IDEF3 process description capture method report, KBSI Inc, AL-TR-1995

[14]   S.T. Polyak, S. Aitken, Manufacturing process interoperability scenario, AIAI-PR-68, 1998

[15]   B. Prasad, Concurrent Engineering fundamentals, Vol. 1, Prentice Hall, 1996

[16]   C. Schlenoff, M. Gruninger, F. Tissot, J. Valois, J. Lubell, J. Lee, The process specification language (PSL) Overview and version 1.0 specification, NISTIR 6459, NIST, 2000

[17]   C. Schlenoff, A. Knutila, S. Ray, Unified process specification language : requirements for modeling process, NISTIR 5910, NIST, 1996

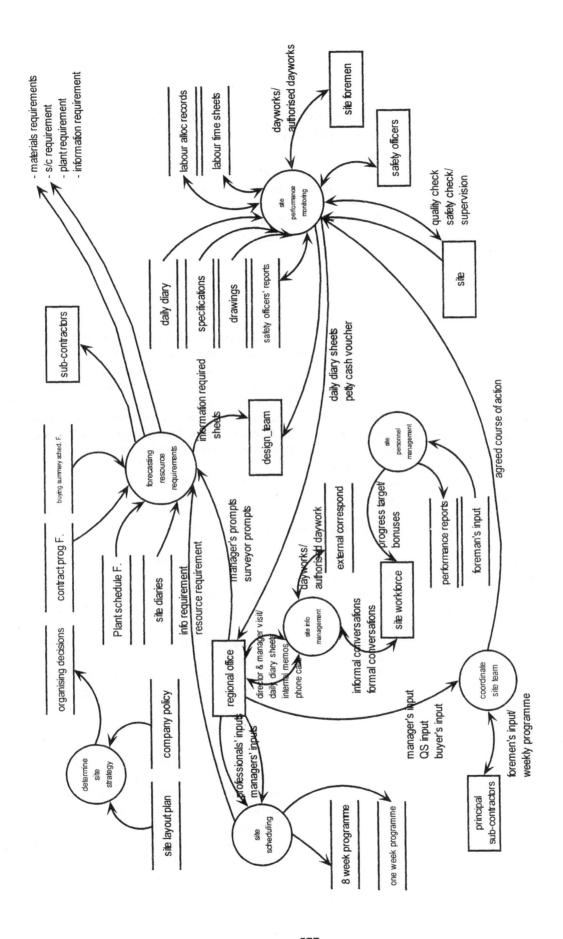

Fig. 6 : DFD representation of the site management process [11]

577

A Comparative Review of Concurrent Engineering Readiness Assessment Tools and Models

Malik M. A. Khalfan

Department of Civil & Building Engineering, Loughborough University, Loughborough, Leics., LE11 3TU, UK.

Chimay J. Anumba

Department of Civil & Building Engineering, Loughborough University, Loughborough, Leics., LE11 3TU, UK.

Abstract

While Concurrent Engineering (CE) is gaining increasing acceptance within many industry sectors, some implementation efforts have not obtained their full potential for reducing costs, reducing time, and increasing efficiency, effectiveness and performance for product development efforts. This is due in part to weak planning to support the implementation. One approach that has been used successfully to improve CE implementation planning is to conduct an organisation readiness assessment prior to the introduction of CE. This helps to investigate the extent to which the organisation is ready to adopt Concurrent Engineering.

This paper focuses on a comparative review of existing readiness assessment tools and models that have been successfully used in the manufacturing and IT sectors. It argues that Readiness Assessment of the construction supply chain is a necessity for the implementation of CE in construction industry. It assesses and discusses the applicability of existing tools and models to the construction industry, and concludes that there is the need of a new CE readiness assessment model for the construction supply chain.

Key Words – Concurrent Engineering (CE), CE Readiness Assessment, Construction Industry.

1 Introduction

1.1 Definition

Concurrent Engineering, sometimes called simultaneous engineering, or parallel engineering has been defined in several ways by different authors. The most popular one is that by Winner et al. [1], who state that concurrent engineering "…is a systematic approach to the integrated, concurrent design of products and their related processes, including manufacture and support. This approach is intended to cause the developers, from the outset, to consider all elements of the product life cycle from conception through disposal, including quality, cost, schedule, and user requirements."

1.2 Different Aspects of CE

There are eight basic elements of CE, which are divided into two aspects as follows [2]:

Managerial and human aspect

- The use of cross-functional, multidisciplinary teams to integrate the design of products and their related processes.
- The adoption of a process-based organisational philosophy
- Committed leadership and support for this philosophy
- Empowered teams to execute the philosophy

Technological aspects

- The use of computer aided design, manufacturing and simulation methods (i.e. CAD/CAM/CAE/CAPP) to support design integration through shared product and process models and databases.
- The use of various methods to optimise a product's design and its manufacturing and support process (e.g. DFM, DFA, QFD).
- The use of information sharing, communication and co-ordination systems.

- The development and/or adoption of common protocols, standards, and terms within the supply chain.

1.3   The Need for CE Readiness Assessment

While Concurrent engineering (CE) is gaining acceptance, some implementation efforts have not realised their full potential for reducing costs, reducing time, and increasing efficiency, effectiveness and performance for product development efforts. This is due in part to insufficient planning to support the implementation [3]. One approach that has been successfully used to improve CE implementation planning is to conduct readiness assessment of an organisation prior to the introduction of CE. This helps to investigate the extent to which the organisation is ready to adopt Concurrent Engineering, and to identify the critical risks involved in its implementation within the company and its supply chain. CE Readiness Assessment has been successfully used for the planning of CE implementation in several industry sectors, notably manufacturing and software engineering, as described below in the next section.

This paper undertakes a comparative review of existing tools and models for CE readiness assessment, with a view to identify the most appropriate model for the construction industry. A brief background to the adoption of CE in the construction industry is provided in section 2. Section 3 details the review, which is followed by a discussion of the most appropriate tool for CE readiness assessment of the construction industry. Further work being done to develop a model for construction is outlined and conclusions drawn.

2   The move Towards CE in Construction

In the context of the construction industry, Evbuomwan & Anumba [4] define Concurrent Engineering as an "…attempt to optimise the design of the project and its construction process to achieve reduced lead times, and improved quality and cost by the integration of design, fabrication, construction and erection activities and by maximising concurrency and collaboration in working practices." This is in sharp contrast with the traditional approach to construction project delivery.

2.1   The Traditional Approach

In the construction industry, based on the client brief, the architect produces an architectural design, which is given to the structural engineer, who on completing the structural design passes the project to the quantity surveyor to produce the costing and bill of quantities. This goes on until the project is then passed on to the contractor who takes responsibility for the construction of the facility. This scenario, which is similar to the 'over the wall' approach [4, 5], is shown in Figure 1.

OVER THE WALL

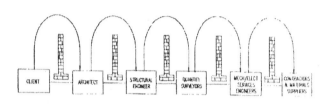

Figure 1: The over the wall approach [4]

Key disadvantages prevalent with this approach include:

- The fragmentation of the different participants in the construction project, leading to misperceptions and misunderstandings:
- The fragmentation of design and construction data, leading to design clashes, omissions and errors;
- The occurrence of costly design changes and unnecessary liability claims, occurring as a result of the above;
- The lack of true life-cycle analysis of the project, leading to an inability to maintain a competitive edge in a changing marketplace;
- Lack of communication of design rationale and intent, leading to design confusion and wasted effort.

To address these issues, there is an urgent need for a shift in paradigm within the construction industry. This should involve the adoption of new business strategies, with the aim of integrating the functional disciplines (see Figure 2) at the early stages of the construction project [4].

2.2   The Application of CE to Construction

There is an urgent need to improve the performance of the construction supply chain participants. This can be achieved during the design process by considering all aspects of the project's downstream phases concurrently. Incorporating requirements from the construction, operation and maintenance phases at an early stage of a project would undoubtedly lead to an overall improvement in project performance. The essential constituents of 'Concurrent Construction' are as follows [6]:

- The identification of associated downstream aspects of design and construction processes
- The reduction or elimination of non-value-adding activities
- The development and empowerment of multi-disciplinary teams

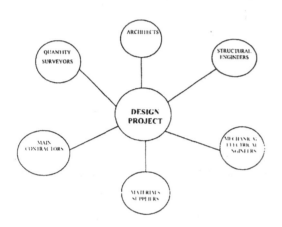

Figure 2: An integrated project team [4]

3   Readiness Assessment Tools and Models

3.1   An Overview of Tools & Models

There are several tools and models, which are being used for readiness assessment of organisations for concurrent engineering. This section compares these models & tools on the following basis:

- Aim of the tool;
- Who has developed this tool;
- What does it cover;
- Its application industry and sectors;
- Status of the tool;
- Usage of the tool either as a CE readiness assessment tool or CE implementation tool; and
- Its appropriateness for the construction industry.

A brief description and comparison of the models and tools (see Table 1) are presented below:

a)   RACE (Readiness Assessment for Concurrent Engineering)

This tool was developed at West Virginia University (United States) in the early 90's and is widely used in the software engineering, automotive and electronic industries. It could be modified for use in the construction and other industries. The RACE-model is conceptualised in terms of two major components: The organisational processes for product development, and the information technology to support the product development process [7, 8]. The Process component is subdivided into ten elements and Technology into six as mentioned in Table 1.

b)   PMO (The Process Model of Organisation)

This model was developed to assess and analyse the processes and technology of an organisation. The process model of organisation (PMO) is a model, which can basically be used for analysing and designing an organisation, its processes and technology in the context of the market in which that organisation operates. The model is used to detect bottlenecks that prevent the organisation to achieve its objectives. Hence, the model is useful in the awareness and readiness stages of the improvement cycle of the product development process [7].

c)   PMO-RACE (A Combination of PMO & RACE)

PMO-RACE is the combination of two models (PMO and RACE) which was developed by the researchers at University of Twente and Eindhoven University of Technology (Netherlands) in the mid 90's. Since the Process Model of Organisations (PMO) can support the identification of key problem areas and the definition of business drivers while the RACE-method is good at determining the performance level of the product development process and supporting the definition of improvement plans once the business drivers have been set, it was suggested that both methods could be combined to support improvement cycles. The combination would deliver 'the best of both worlds' [9].

d)   PRODEVO (A Swedish Model Based on RACE)

PRODEVO was developed at SISU (Swedish Institute for Systems Development) and this development was parallel to the development of PMO-RACE tool. Some of the dimensions and also a couple of the questions are assimilated in the presented tool from RACE model, and to indicate a relation the working name, "Extended RACE", was adopted earlier [10].

e)   CMM (Capability Maturity Model)

CMM was basically developed for software development and evaluation and was developed by the Software Engineering Institute at Carnegie Mellon University in order to manage the development of software for the US government, particularly that which was to be used by the Department of Defence in late 80's [11]. This model can be used as readiness assessment model and, in fact, the RACE model was developed based on ideas from CMM. The CMM model is also currently being used at the University of Sussex in developing benchmarks for process positions across various industries, including the construction industry.

f) SPICE (Standardised Process Improvement for Construction Enterprises)

This tool was developed at the University of Salford, United Kingdom, and is in the form of a questionnaire, which is designed to evaluate the key construction processes within a construction organisation [12]. SPICE is basically intended for evaluating the maturity of the processes of construction organisations and not for CE readiness assessment. It is based on CMM and is presently a research prototype. However, it could be used to assess the process-related aspects of CE implementation.

g) Project Management Process Maturity (PM)² Model

This 5-Level (PM)² Model was developed at University of California, Berkeley in late nineties. The primary purpose of the 5-Level (PM)² Model is to use as a reference point or a yardstick for an organisation applying PM practices and processes. This 5-Level (PM)² Model further suggest an organisation's application expertise and the organisation's use of technology, or it might produce recommendations on how to hire, motivate, and retain competent people. It can also provide and guide necessary processes and requirements for what is needed to achieve a higher PM Maturity level [13].

h) SIMPLOFI Positioning Tool

The tool was designed and developed by the Department of Manufacturing Engineering at Loughborough University. It formed part of the output of the SIMPLOFI (*Sim*ultaneous Engineering through *People, Organisation and Functional Integration*) project in the mid-nineties. The tool focuses on the introduction of one specific product in an organisation. This tool assists those people who are responsible for product introduction within an organisation in answering the question: "I know what product I want to introduce – How do I organise the introduction of this product to achieve this most effectively?" [14].

3.2   Framework for Comparison

The framework for comparison discusses the characteristics of the available tools and models under a number of generic criteria, which include:

- Aspects covered: which discusses the main issues addressed in each tool;

- The status of the tool: which shows the current standing of the model/tool in terms of whether it is a research prototype, commercial tool or currently under development etc.

- Survey method: which identifies how the data collection is carried out – that is either by questionnaires, interviews or both;

- Software availability: this identifies those tools and models which are accompanied by a software that can be used during the readiness assessment;

- Ease of use: an indication of the user-friendliness of the tools/models;

- The Usage of Tools for Readiness Assessment for CE: this identifies the tools and models which can be used for CE readiness assessment; and

- Applicability to the construction industry: since the basic purpose of this comparison is to identify the most suitable tool/model for the construction industry, this criterion assesses the potential use of the models and tools in the construction industry as a CE readiness assessment tool.

3.3   Findings

Table 1 gives an overview and comparison of available tools and models, which are being used to facilitate CE within an organisation. The comparison is based on the framework, which has already been discussed in the previous section.

From the comparative analysis (Table 1), it could be concluded that most of the tools and models discuss improvements in the product development process, and the use of technology to facilitate the development process. Some of the tools and models also cover the organisational environment to support the development process. The status of the tools and models shows that most of the tools and models are under development but some of them are being used on a commercial basis. With regard to software availability, there are only a few tools and models which are accompanied by their own software. Most of the tools and models are easy to use and user-friendly. Most of the tools and models reviewed were developed to assess the product development process within an organisation but can be used as a CE readiness assessment tool after appropriate modification. However, some of the tools and models were basically designed for CE readiness assessment. An assessment of the use of these tools and models within the construction industry shows that none of the tools and models is ideally suitable for use in construction as a readiness assessment tool for CE.

Table 1: Comparison of CE Readiness Assessment & Implementation Tools & Models

Tools/Models→ / Criteria↓	RACE	PMO	PMO-RACE	PRODEVO	CMM	SPICE	(PM)^2	SIMPLOFI
Aspects Covered	**Process** • Customer focus. • Product Assurance. • Leadership. • Team formation. • Strategy deployment. • Agility. • Teams within the Organisation • Process focus. • Management system. • Discipline. **Technology** • Project Architecture. • Application tools • Communication. • Co-ordination. • Information sharing. • Integration.	**Organisational Environment** • Task environment. • General environment. **Processes** • Primary processes. • Control processes: *Strategic level, Adaptive level, & Operational level* • Support processes.	Aspects covered are the same as PMO & RACE because this is the combination of both of these tools.	• Customer & user focus • Process focus • Team & project focus • Life-cycle perspective • Communication	**Process** • Pre-project Phase • Pre-construction Phase • Construction Phase • Post-construction Phase **Information Technology** • Simulation • Integration • Intelligence • Communications • Visualisation • IT support	• Brief Management. • Project Planning. • Project Tracking & Monitoring. • Contract Management. • Quality Assurance. • Project Change Management. • Risk Management. • Organisation Process Focus. • Organisation Process Definition. • Training Programme. • Inter-disciplinary Co-ordination. • Peer Review. • Technology Management.	• Planning to execute a project • Definition of project activities • Cost estimates for the project • Project Management (PM) process • PM-related data collection and analysis • Utilisation of PM tools and techniques • Working as a team • Senior management support	• The structure of teams • Control mechanisms (whether control mechanisms should reside with functions or projects) • The degree to which the process should be parallelised • How specialised people operating the process should be • The degree of automation in the tools used.
Status of Tool/Method	Commercial	Development Ongoing	Development Ongoing	Development Ongoing	Commercial	Research Prototype	Development Ongoing	Commercial

	Tool 1	Tool 2	Tool 3	Tool 4	Tool 5	Tool 6	Tool 7	Tool 8
Survey Method	Questionnaire & Interview.	Questionnaire & Interview.	Questionnaire & Interview.	Questionnaire.	Questionnaire & Interview.	Questionnaire.	Questionnaire.	Questionnaire.
Software Availability	Yes, also uses other software (e.g. SPSS).	Can use any modelling software.	Yes	None.	Yes, but also use other software e.g. SPSS.	None.	None.	Yes.
Ease of Use	Yes, but technological aspect is complicated to answer and is only for specialists.	Yes, but seemed to be incomplete, that's why merged with RACE later on.	Yes, and it seems to be completed after the combination of PMO & RACE.	Yes	Yes	Yes, MCQs are developed with additional space for comments.	Yes.	Yes, user-friendly software.
Can be used for Concurrent Engineering Readiness Assessment?	Yes, basically made for this purpose.	Basically used for analysing & designing organisations.	Yes, mainly for readiness assessment but also used for CE implementation process.	Basically developed for assessing concurrent engineering (CE) process.	Yes, but basically used for CE Implementation process.	Basically used for Process Improvement.	Basically used as a yardstick for an organisation applying PM practices and processes.	Basically used to assist those, who are responsible for product introduction within an organisation.
Appropriateness for use in Construction	Yes, but requires some modifications.	Yes, but basically used for analysing and designing an organisation, its process and technology.	Yes, but RACE model requires modification before applying to construction.	Yes, but it requires changes to address construction specifically.	Yes, but basically developed for software industry, therefore it requires changes before applying to construction.	Yes, but this tool is basically made for process improvement within construction projects.	Yes, but this tool is basically developed to determine and to position an organisation's relative PM level with other organisations.	Yes, but this tool focuses on the introduction of one specific product in an organisation. Therefore, in any construction organisation, it can be used for a specific project and it would give the position of the project and not the position of the organisation.

4   Which Assessment Model?

After analysing the comparison matrix (see Table 1), it could be said that RACE would be the best to use as the Readiness Assessment Tool for Concurrent Engineering in the construction industry because of the following reasons:

- Aspects covered in RACE model such as customer focus, team formation, management systems, communication & integration systems, etc., can be used readily for CE readiness assessment in the construction industry, after some modification, due to the similar structure and requirements of the construction industry;
- Commercial usage of RACE model makes it more reliable;
- RACE model questionnaire addresses and assesses the critical business drivers in the construction industry; and
- Since RACE is basically a CE readiness assessment model, therefore, it is more appropriate than other tools and models, which were developed to assess the project/product development process within an organisation.

However, it requires adaptation and modification for this purpose because, basically, this tool was developed for readiness assessment for concurrent engineering in other industries such as software engineering. Thus, it needs to be tailored to the requirements of the construction industry and the people working within the industry.

The following are some of the reasons which indicate that RACE in its current form is not suitable for the construction domain and requires modification for use in assessing the construction industry:

- RACE is basically designed for assessing the readiness of other industries such as software, automotive, manufacturing, and electronic industries, all of which have different characteristics to construction;
- Aspects covered focus on the processes in the above mentioned industries and require changes to assess the construction process;
- The structure of teams within above mentioned industries are different from typical construction project teams;
- The level of technology usage in the afore-mentioned industries is different from that in the construction industry;

- The products of the other industry sectors satisfy a large number of customers whereas a construction project is one-off in nature, typically fulfilling the needs of a particular client or organisation;
- Level of integration, communication, co-ordination, and information sharing are different for construction and above mentioned industries; and
- Managing a manufacturing product and a construction project require different levels of management skills.

It is evident from the above that there is the need for a model, which is basically designed for the construction industry and can be used to assess the readiness of the industry for the adoption of CE.

5   Further Work

Further work towards the implementation of CE in the construction industry is focusing on the following:

- Development of the CE Readiness Assessment Model for Construction Industry;
- Development of a CE Readiness Assessment Software;
- Detailed survey and assessment of key sectors of the construction supply chain using the prototype software; and
- Formulation of CE implementation strategies for the construction industry.

6   Summary & Conclusions

This paper has presented a comparative review of Concurrent Engineering (CE) readiness assessment tools and models. In the early part of the paper, the general principles of CE were introduced and the move towards its adoption in construction discussed. A comparative review of existing tools and models for CE readiness assessment was then presented which includes an overview of the tools and models, comparison framework, comparison table, findings from the table and conclusions drawn from the comparison. The limitations of the RACE model, in its use for assessing the construction industry, were then highlighted. The need for a CE readiness assessment model for the construction supply chain was also stated and the further work being done in this regard outlined.

The following conclusions can be drawn from the discussion presented in this paper:

- CE is an new approach, which can make project/product teams less fragmented, improve project/product quality, reduce project/product time-scales, and reduce total project/product cost;
- It is necessary to carry out CE readiness assessment of the construction industry before CE implementation so as to ensure that maximum benefit is achieved;

- There is also a need to develop an appropriate CE readiness assessment tool or model for the construction industry as existing models are not appropriate in their present form.

References

[1] Winner, R. I.; Pennell, J.P.; Bertrend, H.E. & Slusarczuk, M. M. G., 1988, The Role of Concurrent Engineering in Weapons System Acquisition, IDA Report R-338, Institute for Defence Analyses, Alexandria, VA.

[2] Chen, G., 1996, The Organisational Management Framework for Implementation of Concurrent Engineering In the Chinese Context, Advances in Concurrent Engineering, Proceedings of 3rd ISPE International Conference on Concurrent Engineering: Research & Applications, University of Toronto, Ontario, Canada, 26-28 August 1996, pp.165-171.

[3] Componation P. J. & Byrd Jr., J., 1996, A Readiness Assessment Methodology for Implementing Concurrent Engineering, Advances in Concurrent Engineering, Proceedings of 3rd ISPE International Conference on Concurrent Engineering: Research & Applications, University of Toronto, Ontario, Canada, 26-28 August 1996, pp.150-156.

[4] Evbuomwan, N. F. O. & Anumba, C. J., 1998, An Integrated Framework for Concurrent Life-cycle Design and Construction, Advances in Engineering Software, 1998, Vol. 5, No. 7-9, pp.587-597.

[5] Prasad, B., 1997, Seven Enabling Principles of Concurrency and Simultaneity in Concurrent Engineering, Concurrent Engineering in Construction, Anumba, C. J. & Evbuomwan N. F. O. (Eds.), Proceedings of 1st International Conference organised by The Institution of Structural Engineers Informal Study Group on Computing in Structural Engineering, London, 3 & 4 July 1997, pp.1-12.

[6] Love, P. E. D. & Gunasekaran, A., 1997, Concurrent Engineering in the Construction Industry, Concurrent Engineering: Research & Applications, Vol.5, No.2, June 1997, pp.155-162.

[7] Wognum P. M.; Stoeten, B. J. B., Kerkhof M & de Graaf, R., 1996, PMO-RACE: A Combined Method for Assessing Organisations for CE, Advances in Concurrent Engineering, Proceedings of 3rd ISPE International Conference on Concurrent Engineering: Research & Applications, University of Toronto, Ontario, Canada, 26-28 August 1996, pp.113-120.

[8] CERC Report, 1993, Final Report on Readiness Assessment for Concurrent Engineering for DICE, Submitted by: CE Research Centre, West Virginia University, June 1993.

[9] de Graaf, R. & Sol, E. J., 1994, Assessing Europe's Readiness for Concurrent Engineering, Proceedings of Conference on Concurrent Engineering: Research & Application, 1994, pp.77-82.

[10] Bergman, L. & Ohlund, S., 1995, Development of an Assessment Tool to Assist in the Implementation of Concurrent Engineering, Proceedings of Conference on Concurrent Engineering: A Global Perspective, 1995, pp.499-510.

[11] Aouad, G.; Cooper, R.; Kagioglou, M.; Hinks, J. & Sexton, M., 1998, A Synchronised Process/IT Model to support the co-maturation of processes and it in the Construction sector, Time Research Institute, University of Salford.

[12] SPICE Questionnaire, 1998, Key Construction Process Questionnaire, Ver. 1.0, Salford University, July 1998.

[13] $(PM)^2$ Model web site: http://www.ce.berkeley.edu/~yhkwak/pmmaturity.html

[14] Brookes, N. J. ; Backhouse, C. J. & Burns, N. D. (2000), Improving Product Introduction Through Appropriate Organisation: The Development of the SIMPLOFI Positioning Tool, In press.

Issues in Management of Distributed Concurrent Engineering Design in Object-Oriented Databases*

A. Al-Khudair[1], W. A. Gray[1], J. C. Miles[2]

[1]Department of Computer Science, University of Wales-Cardiff, UK

[2]Cardiff School of Engineering, University of Wales-Cardiff, UK

{A.I.Khudair | W.A.Gray}@cs.cf.ac.uk , MilesJC@cf.ac.uk

Abstract

This paper will discuss a novel distributed support environment for Concurrent Engineering (CE). We will identify the issues to be addressed and look in detail at how to support the versions of design artefacts and their configuration management in an Object-Oriented Database (OODB) system. We aim to support heterogeneity not only for the design tools but also for different platforms of distributed OODB system, rather than changing the preferred design tools of individual designer.

1 Introduction

Concurrent Engineering (CE) has received considerable attention in recent years due to its efficiency in time and cost reduction in the product development phases (i.e. Requirement Definition, Conceptual Design, Detailed Design, Development, Manufacturing, Marketing, and Support). CE achieved major improvements by conducting these phases *concurrently* rather than *sequentially*. This leads to a consideration, during the upstream phases, of all the elements (or phases) of the product development from conceptual design to marketing [5,6] .

During the CE process, many changes are made to the design artefact and a large amount of data and other information pass between the participants. The management of both simultaneous tasks as well as keeping the history of design changes (versions) is still a difficult problem which may not be solved using CAD systems alone [10]. This is because they lack the support of a powerful management system that can integrate and keeps track of all the phases and states of a large and complex design artefact [15,16]. Hence, a management tool

(i.e. database system) is essential for keeping track of the evolution and change of design artefact as well as improving the communication and cooperation in a distributed CE design environment.

Using database systems to support engineering environments is an active research area among the database and engineering communities [1,10,15,16]. The notion of *engineering database systems* was introduced to capture the requirements of engineering applications. Most of the proposals and systems focus on the database facilities which enable the management of the engineering design process [5,7,10,11,15,16]. CAD systems are used to manipulate the geometrical aspects of the design object whereas the database system is used as a *kernel* to keep track of the overall hierarchy of design objects (or structure), design evolution (or versioning), and the relationships between design objects. Hence, CAD systems should be built on top of a database system.

Conventional database systems, supporting record-based applications, are considered unable to satisfy the requirements of engineering applications. For this reason, advanced database systems that can handle such requirements are needed. OODB systems are considered capable of satisfying the requirements of CAD systems and other engineering domains since they possess rich modelling and manipulation features [1,8,9,10,12,14].

This paper discusses the main issues to be considered in the management of a distributed CE design environment. The discussion shows also the support of these issues in the commercial OODB systems based on the evaluation conducted in [3]. The remainder of this paper is organized as follows. In Section 2, the basic object-oriented model is reviewed. In Section 3, the basic concepts of CE are reviewed. In Section 4, the issues related to the management of distributed CE design in OODB system are discussed. In Section 5, performance considerations are addressed. In Section 6, concluding remarks are given.

---

* This work is part of the DESCRIBE project and is partially supported by EPSRC.

2 Object-Oriented Databases Overview

A data model is a logical organization of the real-world objects (entities), constraints on them, and relationships among them. A data model that captures object-oriented concepts is an *object-oriented data model* [9]. An *object-oriented database* is a collection of objects whose behavior, state, and relationships are defined in accordance with an object-oriented data model. An *object-oriented database system* is a database system which allows the definition and manipulation of an object-oriented database.

In the object-oriented paradigm, all conceptual entities are modelled as objects, and an object is defined by two parameters *structure* and *state* . The structure of an object provides the *structural* and *behavioral* capabilities of that object, which is defined by a set of instance variables (or attributes), methods, and/or rules (or integrity constraints). The state of an object assigns data values to the instance variables of the object and the methods that operate on them. Every object has its own *unique* identifier, called object identifier (OID), that distinguishes it from other objects. (In this paper, we will use the terms *instances, objects,* and *design objects* interchangeably). The amount of information may be unmanageably large if every object carries its own structure (i.e. attribute names and methods). Therefore, a set of objects sharing the same structure are grouped together in a *class* (or *class definition*) [14]. A *database schema* is a set of class definitions connected by the superclass/subclass relationships that is called *class-hierarchy*. It is represented by a *class lattice* (or a Directed Acyclic Graph, DAG).

In the class-hierarchy, a superclass is a *generalization* of its subclasses, while subclasses are a *specialization* of their superclass. Any class in the subclasses list of a superclass *inherits* all the attributes and methods of the superclass and may have additional attributes and methods which express its own need. An instance of a subclass is also a logical instance of all its superclasses. Messages are sent to an object to get the values of the attributes and the methods encapsulated in it. There is no way to access an object except through the public interface specified for it.

3 Concurrent Engineering Design Overview

In general, the concurrent design of an artefact passes through two stages before the final state of a design can be accomplished. These stages are:
• **Decomposition:** In this stage, the design artefact is *decomposed* into autonomous subparts *recursively* until reaching a state of primitive design units.
• **Aggregation:** In this stage, the design subparts are

reconstructed recursively to form a higher level subpart until reaching a complete design artefact.

Figure 1 shows these stages where a design artefact is decomposed into three parts: A, B, and C designed concurrently.

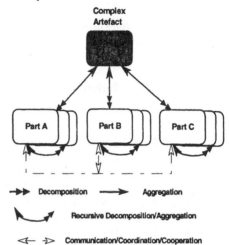

Figure 1. Concurrent engineering design process

Boundaries between the parts (or components) of an artefact (or a product) should be taken into consideration during the design of each part in order to define a clear interface between them. Without careful interface definition, inconsistencies may occur [6]. (In this paper, we will use the terms *part* and *component* interchangeably).

A CE design environment can be spread over geographically distributed sites. OODB systems should be built in using distributed architecture to support such requirement. Furthermore, this distribution should be transparent to the users.

4 CE in Object-Oriented Databases

In this section, we discuss the architecture of a typical distributed CE design environment. Then, the main issues related to the support of this environment in an OODB system are reviewed. These issues are discussed in terms of four phases of support : mapping, semantic extensions, collaboration, and intelligence. In the mapping phase, the objective is to model, in the OODB system, the fundamental activities which are normally carried out during the design process. The semantic extensions phase incorporates advanced issues , such as version and configuration management, into the OODB system. The collaboration phase models the normal cooperation and communication between multidisciplinary teams. The intelligence phase attempts to gather knowledge, which may be used during the product life cycle, that can assist the database system throughout the design stages. Note that these phases need not be applied in a strict serial order. They are organized in terms of their *complexity*. Therefore, an aspect of a successive phase may be incorporated even if the current phase is not complete.

The corresponding support of each phase in the commercial OODB systems is shown particularly in O_2, Objectivity, and VERSANT since they are available in our machines [24,25 ,26].

4.1 Architecture of the Distributed CE Design Environment

An OODB system supporting a distributed CE design environment consists of *local* and *global* servers. The local servers are used for distributed CAD clients whereas the global server represents a *common* view of the engineering data. These servers can run on heterogeneous platforms [3]. Therefore, heterogeneity is supported not only for the design tools (i.e. CAD systems) but also for different platforms of distributed OODB system. This setup is necessary to avoid changing the preferred design tools of individual designer. It should be noted that this distribution is *transparent* to CAD clients.   . The advantage of the distributed architecture is to overcome the contention for CPU and I/O services as well as reducing the remote access delay. These cause performance degradation in centralized systems [17].

Figure 2 shows the basic architecture of the distributed CE design environment. In the figure, the following components are identified: CAD systems, Mapping Protocol, Local OODB Servers, Local Storages, Global OODB Server, and Global Storage. The figure also demonstrates the distributed *client/server* architecture. The following is a discussion of each component.

●**CAD Systems:** The geometrical and topological aspects of the design objects are manipulated using these systems. In CE environment, heterogeneous CAD systems may be used under different hardware and software platforms. Furthermore, a high degree of distribution is adopted. A change to these systems is normally undesirable. Instead, the management tool (i.e. OODB system) should be able to communicate with such diverse platforms.

●**Mapping Protocol:** The aim of this protocol is to *map* the design objects that are founded in CAD systems into the OODB server (i.e. defining an interface between them). The protocol includes the necessary mechanisms to exchange the engineering data as well as maintaining the consistency between CAD systems and the OODB server. Engineering data exchange standards are used within this protocol to define a *uniform* interface to the OODB server. STEP standard represents a desirable means of engineering data exchange. Other exchange format and

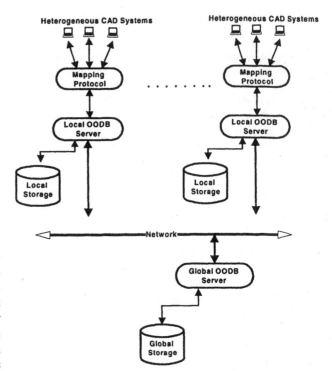

Figure 2. The architecture of the distributed CE design environment

standards such IGES and DXF can also be used [19]. A programming language interface is used to communicate with the OODB server. C++ or Java binding can be used to accomplish this task. The description of the mapping mechanisms is beyond the scope of this paper.

●**Local OODB Servers:** These distributed servers are used to store and manipulate design objects taking full advantage of OODB system features. The *persistence* of design objects is supported via these servers. The engineering data which resides in a particular local server can be accessed from local servers elsewhere through an external interface.

●**Local Storage :** This distributed storage is used to store the engineering data in the physical medium (i.e. the disk). It is attached to each OODB server which coordinates the access to it.

●**Global OODB Server:** This server is used to store and manipulate the engineering data that is common to distributed CAD systems. Global integrity constraints as well as *central locking* are also maintained in this server [18,24]. Design objects are migrated from the local OODB server to the global OODB server if there is a need to share them with other CAD systems. Note that the global OODB server does not need a direct interface with CAD systems. Its data can be accessed via local OODB servers.

●**Global Storage :** This storage is used to store global engineering data in the physical medium (i.e. the disk). It is attached to the global OODB server which coordinates access to it.

This architecture *integrates* distributed heterogeneous CAD systems each of which may contain a set of components of the complex artefact, thus facilitating a *unified* view of a single complex artefact.

4.2 Mapping Phase

Design objects are originally founded in a CAD system and then mapped, through the mapping protocol, into the database system. Once the complex artefact is mapped, it will take full advantage of OODB features. The following discussion assumes that the design objects are already mapped into the OODB system.

4.2.1 Dynamic Schema Evolution

The hierarchy of the design objects, of a complex artefact, is represented using an OODB schema. This schema sets up the core architecture of the design artefact (i.e. mapping the real world design into the database). Typically, the schema consists of a set of classes connected by superclass/subclass relationships. Once the design hierarchy is defined, the design schema may be subject to change by the designers. *Dynamic schema evolution* permits this change to take place *on-line* without the need to bring the system down. Most of the commercial OODB products support this facility [3].

4.2.2 Propagation of Schema Changes on Design Instances

Perhaps the most important issue in the mapping phase is how to propagate schema changes to its instances. The modification of the existing schema may be unknown to the instances that were created before incorporating the change to it. Thus, there can be a problem when a CAD system accesses the new schema version while the instances are still under the old version definition. A conversion mechanism is, therefore, required to bring the instances into the new state of the schema (i.e. change propagation). There are, however, many approaches that can be used to manage propagation of changes to instances. Some are described as follows [2,14,20]:

■ **Immediate update (or conversion):** Instances are immediately affected by the change in the class. All instances of the class are changed in order to conform with a new definition. All instances of the class are unavailable for use by others for the duration of the conversion operation. Although this approach is efficient once the conversion has been made, the disadvantage is that the performance of the system is affected by frequent

and immediate updates. This approach is adopted in GemStone OODB system [3].

■ **On-use (or Lazy) update (or incremental conversion):** Updates to instances are deferred until the instance is used. The disadvantage of this approach is that it requires a permanent propagation mechanism throughout the system's lifetime. This is the approach used in ITASCA and VERSANT OODB systems [3].

■ **Screening:** Instead of converting instances, a filter may be placed between the accessing routines and the instance versions that are accessed. The filter acts just as a database view in that it gives the user of an object the "illusion" that the object is of the class version that the user expects.

■ **Write-once class:** Disallow class modification when an instance of a class is created. Instead, a new version of the class must be defined to incorporate the change and the old instances are copied to it.

■ **Schema mapping:** Defer updates to instances indefinitely, or until a reorganization is requested explicitly, maintaining a mapping between the current representation of classes and all previous versions.

O_2 and Objectivity OODB systems support both immediate and on-use updates. The user can choose the approach that matches his needs [3].

4.2.3 Computational Support in OODB

Traditional database applications include simple mathematical computation. In contrast, engineering applications require a complex mathematical computation. For example, a change to a part of a design may result in the need for complex mathematical computations to find the new values of the properties of this part. In order to match this requirement, the database system should support a *computationally complete* programming language [1]. Most of the commercial OODB products support such language [3]. Alternatively, complex computations can be done by an external tool, such as a structural analysis package, and then wrapped within the design object [8].

4.3 Semantic Extensions Phase

The basic modelling facilities of the mapping phase may not be sufficient to capture the design process. Semantic extensions to the basic model are essential in order to enable the database system to handle the activities normally conducted during the design process. These extensions are versions and composite objects which introduce VERSION-OF and PART-OF relationships [9]. Versioning is the process of keeping multiple versions of the same object (i.e. maintaining the history of a design changes). A composite object is an object which relates to other objects by the

PART-OF relationship. This type of object is used to link the components of a complex artefact (i.e. configuration).

4.3.1 Versioning in Object-Oriented Databases

Versions have been used in the literature for various purposes such as concurrency control, fault tolerance (or recovery), performance enhancement in distributed systems, and broadcasting [15]. In the CE environment, versions can be used, in addition to the above purposes, to keep track of the design evolution which enables design *reuse*. Therefore, these versions represent a record of the changes that are added to the design object throughout its life time. This will add a *temporal dimension* to the CE environment. Furthermore, a *rollback* to a previous stable state of the design can be made whenever the new changes introduce problems. Keeping a record of design changes also represents a valuable documentation and auditing aid.

Changes can be managed in object-oriented databases to both parameters, structure and state of an object . In the existing OODB systems, changes to the state of an object are maintained via *version management* [2,7,15]. Also, structural changes are supported in most object-oriented database systems. Such changes to a class are referred to as *schema evolution* in the literature [14,20]. The notion of version management can also be extended to these structural changes. Thus, history management in object-oriented databases can be divided into the following aspects [2,20]:

■ **Object Versioning**: The concern here is to maintain the history of a *state* of an object (or instance), that is, to maintain the history of *data values* of an object, i.e., the history of the object structure is not maintained.

■ **Class Versioning**: The concern here is to maintain the history of *class definition*, that is, to ensure that the class definition, which existed before the change, is retained.

■ **Schema Versioning**: The concern here is to maintain the history of whole schema (entire class-lattice) rather than individual classes.

Although class and schema versioning are an interesting field of research, the scope of this paper is limited to object versioning. A discussion of these advanced issues can be found in [2,9].

Object versioning is the creation of a new version of the object. That is, the object state before the change is retained as well as the new state, allowing multiple versions of object states to co-exist. In general, the mechanism for object updates in a system with version management support can be shown as follows:

$$V_i \xrightarrow{\alpha} V_{i+1} \xrightarrow{\beta} V_{i+2}$$

Where V is object version, i is version number, and α and β are changes to the object. Note that α and β are changes to a current version of object data (or object state).

A versioned object consists of a hierarchy of versions called a *version-derivation graph* connected by the parent/child relationships. If a new version of an object introduces cycles in the version-derivation graph, then the version is rejected. New versions of an object need not be generated in strict linear sequence. It is possible that two or more versions of an object are derived from the same parent version, i.e., *alternative versions* which can be used to experiment with different alternatives of a design artefact. It is also possible to have an object version derived from two or more existing versions of the object, i.e., *version merging* which can be used to produce a version of a design artefact that integrates different features scattered in the parent versions which are possibly developed by different designers. This is shown in Figure 3 where versions v3 and v4 are derived from version v2, whereas version v5 is derived from versions v2, v3, and v4. Every version of an object has its own *unique* identifier that distinguishes it from other versions in the scope of the corresponding object.

Object versioning is supported in many commercial OODB systems such as GemStone, ITASCA, O_2, Objectivity, and VERSANT [3]. The rest of this section is a review of the main issues related to versioning.

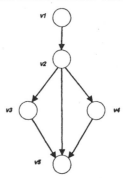

Figure 3. A graph of versions derivation

• **Creating A New Object Version:** We differentiate between the creation of an object itself and the creation of a version of this object. The creation of an object is the process of establishing new knowledge in the database and adding a new object as an instance of a specific class. The creation of a version of an object is the process of *augmenting* an existing object (or an object version) and hence creating a new version. Two types of relationships are introduced, the VERSION-OF relationship and the DERIVED-FROM relationship [9]. The former is used to represent the relationship between an object and its versions. The latter is used to represent the relationship between two successive versions of an object (i.e. the old version and the new version derived from it). When a change is necessary to a design artefact, a new version of the design is created to hold the required changes while preserving the old versions of the design, thus allowing the maintenance of a complete history of the design changes. Figure 4 shows the creation of a new object version from an existing version "Car". Both versions continue to exist and they can be accessed at any time. The new version is included in the version-derivation graph and connected to its parent(s) through the DERIVED-FROM relationship. If a new version of an object introduces cycles in the version-derivation graph, then the version is rejected.

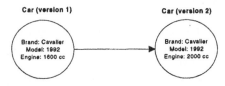

Figure 4. The Creation of a new object version

There are two types of bindings (or references) to an object version: *static* and *dynamic*. In *static binding*, the reference is identified for a specific object version. In *dynamic binding*, the object name is specified without a version number. In this case, the default version number is selected. The *default version* number is the latest version created or it can be specified by the user. Clearly, dynamic binding is useful when no object version number is known in advance.

• **Deleting Object Version:** When it is necessary to delete an object from a version-derivation graph, the following steps are suggested: (i) Delete the relationship to the object (VERSION-OF relationship), (ii) Delete all the relationships to other versions if any (DERIVED-FROM relationship), (iii) Delete the version from the version-derivation graph. A version may be referenced by other versions and/or objects. Therefore, the deletion of a version may cause a *dangling reference*. Moreover, the version-derivation graph may be disconnected by a deletion of a version. This can be reconciled by connecting each version derived from the deleted version to the immediate predecessor version of the deleted version. This may introduce a confusion if the deleted version was driven from two or more versions (i.e. version merging) and from the deleted version, two or more other versions were driven. The problem is, to which parent of the deleted version the disconnected versions should be linked. Therefore, some OODB systems do not remove the deleted version from the version-derivation graph. Instead, *logical* deletion is done. This approach is adopted in VERSANT OODB system [3]. Some restrictions may be imposed on the deletion of an object version like preventing deletion when the version has a child version that is directly derived from it.

• **Storage Issues:** In a large and complex artefact which has different versions at different levels of the design (i.e. configurations), it is essential to take into account the storage requirements of the design. A design normally consists of a complex data (i.e. text, image, drawings etc.) which may consume a large amount of storage space. One solution is to store only the difference between design versions which is called the *delta*. Some proposals, such as [15], do not consider storage issues, rather, versions are stored completely. Some OODB systems, such as VERSANT, do not use version deltas, rather a complete version is stored no matter how similar it may be to its predecessors [26]. Although keeping only the delta of design versions can save a considerable amount of storage space, the *cost* of *reconstructing* a design version from a delta state to a complete design state may incur system overhead particularly in a frequent version-retrieval environment.

Clustering is another important issue which allows the grouping of design versions in physical storage (i.e. the disk) according to identical properties between them. Hence a fast access to these versions can be obtained [17]. This is called *physical clustering*. Logical clustering, on the other hand, is to group versions of similar properties "logically" regardless of their physical storage location [15]. Physical clustering is supported by GemStone, ITASCA, O_2, Objectivity, and VERSANT OODB systems [3].

4.3.2 Configuration Management

Complex design artefact is normally decomposed into subcomponents to facilitate the design activities. Each subcomponent may, in turn, be subject to further decomposition recursively. To map this requirement into the object-oriented model, a composite object may be used. A composite object is an object which is composed of a set of other objects rather than primitive domains. A PART-OF relationship is defined to link a composite object with its components. A *configuration* is the composition or grouping of objects into a higher level object. There are semantic differences between a configuration and a composite object. However, in our discussion a configuration is considered as a large scale composite object.

Configurations work in a manner similar to that of the normal object except that they have higher level abstraction. The issues discussed within normal object versioning can be extended to configurations. The notion of configuration in some commercial OODB systems, such as O_2, refers normally to a collection (or set) of versions belonging to the same object which is different from our discussion here. However, the notion of a composite object and/or introducing new classes can be used in these systems to model configurations [3,8]. The rest of this section discusses the main issues related to configuration management.

• **Creating Configurations:** In order to link the components of a complex artefact as a cohesive unit, a configuration structure must be used. A PART-OF relationship is used to connect a configuration to its components. Each configuration is assigned an identifier called a *configuration identifier* CID which uniquely distinguishes it. An example is a car configuration which can be used to represent the components (or attributes) of a car such as Engine, Body, and Weight. The Weight attribute has a primitive domain which is Number class whereas the Engine and Body attributes have non-primitive domains which are objects.

• **Deleting Configurations:** The deletion of a configuration may incur the deletion of all its components (or objects) whose existence depend on it whether they are shared or not. If a component is independent (i.e. its existence does not depend on the configuration existence), then it is not deleted because it may be referenced by other configurations and/or objects. Note that the deletion of a configuration becomes more complicated when versioning of a configuration of a design is adopted. Further, a deletion of a particular component of a configuration by other users may lead to a *dangling reference*. Hence, the deletion of a component should be avoided if a configuration still references it. To assist this feature, each component of a configuration may contain a *reverse reference* showing to which configuration it belongs. It is preferable in a design database to keep deletions to a minimum. Versioning, which is discussed next, may be considered as a desirable alternative to deletion by introducing a new version of the configuration which does not include the deleted component while preserving the references to the "deleted component" in the previous version. This is called *virtual deletion*.

• **Configuration Versioning:** *Configuration versioning* is the creation of a new version of the configuration. That is, the configuration state before the change is retained as well as the new state, allowing multiple versions of a configuration state to co-exist. The evolution of a configuration is represented as a hierarchy called a *configuration-derivation graph*. If a new version introduces cycles in the configuration-derivation graph, then the version is rejected. New versions of a configuration may not be generated in strict linear sequence. It is possible that two versions of a configuration are derived from the same old version, i.e., *alternative configurations*. It is also possible to have a configuration version derived from two existing versions of the configuration, i.e., *configuration merging*. Every version of a configuration has its own unique identifier

that distinguishes it from other versions in the scope of the corresponding configuration.

The concepts discussed for object versioning in Section 4.3.1 may be extended to versioning of a configuration. Of course, keeping track of different versions of a configuration is more sophisticated than the versioning of a simple object. Therefore, we propose three types of configuration management:

1 Simple Configuration: The configuration and its components are not versioned.

2 Partially-Versioned Configuration: Either the configuration is not versioned while its components are versioned or the configuration is versioned while its components are not versioned.

3 Totally-Versioned Configuration: The configuration itself is versioned and its components are also versioned.

The totally-versioned configuration facilitates the modelling of the activities in a real world design such as experimenting with alternative configurations using different combinations of components versions to reach a configuration that best matches predefined design requirements. This manipulation is hard to achieve without the explicit support of versioning of configurations because old configurations may be overwritten by the new one. Moreover, without careful management of the evolution of a configuration and its components (i.e. totally-versioned configuration), major inconsistencies may arise between the components of a design artefact. Therefore, it is essential to identify which component (or object) version is compatible with which configuration version [13]. The binding of a configuration to its components can be *static* or *dynamic*.

• **Temporal Integrity:** One of the important issues in the management of multiversion design configurations is to preserve the integrity in each state of the design evolution. This is called *temporal integrity*. This type of integrity may be violated if a change affecting a past state of a design, which was preserving integrity at that time, is introduced to match a current need. It is essential to restrict the design changes which result in such violation. More discussion about this issue may be found in [13].

4.4 Collaboration Phase

CE involves different multidisciplinary teams working on different aspects of the design. The interaction between designers is crucial to coordinate design efforts, enhance product quality, and exchange ideas. One of the drawbacks during the product development is the lack of cooperation [23]. A better cooperation can be achieved using a well defined communication system, workspace management, change management, and cooperative transaction management which are adopted in a distributed environment. These issues are addressed here.

4.4.1 Communication System

A database system supporting CE should have a communication platform to facilitate the interaction between designers. This communication may be conducted apart from the database system in the form of meetings, individual discussions, telephone calls, etc. It can be also utilized and incorporated as part of the database system which can help to rationalize the design changes.

We propose three modes of communication in a collaborative CE design environment where different multidisciplinary teams are working together (we use the term participant here instead of designer to emphasise the multidisciplinary nature of CE):

- **Intra-disciplinary Communication**: Where participants from a particular discipline (e.g. electrical engineers) are interacting with each other in terms of the *same discipline perspective*.

- **Inter-disciplinary Communication** : Where participants from a particular discipline (e.g. electrical engineers) are interacting with other participants who are from other discipline (e.g. mechanical engineers) in terms of the *different disciplines perspective*. That is, each participant is viewing the design from the point of view of his discipline.

- **Global Communication** : Where all multidisciplinary participants (i.e. from design, manufacturing, marketing, etc.) are interacting together in terms of *each discipline perspective*. That is, each participant is viewing the design from the point of view of his discipline.

Note that these communications may happen with the same component, different components, or the whole configuration. Furthermore, they can be held in synchronous or asynchronous mode. The database system should be able to consider these communications either explicitly or through an external tool. Further discussion about communication in a cooperative engineering design can be found in [6,10].

4.4.2 Workspace Management

A workspace is part of the database which is used to store and manipulate designs at different levels of authorization. A workspace can be *private* or *shared*. A private workspace is owned by specific designers who are utilizing it for the development of a design. Only the owners of a private workspace can read, modify, or delete its contents. A shared workspace is common among different designers who can read or deposit their designs in it (i.e. not common to other designers of the same group). A shared workspace can be shared among all designers or a subset of them. This introduces *public* and

project workspaces. A public workspace is open to all designers to read or deposit designs in it. However, there should be a verification mechanism which allows only mature (or complete) designs to be deposited in it [16]. A project workspace represents an interaction area between the designers of a specific project. Thus, the database can be seen as a hierarchy of public, project, and private workspaces. There are different authorization levels to access each type of workspace. A private workspace is only accessible to its owners and access is granted by the designer who creates it. A project workspace is only accessible to the designers in the same project and access is granted by a project manager. A public workspace is accessible to all designers and access is granted by a database administrator. We adopt in this discussion the model proposed in [14]. A three-layered workspace hierarchy is adopted in many other proposals with different names (or terminology) such as [16] where these layers are called *private, group,* and *archive*. Some commercial OODB products support private and shared workspaces such as ITASCA and VERSANT [3].

The concept of dividing the database into private, project, and public workspaces has introduced new types of versions which are not discussed in Section 4.3.1. These types are sensitive to their location within the three-layered hierarchy. They are *transient, working,* and *released* versions [14]. A transient version resides in a private workspace where it can be updated or deleted by the designer who creates it. A working version resides in a private or project workspace where it is considered stable and cannot be updated but it can be deleted by its owner. A released version resides in a public workspace where it is neither updated nor deleted. These types of versions are supported in some commercial OODB systems such as ITASCA and VERSANT [3]. *Checkin* and *checkout* operations are introduced to deposit and retrieve design objects to/from workspaces. In the engineering context, the checkout operation incurs a *long-duration transaction*.

4.4.3 Change Management

A design may be subject to numercus changes before reaching a stable state. The management of these changes needs an overall mechanism which can trap the concerned changes and then notify the affected designers in order to undertake subsequent actions. Without a powerful notification mechanism, major inconsistencies between design objects may occur. There are two aspects of change management. These are *change notification* and *change propagation* [14,16]. In the former, the designers who may be affected by the change are notified. This notification can be *immediate* (by sending a message to the affected users) or *deferred* (by updating flags in the design object itself, and designers are aware of the changes only when they access the design object). In the latter, a deeper action takes place to incorporate the change in the structure of a design. Katz [16] adopts a technique of change propagation which incorporates new versions into configurations. Change management is supported in most of the commercial OODB systems such as ITASCA, O$_2$, Objectivity, and VERSANT [3].

4.4.4 Long and Cooperative Transactions Management

Long-duration manipulation of design objects as well as cooperation between multiple designers are typically inherent in most engineering design environments. For this reason, an advanced transaction model supporting long and cooperative transactions is needed. A long duration transaction is a transaction which may take minutes, hours, days, or even months to complete. It is initiated by a checkout operation into a private database. The *atomicity* property of the traditional transaction model implies that this type of transaction can be either performed completely or not at all. Hence, if a long transaction cannot be completed, then all the work that has been done, possibly in months of time, must be re-initiated which may be very expensive. This problem can be alleviated by considering a long transaction as a set of short-duration transactions each of which preserves the *atomicity* property [9]. Hence, recovery can be made up to the point where the latest short transaction was committed. Another problem associated with long transactions is the long-duration lock. A long transaction locks the design object when it checks it out in the update mode. Other transactions trying to access the same design object in a conflicting mode must be blocked until the long transaction holding the lock completes. This problem can be solved using various techniques such as versioning of objects, group transaction, soft locks, and change notification [9,22]. Long transaction is supported in most of the commercial OODB systems available today such as Objectivity and VERSANT [25,26].

Cooperative transactions allow several users to participate in the same transaction. Moreover, two or more transactions can interact and share results with each other while they are in progress [21,22]. The isolation property of the traditional transaction model restricts the interaction between transactions. Therefore, this property must be relaxed in order to support cooperative transactions [22]. This may sacrifice the consistency of the database. However, consistency issues may be resolved in a level higher than the database level (i.e. human interaction). To the best of our knowledge, the only commercial OODB product that has the means to support cooperative transactions is ITASCA [3]. More detailed discussion of cooperative transaction can be found in [21,22].

4.5 Intelligence Phase

This phase is considered the most advanced phase. It includes the incorporation of Artificial Intelligence (AI) approaches to make the system supporting CE able to interact with the design process. Incremental knowledge acquired during product development may be utilized in the design phase. Expert systems or more advanced AI approaches such as first–order predicate logic and object-oriented programming, constraints, and constraint networks can be used to represent CE domain knowledge [7]. AI issues within the design process have received a lot of attention in the literature. We address here only constraint management. More details of incorporating AI approaches in the CE design process can be found in [7].

4.5.1 Constraint Management

Constraints which may affect other parts of the design may be added, deleted, or modified during the design process. Further, conforming to a specific constraint may lead to a conflict with other constraints. Design constraints can be modelled in terms of three modes. The first is by using constraint support that is available in engineering standards such as STEP part 47 (Shape Variational Tolerances) [28]. The second is by encapsulating constraints inside the design object itself as methods that are triggered whenever a change happens to the design. This checks whether the new state of the design matches the constraint or not. The third is by using a more advanced AI systems such as constraint networks systems[7]. A combination of these modes can also be adopted.

Immediate constraint checking may incur system overhead particularly in a change-intensive environment. Therefore, *deferred* constraint checking may be used to utilize the system's idle times. However, deferred checking may leave the data in an inconsistent state until the checking begins [18].

5 Performance Considerations

Perhaps one of the most important issues in computing environment is how to ensure a good performance. Performance is normally sensitive to the application domain rather than how fast the OODB system may be [24]. Engineering applications contain complex data and relationships which require a high degree of performance. The distributed architecture of OODB systems plays an important role in performance enhancement. Although OODB systems show a better performance compared to relational databases particularly in engineering applications [1,12], a benchmark may be used to examine the performance of the OODB system which is intended for use in the distributed CE design environment. Performance *tuning* (or optimization) can also be adopted to increase the degree of performance. This feature may be supported using tools available in OODB systems to allow a customized performance tailored to the needs of the engineering application. This is supported in some OODB systems such as O$_2$, Objectivity, and VERSANT.

6 Conclusions

In CE design, many tasks are performed simultaneously. Although CAD systems can support the development phases of a design artefact, they lack the facilities to integrate and manage extensive concurrent design activities. Managing these activities can be accomplished through OODB systems because they have the modelling and relationship support necessary for CE.

In this paper, we review the main issues concerning the management of distributed CE design in OODB system. The issues are discussed in terms of four phases of CE design support. The discussion shows also the support of these issues in the commercial OODB systems. We are currently developing a management system to support design evolution in a distributed CE environment. This system will enable design *reuse* as well as the traceability of the history of design objects, hence adding a *temporal dimension* to the CE environment. We are also continuing an earlier work in this area where an O$_2$ OODB system is used as back end and AutoCAD is used as front end in the design environment[8].

References

[1] S. Ahmed, A. Wong, D. Sriram, and R. Logcher, "Object-Oriented Database Management Systems for Engineering: A Comparison," *Journal of Object-Oriented Programming*, Vol. 5, No. 3, 1992.

[2] A. Al-Khudair, A. Shah and H. Mathkour, "Issues in Management of Class-History in Object-Oriented Databases," *The Third Golden West Intl. Conf. on Intelligent Systems*, Las Vegas, USA, 1994.

[3] A. Al-Khudair, "Prominent OODBs Support for Distributed Concurrent Engineering Design: Issues and Comparisons," *Internal Report, Department of Computer Science, University of Wales-Cardiff*,UK, Feb. 1999.

[4] A. Al-Khudair, A. Shah and H. Mathkour, "Design of Class History Management System for Object-Oriented Databases," *VIII International Symposium in Informatics Applications*, Chile, 1994.

[5] M. Hague, and A. Taleb-Bendiab, "Tool for the Management of Concurrent Conceptual Engineering Design," *Concurrent Engineering: Research and Applications*, Vol. 6, No. 1. 1998.

[6] B. Prasad, Concurrent Engineering Fundamentals, Prentice Hall,1996.

[7] P. O'Grady and R. Young "Issues in Concurrent Enigineering Systems,", *Journal of Design and Manufacturing*, Vol. 1, PP. 27-34, 1991.

[8] I. Santoyridis, T. Carnduff, W. A. Gray and J. C. Miles, "An Object Versioning System to Support Collaborative Design within a Concurrent Engineering Context," *BNCOD15, London, UK*, 1997.

[9] W. Kim,.Introduction to Object-Oriented Databases, MIT Press,1990.

[10] R. Fruchter, K. Reiner, L. Leifer and G. Toye, "VisionManager: A Computer Environment for Design Evolution Capture," *Concurrent Engineering: Research and Applications*, Vol. 6, No.1,1998.

[11] R. Ahmed and S. Navathe, "Version Management of Composite Objects in CAD Databases," *Proc. Of the 1991 ACM SIGMOD International Conference on Management of Data*, Denver, Colorado, USA, 1991.

[12] D. Barry., The Object Database Handbook, Wiley Computer Publishing, 1996.

[13] J. Rykowski, and W. Cellary, "Using Multiversion Object-Oriented Databases in CAD/CIM Systems," *Lecture Notes in Computer Science*, No. 1134, pp. 1-10, 1996.

[14] H. Chou and W. Kim, "Versions and Change Notification in an Object-Oriented Database System", *The 25th ACM/IEEE Design Automation Conference*, 1988.

[15] K. Dittrich and R. Lorie, "Version Support for Engineering Database Systems," *IEEE Transactions on Software Engineering*, Vol.14, No. 4, 1988.

[16] R. Katz, R., "Toward a Unified Framework for Version Modeling in Engineering Databases," *ACM Computing Surveys*, Vol. 22, No. 4,1990.

[17] M. Özsu and P. Valduriez, Principles of Distributed Database Systems, Prentice-Hall, 1999.

[18] S. Yoo and H. Suh, "Integrity Validation of Product Data in a Distributed Concurrent Engineering Environment," *Concurrent Engineering:Research and Applications*, Vol. 7, No. 3, 1999.

[19] D. Spooner and M. Hardwick, "Using Persistent Object Technology to Support Concurrent Engineering Systems," *Concurrent Engineering: Methodology and Applications*, P. GU and A. Kusiak (Ed.), Elsevier Science Pub., 1993.

[20] J. F. Roddick, "A Survey of Schema Versioning Issues for Database Systems", *Information and Software Technology*, Vol. 37, No.7, 1995.

[21] G. Kaiser, "Coopertive Transactions for Multiuser Environment," *In: Modern Database Systems: The Object Model, Interoperability and Beyond* (W. Kim ed.), ACM Press, NewYork, USA, 1995.

[22] W. Wieczerzycki, "Multiuser Transactions for Collaborative Database Applications, " *9th International Conference on Database and Expert Systems Applications DEXA'98*, Vienna, Austria, 1998.

[23] D. Clausing, "World-Class Concurrent Engineering", *In Concurrent Engineering: Tools and Technologies for Mechanical System Design* (E. Haug ed.), Springer-Verlag, Berline, 1993.

[24] O$_2$ ODBMS, Release 4.5, Ardent Software, Inc. , 1998.

[25] Objectivity ODBMS, Release 5.2, Objectivity, Inc., 1999.

[26] VERSANT ODBMS, Release 5.2, Versant Ltd.,1998.

Integration Technologies to Support Organisational Changes in the Construction Industry

M. Sun, G. Aouad

School of Construction and Property Management
University of Salford
Salford, M7 9NU, UK

Abstract

The recent Egan report on the UK construction industry has concluded that "integrated construction process" holds the key to further improvement in efficiency and productivity. In the last decade or so, considerable R&D efforts have been devoted to the integration issues in numerous research initiatives. A range of technologies of different levels of maturity have emerged from these studies, including data and process modelling, data management, data exchange using STEP standards, integrated project database, object oriented implementation, concurrent engineering systems, etc. The paper seeks to explore how these integration technologies can be used to support organisational changes necessary if the performance improvement is to be achieved.

1 Introduction

Construction is one of the largest economic sectors in all the developed economies. For example in the UK, the industry had an output of some £58 billions in 1998, equivalent to roughly 10% of GDP, and employs around 1.4 million people [1]. The performance of the construction industry has consistently lagged behind other industrial and service sectors [2]. In recent years as a result of a business down turn and increasing international competition, performance improvement is seen as a matter of life and death for many construction companies in the UK. Latham [3], Egan and other studies have identified a number of factors preventing the construction industry from achieving necessary improvements which have already been achieved in other manufacture and services industries:

Fragmented supply chain. The construction industry consists of hundreds of thousands of firms, over 90% of them are small and medium size with less than 10 employees. Most construction projects undertaken by our industrial partners and others often involve an ad hoc team of 15-20 of these firms located at disperse places. Similarly, the majority of the clients are one off clients. Fragmentation is a key feature of the industry structure and client base. The supply chain relationship is not a stable one.

Lack of industry standards for information exchange. With the increased use of computers, more and more information in construction is now generated in digital form. However, due to the lack of standards for information storage and communication, electronic information exchange between programs, tasks and enterprises is still not common.

Poor cross-disciplinary communication. A typical construction project consists of many sub-processes often carried out by different professionals at different locations. Poor cross-discipline communication has widely been identified as a bottleneck for performance improvement, and it often re-enforces the confrontational and blaming culture so common in the construction industry.

Lack of process transparency. Due to geographical and organisational barriers, many processes during design and construction are often repeated by different project partners. The lack of transparency means knowledge not being shared thus resulting in waste and inefficiencies.

Poor knowledge management at industry, enterprise and project levels. Construction is a project-based process. A false belief is that because each project is unique, knowledge gained in one project can not be re-used in others. Egan concluded based on other studies "not only are many buildings, such as houses, essentially repeat products which can be continually improved but, more importantly, the process of construction is itself repeated in its essentials from project to project. Indeed, research suggests that up to 80% of inputs into buildings are repeated." Better knowledge management would enable the capturing and re-use of knowledge at project, enterprise and industry levels.

The solution to the above problems lies in the integration of the processes throughout the construction supply chain. The potential benefits through integration in construction have been well argued by many authors, these include saving time and project costs, eliminating unnecessary processes, promoting partnership and teamwork culture, increasing client satisfaction, etc. Recent advances in Information and Communication

Technologies (ICTs) offer a platform to develop integration solutions throughout the supply chain in construction. Indeed, ICTs have been transforming every aspect of business activity as we move into the age of "digital economy". They allow businesses to re-engineer both their internal processes and their supply chain relationships to strip out waste, improve quality and give better customer service.

2 Drivers for Changes in Construction

2.1 Government Initiatives

Two influential reports on the UK construction industry have been published in the 1990s. The Egan Report on Rethinking Construction stresses the need for the industry to make substantial changes in its culture and structure, as a driver for improvements in efficiency, quality and safety. The report forwards the earlier recommendations of the Latham Report, which highlighted the need for the industry to adopt a team approach and invest in high-quality training, in order to strive towards a 30% reduction in costs. Both reports draw on the experience of other industries that have successfully adapted organisational cultures, introduced new business processes, achieved significant cost reductions and generated major gains in productivity and quality.

2.2 International Competition

Further, the pressures of globalisation are increasing. The 1990s has witnessed some significant political and economic changes. For example, the emergence of the EC single market means a erosion of trade barriers between the UK market and the rest of western Europe, as a result of which the UK economy has become a 12th part of a larger economy. The 1993 signing of the GATT agreement further removes entrants' barriers into the UK construction market and some of the overseas markets where the UK firms had hitherto enjoyed significant market share. Elsewhere there are cross-border trading organisations such as the Association of South East Asian Nations, North American Fair Trade Agreements and Economic Community of African States. The implication of the emergence of these cross-border trading organisations is a threat to the growth of the UK construction industry. First, the emergence of the European single market means that the UK construction market which accounts for 85 per cent of the output by UK construction industry is now opened to competitors from the rest of the community. Second, those regions beyond, where the UK contractors had hitherto enjoyed easy access, will now become very much more competitive and difficult to penetrate due to tighter entry regulations. Third, there are other threats

from those developing nations because of cheap labour, proximity and language barriers. This probably explains the reasons for continuous falls in orders obtained by British firms in regions such as Latin America, Africa, and the Middle East.

2.3 Technological Development

We are experiencing an 'Information Revolution', which is fundamentally changing the way we live, learn, work and play. The driving force behind this revolution is the convergence of computing and telecommunication technology. Today, computer performance is measured in the billions of instructions per second and "palm-top" computers far exceed the performance of the 1960s mainframes at a fraction of the price. Wireless technology and the integration of data, voice, video and graphics capabilities over fast, yet cost-effective, networks now allow "any time, any place, and any form" communication and information sharing. These dramatic improvements in IT price and performance will lead to equally dramatic changes in organisational strategy, structure, processes, distribution channels and work. The impact of IT on modern society is profound. Computers are already everywhere in offices, stores, banks, homes and even coffee shops. At the macro economic level, IT has enabled the globalisation of manufacturing and services, and has subsequently brought about large-scale changes in the industrial makeup of all the advanced industrial nations. We have witnessed the rapid growth of some industries such as computers, communications, software and financial services by enabling new products, services, and efficiencies, while other more traditional industries have stalled or even contracted by comparison. At the individual level, IT literacy is becoming an essential requirement for most professional jobs. With the imminent arrival of digital television, on-line shopping and banking, IT will become a basic skill every citizen of the community needs to have.

For the construction professionals, the initial surge of enthusiasm for computer applications started in the early 1960s. There was an optimistic view of the computer's potential as a supporting tool for design and construction and the time needed to develop this potential. In the 1970s and early 1980s, the initial excitement was replaced by a greater realism about what computers could offer [4]. The change of opinion was caused by a combination of the high capital costs of computing hardware and the limitations of the available computer software at the time. Since the mid 1980s, the penetration of computers in the construction industry has been accelerating thanks to the rapid development in computer hardware especially Personal Computers (PCs). The increase in computer ownership and IT applications in construction practices were confirmed by a series of surveys by Royal institute of British

Architects (RIBA) [5], Construction Industry Computing Association (CICA) [6] and other professional organisations in the UK.

3   Integration Strategies

There is a growing demand for computer integrated construction. The essence of integrated construction is the ability for different professionals to share project information by either accessing a central data repository or by exchanging data electronically. The integration principle is very appealing to the industry itself. Its importance is reflected by the growing number of funded research initiatives in Europe and UK. Amor [7] and Eastman [8] reviewed a long list of recent research projects, including COMBINE, COMMIT, ICON, OSCON, SPACE, ToCEE, RISESTEP, CIMSteel, CONCUR, GEN, VEGA, RATAS, ISO-STEP, IAI/IFC, etc. The on-going integration efforts can be divided into three categories according to the breadth and depth of process and data integration (Figure 1) [9]:

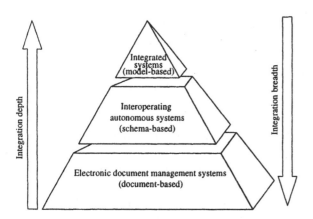

Figure 1 Three integration strategies

Electronic document management systems. This is a broad but shallow form of integration. All AEC applications are independent from the management system. The system only manages the outputs, or documents, of AEC applications. Each document is treated as one entity. The system has no knowledge of the detailed contents of the document.

Inter-operating autonomous systems. This is about data exchange between otherwise autonomous AEC applications. In this approach, each application has its own internal data format, however, all applications subscribe to a common data model. Data can be exchanged between application using the data model as a reference through some neutral formats, such as ISO-STEP physical files. The exchange may be conducted

directly between two applications or through a shared database.

Fully integrated concurrent engineering systems. This is the highest level of integration. In a fully integrated system, the system needs to have knowledge of data as well as processes. Project information is stored in a unified repository. The system controls the sequence in which AEC applications are used and multiple applications can interact with the project database simultaneously.

2   Integration Technologies

With the increased use of construction software, the traditional cross discipline communication is increasingly manifested as an issue of data exchange and data sharing between different software applications. Unfortunately, because most of the software packages are developed by different vendors, they all have their own particular data format. To achieve data exchange, a data mapping mechanism is required between two applications involved in the exchange. Given the large number of applications used throughout the building life cycle, it is impractical to set up one-to-one mapping between all of them. A more efficient solution is to use a neutral data format as a medium for the data exchange [10]. This neutral data format is an integrated data model which captures the full semantics of a building system and its components. In this approach, each application only works with a subset of the model. This subset is often described as an aspect model. Data exchange can be achieved between all the aspect models by mapping through the integrated data model.

2.1   Building Data Model Standards

Historically, the initial requirement for a standardised data model came from the need for different versions of CAD application to share their graphic files. IGES (the Initial Graphics Exchange Specification) of the United States was developed for this purpose [11]. However, graphical and geometrical data is only part of the information required in a building project. IGES is not able to support the exchange of other types of data such as construction, thermal, light, etc. Therefore a new project, PDES (Product Data Exchange Specification), was proposed in the US in the early 1980s to overcome these limitations. In the same period, similar efforts have been made in other countries, for example, the SET (Standard d'Echange et de Transfert) in France and the VDAFS (Verband der Deutschen Autombilindustrie Flaechen Scnittstelle) in Germany. In 1983, all these initiatives have been co-ordinated into a major international program under the umbrella of the International Standard Organisation, Standard for Exchange of

Product data (STEP) [12]. STEP seeks to define not only standard data models to facilitate information exchange but also standard methodology for data modelling and data exchange. At present the STEP standard is still evolving. It has a very ambitious aim of defining standard data models for all manufacturing sectors, such as aerospace, automobile and AEC industries.

Since the mid-1980s, there have been a number of studies into construction specific data modelling [13, 14]. Several well-known building data models have been developed by the research community, for example the RATAS model [15], the Building Core Model [16], the Integrated Data Model of COMBINE project [17], the Integrated Product Model of COMBI [18] and the Logical Product Model for CIMSteel [19]. All these models adopted an object oriented paradigm which describes the building system using objects or entities and their inter-relationships.

The lack of integration between commercial different AEC applications has become a barrier for the wide use of these systems. To address this issue twelve companies involved in the AEC and Facilities Management (FM) Industry, many of whom are software developers, started the International Alliance for Interoperability (IAI) in 1995 [20]. The aim of IAI is to achieve software interoperability in the AEC/FM industry. The mission is to define, promote and publish a specification for sharing data throughout the project life cycle, globally, across disciplines and across technical applications.

In essence, IAI is another data modelling initiative which seeks to specify how the 'things' that could occur in a constructed facility (including real things such as doors, walls, fans, etc. and abstract concepts such as space, organisation, process etc.) should be represented electronically. These specifications represent a data structure supporting an electronic project model useful in sharing data across applications. Similar to the STEP and the building models mentioned above, IAI also adopted the object oriented approach. The specification of each type of real world object, such as doors, walls, windows, is called a 'class'. The IAI model will be a collection of classes, thus is termed 'Industry Foundation Classes' (IFCs). IFCs, once developed, will enable interoperability among AEC/FM software applications. Software developers can use IFCs to create applications that use universal AEC/FM objects based on the IFC specification. All IFCs compliant software will be able to share project data in an electronic format.

2.2 Integrated Project Database

In parallel to the above mentioned data modelling activities, there are many research projects world wide which investigate the integration of AEC applications through an implemented project database. The aim of an Integrated Project Database (IPDB) is to provide a consistent and reliable storage of the project information, and serve as a data exchange hub for different tasks during the construction process [21]. The requirements for an integrated project database can be summarised as follows [22]:

- *Persistent building model*. Building design is a long process during which information about the building is enriched gradually. The first requirement for the IPDB is to hold a persistent model for different states of the building.
- *Data exchange interface*. The second requirement for the IPDB is to support data exchange interface with third party construction software through the use of product data technology.
- *View Integration*. Each software package deals with a particular aspect of the construction problem, thus, it has a partial view of the building model. These partial views need to be merged into a single coherent building model in the central data repository. This determines the requirement to maintain a consistent building model and to support the growth or population of this model through the integration of partial views.
- *Change management*. In a real world construction project if the architect feels the need to make changes to the building layout, he or she would inform the HVAC engineer to halt the heating and cooling system work, since they will have to be re-designed as a result of the building geometry changes. In a computer based design support system, a degree of concurrent engineering management is expected. This determines the requirement of the IPDB to provide the underlying support for change management and conflict prevention.
- *Design versioning*. An essential feature of any design process is the requirement to explore alternative design solutions. Design versioning should allow each discipline to take a version of the design at any stage. These versions will develop independently and co-exist. They also need to be merged to produce one coherent version of the design.
- *Project history*. An IPDB should be able to record the project history trail and allow professionals to re-examine the decision making process and to rollback to an early consistent state of the project.

Amor [23] presented a survey of six major EU and UK research projects which involve the development of IPDBs. He revealed that there is a consensus on the conceptual approach and the object oriented database implementation of the IPDB amongst these projects. However, the scope of these existing IPDBs is still limited and a cradle-to-grave IPDB will not appear in the near future. The IPDB scalability and the interface with application software still pose issues for further research.

2.3 Data Exchange using STEP

In a computer integrated construction environment, data exchange refers to the exchange of data between AEC applications, often using neutral format files. A sending application translates data from its internal format and encodes it into an established neutral format. This file is then transferred to the receiving application where the data is translated into the internal format of the receiving system. DXF is a well-known file format for exchange graphic file between CAD systems. Recently, the STEP physical file format has emerged as the neutral format for exchanging full product data [12].

A STEP file is a text file that contains values of data. Its data structure complies with a conceptual data model which defines the explicit standardised data specification for interpreting the STEP data. It is this model that provides a documented explanation of the context (scope) and meaning (relationships) of the data to be exchanged. It is used, along with an encoding algorithm, to read and write STEP physical files that contain both the data and its associated context, thus enabling effective and flexible communication between computing systems.

2.4 Virtual Workspace

Information sharing is only one aspect of an integrated construction process. Another important aspect is the human interaction between different professionals in a construction project. It has been widely acknowledged that some of the best ideas are developed "on the back of an envelope" in the canteen or over a cup of coffee when people have the opportunity to interact with each other. The BAA's Genesis project has demonstrated the benefits for bringing the project team into one office location and to operate in an integrated fashion. However, given the organisational structure of the construction industry, it is not always practical to get all participants of the project team to one physical location for the duration of the project. Recent advances in the Internet and Computer-Support Collaborative Work (CSCW) technologies make it possible to develop a shared virtual workspace to support close teamwork and project integration between construction organisations and professionals physically located at dispersed places.

5 A Distributed Collaboration System Supporting Partnering

The following introduces an on-going project, GALLICON, which seeks to apply the integration

technology to support the organisational changes necessary in the construction industry.

5.1 Business Background

The AEC industry practice is going through a transformation in recent years [24]. Traditionally, the widely held view of the construction process is that it is different from manufacturing because every product is unique. The conventional processes assume that clients benefit from choosing a new team of designers, constructors and suppliers competitively for every project they do. As a result of the repeated selection of new teams, there is little incentive for construction companies involved to invest and develop teamwork skills and innovation, both of which are vital to further efficiency improvement. Critically, it has prevented the industry from developing products and an identity - or brand - that can be understood by its clients.

The problem has been recognised and efforts are being made to improve the situation. One of the solutions emerged is the partnering arrangement between clients and contractors replacing the conventional competitive tendering. The key to the partnering approach is that a client and a contractor form a long term stable relationship whereby the client's new projects will be automatically awarded to the partnering contractor. Such a arrangement will enable the contractor and its sub-contractors to improve the repeated process and offer the client best value for money.

The main industrial sponsors of the GALLICON project, a contractor and a major client in the water treatment industry, have such a partnering system in operation since June 1997. The partnering is extended to the repeated use of the same cost consultants, design consultants, mechanical and electrical engineers. So far, some 16 projects have been completed under the partnering agreement with evident benefits for all partners involved. To further improve the operation, the partners seek to explore integrated project database technology for the communication and information sharing tasks.

5.2 System Architecture

Figure 2 shows the system architecture of the GALLICON prototype. It consists of an integrated project database and other shared project resources, a Process Manager and interfaces to four third party AEC software packages. The database provides a central consistent repository for project information. The included AEC applications communicate with the database through purpose-built data mapping interfaces. In line with the process driven integration strategy, the Process Manager module is at the core controlling the operation of the system.

A typical scenario of using the GALLICON prototype can be described as follows. At the start of a new water treatment project, the project manager will set up the integrated project database and give all participants appropriate levels of permission of access. A process model is also produced based on a predefined template. The model will used to control the operation of the system. The designer starts the design of the water treatment plant using a CAD tool. Project information, layout and specification, is transferred to and stored in the integrated project database. The Process Manager monitors the progress of the design activity. At certain stage, a cost estimate is needed. The Process Manager will inform the cost estimator who will perform the task using a cost estimating software and project data downloaded directly from the database. The cost estimator is able to see the exact status of the project from the graphic process model and existing design using VRML viewer. The designer and cost estimator can work collaboratively to modify the design and components specifications. The visualisation tools will facilitate the inter-disciplinary communication. When the design is near completion, the planner is given permission by the Process Manager to start planning. Again, project data are stored in the database

and readily available to the project planning software. The benefit for the client is that they can monitor the project decision making process, know the design they are going to get, how much it costs and how the on-site operation is scheduled to completion. All these are provided to the client by the process manager and the VRML visualisation tool without the need to use specialised CAD, cost estimating or planning software.

5.3 Distributed Collaboration

The GALLICON prototype intends to support the data integration between the software packages for CAD design, cost estimating and project planning as well as the collaboration between the remotely located professionals who use the software. To achieve effective communication between the team members and realise the potential of the integrated solution, it is essential to keep everyone well informed on the project status at any given time. In other words, all participants need to know when they are required to perform a task and what project information is available for their decision making. It is well known that graphic is the most effective method for human communication. The use of graphic visualisation in GALLICON is two-fold.

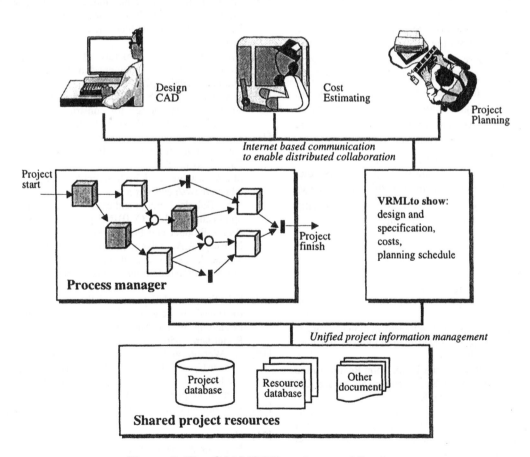

Figure 2 The GALLICON system architecture

First, it has a graphic representation of the process model which shows all the tasks and task dependencies of a water project. Using an appropriate colour code, i.e., grey shade for the completed tasks, the process manager will be able to inform all project participants the progress and current status of the project. The second aspect of visualisation is the VRML representation of project design, specification, costs and planning schedules. An Internet based implementation was adopted for the project manager and VRML viewer so that the users can access them using standard World Wide Web (WWW) browser such as Microsoft Internet Explorer or Netscape. The follow scenario illustrates how the GALLICON prototype is intended to work.

6 Conclusions

Supply chain partnering and integrated construction process are widely recognised as vital to the performance improvement in the construction industry. With the increased use of computer software applications throughout the supply chain, the demand for integration is often manifested as the requirement for integration between these applications. The objective of integration is to achieve coherent management and electronic sharing of information and knowledge during construction projects.

Integrated construction process requires an integrated IT system that will enable the project team members to work together and share project information seamlessly. In recent years, many EU and UK research projects investigated various technical aspects of such a system. A range of technologies have emerged from these studies, including data and process modelling, data management, data exchange using STEP standards, integrated project database, object oriented implementation, concurrent engineering systems, etc. The GALLICON system described in this paper shows how these technologies can support the partnering practice between construction companies.

References

[1] J. Egan, Rethinking Construction, Department of Environment, Transport and Regions (DETR), UK, 1998

[2] P. Barrett and M. Saxton, The transformation of 'out-of-industry' knowledge into construction industry wisdom, CRISP report, 1999

[3] M. Latham, Construction the Team. Joint Review of the procurement and contractual arrangements in the UK construction industry, Final Report, 1994

[4] V. Bazjanac, The promises and the disappointments of Computer Aided Design, in Negroponte N., (eds) Computer Aids to Design and Architecture, Mason/Charter, London, 1975

[5] Howard R., "Computer survey results compared", RIBA Practice 4, April 1988

[6] CICA, The CICA CAD systems sales surveys, 1981 - 1996. Construction Industry Computing Association, Cambridge, UK, 1996

[7] R. Amor, A UK survey of integrated project databases, in Proceeding of CIB W78 Information Technology in Construction Conference, Stockholm, Sweden, June 3-5, 1998, pp67-75

[8] C. Eastman and G. Augenbroe, Product modelling strategies for today and the future, in: Proceeding of CIB W78 conference, Stockholm, Sweden, June 3-5 1998, pp191-207

[9] M. Sun and G. Aound, Control Mechanism for Information Sharing in an Integrated Construction Environment, in Proceeding of The 2nd International Conference on Concurrent Engineering in Construction – CEC99, Espoo Finland, 25-27 August 1999, pp121-130

[10] H.C. Howard, Levitt R.E. and et al, "Computer integration: reducing fragmentation in AEC industry", Journal of Computing in Civil Engineering, No. 3, 1989, pp18-32

[11] B. Warthen, "A history of AEC in IGES/PDES/STEP", Computer Integrated Construction, Vol.1 Issue 1, 1989, pp4-8

[12] ISO TC 184, Industrial Automation Systems - Product Data representation and Exchange, 1993

[13] W. F. Danner, A global model for building-project information: Analysis of conceptual structures, U.S. National Bureau of Standards, PB88-201546, April 1988

[14] B.C. Bjork, "Basic structure of a proposed building product model", Computer Aided Design, vol. 21 number 2, 1989, pp71-78

[15] E. Enkovaara, M. Salmi and A. Sarja, RATAS project, Computer aided design for construction, published by Building Book Ltd., Helsinki, Finland, 1988

[16] J. Wix, Computerised exchange of information in construction, Building Construction core model, 1996, Paper is available at: http://www.bre.co.uk/bre/research/cagroup/consproc/itra/bccm/news/news5.htm

[17] A.M. Dubois, J. Flynn and et al, "Conceptual modelling approaches in the COMBINE project", in Scherer R.J. (Eds.), Product and Process Modelling in the Building Industry, A.A. Balkema, Rotterdam, pp555-565, 1995, The model itself is available at: http://dutcu15.tudelft.nl/~combine/idm/index_schema.html

[18] R.J. Scherer, "EU-project COMBI – Objectives and overview", in Scherer R.J. (Eds.), Product and Process Modelling in the Building Industry, A.A. Balkema, Rotterdam, pp503-510, 1995

[19] A. Watson and A. Crowley, "CIMSteel integration standard", in Scherer R.J. (Eds.), Product and Process Modelling in the Building Industry, A.A. Balkema, Rotterdam, pp491-493, 1995

[20] IAI, Introduction to IAI, IAI CD release version 2.0, March 1999

[21] Construction IT bridge the gap, Department of Environment report, UK, 1995

[22] M. Sun and S. Lockley, "Data exchange system for an integrated building design system", Automation in Construction, 6(1997), pp147-155

[23] R. Amor, A UK survey of integrated project databases, in Proceeding of CIB W78 Information Technology in Construction Conference, June3-5, 1998, Stockholm, Sweden, pp67-75

[24] Consultative committee on Construction Industry Statistics, The state of the Construction Industry, Issue 2, July 1994

A Method for Designing Database Schema using Multi-Context Map –Property Centric Modeling for Discrete Manufacturing System

Isao Yamada, Satoshi Kumagai

Research & Development Headquarters, Yamatake Corporation,
1-12-2,Kawana, Fujisawa, Kanagawa, 251, Japan

Abstract

In order to satisfy variety of customer needs, manufactures supply many kinds of products. This causes process modification. In order to be adjustable to the process modification, the information related to a process, or process properties, should be specified for each process to be managed by manufacturing system.

In manufacturing system, process properties stem from materials, which includes products and are typically associated with them because each process focuses on different information about material. This causes variation in properties even from the single source of material. On the other hand, the identity of individual materials should be maintained through all process for designing database schema for manufacturing management system.

This paper presents a method of designing database schema for manufacturing system based on the concept of "Property Centric Modeling", which is flexible for the process modification. It resolves the issue of identification and recognition about physical material in the information world.

1 Introduction

Things involved with manufacturing systems are not only physical materials such as facilities, raw materials, and intermediate/final products, but also information relating to them. Although physical materials (hereafter material) as well as their related events are clearly distinctive in the physical world, they are not always specified in the information world, or within the manufacturing management system. In other words, materials within the physical world are individual per se regardless of views on which manufacturing management system stands. We refer to such characteristics of material as individuality".

In this paper, individuality is distinguished from "uniqueness ". Uniqueness depends on a particular view of manufacturing management, so as to enable identification of material from the view. For example, each part to be assembled in the same lot has individuality in the physical world; however, that part does not always have the uniqueness of information world because it might not necessary be uniquely addressed in a context of manufacturing management system.

We consider that information world consists of properties, which are associated with material and propose a method of "Property Centric Modeling". Each property in the information world can be regarded as a mapping which provides a correspondence from a physical material to the information associated with it. In other word, for any property, at least one physical material associated with it exists in the physical world. In general, manufacturing management systems focus on a set of properties for each process to manage progress of products, quality of products, etc. Specifying the process related information for each process facilitates the manufacturing management system to follow the process modification.

2.Property Centric Modeling

Information maintained in the manufacturing management systems usually relates to the properties of a material, which include drawings, products, and row materials. Since each material is clearly distinguished from the others in the physical world, its individuality is trivially assured in the physical world. In designing manufacturing management systems, materials existing in the physical world are often mapped to sets of properties of information world. The manufacturing management systems are required to be identifying one thing, which might be material itself in the same group for a particular purpose. This is actually done by taking various sets of properties into account. However, manufacturing management systems do not always uniquely identify every set of properties associated with the material.

In Property Centric Modeling, we will distinguish individuality and uniqueness in terms of the relationship between the physical world and the information world.

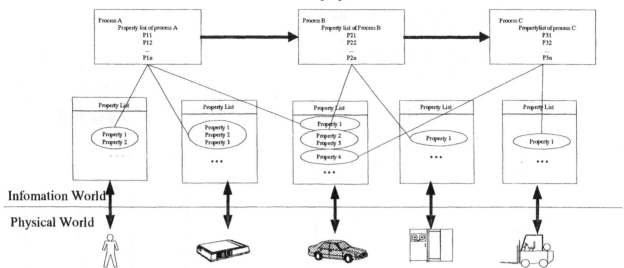

Figure1: Process and Property

2.1 Individuality and uniqueness of information in the manufacturing system

We regard the manufacturing management systems dealing with the information world as set of properties. Any property sometimes has a particular correspondence with physical material such as drawings, products, and row materials. As mentioned before, any material has individuality, because each material can be trivially distinguished from the others in the physical world. Consequently, a particular collection of properties associated with that material has individuality in the information world if the collection has one-to-one correspondence with that material.

We can regard a set of properties existing only in the manufacturing management system as an individual thing as illustrated in Figure 2, if the set has one-to-one correspondence with the material. That set of properties can be regarded having "individuality" as well. For example, the properties, which are associated to information, such as message among processes, could have individuality through the association with a material. The association is actualized as an event, which is related to a material

Therefore, in Property Centric Modeling, individuality of thing which could be a material and/or a set of

properties is defined as follows: Thing has individuality if and only if it can be distinguished form the others.

Manufacturing management systems are not always required to uniquely identify every material and set of properties. The number of identification or label to be hold in the manufacturing management system is limited for a particular objective. The property sets that have individuality without unique label in the system can be regarded as resources for the system. For example, parts that are managed by a group or lot do not need to be identified in the manufacturing systems. It is enough for the system to assign properties such as lot number.

Therefore, in Property Centric Modeling, uniqueness of thing which could be a material and/or a set of properties is defined as follows: Thing has uniqueness if and only if it can be properly labeled in the manufacturing management system.

2.2 The classification of properties

The properties of the information world can be divided into several groups. In the following section, a classification of properties of the method is presented for implementing the Property Centric Modeling.

2.2.1 Elementary properties

Elementary property is minimal unit of property in Property Centric Modeling and is represented in a combination of attribute and its value. Elementary property is often recognized outwardly as state changes caused by manufacturing process.

Manufacturing processes cause varieties of changes to properties. For example, shape property of a part to be assembled is changed from flat-shape to box-shape in a press process. Each change of property is made by the process function. We consider the shape property as a process's property, because this property change depends on the process function.

property is one of properties in the group, which will guarantee uniqueness of the group among the others. Singular property must remain unchanged during the manufacturing processes. In case of multiple candidates

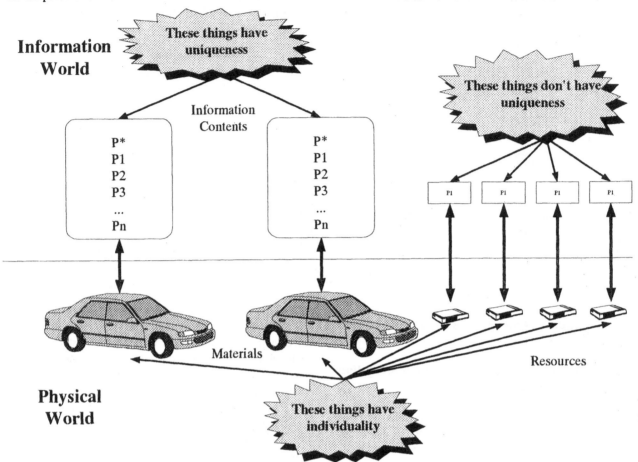

Figure2 Individuality and Uniqueness

(1) Process function property

Process function property is associated with a particular manufacturing process. The value of the attribute is modified through an occurrence of the process. Therefore, process function property is encapsulated in the process. Some process function properties are generated in the course of the process function. In these cases, the initial value is null and set to a fixed value at the completion of the process.

(2) Singular property

In Property Centric Modeling, various combinations of elementary property combination are considered, so as to be implemented in the manufacturing management system. Each group of property will need to be uniquely addressed by the manufacturing management system. Singular

for singular property in the same group, only one of them must be specified.

2.2.2 Composite property

Group of elementary properties is classified as below:

(1) Recognized property by the process performer

In each process function, there is someone or something that plays a central role in the process. That is called process performer of the process function in Property Centric Modeling. Process performer can be an operator and/or facility in the process.

Recognized property by the process performer is grouped properties by the process performer. Since such grouping tends to depend on the process and/or intention of the process performer, it is effective within the scope of the process.

(2) Information contents of material

Certain set of properties can be grouped under the name of a particular material. It will include recognized property by the process performer because the process performer would regard them belonging to the material. An information contents of material is aggregation of recognized property by the process performer for all of the manufacturing processes. The name of information contents should be equal to the material's name.

	Elementary Property	Composite Property
Single process	Process-Function-Property	Recognized-Property By the process Performer
All of the manufacturing processes	Singular-Property	Information contents of material

Table 1 The classification of properties

In Property Centric Modeling, the granularity of properties is recognized in reference to the manufacturing process. This is summarized in Table1.

3. Procedure of capturing the properties

In general, it is not easy to enumerate properties without any references. Focusing state changes in the manufacturing process enables to easily capture the properties because state can be regarded as a property with a specific value. In this procedure, properties within the scope of the state changes in the process functions are taken into account. The states to be considered are those recognized by process performer. When a material passes from one process to another, the process performer acknowledges the state changes in reference to the material.

In the followings, format of property is provided in two ways. When property is for describing a process, it is noted " attribute name" of the property. For describing the state of the process, it is noted "attribute name: value". In addition, it should be noted that what manufacturing management system manages is just label to group of individual material in the physical world because each individual material in the group can be assumed to be the same for the system. Therefore attributes and their values to each in the same group are not specified as long as the material's state is in the same state.

(STEP1) Create property list relating to a process
State changes between before and after a particular

process are to be considered.
(STEP2) Capture singular properties
Choose the properties whose attribute value does not change through all of the manufacturing process as singular properties. In practices, the property whose value remains unchanged between the pre-process state of the first process and the post-processed state for the end process in the manufacturing processes should be considered at hand. Make clear the mapping between the singular property and actual identification used in each process.
(STEP3) Capture process function properties
Extract the properties whose attribute values are modified in the process.
(STEP4) Grouping the listed properties
For each process, define the set of properties as recognized properties by the process performer.

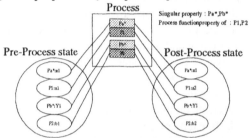

Figure 3 Grouping the listed properties

(STEP5) Associate the recognized property for materials
Associate the grouped recognized properties in the information world for a material in the physical world. All of the grouped properties are not always associated with a particular material. Those properties are transferred as messages between processes.

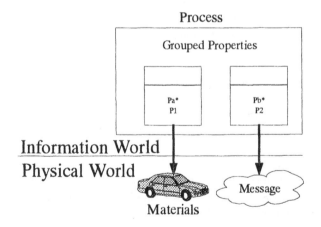

Figure4 Map the recognized objects for the materials

(STEP6) Define information contents of material
The previous steps are executed, focusing on each

process. To generate information contents for a particular material, all of the recognized properties should be aggregated to the material. By comparing upper and lower bounds of value of those properties, some of the properties whose labels are the same can be merged

4.Procedure of constructing database schema

The Property Centric Modeling can specify the objects to be managed as information contents and their attributes in the systems. In this chapter, we apply the Property Centric Modeling to the construction database schema.

(STEP1) Create a function model
Create a function model for the manufacturing system with IDEF0/MCM.
(STEP2) Create a process list
In order to specify the area where support system focuses, create process list for all of the manufacturing processes in the manufacturing system.
(STEP3) Create a process connection list
Specify the connections among processes, based on the material flow in the manufacturing system.
(STEP4) Execute the Property Centric Modeling
(STEP5) Define data model
Define data model of the manufacturing management system in the form of Entity-Relationship diagram. Each entity is composed of the properties obtained in the steps above mentioned. The data model consists of the following sub-models of entities – Material model, Process model, and Inventory model and their relationships.

-Material model
Material models are defined as a collection of information contents of materials.

-Process model
Process models are defined as a collection of process function properties, whose attributes should include start-time and end-time of process function. At least one singular property should be included.

-Inventory model
Inventory model is defined as a collection of grouped properties, whose attributes should include quantity of material and associated timestamp.

5.Exsample

5.1 The sample production system

A sample of automated manufacturing system is presented in Figure 5 [3]. The purpose of the manufacturing system is to bond a silicon sensor chip on the mount. The assembly line is consists of multiple functional modules and conveyers connecting them. Here, we employ Multi-Context Map [1][2][11][12] (hereafter MCM) for describing the system. MCM can specify both information and material flows in manufacturing, as well as resources involved. Figure 6 illustrates the notation of the building block of MCM, which is called Context Map (CM). In each CM, thick solid line is used for material, and thin one for information or event flow. Perspective arrows, which are attached to the upper edge of the box, imply associated process performers.

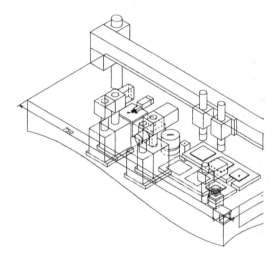

Figure 5 The Outline of Automated Assembly Line

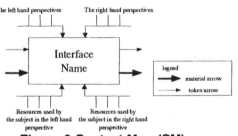

Figure 6 Context Map (CM)

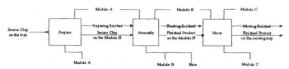

Figure 7 The sample manufacturing system (STEP1)

5.2 Prototype system

We developed a prototype system, which implements the method. The prototype supports each step of the procedure as the followings.

Manufacturing processes are enumerated for (STEP2) of Section 4 as shown in Figure 8. The window allows to entry each manufacturing process name and its associated facilities and/or operator. Input and output materials to the process are also specified.

For the material flow specified, connection between processes is defined as in Figure 9. Process function properties are captured in the window in Figure 11. Those properties are associated with a material and specified as recognized properties by the process performer in the window of Figure12. All of the properties are aggregated for each material and entities as information contents of the material are generated as in Figure 13,14,15.

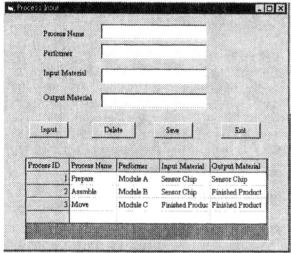

Figure 8 Create a process list (STEP2)

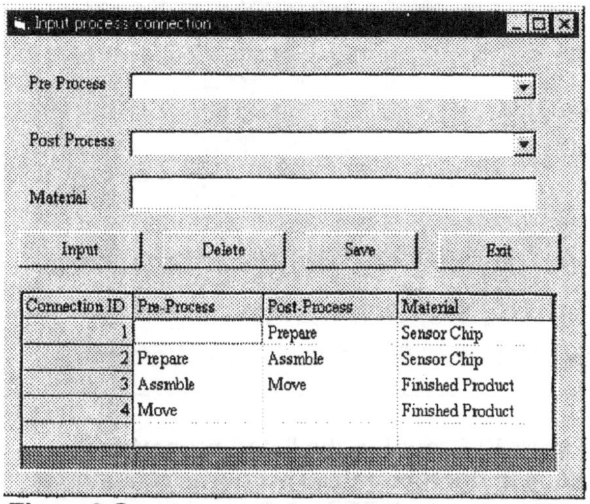

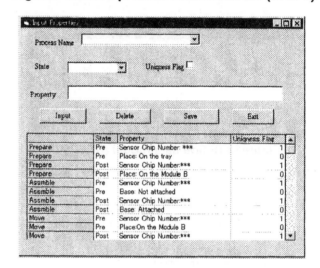

Figure9 Create a process connection list (STEP3)

Figure10 Create property list relating to a process & Capture singular properties (STEP4-1, 2)

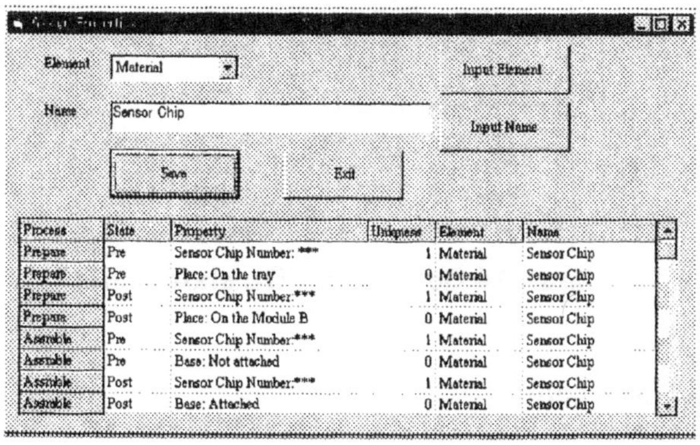

Figure11 Capture process function properties (STEP4-3)

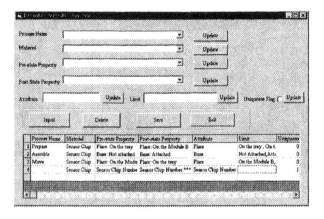

Figure12 Grouping the listed properties & Associate the recognized property for material (STEP4-4, 5)

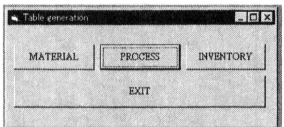

Figure13 Define data model (STEP5)

```
create table Prepare(
    Sensor_Chip_Number      int,
    In_time                 timestamp,
    Out_time                timestamp)

create table Assemble(
    Sensor_Chip_Number      int,
    In_time                 timestamp,
    Out_time                timestamp)

create table Move(
    Sensor_Chip_Number      int,
    In_time                 timestamp,
    Out_time                timestamp)

create table Sensor_Chip(
    Sensor_Chip_Number      int,
    Place                   char(20),
    Base                    char(5))
```

Figure14 Generated SQL statements

6.Discussion

6.1 Comparison with other methods

(1) Comparison with data centered approach

In general, design method for database schemas uses data centered approach [4][5][8]. In data centered approach, since data design and process design are conducted separately, data model sometimes is not clearly associated with process model. In order to comply with modification of manufacturing and/or data processes, it is necessary to specify entities and relationships associated with them accordingly. In Property Centric Modeling, since processes and their related information are clearly associated by process function properties, data model and process model.

(2) Comparison with Object-Oriented Modeling Language/Method

In Object-Oriented Modeling Language/Methods [6][9] such as UML and OMT, real world is recognized as a set of objects. For this reason, it is important for designing useful reusable system to define the objects. However, object design depends on analysts. Object consists of data and method, which relates to the data. In well known Object-Oriented Modeling Method, analyst makes a frame for an object tacitly and analysis the data/method for a focused context. On the other hand, in Property Centric Modeling, data and methods are captured as properties. Then the properties are categorized as objects to manage. Thus, Property Centric Modeling provides a method for object design.

6.2 Defined viewpoint

Viewpoint should be defined, because the recognized information depends on the viewpoint. One thing might be recognized as a product from supplier's view and also as a part to be assembled from a consumer's viewpoint. Viewpoint is not made clear in general data modeling method. In Property Centric Modeling, the information is captured as the material's properties or ontology [10] of material by making clear the viewpoint.

7.Future Works

7.1 Application to Multi-Products production system

Actually the same production line is used for manufacturing different products. In implementing a production management system using the method proposed, one table of data to be managed is assigned to each process. In the example presented in Figure 16, three tables should be created for the process 'A', because it is presented as those tables which are associated with part X, Y, and Z in the method. In case of changing the characteristics of products, all of the tables for the process need to reconsider.

For Multi product line, it is reasonable to make the table for each product firstly, then integrate all of the tables for the processes. This facilitates the incremental design for the product modification.

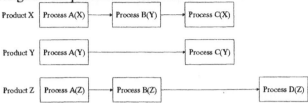

Figure16 Multi-Products production system

7.2 The properties required for the total production management

Enterprise activity involves various properties, which are not always having explicit relationships with material and its manufacturing processes. Followings are the other candidates of properties for modeling production in terms of enterprise engineering.

(1) Inherited property

Manufacturing process is often associated with property which is created and/or modified, and passed by previous processes. This property is referred to as inherited property.

(2) Reference property

Properties are sometimes provided by the manufacturing processes, which are beyond the scope of target production management system. Those properties are external of the field of interest and stay intact through the manufacturing processes and called reference properties.

(3) Physical property

This property represents the property, which is physical characteristic of material.

(4) Performer property

This property represents the property, which should be owned by performer of process function. This category is useful for modeling enterprise activities from a particular perspective, such as a person, organization, and facility.

(5) Derived property

This property can be explained by using the other properties. For example, cost processed amount.

8.Concluding Remarks

In this paper, we introduced 'Property Centric Modeling' method, and applied it to the design of database schema for production management system. We think that our method can be applied for various applications. Product design suitable for process operations could be supported by taking the properties, which should be acknowledged, between process and product into account.

Acknowledgements

We would like to express our thanks to Prof. Kiyoshi Itoh at Sophia University who gave us valuable advice.

References

1. Akiko Hasegawa, Satoshi Kumagai, Kiyoshi Itoh : Collaboration Task Analysis by Identifying Multi-Context and Collaborative Linkage, CERA Journal , to appear (2000)

2. Satoshi Kumagai, Kiyoshi Itoh : A Design Method of Concurrent Engineering Process using Multi-Context Map, JSSST FOSE'97 (Foundation of Software Engineering),pp.51-58, 1997, in Japanese.

3.Yoshitaka Tomita: An Application of IDEF0 Method for Automated Assembly Line Building,

4.Richard J. Mayer: IDEF1 Information Modeling, 1994

5.Thomas A. Bruce: Designing Quality Databases with IDEF1X Information Models, 1992

6.James Rumbaugh: Object-Oriented Modeling and Design, 1991

7.Biren Prasad: Concurrent Engineering Fundamentals, 1997

8.James Martin: Information Engineering, 1989

9.http://www.ogis-uml-university.com/download/download2.html

10.Perakath C.Benjamin, Christopher P.Menzel, Richard J.Mayer, Floarence Fillion, Michael T.Futrell, Paula S.deWitte, Madhavi Lingineni, IDEF5 Method Report,1994

11.Satoshi Kumagai, Akiko Hasegawa, Ryo Kawabata, Kiyoshi Itoh: Building Workflow for Collaboration Task using Multi-Context Map, Journal of Integrated Design and Process Science, to appear, (1999).

12.Satoshi Kumagai: A Study on Analysis Methods in Collaboration Domain, Doctoral Thesis, Sophia Univ., Japan, 2000.

Process Management in the Interdependent Activities of a Building

T. Kaneta

Department of Architecture and Architectural Systems, Kyoto University, Kyoto, Japan 606-8501

H. Nagaoka and S. Furusaka

Department of Architecture and Architectural Systems, Kyoto University, Kyoto, Japan 606-8501

Abstract

This paper presents the first step of an overall approach to the integration of design and construction in building construction projects. Its approach relies on the information technology to support members of design and construction teams throughout the project's life cycle. To manage the information correctly in execution of jobs, the authors covered all the items to be considered in view of project management. Characters of the supporting tools are discussed around integrated object oriented model. As the primary tool, process model is described with its prototype using Object Pascal.

1 Introduction

Integrating information is now indispensable in building design and construction from the programming to construction phase, because it have improved the collaboration of the multiple disciplines such as architectural design, structural design, cost planning, and scheduling. It also enables the earlier manufacturing of the items that would be procured to the construction site. The drawings and documents will be made more precisely, which improves the productivity of the project.

As Oxman points out, Concurrent Engineering (CE) opens the possibility for simultaneous work on complex collaborative, engineering tasks [22]. Indeed, there are some frameworks for concurrent engineering in manufacturing field [5, 19, 20, 23], but in building design and construction the complexity of collaboration is significantly higher and goes beyond the sharing of data [22]. As Betts mentions, efficient interfaces between the participants of construction projects should be produced [1].

The authors have analyzed the process of building design and construction in a project based on the concept of CE in the past studies [15, 16]. As for the activity in the process, it is important to clarify what is necessary to

execute it efficiently, paying attention to another activity in the different discipline.

Though several useful process models have been already proposed [12, 13, 24], few methods [8, 10] are shown to implement the models in the project and to combine them with the product models that have multiple views. Most of the proposed system are focusing construction phase but not total process [6, 27, 28, 29].

In this paper, the procedure for integrating the process and product model is proposed. Then, overall support system for process management is shown. Finally, the prototype of the system is developed through object-oriented design.

2 Activity Execution Procedure in a Project

2.1 Modeling the Activity Execution Procedure

The effort paid by the architect and the engineers to secure the preciseness among their activities is captured to examine what kinds of support are required [14]. The activity execution procedure is shown in Figure 1. It can be divided into three cases as follows:

a) Normal Procedure

An activity, which has a serial or interdependent relation to another activity, has normal procedure. After input and additional information are acquired, subroutine "Normal Design" is executed, and output information is processed.

b) Changing Procedure

In case that executing and modifying the same activities again, the activity to re-execute is specified at first. Then, "Normal Design" is executed, evaluating multiple alternatives.

c) Iteration Procedure

In case that temporary output information remains in the former activity, it should be iterated to fix the output information. After uncertainty is cleared, the former activity is executed through "Normal Design".

Afterwards, the temporary information is examined and substituted.

2.2   Subroutine of the Procedures

Above each procedure has a subroutine, which can be analyzed to contain checklists. There are eighteen checklists in subroutine "Normal Design" as shown in Figure 2. The architect or engineers should examine the input information. If it is confirmed that there is logically no contradiction in the input information from the upstream activities, they should manage the relation to the interdependent activities, repeating item (10-14). Then the relation to the downstream activities is examined in items (14) to (17). At this time, the preciseness of the output information should be guaranteed. The output information is stored in the indicated place.

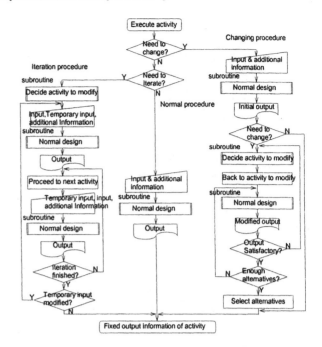

Figure 1 Activity execution procedure

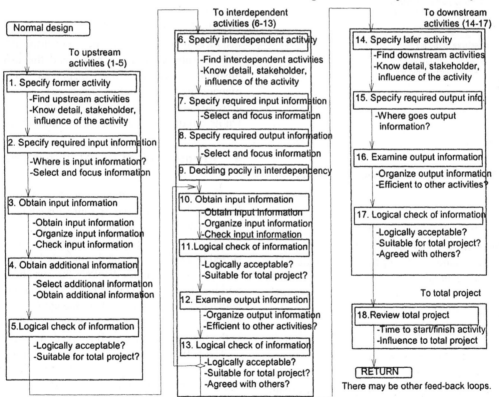

Figure 2 Subroutine "Normal Design"

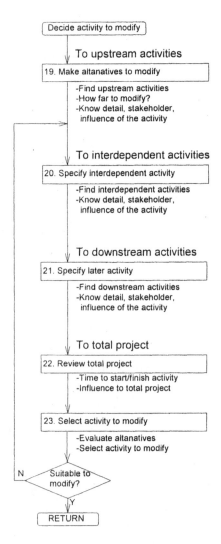

Figure 3 Deciding the activity to modify

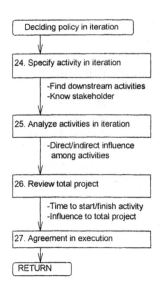

Figure 4 Deciding the policy in iteration

In subroutine "Deciding the Activity to Modify", there is five checklists shown in Figure 3. As the activity to modify is selected among the upstream activities, the candidates should be indicated (19). Next, the influence on interdependent activities (20), downstream activities (21), and the total project (22) should be examined when the output information is changed.

In subroutine "Deciding the Policy in Iteration", there is four checklists (24-27) as shown in Figure 4.

The procedures and checklists might vary in each project. However, they cover all the points: upstream, interdependent, downstream, and total project.

3 Method to Support Activity Execution

The twenty-seven items in the checklists were classified and examined as shown in Table 1. For instance, item (1) targets the input information, and so it will be supported if the name and the influence of the upstream activities are displayed. Thus, all the management information necessary for executing the activities can be developed for entire 300 activities in the whole building design and construction process.

4 Framework of the Management System

The framework of the management system is shown in Figure 5. It is composed by four subsystems of the process model, the activity execution procedure, the product model, and the engineering programs. All information is stored in an object-oriented, integrated model.

The process model displays the structure of the entire building design and construction with the AllJobs class, which can make it clear where the activity is located in the entire process. Activities that have considerable influence to the focused activity are displayed in the network chart.

The activity execution procedure displays management information to execute the activity with the OneJob class. The name and the outline of the focused activity are displayed. Moreover, the table that contains the activities related to the focused activity is also displayed. The number of the business, the name, and the degree of the influence are shown in the table.

Table 1 Method to support activity execution

	Classification	Method to support	Objective items
1 Input information	Name of former activity	Indicate number, name, influence of activities	1, 19
	Required input information	Indicate list and class name of input information	2
	Contents of input information	Install input information from system	3
2 Additional information	Name of additional information	Indicate list of additional information	4
	Contents of additional information	Install additional information	4
3 Interdependent	Name of interdependent	Indicate number, name, influence of activities	6, 20
	Interdependent input	Install input information from other activities	7, 10
	Interdependent output	Store output information to other activities	8, 12
	Interdependent manual	Remind users by sending messages	9
4 Iteration	Name of iteration	Indicate number, name, influence of activities	24, 25
	Iteration input	Indicate input information to be examined again	24
	Iteration output	Indicate output information to be examined again	24
	Iteration manual	Remind users by sending messages	27
5 Output information	Name of later activity	Indicate number, name, influence of activities	14, 21
	Required output information	Indicate list and class name of output information	15
	Contents of output information	Store output information to system	16
6 Total project	Logical check	Leave to users' own decision making	5, 11, 13, 17, 23
	Review	Evaluate by scheduling program	18, 22, 26

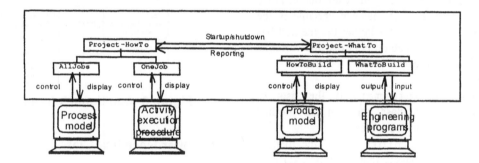

Figure 5 Framework of the management system

The product model manages input and output information with the WhatToBuild class for structure/material information and the HowToBuild class for construction information. Design information in the drawing and documents are related outside of this system. The characters requested to the product model are as follows:

a) Career Growth of the Model

The product model should grow up as the amount of information increases with the progress of the process. Therefore, it is assumed to treat outline information in the first stage of the process, and detailed information after mid-term of the process. They should be replaced as time goes. The former is stored in super-class, and the latter in sub-class of the system. The sub-class inherits the attribute of its super-class, and so they can share the instance variables and the methods.

b) Flexibility for Changes

To correspond to the update of information on the changes, the product model should be flexible to the change. In middle term of the process, the information decided in the first stage of the process can be updated. The class structure of the product model can be also modified. Displaying last re-written time on the screen, the change will be recognized easily.

The engineering programs are subsystems by which a technological calculation such as cost estimation [11] are processed.

5 Object Model of the Management System

5.1 Structure of the Object Model

The classes of the management system are shown in Figure 6. According to the OMT method [26], a rectangle indicates a class that can be divided into class name, instance variables, and methods. Multiple relationship, for example, in case a space contains three spaces, are marked with black circles.

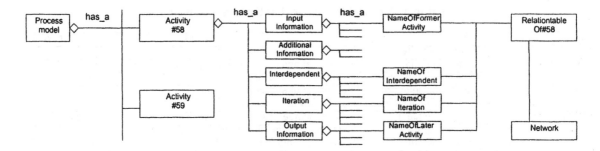

Figure 6 Classes of the management system

5.2   Input Information Class

As shown in Figure 7, there is no instance variable in InputInformation but there is one method. This method sends messages to three classes.

In NameOfFormerActivity there are one instance variable and two methods. The instance variable is the Boolean type (0 or 1) indicating whether the method has been executed before. One method controls this variable. Another method creates an instance of RelationTable.

In RequiredInputInformation the lists of input information, that is, class name in the product model are defined as instance variables.

In ContentsOfInputInformation the method to create an instance for classes that contain CAD data will be defined. The development of this method is left to the future study.

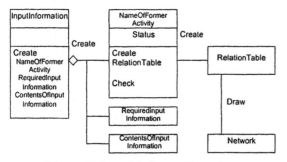

Figure 7 Input information class

5.3   Additional Information Class

As shown in Figure 8, there is either no instance variable in AdditionalInformation but there is one method. This method sends messages to two classes.

In NameOfAdditionalInformation there are one instance variable and two methods. One of the methods displays the list of additional information.

In ContentsOfAdditionalInformaion there are methods that install the additional information, such as application form to the government.

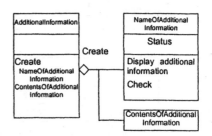

Figure 8 Additional information class

5.4   Interdependent Class

The interdependent class exists only if there are interdependent activities to the focused activity. Interdependent has four sub-classes as shown in Figure 9.

NameOfInterdependent has one instance variable and two methods. The instance variable, the Boolean type, indicates the execution of the method. One method controls this variable. Another method creates an instance of RelationTable.

In InterdependentInputInformation the method to display the input information from the interdependent activity, for instance, the alias of the CAD data, is defined.

In InterdependentOutputInformation the method to send the output information to the interdependent activity is defined.

InterdependentManual contains a method that reminds the user of the way to manage the interdependent activities by displaying messages.

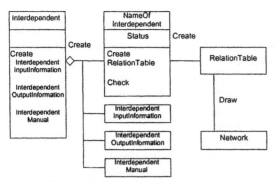

Figure 9 Interdependent class

5.5   Iteration Class

The iteration class exists only if there are iteration activities to the focused activity. Just like `Interdependent`, `Iteration` has four sub-classes as shown in Figure 10.

`NameOfIteration` has one instance variable and two methods. One of the methods creates an instance of `RelationTable`.

`IterationInputInformation` specifies the one that requires iteration among input information from the upstream activities. For example, when the whole input information is displayed by class name of the product model, this method selects and displays the class that includes information to be examined again. The user can recognize the temporary information by this selection.

`IterationOutputInformation`, in reverse, specifies the one that requires iteration among output information to the downstream activities. The method to select only the classes which iteration is required.

`IterationManual` contains a method to show the way to manage the iteration activities.

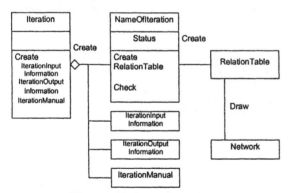

Figure 10 Iteration class

5.6   Output Information Class

The structure of output information class , shown in Figure 11, resembles that of input information.

In `NameOfLaterActivity` a method is defined to mark the checkbox to show this activity is properly executed. Another method creates an instance of `RelationTable`.

In `RequiredOutnputInformation` the lists of output information produced and stored through the activity are defined with class name in the product model as instance variables.

In `ContentsOfOutputInformation` the method to create an instance for classes that contain CAD data will be defined.

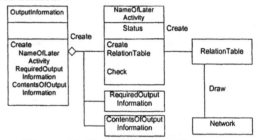

Figure 11 Output information class

5.7   Relation Table Class

The class structure around `RelationTable` is shown in Figure 12. As mentioned, the instance of `RelationTable` can be created by four classes above. `RelationTable` sends message to `Network` to draw the network.

`RelationTable` contains number, name, and influence of the activities of upstream side, interdependent, downstream side, and iteration as instance variables as shown in Figure 13. In this paper, the degree of influence between the activities is evaluated subjectively assuming to have four values of {0.0, 0.3, 0.7, 1.0}. The activities included within the iteration circuit are deemed to be 1.0.

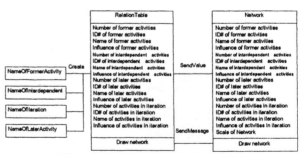

Figure 12 Relation table class

5.8   The Product Model

As Ford et al. proposed, in modeling at a lower level of abstraction there is the need to make compromises when attempting to incorporate the differing information needs of the various users of information [9]. In this paper, the product model was divided roughly into seven kinds, referring past studies [21, 25]. The structure is shown in Figure 13. The detail is also shown in Figure 14.

`Building` is an aggregation class of the physical material information. `Site` is an aggregation class that stores information concerning the site of the project. `ArchitecturalSpace` is an aggregation class that stores concepts and images of each space. `ConstructionMethod` has sub-classes such as

`ReinforcedConcreteStructure` and `SteelStructure`. They are aggregating classes that store the information about procurement and construction method. `Schedule` and `Cost` are aggregating classes of the information for project management. The condition of the total project is stored in an aggregating class `ProjectCondition`. They are defined for a typical building project.

Common instance variables among multiple classes are indicated in Italic. For example, the instance variable *BuildingArea* of `Schedule` is also the instance variable of `Building`. These instance variables are attributes of each class. To improve the productivity, the users have to access only one class, then other classes acquire the updated value by referring that class.

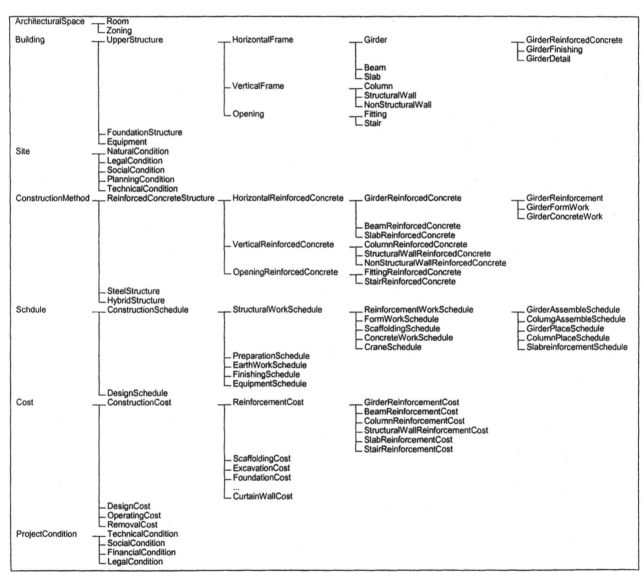

Figure 13 Class structure of the product model

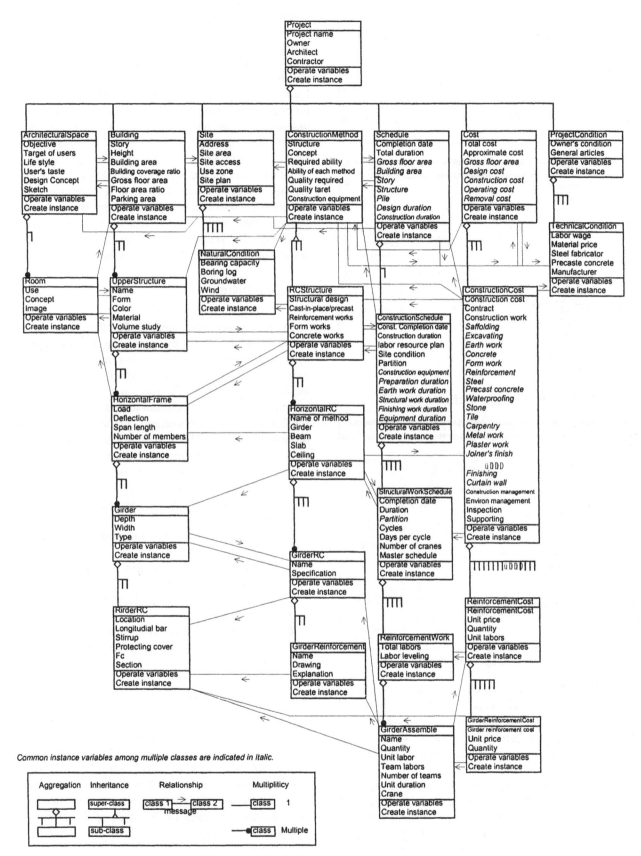

Figure 14 Relationship among classes

619

6   Example of the System Implementation

The prototype of the above-mentioned object model was designed in the object-oriented language, Object Pascal. The development was on the environment offered by Delphi [3], and Paradox [4] for the background relational database.

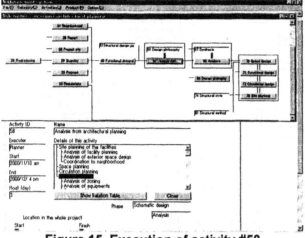

Figure 15   Execution of activity #58

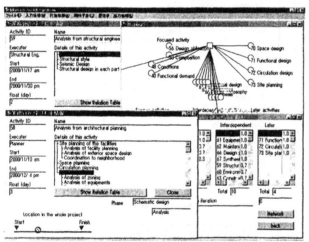

Figure 16   Interdependent activities

A small restroom in a park was adopted as an example of a building project. The execution of the activity #58 "Analysis from Architectural Planning" is shown in Figure 15. The step here is an example of managing interdependent activities. An architect or a consultant is in charge for the activity #58.

The information for executing this activity is shown in Figure 15. For example, it is necessary to recognize the interdependent activities at the time of "Fix the activities the design results interfere" in "Normal Design" in Figure 2. The architect can find the activities related to the

activity #58 by referring the relation table and network as shown in Figure 16. Activity #59 exists among such activities.

If the user opens the window of activity #59, the structured details of activity #59, such as "Analysis of the load of the roof", are displayed. The class name and instance variables of the product model that becomes the destination of the output information are defined in this model. For example, `ArchitecturalSpace` has relationship with `RequiredOutputInformation`.

As the users are supported by such a system, the effort for the management can be reduced. It is possible to refer the interdependent class, the architect is reminded to communicate with the structural engineer who is executing the activity #59. They can collaborate in architectural planning and structural design.

7   Conclusions

In this paper, the method for the executing the activities is discussed by proposing the management system that integrates the process model, the activity execution procedure, the product model and the engineering programs. To apply the idea of CE into a building project, it is important to secure preciseness and flexibility in the information produced from multi-discipline activities. This system is described the first step of the information-integrated paradigm.

The obtained findings are summarized as follows:

1. The execution procedure of a general activity is modeled, and the checklists in the procedure are proposed.
2. The framework of the management system was shown.
3. The prototype of the system was developed using object-oriented language. The integration of the process model and the product model was shown in the prototype.
4. Using the visibility and inheritance of object-oriented design, the career growth and the flexibility were enabled.

Acknowledgement

The authors wish to acknowledge the help and advice received from Dr. S. Hokoi, Dr. M. Yamazaki, and Dr. Y. Zhang of Kyoto University, Mr. E. Miyake of Ritsumeikan University, and Mr. K. Kimoto of Konoike Construction Co., Ltd. This research was supported by the Grant-in-Aid for Scientific Research of the Ministry of Education, Science, Sports and Culture, the Japanese Government.

References

[1] M. Betts and T. Wood-Harper, "Re-Engineering Construction: A New Management Research Agenda," Constr. Mgmt. and Economics, Vol. 12, pp. 551-556, 1994.

[2] G. Booch, Object Oriented Design with Applications, The Benjamin/Cummings Company, Inc., 1991.

[3] Borland Delphi 2.0J for Windows 95 & Windows NT

[4] Borland Paradox 7J for Windows 95 & Windows NT

[5] D. E. Carter, B. S. Baker, Concurrent Engineering: The Product Development Environment for the 1990s, Addison-Wesley Publishing Co., Inc., 1992.

[6] K. Currie and B. Drabble, "Knowledge-Based Planning Systems: A Tour," Int. J. Project Mgmt., Vol. 10, No. 3, pp. 131-136, 1992.

[7] J. E. Diekmann and H. Al-Tabtabai, "Knowledge-Based Approach to Construction Project Control," Int. J. Project Mgmt., Vol. 10, No. 1, pp. 23-30, 1992.

[8] M. A. Fischer, F. Aalami, C. Kuhne, and A. Ripberger, "Cost-Loaded Production Model for Planning and Control," Proc. 8th Int. Conf. on Durability of Building Materials and Components, CIB W78 Workshop, Vancouver, Canada, pp. 2813-2824, 1999.

[9] S. Ford, G. Aouad, J. Kirkham, P. Brandon, F. Brown, T. Child, G. Cooper, R. Oxman, B. Young, "An Information Engineering Approach to Modelling Building Design," Automation in Constr., Vol. 4, pp. 5-15, 1995.

[10] T. Froese and J. Rankin, "Construction Methods in Total Project Systems," Proc. Int. Computing Congress, ASCE, Boston, MA, pp. 383-394, 1998.

[11] S. Furusaka, T. Kaneta, M. Yamazaki, Y. H. Yi and S. Ishibashi, "New Approximate Estimation System of Building Using Neural Network," Proc. Joint Triennial Symposium CIB W65 and W55, Cape Town, South Africa, pp. 870-878, 1999.

[12] S. Globerson, "Impact of Various Work--Breakdown Structures on Project Conceptualization," Int. J. Project Mgmt., Vol. 12, No. 3, pp. 165-171, 1994.

[13] T. Heath, D. Scott and M. Boyland, "A Prototype Computer-Based Design Management Tool," Constr. Mgmt. and Economics, Vol. 12, pp. 543-549, 1994.

[14] Japan Institute of Architects, Architects' Professional Services, JIA, Tokyo, Japan, 1992.

[15] T. Kaneta, H. Nagaoka, S. Furusaka, K. Kimoto, and H. Okamoto, "Structural Analysis of Design and Construction Process of Building - towards Concurrent Engineering," Proc., 1st Int. Conf. on Concurrent Engineering in Construction, London, UK, pp. 77-86, July 1997.

[16] T. Kaneta, H., Nagaoka, S. Furusaka, K. Kimoto, and H. Okamoto, "Process Model of Design and Construction Activities of a Building," Computer-Aided Civil and Infrastructure Engineering, Vol. 14, pp. 45-54, 1999.

[17] T. Kaneta, H. Nagaoka, S. Furusaka, K. Kimoto, and Y. Zhang, "Configuration Management in Design and Construction of Building Using Concurrent Engineering," J. Archit. Plann. Environ. Eng., AIJ, Tokyo, Japan, No.492, pp. 163-170, 1997.

[18] T. Kaneta, H. Nagaoka, and S. Furusaka, "Integrated Modeling of Information in Concurrent Building Construction Project," J. Archit. Plann. Environ. Eng., AIJ, Tokyo, Japan, No.497, pp. 171-178, 1997.

[19] F. Kimura, "A Computer-Supported Framework for Concurrent Engineering," Proc. JSPE/IFIP TC5/ WG5.3 Workshop, Tokyo, Japan, pp. 345-359, 1993.

[20] S. H. Lim, N. Juster and A. Pennington, "An Information Support System for Enterprise Integration," Concurrent Engineering: Research and Applications, Vol. 5, No. 1, pp. 13-25, 1997.

[21] G. A. Nederveen, F. P. Tolman, "Modelling Multiple Views on Buildings," Automation in Constr., Vol. 1, pp. 215-224, 1992.

[22] R. Oxman, "Data, Knowledge and Experience in Multiuser Information Systems," Constr. Mgmt. and Economics, Vol. 13, pp. 401-409, 1994.

[23] B. Prasad, Concurrent Engineering Fundamentals: Integrated Product and Process Organization, Prentice-Hall, Inc., Upper Saddle River, NJ, 1996.

[24] M. W. Radtke and J. S. Russell, "Project-Level Model Process for Implementing Constructability," J. Constr. Engrg. and Mgmt., ASCE, Vol. 119, No. 4, pp. 813-831, 1991.

[25] K. F. Reinschmidt, F. H. Griffis, P. L. Bronner, "Integration of Engineering, Design, and Construction," J. Constr. Engrg. and Mgmt., ASCE, Vol. 117, No. 4, pp. 756-771, 1991.

[26] J. Rumbaugh, et al., Object-Oriented Modeling and Design, Prentice-Hall, Inc., 1991.

[27] V. E. Sanvido, B. C. Paulson, "Site-Level Construction Information System," J. Constr. Engrg. and Mgmt., ASCE, Vol. 118, No. 4, pp. 701-715, 1992.

[28] O. Shaked and A. Warszawski, "Knowledge-Based System for Construction Planning of High-Rise Buildings," J. Constr. Engrg. and Mgmt., ASCE, Vol. 121, No. 2, pp. 172-182, 1995.

[29] G. Winstanley and J. M. Kellett, "A Computer-Based Configuration and Planning System," Int. J. Project Mgmt., Vol. 11, No. 2, pp. 103-110, 1993.

CHAPTER 11

Knowledge-Based Concurrent Engineering

Taking KBE into the Foundry

C. N. Bancroft, S. J. Crump, P. J. Lovett, D. Bone and N. J. Kightley
Knowledge Engineering and Management Centre, School of Engineering, Coventry University,
Coventry, UK. CV1 5FB
Telephone: +44 (0) 24 76888999; Email: c.bancroft@coventry.ac.uk

Abstract

This report on research and software development at the Knowledge Engineering and Management Centre, Coventry University, updates a paper presented at the CE'99 conference. It recaps briefly on Knowledge-Based Engineering (KBE) and the aims of the REFIT project. The progress of the software being built is described, including new collaborative projects with Butler Manufacturing and Armitage Shanks. Additions to the KBE system development methodology that is under construction are discussed, with an emphasis on the work on knowledge elicitation and representation. Some wider issues are considered, specifically the relationship between KBE and knowledge management, and problems that can occur in KBE projects. The paper also discusses the transfer of knowledge and skills relating to KBE to the foundry industry, and a new multi-media training package being constructed for this purpose.

1   Introduction

This section recaps briefly on the material that was presented at the CE'99 conference, and is described in [1].

KBE entails the capture and embedding of specialised engineering knowledge in a computerised application. This can then be used repeatedly, without the physical presence of the experts who have supplied the knowledge. The automation of repetitive processes saves time and frees the experts for more creative activities. KBE can be particularly effective in the support of concurrent engineering in product and tooling design. Knowledge relating to a wide range of disciplines can be incorporated in a design package, and reflected in designs, thus eliminating problems caused by their discovery at a later stage. Consequently, KBE development software often incorporates geometrical modelling features, or can be linked to one or more CAD packages.

The REFIT (Revitalisation of Expertise in Foundries using Information Technology) project is funded by the ADAPT initiative of the European Social Fund, to improve the competitive position of small foundries within the UK West Midlands. Nevertheless, its relevance is not limited to this sector of industry, and the results are expected to be applicable to most Small to Medium-sized Enterprises (SMEs) in the field of engineering and beyond.

REFIT aims to encourage and support the uptake of KBE by SMEs, using two methods. Firstly, "technology demonstrator" applications are being built, in co-operation with industrial partners, for use in an industrial environment. These will show that KBE applications offer practical solutions to real-world problems, and will demonstrate the competitive advantage they can provide. Secondly, a methodology for KBE system development is being compiled, to counter the lack of guidance for practitioners in this area.

2   Changes at the Centre

Since the REFIT project began, the Knowledge Based Engineering Centre at Coventry University has become the Knowledge Engineering and Management Centre. The change of name reflects its growing expertise in the area of Knowledge Management (KM) as well as the application of KBE solutions to engineering problems.

Although KBE was recognised as a discipline before KM, the former can be regarded as a specialised form of the latter. KM is chiefly concerned with the efficient collection, storing and sharing of knowledge between humans, (albeit often using computerised tools). KBE embeds that knowledge within a computerised application, which then uses it to process engineering information.

There are clear benefits to be gained from a cross-fertilisation of ideas between the two disciplines, and this extra dimension within the Centre presents exciting opportunities for their exploitation.

3   The Methodology – a Brief Update

It is necessary at this point briefly to describe the methodology for KBE system development that is being compiled in parallel with the demonstrator development work. It will be referred to in succeeding sections that discuss the applications being built. The methodology now has a name: KOMPRESSA (Knowledge-Oriented Methodology for the Planning and Rapid Engineering of Small-Scale Applications). It has been divided into 9 activity groups as follows, each containing activities, guidelines and techniques:

(1) Initial Investigation
(2) Application Classification
(3) Requirements Analysis
(4) Tool Selection
(5) Design
(6) Implementation
(7) Validation, Verification and Testing
(8) System Realisation
(9) Maintenance

Activity Groups 1, 3 and 5 are complete (subject to further evaluation and modification during the remainder of the project), and are described more fully in Section 9. Activity Group 5 (Design) is the largest, and includes the topics of knowledge elicitation and knowledge modelling, which are heavily based upon the experience gained with the industrial collaborators in the project.

One of the main aims of the methodology is to enable as much development work as possible to be carried out by employees of the organisation for whom the application is intended.

4   Butler Manufacturing

This company is a SME, with approximately 200 employees, based in the West Midlands in the UK. It produces gravity-fed and hard sand aluminium castings, and is currently enhancing its manufacturing capacity and introducing the low-pressure process. The product range includes engine components and charge air coolers for the automotive trade – precise, safety-critical components that require a high level of expertise.

Butler Manufacturing and REFIT began co-operating on the development of a KBE system to support tooling

design in January 1999. As with all foundries the company must design the tools required to produce castings according to customer requirements, which may be specified in the form of drawings or as electronic files in proprietary CAD formats. Tooling designs may be 3D computerised models, or 2D files or drawings, to suit the needs of the process to be employed for tooling manufacture.

In the case of the low-pressure equipment used at Butler the machine configuration that is needed to manufacture a casting consists of 15 fundamental components (see Figure 1). This includes the machine on which the tooling is placed, the tooling itself and the various pieces of equipment necessary to attach the tooling to the machine. The set of 15 models comprising the overall machine configuration for a given product design is known as a general assembly. Software is being developed in Pro/ENGINEER for the automatic generation of a general assembly based upon a crude approximation to a specific casting geometry. This is achieved by the use of a generic model of the general assembly with embedded knowledge about how it is affected by the geometry of a casting. As casting geometry is created, it is automatically reflected in the die design, which in turn is reflected in the remaining parts of the assembly, individual models and associated drawings.

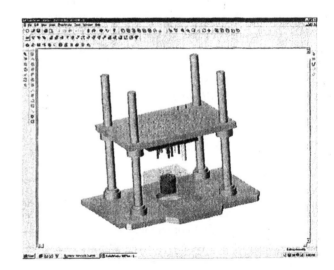

Figure 1. SolidWORKS model of casting tooling

The benefits to the company of this application are expected to be considerable. Most obvious are the time saved in producing the original tooling design, the liberation of time for the design office, and the provision of a degree of repeatability in the design of equipment for the manufacturing process. Furthermore, the capture of engineering best practice and its automatic application to new designs maximises the returns on the company's intellectual capital.

5 Armitage Shanks

Armitage Shanks Limited Brassware division began collaborating with REFIT in 1998. Candidate application areas included business strategy, product design, cost estimation, and training. In particular, a need was identified for the capture of knowledge from specific functional areas, to be made readily accessible throughout the company. A previous paper [1] described the need to transfer knowledge from the design office to the shop floor, and the Component Identification Tool was built for this purpose. It is a graphical database application that provides a breakdown of the cast components in any product. The next step was to enable knowledge from the shop floor to be communicated to the designers.

Work is now in progress on a second project that includes the capture, maintenance and recycling of "black art" knowledge from the foundry. The result will be a PC-based multimedia tool to assist in the training of new operatives and to educate designers and toolmakers in shop floor practices. It is hoped that this will reduce or eliminate recurrent problems related to design features.

The foundry has four functional areas: the core shop, metal management, the furnace room and finishing, all of which will be subject to the knowledge capture process. A KOMPRESSA unified knowledge model (described in Section 9.3) has been generated for the core shop, following a series of interviews with the senior core setter. Many of the components of this model take the form of process diagrams, of the type shown in Figure 2.

The training package for the core shop first introduces the user to the basic concepts of core making. This uses simple animated diagrams that are designed to be accessible to users at all levels. Specialist modules then explain in detail how a particular machine is set up and operated. This will depend on the type of user, with maximum detail for core setters and minimum detail for machine operators. Extensive use is made of photographs, with video clips to show critical operations, as illustrated in Figure 3.

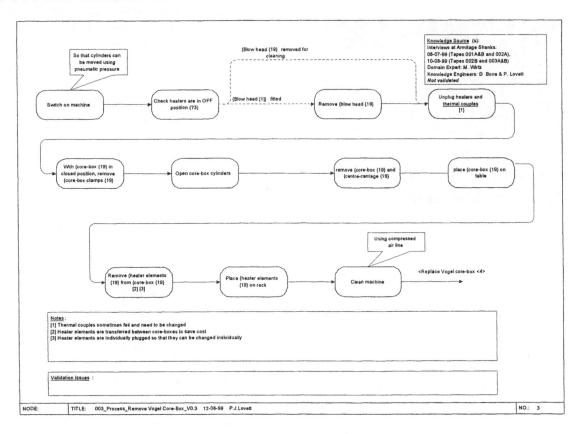

Figure 2. Model of a core shop process

Video recordings can both reveal and demonstrate "tricks of the trade" that might otherwise be missed. Thus tacit knowledge that is difficult or impossible to describe verbally can be transferred without awareness on the part of the knowledge provider or recipient.

A recent reorganisation has led to the transfer of core making machines between factories. Company personnel have stated that had the software and its documentation been available at this point, the machines would have been installed more easily and they would have been operational earlier.

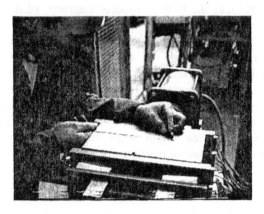

Figure 3. Stills from a video recording

6   PDC

PDC and REFIT began work in 1998 to produce a "Standards Drawings Generator", which was described in [1]. The user interface has now been greatly simplified, and a first copy of the software has been supplied to the company on CD for evaluation. Dramatic savings in the time taken to produce engineering drawings of brake discs have been demonstrated. Further enhancements are envisaged, including extending the range of drawings that can be produced, displaying tolerances, and increasing the quantity and quality of the on-line help.

7   Other Development Projects

REFIT is also working with two other companies. The Bridge Foundry project was described in [1]. The team is also collaborating with Rolls-Royce to develop a KBE system to support turbine blade manufacturing strategy. This will be the subject of a future publication.

8   Problems in KBE System Development

The following section recounts the experiences and opinions of the REFIT team relating to some of problems that have occurred within the project, and that can occur during the development lifecycle of a KBE application. Tips are given that may lessen the impact of such problems in future projects of this type.

8.1   Plan Hard, Plan Long and Keep Planning

REFIT began life with an ambition to give the foundry industry a competitive edge to evolve and survive by transferring KBE skills to foundry personnel. Industrial support for a project proposal was sought and obtained via a committee then known as Foundry 2000. Different individuals were responsible for drawing up the proposal and for managing the project, and in hindsight, the transition between these activities could have been better co-ordinated. In particular, there were many unknowns when the proposal was compiled, but these were not communicated effectively, and gave rise to the need for project changes at a later date. Project managers would be well advised to:

- Assume a proposal contains little planning
- Pay great attention to the transition between proposal and project management
- Seek out willing and experienced advisors
- Make detailed project planning part of the delivery of projects.

8.2   Selection of Project Partners

The nature of the funding means REFIT is a high-risk project. Target beneficiaries are SMEs facing a battle to survive. However, large companies were included as agents of change. REFIT started with industry-recommended project partners. Too little thought was given to profiling companies and individuals according to our own needs. This was true for the larger organisations as well as SMEs. Problems faced included:

- Company closure
- Redundancies
- Company Restructuring
- Staff movements, both internal and external
- Conservative, protectionist and dominant attitudes.

To reduce the likelihood of problems, partner selection should combine some formal assessment of suitability with past experience and subjective judgement.

8.3 Engaging Partners

One result of failing to integrate planning and management at the outset of the project was a lack of awareness that proposed partners had not committed to the project. However this soon became apparent. All partner relationships had to be developed before formal commitment was obtained. Relationships were developed through the following mechanisms:

- Education and awareness courses
- Company visits to commercial applications
- Trial or induction periods on the project
- Training by external parties
- Recommendations from external parties
- The identification of potential project champions
- Meetings and discussion groups with interested parties
- Supporting internal presentations made by potential project champions.

This led to strong partnerships in those that remained and a high level of commitment from project champions.

8.4 Maintaining and Managing Relationships

A reticence to share knowledge and the resistance and/or refusal to adapt to and use new technology are often described as barriers to the successful application of IT and knowledge management. This has not been apparent in the case of REFIT and its partner organisations. Readers may regard this as surprising given that REFIT's partners are in a fluid high-risk sector and have faced change to varying degrees. One possible explanation is that participating individuals see KBE as an investment in their own future. Advice on maintaining and managing relationships, based on the experience of the REFIT team, would include:

- Acknowledge that the clients are the experts and the developers are enablers.

- Stress that KBE supports a person's job function rather than replacing him.
- Agree the protection of the client's corporate knowledge.
- Be open and honest about issues such as cost, limitations, etc.
- Project manage: agree objectives, specifications, roles and responsibilities, milestones etc.
- Maintain frequent communication (verbal and textual).

8.5 Application Development Issues

Tacit knowledge has proved difficult to elicit. Within the REFIT partner group tacit knowledge has been referred to as tricks of the trade, rules of thumb, gut feeling, etc. Difficulties can be caused by:

- The inexperience of the knowledge engineer, and
- his unfamiliarity with the expert's domain
- A conceptually difficult domain area
- A small number of domain experts, making knowledge validation more difficult

The team have become increasingly adept at handling such problems as they have gained experience in the field. The solutions that have been discovered have been documented in KOMPRESSA, so that others can benefit from them.

8.6 Running Successful KBE Projects

The advice [2][3], on running successful projects or remaining competitive as an organisation is to facilitate change, in particular meeting the changing needs of the customer whilst minimising the "cost of change". Key attributes of those involved are flexibility and adaptability. Whilst in agreement with these sentiments, avoiding unnecessary change is seen by REFIT as the single most realistic way of maintaining project trajectory. By reducing and controlling "fire fighting", an organisation can facilitate the management of expectations of its customers and afford the occasional introspective reflection. Based on existing documentary evidence from project partners and the funding body support unit, it is projected that REFIT will meet the expectations of all supporting parties. The cornerstone of this success has undoubtedly been a shared belief of the validity of the work undertaken. All team members and indeed partners have demonstrated a willingness to learn and develop. These key attributes provide individuals, teams and organisations with knowledge, both theoretical and practical, that can then be drawn upon to reduce unnecessary mistakes, solve problems and facilitate the initiation, acceptance and management of change as necessary.

The theme of REFIT as an organisation, its achievements, failures and lessons learnt is to be the basis of a future publication.

9 KOMPRESSA – In Further Detail

As already indicated, KOMPRESSA aims to transfer much of the responsibility for KBE system development away from information systems specialists to employees of the client company. Ideally, only the coding would need to be carried out by external agents, although the extent to which this will be practical is yet to be seen. One reason for this aim is financial – REFIT is intended to improve the competitive position of foundries, so minimising expense is vital. A second reason is that managers and engineers within the organisation may be in the best position to carry out the task. (This is further discussed in Section 9.3). An attempt has therefore been made to make use of techniques and concepts to which managers and engineers can relate, and with some of which they will be familiar (such as IDEF0 [4], cause and effect diagrams [5], and decision trees [6]).

The rest of this section examines Activity Groups 1, 3 and 5 in terms of their purpose, how it is achieved, and some of the more important techniques they contain. KBE system development is not a linear process, and it should not be assumed that activities will be performed in the order they are described. Many will take place in parallel, many will not be relevant to all projects, and there will be a great deal of iteration. Advice on these aspects will be provided in Activity Group 2: Application Classification.

9.1 Activity Group 1: Initial Investigation

The activities in this group are concerned with administrative and business issues as much as technical ones. It is important to gain a thorough understanding of client organisations and the motivation for their KBE projects, which will ultimately be financial rather than technological. It should be noted that the terms "client" and "developer", as used within this text, *may* refer to distinct organisations. This will be the case if development work is being performed by external consultants. Alternatively, they may refer to different people or groups of people within a single organisation – the distinction between the terms being the development role rather than who is employing whom. The extent to which many of the activities in the Initial

Investigation will be necessary may depend on which of the above situations applies.

Topics that are addressed in this part of the methodology include:

- Making contacts and establishing a good working relationship
- Conducting interviews
- Identifying a project champion
- Either identifying a suitable application area, or justifying an existing proposal
- Problems that can occur with KBE system development, and reasons why such projects can fail
- A modelling technique (Activity Diagramming) to assist in some of the above areas, and for organisational analysis and documentation. It uses multiple viewpoints, including information and information flows, lines of communication and responsibility, and knowledge sources and bottlenecks. It is based upon an adaptation of the IDEF0 method, which will be familiar to many of its intended users. An example is shown in Figure 4.

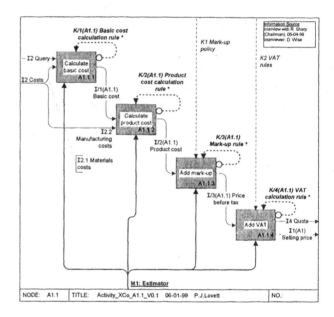

Figure 4. An activity diagram

9.2 Activity Group 3: Requirements Analysis

The activities within this group do not differ greatly from those of more traditional (i.e. not knowledge-based) information systems. They are included because KOMPRESSA is intended to be suitable for use by those with little or no software development experience.

Essentially, requirements analysis is concerned with ensuring that the application to be built will fulfil the needs of the clients (and it may therefore also be necessary to educate them in the capabilities of KBE). The aim is to avoid the danger of providing what may be a first class solution to the wrong problem. The "acid test" of whether the users' requirements have been met is the extent to which the finished system is used in practice. Reviews throughout system development will confirm that the requirements are being adhered to.

The techniques to be applied within this activity group are, whenever possible, either the same as, or extensions of, those already introduced, and new techniques follow established principles. This philosophy is applied throughout the methodology.

Requirements have been divided into the following categories:
- Functionality
- User interface
- Information
- Knowledge elicitation
- Performance

The only aspect that differs significantly from traditional information systems is that of knowledge elicitation requirements analysis. This involves identifying major sub-divisions of the knowledge to be scrutinised, and the sources or owners of that knowledge. (Sources may be human, textual or take some other form). An initial strategy for obtaining the knowledge must then be devised, and availability of the sources must be assured.

The representation methods for user requirements are largely graphical, with supplementary text. This allows the overall structure of the content to be clearly depicted, without sacrificing detail. In addition, hierarchical decomposition is used extensively to provide high-level abstractions of complex subjects. This is true of the methodology as a whole, and will be illustrated in the examples provided in section 9.3.

The methodology stresses the need for the contents of the requirements specification (which is the document produced as a result of the analysis) to be formally agreed by all interested parties. This includes at a minimum the developers, the clients, and the prospective users.

9.3 Activity Group 5: Design

Three aspects of design are embraced within the activity group:
- Functional design
- User interface design

- Knowledge capture

Information requirements are satisfied by knowledge capture activities, and performance requirements are reflected in all aspects of design, and also by implementation issues, dealt with in Activity Group 6. This is illustrated in Figure 5. The three aspects of design are highly inter-dependant, and mechanisms are provided to link the models used for their representation.

Techniques and guidelines are of course provided for functional and user interface design. However, this section will concentrate on knowledge-related activities, which distinguish KBE from conventional software.

An important feature of knowledge based systems, including KBE, is the extraction ("elicitation") of knowledge from individuals ("domain experts") and other sources, and its representation. Building a KBE system will always be a co-operative exercise between domain experts and specialists in knowledge-related technologies ("knowledge engineers"). This situation presents a fundamental problem – the knowledge engineer has the skills to embed knowledge in a software package, but the subject matter will be outside his normal sphere of experience. On the other hand, while the expert understands the knowledge, he does not know how to present and prepare it for inclusion in a computerised application. To bridge this "knowledge gap", one or the other must learn about the other's job, thereby creating some common ground. Usually most movement is made by the knowledge engineer.

KOMPRESSA is based on the view that there are considerable benefits to be gained if individuals from the knowledge-holding organisation play a larger part in knowledge elicitation and modelling than has erstwhile been the case. Firstly, this approach is likely to be less costly. Perhaps more importantly, it may be easier for individuals with some knowledge of the domain area to learn techniques for knowledge capture than for an external knowledge engineer to comprehend technical subject matter with which he is entirely unfamiliar.

Many types of knowledge are used in engineering design [7] – hence KOMPRESSA provides a correspondingly wide range of elicitation techniques (28), and modelling techniques (17), together with advice on selecting those most suitable for the task in hand. The models are more than a means of documenting knowledge. They are used as a communication aid in elicitation sessions, and incorporate a validation mechanism. Figure 6 shows an example of one type of knowledge model provided by KOMPRESSA – A component diagram. This type of model can be used to represent physical objects, or less tangible things such as organisations, problems,

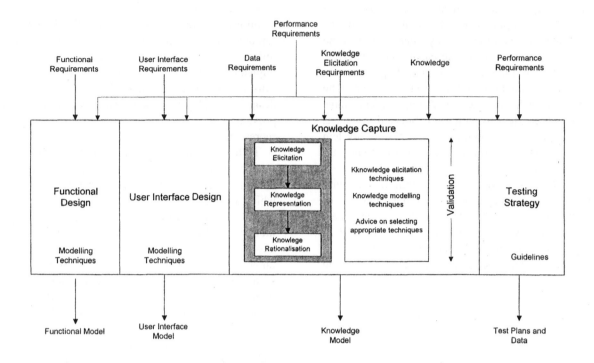

Figure 5. Design activities

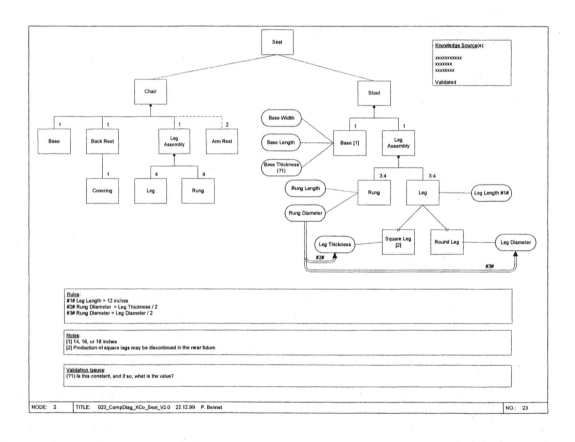

Figure 6. A component diagram

632

responsibilities, etc., and the elements of which they are composed. It can also be used to depict characteristics of these elements, and rules relating to them.

Figure 6 also illustrates two important mechanisms used in all models. The box in the top-right corner is used for recording knowledge sources, and the box at the bottom records outstanding validation issues.

One of the problems that is sometimes encountered when building KBE systems is the large volume of knowledge that can be generated, and which may be difficult to navigate unless it is well structured. For this reason individual knowledge models are linked via references, and a "unified knowledge model" (see Figure 7) is created. This provides an overview of the knowledge for the whole project. Each node on the tree can refer to any of the 17 types of individual knowledge model.

10 NO DEFECTS

An industry-wide awareness of REFIT's aims, activities and achievements is cultivated through close collaboration with key trade bodies within the foundry sector. The transfer of KBE knowledge and skills to technicians, engineers and managers within the industry is achieved using three key mechanisms:
- KOMPRESSA – as described in Sections 3 and 9.
- Collaborative focus groups – working together on shared problems with the guidance of the REFIT team.
- NO DEFECTS.

NO DEFECTS is a 12-month project to develop and deliver a foundry-specific training programme to employees within the industry. The programme, entitled "An Introduction to Knowledge-Based Engineering" aims to equip participating individuals and organisations with both the tools and knowledge to become practitioners of KBE and reduce the need for contract support. The programme benefits from the aid and involvement of the Institute of British Foundrymen and complements their vision for continuous improvement within the UK casting industry.

There are three main elements of the programme: development, evaluation and delivery. Programme development draws on industrial and academic expertise and experience. Core training material is taken from REFIT's findings and experience and is strongly linked with the KOMPRESSA methodology. This material is released to a group of evaluators whose role is to review the material and structure for relevance and suitability for the target audience. A period of face-to-face training for up to 50 foundry employees will take place. Feedback

from those trained will be used to shape the finalised syllabus and course material. Each employee will spend 40 hours of contact time being trained. Following this period the programme will be amended and will then be made available on CD ROM along with support for its use. The initial target is to train around 250 employees using this method.

In order to support the course material and reinforce learning the programme includes a practical element that enables students to gain first-hand experience with KBE software. A broad outline of the syllabus is as follows:
- An Introduction to Knowledge Based Engineering
- REFIT, Methodologies and KOMPRESSA
- First Steps
- Setting the scope of a project
- Design
- Knowledge capture
- Programming
- Testing
- System realisation
- Maintenance and project review

11 Future Work

REFIT terminates at the end of 2000. Before then it is planned to:
- Complete the 6 technology demonstrator applications.
- Complete the first version of KOMPRESSA, paying particular attention to the areas of KBE system testing and maintenance.
- Complete and deliver the NO DEFECTS training package.

The results of the research and practical experience emanating from the REFIT project will be applied to future work at the KEM Centre (both research and business support). KOMPRESSA itself would benefit from further testing and evaluation, and opportunities will be sought to apply and review the methodology.

Training activities will continue, based on NO DEFECTS.

12 Summary and Conclusions

KBE presents real opportunities for small foundries to obtain advantage an intensely competitive market. However, there is a reluctance to invest time and money in a technology that is still considered to be unproven for smaller organisations. REFIT is helping to provide the evidence that such systems are viable, in the form of technology demonstrators that operate in an industrial environment.

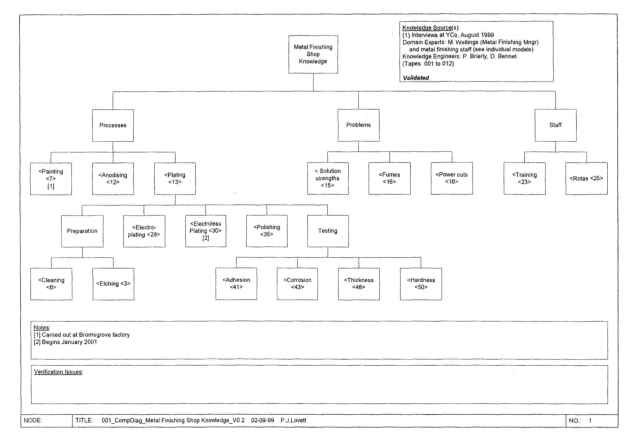

Figure 7. A unified knowledge model

The development of KBE systems is not a trivial exercise, and the provision of methodological support (KOMPRESSA) should both encourage practitioners to undertake the task, and improve the quality of finished systems. Those companies collaborating with REFIT in the project have already benefited from the transfer of the necessary skills. It is hoped they will form a core of informed enthusiasts, to encourage the spread of KBE and the rewards it can provide.

The NO DEFECTS software and training program will enhance this process, and will extend the opportunity to learn about KBE to many more of those working in SMEs in the foundry industry. Other sectors will also be encouraged to take advantage of the new technology, which has much to offer throughout engineering.

Acknowledgements

The authors would like to acknowledge the contribution and support of all the members of the REFIT consortium, particularly that of the lead partners: Butler Manufacturing, Armitage Shanks Ltd., Brassware Division, Precision Disc Castings Ltd., The Bridge Foundry Company Ltd. and Rolls-Royce PLC. We thank the trade bodies IBF, BICTA and BMCA for their co-operation and assistance with the NO DEFECTS project.

References

[1]   C. N. Bancroft, S. J. Crump, P. J. Lovett, D. Bone, A. Ingram and N. Kightley, "REFIT – Knowledge Based Engineering Applied to the Foundry Industry", Proceedings of CE'99 (Advances in Concurrent Engineering), 327-335, 1999.

[2]   K. H. Payne et. al., "MACRO and the Future of Integrated Product Development", Proceedings of CE'99 (Advances in Concurrent Engineering), 233-240, 1999.

[3]   A. Taylor, "IT Projects: Sink or Swim", The Computer Bulletin, January 2000, British Computer Society, 24-26, 2000.

[4]   Federal Information Processing Standards Publication 183 – Integration Definition for Function Modelling (IDEF0), National Institute for Standards and Technology, 1993.

[5]   Ishikawa, K (translated by Lu David J, "What is Total Quality Control? The Japanese Way", Prentice-Hall, Englewood Cliffs, N.J., 1985.

[6]   D. J. Flynn, "Information Systems Requirements: Determination and Analysis", McGraw-Hill, 1992.

[7]   P. A. Rodgers and P. J. Clarkson, "An Investigation and Review of the Knowledge Needs of Designer in SMEs", The Design Journal, 1(3), pp 16-29, 1999.

Knowledge Needs and New Products Development within a Concurrent Engineering Environment

J.-L. SELVES*, Z.Y. PAN*, E. SANCHIS*

* Laboratoire Gestion et Cognition . IUT Ponsan – Université Paul Sabatier.
115, route de Narbonne. 31077 - TOULOUSE – Cedex. FRANCE

Abstract

Related to new products development processes the concurrent engineering approach is characterised by a project team that tries to develop new products using co-operative work. It responds more quickly to changing market conditions and can be improved and fertilised with the information and communication technologies. Thus, a process to capture, organise and use multilevel knowledge is defined and revised using information and communication technologies. Knowledge is partitioned into modules or elements that are composed of documents associated with multilevel knowledge and, a software agent. These elements are available in resource centers or web sites and project members could ask them to build solutions. All these operations and resources constitute a knowledge management system adapted to co-operative problem solving and produce a project knowledge card, which allows controlling new products development processes

1 - Introduction

New products development processes led us to consider the conception project management which is associated to two goals: on one hand a goal of organisational and technical search for solutions, on the other hand a goal of making which clarifies procedures and necessary means to set-up the project. These goals are carried out within the framework of a process, which can be characterised by approach, organisation, resources and methods.

1.1 Approach

Two models include most approaches related to new products development processes: the traditional sequential approach and the concurrent engineering approach [1]. The traditional approach, where the project went sequentially from phase to phase [3] (concept development, feasibility testing, product design, development process, pilot production and final production) and from functional team to functional team, was largely used in manufacturing industry but may conflict with today's competitive requirements: speed and flexibility. The last approach, where the overlaps extends several phases, is characterised by a project team that tries to develop new products using co-operative work and can respond more quickly to changing market conditions. Since 1990, it is considered as the best model in most development projects [2, 3, 4,14].

1.2 Organisation

In term of organisational structure of the team project, there are four types of structure: functional structures, lightweight team structure, heavyweight team structure (matrix structure), autonomous team structure [23]. In the context of concurrent engineering, matrix structure and autonomous heavy structure are the most often adopted.

1.3 Resources and methods

Now let us consider resources and methods. Till the end of the years eighty, the researchers tended to consider that the team of conception, process data and information to model systems to be conceived on an individual basis.The project team was considered as a system of search and production of information and the driving force of the search for solutions is the intelligence of the team. Thus, in the years ninety the researchers moved towards the organisational learning, and, by introducing temporal dimension into the process of creation of knowledge [22] propose an ideal model of the process which includes 5 steps: the sharing of tacit knowledge, the creation of concepts, the justification of concepts, the construction of an archetype, the extension of the knowledge in the various levels of the organization. With this approach one can consider the team of project as an organization of creation of knowledge. The driving force of the search for solutions is the capacity of learning and the creativity of the team.

Thus, in this article we shall study quite particularly the knowledge, which constitute the main resource in projects and the methods to acquire, to share, to use and to increase this knowledge [15, 16, 17]. This within the framework of the concurrent engineering, that is, with a multidisciplinary team working in a cooperative way through cycles of negotiation. At first we shall describe, from a real case, the new product development process within concurrent engineering, then we shall interest in the modelling of this process, the use of knowledge in problem solving and finally we shall propose a solution of knowledge management fertilised with the information and communication technologies [18,19,21].

2 - A Typical Approach for Concurrent Engineering (CE)

Within a CE environment a multidisciplinary project team works to reach goals like quality, low price and maximum speed [5]. Everyone must understand the other's person position and try to talk to each other with a "multidisciplinary language". The team members operate with two kinds of knowledge: the necessary knowledge from across all areas included in the project for evaluation and negotiation and, knowledge accumulated on an individual basis and a narrow area of focus for studying in depth new components [6]. To understand a development process it seems natural to be interested in the car industry because this one was in the base of the project management and we will study quite particularly, in the case of the TOYOTA company because it is considered as a model for the concurrent engineering [24]. The vehicle development cycle is organised around key milestones [24]:

- Evaluation report which includes customer reactions to the current product and predicted market conditions
- Vehicle concept or the qualitative description of the target customer, vehicles characteristics and important competitive features.
- Vehicle sketches or various styling alternatives which includes several functional groups studies (interior body design,...) and supplier presentations.
- Concept approved. It consists to make from five to twenty different 1/5 scale clay models.
- K4 approved. K4 are functional documents that contain many sketches and drawings with dimensions. Once each functional group reach agreement, the K4 is widely circulated for approval.

- First prototype based on K4 documents, internal CAD data and component prototypes from suppliers.
- Second prototype based on test results from the first prototype
- Production trial

The architecture is modular and each module corresponds to a subsystem studied by a functional group. That allows the members of the team project to work under a matrix organisation. For each module, the designers make their search for several solutions, which are explored and experimented first at subsystem level, and if, needed at the system level. The coherence of interfaces is insured with a checklist of the constraints, which indicates limits not to be exceeded. A schema is proposed by the authors to represent Toyota set-narrowing process (fig 1)

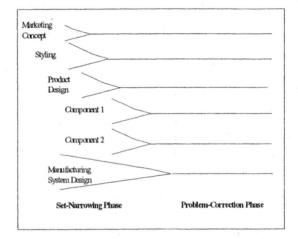

Figure 1 Toyota Set-narrowing Process [24]

In this process, first of all, two types of knowledge are confronted: knowledge on the customer (evaluation of the previous products, current and future needs) and knowledge on the product, to define goals to reach. Then negotiations are engaged to determine, according to functions and performances, the characteristics of the modules, which allow the emergence of the concept. From the concept, the various functional groups and the subcontractors helped by documents of specifications and constraints work to verify feasibility and to build a prototype. Finally more precise documents on various possible solutions are published and are widely spread in the company to be approved and to lead to the final prototype. We find indeed here the five phases defined by I. Nonaka and H. Takeuchi [22] and quoted in the introduction and we can distinguish four types of knowledge, which move during the project:
- The targets which become technical specifications.
- The constraints (manufacture, previous projects, know-how,) which limit the space of solutions.

- The graphic representations (CAD) which determine spatial configuration and correspond to propositions or solutions.
-The results of tests made with real representations or prototypes. They allow evaluating solutions.

At this point, we have to note that the originality of this product development process is related to the parallel search for solutions based on an important number of alternatives. Each alternative being studied in a very detailed way. This is made possible by a voluntary policy to delay at most the choice of alternatives.

3 - Methods used in Concurrent Engineering

Concurrent engineering belongs to collective phenomena setting up a self-intelligence which one does not explain and understand. Thus, about Toyota the author's [24] often speaks of "absurd behavior" and "against the logic" and in the same time of " the best results". It is not a question of analyzing a process by cutting it in tasks like the classic engineering, but to build solutions at various levels using emergence, that is to configure in an optimum way a complex set of components or modules to reach a goal or to realize a function. Planning or rational global approaches are not able to solve this type of complex problem. Only members of a team in interaction, each of them with his intelligence finding a solution which becomes integrated in an optimum way into a global structure, allows to move ahead in the resolution of this type of problem.

H Takeuchi and I. Nonaka [3] compare the practices of the concurrent engineering in the process of development of new products with those that are used in a game: rugby. There is a goal: score, and the ball passes from a player to the other one not according to a plan elaborated before the contest but according to actions and decisions which are linked to multiple factors that every player has to estimate (the context, the rules of the game and its personal experience). This set of actions is related to the auto-organization or to the emergence within not programmed situations. Indeed, every player has to made what it is necessary at the right time to become integrated into movements of a team.

The terms of collective intelligence, auto-organization and distributed problem solving are also used to explain behavior builder of the social insect's [25]. This behavior is certainly simpler than those of the members of a team in product development projects, but they were well studied. What allows proposing methods to build an object (the nest) collectively without a preliminary plan. Indeed, G Theraulaz and E. Bonabeau [26] explain behavior builders by means of several types

of logics or mechanisms that exclude plan or guide of building:
-The "gabarit" which consists in building forms by using as guide material or chemical existing forms (chemical gradient of concentration).
-The "stigmergy» where the building activities are directed and are activated by local configurations of material. It is a succession of generally programmed stimuli-answers which are linked in the space and in the time and which leads to a coherent collective construction.
-The "auto-organization" where the structures appear at collective level by using an autocatalytic logic and the physical and geometrical constraints of the environment.

The development of new products, the moving of players and the construction of a nest are processes which come true from situations (current conception, current nest, situation of the players) which activate events and then coordinated actions (new solution, change of the ball, to bring materials) to realize finalized architecture (prototype, score, nest). The specificity of the concurrent engineering results essentially from the coordination and from the coherence of choices or solutions to set-up architecture or structure coherent and optimized. The study of the behavior of the social insects allows us to propose three mechanisms that we can apply to the other processes:
- "Gabarit" is similar to the volume and shape (persons or materials transported) for the building of a vehicle, and, to the available space for the players of rugby.
- The "stigmergic » logic which leads to answers or solutions which belong to a predictable set. For the game every member of the team, in most of cases, chosen a solution among a set of solutions which is known by all the team. These solutions depend on constraints, on the experience or are defined before the contest. The same reasoning may be applied to the conception of product.
- Actions or solutions of emergent and "auto-organized" type. The solution is new for a member of the team and appears from a situation (architecture or structure). Then all the members of the team (development, integration and appropriation) adopt it. This mechanism corresponds to the innovation.

While for the social insects the three mechanisms are generally used separately, they could be used simultaneously for the other processes. We shall notice here that the comments of the authors on Toyota's methods using relatively unstructured development process and multiplying prototypes is necessary to obtain solutions of emergent and auto-organized type.

Nevertheless, difficulty to understand the functioning of these groups where the activity and the intelligence are distributed and where experiment is impossible, has led to create groups completely artificial, multi-agents systems (MAS) in software domain [27], to study interactions between entities and emergent structures. They allow to conceive and to realize organizations of artificial agents capable of acting, of collaborating in common tasks, of communicating, of reproducing and of planning their actions to reach objectives. With this MAS, it is impossible at the moment to simulate processes such as those that are studied in this paper because problems are functionally very distributed and diverse. However, they give us models of entities (the software agents) that can help or replace the human being in the execution of certain tasks related to the concurrent engineering.

Thus, we at first defined types of knowledge (objectives, constraints, solutions, evaluations) which evolve during the progress of the project, then the mechanisms which allow to use them to resolve problems collectively. It is a question now of studying on concrete case the search for solutions or the passage from event to action.

4 - Knowledge and Problem Solving

Chosen case concerns petroleum industry [7] but treated problems could arise in similar terms for other materials or other industrial installations. Indeed, it is essentially the process of problem solving which is examined here by learning on decision model (intelligence, conception, selection) proposed by J.-L. Le Moigne [28] and shown on the Figure 2. This to characterize knowledge used in industrial problem solving.

Let us consider the following problem: an oilfield is situated in the country X and the customer in the country Y. A pipeline transports the petroleum. The customer and the supplier wish to increase exchanged quantity.

In a first phase the dissonance which appears concerns difference between the supplied quantity and the wished quantity. This problem solving is related to the question: which transport ? Or to consider the function transport (intelligence). Several solutions are possible: the ground way, the sea route, the pipeline. Evaluation (possibility of transporting this petroleum, of increasing output in the pipeline, ...) can be realized by taking into account the context (geographic, economic, political), the previous experiences, know and know-how. If selection is impossible, it is then necessary to evaluate in parallel several solutions or to request outside experts (supplement of information). Let us admit that, in this situation, the solution, to increase output in the pipeline, is better than the others. It is then possible to go to the second phase.

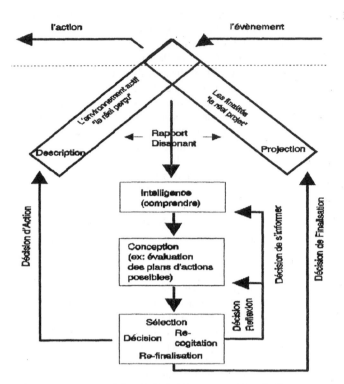

Figure 2   Decision Model Proposed by J.-L. Le Moigne [28]

In the second phase the dissonance which appears concerns difference between the wished output and the current output. Which parameter to modify to increase the output of this quantity? By using knowledge concerning the mechanics of fluids it is possible to propose a calculated modification for the main factors which condition the flow: pressure, temperature, viscosity (chemical modification). Using now the know-how it is possible to inventory the material and the practices to be implemented to modify these factors and to estimate solutions (quality, cost, delay). If this evaluation leads us towards the chemical modification of the viscosity, several methods exist by addition of the another fluid [7] (another crude oil, oil extracts, solvents or additives). The choice can require consequently another phase and/or parallel search for solutions based on an important number of alternatives (including combinations). In this situation, the chemical modification of the viscosity with additives is considered as the best solution.

In the third phase dissonance appears on the viscosity. Which additive to choose and in which quantity so that viscosity is lowered of x %. To know them chemical knowledge should have to allow us to characterize the petroleum and according to these characteristics to choose the additive. This is not possible and this led us to envisage two methods of choice [7]:

- Reasoning by analogy in classifying oils and additives following chemical characteristics. Then according to previous experiments, relating oils and the most effective additives and finally making tests (laboratory or pilots) to confirm efficiency.
- Experimental method or simulation of production steps with reduced oil field equipment's and test of the available additives.

But these methods of choice are expensive and this problem is frequent, it is possible to intend to increase knowledge in this very specialized domain. If it is practicable [8-13], it is necessary then to integrate these new knowledge or innovation [8] into the company by modifying the methods of work and to change the question which becomes: which molecular interaction to decrease and with which type of molecule?

Progress towards a final solution can oblige to take into accounts backlashes or complete changes ("Re-finalization"). For example in the case of a risk of pollution by the additive or a risk on the possibility of buying the additive (possibility of replacing it by an oil extract), ... Steps can be added to the solving process following needs.

It is possible also to study the problem of the manufacture of additives by an industrial company. The first phase would concern the study of the present rival solutions on the market and the fluids to take into account, the second would concern additives (equivalent in the third phase of the previous problem) and would include also the constraints of manufacture, the third could concern the raw material.

The knowledge needs for every phase or level (from macro to micro or from the functional to the detailed material), as well as the evolution of these knowledge is represented on the table 1.

Knowledge	**phase 1**	**phase 2**	**phase 3**
Goals	Functional	Technical	Technical specifications
Constraints	General (economic, technical , ...)	Environment, scientific knowledge,	Scientific knowledge, environment
Solutions	Mainly qualitative orientation	Qualitative and quantitative orientation	Quantitative & qualitative solution
Evaluation	Different studies,..	Studies + Experiments	Tests and experiments

Table 1 - Knowledge Needs for Problem Solving During a Project

5 - Modeling of the Organization and the Production of Knowledge

Thus, the organisation and the production of multilevel knowledge was defined and now using information and communication technologies we propose a model to improve and control new products development projects within a concurrent engineering environment.

5. 1 Modeling of an element of knowledge

We chose to use a uniform model of representation of a knowledge element: the document. This one represents and models in abstracted way knowledge concerning goals, constraints, solutions and evaluations. An initial document is generated in the first phase with corresponding knowledge. Then from this document a first generation of documents will be made by inheriting knowledge of the first document and so until the generation n, which will correspond to the most specialized knowledge. Every document will keep links with the document " father " and the documents "children" to be able to control the

three and regroup documents by function, by component or others characteristics. The knowledge elements so defined will be able to be enriched by the members of the team, the specialists or the subcontractors. For it, in the modeling that we are going to develop, two parts will be considered:
- The element of knowledge such as it appeared during its creation or definition, that is *the original element of knowledge* and we shall note it IKE (INITIAL KNOWLEDGE ELEMENT)
- The set of the enrichments which were brought to it, the history of the copies of the document, its movements, that is, in a global way, the tracks of the interactions of the document within the project. We will call *history* this set of information (History).Thus, we will consider a document as being composed of two elements: the IKE and H (fig. 3).

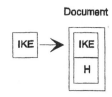

Document

Figure 3 - A Document

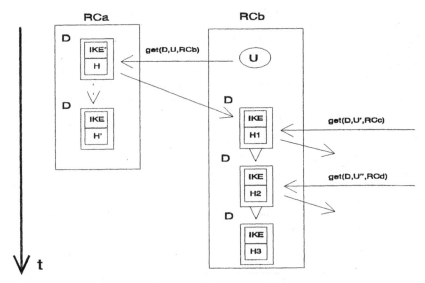

Figure 4 - Support System and Documents

5.2 Modeling of the support system

The support system chosen to implement this process of organization and production of knowledge is built around four component parts:
- The Actors of the project
- The Documents
- The Resource Centers
- The Software agents

The Resource Centers

The knowledge elements are available in the Resource Centers and the members of the project can use them according to their cognitive gaps or their specific needs to build solutions.

Interactions between the first three constituents of the support system are illustrated in the figure 4; the use of the software agents will be clarified in the following section.

A Resource Center RC keeps for the project actors one or several documents. If these documents have been created locally (or in a more precise way the IKE of the document), thus, we shall say that the RC is *Document Manager* (DM) of the document. In the example above, the IKE of the document D was created on RCa and consequently, this last one is DM for D. One supposes that this document was enriched locally and therefore contains a *history* H.

A request is engaged in the following way: an user U dependent on another RC (RCb) asks for a copy of the document D (*get request*). Having transferred the wanted document, RCa updates the historic H ' of D. According to the policy of a RC

for a document, it can send the document in its entirety, that is the IKE provided with complete H part, either, for reasons of confidentiality, the IKE and part of H.

The duplicated document on RCb can also receive requests: the H part of this copy grows then in an independent way of the document D present on RCa. It has for main consequence that a DM cannot know the location of all the copies of the document that it administers.

To be able to collect the various H parts of D, scattered within the organization, the DM which is associated to it (in our example, it is RCa), uses software agents.

The software agents

The software agents implemented to collect the history part possess a dual structure (fig. 5):
- A sub-system which implants what is concerning the fulfillment of the function assigned to the agent; in our application, it is a question of discovering the presence of a copy of a document, of analyzing history and of transporting a copy of this last one towards the DM of the document. The figure below precise the analysis made by the agent
- A sub-system that implants characteristics independent from the function allocated to the agent, the mobility and the faculty to communicate. These properties are also called Attributes [20].

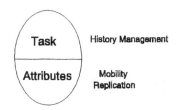

Figure 5 - A Software Agent

We can now clarify the work of the software agents (fig. 6). According to a policy previously defined (from events, at regular interval of time, at the end of every phase of the project for example), the DM of the document D, that is RCa, creates a software agent A1 to analyze the history of the local document D (in the example presented first, he acts of H '). This one contains the set of the RC having requested a copy of D: in our example, RCa emitted an unique copy of D to RCb. The software agent duplicate then as often as there were copies. Every clone of the software agent goes towards a different implied RC. When the clone arrives on the RC, it analyzes history of the copy (H3) and determines if it has also to duplicate: two RC (RCc and RCd) acquired a copy of D, thus two clones A2 and A3 are created and goes respectively to RCc and RCd. The agent A1 transports with him a copy of history H3 and goes to RCa, DM of the document D. When the agent A2 arrives on RCc, the document D does not exist any more (for example, the document not having been considered interesting, was erased), it dies, having no history to be returned to RCa. The agent A3 discovers on RCd that no new copy was made, he then returns towards RCa with the history H4.

The processing of the various histories is made on the site of the DM of the document.

5.3 Control of the process and concurrent engineering

There are two dimensions that allow at any time to control the state of the project and to establish cards of knowledge.

The vertical dimension from links between the documents, which allows, to control the three and to regroup documents by function or by components. The way of doing documents and the links between documents authorize the simultaneous study of several solutions, the overlapping phases and the possibility of studying completely a function, a module or a component independently.

The horizontal dimension which corresponds to the recording of interactions between the members of the project, the specialists or the subcontractors in the part " H " of the document. This dimension should allow to decrease the risk of error and to confirm an innovative idea by favoring its emergence. It can lead to a better control of mechanisms ("gabarit", "stigmergy", emergence and "auto-organization") related to the concurrent engineering and authorizes the creation of cards projects composed of chosen elements having been requested. They will be associated to information concerning the number of requests, the tracks of the interactions, the degree of enrichment and the existence of similar elements.

Figure 6 - Support System and Software Agents

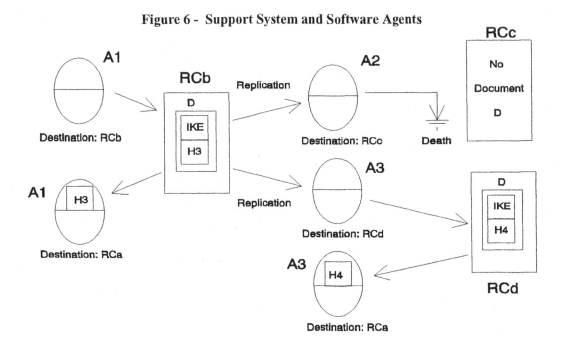

6 - Conclusion

Related to new products development processes the concurrent engineering approach is characterised by a project team that tries to develop new products using co-operative work. It is a system of search and production of knowledge and the driving force of the search for solutions is the capacity of learning and the creativity of the team. Thus, in this paper we have studied quite particularly the knowledge, which constitute the main resource in projects and the methods to acquire, to share, to use and to increase this knowledge. This within the framework of the concurrent engineering.

At first we described, from a real case, the new product development process within concurrent engineering and we distinguish four types of knowledge which move during the project: the targets, the constraints, the solutions or graphical representations, the evaluation of prototypes. We note that the originality of this product development process is related to the parallel search for solutions based on an important number of alternatives and to the coordination and the coherence of choices or solutions. The study of the behavior of the social insects allows us to propose three mechanisms to explain this coherence: the "Gabarit", the "stigmergic» logic and actions or solutions of emergent and "auto-organized" type. Then using problem solving we characterize the knowledge needs for every phase or level (from macro to micro or from the functional to the detailed material) of a project, as well as the evolution of these knowledge.

Thus, a process to organise use and produce multilevel knowledge is defined and revised using information and communication technologies. Knowledge is partitioned into modules or elements. Each of them is composed of a document associated with multilevel knowledge and, a software agent that accomplishes tasks like to go from machine to machine, sense the environment, control birth with duplication and death. These elements are available in resource centres or web sites and project members' could ask them to build solutions for problem solving. Each request produces a duplicate element. The software agents attached to this element transports the document, gives resource centre information about the requester environment and take into account a supplement of information from the requester if needed. All these operations and resources constitute a knowledge management system. It must produce a project knowledge card that will be composed of the requested elements associated with parameters related to the number of request, the identity of the requesters, the quantity of information added.

References:

[1] Christophe **Milder**, *Evolution des modèles d'organisation et régulations économiques de la conception,* Annales des Mines, février 1997, p 35-40

[2] Florence **Charue-Duboc**, *Maitrise d'œuvre, maitrise d'ouvrage et direction de projet,* Annales des Mines, septembre 1997, p 41-48

[3] H. **Takeuchi** and I. **Nonaka**, The new new product development game, Harvard Business Review, January-February.

[4] Donad E. **Carter**, Barbara Stilwell **Baker**, *Concurrent Engineering, The Product Develpment Environment for the 1990s,* Addison-Wesley Publishing Company, 1991.

[5] Frédéric **Gautier**, *Evaluation économique des activités de conception et de développement des produits nouveaux,* Cahier de Recherche du Gregor, 1997.12 (http://www.univ-paris1.fr/GREGOR/).

[6] Behnam **Tabrizi** and Rick **Walleigh**, *Defining Next-Generation Products : An Inside Look,* Harvard Business Review, November-December 1997.

[7] J.-L. **Selves**, Ph. **Burg**, *Décisions techniques et méthodes numériques dans un processus industriel appartenant à l'industrie pétrolière,* CNRIUT'97, Blagnac, 14-16 Mai 1997.

[8] Ph. **Burg**, J.L.**Selves**, J.P. **Colin**, Procédés de détermination des caractéristiques d'un pétrole brut, Brevet N° 9403188 (1994).

[9] Ph. **Burg**, J.L.**Selves**, J.P. **Colin**, *Numerical simulation of crude oil behaviour from chromatographic data,* Analytica Chemica Acta, 317(1995) 107-125

[10] Ph. **Burg**, J.L.**Selves**, J.P. **Colin**, *Crude oil modelling from chromatographic data. A new tool for crude oils classification,* Fuel, 76(1997) 85-91

[11] Ph. **Burg**, J.L.**Selves**, J.P. **Colin**, *Predictions of crude oils kinematic viscosity from chromatographic data.* Fuel, 76(1997)1005-1011

[12], J.L.**Selves**, M.H. **Abraham**, Ph. **Burg** , *A new method for the explanation of liquid properties in terms of molecular interactions,* Fluid Phase Equilibria, (1998).

[13] J.L.**Selves**, Ph. **Burg**, *Prediction of the Water-In-Crude Oils Emulsions Stability from Chromatographic Data,* Journal of Canadian Petroleum Technology, 38(1999)25-28

[14] Jean-Claude **Sardas**, *Ingénierie Intégrée et mutation des métiers de la conception,* Annales des Mines, février 1997, p 41-48

[15] David **Robertson** and Karl **Ulrich**, *Planning for Product Platforms,* Sloan Management Review Summer 1998.

[16] Marc H. **Meyer** and Robert **Seliger** , *Product Platforms in SoftwareDevelopment,* Solan Management Review Fall 1998.

[17] Kim B. **Clark** and Steven C. **Wheelwright**, Managing New Product and Process Development », The Free Press 1993.

[18] Vincent **Giard**, *Besoins technologiques et réseaux,* Cahier de Recherche du Gregor, 1998.05 (http://www.univ-paris1.fr/GREGOR/).

[19] Jean-Louis **Peaucelle**, *La baisse du coût de transaction par le commerce électronique : le moindre temps perdu pour les contacts commerciaux*. Cahier de Recherche du Gregor, 1997.01 (http://www.univ-paris1.fr/GREGOR/).

[20] E. **Sanchis**, *Modular Autonomy for Simple Agents*, Third International Conference on Autonomous Agents - Workshop on Autonomy Control Software - May 1-5, 1999, Seattle (WA).

[21] Daniel **Alban**, *Management du système d'information et politiques relationnelles d'organisation réticulaire*, Cahier de Recherche du Gregor, 1997.01 (http://www.univ-paris1.fr/GREGOR/).

[22] I. **Nonaka** et H. **Takeuchi** , *La connaissance créatrice, la dynamique de l'entreprise apprenante* , DeBoeck Université, 1997.

[23]R. H. **Hayes**, S. C. **Weelwright**, *Dynamic manufacturing, creating learning organization*, The free press, 1988.

[24] Allen **Ward**, Jeffrey K. **Liker** , John J. Cristiano, Duward K. **Sobek**, II, The second Toyota Paradox : How Delaying Decision Can Make Better Car Faster, Sloan Management Review, Spring 1995, p 43-41.

[25]G. **Theraulaz**, F. **Spitz** , Auto-organisation et comportement, édition Hermés, Paris, 1997.

[26] G. **Theraulaz**, E. **Bonabeau**, *La modelisation du comportement bâtisseur des insectes sociaux*, paper in book [25], p 210-234.

[27] J. **Ferber**, Les Systèmes multi-agents. Vers une intelligence collective, 1995, InterEditions, Paris

[28] J.L. **Le Moigne**. *La modélisation des systèmes complexes*, Dunod, Paris,1990.

Intelligent Personal/Team Assistant

J. Pokojski, P. Cichocki, M. Gil

Institute of Machine Design Fundamentals, Warsaw University of Technology, 02-524 Warsaw, Narbutta 84, Poland

Abstract

The paper presents a concept of knowledge based software supporting a long period machine design analysis – intelligent personal assistant. The concept of intelligent personal assistant is based on the maze model of the design process. The functional structure of the whole system and some issues of employing blackboard architecture for knowledge sources integration are shown.

1 Introduction

Every designer during the design process has to make some analysis. The designers do their analysis with the help of some plurality of models and tools. The models which are considered can exist only in the designers' heads or be expressed in a more formal way. If we concentrate on so called formal models the designers build them using often formal methods, formal software.

The human ability of building formal models and examining them is still the key aspect of designing. With every technique of modelling and making analysis some knowledge is connected. The knowledge decides what analysis should be done, what goals they have, how this process should look in a particular case.

The knowledge stored by designers is permanently modified. Each project analysis can give impulses for a new knowledge chunk articulation or modification.

Computer tools used in the design process are developing in the direction of knowledge supported software. Many technologies belonging to artificial intelligence are starting to support classical engineering systems.

Many design problems analysis are based on knowledge which is the result of a long period of design experience [1, 2, 4, 5, 11, 20, 21]. For instance the long period of modelling and analysing dynamic phenomena of some structures with the help of multi-body formalisms [15, 19]. The designer exploits theoretical knowledge together with his own knowledge resulting from his experience. He conducts a kind of dialogue between the problem actually solved, his own available knowledge sources [7, 15, 17, 18, 19] and cases [13, 14] from his experience in the past.

It is not easy to depict what knowledge is needed to solve particular problems in machine design. In any case we need a theoretical basis, experience and the abilities to model successfully, to make correct observations and to create a suitable hypothesis.

Human experts exploit their own memory for storing different kinds of knowledge or evaluation of other knowledge sources; sometimes they have some notes, some schemes.

Experts conduct dialogues with different knowledge sources from this what they remember, what they have as paper notes, what they can restart on their own computers from the past.

The expert knowledge can be acquired and used in some knowledge based system. We can imagine the situation that an expert adds knowledge (created by him) to an existing knowledge based system (containing theoretical knowledge, knowledge acquired earlier) in a dynamic way for his own (expert) purposes. In this case the system fulfils the role of the expert's active notes. This knowledge can contain comments to different models, parameters, achieved results and successful plans etc. The stored knowledge can be used by the expert himself and the people collaborating with him.

The paper presents concepts and pieces of the implementation of a knowledge support system of machine design analysis called intelligent personal assistant.

The concept of the intelligent personal assistant, from its functionality point of view, reminds the concept of a word processor.

The word processors are mentioned as „type-writing machines" with the possibility of easy storage and reusing of former texts.

We can observe that people who started to use word processors, after some time, stop writing by hand. The possibility of text reusing is the main argument for that fact.

If a user makes some documents for himself he knows well that everything what is done should be done on a quality level accepted by him (her). Because most of the documents can be reused by the user again. Better-quality-documents mean less work in future, on the stage of document reusing.

Lets assume we want to collect not only text documents but also want to collect the user's personal opinions, personal knowledge chunks, having different representations. The users can have different associations connected with different pieces of knowledge or information.

The users use in their work different computer tools. These tools have some structure, parameters, model of user's profile etc. Often tools can work as a sequence of tools when there is some know-how to organise this process an to build data exchange connections. All the above information can have practical value for the user.

It could be useful if we could store all this knowledge and information and could later reuse it or compose its new version just similar like in case of text documents processing.

In the next chapters there are shown two aspects of intelligent personal assistant: 1) structure of functioning, 2) attempts of employing blackboard architecture as integration tool.

2 The structure of the personal assistant

The backbone of a personal assistant layer is based on relational data base technology. The user's interface for this application is prepared in database environment with the use of MS Visual Basic. Personal assistant integrates:

-managing all kinds of data arising during machine dynamic analysis, including index and search system,

-extended connections between data,

-project plan editor supporting the maze model of the design process,

-artificial intelligence methods:

- expert system used for supporting decisions, for managing tasks in the project node and for preparing more complicated dialogs,

- case base reasoning used for comparing and reusing paths in the maze model,

- blackboard architecture as a tool to integrate different knowledge sources.

One of the issues of a personal assistant is to integrate all kinds of data arising during long period analysis. As possible data format we have considered:

-data stored in files saved on computer disks, which can be linked to a personal assistant or stored in it,

-descriptions and placing of paper documents (books, articles, project specific documents), or links to web documents,

-specifications of project nodes including analysis which is based on expert systems,

-various data types predefined in a system or defined by a user.

This system has a flexible and open architecture, and new data types can easily be added. The main ideas of this structure are shown in figure 1.

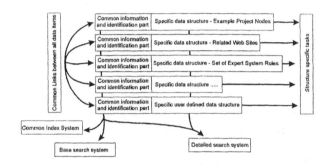

Figure 1 Personal assistant data structure

The idea of the data structure is similar to ideas which we know from object programming.

2.1. Nodes

In our work we assume, that the problem can be divided into small sub-problems connected with each other. These sub- problems are presented in the system as nodes [4, 7, 18]. The node is a specific data structure which can be linked with other nodes, a set of tasks and pieces of information. In fig. 2 a sample node is presented.

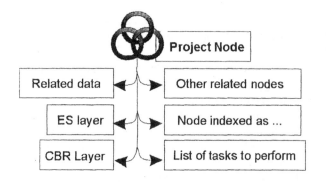

Figure 2 Sample structure of the project node

When the user focuses on a particular node, then a set of related data is presented to him and a small set of related expert system rules is loaded. Then he can use the data or perform tasks from a list. As a result he produces a new description of the problem in the form of object – attribute – value. This is the basis on which rules can operate. In this way an expert system can suggest which tasks should be performed earlier, can show alternative paths in the maze model, or can suggest other possible solutions.

New data arising during the interaction with the system can be stored in a personal assistant and can be linked with a specific node. The possible data types are expert system rules. The user can add new remarks in this form and run them later – all of them or only some sets from specific sessions. In this way he can reconstruct his own suggestions from the past.

Nodes and links between the nodes are presented to the user in a maze editor. In this editor we present the user a sequence of completed nodes, a current node, and possible next steps. The completed nodes are analysed to find out circular paths. The editor always focuses on the current node and presents the next possible nodes in the form of a tree. The nodes have hierarchic order so the problem can be considered on different levels – from the most general to the most detailed.

2.2. Indexing system

Whenever we have a big number of data, a specific index system is necessary. It has to generalise and order information. Generalising helps to find out related data when we don't exactly know what we are looking for. In our program the index system consists of two sub indexes. The first one – base – includes a general dictionary with common used terms placed in a multilevel hierarchy. The second one additional index includes different descriptions

of terms from the base index. Sometimes expressions from the additional index are not fully correct in technical sense but commonly used.

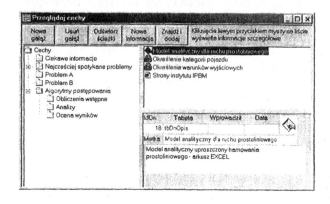

Figure 3 Index browser

These two indexes are converted for phrases for the sake of easy reading and understanding a context. This approach is useful for automating assigning information to the categories of the index. We have tested this system with indexes containing 20 thousands of entries placed on many levels. It seems to work well. The sample browsing through the index is presented in fig 3.

2.3. Searching system

As an addition to browsing information related with nodes we developed a search tool which is able to help the user to find information on the base of full text search. This tool allows the user to set a base – collection of data and search it for all or some of the entered words.

Figure 4 Information browser

2.4. CBR applied for user session

During the interaction with a system a sequence of nodes completed or visited by the user is recorded. These sequences and sets of design variables from different sessions can be compared. We can present a comparison of different coefficients of similarity. The highest coefficient of similarity means, that the user visited the same nodes a second time. When we want to compare sequences less accurate we eliminate some nodes from the path or pass over circuit sequences, and compare them again. In this way we can find out the best and most frequently used paths from the maze and suggest them in future sessions. We can also identify which sets of information are frequently used in given task and next time present them in advance.

3. Blackboard architecture

Blackboard architecture has its beginning in speech understanding systems (Hearsey II) [9, 10]. It was often used as a solution for solving large multi-dimensional problems where not only domain knowledge is important but also control knowledge.

Blackboard architecture [2, 9, 10, 12, 20, 21] consists of the blackboard, a set of knowledge sources and a control mechanism. The blackboard is a data base. As a global data base it is visible for all knowledge sources. It stores data and hypotheses. The blackboard can consist of a hierarchy of levels. This concept of a set of knowledge sources reflects all knowledge dealing with the particular problem solving. Knowledge sources can share the information available on the blackboard. They create and modify the content of the blackboard. Knowledge sources can be activated after having satisfied the prescribed conditions. The control mechanism can decide about a selected control strategy.

During the last 20 years some of blackboard architectures have been developed which differ especially from the control mechanism point of view. Blackboard architecture was tested in many real life problems [9, 10, 20, 21].

In [12, 20, 21] blackboard architecture is used as an integration platform of design tools. In [12] experimental application is presented in which blackboard architecture was used to integrate environment for cooperative engineering design. The main intention of the authors was to integrate engineers cooperating in one large project.

Blackboard architecture was invented as a structure reflecting the human way of solving problems. Most of the applications exploit this concept as a tool for solving problems with external knowledge sources. For instance the integration of different aspects of the design process.

In our work blackboard architecture is used as integration platform for different aspects of design analysis which are controlled via a personal designer.

Traditionally, the design process is perceived as a serial process in which the design is passed through different stages. Real experience differs from this classical view. The design process is often parallel, multithreaded and not fixed like the one-way algorithm. Moving from one design stage to another is rather like exploring a labyrinth. In such a case, the sequence of stages is not in an a'priori established order, but rather free search, supported only by the data and knowledge provided by the designers' team.

We were taken on development of the breaking system for mobile cranes [1,2] together with the Industrial Institute of Construction Machinery (PIMB, Kobyłka). During the project development we focused on the enhancement of the design process. We have tried to

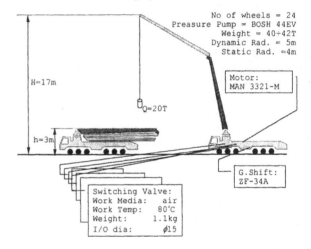

Figure 5 Schematic diagram of the truck crane

reach two main goals: 1) improve the performance and quality of the design as a whole; 2) make the work for the particular designer easier and more convenient thus turning their efforts into creativity.

When we take into account bigger projects, like the design of a mobile crane, we have to reject the simple linear model of the design and adopt a different one. Such complex designs, where collaboration of specialists from different domains work together, and the design consists of many sub-processes, can be modelled with the maze [4, 7, 18]. The maze model can be built with the support of blackboard architecture.

The blackboard is used as a central repository for all shared information. The information on the blackboard represents facts, assumptions, and deductions made by the system during the course of solving a problem. Each participant brings a unique set of knowledge to bear, and each may employ an original problem-solving strategy. Each expert views the information on the blackboard, and if possible, tries to contribute to the solution. A facilitator controls "the chalk", mediating among the experts competing to write/delete on the blackboard. The function of the facilitator is to determine which expert, at a given point in time, has the most important insight or information to contribute to the solution of the problem.

To implement such environment we have chosen a commercially available blackboard development system GBB (Generic Black-Board [3]) from Knowledge Technologies Inc. This is a domain independent framework, similar to the well known BB1 [9,10,20]. We have defined knowledge sources to be capable of processing two functions: a) provide knowledge as experts opportunistically provide knowledge forming a final solution; b) support performing repeated tasks – we call them agents rather than knowledge sources, because they are not knowledge based.

Blackboard consists of the information or entries generated by the knowledge sources during an ongoing process. In our case it could be organised into some number of levels. Each level contains objects or attributes that are important to the design. The hierarchy of levels on the blackboard depicts the hierarchy of the product being designed. In our case it is a mobile crane like that shown in fig. 5. We can divide our project repository into the following levels: basic properties, mechanical properties, parts, sets, systems, and so on (fig. 6). The low level objects represent atomic data, like the weight of a particular element, its dimensions and other reacquired specifications. The top level data represent global solutions, for example the breaking system: pneumatic, split circuitry, with two air storage tanks, with two pneumatic pistons actuators per wheel, etc. Normally, knowledge sources are designed and specific to a certain level in the blackboard, i.e. the activation of the KS (Knowledge Source) depends on data generated on some level of the blackboard, and the actions of that KSs modify entries in the same level. Although there are also KSs, whose actions generate a new object on the upper or lower level .

When starting the application simple KSs (we call them agents) acquire requirements and input data for the project. There are many objects which are generated on the lowest level. Calculating KSs which find their substrates on the blackboard start to activate a generation of a more accurate description of the designed object. Then selecting KSs start to run and to introduce parts which are reacquired by the project and are satisfied by

basic properties. This process opportunistically goes on and the vision of the product emerges.

According to the original blackboard architecture there must be a control mechanism. The simplest one can consist of an agenda and monitor. The monitor watches the blackboard and places any activation of the blackboard to the agenda. The KSs with the highest priority are executed. The order of the execution is very important. The resolution strategy focuses on the object or level of the blackboard and orders the KSs based on how they can address the focused object. Currently we are working with a simple agenda based control mechanism.

The system being created provides major benefits to CAD by improving the speed and effectiveness of the product development.

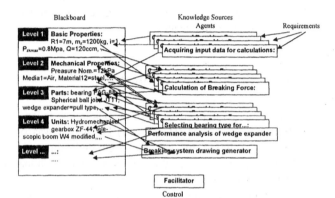

Figure 6 Blackboard levels and corresponding knowledge sources

4. Conclusion

The presented concepts and parts of implementation are the basis for dedicated knowledge based environment being developed.

References

[1] Gil M.: Concepts and Realisation of Computational System Supporting the Design of Mobile Crane Breaking System. Master Thesis, Faculty of Automobiles and Heavy Machinery, Warsaw University of Technology, pp. 1-38.

[2] Blackboard Architecture, http: //www.stc. westinghouse.com

[3] Blackboard Technology Group, Inc. , GBB manuals, 1998.

[4] Boulanger S., Gelle E., Smith I.: Taking Advantage of Design Process Models. IABSE COLLOQUIUM, Bergamo, pp. 87-96, 1995.

[5] Cichocki P.: Concepts and Realization of Database for the Design of Mobile Crane Breaking System. Master Thesis, Faculty of Automobiles and Heavy Machinery, Warsaw University of technology, pp. 1-30.

[6] Cichocki P., Gil M., Pokojski J.: Heating System Design Support. In Lindemann U., Birkhofer H., Meerkamm H., Vajna S.(eds.): Proceedings of the 12th International Conference on Engineering Design. Tu Munchen, 1999, pp. 60-68.

[7] Cichocki P., Gil M., Pokojski J.: Heating System Design Support. In Smith I. (Ed.) : Artificial Intelligence in Structural Engineering. Information Technology, Collaboration, Maintenance, and Monitoring. Springer-Verlag, Lecture Notes in Artificial Intelligence 1454, 1998, pp. 60-68.

[8] CLIPS 6.0. NASA Software technology, manuals, 1995.

[9] Craig I. : Blackboard systems. Ablex Publishing Corporation, Norwood, New Jersey, 1995.

[10] Englemore R., Morgan T.: Blackboard Systems. Addison-Wesley Publishing Company, 1988.

[11] Fenves S., Turkiyyah G.: Knowledge-Based Assistance for Finite-Element Modeling. IEEE Expert Intelligent Systems & Their Applications, pp. 23-32, June 1996.

[12] Lander S., Corkill D., Staley S.: Designing Integrated Engineering Environments: Blackboard-Based Integration of Design and Analysis Tools. Concurrent Engineering: Research and Applications, Vol. 4, No. 1, pp.59-70, March 1996.

[13] Maher M., de Silva Garza A.: Cased-Based Reasoning in Design. IEEE Expert Intelligent Systems & Their Applications, pp. 34-41, March-April 1997

[14] Maher M., Pu P.: Issues and Applications of Case-Based Reasoning in Design. Lawrence Erlbaum Associates, Publishers, 1997.

[15] Pokojski J.: Knowledge Based Support of Machine Dynamics Analysis. In Chawdhry P.K., Ghodous P., Vandorpe D. (Ed.): Advances in Concurrent Engineering CE99, Sixth ISPE International Conference on Concurrent Engineering: Research and Applications, Bath, UK 1999. Technomic Publishing Co. Inc., Lancaster, USA, 1999, pp. 336-344.

[16] Pokojski J.: An Integrated Intelligent Design Environment on the Basis of System for Flow Dynamics Analysis. International Conference on Engineering Design, Praga, pp. 1333-1338, 1995.

[17] Pokojski J.: Expert System technology in Machine Design. Advances in Engineering, Proceedings of the IX German- Polish Seminar, Koln, pp. 118-125, 1997.

[18] Pokojski J.: Product Model Transformations in Maze Model of Design Process. Computer Integrated Manufacturing, Proceedings of Int. Conf. CIM 99, Zakopane 9-12.03.1999, WNT, Warsaw, pp. 121-128, 1999.

[19] Pokojski J., Cichocki P., Gil M.: Intelligent Personal Assistant for Machine Dynamics Problems. In Borkowski A.(Ed.) :Artificial Intelligence in Structural Engineering. Proceedings of the 6th EG-SEA-AI Workshop Wierzba 99, WNT, Warszawa, 1999, pp. 159-164.

[20] Sriram R.: Intelligent Systems for Engineering. Springer-Verlag, 1997.

[21] Tong Ch., Sriram D. (Ed.): Artificial Intelligence in Engineering Design. Vol. 1,2,3. Academic Press, 1992.

A knowledge-based environment for modelling and computer-aided process planning of rapid process manufacturing

A. Deglin, Prof. A. Bernard

CRAN (Research Center for Automatic Control of Nancy)

Université Nancy I - BP 239   54506 Vandoeuvre les Nancy Cedex – France

Tel: +33 3 83 91 27 29. Fax: +33 3 83 91 23 90. E-mail: alain.bernard@cran.u-nancy.fr

Abstract

This paper introduces a knowledge-based environment dedicated to the choice of rapid manufacturing processes.

Rapid manufacturing processes are not limited to layer-manufacturing machines, but they also integrate CAD, reverse engineering, indirect methods for metallic and plastic part manufacturing, etc...

Due to short delays, people have no time to test and compare different solutions of rapid manufacturing. Even if people have time, tests are time and money consuming. It is also very difficult for someone to know all about industrial technologies, and to be able to evaluate a multi-criteria choice in a short time.

The aim of the proposed knowledge-based environment is to propose, from a detailed functional specification, different alternatives of rapid manufacturing processes, which can be ordered and optimised when considering a combination of different specification criteria (cost, quality, delay, aspect, material, etc...).

At present a first version of the conceptual model has been implemented on KADVISER platform and validation tests are ongoing based on industrial case studies. This project will be achieved on end of June 2000. So, the first practical results will be available for presentation at the beginning of next July.

Key-words: knowledge-based system, rapid manufacturing, optimisation..

1 Introduction

Currently, in the field of rapid prototyping, a great number of new technologies joined already more conventional ones, and they did not have of cease to progress. This is why today the possibilities are multiple and varied to obtain a prototype, tools, a series of part... But of course, considering the number of processes, materials, machines, subcontractors...,there are means more or less judicious to obtain the desired results [1] [2].

In this maze of solutions, this meander of processes, and the possible subcontractors, a system of assistance to the choice of the various processes of rapid manufacturing will be able to thus allow many services of the companies to share the knowledge and the solutions delivered by the systems. For example the engineering and design departments, to produce a prototype model of design, the manufacturing engineering to validate a process, but also the Research and development to validate a concept [6].

Life cycle of products is very short. So, companies have to adapt their development and industrialisation organisation in order to reduce time-to-market, based on numerical information that has become the reference for the product. In fact, new challenges concern the capability to manufacture the just necessary number of products. The main consequence is flexibility for tool manufacturing with low-price and consumable tools, instead of very cost-consuming tools. This is possible because of numerical information, used along the complete development of the product, and due to new materials. But, consequently, it is necessary to validate product and process concepts very early during the design stage. Some recent developments in rapid manufacturing allow such possibilities [3] [20].

These examples show that the dynamic evolution of technologies is not easy to take into account in real time. This is why it is strategic to find the just necessary process for given specifications.

What is proposed in this paper is an approach for knowledge and know-how modelling, and for computer-aided process planning (CAPP) from given specifications (type of part, material, delay, quality, colour, etc...).

2 Rapid manufacturing process

The experience shows that quite all-industrial products are concerned by rapid manufacturing technologies. It is mainly due to the variety of technologies and more especially of materials. As said before, there are various fields of applications to favour less time consuming for product development.

The other main evolution is related to the integration of layer-manufacturing technologies with CAD and numeric models. Such CAD systems and environment are not so expensive and are user-friendlier. Due to the stability of the technologies, the results of rapid manufacturing processes are really 3D physical images of CAD models of parts or tools, more functional and accurate, in many different materials [3] [4].

In spite of last arguments in favour of the use of rapid manufacturing processes, this domain has a permanent evolution. It is very difficult to appreciate all the real capabilities offered for industrial applications. But it is really more and more interesting to use such means because the number of parts that have to be manufactured at one time seems to be lower and lower. This fact induces a new tendency and favors the development of rapid tooling technologies, in order to obtain some real industrial processes which are alternatives to actual traditional processes. The main originality is to develop materials and technologies that allow obtaining economical and consumable tools which life time is limited. And due to the integration of CAD reference models with these new processes, it is possible to produce as many economical tools as needed, even if this is in one, ten years or more [17].

In order to be able to decide what is the best process at a time, taking into account the industrial context and the particular specifications of the product, it is strategic to capture and model the knowledge related to all the technologies. If so, it would be possible to distribute this knowledge to all the services which need it (design, manufacturing, industrialisation, control, …). Of course, the volume and the dynamic aspect of this knowledge have to be taken into account when choosing conceptual models for data modelling and data processing. One of the main difficulties is that data will be related to both knowledge and know-how.

All these arguments show the necessity of a computer environment dedicated to CAPP for rapid manufacturing. This system will allow to obtain alternatives of rapid manufacturing processes and also to choose the best one optimised from multiple-criteria or based on a particular main criterion.

The interest for a large company to use such environment is to allow all the services having the global technological information at the same time. According to us, it should also be a tool for general technological knowledge representation.

Concerning a small or medium company, this information system has a strategic interest because this environment should be accessed through a Web application. In particular, that will allow evaluating alternatives of processes based on real means and companies (subcontractors and all their technologies, their capabilities in terms of cost, delay, quality, etc…).

The last aspect is of course that this system will also contribute to help training sessions in order to highlight technologies and the main choice criteria between technologies and more generally between manufacturing processes.

3 Related research

Concerning work relating to the subject, it is important to remark that the topic is very recent. We can notice the initiative of the University of Singapore [19], the based of the approach, an Rapid Prototyping and Manufacturing material/machine selector has been developed to assist the user select the right Rapid Prototyping and Manufacturing material/machine for specific part fabrication. Using rule-based reasoning and database, the selector first finds the applicable Rapid Prototyping and Manufacturing material/machine. From the list of applicable Rapid Prototyping and Manufacturing systems, the optimal combination of material, machine and part orientation is then found out based on the minimisation of cost.

The University of Northwersten [18] proposed a RP process selection for tool making, witch involves the selection of a particular RP process and material, is a sub-decision that must be made as part of the tool path selection process.

It is also significant to speak about other works in other technological fields and in particular in the field of machining process planning.

Our work is particularly based on our experience thanks to two past projects on machining process planning. These projects carried out by Richard and Derras used the same expert software support that in our case [7] [13]. But, what has to be underline is that we plan to retard our approach to other technical fields because of the generality of the proposed concepts. The main problems will be the quantity of information that will have to be capitalised and the computing reasoning time based on this information.

Knowledge is defined in a general way as being the unit of the know, the experiments, the rules and the expertise [9].

Three categories of knowledge have been distinguished:

- Knowledge relating to knowing: it's descriptive, static, directly usable and is acquired while being informed, information is thus the privileged vector of this type of knowledge.

- Knowledge relating to making: it's dynamic and generally correspond to methods or procedures, the privileged vector of this type of knowledge is the training.

- Knowledge relating to understanding: it results from the enrichment of knowledge relating to knowing and

making obtained through the experience sharing lived by different people, in more or less close contexts. This knowledge is not directly transferable, the communication is the privileged vector of this type of knowledge.

The capture and modelling of knowledge suppose a step of identification of relevant pertinent knowledge in order to allow its management.

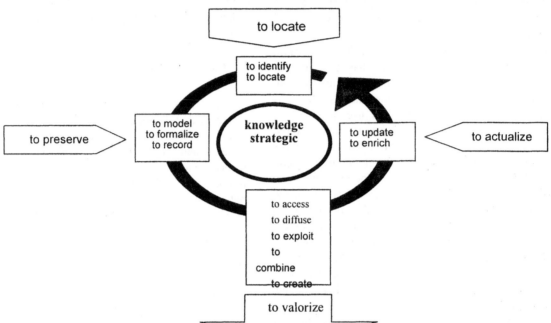

Figure 1 Major functions of a knowledge-based system

Knowledge formalisation and capitalisation, are significant elements, which for these ten last years has worried all the industrialists.

Today, the field of rapid prototyping, and even more rapid manufacturing come within domains where or each actor represents his technologies, and its competencies represents only one small part of the complete field. The interest for a company to model technologies and the knowledge of the field, is to pre-empt the future needs, to capture and model in a field without being tributary of the expert, who can leave to the retirement, competition or to even be sick... In fact makes it capitalise the field of rapid manufacturing by means of computer, makes reliable, makes safe, and perennialises the knowledge and know-how [10].

With the objective of computerisation, it is necessary to give a physical and static representation of the field in order to make it possible to limit and know the various values to be handled and intervening in the field. It is the first work of analysis, it makes it possible to include/understand, then to generalise the mechanisms in order to draw a certain number of concepts, principles and

heuristic governing the field. Once this work is carried out, it is used as a basis for the system.

The processing of knowledge capture and modelling can be broken up into several categories:

-   The acquisition of knowledge is specific to the fields and to the problems to be treated. This knowledge, primarily symbolic, often expresses relations between objects, but also formal or numerical calculations on these objects.

-   The representation of knowledge according to models; as general as possible, as independent of the processing as possible and sufficiently semantic so that the acquisition of knowledge is easy.

-   The implementation of the reasoning on this knowledge which results in models of handling of knowledge more or less general, related to the models of representation translating the principles of reasoning by the absurdity, recurrence, or by analogy.

-   The control of the reasoning and its implementation on specific problems to reduce the space of search for solution.

- The explanation of the reasoning in two objectives:
    - to leave a detailed trace of the reasoning undertaken with the developer of the system
    - to give a concise explanation to the user of the system
- The revision of knowledge must determine up to what point one can modify in the established knowledge as a preliminary and consider an automatic revision of knowledge (maintenance of coherence)

5 Representation of knowledge

The problem of the representation of knowledge consists in finding a correspondence between an external world and a symbolic system. Knowledge that the computers usually handle is of a numerical nature. The expert system of assistance to the choices of the various processes of rapid prototyping, has as a principal objective, to answer a need capturing and modelling of the knowledge and know-how in the field of rapid manufacturing. This is like restoring a relevant orientation in term of process solution to the case imposed to the various actors [11] [12]. The schedule of conditions of the application emphasises two great phases of use of the environment to be set up:

- The capture and modelling of knowledge, which represents the whole of the parameters, the behaviours and the variables to be integrated in the expert system, in order to be representative of the field. This has to be introduced into the system by the experts and by the knowledge engineers.
- Extraction and exploitation of knowledge, represents the service accessible by the users, this service results in a solution in terms of processes, technologies, subcontractor... according to the various specification of the part that the user defines.

Modelled knowledge should represent the processes and the technologies used in rapid manufacturing. The models representing the processes and the technologies are representative of the significant aspects and criteria of decision. As we can see it in the cognitive model (figure 3), the criteria indicated by the user in the schedule of specification have similarities with the criteria of the processes figure 2 and meta-technology figure 4, so then allowing a correlation making it possible to generate a solution.

Knowledge is of type capacitating, behavioural and environmental. In other terms capacitate material, type of machine, minimal thickness of realisation, dimensions... Knowledge of the behavioural type results in the difference between the input and output states of the process. Environmental knowledge defines the context of manufacture, the parameters of delivery (time, cost, quality,...).

The modelling of knowledge will be carried out within a system, which will make it possible to store various knowledge classified by processes and by technologies [14].

The representation of the cognitive model of the expert system dedicated to the assistance to the choices of the various processes of rapid manufacturing is given on figure 2.

The cognitive model of process represents the structure making it possible to the experts to model examples of processes (case studies) and technologies used in the field of the rapid manufacturing of products.

The cognitive model of the specification model is presented on figure 3. This makes it possible to the user to inform the various constraints associated with its need, and on which the system will base its reasoning.

The model of specifications is a significant element of structure, because it allows the user transcribing its need of a formal manner, and allows then the system working on the stored of objects (technologies and process). One of the possible solutions for reasoning should be the comparison of the user specifications with the stored technologies, processes and meta-technology which allow of define the procedure of solution.

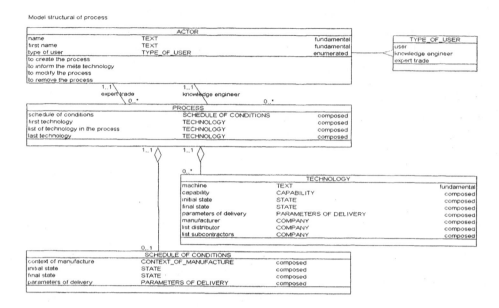

Figure 2 Part of the cognitive model of process model

The third major element in the system is the model of knowing of technical manufacture and represented by the model of meta-technology (given on figure 4) and by the model of technology. We understand by meta-technology, the generic class of technology having for common point a given technology of fabrication. For example:

- stereolithography meta-technology gathering the technology of the sla250 type, sla 3500...
- powder sintering meta-technology, corresponding to the technology of EOSINT P 350,
- DTM Sinterstation 2500,...

Figure 4 represents the basic elements of the meta-technology model.

The meta-technology and the technology are elements that constitute the knowing, these two models are defined and introduced in the system only by people having the technical knowledge; these are thus the experts as well as the knowledge engineers who have the privilege to implement the base of new technologies.

6   Utilisation of the CAPP system

For the utilisation of the CAPP system we set up two types of reasoning allowing to bring different functionalities, closely connected to give a real efficiency in order to satisfy the various user's needs. As shown on figure 5, the two types of reasoning are:

- the case based reasoning, this type of reasoning makes it possible to model, manage, adapt the processes modelled based on case studies, this having for principal advantage of extracting from the process solutions having been validated in the past. The two main disadvantages of this technique applies in some, very vast field such as the field of rapid manufacturing that the coherent number of process case studies must be very significant in order to allow the system finding a solution corresponding to the user specification. In addition, the innovation is difficult to integrate due to the lack of experience.

- the bottom up generation of process, this type of method of reasoning is complementing to the case base reasoning, therefore have the disadvantage of suggesting some solutions not extracted from case studies and not validated by concrete industrial application, and the advantage of allowing to integrate some innovation. When one applies this method, he has the succession of " linear " processes to make very quickly some simplifications in the tree structure made up of technologies which constitute the processes [16].

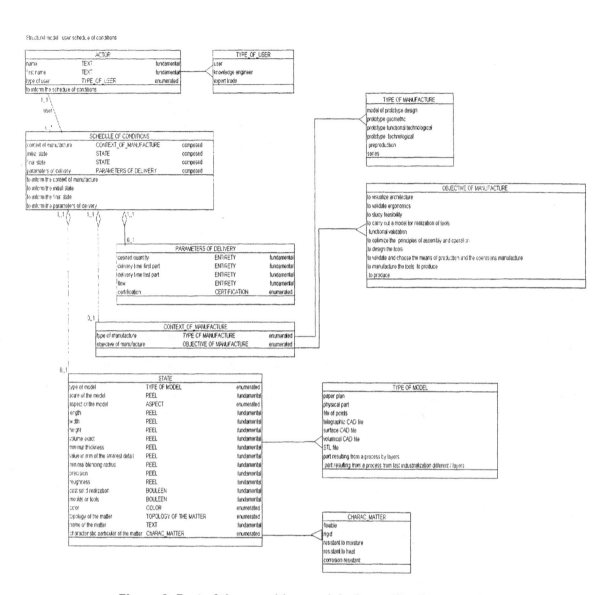

Figure 3 Part of the cognitive model of specification model

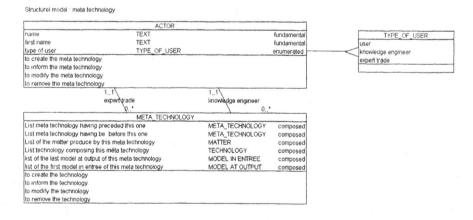

Figure 4 Part of the cognitive model of meta-technology model

655

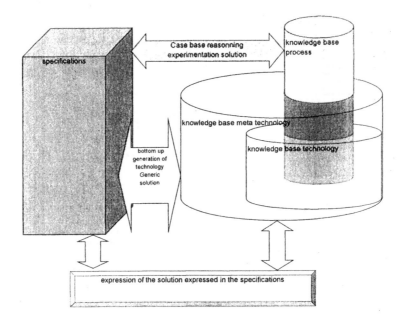

Figure 5 Model representing the types of reasoning applied in the CAPP system

Based on these kinds of reasoning the user is able, thanks to the user interface, to fill the various fields of the specification form. All these parameters do not require to be informed because some are not known and are not of primary importance for the treated case.) This form includes all the parameters relevant for the definition of the user needs.

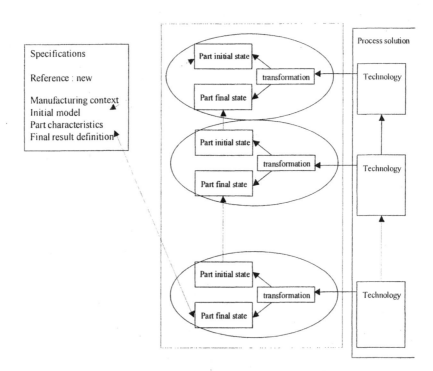

Figure 6 Basic point of one of the modelling of the generation of process [5]

During the description of the various parameters of the user specification, the system makes the first reasoning, making it possible to simplify the tree structure of the various meta-technologies as well as technologies. To allow a rapid and efficient simplification and a fast defined reasoning, an example of criteria making it possible to create the first outline of process solution is:
- Type of model in input
- Topology of the material for the final desired part

The type of model in input makes it possible to define all the processes and the meta-technology having the same type of model in input as that by the user.

The topology of the matter of the final part makes it possible to define the meta-technology as well as the processes potentially capable to produce a part in a coherent state compared to the specification, or potentially able to be the final process of the required solution.

Then the system analyses if a process has the two both elements (topology of the material and type of model of input). In this case, the process is made up only of one meta-technology, and in consequence of one technology. Then, if the system does not find any type of process having the topology of the material and the model in input, it generates a solution by bottom up output generation of process. It starts from the final meta-technology and see for a meta-technology such as its material topology is coherent with the one in input of the final meta-technology. This is repeated for all the final processes, and then gradually in order to determine the complete solutions archived when finding the first possible processes, corresponding to the type of model in input given by the user.

It is clear that the various criteria making it possible to select the proposed solutions between the various processes constituting each one a potential solution are more numerous than both used for the example. To allow a diagrammatic vision of the process determining the solution, one has to refer to figure 6.

7 ACPIR CAPP application

Some part of screens are presented bellow, printing from the CAPP system (figures 7,8,9) in order to give a more precise idea of the system requirements, as well as its ease of use.

To model the knowledge and know-how of the field, we chose to use an industrial expert environment, a system containing knowledge, Kadviser [8], which allows from its intrinsic construction:
- an object based representation
- the modularity of the base of knowledge
- a communication with external applications , using DBMS standard (Data Base Management System), geometrical modeller...
- an easy and user-friendly interface with the various actors using this system.

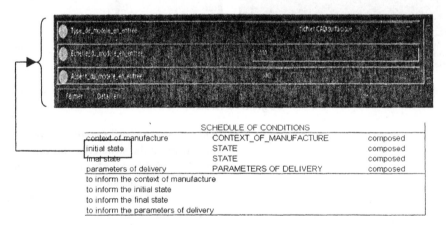

Figure 7 Example of screen for the definition of the model in input and part of the cognitive model

Kadviser is a generator of expert system including an inference engine for constraint propagation based on the mathematical logic of command 1, allowing the non-monotonous reasoning by management of assumptions on the objects which are handled through typified and quantifiable variables. The mechanism of inference and the principle of resolution are non-monotonous and based on a non-configurable logic by the user. Because of this mode of reasoning, Kadviser knows the opposite of any rule and includes the deductive non-complete. Its capacities of simulation of reasoning, as well as the potential for the processing of overstrained problems (without precise solution), under constrained (multiple precise solutions) or even complete (single precise solution) confer the possibility of treating expert applications in different fields.

The global application menu is used as command interface with to the user, making it possible for him to interact with the CAPP system (specification definition, solution collection). The model in input makes it possible to define which are the various processes, which have the same starting state.

PARAMETERS OF DELIVERY		
desired quantity:	ENTIRETY	fundamental
delivery time first part	ENTIRETY	fundamental
delivery time last part	ENTIRETY	fundamental
flow	ENTIRETY	fundamental
certification	CERTIFICATION	enumerated

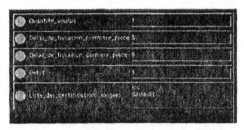

Figure 8 Example of screen expressing the parameters of delivery of the need and part of cognitive model

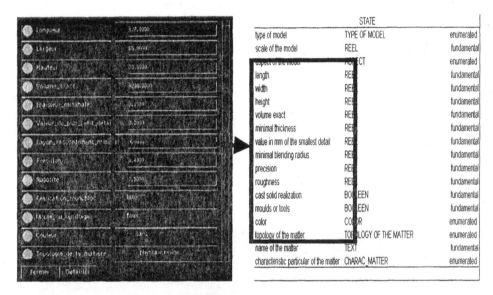

STATE		
type of model	TYPE OF MODEL	enumerated
scale of the model	REEL	fundamental
aspect of the model	ASPECT	enumerated
length	REEL	fundamental
width	REEL	fundamental
height	REEL	fundamental
volume exact	REEL	fundamental
minimal thickness	REEL	fundamental
value in mm of the smallest detail	REEL	fundamental
minimal blending radius	REEL	fundamental
precision	REEL	fundamental
roughness	REEL	fundamental
cast solid realization	BOOLEEN	fundamental
moulds or tools	BOOLEEN	fundamental
color	COLOR	enumerated
topology of the matter	TOPOLOGY OF THE MATTER	enumerated
name of the matter	TEXT	fundamental
characteristic particular of the matter	ChARAC_MATTER	enumerated

Figure 9 Example of screen expressing the final need and the characteristics of the part and part of cognitive model

10 Conclusions

In conclusion, the ACPIR CAPP expert system for the assistance to the choices of processes of rapid manufacturing makes it possible to model the knowledge and know-how of the field. Its main objective is to make it possible to the various users of the system to find, whatever their need, a solution through the system.

The first phase carried out at the time of this project was the census of technologies, the processes, the criteria, the procedures and behaviours, the uses and habits of this

After the implementation of the models and the programming in the base of knowledge, the phase of development of the rules was engaged, this making it field. This stage of " cognitic " allowed highlighting and implementing the structures and functions to be integrated into the CAPP expert system. In parallel search and analysis of a great number of fast case studies in the field of rapid manufacturing has been carried out.

With this stage followed the phase of formalisation and object modelling of the specifications representative of the need, processes representative of the case studies, and meta-technologies and processes representing technology used for the various stages of prototype part manufacturing.

possible to make some simulation, and to valid the coherence of the system.

After the programming phase, we began the experimental phase, which made it possible to validate the modes of reasoning, and behaviour of the " theoretical " system with respect to reality and practice.

It is to specify that we assume the work carried out in the application field of rapid manufacturing is important applicable to any type of technology, and thus more generally the industrial production.

11 Acknowledgements

The authors would like to acknowledge different persons. T. Fouquerel (from Aérospatiale) for the industrial validation of the idea. The student who worked first on this project, [15]. The different experts (especially T. Deschamps from Resine Technologie and Y. Seeleuthner from Ateliers Cini) for their availability during the interviews. Finally, Kade-Tech company for their help in Kadviser (Knowledge based system) use.

References

[1] AFPR, *Proceedings of the 7th European Conference on Rapid Prototyping*, November 1998, Paris, France.

[2] A. Bernard and G. Taillandier. *Le prototypage rapide*, 1998, N° ISBN 2-86601-673-4, Editions Hermès, Paris, France.

[3] A. Bernard (Coordinator), *Développement rapide de produit*, International Journal of CADCAM and computer graphics, Vol. 13, n°4-5-6, December 1998, N° ISSN 0298-0924, Editions Hermès, Paris, France

[4] A. Bernard, *State of the art on reverse engineering*, Time-Compression Technology (TCT'99), October 1999, Nottingham, UK.

[5] A. Bernard, *Computer-aided process planning for rapid prototyping* Solid Freeform Fabrication Symposium at Austin 1999, pp 39 – 45.

[6] A. Deglin, *Analysis of needs for CAPP system development in rapid manufacturing field*, internal report, CRAN, October 1998, France.

[7] C. Derras, Formalisation de l'imprécision informationnelle et des incertitudes décisionnelles des connaissances expertes pour la génération de processus de fabrication. Thèse de docteur 1998

[8] Technique documentation on kadviser, kadviser is a trademark of Kade-tech. 17 chemin du Petit Bois - 69130 Lyon-Ecully - FRANCE tél : + 33 4 72 86 11 00 - fax : + 33 4 78 33 43 12

[9] Kassel G. *Contribution à la représentation des connaissances pour les Systèmes Experts de Seconde Génération : projet AIDE, Habilitation à diriger les recherches*, Rapport de recherche Heudiasys 95/55. Université de Compiègne, 1995.

[10] J.L. Perpen, *Définition et réalisation d'une application orientée objet pour la maitrise du processus de coupe*, Thèse Université de Bordeaux, 2000, France.

[11] F. Rechenmann, SHIRPA : *système de gestion de bases de connaissances centrées objet*, INRIA/ ARTEMIS, Grenoble, France, 1988.

[12] F. Rechenmann, *les catégories de connaissances et leur modélisation*, Actes 2nd rencontres Théo Quant sur "décision spatiale" , Besançon, France, octobre 1995.

[13] P. Richard, *Contribution à l'automatisation de la préparation de la fabrication pour les systèmes de production manufacturiers*, Mémoire C.N.A.M. 1997.

[14] SME, *Proceedings of Rapid Prototyping and Manufacturing (RP&M'99)*, April 1999, USA.

[15] V. Trousselard, *First version of a CAPP system for rapid manufacturing process generation*, final report, ESIAL, June 1999, France

[16] F. Villeneuve, *Génération ascendante d'un processus d'usinage*, Thèse Ecole Centrale Paris 1990.

[17] T. Wolhers, Rapid Prototyping and Manufacturing *State of Industry report*, 1999

[18] Wanlong Wang, James G.Contey, Henry W. Stoll, Rui Jiang, *RP Process Selection for rapid Tooling in Sand Casting*, Solid Freeform Fabrication Symposium Austin 1999, pp 19 - 27

[19] F. Xu, Y. S. Wong, H. T. Loh, *A Knowledge-based Decision Support System for RP&M Process*, Solid Freeform Fabrication Symposium Austin 1999, pp 9 -18

[20] S. G. Zhang, A. Ajmal and S. Z. Yang, *Reverse engineering and its application in rapid prototyping and computer integrated manufacturing*, Proceedings of Computer Applications in Production and Engineering (CAPE'95), May 1995, Beijing, Chine, pp. 171 - 178.

Determining Performance with Limited Testing when Reliability and Confidence are Mandated

R. M. Dolin, C. A. Treml

Los Alamos National Laboratory, ESA-EA, Los Alamos, NM 87545

Abstract

In an environment of increased cost, regulation, and accountability, the ability to do unlimited testing to verify the performance parameters of a product are diminishing. At the same time, liability and maintenance concerns are requiring that clearly established operating conditions be defined. Because of this, product developers often encounter situations where reliability, and the confidence in that reliability, are mandated. Furthermore, only a limited number of tests can be performed to validate these mandated conditions. A methodology is presented for establishing performance parameters in these situations. The methodology is based on Bayesian Hypothesis Testing theory. It is shown how an iterative hypothesis can be defined. Within the a Bayesian framework, equations are derived for computing a necessary prior, determining the minimum number of tests needing to be performed, establishing how many failures can be observed for a given number of tests, and inferring an overall reliability distribution given a mandated confidence.

1 Introduction

As costs and regulations continue to rise, the ability to perform unlimited testing is diminishing. At the same time, liability and maintenance concerns have increased sensitivity to product performance reliability. In some situations product reliability is mandated *a priori* and the necessary confidence in that reliability is also stipulated. Furthermore, in these situations, verifying a product conforms to mandated conditions must often be done with a predetermined limited number of experiments.

This paper presents a methodology for establishing a product performance hypothesis via limited testing when reliability and confidence are mandated. Using classical Bayesian based binomial hypothesis testing formulas we develop a methodology for determining a necessary hypothesis given mandated conditions. We also show how this approach can be used with historical information to define a necessary prior probability. Finally, once a hypothesis has been established based on the results of

limited testing we derive the equations for inferring an overall reliability distribution.

An example problem is presented whereby this methodology is used to predict the overall performance characteristics of an inventory based on the results of twenty-five performance tests randomly selected from the inventory with product reliability and confidence mandated. A necessary prior probability is determined based on the results of ten historical tests. Finally, after the hypothesis has been determined, i.e., the performance characteristics defined, a reliability distribution is generated to infer reliability over a range of performance values.

2 Bayesian Binomial Hypothesis Testing

In hypothesis testing, one or more propositions are stated and probabilistic inference is used to determine what hypothesis is correct and under what circumstances it is valid. Beyond simply determining the truth or falsity of a proposition, hypothesis testing can be utilized in other ways. For example, probabilistic inference can be used to determine how many products from some inventory need to be tested in order to achieve a prescribed level of probable confidence that the remaining products in the inventory have some characteristic defined by the hypotheses.

A framework for hypothesis testing can be developed using Bayes' theorem. This framework is used to estimate the number of tests required to achieve some mandated level of reliability and confidence in an inventory. The approach used to develop our methodology is based Jaynes' Bayesian based binomial hypothesis testing.[5] Rather than approaching uncertainty from the classic frequentist standpoint, we treat it as a measure of ones current state of knowledge.

Using the concept of evidence developed within the Bayesian framework, we measure the impact new information has on the current state of knowledge. In this framework, evidence is dependent on prior knowledge as well as on whatever knowledge is gained insitu. It is shown by example, that evidence can be used to not only establish a hypothesis in mandated situations, but also to determine how many tests are necessary to validate the problem's premises.

2.1   Hypothesis Testing Formula

The Bayesian binomial hypothesis test formula can be stated in terms of probability or evidence. Evidence is defined in units of decibels and is a linear function of probability. In general, evidence is a more convenient system of units in which to represent knowledge. The general Bayesian based binomial hypothesis testing equation for confidence given as

$$e(A|D,X) = e(A|X) + (w_s \cdot n_s) + (w_f \cdot n_f).$$

Where

$e(A|D,X)$ = Evidence supporting hypothesis **A** given test data D and prior knowledge X (the posterior confidence).

$e(A|X)$ = Evidence supporting hypothesis **A** given prior knowledge (the prior probability).

w_s = Weighting factor for success (ratio of probabilities for success of hypothesis **A** over hypothesis **B**).

w_f = Weighting factor for failure (ratio of probabilities for failure of hypothesis **A** over hypothesis **B**).

n_s = Number of successes encountered during testing.

n_f = Number of failures found during testing.

2.2   Hypothesis Testing Process

The general hypothesis testing process can be broken into the following steps.

1) State the hypotheses. For example,

A: Products are at least *R* reliable for loads up to and including *L.*

B: Products are at least *r* reliable for loads up to and including *l.*

2) Combine pre-existing data and information into a Prior Probability.

P(A|X) = Probability that hypothesis A is true given any prior knowledge X.

3) Use the product rule of probability theory to apply Bayes theorem.

$$P(D,A|X) = P(D|A,X) \cdot P(A|X)$$
$$= P(A|D,X) \cdot P(D|X)$$

$$P(A|D,X) = P(A|X) \frac{P(D|A,X)}{P(D|X)}$$

Here, P(A|D,X) is the probability hypothesis **A** is true given current data D and prior knowledge X. This is often called the posterior probability or simply posterior. It is a measure of confidence in the primary hypothesis. The P(D|A,X) is often called the likelihood probability, or just the likelihood.

4) Define the confidence formula that measures the degree of belief in hypothesis A being true given all currently available information.

3   Mathematical Derivation for Mandated Situations

For a given hypothesis, we can define weighting functions specifying the contribution each test result has on our confidence in the validity of the hypothesis. Rewriting the hypothesis testing equation for confidence yields,

$$\text{Confidence} = [\text{Prior} + w_s \cdot n_s + w_f(n_t - n_s)].$$

While confidence is usually expressed in units of percentage, it is mathematically convenient to work in units of decibels, which are $\log_{10}$ based. Confidence (C) expressed in decibels is $10\log_{10}[C/(100-C)]dB$. This is generally referred to as Evidence, and is written

$$E = [P + w_s \cdot n_s + w_f \cdot (n_{tests} - n_s)]. \qquad (1)$$

Where, E, P, w_s, and w_f, are all expressed in decibels and defined as follows:

E = Evidence supporting the validity of the hypothesis. It's a measure of confidence, e.g., if E = 12.94dB, then there is a 95% confidence the hypothesis is true.

P = Degree of belief the hypothesis is true prior to testing. For binomial testing, the minimum confidence is 50%, usually after assessment of available knowledge, this adjusts upward.

w_s = Weighting factor for success. Amount confidence increases with each successful test.

w_f = Weighting factor for failure. Amount confidence decreases with each failure.

n_t = Number of tests performed.

n_s = Number of successes in the n_t tests.

In a limited testing environment, an upper bound on n_t is known. The number of failures, m, allowed in n_t tests while maintaining k confidence in the hypothesis is either determined based on the hypothesis or specified *a priori* based on other criteria and used to define the hypothesis. Writing the evidence equation (1) in terms of the allowable failures variable m, yields

$$E = [P + w_s(n_t - m) + w_f \cdot m]$$

from this it follows:

$$m = \frac{P + w_s \cdot n_t - E}{w_s - w_f}.$$

This formula specifies the m failures in a series of n_t tests that are allowed while maintaining k confidence in the hypothesis. In other words, given an hypothesis **H** and prior belief P we can determine how many failures, m, can be absorbed while maintaining a mandated confidence (expressed as evidence E.)

Suppose the performance characteristics of an inventory are to-be-determined based on the results of a limited test series. The inventory is mandated to operate at a to-be-determined performance level such that the products are R reliable with k confidence. If the failure loads observed in testing are ordered from highest to lowest, the m lowest failure loads can be disregard because they represent the allowable failures that can be absorbed. The $(m+1)^{st}$ lowest failure load then becomes the established performance load, L, specified for the conditions of the hypothesis. For example:

$m=0 \Rightarrow$ The lowest failure load found in testing is the established performance load since we can accept no failures. (The hypothesis is subsequently modified to reflect this.)

$m=1 \Rightarrow$ We can absorb one failure in the test series so the second lowest failure load observed in testing becomes the performance metric the hypothesis is modified to reflect.

$m=2 \Rightarrow$ We can absorb two failures in testing so the third lowest failure load observed in testing becomes the performance metric the hypothesis is modified to reflect.

etc…

3.1 Failure Load Intervals

There are many uncertainties involved with testing for failure. The sensitivity of the test equipment, the repeatability of an experiment, and the consistency of the test samples themselves just to name a few. Because of this load intervals are often used when establishing product performance. Load intervals assume that failures within a given range of loads represent equivalent failures. When load intervals are used, the hypotheses is modified to capture the allowable ranges. It's the responsibility of decision-makers to define the ranges and determine their distinctions.

The measured failure loads for a series of tests are discretized based on problem fidelity. A possible set of load ranges that might be used to discretize what is often viewed as continuous measurements for failure loads, L_f is shown in Table 1. The sample discretization is shown for the results of a series of 25 tests.

Load Interval	Load Range	# failures
L_1	$L_0 < L_f \leq L_1$	1
L_2	$L_1 < L_f \leq L_2$	1
L_3	$L_2 < L_f \leq L_3$	1
L_4	$L_3 < L_f \leq L_4$	4
L_5	$L_4 < L_f \leq L_5$	2
L_6	$L_5 < L_f \leq L_6$	3
L_7	$L_6 < L_f \leq L_7$	6
L_8	$L_7 < L_f \leq L_8$	4
L_9	$L_8 < L_f \leq L_9$	2
L_{10}	$L_9 < L_f \leq L_{10}$	1

Table 1: Load Range Discretization

Usually, it is convenient to view this discretization in the form of its corresponding histogram for the number of failures per load interval. Figure 1 shows the histogram of failure loads corresponding to table 1.

3.2 Determining Failure

Often confidence and reliability are mandated by technical, performance, administrative, or regulatory requirements. When its known prior to testing that with k confidence, $R\%$ of the time a test must be successful, the products definition of failure can be determined. Since

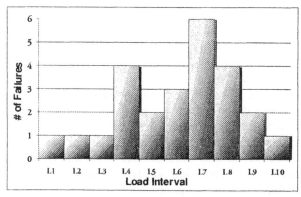

Figure 1: Failure Histogram

only *k* confidence with *R* reliability is sought, failure loads of individual products can be omitted when establishing the definition of failure.

This reverses the logic of standard hypothesis testing making the hypothesis itself the iterative solution variable. In the language of binomial hypothesis testing, the stated hypothesizes become

A: At least *R%* of products survive loads at or above load *L*..

B: At least *r%* of products survive loads at or below load *l*.

Hypothesis **A** is the desired level of reliability. Hypothesis **B** is the lowest acceptable level of reliability.

Suppose it is mandated that the desired reliability of a product inventory is 90%, while the lowest acceptable level of reliability is 80%. Then,

$$P(S|A,D) = R/100 = 0.90$$
$$P(F|A,D) = (100-R)/100 = 0.10$$
$$P(S|B,D) = r/100 = 0.80$$
$$P(F|B,D) = (100-r)/100 = 0.20$$

where $P(S|A,D)$ is the probability of success (however we ultimately choose to define it), given hypothesis A and data D. Likewise, $P(F|A,D)$ is the probability of failure given hypothesis A and data D. Each time a test fails at or above *L* we increase our degree of belief in hypothesis **A** by the amount w_s, where in this example,

$$w_s = 10 \log [0.9/0.8] = 0.5115 \text{ dBs}.$$

Each time a test fails below *l* we decrease our degree of belief in hypothesis A by the amount w_f, where,

$$w_f = 10 \log [0.1/0.2] = -3.0103 \text{ dBs}.$$

The number of successful tests needed to overcome a single failure is then given by,

$$N = |w_f / w_s| = 5.885 \sim 6 \text{ tests}.$$

3.3 Prior Probabilities

Priors are used to capture and apply within an analysis all the knowledge available prior to starting the problem. [4,9] Whenever a hypothesis is stated, there is some degree of believe on its validity. At a minimum in binomial hypothesis testing we must prefer one hypothesis over the other. If not, the hypothesis problem is poorly formed. Hence, the preferred hypothesis has to have a prior probability of least 50%.

One approach to establishing a prior probability is based on subjective metrics. [3,6] For example, given hypothesis A is preferred to hypothesis B, how much more is it preferred, a little or a lot? Based on all the presently available knowledge, e.g., testing, simulation, expert opinion, etc., is it possible to say hypothesis A has a 60% chance of being true? If so, can it comfortably be said that hypothesis A has a 70% chance of being true? What about 80%? Where one stops being comfortable with the likelihood hypothesis A is true is the starting prior for the analysis.

There are many techniques that can be used to determine a prior probability. The approached described above is a combined application of the equivalence and choice rules of prior probability assessment.[2] The power of a prior, and of Bayesian methods in general, is that they allow you to start a problem with your best judgement or belief in the outcome incorporated into the model. Testing is then an exercise to determine how much the actual probability is greater or less than your initial conception. This ultimately leads to a solution with fewer experiments than would be required by classical frequentist approaches since in essence these classical methods start with a zero prior probability.

The downside to picking a poor prior is that you'll need more testing to conclude the truth or non-truth of a hypothesis. However, data eventually overwhelms a bad prior.[7,8] It's just a matter of how much data it's going to take. Less subjective and more mathematically rigorous approaches can also be used to determine the prior probability. These more analytical methods would be preferred when there is quantitative information available, for example, from historical tests or analysis simulation. In many situations quantitative information, when available, is combined with qualitative knowledge, like expert opinion to formulate a prior.

$$m = (E - P - w_s n_t) / (w_f - w_s).$$

3.4 Determining the Minimum Number of Tests

Often it is of interest to know how many tests are required to prove or disprove a hypothesis. This can be for economic reasons like estimating the costs for a deliverable, or scheduling, environmental, or limited supply of resource reasons, just to name a few. To determine the fewest number of tests required to be k confident in a hypothesis, we need to modify the language of the hypothesis. For example, hypothesis A might be modified to say

A: At least $R\%$ of an inventory survives the smallest observed failure load in n_t validation tests.

In other words, the desired reliability, R, is used to define how many physical tests need to be performed using the smallest failure load observed as the load limit L. For the minimum test case, $m=0$, i.e., no failures are allowed. Then,

$$E = P + w_s * n_{mim} \Rightarrow$$
$$n_{mim} = (E - P) / w_s$$

This represents a lower limit on the number of tests needing to be preformed since it assumes no failures are observed ($m = 0$). If for whatever reason we can perform more than this minimum, n_{min} tests, then we can determine how many tests need to be performed successfully? In other words, if we allowed m failures what would the value of n_t be?

Often n_t is either known or mandated. In those situations, we can determine m and from that determine L and ultimately the final language of our hypothesis. For example, suppose the number of tests that can be performed, n_t, is mandated, then

$$E = P + w_s(n_t - m) + w_f \cdot m.$$

We are interested in determining the number of allowable failures, m, that can be observed while maintaining k confidence in hypothesis **A** given that n_t tests can be performed. To accommodate this, hypothesis **A** is modified such that,

A: At least $R\%$ of an inventory survives the $(m+1)^{st}$ lowest observed failure load (L) in n_t validation tests.

Then the number of allowable failures given this hypothesis is,

3.5 Determining the Necessary Prior from Limited Testing

If we are in a limited testing situation and want to determine the necessary prior in order to realize some desired reliability in a threshold load, two cases can be considered. The first case involves allowing no failures and the second case assumes some allowable number of failures. The first case yields the more conservative estimate of a necessary prior.

In *Case 1*, no failures are allowed to maintain k confidence in hypothesis **A**. If no observed failures are allowed, say because n_t is small or because we need a conservative estimate, the following formulation is used to determine the necessary prior.

$$n_t = (E - P) / w_s \Rightarrow$$
$$P = E + (n_t \cdot w_s)$$

In *Case 2*, m failures are allowed while maintaining k confidence in hypothesis **A**. There may be times when some failures are allowed. For example, if a relatively large number of test samples are available or we want to drive the solution to the highest reasonable load interval. Then the following formulation can be used to determine the necessary prior.

$$m(w_f - w_s) = E - P - (n_t \cdot w_s)$$
$$\Rightarrow P = E + m(w_s - w_f) - (n_t \cdot w_s)$$

The necessary prior can be determined in a limited testing environment where either the numbers of samples available for testing is small, or little historical knowledge is available. The necessary prior can be use in situations where it is either desired or required to maintain a minimum confidence in the validity of hypothesis **A**, and if we know the desired or required probability of success, R.

The necessary prior can be used to determine the performance load interval, L, for our hypothesis. In other words, of all the possible load intervals, which one do we have a degree of belief in being true that's equivalent to the computed prior? Whatever load interval that is represents the performance load L in the iterative solution hypothesis.

4   Example: Limited   Test   Environment   with   Mandated   Reliability   and   Confidence

Suppose technical, administrative, or regulatory requirements mandate a load limit for a product be established such that the reliability (i.e., probability of success) is $R=90$ and the confidence in that reliability estimate is $k=90$. A limited test series is to be undertaken, but because resources are limited only 25 test samples are available. Binomial hypothesis theory is to be employed. Because a Bayesian based approach is used, a prior probability needs to be determined.

4.1   Historically Based Necessary Prior

In a series of ten tests performed last year, the lowest observed failure was 1.1k. It is decided a necessary prior can be determined allowing one declared failure. In other words, $m=1$. This decision is based on the reasoning that one failure in ten tests is approximately a 90% reliability. Expert opinion is that this is a reasonable assumption. The equation for determining the necessary prior is

$$P = E + m(w_s - w_f) - N_{tests} \cdot w_s)$$
$$= E + w_s - w_f - 10w_s$$
$$= E - 9w_s - w_f$$
$$= 9.5424 - 9(0.5115) + 3.0103$$
$$= 7.9492 \text{ dBs} = 86.2\%$$

At this point it is not know, which load interval to base the primary hypothesis on. This is because the number of failures that can be tolerated in the current test series is not known. The established product performance failure load is a function of the number of failures that can be tolerated along with the outcome of the tests once performed.

Recall that only 25 test samples are available and both reliability and confidence have been mandated. Hence, the number of failures that can be absorbed while maintaining k confidence in R reliability is

$$m = (E - P - N_{tests} \cdot w_s) / (w_f - w_s)$$
$$= [9.5424 - 7.9492 - 25(0.5115)]/(-3.0103 - 0.5115)$$
$$= 3.1786 \sim 3 \text{ failures.}$$

This means that the product performance failure load will be determined by conducting the 25 tests and recording the failure interval for each test. The results can then be ordered from lowest failure load to highest. Then,

the forth lowest failure load can be used as the load limit parameter L in hypothesis **A**.

> **A**: We are 90% confident that 90% of our inventory survives loads up to and including L.

Another way to state this conclusion in terms of load intervals is

> **A**: With 90% confidence, 90% of the time a randomly selected product from our inventory survives loads up to and including L.

4.2   Building   a   Reliability   versus   Load Distribution

It's often necessary to understand how reliability is distributed over an inventory rather than only knowing the load interval for which some mandated reliability is valid. In addition to knowing for load intervals less than or equal to **L** we are at least k confident in being R reliable, we may also want to know the reliability R_i at other load intervals $\mathbf{L_i}$ as well. Recall, in binomial hypothesis testing there are two hypothesis statements one, of which one must be valid. In general there is not a crisp separation for when one hypothesis is invalid and the other is valid. This margin of uncertainty can be captured by writing both hypotheses in terms of one set of parameters. However, this would require knowing one of the reliability parameters as a function of the other. This can be accomplished by forming an iterative hypothesis equation. For our problem, we state the forward iterative hypotheses as:

> **A**: With k confidence, $R_i\%$ of the time a randomly selected product from an inventory survives loads up to and including load interval $\mathbf{L_i}$.

> **B**: With k confidence, $(R_{i+1})\%$ of the time a randomly selected product from our inventory survives loads up to and including load interval $\mathbf{L_{i+1}}$.

Where $\mathbf{L_i}$ is defined to be the ith load interval for a monotonically increasing set of loads and R_i the reliability for the $\mathbf{L_i}$ load interval given a fixed value for confidence, the necessary prior, and the data from the current test series. From Table 1, the relationships between the load limits L_i and load intervals $\mathbf{L_i}$ are written, $\{\mathbf{L_i}: L_f \in \mathbf{L_i}, (L_{i-1} < L_f \leq L_i)\}$.

If we assume a mandated confidence k, a fixed necessary prior, and utilize current test information, we can solve the inference evidence equation (1) in terms of reliability R_i. Philosophically, fixing the prior is saying the computed reliability distribution is consistently based on the same prior belief. A family of reliability distributions can be generated using different priors to change the belief structure. But that is always true in Bayesian analysis, just as changing the confidence parameter k yields a family of reliability distributions.[1,11] Using our example, we then calculate the reliability distribution according to the mandated confidence (k=0.9), the necessary prior (P=0.862), and the test data shown in Figures 1 and 2.

Starting with the initial evidence equation (1), we can develop a hypothesis expression for determining R_i. First, we rewriting equation (1) in terms of decibels,

$$E = 10\log_{10}\left(\frac{C}{100-C}\right) = 10\log_{10}\left(\frac{P}{100-P}\right) + \cdots$$
$$\cdots + n_s \cdot 10\log_{10}(w_s) + (n_t - n_s) \cdot 10\log_{10}(w_f) \quad (2)$$

where the E is in decibels and C and P are in percentages. The weighting factors for success and failure, w_s and w_f, respectively, defined in terms of reliability are as follows:

$$w_s = \frac{P(S|A)}{P(S|B)}$$

$$= \frac{\text{Prob}(L_f > L_i \text{ given } R_i\% \text{ survive loads up to } L_i)}{\text{Prob}(L_f > L_i \text{ given } R_{i+1}\% \text{ survive loads up to } L_{i+1})}.$$

Where, L_f is the load at which failure occurs and S is the success of hypothesis A, i.e., $L_f > L_i$. It follows that,

$$w_f = \frac{P(\overline{S}|A)}{P(\overline{S}|B)} = \frac{1 - P(S|A)}{1 - P(S|B)}.$$

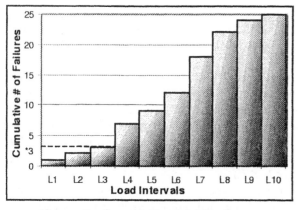

Figure 2: Cumulative Distribution

By the definition of reliability, we can write, $R_i = P(S|A)$. To write an expression for P(S|B), we must utilize the information in the failure histogram from the current test series.

$$P(S|B) = R_{i+1} + [\text{incremental probability of a failure}$$
$$\text{occurring in load interval, } L_i]$$

$$= R_{i+1} + \frac{n_f(i+1)}{n_i} = R_{i+1} + \Delta_{i+1},$$

where $n_f(i)$ is the number of failures recorded in load interval, L_i and $\Delta_i = n_f(i)/n_i$. Now, in terms of reliability and letting log imply $\log_{10}$, equation (2) becomes,

$$\log\left(\frac{k}{k-1}\right) = \log\left(\frac{P}{P-1}\right) + \cdots$$
$$\cdots + n_s(i+1) \cdot \log\left(\frac{R_i}{R_{i+1} + \Delta_{i+1}}\right) + \cdots$$
$$\cdots + m(i+1) \cdot \log\left(\frac{1-R_i}{1-(R_{i+1} + \Delta_{i+1})}\right) \quad (3)$$

Where R_i is the reliability stated in hypothesis **A** and R_{i+1} is the reliability of the binomial alternative in hypothesis **B** and R_i, R_{i+1}, P, and C are represented as probabilities. Also, $m(i)$ is the total number of failures occurring at or below loads L_i (or at or below load intervals **Li.**)

$$m(i) = \sum_{l=1}^{i} n_f(l)$$

In terms of the failures, the total number of successful tests that occur at or below L_i is written,

$$n_s(i) = n_t - m(i).$$

Rearranging equation (3) and letting, $\tilde{R}_{i+1} = R_{i+1} + \Delta_{i+1}$, and $q(i) = n_s(i)/m(i)$ yields,

$$\left[\frac{k(1-P)}{P(1-k)}\right]^{1/m(i+1)} = \frac{R_i^{q(i+1)}(1-R_i)}{\tilde{R}_{i+1}^{q(i+1)}(1-\tilde{R}_{i+1})}. \quad (4)$$

Rearranging (4) and letting $K = \frac{k(1-P)}{P(1-k)}$, yields

$$\tilde{R}_{i+1}^{q(i+1)+1} - \tilde{R}_{i+1}^{q(i+1)} + \left[K^{\frac{-1}{m(i+1)}} \cdot R_i^{q(i+1)} (1 - R_i^{q(i-1)}) \right] = 0 \quad (5)$$

where $R_{i+1} = \tilde{R}_{i+1} - \Delta_{i+1}$. Equation (5) is the forward inference hypothesis equation. Using the example presented throughout this paper, the constant values in (5) are:

$$k = 0.9$$
$$P = 0.862$$

Where k is the mandated confidence and P is the necessary prior based on the mandated confidence and reliability.

In section 4.1, it was shown that $m=3$ failures are allowed in the series of 25 tests, while maintaining $R = 0.9$, reliability for loads up to an including the $(m+1)^{st}$ recorded failure load. Figure 2 shows the cumulative failures in each load interval for the 25 tests. Note that L_3 is the load interval containing $m(3) = 3$ cumulative failures, so L_3 is set as the load limit interval in hypothesis A. From this, the initial reliability used to solve equation (5) becomes, $R_3 = 0.9$.

Using , $R_3 = 0.9$, $P = 0.862$, $k = 0.9$, and the data in Figures 1 and 2, reliabilities for forward load intervals L_4 through L_{10} can be calculated. This is done by first solving for the zeros of the polynomial on $\tilde{R}_{i+1}$ in equation (5), using MATLAB.[10] Then calculating $R_{i+1} = \tilde{R}_{i+1} - \Delta_{i+1}$. The results are shown in Table 2.

In a similar manner, the backward inference hypothesis equation can be formed, using the hypothesis statements:

A: With k confidence, $(R_{i-1})\%$ of the time a randomly selected product from an inventory survives loads up to and including load interval L_{i-1}.

B: With k confidence, $R_i\%$ of the time a randomly selected product from our inventory survives loads up to and including load interval L_i.

From this, it follows that the backward inference equation can be written,

$$R_{i-1}^{q(i-1)+1} - R_{i-1}^{q(i-1)} + \left[K^{\frac{+1}{m(i-1)}} \cdot \tilde{R}_i^{q(i-1)} (1 - \tilde{R}_i^{q(i-1)}) = 0 \right], \quad (6)$$

where $\tilde{R}_i = R_i - \Delta_i$. Equation (6) is a polynomial on R_{i-1}, and can be solved directly for R_{i-1} since all other parameters are known. As with the forward problem, the initial value, $R_3 = 0.9$ was used. The resulting reliabilities for load intervals L_2 and L_1 are shown in Table 2.

Using forward and backward inference hypothesis equations (5) and (6) with the current test data, the mandated confidence ($k=0.9$), the calculated necessary prior ($P=0.862$), and corresponding allowable test failures ($m=3$), the corresponding reliability distribution was generated. The distribution R_i is shown in Figure 3 versus the corresponding load intervals L_i.

Forward Problem				
(i)	R(i)	m(i+1)	nf(i+1)	R(i+1)
3	0.9	7	4	0.74679
4	0.74679	9	2	0.68898
5	0.68898	12	3	0.58595
6	0.58595	18	6	0.35702
7	0.35702	22	4	0.21059
8	0.21059	24	2	0.14429
9	0.14429	25	1	0.11642
Backward Problem				
(i)	R(i)	m(i-1)	nf(i-1)	R(i-1)
3	0.9	2	1	0.93936
2	0.93936	1	1	0.98298

Table 2: Reliability Distribution Results

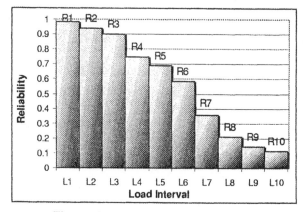

Figure 3: Reliability Distribution

5 Conclusions

Using a Bayesian based hypothesis-testing framework, a methodology was presented for establishing a product's performance parameters in these situations.. It was shown how an iterative hypothesis evolves to satisfy the mandated conditions. Equations were derived for computing necessary priors based on available knowledge. Equations for determining the minimum number of validation tests were also presented. Once these values were known, equations for determining the number of allowable failures given the mandated conditions were derived. It was shown how all of that knowledge can be determined so as to define a hypothesis satisfying the mandated conditions. Finally, a methodology for generating an overall reliability distribution via forward and backward inference was described. An example was presented, whereby a reliability distribution was computed for a fixed confidence over the range of load intervals.

References

[1] Bernardo, J.M., Smith, A.F.M., Bayesian Theory, John Wiley & Sons, New York, 1994.

[2] Clemen, R.T., Making Hard Decisions, An Introduction to Decision Analysis, Second Edition, Duxbury Press, 1996.

[3] Finetti, B. de, "Foresight: Its Logical Laws, Its Subjective Sources," Studies in Subjective Probability, ed. H.E. Kyburg, Jr., H.E. Smokler, Wiley, New York, 1964.

[4] Good, I. J., "Estimation of Probabilities: An Essay on Modern Bayesian Methods," Research Monograph No. 30, MIT Press, Cambridge MA, 1965.

[5] Jaynes, I.T., "Probability Theory: The Logic of Science," Unpublished Manuscript, internet version, 1995.

[6] Jaynes, E.T., "Where Do We Stand on Maximum Entropy, in The Maximum Entropy Formalism", eds. R.D. Levin, M. Tribus, MIT Press, Cambridge, MA, 1979.

[7] Neopolitan, R.E., "Limiting Frequency Approach to Probability based on the Weak Law of Large Numbers," British Journal for the Philosophy of Science, 1989.

[8] Puri, M.L., Ralescu, D.A., "Strong Law of Large Numbers with Respect to a Set-Valued Probability Measure.", Annals of Probability, v. 11(#4), pp. 1051-1054, 1983.

[9] Sheridan, F.K.J., "A survey of techniques for inference under uncertainty," Artifical Intelligence, Rev. 5, pp. 89-119, 1991.

[10] The mathWorks, Inc., Using MATLAB, 1998.

[11] Tiao, G.C., Zellner, A., "Bayes Theorem and the Use of Prior Knowledge in Regression Analysis," Biometrika, vol. 52, p.355, 1965.

CHAPTER 12

CE Perspectives

A Survey of how UK SMEs use Computer Aided Solid Modelling - And an outline for better Implementation

J. Lawson

School of Design, University of East London, 4-6 University Way, London, E16 2RD

A. King

School of Design, University of East London, 4-6 University Way, London, E16 2RD

Abstract

With the globalisation of business the manufacturing sector now trades in a highly competitive market where design efficiency is essential. New technology, and particularly Computer Aided Design, plays an essential part in achieving this goal. However, simply having the technology is no solution: it needs to be used effectively. This paper explains the problems that Small to Medium sized Enterprises (SMEs) face when investing in CAD technology and in particular Computer Aided Solid Modelling (CASM).

The paper concludes with an evaluation and gives recommendations for future research.

1. Computer Aided Solid Modelling (CASM)

Computer Aided Solid Modelling (CASM) is a prime example of a new technology that can, if implemented correctly, bring massive cost and time savings to a company's new product development process. Although various forms of CAD have been used for 30 years, it is the more recent development of CASM that has made its use more widespread in the earlier stages of design [1]. There are two main reasons why CASM has become a viable technology for SMEs to use.

Firstly, the cost of implementing powerful hardware and software systems has significantly dropped meaning that state-of-the-art CASM software is now an option for even the smallest of companies. With this the software has become much more usable due to user-interface developments.

Secondly, CASM software represents much more design information than older 2D Computer Aided Drafting software. Therefore, CASM acts as the "bridge" for the further integration of secondary technologies including Finite Element Analysis, Computational Fluid Dynamics, Rapid Prototyping, and Computer Numerical Control manufacture.

CASM is an ideal foundation to enable further business development.

1.1 Types of 3D CAD

Before CASM techniques can be discussed specifically it is important to understand the other types of 3D CAD software. The three types are Wire-frame, Surface Models and Solid Models.

1.1.1 Wire-frame models: This technique displays an object as a framework of lines as opposed to shaded surfaces. Such models do not give an understanding of surfaces or volumes and because all edges are displayed simultaneously they are visually hard to interpret. However, wire-frame models require relatively small amounts of computer resources.

1.1.2 Surface Models: These give a representation of external surfaces and can be shaded to produce an understandable image with hidden lines removed from view. Surface models only compute an object's faces meaning it is easier to work out complex topography because volumes are not part of this already very complicated calculation

1.1.3 Solid Models: Although these use the most computer resources an object is fully understood by both the user and the software. This is because the enclosed space between surfaces (the volume) is known. Therefore a model is represented in terms of its real-life spatial integrity so that physical properties such as mass can be assigned to enable further analysis.

For further detail on types of 3D CAD see Nottage, J. 1994 [2].

1.2 Different CASM Techniques

Solid Modelling is the most recent and by far the most powerful of the above techniques. However, there are various types of CASM packages that employ different representation methods and techniques. These are discussed below.

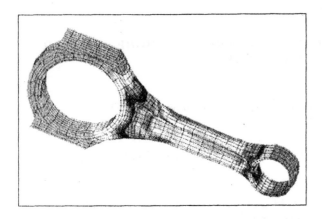

Figure 1 - Solid Model under FEA forces

1.2.1 Constructive Solid Geometry (CSG): CSG represents a solid model through the use of progressively simple 3D shapes. These are used in combination to form the finished, more complex shape. The simplest of these shapes are called primitives and typically include cylinders, cones, spheres and boxes among others. Primitives are combined with each other through the Boolean operations described below:

- Union: Joins two or more shapes together at the area where they intersect.
- Difference: Takes one solid primitive away from the other and keeps the remaining shape.
- Intersection: Keeps only the area of the primitives that intersect.

CSG representations store the solid primitives together with the 'Boolean set operations' in a 'binary tree'. The root node represents the finished object and other nodes represent an intermediate solid, which is in combination of primitives and intermediate solids lower in the tree. The initial primitives are therefore represented in the leaves of the CSG tree. This is a very efficient method or storage, as it gives an historical account of how the final solid was created. Therefore, accessing and altering the model is relatively easy.

1.2.2 Boundary Representation (B-Rep): The B-Rep database stores a solid as a set of vertices, edges, and faces that completely enclose an object's volume. An example is a cube that is constructed from six equal faces that are bound at eight edges.

B-Rep information is explicitly stored meaning it is directly represented rather than implied. For example, an edge defined by an equation and two endpoints is explicitly defined. This means a B-Rep model has a comparatively fast display time because it is evaluated as it is constructed. (A CSG model is evaluated at display times meaning the viewing process is more timely).

However, there are two main limitations with B-Rep representation:
1. Only relatively simple shapes can be constructed.
2. B-Rep modellers are currently limited to Manifold Representation. This means that objects cannot be joined to one another or to themselves at an edge or a vertex.

1.2.3 Pure Primitive Instancing: This technique stores a number of pre-defined geometric shapes in its database. These can then be used in certain design situations where they can be scaled accordingly. For example, a part library consists of useful pre-modelled standard parts such as nuts and bolts. This technique can allow the designer to spend more time on innovation rather than re-drawing existing parts or general shapes.

1.2.4 Cellular Decomposition: In this technique smaller cells that are mutually contiguous and do not interpenetrate, represent a solid volume. This creates an approximation of the object because some cells are partly in and partly outside of the model. The most popular application for this method is Finite Element Analysis (FEA) as each cell or polyhedral can be assigned separate properties and respond differently from the rest of the model.

The main disadvantage with this technique is that models are very hard to build, and are time consuming and resource intensive.

For further detail on CASM techniques see Shahin, T. 1996 [3]

1.3 Parametric CASM

Parametric CASM is a powerful technique that gives the designer a more effective way of working. With Parametric CASM, the geometry of each new feature is implicitly related to the first created. If the first feature is changed, all of the associated geometry is also altered. This is why it is also called Relational CASM as all the dimensions are linked (or related) together.
A typical parametric CASM software package has the following elements:

1.3.1 The Sketcher: This is an environment where 2D profiles can be generated or modified. Standard drawing tools allow complex shapes to be created accurately and at speed. The 2D profiles are then transformed into 3D models.

1.3.2 The SM System: This is where the actual 3D model is created. The most common way of generating a solid is through either an axial or rotational sweep. However, other ways include general surface enclosures (by combining edges and faces to created an enclosure), the combination of primitives using Boolean operations, and using existing solids to produce shells.

1.3.3 The Dimensional Constraint Engine: The DCE is also called the variation or parametric engine. This allows the modification of dimensions and determines the associated changes.
Therefore, if a designer modifies an early part of a Solid Model, the DCE will effect this change and update (and modify) all affected features.

1.3.4 The Feature Manager: This provides a number of common features (such as holes, bosses, fillets etc.) that can be applied to the solid model. They usually represent how an object would be manufactured and are tools that will improve the reliability of an object. These "pre-set" features can also be made relational to existing features. For example, if a hole is defined as a 'through hole', when associated dimensions are changed the hole will always extend through the parent feature. The 'feature driven' philosophy both speeds the design process and simplifies the underlying database.

1.3.5 The Assembly Manager: This enables component parts to be modelled separately and then brought together to assemble the complete product. There are two types of assembly operation. Relative placement 'loosely tacks' parts together so they can become unrelated later. Boolean placement 'permanently glues' parts together forming an irreversible bond.

1.5 CASM Data Transfer

SMEs need to exchange information with their suppliers and customers and therefore need CASM software that will allow fast and accurate data transfer. If the CASM system does not offer reliable data exchange the company will have difficulty in remaining competitive [4].

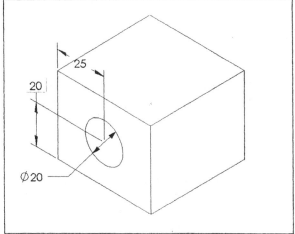

Figure 2 - Parametric component with defined "through-hole".

1.5.1 IGES: Through the IGES (International Graphics Exchange Standard) translation process, the pre-processor of the sending CAD system converts geometry into IGES entities that are close to the native entity in mathematical form or intended content. Similarly the post-processor of the receiving CAD system must convert IGES entities into its own appropriate native geometry.

The IGES translation method is by no means ideal. Similar to the translation of a foreign language the original message can become confused or lost all together. Passing geometry through IGES can leave gaps where surfaces join, erase essential text, and loose all other kinds of associated data.

1.5.2 STEP: STEP (Standard for the Exchange of Product data) is the acronym for the international standard ISO 10303 [5]. The standard is being established to provide a framework for the representation of product data in a format that can be shared by all product databases and CAD systems. STEP will provide a complete, unambiguous, computer-read-able definition of the physical and functional characteristics of a product throughout its life cycle. STEP will also verify a product's physical shape and hence true enclosed geometric forms are maintained.

The advantages include the elimination of translations between native formats and consistent representation for data. The disadvantages include an increase in computer system infrastructure and organisational impacts caused by the shared data paradigm.

2. SMEs and CASM implementation

2.1 The limitations of being an SME

On-going research has identified the main problems associated with being an SME in the manufacturing sector [6, 7, 8]. More interestingly for the purpose of this paper, the research has specified factors that restrict SMEs when they are making investments in technology and have been categorised as follows [9]:

- Lack of capital investment funds
- Lack of staff to investigate the market place
- Lack of access to expert help
- Lack of time to investigate new systems
- Lack of knowledge of available systems

As well as offering a platform for survival, technology investments present the opportunity for growth and company wide improvements. However, it is important that the word 'offers' is stressed, as technology does not automatically bring improvements.

It has been stated that "Poor design process and expensive technology = Expensive poor design process" [10]. Therefore the limitations listed above need to be individually examined and resolved if companies are to get the full potential from technology purchases. One way of achieving this is by following a clear planning and implementation methodology.

2.2 Integrating CASM into the business process

Although CAD implementation guides exist [11, 12, 13, 14], there are presently few that specifically aid companies in successfully implementing and integrating the powers of CASM into their design process.

Buying hardware and CAD is relatively easy. However, the implementation process is difficult. SMEs usually have limited resources (particularly in-house expertise) and the transition between an old and new system can be complex (often in terms of design legacy and adopting new design practice).

2.3 Telephone Survey of SMEs usage of CAD

To identify current usage and implementation issues, the authors conducted a telephone survey of 372 UK companies. The companies were selected from a commercial database [15] and filtered by sector (Mechanical Engineering).

2.3.1 The Telephone Survey: Each company was telephoned between the 26th of January and the 16th of February 2000 during normal working hours (0900-1630). On calling, the most appropriate person within the business was established, and invited to answer questions from a 5-10 minute survey. In order to allow for non-CAD users, the first question determined whether CAD was used, and subsequent questions were tailored to this reply. The survey structure is shown in figure 3.

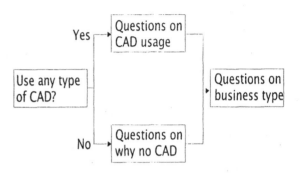

Figure 3 - Telephone Survey Structure

2.3.2 The Survey Results: Of the of 553 UK SMEs contacted, 372 stated if they did, or did not own a CAD system. The remaining 181 companies were either unprepared to give any information, were too busy or were incorrect phone numbers. 232 companies used no form of CAD whilst the remaining 140 were owners of CAD software. Although the benefits of CAD are well established in academia, according to this survey, the majority of UK SMEs within the mechanical engineering sector still do not use it.

Out of the sample of 372 companies, 80 completed the full telephone interviews. Of this number, 70 were CAD owners.

2.3.3 Response from 70 CAD owners: In this survey, an owner is defined as "a company that owns and uses any form of CAD within their design process". This includes both 2D and 3D systems.

- 57 (87% of owners) owned at least one licence of a version of AutoCAD. The remainder own more specialised systems suited to their particular business activity.
- For 63 companies, (90% of owners), the main use of CAD was purely for 2D design. Only 8 (10%) used it for any form of analysis and only 5 (7%) used a combination of CAD and CAM (Computer Aided Manufacture).
- Of the total 88 systems owned, (some companies own more than one), 70 were for 1-5 licenses.

- 54 (77% of owners) have more than 2 employees that can directly use and operate their CAD systems.
- 58 have systems running on either Microsoft Windows 95/98™ and 25 owned systems that operate on Microsoft Windows NT™. Only 5 used any other Computer Operating System.

One objective of the survey was to determine the main benefits that CAD brought to the companies. Figure 4 displays the results of this question.

- The biggest benefit given was the ability to remain flexible. 62 (91% of owners) agreed that this area was a benefit and 21 (40%) thought it was the main benefit. The flexibility of a system is its ability to make quick and often late changes to drawings and designs.
- 10 (19% of owners) agreed that a shortened "time to market" was the main CAD related benefit.
- 5 (10% of owners) stated that reduced operating costs was the main benefit.

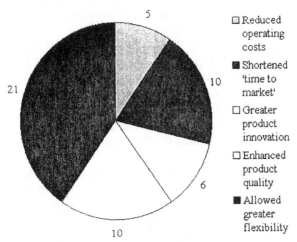

Figure 4 - No. of companies listing main benefits of CAD

The issue of CAD implementation was also addressed in the survey. Companies were asked to state any problems they faced.

- 23 (36% of owners) stated that minimising downtime was a problem.
- 20 (30% of owners) stated training was a problem
- 18 (27% of owners) stated that the initial selection of the CAD system was a problem.
- Other common problems involved data transfer issues, underestimating the costs, and choosing the right vendor to buy it from.

In terms of why the companies invested in CAD systems, the main reason was due to the pressures from industry competitors.

- 36 (53% of owners) agreed that this was the largest influence.
- The next biggest influence was pressure from business customers. 27 (38% of owners) agreed with this.

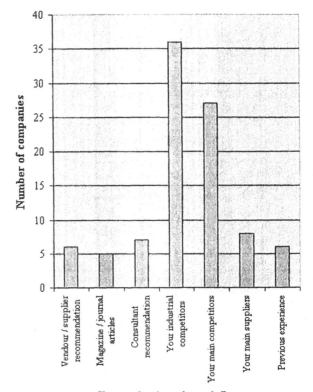

Factors having a large influence

Figure 5 - Factors having a large influence in CAD investments.

In terms of how the companies gained expertise in their CAD system, the following answers were given:

- 18 (26% of owners) had on-site vendor training (13 rated it as either a good or excellent learning method).
- 21 (30% of owners) had off-site vendor training (of which 16 rated it as a good or excellent training method).

2.3.4 Response from non-CAD users: From the 232 companies that did not use any form of CAD, only 10 completed the full survey. When asked about the problems with technology investments:

- 4 said the main problem was associated with having the staff available to help select and implement new systems.
- 4 stated that a lack of time was the main area of restriction.

- 2 stated that obtaining access to knowledge and good advice was the main problem.

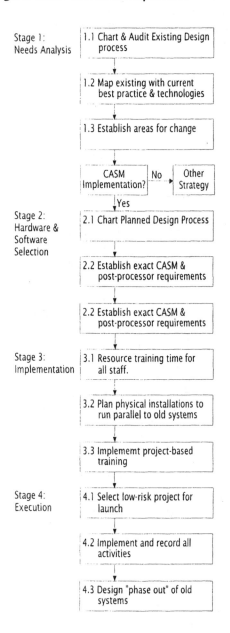

Figure 6: Outline CASM Implementation Methodology

The companies were also asked if they were familiar with the various types of CAD.
- 9 were familiar with the benefits of 2D,
- 7 were familiar with the benefits of 3D CAD,
- 2 were familiar with the benefits of CASM.

The companies were also questioned if they had ever thought about how CAD might be used within their design process.

- 7 would like to use CAD for product drawing and modelling but for various reasons have not got around to it.

2.4 Analysis of the survey

From the results collected a number of interesting issues have been raised. This analysis is summarised below:

2.4.1 Analysis of CAD Owner data:

- Few companies use their CAD systems for further analysis (FEA etc), RP, or CAD-CAM. The widespread use of CASM could significantly raise these figures by giving companies the chance to develop their in-house service. Other results show that few companies are planning to use CAD more extensively within their design process. They are more likely to make upgrades to their existing systems when required.
- Most of the companies are still only using 2D systems. This shows that these companies are not exploiting the benefits of CASM.
- The popularity of "Windows based platforms" that are powerful enough to use mid-market CASM is significant. As there has been a shift towards Windows based systems, CASM is now available to a much larger portion of the market place. Dropping prices combined with the Windows User Interface also makes CASM a more "easy-to-use" option.
- Most companies had more than one person who can use their CAD system/s. If CASM was to be employed successfully and used to its full potential within any sized company the survey suggest that sufficient staff would be available to be retrained.
- Results clearly state that the biggest benefit CAD gives to companies is the ability to remain flexible. The ability to make quick and often late changes to drawings and designs is a very powerful tool and one that should not be underestimated. However the flexibility of CASM is far greater than that of 2D CAD systems.
- The significant majority of engineering SMEs use AutoCAD as their main system. Implementing a 2D system into the design process does not have the same impact as integrating CASM as CASM can (and ideally should) be developed across a greater part of the design process. It is therefore worth taking into consideration that implementation problems are likely to be much more extensive when installing CASM. The flaws of leaving technology investments until the last

moment is discussed in Chris Austin's 1991 book entitled, 'Planning for CAD Systems' [16], where he calls the scenario 'Future Shock'.

- The survey indicates that the main reason for investing in CAD systems is that it can be used as a competitive tool. Also, it is common that owning a certain type of CAD system (i.e. AutoCAD) is an industrial requirement or standard. As the use of CASM becomes more widespread (due to newer lower-cost packages) this is also likely to become an industrial requirement.

- Many respondents said that intense training was a very effective way to learn, although applying this acquired knowledge into their own projects was difficult. Therefore, implementation should involve training with specific company 'benchmark' projects rather than just the conventional methods.

2.4.2 Analysis of Non-CAD Owner Data:

- Surprisingly there were no companies in the survey that considered the financial investment to be the main problem when investing in new technology. This indicates that companies are in a position to invest in CASM technology.

- A thorough CASM implementation methodology could ease problems SMEs face when investing in new technology. Providing it was updated it would act as an invaluable guide and knowledge source. Staff and time availability issues would also be indirectly addressed.

- The large majority of SMEs not owning CAD systems are unfamiliar with CASM. An implementation methodology should address this issue.

3. An Outlined CASM Implementation Methodology

globalisation of business such a powerful tool should not be overlooked.

Further work will include the development of an implementation methodology that is widely available to the UK manufacturing sector. The initial outlined methodology will be tested and developed by working with three east London businesses. In addition to the methodologies contents the way it is structured and incorporated is also of vital criteria.

The long-term aim is to create an electronic database that can either be accessed on-line or as an individual piece of software. In addition to the full methodology a package selection guide will be included. The company's requirements and investment

There are two main issues that have been addressed throughout this paper. One is the benefit of CASM and the other is the result from the survey showing that CASM is not widely used in SMEs. These issues lay the foundation for further work and the development of a method for CASM implementation.

The benefit of CASM technology is widely documented and is often discussed within academia. What is not well documented is how companies, and in particular SMEs, should select and integrate this into their design process. Much is left to experience, and learning through costly mistakes. An implementation methodology that is fully developed and tested would aid SMEs to integrate CASM efficiently, and more importantly effectively, into the design process.

The authors' survey has established the problems and the benefits with present CAD systems. These results are of vital importance as CASM technology and its implementation will involve similar issues but often on a much larger scale. The following outlined methodology takes into consideration all of these factors and would limit these potential consequences.

Before companies invest in CASM they need to clearly establish and understand a set of aims and expectations. Therefore the first stage would be to identify the company's needs. The outline methodology is shown in figure 6.

4. Summary and Recommendations

This paper has explored how SMEs in the manufacturing sector should approach and reap the full benefits of CASM. With the help of an implementation methodology companies can dramatically develop their design process and strive towards ultimate design efficiency. With the

criteria would be manually inputted and the output will provide the package solution/s that are most suited. Hyperlinks could then take the user directly to the relevant web pages where more information can be found. This documentation technique will be easy to update and manage. Paper techniques would be less preferable as they are harder to access and they quickly become obsolete. An electronic database would be an invaluable resource and would help the development of CASM within the SME manufacturing sector.

5. Conclusions

- CASM packages are now very sophisticated, yet easier to use, and are priced to allow more widespread use. As the CASM market grows, SMEs will need implementation guidelines to follow.
- The main limitations of SMEs are known and have been further defined through the authors' survey.
- A CASM implementation methodology is required to aid companies in getting the most from their systems and to avoid mistakes.
- The contained outlined methodology will supply the basis for further work and development.
- The final methodology will be available in an electronic format either via the Internet or as a separate piece of software. It will also include a CASM package selection guide.

References

[1] P. Brooker, "At last! The CAD Revolution is here," Engineering Designer, November / December pp 4-6, 1994.

[2] J. Nottage, "Computer Aided Geometric Design Principles," Engineering Designer, November / December, pp 8-12, 1994.

[3] T. Shahin, "Automation of Feature-based Modelling and Finite Element Analysis for Optimal Design," Brunel University, pp 5-49, 1996.

[4] P. Martin, "Buying or Upgrading PC-based CAD Systems to 3D," Engineering Designer, May / June, pp 16-17, 1997.

[5] NIST, "ISO 10303 Parts 1-518," National Institute of Standards and Technology, USA, 1991.

[6] P. Drake, et al, "Slipping Systems into SMEs," Manufacturing Engineer, October, pp 217-220, 1998.

[7] H. Marri, et al, "An Investigation into the Implementation of Computer Integrated Manufacturing in Small and Medium Enterprises," International Journal of Advanced Manufacturing Technology, No. 14, pp 935-942, 1998.

[8] S. Woolgar, et al, "Abilities and Competencies required, particularly by small firms, to identify and acquire new technology," Technovation, Elsevier Science Ltd, Vol. 18, No 8/9, pp 575-584, 1998.

[9] J. Bennett, et al, "Technology transfer for SMEs," Manufacturing Engineering, June, pp 139-148, 1998.

[10] D. Muir, B. Hillier, "Design Profit into Products," Time-Compression Technologies, Vol. 7, No. 4, pp 20-25, 1999.

[11] M. Hashemipour, et al, "A computer-supported methodology for requirements modelling in CIM for small and medium size enterprises: a demonstration in Apparel Industry," International Journal of Computer Integrated Manufacturing, Vol. 10, No. 1-4, pp 199-211, 1997.

[12] M. Hashemipour, S. Kayaligil, "Identifying integration types for requirement analysis in CIM development," Integrated Manufacturing Systems, Vol. 10, No. 3, pp 170-178, 1999.

[13] R. Kaswen, "Implementation of CIM Technologies," International Engineering Management Conference, IEEE, pp 100-103, 1990.

[14] D. Miles, "Implementing Design Technology ...an executive guide," Computer Suppliers Federation, 1998.

[15] BT Plc, "Yell," Yellow Pages, www.businesspages.co.uk. 1998.

[16] C. Austin, "Planning for CAD systems," McGraw-Hill International, 1991.

Introducing and Implementing Concurrent Engineering in the Mexican Automobile Industry

A. Al-Ashaab, T. Valdepeña, D. Quiroz, Ma. P. Jaramillo, L. E. Peña, J.L. de Leon and I. Silva
Concurrent Engineering Research Group
CSIM/DIA, ITESM Campus Monterrey
E. Garza Sada 2501 Sur. C.P. 64849
Monterrey, N.L. Mexico

Abstract

*The Mexican Automobile industries are playing an important role in shaping the economy of the country. In order to support this sector in having a customer oriented approach that would produce a quality product at less cost and in shorter time, the concurrent engineering research group, of the ITESM Campus Monterrey, is making efforts in introducing and implementing Concurrent Engineering in this important sector. These efforts are divided into several stages, they are **introduction**, **detailed diagnostic study**, **planning**, **CE pilot project**, **evaluation** and **expansion**. The mechanism of approaching the industries and the activities performed are reported and several examples gained from the authors' experiences are presented. The evolving issues are discussed then a conclusion is drawn.*

1. Introduction

The Mexican automobile industry is the fastest growing manufacturing sector in the country. This is due to the fact that many big OEM (Original Equipment Manufacturer) car industry such as GM, FORD, Chrysler, VW, NISSAN as well as heavy track industry and Auto bus such as Mercedes, Kenworth, Volvo, Navistar, Dina, John Deere and Caterpillar have been (and making more) investing in Mexico. This made the Mexican industry to grow as 1st tier provider and making industrial group that give complete sub-assembly to the OEM. However, this increase pressure on the industries in order to sustain and improve their market share. The situation has led the Mexican companies to look for a customer-oriented approach to ensure delivering a product that gives total customer satisfaction. This satisfaction is in terms of quality, cost, time, services and the ecological aspect of the product. As such, the Concurrent Engineering (CE) philosophy has attracted the auto part manufacturing companies in Mexico to aid achieving their aims.

In order to support this sector in this need, the concurrent engineering research group is making efforts in introducing and implementing Concurrent Engineering in the Mexican industries especially the automobile sector that are responded positively. These efforts are divided into several stages (see figure 1); they are **introduction**, **detailed diagnostic study**, **planning**, **CE pilot project**, **evaluation** and **expansion**. The mechanism of approaching the industries and the activities performed are reported and several examples gained from the authors' experiences are presented. The evolving issues are discussed then a conclusion is drawn.

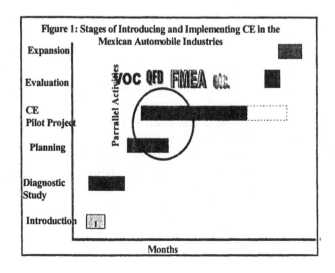

Figure 1: Stages of Introducing and Implementing CE in the Mexican Automobile Industries

2. Concurrent Engineering and Mexican Automobile Industry

As all the auto part manufacturers, the Mexican automobile industry had to have the SQ 9000 certification as OEM's requirements. The APQP (Advanced Product Quality Planning and Control Plan) applications are an important part of this quality system where product quality is the responsibility of multidisciplinary team not only the quality department (APQP 1995).

At the start most the effort had been on the documentation part of the QS 9000 without paying attention to the actual application of the tools and methods such as QFD, DFMA, FMEA of the design, Value Engineering and the real multidisciplinary teamwork of the CE applications. The reason was most the Mexican industries where providing manufacturing services to the OEM (i.e. no product design and product engineering activities). However, in time the Extended Enterprise became an important issue in the manufacturing strategy of the OEM. As a such, the first tier providers where expected to participate effectively in the process of product development from its concept development until delivery and services, as well as its disposal management. That was the point where more effective approach to product development was needed and the Concurrent Engineering (CE) application has been the answer to the Mexican automobile industry.

3. Approaching the Mexican Automobile Industry.

The effort of promoting CE applications in Mexico started back in 1994 when the first author gave CE courses and many seminars (while he was working in the ITESM campus Morelos in Cuernavaca) to universities as well as industries. The interests led to form a formal CE research group in 1995 then the foundation of the Mexican society of Concurrent Engineering in 1996 (SMIC 2000) as well as Spanish CE publications. The INA "the Mexican National Auto Part Industry Association" (INA 2000) provided us with a directory of all the industry where information was sent to them. Several companies replied to us wanted more information, courses and seminar. However due to the QS 9000 certification activities (mainly in 1997) no formal project was made.

In 1998 another effort started formally in the north city of Monterrey where many big industry are around. A formal letter was sent to several general directors and managers of product engineering. Formal CE presentation and 16 hr short course where given in many cases that aided in having a better understanding of the subject.

The above gave an opportunity to make in depth field study of the current practice of product development and identifying many opportunities of improvement. As results, a practical methods of introducing and implementing CE have been developed, which is presented in, derail in the following section.

4. Practical Approach to Introduce and Implement CE in the Mexican Automobile Industry

The objective of the effort is to implement Concurrent Engineering (CE) in the Mexican Automobile Industry to improve the process of product development and achieve customer satisfaction in term of product quality, shortening the lead time and cutting the production cost then ensuring high business profitability. The aims are:-

- Working with the general manager and the senior functional managers to plan for a change that is going to improve the company performance and ensure business profitability
- Creating the required knowledge and culture of Concurrent Engineering in the company by means of training courses in different levels
- Developing a methodology for new product introduction and development

- Ensuring 100% complying with customer requirements and specification
- Reducing the product development cost and time by at least 10 %

The following subsections present in detail the stages of introducing and implementing CE in the Mexican automobile industries.

4.1 The CE Introduction Stage

The *introduction stage* is the formal introduction of CE to the company as well as getting the understanding and the support of the direction. This is because without that support, it is hard to have a complete successful implementation. In the same time a project co-ordinator is assigned to help in the CE development project. It is the authors practice to ask the general director to start the work by talking to the engineers about the importance of the CE project and their contributing in given the correct data as well as the detail discussions in order to have results that reflect the company needs.

In order to measure the performance of a company a questionnaire have been developed and applied to several engineers. The *Performance Measurement* is related to organizational aspect, Information, Human Resources and Technology (Al-Ashaab and Molina 1999). The results are projected in a radar diagram, see figure 2, which indicate to the company the aspects they are good in and other that need more attention and improvement. In general there is good teamwork practice and a good customer focus. The main issue is the actual planning and the implementation of the tools and methods as well as the integration with the providers mainly the toolmakers. In the other hand, more attention should be paid to the human resources issue, especially the empowerment of personals of the company.

A 16hrs-training course of CE usually is given to a number of engineers and personal of the company. This to learn about the principal and fundamental of CE as well as it related tools and methods and the supportive information system.

The deliveries of the introduction stage are having a sound understanding of CE and the identification of the opportunities of improvement as results of the performance measurement.

4.2 The Detail Diagnostic Study

Parallel to the introduction stage a detail diagnostic study is usually carried on to analyze the current practice of product development from different point views. These are, Organizational aspect, Information, Human Resources and Technology (Al-Ashaab and Molina 1999).

The diagnostic study includes the relation with the clients/market, providers as well as the partners with the company extended enterprise framework.

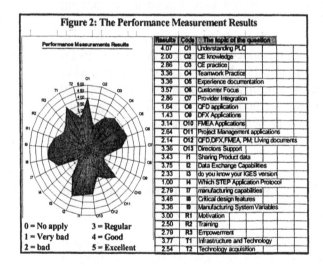

Figure 2: The Performance Measurement Results

Results	Code	The topic of the question
4.07	O1	Understanding PLC
2.00	O2	CE knowledge
2.86	O3	CE practice
3.38	O4	Teamwork Practice
3.38	O5	Experience documentation
3.57	O6	Customer Focus
2.86	O7	Provider Integration
1.64	O8	QFD application
1.43	O9	DFX Applications
3.14	O10	FMEA Applications
2.64	O11	Project Management applications
2.14	O12	QFD,DFX,FMEA, PM, Living documents
3.36	O13	Directions Support
3.43	I1	Sharing Product data
3.75	I2	Data Exchange Capabilities
2.33	I3	do you know your IGES version
1.00	I4	Which STEP Application Protocol
2.79	I7	manufacturing capabilities
3.46	I8	Critical design features
3.38	I9	Manufacturing System Variables
3.00	R1	Motivation
2.50	R2	Training
2.79	R3	Empowerment
3.77	T1	Infrastructure and Technology
2.54	T2	Technology acquisition

0 = No apply   3 = Regular
1 = Very bad   4 = Good
2 = bad   5 = Excellent

The key activities in this stage are: -

- Interviewing different engineers and personal of the company who are involved directly in the process of product development
- Identify the company experience and the training requirement of all the tools and methods that support the CE implementation, such as QFD, DFx,

FMEA, Project management, CAD/CAM/CAE, data exchange (IGES/STEP) and VRML as well as the different information system

- To have a sound understanding of the quality system, ISO/QS 9000, and how this could be improved by the CE implementation
- Identify the critical design features of the product.
- Identify the variable of the manufacturing system that influence product quality as well as the source of the variations
- Doing a close workshop to develop methodology for the new product introduction and development.

The delivers are a detail technical report evaluating the current practice of product development and the identification with evident the opportunities of improvement. The product life cycle representation is standardized using IDEF0 notation (IDEF 2000). This would give a good understanding of the company product development requirement then a methodology for new product introduction and development (NPI/D) is proposed. Finally a workshop is organized to review, modify and finalize the proposed NPI/D methodology. The following sub-section presents a panorama of a possible NPI/D methodology.

4.2.1 New Product Introduction and Development Methodology

International companies have developed through their long time experience a standard methodology for their new product introduction and development. Example of that Whirlpool developed a Customer-to-Customer product development (C2C), United Technologies developed their Integrated Development System (IDS), GE developed their design for Six Sigma (DFSS), and Caterpillar developed their NPI called CPPD (Concurrent Product and Process

Development). This is to standardize the process of product development based on the company experiences ensuring a highly customer oriented approach. As such the authors believe in the importance of developing a customized NPI/D methodology to each company or industrial group and has been always one of the main deliveries of the diagnostic studies.

The NPI/D methodology development is based on the principle of Concurrent Engineering, APQP requirement and the company best practice and experience. This support the integrated product development that cover details related to the OEM and final car user requirements as well as the integration with the key providers

In the NPI/D methodology, see figure 3 as a general example, the emphasis is on the full understanding of the customer needs. This is achieved by making a formal study of the voice of the customer then representing these requirements into product specification by using the QFD. Hence, the CE team will have the understanding the requirements and what are needed to achieve them.

The methodology, also, emphasis on the use of project management as formal techniques in planning and administrating the product development project. The following revision is done carefully and explained in detail to whoever involved in product development (not on the CE team):-

- The critical design features
- The detail product technical specifications
- The manufacturing system variables that effect product quality
- Tool specifications

The above is needed to make sure there is a common understanding among the personal in the different departments of the company as well as the providers.

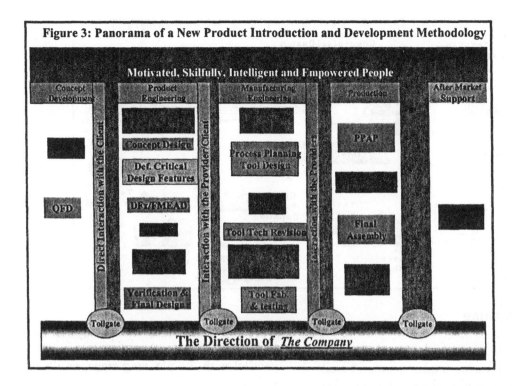

Figure 3: Panorama of a New Product Introduction and Development Methodology

Figure 3 illustrates a view of the possible product development methodology. The CE team (or also called the APQP team) is selected (with a project leader) and empowered by the direction. The product life cycle is divided into several key stages and each stage has got several activities to be performed by the CE team with the involvement of the customers and key providers. The work is evaluated by the direction through tollgates.

It is important to mention here that the typical cross-functional team in the Mexican Automobile industry is the Heavyweight team structure due to the fact that most Mexican companies are small-medium enterprises, therefore, they lack of human resources. However in order to keep the team to team relationship with the OEM as well as the key providers across borders many companies are using the video conference as a mechanism for communication.

4.2.2 CE Tools and Methods Integration

There are many tools and methods that support CE applications and the product development, such as QFD, DFMA or DFx, FMEA, Project Management, etc. One aspect the authors emphasis during the training courses and then later on in the implementation is the integration aspect of these tools and methods. Figure 4 illustrates a view of the meaning of this integration. The issue is the link and interaction between activities through their results. For example, the design should be driven by the customer through its requirements that are captured in the QFD. The potential product design failure is analysed in the FMEAD. The analysed functions should be related to the "what" of the QFD. This is because a function that is not part of the customer requirements then it should be considered carefully to be eliminated, and so on with other activities. This aid making the results as living documents which must be updated whenever a change is required on the product specifications and design as well as the manufacturing system.

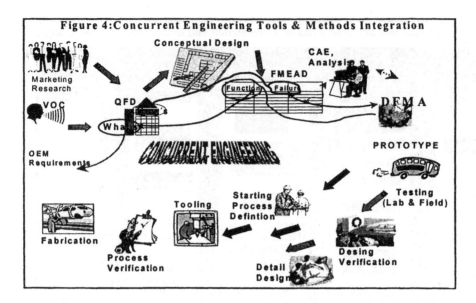

Figure 4: Concurrent Engineering Tools & Methods Integration

4.3 The CE Planning

This stage is to plan in detail for the Concurrent Engineering implementation based on the developed methodology. This includes the development of all documents (format, customized manuals for the use of the tools such as FMEA, DFx, check list, etc) to support the multi-disciplinary activities as well as the tollgates review documents in the key stages of the product development. In the same time arrange any specific training courses required by the CE team. At the end of this CE planning stage the company is going to have all the required knowledge, tools and methods as well as their documentations for the CE implementation.

One of the problems encountered is most the companies do not have a good record on the cost of the product development. This is important in order to help calculating cost reduction. However, cutting the product development time and reducing and elimination the re-work are the most important issues that companies are looking for.

4.4 CE pilot project

This stage is the actual execution of the CE plan (defined in the pervious section). The pilot project is a real project of the company

where the planed activities are applied. Several activities of the pilot project are done in parallel with the planning stage, as illustrated in figure 1.

4.5 Evaluations and Expansion

These stages are the final of the work. The first is to evaluate the actual CE implementation. It is to make sure that what have been planned is under the right executions. This acts as lesson learning for continuous improvement. Finally the company is ready to expand the CE implementation to other projects.

4. Conclusion

The Mexican Automobile industries are indeed considering, seriously, the formal implementation of the Concurrent Engineering. This is walking hand to hand with their QS 9000 implementations ensuring the fabrication of a high quality product to their OEM and the final user. In the same, is giving them the role of the active partner who participates in product development. In doing that a practical approach of the CE introduction and implementation is important to speed us the process. The paper presented a traditional CE approach based on the cross-functional teamwork and the use in an integrated way the tools and methods that support the implementation. This is done through the 6 stages: *introduction, detailed diagnostic study, planning, CE pilot project, evaluation* and *expansion*. The stages of the approach and the activities related to them have proved its practicality within the Mexican industry and the work is going to complete the full implementation.

The authors found that the CE activities could be supported with the development of a knowledge based engineering (KBE) applications. In one of the studies we found that the manufacturing engineering could be cut by 90% through the KBS applications so as the plastic product development, work is under way in developing prototype systems (SPEED 2000). More attention should be paid to the Product Data Technologies to ensure good CAD/CAM/CAE and other product life cycle data integration. The industries depend only on the commercial tools without having a good understanding of the technology.

The CE efforts is in the right direction and with more commitments and understanding the result will ensure a competitive advantages to the Mexican automobile industry in having a customer oriented approach that support the development of better, faster and cheaper product.

Acknowledgement

The authors would like to express their gratitude for the Mexican industry that kindly provided us with access and information to realise our work. We would like to mention in particular, Ramirez Group, Carplastic of the Visteon Automotive, DESC/UNIKO Group, and Cardanes of the SPICER GROUP, CIFUNSA and DITEMSA of the GIS GROUP as well as the British Council office Mexico City.

References

Al-Ashaab A. and Molina A, 1999, Concurrent Engineering Framework: A Mexican Perspective", CE99 Bath, UK.

APQP 1995 "Advanced Product Quality Planning (APQP) and Control Plan", Reference Manual, Chrysler Corporation, Ford Motor Company, and General Motors Corporation.

INA 2000, http://www.ina.com.mx, the home page of the Mexican National Auto Part Industry Association

SMIC 1999, CE Information Centre, http://w3.mor.itesm.mx/~smic/ceic. The Mexican Society of Concurrent Engineering.

SPEED 2000. SPEED Project, Supporting Plastic Engineering Development, http://tamayo.mty.itesm.mx.

A Simulation-Based Framework for the Implementation of Concurrent Engineering

D. J. Rankin and P. K. Chawdhry

Department of Mechanical Engineering, University of Bath, Bath BA2 7AY, United Kingdom

Abstract

Successful implementation of Concurrent Engineering (CE) is key to competitive product development. This paper highlights the need for an accurate implementation mechanism for CE. A framework has been developed with the aim of fulfilling this need. Key factors for successful implementation of CE have been identified. Scenario-based simulation models have been developed for four key success factors, involving over 30 inputs and 20 outputs. Models were simulated in MS Excel to enable scenario-based assessment of CE implementation in a specific enterprise. The simulations were validated in industry through interviews and informal assessments. The need for models for a dynamic simulation has been highlighted. It is expected that such a simulation would include the time and cost factors associated with CE implementation, and thus, would be a comprehensive management tool for decision support in CE process design.

1 Introduction

Since its introduction in the 1980s, CE has generated much interest in the techniques and methodologies for concurrent product and process development. Recent years have seen the interest shifting towards the practical implementation of these techniques in the industry. Many papers highlight the merits of CE and review successful case studies. However, little *solid* information has been documented concerning the implementation of CE. A few assessment tools and implementation methodologies have been developed, such as the RACE[1] model, and the PACE[2] project. However, these are often complex and time consuming to comprehend by a practitioner, and do not produce a definite strategy for enterprises wishing to proceed with CE implementation. In addition, some implementation methodologies claim to be suitable for all sizes of enterprise, nevertheless the majority are still non-specific to a company's particular requirements and goals. Additionally, knowledge of the CE philosophy is often required in order to understand these methodologies, and this knowledge may *not* be present within the client's enterprise. Therefore, there is the need for a strategic CE implementation tool, which could be tailored to meet the needs of *any* size of company and be used by the management with ease.

2 Background and Related Work

An ESPRIT Working Group[3] discusses the evolution of assessment models for CE and earlier Quality Programs. The progression from 1D models onto 2D models is highlighted, followed by an in-depth discussion of the 'Readiness Assessment for CE'. The paper illustrates the double-edged concept of the race model: the organisational and technology issues. The main aim of the RACE model is to identify bottlenecks and remedial initiatives. These initiatives are ranked in order of importance in accordance with their relationship with the business drivers. The article concludes that assessment models continuously change and are flexible to adapt to each particular organisation. It also highlights the difficulties encountered when attempting to prove why one assessment model is any better than another.

Componation & Byrd Jr.[4] introduce readiness assessment practices and the increasing need for such assessments for implementing CE. Bergman & Ohlund[5] define the role of tools in the change process as including: the formulation of a change strategy and the monitoring of the change process. The implementation strategy discussed by Paashuis & Boer[6] is an 'Operations Strategy Framework' which is occupied with CE integration mechanism decision alternatives. Clegg, Cole & Wolak[7] commence by defining the barriers that exist to CE implementation, and focus upon the need for 'corporate knowledge' for a successful implementation. Whereas, Nielsen[8] focuses upon the need for multifunctional teams within a Concurrent Engineering environment.

Many authors in the past have adopted different approaches to CE and, more specifically, its implementation. However, there are few approaches which combine the significance of the 5 key areas of CE:

- People
- Processes
- Organisation
- Tools & Techniques

- IT Infrastructure and Data Management

There is a requirement for a CE implementation support tool which acknowledges the necessity of all of the key features and the ways in which they inter-relate.

A common methodology for implementing CE is the 'Integral Approach'[9] which entails an assessment of the status of the Product Development Process, to establish its progression. Usually, a 'roadmap' is then developed to lead the change from the current process to the new. It is suggested that this approach be used in conjunction with the framework outlined below.

3 Methodology

In the course of an extensive literature analysis preceding this work, key factors for successful CE implementation were identified [11] (Figure 1). All of the above factors became the *Objective* of a CE implementation model, which aimed to measure the drivers and indicators with respect to each of the key factors.

Figure 1. Implementation Objectives

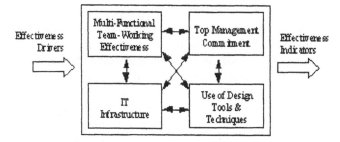

3.1 Hypotheses

The following hypotheses were formed for this research:

1. There exists a finite and measurable set of controllable inputs which can affect the CE implementation process.
2. There exists a set of external (non-controllable) influences upon the CE implementation process.
3. There exists a finite set of measurable outputs for the CE implementation process.
4. The inputs to the CE implementation process affect the outputs of the process.
5. The degree to which the inputs influence the different outputs varies between input-output pairs.

6. It is possible to assign weightings for each relationship within the CE implementation process.

4 Models

Figure 2 shows the Multi-Functional Team-Working. The starting point for each Model was to identify the measurable Outputs of the Objective. Subsequently, the factors required to *achieve* the objective, the Direct Inputs, were defined. Once the 'Direct Inputs' to the Model had been recognised, it was possible to consider the 'Indirect Inputs'. These were the factors which give rise to the Direct Inputs, whether it be individually, or in combinations. This process was repeated until the Indirect inputs became 'Primary' inputs, i.e., they could be either measured, or reduced no further.

For example, Figure 2 indicates that there are three Indirect inputs to the Shared Skills box, which are 'Corporate Knowledge Base', 'Continuous Learning / Training' and 'Appreciation for Knowledge'. In turn the 'Corporate Knowledge Base' box can be seen as having two primary inputs: 'Corporate Knowledge Base Policy' and 'IT Infrastructure'. The 'Continuous Learning / Training' box has two inputs, one is a Primary input (Goal Driven Objectives) and the other is an Indirect input (Personal Initiatives), which subsequently has its own Primary input (Human Resources Learning Policy). The third input to the Shared Skills box (Appreciation for Knowledge) is also an Indirect input, with its own Primary (Team Culture).

In short, the Models (Figures 2-5) were created by 'working backwards' from their Objective until such a point where all of the outputs and inputs were measurable. No further explanation of the content of the models will be given as they are relatively self-explanatory. Further details concerning the model development are given in [11]. The models were implemented to form a simulation-based framework.

5 Simulations

A computer-based simulation was developed to form the basis of the Implementation Framework. This was due to the following reasons:

- In industry, it is perceived that 'methods implemented on a computer have a stronger positive influence than those on paper-based systems'[10].
- A state-of-the-art simulation was more likely to achieve acceptance in both academic and indus-

trial circles; a computer-based system appeared a practical method to satisfy this requirement.

- A computer-based simulation would be easier to use to assess various scenarios of CE implementation and to carry out 'what-if' analysis to aid management decision processes.

5.1 Features of Simulations

The features required in order for the framework to be practical were identified as:

1. Allowance for the user to input values to ascertain their status in terms of CE implementation.
2. Dynamic in nature such that the user could specify a target situation and be informed of paths to take in order to reach such a point.
3. Provision of timescales and costs associated with processes such that the user can make informed decisions as to which path to take.
4. Not excessive in length, in terms of time for training and use.
5. Easy to understand such that no prior knowledge of the CE is required.
6. Simple to use, including a comprehensive and familiar user interface.
7. Adaptable to suit the requirements and goals of *any* size of enterprise.
8. Able to run without the need for specialised hardware or software.

5.2 Development of Simulations

A static, simulated model for each of the four objectives was developed. At this stage, the simulation is termed 'Static' as it does not predict any changes over time, only the as-is results for the scenario entered. The simulations were each constructed in a Microsoft Excel Spreadsheet. The flowcharts of the Models were re-documented in Excel, with an 'Entry Box' being placed below each Primary Input and Output, Figures 6 and 7. Note the following concerning the simulations:

- The Objective from the Model has been omitted and replaced by a series of arrows. These arrows link the Direct Inputs and the Measurable Outputs, and represent the fact that the input has a direct effect upon the output.
- A score for the overall Objective of each Simulation is illustrated.
- Weightings assigned to various relationships are shown, with, for example '*0.8' written along a

given arrow indicating that the input in question contributes towards 0.8 (or 80%) of the output.

5.3 Assignment of Weightings

Weightings were assigned in several areas of the Simulations to relate the following:

- Primary inputs to the Indirect and Direct inputs
- Direct inputs to the Objective
- The Objective to the Measurable outputs
- Direct inputs to the Measurable outputs

For example, the Entry Box below the 'Shared Skills' Input reads 100%. From the weightings indicated, it can be deduced that 25% is due to the Corporate Knowledge Base, 50% due to Continuous Learning / Training, and 25 % due to Appreciation for Knowledge. In turn, each of these inputs is related to the Primary inputs. The weightings were determined by analysing the literature and background information [11].

5.4 Use of the Simulations

In order to utilise the simulations, the user loads the models in MS Excel. For a given model, the user enters a value between 0 and 10 in each of the Entry Boxes for the Primary inputs. The intention is to draw up a list of criteria to be met in order to attain each 'Score' such that the score selection would not be subjective. This is to be performed for *all* of the Primary inputs. Having entered the values, the spreadsheet calculates a score for: the Indirect inputs, which are constructed from more than one Primary input, followed by the Objective and, each of the measurable outputs. The output score is in the form of a percentage.

6 Results and Validation

In order for the Models and Simulations to be reliable, their validation was required. Interviews were performed within Industry to assess whether or not the models gave results that were both accurate and useful. The objectives set for the interviews were to establish whether the static simulation was accurate, easy to understand and use. The potential uses of a Dynamic Simulation were also investigated.

Three interviews were conducted in total. The job titles of the Interviewees were Chief Engineer, Engineering Manager and Joint Chief Executive (previously R & D Director). All of the Interviewees were generally positive regarding the Models and Simulations. Overall, the work was well understood and benefits of using the tool were appreciated. The main points highlighted during the course of the interviews can be summarised as follows:

- The Simulations were generally accurate.
- A Static Simulation would be very useful: as an assessment model prior to implementing CE; in mapping progress of Continuous Improvement or CE Programs; in Benchmarking Suppliers.

It was also noted that a Dynamic Simulation would be very useful in implementing Concurrent Engineering, and would be a further practical aid to decision-making.

7 Conclusions and Further Work

A simulation-based framework was developed for CE implementation. The framework consists of 4 key implementation objectives and corresponding models. These models were simulated in MS Excel and validated in Industry. The static simulations could be utilised as follows:

- Scenario-based assessment of readiness for concurrent engineering
- Map progress of CE implementation over time within a specific enterprise
- Benchmarking of suppliers against one another

The benefits of the simulations are summarised:

- Suitable for any size of company
- No prior CE knowledge required for use
- User-friendly interface
- Management tool for decision support

Future work would include development of models for other key factors. Another useful step would be to develop a Dynamic Simulation for the implementation of CE. This was to be based upon the Static Simulation presented here. A Dynamic Simulation would firstly require the incorporation of cost factors. The user would be required to enter their current status, as for the Static Simulation, and subsequently, a target situation. The costs incurred in order to meet such a target would be presented to the user. The integration of time factors would further enhance the simulation and provide its dynamic nature. In terms of the target situation, the simulation should then present the user with, not only the resources required, but also the timescales required to implement the changes.

The simulations can be used to carry out what-if analysis by the decision-maker. For example, if a company, having entered its current situation in the Static Simulation, noted that the scores were low in a particular area, say 15% in 'IT Infrastructure and Data Management

Effectiveness'. The user could then input a target score, e.g. 90%, and the Dynamic Simulation should suggest several different scenarios for implementing changes to reach the target, including the associated timescales and costs involved.

In conclusion, the research has demonstrated the feasibility to develop models for CE implementation that are capable of predicting its effectiveness. This has laid the basis for future work on a Dynamic Simulation as well as the development of more extensive models.

References

1. Van de Kerk, Klaassen, Peek, Slagt, Schouwenaars & Schrijver, "RACE Web Site", www.tm.tue.nl/race/welcome.html
2. Driva & Pawar, "Overview of PACE from Conceptual Model to Implementation Methodology", Proceedings of PACE'97. pp 13-25.
3. ESPRIT Working Group, "Assessment Models", Sol. CIM at Work CIMMOD and CIMDEV, Aug. 95 //www.tm.tue.nl/race/models/index.htm.
4. Componation & Byrd Jr, "A Readiness Assessment Methodology for Implementing Concurrent Engineering", Proceedings of CE96. pp 150-156.
5. Bergman & Ohlund, "Development of an Assessment Tool to Assist in the Implementation of Concurrent Engineering", Proceedings of CE95. pp 499-510.
6. Paashuis & Boer, "Organising for Concurrent Engineering; An Integration Mechanism Framework", Integrated Manufacturing Systems, Vol. 8, No 2, pp 79-89. 1997.
7. Clegg, Cole & Wolak, "A Support Tool for Practising CE", Proceedings of CE95. pp 665-675.
8. Nielsen, "Teambuilding – The Stepping Stone to Success", Proceedings of PACE'97. pp 69-79.
9. Thoben & Weber, "Designing Information and Communication Structures for Concurrent Engineering – Findings from the Application of a Formal Method", ICED '99. pp 989-994.
10. Evbuomwan, Sivaloganathan & Jebb, "A State of the Art Report on Concurrent Engineering", Proceedings of CE94. pp 35-44.
11. D. J Rankin, "A Framework for the Implementation of Concurrent Engineering", Department of Mechanical Engineering, University of Bath, UK, May 2000.

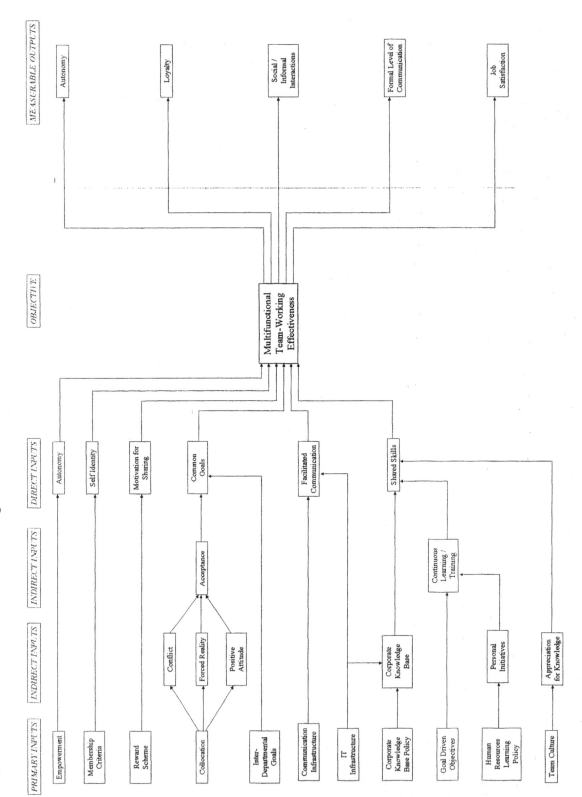

Figure 2. Model – Multi-Functional Team-Working Effectiveness

Figure 3. Model – Top Management Commitment to CE Process

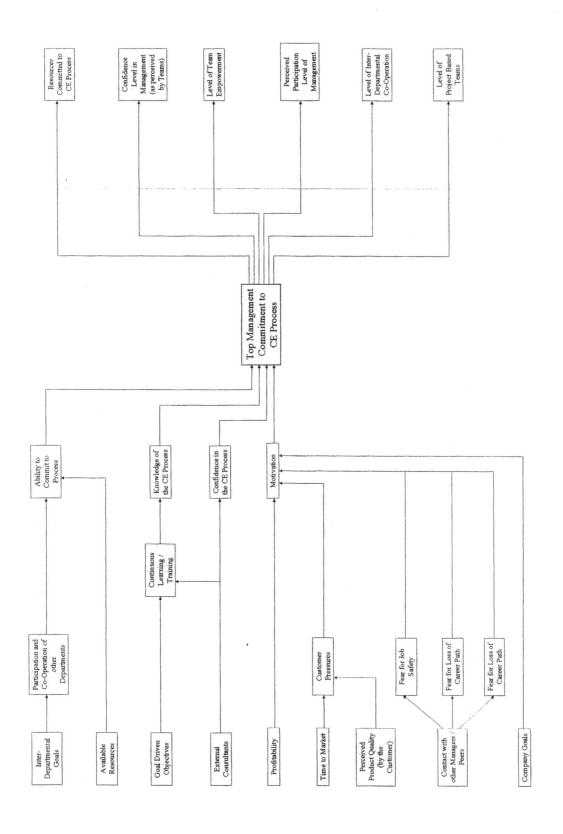

Figure 4. Model – Effective Use of Design Tools & Techniques

691

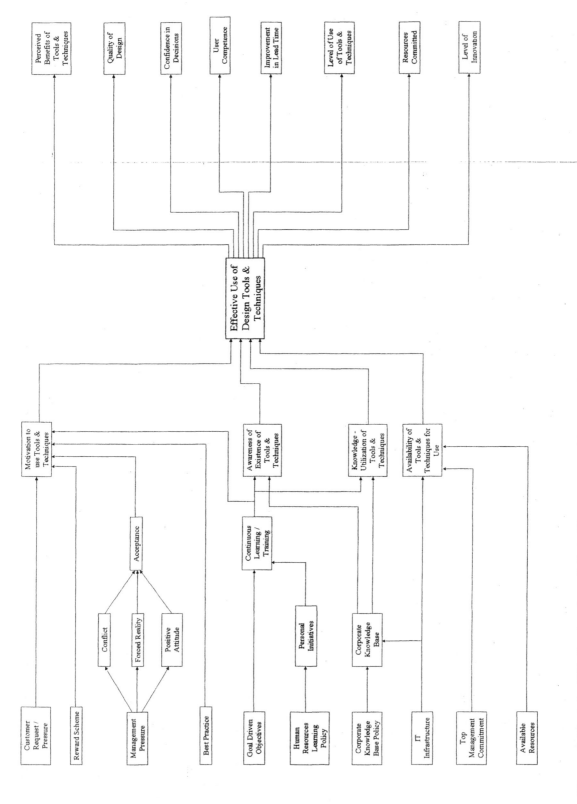

Figure 5. Model – IT Infrastructure & Data Management Effectiveness

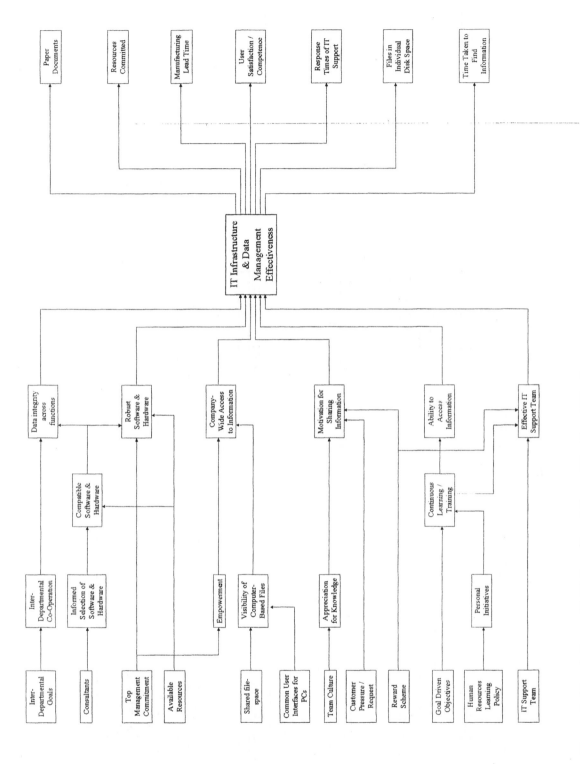

Figure 6. Simulation – Multi-Functional Team-Working Effectiveness

693

(The input and output scores correspond to an idealised implementation)

694

Figure 7. Simulation – Top Management Commitment to CE Process

(The input and output scores correspond to a validation case representing a typical company)

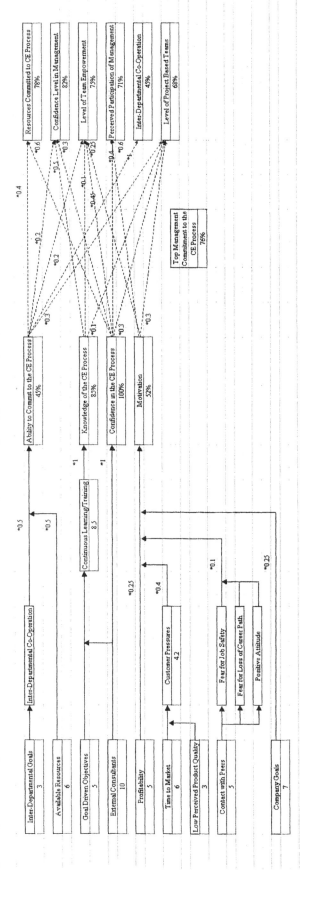

695

CHAPTER 13

Training and Education in CE

CSCW in Industry and Education:
Transferring Knowledge in Engineering Design

S P MacGregor
DMEM, University of Strathclyde, Glasgow, Scotland G1 1XJ

A I Thomson & W.J. Ion
as above

Abstract

This paper describes the findings of three recently completed research programs conducted at the University of Strathclyde. Each of the projects involves the implementation and usage of Computer Supported Co-operative Working (CSCW) technology within the design process. Two of the projects described are education based whilst the other was conducted in collaboration with industry based on live industrial projects. The paper presents the findings of each of the projects and then explores opportunities for cross-sectoral lessons.

1 Introduction and background

In recent years, extensive research and experimentation has been conducted in the use of Computer Supported Collaborative Working (CSCW) tools within both the industrial and educational arena [1-8, 9-14]. Opportunities exist for lessons to be learned through the identification of similarities and differences in the findings of such research. Allowing, recognition of generic findings and possible transferable lessons.

This paper presents the findings of three recent research projects carried out at the University of Strathclyde. Two of the projects, ICON [9, 12] & ICON2 [10, 11, 12] investigated the usage of CSCW tools by disparate engineering design students. Whilst, a Design Council funded research project investigated the introduction and usage of shared workspace technology within the design process of three companies and their supply chain [16,17].

This paper presents an overview of the approach and findings of the ICON and the Design Council funded "Integration of Design Specialists Through Shared Workspaces" projects. This is followed by a comparative reflection of the results allowing identification of generic findings together with opportunities for cross-sectoral lessons.

2 ICON

2.1 The ICON concept

ICON (Institutional Collaboration Over Networks) began in June 1997 with a week-long collaborative design project involving four pairs of students. The students came from the Product Design Engineering course at Strathclyde University and the same course run jointly at Glasgow School of Art/Glasgow University. The students tackled different design briefs and were asked to provide a solution and present their findings at the end of the week. They were restricted to the use of network technologies such as audio/video conferencing and chat tools and were prohibited to communicate by any other means. Although difficulties were encountered through the week, the project was considered a success in that it proved virtual collaborative design projects were feasible. The difficulties encountered were also examined and resultant changes made to the system for the second project, ICON2, which ran for eight days in September 1998. This iterative process, illustrated in figure 1, resulted in modifications being made to the system in terms of the methodological approach and project organisation, in the pursuit of ensuring future success. By identifying the barriers to effective communication and work, benefits of adopting CSCW in a design context can be maximised and remedial strategies and guidelines can be developed for implementation by the design community.

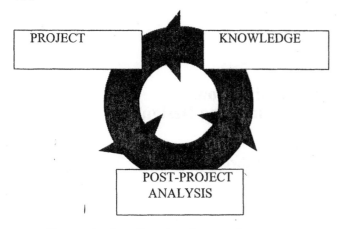

Figure 1 - Iterative development process

The underlying philosophy to the ICON projects involved improving the accessibility CSCW tools in order that policy makers in academia could implement similar projects easily, efficiently and with as little as possible start up. This included using as much freeware as possible and avoidance of ISDN.

The main objectives of the ICON projects can be outlined as follows:

- To ascertain the practicality of facilitating Internet based remote collaborative design projects for students;

- To introduce students to remote design environments and technology currently being adopted in industry and further, to investigate any obstacles encountered in implementing and utilising this technology in design education;

- To investigate the impact of the technology on the effective teaching and practice of the design process;

- To study the ability of the technology to overcome cultural differences between the participating institutions and to generally promote inter-institutional collaboration;

- To encourage diversified learning. The product design engineering degrees taught by each of the participating departments incorporate many common elements, however each of these departments presents its course from a different pedagogical perspective. ICON aimed to encourage the students to share their knowledge base in an attempt to widen both their individual and institutional skill sets.

2.2 Methodology

In the six weeks prior to the first ICON project three briefing sessions were arranged to introduce students to the project methodology, the technology and participating staff and students. Evaluation of the project took place through the implementation of pre/post project questionnaires and interviews, project diaries and video recordings of the final presentations.

The design interface for the project came in the form of a project website incorporating design briefs, project schedules and technical support. The briefs, which were disclosed on the morning of the first project day varied for each team and were allocated randomly. The briefs were

- Design a portable electronic scanner/sketch aid/notetaker for students

- Design a powered weeding device for the 60+ market

- Design a portable facial hair remover

- Design a hot frothy chocolate machine

True to the ICON philosophy participating students were restricted to using the tools and technologies made available to them. The computers used for ICON were cross platform PC/Mac, due to the existing facilities at each institution. CuSeeMe provided a single video link between the institutions with audio conferencing being provided for all teams. Microsoft office, BSCW (Basic Support for Co-operative Work), Netscape Navigator, Paintshop Pro, Adobe Photoshop, AutoCAD, and Peoplesize were provided to aid the completion of the briefs.

2.3 Findings

In general, the first ICON project was a success. Although technical difficulties were encountered throughout the week the students enjoyed the experience and reported a significant improvement in their computer competence. The students were content with the project, mainly due to low levels of expectation. Despite the positivity participants did not envisage such methods of working as constituting a replacement for conventional practices. Perceptions and levels of expectation were found to be central to the success of implementing such projects and this aspect will be discussed later in the paper.

Participating students made good use of the tools, BSCW being the most successful. The use of audio was found to be a critical factor in effective communication, concurring with other research [13].

The following guidelines were distilled from the first ICON project. These were used as the basis for designing the system for ICON2 project,

- Agree on the technologies in advance and ensure that they can be used at each site;

- Wherever possible, standardise the hardware and software. Many difficulties were experienced in this project because communications technologies had to be cross platform i.e. PC & Mac;

- Allow enough time to set up and test the technologies. Attention to details such as student shared workspace registration and configuring email addresses on machines to be used for the project may also be necessary;

- Provide a variety of technologies so that if one communication channel fails another may be adopted;

- Allow students time to familiarise themselves with the technology before the project commences;

- Ensure staff with the necessary expertise are available to assist students having technical problems;

- Be realistic about what can be achieved. Eight students, four from each institution, was considered a manageable size for this project.

3 ICON2

3.1 ICON2

Similar to ICON, ICON2 involved the partnership of four pairs of students from each academic institution. Three teams were restricted to the use of network technologies to complete the brief whilst, a control team was also nominated who could meet as they wished and were allowed to use any means of communication with the notable exception of audio and video conferencing. The project lasted for eight days and was split into two main phases. Phase 1 comprised a sacrificial project that allowed the students to get used to the technology at the same time as conducting research for phase 2. The required deliverables from this part of the project comprised a Product Design Specification and a Theme Board. Phase 2 started with the participating students being presented with a design brief. This was as follows:

A portable syringe driver is the "Walkman" of the medical services industry. The small, now pocket-sized devices are connected by a thin flexible tube to an intravenous cannula in the patient. The objective of the project is to investigate and propose a new visually, ergonomically and technically advanced appropriate design which meets the complex procedural medical, clinical user and patient demands of such a product.

Each team was to develop a product to meet the brief and to present their chosen concept using a CAD produced product layout drawing together with a rendered presentation graphic. The Clyde Virtual Design Studio (CVDS) [14], which was developed in the wake of ICON, was used as the basis for tackling the brief. The CVDS integrated the set of tools/facilities that were required for the successful completion of the project. The CVDS consists of four main components:

Data management - *storage space for project work:* In the form of BSCW which was downloaded free from the Internet and configured for use with the CVDS.

Communications suite - *local audio, video, chat and whiteboard facilities:* Microsoft Net Meeting, Netscape Communicator and Ewgie chat were provided.

Local applications - *relevant software programs:* AutoCAD R13, 3D Studio MAX, Microsoft Office, Paintshop Pro, Adobe Photoshop and PeopleSize were provided.

Reference area - *General and project specific links and information.*

As with the first ICON project an effort was made to record as much information as possible. For ICON2 the following methods were employed:

- Daily project videos;
- Pre/post project questionnaires
- Support staff logs
- On-line diaries
- One to one post project interviews
- Presentation video

3.2 Findings

The main findings on ICON2 can be classified in the following areas:

- System and tool usage
- Benefits
- Barriers

System and tool usage

There is much debate with regards to the effectiveness of synchronous V's asynchronous working practices [2, 6 & 9]. The overall split for all the teams in ICON2 was 68% asynchronous V's 32% synchronous. This level of synchronous work is relatively and may be attributed to the fact that the project was of short duration and that the students didn't know one another personally. Therefore, a high level of synchronous work was required in order to ensure the design was moving in the right direction, especially with tight deadlines looming. Other factors may include the ability of the tools to facilitate synchronous communication and the perseverance of the students, an attitude that was, perhaps, expected of them.

Each of the three networked teams displayed very different patterns in communicating. Figure 2 shows how team 3 collaborated over the course of the project.

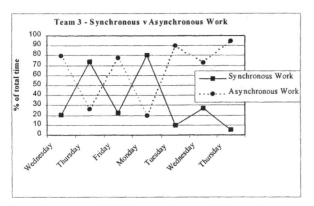

Figure 2 team 3 daily synchronous/asynchronous split

Work Modes and Habits

From examination of the personal log and on-line diaries it became apparent that the participants generally, displayed the same work mode at each of the various stages. By examining this it may be possible to maximise benefits and ease teething trouble in future projects. To this end, figure 4 is proposed. This figure is basically a summary of the work practices and attitudes of the ICON2 participants. The students picked up the technology relatively quickly and used only what they found useful after a while. Other results included a decrease in the need for technical support although most students agreed that constant support would always be welcomed. One aspect of the model that could be improved is the participants' tendency to become less

tolerant as the project goes on. As one evaluator mentioned, " . . . *the more it goes well, the more annoying it is when the odd thing goes wrong.*" These frustrations can be attributed, partly, to the tight time-scale of the project.

Stage	Induction	Familiarisation	Expert
Attitude	Keen	Content	Frustrated/ Stressed
Tools Usage	Using a wide variety	Narrowing usage	Whatever appropriate to deliverables
Attitude to tools	*"open to anything"*	*"open to what works"*	*"nothing works"*
Level of back-up	Tools instruction needed	Software support only	Moderate software support

Less Tolerance

Time

Figure 3 work-modes and attitudes during Icon2

Additional training and experience of using such a system should result in minimal frustrations and less detraction from the design process. We may not be able to change the fact that participants become less tolerant as the project progresses (this may not be a bad thing, participants will not settle for mediocre performance) but if we can manage the project effectively this may not matter as much. Additionally, other measures such as on-line tutorials may substitute the need for software support.

Benefits

The main benefits of the ICON2 project can be summarised as follows:

- A steep learning curve for the students that resulted in significant improvements in their computer skill and confidence levels;

- Greater experience and knowledge for staff for eventual input into a new system design;

- Widening of scope of students knowledge and experience through working with students from another institution;

- Strengthening of inter-institutional links;

- Student perception of certain elements of the design process being improved through practical project;

Barriers

The barriers encountered in ICON2 can be categorised as:

- Technology
- Educational Issues
- Social/Cultural
- Project Management
- Psychological/physiological

For a detailed discussion regarding the barriers, refer to [15].

3.3 Conclusions/recommendations

Analysis of findings allows certain guidelines to be produced for future virtual collaborative design projects.

- A project manager or champion should be appointed who should be aware of the issues and possible outputs of the project;

- Any additional input to projects should be cleared by the project manager and its effects closely considered;

- Future collaborative design projects should closely resemble conventional semester long modules so that the technology is the only differing factor in the project;

- Students should be trained adequately in all software that is to be used in project. At least an introduction to the packages should be completed;

- Participating students should be trained in adopting the correct behaviour for such projects;

- Periodic contact between colleague(s) should be encouraged to ensure working towards a common goal;

- Depending on the specifics of the teaching methodology consideration should be made to letting design partners' meet physically before and/or after the project. Working on the start of the project in person and presenting findings afterwards are options;

- Encourage high level synchronous collaboration after long periods of non-contact. In a conventional project this will usually mean the start and end of sessions and the start of important phases;

- Materials should be developed which can support the detail design phase of the design project including online specifications and design methods;

- Design briefs should be similar but not identical to encourage the correct levels of inter-institutional and collocated collaboration;

- On site support at participating institutions is advised for the start of any project until students become familiar with technology;

- Software support should be considered for stages of the project when students are required to produce an output. Online support is an option;

- Adequate space should be provided for students to carry out other activities at their workstation, such as sketching and writing;

- An awareness of stages of behaviour in project participants should be evident in support staff;

- An awareness of possible effects of different personalities and how to respond is also advised;

- Students should take appropriate exercise to combat any physiological problems associated with long periods in front of a computer;

Another important need for such projects is that of *setting up* future projects, ensuring that any preparation is completed as economically and effectively as possible. The following stages have been identified:

IDENTIFICATION - *methodology, resources*
INSTALLATION - *technology*
PREPARATION - *briefs, students*
RUNNING & MONITORING - *support*
EVALUATION - *input to next iteration*

4 Integration of design specialists through shared workspaces

4.1 Introduction and background

This recently completed research, funded by the UK Design Council, investigated the introduction and usage of shared workspace technology within the design process of a number of small and medium sized enterprises (SME's).

Within the context of this paper the term shared workspace, has been used to describe a computer based collaborative working system typically consisting of the following functionality:

video and data conferencing ;
real time application sharing ;
shared whiteboards;
file transfer.

Development and use of such technology to date has been dominated by large multinational companies. Ford, for example, have used the latest collaborative technologies to allow their seven design centres, each of which specialises in different aspects of design, to communicate effectively across great distances and different time zones. Benefits demonstrated by such projects include:

- improvements in the flow of work allows companies to move and react faster;

- product development lead time and costs are reduced while maintaining or improving quality;

- time to market is reduced

- relationships and efficiency of communications throughout the supply chain.

It is clear that current desktop data and video conferencing technology offers the possibility for companies to collaborate effectively and at relatively low cost over Networks using personal computers. Furthermore, the benefits that collaborative working technologies can bring to the new product development process are apparent. However, their widespread use has been restricted by a number of organisational and technological issues. The main aim of this research is to address the key issues relating to the implementation and adoption of these technologies within companies and their supply chain.

4.2 Project approach

The research approach adopted is best described as a series of industrial case studies involving a number of companies from a range of industries specifically Product Design, Construction and Electronics Manufacture.

The general methodology adopted within each of the companies was to run successive case studies each building upon and testing the findings of the previous. Therefore, each case study followed a different methodology focusing on slightly different aspects. The first company case study commenced in May 1997 with a series of trials being carried out in Hulley & Kirkwood a mechanical and electrical building services consultancy. This was followed by a product design company, Devpro

starting in December 1997. Finally, the Keltek electronics case study began in June 1998.

A variety of data collection methods each aimed at capturing specific types of information were devised and employed during the case studies specifically:

- Initial Structured Interview
- Post Demonstration Questionnaire
- Diary of Observations:
- On-Line Logging System
- Weekly Questionnaires:
- Final Interview.

4.3 Findings

The results of this research can be classified within the following areas:

- Barriers to the introduction and usage of the shared workspace;
- Typical system usage;
- Benefits obtained through system usage;
- Perceptions of the company throughout the introduction and usage of the shared workspace;

Barriers to the introduction and use of shared workspaces

Barriers identified during the introduction and use of the shared workspace can be classified under the following main areas:

- Management;
- Psychological / Perceptions;
- Technology Related;
- Training.

A full description of each of the barriers identified is provided in [16]

Benefits of shared workspace usage

The main benefits achieved within the case study companies can be summarised as:

- Reduction in the time taken to carry out a variety of design activities due to less re-work, ambiguity, file transferring and paper chasing;

- Improved design quality;

- The companies design costs are reduced therefore, reducing the cost of the services to their clients giving them a market advantage over their competitors;

• Due to improved communication clients invest less time on design, reducing their costs further;

• Companies adopting the technology feel they are getting a better response from remote parties using the shared workspace than they would get adopting a conventional communication tools.

Typical system usage

In general, the shared workspace technology was not used to replace travel. Despite the fact that prior to the introduction of the technology companies felt one of the greatest benefits they would achieve would be a reduction in frequency of travel and savings relating to this. In reality, there were certain activities that each of the companies felt could not be carried out remotely therefore, the reduction in travel and related costs was found to be negligible. Typically, the system was used to enhance design activities that were previously carried out using conventional asynchronous communication media such as the application of e-mail, telephone, fax etc. to carry out the following activities:

• introducing and discussing design changes;
• clarifying design details;
• presenting designs to clients for approval;
• discussing project progress.

The shared workspace was used differently in each of the participating companies. Hulley and Kirkwood initially found it difficult to co-ordinate times for synchronous use of the system. After a few months of barely using the system they devised a method of employing the system in an asynchronous manner, which was found to be advantageous. In contrast, Keltek adopted the shared workspace in a synchronous manner from commencement of usage. The main difference between the companies being that the key system users in Keltek were predominantly located at their desk whilst, the key system users adopting the system asynchronously spent a considerable percentage of their time out of the office.

For synchronous use application sharing was found to be the most useful tool, being employed almost 100% of the time. On the other hand, Hulley and Kirkwood who adopted an asynchronous mode of working found the whiteboard to be an extremely useful tool employing it more than 90% of the time. Usage of the shared workspace went through 'cycles' i.e. short periods when it was used synchronously almost on a daily basis followed directly by longer periods when it may not be used for several months whilst designers work alone of with co-located team members.

Company perceptions

Company perceptions changed dramatically throughout the project. Initially, prior to the introduction of the technology all of the companies were enthusiastic at the prospect of using the shared workspace. Initial impressions were that the shared workspace would have a positive impact on communication between remote design team members. Preconceived benefits include:

• Reduced travel, savings in flight tickets and a reduction in the time spent travelling;
• Improved communication within the distributed design team;
• Closer working relationships within the company and their supply chain;
• Better quality products;
• Reduction in the time taken to execute interactive processes, fewer redesigns.

A minority of prospective system users had reservations about using the shared workspace instigated by concerns that their IT skills may not be adequate.

Once the system was introduced in each of the companies initial enthusiasm was thwarted by a number of barriers, primarily relating to:

• The location of the shared workspace;
• Confidence in using the system;
• Technical issues;
• Fire fighting.

Although over 90% of prospective system users had initially, expressed an enthusiastic interest in using the shared workspace when the system was installed less than five percent of them retained this enthusiasm. In instances where individuals overcame initial barriers and used the system to their advantage it was found that initial enthusiasm returned. The companies who adopted the system within their design process have either purchased addition systems or have made plans to do so in the near future.

Guidelines

The main output of this research was the development of guidelines for the effective introduction and usage of the shared workspaces within the design process. These guidelines follow four basic stages:

Recognition of need
Preparation
Introduction
Adoption

Recognition of need:
Identify need to improve communication;
Investigate means of improving communication;
Identify shared workspaces as solution;
Ensure the information regarding the cost of shared workspace technology is factual;
Ensure that the remote parties with which your company frequently communicates with have compatible software;

Preparation:
Select and appoint a committed shared workspace champion; ¡
Ensure IT staff are involved as early as possible and that sufficient time is set aside in their schedule for system installation;
Review information sharing requirements;
Specify hardware and software;
Any future plans for changing operating system should be considered when purchasing the shared workspace system;
Review information current information sharing practice;
Identify key system users;
Procure equipment - one shared workspace should be purchased for each key system user;
Review current network infrastructure and arrange update where necessary to ensure seamless access to the network for each key system user;
Set-up and test equipment;
The shared workspace should be located where the key system user has readily available access, preferably it should be installed on the PC of the key system users PCs';
Set up should ensure the system on which the shared workspace is installed mirrors standard company set up;
Ensure audio is of a high standard;
Ensure sufficient RAM is available to support the shared workspace software and the operation of additional software;

Plan training:
Proper training is essential as many prospective system users perceive the shared workspace to be difficult to use and as a result are reluctant to use it. Effective training will help overcome this;
System users should be trained both to set-up and use the system properly as lack of knowledge and inexperience in usage and setting up can lead to crashing which in turn can result in lack of confidence amongst system users;

Introduction:
Identify pilot project and key system users;
Initially, until confidence is gained, the shared workspace should be used on projects which are not on the critical path;
Use system within pilot project;
Until users become familiar and confident system users the local video window should be made small or completely invisible as many people find looking at their own image extremely off putting;

Adoption:
Extend use;
Monitor benefits;
Develop infrastructure;
Train new staff.

5 Comparative findings and conclusions

5.1 Common areas

Through comparing the findings for the educational and industrial projects described in this paper it is clear there are areas of similarity between the project findings. These are discussed in the following paragraphs.

Barriers: the majority of barriers identified in both projects can clearly be classified under common headings, specifically:

Technology;
Social/ cultural;
Management;
Psychological/physiological;
Training.

Usage:

• In both the educational and industrial based projects it was found that the technology was employed in different ways by each "team";

• Majority of communication still asynchronous despite user preconceptions that synchronous communication would increase dramatically. This finding is common to other research projects [5].

• Industrial case studies show that users go through short phases of intensive synchronous usage followed directly by longer periods of asynchronous communication. A similar pattern is apparent in the ICON2 project illustrated in figure 2;

• Both projects show that CSCW tools cannot replace face to face meetings, in their present form;

Technology:

• A lack of sufficient training/preparation can lead to a severe lack of confidence in the system and ultimately, non-use;

• Good quality audio was found to be absolutely critical in both educational and industrial usage with video taking the role of a secondary communication link. This result is common to other research findings [13];

• Both projects show that a lack of hardware / software standardisation can render the system unusable;

Management:

• As CSCW involves the collaboration of people and institutions conflicting objectives can sometimes mean that ventures fail to become reality. Both the educational and industrial projects showed that effective management is crucial to compromising on different agendas such as policy, timetabling and resources;

• A project champion is essential in both industrial and academic projects to ensure successful management of the system.

5.2 Transference of lessons between sectors

The majority of findings from both projects are common. However, both projects provide scope for lessons to be transferred from industry to education and vice versa.

Industry to Education:

• More barriers were identified in the industrial based projects, due mainly to the almost artificial "sheltered" nature of the ICON projects. These additional barriers have been developed in to guidelines which could prove useful in future educational projects;

• Industrial case studies can be used as stand alone teaching material within educational environments in order to provide students with a realistic overview of "the real world";

Education to Industry:

• Both ICON projects were well managed and prepared for well in advance of commencing. As a result, system usage was smoother than within the companies. Industrial case studies show that companies are keen to commence usage in order to achieve the perceived benefits and tend to gloss over the preparation and management stage, often to the detriment of successful technology implementation;

• In ICON2 a variety of technologies were provided to ensure a back-up was available in case of failure of other communication media. In addition, the continuous availability of technical support eased potential problems. This approach would prove beneficial in industry where key system users become very frustrated when communication technology fails, often becoming annoyed to the point where they may not use the system again ;

• The ICON projects showed that students adopted high levels of synchronous work to cope with short project deadlines. Results from industry showed key system users tended to "back off" employing technology when tight deadlines were looming often resorting to conventional asynchronous modes which are more time consuming.

• Throughout the ICON projects, particular attention was paid to the learning process that the students went through in the course of the project, both in terms of the new technology and the core material. As CSCW technologies are new to most people in education and industry, the latter can learn from the former in methods that maximise quick and efficient uptake of the new systems - all engage in the learning process.

5.3 Conclusions

It is evident from the research findings presented in this paper that lessons can be transferred between education and industry sectors. Furthermore, common guidelines can be developed in the form of procedural stages to facilitate the successful implementation and usage of CSCW technology within both arenas.

References

[1] C.E.Siemieniuch & M. Sinclair, "Real-time collaboration in design engineering: an expensive fantasy or affordable reality ?, Behaviour & Information Technology", Vol. 18, No. 5, 361-371, 1999.

[2] S. Nidamarthi, R.H. Allen, S.P. Regalla & R.D.Sriram, Observations from multidisciplinary, internet-based collaboration on a practical design project", International Conference On Engineering Design ICED 99 Munich, P709-714, August 24-26,1999

[3] Sandkuhl, K & F, Fuchs-Kittowski, "Telecooperation in decentralized organisations: conclusions based on empirical research, Behaviour & Information Technology", Vol. 18, No. 5, 339- 347, 1999.

[4] R, Fructer, 1996. "Interdisciplinary Communication Medium in Support of Syncronous and Asyncronous Collaborative Design". *Proceedings of the First International Conference of Information Technology in Civil and Structural Engineering Design (ITCSED).* University of Strathclyde, Glasgow. 14-16th August 1996.

[5] A. May, C. Carter, S.M.Joyner, W. McAllister, A. Meftah, P. Perrot, P. Pascarella, H. Chodura, M. Doublier, P. Carpenter, P. Caruso, C.Doran, V D'Andreas, P. Foster, J. Pennington, V. Sleeman & R. Savage, "Team based european automotive manufacturing: final results of demonstrator evaluation" (Rover/ Team/ WP5/DRR007) Loughborough, Leics LE11 1RG, UK: HUSAT Research Institute, Loughborough University, pp1-75

[6] A.Milne & L.Leifer, "The Ecology of Innovation in Engineering Design", International Conference On Engineering Design ICED 99 Munich, Vol 2, P935-940 August 24-26,1999

[7] Petrie, C. "Madefast" Home Page. http://www.madefast.org/

[8] "Suppliers and Manufacturers in Automotive Collaboration (SMAC)" Home Page, http:// greenfinch. analysis. co. uk /race /p17/present/smac/default.htm

[9] M. Sclater, N. Sclater, L. Campbell, "ICON: evaluating collaborative technologies", Active Learning, vol 7, CTI, Dec 1997.

[10] S. P. MacGregor, W. J. Ion, "Introducing and developing virtual design environments - an effective platform for collaborative design projects in academia", Proceedings of EDE '99, University of Strathclyde, September 1999, pp243-247.

[11] S. P. MacGregor, "Investigation and development of guidelines for the implementation of collaborative virtual design projects in academia", MEng dissertation, DMEM, University of Strathclyde, 1999.

[12] L. Campbell, I. Ali-MacLachlan, A. I. Thomson, W. J. Ion, A. S. MacDonald, "Institutional collaboration over networks - ICON, A comparison of two collaborative design projects", Proceedings of CADE '99, University of Teeside, April 1999.

[13] J.S.Kirschman & J.S. Greenstein, "The use of computer supported cooperative work applications in student engineering design teams: matching tools to tasks", International Conference On Engineering Design ICED 99 Munich, P1299-1302, August 24-26,1999

[14] W. J. Ion, A. I. Thomson, D. J. Mailer, "Development and evaluation of a virtual design studio", Proceedings of EDE '99, University of Strathclyde, September 1999, pp163-172.

[15] N. Sclater, H. Grierson, W. J. Ion, S. P. MacGregor, "Online collaborative design projects: overcoming barriers to communication", to appear in special issue on virtual universities in the International Journal of Engineering Education, 2000.

[16] A.I.Thomson & W.J.Ion, "Integration of Design Specialists Through Shared Workspaces" Design Council Final Report, 1999.

[17] A.I.Thomson & W.J.Ion, "Intorducing and Using Shared Workspaces in the Design Process: From Perceptions to Effective Widespread Usage, International Conference On Engineering Design ICED 99 Munich, P1829-1832, August 24-26,1999.

Implementation of a Demonstrator to Illustrate Project Management Techniques within a General Background Engineering Syllabus

Bernard PHILIPPE
Assistant Professor in Automation

Véronique SADAUNE
Assistant Professor in Electronics

École Nationale Supérieure d'Ingénieurs en Mécanique Énergétique Université de Valenciennes France

Abstract

The paper aims at presenting a demonstrator used to illustrate an original approach allowing the training of general background engineers within a team project type organisation.

Within the third year, it was then decided, four years ago, to lead an innovative experiment within the School. This one has been on a yearly basis and consists in bringing knowledge, no longer via a vertical approach, but via a horizontal approach (project team).

The general idea is to propose students to design a system defined by the specifications (translated into Quality, Cost, Delay) defined either by an industrialist or by the School teaching staff. To illustrate our approach, it was decided to develop a demonstrator as an autonomous mobile platform moving around the School.

Key words

Mechatronics, Project team, Concurrent engineering, Intelligent mobile platform.

1 Introduction

The " École Nationale Supérieure d'Ingénieurs en Mécanique Énergétique de Valenciennes" aims at training general background engineers. (Research & Development). Due to its geographical situation (car and rail constructors and parts manufacturers), the main competence field of ENSIMEV is ground transportation.

The pluridisciplinary training is organised around four main axes: Mechanical Design and Manufacturing, Fluids Mechanics and Energy, Mechatronics and Advanced Materials for Mechanics and Energetic. Within the third year, the students can investigate one of these axes within the framework of an option.

With the aim of helping young graduates to fit in their new professional environment, it is important to sensitise students to team-based project management [1]. Indeed, for the sake of time saving and of cost reduction, firms increasingly work with concurrent engineering concepts [2][3] (work in team project), and even cooperative engineering integrating virtual design office concepts.

Within the third year, it was decided, four years ago, to lead an innovative experiment within the School through project management. A working group included, on the one hand, 4 to 8 students (that is to say a potential of 800 to 1600 working hours) and, on the other hand, by a tutor (pedagogical project manager), a manager consultant (for project management) and experts (teachers or engineers having a recognised scientific knowledge) has been constituted.

In order to allow to synthesise the whole project process (concurrent engineering), it was decided to develop a demonstrator as an autonomous mobile platform. This one, validated this year, will be used as an aid for management project course and will be used for the project manager formation. It will illustrate for each step (analysis, activity research, time and cost management, technological choice of components and dimensioning, interaction between the different technological solutions, etc...), the concepts, the methods and the tools implemented to structure and manage the project.

2 Problem positioning

The industrial world is in perpetual evolution (globalisation, cost and delay reduction, ...) [7] [8]. It is then imperative, within the framework of scientific and technical teaching, to prepare students to the actual practices awaited by the firms. This point of view needs to question the education given and to define a novel pedagogical approach [9][10][11]. Several experiments was lead in field as varied as training period, integrated degree courses, tutorial projects, distance learning, use of the novel technologies (virtual practical model, videoconference courses, learning educational tools from CD ROM, ...) [12][13][14][15].

Four years ago, the introduction of project team marked an important in the new industrial needs taking into account. Each year, an evaluation is done and shows that only 70% of the awaited results are achieved, table 1, 30 % remaining to the responsibility of the teaching tutor and of the resource men (the engineers and the technicians of the School). A finest analysis of this situation (table 2) points out it is mainly due to a bad evaluation by the students of the difficulties linked to project management (lack of a global view of the project, impregnation of the management techniques, ...).

Criterions	%
Students satisfaction	100
Interest for the subject	95
Obtained results / Expected results	70

Table 1: Result of the assessment

Problems encountered	%
Organisation (project management)	30
Functional analysis (methodology)	25
Technological knowledge	20
Autonomy	10
Humans relations (internal conflicts)	10

Table 2: Repartition of the problems encountered

To correct this problem and, in the future, to obtain 100% of the expected results in the projects realisation, it was then decided to lead an innovative experiment. This one will be on a yearly basis and will consist in a horizontal approach (courses monitored according to the needs) of the system design.

To validate and so showing the efficiency of this novel approach, it has been decided, within the framework of a project team, to conceive a demonstrator to illustrate the methodology proposed.

The subject chosen had to be sufficiently complete to imply a maximum of disciplinary taught in the School (mechanics, electronics, automation, manufacturing, ...)

The general idea was to propose to students to design, by themselves, a mechatronics system. This one is a mechanical system with electrical or hydraulics actuators and sensors added, associated to a control device to improve its static and dynamic result [6]. It was defined by the specifications (translated into Quality, Cost, Delay) given by the teaching staff. The work asked was to supply a detailed pilot study corresponding to the choice of the different industrial components and their integration. The School technical staff will carry out the manufacturing. The final choice for the project has gone to the conception and the realisation of an autonomous mobile platform operating supplies transport in the School.

The next paragraph presents the different steps followed to obtain the demonstrator and to validate the project management approach set up.

3 Project sequence

The whole approach proposed turns on the 10 phases described on the table 3.

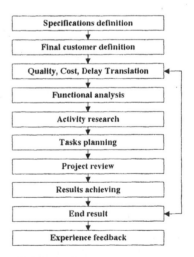

Table 3: Course sequence

Two methods, associated with 2 specific tools available within the School, have been chosen and used, after a short training period, to obtain a homogeneous structuring of the whole approach.

The first one, called S.A.D.T. (System Analysis and Design Technique)[4], associated with the IDEF0 tool [5], provides the decomposition in levels of abstraction. It structures the reflection from linguistic units (actions) subjected to some constraints, supported by mechanisms and through which circulate information (signal, data, ...).

The second one, commonly used in the industry for production management, associated with MS-PROJECT software, allows to introduce the notion of time. It organises, as a planning, the whole necessary tasks in the project progress.

3.1 Specifications and final customer definition

The specifications description relative to the project was stated by the teaching team (constituted here by an assistant professor in automation, an assistant professor in electronics and a qualified teacher in electrical engineering) associated with the engineers and the technicians of the School.

The autonomous mobile platform specifications were the following:

- Total rolling weigh: 150 kg ;
- Autonomy: 1 hour;
- Size: 700*500 mm ;
- Maximal speed: 1 m/s;
- Study duration: 200 hours/person;
- Budget: $6000;
- Pedagogical constraint:
    - To use industrial components available in shop;
- Number of persons implies:
    - 6 third year students;
    - 3 teaching tutor;
    - 1 mechatronics engineer
    - 1 technician.

After the presentation of the project at all the third year students, a team project was set up and a student project manager was named.

This takes place during a forum following by interviews where, after the presentation of all the available projects, the students choose a particular subject. The teaching tutor, on the basis of the following criterions, constitutes the team project:
    - motivation for the project
    - scientific and technical competence
All the accepted project team elects the student project manager. The teaching tutor arbitrates in case of problem.

3.2 QCD Transcription

The student team and the teaching team assume the specifications transcription in terms of quality, cost, delay. Here:

Quality (targets to reach):
- to present a detailed pilot study of the autonomous mobile platform (according to the specifications given higher)
- to validate the different steps of the approach project proposed

Cost: $6000

Delay: 15 weeks, one and a half day per week

From this moment the work of the team project really begins under the responsibility of the student project manager. The teachers and the resource men (the engineers and the technicians) are yet present to guide the students in the study progress, and to answer to scientific and technical questions.

3.3 Functional analysis

This point allowed to reveal of the main functions and tasks to well lead the project. A SADT approach, illustrated Figure 1, pointed out five main functions (Control, Power, Energy, Sensors and Man/Machine Interface). It consisted in isolating the problem from its context, then in translating the initial specifications in terms of expressed and not expressed needs, of functions, of constraints, of information flow,...

After two working weeks, the student project manager presents the obtained analysis results to the tutor who validates the whole decomposition.

3.4 Activity research

The next step consisted in research the activities (tasks) to see the whole project through. A SADT type approach, figure 2, was also adopted here. It allowed to isolate and to define the resources generated by each basic project activity. The purpose was to underscore, from the beginning, the man, material, documentation and complementary formation requirements.

To provide the quality all along the project (traceability), a document codification was organised. It was common for all the project teams and permits to identify the document type and the project referenced. As an example, FA99 PLP ENS 01 is for:

- FA: Functional Analysis;
- 99: Project year;
- PLP: Project Team;
- ENS: Internal Project;
- 01: Number.

3.5 Tasks Planning

A duration and a means was allocated to the previously isolated actions (figure 2), which became tasks. The whole task was planned to meet the time allowed as it is shown figure 3. The available project control tool used is MS-PROJECT software. It is important, at this stage, to organise the work load of each people involve in the project while taking care about keeping the overall coherence in the project progress (task synchronisation).

More, this step permitted to identify the tasks that conditioned the overall project delay and which ones having a more flexible time positioning.

In our case, an other difficulty consisted in managing the means availability conflicts (human and material one) in the team project, taking into account, on a more general level, the whole team-based project management (15 team projects work in parallel in the School). A synthesis reunion was organised three weeks after the beginning of the team project to solve these possible problems (CAD licences, workshop, technicians, …).

3.6. Project review

This phase represents more than 80% of the project time.

The previous problem analysis (§ 3.3) and the task planning (§ 3.4) ended in the allocation of one function to one student, according to his competence or to his centre of interest. Each student has to work in parallel (concurrent engineering concept) on the functional decomposition of each subset. A settling point has synchronised the sequence of the project.

It is necessary, here, to define the technology used.

This step consisted of doing bibliography researches on the whole technological components able to realise the basic function isolated previously. For this work, students need experts to help them in their component choice (magnetic or optic sensors, asynchronous or continuous motor, pneumatic or electric energy, Programmable Logical Controller or micro-controller, etc…). The answer has to be adapted to each specific requirement, as bibliography reference, suppliers' technical documentation or courses.

At this study level, the presence of the mechatronic engineer and the technician is essential, because they can help the students in the choice of the technologies usable and compatible with the available resources within the School.

3.7. Results achieving and end result

The last phase permitted to do the final settling point on the sequence of the project in the aim to verify the overall coherence of the proposed solution. From a practical point of view, it has led to the writing of a report. The location of each component, the plans, the nomenclature, … were also realised to give to the technical staff the specifications required to manufacture the autonomous mobile platform drawn figure 5.

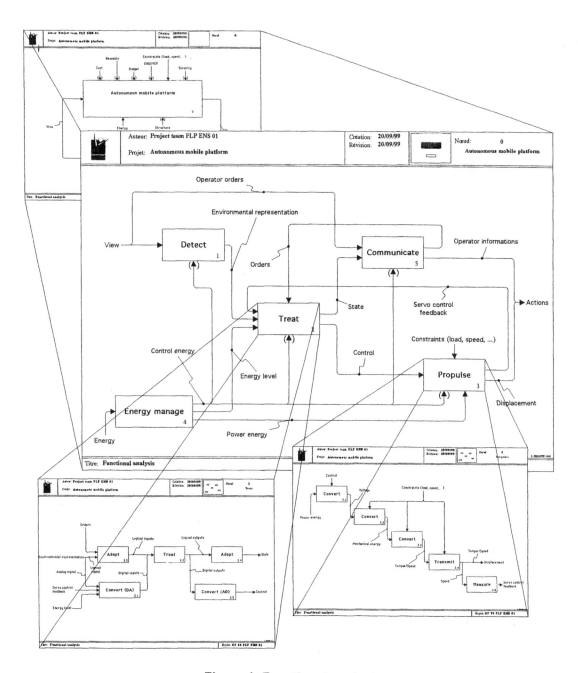

Figure 1: Functional analysis

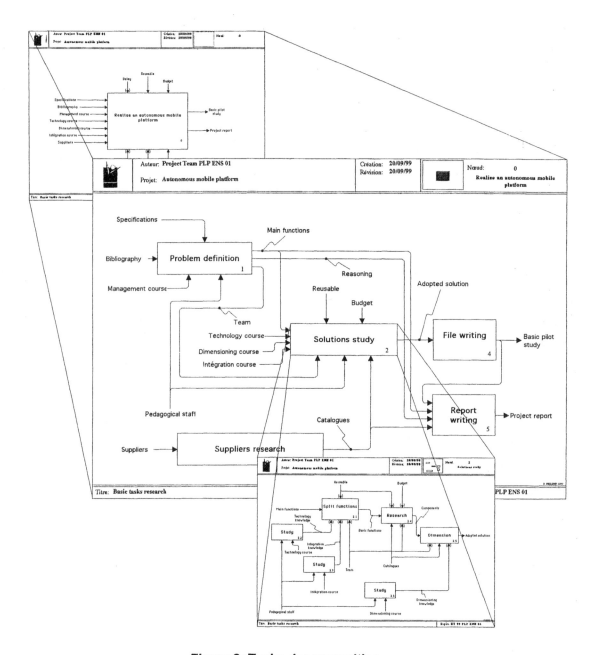

Figure 2: Tasks decomposition

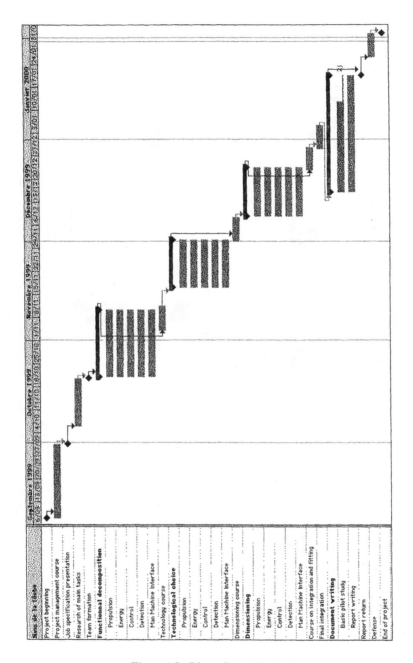

Figure 3: Planning adopted

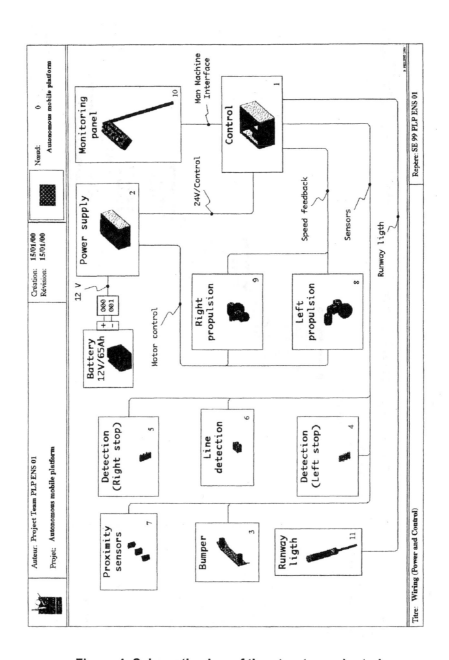

Figure 4: Schematic view of the structure adopted

Figure 5: General view of the adopted structure

The student team presents and defends their detailed pilot study during a viva. This one must justify the organisation and the project management, like whole the technological choices made. A comparison between the results obtained and awaited is realised. In the present case, 100% of the targets were reached. Indeed, the detailed pilot project was presented within the time allowed and the whole work validates the project approach developed.

The last stage (experiment feedback in table 3), which consisted in getting learning with the idea to realising, in the future, similar project, has not been performed because, for the moment, within the framework of the project teams developed in our School, the subjects ere each year different.

4 End result

The experiment led this year has given most satisfactory result.

From the detailed pilot study provided by student, the technical staff of the School has manufactured the autonomous mobile platform now operational, Figure 6a and 6b, and which serves as a demonstrator.

Figure 6a: Side view photography of the autonomous mobile platform

The approach proposed is valid, from a knowledge assimilation point of view, because the students are now able:
- to structure a project
- to discuss with suppliers
- to exploit technical documentation
- to have a better global view of the project
- to apprehend different technology

An inquiry, applied to the students, has pointed out that 86 % are satisfied by the approach proposed. They find a greater motivation because they can use the acquired knowledge, either directly during their training period, or in their extra school technical type activities (clubs and associations).

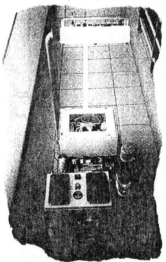

Figure 6b: Front view photography of the autonomous mobile platform

Yet some problems have remained:
- conflict of persons (between the student)

- difficulties of communication within the team (and with the teaching team)
- lack of critical distance on some technical sides (choice and dimensioning)

With regard to the effects on the pedagogical team, we can quote:
- a greater interest of students for the technical fields (because a specific needs exist within the framework of the project)
- the organisation of a dynamics between teachers of different speciality (federation around a common subject)
- the achieving of a demonstrator that will be install next year in the project management technique training
- a better view for students on the project manager job and on the importance of communication within a team

We can note that to carry this type of teaching to a successful conclusion, a versatile team capable of treating all the project aspects is necessary. This requires a large contribution from everyone both at preparation level and at project supervision level. It is important to be able to find an answer to the student theoretical and practical questions as they advanced in study.

5 Conclusion

The students have experimented a global approach necessary to the design of a mechatronics system (autonomous mobile platform). Considering the results reported in the previous paragraph, it seems justified to introduce a specific course aiming at bringing students the knowledge required to structure and to manage a project. This course, entitled "Applied Mechatronics Design", will be structured around the points stated in the § 3, and will use, to illustrate each step, the technological aid constituted by the demonstrator realised.

To be beneficial, the course proposed should be programmed, at later, at the beginning of the third year. At the ENSIMEV level, it is the only conceivable solution for the moment (given the current training organisation). Nevertheless, within the framework of School development, a new

pedagogical program is under study. It schedules this course at the end of the second year. This will allow students to be able to begin the third year project team in optimal conditions aiming at the achieving of 100% of awaited results.

References

[1]   J.P.Lewis, "Team-based project management", Amacom (October 1997)

[2]   J.Turino, "Managing concurrent engineering", Kluwer Academic Publisher (March 1992)

[3]   J.P.Lewis, "What every engineer should know about concurrent engineering", Maecel Dekker (30 June 1995)

[4]   D.A.Marca, C.L.McGowan, "SADT: Structure Analysis and Design Techniques", Mc Graw-Hill Software Engineering Series

[5]   D.A.Marca, C.L.McGowan, "IDEF0-SADT business process & enterprise modelling", Eclectic Solutions Corporation (October 1993)

[6]   G.Schweitzer, "Mechatronik an der ETH Zürich", Bulletin SEV/VSE, 1/1989, pp.10-4.

[7]   D.I.Cleland,"Project management: the strategic trajectory", proceeding of the 11th Internet World Congress on project management, Florence,Vol.1, pp.313-323 (June 1992)

[8]   V.A.Mabert, J.F. Muth, R.W.Schmenner, "Collapsing new product development times: six cases studies", Journal of product innovation management, n°9, pp.200-212 (1992)

[9]   D.J.Hargreaves, "How undergraduate students learn", European Journal of Engineering Education, 21(4) (1996)

[10] G.Heitmann, "Project-oriented study and project-organized curricula", European Journal of Engineering Education, 21(2) (1996)

[11] R.J.Lenschow, "From teaching to learning: A paradigm shift in engineering education and life long learning", European Journal of Engineering Education, 23(2) (1996)

[12] I.Qasem, H.Mohamadian, "Multimedia technology in engineering education", Proceedings IEEE Southeastcon, Vol.1, pp.46-49 (1992)

[13] L.Sauvé, "On-line teaching and learning: Some avenues", Proceeding of the 19th world conference on open learning and distance education of ICDE (CD ROM). The new educational frontier: Teaching and training in a networked world, Vienna (June 1999)

[14] J.M.Thiriet, M.Robert, "Development of virtual educational tools for teaching and through teaching", Proceedings of the 9th annual IAEEIE conference, pp.161-166, Lisboa (May 1998)

[15] L.Wilkerson, W.H.Gijselaers, "Bringing problem-based learning to higher educational theory and practice", n°68, Jossey-Bass Publisher (Winter 1996)

Communicative and cooperative manufacturing systems engineering

M. Guillemot, G. Louail, D. Noterman

Equipe d'Etudes Technologiques du Laboratoire de Génie Electrique et Ferro-électricité
Institut National des Sciences Appliquées de Lyon – 69621 Villeurbanne Cedex

E. Bideaux

Laboratoire d'Automatique Industrielle
Institut National des Sciences Appliquées de Lyon – 69621 Villeurbanne Cedex

Abstract

In the next century, the new technologies and the economical needs will issue in changing the work organization and the management of the enterprises. This will further new concepts to come out as the "communicative enterprise". The goal of the project described in the paper is to create new, interactive tools and pedagogical methodologies in order to initiate the engineering students to the "cooperative" work and to teach them the necessary means (tools, methodologies). The application field concerns the development and implementation of automated manufacturing systems.

1 Introduction

It is obvious that the competitiveness of the enterprises implies nowadays the reduction of the design delays [5,6]. In a first approach, this means the reduction of the delay in between the arising of a new product concept and the effective manufacturing of this new product. In a second approach, this implies to put into practice a concurrent engineering structure, which allows designing the product as a project managed by an interdisciplinary team [7,8]. The close collaboration in between the members of the interdisciplinary team is then crucial for a successful outcome. The main difficulties emerging from this new design approach are related to the cooperation, the coordination and the management of the information within the project team [1,2].

The object of this paper is the statement of an educational project, which consists in a teaching-training course for post-graduate students. Communicative and cooperative work is approached in the context of the manufacturing system engineering. To build up this educational project, we took into account:

- The development of a work methodology based on the diagram of the V cycle, using shareable software at every stage if possible. The whole project is gathered together in a document base provided by the GroupWare Lotus Notes.

- The implantation of communicative structures on distant sites (Institut National des Sciences Appliquées de Lyon (INSA), Université Claude Bernard Lyon I (UCBL), Ecole Nationale Supérieure des Mines de Saint-Etienne (ENSMSE)). Each site is fit out with communication tools (Visio-communication, network, network devices, etc) that allow to communicate in between the sites with PLC or PC's.

- The implementation of specific software that provides the simulation of industrial operative parts consisting in either continuous processes, or manufacturing systems [3,4].

- The initiation of a reflection about the contribution of this new technologies concerning the pedagogy in our university.

An educational group associating lecturers of the electrical department and the mechanical engineering department in the INSA de Lyon initiated all these works. The project is nowadays included in a global study driven by the AIP/RAO and is supported by regional (Région Rhône-Alpes) and private funding (Schneider Electric).

2 Project organization within concurrent engineering

The design of the manufacturing process and its automation is crucial for the competitiveness of the enterprise. It can be considered as a product and it life cycle may be represented by a V cycle as shown on figure 1.

Méthodology

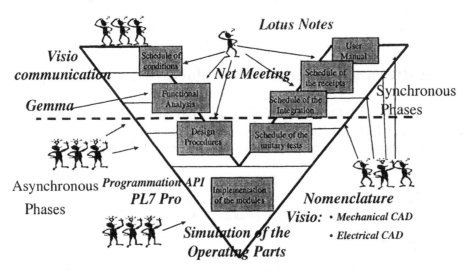

Figure 1: Diagram of the V cycle of an automatism project.

The V cycle shows two distinct stages:

- The first one consists in synchronous phases, which needs the simultaneous presence of the different active partners even if they are located at distant places.
- The second one consists in asynchronous phases. During these phases the actions or the tasks may be autonomously realized.

During all these phases, that constitutes the V cycle, the information and the technical data have to be managed and accessible by all the participants in the project. The federating tool of the project is thus a GroupWare like Lotus Notes. It allows also managing the project by setting the schedule of the different phases and of the meeting for the resynchronization.

The concurrent engineering define then two situations of work.

Concerning the synchronous phases, the partners of the project have to realize an analysis of the schedule of conditions, the functional analysis and then the task scheduling. In order to achieve this first stage, the Visio-communication tools present interest in allowing quick and efficient meetings. The videoconferencing using video and sound projection of the different sites (e.g. Meeting Point) and the application sharing (e.g. NetMeeting) have been used efficiently.

The technical part of the project is realized during the asynchronous phases. The physical and software tools usually used during this step have to allow remote control and to be shareable by the different partners. In this case, the communication is rather a point-to-point work and its objective is the partial resynchronization of contiguous parts of the project, the exchange of parameters or data, or the validation of a subsystem. The Visio-communication is not crucial for this step, but it proved to be useful for a better understanding.

Therefore we define an architecture for the classroom which enables the work in the two situations. The first half of the classroom is dedicated to the synchronous phases and integrates the necessary equipment for the videoconferencing of a large meeting. This equipment allows communication via Ethernet or RNIS as shown on figure 2. The second half consists in PC workstations with dedicated software and enabling point-to-point communication. Up to now, this architecture (figure 3) has been implemented in two places at the INSA (at the Electrical Engineering and the Mechanical Engineering Departments), and is presently put into service at the UCB and the ENSMSE.

Figure 2: Architecture of the classroom for the synchronous tasks.

Figure 3: Architecture of the classroom for asynchronous tasks.

3 Pedagogy around an automatism project within concurrent engineering

In the next parts, we present the methodology and the pedagogy we applied with students to develop this concept within automatism projects.

3.1 The first synchronous phases

Once the main objectives of the automatism project have been presented, all the participants of the project have to meet to establish the schedule of conditions for the proposed system.

As a meeting at a single place may be difficult, especially if the different groups are working in distant sites (and it is usually the case in the industry), we propose then to use the part of the classroom, which is dedicated to video-communication. This conference zone (figure 2) is equipped with a large angle camera, which allows either a global view of the participants or a zoom on a specific intervening person. A directional microphone permits the recording of the discussion. The software MeetingPoint performs the exploitation of this equipment, particularly the synchronization of the sound and of the images and their transmission to the connected sites, that is the location of the other groups. By using also NetMeeting, we dispose of some important tools as: a white board, a chat tool, a file exchange, and an application-sharing tool.

All this equipment is enough flexible to efficiently achieve the schedule of conditions for the proposed system. In the same synchronous phase, it is then necessary to realize the functional analysis. This important stage of the automatism project consists in the description and the schedule of the different stopping and functioning

modes for the system. We use to achieve this phase a graphical analysis tool, called GEMMA that is implemented in the pedagogic software CADEPA (figure 4). During this phase the remote sites share CADEPA in order to set up the functions and their schedule and the tasks are shared out among the different participants.

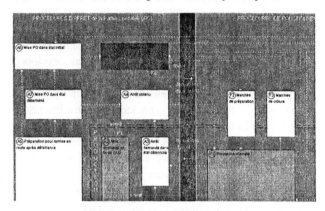

Figure 4: Functional analysis

3.2 The asynchronous phases

The first task of each group of participant working on the same part of the global project is to set up the design procedures. The implementation of the different specifications may start afterwards.

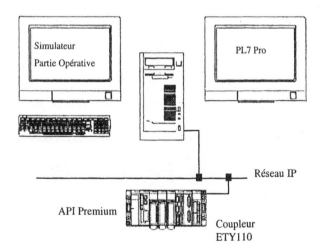

Figure 5: Architecture of the user workstation

The implementation of the control is realized on PLC (TSX Premium from Schneider Electric) using the package PL7PRO (SE). This package offers 4 different development languages, which are conform to the IEC-1131-3 norm:

- the graphical language (Grafcet),
- the ladder language (LD),

- the structured language (ST),
- the instruction list (IL).

At this step, the participants use the second part of the classroom and develop the previously defined specifications on a PC workstation (figure 5). As each workstation is connected with a PLC, the developments may be directly tested. To complete this possibility, it appears crucial to use a real tool for testing the PLC developments, and therefore we developed a simulator for the operating of the process. This software was developed in the Electrical Engineering and Mechanical Engineering Departments within the framework of student project. It is an extension of the supervision package Monitor Pro (from Schneider Electric) and provides different libraries of components (actuators, conveyor belts, sensors, signaling units, control units, etc.) which allow to build up any kind of manufacturing process (figure 6 and 7) within:

- a screen representing the synoptic of the process,
- a screen representing the operator desk.

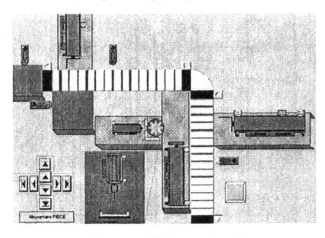

Figure 6: Synoptic of the industrial process

Figure 7: Operator desk

The development of the process representation is a copy and paste from the libraries towards the synoptic screen or the operator desk screen.

When the participants reach the debugging step, the application representing the process is loaded (or developed by the users) and is immediately put in communication with the PLC via Ethernet (or another local network). The simulator provides an animation corresponding to the state of the PLC outputs and may be used to act on the inputs of the PLC.

As the simulator exist in one user and web server version, it may be used via application sharing or remote control.

In the same asynchronous phase, the participants have to implement other parts of the automatism project as:

- The nomenclature (figure 8) using Excel (Microsoft) and hypertext links towards supplier's catalogs or data bases,

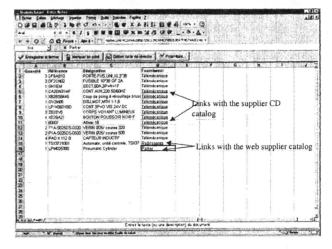

Figure 8: Nomenclature using Excel

- The mechanical and the electrical CAD (power and control parts) using a single software (Visio) or dedicated software (figure 9).

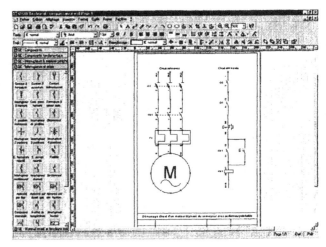

Figure 9: The mechanical and electrical CAD VISIO

Because Excel and Visio have direct links towards Lotus Notes, all the global information may be managed in real time.

3.3 The integration – The last synchronous phases

The last synchronous phases start once all the asynchronous phases have been achieved. In the last stage, the participants have to realize the final integration tests, the schedule of receipts and operator manual before the confrontation with the principal.

The integration is done using simultaneously the both parts of the classroom; all the participants transfer their developments on the same NT server and the global application is integrated on a single machine. The final integration tests can then be achieved.

All the previous constraints have imposed the following architecture for each classroom (figure 10):

- One application server (NT Server) supporting the database, Lotus Notes, the simulator and the video conferencing equipment (camera, video projector, RNIS connection, etc.).
- 6 client machines (12 in July 2000) with the application software, a videoconferencing kit and 2 displays :
  - one to refer to the PLC developments,
  - two when using the simulator (process and operator desk).
- 6 PLC (12 in July 2000) with an Ethernet coupling device.
- All this equipment is connected on the Ethernet network and the remote control is possible from any place.

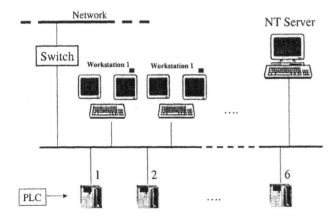

Figure 10: Architecture of the classroom

4 Conclusions and perspectives

This physical and software architecture allow us to manage different pedagogic experiences efficiently:

- between the Electrical Engineering and Mechanical Engineering Departments, with group of 30 students, we provide an initiation to concurrent engineering. 240 students from the INSA will have follow this 3 to 6 hours lecture.
- two other sites (UCB Lyon I and ENSME.SE) are presently installed and will enable us to drive a remote cooperative project in between the INSA and these two sites with 2 students (40 hours).

In a first analysis, these experiences point out that the role of the lecturer is basically modified in a function of group animation or project coordinator.

At short term, new development seems interesting on the basis of the present architecture:

- the integration of the process simulator directly in the PLC. This is actually possible with the coming out of the Web server functionality on the Ethernet coupling devices.
- to complete the process simulator with continuous processes. This development has been tested using Monitor Pro as the user and the PLC interface and Matlab (Mathworks) as the model solver (figure 11). It allows studying the tuning of the controller, consisting in functional blocs in the PLC.

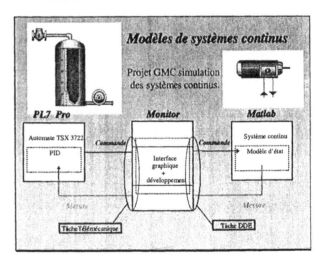

Figure 11: Simulator with continuous process

To conclude, we want to underline that the behavior is changed by these new pedagogic methods applied with concurrent engineering. The association with competencies in the domain of human resources seems finally a crucial point in order to achieve a better integration of the human behavior in the project management.

ACKNOWLEDGEMENTS

This work has been performed within the "Contrat d'Objectifs 1998" program by Région Rhône-Alpes.

References

[1] Claude Foulard, 'L'entreprise communicante', Hermès 1998.

[2] Melissa Saadoun, 'Le projet groupware, Des techniques du management au projet groupware', Eyrolles, 1996.

[3] Yu, B., Harding, J.A. , and Poppplewell, K., 'A Reusable Entreprise Model', International Journal of Production and operation Management.

[4] Poppplewell, K., 'Factory Modelling and Simulation of Manufacturing Systems', Proceedings of the first International Conference on Responsive Manufacturing', Nottingham, Sept. 1997, pp. 71-82.

[5] Harding, J.A. , Yu, B., and Poppplewell, K., 'Factory Design to Further Competitive Advantage', Proceedings of the World Innovation & Strategy Conference, Chapman, R.L. and Hunt R. (Eds), Sydney, Australia, Aug. 1998, pp. 437-442.

[6] Gupta, A, Wilemon, D., 'Accelerating the Development of Technology-Based New Products', California Management Review, Winter 1990.

[7] 'Simultaneous Engineering – Integrating Manufacturing and Design', Editor C. Wesley Allen, SME Dearborn, MI, 1990.

[8] Nevin, J., Whitney, D., et al, 'Concurrent Design of Products and processes, a Strategy for the Next Generation in Manufacturing', McGraw Hill Publishing, New York, 1989.

CHAPTER 14

Manufacturing

Automatic Determination of the Configurations

of a Mechanical Assembly

Ch. Forster

Laboratoire Roberval, U. T. C., Compiègne, France 60 200

J. - P. Boufflet

Heudyasic, U. T. C., Compiègne, France 60 200

F. Lecouvreur

Laboratoire Roberval, U. T. C., Compiègne, France 60 200

Abstract

We propose herein a mathematical modelization for mechanical assemblies. Our aim is to automatically determine the all-configurations of this assembly, which are obtained by the "sliding" of parts in a given direction. This model will be later useful to build and to solve the tolerance chains occuring in the unidirectional tolerancing process.

The proposed model is based on the graph theory. A graph, called, in the following, the part-graph, is associated to the assembly. The vertices represent the parts, and the edges indicate the nature of the contacts between the parts. The part graph contains not only the actual configuration of the assembly, but all the possible configurations. A very close link is made between the mechanical and technical data, and the mathematical properties of the graph. These properties permit the verification of the coherence of the studied assembly.

A configuration of the assembly is represented in this model by a partial graph, issued from the part-graph. This partial graph has particular properties too. These properties allow the automatic determination of the all-configurations of the assembly, wich may be more rapid if particular cycles or cutpoints exist in the part-graph.

1 Introduction

Tolerancing is a very complex stage of the design process. It consists in limiting the admissible tolerances (variations) of the dimensions and the geometrical defaults of the parts of a mechanical assembly, whereby it ensures performance, interchangeability, and low cost manufacturing.

We have focussed on the unidirectional tolerancing, which concerns only the dimensions (not the geometrical defaults) in one given direction. Our aim is to develop an automatic method of tolerancing. This method will be implemented in a CAD system, following the works of [1], [2], [3] and [4].

The first step of this method, that we have identified, is to automatically determine the all-configurations of the mechanical assembly. We present in this paper the details of this automatic determination. In the first part, we introduce definitions. In the second part, we present the mathematical model that we associate to a mechanical assembly. Then, in the second part, we present the mathematical model associated with the mechanical assembly, the part-graph. A partial graph, extracted from this part-graph, is used to represent the actual configuration. And we define some particular cycles of the part-graph, which are useful for the next part. In the third part, we develop the automatic determination of the all configurations of the mechanical assembly. We finally explain how particular cycles and cutpoints accelerate this search.

2 Definitions

Let us consider a mechanical assembly, of which the parts have been created and assembled with a general purpose CAD system. Parts are supposed to be geometrically perfect and rigid.

The presence of "clearance" permits parts to move and to define a new configuration of the mechanical assembly. As in continuum mechanics, the word « configuration » signifies the domain of the space used by all the parts of the mechanical assembly at a given time. In unidirectional tolerancing however, we only deal with the configurations in which the parts are in contact, i.e. when they are in extreme positions. We define therefore only a finite number of relevant configurations, without reference to time.

To localize these clearances, or, *a contrario*, to underline the contacts between the surfaces, the designer defines which are the pairs of surfaces in contact or able to enter into contact, and the nature of the contact :

- permanent contact (CP), for instance for fixed parts ;
- closed contact (CF) for two surfaces in contact which may be separated ;
- opened contact (CO) for two separated surfaces able to enter into contact.

We consider the tolerancing in the direction $\vec{d}$, in which the designer will impose the functional constraints. So we are interested in the positional changes of the parts in this direction. This supposes that in the other directions, contacts between parts are maintained.

For each part, one considers the surfaces (or the portions of surfaces) that are parallel to the direction $\vec{d}$. It is easy to define the coordinate of these surfaces relative to an axis parallel to the direction $\vec{d}$, with an arbitrary origin.

For example, we consider an assembly of three prismatic parts arbitrarly numbered ①, ②, et ③ (fig. 1). We are interested in the tolerancing in the « horizontal » direction $\vec{d}$, i. e. in this first step, in finding all the configurations of the assembly, obtained by the "sliding" of the parts in the direction $\vec{d}$. The j-th surface of the i-th part in the sense given by $\vec{d}$ will be denoted $\pm i/j$. Surfaces are oriented positively or negatively (relative to $\vec{d}$) by their outward normal. The coordinate of the j-th surface of the i-th part, relative to an arbitrary axis parallel to $\vec{d}$ is denoted x_j^i.

3 Mathematical modelization of a mechanical assembly

3.1 Part-graph associated with an assembly

The mechanical assembly is then modelled by a graph ([5]) of which the vertices are the parts, and which is called, in the following, the « part-graph ». Two vertices associated with two parts are linked by an edge if, and only if, the parts are, or may, enter into contact. It is worth noting the significance of the edges, which express not only the contact between two parts, but also the ability to enter into contact. The part-graph does not contain only the configuration chosen by the designer ; it may contain all the configurations obtained by sliding

the parts. One could build a graph with the surfaces instead of the vertices, but this would introduce a supplementary and unprofitable complexity.

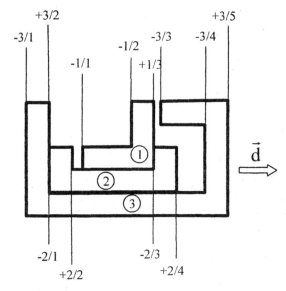

Figure 1 Treated example

The number of vertices, denoted n, corresponds to the number of parts of the assembly. Each vertex is arbitrarly numbered with a number between 1 and n.

The obtained graph is not ordered, because the binary relation «to be, or to be able to enter into contact» is symmetric. It is a 2-graph, i. e. there cannot exist more than two edges between two vertices ; indeed, in unidirectional tolerancing, one part can be positioned by another part only with two surfaces, one for the positive sense, the other for the negative sense.

The part-graph is connected, i. e. it is always possible to find a walk between two vertices. A walk is such a sequence of edges that the final vertex of one edge is the first vertex of the sequential edge. Here, we only consider elementary paths (each vertex is encountered only once), which we shall call « walk », for simplicity.

For each part being positioned, in a sense relative to $\vec{d}$, by at least one other part, it is always possible to find a sequence of surfaces in contact from a first part to a last part.

The number of edges is denoted m. Each edge is numbered in an arbitrary manner with a number between 1 and m.

In addition, each edge has two values. The first one is a pair $\{\pm i_1/j_1, \pm i_2/j_2\}$ which represents the concerned surfaces (and consequently, the parts). The second value is a symbol $\xi \in \{CO,CF,CP\}$, which represents the nature of the contact between the two surfaces. These values may be written in a table, and describe unambiguously the actual configuration.

It is worth noting the significance of the edges, which express not only the contact between two parts, but also the ability to enter into contact. The part-graph does not contain only the configuration chosen by the designer ; it may contain all the configurations obtained by sliding the parts, i. e. by changing the values of ξ.

When applying this model to our example, one obtains the following part-graph (fig. 2), and the following edge description (table 1).

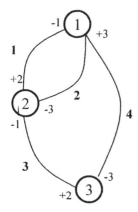

Figure 2 Part-graph of the treated example

Edge	Surfaces	ξ
1	{+2/2,-1/1}	CO
2	{-2/3,+1/3}	CF
3	{-2/1,+3/2}	CF
4	{-3/3,+1/3}	CO

Table 1 Edge values of the treated example

In addition, one may notice of the vertices and the edges, that each vertex is linked with at least one edge valued CF or CP ; if it is not the case, the part is not well positioned in the assembly (relative to $\vec{d}$). If a vertex is linked to the rest of the graph by only one edge, this edge must be valued CF (this corresponds to a part allowed to disconnect from the assembly) or CP. Between two vertices, one can have :

- nothing ;
- an edge valued CO, CF or CP ;
- two edges, necessarily valued or {CO, CF} or {CP, CP}. One cannot have {CO, CO} (one of the CO cannot be closed) ; {CO, CP} (the CO cannot be closed) ; {CF, CF} (contact redundancy) ; nor {CF, CP} (the CF cannot be opened). The case {CP, CP} corresponds to fixed parts without clearance and has to be treated like the case {CO, CF}.

For each edge, the surfaces must have opposite outward normals, because we are dealing with contact between rigid bodies.

The above remarks, based on mechanical and technological fundaments give useful properties for the part-graph. After the contacts have been imposed, the most of these properties are easy to verify; and they permit communication with the designer, in order to control the initial assembly. When implementing the method, these properties facilitate the use of the most efficient algorithms of graph theory.

3.2 Partial graph associated with a configuration

According to the definition we gave before, a configuration corresponds to a combination of contacts between parts, such that each part or group of part is positioned in one sense, or the other, by contact with other parts.

The initial configuration can be modelled by the partial graph obtained by suppressing in the part-graph all the edges of type CO (fig. 3). There exists a partial graph for each configuration contained in the part-graph, and these partial graphs are very useful in constructing the tolerance chains. The number of edge in a partial graph will be denoted m', with m' ≤ m.

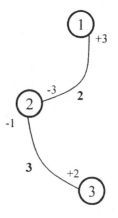

Figure 3 Partial graph associated with the initial configuration of the treated example

If the partial graph corresponds to a "correct" initial configuration, it verifies the following properties.

The partial graph is connected, and consequently, the part-graph is connected too.

It is now a 1-graph, i. e. there cannot exist more than one edge between two vertices ; if there are two edges (necessarily valued CF on this partial graph), the two parts are in simultaneous contact by two different pairs of surfaces, which are contact redundant. In addition, having no loop, this partial graph is a simple one (a simple graph is an 1-graph without loop).

Finally, in this graph there is no closed walk. A closed walk would consist in two walks of edges valued CF, or CP, between two parts of the closed walk, i. e. two sequences of parts having equal length, which is impossible in the case of rigid bodies. One can deduce that each edge of this partial graph is a cutline. A cutline is an edge that transforms a graph in two subgraphs when being suppressed. The proof is trivial : if one suppresses one edge and if the graph is connected, there exists a walk linking the two vertices associated with the edge. It means also that there was a cycle on the graph, made of the suppressed edge and the walk. This is in opposition with the previous remark according to which there is no cycle in a graph associated with a configuration.

Consequently, at the beginning, it is necessary to verify that the partial graph associated with the initial configuration is connected, and that there is no cycle. Classical algorithms of the graph theory are available for these verifications.

3.3 Study of the cycles of the part-graph

In the part-graph, a cycle is a walk of which the initial and the final vertices are identical. It corresponds to a group of parts that are positioned each by the others. Taking account of the fact that the part-graph is not a 1-graph, a cycle cannot be represented by the sequence of the encountered vertices, but by the sequence of the encoutered egdes.

In the chosen example, there are three cycles :

- the cycle {1,2}, which indicates that the positions of the parts ① and ② are dependent ;
- the cycles {1,3,4} and {2,3,4}, giving the position of the part ② relative to ① and ③.

It is worth noting that the configurations associated to the cycles {1,3,4} and {2,3,4} do not exist simultaneously. There is a commutation between the values CO and CF in the cycle {1,2}. One could ask the designer to indicate that these values are dependent, but such a method would be difficult to apply for complex assemblies. We shall see that it is possible to sort the cycles of the part-graph, and that this sorting permits to find the previous commutations.

After having analysed the different cycles which may exist in the part-graph, one defines two sorts of cycles, caracterized by the variation of the sense of the outward normal of the surfaces in contact during the "traversal" of the parts.

Cycle of type I :

Along such a cycle (fig. 4), the traversal of each part introduces a change in the sense of the outward normal from the initial surface (by which one "enters" in the part) and the final surface (by which one exits the part), e. g. the cycles {1,2} and {1,3,4}). In this cycle, each part is positioned in one sense, or the other, by two different surfaces with opposite normals.

A particular case is that there exits only one cycle of type I between the concerned parts, e. g. {1,2}. This corresponds to a group of parts which are positioned each relative to the other; by the same surfaces in the two senses. That is the standard case studied in unidirectional tolerancing.

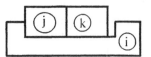

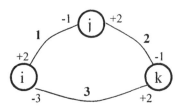

Edge	ξ
1	CF
2	CF
3	CO

Figure 4 An example of particular cycle of type I

Cycle of type II :

Along such a cycle (fig. 5), there exists at least one part from which the traversal introduces no change in the orientation of the outward normal of the initial and final surfaces (e. g. the cycle {1,3,4}). This conservation of the normal orientation indicates that the concerned part is in contact with the rest of the cycle, only one or two surfaces having the same orientation. In order to close the cycle, and to have edges between surfaces with opposite normals, it must exist another traversal without a change of normal in the cycle (or the concerned part is not correctly positioned, as mentionned above). In such a cycle, however, traversals without normal change always appear in pair number.

In addition, let us consider two parts of such a cycle, for which no change of normal occurs during their traversal. There exist two walks between these two parts (two "half" cycles). These walks correspond to two groups of parts existing between the two parts. These two groups can not have the same length in the direction $\vec{d}$, because the parts are rigid. Consequently, there exists at least one egde valued CO in the two walks.

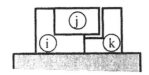

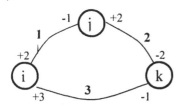

Edge	ξ
1	CF
2	CO
3	CF

Figure 5 An example of cycle of type II

The search for all the cycles of the part-graph leads to very expensive computation. In fact, it is sufficient to find the particular cycles of type I. Only these cycles simplify the search for the all-configurations of the assembly, and later, the determination of the tolerance chains.

Finding these cycles, it is worth noting that these cycles belong to the fundamental cycles of the part-graph. The number of fundamental cycles is given by the cyclomatic number ν,

$$\nu = m - n + p$$

where p denotes the number of connected components of the graph. The part-graph being connected, $p = 1$ so that :

$$\nu = m - n + 1$$

One has to search for the fundamental cycles for the part-graph, and to study the ν cycles in order to find the particular cycles of type I. Classical algorithms of the graph theory are available ([5]).

Concerning the cycles of type II, it has been mentioned above that one has to verify the compatibility of groups of parts lying in parallel between pairs of surfaces, particularly by controlling the presence of edges valued CO. This is very difficult in the part-graph, and we shall see that there exists a general and efficient means to control this compatibility on the partial graph.

4 Determination of the all-configurations of an assembly

4.1 General method

When parts slide from the initial configuration, one modifies contacts between parts, and one obtains other configurations. In order to find all the configurations of the assembly, we use the following proposition :

« the number of edges of type CO remains constant from one configuration to another »

To demonstrate this proposition, we use the partial graph associated with a configuration. Let us consider a given configuration, its partial graph and two parts linked by an edge valued CF. When separating these two parts to obtain a new configuration, one suppresses, in the partial graph, the edge between the two vertices associated with the two parts. This edge is a cutline, as established previously. The partial graph is divided into two distinct connected subgraphs, corresponding to a part or a group of parts disconnected from the rest of the assembly. In order to return to a satisfying configuration where every part or group is positioned, one has to add an edge to the partial graph, i. e. to « close » a contact elsewhere. The change of a CF to a CO implies necessarily a change CO $\rightarrow$ CF, such that the number of CO remains constant.

This proposition is fundamental to the construction of the algorithm able to find all the configurations of the assembly :

- let m_{CO} be the number of edges valued CO in the part-graph, and m_{CF} the number of edges valued CF ; one searches for all the possible combinations changing m_{CO} edges valued CF in CO, that is $C_{m_{CF}+m_{CO}}^{m_{CO}}$ combinations ;

- one controls each combination to know that it corresponds to a configuration, i. e. that the associated partial graph be connected.

The three following points are noteworthy.

Firstly, one has to control if the partial graph associated with the initial configuration given by the designer is connected, in order to compute the correct number of CO.

Then, that the number of CO remains constant from one configuration to another, does not imply that every combination convenes.

Finally, one has to verify that no geometrical interference occurs with the change CO $\rightarrow$ CF. For the systematical computation of the combinations may move or may remove CO valued edges in groups of parts lying in parallel in cycles of type II. And this would lead to incompatible lengths or to geometrical interferences. It is absolutely necessary to verify in each partial graph associated with a combination, that surfaces linked by a CO valued edge in the part-graph, do not interfer. One has to find the shortest directed path in the partial graph between these two surfaces. The partial graph being a simple graph, there is only one path and the first founded one is correct. The search of the path is made with the "deep first search" algorithm. When having determined this path, intermediar parts and surfaces are known, and the use of the coordinate of each surface permits to compute the oriented distance (in the direction $\vec{d}$). If the distance is positive, there is no interference; if the negative case, a part penetrates in the other one.

With this verification, one eliminates the bad combinations, and one avoids the systematic control of the cycles of type II on the part-graph, as mentioned above.

In the chosen example, the partial graph of the initial configuration is connected (fig. 3), and gives a satisfying combination. One reads, in the table 1, that there are 2 CO et 2 CF. There are $C_{2+2}^2 = 6$ possible combinations, which are given in table 1.

Arête	I	II	III	IV	V	VI
1	CO	CO	CO	CF	CF	CF
2	CO	CF	CF	CO	CO	CF
3	CF	CO	CF	CO	CF	CO
4	CF	CF	CO	CF	CO	CO

Table 2 Possible combinations by « displacement » of the CO for the treated example.

The combination number I does not convain, because there exists a penetration of the part ① in the part ② (this interference is detected because the distance between the surfaces $+1/3$ et $-2/3$, computed with the partial graph, is negative). We reach the same conclusion for the combination VI (the part ③ is floating). Finally, only 4 combinations give satisfying configurations. The partial graphs are given in figure 6, and the admissible configurations different from the figure 1 are represented in the figure 7.

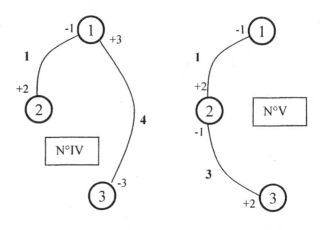

Figure 6 Admissible partial graphs for the treated example

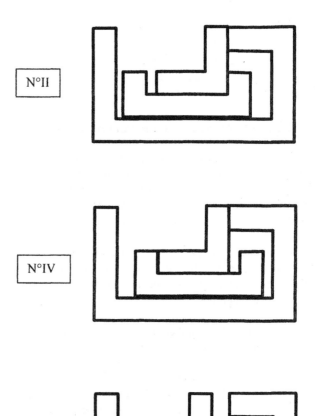

Figure 7 Admissible configurations for the treated example

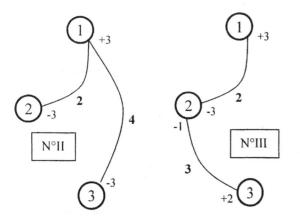

4.2 Simplification in the case of particular cycles of type I

One could notice that there exists a particular cycle of type I in the part-graph (fig. 2), with 1 CO valued edge, and with 1 CF valued one. This corresponds to 2 combinations. In the rest of the part-graph, there are 1 CO valued edge, and 1 CF valued one, i. e. $C_{1+1}^1 = 2$ supplementar combinations. The total number of combinations is therefore 2*2 = 4, which is less than the previous result. In fact, having distinguished these two subgraphs, one expresses explicitly that the first CO valued edge concerns only the cycle of type I, and that the second CO valued edge concerns the rest of the graph. One removes also the two combinations in which the first CO valued edge "migrates" into the rest of the graph, and *vice versa*.

This is the reason of the search for the particular cycles of type I in the part-graph. They can be treated separately from the rest of the graph, and they reduce the number of studied combinations for the rest of the graph.

4.3 Simplifications in the case of cutpoints

It is possible to get a similar reduction when cutpoints exist in the part-graph. A cut point is a vertex that divides the graph in two disconnected subgraphs when being suppressed. This corresponds to a part connecting to « independent » groups of parts.

In this case, the CO valued edges of one subgraph do not act upon the other subgraph. The combinations are constructed and tested for each subgraph separately. The combinational calculcus of the all-combinations is transformed into the product of two lower combinational numbers for each subgraph. It is then easy to compile the configurations obtained for each subgraph, in order to build the configuration relative to the whole assembly.

5 Conclusion

After having studied the syntax and the semantic of the functional constraints in the unidirectional tolerancing, we have pointed out the necessity of finding all the configurations of a mechanical assembly before imposing functional constraints.

Firstly, we propose an efficient mathematical model based on the graph theory, in order to represent a mechanical assembly and its configurations. To the assembly is associated a part-graph, of which the vertices are the parts, and of which the edges give the presence, or the possibility, of contact. The study of the properties of the part-graph permits the validation of the design of the assembly, from the point of view of the positioning of the parts or groups of parts.

Within this model, a configuration of the mechanical assembly is represented by a partial graph issued from the part-graph. Taking into account the properties of this partial graph, one can determine in a systematic way, all the configurations of the assembly.

Finally, we indicate how the presence of particular cycles and cutpoints in the part-graph can accelerate in a efficient manner the search of the all-configurations.

In this work, we have limited ourselves to presenting the general method and to proving its ability to find the all-configurations of an assembly. Whenever it was possible, we insisted on the information contained in this method, which permits a dialog with the designer, in order to validate each step. The related algorithms of the graph theory are standard, and will be detailed later.

The next step of our work deals with the automatic construction and solving of the tolerance chains and with the connection to a general CAD system, whereby may be a general and efficient automatic method for unidirectional tolerancing.

Acknowlegdments

The authors thank very gratefully the ANVAR (National Agency for Research exploitation), which has sponsored this work.

References

[1] R. Södeberg, « CATI : a computer aided tolerancing interface », Advances in Design Automation - ASME 1992, DE-Vol. 44-2, Vol. 2, 1992.

[2] M. N.Mliki, « Contribution à l'étude et à la réalisation d'un système de cotation fonctionnelle assistée par ordinateur », PhD Thesis, Marseille III, 1992.

[3] D.Mennier, M. N. Mliki, « CotAO : intégration de la cotation fonctionnelle assistée par ordinateur dans le processus de conception-fabrication », GSI4, Marseille,1993.

[4] B. Ramani, S. H. Cheraghi, J. M. Twomey, "CAD-based integrated tolerancing system", Int. J. Prod. Res., Vol. 36, N° 10, 1998.

[5] M. GONDRAN, M. MINOUX, Graphes et algorithmes, Edts. Eyrolles, Paris, 1995.

On-line Control Based on Genetic Algorithm for Flexible Manufacturing Systems

Tiberiu Stefan Letia

Department of Automation, Technical University of Cluj-Napoca, Cluj-Napoca, RO 3400

Honoriu Valean, Adina Astilean, Calin Gruita

Department of Automation, Technical University of Cluj-Napoca, Cluj-Napoca, RO 3400

Abstract

The paper presents a method for solving the control problem of a class of flexible manufacturing systems (FMS).

The controlled processes are modeled by: P/T Petri nets and controlled timed Petri nets. The FMS's requirements are described by sequences of timed events. The input strings of events are considered to be asynchronous.

An architecture structured on multiple layers is proposed to implement the control system. The upper level (called supervisor) uses a hybrid genetic algorithm (HGA) for finding the control sequences which lead the non-deterministic discrete event system to a desired behavior.

The Petri nets models are used by HGA for searching the sub-optimal solution of the control sequences. Three types of chromosomes are proposed to adapt the HGA to the different models.

The lower level implements simple controllers which execute the sequences sent by the supervisor. In this way the non-deterministic discrete event process is transformed into a deterministic one. Another proposed way is to transform the system into a synchronous one.

1 Introduction

Manufacturing environments that deal with a large variety of product types produced on general purpose machine tools, stoves, tanks, conveyors and robots (capable of performing a variety of operations and integrating by an automatic transport facility for transferring parts) are considered in this paper.

The FMS capable to process many types of parts require rapid job switching to reduce idle time. The extension of flexibility leads to large variations of task execution times and non-deterministic arrival times for parts and subassemblies.

Planning and scheduling are used for projects consisting of several tasks with timing constrains associated with them. The control of FMS needs to find sequences of events which correspond to the imposed behaviors taking into account the real-time requirements.

Cooling [3] categorizes the computer systems in three general groups:

Batch – in case we don't mind when the computer results arrive;

On-line – when we would like the results within a fairly short time (typically a few seconds);

Real-time – when we need the results within a definite short time scale (typically milliseconds to seconds).

In the case of flexible manufacturing systems a mixture between on-line system and real-time system is met. Usually the planning (achieved by the upper level) is implemented in an on-line manner, meanwhile the scheduling fulfilled by the lower level meets the real-time constraints.

Generally, manufacturing scheduling problems are NP-hard. i.e. no algorithm exists that can find optimal solutions to these problem in polynomial time [9]. Heuristic methods are used for solving this kind of problems, but typically they become intractable (i.e. take more than polynomial time) when additional constraints are added or the problem size grows.

In the last few years, the focus has been shifted to a more realistic scenario of decision support system [9]. It uses genetic algorithms and neural nets to help managers in decision making.

There are many methods to be used for the specification of the flexible manufacturing systems and corresponding control algorithms. Among them, different categories of Petri nets (condition-event, place-transition, stochastic, timed, high-level, controlled timed Petri nets) are used for different applications which require the modeling, analysis, synthesis, control of the FMS [4].

Planning involves selecting among alternative routes to meet the required characteristics of the final product.

Scheduling involves the setting of start/finish times for the individual processing tasks at the subordinate entities of the controller.

Execution verifies the physical preconditions for scheduled tasks and carries out the dialogue with the subordinate controllers required to physically perform the tasks.

The main concern is that, since the production requirements change frequently, the plans and schedules must also be changed. The estimated routes, orders of the activities and the processing times are found out by the computing of a next state achieved by the controlled system after a pre-calculated moment of time.

2 The architecture of the on-line control system

The architecture of the on-line control system is presented in figure 1, where
- PSK signifies the Problem Specific Knowledge;
- CTPN is the Controlled Timed Petri Net and
- HGA is the acronym of the Hybrid Genetic Algorithm.

resource (M_1, M_2,..., M_n) or a conveyor (C_0, C_1). The parts are received from the C_0 conveyor. A sensor signals the arrival of a part on the conveyor in a position where the robot is able to pick it up. The processed parts are delivered to the conveyor C_1.

Each part has attached a set of *technical attributes* which defines the required manufacturing or processing operations and are used to create the technology for the part.

3 The models of the controlled system

The controlled system usually has non-deterministic

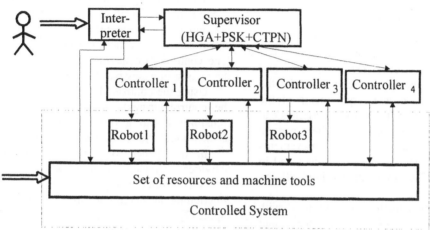

Figure 1 The architecture of the FMS

The considered three levels control architecture allows the reducing of complexity limits responsibility and distributes the functions. *The Interpreter module* is awaked when a new part (or container) arrives, or is signalled and has to be manufactured. The interpreter module takes the sequence of the requested activities (implied by technology of the part or container which is memorised into a technology library) and verifies if the hardware resources composing the FMS can fulfil it. If all the activities can be achieved in the FMS, the interpreter asks the supervisor to plan the new coming part into the system; otherwise it rejects the request.

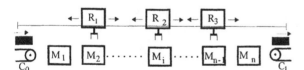

Figure 2 The structure of a part of the controlled system

Figure 2 presents a part of a FMS (proposed for testing the control system) composed of three robots (R_1, R_2, R_3) which transport the parts through various locations (tanks, stoves, machine tools, conveyors etc.) throughout the section, taking into account the specified requirements. The robots are able to glide to the left or right, according to the commands and can pick up or put down a part or a container with a set of parts. The sensors signal the positioning of the robots in front of a

characteristics. They can be seen in the multiple possible choices between activities which can be executed or the delays involved by the execution of some activities. This non-determinism is modelled by multiple transitions which can be executed from a state (i.e. marking) or the variable delays of transitions.

The existence of the multiple choices makes possible the flexibility implementations, but it requires the control system to transform the entire system into a deterministic one.

The non-deterministic delays characterise an asynchronous behaviour which makes harder the solving of the control problem, especially when multiple simultaneous sequences of events involving the evolution of the FMS are possible. The solution of the control problem involves the transformation of the asynchronous system into a synchronous one by adding controllable transitions. In this way the asynchronous events become synchronous (at least partially).

The requirements of the behavior of the FMS could contain only sequences of activities (transitions) or in the general case, the minimum and maximum delays between them could be added. In the first case when only one sequence of events or transitions is possible, the asynchronous approach (i.e. the event driven approach) can be used. That leads to obtain the maximum speed (i.e. the maximum system throughput). The last case implies the synchronous approach (i.e. the time driven system). That could lead to a loss of the system throughput.

4   Specification of the system required behaviour

The parts or the containers with parts have to be processed as technologies require. That means the system has to perform some specified activities. A simple method is proposed to describe the manufacturing requirements of the parts (product or containers). As example, a part technology can be described as follows:

$$r_k = a_0 \cdot a_2[t^2_1, t^2_u]) \cdot (a_5[t^5_1, t^5_u] + a_7[t^7_1, t^7_u]) \cdot \ldots \cdot$$
$$\cdot a_{(n-2)}[t^{n-1}_1, t^{n-1}_u] \cdot a_n.$$

where a_i represents the asked activity and $[t^i_1, t^i_u]$ the limits of the necessary and accepted duration. The symbol "•" represents the concatenation operator and "+" corresponds to an alternative. Sequences like the above one describe the real-time requirements of the FMS behaviour for a period of time. Some activities can be made on a particular machine, but others can be performed on more machines. The movements (performed by robots or conveyors) are not included in the required sequences. They are computed implicitly by supervisor.

Usually a specification method is introduced with the aim of the *verification* of the correctness and completeness as well as system *validation* which consists in controlling the condition of liveness (e.g., the absence of deadlocks) and the meeting of timing constraints (e.g., deadline, timeout, etc.) [2]. Using a collection of specification methods, like that proposed in the paper, we get the advantage to take the most suited

description for every aspect of the system, but also the disadvantage of losing the power of the methods for verification and validation. As a matter of fact, some specification tools propose the simulation of the system for verification and validation. A similar way is used in the present paper, connecting the verification and validation with control sequences synthesis.

In the FMS control problem, it is necessary to find sequences of events which lead the system to requested behaviour and to demonstrate that it meets the real-time requirements. Some efficiency aspects of the system behaviour have to be evaluated and taken into account too.

The searching of a control solution (sequences of events), the verification and validation of the system behaviour is implemented in the software component called supervisor which is executed when the predicted (and controlled) behaviour has to be changed following the request of a new part to be processed.

Using the controlled timed Petri nets (together with the control strings), the non deterministic system is transformed into a deterministic one. This way, the asynchronous system becomes a synchronous one.

In this paper, only the deterministic delays of activities (or transitions) are considered.

A *controlled timed Petri net* is a timed Petri net functioning at maximum speed with external control inputs [7].

A *transition is state enabled* (from the point of view of the current marking) if and only if each of its input places are marked with at least one available token or generally at least with the number of tokens specified by the incidence matrix.

A *controllable transition is executable* for a given

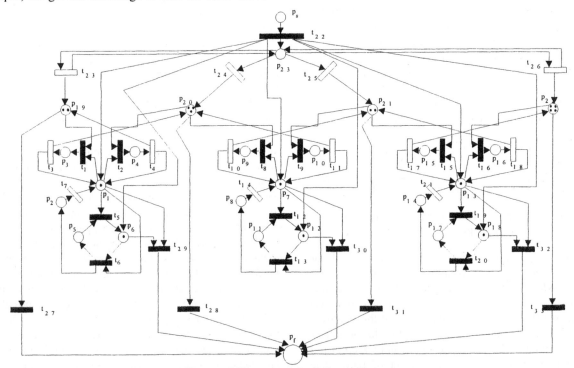

Figure 3 The model of the actions

marking if it is both state enabled and control enabled from the point of view of the control vectors.

Controllable transitions are used to impose the system a deterministic behaviour. A controllable transition may be disabled by the supervisor which ensures that the behaviour of the system is within a legal behaviour. An uncontrollable transition corresponds to an event which may not be prevented from occurring by the supervisor [7]. Using the controllable transitions the problem of selection of the alternatives is solved thus the confusions are removed. In this way, the non deterministic system is transformed into one with a predicted behaviour.

For the mentioned testing part of the FMS, the model for the actions of the three robots allows that the possible executions in the three subsystems occur concurrently, each of the subsystems being separately expressed as a controlled timed Petri sub-net. In figure 3 the model of the actions accomplished by robots is presented. The significance of places and transitions (for the robot R_2) is the following: t_8 – start the left shifting; t_{10} – finishes left shifting; t_9 - start the right shifting; t_{11} – right shifting is finished; t_{12} - start the unloading of the part (or container); t_{14} - finishes the loading (unloading) of the part (or container); t_{13} - start the loading of the part (or container); $t_{22}, t_{24}, t_{25}, t_{28}, t_{30}, t_{31}$ - link transitions; p_s - start place; p_f - final place; p_7 - the robot waits the commands; p_9 - the robot accomplishes a left shifting; p_{10} - the robot accomplishes a right shifting; p_8 - the robot loads (unloads) a part (or container); p_{20}, p_{21} – the places conditioning the moving of the robot to the right or to the left; p_{12} - the robot is unloaded; p_{11} - the robot is loaded.

The Petri net does not contain any information about the presence of the parts on the resources. However, this information exists in PSK.

The controllable transitions (dark filled) are allowed or not to be executed using a string of control vectors which specifies the permissions and the corresponding delays.

5 The supervisor

The supervisor has the main goal to find out the sequences which must be performed by the different components of the FMS and to send the controller the control sequences (control vectors) which fulfil the desired behaviour of the system. The HGA are proposed to be used for control sequences synthesis.

As Murata et al. [8], the use of genetic algorithms with two objectives is proposed for searching the planning solution. One of the objectives concerns the R-T constrains fulfilment, while the other corresponds to the efficiency of the system. The efficiency of the control system is expressed by an algebraic formula containing duration of the finishing of the parts requested to be processed, the costs of some activities and also some penalties for the cases when a part doesn't cover all the required sequences of activities or some other non required activities are made. The R-T

objective contains penalties of non meeting of the R-T requirements.

The supervisor tries to plan (or to add to the previous plan) the new asked sequence of activities using the following algorithm:
1. The supervisor waits for being signalled by interpreter module. It gets the new sequence of activities to be planned from interpreter module.
2. The supervisor anticipates (according to the current plan) the controlled system state after T time units (t. u.) necessary for the new planning activity.
3. The supervisor uses the HGA to get the new planning for the time horizon during T t. u.
4. If it fails to get a feasible planning from the point of view of R-T constraints (during T t. u.), the supervisor repeats the activities starting from step 2; else it sends to controllers the new permission strings.

If the supervisor fails to the first attempt to get the new system trajectory, the introducing into the system of the new part is delayed by another T t. u. The cause of the fail can be the overloading of the controlled system (there are too many parts or containers to be manufactured) or just the low convergence of the algorithm.

During the time the supervisor searches for a new plan, the controllers lead the discrete event processes on the previously predicted trajectories and lead the robots or command the tools to fulfil the supervisor planning. That means they get the control strings (i.e. the sequences of transitions to be executed) from the supervisor.

Three types of chromosomes have been experimented for solving the planning problem as they are represented in figure 4. The string (figure 4.a) is suited for a deterministic model and a system without timing requirements. In this case, the chromosome provides the order of execution for the controllable transitions. The chromosome in figure 4.b is used for an event driven implementation, meanwhile the matrix from figure 4.c is more appropriate for the non-determinism in the transition delays and with hard real-time requirements (upper and lower limits). The results of using the first type of chromosome are given in [1].

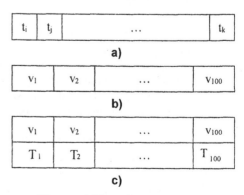

t_i	t_j	...	t_k

a)

v_1	v_2	...	v_{100}

b)

v_1	v_2	...	v_{100}
T_1	T_2	...	T_{100}

c)

Figure 4 The chromosomes

v_1, v_2, ..., v_{100} are the genes and they correspond to the control vectors which are used to solve the choice non-determinism. They are natural numbers in the domain {1,2, ..., maximum number of controllable transitions}. The controller must convert the natural number coded control vector into a binary one, before applying it to the controlled system in equal periods of time.

In the matrix coded chromosome the elements T_1, T_2,..., T_{100} correspond to transition delays from the moment the controllable transitions are enabled (from the point of view of the markings). Whenever a controllable transition is enabled, a control vector is used to solve the problem. The last type of chromosome involves the application of every control vector to the duration of time given by the time associated to it (i.e. T_i).

6   The lower level control algorithm

Three different ways for implementing the controllers are used.

In the first case (when the individual corresponds to the structure of figure 4.a) the controllable transitions are delayed according to the order given by the chromosome.

The other two ways are using the control vectors (figures 4.b and 4.c) allowing or not the execution of the controllable transitions.

The controllers detect the points when the controlled transitions are enabled from the point of view of the marking. They take one by one the control vectors from the strings sent by supervisor and analysing the enabled transitions, permit or not the executions of the controllable transitions.

Therefore, there are two different ways of using the control vectors. One way is time-driven when the controllers are synchronised with the time and another one is event driven when they are synchronised by the enabling events of the controllable transitions. The decisions of using one or another of the methods is determined by the variations of the duration of the activities performed by the components of the FMS.

7   The simulation results

The static priority scheduling was chosen for real-time scheduling of software tasks [4].

For simulation, populations of 100 chromosomes were used. Each chromosome contains 100 genes, corresponding to the control vectors. The tests consist in 5 trials of 500 generations, using the genetic operators as presented in figure 5.

The maximum fitness value is 25000. This value is decreased by a penalty function, penalizing the chromosomes which are not respecting the requirements.

The simulation results (fitness evolution) are depicted in figure 6a, b, c, d and e.

Figure 7 presents the average computing time per generation of 100 chromosomes for the 5 trials.

8   Conclusions

A mixture of specification methods can be better used to describe different parts of a complex system structure or behaviour. Intelligent methods can be used for system behaviour verification and validation replacing partially the human laborious activities.

As it is known, an asynchronous discrete event system can be transformed into a synchronous one in some circumstances. This permits to find out the control sequences as well as the R-T verification of the proposed scheduling method.

The relation between R-T scheduling and system throughput can be easily seen. The scheduling method influences the manufacturing throughput. Also, the time requested for planning can decrease the system throughput. In its turn, the time horizon for searching a plan influences the planning duration. The time horizon length is determined by the length of the sequences of activities. The planning duration depends on:

- the processor used for implementation
- the implementation of the HGA (a parallel implementation can be used for decreasing the execution duration)
- the method used by HGA for coding, evaluation of fitness and so on

References

[1] A. Astilean, C. Gruita, T. Letia, "Control Sequences Synthesis Method for Deterministic Discrete Event Systems", Proc. of 12'th International Conference on Control Systems and Computer Science, Bucharest, Romania, vol. 1, pp. 346-351, 1999.

[2] G. Bucci, M. Campanai and P. Nesi, "Tools for Specifying Real-Time Systems", In Real-Time Systems, 8, Kluwer Academic Publishers, Boston, pp. 117-172, 1995.

[3] J. E. Cooling, "Software Design for Real-Time Systems", Thompson Computer Press, 1991.

[4] F. DiCesare, G. Harhalakis, J. M. Proth, M. Silva, F.B. Vernadat, "Practice of Petri Nets in Manufacturing", Chapman & Hall, 1993.

[5] C.J. Fidge, "Real-Time Schedulability Test for Preemptive Multitasking", In Real-Time Systems, 14, Kluwer Academic Publishers, Boston, pp. 61-93, 1998.

[6] A. Giua, F. DiCesare, M. Silva, "Supervisors for Generalised Mutual Exclusion Constraints", In Proc. of The 12'th World Congress IFAC, Sydney, vol. 1, pp. 267-270, 1993.

[7] J. Long, B. Descotes-Genon, M., "Control Synthesis of Flexible Manufacturing Systems Modelled by a Class of Controlled Timed Petri Nets", In Proc. of The 12'th World Congress IFAC, Sydney, vol. 1, pp. 245-248, 1993.

[8] T. Murata, H. Ishibuchi and K.H. Lee, "Application of Two-Objective Genetic Algorithm to ShopFloor Scheduling Problems with Interval Processing Time", Proc. the EUFIT'96, Ed. Elite Foundation, Achen, Germany, pp. 443-447, 1996.

[9] J.S. Smith, W. Hoberecht, S. B. Joshi, "A Shop Floor Control Architecture for Computer Integrated Manufacturing", DDM 9009270, DDM-9158042.

Trial No. (Figure No.)	Mutation %	1 Point Crossover %	2 Points Crossover %	Pattern Crossover %	Best fitness value
1	70	10	10	10	18150
2	10	70	10	10	15980
3	10	10	70	10	16230
4	10	10	10	70	16750
5	25	25	25	25	20120

Figure 5 Test data

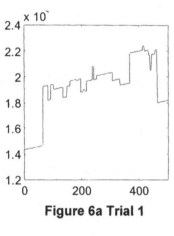

Figure 6a Trial 1

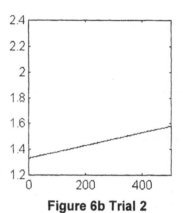

Figure 6b Trial 2

Figure 6c Trial 3

Figure 6d Trial 4

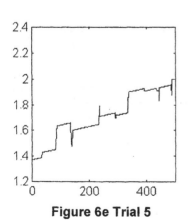

Figure 6e Trial 5

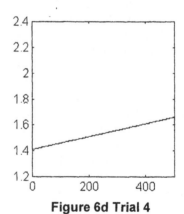

Figure 7 Average elapsed time / trial

Evaluation of product development cycle cost and simultaneous engineering

Michel Aldanondo, Paul Gaborit

D.R.G.I - Ecole des Mines d'Albi Carmaux - Route de Teillet - 81013 Albi CT Cedex 09 - France

Philippe Clermont

L.C.P.S.I - Ecole Nationale d'Ingénieurs de Tarbes - Avenue d'Azereix - 65 016 Tarbes Cedex - France

Abstract

Our communication deals with simultaneous engineering evaluation with the help of discrete simulation. Our purpose is to show how the degree of activity overlapping influences the total cost of the product development cycle when facing disturbances.

1 Introduction

Among all the methods and approaches set up in the field of Concurrent Engineering in order to improve company competitiveness, some of the most frequently used are Multi-disciplinary Team (MT), Design For Manufacturing (DFM) and Total Quality Management (TQM) as explained in Lawson survey [1]. These three elements allow to achieve Simultaneous Engineering (SE) which targets to overlap engineering activities in order to reduce the product development cycle (PDcycle).

Without perturbation, it is clear that SE allows reducing the PDcycle while keeping cost approximately the same. This is not so clear when disturbances or perturbations affect the PDcycle. In fact, a PDcycle activity can become not feasible and the result of a previous already completed activity might need to be reconsidered. This lead to a kind of iterative behavior of the PDcycle and can provide cost increases.

As perturbation occurrence is frequent during the PDcycle, we propose to study how overlapping affects the PDcycle cost when perturbations are present.

This communication is divides in two main sections. The first one deals with modeling of PDcycle and perturbation reactions. These elements are set up in an event driven simulator, and the second section presents some simulation results allowing us to derive some conclusions about overlapping interest versus PDcycle cost.

We must warn the reader that we are going to propose a model of the PDcycle and conduct simulation on this model. We want to identify the first main tendencies of overlapping interest and our conclusions will be directly linked with our model assumptions.

2 Product Development cycle modeling

2.1 Modeling elements

2.1.1 – Activities, steps and overlapping

We consider that the PDcycle can be represented by a set of successive activities (Ai). We aim to study the activity overlapping effect on PDcycle cost. Therefore we define the overlapping between to successive activities (Ai, Ai+1) as : beginning date (Ai+1) - ending date (Ai).

In order to take into account overlapping in discrete simulation, we divide each activity in successive steps (Si,j) characterized by a cost Ci,j and a duration Di,j. We associate overlapping with the duration sum of a given set of the last steps of Ai. Therefore, overlapping can be modeled with an anteriority constraint between the end of one of the step Si,j of activity Ai and the beginning of the first step Si+1,1 of activity Ai+1 as shown in figure 1. In this representation, each constraint represents an information flow between the two activities.

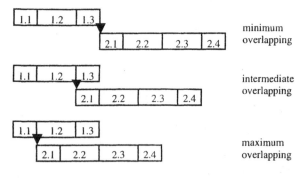

Figure 1 Overlapping and constraints

In the example of figure 1 when overlapping is maximum, it is obvious that other information flows exist between the step couples S1,2-S2,2 and S1,3-S2,3. In that case, if the duration's of the various steps are not equal, discontinuities ("holes" in the planning) can appear during the progress of the downstream activity (A2). In order to avoid this kind of synchronization problem, we assume that all the step durations are equal as represented in figure 2. This hypothesis is valid if we consider that the decomposition activity/step is made according to synchronized project reviews.

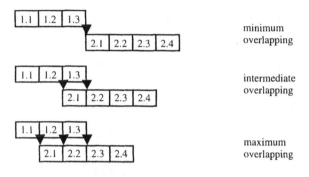

Figure 2 Synchronized activities

Therefore our model considers activities composed of steps of equal duration. An overlapping level corresponds with a set of anteriority constraints and a can be quantified by an overlapping degree (Dvrlp) equal to the number of steps of the downstream activity Ai+1 that can be done before the end of the upstream activity Ai. When the overlapping degree is modified, the way people used to work and the information flows are modified. We will not go further in overlapping modeling and advice the reader to consult the work of Krishnan [2].

2.1.2 – Perturbations

Perturbation occurrence is not modeled, because we are just interested in the perturbation consequences in terms of unfeasible steps. Thus we characterized each step by a probability of success (Psuc(Si,j)).

2.1.2 – Perturbation reaction

When a step is declared unfeasible (Sfail), we consider that (i) it exists, upstream in the PDcycle, an already completed step which can bring a solution to this perturbation, (ii) if this step is achieved again, the previous unfeasibility might disappear. This "solution" step (Ssol) can not be initially defined, because on one hand it depends of the overlapping degree and on the other, it is relevant to the PDcycle and to the activity

unfeasibility. Therefore a statistical rule provides for each step located upstream of step Sfail its probability to be step Ssol. This matches two different cases :

- Case 1. During the PDcycle, a wrong decision has been taken by step Ssol, and this decision is the cause of the failure of Sfail. With information provided by step Sfail to step Ssol, people in charge of step Ssol can take the "good" decision and prevent the unfeasibility of step Sfail. This is in fact an internal perturbation.

- Case 2. All the decisions taken during the steps located upstream of step Sfail are correct, thus the perturbation is external (market shift, customer demand modification...). It is necessary to identify an upstream step Ssol able to : take into account the perturbation, reconsider its previous decision and provide elements allowing Sfail to be completed smoothly. Elements concerning this case have been already reported by Aldanondo in [3].

We choose, for the statistical rule, identifying step Ssol, to take into account a degree of implicit validation. Implicit validation is based for a given step on the number of downstream steps successfully achieved. As an example for a PDcycle composed of 4 successive steps, if the three first steps (1,2 and 3) are successful and step 4 is unfeasible; we define full implicit validation as :
- prob(Step3=Ssol) = 3times Prob(step1=Ssol),
- prob(step3=Ssol) = 2times Prob(step2=Ssol).
Full implicit validation statistical rule would compute Prob(step1=Ssol) = 2/11, prob(step2=Ssol) = 3/11 and prob(step3=Ssol) = 6/11. A coefficient of implicit validation allows to modulate linearly between full implicit validation and equally probable distribution where Prob(step1=Ssol) = Prob(step2=Ssol) = Prob(step3=Ssol) =1/3.

Once the step Ssol is processed again, it is necessary to reconsider, totally or partially, all the already processed steps located downstream of step Ssol. For this, we decide to include a learning effect. This mean that each time a step is processed again, its duration and its cost are decreased with a learning coefficient. For example a learning coefficient of 0.5 used for a step with duration equal to 1 would provide :
- first time : duration = 1,
- second time : duration = 0.5^1,
- third time : duration = $0.5^2 = 0.25$ and so on.

2.2 Previous works and simulator set up

In order to set our modeling elements in an event driven simulator, we have studied existing modeling approaches and simulation solutions. Two kinds of contributions can be found. The first kind is base on a functional approach as described by Kusiak [4] eventually improved with some behavioral elements as those described by Ramat [5] or Elmaghraby [6]. The second kind is based on behavior model as Petri Nets as the work of Marier [7]. Recently, first elements concerning a work targeting the same goals as us have been reported in [8].

The closest approach to our needs is the work of Ramat who proposed a model called RAIH. This model is a functional breakdown structure with probability of success affected to each step. We take this model as a basis and add some elements to match our needs and set up all this in a specific simulator. Aiming first general tendencies, we have limited the simulation parameters as follow :
- cost and duration are equal to one for any step belonging to any activity,
- the number of steps per activity is the same for all activities,
- the probability of success for any step belonging to any activity is the same.

2.3   Cases and indicators

Different PDcycle structures have been investigated. We mean by structure the number of activities and the number of steps per activity. Cases go from three activities composed of three steps (3x3) up to six activities composed of six steps (6x6) including intermediate asymmetric decompositions. For each structure all the possible overlapping degrees are simulated.

The simulator computes two indicators the total PDcycle time, Ctm (Dvrlp), and the PDcycle cost, Ccs (Dvrlp). As we wish to compare various cases of problem we need an indicator which is independent of the PDcycle parameters (number of activities and number of steps per activity). In this purpose, we calculate a ratio that we call the PDcycle cost gap, Gcs(Dvrlp). It quantifies a percentage gap between the cost provided by a given overlapping degree and the cost provided with maximum overlapping. It is defined by :

$$Gcs(Dvrlp) = 100x[Ccs(Dvrlp)–Ccs(max)] / Ccs(max)$$

Thus, a positive gap means that the PDcycle with a given overlapping degree is more expensive than the one with maximum overlapping.

3   Experiments and results

3.1 – Overlapping and probability of success

For this first experiment, we have considered the following case :

- structure 4x4 or 4 activities of 4 steps,
- no learning effect and no implicit validation,
and the variations of :
- probability of success : 0.9 0.925 0.95 and 0.975,
- overlapping degree : 0 1 2 and 3.

With this data, we are going to study PDcycle time and PDcycle cost gap. Each curve characterized a given probability of success.

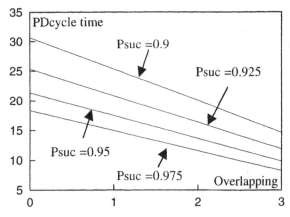

Figure 3.1 PDcycle time

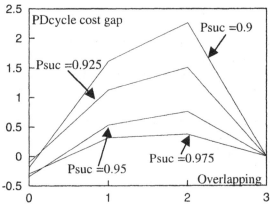

Figure 3.2 PDcycle cost gap

The results of figure 3 show the PDcycle time (figure 3.1) and the PD cost gap (figure 3.2).

The interest of overlapping is clearly shown in term of cycle time but it is not the same thing for the cost. In that case, null or maximum overlapping have a same cost order of magnitude, but intermediate overlapping are always more expensive than extreme overlapping situations whatever the probability of success is.

When the probability of success decreases the cost gap decreases. This is normal because the case tends towards a PDcycle with no disturbance.

The following experiment will try to show if extreme overlapping always produce the same cost order of magnitude for other structures.

3.2 – Structures and probability of success

In that case, we are going to study only the PDcycle cost gap between maximum and minimum overlapping :

$$Gcs(min) = 100x[Ccs(min)-Ccs(max)]/Ccs(max)$$

The experiment is characterized by :
- no learning effect and no implicit validation,
and the variations of :
- probability of success : 0.9 0.925 0.95 and 0.975,
- case structure : 3x3 4x4 5x5 and 6x6.

Each curve represents the cost gap evolution of a given structure according to the probability of success.

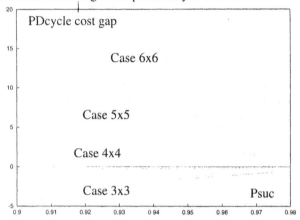

Figure 4 Cost gap (max/min) and structures

The results of figure 4 show that when the number of activities and the number of steps are increasing,

maximum overlapping becomes interesting in term of cost whatever the probability of success is.

The structure 4x4 is a kind of neutral case (as seen in the previous experiment), while null overlapping is always cheaper for 3x3 structure and always more expensive with 5x5 and 6x6 structures.

The interest of maximum overlapping when structure becomes large must be noticed (5% to 16% for 6x6 case).

The following experiment will therefore study for a given probability of success all structures with all possible overlapping.

3.3 – Structures and overlapping

As the previous experiment, we are interested in cost gap analysis for various overlapping degrees :

$$Gcs(Dvrlp) = 100x[Ccs(Dvrlp)-Ccs(max)]/Ccs(max)$$

The experiment is a group of three experiments characterized by :
- probability of success equal to 0.95.
- case structure : 3x3 4x4 5x5 6x6 and all intermediate structures,
- overlapping : all possible degrees.

3.3.1 – Basic experiment

This first one does not take into account learning effect and implicit validation. Each curve of figure 5 represents the cost gap evolution of a given structure according to the overlapping degree.

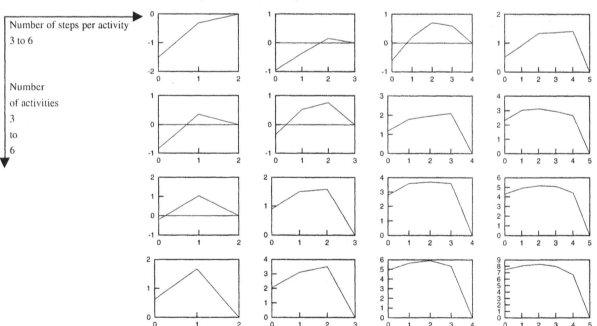

Figure 5 - Cost gap, structures and overlapping

Each line has the same number of activities and each column has the same number of steps. This explains that overlapping possibilities change according to the column number. Upper left is 3x3 structure and lower right is the 6x6 structure. This set of curves allows deriving the following conclusions :

(1) When the number of activities and the number of steps per activity increase, maximum overlapping provides the cheapest PDcycle cost.

(2) For the largest structures (5x5 - 6x5 - 5x6 and 6x6) the cost of intermediate and null overlapping are of the same order of magnitude.

(3) For the smallest structures (3x3 - 3x4 and 4x3) null overlapping is always cheaper than maximum overlapping.

(4) If we consider now PDcycles with approximately the same number of steps (nbstp), we can identify seven groups :
- group 1 : 3x3 : nbstp = 9
- group 2 : 4x3 - 3x4 : nbstp =12
- group 3 : 5x3 - 4x4 - 3x5 : 15 ≤ nbstp ≤ 16
- group 4 : 6x3 - 5x4 - 4x5 - 3x6: 18≤ nbstp ≤ 20
- group 5 : 6x4 - 5x5 - 4x6 : 24 ≤ nbstp ≤ 25
- group 6 : 6x5 - 5x6 : nbstp = 30

- group 7 : 6x6 : nbstp =36

Inside each group we can note that the shape of the curves and the maximum PDcycle cost gap are very close. Therefore, we tend toward the conclusion that the number of activities and the number of steps per activity are not the key parameters for our study. The important factor is in fact the total number of steps of the PDcycle.

This set of conclusions is based on a simple model of the PDcycle. Therefore, in the next two experiments, we are going to study if these conclusions remain valid when taking into account learning and implicit validation.

3.3.2 – Learning effect influences

This second experiment, figure 6, takes into account learning effect with two settings of the learning coefficient : 0.75 and 0.5. For each structure three curves are presented : the basic experiment (solid line), learning coefficient 0.75 (dotted line) and learning coefficient 0.5 (point line).

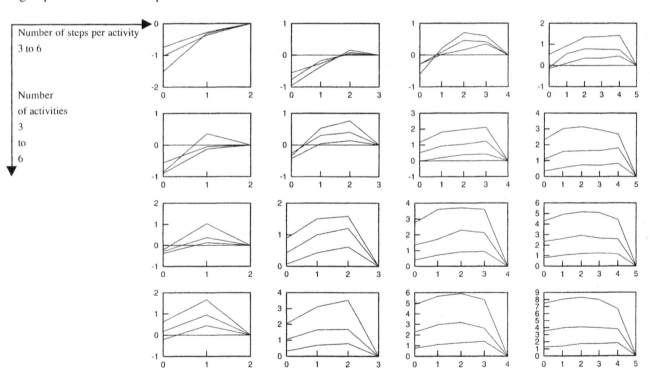

Figure 6 Learning effect on cost gap

The conclusions of the basic experiment remain all valid. The only observed effect is that learning effect acts like a damper. All the values of the PDcycle cost gap are attracted towards the null value (positive values tend to decrease and negative values tend to increase). This behavior is normal because, as cost decreases when rework, the effect of perturbations gets smaller and the cost gap tends to the null value.

3.3.3 – Implicit validation influences

This second experiment takes into account implicit validation with two settings of implicit validation : 0.5 or medium between full implicit validation and no implicit validation and 1 or full implicit validation. As before, figure 7 presents three curves: the basic experiment (solid line), medium implicit validation 0.5 (dotted line) and full implicit validation 1 (point line).

The conclusions of the basic experiment remain globally valid, but the effects of implicit validation are more complex. Implicit validation does not always act like a damper.

For large structures, group 5, 6 and 7 the effect is like a damper with a stronger influence on low overlapping.

For structure of other groups and for some overlapping, the PDcycle cost gap tends to increase with implicit validation (3x4 - 3x5 - 4x4 for intermediate overlapping) and for other cases it tends to decrease (5x3 - 6x3 -5x4 for minimum overlapping). It can be also pointed out that for cases 3x4 - 3x5 and 4x4: medium implicit validation (dotted line) increase the PDcycle cost gap then with full implicit validation (point line) the gap decreases.

Finally, the shapes of the curves for each group when the total number of steps of the PDcycle gets smaller (groups 1, 2, 3 and 4) are not the same. Therefore implicit validation effect is function of both the number of activities and the number of steps per activity.

We do not have at the present time any element allowing us to explain this kind of bumpy behavior. Other experiments are needed to analyze more accurately this effect.

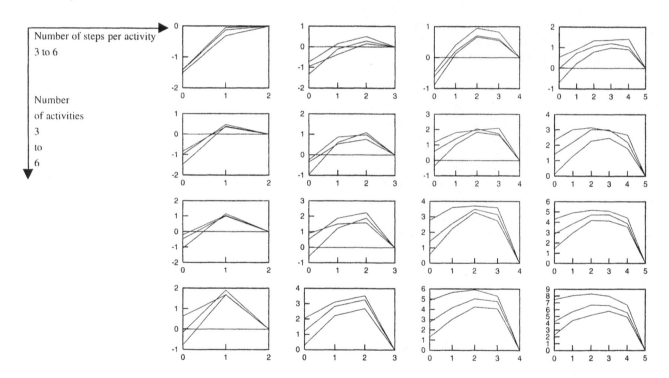

Figure 7 Implicit validation effect on cost gap

4   Conclusions

From these experimental results and other explained in the Ph.D. thesis of Ph.Clermont [9] it is possible to derive the following set of conclusions :

Simultaneous Engineering set up always provides cycle time decreases but does not provide systematically cost improvements.

Intermediate overlapping values (between minimum and maximum) are generally not interesting in term of cost.

When the number of activities and the number of steps increase, the interest in term of cost of maximum overlapping is growing.

The number of activities and the number of steps per activity are not the key factors for overlapping interest when facing disturbances. It is the total number of steps of the PDcycle, which is the key factor.

Learning effect and implicit validation does not modify the general ideas of the previous conclusions. Learning effect just moderates the PDcycle cost gap in absolute value. Therefore, when taking learning into account, the cost differences are smaller. The effect of implicit validation is more delicate to predict.

Of course, we have been reasoning on a model of the PDcycle which is far from reality and did not take into account the cost benefit of putting a product earlier on the market. We did not also consider that, when overlapping is increasing, the costs of the upstream steps are increasing and the probabilities of the downstream steps are increasing. But we were looking for the first tendencies of cost evolution versus overlapping when facing disturbances and we think that the provided data can be considered as a good basis.

References

[1]   Lawson M. and Karandikar H.M., "A survey of Concurrent Engineering", Concurrent Engineering Research and Applications, Vol 2 pp 01-06, 1994.

[2]   Krishnan V., Eppinger S.D. and Whitney, "Accelerating Product Development by Echange of Preliminary Product Design Information", journal of Mechanical Design, Vol 117, pp 491-497, 1995.

[3]   Aldanondo. M and Clermont. Ph, "Cost and Cycle Time reduction in C.E. Local Reactivity : concepts and case study", Proceedings of Concurrent Engineering 1997 pp 245-251. Detroit USA..

[4]   Kusiak A. and Wang J., "Qualitative analisys of the design process", Proceedings of the Intelligent Concurrent Design: Fundamentals, Methodology, Modelling and Practice, ASME, Vol. 66, pp. 21-32, 1993.

[5]   Ramat E., Lenté C. et Tacquard C., "Incertitude et projets d'innovation, le modèle RAIH", European Journal of Automation, RAIRO-APII-JESA, Vol. 31, n°4, pp. 615-643, 1997.

[6]   Elmaghraby SE - "Activity Network : project planning and control by network models", Wiley New-York 1977.

[7]   Marier S. - "Modélisation et évaluation des performances des processus industriels semi-structurés" PhD thesis Institut National Polytechnique de Grenoble - December 1996

[8]   Nadia Bhuiyan and Vince Thomson, "Simulation of concurrent engineering process", Proceedings of IEPM conference vol n°2 pp 221-221, Glasgow Scotland July 1999.

[9]   Clermont Ph., "Apport de réactivité dans le cycle de développement du produit. Formalisation d'une démarche", PhD thesis University of Bordeaux I, January 1998

The co-operative mechanical design and the following of the work-piece geometry through the manufacturing process, a software application in turning.

V. Wolff

CASM, INSA Lyon, Villeurbanne, FR 69100

J. F. Rigal

CASM, INSA Lyon, Villeurbanne, FR 69100

Abstract

Our paper is a contribution to research in Concurrent Engineering applied to cooperative mechanical design. After a bibliographical analysis, the concept of " points of view " and links between the various " know how " are adopted as bases for working on simultaneous engineering in product design and manufacturing. The manufacturing process planning is a central element of the relation between the design and the manufacturing engineering, and more, tolerancing is a central activity. Our contribution relates to the research for tolerancing models which are able to follow the work-piece geometry through the stages of a manufacturing process. A model is defined to qualify the turning process. A geometry of a test work-piece and an experimental procedure are defined. A software development of this approach has been carried out. The results of simulation are compared with the results obtained on a real lathe. The majority of the dimensions of the machined parts are in the interval of tolerance calculated by simulation.

1 Introduction

Concurrent Engineering applied to cooperative mechanical design is now in an industrial development phase [SOH92]. The current developments gathered under the term of Co-operating Design, cover primarily work relating to the technical data management [TOL92]. They are now organized around models based on the concept of features. [ARI 92]

A bibliographical analysis allows us to point to a small number of notions. The concept of " points of view " [TICH 97] seams to be a convenient idea in the cooperative design method. It is based on the observation of the mechanical design process. Every one, during the procedure, don't need to have the knowledge of all data.

For example, if the term of " hole " is used, each actor of the CAD-CAM process has a different meaning of this feature. (Fig. 1)

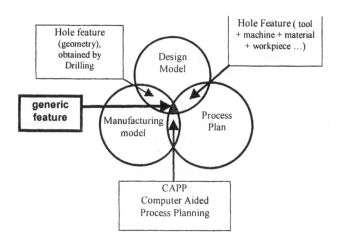

Figure 1   Different definitions of the hole feature

A second aspect is the links between the various " know how" (based on each speciality).

These two concepts are adopted as bases for working on simultaneous engineering in product design and manufacturing.

The following conclusions are highlighted and retained :
- the data-processing system of a product life cycle must be flexible to integrate the innovations such as the new knowledge or the new processes (Fig. 2)

- for efficiency, the concurrent engineering the system must integrate a maximum of information on the physical reality of the stages of the product life cycle.

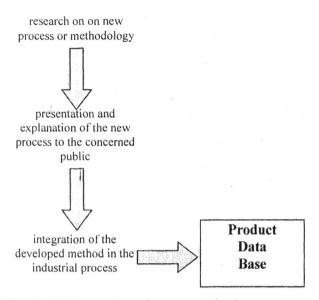

Figure 2   Integration of a new method or process in an industrial process.

2   Manufacturing Process Planning and Tolerancing

The manufacturing process planning is a central element of the relation between the design and the manufacturing engineering, and especially, tolerancing is a central activity [ANS96, WILL92]. This activity is one of the major relations between design and manufacturing services. Those elements are often in a contractual matter. They are controlled and defined by standards.

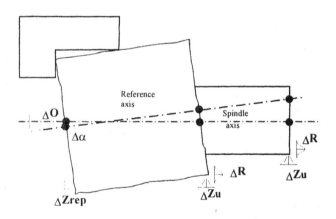

Figure 3   Tri dimensional model of turning

Our contribution to the developments in simultaneous engineering relates to the research for tolerancing models

able to follow the work-piece geometry through the stages of a manufacturing process [RIG 96].

The "Delta L" tolerancing method, initially developed by Pr. BOURDET [BOU 73] , provide a methodology to simulate the part manufacturing process. It deals with the uni-dimensional problems. A specific extension of this method for the revolution work-piece obtained in turning is proposed, according with the Concurrent Engineering concepts. Our proposition try to take into account the tri-dimensional reality. (Fig. 3)

3   A 3D model for revolution work piece

A model is defined in the form of 4 couples of values (P, ΔP) to qualify the turning process (Fig. 4). The relations between the tolerance standards and these 4 proposed values have been analyzed [WOL 99]. A geometry of a test work-piece and an experimental procedure are defined.

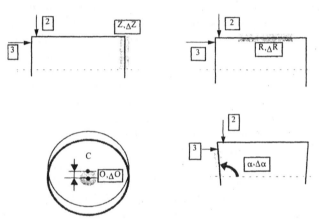

Figure 4   The proposed model for tolerancing rotational parts.

Z, ΔZ is a parameter traditionally named ΔL in the Delta L method of Pr. BOURDET. It represents the dispersion along the Z axis.

R, ΔR is similar to the precedent. The difference is the concerned axis. It characterises part's diameter (2xR).

α, $\Delta\alpha$ permit to characterise the orientation of surfaces. It is possible to define two values for α in the cases of parallelism ($\alpha = 0$) and perpendicularity ($\alpha = \pi/2$).

O, ΔO is a parameter which determine the position of the centre of each section. The value is always zero because we consider turning operations. The only characteristic to consider is ΔO.

Same as the "Delta L" method, we have to find a path to join any surface machined by a turning operation to the two setting surfaces. In this case, two path are possible (Fig. 5). A small number of work-pieces (5 only) are used to obtain the characteristics of a lathe.

Surface	Sequence 10	Sequence 20
1	R	
2	U	
...		
h	R	
i	U	R
j	U	R
k		U
...		

surface 1 = setting surface in sequence 10
surface k = machined surface in sequence 20

Figure 5 path between setting surface and machined surface

A number of relations (Fig. 6), using the P ΔP parameters, were defined to calculate the tolerancing intervall expeted with the mafufacturing process. These relations were defined to be nearest as possible to the ISO tolerancement [MAT95].

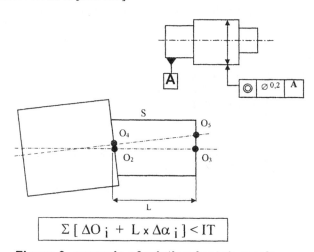

$$\Sigma\,[\,\Delta O_i\,+\,L\,x\,\Delta\alpha_i\,]\,<\,IT$$

Figure 6 example of relation for tolerancing.

4 Application

A software development of this approach was carried out. Various calculation modules and input data procedures are presented as well as a CAD/CAM interface allowing the automatic transfer of the ideal geometry through the IGES standard. An application is proposed through an example (Fig. 9).

Numero	Nature	Cd Dpt(x)	Cd Dpt(y)	Cd Arv(x)	Cd Arv(y)	Cd Centre(x)	Cd Centre(y)	Type
1	ARC	37.	9	37	9.	37.	0.	Plane
2	ARC	18.33	20.35	18.33	20.35	18.33	0.	Plane
3	ARC	842171e-014	23.5	842171e-014	23.5	.842171e-014	0.	Plane
4	LIGNE	45	9	37	9	....	....	Cylindrique
5	ARC	45.	9	45	9.	45.	0.	Plane
6	LIGNE	45.	-9.	37	-9.	....	....	Cylindrique
7	LIGNE	37	11.35	35.22083	11.35	....	....	Cylindrique
8	ARC	37.	11.35	37.	11.35	37.	0.	Plane
9	LIGNE	37.	-11.35	35.22083	-11.35	....	....	Cylindrique
10	LIGNE	35.22083	11.35	18.33	20.35	....	....	Quelconque
11	ARC	35.22083	11.35	35.22083	11.35	35.22083	0.	Plane
12	LIGNE	35.22083	-11.35	18.33	-20.35	....	....	Quelconque
13	LIGNE	18.33	23.5	0.	23.5	....	....	Cylindrique
14	ARC	18.33	23.5	18.33	23.5	18.33	0.	Plane
15	LIGNE	18.33	-23.5	0.	-23.5	....	....	Cylindrique
16	LIGNE	93.25	5	85	5.	....	....	Cylindrique
17	LIGNE	93.25	-5.	85.	-5	....	....	Cylindrique
18	LIGNE	82.5	11	67	11	....	....	Cylindrique
19	LIGNE	82.5	-11.	67	-11.	....	....	Cylindrique
20	LIGNE	67	11	45	9.	....	....	Quelconque
21	ARC	67.	11	67.	11	67.	0.	Plane

Figure 7 software application, IGES interface

The surfaces numbered in the figure 7 are used to define the setting surfaces and the machined surfaces. There are two sequences in the example, and the 18,33mm dimension is calculated by simulation. The data acquisition interface is shown in the figure 8.

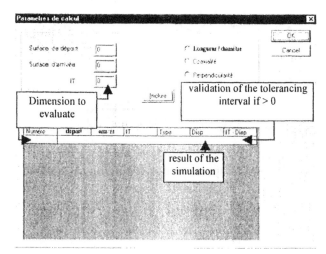

Figure 8 interval calculated by simulation

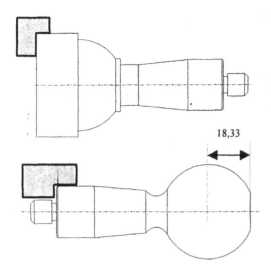

18,33

Figure 9 application to an example

The results of the numerical simulation are compared with the results obtained on a real lathe and for the type of manufacturing operations implemented. An example of results is given in figure 10. We used small series of 15 parts. The comparison was fine. The majority of the dimensions of the machined parts were in the interval of tolerance calculated by the simulation.

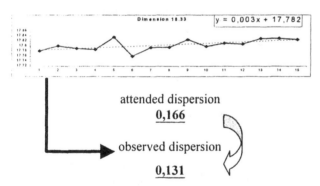

Dimension 18.33   y = 0,003x + 17,782

attended dispersion
0,166

observed dispersion

0,131

Figure 10 comparison between simulation and realization

5 Conclusion

During tests, for certain dimensions, it appeared that the characteristic values of an operation could vary in the course of time. For an always reliable simulation, the events modifying the physique of the system should be taken into account dynamically. This would be a guarantee in the case of production of precision parts. The proposed method, because of its simplicity of

implementation, makes it possible to integrate variations directly. Thus, further software development relates to the debugging of measurements procedures and to the record of the evolutions of the machine tool dispersions obtained in the course of time, during a real production.

These conclusions underline the effectiveness of the method and the software tools proposed for an application in the complex process of cooperative design and manufacturing engineering of the mechanical products.

References

[ANS 96] ANSELMETTI B. , "Intégration des fonctions conception et fabrication en CFAO: traitement de la cotation", CIMAT'96, Grenoble, 1996.

[ARI 92]   ARIKAN M.A.S. , TOKUK O.H. , "Design by using machining operations." , Annals of CIRP , vol. 41/1/1992 , p 185-188.

[BOU 73] BOURDET P. , "Chaîne de cotes de fabrication." , ITET , Paris 1973.

[MAT95], MATHIEU L. , "Le tolérancement normalisé, langage du mécanicien " Colloque Tolérancement et chaines de cotes, ENS Cachan, 02/1995, 18 p.

[RIG96] Rigal J.-F., WOLFF V., ROUMESY B., RAYNAUD S., , "Un modèle géométrique pour le suivi de la cotation en cours de fabrication" IDMME'96, Nantes avril 1996, pp. 897-905.

[SOH92] SOHLENIUS G. , "Concurrent engineering", Annals of the CIRP, Vol. 41/2/1992, pp. 645-655.

[TICH 97] TICHKIEWITCH S. , VERON M. , "Methodology and product model for integrated design using a multi view system." , Annals of CIRP , vol. 46/1/1997 , p 81-84.

[TOL92] TOLLENAERE M. , "Quel modèle produit pour concevoir, CNES, Grenoble, 1992, (on line) www.3s.hmg.inpg.fr/ci/tollenaere/publications/CNES92.html

[WIL92] WILHELM R., LU S. C.Y. , "Tolerance synthesis to support concurrent engineering" Annals of the CIRP, Vol. 41/1/1992, pp. 197-200.

[WOL 99] Internal Report , march 99 , Laboratory CASM, pp. 98.

CHAPTER 15

Enterprise Engineering Applications

Opal: an Integration Platform
for Concurrent Engineering Environments

B. Finel[2], D. Roy[1,2], F.B. Vernadat[1,2]

MACSI Project, INRIA-Lorraine[1]
LGIPM, ENIM/University of Metz[2]
Ile du Saulcy, F-57045 Metz cedex 1, France
vernadat@loria.fr

Abstract

OPAL is a software platform or middleware component for integrated information and process management dedicated to distributed, heterogeneous design and manufacturing environments. Its principles are appropriate for the extended or the virtual enterprise as well. Its main originalities are (1) to make easier integration of modern business application systems with stand-alone legacy systems (thanks to a sound set of encapsulation services), (2) to control and monitor execution of business processes (using a workflow engine) and (3) to provide user access to any kind of engineering data or technical documents (using advanced hypermedia facilities) wherever these items are located in the public workspace of the enterprise net. OPAL has been developed as an ESPRIT project as part of the AIT Initiative. It takes advantage of modern technologies for business process management and information sharing including CORBA-compliant object request brokers, object-oriented technology, neutral product and process data exchange formats (STEP), workflow engines and hypermedia technology (HTML and Internet).

Keywords

Enterprise integration, Integration platforms, Middleware components, Encapsulation modules, Design and manufacturing environments, Car industry, AIT, OPAL.

Introduction

Enterprise Integration (EI) deals with increasing interoperability among people, machines and applications to enhance synergy within an enterprise to better achieve business objectives [6].

OPAL, the European ESPRIT Project 20377, is aimed at developing a solution to this recurrent problem. This holistic goal in computer science must conciliate several incompatible objectives, e.g. to allow access to information items whatever their nature may be (alphanumeric, numeric, strongly formatted, hypertext, etc.), stored within various information systems of the enterprise, at different locations in various formats. By enterprise, we mean the networked enterprise (be it extended, virtual or forming a network). This access must be made easy and is performed by means of a user-friendly or even intuitive interface with high performances. In this paper, the OPAL architecture is first presented. Then, solutions proposed for encapsulation of business applications and legacy systems are explained. Finally, a pilot application concerning engineering change order (ECO) management in the car industry is discussed.

OPAL differs from generic integration platforms such as CIMOSA-IIS [9], AIT-IP[4], and NIIIP [5] in the sense that rather than provides new services, it is built from stats-of-the-art existing technologies and standards.

OPAL architecture

The global system architecture of OPAL is depicted by Fig. 1 [3, 7]. It consists of a central Repository mediating between a development environment and an execution environment based on a workflow engine, a middleware layer (object request broker) and central services for system-wide message and object exchange and interoperability.

The OPAL Repository stores necessary business process, data and organisation models to control business support applications. These models are stored in the Repository using OPAL Classes in a common object-oriented structure to neutralise heterogeneous business data.

The execution environment is used to run business applications and processes. Each application package to be used (e.g. CAD, PDM or MRP system) must be encapsulated, either using its proprietary application program interface (API) if it has one or using dedicated OPAL classes in the case of legacy systems, to become an agent of the integrated system (via the OPAL encapsulation services). Once encapsulated, the application can then be interfaced to and accessed by other modules via the middleware layer (based on ORBIX from IONA, a CORBA-compliant object request broker) and the central services (catering for naming services, time services, communication services and information services).

other (OPAL) components to have read and possibly write access on objects of the systems, to invoke methods for these objects, etc.

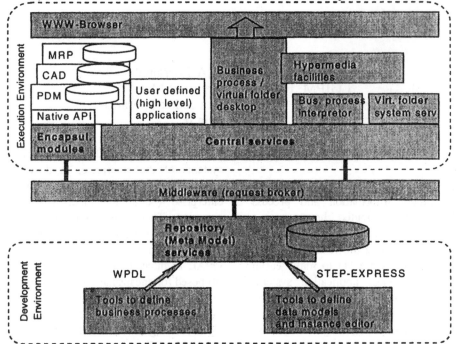

Figure 1. Overall architecture of the OPAL platform

Because this information infrastructure is dedicated to integrated engineering and manufacturing data management, it also provides so-called virtual folder system services to deal with exchange of compound product or engineering data. The virtual folders are used to bind and transmit technical documents (e.g. CAD drawings, bills-of-materials, product documentation, simulation data, text files, etc.) put together as one document and to be exchanged by users or application systems across the system. Hypermedia facilities and web browsers provide users with the ability to navigate through the product data to which they have access, wherever they are located in the system. Users are provided with a user terminal, called the user desktop,from which they can access technical data without to have to know where they are stored and under which format but they can also start new business processes or check the status of on-going processes.

To access to legacy systems, OPAL provides some special interfaces called the Encapsulation modules.

Encapsulation modules

The encapsulation modules are the link between the OPAL system and the integrated data administration systems. Each of these systems to be integrated in the OPAL system must have a suitable interface. Modern systems are no more closed applications but offer different types of interfaces. Regarding the needs of the OPAL architecture it is necessary to have access to the data administration systems by means of an application programming interface (API) which enables

As a matter of fact each system API is different from other ones. Even if the technique used appears to be identical (e.g. using DDE, OLE), the way to access the objects is usually completely different. This is the reason why encapsulation modules are necessary to integrate a data administration system via its specific API in the OPAL system.

It is the task of the encapsulation modules to hide the technical differences of the system specific interfaces and to offer a homogeneous syntax on the second level of the OPAL interface. This means that for each data administration system integrated in the OPAL system, an encapsulation module has to be developed.

The three major functions used to build encapsulation modules are [8]:

EMOpenSession

- Opens a connection to the data administration system if the user name and password are correct. Otherwise a suitable error code is returned.

- Output argument is an internal (system specific) session handle which has to be usedfor any further callof an encapsulation module function to uniquely identify this connection.

EMClosesession:

- Closes the connection identified by the given internal (system specific) session handle.

EMGetInstanceByIdentifier

- Establishes the correct user context with the given internal (system specific) session handle if necessary.

- Creates an object container for the output argument 'Result' according to the given argument 'OPALClassName'.

- Retrieves the object belonging to the 'Identifier' parameter and converts it from the system internal structure to the external (OPAL) structure as defined in the 'MapTbl' parameter. Inserts the values of the instance into the object container 'Result'.

To offer to the end-user a friendly and useful interface to the legacy systems, a hypermedia facility interface is provided by OPAL [2].

processor which is able to interpret the pairs of attribute-names and values to update existing objects, to create new objects or to establish relationships between two existing objects.

The pilot users have identified requirements for an HTML based user interface:

- For each step in the process a new screen has to be produced which shows only the necessary attributes. Views will be used to filter the content of complex objects.

Method name	EMOpenSession	
Input arguments	in string UserName	System specific user name
	in string Passwd	System specific user password
Output arguments	out string SessId	Internal session handle
Return value	OPALErrorRec	Error structure
IDL Declaration:		

OPALErrorRec EMOpenSession, (in string UserName, in string Passwd, out string SessId);

Method name	EMCloseSession	
Input arguments	in string SessId	Internal session handle
Return value	OPALErrorRec	Error structure
IDL Declaration:		

OPALErrorRec EMCloseSession, (in string SessId);

Method name	EMGetInstanceByIdentifier	
Input arguments	in string SessId	Internal session handle
	in string OPALClassName	OPAL class name
	in OPALReference Identifier	Identifier of the object
	in OPALMapTbl MapTbl	Attribute mapping table with name of OPAL class and system specific class and attributes in system specific class.
Output arguments	out OPRep Result	Object referenced by the Identifier
Return value	OPALErrorRec	Error structure
C++ Declaration:		

SessId, in OPALReference Identifier, in OPALMapTbl MapTbl, out OPRep Result);

Hypermedia Facilities

The Hypermedia facilities (developed by INRIA) mainly consist of the following components:
- One wrapper "ocwrapper" which maps the OPAL object container structure to an ordered labelled input tree. This tree is given in a string representation.
- One wrapper to generate HTML output based on an internal tree structure.
- A declarative language YATL to specify the conversion rules.
- A generic tree translation program which converts an input tree structure to an output tree structure.

For the PSA pilot the Hypermedia facilities have been used to create form-based HTML pages with controls such as textfields or choice-boxes to allow user input. The input of such pages is sent to a CGI, pre-

- Each attribute in each screen is either an input element or a read-only value.
- For attributes representing relationships it should be possible (i) to search for existing objects, (ii) to create a new object or (iii) to type in the id of an existing object and set a relationship to this object.
- It should be possible to use labels in the HTML page in different languages which map to the names of the attributes.

Taking these requirements into account a new kind of declarative files have been introduced, called mapping files where all the additional information can be specified. When dynamically creating a HTML page, the existence of a mapping file for the given class is tested and the specifications defined in that file are used.

Within the code provided by INRIA only the ocwrapper, which maps the OPAL, object container structure to an ordered labelled input tree is modified.

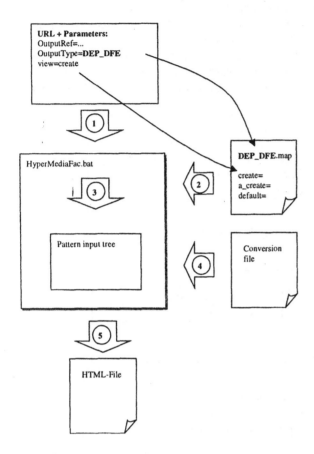

Figure 2. Steps used to dynamically create a HTML page

This module looks for the existence of a mapping file for a given class, parses this file and uses the information from the file to produce some more sophisticated output. Having such concentrated input the conversion files have also been extended to be able to match the various input (i.e all the elements usefull to the end-user at the current stage) and to generate corresponding HTML output.

The main extension to the Hypermedia facilities was the introduction of a further specification file, which is used during creation of the pattern input file (Figure 2). With this file, a richer input allows to write more sophisticated conversion files which are used during creation of the HTML files.

The ability to update objects is provided by a CGI-processor, which is called when submitting the various OK, SET, CREATE or QUERY buttons, which are created on the HTML form pages.

The next section presents, an industrial application used as a pilot for the OPAL architecture. The application case or industrial pilot considered by partners of the project is the automation of a business process at PSA Peugeot Citroën which, without being too much complex, involves both computer based applications and human operations.

The business process, called DEP (Demande Evolution Produit) in the PSA jargon, concerns a process for product/process change request. The application scenario is briefly described in the next section. It will be used to show progress achieved in the company with OPAL [1].

Industrial pilot

Description of the DEP process

To understand the flows of exchanges (mostly work flows and document flows) in the DEP application, it is necessary to briefly point out the organisation, which was in place when the project was launched. This structure, which has been improved since then, can be summarised as follows:

- The Design Department carried out the design of the parts (products).
- The Process Planning Department carried out the design of the facilities (manufacturing processes).
- The Factories manufactured and assembled the parts.

This sequential flow of tasks was valid for the whole set of process plans of the PSA group and for all production sites. Thus, a request for modification issued by one of the production sites had to be the subject of a global validation to be taken into account. Indeed, the manufacturing facilities were common to the Production Units and products were manufactured according to the production plan of the company.

In concrete terms, a request for evolution of a product part or of a process required a significant flow of information items to be processed, which, before OPAL, existed in many varied forms, e.g. telephone calls, fax documents, internal electronic mail messages, file exchanges, consultation of drawings, etc. All this without mentioning commercial negotiations which also had to take place.

DEP application

The DEP application comprises several information systems as illustrated by Figure 3:Internal electronic mail: IBM System 3270 (Memo).

- Product data management system: SHERPA under Unix (BORNEO database).
- Business application of the design bills-of-materials: Common service of the Design Department (SCE) which works under MVS.
- Servers operating under Windows NT.

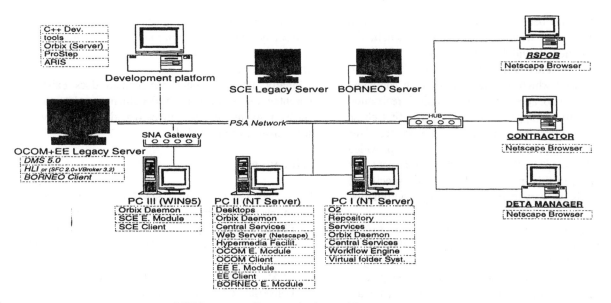

RSPOB represents the person in charge of the request.

Figure 3. PSA pilot application

DEP scenario

The illustrative application scenario selected is a request for evolution of a part such as crankshaft bearing lubrication in a car engine (Figure 4).

Analyses of reports on day-to-day technical problems about certain car engines highlighted a problem of lubrication of crankshaft bearing. A solution could consist in increasing the lubrication hole diameter.

initialises a DEP request. The responsible technical person, identified by the issuer, has access to the sets of information relating to the parts and can always consult drawings, if he wishes, through his Web interface and Encapsulation modules. The person in charge carries out the consultation controls.

The DEP status then moves from the INIT state to the AVIS (advice) state. After reception and consolidation of consultations, the technical person

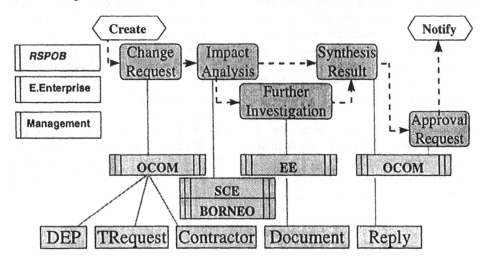

There are four Legacy Data Models to encapsulate:

- Product Management : *«SCE»*
- CAD Vault : *«BORNEO»*
- Change Management : *«OCOM»*
- Extended Enterprise Access : *«EE»*

Figure 4. Workflow Application Pilot

diameter.

Using the OPAL environment by means of the hypermedia facilities, the issuer of the request

modifies, after approval by his superior, the modifications. The DEP status then moves from the AVIS status to the SIGNed status. After the modification has been notified to all actors concerned (status DIFF) and without reply on their behalf, the request is signed off and thus completed.

Further to the description of this simplified application case which is specific to a given company, it is necessary to appreciate the technical realisation which leads to this apparent simplicity. The first phase concerns the modelling of the business processes. This phase determines the level of Quality of the provided service. Of course, the necessary requirement for effectiveness of this modelling phase is not specific to OPAL. This is very common any time automation mechanisms have to be implemented.

The second phase of modelling relates to the modelling of data involved in processes. In the case of OPAL it is significant to note that it is not necessary to model, for example, the whole system of management of the technical data, but only the part relevant for the case treated. Thus, this modelling phase can apply to the 20% of the system, which accounts for the 80% of the requests from the users. The use of an architecture such as OPAL, using standards such as CORBA, STEP or C++, is of course a significant success factor for the re-use of previous work.

Conclusion

The results obtained within the framework of this collaborative project are satisfying in more than one way. First of all, OPAL made it possible to show that the use of advanced but existing information and communication technologies could successfully lead to an operational solution for enterprise integration. Secondly, the encapsulation techniques for application systems, under the constraint of absolutely avoiding to modify existing applications, appear completely workable although usually complicated and sometimes tricky.

As mentioned earlier, OPAL was tested on a simplified but significant industrial application scenario. Performances in terms of behavioural aspects and response times were not critical in this application.

They have not been tested for heavy load of the system (i.e. many requests to be processed in parallel).

Security aspects, which are absolutely mandatory in the case of industrial deployment, have not been sufficiently considered.

Finally, no consistency mechanism does exist to guarantee the validity of a document or data. This aspect remains to be investigated and relevant mechanisms to be provided.

References

[1] B. Finel, J.M. Glaudeix, D. Roy, F.B. Vernadat, "OPAL: an integration approach for engineering environments", EDA'99 INt. Conf. on Engineering Design and Automation, Vancouver, CDN, August 1-4, CD-ROM, 1999.

[2] OPAL, OPAL WP4 workflow application pilot, Deliverable 4.2, December, 1998.

[3] B. Finel, Y. Harani, F.B. Vernadat, "OPAL: An integration platform for design and manufacturing activities", Proc. 2nd International Conference on Engineering Design and Automation, Maui, Hawaii, August 9-12. CD-ROM, 1998.

[4] AIT Consortium, AIT Integration Platform, ESPRIT Project EP 22148, Deliverable, December. http://www.ait.org, 1997.

[5] R. Bolton, A. Dewey, A. Goldschmidt, P. Horstmann, NIIP – The National Industrial Information Infrastructure Protocols for industrial Enterprise Integration: Enabling the Virtual Enterprise, in Enterprise Engineering and Integration: Building International Consensus (K. Kosanke and J. G. Nell, eds), Springer -Verlag, Berlin. pp 293-306, 1997.

[6] F.B. Vernadat, Enterprise Modeling and Integration: Principles and Applications, Chapman & Hall, London, 1996.

[7] OPAL. OPAL Architecture Definition. OPAL EP 20377, Deliverable 3.1.1, December, 1996.

[8] OPAL. OPAL Prototypes of encapsulation modules. OPAL EP 20377, Deliverable 1.3.2, December, 1996.

[9] ESPRIT Consortium AMIPE,CIMOSA: Open System Architecture for CIM, 2nd edition, Springer-Verlag, Berlin, 1993.

Model of Product / Production System Evolutions by Coloured Petri Net

A.Collaine

LRPS, ENSAIS, 14 boulevard de la victoire, 67084 STRASBOURG Cedex

Philippe Lutz*, Jean-Jacques Lesage**

* LRPS, ENSAIS, 14 boulevard de la victoire, 67084 STRASBOURG Cedex
** LURPA, ENS-Cachan, 61 avenue du Président Wilson, 94235 CACHAN Cedex

Abstract

Today, the enterprises have to continuously evolve. This supposes to propose new products derived from old products and to adapt the production system for them, in a concurrent engineering context. Enterprises must then answer to the following questions: how to evolve? Could the enterprises support these evolutions with competitive delays and cost?

Our contribution aims to help the enterprises to do evolutions choices by proposing two models for product and production system evolutions.

The first one is a generic model using evolution parameters and describing evolution links between them.

The second one, implemented for particular cases, is based on Coloured Petri Net and describes the dynamic of evolutions. It assigns the parameters on the place and uses arc expressions and colours to represent the evolutions. We present this model by constructing a basic example.

Our conclusion explains the interest of the coloured Petri net model in the simulation of evolutions.

1 Works context and objectives

Some ten years ago, the enterprises extended without any delays, costs and quality constraints. They could have their own tempo. Today they have to satisfy more demanding customers; they have to propose large diversity of products, to produce in more shorter delays, with a good quality and competitive costs.

In this ultra-competitive context, the enterprises could live only while proposing new products, while innovating and while making evolve old products. In the same way, they have to make production systems evolving, like their products. The enterprises must design production systems for the new products and adapt existing production system for evolving products. These modifications must be done in a concurrent engineering context : products and production systems must evolve simultaneously.

Many tools exist in order to help the manufacturers to design production system. But it is not the same for the redesigning and evolutions of production systems that are the consequence of products evolutions. However this problem is more and more real : lots of "new" products on the market are in reality old products evolutions ; they aren't design *ex nihilo*. In fact, the classification of product evolutions in grade from "minor improvement" by "discovery" that is synonym of real new products, shows us that new concepts and discoveries represent just 5 % of these evolutions [1][2].

Let us consider the case of a product evolution. This product evolves from a N generation (product P_N) to a N+1 generation (product P_{N+1}). In the production system life cycle and especially in its exploitation, the enterprise must make it evolve in a causal answer to the product evolution. The production system must then evolve from a N generation (production system S_N) to a N+1 generation (production system S_{N+1}) (Figure 1).

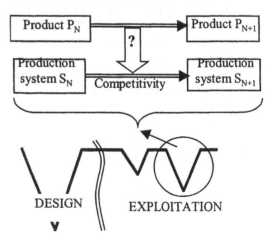

Figure 1 : context of enterprises evolutions

In order to be competitive, the enterprise has to define feasible solutions and to analyse their pertinence. Then the problem is to answer the following questions : is the company able to make evolve the production system? How to make it evolve? How to validate the relevance of the evolutions? On the analysis of which criteria the choices of evolution can be based?

In this context, our work aims to help the enterprise answer the two last questions. Then, we propose a methodological framework, tools and models with the objective to supply the decisions-makers with performance indicators and evaluators. Then they could make their choices.

The proposal method, not presented in this article, is to be set up for particular cases of companies of manufactured products. It bases on the classification of evolutions in four generic schemes. These schemes use the definition of evolutions modes in association with static evolution model. With the classification and for a particular study case, we propose the construction of performances indicators and evaluators, supported by a second evolutions model : the dynamical model.

In the second chapter of this article, we present the definitions of the four evolutions modes and seven parameters of evolutions. The parameters of evolutions represent the product and its production system. For each evolutions mode, we have constructed a static model. It describes the evolutions links between the parameters. These models are proposed in the third part. They include all the evolutions cases excepted when these evolutions suppose to design a new production system. The whole {evolutions modes, static models} is a general framework. It permits to target *a priori* the type of evolutions in a particular case and to locate in which generic scheme this case is.

In the fourth chapter of this article, we present the dynamical model by constructing a simple example. This second model completes the links representation of the first model by add the evolutions dynamics. It bases on the Coloured Petri Net [9]. For a study case, it gives a picture of evolutions generated by the product evolutions. Then, we represent by this model which parameters have evolved, how many times and in result of the evolutions of which parameters.

2   Evolutions representation : modes and parameters

2.1   System evolutions modes

Melese proposes a macroscopic classification of the working modes of an organisation [3]. The four proposed modes represent the nivel of organisation adaptability to face to strategical enterprises modifications. We decline these adaptability modes in a microscopic level and define four evolutions modes that represents the production system evolutions resulting of product evolutions.

The first mode named "*stationary efficiency*" is "*the produce of the enterprise machine in a stable environment: products, markets, means, budgets are known*". This evolution mode translates the stabilised working of the organisation. At the production system scale, the "*stationary efficiency*" represent its normal working. In this mode, a product evolution involves just minor production system evolutions. And they have any consequences on over elements of the production system.

The "*operational adaptability*", "*capacity to answer to modifications of levels of different variable*" is the second mode. The production system is supposed to have parameters that will evolve and make over parameters evolve. But these evolutions don't represent important overturning. For example just the production means are modified.

The third mode, named "*strategic adaptability*" is "*the capacity to answer to parameters types modifications*". At the production system scale, it includes major evolutions of it (for example proceeding modifications) that could modify completely the manufacturing.

The fourth mode is the "*structural adaptability*". It represents "*the capacity for the system to modify its structure to answer to structural modifications of the environment*". In this mode, we consider that the product evolutions are very important and assign the design of a new production system.

In our following work, we are not interested in the first and fourth modes. In fact, the first one has any consequence and the fourth one is on the contrary synonymous of fundamental redesigning.

2.2   Generic   parameters   of   product   and production system

Our objectives are to describe the causality links between product evolutions and production system evolutions. For this reason, we try to represent product and production system by parameters. The parameters research was made in existing models; we have extract of these models the typical aspects of the product and of the production system, independently of the trade; moreover, we separate product and production system in order to can represent the links, after the parameters definition.

- **Product parameters**

In the literature, we found lots of product models. They permit to represent different aspects of a product, in conditional of the steps of its life cycle. Geometrical models, structure oriented models, feature based models, trade oriented models, knowledge oriented models ... for all using, a model product exists.

Our objective is to propose a modelling from with we can describe the evolutions links with the production system.

A simple geometrical model can't be sufficient. In fact, it details the product geometry without showing manufacturing particularities (for example, group of surfaces for manufacturing)

A feature or object oriented model permit to represent the links between the product and its production system. But our work is interested in all trades of manufactured products (assembly and tooling, plastic or metallic products...)

A feature or object oriented model is more interesting. It is in fact adapted in the representation of the links between product and production system. But our work concerns all the mechanical trades (assembly or tooling, plastic or metallic products...). A feature or object oriented model for all mechanical trades doesn't exist.

For these reasons a functional appears to be interesting. It includes the links between the product and specifications, who can the evolutions come from. It permits to identify quickly and represent the evolutions links between the product and its production system.

In a first time we propose to represent the product by the two parameters: services functions and technical functions [5]. In a second time, we add to this representation the manufacturing functions. These functions are the technical aspects of the product, linked with the manufacturing; it is for example geometrical elements of the product that are used to lie the product on the machine or for the tools accessibility...

If the functions are in the same time manufacturing functions and technical functions, we use the technical functions (we give the priority at the using of the product).

To conclude, the product parameters are:
- services functions,
- technical functions,
- manufacturing functions.

- **Production system parameters**

The objectives of the production system parameter are the representation of the working and the consequences of the product evolutions. That's why we don't interested in planning, control ... models.

The other models are mainly based on functional aspects and activities aspects.

The functional view described in [6] proposes a whole of generic functions. These functions are valid for all production system. They represent the production system functions in declining the four functions: TRANSFORM, VERIFY, TRANSPORT, STORE. This model is complete but it doesn't permit to distinguish means and activities. It's why it is insufficient for our work.

Lots of model based on the description of the production system by activities exist: ISO-GAM, IDEF 0, IDEF 3, CIM-OSA, GRAI ...[7]. These models often propose a flow representation. These flows are transformed in the activities by the means. Then, in the context of our work, we see it is important to represent the tasks, their chain and the means used for their production.

Activities models and functional aren't alone satisfying. For the activities, we must interested in several description levels. In assembly, the production system description uses two levels: the range describes the product states; the tasks describe the production operations with the means [8]. To describing tooling, the first step is the process definition; the second step is a complete description by the manufacturing range that is describing the all operations and means. So we propose to describe the production system by two different levels for the task description: the logical range (for the process) and the physical range (elementary operations). We propose soon to represent separately the means and their organisation.

We have then four parameters for the production system description:
- the logical range (GL),
- the physical range (GP),
- the means (MO),
- the organisation (IM).

3   Evolutions static model

In the two evolutions modes (operational adaptability and strategic adaptability), we consider lots of example of evolutions. Their analysis permit us to build the parameters causality links. That's : for a parameter evolution, which parameters evolve. We represent these links in the two matrixes below.

Operational adaptability						
	FS	FT	FR	GL	GP	MO	IM
FS	1	2	0	0	0	0	0
FT	0	1	0	1	1	2	2
FR	0	0	1	0	1	1	1

GL	0	0	0	0	1	2	0
GP	0	1	1	1	1	2	1
MO	0	1	1	0	1	1	1
IM	0	0	0	0	0	0	0

Strategic adaptability							
	FS	FT	FR	GL	GP	MO	IM
FS	1	3	3	2	2	2	2
FT	0	1	2	2	2	2	2
FR	0	1	1	0	2	2	2
GL	0	1	2	1	3	3	3
GP	0	1	1	1	1	3	3
MO	0	1	1	1	2	1	3
IM	0	0	0	1	1	1	1

If is the matrix line number and j the matrix column number:

- c_{ij}=0 means : if the parameter of the line i evolves, the parameter of the column j doesn't evolve ; on the contrary c_{ij} =1,2,3. The number 1, 2 or 3 represent the importance of the evolution;
- c_{ii}=1 means : the parameter of the line i can set of the evolutions.

These matrix models are wrapping models. In a particular study case, the matrix is building by questioning. It permits to situate in which evolutions schemes the case is. Then the dynamical modelling and the simulation can be done.

4 Evolutions dynamical model by Coloured Petri Net

4.1 Specifications of the dynamical modelling

The static model permit, for the two evolutions modes, the representation of the causality links between the parameters evolutions. The use of the evolutions modelling in a study case must permit to situate in which evolutions cases we are (interest of the static model) ; it must permit more to identify all the evolutions propagation. It's why we propose the dynamical model. It uses the Coloured Petri Net [9].

The proposal dynamical model aims to represent a study case. This study case includes for product different evolutions cases. We name these cases evolutions classes.

The constructed net must permit to identify :

- which parameters evolve,

- how many time they evolve,
- in consequence of which parameters of evolutions they evolve,
- in which evolutions class they evolve?

We wish soon that our representation permitted by its schematism to simply identify the evolutions links.

4.2 Net definition

- **Places and transitions definition**

We assign a place for an evolution parameter. The transitions are then the evolutions links between the parameters (we can see them in the particular study case matrix).

We note $T_{Pi,Pj}$ the transition between place Pi and place Pj.

- **Evolutions representation**

In order to have a legible model, any additional arc or transition or place can be created for the dynamical representation. So the own arc expressions are characteristic of the evolutions.

We note $f_{Pi,Pj}$ the upper arc expression of the transition $T_{Pi,Pj}$ and $g_{Pi,Pj}$ the down arc expression of this transition.

With the circulating tokens, we must identify which parameters are evolving and in which evolutions class. The token colour is named COLOR EVOL. This colour arranges the two colours COLOR PARAM and COLOR CLASSEVOL. COLOR PARAM represent the parameter of evolutions (parameter which is evolving); this colour includes then seven elements (an element for an evolution parameter) and more the element FINAL which represent each ended evolution parameter (without consequences). COLOR CLASSEVOL represents the evolutions class in which the parameter is evolving.

Each arc expression is coloured by the colour of the evolutions class.

Then, the constructed coloured Petri net is :

CP-net = {Σ, P, T, f, g, M_0};
   with :

       Σ : colour set,
       P : set of places,
       T : set of transitions,
       f : upper arc expression,
       g : down arc expression,
       M_0 : initialisation function = initial marking.

Σ = { COLOR PARAM = with FS/ FT/ FR/ GL/ GP/
  MO/ IM/ Final ,
  COLOR CLASSEVOL = with I/ II/ ... N},
  COLOR EVOL = product PARAM *
  CLASSEVOL},
P = {FS, FT, FR, GL, GP, MO, IM},
T={ $T_{FS,FT}$, $T_{FS,FR}$ ...}.

4.3 Example of net building for a simple case study

Let us consider the example of a single row tapered roller bearing. This product is manufactured by the enterprise Timken in Colmar (France). Its components are : Cup (outer race), cone (inner race), rollers and cage (Figure 2).

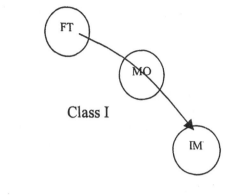

Class I

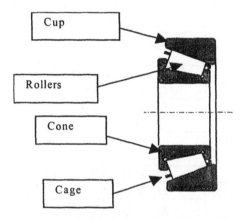

Figure 2 : Components of a single row tapered roller bearing by enterprise Timken

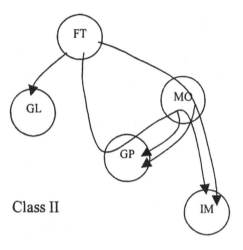

Class II

Figure 3 : propagation evolutions scheme.

For each customers request, a sale-engineer define if the product is an existing product (in the catalogue) or if it is an evolution case. We analyse all the evolutions cases for the studied product. There are many evolutions. In this article, we propose just two different cases, which represent two evolutions class. We present these class below with the figure 3.

The first class is noted I. It represents cone face radius value and tolerances modification (FT evolution). This evolution involves a screw cutting equipment change (MO evolution). This new equipment changes the equipment storage (IM evolution).

The second class (II) affects a bearing without cup. Rollers and cage are the same but the cone has new measurements. These are evolutions off technical functions (FT evolutions).

The absence of cup involves evolutions of logical range (GL), physical range (GP) and means (MO) (removal of elements of GL, GP and MO linked with cup manufacturing).

The evolutions of cone change the technical range (GP) and means (MO) (new equipment, new uses machine for screw cutting, position of marking operation). After these evolutions, we can see evolutions of organisation (IM), means (MO) and technical range (GP) (equipment storage, new equipment, tools path for cutting).

Then, in class I, FT evolution involves MO evolution, that involves IM evolution. In class II, FT evolution involves GL, GP and MO evolutions. GP evolution involves MO evolution; MO evolution involves GP and IM evolution.

- **Constructed net**

The constructed net consists in five places for the five parameters of evolutions : FT, GL, GP, MO and IM. These places are linked by evolutions links named below (see figure 4).

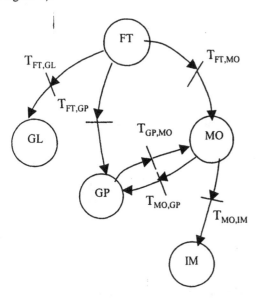

Figure 4 net structure

In order to represent on the net the two evolutions class, we use the colour : COLOR CLASSEVOL = with {I,II}.

Interest now in the $T_{FT,GP}$ transition (evolutions between FT and GP). This evolution path doesn't exist in the class I. The arc expression $f_{FT,GP}$ and $g_{FT,GP}$ don't have the same signification in the class I and class II. We must then write each arc expression for the two colours of COLOR CLASSEVOL (I and II).

Evidently, $f_{FT,GP}(I)=\varnothing$ and $g_{FT,GP}(II)=\varnothing$

The arc expression $f_{FT,GP}(I)$ represents the consumed tokens of the FT place in order that the transition $T_{FT,GP}$ occurs. It is possible if the place FT has a token authorising evolution of parameter GP (in class II). This token is (II,GP). Then $f_{FT,GP}(I)=1'(II,GP)$. Let us interesting in $g_{FT,GP}(II)$. This arc expression represents the realised evolution and the following evolutions in representing them by tokens in the GP place. $g_{FT,GP}(II)$ must permits MO evolution. Then $g_{FT,GP}(II)=1'(II,MO)$.

In order to distinguish last evolutions we propose to represent these evolutions by the parameter FINAL instead of the name of the evolution parameter. Then in class II, the down arc expression of $T_{FT,GL}$ transition is $g_{FT,GL}=1'(II,FINAL)$.

When we construct the all arc expressions with these principles, we obtain the arc expressions presented in the below table.

Trans.	Arc expressions fij
$T_{FT,GL}$	$f_{FT,GL}(I) = \varnothing$
	$f_{FT,GL}(II) = 1'(II,GL)$
$T_{FT,GP}$	$f_{FT,GP}(I) = \varnothing$
	$f_{FT,GP}(II) = 1'(II,GP)$
$T_{FT,MO}$	$f_{FT,MO}(I) = 1'(I,MO)$
	$f_{FT,MO}(II) = 1'(II,MO)$
$T_{GP,MO}$	$f_{GP,MO}(I) = \varnothing$
	$f_{GP,MO}(II) = 1'(II,MO)$
$T_{MO,GP}$	$f_{MO,GP}(I) = \varnothing$
	$f_{MO,GP}(II) = 1'(II,GP)$
$T_{MO,IM}$	$f_{MO,IM}(I) = 1'(I,IM)$
	$f_{MO,IM}(II) = 1'(II,IM)$

Trans.	Arc expressions gij
$T_{FT,GL}$	$g_{FT,GL}(I) = \varnothing$
	$g_{FT,GL}(II) = 1'(II,FINAL)$
$T_{FT,GP}$	$g_{FT,GP}(I) = \varnothing$
	$g_{FT,GP}(II) = 1'(II,MO)$
$T_{FT,MO}$	$g_{FT,MO}(I) = 1'(I,IM)$
	$g_{FT,MO}(II) = 1'(II,IM) + 1'(II,GP)$
$T_{GP,MO}$	$g_{GP,MO}(I) = \varnothing$
	$g_{GP,MO}(II) = 1'(II,IM) + 1'(II,GP)$
$T_{MO,GP}$	$g_{MO,GP}(I) = \varnothing$
	$g_{MO,GP}(II) = 1'(II,FINAL)$
$T_{MO,IM}$	$g_{MO,IM}(I) = 1'(I,MO,AE)$
	$g_{MO,IM}(II) = 1'(II,FINAL)$

- **First marking**

The first marking is constituted by tokens placed in the place of the initial parameter, which evolves. They include for all evolutions class tokens, which represent following evolutions. For class I the first parameter which evolves is FT. This evolution involves MO evolution. Then, we place an initial token (I,MO) in the FT place.

With these principles the all initial marking is :
M_0 (FT,GL,GP,MO,IM) =
$(1'(I,MO)+1'(II,GP)+1'(II,MO)+1'(II,GL) ; \varnothing ; \varnothing ; \varnothing ; \varnothing)$

• **Simulation analysis**

The all net simulation permits to see nine different state of the net by the marking evolution. Each state permits to see which parameters have evolved : they have represent by tokens in their place. If the token have colour FINAL, that's mean there is no following evolution. On the contrary, we see the following evolution by the name of the parameter.

5 Conclusions et perspectives

The present article proposes two models in order to represent causality links between product and production system evolutions.

The first one distinguishes two evolutions modes. For each all, it defines the static aspects of evolutions in representing causality links between parameters of evolutions. The static model proposes a generic framework. It permits to categorise a study case.

The dynamical model uses the Coloured Petri net. It is building only for study case. It permits to represent the evolutions spreading. The above model can evolve. In fact, it actually permits to distinguish the evolutions classes but not to compare them. This problem can be solving in adding arc expressions. In the same way, arc expressions can complete the modelling for informing better the simulation (in function of the performance indicators).

Our future work aims to precise the dynamical model. It must be more adapted to the simulation. For each evolutions modes, we want to define the necessary performance indicators and use the dynamic model in order to evaluate these indicators. We will to precise the consequences of evolutions on the information system.

References

[1] D. Cavalucci, P. Lutz, , "TRIZ, une nouvelle approche de résolution des problèmes d'innovation", revue française de gestion industrielle, Vol 16, N°3, 1997.

[2] G. Altshuller, "And suddenly the Inventor appeared," Second Edition, Worcester, MA, Technical Innovation Center, 1996.

[3] M. Melèse, Approche systémique des organisations, Ed. Hermès, 1989.

[4] F-L. Krause, F.Kimura, T. Kjellberg, S.C-Y. Lu, A Book, "Product modelling," Annals of the CIRP, Vol. 42, pp.695-703, 1993.

[5] NF X50-501.

[6] A. Livet, M.Barth, R. DeGuio "Activity Based Reference Model for Production System Design and Evaluation", 3rd International Conference on Engineering, Design and Automation, Vancouver, B.C., Canada, August 1-4, 1999.

[7] A. Collaine, "La conception des systèmes de production multiproduit," Mémoire de DEA PA, LURPA, Ens-Cachan, 1996.

[8] C. Perrard, P. Lutz, J-M. Henrioud, A. Bourjault, "The MARSYAS Software : an efficient tool for the rational design of assembly systems ", Proceeding of the 14th International Conference on Assembly, Adelaïde, Australia, pp. 77-85, 1993.

[9] K. Jensen, J.W. De Bakker, W.-P. De Roever, G. Rozenberg, "An introduction to the theorical aspects of Coloured Petri Nets," A decade of concurrenve, Lecture notes in Computer Science, Vol. 803, pp.230-272, 1994.

Method for identifying work groups in a complex design process

Zahra Idelmerfaa, Houssein Adoud, Eric Rondeau

CRAN ESA CNRS 7039

Université HP, Nancy I, B.P. 239

54506 Vandoeuvre-lès-Nancy, France

e_mail: zahra.idelmerfaa@cran.u-nancy.fr

Abstract

Today, collaborative engineering is seen as a structure of considerable interest in the manufacturing engineering community [1]. However, this structure does not simplify the design process, rather it adds a tremendous amount of inter-tasks coupling which makes the overall work considerably more difficult [2]. Consequently, it is important to provide methods that allow to structure and to effectively control the work process. In this paper, the authors propose a method for analyzing a complex design process in order to structure it into a set of work groups and to schedule them. The first step is based on Steward and Eppinger's work [2], [3]. It aims at modeling inter-task dependencies by means of a matrix called Design Structure Matrix and then at reorganizing this matrix in order to optimize the information flow. However this method can define work groups collecting a large number of actors and consequently can lead to a complex management of these groups. The second step is then to break the initial work groups to obtain smaller groups. For this, we propose to apply a spectral algorithm [7] which takes into account criteria relative to workload balancing. The interests to apply this method to complex collaborative work are studied in this paper.

1 Introduction

To improve the design process, the new work organization tend to take into account as soon as possible the whole of relevant information not only on the product but also on the way of realizing it, i.e. on the design process. These approaches need work structures which allow to collect a great number of skills and thus of actors within the design project, but especially they require to make cooperate these actors in the most effective possible way.

However, this kind of organization does not simplify the design process, rather it adds a tremendous amount of inter-tasks coupling which makes the overall work considerably more difficult [2]. Consequently, it is important to provide methods that allow to structure and to effectively control the work process.

In this context, this study aims at proposing an approach for the analysis of a complex design process in order to deduce the optimal organization from it. The definition of this organization is based on the knowledge deduced from the analysis of previous projects. The objective is then to re-use this knowledge in similar future projects.

Our contribution is to define a decision-making support relative to the organization of the actors of a project. The approach consists to identify and schedule work groups by means of the analysis of the exchanges between the actors during the design process. This analysis allows to control if the different groups can work effectively, otherwise to restructure them according to several criteria.

2 Phase of preliminary analysis

2.1 DSM method [2],[3]

The first phase of the proposed approach aims at determining an initial organization of the design tasks. It allows to identify and schedule in a design project :
- the tasks that can be sequenced so that each one can be executed only after it receives all the information it requires from its predecessors (Series tasks),
- the tasks that do not depend on others tasks (Parallel tasks),
- the tasks that are interdependent and must be executed simultaneously (Coupled tasks).

This phase is based on the results of work, relative to the management of complex systems, of Eppinger, Steward and Smith [2][3][4].

In this work, the design process is modeled by means of a matrix called DSM (Design Matrix Structure). In a DSM matrix, each design task is represented by an identically labelled row and column of the matrix. The elements "1" within each row identify which tasks must contribute information for the proper achievement of the design process. The rows indicate a chronology of the tasks. The matrix is initially not structured (unspecified chronology) and does not show any coupled task.

	1	2	3	4	5	6	7	8	9	10	11
1	x		1			1					
2		x			1			1			1
3	1		x			1					
4				x						1	
5					x	1	1		1		
6	1		1			x					
7		1	1		1		x			1	
8		1		1		1		x			1
9			1		1				x	1	1
10				1					x		
11	1								1		x

Figure 1 Example of an initial DSM matrix

For example, the initial matrix depicted on figure 1 represents the exchanges between eleven actors. The diagonal elements in the matrix are only included to distinguish the diagonal and the upper and lower triangles of the matrix. The upper triangle visualizes unknown information and the lower triangle known information.

Thus, the element "1" in row a_3 are in columns a_1 and a_6 indicating that the actor a_3 requires information to be transferred from the actors a_1 and a_6. The information to be transferred from the actor a_1 are known because the actor a_1 has begun his task. On the other hand, the information transferred from the actor a_6 are unknown and must be estimated by the actor a_3 because a_6 has not yet started his task.

The interest of DSM is to allow groups to visualise the relationship among their various activities. This knowledge is then used to reorganise the tasks. The objective is to find a sequence of the design tasks which allows the matrix to become lower triangular. The proposed method is a partitioning process.

The partitioning process consists of rearrangement of the initial matrix by interchanging rows and swapping the corresponding columns to achieve a more organized design sequence, i.e. which allows the different actors to work on valid information. The new arrangement of the matrix is defined by research of circuits in the matrix. The method is based on the graph theory and consists in identifying the strongly related components inside the matrix. The actors of a same strongly related component are then collected within a coupled task, i.e. a work group where the actors must closely cooperate.

The objective of the partitioning of the matrix is thus to eliminate the maximum of unknown factors. In the example, three strongly related components constituted respectively of the actors {a1, a3, a6}, {a4, a10}, and {a2, a5, a7, a8, a9, a11} are identified. They are gathered in the matrix in three coupled tasks which are then scheduled in order to remove the "1" on the upper triangle of the matrix and to obtain a partial order.

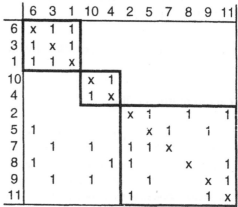

Figure 2 Partitioned DSM Matrix

The partitioned matrix described on figure 2 indicates that the two coupled tasks implying the actors {a1, a3, a6} and the actors {a4, a1} can be executed in parallel because they do not have any interaction. Then, as soon as these two tasks are performed, the third coupled task implying the actors {a2, a5, a7, a8, a9, a11} will be able to intervene since it will have all information it requires, transferred from the first two tasks.

2.2 Limits of the method

The major interest to define work groups by a research of the strongly related components in the exchange matrix is to allow to schedule the tasks to be performed. However, the decomposition of the design process by means of this method does not allow to control the size of the groups i.e. the number of actors within a task. Thus, the application of this method to a complex work organization can lead to the constitution as well of groups of very small size as to of groups of significant size. We are primarily interested in the next of the study in the case of groups of significant size. Indeed, to make collaborate at the same time, in a same work session, a too great number of people can strongly slow down the performance of this group. It is thus important to be able to manage this aspect. For that, a second phase of analysis must be added to reorganize a group containing too many participants. The objective is then to define sub-groups of actors inside a group by means of a phase of refinement.

In the following paragraphs, we present this phase of refinement. Firstly, a method called tearing method is presented. It is an extension of the DSM method [4] and proposes to classify each cooperation and after a partitioning process, to eliminate the least significant cooperations in order to refine the definition of the groups. The second method, more adapted for our study, is based on a spectral algorithm. The application of this method and its extension for the analysis of complex work organizations are secondly presented.

3 Partitioning method for coupled tasks

3.1 Tearing method [5][6]

For this second phase, Eppinger proposes to use the technique of "tearing" which allows to divide a coupled task in sub-tasks. This approach aims at defining different levels in the exchanges between the actors. It distinguishes strong and weak cooperations between the actors. The strong cooperations are represented in the matrix by "1" and the weak ones by "+". Then, a partitioning process which does not consider the weak cooperations is achieved in order to identify the actors who must cooperate in a priority way.

Thus, if we consider that the third group of the exchange matrix on figure 3 contains too many actors and must be reorganized in several sub-groups, we must initially distinguish in the matrix the weak cooperations by "+". In the example, three weak cooperations are identified. Then, a new partitioning phase identifies two strongly related components, i.e. two sub-tasks, one collects four actors (a2, a8, a9, a11) and the other one two actors (a5,a7).

Therefore, this technique allows to divide a complex and important coupled task into several sub-tasks which will be easier to manage. It also allows to define a scheduling between these new sub-tasks. But, this scheduling is limited since the first partitioning phase

relative to the global coupled tasks must also be taken into account.

	2	5	7	8	9	11
2	x	1		1		1
5		x	1		1	
7	1	1	x			
8	1			x		1
9		1			x	1
11	1				1	x

	2	5	7	8	9	11
2	x	+		1		1
5		x	1		1	
7	1	1	x			
8	+			x		1
9		+			x	1
11	1				1	x

	2	8	9	11	5	7
2	x	1		1	+	
8	+	x		1		
9			x	1	+	
11	1		1	x		
5			1		x	1
7	1				1	x

Figure 3 Application of tearing to a coupled task

However, the "tearing" phase is difficult to implement because it is necessary to study and compare each cooperation in order to classify it in weak or strong cooperation. This heuristic processing is long and tiresome. In addition, it appears that the cooperation typology could be more effective if it is achieved at the beginning of the study in order to be taken into account in the global design process organization. Another remark is that the partitioning operation after the "tearing" phase defines sub-groups but does not allow to predefine their number and thus does not allow to control the size of these groups. The second partitioning phase can again lead to identify sub-groups which contain too many actors. It would then be necessary to apply again a "tearing" phase and thus to redefine a new level of cooperation, ... and so on.

That's why, we propose to apply an algorithm based on the following conditions : If a group defined by the preliminary partitioning method contains a too great number of actors, it is necessary to be able to divide it into n sub-groups collected on average i actors. In this processing, two criteria must be taken into account :
- Criterion 1: having a minimal number of inter-group cooperations,
- Criterion 2: defining a priori the average number of actors in each group.

The application of the second criterion is important to efficiently manage the design process. It allows to globally ensure the workload balancing between each group.

Indeed, low-size groups converge more easily towards a solution but quality or ingeniousness of this solution is not generally the best one, while big-size groups behave in a completely opposite way.

The analysis of these criteria conforms to the problems defined in researches on the processing of software applications based on parallel architectures. In theses applications, the objective is to try to balance the loads of software processing on multiprocessor architectures. In the following paragraph, we present one of these algorithms and we apply it for the organization of complex tasks into sub-tasks in respect to the previously defined criteria.

3.2 Spectral algorithm

3.2.1. Introduction

The main problem of partitioning a graph can be defined as follows :

Given a graph $G =(V,E)$, V the set of nodes, E the set of edges and $|V| = n$, the partitioning of this graph corresponds to the decomposition of the set of nodes V in k sub-groups $V_1, V_2,...., V_k$ such as :

- $$\sum_{i=1}^{k} Vi = V \ and \ V_i \cap V_j = \phi \ \ i \neq j$$

- the number of edges connecting these groups must be minimized,

- the weights of nodes in each group must be balanced.

To solve this problem, several types of algorithms were developed such as the spectral algorithms [7], the geometrical algorithms [8], or the multilevel algorithms [9]. We are interested in this paper in the spectral algorithm and show the interest of the application of this algorithm for the organization of groups inside coupled tasks.

3.2.2. Description of the spectral algorithm [7]

This algorithm uses the following procedure to partition a graph into two subsets :

- Building the Laplacienne matrix L of the graph, $L=D-A$ where A is the matrix such as : $A=[a_{ij}]$ and,

$$a_{ij} = \begin{bmatrix} w(v_i, v_j) & if (v_i, v_j) \in E_m \\ 0 & otherwise \end{bmatrix}$$

(w represents the weight of the edges)

and D is a diagonal matrix such as $D =[d_{ij}]$ and

$$d_{ij} = \begin{bmatrix} \sum w (v_i, v_j) & if \ i = j \\ 0 & otherwise \end{bmatrix}$$

- Calculating the eigenvalues of the Laplacienne matrix L;
- Identifying the second smaller eigenvalue λ_2 and finding the eigenvector y corresponding to this eigenvalue;
- Calculating the average M of the values yi;
- Dividing the set of nodes V into two subsets P_1 and P_2 with the following criteria :

If $yi >=M$ then $V_i \in P_1$
If $yi < M$ then $V_i \in P_2$.

3.2.3. Symmetrization of a graph

The application of the spectral algorithm to an exchange matrix cannot be direct. Indeed, this algorithm is applicable to non-oriented graphs whereas the exchange matrices which we study handle oriented graphs. It is thus necessary to specify an intermediate phase named symmetrization procedure of exchange graphs. This procedure is done as following (figure 4) :

- Step 1: in case of a single arc between two nodes, add an artificial arc that has the opposite direction of the already existing arc, and let the weight of the new arc be zero;
- Step 2: replace each pair of arcs between two nodes by an edge such that the weight of this edge is the sum of the two arcs.

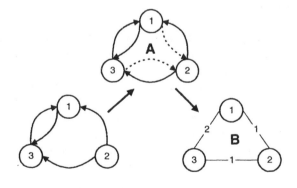

Figure 4   Symmetrization phase of a graph

3.2.4. Application to an exchange matrix.

We take again the example of the third group obtained after the preliminary partitioning phase (actors {a2, a5, a7, a8, a9, a11}). We wish to divide this group into two sub-groups. Figure 5 shows the result of the symmetrization procedure (graph G).

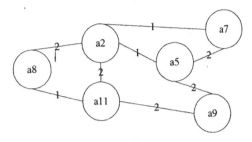

Figure 5   Application of the symmetrization procedure to a coupled task.

The eigenvectors of the Laplacienne matrix L of the graph G are :

	vp_1	vp_2	vp_3	vp_4	vp_5	vp_6
a_2	0,5354	0,2884	0,1749	0,2039	0,6255	-0,4082
a_5	0,5657	-0,4383	-0,0606	-0,3935	-0,4034	-0,4082
a_7	-0,4299	0,227	0,523	-0,5617	0,0319	-0,4082
a_8	-0,2671	-0,466	0,3511	0,6364	-0,1287	-0,4082
a_9	-0,3686	-0,2417	-0,6807	-0,138	0,3958	-0,4082
a_{11}	-0,0355	0,6306	-0,3143	0,2529	-0,5213	-0,4082

The eigenvalues of this matrix are : $\lambda_1=6,88$; $\lambda_2=5,60$; $\lambda_3=2,90$; $\lambda_4=1,96$; $\lambda_5=8,67$; $\lambda_6=0$.

The second smaller eigenvalue is λ_4. We thus calculate the average M on vp4, i.e. M=0. We deduce that {a2, a8, a11} constitutes a first group (positive values of vp4) and that {a5, a7, a9} constitutes the second group (negative values of vp4).

The analysis of this decomposition shows that the obtained organization gives an optimal solution according to all criteria since :
- the number of actors in the two groups is balanced (3 in each group),
- the interaction between the groups is minimized,
- the exchanges inside the groups are homogeneous, indeed the sum of the edge weights is five in the first group and four in the second one.

3.2.5. Integration of valued cooperations

The interest of the tearing method is that it allows to specify weak and strong cooperations. We show in this paragraph that this concept can easily be integrated in a spectral algorithm. It is necessary for that to modify the symmetrization phase in order to introduce into non-oriented graph the different cooperation levels. The operation consists to associate to each arc a weight proportional (or with another criterion) to the cooperation level. The difficulty is obviously to determine the relation between the weight of an arc and the valued level of a cooperation. We propose in a first approach to define a ratio of 2 between a weak cooperation and a strong cooperation. The symmetrization steps in case of a two-level cooperation are :

- Step 1: in case of a single arc between two nodes, add an artificial arc that has the opposite direction of the already existing arc, and let the weight of the new arc be zero;
- Step 2: each arc corresponding to a strong cooperation is affected of a weight "2", and each arc corresponding to a weak cooperation is affected of a weight "1";
- Step 3: replace each pair of arcs between two nodes by an edge such that the weight of this edge is the sum of the two arcs.

Figure 6 represents the non-oriented graph obtained after the symmetrization phase. The distribution of the actors generated by the spectral algorithm gives the same result as previously, i.e.
- group 1 : {a2, a8, a11},
- group 2 : {a5, a7, a9}.

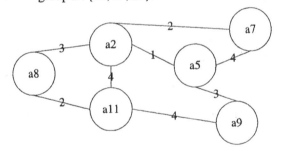

Figure 6 Valued symmetrization of a coupled task

3.3   Comparison of the two methods

3.3.1. From the example.

The decomposition obtained by the tearing method has given a group composed of {a2, a8, a9, a11} and another one of {a5, a7}. We note that this solution is unbalanced in term of actors (respectively 4 and 2), in term of exchanges inside the groups (respectively 4 and 13) and finally that the exchanges between the groups are equal to

6. The organization generated by the spectral algorithm leads to a balance solution in term of number of actors in each group (3 in each group), in term of intra-group exchanges (9 for group 1, and 7 for group 2) and gives an inter-group exchange equal to 7, i.e. slightly higher than that defined by the tearing method.

3.3.2. In general manner (figure 7).

The tearing method breaks the coupled task into several series sub-tasks. In opposition, the spectral algorithm breaks the coupled task into parallel sub-tasks in concentrating the strong relationships between the actors inside each sub-task. The advantage of the spectral algorithm is to maintain the concurrent activity of a coupled task since the whole actors continues to work together in the same temporal period. This decomposition of the coupled task only enables to limit the major information flow necessary to work in a sub task inside this sub task and to avoid to diffuse the all information to all the actors of the coupled task.

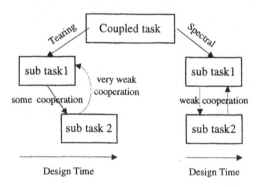

Figure 7. Decomposition of a coupled task

4 Conclusion

We propose in this article a two-step method to analyze and organize the cooperative work in complex systems. The first level is based on the research of strongly related components of an oriented graph which represents the exchanges between the actors of the design process. The interest of this approach is to determine the scheduling of work groups. But it does not enable to define a priori the size of these groups. When these groups gather a too great number of actors, it can be very difficult to manage them. Therefore, we propose a second analysis level which allow to divide an important group into several sub-groups. This decomposition is achieved by using a spectral partitioning algorithm which takes into account criteria relative to the load balancing of the groups such as the number of actors

in each group, the number of exchanges inside a group, and the number of exchanges between the groups.

To note, the DSM method coupled to the spectral algorithm to decompose the large coupled tasks can be used to specify the point "a" (relative to the design process steps elaboration) of the design plan described in the new recommendation of the ISO 9001:2000.

In the future, other mechanisms would also be studied and developed. It concerns for example the aggregation of groups if the first level of partitioning leads to small sizes of groups. The objective will be to find criteria which enable to collect in a better way these different groups in order to improve the concurrence between the tasks.

References

[1] Kusiak, A., Wang J., « Decomposition of the design process », Journal of Mechanical Design, Vol. 115, pp. 687-694, 1993.

[2] Eppinger S.D., « Model-based approaches to managing concurrent engineering », Journal of Engineering Design, Vol. 2, n°4, pp. 283-290, 1991.

[3] Steward D.V., « Planning and managing the design of systems », Portland international conference on management of engineering and technology, pp. 189-193, 1991.

[4] Smith R.P., Eppinger S.D., « Identifying controlling features of engineering design iteration », Management Science, vol 43, n°3, 1997, pp 276-293, 1997.

[5] Steward D.V. « The design structure system : a method for managing the design of complex systems », IEEE Transaction on Engineering Management, Vol. EM 28, n°3, pp. 71-74, 1981.

[6] Steward D.V. « Partitioning and tearing systems of equations », Journal of SIAM, serie B, Vol. 2, n°2, pp 345-365, 1965.

[7] Pothen A., Horst D., Kang-Pu L., « Partitioning sparse matrices with eigenvectors of graphs», SIAM Journal of Matrix Analysis and Applications, Vol. 11, n° 3, pp. 430-452, 1990.

[8] Heath M. T., Raghavan P., «A Cartesian parallel nested dissection algorithm», SIAM Journal of Matrix Analysis and Applications, Vol. 16, n°1, pp. 235-253, 1995.

[9] Karypis V., Kumar, « Multilevel graph partitioning and sparse matrix ordering», International Conference on Parallel Processing, 1995.

A Drawing Version Control System for the Japanese Construction Industry

KATAOKA, Makoto

Building Construction Division, Shimizu Corporation, Tokyo 105-8007, Japan

HIRASAWA, Gakuhito

Computer Integrated Construction, CSTB, Sophia Antipolis, F-06904

Abstract

A unique relationship among architects, engineers and general contractors characterizes the Japanese construction industry. General contractors provide coordination for all of the subcontractors working on a particular job, and they are responsible for building quality, safety, and cost control. In this paper we discuss issues surrounding the management of drawings in the Japanese construction industry, highlighting how our Building Data Management System provides improved drawing version control and information management. Our research focuses on information control during construction with specific emphasis on the management of drawings. The drawing version control system we have developed enables users to manage drawings easily and retrieve important information quickly. The system can trace the history of drawings and the relationship between them. The concept of concurrent engineering using this system and other technical matters are discussed. Feedback from test users of BDMS is also introduced and examined.

1 Introduction

1.1 The Japanese Construction Industry

The Japanese construction industry is unique. The relationship between general contractors and A/Es (architects and engineers) is unlike its European and U.S. counterparts. A/Es in Japan do not produce detailed drawings as is found in Europe and the U.S. The detailed drawings required for construction, so-called shop drawings, are instead produced by general contractors and their subcontractors. A/Es only produce preliminary drawings. Most A/Es do not seem to have the expertise required to draw detailed working drawings for complicated buildings. In Japan, general contractors employ engineering and research staff along with basic construction personnel. The Japanese general contractors' ability to provide high-tech construction engineering enables A/Es to concentrate on preliminary design.

Therefore, construction managers sometimes claim that A/Es' drawings are not useful for actual construction, and indeed no operative refers directly to the working drawings drafted by A/Es. It is the general contractors and subcontractors who generate most construction drawings. If the A/Es' drawings have faults or inconsistencies, and the contractors build it as specified in these drawings, the contractors are likely to be held accountable for the failure. For this reason, general contractors make great efforts to eliminate all discrepancies of construction documents and coordinate subcontractors' work.

1.2 Coordination Method

A general contractor's main responsibility at a construction site is the coordination of numerous works. There are various jobs required in any given building project. A simple reinforced concrete building project, for example, has excavation, foundation, temporary facility work, steel reinforcement, forming, concrete casting, plumbing, tiling, partition making, clothing wall, and so on. Different subcontractors do each job separately. Building construction projects, unlike factory production, have to be operated on-site, where workers move according to its schedule. The state of the construction site changes day by day. If the schedule is not planned properly, efficiency, safety, quality, and cost invariably suffer. To improve productivity and quality, construction managers make great efforts to coordinate the works. In coordination, drawings are the most useful tools, as they are in some other industries.

As mentioned earlier, general contractors have to manage thousands of drawings drafted by many subcontractors and themselves. As an example, A/Es make steel frame plans in their working drawings but the location of sleeves, where ducts of the plumbing work intersect, are not specified. A general contractor assigns a steel frame fabricator to prepare shop drawings and a plumbing subcontractor to provide a detailed plan for the duct allocation. These subcontractors aim for precise design, however the A/Es' drawings do not necessarily contain sufficient information for decision making. As a result, the subcontractors' drawings may conflict with each other, which can cause major problems. The general contractor has to detect the problem in advance and order either or both of the subcontractors to change plans in such a case. Errors of this nature are difficult to eliminate completely, as the drawings are produced independently. The coordination of these drawings is often conducted by superimposing drawing data with CAD systems, where it is easier to find errors. Construction managers often use superimposed drawings when they check for interference among air-conditioning equipment, plumbing equipment, electrical equipment and building elements such as columns, girders, and openings. They call the superimposed drawings "coordination drawings" or "equipment plotting drawings," which are perhaps not used in Europe or U.S. Coordination drawings are widely utilized in Japanese construction sites.

1.3 Drawing Issues

Drawings of a building express many forms of information such as shape, size and material. Since they also imply information concerning construction method, sharing drawings means sharing works. It is thus important to manage drawings in order to accomplish concurrent engineering in the construction industry.

To manage drawings drafted by different groups of architectural engineers, version control is a significant issue. Large projects may involve over a thousand drawings, and even more if extra orders for change are added. It is difficult to monitor so much information effectively—almost impossible if a controlling system

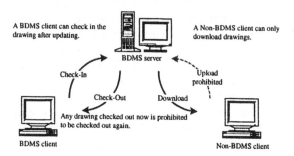

Figure 1 Check-Out and Check-In

does not exist. Draftspersons should inform managers and their colleagues of changes in their drawings so that everyone involved has the same information. Even if they use advanced CAD software packages, such rules are essential.

Some CAD systems have a function that enables users to include other drawings into one drawing, and is useful for coordination purposes as the included drawings can be automatically updated when the original drawing files are revised. Operators, however, may not notice the changes and may fail to check the drawings. Therefore, some informing functions are necessary.

2 System Concept

Some operating systems, which can be inexpensively acquired, have a useful file sharing system. Many organizations currently impose rules for sharing drawing files on a computer network. In addition to such rules, we have considered more elements for effective utilization of network systems. Some of the vital purposes of drawing management in construction sites are as follows:

- Sharing drawings and quick communication
- High efficiency of drafting
- Consistency among drawings
- History of decision making
- Elimination of falsification
- Schedule management
- Cost management according to change order
- Delivery management of drawings

We have developed a system called the Building Data Management System (BDMS) for these purposes. Below, we address some of the advantages of BDMS.

2.1 Exclusive Updating

Most people usually write down additional information on the printed drawings after the drawings are completed and printed. In particular, the drawings approved by supervisors with their signature, or the signature stamps commonly used in Japan, cannot be printed again. In this case, there is the only one original for each drawing, and there is no possibility for two or more persons to revise the drawing simultaneously.

On the other hand, draftspersons who use a CAD system often copy a file for updating in order to allow other users to refer to the original file. CAD systems may allow multiple users to open the same file at the same time, but this can be problematic because users may refer to uncompleted temporary data while another user can save changes halfway. Copying a file for updating may seem sensible, but if two users copy the same file and return it after updating, only the changes of the last file saved are reflected. Consequently, the following conditions are necessary:

E All changes must be included.
E Integrity of drawings should be maintained while being updated.

We implement these functions with an exclusive control, which allows only one user to copy and update a drawing, and which allows other users only to read the file, not to replace it, until the file is completely updated. We use the terms check-out and check-in to mean copying a file from the server to revise and replacing the file back into the server respectively (See Figure 1).

2.2 Version Control with Branching Alternatives

CAD operators may restart to design from an earlier point. For CAD users to retrieve the old versions of drawings, the system stores the every version of drawings. The files of all versions, stored in the BDMS server, have the names of sequential numbers as their IDs. BDMS stores the derivation information that shows the derived drawing and its parental drawing (See Figure 2). The parental drawing attribute enables the users to trace the old drawings, and a tree hierarchy can be realized with the attribute. Construction managers consider a wide variety of options in the construction stage. This function gives the managers the history of decision making as it takes place.

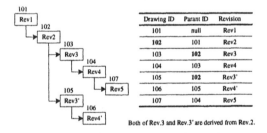

Drawing ID	Parant ID	Revision
101	null	Rev1
102	101	Rev2
103	102	Rev3
104	103	Rev4
105	102	Rev3'
106	105	Rev4'
107	104	Rev5

Both of Rev.3 and Rev.3' are derived from Rev.2.

Figure 2 Management of alternative history

2.3 External Reference

Some CAD systems have functions to refer to other

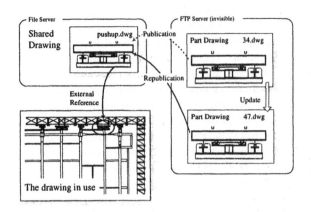

Figure 3 Management of external reference

drawings. There are two forms of such functions. One is to copy and include data of the referred drawings into the drawing in use. The referred data is stored in the file but will not be automatically updated even when the original drawing is modified. The other function is that the drawing in use can refer to external files, referring the file path—the places and the names of the external files. In this case, the referring file will be updated automatically when the referred files are changed because the referring file reloads data when opened or refreshed. CAD operators are expected to choose which function is optimal for their work. Assuming no change on referred files, the latter function includes the former one. We, therefore, make use of the latter function to implement an automatic system of recording the external referential relationship between drawings.

All drawing files stored in the BDMS server are invisible for CAD users, and they have the names of sequential numbers for easy operation in the server. The approved drawings are, therefore, available to the public, given a new plain name so that the users can easily search and refer to it (See Figure 3).

If changes are made to one referred drawing using external reference system, the changes are copied to the referring drawings automatically, but without notification. However, notification is often necessary, and it is preferred if it is automatic. If CAD users do not know what has been changed, errors can occur. Also, drawings using external reference should be checked for conflicts after data has been changed. BDMS stores data of referential relationship in order to notify its users of possible inconsistencies when referred drawings have been changed. Comparing the referential data and updating timestamps, BDMS can identify potential problems and recommend users review files to ensure consistency.

2.4 Printing Management

As explained earlier, the differences between printed drawings and electronic data can cause serious problems. Changes added into a printed drawing should be added into the data stored electronically; otherwise, drawings will not be correct. Organizations that use CAD software packages, of course, attempt to keep printed drawings and electronic data consistent. BDMS helps its users to do so. The customized CAD system employed by BDMS records the user ID, the drawing ID, the time and the purpose of printing when the user prints out a drawing. There are various purposes for printing out drawings such as submissions for approval and drafts for review. BDMS prompts the users to enter the purpose when they wish to print out a drawing, thus the users can input the purpose of printing. BDMS also prompts users to print drawings following updates. Users can reprint they find it important to do so. This system can also be used for the management of drawing delivery, if the users put in delivery records. Consistency between actual print and data stored in the database is one of the most important issues.

2.5 Safe Saving

CAD users often save backup files while they are editing, and these files are often renamed. If users save a file with a different name as the BDMS server knows, the system has to give a correct name to the file. Some actions by users are automatically recorded while editing it. If the user fails to save the file at the end, the database conflicts with the drawing data. The BDMS client keeps the original name of the file and forces the CAD system to save the modified drawing with the original name. The users, in this case, do not have to do anything but exit the CAD system. BDMS also saves the DWF file of the drawing for its users to browse both with WWW browsers and BDMS clients.

2.6 Problem Detection

We have discussed important conditions of drawing version control. The data must be consistent and ready to be referred to. The system is expected to notify the users when problems occur. BDMS notifies the users when it detects problems in the following ways.

One is the case of falsification or deletion of the shared drawing files. Multi-user operating systems can grant higher levels of access. BDMS can detect such problems of falsification or deletion since shared drawings must not be falsified. BDMS also detects cancellation of publication of drawings that are referred to as external references. The drawings that refer to the cancelled drawings are probably defective, thus BDMS points out the problem to its users.

Updates in referred drawings are also detected. As illustrated earlier, the drawings should be checked for conflicting information, especially drawings with external references. Printed drawings are checked for reference changes in the same way.

3 Implementation of BDMS

We implemented the first version of BDMS in 1998. BDMS has been revised and improved since. We designed the database schema before implementation. The schema has naturally changed for implementation of more advanced functions. This section shows the database schema and configuration of BDMS.

3.1 Schema of Database

The relational database schema we designed is shown in Figure 4, expressed in IDEF1X.

The main entity is DRAWING. The entity has the drawing ID as a primary key. The drawing name, the name of the file, and the project ID are also attributes. Every drawing belongs to one of projects defined in the PROJECT entity. To store a tree structure in the database, we added 'parental drawing ID' as a recursive attribute pointing the drawing from which it was derived. By tracing the pointers, the system makes a hierarchy of

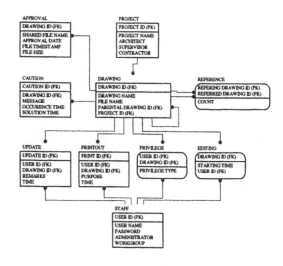

Figure 4 Schema of BDMS database

drawings. The remarks noted in updating can occur multiple times for each drawing. Thus we created the entity UPDATE, which records one remark or more for each drawing. The PRINTOUT entity stores many lines of the printing history for each drawing as well.

Each record of the REFERENCE entity demonstrates the relationship between two drawings that have an external reference relationship. The entity has two kinds of drawing IDs as attributes. One is the referring drawing ID and the other is the referred one. The entity also preserves the number of objects. By tracing the relationship of REFERENCE, users retrieve information about which drawings are influenced in chains by a certain drawing.

The STAFF entity and the PRIVILEGE entity are used for certification of users and restriction of writing and reading data. We assume that BDMS is used in organizations consisting of many groups, where certification and restriction are important issues. The primary key of the PRIVILEGE entity is a set of user IDs and drawing IDs. The three types of privilege are 'read and write', 'read only' and 'no access.'

The entity EDITING stores temporary data, and each record shows the user and the drawing in use. The drawing recorded in this entity cannot be used for updating by other users to prevent inconsistencies. The record is removed after updating.

The approved drawings are recorded in the entity APPROVAL. When a drawing is approved, the record is made with the approval date. It also stores the size and the timestamp of the file for BDMS to detect falsification and deletion.

The CAUTION entity stores problems that BDMS detects. The attribute solution time is null while the problem is unsolved. When the solution time is filled, BDMS no longer displays the problem. Because the record is kept in the database, users can retrieve the history of conflict occurrence and solutions.

3.2 System Features

BDMS is a type of client/server system. The server consists of a file server, a relational database management system, an FTP server, and a WWW server. The client system consists of a BDMS client system, a customized CAD system and a WWW browser. The BDMS client is used for check-out, check-in and controlling CAD. Users can access various information with web browsers. Users can also refer to the remarks and the latest drawings both with the BDMS client and with web browsers. Many programs for administrative management are implemented in WWW services as well. The BDMS administrators can maintain the database with their local

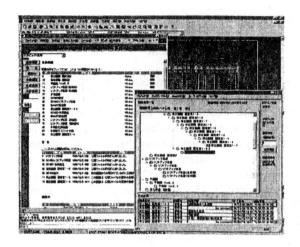

Figure 5 BDMS client system

machines.

When a user selects a drawing from the list shown in a tree form in the BDMS client system to review it, BDMS retrieves the file of the drawing by FTP and writes an editing record into the database. When the user quits editing the drawing, BDMS transfers the file into the server and removes the editing record from the database. The BDMS client also prompts the user to enter some remarks concerning the revision of the drawing.

The users of BDMS do not have to recognize new functions or unfamiliar manipulation for the system to store drawing management data. We customized AutoCAD R14 and AutoCAD 2000 with ObjectARX, and we use a notification function, which enables us to make event-driven programs. On booting up AutoCAD, BDMS attaches reactor programs to AutoCAD. When users select the menu 'external reference' on the CAD system, one of the reactor programs accesses the database. When users delete 'external reference' objects from the drawing, another reactor program functions. Such event-driven programs can be implemented in the CAD software packages with customizing and notification features. BDMS can employ any relational database management system that gives stored procedure and trigger functions. We employed Microsoft's SQL Server.

4 Practical Issues

4.1 Conventional System

BDMS is now being used in two organizations. One of them is a production planning section of a general contractor. The production planning section makes many drawings according to requests from construction sites. As the section consists of construction engineers, they consider the combination of architectural drawings, structural drawings, and construction conditions to produce detailed drawings. Labor and cost of the management at the construction sites can be deduced with such an organization. This kind of department works for many projects concurrently, and there are many small groups of engineers. To avoid confusion, they used to have some rules for sharing drawing files.

Our users used to share drawing files in the same file server and stored the files in fixed folders allocated to each project. The names of files were difficult to understand because they were defined by restricted rules. For example, the first two letters were abbreviation of the name of the project and the following two letters categorized the type of the drawing. The file name also had the serial number of the drawing. An example file name could be "HBAH-305.DWG," which means the detailed plan of the 3rd floor of Honda Building, drawing number 05. The contents of the drawing were difficult to guess even if we knew what the abbreviation meant.

The drawing files of each project were stored in the project folders, and users were able to copy the drawing files onto the removable media to transfer to their subcontractors. They only stored the latest files, although some draftspersons saved backups in case of emergency. The files in the folder, including other documents such as schedules and quality control sheets, were transferred to removable media after completion of construction.

4.2 Feedback

Users complain that they cannot check out two drawings or more when they want to hand over many drawings to the subcontractors. This restriction is placed to avoid deadlocks. Moreover, even if they check out many drawings, the drawings cannot be checked in after the drawings are returned from subcontractors because the places in the tree structure are not retained. The information about the check-out is kept only in the client computers. Hence, we think that check-out information should be hidden in the drawings. With the information managed by BDMS stored in each drawing, they can be restored into the server.

The second issue is on legal approval on the network. We should improve the approval function because BDMS puts only the approval date as an attribute of the approved drawings, although users do not need to make drawings approved on the network at the moment. In some large projects, the client, the A/Es, and the contractors are already connected on the network. Recent studies on application of electronic systems may enable us to

approve drawings in a much safer way on the network in the near future.

The last issue is the drafting custom. Our test users have not yet used either external reference or alternative design. That is because they are not familiar with the inclusion of others' drawings. However, they express their interest in using these features to ease the coordination of drawings. To accomplish this purpose, they have to make some effort. They should attempt to change their thought processes when producing building data, not to make the drawings. Recognition of the difference would make users achieve it. Changes in the way drawings are drafted will change business flow.

5 Summary

We have discussed issues surrounding construction drawings in Japan, and we have presented our Building Data Management System as a quality solution. We focus not only on drawing management but also on management of all kinds of information related to decision making. Questions for future study include how to manage other kinds of documents such as specifications, schedules and change orders. The change orders, which are vital information for construction management, are currently difficult to control in BDMS. We have to study what and how architects and engineers order, and what kinds of documents are related to each other. After that we will try to implement a new system that controls not only drawing data but also architectural objects as management units. BDMS can be further extended to integrate construction management information.

Reference

[1] Hirasawa, Gakuhito and Makoto Kataoka, "Development of a Drawing Version Control System, used to support alternative designs and ensure the integrity of external references," CONCURRENT ENGINEERING IN CONSTRUCTION: CHALLENGES FOR THE NEW MILLENNIUM, CIB Publication 236, pp.161-166, 1999.

Product Duality in Concurrent Engineering

Ph. Chazelet

Laboratoire d'Automatique de Besançon, UMR CNRS 6596, Université de Franche-Comté, ENSMM, Institut de
Productique, 25 Rue Alain Savary, 25000 Besançon, France
e-mail : chazelet@ens2m.fr

F. Lhote

Laboratoire d'Automatique de Besançon, UMR CNRS 6596, Université de Franche-Comté, ENSMM, Institut de
Productique, 25 Rue Alain Savary, 25000 Besançon, France
e-mail : flhote@ens2m.fr

Abstract

The aim of the paper is to bring a new point of view on the purposes of the concurrent engineering and to extend them to another scientist context. We will show a product-orientated point of view on the concurrent engineering. We will explain the main following hypothesis: the concurrent engineering has not just for goal to design, in a simultaneous way, the product and its manufacturing system but, more fundamentally, to design, in the same time, the functional aspects and the physical aspects of the product, which is both a generic object and an individual object.

After a definition of what are the generic product and the individual product and a presentation of their differences, we will explain how the concurrent engineering allows treating globally this duality of the product. This representation of the concurrent engineering highlights the main problem of the nowadays customer-orientated factories. The factory, seen as a double value chain, creates two kinds of products (generic and individual) and so exchanges them with two kinds of customers or two recipients (the supposed final customer of the individual product and the known final customer of the generic product).

There is a danger, for the factory, to treat, in a sequential way, the design of its two products and the satisfaction of its two customers. The concurrent engineering gives a good answer to this problem. However, for implementing the concurrent engineering in this context, we will suggest to develop and to use two complementary concepts following the previous ones: the intermediate customers and the intermediate objects. We will propose a typology of these intermediate customers and we will show the interest to define and to use the intermediate objects that are exchanged between these actors along the value chain of the factory.

1 Introduction

The world of the factory and its socio-economic environment, mainly the market, undergo important and continuous evolutions. One of the most important is the rise of the design at every level:

- at the macro level, with the development of the functional aspects of the needs of the *market*, the development of the products diversity, the decrease of the products lifetime and the change of the competition conditions (supply much bigger than demand),
- at the meso level, with the increase of the design *process* and the bigger and bigger importance of the upstream phases of the design. We know now that these phases are responsible of the main part of the development costs [1] [2],
- at the micro level, with the development of the researches in order to improve performances of design activities seen as problem solving activity [3]. Moreover design activities are founded in other processes than the only design process (for instance, in the manufacturing process).

The concurrent engineering theories suggest, in order to improve the design process, to design in a simultaneous way the product and its manufacturing system. However, the concurrent engineering takes only into account a little part of the design process. In fact, it chooses to start from the design activity to represent the design process. So, in order to have a concrete view of one process, it is necessary to build a representation based on the result of this process [4]. Actually, an object-based representation of the design process i.e. a *product-orientated* point of view allows to define precisely and completely the design process and to enlarge our vision on upstream activities of the factory in order to improve theirs efficiencies. We have adopted yet the same systemic approach in order to suggest an enterprise modelling [5].

To our mind, the concurrent engineering has not just for goal to design, in a simultaneous way, the product and its manufacturing system but, more fundamentally, to design, in the same time, the functional aspects and the physical aspects of the product, which is both a generic object and an individual object.

2   Generic Product and Individual product

2.1   The Design-Implementation Matrix

The concurrent engineering is based on the following observation: the organisation of factory, inherited from F. W. Taylor and H. Fayol, develops sequentially the tasks (sequentiality between the product design tasks and the manufacturing system design). This organisation was adapted to a socio-economic context where the supply was lower than the demand. But, the context has changed. Now, the supply is superior to the demand and the factories are trying to develop differentiation on the quality-cost-time criteria. The aim of the concurrent engineering is to improve the quality, to decrease the costs and to manage the time. So the concurrent engineering takes into account the product design tasks and the manufacturing system design (Figure 1: The Product-Process Matrix) and suggests to design simultaneously the product and the manufacturing system.

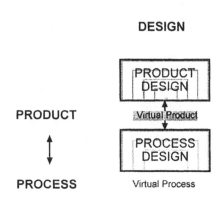

Figure 1 The Product-Process Matrix

The observation on which is lying the concurrent

engineering is good but too limited. We want to improve this observation and, for this goal, we are trying to find a more global and larger vision of the design-implementation process with a *systemic point of view*. That is why we will enlarge our vision towards the upstream of the product design (Figure 2: The Service-Product-Process Matrix) and towards the downstream of the manufacturing system design (Figure 3: The Design-Implementation Matrix).

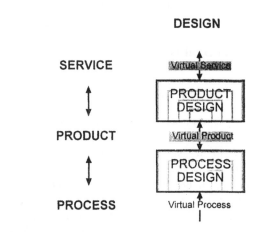

Figure 2 The Service-Product-Process Matrix

To model the design-implementation process as a matrix is to obtain a dynamical representation of the process as a whole, and not divided into parcels. The process is modelled in its evolution. This view shows its finality: the process moves on from (virtual) service to (real) service (Figure 3: The Design-Implementation Matrix).

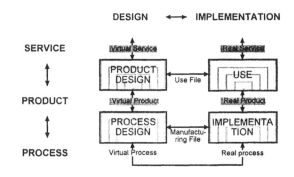

Figure 3 The Design-Implementation Matrix

The matrix shows that there are two worlds:
- the *virtual* world ruled by the design activities,
- the *real* world where the implementation activities are prevailing.

Both of them are cut crosswise by three levels:
- the level of *service* which represents the immaterial,

intentional and functional aspects of the product,

- the level of the *product* which represents the material, physical and technical aspects of the product,
- the level of the *manufacturing system.*

Classically the concurrent engineering establishes a link between the two last levels but do not establish one neither between the two first levels nor between the two worlds.

Our aim is to tackle the fields not yet explored by the concurrent engineering and to create the links between them.

2.2 Product Duality

2.2.1 The Product, Support of Service

A product is an intentional object, aimed towards recipients for whom it offers functions of service incorporated to a material support [6]. The main difference between the material and the immaterial, though both appear in the manufactured product, even if the first includes the second, is at the level of the notion of " intentionality " [7]. It is in the immaterial side incorporated into the object that we find its finality. When there is no goal in its material side. The object has intrinsically an intention; i.e. it expresses a goal defined at the time of its design by a human spirit. What is doing the value of the product is the intentional content, which is printed on the physical support.

The expected service of a manufacturing product has two characteristics:

- it uses a physical support,
- it moves from a man to another.

In this way, manufacturing product constitutes:

- a potential of service. As material object endowed with functions of services, the product, when it is not in use, contains a potential service. Service which becomes real only with the implementation of a use action ;
- a support of service. A material object in which we incorporate functions of service constitutes it. Submitted to a use action, it will carry out the provision of service for which it is intended (Figure 4: The Product, Support of Service).

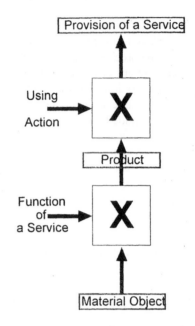

Figure 4 The Product, Support of Service

All the products submitted to a use action develop an activity pointed towards a goal and linked to services, in order to satisfy one or several recipients. The product is then an intentional material object. It incorporates, on the one hand, a physical support (structure) moving in a world of pure causality: the use of material object is a cause that produces effects. And on the other hand, finality (function of service) moving in a world of intentionality: the functions of service are a means to an end.

2.2.2 The Concept of Product: Physical Point of View

From a physical point of view, the product is a material object, which undergoes transformations. Natural resources are firstly extracted from the environment. So they constitute raw materials which, after a first transformation, become intermediate products and after a second one, a final product. Which, once distributed, will become a sold product. Its consumption will lead to waste, which, once recovered and extracted, will constitute raw materials again. The use of sold product will require, for its part, a maintenance, which will transform it into maintained product (Figure 5: The Concept of Product: Physical Point of View) [8].

design/manufacturing of a new product as the integration of two flows (Figure 7: Global View of the Material and Immaterial Flows):

- A "virtual" flow or intentional flow which goes from product concept to *generic product*.
- A "real" flow or material flow which goes from raw material to *individual product*.

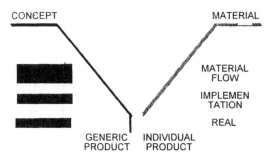

Figure 7 Global View of the Material and Immaterial Flows

These two flows are generally considered as two flows that follow each other sequentially. The flow of design, leading to the generic product, precedes the flow of implementation that leads to individual products.

The classical model is a *sequential* representation, both of the two flows (or processes) and of the activities that compose them. No link is made between activities of two processes. It seems that these are completely autonomous and meet each other only in the occasion of the transition from one to another (Figure 8: Classical View).

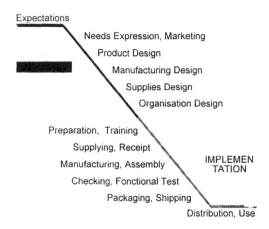

Figure 8 Classical View`

The concurrent engineering has chosen another representation of virtual and real flows. The activities of two flows are linked one by one along each of the two branches of the *cycle in V*. It exists one-to-one links between activities of design and activities of manufacturing (Figure 9: V Cycle View). Flows become

closer but the sequential characteristic of the classical representation is still there.

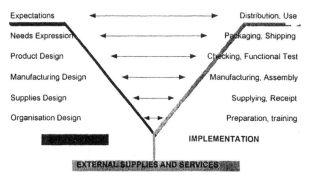

Figure 9 V Cycle View

The one-to-one links between activities of design and activities of manufacturing are quite rich to exploit, following an approach of concurrent engineering, on condition that the flows of design and manufacturing are modelled in a parallel way, and no more in a sequential way.

2.3.2   Intentional Flow and Generic Product

All along the flow of design, activities follow activities. Actors develop these activities. The aim of the activity, of these actors, their task is not to implement a service, a product or a manufacturing system, but to design a service, a product or a manufacturing system. In other words, they do no supply a real service, a real product or a real manufacturing system, but a virtual service, a virtual product or a virtual manufacturing system for which users are yet in a virtual state. Service, product or manufacturing system will move on from a potential state, at the beginning of the flow, to a virtual state, at its end. We are in the virtual phase of the matrix.

What is the status of the product ("product" level of the matrix) in this phase?

Along this intentional flow, we follow a immaterial product and its evolutions (transformations). This immaterial product will be design only one time, then, eventually re-design afterwards (much later), if circumstances or market require it. That is why we will talk about *generic product* since it is common to all the users. This product is not an individual property of every user. It is not individualised even if it tries to fulfill individually every user. It is a generic entity quite different from the individual product (Figure 10: Life Period of the Generic Product).

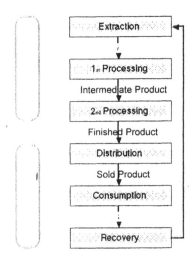

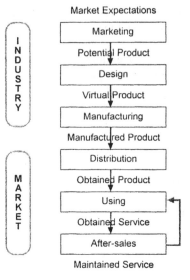

FUNCTIONAL POINT OF VIEW :
PRODUCT = SUPPORT OF THE
FUNCTIONS OF A SERVICE

Figure 5 The Concept of Product: Physical Point of View

Figure 6 The Concept of Product: Functional Point of View

The product has material and physical aspects. But the product has also immaterial and intentional aspects. The product is constituted by a physical support [9], i.e. by physical components, assembled in order to become an autonomous product. We find this materiality of the product even in the components, which can not be broken down, called the " features ". But the product is also an object which, in use, reveals the intentional and immaterial aspects adapted to the service it contributes to achieve when it is the object of a use action.

2.2.3 The Concept of Product: Functional Point of View

From a functional point of view, the product is a support of functions of services. These functions of services are perceived by marketing. There are the market expectations. Marketing interprets them towards design under the form of a potential product. Which will be transformed into initial product when moving from design to manufacturing. Manufacturing gives an achieved product to the distribution. This one gives an acquired product to the user, who will obtain finally a service after use, service which can be maintained with a logistic after-sales help and can allow new uses. (Figure 6: The Concept of Product: Functional Point of View) [8].

As a medal, two inseparable obverses constitute the product:
- the first one represents the physical product.
- the second one represents the intentional product.

These two aspects of the product, these two "products", are inseparably joined in the final product. Therefore we propose to use methods of concurrent engineering in order to further the connection between design activities linked to functional aspects and design activities linked to technical aspects of the product. The aim is to re-create, within the activity, the uniqueness of the product and to avoid the schizophrenia peculiar to the splitting and to the sequentiality of these activities.

The concurrent engineering has not just for goal to design, in a simultaneous way, the product and its manufacturing system but, more fundamentally, to design, in the same time, the functional aspects and the physical aspects of the product.

2.3 Flow Duality

The service, the product and the manufacturing system are successively designed and then achieved. So two worlds exist: the one of the design (virtual) and the one of the implementation (real); the finality of the factory being to achieve the transition from virtual to real.

2.3.1 Global Representation

As a general rule, we can represent the process of

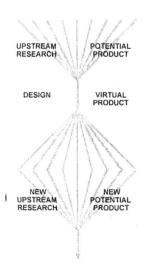

Figure 10 Life Period of the Generic Product

2.3.3 Material Flow and Individual Product

Actors of the different activities, which take place along the flow of implementation, have not for task to design a service, a product or a manufacturing system. In other words, they do no supply a virtual service, a virtual product or a virtual manufacturing system, but a real service, a real product or a real manufacturing system for which the user is identified. Service, product or manufacturing system will move on from a virtual state, at the beginning of the flow, to a real state, at its end. We are in the real phase of the matrix.

What is the status of the product ("product" level of the matrix) in this phase?

Along this physical flow, we follow a material product and its evolutions (transformations). This material product will be produce and re-produce in several copies. This product will be unique for each individual user. That is why we will talk about *individual product* since it is peculiar to every user. It is an individual entity quite distinct, by nature, from the generic product (Figure 11: Life Period of the Individual Product).

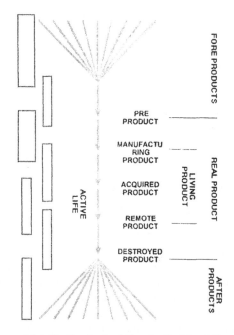

Figure 11 Life Period of the Individual Product

The physical, material and real aspects, at the one hand, and the intentional, immaterial and virtual aspects, from the other hand, are simultaneously present in the product. But if these two kinds of aspects meet each other only at the end of the process of design/implementation of new products, there are significant risks (as there were when the product and the manufacturing system were design separately) to obtain a poor result, in economic conditions far from efficient. Actually, the final goal is really to obtain a product where functional requirements meet technical requirements in an optimal way. So it is preferable that this junction takes place as soon as possible, in order to avoid, at the end, a product-service quite different from the expected product-service. This will to bring together functional and technical within the product is materialised at the level of activity and is manifested by a strong need of concurrency.

Moreover, the integration and the connection of the activities related to the design of the generic product with those related to the implementation of the individual product, are necessary, insofar as one places oneself in a philosophy of concurrent engineering type, i.e. of bringing together the activities of the upstream with the activities of the downstream. The sequentiality of the two flows as the one of the inside various activities and the lack of cross connections between activities of the two flows have critical consequences in terms of performance that the methods of the concurrent engineering can contribute to reduce.

This setting in interaction of technical and functional, on the one hand, and of the design and the implementation, on the other hand, needs links. The company finds them through the intermediate objects, which constitute temporary states of the final product, and through the intermediate recipients that create and use them.

3   Intermediate Recipients and Intermediate Products

3.1   Intermediate Recipients

There are many actors, inside and outside of the company, which come into contact with the technical product or the functional product as with the generic product or the individual products. These actors carry out design and/or implementation activities on these products, in their final or temporary states. The contact with the product, in all its forms, can be more or less direct. All these actors have a common point: the product and the finality of this product (service which it will deliver).

An exhaustive typology (Figure 12: Actors Typology) of all these actors shows their number and especially the variety of the connections which they maintain with the product or its avatars. To represent them, it seems necessary to classify them by category. The selected criteria of categorization are the following:

- the position of the actor inside or outside of the company,
- the activity of design or implementation that he carries out on the object,
- the degree of direction or indirection that he has with regard to the object.

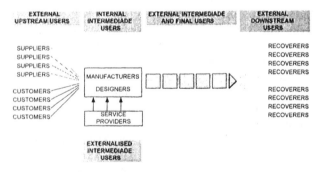

Figure 12 Actors Typology

The set of the actors and the links they weave between them constitutes, according to the words of M. Porter [11], a *chain of value* insofar as they gain to form part of the process and as they enrich this one. Each actor profits, in his activity, of the value brought to the product by the actor who precedes him along the process and adds value to the product before transmitting it to the following.

All the actors are users, more or less temporary, of the product that they receive. They use, more or less partially, this product of which they are the recipients. Each actor is the consignee of the product that he receives, and of this one only. More precisely, the state of the product that is intended to him must correspond, in a particular and precise form, with his requirements in terms of use. To take into account the status of user and recipient of product of all the actors of the chain of value, is to highlight the fact that each actor has to lead his activity by integrating the operational requirements of the product recipients (he modifies the product because of his activity) who are downstream from the process. It is about a factor of organisational coherence. It is in this spirit that we can see the development of that we can call the Design for X. The control of this requirement contributes to the *effectiveness* of the process.

It is necessary to add that the more significant the number of actor is, the more the risk, already mentioned, to see the functional requirements disappearing to the profit of the technical requirements is high. Another obligation is thus made to each one. That is to transmit, at the same time that the technical object, the intentionality specific to the product. The activity that each one carries out aims to contribute to provide, with the final recipient, a precise service. The control of this requirement contributes to the *efficiency* of the process.

Each actor of the process, from the provider of raw material to the recoverer of subsets of the product is thus a product user. The product is designed (finality of the product) to provide a service to the end-user, product recipient. The many users who move throughout process are thus only intermediate recipients of the product. The product that these *intermediate recipients* will handle is an *intermediate product* compared to the final product that the final recipient will use.

3.2   Intermediate Products

The intermediate recipients of the product receive and use intermediate objects. These products are known as intermediate because they correspond to a partial and temporary state of the final product. The activity of the intermediate recipients of these intermediate objects consists in progressively modifying this provisional state by giving it more reality as process develops.

The intermediate product that receives each intermediate recipient constitutes a *vehicle of information* for this one. Indeed, the intermediate recipient will use the information contained in the intermediate product to carry out the task that was assigned to him and to modify, from this information, the state of the product by adding value and information to it. The information provided to each intermediate recipient and contained in each intermediate product must be enough relevant to allow the following intermediate recipient to carry out his activity.

The intermediate product is more or less material depending on whether one is:

- on the virtual process of design which will lead to the generic product or
- on the real process of implementation which will lead to the individual products.

It is, also, more or less material depending on whether one places oneself upstream or downstream from the process of design-implementation of new products.

In this last part, we have spoken essentially about products. Our matter can be extended, in the same way, to the services and the manufacturing systems. There are, in the same way, intermediate services and intermediate manufacturing systems. We will use successively the terms of objects and intermediate objects to deal with the whole of the final and temporary states of the service, the product and the manufacturing system.

Along the design process the intermediate objects are transmitted from an intermediate recipient to another one. They constitute a vehicle of information, a basis of work, for the following recipients. This information is completed by documents. The marketing study, the market study, the specifications of the product, the specification of the manufacturing system... are as many intermediate objects (Figure 13: Design Intermediate Objects).

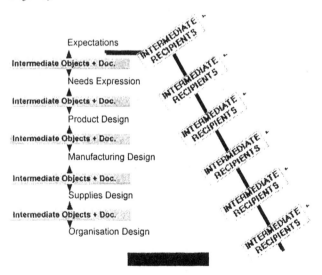

Figure 13 Design Intermediate Objects

In the same way, there are intermediate objects along the process of implementation. These intermediate objects fulfill the same functions as those that are on the design process. The information that they transmit is also supported by documents. The various states of the manufacturing system, the product and the service are as many intermediate objects (Figure 14: Implementation Intermediate Objects).

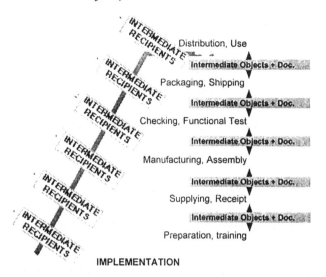

Figure 14 Implementation Intermediate Objects

It is through the intermediate objects, where is located the information, and the working documents which come with them that is transmitted the intentionality. The loss of intentionality remains a permanent danger throughout these transmissions of intermediate objects that are as many translations of the intentionality and thus as many risks of betrayal. This is why documents seeking to express this intentionality come to support the intermediate object. The transmission of these documents follows the process. Moreover, transversely, interactions exist and documents of another nature are transmitted in the same objective. These one-to-one transverse links between actors of the design and actors of the implementation involved in parallel phases of the development of the process are based on a community of interest (Figure 15: Design-Implementation Cross Connections).

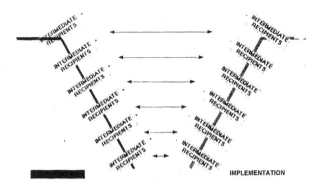

Figure 15 Design-Implementation Cross Connections

4   Conclusion

The design takes an increasingly significant place for the company on the level of its markets as much as of its processes and its activities. To answer this new strategic stake the assistance of the concurrent engineering can be very useful. Indeed, the concurrent engineering proposes to bring closer, as soon as possible, the downstream activities and the upstream activities. Transposed to the new context, it is a question of improving the process of design by:

- an optimized link between the activities of design and the activities of implementation [10],
- a connection between the design of the functional aspects and the design of the technical aspects of the product.

The implementation of this concurrent engineering, adapted to this new context, goes via the well understood use of the intermediate objects by the intermediate recipients.

References

[1]   I. M. Author, "Some Related Article I Wrote," Some Fine Journal, Vol. 17, pp. 1-100, 1987.

[2]   A. N. Expert, A Book He Wrote, His Publisher, 1991.

[1]   J. Chevallier, Produits et analyse de la valeur : AV et analyse fonctionnelle : deux clés pour un produit performant, CEPADUES, 1989.

[2]   S. Bellut, La compétitivité par la maîtrise des coûts : conception à coût objectif et analyse de la valeur, AFNOR, 1990.

[3]   F. Lhote, Ph. Chazelet and M. Dulmet, "The Extension of Principles of Cybernetics towards Engineering and Manufacturing", Annual Reviews in Control, Vol. 23, pp. 139-148, 1999.

[4]   Ph. Lorino, Le contrôle de gestion stratégique : la gestion par les activités, Dunod, 1991.

[5]   Ph. Chazelet and F. Lhote, "From Finalities Representations to Enterprise Modelling", Advanced Summer Institute 1999 of the ICIMS Network, September 22 – 24, 1999, Leuven Belgium.

[6]   Ph. Chazelet and al., "La dualité technico-sociale des produits industriels et ses implications", Journées de travail du GRP, 16-17 octobre 1997, ENS Cachan.

[7]   J. R. Searle, Intentionalité : un essai de philosophie des états mentaux, Les éditions de minuit, 1985.

[8]   Ph. Chazelet and al., "L'intentionnalité dans le processus de projection/production de nouveaux produits manufacturiers", Projet ACNOS, Journées des 17 et 18 décembre 1997, Strasbourg.

[9]   M. Minsky,. La société de l'esprit, Interéditions, 1986.

[10] F. Lhote and al., "La conception des systèmes de production", PREDIT-ICPV, Programme de recherche ACACIA : Analyse et caractérisation de l'Activité de Conduite d'Installations Automatisées, pp. 59-70, Avril 1995.

CHAPTER 16

Applications in CE

Concurrency in Large Made to Order Products: a Case Study in the Design of Offshore Oil and Gas Production Facilities.

Dr. P. Williams, Dr. P. Norman, Prof. I. Ritchey

Engineering Design Centre, University of Newcastle, Newcastle upon Tyne, NE1 7RU, United Kingdom

Abstract

An initial investigation, by the Engineering Design Centre at Newcastle University in partnership with a leading offshore Engineering Procurement and Construction (EPC) contractor in the offshore oil and gas industry, into the EPC contractor's design processes was carried out. The academic and business communities have to date paid less attention to the development of engineering methods in the design and construction of offshore facilities than in volume production sectors. A mature North Sea, and concerns over future price stability have made change an imperative. The case study presented here analyses methods and results deriving from a series of interviews with the contractor's lead engineers. Several themes will be developed throughout the paper: concurrency and it's relationship to the unplanned costs as well as potential solutions to construction rework and abortive work.

1 Introduction

This paper presents an initial investigation into the design practices of a leading UK offshore facility design and construction contractor, originating out of the early stages of a technology transfer program between the industrial contractor and the Newcastle Engineering Design Centre. The company had developed its engineering, procurement and construction business within the UK from a series of acquisitions throughout the 1990s. It had undergone several periods of internal restructuring in an attempt to integrate its IT and business systems across the different functional units (design, construction, commissioning et cetera). Many of the systems in place were legacies from previous takeovers. The technology transfer program was one component of a wider business improvement and reorganisation scheme.

The highly cyclical nature of the oil and gas industry, in which to a large extent the profitability of many oil companies is dependent upon the OPEC countries limiting their supply, has placed a great pressure to reduce costs at all levels throughout the oil and gas industry sector. This is to ensure sustained growth and deliver shareholder value [1]; yet the recent low oil prices and continued doubt over the future stability of oil prices have made investment in the industry problematic [2] especially in high cost production areas such as the United Kingdom and Norway. In a recent report by the Oil and Gas Industry Task Force [3] it was also recognised that there is a general perception that the UK does not attain world class standards in the design and engineering of the offshore facilities. For example bench-marking against the Gulf of Mexico has shown that some facilities can cost up to forty percent more in the UK than GoM. Even taking account of the North Sea's harsh environment and the different regulatory conditions there is considerable room for improvement. The principal area of concern was cultural rather than technical.

The initial drive in the mid part of this century to examine and develop better design methods was a result of the post second world war economic boom and the beginning of the mass consumer market [4]. In more recent times, and especially in the United States, a re-evaluation of the design and manufacturing methodologies in the late 1980s and early 1990s was spurred on by the decline in the USA's productivity, and the fact that by comparison to Japan's economic health throughout the 1980s America's looked poorly.

Since the 1980s there has been a wealth of procedural and cultural initiatives, most of which have originated in the automotive and aerospace businesses, along with their inevitable acronyms and 'buzzword' phrases. For example, Total Quality Management (TQM), Integrated Product Development (IPD), Concurrent Engineering (CE) [5], Virtual Enterprise (VE), Quality Function Deployment (QFD) [6] et cetera. Some of these have migrated to the Large Made To Order (LMTO) sector, yet there have been fewer sector specific initiatives. Jagannathan et al. [7] define concurrent engineering as: "the process of forming and supporting multifunctional teams that set product and process parameters early in the design phase". While Dean and Unal [8] define CE as: "Concurrent engineering is getting the right people together at the right time to identify and resolve design problems. Concurrent engineering is designing for assembly, availability, cost, customer satisfaction, maintainability, manageability, manufacturability, operability, performance, quality, risk, safety, schedule, social acceptability, and all other attributes of the product". Most of the research has focused on production line products (automotive, aerospace and defence industries for example). All of the above design philosophies attempt to combine the various disciplines involved in the development and production of a product at the very early stage [9]. This is to maximise the opportunity for reducing costs, as it has long been recognised that the greatest opportunity for reducing overall product costs is in the concept and early design stages. British Aerospace has estimated that the first five percent of the product development dictates eighty percent of the final cost [10].

There has been relatively less treatment of design methods within the large MTO sector by the academic or wider business arenas than in other industry divisions. This may be a consequence of the very small core in house design capabilities maintained by most large corporations. During times of market activity organisations import temporary labour through agencies or subcontracted consultants [11,12] hampering the development of an optimum design structure. Greater emphasis has been placed on technological solutions, and especially information technology. Hence these procedural and systems undertakings have been accompanied by an equally extensive array of information technologies that attempt to enable the procedural developments to be implemented in design environments dominated by Computer Aided Engineering (CAE). These enabling technologies include developments such as the international standards for data representation exchange protocols [13,14], collaborative design agents [15] and matrix scheduling techniques [16,17].

There are several large software manufacturers purporting to offer design solutions based around concurrent engineering principles, combining multi-disciplinary elements (such as the layout, structural, HVAC, and process engineering tools) into a single unified design environment. Well known examples are:

- AutoCAD (AutoDesk) www.autodesk.com
- Plant Design Management System (PDMS) (Cadcentre) www.cadcentre.co.uk
- Plant Design System (PDS) (Intergraph) www.intergraph.com
- Pro Engineer (Parametric Technology Corporation) www.ptc.com

One important factor in the engineering of offshore installations is the pressure to reduce the time between project conception and realisation as this will reduce costs significantly. For example costs can be reduced for large CAPEX (capital expenditure) projects by reducing the time between sanction and implementation; doing so decreases the finance cost, i.e. reducing the time cost of borrowing. The company used in this research estimated these savings to be $1/4\%$ of the gross project value per month of sanction time eliminated. For example, on a £100m two year project a reduction of the sanction time by one month savings of £125,000 per year would be obtained. Such factors make concurrency in the design and manufacture of large made to order (LMTO) products within the oil and gas industry a highly attractive option: reducing costs and still meeting the clients schedules. The need for change within the oil and gas sector at all levels is widely recognised and has recently been reiterated again in an article in the SPE review [18].

2   Capturing the Design Process

2.1   Interview Procedure

A series of nine semi-structured interviews were carried out over a period of three days at the companies design offices in London. The aim of this investigation was to:

- Develop an understanding of the current design procedures with their associated problems and limitations.
- Determine whether the current work practices reflect the best industry standards and utilise the design tools in the most efficient and innovative way.

From this analysis suggestions for improvements will be made. The Interviewee selection was predetermined by the author and the vice president of engineering at the company in order to obtain the participation of the head engineers from each discipline. The interviews were conducted by the author and a company representative, and divided into two sections. The first section was used to determine the current design management structure, which was achieved through the use of an IDEF0 [19] activity model representation.

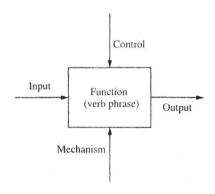

Figure 1   An IDEF0 activity modelling element.

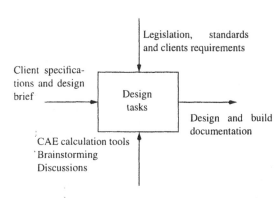

Legislation, standards and clients requirements

Client specifications and design brief

Design tasks

Design and build documentation

CAE calculation tools
Brainstorming
Discussions

Figure 2 An activity modelled using IDEF0.

A pre-prepared single IDEF0 element was drawn on a white-board with the component labels (see figure 1); an explanation defining the meaning and function of each component (input, output, control, mechanism and activity) which constituted the IDEF0 element was then given with examples.

The Interviewee was then asked to fill out the IDEF0 diagram element based on the current work group discipline arrangement, paying particular attention not only to the information type and its definition but also to its source or destination. The results from this stage were then printed off and subsequently transcribed into an electronic format and collated into a single IDEF0 diagram encompassing all of the engineering disciplines. The following section of each interview was based around the questions outlined below, but the interviewees were allowed to expand on areas and diverge if it was deemed that the information would contribute to the overall aims of the investigation.

- How do you view the design methodology as practised by the company, and your discipline's role within the overall process?
- Describe a typical design activity.
- What are the limiting factors, both technological and procedural, on your group's activities which for example prevent an increase in productivity?
- Are there any specific feedback loops or inter discipline interactions which are problematic?

The points and issues originating from the above questions were bulleted during the interview and again printed directly from the white board after the interview was terminated. Throughout the interviews it became obvious that the issues and comments raised fell into two broad categories: procedural concerns and those related to IT. Most of the information obtained from this second section was fairly company specific and therefore it was deemed inappropriate to publish it for public consumption. However more general points could be made; these are detailed in table 1 below.

Table 1: Concerns and issues raised

IT	Procedural
Problems	**Problems**
Different products with the same function which are non compatible.	Single organisational issues: multiple reporting accountability and responsibility lines
Communication difficulties between engineering related but functionally different information technologies.	Lack of feedback and implementation of lessons learnt.
Not exploiting the available tools efficiently or innovatively.	The management of changes introduced through the development of vendor package data.
Desires	Poor planning for changes, and the management of change costs.
A need to expand in-house systems for more functionality.	Improving first guess estimates for early procurement so that changes are less likely.
Reduction to a minimum and the imposition of company wide IT systems.	Late delivery of vendor package data.
	Exceptionally demanding scheduling: specifying long lead items before the process has been fully elucidated.
	Desires
	Improvement in the evaluation of the impact that changes have on downstream activities and the tracking of changes introduced.
	Development of standards (for example, for the control, electrical and instrumentation design) which are set early in project development.

The total duration of each interview ranged from sixty to ninety minutes.

2.2 Feedback Loops and Work Flow

A clearer representation of the feedback loops and scheduling within the six engineering disciplines under consideration can be obtained from the construction of a trivial Design Structure Matrix (DSM), see table 2. The IDEF0 activity model was used to determine the data flows from which the DSM could be constructed. The Design Structure Matrix system was originally developed by Steward to improve the planning of complex systems with a large number of interdependencies [17]. The matrix outlined below is the most basic element of the DSM system and is termed a precedence matrix; this can be used to represent the data interdependencies between activities [20]. An activity is assigned a row and column, and in each the activity has the same relative position to the other tasks in the matrix. The activities are also listed in order of execution, and by reading across a row the inputs needed to perform the respective activity can found. For example EI&T require data from the process, piping and mechanical disciplines. Likewise reading down a column reveals the tasks which receive output data from the activity corresponding to the column. By only including the discipline activities represented on the IDEF0 scheme a false impression of

the work flow and the number of feedback loops is created. The scheme represented by table one would be more typical of a company with little sub contracted design work. This does not reflect the case study company's large dependency on process packages (multiple pieces of process equipment designed and manufactured to form a single unit by a subcontractor, for example a compression package). The introduction of the Package vendors results in the creation of two more feedback loops, as demonstrated in table 3.

Table 2: Precedence matrix of the engineering disciplines.

Engineering Discipline	Process	Piping and layout	Mechanical	EI&T	Structural	Naval
Process		1				
Piping and layout	1		1			
Mechanical	1	1		1		
EI&T	1	1	1			
Structural	1	1	1			
Naval	1	1	1		1	

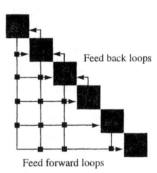

Table 3: Precedence matrix including the package vendors.

Engineering Discipline	Process	Piping and layout	Mechanical	Package Vendors	EI&T	Structural	Naval
Process		1		1			
Piping and layout	1		1				
Mechanical	1	1		1	1		
Package Vendors		1					
EI&T	1	1	1	1			
Structural	1	1	1				
Naval	1	1	1			1	

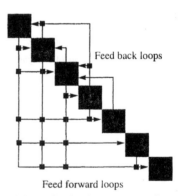

One crucial feedback loop is the data supplied by the package vendors back to the process group. If there is any delay or an unexpected difference between the initial estimates and the data subsequently returned this could potentially have a huge detrimental effect on the design project as the changes introduced ripple out to all the other disciplines. The detailed design process continues throughout much of the time that the facility is being constructed. Changes to an erected structure cost considerably more than an equivalent piece of work

performed from fresh, due to work disruption, setting up area access, material wastage et cetera. This analysis is confirmed by the concerns raised during the interviews: a large number of comments were directed at the way in which changes were generated and managed. Examination the documents used to approve design changes show that for a typical CAPEX project only twenty percent of post construction changes could be booked to the clients account.

After the pre-qualification selection (which will set the minimum standards for quality, safety and environmental management) it is assumed by the contractor that he will be able to meet the client's scheduling constraints. Once the contract has been won the main project driver switches to that of meeting the client's scheduling. This simple driver shift is crucial to understand the way in which the engineering contractors manage their design procurement and construction business. The traditional work flow model typified by figure 3 below is no longer an accurate or an appropriate representation of the way in which large made to order products in the offshore oil and gas industry are designed and fabricated. The detailed process and conceptual engineering design activities (where the piping and instrumentation diagrams, heat and mass balance, line lists, preliminary equipment specifications et cetera are fully developed) are commonly termed as the Front End Engineering Design (FEED).

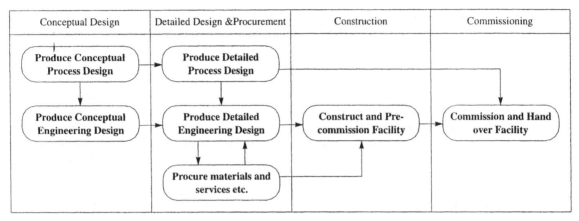

Figure 3 Extract from 'Process Plant Engineering Activity Model' [21].

Modern engineering schedules have dictated that the time allowed for the FEED, detailed engineering design and construction stages have been compressed, which has resulted in break down in the distinction between these three work phases at the macro level. These activities have been deconstructed into smaller work units or modules which can be performed in parallel. For example:

- The generation of process and instrumentation diagrams (P&ID) is performed in parallel with the erection of structural steel work, allowing the installation of pipe section by section soon as the relevant P&ID has been produced.
- Procurement of long lead items is performed early on in the front end engineering design stage, before the detailed process design has been fully elucidated.

An internal company audit into the delays in fabrication emphasised the importance of design changes as the major cause of project overruns (see table 4). From which it can be deduced (assuming eighty percent of design changes are internal) that on average forty percent of all delays are a result of design development changes, rather than alterations originating with the client, which is the largest contributor to fabrication inefficiencies.

One obvious outcome of an economic environment with low market activity and a high number of engineering and construction contractors looking for work is that their clients are in a strong position to pass on to the contractor a large amount of the risk in the development of offshore facilities. This can lead to the contractor buying work with low bids hoping to survive

Table 4: Reasons for fabrication delays [22].

50%	Design change
25%	Bulk materials delivery failures
8%	Late equipment
8%	Design - fabricator interface
9%	Other

until market activity increases or rival firms close thereby removing capacity from the system. The emphasis on transferring risk to the contractor is reflected by the oil companies' shift towards functional specifications and distancing themselves from involvement with the front end engineering and detailed design stages. Hence only the conceptual design phase has not been integrated into the concurrent engineering cycle as this has remained in the control of the operating client companies.

3   Discussion

3.1   Developing Standard Modules

Integrating the front end engineering with the detailed design and procurement, as commented on above, can reduce the sanction and development time. Yet if the greater number of feedback loops cannot be managed successfully this will result in a financial loss for both client and contractor. The client will be less likely to meet production targets, due to late delivery, and the contractor will incur large reworking costs.

Improving the certainty of vendor package data early on in the design process would reduce or eliminate many of the negative effects of the package vendor to process and piping (via mechanical) feedback loops. A comparison study between the GoM and UKCS highlighted the different package engineering philosophies. In the GoM a range of effective standard skid mounted packages has evolved to meet most of the operators' needs. A large number of similar skid mounted units are required for both onshore and offshore operations [23]. A culture has emerged where there is no desire to over engineer. Previous engineering solutions are modified rather than completely redeveloped for each new facility. Standard procedures and "cook book solutions" have been used to minimise the engineering development time. Engineering costs are typically four percent of the total topside costs versus ten for the UKCS. This arrangement is likely to have developed because of the diversity of the GoM oil field sector. There are numerous fields, small operators, designers, suppliers, fabricators and installation contractors were there is a strong competition and a desire for minimum cost solutions. This diversity is supported by the lower CAPEX entry costs into shallow water (in many cases less than 40ft) GoM development, where the environmental conditions are far more benign than in the UKCS. As would be expected, the GoM costs escalate for operating in deeper waters where there is no market to supply readily available standard units, and equipment has to be integrated into a multi deck structure.

CRINE (Cost Reduction In the New Era) was an initiative supported and funded by the UK oil and gas exploration and production industry (UK Offshore Operators Association, the UK Department of Trade and Industry, UK & international trade associations as well as many commercial companies) which has attempted to increase the global competitiveness of its participants by reducing the capital costs of developing oil and gas fields by some 30%. This organisation has made many recommendations: one of the most important, since the mid 1990s, was the introduction of standard functional specifications [23] as "Prescriptive company specifications were also considered to inhibit both innovation and standardisation". CRINE functional specifications were developed as frame works for the procurement of vendor packages (i.e. excluding commodity items: valves flanges and other fittings) such as control systems, chemical injection packages, filtration packages et cetera, not as ready for procurement specification, yet this development has been poorly received [24]. The standard functional specifications aim to replace operators' and contractors' own detailed requirements. Suppliers would then have the opportunity to develop their own proprietary equipment around ISO standards [25-27]. In the development of CRINE's own strategy for the implementation of functional specifications they have attempted to incorporate the relevant ISO standards.

One example given above of concurrent engineering work-flow was the installation of piping as soon as individual sections were routed and codified by a P&ID. Also noted previously was that even small changes post-fabrication, because of access restrictions (scaffolding may need to be re-erected), material wastage et cetera, make rework highly undesirable. If the functional specifications for packages incorporated standard utilities, control, power, and process interfaces along with firm dimensional restrictions, this would decrease the likelihood of layout changes, as the package modules could be treated as 'black boxes' and arranged accordingly. Hence the structural steel work and piping could be erected and installed prior to the delivery of the subcontracted package (as with current practice), but with a higher degree of confidence that rework will not be required. The use of standard functional solutions would remove much of the necessity for the mechanical engineering discipline group to specify the vendor package details; this function could be performed by the process group in parallel with the development of the equipment list. This would also have the effect of shortening the package vendor-process feedback loop (see table 3) increasing the chance of capturing errors pre-fabrication. Reducing the effects of the feedback loops through the use of standards was one topic raised in the interviews by the lead engineers who commented that early specification of project standard for the control system would speed up the control design. The E&IT discipline group would reduce its dependence upon data from the package vendors: again increasing first guess certainty decreases the effects of the feedback loop (E&IT to mechanical disciplines, see figure 3).

The concept of using modularity combined with specifications which impose and police interfaces as a method of structuring systems and managing complexity is not new. This technique has been used for some time to develop and manage the design of software [28]. With the domination of the engineering design processes by electronic formats and tools there has been increasing interest in the transfer of concepts and methodologies from software engineering into the design and fabrication in more traditional engineering disciplines.

3.2   Is There The Will and Resources for the Investment in Change?

Without price predictability, to which market activity forecasts are strongly related, EPC business will be reluctant to commit the investment required to develop a catalogue of pre-optimised solutions and maintain development outside the scope of a particular client contract. There is continued uncertainty in the stability and magnitude of the price rebound from lows of 10US$/bbl. Yet the current oil price movements have not affected to any great extent oil price projections beyond

2005, which continue to have a wide range from the mid teens to the low thirties (1997 US Dollars per Barrel) [29]. But it should be noted that regardless of the price scenario produced by the United States Energy Information Administration (EIA) oil consumption will rise considerably, on average 1.8 percent per year over the next twenty years, with the developing world contributing to two thirds of this expected growth. Moreover,

'...[in the near term], deep water exploration and development initiatives are generally expected to be sustained world wide with offshore West Africa emerging as a major future source of oil production. Technology and resource availability can sustain increments in oil production capacity at prices ranging between $18 and $22 per barrel. The current oil price regime will, however, slow the pace of development in some highly prospective areas, including especially the Caspian Basin region...[29]'

Hence, two conclusions can be drawn: oil consumption will increase with the industrialisation of the developing world; and, the major new offshore developments will be out side the UKCS and GoM. Therefore UK EPC contractors will have to compete for work in an international arena. In a recent press release Kvaerner Oil and Gas stated that it would reduce its reliance on the North sea from 80 per cent to 50 per cent of all business activity.

As insurance against price instability and in an attempt to ensure future profitability there has been considerable consolidation in oil companies. Similar consolidation might be required in the EPC sector before standardised modules and engineering solutions can be developed. Though the highly cyclical nature of the upstream oil and gas industry and the EPC contractors' low volume output have contributed in deterring major design process alterations, such as a shift towards modular based designs, the UK Oil and Gas Industry Task Force's Competitiveness Working Group claimed, as previously indicated, that the principal impedances to improving competitiveness were cultural [3]. To improve the UK offshore industry's international competitiveness a further joint industry and UK governmental organisation, LOGIC (Leading Oil and Gas Competitiveness), was established in September 1999 to take forward and implement CRINE's recommendations in several areas:

- Improve supply chain management via consultancy support and training.
- Co-ordinate projects aimed at supporting and promoting collaboration across industry in the execution of non competitive activities.
- Benchmark capital and operating expenditure on a global basis.
- Identify leading practices and transfer learning.

The establishment of LOGIC, as a replacement for CRINE, was the prime deliverable of the Competitiveness Working Group. Though the Competitiveness Working Group highlighted that a standardised approach to design which maximised repeatability could potentially produce up to a forty percent reduction in the overall cost of topside facilities, the development or implementation of a program which could bring about greater opportunities for repeatability or increased economies of scale was omitted. It was recognised that:

'...[it was] outside the scope of this study to identify ways in which these could be changed in the urgent time scale of actions required by the Task force. [30]'

In summary the economic environment, the joint industry and governmental organisations' focus on supply chain management and the culture of co-operation, have meant that examination and development of improvements to the design and manufacturing process of the offshore facilities have remained relatively unexplored.

4   Summary and Conclusions

Increased use of modules. The adoption of functional specifications enabling EPC contractors to move towards pre-engineered solutions and a 'product catalogue', within a frame work for regulating module interfaces, looks to be a highly promising way forward in an attempt to reduce facility development and build costs. The depth of a module's pre-engineering is dependant upon the individual economic conditions, but all modules should adhere to the interface regulations. Controlling the interfaces can allow for optimisation of the design information flows and subsidiarity in change management. This clearly identifies alterations that are local and hence do not ripple out beyond the module in question.

However, the recommendation for utilising both functional specifications and standard module interfaces can seem contradictory as standards impose conditions which are not purely functional. The relationship between an EPC contractor and a client operator could be based around 'pure' functional specifications. This could be achieved as the number of dependencies which impose upon an offshore installation and the way in which it interacts with its wider environment are obviously far fewer than the number of dependence relationships for an individual module within an integrated production platform itself. Functional specifications can allow the EPC contractor to develop optimum designs which can suit a range of clients, permitting reuse and the development of a product catalogue. But the reliance on subcontracted packages restricts the advantages of adopting a purist functional specification approach to the subcontractors. Reducing the effects of the dependencies represented by the feedback loops highlighted earlier in this paper can be achieved by developing a hybrid approach where the functional specifications operate within a framework standardising the interfaces between modules. This combination would improve the efficiency of the design process by increasing the certainty of the initial estimates in a feedback loop: standard module interfaces have been agreed upon and similar products arrangements form part of the EPC contractors solutions catalogue. Further research is required to:

- Determine why up until now the CRINE functional specifications have been poorly received, and whether industry wide standards would be more appropriate than EPC contractors developing alliance relationships with their engineering package suppliers.
- Develop a methodology that could identify a product or module which pre-engineering and standardisation would suit better rather than directing resources into developing technological solutions to speed up the design and realisation of a module.
- Quantify the economic relationship between the extended functionality obtained from a standard module and the loss imparted to the client from not having the most fit for purpose design. Once this relationship has been identified it could then be used as a criterion for searching the design space to identify the most economically suitable solution.

Skills. With the common use of multi-functional software tools (integrating the process, structural and piping designs) and with the need to respond rapidly, once a project has gained approval for commencement by increasing manning levels, a shift is needed towards multi-skilled personnel who can take advantage of CAE environments which span the traditional engineering discipline boundaries. It is well beyond the scope of this paper to examine the agency and staff labour market within the oil and gas industry. Nevertheless without structures in place that enable life long learning capital intensive industries relying on a flexible use of agency staff will continually find skill shortages as the pace of technological advancement is not matched by the skill sets available.

Acknowledgments

The authors would like to express their thanks to: all our colleges and industrial partners at the Newcastle Engineering Design Centre who have provided invaluable comments and assistance (for further information on either please read our website at www.edc.ncl.ac.uk); the TCD for their support and funding; M. Maglio for her help in proof-reading this paper.

References

[1] Lukman, R., "Current Challenges and Market Outlook: An OPEC Point of View", 10th Montreux Energy Roundtable, Montreux, Switzerland, 19th April 1999.

[2] Lukman, R., "OPEC's view on Policy Options for the Oil Industry in the Middle East and North Africa in a Globalized Economy", Afternoon Ministerial Session of the 9th Annual Conference of the Centre for Global Energy Studies, London, United Kingdom, 22nd - 23rd April 1999.

[3] OGITF Oil and Gas Industry Task Force, "The Oil and Gas Industry Task Force Report: A template for Change, Department of trade and Industry", September 1999, DTI/Pub 4465/8k/10/99/AR. URN 99/924.

[4] Birmingham, R., Cleland, G., Driver, R., Maffin, D., *Understanding Engineering Design: Context Theory and Practice*, (Hertfordshire, United Kingdom: Prentice Hall, 1997).

[5] Blackhouse, C. J. and Brookes, N. J. (Ed.), *Concurrent Engineering: What's Working Where*, (Hampshire, United Kingdom: Gower Publishing, 1996).

[6] Dean, E. B., "The Many Dimensions of Program management", 14th Annual Conference of the International Society of Parametric Analysits, Munich, Germany, 25-27th May 1992.

[7] Jagannathan, V., Cleetus, K. J., Kannan, R., Matsumoto, A. S., and Lewis,J. W., "Computer Support for Concurrent Engineering: Four Strategic Initiatives," Concurrent Engineering, September/October, pp. 14-30.

[8] Dean, E. B. and R. Unal, "Elements of Designing for Cost", presented at The AIAA 1992 Aerospace Design Conference, Irvine CA, 3-6 February, AIAA-92-1057.

[9] Ramana, Y. V., et al, "Computer Support for Concurrent Engineering" IEEE Computer, Special Issue Vol 26 No. 1 January 1993 pp 12-16.

[10] Krajewski, L. E. and Ritzman, L. P., *Operations Management Strategy and Analysis*, (Reading, Mass., USA, Addison Wesley, 1993).

[11] Edwards, R. *Contested Terrain: the Transformation of the Workplace in the 20th Century*, (London, UK, Heinemann, 1979).

[12] Kennedy, P., "An Analysis of Design Production Using Social Science". Published in: Tiala, P., Smrcek, L., (Ed.) *Proceedings of the First International Conference on Advanced Engineering Design*, Prague, 1999.

[13] ISO 10303-1:1994 "Industrial Automation Systems and Integration - Product Data Representation and Exchange, Part 1: Overview and Fundamental Principles", TC184/SC4.

[14] ISO/WD 15926-1 (E), "Industrial automation systems and integration - Integration of life cycle data for oil and gas production facilities - Part 1: Overview and fundamental principles", TC184/SC4.

[15] Florida-James, B. et al, "An Agent Mechanism for Version Support in Engineering Design", Proceedings of the 1998 ASME Design Engineering Technical Conference, Atlanta Georgia, USA, 13th -16th September 1998.

[16] Rogers, J. L., "Reducing Design Cycle Time and Cost Through Process Resequencing", Proceedings of the International Conference on Engineering Design (ICED'97), Vol. 1. pp. 193-198.

[17] Steward, D. V., *Systems Analysis and Management: Structure, Strategy and Design*, (New York, USA, Petrocelli Books, 1981).

[18] Clutterbuck, P., "Oil must learn from other industries", *SPE review*, Issue 122, November 1999, pp 6-7.

[19] National Institute of Standards and Technology, Computer Systems Laboratory, *Integration Definition for Function Modelling (IDEF0)*, (Springfield VA 22161, USA, National Technical Information Service US Department of Commerce, Federal Information Processing Standards 183, December 1993).

[20] Scott, J. and Sen, P., "Reducing Product Development Lead Times Using The Design Structure Matrix System",

[21] Harrow, P. "Process Plant Engineering activity Model", PISTEP, June 1994.

[22] Internal company data collated in mid 1998.

[23] Hall, J., "Gulf of Mexico Comparative Study" CRINE 14th May 1999.

[24] CRINE "Guidance Notes for Effective use of Functional Specifications" CRINE, February 1999.

[25] ISO 13879:1999, "Petroleum and Natural Gas Industries - Content and Drafting of a Functional Specification", TC67.

[26] ISO 13880:1999, "Petroleum and Natural Gas Industries - Content and Drafting of a Technical Specification", TC67.

[27] ISO/DTR 13881, "Petroleum and Natural gas Industries -- Classification and Conformity Assessment of Products, Processes and Services", TC67.

[28] McDermid, J. et al, "Towards Industrially Applicable Formal Methods: Three Small Steps and One Giant Leap" Second International Conference on Formal Engineering Methods, Brisbane, Australia 9-11 December 1998. Published in: Staples J., Hinchey, M. and Lui, S. (Ed.) *Formal Engineering Methods*, IEEE Computer Society pp 76-88.

[29] Energy Information Administration, *International Energy Outlook 1999* (Washington DC, USA, Office of Integrated Analysis and Forecasting, March 1999).

[30] Oil and Gas Industry Task Force, *Task Force Supplementary Papers - Competitiveness. A Comparative Study of the UK and Gulf of Mexico Topsides Facility Design and Fabrication Practice* (Aberdeen, UK, Infrastructure and Energy Projects Directorate (IEP), 1999)

Formal Definition of Premises for Concurrent Design Process of Large Complicated Systems and Their Consequences

Kikuo Fujita

Department of Computer-Controlled
Mechanical Systems
Osaka University
Suita, Osaka 565-0871, Japan

Shin'ichi Kikuchi

Department of Computer-Controlled
Mechanical Systems
Osaka University
Suita, Osaka 565-0871, Japan

Abstract

This paper discusses concurrent design process of large complicated systems toward the establishment of formal fundamentals that are necessary in the development of agent-based distributed design systems. Under the assumption that an artifact is modeled with a set of lumped mass systems and that the number of variables is fairly huge, the necessity of hierarchical and horizontal decomposition of design tasks is formulated under the time-boundness of design process, granularity levels in modeling and consequent fidelity degrees of respective models. This leads the understanding that concurrent design process is under goal cascading across various aspects toward total compromise of overall design requirements. This paper concludes with the definitions of three typical situations in goal coordination and required mechanisms in the implementation of agent-based concurrent design systems.

1 Introduction

Design problem of large complicated artifacts[1] such as ships, aircraft, automobiles, etc. is solved by a large engineering team through long design process. The state-of-the-art of digital technologies are struggling to improve and enhance the performance of such design process through geometric modeling, seamless integration with computer-aided engineering, total product data management, etc. Behind this movement, concurrent engineering has been a main stream of design engineering research in the last decade. While its focus is widely spread over from life-cycle engineering to computer-supported cooperated work, concurrent engineering concept provides some backgrounds on those digital technologies. However, the outcomes of computer based concurrent engineering technologies are still subsidiary in the aspect of design problem solving. For instance, geometric modeling and consequent integration with engineering analysis contribute to facilitate design consideration across various aspects such as different disciplinaries, life-cycle phases, but the sum of them cannot perform design itself due to the lack of synthesis viewpoint. Thus, concurrent design problem solving must be settled as a next major stream of concurrent engineering research.

Toward this direction, Fujita *et al.* developed an agent-based distributed design system architecture [1] and proposed task distribution framework [2]. Their standpoint is that computer-supported concurrent engineering systems require three steps of development; distribution of tasks, parallelization of tasks, and coordination among tasks. The former provided a basis for distribution with an application to basic ship design. The latter tacked on its extension to the other steps through aircraft design example. While applications are indispensable to deal with design problems of large complicated artifacts, any formal background must be essential to generally establish rational means for computer-supported concurrent engineering.

This paper discusses the concurrent design problem solving of large complicated artifacts under the above standpoint. In order to go into synthesis aspects apart from application contents, the discussion is explored under the assumption that an artifact is modeled with a set of lumped mass systems. Further, it is assumed that the most rational design over such a representation is not possible and any relatively rational design result must be sought within a given time limit. These assumptions lead the formal necessity of concurrent design process and a form of distribution and coordination among divided design tasks

---

[1] The paper title uses a term of *systems* instead of artifacts. However, since the meaning of systems includes natural systems, computer systems and so forth, we distinguish artificial systems that are subject of engineering design and the others with the term of *artifacts* throughout this paper.

through goals. This paper concludes with representative coordination situations under cascaded goals.

Besides, the design problem solving has been already studied widely with design examples of small devices. However, the size of design problems must be essential on concurrency, coordination and so forth in design problem solving.

2 Simplified Representation of Design Problem of Large Complicated Artifacts

2.1 Formal investigation into concurrency in design process

'Theories versus applications' is a measure for research categorization. While theories tend to be general, they may not be directly effective in applications. Since concurrent engineering was rather initiated with the necessity in application world, the tools and methods in concurrent engineering aim for effectiveness in applications apart from theories. However, for instance, theories in engineering research have provided better understanding of design, even though their outcomes do not have any direct effects in applications. Further, mathematical theories can have some appropriate nature to access the essential fundamentals in a specific class of problems.

While the issues related to concurrent engineering are widely spread, concurrency of design process is demanded by the size of an artifact and associated design problem. In general, larger problems must take much time and effort to solve. Various methods and algorithms have been developed to shorten such time and to reduce such effort. But, concurrent engineering must be stand on the viewpoint that we cannot find the best design solution within a bound time even with excellent methods and algorithms, and that instead we need to find a better solution with 'limited rationality.' Since concurrency and consequent parallelism of design tasks can put much effort in a fixed time scale, the structure of task division and coordination among them must be essential toward better limited rationality.

This paper investigates these issues to theoretically clarify premises for concurrent design process and necessary prerequisites of computer-supported concurrent design systems for large complicated artifacts.

2.2 Representation as a set of lumped mass systems

Toward the above direction, this paper assumes that the representation of an artifact can be as follows:

All contents for designing an artifact can be represented with a series of real continuous variables, the number of which is very huge. The relationships among them, that are functions, are fairly smooth.

This indicates that the sizes of a design problem and its divided tasks are countable with the number of variables related, and that partial influence of design changes to the other related variables is predictable somehow. While this situation seems to be too simple, the huge system of variables and relationships generates a form of enough complicated design problems in the aspects of size-oriented difficulties in design problem solving and their relaxation methods to obtain possible solutions.

Besides, the above assumption is also valid to eliminate any particular contexts such as geometric representation, contents of engineering analysis, product life-cycle issues and so forth from the consideration.

2.3 Types of design-related variables and their relationships

The design can be stated as a problem to find an imaginary representation of an artifact that will accomplish required performance under the real environment. Under this statement and the above assumption, the design-related variables can be categorized as follows[2]:

- *Decision variables* $\cdots$ Variables representing how an artifact is apart from the environment. Since design must directly determine such variables, they can be referred as *decision* variables[3]. They are denoted as x in the following.

- *Performance variables* $\cdots$ Variables representing what and how an artifact *performs* in the environment. They are denoted as z.

- *Intermediate variables* $\cdots$ When assuming design can be divided into subparts, the relationships between decision variables and performance variables must be correspondingly decomposed into subsets. This requires *intermediate* variables between both. They are denoted as y.

When following to the standpoint that analysis activity means to simulate the performance of an artifact, analysis

---

[2] This categorization partially follows to the definitions by Kusiak and Wang [3] and by Sreeram and Chawdhry [4], while our past research [2] used the categorization into design items, intermediate items and objective items.

[3] They correspond to design variables in a sense of design optimization.

can be represented as an activity to obtain z against given x by solving the following form of equations:

$$z = f(x) \qquad (1)$$

Further, since intermediate variables are between decision variables and performance variables, the above directionality from decision to performance brings the following interpretation of Eq. (1) with intermediate variables:

$$\left. \begin{array}{rcl} y_i &=& f_{y_i}(x_{y_i}, y_{y_i}) \\ z_j &=& f_{z_j}(x_{z_j}, y_{z_j}) \end{array} \right\} \qquad (2)$$

Where, the suffixes in these equations do not mean exact corresponding relationships but distinction among terms.

Besides, Equations (1) and (2) imply that the function $f(x)$ can be composed of a set of functions like f_{y_i} and f_{z_j}.

3 Interpretations of Some Design Theories and Models under Simplification

The meanings of some theories and tools on design synthesis under the above assumption are instructive to reveal the direction of discussion.

3.1 Design process perspective

Yoshikawa's general design theory [5] was a pioneering work that formally models the design problem. The theory is based on the mapping relationship from functions to an entity, and design is defined as an inverse problem to seek an entity that can perform given functions. Since his concern is conceptual design, the theory employs topology space as mathematical framework, while the assumption in this paper has no concerns on conceptual design or topology. However, the shape of an inverse problem is significant in the form of Equation (1). In his standpoint, the shape indicates that perfect design requires perfect knowledge on all artifacts including ones that would be invented in the future, and that it might be practically impossible when considering the wide possibility of artifacts. By transferring these observations to a set of lumped mass systems, if we could have the explicit representation of $x = f^{-1}(z)$, we could solve the problem with less time and cost. But, it is obviously impossible under a huge system of variables as a nature of mathematical equations as well as design concepts.

Simon explored the sciences of the artificial [6], in which he provided two concepts, *satisfying* and *nearly decomposable systems*, as fundamental characteristics in designing artifacts. The former concept means that the optimal solution is impossible in designing an artifact and

that a solution is a result of compromise among various criteria. The latter means that the contents of an artifact can be decomposed into subparts, but interactions among them can never be neglected, but that a better organization of decomposition can bring a better solution since it can partially resolve the complicatedness natively associated with an artifact. These issues are straightforwardly interpreted as follows: While a design solution must be restricted under limited rationality, better organization and utilization of Eq. (1) through the form of Eq. (2) can bring a better solution.

3.2 Computational framework perspective

Many computational design schemes have been developed under the representation with variables and relationships among them.

Constraint-directed reasoning techniques have been applied to design problem solving in this direction. Such researches proved the effectiveness of constraints in design through bi-directional reasoning mechanism over variables, since design is an inverse problem. While it seems to be straightforwardly applicable to the system defined with Eq. (1), the size-oriented difficulty has been out of focus. In the other words, while constraints paradigm can handle a relatively small set of variables, it is doubtful that it can simultaneously handle a huge set of variables.

Another category of computational schemes is game-theoretic negotiation and similar ones. Such schemes are essential and important for coordination among agents in concurrent design systems. They are supported by well-established theories and their validity is ascertained through applications. However, they still have distances to real applications in the aspect of problem size as well.

What is missing in present computational design frameworks? The answer may share some background with other computational methods across different fields. As the above two cases indicate, it must be related to the size-oriented difficulty that can never be overcome natively even with any means. Studies into emergent system theory focus on the interaction between an agent and its environment. When the size of a problem is measured with both agent and environment, it may be grown to be infinite even for a very small problem, since the boundary of a system cannot be defined under emergence. Emergent system theory is also efficient to understand the design problems that are difficult to understand with a traditional sense [7]. Further, limited rationality has become a shared key concept in artificial intelligence beyond the shortcoming in the past paradigms [8].

An underlying common concept of these movements is the explicit consideration on *time-boundness* in problem solving. This must be a significant issue for concurrent engineering research as well.

4 Types and Components of Design Knowledge under Time-Boundness

4.1 Phase based categorization of knowledge

Phase based understanding of design process is often useful. Dym defined the four task phases of design problem solving; *design*, *verify*, *critique* and *modify* [9]. Following to the categorization of design-related variables, his definition of tasks can be interpreted as follows:

- *Design* means to set the decision variables x.

- *Verify* means to obtain the performance variables z.

- *Critique* means to make some judgment on the obtained performance variables z.

- *Modify* means to adjust the decision variables x based on the result of critique.

Further, the directionality of the function $f(x)$ in Eq. (1) leads the following categorization of design knowledge over the above task definition:

- *Forward-chaining knowledge* that corresponds to $f(x)$ used in verify phase. This is the knowledge of analysis as aforementioned.

- *Backward-chaining knowledge* that corresponds to $f^{-1}(z)$ in design and modify phases. The nature of design activity as an inverse problem indicates that even under the explicit representation of $f(x)$ the explicit representation of $f^{-1}(z)$ is not available except the cases where the number of argument variables is quite small and the form of $f(x)$ is extremely simple. This is a reason for the iteration of the above task phases.

- *Critique knowledge* that does not directly link with $z = f(x)$ and that are mainly used in critique phase to measure whether z are superior or inferior toward compromise within them. This is necessary because design requirements that are given as specifications are temporary and they must be refined or relaxed through the reflexion of gotten z so as to find a feasible and satisfactory solution.

While analysis knowledge concerns the first one, synthesis knowledge includes all three types. For instance, design optimization, which is a very simple form of computational design synthesis, consists of optimization algorithm and analysis code. The former corresponds to backward-chaining knowledge, and the latter corresponds to forward-chaining knowledge. However, since optimization does not concern on the value of a design result, its framework does not include so-called critique knowledge.

4.2 Beyond real mapping from decision to performance

First, the discussion focuses on the fidelity of forward-chaining knowledge among the above categories.

Research into analysis is seeking better understanding of physical phenomena. 'Better' means here high fidelity as much as possible. The hidden side of this dogma implies that analysis can predict performance of an artifact but it cannot perfectly show its real performance in a real environment. If real performance would be sought, the number of decision variables for representing an artifact would go up to the infinite, and time and cost to manipulate the representation would go to the infinite as well. That is, perfect exactness is impossible in engineering models.

Beyond the impossibility of real mapping from decision to performance, the utility of any forms of analysis is highlighted as forward-chaining design knowledge. That is, the above meaning of 'better' is not essential in designing; knowledge as a poor understanding is handy and cheap, while knowledge as a better understanding is heavy and expensive, in its utilization. Further, when consider the situation where utilizing any knowledge in a fixed time-scale, the former can go into wider and global issues, but the latter can go only into narrow and local issues. Thus, every grade of fidelity in knowledge is useful or necessary for designing an artifact.

4.3 Granularity in decision variables and design knowledge

The concept of granularity [10] makes the above point more clear. When viewing an artifact, we can see it only through representative features, or we can look its narrow precise details as well, by switching views [12].

In the form of Eq. (1), while a whole of x represents an artifact, some elements are representative and some others are subsidiary. That is, various adjusted subsets of decision variables can redundantly represent an artifact under different views. When such subsets are denoted as $x_{g_1}, x_{g_2}, \cdots, x_{g_i}, \cdots$ from coarse view to fine view, the relationship among them is defined as follows:

$$S(x_{g_1}) \subset S(x_{g_2}) \subset \cdots \subset S(x_{g_i}) \subset \cdots \tag{3}$$

Where, $S(\bullet)$ means the set of elements listed in a vector $\bullet$.

According to the granularity levels in decision variables, forward-chaining knowledge $z = f(x)$ is varied based on to which granularity level of decision variables it is applied. When denoting subsets of knowledge that are applicable over x_{g_1}, x_{g_2}, $\cdots$, x_{g_i}, $\cdots$, respectively as $f|_{x_{g_1} \to z}$, $f|_{x_{g_2} \to z}$, $\cdots$, $f|_{x_{g_i} \to z}$, $\cdots$, the relationship among their utilization costs is expected as follows:

$$C\left(f|_{x_{g_1} \to z}\right) < C\left(f|_{x_{g_2} \to z}\right) < \cdots < C\left(f|_{x_{g_i} \to z}\right) < \cdots \tag{4}$$

Where, $C(\bullet)$ means the cost for operation of $\bullet$.

The tendency of Eq. (4) is partially caused by the relationships on the number of related variables shown in Eq. (3). However, what is more importantly it is rather caused by the contents of $f(x)$. For instance, while the most coarse form of $f|_{x_{g_1} \to z}$ could take a form of algebraic equations, any fine form of $f|_{x_{g_i} \to z}$ requires any numerical simulation or else.

This difference is more significant in the contents of backward-chaining knowledge, since algebraic equations may be solved into the inverse forms but numerical simulation or else cannot be solved in similar ways. This means that explicit representation of backward-chaining knowledge must be very limited under the contents of forward-chaining knowledge.

4.4   Fidelity levels in performance variables

The granularity in decision variables and forward-chaining knowledge indicates that a specific performance variable can be computed through several different ways and that it is multiplely represented with different levels of fidelity correspondingly. The relationship among such fidelity levels of a performance variable z_j, ($z_j \in S(z)$), is expected as follows in correspondence with a series of forward-chaining knowledge:

$$\Phi\left(f|_{x_{g_1} \to z_j}\right) < \Phi\left(f|_{x_{g_2} \to z_j}\right) < \cdots < \Phi\left(f|_{x_{g_i} \to z_j}\right) < \cdots \tag{5}$$

Where, $\Phi(\bullet)$ indicates how the value of $\bullet$ is trustworthy.

In the reality in real environment, a performance variable must be verified and critiqued with high fidelity finally in design process. However, verification of performance variables with less fidelity must be necessary due to its less cost shown in Eq. (4).

4.5   Dominatedness in design knowledge

Behind forward-chaining design knowledge and various differences among the operations of its segments, how

backward-chaining design knowledge is formed and acts is essential to reveal the characteristics of design process of large complicated artifacts.

Consider the situation where a performance variable z_j is supported by two forward-chaining knowledge segments $f|_{x_{g_{i-1}} \to z_j}$ and $f|_{x_{g_i} \to z_j}$ under the relationship $S(x_{g_{i-1}}) \subset S(x_{g_i})$ on granularity level. Natively, it is expected that $f^{-1}|_{z_j \to x_{g_{i-1}}}$ and $f^{-1}|_{z_j \to x_{g_i}}$ could be formed somehow, if their operations could use infinite cost. However, backward-chaining knowledge as the exact inverse of forward-chaining knowledge is impossible as aforementioned. Thus, their executive approximations, which are empirically established, are denoted as $\widetilde{f_{g_{i-1}}^{-1}}|_{z_j \to x_{g_{i-1}}}$ and $\widetilde{f_{g_i}^{-1}}|_{z_j \to x_{g_i}}$, respectively. 'Empirically' means here that an executive approximation can recommend some suitable values for decision variables rather than their best values for accomplishment of desired performance.

Further, under observing the form of actual design knowledge in some applications [13, 14, 15], when denoting $S(\widehat{x_{g_i}}) = \left\{ x_\ell \,|\, x_\ell \in \overline{S(x_{g_{i-1}})} \cap S(x_{g_i}) \right\}$, $\widetilde{f_{g_i}^{-1}}|_{z_j \to x_{g_i}}$ is divided into $\widetilde{f_{g_i}^{-1}}|_{z_j \to x_{g_{i-1}}}$ and $\widetilde{f_{g_i}^{-1}}|_{z_j \to \widehat{x_{g_i}}}$. Rather, $\widetilde{f_{g_i}^{-1}}|_{z_j \to \widehat{x_{g_i}}}$ is possible, but $\widetilde{f_{g_i}^{-1}}|_{z_j \to x_{g_{i-1}}}$ is impossible. In the other words, the utilization of $\widetilde{f_{g_i}^{-1}}|_{z_j \to \widehat{x_{g_i}}}$ requires the values of $x_{g_{i-1}}$ as its prerequisite.

Under these relationships[4], the utilization of $\widetilde{f_{g_{i-1}}^{-1}}|_{z_j \to x_{g_{i-1}}}$ *dominates* the utilization of $\widetilde{f_{g_i}^{-1}}|_{z_j \to \widehat{x_{g_i}}}$. That is, the decision on $x_{g_{i-1}}$ must precede the decision on $\widehat{x_{g_i}}$. This also leads the necessity of redundant design knowledge across different granularity levels.

4.6   Time-boundness of design process

By summarizing the points extracted in this section, the design process is illustrated as shown in Fig. 1. That is, since dominate decision variables must be determined beforehand against the others under the dominatedness of design knowledge, in the early stage a model with coarse granularity is used, and fidelity is still less, but the less cost per model operation permits much design iteration. This enables to globally explore potential possibility of design solutions, while the satisfaction of design requirements must be ambiguous due to less fidelity under coarse granularity. On the other hands, the design must be finalized with satisfying requirements with enough fidelity

---

[4] Knill *et al.* [11] proposed an efficient method for response surface approximation by combining two-layer models on aerodynamics toward its application to optimal configuration design of high-speed civil aircraft. Their method shares some underlying concepts on dominatedness and fidelity with the discussion of this paper.

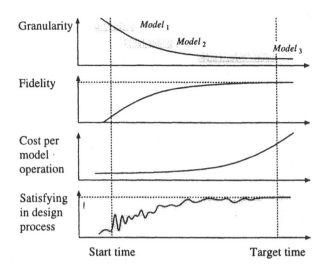

Figure 1   Time-boundness of design process

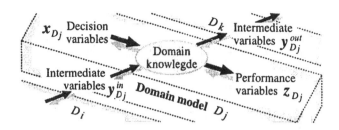

Figure 2   A domain and its surroundings

- *The global model* with principal particulars $\cdots$ The issues related only with principal particulars. This acts as the root of design, and each design problem has a unique one.

- *Domain models* with other variables $\cdots$ The other issues except the global model. Since these are the majority in the whole issues on their volume, they must be divided into appropriate subsets under the necessity of organizational decomposition of design problem solving.

Where, a model means here a set of variables and associated knowledge.

at the final stage, i.e., at the given target time due to *time-boundness*. Thus, gradually, granularity level must be getting fine, and fidelity must be increased, respectively. However, since cost per model operation is drastically increased accordingly, global shape of a design must be almost frozen, for instance in the middle stage, as early as possible with keeping the possibilities of small design changes to cure the deviations drawn from granularity shift and consequent fidelity increase.

This scenario is nearly equivalent to the preferable situation of design process recommended under concurrent engineering. To formally argue or computationally implement such a scenario, organizational structure of design process and coordination mechanism among divided tasks in problem solving are indispensable [1, 2].

5   Decomposition of Design Process

5.1   Two types of representations and variables

Types of design knowledge and its chunks are further categorized toward definition of design process structure. According to specific design problems [13, 14, 15], it is general that a few representative variables, which are called as *principal particulars*, can roughly express the concept of overall design and that they totally dominate other succeeding design. In the case of ship design, length, breadth, draft, depth, block coefficient, propulsion power, lightweight, deadweight, meta-center height, etc. are such variables [13]. Under this point, design related issues can be split into the following two types:

5.2   Horizontal decomposition of domain models

As for domain models, since each segment of forward-chaining knowledge is generated within a corresponding disciplinary, domain models can be decomposed into subsets based on disciplinarity with loose coupling among them. This decomposition scheme is well conformed to Simon's concept of nearly decomposable systems [6].

Autonomy of each divided domain model demands that it includes some decision variables and some performance variables, because both of them are indispensable for an independent unit of design activities. Further, its loose coupling with the others demands intermediate variables at its boundary. As a result, the shape of a domain model D_j can be configured as shown in Fig. 2. That is, it has the decision variables x_{D_j} that are to be determined and refined by its own responsibility and the performance variables z_{D_j} that are locally deduced under the major influence of its own design variables[5]. Further, the intermediate variables in the decision variable side $y_{D_j}^{in}$, which are distinguished as the intermediate input variables, and ones in the performance variable side $y_{D_j}^{out}$, which are distinguished as the intermediate output variables, relate

---

[5]   Under a series of domains, D_j ($j = 1, 2, \cdots, J$), it is assumed that all decision variables x can be exclusively decomposed into a set of x_{D_j} and that all performance variables z can be exclusively decomposed into a set of z_{D_j}.

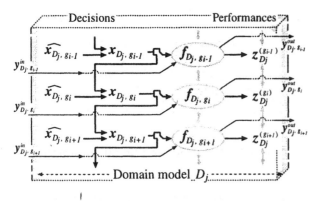

Figure 3 Hierarchical linking among models

to other domain models under the relationships of Eq. (2) as the shape of coupling[6]. In the outside of this domain, each element of $y_{D_j}^{in}$ corresponds to an intermediate output variable of another domain, and each element of $y_{D_j}^{out}$ corresponds to an intermediate input variable of other domains, respectively.

Besides, types of domains are categorized into subentity driven domains and subperformance driven domains[7]. The former one mainly concerns the determination of subentity's contents, and the latter one mainly concerns the performance that is under the balance among several subentities. In the case of aircraft design [2], domains of main wing, fuselage, etc. are the former, while domains of cruise range, propulsion, etc. are the latter.

5.3 Hierarchical decomposition of design process

Beyond a series of domains, granularity levels and dominatedness in design knowledge indicate that each domain is further decomposed into several layers within it. When applying the relationships among $x_{g_{i-1}}$, x_{g_i} and $\widehat{x_{g_i}}$ onto Fig. 2, the circumstance of design process for a domain model D_j is illustrated as shown in Fig. 3 in the aspect of forward-chaining structure.

That is, after a set of dominate decision variables $x_{D_j,g_{i-1}}$ is given and correspondingly $z_{D_j}^{(g_{i-1})}$ is computed as an estimation of z_{D_j} through forward-chaining knowledge $f_{D_j,g_{i-1}}$ in the granularity level g_{i-1}, the focus is moved to the next level g_i. There, first the supplementary decision variables $\widehat{x_{D_j,g_i}}$ are determined somehow for instance by

---

[6]   The definition of intermediate variables is here materialized as the variables at the borders among domain models, while the original definition means something between decision and performance variables.

[7]   While the previous research [2] used entity domain agents and function domain agents under entity-function structure, the categorization used in this paper is more appropriate under the autonomy of agents.

referring $z_{D_j}^{(g_{i-1})}$, $x_{D_j,g_{i-1}}$, etc., x_{D_j,g_i} is formed with it and $x_{D_j,g_{i-1}}$. Then, design knowledge f_{D_j,g_i} is applied over x_{D_j,g_i} and the related intermediate input variables y_{D_j,g_i}^{in} to compute $z_{D_j}^{(g_i)}$ and the related intermediate output variables y_{D_j,g_i}^{out}. The resulted $z_{D_j}^{(g_i)}$ is expected to have higher fidelity than $z_{D_j}^{(g_{i-1})}$. Following these, the operations in the next level g_{i+1} succeed.

The chain of these design operations is recursively nested from coarse to fine levels. The coarsest one is linked with the global model, and the piling bottom is stopped at any appropriate level.

6 Goal Cascading Across Decomposed Design Tasks

6.1 Ill-posedness on design requirements

While the discussion of the previous section mainly follows the structure of forward-chaining knowledge, the backward-chaining knowledge is rather essential.

Design is often referred as a typical ill-posed problem apart from well-established procedures for solving any problems. What is called *design spiral* [16] well illustrates the situation of such ill-posedness under the necessity of the critique phase in design process. That is, while a set of design requirements are given as preconditions, they must be refined through design iteration or compromise among performance variables, for instance, if the satisfaction of conditions is impossible due to over-specification, or since enough understanding of native requirements requires reflexion from any embodied design result. Further, such refinement is executed across granularity levels under the aforementioned operationality of design knowledge. These points indicate that each performance variable must be managed not only across the granularity levels but also toward compromise against the others.

6.2 Goals toward compromise under operationality versus ambiguity

Since the critique of computed values of performance variables must be iterated to find appropriate compromise among/within design conditions, it is desirable to easily operate such contents under critique knowledge. Operation includes here to compare the requirements and results, to refine requirements, to provide information to backward-chaining knowledge, etc. However, ease in operation leads ambiguity in performance prediction under the granularity and fidelity levels. Thus, the design process is formed as shown in Fig. 1 under time-boundness, and the operation must further include to predict their final relationships, to provide intermediate objectives toward modification of decision variables, satisfaction of requirements and so forth, etc.

The management of the above ambiguity requires managing the range of both performances and requirements rather than their exact values. Further, since design must be finalized with a series of values, such ranges must be gradually squeezed through design process. Under these characteristics, this paper calls the surrounding relationship on a performance variable *a goal*[8].

Under the goal concept, each performance variable z_ℓ at least has the following attributes for its management:

- Predicted values in respective granularity levels. Each is denoted as $z_\ell^{(g_i)}$.

- Their fidelity degrees in computing $z_\ell^{(g_i)}$. Each relates to the fidelity of forward-chaining knowledge $\Phi\left(f|_{x_{g_i} \to z_\ell}\right)$. This may be expressed as a range, $\check{z}_\ell^{(g_i)} = \left[\check{z}_\ell^{(g_i)L}, \check{z}_\ell^{(g_i)U}\right]$, under the assumption of design space smoothness.

- Preference of the variable defined in design requirements. This is denoted as $\check{z}_\ell$, which may be able to be also represented as $\check{z}_\ell = \left[\check{z}_\ell^{L}, \check{z}_\ell^{U}\right]$.

Where, $z_\ell^{(g_i)} \in \check{z}_\ell^{(g_i)}$ must be true, while $z_\ell^{(g_i)} \in \check{z}_\ell$ and $\check{z}_\ell^{(g_i)} \subset \check{z}_\ell$ hopefully but not necessarily stand up respectively.

6.3 Squeezing goals across granularity levels

Figure 4 shows the relationships across goals through granularity levels. That is, in a coarse level of granularity g_{i-1}, after preferences are first temporarily defined as $\check{z}|_{t=\tau-1}$, decisions are made on $\widehat{x_{g_{i-1}}}$ to form $x_{g_{i-1}}$, and the performances are predicted as $z^{(g_{i-1})}$ through $f_{g_{i-1}}$. Under this situation, any critique knowledge is applied to the contrast between $\check{z}|_{t=\tau-1}$ and $z^{(g_{i-1})}$. Then, the preferences are squeezed, i.e., narrowed into $\check{z}|_{t=\tau}$ since the result of decisions in this level guarantees promises in the next granularity level, or the preferences are refined if necessary in the current level. This sequence is recursively iterated from $\check{z}|_{t=\tau}$ to $\check{z}|_{t=\tau+1}$ as well as from $\check{z}|_{t=\tau-1}$ to $\check{z}|_{t=\tau}$. Furthermore, the following equations hopefully but not necessarily stand up for each z_ℓ respectively toward a final

---

[8] This term is borrowed from goal programming [17] that is a technique for multi-objective decision making. While Ward *et al.* [18] modeled concurrent engineering process with set theory, their viewpoint is almost similar to the purpose of goal concept.

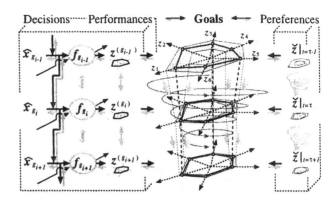

Figure 4 Squeezing goals

design result:

$$\cdots \supset \check{z}_\ell|_{t=\tau-1} \supset \check{z}_\ell|_{t=\tau} \supset \check{z}_\ell|_{t=\tau+1} \supset \cdots \quad (6)$$

$$\cdots \supset \check{z}_\ell^{(g_{i-1})} \supset \check{z}_\ell^{(g_i)} \supset \check{z}_\ell^{(g_{i+1})} \supset \cdots \quad (7)$$

Besides, since promises from coarse to fine levels might result in misleading of consequent design exploration, the refinement of preferences might be retrogressed to early stages of design process, i.e., more coarse granularity levels. This means that focus for squeezing preferences is iteratively shifted up-and-down across levels.

6.4 Role of intermediate variables across domains

As shown in Fig. 2, design knowledge and process are split into a set of domains, which is a native source of concurrency in design. When multiplying the situations of Figs. 2 and 3 onto the situation of Fig. 4, intermediate variables must be managed as similar to performance variables. The reason is that intermediate output variables computed in a specific domain behave as performance variables inside the domain, but their preferences are recommended by the other domains since they are intermediate input variables in there. This means that the goal concept is expanded to intermediate variables.

In the other words, the concurrency among domain models is realized through the goals of intermediate variables, that is, the goals can act as plays among domain models. Since domain models are linked each other, the dependency among domains requires, for instance, that forward-chaining operation in a certain domain requires the values of its intermediate input variables, and that backward-chaining operation in a certain domain requires the preferences of its intermediate output variables. Goals enable to temporarily release prerequisites for respective operations so as to overcome this restriction toward parallel operations under the native concurrency.

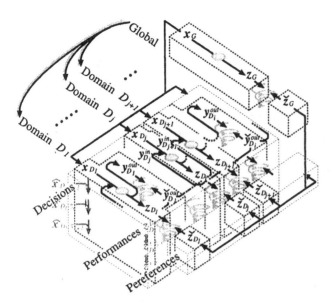

Figure 5   Infinitely cascaded goals

6.5   Infinitely cascaded goals

The overall situation of goals on performance variables and intermediate variables is modeled as a series of *cascaded goals* as shown in Fig. 5. That is, first after design condition is globally given, the corresponding goals $\check{z}_G$ are assumed, and the representative decision variables x_G are determined so as to promise successful design. Then, x_G and $\check{z}_G$ are distributed to respective domain models. Under this, for instance, in a domain D_j, $\check{z}_{D_j}$ is given from $\check{z}_G$, and x_{D_j} are going to be determined. This operation requires some interactions with other domains through $y_{D_j}^{in}$ and $y_{D_j}^{out}$ under the situation of Fig. 2. The preferences $\check{z}_{D_j}$ must be forwarded to each element of $y_{D_j}^{\check{in}}$ so as to promise successful determination of decision variables x_{D_j}. The forward of each element of $y_{D_j}^{\check{in}}$ to other domains D_i, $(i \neq j)$ results in preference establish of one of $y_{D_i}^{out}$ in its corresponding domain. On the other hands, since the domain D_i has its own performance variables $\check{z}_{D_i}$, the operations in D_i must be performed under preferences on both $\check{z}_{D_i}$ and $y_{D_i}^{\check{out}}$. This indicates that the original domain D_i has preferences on $y_{D_j}^{\check{out}}$ as well. Furthermore, every domain is under the same situation, and it means that the linkage among domains through goals of intermediate variables is natively infinitely cascaded. Beyond such infinity, the time-boundness of design process implies the necessity of goals as plays and the design process situation shown in Fig. 1 in the aspect of gradually satisfying goals in accordance with the design progress from coarse to fine granularity levels, from less to higher fidelity, etc.

7   Goal Driven Coordination

7.1   Global compromise versus local compromise

The goal concept provides a means for coordination among decomposed design tasks between the global model and domain models, among domains and across granularity levels. The discussion crystallizes several representative situations of coordination as follows:

- *Inter-coordination* $\cdots$ Since concurrent design process is fundamentally enabled by domain based decomposition, and since they are linked with intermediate variables, $y_{D_j}^{in}$ and $y_{D_j}^{out}$, their autonomous design operations require information exchange. Further, since such information includes goal information, they should be locally coordinated among specific domains.

- *Global-coordination* $\cdots$ While the above local coordination in respective domains is indispensable for flexibility, the global viewpoint is necessary toward global optimality or conflict resolution behind it. Thus, the overall design should be globally coordinated through any supervising function. The global model is useful for this purpose as well as initialization of principal particulars.

- *Inner-coordination* $\cdots$ Since each domain D_j has a pile of decision variables $\widehat{x_{g_i}}$ and corresponding redundant models across granularity levels g_i, the shift among levels must be managed. Thus, how and when the determination of each $\widehat{x_{g_i}}$ is frozen or refined should be managed. This requires the coordination among hierarchically redundant models. For instance, under specified $\check{z}_{D_j}$ and $y_{D_j}^{\check{out}}$, even when $y_{D_j}^{in}$ is changed or either $\check{z}_{D_j}$ or $y_{D_j}^{\check{out}}$ is recommended to be modified from the others, it may be appropriate that its influence is absorbed by the local refinement of x_{D_j} within locally internal tradeoff. Under fixed $x_{D_j, g_{i-1}}$, the refinement for y_{D_j, g_i}^{in} is executed within $\widehat{x_{D_j, g_i}}$ so as to reduce the global propagation of refinement influences. In this meaning, $\widehat{x_{D_j, g_i}}$ is recognized as plays under $x_{D_j, g_{i-1}}$.

While design task division can bring their autonomy in concurrent design process, all of these coordination situations are indispensable to lead superior process and design result.

7.2   Mechanisms for coordination

Implementation of the above coordination situations requires some mechanisms in the following directions:

- Fundamentally, distributed tasks require any negotiation scheme to coordinate their own tasks and the others. While negotiation is a central issue in distributed problem solving, one for design of large complicated artifacts needs to face with granularity levels and consequent fidelity degrees.

- Granularity level on which each design stage focuses must be arranged under the time-boundness. It must be dynamically and robustly scheduled under the uncertainty due to coarse granularity toward efficient design progress, autonomy of tasks and so forth.

- Since dynamical scheduling must meet with the relationships of Eqs. (6) and (7), it is important to reflexively predict ambiguity, e.g., how $f(x_{D_j,gi-1})$ and $f(x_{D_j,gi})$ are different each other, etc., under a certain granularity and consequent model fidelity to properly propagate preference levels across goals. Further, it is necessary to predict cost per model operation behind them as well.

While these issues remain as future works, they are derived from the formal premises of concurrent design process.

8 Concluding remarks

This paper discussed the underlying structure of concurrent design process of large complicated artifacts in order to provide formal bases to agent-based concurrent design systems. The discussion was based on an assumption of simplified but formal representation of an artifact as a set of lumped mass systems, and it revealed some essential situations in concurrent and coordinational design process. It is planed to develop an agent-based distributed design system based on them as future research with establishing concrete mechanisms of goal-driven coordination.

Acknowledgments

This research is partially supported by "Research for the Future" Program of the Japan Society for the Promotion of Science (JSPS) under "Methodology for Emergent Synthesis" Project (project number 96P00702).

References

[1] Fujita, K. and Akagi, S., "Agent-Based Distributed Design System Architecture for Basic Ship Design," *Concurrent Engineering – Research and Applications*, Vol. 7, No. 2, pp. 83-94, 1999.

[2] Fujita, K. and Kikuchi, S., "Task Distribution and Coordination Framework for Engineering Systems Design based on Entity-Function Structure," *Advances in Concurrent Engineering CE99 — Proceedings of 6th ISPE International Conference on Concurrent Engineering*, pp. 199-208, 1999.

[3] Kusiak, A. and Wang, J., "Dependency Analysis in Constraint Negotiation," *IEEE Transactions on Systems, Man, and Cybernetics*, Vol. 25, No. 9, pp. 1301-1313, 1995.

[4] Sreeram, R. T. and Chawdhry, P. K., "A Unified Scheme for Conflict Negotiation in a Multi-Agent Decision Process," *Advances in Concurrent Engineering CE99 — Proceedings of 6th ISPE International Conference on Concurrent Engineering*, pp. 129-141, 1999.

[5] Yoshikawa, H., "General Design Theory and a CAD system," *Man-Machine Communication in CAD/CAM, Proceedings of The IFIP WG 5.2-5.3 Working Conference 1980*, pp. 35-57, North-Holland, 1981.

[6] Simon, H. A., *The Sciences of the Artificial (Third Edition)*, The MIT Press, 1996.

[7] Fujita, K. and Akagi, S., "Toward Engineering Design Synthesis of Engineering Systems," *Workshop on the Methodology of Emergent Synthesis, WMES '98*, pp. 41-48, 1998.

[8] Russell, S. and Wefald, E., *Do The Right Thing — Studies in Limited Rationality*, The MIT Press, 1991.

[9] Dym, C. L., *Engineering Design — A Synthesis of Views*, Cambridge University Press, 1994.

[10] Hobbs, J. R., "Granularity," *Proceedings of the Ninth IJCAI*, pp. 1-4, 1985.

[11] Knill, D. L., Giunta, A. A., Baker, C. A., Grossman, B., Mason, W. H., Haftka, R. T. and Watson, L. T., "Response Surface Models Combining Linear and Euler Aerodynamics for Supersonic Transport Design," *Journal of Aircraft*, Vol. 36, No. 1, pp. 75-86, 1999.

[12] Fujita, K. and Akagi, S., "A Framework for Component Layout and Geometry Design of Mechanical Systems: Configuration Network and its Viewing Control," *Proceedings of the 1995 ASME Design Engineering Technical Conferences*, DE-Vol. 82, Vol. 1, pp. 515-522, 1995.

[13] Tomita, T., *SENPAKU-KIHON-SEKKEIRON*, Maruzen Publishing Service Center, (*In Japanese*), 1982.

[14] Torenbeek, E., *Synthesis of Subsonic Airplane Design*, Delft University Press, 1976.

[15] Raymer, D. P., *Aircraft Design: A Conceptual Approach*, AIAA education series, 1989.

[16] Buxton, I. L., *Engineering Economics and Ship Design*, British Ship Research Association (BSRA) Report, 1971.

[17] Lee, S. M., *Goal Programming for Decision Analysis*, Auerbach Publishers, 1972.

[18] Ward, A., Sobek, D. K. II, Cristiano, J. J. and Liker, J. K., "Toyota, Concurrent Engineering, and Set-Based Design," *Engineered in Japan — Japanese Technology-Management Practices*, pp. 192-223, Oxford University Press, 1995.

Candidate Piping Routes Generation by Genetic Algorithm

Teruaki Ito
Department of Mechanical Engineering
The University of Tokushima
2-1 Minami-Josanjima
Tokushima 770-8506, Japan

Abstract

The paper describes an approach to generate candidates piping routes by genetic algorithm (GA) to support interactive planning of a piping route path for layout design. The author has proposed a hybrid approach to piping route path design, and reported some of the basic features of GA-based piping route design. In this paper, the basic ideas and their extensions are described in further details. The paper covers the definition of genes to deal with pipe routes, the concept of spatial potential energy, the method to generate initial individuals for GA optimization, the concept of zone in route generation using GA, evaluation of crossover methods, and definition and application of fitness functions.

1 Introduction

Given the initial specifications for a product, a designer must create the description of a physical device that meets those requirements. The final design must simultaneously meet cost and quality requirements, as well as meet the constraints imposed by activities such as manufacturing, assembly, and maintenance. In addition to this basic approach to product design, the competitiveness of companies is closely related to their capability to timely create innovative products to satisfy customer requirements which are becoming increasingly individualized and diverse. Concurrent Engineering (CE) (Prasad, 1993) is one of the essential concepts meeting those challenges. In addition to the key idea to develop high quality products and offer them at a lower price and in significantly shorter time to the competitive global market, Concurrent Design (CD) (Finger, 1992) concerns with more sophisticated designs using considerations of various phases of design simultaneously and cooperatively. Several methodologies of CD including multi-agent architecture, engineering data management, virtual manufacturing, etc. have already been applied to design

support systems, and their availability and effectiveness have been reported.

In the mean time, the radical notion that interactive systems are more powerful problem-solving engines than algorithms is making the new paradigm for computing technology built around the unifying concept of interaction (Wegner, 1997). As for designers who use design support systems as a tool to work on design tasks, thoughts of designers are effectively activated by nice system interactions (Hancock, 1989). The interactions may stimulate the brain of designers to produce some innovative ideas. Using the basic categories of interaction style including key-modal, direct manipulation and linguistic (van Dam, 1997), various approaches are studied to build interactive systems (Newman, 1995).

The study focuses on these two paradigms regarding concurrency and interaction in design issues, and proposes a genetic algorithm (GA) approach (Goldberg, 1989 & 1994) to pipe route planning (Takakuwa, 1978) in layout designs. Layout design is critical in engineering designs of various production systems, chemical plants, power plants, factories, etc. Its main aim is to effectively utilize the functionality of component equipments and to appropriately satisfy the spatial constraints. Pipe route planning is one of the key issues in these layout designs. Our goal is to create a computer-based design system with a powerful problem-solving engine that will enable a designer to make concurrent rather than sequential consideration of requirements regarding pipe route planing in a collaborative and interactive manner, and to evaluate the impact of design alternatives in terms of various perspectives.

The author has proposed a hybrid approach to piping route path design (Ito, 1998), and reported some of the basic features of GA-based piping route design. In this paper, the basic ideas and their extensions (Ito, 1999) are described in further details. In the following, the paper first review pipings around us and its route planning. It then provides an detailed description of our methodology based on GA, which includes the definition of gene, the concept of spatial potential energy, the method to generate initial individuals, the concept of zone in route generation, evaluation of crossover method, and the definition and application of fitness functions. Lastly some results of simulation based on our methodology is presented.

2. Pipings around us and its route planning

Simply speaking, the goal of pipings is to transfer a certain volume of liquids from a starting point to a destination point within a certain time interval through an appropriate route. Pipings are very popular in our daily lives and we can see them in many places, for example, in electric power plants, chemical plants, factories, buildings, sewage, automobiles, etc.

As for design of pipings, the design should be considered from various view points. First of all, it is very important to determine the appropriate flow speed from economic sense of view. The inner diameter of piping can be calculated based on the flow speed to minimize the overall pressure losses. In plant piping designs, pipings are categorized to several classes: process lines for insides of equipments, yard lines for outside of equipments and utility lines for water, steam and fuels. Pipings of the same categories in plant pipings should be arranged to be placed together. From the view point of mechanical design, on the other hand, pipings are categorized differently, namely, in terms of the place to be installed: on the ground, in the air and under the ground. Maintenance is also another important factor to be considered in designing a pipe route. Although piping should be accessed easily for maintenance, but they should not be placed on the walk way. Equipments should be placed in such a way as pipings are placed in parallel or rectangular, basically avoiding an diagonal path. Each supporting should possess the durability to support the pipings. For example, materials of the supporting should have good heat resistance so that they can minimize shape deformation.

As far as piping design is concerned, there are so many things to consider from various perspectives as mentioned above, and piping designs, therefore, are a difficult and time consuming task. One of the most difficult part in piping design would be to conduct pipe route planning, which is to design the appropriate route for piping connecting the starting and goal points. If the optimal pipe route is designed, it could be said that the most difficult part of piping design is completed. But it is very difficult even for a skilled designer to design the optimal pipe route especially under very complicated spatial constraints, which is just like going through a labyrinth, while keeping an appropriate space between the pipe and the surrounding walls or the equipments metaphorized as obstacles.

As a general approach to piping design, a designer initially creates an appropriate design model, and interactively and iteratively modifies the model in a trial-and-error manner until an appropriate design specification is completed. Our approach focuses on this interactive, iterative and concurrent session of designing. During the designing session using our methodology, various pipe route candidates are generated based on the starting and goal points, and/or several subgoals if necessary, all of which are specified by the designer using a simple operation of pointing device, or mouse, which would not interrupt the designer's thoughts for piping design. We use the optimization method of GA in our approach, however, the goal of our approach is not just to find the best pipe route, but rather to present the designer several appropriate pipe routes which could be a clue for the designer to find the best pipe route. The designer would collaboratively proceed pipe route planning, referring to the proposed pipe route. We will describe our GA approach in more detail in the remaining sections.

3. Definition of chromosome and Spatial potential energy

As an optimal search method for multiple peak functions, GA stemming from the generation of the evolution of living things is applied to various optimization problems and its validity has been verified so far (Yamamura et al., 1994).

In order to use GA in pipe route planning as one of the optimization problems, a route from a starting point to a destination point is represented by a character string and is regarded as a design parameter.

In our approach, a working space for pipe route planning is represented by a model, and the space is divided into the cells of MxN. A route is represented using a combination of cells connecting a starting cell and a destination cell. To represent direction of a route path, a set of unit vector $\{r, u, l, d, o\}$ is defined, each vector represents right, up, left, down and stop, respectively, and character string of $\{1, 2, 3, 4, 0\}$ corresponds to each vector. Using information on the cells which compose the route, each individual is coded. For example, the gene type for a route is expressed using symbols including $\{1, 2, 3, 4, 0\}$, where zero means the current point already reached the destination cell.

In pipe route planning, high priority is given to the shorter route path. In addition to this, a route must go along the wall and obstacles as closely as possible, avoiding a diagonal path, the most appropriate route is designed. Using the concept of spatial potential, the degree of access to the wall or the obstacles is quantitatively calculated, and used as a part of objective function for the generation of a pipe route using GA.

To determine the distribution of spatial potential, the working space is divided to MxN(M=1,2,...; N=1,2,...), those cells which contain any portion of the obstacles are given the potential value Pn, each cell is given the potential value of P1, P2, ..., Pn-1 according to the distance from the obstacle cell. Those cells which are located next to the wall are given the potential value P0, which means that the route path is more favorable if it goes along the wall. Only the positive values are used as a potential energy. The higher value means that the cell is far from the wall or the obstacles. In this way, potential energies of each cell are determined.

4. Studies on crossover method

4.1 Characteristics of genes

Determination of crossover methods depends upon the application. For example, uni-crossover is suitable for the case that each gene has individual information, blend crossover is suitable for the case that genotype has a continuous value. In the case of a piping route path, genotype shows a continuous value. A route must connect a starting and a goal points. In addition to that, considering the layout of equipments, physical conditions for temperature, maintenance and so on, the most appropriate route among various candidate routes is designed. In other words, how to reach the goal point is the important thing to consider.

4.2 Applied crossover operations

A route must connect the two points, namely, the starting and goal points. The length of genotype is variable and not fixed, which means that a genotype is elastic like a rubber band. A route path is like a rubber band which is fixed with a pin in each of the terminals, and it smoothly expands.

Sometimes, genes having different length must be crossed over. To cover the difference in length, we do as follows. Set the length of parent 1 to l1, that of parent 2 to l2 where ,. Then set the half of l1 to lb. lb is multiplied by a random number between 0 and 1 and gives lp. The portion between 0 and lp and the portion between lb and 1 are exchanged. Since the length of genes may differ each other, additional vector array may be inserted if the length of generated gene is too short to connect each end of the parent gene.

To avoid generating genes including obstacles, potential value of each cell is checked to see if the cell in

on the obstacle. If it is, a vector is repeatedly generated until it does not on the obstacle. In this way, we could obtain those genes which does not contain obstacle cells in the earlier generations. As a result, a wide range of search area is considered in GA.

From the results, two-points crossover tries to avoid obstacles, aggressively find a new route, and can find a better route during an earlier generations. On the other hand, uni-crossover can inherit the characteristics from each parents at the same ratio, do not change the genotype, easily to conduct crossover operations even between parents with different lengths, and can generate various kinds of vector arrays. The study mainly applied two-points crossover operation but also applied uni-crossover operation as a control.

5. Introduction of tendencies in directions

5.1 Concept and definition of tendencies in direction

A gene goes from a starting point towards a goal point. The most important thing is that a gene must reach the goal point without fail, which means that a gene possesses the characteristics that it moves towards the goal. Otherwise, a gene randomly goes inside the search area, which is time consuming and a waste of time.

We define the concept of "zone" to give chromosome the tendencies in the direction of a route path from a current position towards the goal point. To determine a zone, coordinates of current cell and goal cell are used. When coordinates of a goal is (X, Y) and those of current cell is (myx, myy).

Then, priority vector is set to each zone. If the priority is set too high, however, all chromosomes have the same tendency in direction and a route path cannot be appropriately generated. The priority is set in a trial-and-error manner to generate chromosomes having variety of route paths. In this case, careful considerations are also taken so that variety of route path can be generated, otherwise all the route paths would only go straight in the right direction.

5.2 Simulation for generations of individual

Using the concept of tendencies in direction for a pipe route, we could make pipe routes under control and give them the tendencies so that the paths are likely to go towards the goal point. Consequently, initial individuals are effectively generated. The method is also applicable when an additional portion is patched to cover the shortage of route path in crossover operations.

Using priority vectors only would not be a good method to generate a variety of chromosomes. That would generate those chromosomes which have the same tendencies in terms of direction so that the route paths converged by GA operations is likely to be a local optimization. Although it was difficult to find appropriate priority values, we made it in a trial-and-error manner and to applied the concept to generate initial individuals.

6 Fitness function

6.1 Elements of fitness function

In general, pipe route planning is considered from various perspectives. Some of the typical perspectives to the planning would be (a) shorter length of route path, (b) arrangement of the pipes under the same categories, (c) guarantee the maintenance spaces, and so on. Our study considers the shorter length of a route path and, from this respect, elements for fitness function were studied. We also considered that the route should be as straight as possible but no diagonal path is permitted. The number of turning points in the path should be considered. The path should go along the surrounding wall in the work space or obstacles placed in the work space as close as possible. The potential energy is set lower in these cells. Considering these elements, fitness function was defined as described in the following section.

6.2 Definition of fitness function

(6.1) shows the fitness function applied in our approach.

$$f(x) = f_0 + f_1 + p_{max} + C + W \cdots\cdots\cdots (6.1)$$

(6.2) accumulates the number of cells and gives the length of a route path.

$$f_0 = \sum_{k=1}^{N} x_i \cdots\cdots\cdots\cdots\cdots\cdots (6.2)$$

Length of a route path is evaluated in (6.2), but it is not enough to cover a variety of route paths. For example, different routes having the same fitness value have to be considered. Since the shape of the route is different in these two paths, we applied the function (6.3), which accumulates the total potential values for the route and (6.4), which considers the maximum value of potential energy in the cells of the route.

$$f_1 = \sum_{i=0}^{N} p_i \cdots\cdots\cdots\cdots\cdots\cdots\cdots (6.3)$$

Every time the direction of a route is changed, a certain weight is added as C in (6.1). If a route path is on any

$$f_3 = p_{max} \cdots\cdots\cdots\cdots\cdots\cdots\cdots (6.4)$$

obstacles, or the path contains cells including obstacles, a large weight is added as shown in (6.5).

$$W = p_{obstacle} \times A \cdots\cdots\cdots\cdots\cdots\cdots (6.5)$$

7 Approach to pipe route planning

7.1 Conditions for pipe route planning

Although pipe route planning depends upon various kinds of factors as mentioned before, the study emphasize the minimum length of the route path and determined the conditions as follows.

No significant problem was observed in genotype based on vector representation, the definition is applied without any modification. As for potential energy values in the work space, 4 classes of categories were defined. Cells on obstacles, cells surrounded by spaces, cells around obstacles and cells around the wall were classified as P0, P1, P2 and P3, respectively. Conditions are different between pipings indoor and pipings outdoor, the study focused on pipings indoor and set up the potentials as P0>P1>P3>P2.

As for crossover methods, extended two-points crossover and uni-crossover were applied.

In two-points crossover operation, individuals generated by two-points crossover are sorted by the value of fitness function, top 50% of them were randomly processed two-points crossover operation to generate offsprings, and the bottom 50% were untouched and exchanged with the offsprings. In uni-crossover operation, Individuals generated by uni-crossover were sorted by the value of fitness function, top 70% were extracted, 85.7% out of the 70% were randomly selected to be processed crossover operation, and the bottom 30% were untouched and exchanged with them. In both crossover methods, fitness function (6,1) was applied.

Based on the genotype, crossover methods, selection methods and fitness function, various parameters were studied to determine appropriate values and settings.

7.2 Reconsideration of generation method for initial individuals

Using a starting point and a goal point, initial individuals are randomly generated before GA procedure. The zone defined in section 5 are used here.

At first, the zone of a current cell is determined using coordinates of a starting and a goal points. A roulette based on the ratio of priority vector in the zone is set up, an arbitrary point on the roulette is determined using a random number generator between 0 and 1, and the first gene is selected. A current point is forwarded using the selected vector, and the coordinates of the updated current cell are obtained. Then the current zone is set up based on the updated current cell and the goal cell. The same procedure is repeated until the current point reaches the goal cell. In this way, initial individuals are generated.

In the mean time, to find the most appropriate route, the initial route should be as random as possible in the working area. Without considering obstacles, routes are generated only by the specified cells for the starting and destination points. But most of the routes in the initial individuals have the tendency to go straight to the goal point. The routes using these initial individuals did avoid obstacles and reached the goal point in the end. Judging from the route path length, the number of turnover and overall observations, the most appropriate route is not always found. The next section describes our approach to solve this problem.

7.3 Introduction of intermediate points

Most of the routes in the initial individuals have the tendency to go straight to the goal point. To generate initial individuals more randomly, we defined intermediate points to be passed in the route and applied in our approach.

Making a bisector cutting through the line which connects the starting cell and the destination cell, an arbitrary cell on the bisector is selected. Making an arbitrary line passing through the bisector cell from the starting cell to the destination cell, initial individuals are generated referring to this line. Since the arbitrary cell is randomly selected, the route paths cover the overall working space.

7.4 Evaluations on crossover methods

Using the initial individuals generated in the method mentioned in section 7.2, simulation of piping route path planning were carried out. As results, the followings were observed.

(1) Uni-crossover converges individuals at the earlier generations.

(2) Uni-crossover does not always exclude those individuals which include obstacle cells.

(3) Two-points crossover is superior to uni-crossover.

Although two-points crossover generated appropriate route paths, some of them seem to be a locally optimized route paths. To avoid that, we applied dynamic selection ratio based on the minimum fitness value, average fitness value and the number of cells on obstacles. The first selection ratio of 40% is used until all of the individuals become obstacle free. Then the ratio is set down to 3% and to study all the possible routes. When the convergence status becomes a certain level, the ratio is set back up to 40%. If the difference between the average fitness value and the minimum fitness value is below 5, we assumed that the convergence is going to terminate. In this way, we excluded those individuals including obstacles in the earlier generations, we tried to take time to find the most appropriate route path without converging to a locally optimized route. When individuals are likely to converge to an appropriate route path, convergence speed is accelerated. In addition to the distinction using the fitness value, a certain number is subtracted from the fitness value to distinguish each individual more effectively.

7.5 Coordinates conversion

The above mentioned process is described based on a specific direction. In actually, we have to deal with various directions. A route path may be in any of the direction to down-right, down-left, up-left or up-right. Since it is not effective to consider each direction separately, we introduced a temporary coordinates system into which actual coordinates of routes are mapped to conduct GA operations. Combining rotation at 90 degrees and parallel transfer operations, any route can be converted to temporary coordinates. A route is mapped to a temporary coordinates, given GA operations to optimize the path, and returned to the original coordinates system.

7.6 Route adjustment using subgoal

Routes generated based on the starting and goal points are not always appropriate in terms of pipe route planning. For example, even if a path is not the shortest of all the candidates routes, it might be an appropriate path because it goes through a certain point. We have realized the setting of subgoal in GA route planning. Once a subgoal is set up in addition to the starting and goal points, the first portion of route connecting the starting and subgoal points is automatized and generated. When the portion is generated, the remaining portion is designed and added to the first portion. More than one subgoals can be specified.

In this way, a designer can specify the points to be passed in the route, which will support the designer's thought in finding the appropriate route.

8. Concluding remarks

The paper described the method to conduct pipe route planning using GA. The definition of genes to deal with a pipe route, the concept of spatial potential energy, the method to generate initial individuals, the concept of zone in route path generation using GA, evaluation of crossover methods, definition and application of fitness functions were described.

We have developed a prototype system to conduct pipe route planning using GA and evaluated the results of simulation to show the validity of our approach. Figure 1 is a result of route path generation simulation and shows convergence of fitness values in each individual. The vertical axis on the left shows fitness value in each individual, which is shown as points on the graph, where the horizontal axis show the generations. The vertical axis on the right shows the ratio of individuals with the lowest fitness value out of the total individuals. Some of the routes with the lowest fitness value, or the most appropriate route at the specified generation are also shown in the graph. The figure shows that individuals are converged to the appropriate route after the 245th generations.

References

Finger, S., Fox, M.S., Printz, F.B. and Rinderle, J.R., "Concurrent Design", Applied Artificial Intelligence, 6, pp.257-283, 1992.

Goldberg, D.E., Genetic Algorithm in Search, Optimization and Machine Learning, Addison-Wesley Publishing Co., 1989.

Goldberg, D.E., "Genetic and Evolutionary Algorithms Come of Age", Communications of the ACM, 37(3), pp.113-119, 1994.

Hancock, P.A. and Chignell, M.H. (eds.), Intelligent Interfaces: Theory, Research and Design, Elsevier Science Publishers B.V., 1989.

Ito, T. and S. Fukuda "Hybrid Approach to Piping Route Path Design Using GA-Based Inspiration and Rule-Based Inference", Concurrent Engineering: Research and Applications, Vol.6, No.4, pp.323-332, 1998.

Ito, T. and S.Fukuda, "A genetic algorithm approach to piping route path planning," Journal of Intelligent Manufacturing, Vol.10, No.1, pp.103-114, 1999.

Newman, W. and Lamming, M., Interactive System Design, Addison-Wesley, 1995.

Prasad, B., Concurrent Engineering Fundamentals, Vol.1, Prentice Hall, 1996.

Takakuwa, T., Analysis and Design for Drainpipe Networks, Morikita Shoten, 1978. (in Japanese)

van Dam, A., "Post-WINP user interfaces", Communications of the ACM, 40(2), pp.63-67, 1997.

Wegner, P., "Why interaction is more powerful than algorithms", Communications of the ACM, 40(5), pp.80-91, 1997.

Yamamura, M. and Kobayashi, S., "Toward Application Methodology of Genetic Algorithms", Journal of Japanese Society for Artificial Intelligence, 9(4), pp.506-511, 1994 (in Japanese).

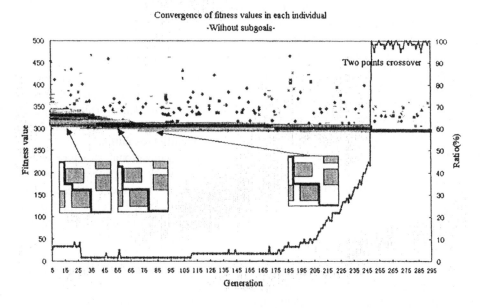

Figure 1. Convergence of fitness values in each individual

Research on Coordination in Concurrent Engineering

Wensheng Xu, Guangleng Xiong, Tianqing Chang

Department of Automation, Tsinghua University
Beijing, 100084, P.R. China

Abstract

The coordination of Concurrent Engineering (CE) product development process is studied. This paper analyzes the concept, objectives and functions of coordination management in CE product development process (CEPDP), presents a framework of coordination management, and proposes the key techniques in CEPDP coordination management.

1 Introduction

In CEPDP, three problems will occur frequently: 1) the establishment and management of workflow; 2) the settlement of the relations which interact with and influence each other in the product development process; 3) the resolution of conflicts in Integrated Product development Teams (IPTs). To solve these problems, efficient coordination tools, systems and relevant key techniques are required. Coordination is an important problem in CEPDP. Many researchers have done much research work in this area[1,2]. This paper introduces some research work on coordination in our project "Concurrent Engineering" (No. 863-511-9504-002) and "Conflict Management in Concurrent Engineering" (No. 863-511-941-009) which are supported by the CIMS Subject of the State High-Tech Development Plan of China (863/CIMS).

2 Analysis on CEPDP coordination

2.1 Definition of CEPDP coordination

Coordination is a broad sense concept in management science. It is widely accepted that coordination is to adjust relations among activities for the purpose of realizing the main objective of the system[2]. CEPDP is a complex process, which has clear and definite objectives and large amount of undefined parameters. In this process, the requirement for coordination is very strong and its function is very evident. We propose that the definition of

CEPDP coordination is: in the organizational structure of IPTs, to collect decision-making information related to the process, to identify existing mistakes in the development process and potential mistakes that would be only exposed in the later stages of the development process, to adjust dependent relations among and within IPTs, and to provide effective decision-making schemes and conflict resolution schemes, so as to ensure the completion of the main CE objectives[3].

2.2 Tasks of CEPDP coordination

CEPDP coordination should carry out the following tasks.

1) To optimize and adjust the relations between activities

In CEPDP, because of the adoption of IPT organizational structure and some DFX tools, conflicts will occur frequently. There are some special activities such as information pre-release in this process. CEPDP coordination should manage these situations properly, organize information and resources effectively to ensure the shortest CE product development lifecycle.

2) Conflict mitigation

The objective of conflict mitigation is to avoid conflicts, so as to minimize the iteration of CEPDP and unexpected incidence. This will greatly improve the CE product development and shorten the CE lifecycle.

3) Conflict resolution

The organizational structure is based on IPTs. Each IPT can work independently, but still there are many relations among them. These relations and different perspectives in IPTs determine the existence of conflicts. Conflicts should be solved properly in order to realize the ultimate objectives of CE.

2.3 Appoachs of CEPDP coordination

CEPDP coordination has two approachs to conflict treatment.

1) Static coordination

In static coordination, we should apply specific relation representation mechanism to express the relations in CE effectively and comprehensively, and avoid conflicts by maintaining these relations.

2) Dynamic coordination

In dynamic coordination, when conflicts occur or essential conflicts exist, it is the requirement of CE to detect these conflicts in time and resolve them properly. Dynamic coordination includes conflict detection and conflict resolution.

2.4 The objects of CEPDP coordination

The functions of CEPDP coordination are embodied by some objects, which include:

1) IPT. IPT is the basic organizational element in CE. The complex relations among and within IPTs should be managed by CEPDP coordination.

2) Time. The temporal relations in CE process mainly involve the temporal points and intervals of the beginning, completion of tasks and pre-release of information. CEPDP coordination should manage these temporal objects to avoid temporal conflicts.

3) Resource. Resource in CE includes designers, equipment, techniques, funds, etc. Resource should be arranged properly to avoid resource conflicts.

4) Parameters. These includes geometrical parameters, weight parameters, electrical parameters, etc. These parameters can be adjusted to meet the functional requirement of products and avoid design conflicts.

3 CEPDP coordination model

From the discussion above, we propose a CEPDP coordination model, which provides a comprehensive framework to solve the coordination problem systematically. Generally speaking, a coordination model should comprise relation representation, problem examination, problem space definition and problem solving. Our CEPDP coordination model is based on constraint satisfaction problems (CSP), as shown in Fig. 1.

This model includes relation model, knowledge model, data collection, conflict resolution, etc.

3.1 Relation model

The inevitable result of developing products and organizing multifunctional teams in CE is that the developers must deal with many dependent and independent relationships which are referred to as constraints. Developers must decide how to formally represent these relations, how to effectively maintain them, how to decrease the frequency of constraint violation and how to present timely notification according to constraint violation information. Therefore, a relation model is needed to formally represent the constraints and the level relationships among constraints. The actual product development process should comply with the relation model[4].

1) Constraint network

A constraint network in a relation model is used to represent all kinds of relations and the relevant variables. A constraint network is the basis of static coordination and dynamic coordination.

2) Constraint recombination

The relations in a constraint network should be regrouped into small sets in order to facilitate constraint calculating and coordination.

3) Constraint propagation

By constraint propagation, values of some variables can be calculated and influence of some variables on other variables can be identified.

4) Consistency checking

Consistency checking of a constraint network can serve as checking scheme feasibility and detecting conflict. The scheme feasibility can be checked through the preprocess function of the constraint network. The preprocess of the constraint network checks for inconsistencies in the constraint network and determines the reasons for the inconsistencies before valuation and propagation.

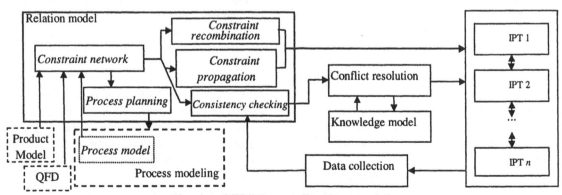

Fig. 1. CEPDP coordination model

Conflicts mean that the developers draw contradictory conclusions for the same condition or for the same purpose. Generally, there are three types of conflicts in the product development process.

- An attribute value proposed by one developer makes it infeasible for another developer to offer a consistent set of values for other attributes.
- Different specialists may have disparate views on approaches or styles used to achieve a design goal.
- Different criteria are used by different specialists to evaluate designs from different perspectives. Typically, these criteria can not be simultaneously and optimally satisfied.

5) Process planning

Product development process can be planned based on constraint network.

3.2 Process model, product model and QFD

CE product development process model is the basis of process management. We propose a kind of integrated multi-view process modeling method[5]. Process model comprises product information view, functional activity view, organizational view and resource view. Process model and product model are sources of constraints for relation model. Process model can provide temporal and logical constraint information for relation model, while product model can provide product structure information for relation model. QFD can provide customers' requirements as constraints for the relation model.

3.3 Knowledge model

In the product development process, the team members have two roles. Firstly, they execute tasks that carry forward the process, for example, executing CAD design tasks. Secondly, as controlling elements, they can indicate the direction for the product development process and conflict resolution process. The decision-making cognitive process is very complex and is difficult to describe precisely and comprehensively. Decision-making activities would occur while some basic knowledge is used. The knowledge can be represented formally to form a knowledge model, which can increase the decision-making effectiveness. A knowledge model has three kinds of knowledge:

1) General decision-making rules.

Such rules are general and are correct in most situations, such as the isolation rule, the trade-off rule, the majority-first rule, etc.

2) Selection rules of decision-making methods.

In general, a problem can be solved using several methods, such as utility theory, fuzzy theory,

mathematical programming, etc. Various decision-making methods have different limitations. Different methods often produce different results, so the proper method must be selected firstly for a given problem. Suitable methods must be recommended for specific problems in the knowledge model. Selection rules can be represented by production rule.

3) Decision-making cases and the successful design cases.

A case can be a successful scheme of solving a problem or an optimized result of a previous design, which includes rich knowledge for past experiences[6]. Case-based reasoning is often adopted by humans and has become a method in the product development process. It can provide the right way for resolving conflicts and provide detailed referenced information. Cases are represented by frameworks.

3.4 Data collecting

With the support of network, some useful product development process data can be collected to provide workflow and design-related information for the consistency-checking module. The data include the resource availability status, the development process progressing status, the design result of some product parameters which are brought forward by the relation model.

The network environment and the application of STEP technique have ensured that most data in the system can be got automatically . Some data also can be got manually.

3.5 Conflict resolution

In the environment of cooperation and coordination, conflict resolution in the development process can be achieved through the following ways:

1) Mathematics-oriented approach

a. Multi-objective decision-making method

Many conflict resolution problems in product development process can be represented as multi-objective decision-making problems. This kind of problem has already had many successful solutions and can be solved directly[7].

b. Fuzzy method

In some cases, high level decision-making can not achieve an exact result if the decision-making criteria for product development are the assemblability, manufacturability, maintainability and reliability. This kind of decision-making problem should apply fuzzy set theory.

2) Artificial intelligence approach

a. Constraint relaxation

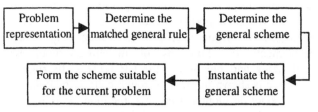

Fig. 2. Diagram of reasoning based on general rules

Constraint relaxation is the method to satisfy the constraint by changing the form of the constraint or changing the domain range of the corresponding design parameters. When performing constraint relaxation, it is needed to test the feasibility of the assumed relaxation value through the calculation function of the constraint network. Any modification may cause new conflicts. So, the perfect resolution of conflicts can only be achieved through reiteration and tradeoff.

b. Case-based reasoning

The basic idea of case-based reasoning is to resolve new problems by indexing and analyzing of similar cases. To make things easier, the cases should be stored in multiple indexing forms. In the process of conflict resolution, developers who are in different domains can index the cases by using different design attributes which they are more familiar with.

c. General rule-based reasoning

The steps of reasoning based on general rules are

shown in Fig. 2. The advantage of this scheme is that it has a large application range, while the shortcoming is that it has a low exactness.

3) Artificial decision-making approach

Developers make decisions based on their intuition and inspiration. This kind of decision-making is not rational and not structured, but is often very useful.

4 The System Architecture of Concurrent Engineering Coordination Management System

To realize the functions of CEPDP coordination model and meet the demand for coordination brought forward by CE, a perfect CEPDP coordination management system is needed to manage all coordination-related factors and objects. The system architecture we provided is shown in Fig. 3. A coordination management system is composed of three modules. They are constraint management system, process management system and conflict management system. The coordination management system exchanges the information with the product model, product development process and IPTs to realize the final coordination.

4.1 Constraint management system

In order to realize and employ the CE relation model, it is needed to establish one constraint management system

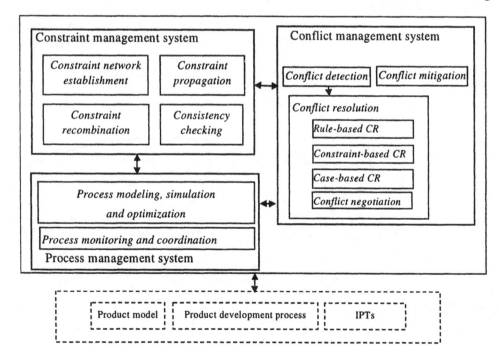

Fig. 3. Architecture of CEPDP coordination management system

to represent and manage the hierarchical constraint relationship in product development process.

1) Levels of constraints

The constraints in product development process can be divided into several levels such as constraints among multifunctional teams, constraints in multifunctional teams, constraints in the design department, constraints among developers who belong to the same task. Classifying the constraint network into a series of constraint sets is an effective way for implementing the constraint network level architecture. Therefore, the constraint network is divided into constraint sets. Fig. 4 is an example of the level architecture in a constraint network.

2) The functions of the constraint management system

The constraint management system has four main functions. They are constraint editing, constraint recombining, constraint propagating and consistency checking. Constraint management system can be effectively applied to check the feasibility of the design schemes, to find out all conflicts in design schemes in time and coordinate the product development process. The developers can estimate the value of some parameters through preprocessing constraint network. They can detect the impact of some decision on other developers, check the correctness of the submitted variable values, find out the conflicts as early as possible through constraint propagation. Thus, the range and extent of some design decisions' impact can be known accordingly.

4.2 Process management system

1) The running structure of the process management system

The radical difference between CE and the traditional product development is the reengineering of the product development process. CE considers the product development process as an integrated and concurrent process. As one important module of CEPDP coordination management system, CE process management system implements the functions of modeling, simulating, optimizing and monitoring the CE process. Its running structure is shown in Fig.5. Through collecting all data in the product development process, the monitoring module detects all process conflicts in time and passing the conflict information to conflict management system. Then conflict management system adjusts the workflow to resolve the conflict according to the conflict information.

2) Process modeling, simulation and optimization

Owing to the complexity of the product development model, it is often impossible to ensure its optimization entirely through analytical methods. Therefore, simulation is necessary for process modeling. When the simulation strategy is decided, it will be convenient to compare the system's performance under different conditions. This can provide decision-making support for improving the product development process.

3) Process monitoring and coordination

The objective of process monitoring is to plan, schedule and monitor the workflow of product development, to carry the correct information and resource to the correct team, and to ensure the product development process converging on the customer requirement. Its detailed functions include the monitoring, planning and scheduling of the workflow. Process model is executed after it is established by the developers using the tool of process modeling, simulation and optimization. Correct execution of the process model can ensure that "The appointed tasks can be accomplished by appointed person in appointed time using appointed resource".

4.3 Conflict management

Conflicts can appear in all design phases and all

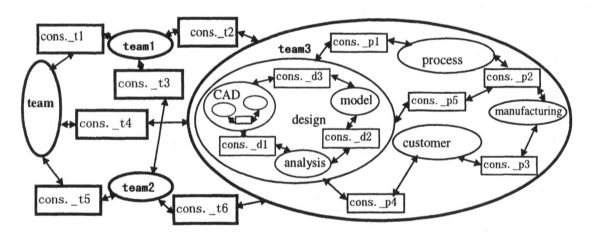

Fig. 4.   Level architecture of a constraint network

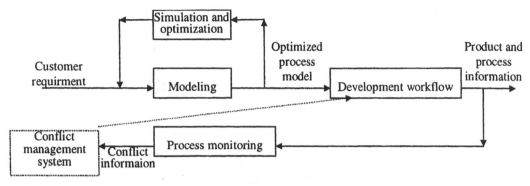

Fig. 5. Running structure of process management system

organizational levels in CEPDP. To some extent, the process of product development is the process of conflict appearing and resolving all the time. A more perfect design comes out with the resolution of the conflicts. To adequately exhibit the benefit of CE, a conflict management system is needed to settle all conflict problems in CE environment and coordinate product development process. Conflict management includes three key supporting techniques -- conflict mitigation, conflict detection and conflict resolution.

1) Conflict mitigation

Conflict mitigation is to mitigate the occuring of some potential conflicts and improve the efficiency of cooperation among and inside the IPTs through some technical and management methods according to the sources and causes of the conflicts. Conflict mitigation techniques include organization dividing and management, design task planning, domain knowledge sharing, product data management, design rationale supporting and etc.

2) Conflict detection

Conflict detection is the technique to monitor and detect the conflicts in CE system. According to the close tie between conflicts and many restrictive and dependent relationships in CE system, we provide the conflict detection technique based on relation model and the consistency processing technique based on constraint network to detect the conflicts in CEPDP.

3) Conflict resolution

Conflict resolution can be divided into two forms -- individual conflict resolution and conflict negotiation. Certain reason and criteria are needed to conflict resolution. We provide the rule-based, case-based and constraint-based conflict resolution techniques and conflict negotiation techniques. To conflict negotiation problem, we put forward the negotiation method in standard satisfactory evaluating space and the negotiation axiom-based conflict negotiation method in CE environment. These conflict resolution techniques are the

summary of engineering practice and they are efficient in CE practice.

5 Conclusion

Coordination is an integrated theory concerning multiple disciplines. It has many theoretical and practical values in CE. This paper provides the system architecture and key supporting techniques of CEPDP coordination management system based on the analysis of coordination problem. Furthermore, key techniques in CEPDP coordination are studied in details in our ongoing 863/CIMS projects.

References

[1] CERC, "Concept of operations for the project coordination board of DICE," Technical Report of Concurrent Engineering Research Center, West Virginia University, Morgantown, West Virginia, 1992.

[2] T. W. Malone, K. Crowston, "The interdisciplinary study of coordination," *ACM Computing Surveys*, vol. 26, no.1, pp. 87-119, 1994.

[3] T. Chang, *CSPs-based coordination theory and methods in concurrent engineering,*[Ph.D. dissertation], Tsinghua University, Beijing, 1999.

[4] T. Chang, G. Xiong, J. Li. "A kind of relation model of product development process in concurrent engineering," *Journal of System Simulation*, vol.8, no.3, pp.19-26, 1996.

[5] Y. Peng, G. Xiong et al., "Methodology of product development process modeling in concurrent engineering," *Journal of System Simulation*, vol.8, no.3, pp.14-18, 1996.

[6] T. Bardasz, I. Zeid, "Case-based reasoning for mechanical design," *Artificial Intelligence for Engineering Design, Analysis and Manufacturing*, vol.7, no.2, pp.111-124, 1993.

[7] T. Chen, *Decision Analysis,* Science Publishers, Beijing, 1986.

Information Technology Support for the Recycling Problem

Karim SAIKALI, Nicole BOUTROS, Bertrand DAVID
Laboratoire ICTT, Ecole Centrale de Lyon, B.P.163
69131 Ecully Cedex, FRANCE, e-mail: Karim.Saikali@ec-lyon.fr

Johnny POQUET
Laboratoire d'Automatique de Grenoble (LAG) ENSIEG-INPG, B.P.46,
38402 Saint Martin d'Hères Cedex, France

Abstract

Recycling of used manufactured products is a major concern for industrialized countries. Common recycling approach, characterized by raw materials recovery, is to limited. Thus another approach, called "noble recycling", is preferred. It consists in extracting recoverable components from wasted products, for further reuse. However, noble recycling is still not enough. The solution is to combine noble recycling with the integration of recycling objectives and constraints at design stage. This is called "green design" or "sustainable design ".

This article describes the two-level approach of the project <<RESTER PROPRE>>, related to this recycling problematic: first level deals with the design of a semi-automated platform for disassembling and recovering manufactured products. We present what are the constraints to satisfy from Information Technology point of view, using different types of information an process management systems; and a multi-agent architecture responsible for coordinating these systems and for the co-ordination of the first and second level. Second level deals with the integration of the recycling objectives at design stage. We present how we plan to use a Product Data Management System as basis integrator between the platform actors and those from design stages.

1 Introduction : the recycling problem

Protecting the environment is one major concern of this new century, especially for industrialized countries. In particular, the recycling of used manufactured products is an important problem that has to be dealt. Indeed, the massive production of such products covers many aspects of the environment protection problem: heavy use of energy and raw materials, growing landfill sites for placing discarded products, etc. For now, the most common approach for products recycling is what we can call "basic recycling", characterized by raw materials recovery. However, this approach quickly reaches its

limits and another one is preferred, that we call "noble recycling". It consists in extracting recoverable spare parts and components from wasted products for further reuse, before basic recycling occurs. This approach constitutes an improvement over the recycling process, but is still not enough. Indeed, we find that noble recycling processes are often too destructive or too expensive to be truly satisfying. The main reason is that recycling objectives are not taken into account at design stage of the products. The solution is then to combine noble recycling with the integration of recycling objectives and constraints at design stage. This integration is called "green design" or " sustainable design " [1]. Indeed, to allow comfortable recycling, it is important to establish a closer link between design and recycling activities. In this way it should be possible to take into account disassembling constraints (as new solutions of reuse) during the design stage; and to have at the disposal at the recycling stage the information on design preferences. This approach transforms the traditional "waterfall" life-cycle of a product into a "looped" life-cycle, by allowing an information feedback between at least design and recycling stages, but also industrialization and manufacturing stages and more generally between product elaboration stages and recycling activities:

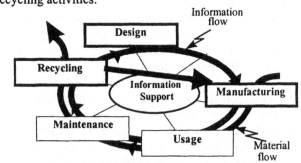

Figure 1 Design for disassembly information and material flows

The new information support must be bi-directional and must allow open access to the information from all stages of the product life cycle. Hence, information and knowledge management and distribution are the basis of this new approach for recycling.

Our paper presents how information technology can help achieving noble recycling and sustainable design objectives. In particular, we discuss the way we use and improve three different but complementary fields in information technology to reach our objectives: Product Data Management, Workflow Management and Multi-Agent architectures.

1.1 The project "RESTER PROPRE"

"RESTER PROPRE" is a French regional project dealing with the recycling problematic of used manufactured products. It integrates the competencies of three laboratories, specialized in different but complementary fields: computer Sciences and new technologies (ICTT-GRACIMP), automatics (LAG), and ergonomics (ERHIST). The main objectives of this project are:

- Study of recycling approaches,
- Design of automatic, manual and semi-automatic disassembly cells,
- Design of a plate-form for recycling (REX),
- Design of a network of recycling plate-forms (implying the study of logistic aspects),
- Study of an information system for information gathering and distribution to the different platform actors,
- Study of an information system for managing the noble recycling process (on the platforms),
- Elaboration of a design stage support for cooperative engineering of sustainable products,
- Design of multi-agent architecture, which is able to manage design, manufacturing and recycling actors in a distributed and cooperative manner.

Our work is based on a real example of recycling, proposed by a non profit association called ENVIE, which purpose is the professional reintegration. First recycling platform, which we studied, was inspired by the organization of an existing recycling unit of ENVIE, which is working on the recycling of so-called "white" goods (refrigerators, washing machines, and cookers).

1.2 REX platform

The recycling platform REX described in Fig. 2 allows manual, semi-automatic and automatic disassembly of products reaching their end-of-life phase. These products

are collected upon request within collect centers that are responsible for sending them to the adequate platform.

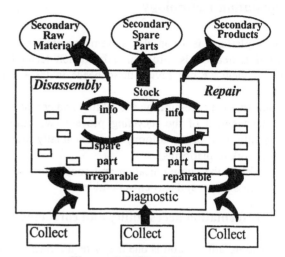

Figure 2 REX platform

The arriving products are analyzed in diagnostic centers which principal task is to decide whether a product is repairable or not. If it is irreparable, it is decided which spare parts are to be extracted. This information is used by the decision system of the disassembly workshop in order to determine the most appropriate disassembly sequences. Note that this step is not always necessary considering that most of the time and for a given product, the same set of components is extracted, or the same repair operations are executed.

We defined three categories of disassembling cells in the workshops : fully automated, semi automated and manual, depending on the complexity of the disassembly activities. In the workshop, end-of-life products are disassembled by extracting spare parts, (reused for repair of other products, component recycling, and secondary spare parts) and by obtaining raw material of high purity (material recycling, secondary raw materials). The repair workshop renews the arriving products by replacing broken components with used spare parts provided by the disassembly workshop (product recycling, secondary products). Finally spare parts and renewed appliances are stocked in warehouses.

At a higher level, we consider a network of cooperating platforms, each of them specialized in some types of products. The recycling process is thus dispatched among these platforms, depending on the nature of the products or parts to be disassembled or recovered and on the availability of each platform in terms of free cells and spare parts.

2 Application needs and constraints on information technology

REX platform is a complex systems generating many constraints and needs. From an information technology point of view we can divide these needs in three main related categories:

1. Information accessibility, availability and correctness,
2. Management of the recycling process (workflows), locally (on one platform) and eventually over a network of platforms,
3. Enabling the communication and cooperation of the process actors and participant entities.

2.1 Information accessibility and correctness

Providing information is one major requirement on the platform. Indeed, working on disassembly without official information seems impossible. Disassembly is an activity, which to be automated, assisted or only shared between several actors, needs to be well known. Thus each organizational unit at each step of the recycling process requires specific and detailed information about the product it handles. For example, the diagnostic phase needs to know everything about the functional structure of a product in order to determine which parts must be extracted. As for the pre-sequencing phase, it needs detailed information about a product structure, its geometry and topology in order to generate the best sequences for disassembly operations. All these information and data exist and are available from the product life cycle stages preceding the recycling phase. Moreover, each information generated by a unit on the platform, and directly related to a product, should be automatically linked to the product definition and data, for further reuse.

As there are many views for a same product and most likely many data sources, the decision was made to use a Product Data Management System (PDMS) for the management of product information. PDMS are information systems used to manage all the information about a product throughout its life cycle. They are able to federate, integrate and reference the data about a product (using a "vault" metaphor), independently from the different information systems and databases used to store it [2][3]. Accessing the correct information is hence possible in a transparent way, choosing the view that fits the user's requirements. We will see in a later paragraph how we use the PDMS functionality to provide the correct informational environment to our platform.

2.2 Management of the recycling process

To increase the interest in noble recycling and sustainable design, it should be proven that these approaches generate interesting results in both environmental and economical aspects. Thus, profitability and productivity constraints have to be taken into account in our solutions, among which a robust management of the recycling process constitutes one step towards our productivity objectives. In order to improve the recycling process efficiency, the process should be automated in most of its parts; and information should be automatically made available just in time at each stage.

Workflow management technology provides the most appropriated solutions to this kind of need; and in particular "process oriented" Workflow Management Systems (WMS). WMS are systems used to define, automate and support the execution of business processes or processes tasks. A "process oriented" WMS is a system that executes and manages a computerized model of a real business process [4]. Existing process oriented WMS enable us to satisfy most of the process management requirements on our platform, mainly : automating the process tasks execution and providing the adequate information to the process actors. However there are still some features, needed in our particular case, which can not be truly provided by existing workflow software. First feature is the *flexibility* of the WMS. We define flexibility here by the capability of adapting a process model during its execution to external changes affecting the process definition. Second feature is *reactivity*, which we consider to be the capability of reacting to and managing exceptions (twists in the process execution). Finally the third feature is the support of automated actors. Indeed, most existing WMS do not well integrate automated actors into a workflow definition (most WMS are "human oriented" and often in best cases, it is only possible to create interfaces using the WMS API). These three features are not currently well supported by existing workflow tools. In this paper we propose an object oriented adaptive workflow framework that allow introducing these capabilities into WMS.

2.3 Enabling the cooperation and communication among the platform entities.

Cooperation and communication are two main keywords in our project; considering that three laboratories of different scientific culture are involved and need to put their work and their competencies together, in order to reach successful results. More generally, any project involving many actors requires strong capabilities of cooperation and communication among the project members, in order to succeed. This is the case of our platform REX, which groups together in a common (virtual) structure many different specialized competencies and systems. Thus, another need to satisfy is to enable all the platform entities to communicate and cooperate. As we previously mentioned, our platform is semi automated which means that human as well as automatic actors coexist. However, traditional groupware softwares (as WMS) do not usually handle automated actors in a group. Hence we need a more flexible architecture, allowing us to handle both human and automated actors. Multi-agents architectures propose well-adapted solutions to this kind of problem, by the creation of "intelligent" collaborative entities (the agents) specialized in specific tasks.

A multi-agent architecture, called AMF (French acronym for multifaceted Agent), was developed a few years ago in our laboratory. AMF is a generic and flexible model that was initially developed for the design of interactive software, but was later successfully applied to the field of groupware (AMF-C) [5]. AMF-C is used at two levels in our project: first level is for enabling the communication and cooperation of platform entities, and the second more specific level is dedicated to the control and supervision of the workshops.

Next paragraphs will develop in more details the preceding concepts and will present the work that was done in our project.

3. Information management : the PDMS

Till now, the "classic" life cycle of a manufactured product started from the design phase and ended at the maintenance phase. However, taken into account the new recycling constraints and laws (European Community treaty of Maastricht contains in article 2 the principle of "durability and non-inflationary growth respecting the environment"), a new stage has to be added to a product life cycle: the recycling phase (fig.1). Considering that PDMS are used to reference and manage all the data of a product throughout its life cycle, our idea was to use this data to provide technical information for the disassembly and repair actors (fig. 3).

In the recycling phase, each participating actor has a distinct view of the product, depending on the type of information it needs. Each type of information corresponds to one or more already existing view(s) on the product, referenced in the PDMS. We hence developed an object oriented information model (in UML) to create an interface between existing views (design, production, etc.) and required views on the platform (diagnostic, workshops, etc.). Object orientation was a necessary approach in order to add evolution capabilities to our information model. Indeed, our "sustainable design" approach implies perpetual evolution of the product definition, which directly affect our information model. Moreover, changes in laws on recycling may require adapting or adding views on a product.

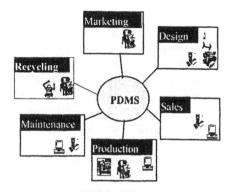

Figure 3 Complete life-cycle information management using PDMS

As an example on how the information model is used, lets consider the fully automated workshop cells. They require detailed data about product components, like their shape and volume, the distance among them, etc. This information is extracted from the design an manufacturing data already referenced by the PDMS and packed into a "workshop cell view" on a product component. Using objects features, it is also possible to add new views to the information model, group together existing views to create new ones, or specialize them, for example in order to create new views on a recent product which could be an evolution of an older one. Moreover, data generated by the platform entities can (and sometimes must) be linked to the recycling views. For example, if we consider that most of the times the platform processes the same products and extracts (or repairs) the same set of components, disassembly operations sequences can be created once by a specific unit of the platform, and stored in the PDMS. The sequences are later automatically associated to the product, reducing thus the time of parts extraction.

The use of the PDMS is however not limited to noble recycling. Indeed, the other main interest of the PDMS is to integrate the information and data that can be generated by the recycling activity, in order to be reused in the design of new products. This information mainly concern the problems that can occur at diagnostic and workshops levels like for instance observations and remarks about a product features, frequent failures and reported breakdowns, difficulties in extracting some components etc. The interest is to capitalize these data and analyze them when designing new "sustainable" products that take into account recycling and disassembly objectives.

3.1 PDMS and interactive learning

As we previously mentioned in paragraph 1.2, we distinguish three types of cells : fully automated, semi automated and entirely manual cells. It is obvious that the granularity of the information handed in the manual case, as well as the way it is presented to the end user, must differ from the fully automated case. We have hence developed an interactive multimedia tutorial, in collaboration with ERHIST, which we describe in the following paragraph.

Providing technical information to human workers requires a larger granularity i.e. less details than for automated actors. For instance, a human worker does not need to be provided with the coordinates of a component, in order for him to extract it. However, human workers require more efforts in the way information is presented. In that case, the PDMS is used as a basis for collecting the required information that must further be filtered and placed into presentation and ergonomic "wrappers". To meet this requirement, we have developed an interactive multimedia tutorial that takes into account ergonomic and learning constraints. The purpose of creating such a tutorial is double : it can be used as a technical reference that can be consulted by experimented workers if they encounter a problem during disassembly or repair operations. The tutorial can also be used as a learning tool for training novice workers. In that case the whole manual disassembly (or repair) process is described, in all its details.

The tutorial proposes two learning modes : first mode is dedicated to novice users. The process is presented activity by activity (in order), where each activity is associated to a component, and is represented as a page (fig.4). A page is divided in a set of elementary operations that can be eventually associated to a tool or a set of tools. Each tool is linked to another page describing the tool and the way to use it. Moreover, a video sequence describing the activity is available to the user.

Figure 4 Multimedia tutorial page

The second mode is dedicated to experimented workers and gives the possibility to access the product components, activities or corresponding tools, independently.

3.2   Automatic generator based on the PDMS

To enhance the tutorial interest and make it a powerful learning tool, it is necessary to cover all the products that are disassembled on the platform. That means that a specific tutorial is needed for each specific product, which is a gigantic task, if done "manually". A first solution is to create one tutorial for each type of products (washing machines, dishwashers, TVs, etc.). This solution is however not interesting enough, for two reasons : first reason is that we might lose some important information specific to a product model. The second reason is that even then, the number of tutorials to develop could be too important. An acceptable solution is to be able to automatically extract the necessary information from databases. Thus, we have designed a prototype of an automatic tutorial generator that we link to the PDMS. Using the data referenced by the PDMS, the generator assembles the information to create pages, using generic presentation patterns. Users are able to choose in a list of products handled by the platform and ask for the automatic generating of the corresponding tutorial.

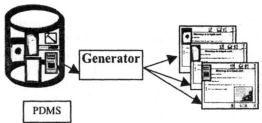

Figure 5 Generator of multimedia interactive tutorials for learning disassembly activities

4.   Adaptive workflow

Many companies evolve in a dynamic environment that is submitted to many unpredictable changes. To remain competitive, a company must be able to quickly react to changes by adapting its business processes to meet the market new requirements and therefore, adapt the model executed by the workflow engine. "Classic" workflow systems however, do not allow handling efficiently enough this dynamic aspect. Flexibility is hence the main target to be reached by the new generation of WMS.

In a Workflow context, we consider that flexibility and adaptability can be obtained using two complementary means :
1. The flexibility of the workflow model and its instances.
2. The flexibility of the system.

1. *Flexibility of workflow model* : F. VERNADAT in [6] and PR. A. SCHEER [7] consider that the flexibility of enterprise management applications necessarily includes an integrated flexible modeling of the enterprise domain : processes, resources, data... This remains true at the level of workflow systems, which are based on business process models of the enterprise. If these models are initially capable of integrating modifications and evolution, these can then necessarily be reflected at run time. JOERIS and al. [10] distinguish two levels of flexibility in WMS : *a priori flexibility* and *a posteriori* flexibility. A priori flexibility consists in creating less restrictive control flow mechanisms in workflow models. In other terms, the workflow definition should, in advance i.e. before execution, propose some alternatives and generic behaviors that allow the workflow user to choose different paths, depending on the situation. A posteriori flexibility is about the "evolution of workflow models", in other terms their capability of dynamically integrate changes at runtime. We mainly concentrate our work on this second aspect of flexibility in the form of a workflow framework based on UML, dedicated to business process modeling and directly exploitable by a workflow engine. More details about the framework will be given in the coming paragraph.

2. *Flexibility of workflow systems* : a flexible (workflow) system is a system that can during run-time, automatically take into account or give the means to react to a deviation of the process model instance. This can be achieved by different means and may be effective at different levels: a *local level*, which means that the processing of the deviation is local and it only concerns one instance of the model (the one that needs to modify the workflow execution); a *global level*, where the processing is reflected on all the running instances of the model but does not affect the initial model itself; and finally a *core global level*, where the handling of the deviation is also reflected on the initial model. Reflecting changes at core global level means that the initial model does not correspond to the real business process anymore. The difference between the global level and the core global level is that in the former, the modification only affects all running instances of the workflow model (to react to a new situation) but does not affect any new instance of the model. That means there is an exceptional situation that is not totally compatible with the process model. However, the model still remains valid for any other case. In the latter case (core global level) it is even impossible to create new valid instances of the initial process model, thus any new instance will include the modifications.

Workflow systems may however have built-in aptitudes to react to modifications of their environment, without being based on a flexible process model (*a priori* flexibility) but in that case, the adaptability may only concern the first two levels described above.

4.1 Adaptive workflow framework

In the field of software design, flexibility and adaptability lead naturally to the concepts of object-orientation. The fact is that object orientation is very successful in bringing solutions to problems of expandability, maintainability and specialization of software applications. The intrinsic object characteristics and especially inheritance, encapsulation and polymorphism make objects particularly attractive for finding solutions to the flexibility problems workflow systems have to face. In the field of process modeling however, and in particular workflow modeling, only a few interesting modeling methods are based on object concept [8][9][10][11]. The fact is that if we take a look at current process oriented workflow systems, none of them proposes modeling techniques based on object formalisms or concepts (even though the system itself may be implemented in an object-oriented language).

Figure 6 adaptive workflow framework

An important part of our work is dedicated to the definition of an object workflow framework (fig.6),

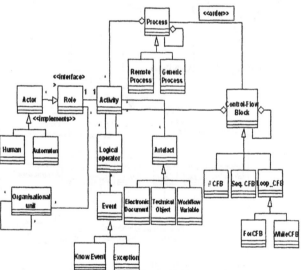

modeled in UML. The purpose of such a framework is to propose a reference and a basis for designing and modeling adaptive workflows. The framework is initially based on the Workflow Management Coalition's reference model [12] over which it introduces some new classes or interfaces. Each class (or interface) of the framework represents a workflow entity for which it provides basic behaviors and attributes.

To model a process using the framework, we propose a simple modeling method composed of three steps : first step consists in defining the requirements of the workflow application. It gives a global view of the main process tasks or sub-processes and the roles responsible of executing them. This requirement phase is modeled using UML use cases [13].

The second step of the method consists in giving a representation of the process dynamic, based on the use cases diagrams. This is done using activity diagrams [13] that propose formalisms that are well adapted to workflow : activities, roles, control flows (sequences, parallelisms, loops), conditions, events and objects used the by activities (data flow).

The two preceding steps only serve as a basis to design the final class model that can be implemented. They should hence be progressively refined until a satisfactory static class diagram is obtained, classes of which are all inherited from the framework components. The following figure describes the three steps of the method :

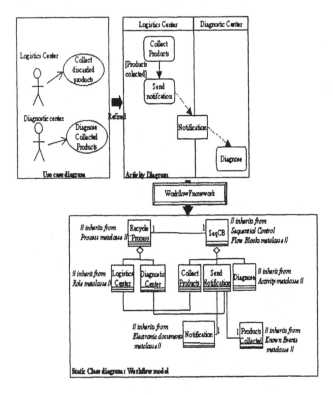

Figure 7 Modeling workflows

Globally, our framework presents four main interests :

1. It can be used as a workflow metamodel. Each class of a workflow model is an instance of a metaclass in the framework. For example if we have a "check payment" activity in a process, then we add a class

"check payment" to the process model and it is an instance of the metaclass "activity" of the framework.

2. The second interest of the framework is that it proposes predefined building blocks. Each class is a business (workflow) object possessing its specific implementation in Java, with its own behavior. Thus we can easily create workflow models by assembling components that directly inherit from these building blocks, without requiring programming skills. It is also possible to adapt the framework and workflow models by adding new features to the children classes or modifying the inherited ones (using overloading and overriding mechanisms). This introduces a great deal of flexibility, but of course requires some programming. However that gives the possibility to customize the metamodel into specific applications, at low design and development costs.

3. All models defined using the framework are reusable and adaptable to new business situations.

4. Many authors and workflow [8][14][15] vendors insist on the fact that an important requirement for modeling language is that they should be easy to use and understand. We believe that our framework fulfills this requirement. Indeed, instead of proposing yet another notation and another programming language (as most WMS do), our framework relies on UML, which is a well spread standardized object modeling notation, and Java which is widely used by many developers. Thus it is not necessary for workflow users to learn new modeling techniques and use new programming languages. Moreover, using the UML facilitates the communication between workflow designers/process modelers (which do not necessarily come from computer sciences domain) and workflow developers, as it can be used as a common modeling notation.

The preceding paragraph introduced the global advantages of the framework, from a general point of view. Next section will present more characteristics of our framework in deeper details and show how they can be used in our platform case.

4.2 Using the framework

On the recycling platform, it is not always known by advanced who will physically process a given task (will it be a robot or a human person). Indeed, the choice of the physical actor sometimes depends on some conditions, like the availability of a given actor, or the complexity of the disassembly or repair operations. Indeed complex operation can not be handled by robots and reciprocally, dangerous operations (like extracting components containing dangerous or poisonous substances) can only be executed by robots. In fact, the only thing that is really known is what competencies (i.e. which *roles*) are needed to fulfill a task. The problem is that, for flexibility sake, we should be able to assign the process tasks indifferently to either human or automated actors in a transparent way, i.e. without changing the activity type. In the framework, we propose a solution to this problem :

An *activity* is a class that contains a description of what has to be done, has its own behavior (like sending a notification to an actor) and is eventually linked to external data' or objects. Each activity is necessarily associated to a *role* that we propose to be an <u>interface</u>, not a class. Hence, a role can be seen as a set of competencies that are proposed by some of the platform entities and each service should be named after an existing activity. A role can be then linked to one or more activities, depending on how many competencies (or services) it proposes.

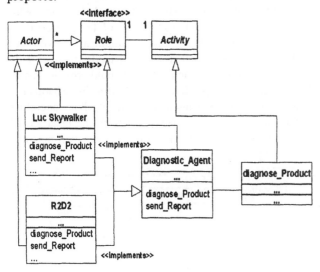

Figure 8 Activities - roles - actors

Roles are associated to actors. An *actor* is a class that corresponds to a real workflow participant. The actor class implements the interfaces, i.e. it provides an implementation for all the role services. This approach is very flexible because is separates "what can be done" (roles) and "what must be done" (activities) form "how it is done" (actors). Hence an activity does not depend on the actor's type because each actor provides its own knowledge on how to process the activity. The advantages are multiple :

- It is not necessary to know the actors by advance (during process modeling) but only the roles, the actors implementation can be provided at run-time

- It is easy to change the way an actor processes an activity only by changing its implementation, without any effect on the process logic and execution
- It is possible to change the actor who will execute a given activity simply by removing the old actor and "plugging" the new one into the process model.

For example in figure 8, the activity "diagnose_Product" of a recycling process should be executed by some "Diagnostic_Agent". This role can handle the execution of the given activity because it is part of its interface. This same role can be taken by two different actors, "Luc Skywalker", which is human or "R2D2", which is a robot. This is possible because both actors implement the method "diagnose_Product" that corresponds to the process activity. However, each actor has its own way of processing the activity.

Another framework application on the platform is how it helps dealing with the objects manipulated by activities. In most cases, WMS are used in office work contexts where workflow activities correspond to office activities. This implies that input and output activities objects are most often electronic documents and data. This is not the case of REX platform where the recycling process activities manipulate real physical objects like dishwashers and TVs, etc.. We introduce a new class called "artifact" that corresponds to any type of object used or produced by workflow activities. This class is refined into three subclasses, respectively : "electronic document", "technical object" and "workflow variable". The interest that we find in the "artifact" class is that in encapsulates the knowledge of how to access the corresponding real object. For example, passing a dishwasher from one unit of REX to another may require launching an external logistic process (find a free carrier, define best trajectory, send notice to carrier, check product arrival, etc.). This other process is independent form the recycling process logic, so it does not have to figure in the recycling process model, for visibility and readability sake. However this "logistic process" may be taken in charge by another workflow engine (or any software engine) and the activity of launching the process could be encapsulated into the dishwasher artifact. The artifact concept reinforces the flexibility of our framework by providing a separation between "which objects the workflow needs" and "how it will get them". It is thus possible to modify the external processes without affecting the process model.

We also use the artifact concept (more precisely the "electronic document" artifact) for querying the PDMS. For each unit of the platform, required information on a product is assembled into packages (the artifacts) which are responsible for querying the PDMS and retrieving the data. The use of artifacts and actors objects requires however to define intelligent behaviors as well as communication protocols between them, the workflow system and any system they communicate with. The coming paragraph will present our approach for dealing with this communication and cooperation problematic.

5.   Communication   and   cooperation management

We previously defined AMF as a multi-agents architecture, initially intended to the design of man-machine interfaces, before migrating to cooperative applications with AMF-C.

AMF organizes each agent in an appropriate number of facets, among which we retrieve two classical components: Presentation and Abstraction facets. The other facets can come either from a finer split of *control* components, from the identification of new characteristics of agents (specific facets), or from the duplication of classical facets (several *presentation* facets corresponding to different views).

AMF expresses interaction control with two kinds of components:

1. Each facet presents several *communication ports* (allowing input, output or both) which can be seen as interfaces of real object methods. These ports avoid having a permanent binding between a function (a service) and its implementation. Moreover, it allows to implementing the function body in heterogeneous languages.

2. The *Control* component is a part of the agent defined by entities called *control administrators* that have three roles:

    - To *connect,* managing logical relations between the communication ports (sources and targets) that are connected to it;

    - To *translate,* transforming the messages which come from the source ports in understandable messages for target ports;

    - To express *behavior,* and so control strategies, using different rules of activation between a source port (A) and a target port (B).

When a facet needs to trigger a distant service, it activates its corresponding output port. This port builds a message and sends it to its associated daemon. Then, the control facet of the owner agent wakes up all the control administrators, which are connected to this source port. If this port is exported (connected to other agents), the activation is recursively transmitted to the parent agents.

Then, each concerned administrator considers its activation conditions (see behavioral role). If these conditions are validated, the message is translated and sent to all the target ports. The activation of these ports runs their associated daemons

These concepts are very similar to the *listener* and *adapter* concepts of *Java Beans* (in fact, the *Java* implementation of AMF uses them). However, AMF relies on a complex engine so that programmer can use predefined components, such as standard administrators, which are real object and not only *interfaces*. Moreover, AMF provides a graphical formalism that expresses the control relationships between ports. More detailed information on AMF can be found in [5].

To enable communication and cooperation among the platform entities, our idea was to associate an AMF agent to each one of them (fig.9). Each agent is provided with specific facets, including facets only dedicated to communication. More particularly, an AMF agent with two facets can represent each actor of the platform : a workflow facet, enabling the communication with the WMS, and another facet, associated to the real actor. If the actor is a machine, then the facet could provide data to the machine and simulate its behavior by establish a dialogue between the agent and the machine operating system. In the other case (human actor), the facet would rather be a presentation facet, for man-machine interfacing.

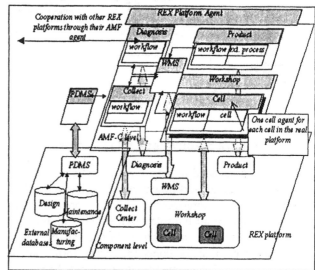

Figure 9 Multi-agents contribution to the platform

This scenario can be reflected to any other entity of the platform, like for instance the workflow artifacts. At a higher level we can associate an AMF agent to the platform, enabling thus the communication among distant and distributed platforms.

6. Summary and Conclusions

In the paper we presented a concrete application on using concurrent information technology to manage and enhance the performances of an industrial system, in our case a recycling platform submitted to profitability and productivity constraints. To achieve our objectives, the main constraints to be satisfied are information accessibility, flexible process management, and cooperation management. We described a cooperation between three types of information systems and architectures : PDMS, WMS and multi-agents architecture; and proposed some improvements on these systems, in order to fit to our platform requirements. A prototype for a multimedia tutorial dedicated to teach product disassembling was also presented and we proposed an architecture for an automatic tutorial generator, that could process any type of product (mostly home appliances). We are also working on a prototype of an adaptive WMS based on the framework. Our work should be integrated with the work of our partners in the project (the design of workshops, the cells and the study of best disassembly sequences) and validated with simulation using "Technomatix".

References

[1] B.T. David, K. Saïkali, N. Boutros, Conception orientée recyclage des produits manufacturiers, 3° Congrès International de Génie Industriel., Montréal, Canada. 1999.

[2] J.M. Randoing, Les SGDT, HERMES, 1995.

[3] M. Maurino, La gestion des données techniques, MASSON, 1995.

[4] T. Koulopoulos, Workflow Market Reference Point 97, The Delphi Group, 1998.

[5] F Tarpin-bernard., B.T. David, "AMF: un modèle d'architecture multi-agents multi-facettes". Techniques et Sciences Informatiques. Hermès. Vol. 18. No. 5. p. 555-586. 1999.

[6] F.B. Vernadat, Enterprise Modeling and Integration - Principles and Applications, Chapman & Hall, 1996.

[7] A.W. Scheer, Business Process Engineering, Springer 1998.

[8] P. Hruby, "Specification of Workflow Management Systems with UML", OOPSLA 98 Workshop on Object Oriented WMS, 1998.

[9] C. Bussler, "Towards Workflow Type Inheritance", DEXA Workshop, 1998.

[10] G. Joeris, O. Herzog, "Towards Object-Oriented Modeling and Enacting of Processes", Technical Report TZI 7/98, Center for Computing Technologies, University of Bremen, 1998.

[12] [WfMC 94] Workflow Reference Model, Workflow Management Coalition (http://www.wfmc.org), 1994.

[13] P.A. Muller, Modélisation Objet avec UML, EYROLLES, 1998.

[14] S. Joosten, "Fundamental Concepts for Workflow Automation in Practice", ICIS'95 conference, 1995.

[15] P. Barthelmess, J. Wainer, "Workflow Modeling", CYTED-RITOS International Workshop on Groupware, Lisbon, Portugal, September 1995.

A Common Base for Data and Plans (BCDP)

A. BRACHON

RATP/ITA 12, avenue du Val de Fontenay – 94120 Fontenay-sous-Bois

D. LE HEN

RATP/SIT 102, esplanade de la Commune de Paris, Noisy-le-Grand, 93160

dlh@ratp.fr

Abstract

RATP is one of the world's biggest multimodal operator of public transport.

The role of BCDP is to manage the fixed installations of RATP, Paris mass transit authority.

1 Introduction

The objective of BCDP (Common Base for Data and Plans) is to place the data related to the fixed installations at the disposal of the RATP units, so that they may :

* handle the patrimonial and technical management of these installations, by giving any chief of establishment the means to fully carry out his mission and, in particular, to control spaces on his responsibility,

* develop and follow-up projects as well administration of existing infrastructures as creation of new infrastructures.

The ambition of the BCDP project is to provide all project team members with consistent and up-to-date company data. Quality and exactitude of these data are guaranteed by updating procedures and project protocol, applied and observed by the participants in the project ; as well for planning and operating as for maintaining works.

The BCDP is structured around 2 main projects:
- the Patrimonial and Technical Reference
- the Project Document Management System (SGDP)

2 The Patrimonial and Technical Reference

The Patrimonial Reference was initialised in 1993. The objective of RATP is to optimise the exploitation of its patrimony (land, real estate, equipment) on technical and financial levels.

General principles:
- a branching describing the places implying a single, transverse coding and common to :
    - graphic data bases
    - and data bases for the maintenance.
- graphic files structured according to the CAD guide principles (cf norm ISO 13567 [1]).

This system will contain the bulk of real estate plans (buildings ,subway stations , RER stations etc).

Functions and services :
  - synthetic reports of occupation (by type of buildings or by stations, by units...) ,
  - consultation and printing of real estate general plans,
  - possibility of downloading graphic files,
  - graphic visualisation after requests related to alphanumeric bases,
  - dynamic management of the safety plans.

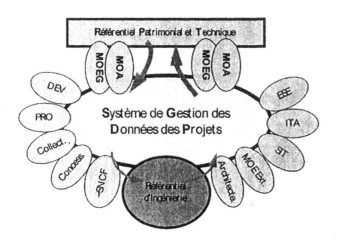

3 Project Document Management System (SGDP)

The efficiency of project management is a fundamental stake for RATP. The optimisation of each project must be carried out according to three dimensions:
-   quality of service,
-   cost,
-   time.

It requires a rigorous method described in the Guide of the Management of Projects.

The SGDP is the suitable tool to organise and manage exchange circuits between all the actors of a same project.

* Projects managed with this method:
- Maison de la RATP (headquarters)
- Championnet (Social Pole)
- Val de Fontenay (Engineering Pole)
- Coeur Transport La Défense
- Revival of the Subway
- Météor Phase II (St Lazare)
- Tramway T1 (Noisy-le-Sec)

The SGDP (often dubbed shortly ADI) is based on PRO-G Exchange (PROSYS).

The objectives of the SGDP are:
- to increase the efficiency of CAD studies production facilities , thanks to rigorous and transverse methods and automated software tools
- to improve technical quality of the documents produced by subcontractors, by working procedures accepted by all and reusable on each project
- to improve the quality of production through a better work organisation :
-   ▪   Identical plans nomenclatures   for all projects, preset plans lists by project,

-   ▪   Automated procedures of files organisations,
-   ▪   Rigourous methods of diffusion, approval and control,
-   ▪   Traceability of the modifications
- to introduce simultaneous engineering:
Methods of pre-synthesis, coordination and synthesis of the conceptual design of the Civil Engineering, Electricity, Fluids and Information Systems.
- to facilitate the constitution of completed work file for later maintenance and exploitation.
- to contribute to reduce the costs and to complete projects on time.

4 Architecture

The software architecture is based on the standards:
-   Oracle (Unix Server)
-   MS – Access and Lotus/Notes
-   ArcView
-   AutoCAD 14
-   Myriad
-   Domino/Notes 4.6 servers (Windows NT 4)
-   Citrix servers (Windows TSE-Metaframe 1.8)

The authorized user can access to BCDP via Netilus, the intranet RATP.

References

[1]   Harmonisation et normalisation des échanges graphiques informatisés dans les projets de construction
Medi@Construct – MEDIACT/M/98004 novembre 1997

Design for Manufacturing: Laser Beam Welding of Centrifugal Machine Components

F. Mandorli, M. Germani, A. Gatto, S. Berti

Mechanical Department, University of Ancona
Via Brecce Bianche I-60131 Ancona Italy
e-mail: m.germani@popcsi.unian.it

Abstract

Design for manufacturing (DFM) is a well known methodology based on the concept that the design of a product needs to be integrated within the design of its manufacturing process in order to reduce design cycle time and improve product value.

Thanks to the continuos innovation in the technological field, more precise and reliable production processes are every time available and designers need to verify their applicability within specific production contexts.

In this paper we present the DFM of centrifugal oil separator components used in the alimentary industry. The motivation for the DFM activity is the introduction of the laser beam welding technology in the production process, due to the need of changing the components material from aluminium bronze to AISI 304 steel.

The use of the laser beam welding technology will reduce the waste of material and the chip-forming machining problems. The specific laser beam used is a Nd:YAG, that will provide a better welding precision, a reduced thermal deterioration and a better effectiveness on high reflecting materials.

1 Introduction

Today's manufacturing systems and the products they produce are very informationally dense and complex. They require vast amounts of specialized knowledge, all focused on single-product problems. Superimposed on this is an exponentially increasing amount of new information and technology coupled with ever-shortening product lifecycles and increasing global competition. New materials, process refinement, new standards and customer requirements are emerging and evolving at ever-increasing rates.

Thus continual and rapid change is one of the leading characteristics of current industrial world. This problem can be handled by concentrating on the new ways to execute the design process. Design for Manufacturing is one of the techniques concerned with this change. More specifically DFM is concerned with understanding how the physical design of the product itself interacts with the components of the manufacturing system and using this understanding to define product design alternatives that help facilitate "global" optimization of the manufacturing system.

The major objective of DFM approach is the identification of the product concept that is easy to manufacture considering the integration of the manufacturing process design and product design. Early consideration of manufacturing issues shortens product development time, minimizes development costs and ensures a smooth transition into production and reduces time to market. [1]

The conditions for a successfull product development in a small and medium enterprise are different from those in large-scale enterprise. DFM is a proven design methodology that works for any size company. In the large-scale enterprises the communication between the personnel involved in product development must be organized in a very effective way through suitable techniques of project management. Thus the manufacturers implement concurrent engineering concepts to improve the communication, such improvement is based on product and process data management software and on expensive CAD/CAE software. Small and medium enterprises have the same aims but they have small liquid assets for starting investment. Instead the small dimension of enterprise allows to deal with product development through an effective collaborative method for a better relationship between designers and manufacturers.

In some medium and small sized Italian mechanical engineering firms are adopted advanced design techniques and methodologies (also Design For X techniques) for the product improvement. In a large number of cases the techniques are applied without a systematic approach and without a real coordination with customers. In this way benefits of advanced techniques application are not very satisfactory [2]. During the few recent years, in Europe, the restructuring of the industry has been quite fast; SME industries have became important for the entire continental economy [3]. In this context researchers are studying suitable methods to improve their manufacturing processes and their design processes [4], [5].

This paper is concerned with the application of a particular design method, involving principles of Design For Manufacturing, for the product development in a SME Italian industry. Starting from new users requirements and from new European standards, we have redesigned, in collaboration with the manufacturer, a machine for the alimentary industry. Our aim has been to provide manufacturing knowledge to the designer in a useful form to integrate, in an effective way, the manufacturing requirements within the product design.

In the following is described the existing machine dwelling upon new specifications resulting from new standards. Further is illustrated the redesign for manufacturing method developed and its application to one of the machine components.

2 Centrifugal oil separator

Present work concerns the redesign of some components of olive oil super-centrifugal separators.

2.2 Tipology of product

The centrifugal separator consists of a decanting vessel that rotates around an axis. The decanting vessel contains the separating mixture.

Centrifugal separators can be classified on the ground of principle used for separation of "heavy" compounds from "light" compounds [6]. There are centrifugal separators based on *decantation* and those based on *centrifugal filters*. In the first type the rotation permits to gather heavy compounds on the external of decanting vessel (called rotor), while light compounds form different layers around axis. Instead in the second type (filtering) the mixture separation is performed by means of a rotating component provided with a filter (called grid) that retains the solid phase.

From manufacturing point of view, centrifugal separators can be classified in "mass" centrifuge and in "high speed" centrifuge (with n between 4500 and 10000

rev/min and with centripetal acceleration of 11000g). A further subdivision for "high speed" centrifuge is based on rotor type; there are centrifuge with cylindrical rotor and disk centrifuge. Present work deals with this last type. In the following figure (Fig.1) you can see the sketch of a disk centrifuge.

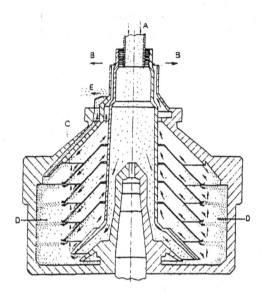

Figure 1 Disk centrifuge

The main characteristic of a disk centrifuge concerns with the presence on the rotating part (rotor) of conical slim sheet metal disks packed one over the other with little gap between them. The disks are supported by an hub, called diffuser. In addition, the diffuser embodies the function of distributing the mixture. The disks are coaxial and integrated with rotor. They have two main tasks: to reduce space for the motion of mixture (reducing swirl effects), and to short the path of particles to obtain the separation.

The machine work can be resumed in the following way. The separating mixture is inserted, coaxially with rotor, within the diffuser at the top of the machine (Fig.1a). The diffuser, provided with suitable form features, drives the mixture within the rotor (Fig.1d) and supplies the needed head for the separation and for the output of final products (oil with various pureness).

When the mixture has reached the gap between disks, the solid particles and the heavy liquid phase flow along the conical surfaces to arrive in the bottom part of decanting vessel.

On the other hand the "light" liquid phase flows towards rotation axis. The value of angle β (35° or 45°), between the generating line of cone and the axis of machine, has been computed to facilitate the sliding of particles and to permit a more easy capture of them. Besides the distance between disks, measured perpendicularly to conical surfaces, permits the countercurrent flow of "heavy" liquid phase without production of swirl.

2.1 New customer requirement

The kind of machine described in the previous paragraph is fairly consolidated both as forms and as performances. However, today exists a new problem regarding the applicabilty in alimentary industry (separation of olive oil). This is the most important application, also from economic point of view, for centrifugal separators. New European community standards advise the manufacturers against the use of certain materials for the production of components designed for treatment of alimentary products.

Such standard has been the input for the redesign process because the centrifugal separators users have immediately required machines that suit the standards. Thus the most important specification for the redesign is the change of materials for some components of machines. In particular we have replaced the current alluminum bronze with a stainless steel (AISI 304 X5CrNi1810 - UNI 6900/71). Such modification weights on the entire product because involves new choices about the productive process and the manufacturing system. Thus we have applied the DFM principles in the redesign process starting from this interesting context.

3 Redesign for manufacturing

The product design process is characterized by a steadily increasing rate of decision making, starting at a minimal level at the beginning of the concept stage and culminating with the product shipment. The early decisions are of great importance in determining the final result and in particular in designing the manufacturing systems and relative costs. When the conceptual phase is considered as a stage separated from manufacturing decisions we have a strictly sequential approach to product design. Such approach, frequently, involves substantial product modifications in the last development stages, when the manufacturing requirements become fundamental. The result is or a low quality product or a very high time to market. The DFM methodologies have been introduced to solve this problem. The tecniques for a pratical application of DFM are various [7]. Every of them can be supported with software instruments, for example manufacturing knowledge based systems or CAD process oriented.

In the present work we have studied a method for the redesign of a mechanical product considering the manufacturing constraints in the early development stages. The application of method is exemplified through the redesign of centrifugal separator. The main goal has been to rationalize the existing information and collect it in a new solution with the support of a commercial parametric feature-based three-dimensional CAD.

In the following figure (Fig.2) we illustrate the developed method.

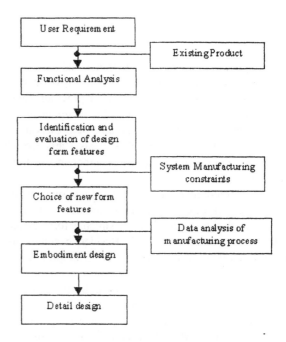

Figure 2 Redesign process

The first step of redesign process is the study of existing solution to collect information on form feature elements and on their function within the specific machine or within the particular component. To execute this task we have used the systematic method of design theory [8]. We obtain a functional structure of machine that helps in the following synthesys effort. Knowledge acquisition phase involves also the machine working and the physical principles used. On the basis of acquired information conceptual solutions are developed to find a compromise between functional requirements and economical and technological constraints. Thus new design form features result from the hypotetical manufacturing process.

Once evaluated different solutions with related possible manufacturing systems, starts the embodiment phase for the chosen solution. In this stage designer needs information on particular technological process chosen to optimize dimensions and technological form features. The last step is the execution of detail design of components with the definition of all the parameters needed for the manufacturing systems.

The described method has been applied for the components of centrifugal separator. In the following paragraph we describe its application to the diffuser.

For the material previously estabilished (AISI 304) it was chosen, between different possible solutions, a laser welding process to obtain the final functional forms. We have studied the process applicabilty for various tipologies of welded joints.

4 Method application

In a new design the aim of functional analysis is to find all needed functions (functional structure) to satisfy the machine tasks, without reference to a previous solution schema. In the case of redesign of the components of an existing machine, with the constraint of preservation of the same interfaces, functional analysis starts from the existing solution to integrate new components in the old machine.

The abstraction obtained from functional analysis permits to criticize the current solution. Thanks to abstraction designer can find not suitable form elements or those not necessary at all. Besides becomes more simple to evaluate different solution principles studying machine from functional point of view.

To exemplify the process we describe the analysis of one of the redesigned components: the diffuser.

In figure 3 the diffuser is in evidence in the circle; in figure 4 are illustrated all the existing form features.

Briefly we list the form elements and related satisfied functions:

The lower conical skirt:

*   Avoid axial translational motion of packed disks;
*   Determine axis of rotation;

The upper external tongues:

*   Determine axis of rotation;

The lower internal tongues:

*   Produce the needed head for fluid motion
*   Determine rotation axis of hub
*   Determine axial position of hub.

Pin between hub and diffuser

*   Torque transmission
*   Reference for assembly (for the high working speed needs a careful mass balancing)

Key positioned in one of the upper tongues:

*   Torque transmission to disks
*   Reference for assembly

The existing diffuser, considering the current form features, is an excellent solution. Such component integrate many fuctions in a very compact way showing an optimal choice of forms layout.

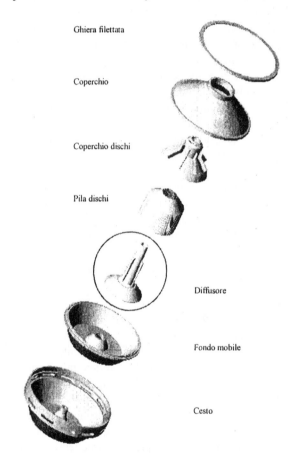

Ghiera filettata

Coperchio

Coperchio dischi

Pila dischi

Diffusore

Fondo mobile

Cesto

Figure 3 3D model of current centrifugal separator

The current material (alluminum bronze) permits to manufacture the diffuser with operations of mechanical material removal starting from a cylindrical bar. Such operations are expensive, with a consistent loss of material. Besides the manufacturing process starting from a high diameter bar involves a low frequency production. Such problems are amplified with new material (AISI 304) because is needed to slow down the feed rate.

One of the possible alternative processes is laser welding. Such solution permits to mantain high quality level and to obtain the same form elements with less material. The result is a consistent cost reduction.

Figure 4 Current diffuser of centrifugal separator

The most important technological constraints concerned with the laser welding process are [9]:

1. To assure the accessibility of laser head on elements to weld;
2. To use a Nd-YAG laser to convey energy by means of optical fibers;
3. To use an 8 axis robot to move the laser head to mantain the appropriate angle between laser beam and surfaces to join.

The first point has been particularly studied because has a great impact on form definition of new part. The new conceptual solution of diffuser is resulted from the evaluation of technological constraints given by new productive process in compromise with design constraints (functional requirements, easy maintenability, respect of interfaces). A preliminary important consideration is related to the impossibilty of manufacture the new diffuser similar to the current. The internal zone of lower conical part is not accessible for a laser head, thus is impossible to weld the tongues. It was decided to divide the diffuser and the impeller in two separate components

to facilitate the manufacturing process. Such solution can be identified with following points:

- Manufacturing of the impeller composed of a conical part of AISI 304 bended sheet metal with tongues positioned on it;
- Manufacturing of tongues in AISI 304 sheet metal;
- Joining of previous components by means of laser welding Nd-YAG;
- Manufacturing of external part of diffuser with conical skirt, the hub and the pipe to support the disks in AISI304.

Having defined the structure of functions we have fulfilled them with some possible alternatives to evaluate the better combination of them. In the following images are illustrated the sketch of conical part and related tongues (Fig.5), and the morfological matrix [8] for definition of functional layout of new impeller (Fig.6).

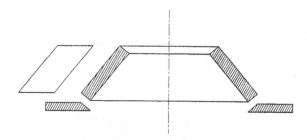

Figure 5 Sketch of parts to join for new impeller

Figure 6 Morfological matrix for the impeller

Once established the morfological layout of new diffuser, we have analyzed in detail the manufacturing process to find the optimal values for the most important manufacturing parameters and to optimize the diffuser dimensions. The working cell used for the tests is of the same type that will be designed for the final production. Such cell is provided with a machine for Nd:YAG laser welding with the following characteristics: laser Continous Wave with effective power of 2000W on the head; focal length of 8 cm; laser beam diameter on focal head; focal length of 8 cm; laser beam diameter on focal point of 0,3 mm. Such machine has an irradiation specific power of $2,8 * 10^4$ W/mm². The focusing head is mounted on a robot assuring the needed precision to laser position.

In the first phase we have studied the compatibility of laser with the particular weld required. Then we have planned systematic experimental tests to optimize the technological cycle (power required, feed rate, angle of incidence of laser) convenient for the specific productive problem.

The welding tests are related to the fastening of tongues on conical skirt with a T joint (see Fig.7), in the table are listed the mechanical properties of material used.

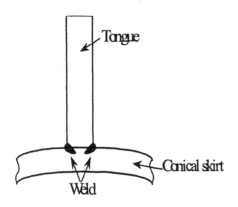

Stainless steel AISI 304		
Tensile Strength (MPa)	Yeld Strength (ε=0,2%)	Hardness (HB)
685	310	88 Max

Figure 7 Sketch of T joint tested

The performances required to welded joint are the following: to be clean, to assure an hermetic seal to avoid the pollution of separating fluids, to strength to centrifugal forces generated by the uniform rotating motion of component (7000 rev/min).

It was evaluated the problem of accessibility of spot laser on point to join for high value of angle between the lateral surface of tongue and the spot. For an optimal angle it was decided the manufacturing of joint with two welds, one for every side of tongue. It was studied the suitable depth to avoid the sovrapposition of the two

welds (Fig.7) and it was calculated the stress on joint with a FEM analysis in working conditions, to optimize the dimensional parameters of welds (for more information see references [10]).

Thus we have spent more time to execution of welding tests with different laser power values (from 500 W to 1700W with a 100W increment), with different feed rate values (2.3 m/min, 3 m/min, 3.5 m/min) and with different incidence angle (30° e 45°). In figure 8 is illustrated a test case with magnification of welded zone. It was analyzed the table of results and it was evaluated two optimal situations for weld parameters values [11]:

- Power 1000 W, feed rate 3 m/min. angle incidence 45°.
- Power 1200 W, feed rate 2,5 m/min. angle incidence 30°.

Between the two groups of values the first is more suitable because the incidence angle permits a better accessibilty in the joint zone. Besides lower power coupled with an high feed rate guarantees a better energy and time savings (better productivity).

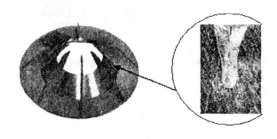

Figure 8 Test case and welded zone analyzed

The new final impeller is represented in Fig.9. It is an assembly of four components, every components satisfies a particular function:

The hub : define axis and transmit the torque;

The conical skirt and the disk : avoid axial flow of mixture and drive the radial flow;

The tongues : supply energy to the fluid

Figure 9 3D model of components of new impeller

The diffuser is composed, in the upper part, of form features needed to direct the flow of mixture, to transmit torque to disks, to interface with the remaining components of machine (for a more complete description of redesign process for the remaining components and for the entire machine see reference [10]). Three-dimensional CAD model of upper part of diffuser and the final assembly are illustrated in the following figures (Fig. 10 and Fig. 11)

The other parts of centrifuge are redesigned in a similar way identifying functions to satisfy and defining related form design features using the principles and the methods of Systematic Design [8].

Figure 10 Hub of new diffuser

Figure 11 Assembly of new diffuser and new impeller

5   Conclusions

The activity of designer is a continual transfomation of information with iterative steps for evaluation and modification of design. The goal is a product that satisfies all customer requirements and all constraints imposed by physical, technological and economical restrictions.

Iteration allows the design to be continuosly improved and optimized over time as better and more complete design information becomes available. More complete information could come in the form of improved understanding gained from a prototype test or revealed through analysis. An unavoidable consequence of design iteration is therefore engineering change. In the early phases of design project the engineering changes are handled fairly easily, on the other hand in the later stages become much more difficult.

DFM is a good technique to avoid modification in the later stages because the designer evaluates the solution simultaneously with the related manufacturing process. We have applied DFM techniques in the redesign process of a machine (centrifugal separator) to satisfy new standard requirements. To minimize errors has been fundamental to use in design phase analysis data related to new techonogical process (laser beam welding). The designers of enterprises involved in development process have found great advantages with the followed DFM technique. The redesign method application has permitted to evaluate new morfological solutions considering simultaneously technological constraints imposed by laser welding. Such approach has reduced development time and has assured the introduction in the market of a good quality machine. The method will be adopted for a new product development starting from an existing solution.

Acknowledgements

The design and development activities have been executed with the collaboration of Meccano S.p.A. (Fabriano- AN – Italy).

References

[1] D. M. Anderson, Design for Manufacturability; Optimizing Cost, Quality and Time to Market, CIM Press, 1990.

[2] E. Manfredi, A. Bonaccorsi, "Design Methods in Practice: a Survey on Their Adoption by the Mechanical Industry", in Proc. of 12th ICED 99, Vol.1, pp.413-417, Munich 1999.

[3] L. Blessing, I. Yates, "Design and Development Capabilities of Small and Medium Sized Enterprises in the UK", in Proc. of 12th ICED 99, Vol.1, pp.119-126, Munich 1999.

[4] E. Hietikko, H. Pakkinen, "Principles and Tools of Concurrent Engineering in Network of Small and Medium Sized Companies", in Proc. of 12th ICED 99, Vol.1, pp.329-333, Munich 1999.

[5] P. Skalak, "Defining a Product Development Methodology with Concurrent Engineering for Small Manufacturing Companies", Int. Journal of Engineering Design, Vol.8 No.4, pp.422-431, 1997.

[6] J.J. Ambler, "The Theory of Scaling Up Laboratory Data for the Sedimentation Type Centrifuge", J. Of Biochimical and Microbiological Techonology and Engineering, Vol.1 No.2, pp.12-26, 1959.

[7] S. B. Bilatos, B. S. Kim, Applications of Design For Manufacturing, ASME Press, 1998.

[8] G. Pahl, W. Beitz, Engineering Design, a Systematic Approach, Springer Verlag, 1993.

[9] Z. Li, "Laser Welding for Lightweight Structures", Journal of Materials Processing Technology, N. 70, pp. 137 - 144, 1997.

[10] O. Bollettini, "Tecnologia di Saldatura Laser: Riprogettazione di Particolari di Super Centrifughe", MS Thesis, University of Ancona, 1998.

[11] C. Zannoni, "Ottimizzazione del Processo di Saldatura con Laser ND:YAG di una Centrifuga in AISI 304", MS Thesis, University of Ancona, 1998.

The Development of a Software Package to Address both Product Design and Manufacturing Process Issues

Richard P. Storrick Jr.

Department of Mechanical and Aerospace Engineering, University of Dayton, Dayton, OH, 45469-0210, USA

John P. Eimermacher

Department of Mechanical and Aerospace Engineering, University of Dayton, Dayton, OH, 45469-0210, USA

Abstract

Considering both design and manufacturing issues up front in the product development cycle ensures the completeness of a design. Software is being developed to facilitate the consideration of relevant issues. The developed program links a computer database with a Graphical User Interface (GUI). Through the GUI a user has the option to choose and answer various questions relating to the design process and the overall project objectives. The answer to a particular question can result in the program presenting other relevant questions to the user for consideration. The responses to all of the questions are stored in a database file. The questions and responses can then be viewed or modified by managers, customers, or another engineers to assist in the design process to achieve the best possible solution.

1 Importance of the DFM software

Implementing a Design for Manufacturing (DFM) approach has numerous advantages that can enhance any type of design or redesign process. DFM can lead to the design being completed in less time with a lower development cost when compared traditional approaches. A quicker and smoother transition can be made into production while having the highest possible quality and reliability that a product can offer. A design that incorporates the DFM approach can also satisfy a customer's current and future needs.

Designing for manufacturing issues early in the design process is an overwhelming task for an individual. In a small to medium company an individual is often called on to undertake the whole design process by themselves. There is such a wide range of issues that need to be addressed that an individual can easily overlook one or more important issues. This task is made considerably easier with a software package that has a list of issues that need to be addressed for the design project to be considered complete.

Through the use of the design software an individual can address all relevant issues that are related to the product development cycle. The above noted software can also be programmed with any necessary reference information that might be needed to understand unknown or vague issues. Full documentation for the reasoning behind the decision can be made and stored for reference by other clients, engineers and managers.

With society moving more toward a legalistic approach to business, it is increasingly important for an engineer to maintain full documentation through out a project. This is becoming increasingly difficult and is made considerably easier with the design software. The design software allows the user to document the reason for each decision. This view at the reasons behind the decisions can eliminate unnecessary steps and avoid reoccurring problems within a company.

2 Design Software

The Design and Manufacturing Issues Program (DMIP) utilizes a Microsoft Access database to store the questions relating to a design or redesign project. The program is then linked directly to Microsoft Access so that manipulation of the database is easily accomplished. The questions are retrieved from the database and displayed to the user through the developed program. After the questions have been considered and answered the responses are then written into the database for future referencing.

Currently, the design software is linked to Microsoft Excel as a primary printing option. To print, the program first interfaces with Microsoft Access to read the entire database and transpose that database to a Microsoft Excel spreadsheet. The spreadsheet can then be manipulated as required by the user.

Overall the design software acts as an interface program that reads questions from a database and displays them to the user. The responses to the questions from the user are then written back to the database for storage. During printing Microsoft Excel is opened and used to print the questions as they are received from the design software. (Fig #1)

843

Figure 1. Program Interface Layout

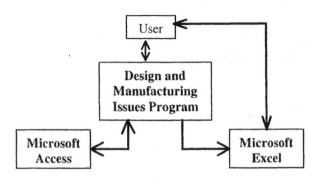

2.1 Form Flow

In general, the entire program is centered on the Main Menu. The only exception to this rule is loading in the project information, which includes creating a new or opening an existing project. The design of the way the windows flow together is intended to allow a user to maintain an easy straight on approach to answering the questions. (Fig #2) The general idea was to have the user return to a central starting point and then proceed outward from that starting point until the chosen option is completed.

When deciding which questions to answer in the question's database the main group needs to be chosen from the Main Menu. Once the main group has been chosen the subcategory will be chosen through the Submenu Window that follows the Main Menu. Only after these two windows have been completed will the user be able to narrow down the entire database to the few questions that currently need addressing.

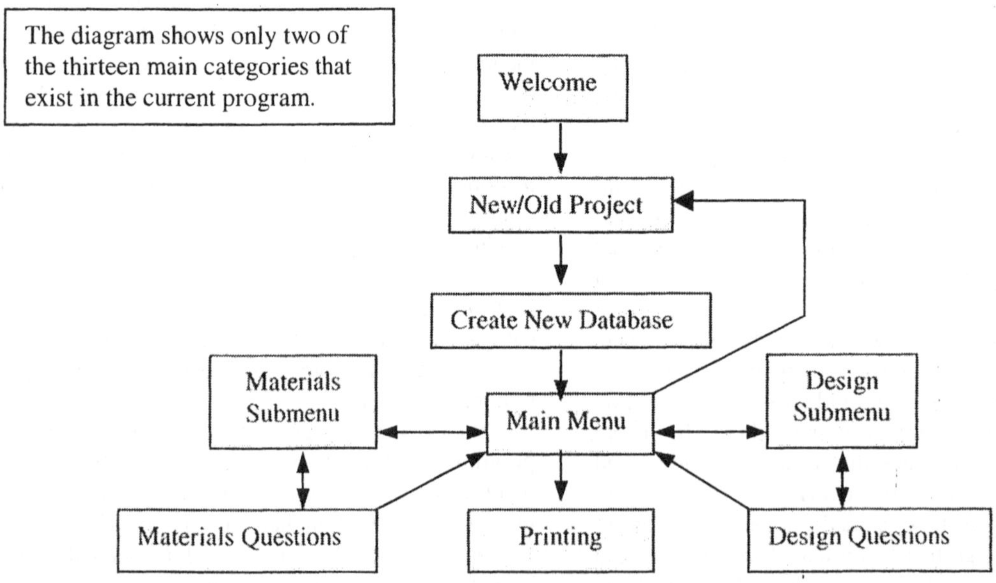

2.2 Window Design

The overall design of each window required the standardization of many aspects to create a pleasant appearance yet maintain an easy to use interface. The main aspects of the program that where necessary to standardize where the color scheme, ergonomics, functionality, simplicity and intuitiveness of each window. These aspects have been designed into each window, but since the requirements of the windows change the Main Menu, Sub Menu and Questions windows used additional concepts to complete their window design.

Throughout the program particular items have been standardized to aid the user in easily using and understanding the program. Peach colored text boxes, which are located in the Questions window as well as the Main Menu and the Print window, indicate message only boxes. The user can not modify these text boxes through the program. Option buttons have been grouped together to simplify the selection of an option within the group. Command functions have similarly been grouped and have always been placed near the bottom of the screen so that the user will be able to easily locate the buttons. Within each of the submenus and subsequent question windows a picture has been placed on the left side of the sub menus and at the top of the questions windows to indicate which main menu and sub category is currently under consideration. These items identify the window that is currently active and helps the user navigate through the program effectively. (Fig #3)

The color selection was chosen very carefully. Because of the nature of the program, it would be easy for a user to lose focus while answering a long list of questions during a redesign project, so a color scheme was chosen to try to

Figure 2. Window Flow Diagram (Simplified)

The diagram shows only two of the thirteen main categories that exist in the current program.

Welcome → New/Old Project → Create New Database → Main Menu

Materials Submenu ↔ Main Menu ↔ Design Submenu

Materials Questions — Printing — Design Questions

keep a users interest. The sky type of background provided a cheerful base that would help the user maintain a positive view of the project during the design process. The titles of the windows needed to blend in with the current color scheme without being overbearing for this reason a burgundy type of color was selected. The color of the message only boxes was selected to indicate the different properties of the text box without over dominating the program window.

Figure 3. Breakdown of the Main Menu

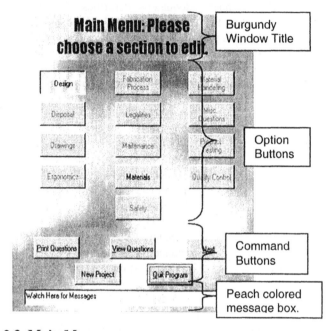

2.3 Main Menu

The main menu is the central window in the DMIP. From this window all aspects of the program can be accessed. The form is separated into two main sections the main grouping buttons and then the command buttons. The main grouping buttons were clustered in the center of the form for ease of use. The command buttons where placed near the bottom of the screen and were spaced out to minimize accidental activation.

The main grouping buttons located in the center of the form determines which of the thirteen main groupings will be selected. The button that is chosen will indicate which sub menu is displayed when the "Next" button is pressed. The main grouping buttons have been designed to only allow one option to be selected at a time and has a default value built in to eliminate the possibility of the user not choosing a grouping to consider. There are currently thirteen main groupings, but if more groupings are necessary, adding the necessary buttons requires the bear minimum of reprogramming of the software. (Fig #4)

Figure 4. The Main Menu

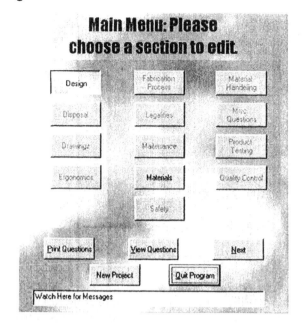

2.4 Sub Menu

Once the main grouping has been selected there can still exist an overly abundant amount of questions to consider. The submenu narrows the number of questions to consider down to a reasonable amount. Each category within a sub menu has anywhere from five to thirty questions. This limits the number of questions that the user can access one screen to a manageable amount.

The preestablished window criteria have been used for most of the window design but due to the unique requirements of the window certain modifications where necessary. Because this window can be one of thirteen that have the same basic function, a picture indicating the main grouping was placed on the left-hand side of the window as an indicator. Again, the number of sub category buttons can increase and decrease as necessary to accommodate different databases with minimal changes to the rest of the program. (Fig #5)

2.5 The Questions Windows

The questions window is where all of the questions are going to be considered and answered. Three questions and all relevant information are displayed at any given time. The window can scroll forward or backward through the question's database, as the user requires. The questions window is also linked to the main menu and the sub menu to allow the user to change main categories or sub categories as wanted.

Figure 5. The Sub Menu

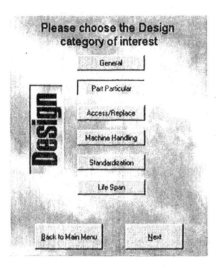

There are three responses for each question. The answers are either Yes, No, or Not Applicable(N/A). Yes indicates the question has been considered and is correct. No indicates the question has been considered and is not correct. Not Applicable exists if the question does not apply to the project. As a result of any of the three answer boxes being checked, the answered box will automatically indicate that the question has indeed been considered. The explain text box is the location that allows the user to indicate the reasoning behind the responses. However, this explanation is limited to 255 characters in length due to the database limitations. (Fig #6)

Figure 6. The Question Window

??? **Design Questions** ???

Can the design be simplified with respect to material selection?

- ☐ Yes
- ☐ No
- N/A ☐
- Answered ☐

Explain

Can the design be simplified with respect to operation (ease of use)?

- ☐ Yes
- ☐ No
- N/A ☐
- Answered ☐

Explain

Can the design be simplified with respect to part specification?

- ☐ Yes
- ☐ No
- N/A ☐
- Answered ☐

Explain

Back   Forward   Sub Menu   Main Menu

3 Database Description

The database that is contained in Microsoft Access is fully integrated into the design program to allow all of the functions of a database to be used with an ease to use GUI. The database questions are grouped into thirteen main groups. (Table #1)   These groups are then subdivided into anywhere from two to seven subcategories that are organized on the Sub Menu Window (Fig #5).   This classification methodology allows for a two letter ID Number to be associated with each question subcategory. This ID Number adequately describes the question location within the entire question database.   The first letter is the grouping designation and the second letter is the subcategory designation.

Table 1. A Listing of the 13 Main Groupings

Design
Disposal
Drawings
Ergonomics
Fabrication Process
Legal Issues
Maintenance
Material Handling
Materials
Miscellaneous Issues
Product Testing
Quality Control
Safety

Each database entry has five different types of data that show the thought that the user has given to each question. The first is the Question; this is what is presented to the user for consideration.   With the question comes the Explanation, this is any supporting documentation that is necessary to support the response to the question.   The three other database entry fields are Yes/No, Not Applicable, and Answered.   These three items are presented through option buttons to the user and have only true or false answers that indicate the response to the question.   The Yes/No data field is the response to the question.   The Not Applicable data field is for those questions that do not apply to the project.   The answered data shows if the question has been answered in any of the three possible responses. (Table #2)

Table 2. Sample of the Questions Database

ID	Question	YesNo	NA	ANSRD	Explaination
dd	Can the design be simplified with respect to material selection?	FALSE	FALSE	FALSE	Explain
dd	Can the design be simplified with respect to operation (ease of use)?	FALSE	FALSE	FALSE	Explain
df	Can the design be simplified with respect to part specification?	FALSE	FALSE	FALSE	Explain
df	Does the design allow the use of standard assembly tools?	FALSE	FALSE	FALSE	Explain
df	Have standard machined features been used wherever possible?	FALSE	FALSE	FALSE	Explain
da	Have alternative design concepts been considered and the simplest and most predicable one selected?	TRUE	FALSE	TRUE	Yes, The simplest design was selected based on cost.
da	Does the design exceed the manufacturing state-of-the-art?	FALSE	FALSE	TRUE	There are designs that are in existance that have a far greater complexity.
da	Is motion or power wasted within the design?	TRUE	FALSE	TRUE	Friction is the primary problem in the current bearing design.

4 Editing the Database

The database is fully adaptable to any market and company. The database currently contains 600 questions that are related to various design and manufacturing issues. Many design related issues are not explored fully within the questions that are currently in the database. It is this reason that the database is fully upgradeable. The database is very easy to modify by adding or removing questions to the database as necessary. This can be accomplished by simply adding a new record to the Microsoft Access questions database located in the reference file. There are only two pieces of information that is necessary to add a new question to the database. These two pieces of information are first, the question itself, and the second is the question ID Number. The question ID Number is determined by the location of the question within the database. Once the desired location of the question within the program is determined the ID Number can easily be figured out and the question can be then typed into the new record.

5 Program Limitations

With DMIP currently being a prototype there exist several applications of the program that have not yet been addressed.

- The DMIP has been developed with the intention of a single user using the program on a single computer. Obviously, this is not the case in industry and the program needs networking capabilities.

- Currently there is no method to alert the user if a question that was answered has any other questions that directly depend on the indicated response.

- There is no cross-referencing available between projects. It is a nice feature that if separate projects have interrelated issues the databases could be linked together to allow easy access to shared question responses.

- If a project has multiple design sections a new project database would need to be created for each section. To eliminate the excessive project files with this type of project a central project database could be linked to satellite databases that would contain only the question information for a singular design project.

- The program is currently a stand-alone program that only asks questions to a user. Further developments are necessary to make the program more applicable to a design environment.

6 Sample Run Through of the Program

To clarify any questions about how the program is to be utilized a sample run through of the program has been included. The sample starts after a user has started the program and has either selected an existing project or has created a new project and has been advanced to the Main Menu for the first time.

The first time the Main Menu is shown to the user the user would more than likely choose to start answering questions. First the main grouping needs to be selected from the Main Menu option buttons. The "Design" button will be selected for the example. After the "Design" button has been selected the "Next" button would be pushed to advance the program to the submenu windows. (Fig #7)

Figure 7. Main Menu (Sample run through)

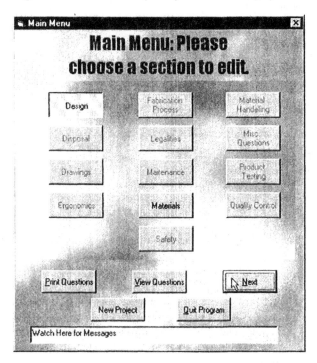

The next window that would appear is the design submenu. Currently the questions database has been limited to about 100 related questions in the design grouping. These 100 questions where placed into one of six submenus with each subcategory containing questions grouped around a specific topic.

To continue with the example the sub menu "Part Particular" will be selected. (Fig #8) This submenu contains questions that pertain to the design of unique part. This whole sub menu will probably not be needed for an entire project but many of the questions found in the submenu can help answer many specialized questions about the project.

After the Part Particular button has been selected and the "Next" button has been pressed the program advances to the Questions Window, where the user answers the questions.

The user is then displayed the three following questions.

- Can parts be made symmetrical for easier assembly?

- Has a base been provided for holding and reference?

- Can multiple parts be combined into a single net shape?

These are the first three questions from the Design Menu under the Sub Menu of Part Particular related questions.

Figure 8. Sub Menu (Sample run through)

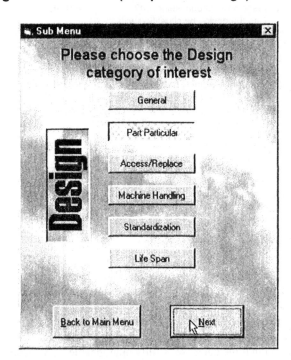

The first question does apply to the project and is determined to be easier to assemble if parts where made symmetrical. Because of this, Yes is selected and the reason for the answer is written in the explain box as "The design of the part was accomplished before a careful design analysis was accomplished." (Fig #9)

The second question was determined to apply and to be No since for the example the overall shape is sphere. This does not allow for a base reference. So No is selected for the second question and the explanation of "The spherical shape of the part does not provide a base for easy reference".

The third question does not apply at all to the project since for the example the spherical shape is composed of two halves with no possibility for redesign. Because the question does not apply at all the N/A option is selected and the explanation of " To halves exist to the design in question and an assembly needs to be placed between the two halves."

At this point the user can press "Forward" to advance to the next set of three questions or go to the Main Menu or the Sub Menu window to choose another question group. Regardless of the user's choice, the question database will update after a selection has been made.

Figure 9.   Questions Window (Sample run through)

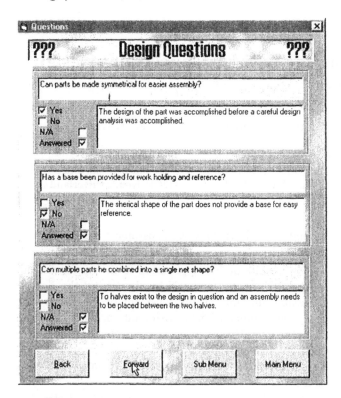

7   Future Work

The Design and Manufacturing Issues Program needs some modifications. Once these modifications have been accomplished the program will then be ready to be linked to the other half of the developing software package. The other half of the software package is an Artificial Intelligence/Expert Systems software program that interconnects all the various raw databases with the user.

The AI/ES software is under construction to maintain all the relationships between various raw databases and the user's project databases. The AI/ES software is not user friendly and would be inappropriate for a user to use. So, instead the DMIP is presented to the user. The DMIP interacts with the user and the AI/ES software translating information between them. This allows the information stored in various raw databases to be organized in an orderly manner that is easily understood and can be easily cross-referenced with other projects by the user.

8 Conclusion

The concept of developing a decision analysis program can greatly aid an individual with a product design process. The program that was developed has limitations. These limitations can be overcome with future programming and development. Even with the current limitations the program is capable of assisting any engineer with any design project.

References

[1] Allen, C.W., King, R.E., Skiver, D.A., Simultaneous Engineering, Springer-Verlag., 1983.

[2] Bakergian, Ramon, "Design for Manufacturability", Vol. 6, Tool and Manufacturing Engineer's Handbook, Society of Manufacturing Engineers, 4th Edition.

[3] Eimermacher, John P., "Design for Manufacturing Handouts" Professor of Mechanical and Aerospace Engineering Department, The University of Dayton, 1999.

[4] Shams, Lutfe, "DFM Checklist Software User Manual", School of Engineering, Mechanical and Aerospace Engineering Department, University of Dayton, May 1998.

[5] Jung, David, Kent, Jeff, Visual Basic Annotated Archives, Osborne/McGraw-Hill, 1998.

[6] Petroutsos, Evangelos, Mastering Visual Basic 6.0, Sybex, Inc., 1998.

[7] Prasad, B., Concurrent Engineering Fundamentals, Prentice-Hall, Inc., 1996.

[8] Smith, Eric A., Whisler, Valor, Marquis, Hank, Visual Basic 6 Bible, IDG Books Worldwide, 1998.

MicroCE: Computer-Aided Support for DFMA Conceptual Design Phase

G. Bornet dit Vorgeat, P. Pu, R. Clavel   A. Csabai, F. Sprumont, P. Xirouchakis   M.-T. Ivorra

MicroEngineering Department
Swiss Federal Institute of Technology
1015 Lausanne, Switzerland

Mechanical Engineering Department
Swiss Federal Institute of Technology
1015 Lausanne, Switzerland

Mecanex SA
Technical Department
1260 Nyon, Switzerland

Abstract

The goal of the MicroCE (Concurrent Engineering Applied to Microengineering Products) project is to provide computer-aided support for the conceptual design phase of the design process. It is a help for the determination of the solution principles of the functional requirements, taking into account mainly technical considerations. The approach of MicroCE considers the conceptual design phase as the solution of a combinatorial problem: the future product is subdivided into elementary functions for which it is necessary to find the best solution from a set of possible ones. The selection criteria are combinations of technical, functional and DFMA (Design for Manufacturing and Assembly) characteristics. The definition of all problem elements (requirements, functions, solution principles, criteria, etc.) is totally dynamic and can evolve over time according to the design projects.

By taking into account a large number of aspects, we improve the general quality of the products. The other main result is the reduction of the development time.

1 Introduction

The main objective of Concurrent Engineering (CE) is the systematic integration, during the design phase of new products, of all elements of the product life cycle from conception through disposal. Concurrent engineering "is designing for assembly, availability, cost, customer satisfaction, maintainability, manageability, manufacturability, operability, performance, quality, risk, safety, schedule, social acceptability, and all other attributes of the product" [7]. This integration is usually realised through computerised tools and cross-functional teams having representatives from internal (e.g., R&D, manufacturing, assembly) and external (e.g., customers and suppliers) stakeholders.

The results of CE application is the improvement of the design process itself by shortening the development time and the improvement of the quality of designed products. Thus, taking into account production considerations during the design phases results in simpler products which are also better adapted to the production resources and therefore more efficiency and lower costs are achieved.

Research in concurrent engineering [9] includes the development of methods and tools facilitating communication and information transfer between the team members [6, 12], and the development of methods and tools focusing on the decision making and problem solving behaviors that occur within teams [2].

The support provided by these CE tools is generally given once the product is already sufficiently defined, during the late embodiment design phase and the detailed design phase of the design process (see figure 1). The main reason is that the methods and tools need sufficient information about the product (geometry, performances, costs, etc.). For example, the manufacturability and assemblability analysis (Design for Manufacturing and Assembly - DFMA) is based on geometrical reasoning about sizes, forms, insertion trajectories, mates, surface qualities, etc. [1, 14].

This implies that the conceptual design phase is less supported by CAD systems, because the uncertainty about the product is high. During this phase, the designers conceptually determine the solution principles achieving the product functions (e.g., how the functional requirements of the product will be achieved). All the same, CAD support of the conceptual design phase can be very beneficial: during this phase between 60% and 80% of the future product costs are defined according to the decisions taken. The economical impact of the conceptual design phase is very important, and taking the right decisions is crit-

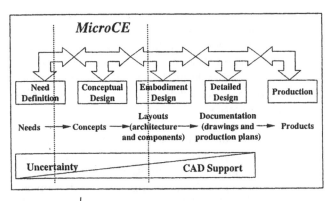

Figure 1: The design process [13]: constituting phases, involved results and CAD support

ical. For the moment, the correctness of the decisions are mainly insured by the experience of team members. This has two consequences:

1. in the case of absence of an expert (retirement, departure, illness) her/his experience is no longer available. This implies that this lost knowledge must be either ignored or rebuilt from scratch: the quality of produced designs is then reduced.

2. the conceptual design phase is not well structured. Results of old projects are not systematically exploited, and similar ideas need to be re-developed each time, with unnecessary time wasting. Checks of the choices are not systematic and then, in the case of error, the corrections are very time (and cost) consuming.

A CAD support, adapted to the conceptual design phase, can avoid this inconvenience. By offering structured knowledge bases about solution principles and design rules (implementing the experts' experience and evolving with the projects), the CAD system guarantees the stability of the information kernel: the necessary information is reusable and still available. By associating analysis tools with these knowledge bases, the CAD system can help the designers in her/his decision making: the quality of the decisions is improved.

In this paper we present MicroCE which is a CAD prototype for such a support of the conceptual design phase, from need definition until early embodiment design. It includes estimations of manufacturability and assemblability of the solution principles. Sections 2 and 3 present respectively the proposed approach in MicroCE and related works. The MicroCE prototype itself is presented and illustrated with an example in section 4. Before the concluding remarks, section 5 presents the main results and the future developments of MicroCE.

2   MicroCE approach

The proposed approach follows the notions developed by G. Pahl and W. Beitz [13] about the engineering design process. They have formalised the different phases of the design process (Figure 1). Design transforms a set of requirements into a physical artifact which realises them. The transformation is a top-down process realised through several iterative refinements based on functional decomposition, solution search and recomposition. Their work is the core of the VDI 2222 standard [15] which defines this process and the associated terminology.

According to this, the conceptual design phase (see Figure 2) is the transformation of the general function of the product into possible concepts realising it. So conceptual design determines first the *general function* describing the main role of the product and covering a set of requirements. For a torque sensor, for example, the general function is simply "measure a torque". The general function is then decomposed into *partial functions*. The decomposition of a given general function is not unique and may depend on the product specifications: a possible (and partial) decomposition for "measure a torque" could be "convert rotation into translation" - to transfer the torque to measure points, "limit overloads" - to protect the sensor integrity, and "guide a solid with one rotational dof (degree of freedom)" - to guide the solid receiving the measured torque. The partial functions can be themselves decomposed into finer partial functions, if necessary. At the end, the obtained partial function are considered to be elementary ones.

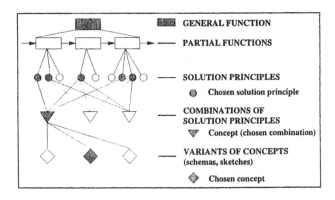

Figure 2: The conceptual design phase [15]

The next step of the conceptual design phase is the determination of the possible *solution principles* for the functions. For each partial function and independently of the problem (realisation of the general function), the designer searches a set of possible solution principles for the function (the principles of how to realise the given function). This set of solution principles should be as large and diverse as possible. These solution principles are then regrouped in *design catalogues* (one catalogue per partial function) which is a data structure allowing their storage, classification and description. The objective of a design catalogue is to facilitate the retrieval of solution principles. Table 1 gives an extract of one particular design catalogue.

Guiding Elem.	Position	Pos	Sketch	Principle
Sliding Bearings	End Guiding	5		Solid slides over the 2 balls
Rolling Bearings	Inside Guiding	6		
Elastic Elem.	End Guiding	7		Thinner section is a hinge between solid and fixed point
Elastic Elem.	Inside Guiding	8		The solid is suspended to a cord in torsion

Table 1: Extract of the design catalogue of the function "Guide a solid with one rotational dof"

Once the necessary design catalogues are created or completed (if already existing), the designer can choose from them the solution principles applicable according to the general function and the specifications. The chosen solution principles are then combined. The number of possible combinations can be huge and it is necessary to select the best ones according to, generally, technical and economical criteria. The combinations satisfying these criteria are called *concepts*. The concepts are then embodied into geometrical layouts which define possible constituent components and their relations in the product.

In MicroCE, the implementation of the conceptual design phase follows the process described above. For this implementation of concept search, we model the conceptual design as a *constraint satisfaction problem* (CSP). Formally, a CSP is defined by the pair $< V, C >$ with:

V the set of the problem variables. Each variable $v_i \in V$ is characterised by its definition domain

which is the set of its possible values. For conceptual design, V is the set of the partial functions. The set of possible values for a partial function is its design catalogue. So a possible value is one solution principle of the design catalogue, which can be represented by a tuple of its describing characteristics.

C the set of problem constraints: $C = \{c_1(V), \ldots, c_m(V)\}$. A constraint is a relation between the definition domains of the variables $v_i \in V$ and represents an interdependence between variables that must be satisfied. For conceptual design, C is the set of specifications and selection criteria. These are used to select the solution principles in the design catalogues and to find the concepts (e.g. choose the correct combinations of solution principles). In MicroCE, they are expressed as arithmetic relations of the characteristics of solution principles. For example, `Guide.Precision + Limit.Position < 4` (see also section 4.1).

The resolution of the CSP consists of search for the states (a possible combination of values) for which the m constraints $\in C$ are satisfied in the search space constituted of all the states. Then the resolution of the conceptual design involves finding concepts in the search space of the combinations of solution principles.

In fact there are two main differences between the formal definition of a CSP and its implementation in MicroCE. The first is that, to represent the concepts, we use tuples instead of using the usual CSP variables. The second is that in classical CSP there are very strong implicit hypotheses: values of the variables must be in a predetermined set while in MicroCE, we have the possibility to determine new solution principles, that is to say, new tuples. The definition domains are dynamic.

The software prototype of MicroCE covers the whole conceptual design phase from the specification definition to the embodiment of concepts. It is composed of 3 modules (see figure 3):

- *Catalogue*: this covers the functional decomposition, the definition of specifications and of criteria, and the management of the design catalogues. In other words, *Catalogue* deals with the knowledge definition and storage.

- *AIAD*: this deals with the constraint resolution. It generates the combinations of solution principles and searches for the concepts. In other words, *AIAD* deals with the combinatorial aspect of the problem.

- *3DLM*: this allows geometrical modeling either of solution principles, or of concepts. It allows their embodiment into abstract layouts (architecture and components), but also allows analysis of the kinematics of these layouts.

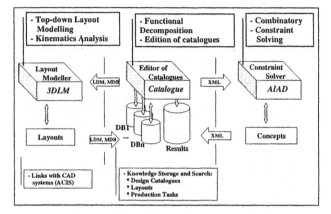

Figure 3: The architecture of the MicroCE prototype

The communication between the modules is controlled by *Catalogue*. In this way it is possible to verify and guarantee the coherence of the information stored in the design catalogues. *Catalogue* and *AIAD* exchange information about functional decomposition and partial functions, solution principles, constraints and concepts through XML files. *Catalogue* and *3DLM* exchange information about layouts of solution principles and concepts. The three modules and the communication are described in more details in section 4.

3   Related work

Research in the domain of engineering design has developed several CAD systems based on the presented conceptual design process, as for example:

- the IKMF of the University of Braunschweig [8] has developed a design catalogue for kinematical joints. The system is mainly composed of databases (built on market and patent surveys) about technical objectives, fundamental solution principles and design tasks. This knowledge is applied for design of liaisons in parallel kinematics machines. All elements (couplings, guides, connections, etc.) are classified and stored according

to the specific requirements of the parallel kinematics domain.

- R. Huber et al. [10] propose a CAD environment for MEMS design (Micro Electro-Mechanical Systems). This environment is an extension to the MEMS domain of the SYSFUND system [16] (Systematisation tool for Functional Design and synthesis) which allows the creation and manipulation of knowledge databases. These design catalogues are based on the FBS product model (Function - Behaviour - State). They allow description of the functional decompositions (Function), description of solution principles (Behaviour) in terms of causality networks between physical phenomena, and description of the links between physical phenomena and components (State).

- S. Carlson-Skalak et al. [4] propose a system for designing pipe networks to deliver cooling fluid to machines, for given pressures and flows. Each element of a network is chosen from a design catalogue. The search algorithm is a genetic algorithm allowing the simultaneous evolution of the network (configuration) and of the components.

In general, the main goal of these systems is to allow description and classification of solution principles (or standard physical components) used in a given and well defined domain. The evolution, over time, of the different knowledge bases is not necessarily easy. We can easily edit (add, modify or suppress) solution principles, but it is difficult to modify the structure of the design catalogues. Depending on the domain, this can imply the swift obsolescence of some information and too great a divergence between knowledge bases and the actual state of know-how in the field.

In the same way, the search algorithms used in these systems are not necessarily very flexible: they are based on the descriptive characteristics of the solution principles, but they do not allow, in general, creation of new search criteria by combining the characteristics. So if the set of applicable criteria cannot evolve and if we still use the same criteria, there exists the risk of finding each time the same solutions and to omit systematically a part of the possible solutions.

Different systems propose a support for the conceptual design based on other approaches, but the goal is still the same: to help the designer to structure his/her ideas in the search for concepts. For example:

- A. M. King and S. Sivaloganathan [11] propose a methodology based on an extension of the Quality

Function Deployment (QFD). QFD is a graphical adaptation of Utility Theory which allows the representation and management of the interdependence between functional requirements and design characteristics. The QFD matrices are completed with considerations about solution principle compatibility for a given concept.

- *A-Design* [3] is a design methodology combining aspects of multi-objective optimisation, multi-agent systems and automated design synthesis. *A-Design* allows the treatment of the ever-changing environment (knowledge, inputs, criteria importance, etc.) of the conceptual design phase.

Those systems are general and very flexible (as MicroCE is), but the adaptation to a particular problem needs a great effort to create the specific knowledge bases (utility functions and evaluation, design catalogues, search criteria). Using them systematically and efficiently in an R&D department should be expected for mid-term. This delay may have as consequence the same inconvenience of obsolescence and divergence for the information contained in the databases.

4   The MicroCE prototype

We present in this section the three modules of the MicroCE prototype in more details.

4.1   The *Catalogue* module

The *Catalogue* module covers mainly the first step of the conceptual design from the general function until the design catalogues and the solution principles. In this process, the most important role of *Catalogue* consists of the definition and maintenance of the knowledge kernel used in the concept search for a given design problem. It allows the management of the design catalogues (creation, edition, saving) and the design problem definition (creation, edition, saving). This definition includes the functional decomposition (combination of design catalogues) and the definition of specifications and selection criteria which are the constraints to satisfy. *Catalogue* allows also the communication between the different modules of MicroCE by translating the data into a file format understandable for the destination module.

The knowledge is stored in databases. *Catalogue* is built on a standard database system (MS Access 97)

which is used for data storage and retrieval. There are two types of databases: the design catalogue databases and the design problem databases.

A *design catalogue database* stores all the information about the possible solution principles for a given function. The information is stored in different record tables according to the possible points of view on the solution principles. So a design catalogue database has three types of record tables, each describing one aspect of the set of solution principles:

- the table of solution principles. This contains the list of all possible solution principles for the function. In this table, a solution principle is only characterised by an identifier, a textual description, a reference to a graphical representation and a reference to a possible *3DLM* layout. It is simply a description of what is the solution principle.

- the tables of descriptive characteristics. This kind of table regroups the characteristics of the solution principles which describe a particular aspect about them: mechanical or electrical performances, dimensions, etc. The DFMA evaluation of the solution principles are given in a table of this kind. These tables allow the user to qualify the solution principles.

- the tables of evaluation scales. These tables give the interpretation of a numerical evaluation of qualitative characteristics. The descriptive characteristics of the second type of table can then be associated with these evaluation scales. The following table gives examples of evaluation scales used to qualify the magnitude of a rotation and the size of an occupied volume.

Notes	Movement	Volume
1	$< 10°$	very small
2	$< 180°$	small
3	$< 360°$	average
4	$> 360°$	big
5	∞	very big

A design catalogue is a dynamic database. It can be modified at any time by adding, editing or suppressing the solution principles it contains, but its structure itself can also be modified by adding, editing or suppressing the tables of characteristics or the evaluation scales. This allows knowledge adaptation over time according to experience accumulated during the projects.

A *design problem database* stores the information about the concept search for a given general function. The information is also stored in different record tables according to its nature:

- a table for the partial functions. This contains the list of partial functions resulting from the decomposition of the general function. The decomposition is done by the user. Each partial function is given by its name and the reference to the design catalogue database.

- a table for the specifications. This contains the list of the parameters of the problem. A parameter is characterised by an identifier, a textual description, a definition and an objective value. The definition of the parameter may be an arithmetical expression of the describing characteristics of solution principles and/or other parameters. Here are some specifications of the sensor example:

Identifier	Description	Expression	Objective
Precision	Estimation of the precision of the measure	Guide.precision + Limit.positn	*Precision* < 4
SetUp	Estimation of the facility of the mechanism set up	Guide.movmt + Transmit.setup	*SetUp* < 10
Measure	Estimation of the measurement range	*Precision* + Limit.load	*Measure* < 8

The expression of the *Precision* specification means that the precision of the measure is estimated by the sum of the evaluation of the guiding precision of the partial function "Guide a solid with one rotational dof" and the estimation of the positioning precision of the partial function "Limit overloads". The objective is that this sum has to be less than 4, what means we want concept with pretty good precision.

- tables for the selection criteria. There is one table per design catalogue. These contain the acceptance conditions for the solution principles concerned. These conditions are given by an identifier and, as for the parameters, an arithmetical expression of the describing characteristics of solution principles. Here are some selection criteria for the solution principles of the function "Guide a solid with one rotational dof":

Identifier	Expression
Constraint_12	DFMA.AlphaSymmetry = 0
Constraint_15	Performance.Movement < 5
Constraint_16	Performance.Volume = 3 Or Performance.Volume = 4

The expression of *Constraint_12* means that the characteristic *AlphaSymmetry* (which is one of Boothroyd's DFA parameters [1]) of the characteristic table DFMA has to be 0°: only the solution principles which are fully symmetrical

around an axis perpendicular to the insertion axis will be selected.

- a table for the concepts. This contains the results of the *AIAD* search. A concept is characterised by an identifier, the solution principles chosen from each design catalogue, and the values for the different parameters. Linked to this concept table there is another table containing references to different models of the concept such as the *3DLM* layouts, CAD models, etc.

So for the problem definition, the user has to create and fill up a design problem database. This can be done either by copying and modifying an existing design problem database for a similar design problem, or by creating a new database. Then she/he has to provide, in order: the list of the partial functions (and the corresponding design catalogues), the list of specifications and the list of selection criteria.

Once these elements are defined, the problem is sent to *AIAD* for solution. The databases are exported in an XML file of which the structure is similar to the problem structure (see following the description of *AIAD*). The results of the search (the concepts) are then sent back to *Catalogue* and stored in the concept table of the design problem database.

Catalogue is then able to create automatically a first geometrical layout from the solution principles of a concept. The resulting layouts are stored in *3DLM* script files (see following the description of *3DLM*). The layout creation is based either by integrating the sub-layouts of the solution principles, or by scaling a parameterised layout of the concepts. The integration is done by merging the sub-layouts into a single layout. The resulting layout is a juxtaposition of the sub-layouts spatially shifted. The scaling is an adaptation to the problem specifications of a predetermined and parameterised layout which models the whole concept.

4.2 The *AIAD* module

The main role of *AIAD* in MicroCE is to find the concepts for a given design problem. It actually solves the CSP of the conceptual design phase. The problem definition is sent to *AIAD* in an XML file by the module *Catalogue*. This file:

- describes the possible functional decompositions into partial functions. The decompositions may be done on several levels.

- gives the list of solution principles which may realize each partial function.

- gives for each functional decomposition, each partial function and each solution principle their describing characteristics with their values.

- gives the formulas that specify the constraints (specifications and selection criteria) that will guide the choices to be made between the possible decompositions and the possible solution principles. The parameters of those formulas are a subset of the describing characteristics.

The defined CSP is a choice problem whose structure is a tree. This structure is similar to that of the XML file.

After having loaded the XML file, *AIAD* searches the solutions of the problem. To do this *AIAD* follows an iterative process from the bottom to the top of the tree: it solves first the constituent sub-problems, corresponding to partial functions, and uses these solutions to then find solutions for the functions of the upper levels. It proceeds like this until the root corresponding to the general function is reached.

If *AIAD* finds too many concepts (the CSP is under-constrained), the user needs to add new constraints. To define these constraints, she/he can evaluate the solutions which were found during the initial search according to user-defined evaluators. She/he can proceed in one or more successive evaluations by using several groups of evaluators. By considering the results of these evaluations, new constraints are deduced from the evaluator definitions.

Moreover, when the specified problem does not admit any solution (the CSP is over-constrained), the user can modify it partially or entirely. She/he can do so by adding new functional decompositions or solution principles and/or by relaxing some constraints (firstly the selection criteria, secondly the specification ones). When *AIAD* has found an acceptable number of solutions, it sends them to the module *Catalogue*.

The graphical user interface (GUI) of *AIAD*, shown in figure 4, is divided into 3 views: the *auxiliary view* (marked **1** in the figure), the *tree view* (**2**) and the *evaluation view* (**3**). In these views, the user can act through contextual menus. When an action needs data input or visualization (for example, for constraints or characteristics) a dialog box is opened.

The only role of the *auxiliary view* is the selection of visualization options for the tree view. The *tree view* shows the trees representing either the whole problem (the left most tree in the view), or the concepts found. The tree view also allows access to the problem elements and their characteristics. The nodes of the trees represent either a function, a functional decomposition, or a solution principle. The arcs of the

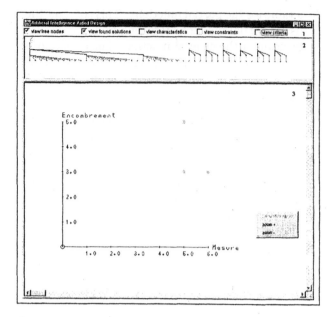

Figure 4: The whole GUI of *AIAD* with its 3 views: (**1**) auxiliary view, (**2**) tree view, (**3**) evaluation view

trees represent either a relation between a function and one of its functional decompositions, a relation between a functional decomposition and one of the partial functions which constitute that decomposition, or a relation between a partial function and one of the possible solution principles which realise that partial function. The use of different colors and shapes allows the identification of the different kinds of nodes and arcs. Moreover the structure of the trees corresponding to the solutions of the problem (the concepts) is the same as that of the problem.

The *evaluation view* shows 2D scatter plots of the set of concepts found. The axis of the plots are user-defined evaluators and the concepts are represented as dots. According to these plots, the designer can select the most promising concepts or refine the problem definition. In figure 4, the evaluator "Encombrement" characterises the volume occupied by a concept and the definition of the evaluator "Mesure" is the same definition of the specification "Measure" of the second table of paragraph 4.1. Since some dots are superposed, we see only 3 dots for the 6 concepts found.

4.3 The *3DLM* module

The *3DLM* module [5] is an integrated layout modeller that provides a tool for defining and maintaining the spatial configuration of mechanical assemblies represented by abstract entities. It is composed of two modules: the Layout Design Module (LDM) and the

Unit Design Module (UDM). Layout and kinematics related parameters can be set up in the LDM while the detailed geometry related data are introduced in the UDM. In MicroCE only the LDM is used, hence the UDM will not be presented here. For more details, see [5].

3DLM is used in MicroCE for geometrical modeling of solution principles and concepts in a first layout and the kinematical analysis of these layouts. A particular solution principle or a concept stored in a database of *Catalogue* can be exported to the *3DLM* module in order to let the designer visualise, analyze or modify the selected 3D model.

It is possible for the designer to deal with the spatial configuration and kinematics related aspects of the assembly. This can be done by dealing with abstract bodies that represent either a single part or a rigid sub-assembly. Assemblies are defined by establishing relationships between the abstract elements using the built-in collection of kinematic constraints (e.g. prismatic joint, revolute joint, etc.). Since the abstract parts lack detailed geometry to attach the kinematic constraints to, reference elements are introduced for the constraints to be attached to the abstract bodies.

The main elements of the *3DLM* are as follows:

- *Layout component*: a layout component is an abstract entity used for representing the spatial position and orientation of layout element. A layout component has no geometry.

- *Design space*: a design space is a simple geometric form attached rigidly to a layout component. This entity represents the shape of the layout element. The layout component and the design space together determine the spatial configuration and the appearance of the layout element.

- *Interface feature*: An interface feature is a marker added to a layout component in order to provide geometric reference for a kinematic constraint. The relative position and orientation of interface features can be changed with respect to the layout component.

- *Kinematic constraint*: The kinematic relationship between layout components are defined by constraints. They always reference two interface features of two distinct layout components.

Once the layout is defined with those elements, it is possible to perform layout evaluations such as kinematic analysis or collision detection by means of a module which is currently under development.

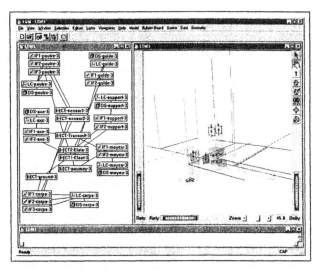

Figure 5: Snapshot of the interface of the 3DLM module. Left: the *graph view* shows the element relationship. Right: the *model view* shows the geometrical model

In the user interface of the *3DLM* module, shown in figure 5, the designer has two views of what she/he is working on. The graph view shows graphically the relationships between the layout elements and allows to access elements properties. The model view shows the geometric models and allows spatial transformation of their elements. Designers have tools to create, delete or modify objects, for example, changing the orientation of an abstract body, or changing the type of a kinematic constraint.

The communication between the *3DLM* and other modules (such as the *Catalogue*) is done by means of script files containing the description of the related configuration. For example the representation of a cubic layout component, called `lc1_adapter`, with three interface features is the following:

```
add_lc lc1_adapter
set_ds cub ds1_adapter lc1_adapter
ds_params ds1_adapter 66 32 62
add_if ld if11_adapter lc1_adapter
if_pos if11_adapter 20 0 -20
add_if ld if12_adapter lc1_adapter
if_pos if12_adapter -20 0 -20
add_if ld if13_adapter lc1_adapter
if_pos if13_adapter 20 0 20
```

Both the *Catalogue* and the *3DLM* modules can read and write descriptive files that conform to a specified syntax. In this way, sub-layouts optimised from a conceptual point of view in the *Catalogue* can be passed to the *3DLM* to realise the spatial arrangement of the components. The *Catalogue* uses default spatial parameters according to built-in rules. The other way to use the two modules is to create a sub-layout in the *3DLM* from scratch and transfer that into the *Catalogue* as a sub-solution, thus giving an extra way of changing the design catalogues.

5   Results

The MicroCE approach has been tested on two design problems furnished by our industrial partner: the design of the mechanical part of the micro-torque sensor and the design of sliprings. The sensor example, as described to the system, represents a search space of 800 possible combinations of solution principles. In the slipring case, the search space consists of 20'736'000 combinations of solution principles. According to the sets of specifications and selection criteria defining each problem, we obtain 6 concepts for the sensor and 8 concepts for the slipring.

According to the fact that the knowledge (solution principles, specifications, selection criteria) on which the concepts are built and searched is the designer's experience, the quality of the concepts obtained are equivalent to what a designer produces for the given problem conditions. But, independently from this, the main result is the time reduction to obtain them. On average, the conceptual design phase (without CAD support) for a slipring is 3 days long. With MicroCE, it becomes 2 hours. This result does not include the design catalogue creation, but includes the definition of specifications and selection criteria adapted to the problem conditions, their tuning during the concept search and the creation of *3DLM* layouts.

This implies two conditions to be efficient during the concept search:

- the biggest part of the knowledge has to be stored in the databases already. The design catalogues needed should already exist and most of the specifications and selection criteria should already be defined. So the design problem definition should be mainly an adaptation of a previous problem definition.

- the user should already know the problem well, first for the problem definition, but also for the concept search to guide the exploration. The designer has only limited help to guide her/him when the problem is under- or over-constrained.

The first condition is not difficult to reach: knowledge reuse is one of the goal of MicroCE. It means that is necessary to create and complete design catalogues, specifications and criteria, not only each time it is necessary, but also independently of the design problems, just for increasing the knowledge bases.

The second condition is more difficult. The reason is that, in its current form, MicroCE offers mainly tools for knowledge input and storage, and CSP solution. The support for decision making is weak. To help the designer in her/his choices (which are the most important criteria? Which are the strong and weak design constraints?), MicroCE should be extended with tools for visualisation and analysis of the current state of the search space. The goal is to provide to the designer a global point of view of the search space and of the different possible options to explore it. With this information, the quality of decisions taken for the space exploration will be improved.

6   Conclusion

The presented approach of MicroCE for the conceptual design support offers several advantages for the user. First of all, the reduction of the duration for concept search. This advantage is completely relevant for the treatment of customer demands. By reducing the delay between the receipt of a demand and the sending of an offer with study of the concepts, it is possible to capture new markets.

Second, the quality of concepts obtained is increased. The designer has facilities for analysis, modeling and knowledge treatment as soon as possible during concept search. The conceptual design is then structured: the reuse of existing solution principles is systematic, the concept exploration procedure takes into account the biggest set of aspects, and the search space is kept as large as possible guaranteeing that no interesting solution is omitted. There is no longer loss of information between projects: the system keeps track of what was done. The support extends the whole project length until the detailed design, guaranteeing the consistency of the decisions taken at each step.

Third, the system is completely adapted to the know-how of the enterprise. The knowledge (functional decompositions, solution principles, specifications, criteria) which is stored in the databases is entirely dynamic and may easily evolve during the projects. This evolution is possible not only for the content of the knowledge, but also for its structure. This limits the problems of information obsolescence and divergence which occur over time.

But, to fully reach these objectives, and according to the promising results, the MicroCE approach should be extended to integrate more analysis tools to provide more information for decision support. This should increase the benefits of the CAD support of the conceptual design phase in terms of development time and product quality and costs.

Acknowledgements

The authors gratefully thank Mr C. Jean-Prost of Mecanex SA and Mr T. Jayet of the Microengineering Department for the numerous discussions about sensor and slipring design. This knowledge is the core of the implemented design catalogues. They also thank Dr Ian Stroud of the LICP for his advice and the reviewing of the draft paper.

The financial support for the MicroCE project is given by the Swiss Federal Institute of Technology. The project is a collaboration between the MicroEngineering Department, the Mechanical Engineering Department and Mecanex SA, Nyon, Switzerland.

References

[1] G. Boothroyd, P. Dewhurst, W. Knight, *Product Design for Manufacture and Assembly*, Marcel Dekker, Inc., 1994.

[2] J. G. Bralla, *Design for eXcellence*, McGraw-Hill, Inc., 1996.

[3] M. I. Campbell, J. Cagan, K. Kotovsky, "A-Design: an Agent-Based Approach to Conceptual Design in a Dynamic Environment", *Research in Engineering Design*, Vol. 11, pp. 172-192, 1999.

[4] S. Carlson-Skalak, M. D. White, Y. Teng, "Using an Evolutionary Algorithm for Catalog Design", *Research in Engineering Design*, Vol. 10, No. 2, pp. 63-83, 1998.

[5] A. Csabai, J. Taiber, P. Xirouchakis, "Design support using constraint-driven design spaces", in *Geometric Constraint Solving and Applications*, B. Brüderlin, D. Roller, Eds, Springer-Verlag, pp. 82-106, 1998.

[6] M. Cutkosky, R. Engelmore, R. Fikes, M. Genesereth, T. Gruber, W. Mark, J. Tenenbaum, J. Weber, "PACT: an Experiment in Integrating Concurrent Engineering Systems, *IEEE Computer Magazine*, Vol. 26, no 1, pp. 28-37, 1993.

[7] B. Dean, R. Unal, "Elements of Designing for Cost", presented at *The AIAA 1992 Aerospace Design Conference*, Irvine CA, AIAA-92-1057, 1992.

[8] H.-J. Franke, D. Hagemann, U. Hagedorn "Systematic Approach to the Design and Selection of Joints for Parallel Kinematics Structures with Design Catalogs", PKM99, *Proceedings of the International Workshop on Parallel Kinematic Machines*, Milano, Italy, pp. 110-118, November 1999.

[9] D. Gerwin, G. Susman, "Special Issue on Concurrent Engineering, *IEEE Transactions on Engineering Management*, Vol. 43, no 2, pp. 118-123, 1996.

[10] R. Huber, H. Grabowski, T. Kiriyama, S. Yoneda, A. Johnson, S. Burgess, "A Design Environment for the Design of Micromachines", ASME, DE-Vol. 83, *Design Engineering Technical Conferences*, Vol. 2, No. 2, pp. 649-661, 1995.

[11] A. M. King, S. Sivaloganathan, "Development of a Methodology for Concept Selection in Flexible Design Strategies", *Journal of Engineering Design*, Vol. 10, no 4, pp. 329-349, 1999.

[12] D. Kuokka, B. Livezey, "A Collaborative Parametric Design Agent, AAAI, *Proceedings of the 12th National Conference on Artificial Intelligence*, pp. 387-393, 1994.

[13] G. Pahl, W. Beitz, *Engineering Design: a Systematic Approach*, Springer-Verlag, 1996.

[14] A. Redford, J. Chal, *Design for Assembly: Principles and Practice*, McGraw-Hill, 1994.

[15] VDI Standard 2222, *Design Engineering Methodics; Setting Up and Use of Design Catalogues*, VDI-Verlag, Düsseldorf, Germany, 1982.

[16] H. Yoshikawa, T. Tomiyama, T. Kiriyama, Y. Umeda, "An Integrated Modelling Environment Using the Metamodel", *Annals of CIRP*, Vol. 43, No. 1, pp. 121-124, 1994.

Author Index

APPLICATION FOR MEMBERSHIP

International Society for Productivity Enhancement (ISPE)
CERA Institute

Personal Information:

Name _____
              First   *Middle*   *Last*

Home Address _____

City _____ State _____ Country _____

Professional and Business Details:

Company/Organization _____

Department _____

Address _____ City _____

Country _____ Tel/Fax/Email: _____

Job Title and Position _____

Field of Specialization and Interest:

____ Manufacturing	____ Inspection and Quality
____ Product Design	____ Computer and Automated Systems
____ Machine Vision	____ AI and Expert Systems
____ Material Handling	____ Robotics
____ Concurrent Engineering	____ Informative Systems
____ Productivity	____ Education and Training
____ Systems Automation	____ Planning and Control
____ CAD/CAM/CIM/FMS	____ Enterprise Integration
____ Technical Memory	____ Management
____ Others (*specify*)	_____

Education:

____ Student   ____ Associate   ____ Bachelor   ____ Master

____ Doctorate   ____ Other (*specify*) _____

Field:

____ Engineering   ____ Science   ____ Technology   ____ Business

____ Management   ____ Other (*specify*) _____

Annual Fee: ISPE CERA Institute Membership Fee: US $35.00 per year

Payment: Please make check payable to "ISPE CERA Institute"

Mailing Address: International Society of Productivity Enhancement (ISPE)
CERA Institute
P.O. Box 3882
Tustin, CA 92781-3882
Email: prasadb@ugsolutions.com